THE ALGEBRA PYRAMID

EQUATIONS AND INEQUALITIES
$2 + 5(7) = 37$ or
$x + 2y > 5$

EXPRESSIONS
$2 + 5(7)$ or $x + 2y$

CONSTANTS AND VARIABLES
$2, 5, 7, x, y$

The algebra pyramid illustrates how variables, constants, expressions, equations, and inequalities relate. At the foundation of algebra, and this pyramid, are constants and variables, which are used to build expressions, which, in turn, are used to build equations and inequalities.

Complex Numbers
Examples: $5 + 2i$, $15 - 3i\sqrt{2}$, etc.

Real Numbers

Irrational Numbers
$-5\sqrt{3}$, $\sqrt{2}$, π, etc.

Rational Numbers $-3, -2.4, -1\frac{4}{5}, 0, 0.\overline{6}, 1$, etc.

Integers $\dots -3, -2, -1, 0, 1, 2, 3, \dots$

Whole Numbers $0, 1, 2, 3, \dots$

Natural Numbers $1, 2, 3, \dots$

Imaginary Numbers
$6i$, $5i\sqrt{3}$, etc.

Arithmetic Summary

Each operation has an inverse operation. In the diagram, the operations build from the top down. Addition leads to multiplication, which leads to exponents. Subtraction leads to division, which leads to roots.

PROPERTIES OF ARITHMETIC

In each of the following, a, b, and c represent real numbers.

Additive Identity
$$a + 0 = a$$

Commutative Property of Addition
$$a + b = b + a$$

Associative Property of Addition
$$(a + b) + c = a + (b + c)$$

Multiplicative Identity
$$a \cdot 1 = a$$

Multiplicative Property of 0
$$a \cdot 0 = 0$$

Commutative Property of Multiplication
$$ab = ba$$

Associative Property of Multiplication
$$(ab)c = a(bc)$$

Distributive Property
$$a(b + c) = ab + ac$$

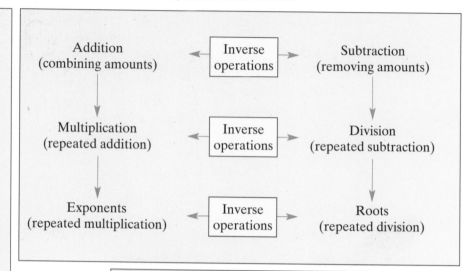

ORDER OF OPERATIONS

1. Grouping symbols
2. Exponents and roots
3. Multiply or divide from left to right
4. Add or subtract from left to right

SOLVING LINEAR EQUATIONS

To solve linear equations in one variable:

1. Simplify both sides of the equation as needed.
   a. Distribute to clear parentheses.
   b. Clear fractions or decimals by multiplying through by the LCD. In the case of decimals, the LCD is a power of 10 determined by the decimal number with the most places. (Clearing fractions and decimals is optional.)
   c. Combine like terms.
2. Use the addition principle so that all variable terms are on one side of the equation and all constants are on the other side. (Clear the variable term with the lesser coefficient.) Then combine like terms.
3. Use the multiplication principle to clear the remaining coefficient.

USING THE QUADRATIC FORMULA

To solve a quadratic equation in the form $ax^2 + bx + c = 0$, where $a \neq 0$, use the quadratic formula:

$$x = \frac{-b \pm \sqrt{b^2 - 4ac}}{2a}$$

FACTORING A POLYNOMIAL

To factor a polynomial, first factor out any monomial GCF, then consider the number of terms in the polynomial. If the polynomial has:

I. **Four terms**, then try to factor by grouping.

II. **Three terms**, then determine if the trinomial is a perfect square or not.

   A. If the trinomial is a perfect square, then consider its form.
   1. If in the form $a^2 + 2ab + b^2$, then the factored form is $(a + b)^2$.
   2. If in the form $a^2 - 2ab + b^2$, then the factored form is $(a - b)^2$.

   B. If the trinomial is not a perfect square, then consider its form.
   1. If in the form $x^2 + bx + c$, then find two factors of c whose sum is b, and write the factored form as $(x + \text{first number})(x + \text{second number})$.
   2. If in the form $ax^2 + bx + c$, where $a \neq 1$, then use trial and error. Or, find two factors of ac whose sum is b; write these factors as coefficients of two like terms that, when combined, equal bx; and then factor by grouping.

III. **Two terms**, then determine if the binomial is a difference of squares, a sum of cubes, or a difference of cubes.

   A. If given a binomial that is a difference of squares $a^2 - b^2$, then the factors are conjugates and the factored form is $(a + b)(a - b)$. Note that a sum of squares cannot be factored.
   B. If given a binomial that is a sum of cubes, $a^3 + b^3$, then the factored form is $(a + b)(a^2 - ab + b^2)$.
   C. If given a binomial that is a difference of cubes, $a^3 - b^3$, then the factored form is $(a - b)(a^2 + ab + b^2)$.

Note: Always check to see if any of the factors can be factored.

Elementary and Intermediate Algebra

TOM CARSON
Midlands Technical College

ELLYN GILLESPIE
Midlands Technical College

BILL JORDAN
Seminole Community College

PEARSON

Addison
Wesley

Boston   San Francisco   New York
London   Sydney   Tokyo   Singapore   Madrid
Mexico City   Paris   Cape Town   Hong Kong   Montreal

Publisher: Greg Tobin

Editor in Chief: Maureen O'Connor

Acquisitions Editor: Jennifer P. Crum

Editorial Project Manager: Kari Heen

Managing Editor: Ron Hampton

Developmental Editor: Lenore Parens

Text and Cover Design: Dennis Schaefer

Production Coordinator: Sheila Spinney

Production Services: Pre-Press Company, Inc.

Assistant Editor: Gregory Erb

Media Producer: Sharon Smith

Software Development: TestGen: Marty Wright; InterAct: John O'Brien

Marketing Manager: Dona Kenly

Marketing Coordinator: Lindsay Skay

Prepress Services Buyer: Caroline Fell

First Print Buyer: Hugh Crawford

Cover Image: © Dale Gulden/IndexStock Imagery

Library of Congress Cataloging-in-Publication Data

Carson, Tom, 1967–

   Elementary and intermediate algebra/Tom Carson, Ellyn Gillespie, Bill Jordan.

    p. cm.

  ISBN 0-201-7296–8

  Includes index.

  1. Algebra. I. Gillespie, Ellyn. II. Title.

QA152.3.C37 2004

512—dc21   2003044330

1 2 3 4 5 6 7 8 9 10   QWC   07 06 05 04

CONTENTS

Elementary and Intermediate Algebra is the fourth book in a series that includes *Prealgebra, Elementary Algebra,* and *Intermediate Algebra.* Written in a relaxed, nonthreatening style, *Elementary and Intermediate Algebra* takes great care to ensure that students who have struggled with math in the past will be comfortable with the subject matter. Explanations are carefully developed to provide a sense of why an algebraic process works the way it does, instead of just an explanation of how to follow the process. In addition, problems from science, engineering, accounting, health, the arts, and everyday life link algebra to the real world. A complete study system with a Learning Styles Inventory and frequent learning strategies is also provided to give students extra guidance and to help them be successful. (See pages xv–xx.)

This text is designed to be versatile enough for use in a standard lecture format, a self-paced lab, or even in an independent study format. A strong ancillary package provides a wealth of supplemental resources for both instructors and students.

Key Features

Study System   A study system is presented in the *To the Student* section on pages xv–xviii. This system is then reinforced throughout the text. The system recommends color codes for taking notes. The color codes are consistent in the text itself: red for definitions, blue for procedures and rules, and black for notes and examples. In addition, the study system presents strategies for succeeding in the course. These learning strategies are revisited in the chapter openers and throughout the body of the text.

Learning Styles Inventory   A learning styles inventory is presented on page xix to help students assess their particular learning style. Learning strategies are then presented throughout the book with different learning styles in mind.

Learning Strategy Boxes   Learning Strategy boxes appear where appropriate in the text to offer advice on how to effectively use the study system and how to study specific topics based on a student's learning style (see pages 7, 125, and 397).

The Algebra Pyramid   An Algebra Pyramid is used throughout the text to help students see how the topic they are learning relates to the big picture of algebra—particularly focusing on the relationship between constants, variables, expressions, and equations (see pages 3 and 390).

Chapter Openers   Like the Algebra Pyramid, chapter openers are designed to help students see how the topics in the upcoming chapter relate to the big picture of the entire course. The chapter openers give information about the importance of the topics in each chapter and how they fit into the overall structure of the course (see pages 1, 191, and 377).

Connection Boxes   Connection boxes bridge concepts and ideas that students have learned elsewhere in the text so they see how the concepts are interrelated and build on each other (see pages 19, 239, and 429).

Your Turn Practice Exercises   Your Turn practice exercises are found after most examples to help give students an interactive learning environment and to prepare them for the exercise sets that lie ahead. This practice step provides immediate feedback so students can gain confidence in what they are learning (see pages 75, 246, and 319).

Real, Relevant, and Interesting Applications   A large portion of the application problems are taken from real situations in science, engineering, health, finance, the arts, or just everyday life. The real-world applications illustrate the everyday use of basic algebraic concepts and encourage students to apply mathematical concepts to solve problems (see pages 114, 203, and 448).

Thorough Explanations   Great care is taken to explain not only how to do the math, but also why the math works the way it does, where it comes from, and how it is relevant to students' everyday lives. Knowing all of this gives students a context in which to remember the concept.

Problem-Solving Outline   A five-step problem-solving outline is introduced on page 107 of Section 2.1 with the following headings:

1. Understand
2. Plan
3. Execute
4. Answer
5. Check

Application examples throughout the rest of the text follow the steps given in this outline, presenting the headings to model the thinking process clearly (see pages 109, 244, and 404).

Warning Boxes   Warning boxes alert students to common mistakes and false assumptions that students often make and explain *why* these are incorrect (see pages 196, 346, and 428).

Of Interest Boxes   Of Interest boxes are positioned throughout the text to offer a unique perspective to content that some students might otherwise consider ho-hum mathematics. Sometimes containing trivia and other times historical notes, Of Interest boxes are designed to enhance the learning process by making concepts fun, interesting, and memorable (see pages 117, 218, and 448).

Puzzle Problems   These mathematical brainteasers, often solved without a formulaic approach, appear at the end of selected exercise sets to encourage critical thinking (see pages 41, 206, and 401).

Collaborative Exercises   These exercises, which appear at least once per chapter, encourage students to work in groups, discuss the mathematics, and use the topics from a particular section or group of sections to solve a problem (see pages 152, 220, and 411).

Calculator Tips   The relevant functions of calculators (scientific or graphing, depending on the topic) are explained and illustrated throughout the text in optional Calculator Tips feature. In addition, an occasional calculator icon in the exercise sets indicates that the problem can be solved using a calculator, though one is not required (see pages 62, 308, and 383).

Review Exercises Since continuous review is important in any mathematics course, this text includes review exercises at the end of each exercise set. These exercises review previously learned concepts not only to keep the material fresh for students, but also to serve as a foundational review for the discussion in the upcoming section (see pages 26, 179, and 453).

Chapter Summaries and Review Exercises An extensive summary at the end of each chapter provides a list of defined terms referenced by section and page number, a two-column summary of key concepts, and a list of important formulas appearing in that chapter. A set of review exercises is also provided (see pages 180, 359, and 454).

Chapter Practice Tests A practice test follows each set of chapter review exercises. The problem types in the practice tests correlate to the short-answer tests in the Printed Test Bank. This is especially comforting for students who have math anxiety or who experience test anxiety (see pages 101, 189, and 374).

Cumulative Reviews Cumulative review exercises appear after Chapters 3, 6, 9, and 13. These exercises help students stay current with all the material they have learned (see pages 264, 541, and 812).

Supplements for Instructors

ANNOTATED INSTRUCTOR'S EDITION (ISBN 0-201-77157-8)

The Annotated Instructor's Edition provides answers to all exercises, including puzzle problems and collaborative exercises, near the corresponding problem. In addition, useful teaching tips are printed in the margin, and a ★ icon indicates especially challenging exercises in the exercise sets. All answers are printed in bright blue to make them easy to locate.

INSTRUCTOR'S SOLUTIONS MANUAL (ISBN 0-201-72975-X)

The *Instructor's Solutions Manual,* written by text coauthor Ellyn Gillespie, contains complete solutions to all even-numbered section exercises, puzzle problems, and collaborative exercises.

PRINTED TEST BANK/INSTRUCTOR'S RESOURCE GUIDE (ISBN 0-201-77162-4)

The *Printed Test Bank* portion of this manual contains a diagnostic pretest for each chapter; four free-response test forms for each chapter, one of which contains higher-level questions; one multiple-choice test per chapter; one free-response midterm exam; two free-response final exams; and one multiple-choice final exam. The *Instructor's Resource Guide* portion of the manual contains mid-chapter check-ups for each chapter.

TESTGEN WITH QUIZMASTER (ISBN 0-321-12193-7)

TestGen enables instructors to build, edit, print, and administer tests using a computerized bank of questions developed to cover all the objectives of the text. Instructors can modify test bank questions or add new questions by using the built-in question editor, which allows users to create graphs, import graphics, insert math notation, and insert variable numbers or text. Tests can be printed or administered on-line via the Web or another network. TestGen comes packaged with QuizMaster, which allows students to take tests on a local area network. The software is available on a dual-platform Windows/Macintosh CD-ROM.

MathXL® www.mathxl.com

MathXL is an online testing, homework, and tutorial system that uses algorithmically generated exercises correlated to the textbook. Instructors can assign tests provided by Addison-Wesley or create and customize their own tests and homework assignments. Instructors can also track their students' results and tutorial work in an on-line grade book.

Students can take sample chapter tests and receive personalized study plans that will diagnose weaknesses and link students to areas they need to study and retest. Students can also work unlimited practice problems and receive tutorial instruction for areas in which they need improvement. MathXL can be packaged with new copies of *Elementary and Intermediate Algebra.* Please contact your Addison-Wesley representative for details.

MyMathLab

MyMathLab is a complete online course for Addison-Wesley mathematics textbooks that provides interactive, multimedia instruction correlated to textbook content. MyMathLab is easily customized to suit the needs of students and instructors and provides a comprehensive and efficient online course-management system that allows for diagnosis, assessment, and tracking of students' progress.

MyMathLab features the following:

- Fully interactive multimedia chapter and section folders from the textbook contain a wide range of instructional content, including videos, software tools, audio clips, animations, and electronic supplements.

- Hyperlinks take you directly to online testing, diagnosis, tutorials, and grade books in MathXL—Addison-Wesley's tutorial and testing system for mathematics and statistics.

- Instructors can create, copy, edit, assign, and track all tests for their course as well as track student tutorial and testing performance.

- With push-button ease, instructors can remove, hide, or annotate Addison-Wesley's preloaded content, add their own course documents, or change the order in which material is presented.

- Using the communication tools found in MyMathLab, instructors can hold online office hours, host a discussion board, create communication groups within their class, send e-mails, and maintain a course calendar.

- Print supplements are available on-line, side by side with the textbook.

For more information, visit our Web site at **www.mymathlab.com** or contact your Addison-Wesley sales representative for a demonstration.

Supplements for Students

Student's Solutions Manual (ISBN 0-201-72976-8)

The *Student's Solutions Manual*, written by Ellyn Gillespie, contains complete solutions to the odd-numbered exercises in each section and to all the section-level review exercises, chapter review exercises, practice tests, and cumulative review exercises.

Videotapes (ISBN 0-321-17886-6)

The videotape series features an engaging team of mathematics instructors who present comprehensive coverage of each section of the text in a student-interactive format. Ellyn Gillespie, a coauthor of the text, presents a major portion of the content in the videos, keeping the presentation of content consistent with the text. The lecturers' presentations include examples and problems from the text, and support an approach that emphasizes visualization and problem solving. A video symbol at the beginning of each exercise set references the appropriate videotape or CD number (see Digital Video Tutor, below).

Digital Video Tutor (ISBN 0-321-17885-8)

The video lessons for this text are also available on CD-ROM, making it easy and convenient for students to watch video segments from a computer at home or on campus. The complete digitized video set, affordable and portable for students, is ideal for distance learning or supplemental instruction.

InterAct Math® Tutorial CD-ROM (ISBN 0-321-12195-3)

This interactive tutorial software provides algorithmically generated practice exercises that correlate at the objective level to the odd-numbered exercises in the text. Each practice exercise is accompanied by both an example and a guided solution designed to involve students in the solution process. Selected problems also include a video clip that helps students visualize concepts. The software recognizes common student errors and provides appropriate feedback.

MathXL® www.mathxl.com

An access card is required and can be purchased through the bookstore or at www.aw.com. See Instructor Supplements descriptions for more details.

MyMathLab

Visit www.mymathlab.com for more information. An access card is required. See Instructor Supplements descriptions for more details.

Addison-Wesley Math Tutor Center (ISBN 0-201-72170-8)

The Addison-Wesley Math Tutor Center is staffed by qualified college mathematics instructors who tutor students on examples and exercises from the textbook. Tutoring is provided via toll-free telephone, toll-free fax, e-mail, and the Internet. Interactive Web-based technology allows students and tutors to view and listen to live instruction in real-time over the Internet! The Math Tutor Center is accessed through a registration number that can be packaged with a new textbook or purchased separately. (Note: MyMathLab students obtain access to the Tutor Center through their MyMathLab access code.)

Acknowledgments

Many people gave of themselves in so many ways during the development of this text. Mere words cannot contain the fullness of our gratitude. Though the words of thanks that follow may be few, please know that our gratitude is great.

We would like to thank the following people who gave of their time in reviewing the text. Their thoughtful input was vital to the development of the text.

Khadija Ahmed, *Monroe County Community College*

Frank Attanucci, *Scottsdale Community College*

Kerry Bailey, *Laramie Community College*

Debra Bryant, *Tennessee Technological University*

Baruch Cahlon, *Oakland University*

Pat Cook, *Weatherford College*

Patrick S. Cross, *University of Oklahoma*

Cheryl B. Davids, *Central Carolina Technical College*

Elias Deeba, *University of Houston—Downtown Campus*

Stephan DeLong, *Tidewater Community College—Virginia Beach Campus*

Laura Ferguson, *Weatherford College*

Margret Hathaway, *Kansas City Kansas Community College*

Carol Murphy, *Miramar College*

Joanne Peeples, *El Paso Community College*

Larry Pontaski, *Pueblo Community College*

Jack Sharp, *Floyd College*

Linda Shoesmith, *Scott Community College*

James Vicich, *Scottsdale Community College*

Linda J. Wagner, *Indiana University—Purdue University Fort Wayne*

Walter Wang, *Baruch College*

We would like to extend a heartfelt thank-you to everyone at Addison-Wesley publishing for giving so much to this project. We would like to offer special thanks to Jennifer Crum and Greg Tobin, who believed in us and gave us the opportunity; to Lenore Parens, who worked so diligently to get the manuscript just right; to Greg Erb, Katie Nopper, and Kari Heen, for keeping us on track; and also to Dona Kenly and Lindsay Skay, for the encouragement and working so hard to get us "out there."

A very special thank-you to Dennis Schaefer, who created the beautiful, student-friendly design; to Ron Hampton, whose keen eyes and editorial sense were invaluable during production; and to Jennifer Carley, Gordon Laws, Katy Faria, Sam Blake, and all of the fabulous people at Pre-Press Company, Inc. for working so hard to put together the finished pages.

To Sharon Smith, Ruth Berry, Mary Ann Perry, and all the people involved in developing the media supplements package, we are so grateful for all that you do. A special thank-you to Laura Hoye, who created the excellent printed test bank. Thank you to Cheryl Davids, John Morin, and John Jacobs for their wonderful job of accuracy checking the manuscript, and to Vince Koehler for accuracy checking and answer writing. A big thank-you goes to Martha Goshaw for her help creating application problems and answer writing.

Finally, we'd like to thank our families (especially Laura, Connor, and Johanna) for their support, encouragement, and love throughout the process of developing this text.

Tom Carson
Ellyn Gillespie Stewart
Bill Jordan

Why do I have to take this course?

Often this is one of the first questions students ask when they find out they must take an algebra course, especially when they believe that they will never use the math again. You may think that you will not use algebra directly in daily life, and you may assume that you can get by knowing just enough arithmetic to balance a checkbook. So, what is the real point of education? Why don't colleges just train students for the jobs they want? The purpose of education is not just job training but also exercise—mental exercise. An analogy that illustrates this quite well is the physical training of athletes.

During the off-season, athletes usually develop an exercise routine that may involve weight lifting, running, swimming, aerobics, or maybe even dance lessons. Athletes often seek out a professional trainer to push them further than they might push themselves. The trainer's job is not to teach an athlete better technique in his or her sport, but to develop the athlete's raw material— to work the body for more strength, stamina, and balance. Educators are like physical trainers, and going to college is like going to the gym. An educator's job is to push students mentally and work the "muscle" of the mind. A college program is designed to develop the raw material of the intellect so the student can be competitive in the job market. After the athlete completes the off-season exercise program, he or she returns to the coach and receives specific technique training. Similarly, when students complete their college education and begin a job, they receive specific training to do that job. If the trainer or teacher has done a good job with hard-working clients, the coaching or job training should be absorbed easily.

Taking this analogy a step further, a good physical trainer finds the athlete's weaknesses and designs exercises that the athlete has never performed before, and then pushes him or her accordingly. Teachers do the same thing—their assignments are difficult in order to work the mind effectively. If you feel "brain-strained" as you go through your courses, that's a good sign that you are making progress, and you should keep up the effort.

The following study system is designed to help you in your academic workouts. As teachers, we find that most students who struggle with mathematics have never really studied math. A student may think, "Paying attention in class is all I need to do." When you watch a teacher do math, however, keep in mind that you are watching a pro. Going back to the sports analogy, you can't expect to shoot a score of 68 in golf by watching Tiger Woods. You have to practice golf yourself in order to learn and improve. The study system outlined in the following pages will help you get organized and make efficient use of your time so that you can maximize the benefits of your course work.

What do I need to do to succeed?

We believe you must have or acquire four prerequisites in order to succeed in college:

1. **Positive Attitude**
2. **Commitment**
3. **Discipline**
4. **Time**

A **Positive Attitude** is most important because commitment and discipline flow naturally from it. Consider Thomas Edison, inventor of the lightbulb. He tried more than 2000 different combinations of materials for the filament before he found the successful combination. When asked by a reporter about all his failed attempts, Edison replied, "I didn't fail once, I invented the lightbulb. It was just a 2000-step process." Recognize that learning can be uncomfortable and difficult, and mistakes are part of the process. Embrace the learning process with its discomforts and difficulties, and you'll see how easy it is to be committed and disciplined.

Commitment means giving everything you've got with no turning back. Consider Edison again. Imagine the doubts and frustrations he must have felt trying material after material for the filament of his lightbulb without success. Yet he forged ahead. Quitting was simply not an option. In Edison's own words, "Our greatest weakness lies in giving up. The most certain way to succeed is always to try just one more time."

Discipline means doing things you should be doing even when you don't want to. According to author W. K. Hope, "Self-discipline is when your conscience tells you to do something and you don't talk back." Staying disciplined can be difficult given all the distractions in our society. The best way to develop discipline is to create a schedule and stick to it.

Make sure you have enough **Time** to study properly, and make sure that you manage that time wisely. Too often, students try to fit school into an already full schedule. Take a moment to complete the exercise that follows under "How do I do it all?" to make sure you haven't overcommitted yourself. Once you have a sense of how much time school requires, read on about the study system that will help you maximize the benefits of your study time.

How do I do it all?

Now that we know a little about what it takes to be successful, let's make sure that you have enough time for school. In general, humans have a maximum of 60 hours of productivity per week. Therefore, as a guide, let's set the maximum number of work hours, which means time spent at your job(s) and at school combined, at 60 hours per week. Use the following exercise to determine the time you commit to your job and to school.

Exercise: Calculate the time that you spend at your job and at school.

1. Calculate the total hours you work in one week.
2. Calculate the number of hours you are in class each week.
3. Estimate the number of hours you should expect to spend outside of class studying. *A general rule is to double the number of hours spent in class.*

4. Add your work hours, in-class hours, and estimated out-of-class hours to get your total time commitment.
5. Evaluate the results. *See below.*

Evaluating the Results:

a. If your total is greater than 60 hours, you may find yourself overwhelmed. It may not occur at first, but doing too much for an extended period of time will eventually catch up with you, and something may suffer. It is in your best interest to cut back on work or school until you bring your time commitment under 60 hours per week.

b. If your total is under 60 hours, good. Be sure you consider other elements in your life, such as your family's needs, health problems, commuting, or anything that could make demands on your time. If you do not have enough time for everything, consider what can be cut back. It is important to note that it is far better to pass fewer classes than to fail many.

How do I make the best of my time? How should I study?

We've seen many students who had been making Ds and Fs in mathematics transform their grades to As and Bs by using the study system that follows.

The Study System

YOUR NOTEBOOK

1. Get a loose-leaf binder so that you can put papers in and take them out without ripping any pages.
2. Organize the notebook into four parts:
   a. Class notes
   b. Homework
   c. Study sheets (a single piece of paper for each chapter onto which you will transfer procedures from your notes)
   d. Practice tests

IN CLASS

Involve your mind completely.

1. **Take good notes**. Use three different colors while taking notes. Most students like using red, blue, and black (pencil).

   - Use the red pen to write *definitions*. Also, use this color to mark problems or items that the instructor indicates will be covered on a test.
   - Use the blue pen to write *procedures* and *rules*.
   - Use the pencil to write *problems* and *explanations*.

   When taking notes, don't just write the solutions to the problems that the instructor works out, but write the explanations as well. To the side of the problem, make notes about each step so that you remember the significance of the steps. Pay attention to examples or issues the instructor emphasizes: they often appear on a test, so make an effort to include them in your notes. Include common errors that the instructor points out, or any words of caution. If you find it is difficult to write and pay attention at the same time, ask your instructor if you can record the lectures with a tape recorder. If your instructor follows the text closely, when he or she points out definitions or procedures in the text, highlight them or write a page reference in your notes. You can then write these referenced items in their proper place in your notes after class.

2. **Answer the instructor's questions.** You should think through every question and answer in your mind, in your notes, or out loud.

3. **Ask questions.** You may find it uncomfortable to ask questions in front of other people, but keep in mind that if you have a question, it is very likely that someone else has the same question. If you still don't feel like asking in class, be sure to ask as soon as class is over. The main thing is to get that question answered as soon as possible.

AFTER CLASS

Prepare for the next class meeting as if you were going to have a test on everything covered so far.

To make the most of your time, set aside a specific time that is reserved for math. Since there are often many distractions at home, study math while on campus in a quiet place such as the library or tutorial lab. Staying on campus also allows you to visit your instructor or tutorial services if you have a question that you cannot resolve. Here is a systematic approach to organizing your math study time outside of class:

1. As soon as possible, go over your notes. Clarify any sentences that weren't quite complete. Fill in any page-referenced material.

2. Read through the relevant section(s) in the text again, and make sure you understand all the examples.

3. Transfer each new procedure or rule to your study sheet for that chapter. Write down important terms and their definitions. Make headings for each objective in the section(s) you covered that day and write the procedures and definitions in your own words.

4. Study the examples worked in class. Transfer each example (without the solution) to the practice test section of your notebook, leaving room to work it out later.

5. Use your study sheet to do the assigned practice problems. As soon as you finish each problem, check your answer in the back of the book or in the *Student's Solutions Manual*. If you did not get it correct, then immediately revisit the problem to determine your error (see the box on troubleshooting). If you are asked to do even-numbered problems, work odd-numbered problems that mirror the even problems. That way you can check your answers for the odd-numbered problems, and then work the even-numbered problems with confidence.

6. After completing the homework, prepare a quiz for yourself. Select one of each type of homework problem. Don't just pick the easy ones! Set the quiz aside for later.

Troubleshooting: For the problems that you do not get correct, first look for simple arithmetic errors. If you find no arithmetic errors, then make sure you followed the procedure rules correctly. If you followed the rules correctly, then you have likely interpreted something incorrectly, either with the problem or the rules. Read the instructions again carefully and try to find similar examples in your notes or in the book. If you still can't find the mistake, go on to something else for a while. Often after taking a fresh look you will see the mistake right away. If all these tips fail to resolve the problem, then mark it as a question for the next class meeting.

7. After making the quiz, study your study sheet. To test your understanding, write the rules and procedures in your own words. Do not focus on memorizing the wording in the textbook.

8. Now it is time to begin preparing for the next class meeting. Read the next section(s) to be covered. Don't worry if you do not understand everything. The idea is to get some feeling for the topics to be discussed so that the class discussion will actually be the second time you encounter the material. While reading, you might mark points that you find difficult so that if the instructor does not clear them up, you can ask about them. Also, attempt to work through the examples. The idea is for you to do as much as possible on your own before class so that the in-class discussion merely ties together loose ends and solidifies the material.

9. After you have finished preparing for the next day, go back and take the quiz that you made. If you get all the answers correct, then you have mastered the material. If you have difficulty, return to your study sheet and repeat the exercise of writing explanations for each objective.

How do I ace the test?

If you have followed all of the preceding suggestions, then preparing for a test should be quite easy.

1. **Read.** In one sitting, read through all of your notes on the material to be tested. In the same sitting, read through the book, observing what the instructor has highlighted in class. To guide your studies, look at any information or documents provided by your instructor that address what will be on the test. The examples given by the instructor will usually reflect the test content.

2. **Study.** Compare your study sheet to the summary in the book at the end of the chapter. Use both to guide you in your preparation, but keep in mind that the sheet you made reflects what the instructor has emphasized. Make sure you understand everything on your study sheet. Write explanations of the objectives until you eliminate all hesitation about how to approach an objective. The rules and procedures should become second nature.

3. **Practice.** Create a game plan for the test by writing the rule, definition, or procedure that corresponds to each problem on your practice tests. (Remember, one practice test is in your book and the other you made from your notes.) Next, work through the practice tests without referring to your study sheet or game plan.

4. **Evaluate.** Once you have completed the practice tests, check them. The answers to the practice test are in the back of the book. Check the practice test that you made using your notes.

5. **Repeat.** Keep repeating steps 2, 3, and 4 until you get the right answer for every problem on the practice tests.

TAKING A TEST

1. When the test hits your desk, don't look at it. Instead, do a memory dump. On paper, dump everything you think you might forget. Write out rules, procedures, notes to yourself, things to watch out for, special

instructions from the instructor, and so on. This will help you to relax while taking the test.

2. If you get to a problem that you cannot figure out, skip it and move on to another problem. First do all the problems you are certain of and then return to the ones that are more difficult.

3. Use all the time given. If you finish early, check to make sure you have answered every problem. Even if you cannot figure out a problem, at least guess. Use any remaining time to check as many problems as possible by doing them over on separate paper.

If You Are Not Getting Good Results

Evaluate the situation. What are you doing or not doing in the course? Are you doing all the homework and taking the time to prepare as suggested? Sometimes people misjudge how well they have prepared. Just like an athlete, to excel, you will need to prepare beyond the minimum requirements. Here are some suggestions:

1. **Go to your instructor.** Ask your instructor for help to evaluate what is wrong. Make use of your instructor's office hours, because this is your opportunity for individual attention.

2. **Get a tutor.** If your school has a tutorial service, and most do, do your homework there so you can get immediate help when you have a question.

3. **Use Addison-Wesley's support materials.** Use the support materials that are available for your text, which include a *Student's Solutions Manual*, the Addison-Wesley Math Tutor Center, videotapes (available on CD-ROM as well), and tutorial software available on CD and on-line. Full descriptions of these supplements are provided on page x of this book.

4. **Join a study group.** Meet regularly with a few people from class and go over material together. Quiz each other and answer questions. Meet with the group only after you have done your own preparation so you can then compare notes and discuss any issues that came up in your own work. If you have to miss class, ask the study group for the assignments and notes.

We hope you find this study plan helpful. Be sure to take the Learning Styles Inventory that follows to help determine your primary learning style. Good luck!

LEARNING STYLES INVENTORY

What is your personal learning style?

A learning style is the way in which a person processes new information. Knowing your learning style can help you make choices in the way you focus on and study new material. Below are fifteen statements that will help you assess your learning style. After reading each statement, rate your response to the statement using the scale below. There are no right or wrong answers.

3 = Often applies   **2** = Sometimes applies   **1** = Never or almost never applies

____ **1.** I remember information better if I write it down or draw a picture of it.

____ **2.** I remember things better when I hear them instead of just reading or seeing them.

____ **3.** When I receive something that has to be assembled, I just start doing it. I don't read the directions.

____ **4.** If I am taking a test, I can visualize the page of text or lecture notes where the answer is located.

____ **5.** I would rather the professor explain a graph, chart, or diagram to me instead of just showing it to me.

____ **6.** When learning new things, I want to do it rather than hear about it.

____ **7.** I would rather the instructor write the information on the board or overhead instead of just lecturing.

____ **8.** I would rather listen to a book on tape than read it.

____ **9.** I enjoy making things, putting things together, and working with my hands.

____ **10.** I am able to conceptualize quickly and visualize information.

____ **11.** I learn best by hearing words.

____ **12.** I have been called hyperactive by my parents, spouse, partner, or professor.

____ **13.** I have no trouble reading maps, charts, or diagrams.

____ **14.** I can usually pick up on small sounds like bells, crickets, or frogs, or distant sounds like train whistles.

____ **15.** I use my hands and I gesture a lot when I speak to others.

Write your score for each statement beside the appropriate statement number below. Then add the scores in each column to get a total score for that column.

	Visual		Auditory		Tactile
	1._____		2._____		3._____
	4._____		5._____		6._____
	7._____		8._____		9._____
	10._____		11._____		12._____
	13._____		14._____		15._____
Total:	_____		_____		_____

The largest total of the three columns indicates your dominant learning style.

Visual learners learn best by seeing. If this is your dominant learning style then you should focus on learning strategies that involve seeing. The color coding in the study system (see page xv) will be especially important. The same color coding is used in the text. Draw lots of diagrams, arrows, and pictures in your notes to help you see what is happening. Reading your notes, study sheets, and text repeatedly will be an important strategy.

Auditory learners learn best by hearing. If this is your dominant learning style then you should use learning strategies that involve hearing. After getting permission from your instructor, bring a tape recorder to class to record the discussion. When you study your notes, play back the tape. Also, when you learn rules, say the rule over and over. As you work problems, say the rule before you do the problem. You may also find the videotapes to be beneficial because you can hear explanations of problems taken from the text.

Tactile (also known as kinesthetic) learners learn best by touching or doing. If this is your dominant learning style, you should use learning strategies that involve doing. Doing lots of practice problems will be important. Make use of the Your Turn exercises in the text. These are designed to give you an opportunity to do problems like the examples as soon as a topic is discussed. Writing out your study sheets and doing your practice tests repeatedly will be important strategies for you.

Note that the study system developed in this text is for all learners. Your learning style will help you decide what aspects and strategies in the study system to focus on, but being predominantly an auditory learner does not mean that you shouldn't read the textbook, do lots of practice problems, or use the color-coding system in your notes. Auditory learners can benefit from seeing and doing, and tactile learners can benefit from seeing and hearing. In other words, do not use your dominant learning style as a reason for not doing things that are beneficial to the learning process. Also, remember that the Learning Strategy boxes presented throughout the text provide tips to help you use your personal learning style to your advantage.

This Learning Styles Inventory is adapted from *Cornerstone: Building on Your Best* by Montgomery/Moody/Sherfield, ©2000. Reprinted by permission of Prentice-Hall, Inc., Upper Saddle River, NJ.

Foundations of Algebra

In Peter Schultz's quote, each of the three people answered the question correctly. Notice that the first two people were rather narrowly focused whereas the third person had a greater vision and appreciation of the finished structure and his or her place in the construction process. Your curriculum and the courses in it are like a building, with the units or chapters like construction materials. Try to keep the big picture in mind as your courses and curriculum build. In the chapter openers of this book, we will discuss how the mathematics in the chapter fits into the bigger picture of the entire course structure to help you see the "cathedral" we are building.

In this chapter, we review the foundation of algebra, which is arithmetic. More specifically, we will review number sets, operations of arithmetic, properties of real numbers, and the evaluation and simplification of expressions. This review is by no means a complete instruction of arithmetic. Since a solid foundation is important to support the rest of our building, if you encounter a topic in this chapter that is a particular weakness for you, then it is important that you consult with your instructor or other more complete sources for extra practice.

"Three people were at work on a construction site. All were doing the same job, but when each was asked what the job was, the answers varied. 'Breaking rocks,' the first replied. 'Earning my living,' the second said. 'Helping to build a cathedral,' said the third."

—PETER SCHULTZ
German businessman

"Build up your weaknesses until they become your strengths."

—KNUTE ROCKNE
Notre Dame football coach 1888–1931

1.1 Number Sets and the Structure of Algebra

OBJECTIVES

1. Understand the structure of algebra.
2. Classify number sets.
3. Graph rational numbers on a number line.
4. Determine the absolute value of a number.
5. Compare numbers.

OBJECTIVE 1 **Understand the structure of algebra.** Learning mathematics is like learning a language. When we learn a language we must learn the alphabet, vocabulary, and sentence structure. Similarly, mathematics has its own alphabet, vocabulary, and sentence structure. In this section we begin the development of the foundation of algebra with an overview of its components and structure. The basic components are **variables** and **constants**.

> **DEFINITIONS**
>
> **Variable:** A symbol that can vary in value.
> **Constant:** A symbol that does not vary in value.

Variables are usually letters of the alphabet, like x or y. Usually constants are symbols for numbers like 1, 2, $\frac{3}{4}$, 6.74, and so on. However, constants can sometimes be symbols like e or the Greek letter π, which each have special numeric values. Variables and constants are used to make **expressions**, **equations**, and **inequalities**.

> **DEFINITION**
>
> **Expression:** A collection of constants, variables, and arithmetic symbols.

Examples of expressions:

$$2 + 6 \quad \text{or} \quad 4x - 5 \quad \text{or} \quad \frac{1}{3}\pi r^2 h$$

> **DEFINITION**
>
> **Equation:** Two expressions set equal to each other.

Examples of equations:

$$2 + 6 = 8 \quad \text{or} \quad 4x - 5 = 12 \quad \text{or} \quad V = \frac{1}{3}\pi r^2 h$$

> **CONNECTION**
>
> Think of expressions as phrases and equations as complete sentences. The expression $2 + 6$ is read "two plus six," which is not a complete sentence. The equation $2 + 6 = 8$ is read "Two plus six is eight." Notice the equal sign translates to the verb *is*, which makes the sentence a complete sentence.

Inequality Symbols and Their Translations	
Symbolic form	Translation
$8 \neq 3$	Eight is not equal to three.
$5 < 7$	Five is less than seven.
$7 > 5$	Seven is greater than five.
$x \leq 3$	x is less than or equal to three.
$y \geq 2$	y is greater than or equal to two.

The algebra pyramid shown illustrates how variables, constants, expressions, equations, and inequalities relate. At the foundation of algebra and our pyramid are constants and variables, which are used to build expressions, which in turn are used to build equations and inequalities.

THE ALGEBRA PYRAMID

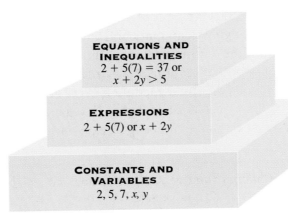

EQUATIONS AND INEQUALITIES
$2 + 5(7) = 37$ or
$x + 2y > 5$

EXPRESSIONS
$2 + 5(7)$ or $x + 2y$

CONSTANTS AND VARIABLES
$2, 5, 7, x, y$

Note: During this course, we will move back and forth between expressions, equations, and inequalities. When we transition from one to the other we will use the algebra pyramid as a quick visual so that it's clear what we are working on.

OBJECTIVE ② **Classify number sets.** Numbers can be placed into different categories using **sets**.

> **DEFINITION** | **Set:** A collection of objects.

Braces are used to indicate a set. For example, the set containing the numbers 1, 2, 3, and 4 would be written {1, 2, 3, 4}. The numbers 1, 2, 3, and 4 are called the *members* or *elements* of this set.

PROCEDURE **WRITING SETS**

To write a set, write the members or elements of the set within braces { }.

EXAMPLE ① Write the set containing the first five letters of the alphabet.

Answer: {A, B, C, D, E}

Numbers are classified using number sets. The set of *natural numbers* contains the counting numbers 1, 2, 3, 4, . . . and is written {1, 2, 3, . . . }. The three dots are called *ellipses* and indicate that the numbers continue forever in the same pattern. The set of *whole numbers* contains all of the natural numbers and 0 and is written {0, 1, 2, 3, . . . }. The set of *integers* contains all the whole numbers and the opposite (or negative) of every natural number and is written { . . . , −3, −2, −1, 0, 1, 2, 3, . . . }. Notice the integers continue forever in both directions.

A number line is often useful in mathematics. The following number line is marked with the integers.

Though we can mark and view only a portion of the number line, the arrows at the ends indicate that the line and the numbers on it continue on forever in both directions. If we traveled along the number line forever in both directions we would encounter every number in the set of *real numbers*.

Some real numbers are not integers. The set of **rational numbers** contains every real number that can be expressed as a ratio of integers.

DEFINITION

Rational number: Any real number that can be expressed in the form $\dfrac{a}{b}$ where a and b are integers and $b \neq 0$.

Note: In the definition for a rational number, the notation $b \neq 0$ is important because if the denominator b were to equal 0, then the fraction would be undefined. The reason it is undefined will be explained later.

For example, $\dfrac{3}{4}$ is a rational number because 3 and 4 are integers. Note that numbers like 0.75 and 75% are also rational numbers because they can be written as $\dfrac{3}{4}$. It is important to realize that the definition of a rational number does not state that the number *must* be expressed in the form $\dfrac{a}{b}$. Rather, if a number *can* be expressed as a fraction, it is a rational number. For example, the number 5 is a rational number because it can be expressed as $\dfrac{5}{1}$.

EXAMPLE 2 Determine whether the given number is a rational number.

a. $\dfrac{2}{3}$

Answer: Yes, because 2 and 3 are integers.

b. 0.4

Answer: Yes, 0.4 is a rational number because it can be expressed as $\dfrac{4}{10}$, and 4 and 10 are integers.

c. $0.\overline{6}$

Answer: A bar written over a decimal digit indicates that the digit repeats without end, so $0.\overline{6} = 0.66666\ldots$, and we say these decimal numbers are non-terminating decimal numbers. Usually we encounter these numbers as quotients in certain division problems, such as when we write certain fractions as decimals. In this case, $0.\overline{6}$ is the decimal equivalent of $\dfrac{2}{3}$. Since $0.\overline{6}$ can be expressed as the fraction $\dfrac{2}{3}$, it is a rational number.

Note: All nonterminating decimal numbers with repeating digits are rational numbers because they can be expressed as fractions with integers in both the numerator and denominator.

YOUR TURN Determine whether the given number is a rational number.

a. 0.56   **b.** -7

Not all numbers can be expressed as a ratio of integers. One such number is π (pronounced "pie"). Because the exact value of π cannot be expressed as a ratio of integers, it is categorized as an **irrational number**.

DEFINITION *Irrational number:* Any real number that is not rational.

Some other irrational numbers are $\sqrt{2}$ or $\sqrt{3}$. (Square roots will be explained in more detail in Section 1.5.) Because an irrational number cannot be written as a ratio of integers, if a calculation involves an irrational number, we must leave it in symbolic form or use a rational number approximation. We can approximate π with rational numbers like 3.14 or $\dfrac{22}{7}$.

ANSWERS

a. yes **b.** yes

The following figure illustrates how the number sets relate in the real number system.

Real Numbers

Irrational Numbers: Any real number that is not rational, such as: $-\sqrt{2}$, $-\sqrt{3}$, $\sqrt{0.8}$, and π.

Rational Numbers: Real numbers that can be expressed in the form $\frac{a}{b}$, where a and b are integers and $b \neq 0$, such as: $-4\frac{3}{4}$, $-\frac{2}{3}$, 0.018, $0.\overline{3}$, and $\frac{5}{8}$.

Integers: $\ldots, -3, -2, -1, 0, 1, 2, 3, \ldots$

Whole Numbers: $0, 1, 2, 3, \ldots$

Natural Numbers: $1, 2, 3, \ldots$

OBJECTIVE **3** **Graph rational numbers on a number line.**   Number lines can be useful tools when comparing numbers or solving certain arithmetic problems. Let's review how to graph a number on a number line.

EXAMPLE **3**   Graph -2.56 on a number line.

Solution Because -2.56 means $-2\frac{56}{100}$, we could divide the space in between -2 and -3 into 100 divisions and count to the 56th mark. Because this is tedious, we will gradually zoom in on smaller and smaller sections of the number line.

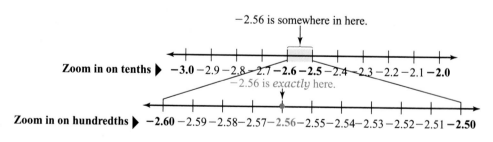

-2.56 is somewhere in here.

Zoom in on tenths ▶ $\quad -3.0 \; -2.9 \; -2.8 \; -2.7 \; -2.6 \; -2.5 \; -2.4 \; -2.3 \; -2.2 \; -2.1 \; -2.0$

-2.56 is *exactly* here.

Zoom in on hundredths ▶ $\quad -2.60 \; -2.59 \; -2.58 \; -2.57 \; -2.56 \; -2.55 \; -2.54 \; -2.53 \; -2.52 \; -2.51 \; -2.50$

ANSWERS

a.

$-1.8 \; -1.79 \; -1.78 \; -1.77 \; -1.76 \; -1.75 \; -1.74 \; -1.73 \; -1.72 \; -1.71 \; -1.7$

b.

$1 \qquad 2 \qquad 2\frac{3}{4} \; 3$

YOUR TURN   Graph on a number line.

a. -1.78

b. $2\frac{3}{4}$

OBJECTIVE 4 Determine the absolute value of a number. The word *value* indicates how much something is worth, or its size. In mathematics, **absolute value** indicates the size of a number.

DEFINITION | *Absolute value:* A given number's distance from 0 on a number line.

For example, the absolute value of 5 is 5 because the number 5 is five units from 0 on a number line. Likewise, the absolute value of −5 is 5 because −5 is still five units from 0 on a number line.

Learning Strategy

In the To the Student section, we suggested that when taking notes, you should use a red pen for definitions and a blue pen for rules or procedures. Notice that we have used those colors in the design of the text to connect with your notes.

The absolute value of a number n is written $|n|$. The examples above would translate this way:

The absolute value of 5 is 5.   The absolute value of −5 is 5.

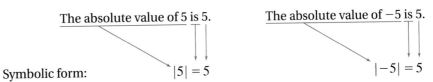

Symbolic form: $|5| = 5$ $|-5| = 5$

What about the absolute value of 0? Since there are no units between 0 and itself, the absolute value of 0 is 0.

RULE | **ABSOLUTE VALUE**

The absolute value of every real number is either positive or 0.

EXAMPLE 4 Determine the absolute value.

a. $|-4.5| =$

Answer: $|-4.5| = 4.5$ because −4.5 is 4.5 units from 0 on a number line.

b. $\left|\dfrac{3}{8}\right| =$

Answer: $\left|\dfrac{3}{8}\right| = \dfrac{3}{8}$ because $\dfrac{3}{8}$ is $\dfrac{3}{8}$ of a unit from 0 on a number line.

YOUR TURN Determine the absolute value.

ANSWERS

a. $2\dfrac{4}{5}$ **b.** 12.8

a. $\left|-2\dfrac{4}{5}\right| =$ **b.** $|12.8| =$

OBJECTIVE 5 Compare numbers. Number lines can also be used to determine which of two numbers is greater. Because numbers increase from left to right on a number line, the number farthest to the right will be the larger of two numbers.

> **RULE COMPARING NUMBERS**
>
> For any two real numbers a and b, a is greater than b if a is to the right of b on a number line.

Consider the number line below where we compare 2 and 8.

8 is farther right than 2.

Because the number 8 is farther to the right on a number line than the number 2, we say that 8 is greater than 2, or in symbols, $8 > 2$.

EXAMPLE 5 Use $=$, $<$, or $>$ to write a true statement.

a. $4 \;\blacksquare\; -4$

Answer: $4 > -4$ because 4 is farther right on a number line than -4.

b. $-2.7 \;\blacksquare\; -2.5$

Answer: $-2.7 < -2.5$ because -2.5 is farther right on a number line than -2.7.

c. $\left|3\dfrac{5}{6}\right| \;\blacksquare\; 3\dfrac{5}{6}$

Answer: $\left|3\dfrac{5}{6}\right| = 3\dfrac{5}{6}$ because the absolute value of $3\dfrac{5}{6}$ is equal to $3\dfrac{5}{6}$.

d. $|-0.7| \;\blacksquare\; -1.5$

Answer: $|-0.7| > -1.5$ because the absolute value of -0.7 is equal to 0.7 which is farther to the right on a number line than -1.5.

ANSWERS

a. $-15 > -21$
b. $-4.1 < 0$
c. $2\dfrac{5}{6} > 2\dfrac{1}{4}$
d. $|-9.5| = 9.5$

YOUR TURN Use $=$, $<$, or $>$ to write a true statement.

a. $-15 \;\blacksquare\; -21$ **b.** $-4.1 \;\blacksquare\; 0$ **c.** $2\dfrac{5}{6} \;\blacksquare\; 2\dfrac{1}{4}$ **d.** $|-9.5| \;\blacksquare\; 9.5$

EXERCISES

For Extra Help

InterAct Math

Addison-Wesley
Math Tutor Center

Video 1/DVT 1

MyMathLab

MathXL

1. Define a set in your own words.

2. Define rational numbers in your own words.

3. Explain the difference between a rational number and an irrational number.

4. Explain why it is incorrect to say that absolute value is always positive.

For Exercises 5–10, answer true or false.

5. Every rational number is a real number.

6. Every real number is a rational number.

7. Every integer is a rational number.

8. Every natural number is a rational number.

9. A number exists that is both rational and irrational.

10. The set of rational numbers contains more numbers than the set of natural numbers.

For Exercises 11–20, write a set representing each description.

11. The days of the week

12. The last ten letters of the alphabet

13. The vowels

14. The states in the United States that do not share a border with any other state

15. The natural numbers that are multiples of 5

16. The even natural numbers

17. The odd natural numbers greater than 7

18. The even integers greater than or equal to 16

19. The integers greater than or equal to -6 but less than -2

20. The integers greater than -2.1 and less than $\dfrac{3}{4}$

For Exercises 21–28, determine whether each of the following is a rational number or an irrational number.

21. $-\dfrac{4}{5}$

22. $\dfrac{1}{4}$

23. $\sqrt{5}$

24. $-\sqrt{17}$

25. π

26. $0.\overline{6}$

27. -0.21

28. -0.8

For Exercises 29–36, graph each of the following on a number line.

29. $-\dfrac{5}{6}$

30. $5\dfrac{1}{2}$

31. $2\dfrac{3}{8}$

32. $-\dfrac{2}{5}$

33. -3.5

34. 7.4

35. 2.45

36. -7.62

For Exercises 37–40, round the given decimal approximation to the nearest tenth, then graph the rounded value on a number line.

37. $\sqrt{29} \approx 5.3852$

38. $\sqrt{80} \approx 8.9443$

39. $-\sqrt{14} \approx -3.7417$

40. $-\sqrt{59} \approx -7.6811$

For Exercises 41–48, simplify each of the following.

41. $|4.5|$ **42.** $\left|2\dfrac{3}{5}\right|$ **43.** $\left|-3\dfrac{1}{8}\right|$ **44.** $|-5.7|$

45. $|-2|$ **46.** $|0|$ **47.** $|-67.8|$ **48.** $|23|$

For Exercises 49–60, use =, <, or > to write a true statement.

49. $-7 \;\blacksquare\; -8$ **50.** $0 \;\blacksquare\; -5$

51. $|2| \;\blacksquare\; |-2|$ **52.** $-3.5 \;\blacksquare\; -3.1$

53. $-8 \;\blacksquare\; 0$ **54.** $|4.31| \;\blacksquare\; |-4.31|$

55. $5.8 \;\blacksquare\; |-5.8|$ **56.** $3\dfrac{5}{6} \;\blacksquare\; 3\dfrac{1}{4}$

57. $-19 \ \blacksquare \ -7$

58. $|4.1| \ \blacksquare \ 4.1$

59. $\left| -6\dfrac{5}{8} \right| \ \blacksquare \ 5\dfrac{3}{8}$

60. $\left| -\dfrac{2}{3} \right| \ \blacksquare \ \left| -\dfrac{4}{3} \right|$

For Exercises 61–64, list the given numbers in order from least to greatest.

61. $-2.56, \ 5.4, \ |8.3|, \ \left| -7\dfrac{1}{2} \right|, \ -4.7$

62. $2.9, \ 1, \ -12.6, \ |-1.3|, \ -9.6, \ \left| -2\dfrac{3}{4} \right|$

63. $0.4, \ -0.6, \ 0, \ 3\dfrac{1}{4}, \ |-0.02|, \ -0.44, \ \left| 1\dfrac{2}{3} \right|$

64. $1.02, \ -0.13, \ -4\dfrac{1}{8}, \ -2\dfrac{1}{4}, \ |-1.06|, \ -2, \ |0.1|$

1.2 Fractions

OBJECTIVES

1. Write equivalent fractions.
2. Write equivalent fractions with the LCD.
3. Write the prime factorization of a number.
4. Simplify a fraction to lowest terms.

In this section we review some basic principles of fractions. It is important to understand fractions in order to understand rational expressions (Chapter 7), which are algebraic fractions. A **fraction**, like $\frac{3}{4}$, is a quotient of two numbers.

DEFINITION | *Fraction:* A quotient of two numbers.

We can use fractions to indicate a part of a whole. For example, the rectangle below has been divided into four equal pieces, three of which are shaded. Thus the shaded region represents $\frac{3}{4}$ of the rectangle.

The top number in a fraction is called the *numerator*, while the bottom number is called the *denominator*.

$$\frac{3}{4} \begin{array}{l} \longleftarrow \text{Numerator} \\ \longleftarrow \text{Denominator} \end{array}$$

OBJECTIVE 1 Write equivalent fractions. Now, let's review how to write *equivalent fractions*, which are fractions that represent the same amount. For example, $\frac{1}{2}$ and $\frac{2}{4}$ are equivalent. We can see that they represent the same amount on a ruler.

We can rewrite $\dfrac{1}{2}$ as $\dfrac{2}{4}$ by multiplying its numerator and denominator by 2.

$$\frac{1}{2} = \frac{1 \cdot 2}{2 \cdot 2} = \frac{2}{4}$$

We could also begin with $\dfrac{2}{4}$ and get back to $\dfrac{1}{2}$ by dividing both the numerator and denominator by 2.

$$\frac{2}{4} = \frac{2 \div 2}{4 \div 2} = \frac{1}{2}$$

RULE **WRITING EQUIVALENT FRACTIONS**

For any fraction, we can write an equivalent fraction by multiplying or dividing both its numerator and denominator by the same number.

This rule makes use of the fact that multiplying or dividing both the numerator and denominator by the same number is equivalent to multiplying or dividing by 1, which does not change a number.

EXAMPLE 1 Find the missing number that makes the fractions equivalent.

a. $\dfrac{7}{12} = \dfrac{?}{48}$

Solution $\dfrac{7 \cdot 4}{12 \cdot 4} = \dfrac{28}{48}$ Multiply the numerator and denominator by 4.

b. $-\dfrac{30}{42} = \dfrac{?}{7}$

Solution $-\dfrac{30 \div 6}{42 \div 6} = -\dfrac{5}{7}$ Divide the numerator and denominator by 6.

Note: When a fraction is negative, the sign can be placed to the left of the numerator, to the left of the fraction line, or to the left of the denominator.

$$-\frac{30}{42} = -\frac{5}{7} = \frac{-5}{7} = \frac{5}{-7}$$

YOUR TURN Find the missing number that makes the fractions equivalent.

a. $\dfrac{-9}{16} = \dfrac{?}{32}$ **b.** $\dfrac{54}{72} = \dfrac{3}{?}$

OBJECTIVE 2 Write equivalent fractions with the LCD. It is sometimes necessary to write fractions that have different denominators as equivalent fractions with the common denominator. A common denominator can be any number that is a *multiple* of the denominators.

ANSWERS

a. $\dfrac{-9}{16} = \dfrac{-18}{32}$ **b.** $\dfrac{54}{72} = \dfrac{3}{4}$

Multiple: A number that is evenly divisible by a given number.

The numbers 2, 4, 6, 8, 10, ... are multiples of 2 because they are divisible by 2.

The numbers 3, 6, 9, 12, 15, ... are multiples of 3 because they are divisible by 3.

Consider the fractions $\frac{1}{2}$ and $\frac{1}{3}$. Note that we could use 6, 12, 18, and so on, as common denominators. However, it is usually preferable to use the smallest or **least common multiple** for the common denominator. Such a denominator is called the **least common denominator**.

Least common multiple (LCM): The smallest number that is evenly divisible by each number in a given set of numbers.

Least common denominator (LCD): The least common multiple of the denominators of a given set of fractions.

Note that the LCD for 2 and 3 is 6. We can now write the equivalent fractions with 6 as the LCD.

$$\frac{1}{2} = \frac{1 \cdot 3}{2 \cdot 3} = \frac{3}{6} \quad \text{and} \quad \frac{1}{3} = \frac{1 \cdot 2}{3 \cdot 2} = \frac{2}{6}$$

EXAMPLE 2 Write $\frac{5}{8}$ and $\frac{1}{6}$ as equivalent fractions with the LCD.

Solution The LCD of 8 and 6 is 24.

$$\frac{5}{8} = \frac{5 \cdot 3}{8 \cdot 3} = \frac{15}{24} \quad \text{and} \quad \frac{1}{6} = \frac{1 \cdot 4}{6 \cdot 4} = \frac{4}{24}$$

YOUR TURN Write each pair as equivalent fractions with the LCD.

a. $\frac{4}{9}$ and $\frac{5}{6}$

b. $-\frac{3}{4}$ and $-\frac{5}{8}$

OBJECTIVE 3 Write the prime factorization of a number. We now consider simplifying fractions. When simplifying fractions, we work with **factors**, which are numbers or variables that are multiplied together.

Factors: If $a \cdot b = c$, then a and b are factors of c.

ANSWERS

a. $\frac{8}{18}$ and $\frac{15}{18}$

b. $-\frac{6}{8}$ and $-\frac{5}{8}$

In the multiplication statement $20 = 2 \cdot 10$, the 2 and 10 are factors. Notice that a factor of a given number divides the given number evenly. The numbers

1, 2, 4, 5, 10, and 20 are all factors of 20 because they all divide 20 evenly. When a number is expressed as a product of its factors, it is said to be in *factored form*. Factored form is sometimes referred to as a *factorization*. Below are some factorizations for 20:

$$20 = 1 \cdot 20 \quad \text{or} \quad 20 = 2 \cdot 10 \quad \text{or} \quad 20 = 4 \cdot 5$$

Some natural numbers only have two distinct factors, 1 and the number itself. These numbers are called **prime numbers**.

DEFINITION

Prime number: A natural number that has exactly two different factors, 1 and the number itself.

Of Interest

Eratosthenes (276–196 B.C.) developed a method for finding prime numbers that he called a *sieve* because it sifted out the composite numbers leaving only prime numbers. The method involves writing odd natural numbers beginning with 3, and then marking out all odd numbers divisible by 3. Then write 5 and mark out all odd numbers divisible by 5, and so on.

3, 5, 7, 9̶, 11, 13, 1̶5̶,
17, 19, 2̶1̶, 23, 2̶5̶, 2̶7̶,
29, 31, 3̶3̶, 3̶5̶, 37, 3̶9̶,
41, 43, 4̶5̶, . . .

Note that 0 is not a prime number because it is not a natural number (0 is a whole number). Also, 1 is not a prime number because the only factor for 1 is itself and, by definition, a prime number must have *two different* factors.

The first prime number is 2, because it is the first natural number greater than 1 that has exactly two factors, 1 and 2. The number 4 is not a prime number because it is divisible by a number other than 1 and 4, namely 2. Below is a list of prime numbers:

2, 3, 5, 7, 11, 13, 17, 19, 23, 29, 31, 37, 41, 43, . . .

Note: A natural number that has factors other than 1 and itself is called a *composite number*. Some examples are 4, 6, 8, and 9. Notice that 0 and 1 are not composite, so they are neither prime nor composite.

DEFINITION

Prime factorization: A factorization that contains only prime factors.

When simplifying a fraction it is useful to write the **prime factorization** of the numerator and denominator.

For example, the prime factorization for 20 is $2 \cdot 2 \cdot 5$.

We can find the prime factorization of a number using any factorization of the number as a starting point. Consider 20 again and suppose we factored it as $4 \cdot 5$. Since 4 is not a prime number, we can factor 4 as $2 \cdot 2$. It looks like this:

$$20 = 4 \cdot 5$$
$$= 2 \cdot 2 \cdot 5$$

We will use factor trees to find prime factorizations. The idea is to draw two branches beneath any composite number and write two factors of that number at the end of the branches. Following are two ways we could use a factor tree to find the prime factorization of 20. Notice that when a branch ends in a prime factor we circle the prime factor. This is optional, but many people find it helpful in making the prime factors stand out.

EXAMPLE 3 Find the prime factorization of 360.

Solution Use a factor tree.

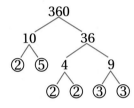

Factor 360 to 10 and 36. (Any two factors will work.)

Factor 10 to 2 and 5, which are primes. Then factor 36 to 4 and 9.

4 is then factored to 2 and 2, which are primes, and 9 is factored to 3 and 3, which are primes.

Answer: $360 = 2 \cdot 2 \cdot 2 \cdot 3 \cdot 3 \cdot 5$

YOUR TURN Find the prime factorization.

a. 84 **b.** 280

OBJECTIVE 4 Simplify a fraction to lowest terms. We can use prime factorizations to simplify fractions to **lowest terms**.

DEFINITION *Lowest terms:* Given a fraction $\dfrac{a}{b}$ and $b \neq 0$, if the only factor common to both a and b is 1, then the fraction is in lowest terms.

For example, $\dfrac{3}{4}$ is in lowest terms because the only factor common to both 3 and 4 is the number 1, whereas $\dfrac{6}{8}$ is not in lowest terms because 6 and 8 have a common factor of 2.

To develop the method of reducing fractions using primes, we must understand that if the numerator and denominator are identical and not both zero, then the fraction is equivalent to 1.

$$1 = \frac{1}{1} = \frac{2}{2} = \frac{3}{3} = \frac{4}{4} = \frac{5}{5} = \cdots$$

This suggests the following rule.

ANSWERS
a. $84 = 2 \cdot 2 \cdot 3 \cdot 7$
b. $280 = 2 \cdot 2 \cdot 2 \cdot 5 \cdot 7$

SIMPLIFYING A FRACTION WITH THE SAME NONZERO NUMERATOR AND DENOMINATOR

$$\frac{n}{n} = \frac{1}{1} = 1, \text{ when } n \neq 0.$$

This rule can be extended to common factors within a fraction. It allows us to eliminate factors that are identical in the numerator and denominator of a fraction.

ELIMINATING A COMMON FACTOR IN A FRACTION

$$\frac{an}{bn} = \frac{a \cdot 1}{b \cdot 1} = \frac{a}{b}, \text{ when } b \neq 0 \text{ and } n \neq 0.$$

These rules allow us to write fractions in lowest terms using prime factorization. The idea is to replace the numerator and denominator with their prime factorizations and then eliminate the prime factors that are common to both the numerator and denominator. Consider the fraction $\frac{10}{15}$. Replacing the 10 and 15 with their prime factorizations, we have:

$$\frac{10}{15} = \frac{2 \cdot 5}{3 \cdot 5}$$

$$= \frac{2 \cdot 1}{3 \cdot 1}$$

$$= \frac{2}{3}$$

Note: The common factor of 5 is replaced by 1. In Example 4 we will leave this step out.

This process of eliminating common factors is called *dividing out* the common factors and is sometimes written this way:

$$\frac{10}{15} = \frac{2 \cdot \cancel{5}^{1}}{3 \cdot \cancel{5}_{1}} = \frac{2}{3}$$

Though there are different styles of showing which common factors are eliminated, we will simply highlight the common factors that are eliminated.

SIMPLIFYING A FRACTION TO LOWEST TERMS

To simplify a fraction to lowest terms:

1. Replace the numerator and denominator with their prime factorizations.
2. Eliminate (divide out) all prime factors common to the numerator and denominator.
3. Multiply the remaining factors.

EXAMPLE 4 Simplify to lowest terms.

a. $\dfrac{28}{70}$

Solution $\dfrac{28}{70} = \dfrac{2 \cdot 2 \cdot 7}{2 \cdot 5 \cdot 7} = \dfrac{2}{5}$ Replace the numerator and denominator with their prime factorizations, then eliminate the common prime factors.

b. $\dfrac{156}{210}$

Solution $\dfrac{156}{210} = \dfrac{2 \cdot 2 \cdot 3 \cdot 13}{2 \cdot 3 \cdot 5 \cdot 7} = \dfrac{26}{35}$

CONNECTION

You may remember a method for simplifying fractions in which you divide out the greatest common factor. For example, to simplify $\dfrac{28}{70}$, we recognize that 14 is the greatest common factor of 28 and 70 and divide both by 14 to get lowest terms: $\dfrac{28}{70} = \dfrac{28 \div 14}{70 \div 14} = \dfrac{2}{5}$. This method is a perfectly good way of simplifying fractions. However, we show the prime method to prepare for simplifying rational expressions in Chapter 7.

YOUR TURN Simplify to lowest terms.

a. $\dfrac{36}{48}$ b. $\dfrac{126}{315}$ c. $\dfrac{312}{378}$

On a scientific calculator, fractions are entered using the $\boxed{a^{b/c}}$ key. To reduce a fraction, enter it and then press $\boxed{=}$. For example, to reduce $\frac{156}{210}$, type: 156 $\boxed{a^{b/c}}$ 210 $\boxed{=}$ or, more specifically,

$\boxed{1}$ $\boxed{5}$ $\boxed{6}$ $\boxed{a^{b/c}}$ $\boxed{2}$ $\boxed{1}$ $\boxed{0}$ $\boxed{=}$ *You should now see 26/35.*

On a graphing calculator, enter a fraction using the $\boxed{\div}$ key, then to reduce the fraction, press the $\boxed{MATH}$ key and select 1:Frac, then press $\boxed{ENTER}$. For example, to reduce $\frac{156}{210}$ on a graphing calculator, type:

$\boxed{1}$ $\boxed{5}$ $\boxed{6}$ $\boxed{\div}$ $\boxed{2}$ $\boxed{1}$ $\boxed{0}$ $\boxed{MATH}$ $\boxed{1}$ $\boxed{ENTER}$

1.2 EXERCISES

For Extra Help

InterAct Math

Addison-Wesley
Math Tutor Center

Video 1/DVT 1

MyMathLab

MathXL

1. What are two ways to generate fractions equivalent to a given fraction?

2. Is $2 \cdot 2 \cdot 4 \cdot 7$ a prime factorization? Explain.

3. Is it possible for two different numbers to have the same prime factorization? Explain.

4. Is every whole number a composite number?

5. How do you find the prime factorization of a given number?

6. How do you know if a fraction is in lowest terms?

For Exercises 7–10, answer true or false, then explain.

7. Every prime number is a whole number.

8. Prime numbers are rational numbers.

9. Every composite number has at least one factor that is a prime number.

10. Two different numbers can have the same prime factorization.

For Exercises 11–14, identify the fraction represented by the shaded region.

11.

12.

13.

14.

For Exercises 15–18, write the length of each line segment in lowest terms.

15.

16.

17.

18.

For Exercises 19–26, find the missing number that makes the fractions equivalent.

19. $\dfrac{3}{10} = \dfrac{?}{40}$

20. $\dfrac{5}{8} = \dfrac{?}{16}$

21. $-\dfrac{7}{12} = \dfrac{?}{60}$

22. $\dfrac{2}{-5} = \dfrac{6}{?}$

23. $\dfrac{8}{60} = \dfrac{2}{?}$

24. $\dfrac{-27}{30} = \dfrac{-9}{?}$

25. $\dfrac{4}{7} = \dfrac{?}{28}$

26. $\dfrac{6}{8} = \dfrac{?}{4}$

For Exercises 27–34, write the fractions as equivalent fractions with the LCD.

27. $\dfrac{4}{5}$ and $\dfrac{2}{3}$

28. $\dfrac{5}{7}$ and $\dfrac{3}{8}$

29. $\dfrac{4}{9}$ and $\dfrac{7}{6}$

30. $\dfrac{5}{8}$ and $\dfrac{7}{12}$

31. $-\dfrac{11}{18}$ and $-\dfrac{17}{24}$

32. $-\dfrac{9}{20}$ and $-\dfrac{7}{15}$

33. $\dfrac{1}{16}$ and $\dfrac{15}{36}$

34. $-\dfrac{13}{21}$ and $-\dfrac{9}{14}$

For Exercises 35–42, write the prime factorization for each number.

35. 36

36. 42

37. 250

38. 810

39. 44

40. 33

41. 64

42. 48

For Exercises 43–50, simplify to lowest terms.

43. $\dfrac{72}{90}$

44. $\dfrac{57}{76}$

45. $-\dfrac{120}{160}$

46. $\dfrac{42}{91}$

47. $-\dfrac{78}{104}$

48. $-\dfrac{39}{52}$

49. $\dfrac{60}{84}$

50. $-\dfrac{105}{189}$

For Exercises 51 and 52, determine whether the simplification is correct. If not, explain why.

51. $\dfrac{\cancel{2}^{\,1} + 3}{\cancel{2}_{\,1}} = \dfrac{1 + 3}{1} = 4$

52. $\dfrac{84}{240} = \dfrac{\cancel{2}^{1} \cdot \cancel{2}^{1} \cdot \cancel{3}^{1} \cdot 7}{2 \cdot \cancel{2}_{1} \cdot \cancel{2}_{1} \cdot \cancel{3}_{1} \cdot 3 \cdot 5} = \dfrac{7}{30}$

For Exercises 53–56, write each fraction in lowest terms.

53. A student scores 294 points out of a total of 336 points in a college course. What fraction of the total points is the student's score?

54. The nutrition label on a package of frozen fish sticks indicates that a serving of fish sticks has 250 calories with 130 of those calories coming from fat. What fraction of the calories in a serving of fish sticks comes from fat?

55. At a company, 300 of the 575 employees have optional life insurance coverage as part of their benefits package. What fraction of the employees have optional life insurance coverage?

56. Laura uses 120 square feet of a room in her home as an office for her business. The total living space of her home is 1830 square feet. For tax purposes, she needs to compute the fraction of her home that is used as an office. What is that fraction?

For Exercises 57–60, use the following bar graph, which shows the number of hours that Carla spends each week in each activity. Write all fractions in lowest terms.

57. What fraction of a week does Carla spend working?

58. What fraction of a week does Carla spend sleeping?

59. What fraction of a week does Carla spend in all of the listed activities combined?

60. What fraction of the week does Carla have for free time, which is time away from all the listed activities?

For Exercises 61–68, use the following table that shows the number of households per 1000 that were victims of property crime in each year. Write all fractions in lowest terms.

National Crime Victimization Survey, Property Crime Trends, 1973–2000

Year	Total property crime*	Burglary	Thefts	Motor vehicle thefts
1973	520	110	391	19
1974	552	112	421	19
1975	554	110	424	20
1976	544	107	421	17
1977	544	106	421	17
1978	533	103	412	18
1979	532	101	413	18
1980	496	101	378	17
1981	497	106	374	17
1982	468	94	358	16
1983	428	84	330	15
1984	399	77	307	15
1985	385	75	296	14
1986	373	74	284	15
1987	380	75	289	16
1988	378	74	287	18
1989	373	68	287	19
1990	349	65	264	21
1991	354	65	267	22
1992	325	59	248	19
1993	319	58	242	19
1994	310	56	235	19
1995	291	49	224	17
1996	266	47	206	14
1997	248	45	190	14
1998	217	39	168	11
1999	198	34	154	10
2000	178	32	138	9

*Victimizations per 1000 households

Source: U.S. Department of Justice, Bureau of Crime Statistics

61. What fraction of households were victims of property crime in 1973?

62. What fraction of households were victims of property crime in 1992?

63. In 1987 what fraction of households were not victims of property crime?

64. In 1994 what fraction of households were not victims of property crime?

65. a. What year had the highest number of property crimes?

b. What fraction of households were victims in that year?

66. a. What year had the lowest number of property crimes?

b. What fraction of households were victims in that year?

67. In 1974, what fraction of the total property crimes were burglaries?

68. In 1985, what fraction of total property crimes were motor vehicle theft?

For Exercises 69–76, write each fraction in lowest terms.

69. What fraction of a year is 30 days? Use 365 days for the number of days in a year.

70. What fraction of an hour is 8 minutes?

71. What fraction of a minute is 40 seconds?

72. What fraction of a foot is 4 inches?

73. In the second session of 2002, the U.S. House of Representatives had 211 Democrats, 222 Republicans, and 2 Independents. What fraction of the House of Representatives was not Republican? (*Source:* Official list of United States House of Representatives)

74. In 2002, the U.S. Senate had 50 Democrats, 49 Republicans, and 1 Independent. What fraction of the Senate was not Democrat? (*Source:* Official list of United States Senators)

Of Interest

It takes Earth one year to complete one revolution around the Sun, which is actually $365\frac{1}{4}$ days. Since it is impractical to have $\frac{1}{4}$ days on the calendar, every four years, those $\frac{1}{4}$ days add up to a full day, which is why we have a leap year with 366 days.

75. One molecule of lactose contains 12 carbon atoms, 22 hydrogen atoms, and 11 oxygen atoms. What fraction of the atoms that make up the molecule are carbon?

76. One molecule of glucose contains 6 carbon atoms, 12 hydrogen atoms, and 6 oxygen atoms. What fraction of the atoms that make up the molecule are other than carbon?

REVIEW EXERCISES

1. Is $5x + 2$ an expression or an equation? Explain.

2. Write a set containing the names of the four planets closest to the Sun.

3. Is 0.8 a rational or irrational number? Explain.

4. Graph 3.7 on a number line.

5. Simplify: $|27|$

6. Use $<$, $>$, or $=$ to make a true statement: $|-6|$ ▨ 6

1.3 Adding and Subtracting Real Numbers; Properties of Real Numbers

OBJECTIVES

1. Add integers.
2. Add rational numbers.
3. Find the additive inverse of a number.
4. Subtract rational numbers.

OBJECTIVE **1** **Add integers.** We now turn our attention to adding and subtracting numbers. We begin with addition. First, consider the parts of an addition statement. The numbers added are called *addends* and the answer is called a *sum*.

$$2 + 3 = 5$$

addends   sum

There are three important properties of addition that are true for all real numbers. First, the sum of a number and 0 is that number. Because adding 0 to a number does not change the identity of the number, 0 is called the **additive identity**.

$$2 + 0 = 2$$

Second, the order of addends can be changed without affecting the sum. This is known as the **commutative property of addition**.

$$2 + 3 = 5 \text{ or } 3 + 2 = 5$$

Third, when adding three or more numbers, we can group differently without affecting the sum. This is known as the **associative property of addition**.

$$2 + (3 + 4) \qquad \text{or} \qquad (2 + 3) + 4$$
$$= 2 + 7 \qquad\qquad\qquad = 5 + 4$$
$$= 9 \qquad\qquad\qquad\qquad = 9$$

Following is a summary of these properties.

Properties of Addition	Symbolic Form	Word Form
Additive Identity	$a + 0 = a$	The sum of a number and 0 is that number.
Commutative Property of Addition	$a + b = b + a$	Changing the order of addends does not affect the sum.
Associative Property of Addition	$a + (b + c) = (a + b) + c$	Changing the grouping of three or more addends does not affect the sum.

Adding Numbers with the Same Sign

With these properties in mind, let's consider how to add signed numbers. First we learn how to add numbers that have the same sign. It is helpful to relate adding signed numbers to money. For example, adding two positive numbers is like adding credits, whereas adding two negative numbers is like adding debts.

$$15 + 10 = 25$$

This addition is like having $15 in an account and depositing another $10 into the same account, bringing the total to $25.

$$-20 + (-8) = -28$$

This addition is like having a debt balance of $20 on a credit card and charging another $8 on the same card, bringing the balance to a total of $28 in debt.

We can also illustrate each of the above cases using a number line. The first addend is the starting point and the second addend indicates the direction and distance to travel on the number line so that we finish at the sum.

These two cases are similar and suggest the following procedure.

PROCEDURE ADDING NUMBERS WITH THE SAME SIGN

To add two numbers that have the same sign, add their absolute values and keep the same sign.

EXAMPLE **1** Add.

a. $15 + 13$

Solution $15 + 13 = 28$

b. $-14 + (-23)$

Solution $-14 + (-23) = -37$

Note: In terms of money, part a illustrates two credits, whereas part b illustrates two debts.

YOUR TURN Add.

a. $32 + 19$

b. $-62 + (-13)$

ANSWERS

a. 51   **b.** −75

Adding Numbers with Different Signs

Now consider adding two numbers with different signs, which is like making a payment toward a debt. Consider the following cases:

$$-15 + 3 = -12 \qquad\qquad -9 + 20 = 11$$

This addition is like having a debt of $15 and a payment of $3 is made toward that debt. This decreases the debt to $12.

This addition is like having a debt of $9 and a payment of $20 is made toward that debt. Because the payment is more than the debt, we now have a credit of $11.

Using number lines:

Note that the commutative property of addition indicates that in the above examples, the addends can be rearranged without affecting the sum.

$$3 + (-15) = -12 \qquad\qquad 20 + (-9) = 11$$

These examples suggest the following procedure.

PROCEDURE   ADDING NUMBERS WITH DIFFERENT SIGNS

To add two numbers that have different signs, subtract their absolute values and keep the sign of the number with the greater absolute value.

EXAMPLE 2   Add.

a. $22 + (-14)$

Solution $22 + (-14) = 8$

b. $-13 + 20$

Solution $-13 + 20 = 7$

c. $12 + (-18)$

Solution $12 + (-18) = -6$

d. $-24 + 5$

Solution $-24 + 5 = -19$

Note: In terms of money, parts a and b illustrate situations in which the credit is greater than the debt. Therefore, the result is a credit, or positive.

Note: In terms of money, parts c and d illustrate situations in which a payment toward a debt is not enough to pay off the debt. Therefore, the result is still debt, or negative.

ANSWERS
a. 9   b. −8
c. 6   d. −15

YOUR TURN   Add.

a. $15 + (-6)$ **b.** $27 + (-35)$ **c.** $-13 + 19$ **d.** $-29 + 14$

Adding Fractions with the Same Denominator

Recall that the set of rational numbers contains any number that can be expressed as a ratio of integers. This set includes the integers themselves because every integer can be expressed as a ratio with 1 as the denominator. Let's now turn our attention to the rest of the rational number set, namely fractions and decimals. Consider the intuitive fact that adding two quarters equals a half (in money this would be a half-dollar). The circle below shows this sum.

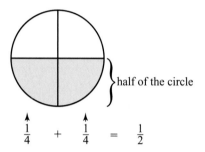

$$\frac{1}{4} \quad + \quad \frac{1}{4} \quad = \quad \frac{1}{2}$$

Notice if the numerators are added and the denominator remains the same, the result simplifies to $\frac{1}{2}$.

$$\frac{1}{4} + \frac{1}{4} = \frac{1+1}{4} = \frac{2}{4} = \frac{1}{2}$$

Conclusion   When adding fractions that have the same denominator, we can add the numerators and keep the denominator the same.

EXAMPLE **3**   Add.

a. $\frac{5}{8} + \frac{1}{8}$

Solution $\frac{5}{8} + \frac{1}{8} = \frac{5+1}{8} = \frac{6}{8}$

Because $\frac{6}{8}$ is not in lowest terms, we must reduce.

$$\frac{6}{8} = \frac{2 \cdot 3}{2 \cdot 2 \cdot 2} = \frac{3}{4} \qquad \text{\textbf{Divide out the common factor, 2}}$$

b. $-\frac{3}{10} + \left(-\frac{1}{10}\right)$

Solution Because the fractions have the same sign, we add and keep the same sign.

$$-\frac{3}{10} + \left(-\frac{1}{10}\right) = \frac{-3 + (-1)}{10} = -\frac{4}{10}$$

Since $-\dfrac{4}{10}$ is not in lowest terms, we must reduce.

$$-\dfrac{4}{10} = -\dfrac{2 \cdot 2}{2 \cdot 5} = -\dfrac{2}{5} \qquad \text{Divide out the common factor, 2.}$$

c. $\dfrac{7}{9} + \left(-\dfrac{4}{9}\right)$

Solution Because the two addends have different signs, we subtract the absolute values and keep the sign of the number with the greater absolute value.

$$\dfrac{7}{9} + \left(-\dfrac{4}{9}\right) = \dfrac{7 + (-4)}{9} = \dfrac{3}{9}$$

Because $\dfrac{3}{9}$ is not in lowest terms, we must reduce to lowest terms.

$$\dfrac{3}{9} = \dfrac{3}{3 \cdot 3} = \dfrac{1}{3} \qquad \text{Divide out the common factor, 3.}$$

YOUR TURN   **Add.**

a. $\dfrac{3}{8} + \dfrac{1}{8}$ **b.** $-\dfrac{5}{9} + \left(-\dfrac{2}{9}\right)$ **c.** $-\dfrac{7}{12} + \dfrac{5}{12}$

Adding Fractions with Different Denominators

If the denominators are different, then we must first find a common denominator. Recall that in Section 1.2 we wrote equivalent fractions with the LCD. Consider the addition $\dfrac{3}{4} + \dfrac{1}{6}$. The LCD of 4 and 6 is 12. We now write each fraction as an equivalent fraction with the LCD.

$$\dfrac{3}{4} + \dfrac{1}{6} = \dfrac{3(3)}{4(3)} + \dfrac{1(2)}{6(2)}$$
$$= \dfrac{9}{12} + \dfrac{2}{12}$$
$$= \dfrac{11}{12}$$

To write $\dfrac{3}{4}$ as an equivalent fraction with 12 as the denominator, we multiply the numerator and denominator by 3.

To write $\dfrac{1}{6}$ as an equivalent fraction with 12 as the denominator, we multiply the numerator and denominator by 2.

PROCEDURE ADDING FRACTIONS

To add fractions:

1. Find the LCD.
2. Write each fraction as an equivalent fraction with the LCD.
3. Add the numerators and keep the LCD.
4. Simplify.

EXAMPLE 4 Add.

a. $-\dfrac{1}{6} + \left(-\dfrac{5}{9}\right)$

Solution The LCD of 6 and 9 is 18.

$$-\frac{1}{6} + \left(-\frac{5}{9}\right) = -\frac{1(3)}{6(3)} + \left(-\frac{5(2)}{9(2)}\right)$$ Write equivalent fractions with 18 in the denominator.

$$= -\frac{3}{18} + \left(-\frac{10}{18}\right)$$ Add numerators and keep the common denominator.

$$= \frac{-3 + (-10)}{18}$$ Because the addends have the same sign, we add and keep the same sign.

$$= -\frac{13}{18}$$

b. $-\dfrac{4}{5} + \dfrac{2}{3}$

Solution The LCD of 5 and 3 is 15.

$$-\frac{4}{5} + \frac{2}{3} = -\frac{4(3)}{5(3)} + \frac{2(5)}{3(5)}$$ Write equivalent fractions with 15 in the denominator.

$$= -\frac{12}{15} + \frac{10}{15}$$ Add numerators and keep the common denominator.

$$= \frac{-12 + 10}{15}$$ Because the addends have different signs, we subtract and keep the sign of the number with the larger absolute value.

$$= -\frac{2}{15}$$

YOUR TURN Add.

a. $\dfrac{5}{6} + \dfrac{1}{4}$ **b.** $-\dfrac{1}{15} + \left(-\dfrac{5}{12}\right)$ **c.** $\dfrac{7}{8} + \left(-\dfrac{1}{5}\right)$

ANSWERS

a. $\dfrac{13}{12}$ **b.** $-\dfrac{29}{60}$ **c.** $\dfrac{27}{40}$

Adding Decimal Numbers

Let's now review adding decimal numbers, which are also a part of the set of rational numbers. Recall that to add decimal numbers, we add like place values.

EXAMPLE 5 Angela has a balance of $578.26 and incurs the following debts:

$$\$25.67, \$92.45, \$78.50, \$15.00$$

What is Angela's new balance?

Solution First, we calculate the total debts.

$$-25.67 + (-92.45) + (-78.50) + (-15.00) = -211.62$$

Now we reconcile the total debt with the current balance.

$$578.26 + (-211.62) = \$366.64$$

YOUR TURN

Daryl has a balance of $-\$452.75$ on a credit card. In one month he charges $25.80, $65.75, and $35.21 using the same credit card. If the finance charge is $7.47 and he makes a payment of $120, what is his new balance?

OBJECTIVE 3 **Find the additive inverse of a number.** What happens if we add two numbers that have the same absolute value but different signs, such as $5 + (-5)$? In money terms, this is like making a $5 payment toward a debt of $5. Notice the payment pays off the debt so that the balance is 0.

$$5 + (-5) = 0$$

Because their sum is 0, we say 5 and -5 are **additive inverses**, or opposites.

DEFINITION

Additive inverses: Two numbers whose sum is 0.

EXAMPLE 6 Find the additive inverse of the given number.

a. 9

Answer: -9 because $9 + (-9) = 0$

b. -7

Answer: 7 because $-7 + 7 = 0$

c. 0

Answer: 0 because $0 + 0 = 0$

ANSWER

$-\$466.98$

To enter a negative number or find an additive inverse on a graphing or scientific calculator, use the ⟨(−)⟩ key. Note that this key is different from the subtraction key.

To find the additive inverse of 9, type: ⟨(−)⟩ ⟨9⟩

To find the additive inverse of −7, type: ⟨(−)⟩ ⟨(−)⟩ ⟨7⟩ ⟨ENTER⟩

On older scientific calculators, the ⟨+/−⟩ key is used to find the additive inverse of a number. You enter the value first, then press the ⟨+/−⟩ key. Each time you press the ⟨+/−⟩ key the sign will change.

To find the additive inverse of 9, type: ⟨9⟩ ⟨+/−⟩

To find the additive inverse of −7, type: ⟨7⟩ ⟨+/−⟩ ⟨+/−⟩

YOUR TURN **Find the additive inverse of the given number.**

a. −12 **b.** 0.67 **c.** $-\dfrac{3}{8}$

OBJECTIVE **4** **Subtract rational numbers.** Now let's turn our attention to subtraction. The parts of subtraction are:

$$8 - 5 = 3$$

| Minuend | Subtrahend | Difference |

Recall that when we add two numbers that have different signs, we actually subtract. This implies that subtraction and addition are related. In fact, every subtraction statement can be written as an equivalent addition statement. Notice that the subtraction statement $8 - 5 = 3$ and the addition statement $8 + (-5) = 3$ are equivalent: they are both ways of finding the balance given an $8 credit with a $5 debt. Note that in writing $8 - 5$ as an equivalent addition $8 + (-5)$, we change the operation sign from a minus sign to a plus sign and also change the 5 to its additive inverse, -5.

Any subtraction statement can be written as an equivalent addition statement using this procedure. Consider $8 - (-5)$, which represents an $8 credit subtracting a debt of $5. What does it mean to subtract a debt? If a bank records a debt and then realizes it made a mistake, then it must reverse that debt. Subtracting, or reversing, a debt means that the bank must record a credit (or deposit) to correct the mistake. In math terms, this means that $8 - (-5)$ is equivalent to $8 + 5$, which equals 13. Notice this still follows the process discussed above.

ANSWERS

a. 12 **b.** −0.67 **c.** $\dfrac{3}{8}$

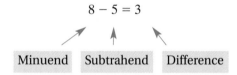

Change the operation from minus to plus. ——— $8 - (-5)$ ——— Change the subtrahend to its additive inverse.

$$= 8 + \quad 5$$
$$= 13$$

REWRITING SUBTRACTION

> To write a subtraction statement as an equivalent addition statement, change the operation symbol from a minus sign to a plus sign, and change the subtrahend to its additive inverse.

EXAMPLE 7 Subtract.

a. $-15 - (-6) =$

Solution Write the subtraction as an equivalent addition.

Change the operation
from minus to plus.

$$-15 - (-6)$$

Change the subtrahend
to its additive inverse.

$$= -15 + 6$$
$$= -9$$

Note: We can think of this as subtracting a debt of $6 from a debt of $15. Subtracting a debt from a debt is equivalent to making a deposit or payment against the existing debt balance, so this decreases the amount of debt.

b. $-\dfrac{2}{3} - \dfrac{3}{4} =$

Solution Write an equivalent addition, then find a common denominator and continue the process of adding fractions.

$$-\frac{2}{3} - \frac{3}{4}$$

Change the operation
from minus to plus.

Change the subtrahend
to its additive inverse.

$$= -\frac{2}{3} + \left(-\frac{3}{4}\right)$$

$$= -\frac{2(4)}{3(4)} + \left(-\frac{3(3)}{4(3)}\right)$$

Write equivalent fractions with the common denominator, 12

Note: If we begin with a debt and subtract a positive amount, we go deeper in debt.

$$= -\frac{8}{12} + \left(-\frac{9}{12}\right)$$

$$= -\frac{17}{12}$$

📟 *Calculator*
TIPS

To enter a problem like Example 7 (b) into a scientific calculator, type:

$\boxed{(-)}$ $\boxed{2}$ $\boxed{a^{b/c}}$ $\boxed{3}$ $\boxed{-}$ $\boxed{3}$ $\boxed{a^{b/c}}$ $\boxed{4}$ $\boxed{\text{ENTER}}$

On a graphing calculator, type:

$\boxed{(-)}$ $\boxed{2}$ $\boxed{\div}$ $\boxed{3}$ $\boxed{-}$ $\boxed{3}$ $\boxed{\div}$ $\boxed{4}$ $\boxed{\text{ENTER}}$

To see the answer in fraction form, press $\boxed{\text{MATH}}$ *and select the 1:Frac option from the menu, then press* $\boxed{\text{ENTER}}$ *.*

c. $5.04 - 8.01 =$

Solution Write an equivalent addition statement.

Note: We begin with a positive balance of $5.04 and subtract $8.01. Because we are subtracting more than we have, the result is negative.

$$5.04 - \mathbf{8.01}$$
$$= 5.04 + (\mathbf{-8.01})$$
$$= -2.97$$

Note: To actually calculate the difference, the number with the greater absolute value must be placed on top.

$$\begin{array}{r} 8.01 \\ -\,5.04 \\ \hline 2.97 \end{array}$$

We then use the equivalent addition to figure whether 2.97 is a positive or negative difference. Because the number with greater absolute value is negative, this result is negative.

YOUR TURN Subtract.

a. $25 - (-17)$ **b.** $-\dfrac{3}{8} - \left(-\dfrac{1}{3}\right)$ **c.** $0.06 - 4.02$

EXAMPLE 8 On Mars, the daytime high temperature can reach $-15°C$, while at night the temperature can plummet to $-140°C$. What is the difference between the high and low temperatures?

ANSWERS

a. 42 **b.** $-\dfrac{1}{24}$

c. -3.96

Solution Difference indicates that we subtract the temperatures.

$$-15 - (-140) = -15 + 140$$
$$= 125$$

Write as an equivalent addition.

YOUR TURN

In an experiment, a mixture begins at a temperature of 5.8°C. The mixture is then cooled to a temperature of $-28.6°C$. Find the difference of the initial and final temperatures.

ANSWER

34.4°C

1.3 EXERCISES

1. Explain the difference between the commutative property of addition and the associative property of addition.

2. Explain why 4 and -4 are additive inverses.

3. Why is 0 called the additive identity?

4. In your own words, explain the process of adding or subtracting two fractions.

5. In your own words, explain how to add two numbers that have the same sign.

6. In your own words, explain how to add two numbers that have different signs.

7. Explain how to write a subtraction as an equivalent addition.

8. Explain why $6 - 5$ and $5 - 6$ have different answers whereas $-6 - 5$ and $-5 - 6$ have the same answer. (*Hint:* Think about the properties of addition.)

For Exercises 9–20, indicate whether the given equation illustrates the additive identity, commutative property of addition, associative property of addition, or additive inverse.

9. $5 + 6 = 6 + 5$

10. $-5 + 0 = -5$

11. $0.8 + (-0.8) = 0$

12. $-\dfrac{2}{7} + \dfrac{3}{4} = \dfrac{3}{4} + \left(-\dfrac{2}{7}\right)$

13. $-8 + (7 + 3) = (7 + 3) + (-8)$

14. $(-4.6 + 4.6) + 9.5 = 9.5$

15. $(3 + 5) + 2 = 3 + (5 + 2)$

16. $6 + (2 + 3) = (6 + 2) + 3$

17. $-\dfrac{4}{9} + \dfrac{4}{9} = 0$

18. $-2 + (3 + 5) = (-2 + 3) + 5$ **19.** $\left(\dfrac{1}{2} + \dfrac{3}{5}\right) + \dfrac{1}{6} = \left(\dfrac{3}{5} + \dfrac{1}{2}\right) + \dfrac{1}{6}$

20. $-12.5 + (-6.8) + (-9) = -12.5 + (-9) + (-6.8)$

For Exercises 21–42, add.

21. $6 + (-12)$ **22.** $-5 + (-7)$ **23.** $-8 + 4$ **24.** $-16 + 19$

25. $27 + (-13)$ **26.** $29 + (-7)$ **27.** $\dfrac{1}{6} + \left(-\dfrac{5}{6}\right)$ **28.** $-\dfrac{3}{5} + \left(-\dfrac{1}{5}\right)$

29. $-\dfrac{4}{9} + \left(-\dfrac{2}{9}\right)$ **30.** $\dfrac{5}{8} + \dfrac{1}{8}$ **31.** $\dfrac{3}{4} + \dfrac{1}{6}$ **32.** $\dfrac{1}{4} + \dfrac{7}{8}$

33. $-\dfrac{1}{12} + \left(-\dfrac{2}{3}\right)$ **34.** $-\dfrac{2}{5} + \left(-\dfrac{3}{20}\right)$ **35.** $-\dfrac{5}{6} + \dfrac{4}{21}$ **36.** $-\dfrac{5}{16} + \left(-\dfrac{3}{12}\right)$

37. $0.21 + 0.05$ **38.** $0.06 + 0.17$ **39.** $-0.18 + 6.7$ **40.** $-15.81 + 4.28$

41. $-0.28 + (-4.1)$ **42.** $-7.8 + (-9.16)$

For Exercises 43–54, find the additive inverse.

43. 5 **44.** -6 **45.** 0 **46.** 7

47. $\dfrac{5}{6}$ **48.** $-\dfrac{6}{17}$ **49.** -0.29 **50.** 2.8

51. $-x$ **52.** b **53.** $\dfrac{m}{n}$ **54.** $-\dfrac{a}{b}$

For Exercises 55–72, subtract.

55. $4 - (-3)$ **56.** $-13 - (-6)$ **57.** $-8 - (-2)$ **58.** $6 - (-7)$

59. $6 - 15$ **60.** $8 - 20$ **61.** $-\dfrac{4}{5} - \left(-\dfrac{2}{7}\right)$ **62.** $\dfrac{3}{8} - \left(-\dfrac{5}{6}\right)$

63. $\dfrac{7}{10} - \left(-\dfrac{3}{5}\right)$

64. $-\dfrac{1}{2} - \left(-\dfrac{1}{3}\right)$

65. $-\dfrac{1}{5} - \left(-\dfrac{1}{5}\right)$

66. $-\dfrac{2}{5} - \left(-\dfrac{4}{10}\right)$

67. $4.01 - 3.65$

68. $-7.1 - (-2.3)$

69. $-6.1 - (-4.5)$

70. $8.1 - 4.76$

71. $0.07 - 5.82$

72. $0.107 - 5.802$

For Exercises 73–80, add or subtract.

73. $-18 + |-12|$

74. $|-24| + 11$

75. $-42 - |-14|$

76. $31 - |-54|$

77. $|-2.4| - |-0.78|$

78. $-0.6 - |9.1|$

79. $\left|-\dfrac{3}{8}\right| - \left|\dfrac{5}{6}\right|$

80. $\left|-\dfrac{4}{5}\right| - \dfrac{3}{4}$

For Exercises 81–90, solve.

81. The Disney corporation's fourth-quarter financial report for 2000 contains the following information. Find the net profit or loss.

Fourth-quarter report for the year ending September 30, 2000

Income (in millions)	Expenditures
Revenue = \$25,020	Amortization of intangible assets = \$442
Gain on sale of Eurosport = \$93	Corporate and other activities = \$91
	Interest expense = \$523
	Income taxes = \$1458
	Minority interests = \$127
	Loss related to Internet group = \$785

82. Below is a balance sheet of income and expenditures for a small business. Find the net profit or loss.

June 2000 balance sheet

Income	Expenditures
Revenue = \$24,572.88	Materials = \$1545.75
Dividends = \$1284.56	Lease = \$2700
	Utilities = \$865.45
	Employee wages = \$21,580.50

83. In engineering, the resultant force on an object is the sum of all the forces acting on the object. The diagram below shows the forces acting on a steel beam. Find the resultant force. What does the sign of the resultant force indicate?

820.7 N   915.6 N

−2004.5 N

Of Interest

N is an abbreviation for the metric unit of force called the newton. The unit is named after Sir Isaac Newton (1642–1727) in honor of his development of a theory of forces.

84. A family's taxable income is their income less deductions. Following is a table that shows the Brendel family's income and deductions for the 2000 tax year. What is the Brendel family's taxable income?

Income	Deductions
Mr. Brendel = $31,672.88	Mortgage interest = $6545.75
Mrs. Brendel = $32,284.56	Charitable donations = $1200
Dividends = $124.75	Medical expenses = $165.45
Miscellaneous income = $2400	Exemptions = $10,800

85. On January 1, 2002, the Dow Jones Industrial Average closed at 10,021.5. On June 1, 2002, the DJIA closed at 9925.25. Find the difference in closing value from January 1 to June 1.

86. On May 3, 2001, the closing price of EMC Corporation's stock was 40\frac{1}{8}$. On May 2, 2001, the closing price was 42\frac{1}{4}$. Find the difference in closing price from May 2 to May 3.

87. On May 3, 2001, the closing price of Microsoft Corporation's stock was 68\frac{17}{32}$. If the price had a change of $-$1$\frac{7}{32}$ from the previous date's closing price, what was the closing price on May 2?

88. On May 2, 2002, an investor found that the NASDAQ closed at 1644.82, which was a change of -32.71 from the previous day's closing value. What was the closing price on May 1?

89. The temperature of liquid nitrogen is $-208°$C. An apple is placed in the liquid, and the apple then raises the temperature of the liquid to its boiling point of $-196°$C. Write an equation that describes the difference in the temperatures, then calculate the difference.

90. Absolute zero is the temperature at which molecular motion is at a minimum. This temperature is $-273.15°$C. A piece of metal is cooled to $-256.5 °$C. Write an equation that describes the difference in the temperatures, then calculate the difference.

Of Interest

Absolute zero refers to 0 on the Kelvin scale, which is the scale most scientists use to measure temperatures. Increments on the Kelvin scale are equal to increments on the Celsius scale, so a change of 1K corresponds to a change of 1°C.

Use the following line graph to answer Exercises 91 and 92. The graph shows the mean composite ACT score of entering college freshmen from 1971 to 1994.

91. Write an equation that shows the difference in the mean composite score from 1970 to 1980.

92. Write an equation that shows the difference in mean composite score from 1980 to 1994.

OFF TO COLLEGE

Source: U.S. Department of Education

Use the following bar graph to answer Exercises 93 and 94. The graph shows the median annual income of year-round full-time workers 25 years old and over, by level of education completed and gender, from 1989 to 1998.

93. In what year was the difference between salaries for males and females with a bachelor's degree the greatest? How much was that difference?

94. In what year was the difference in salaries greatest between a person with a high school degree and a person with a bachelor's degree? What was that difference?

MEDIAN ANNUAL INCOME OF YEAR-ROUND FULL-TIME WORKERS

25 YEARS OLD AND OVER, BY LEVEL OF EDUCATION COMPLETED AND SEX: 1989 TO 1998*

*In constant 1998 dollars

Source: U.S. Department of Education

Fill in each square with a number 1 through 9 so that the sum of the numbers in each row, each column, and the two diagonals is the same.

Dollars and Sense

Your group is to prepare a simple monthly household budget. The budget should be for a "typical" student or for a new employee in your field of study. (However, you should not use any of your personal information.) Remember to be realistic about both income and expenditures.

1. Discuss and agree upon an amount of income you think an average working student makes each month (or use a starting salary for a person in your field of study).

2. Estimate average expense amounts for each of the following. If your group feels that an expense does not apply, then explain why.

Taxes	Transportation/Gasoline
Housing	Child care
Electricity	Loan repayments
Water	Hair care
Telephone (home and cellular)	Clothing
Cable TV	Magazine/Newspaper subscriptions
Internet service	Allowance
Food	Movies
Household supplies	Other entertainment (video rentals,
Medicine	concerts, eating out)

3. Using your estimates in Exercises 1 and 2, calculate the monthly net.

4. Discuss and list ways that a person could cut expenses to improve his or her monthly net.

REVIEW EXERCISES

1. Write a set containing the last names of the first 4 presidents of the United States.

2. Evaluate: $|-7|$

3. Find the missing number that makes $\dfrac{40}{48} = \dfrac{?}{6}$ true.

4. Use $<$, $>$, or $=$ to make a true statement: $-\dfrac{4}{5}$ $-\dfrac{8}{20}$

5. Find the prime factorization of 100.

6. A company report indicates that 78 employees out of 91 have medical flexible spending accounts. What fraction of the employees have medical flexible spending accounts? Express the fraction in lowest terms.

Multiplying and Dividing Real Numbers; Properties of Real Numbers

OBJECTIVES

1 Multiply integers.

2 Multiply more than two numbers.

3 Multiply rational numbers.

4 Find the multiplicative inverse of a number.

5 Divide rational numbers.

OBJECTIVE 1 **Multiply integers.** In a multiplication statement, *factors* are multiplied to equal a *product*.

$$2 \cdot 3 = 6$$

Factors   Product

In Section 1.3, we discussed properties of addition. Multiplication has similar properties to addition. The table below shows the properties of multiplication.

Properties of Multiplication	Symbolic Form	Word Form
Multiplicative Property of 0	$0 \cdot a = 0$	The product of a number multiplied by 0 is 0.
Multiplicative Identity	$1 \cdot a = a$	The product of a number multiplied by 1 is the number.
Commutative Property of Multiplication	$ab = ba$	Changing the order of factors does not affect the product.
Associative Property of Multiplication	$a(bc) = (ab)c$	Changing the grouping of three or more factors does not affect the product.
Distributive Property of Multiplication over Addition	$a(b + c) = ab + ac$	A sum multiplied by a factor is equal to the sum of that factor multiplied by each addend.

Multiplying Numbers with Different Signs

To determine the rules for multiplying signed numbers, consider the following pattern. Pay attention to the product as we decrease the second factor.

$2 \cdot 4 = 8$

$2 \cdot 3 = 6$ ⟵ As we decrease the second factor by 1, the product decreases by 2, from 8 to 6.

$2 \cdot 2 = 4$ Notice that the same pattern continues with the product decreasing by 2 each time we decrease the factor by 1.

$2 \cdot 1 = 2$

$2 \cdot 0 = 0$ Notice that when we decrease the 0 factor to -1, we continue the pattern and decrease the product by 2

$2 \cdot (-1) = -2$ ⟵ so that it is -2.

$2 \cdot (-2) = -4$ Decreasing the factor by 1 again decreases the product by 2, so that we get -4.

This pattern suggests that multiplying a positive number by a negative number equals a negative product. Note that by the commutative property of multiplication we can exchange the factors and conclude that multiplying a negative number times a positive number also equals a negative product.

$$2 \cdot (-2) = (-2) \cdot 2 = -4$$

RULE **MULTIPLYING TWO NUMBERS WITH DIFFERENT SIGNS**

When multiplying two numbers that have different signs, the product is negative.

EXAMPLE 1 Multiply.

a. $9(-8)$

Solution $9(-8) = -72$

b. $(-12)3$

Solution $(-12)3 = -36$

Warning: Make sure that you see the difference between $9(-8)$, which indicates multiplication, and $9-8$, which indicates subtraction.

YOUR TURN **Multiply.**

a. $13 \cdot (-7)$ **b.** $-5 \cdot 9$

Multiplying Numbers with the Same Sign

Now let's examine multiplying numbers with the same sign. Consider the following pattern to determine the rule.

$(-2) \cdot 4 = -8$ From the rule we already established, a negative number times a positive number is a negative product.

$(-2) \cdot 3 = -6$ As we decrease the positive factor, what happens to the product?

$(-2) \cdot 2 = -4$ ← Notice the product is actually increasing!
$(-2) \cdot 1 = -2$ From -6 to -4 is an increase of 2.
$(-2) \cdot 0 = 0$
$(-2) \cdot (-1) = 2$ ← To continue the same pattern, when we decrease the 0
$(-2) \cdot (-2) = 4$ factor to -1, we must continue to increase the product by 2. This means the product must become positive!

This pattern indicates that the product of two negative numbers is a positive number. We have already seen that the product of two positive numbers is a positive number, so we can draw the following conclusion.

ANSWERS

a. -91 **b.** -45

MULTIPLYING TWO NUMBERS WITH THE SAME SIGN

When multiplying two numbers that have the same sign, the product is positive.

EXAMPLE 2 Multiply.

a. $-6(-8) =$

Solution $-6(-8) = 48$

b. $(-9)(-7) =$

Solution $(-9)(-7) = 63$

YOUR TURN **Multiply.**

a. $-6(-9)$ **b.** $(-3)(-13)$

OBJECTIVE 2 **Multiply more than two numbers.** Suppose we have to multiply more than two numbers, such as $(-2)(-3)(5)$. The associative property of multiplication allows us to group the factors any way we wish.

$$(-2)(-3)(5) = 6(5) \qquad \text{Multiply the first two factors.}$$
$$= 30$$

or

$$(-2)(-3)(5) = (-2)(-15) \qquad \text{Multiply the second and third factors.}$$
$$= 30$$

Notice that no matter which way we choose to multiply the factors, the result is the same. Also notice that the result is positive because there are two negative factors involved in the multiplication. What if we have three negative factors as in $(-2)(-3)(-5)$?

$$(-2)(-3)(-5) = 6(-5) \qquad \text{Multiply the first two factors.}$$
$$= -30$$

or

$$(-2)(-3)(-5) = (-2)(15) \qquad \text{Multiply the second and third factors.}$$
$$= -30$$

ANSWERS

a. 54 **b.** 39

With three negative factors, the result is negative. This suggests the following rule.

MULTIPLYING WITH NEGATIVE FACTORS

The product of an even number of negative factors is positive, whereas the product of an odd number of negative factors is negative.

EXAMPLE 3 Multiply.

a. $(-1)(-1)(-2)(6) =$

Solution Because there are three negative factors (an odd number of negative factors), the result is negative.

$$(-1)(-1)(-2)(6) = -12$$

b. $(-1)(-3)(-7)(2)(-4) =$

Solution Because there are four negative factors (an even number of negative factors), the result is positive.

$$(-1)(-3)(-7)(2)(-4) = 168$$

YOUR TURN **Multiply.**

a. $(-3)(4)(-5)(2)$ **b.** $(6)(-5)(-1)(2)(-1)$

OBJECTIVE 3 Multiply rational numbers.

Multiplying Fractions

On a ruler, note that half of a fourth is an eighth.

When preceded by a fraction, the word *of* means multiply, so the above sentence would translate as follows:

Half of a fourth is an eighth.

$$\frac{1}{2} \cdot \frac{1}{4} = \frac{1}{8}$$

Notice that the numerators multiply to equal the numerator and the denominators multiply to equal the denominator. This suggests the following rule for multiplying fractions.

RULE	**MULTIPLYING FRACTIONS**

$$\frac{a}{b} \cdot \frac{c}{d} = \frac{ac}{bd}, \text{ where } b \neq 0 \text{ and } d \neq 0.$$

Multiplying and Simplifying

In mathematics, it is expected that we always simplify results to lowest terms. Consider $\frac{3}{4} \cdot \frac{5}{6}$. We can multiply first, then simplify.

$$\frac{3}{4} \cdot \frac{5}{6} = \frac{15}{24} = \frac{3 \cdot 5}{2 \cdot 2 \cdot 2 \cdot 3} = \frac{5}{8}$$

Note: After multiplying, we divide out the common factor, 3.

Or, as an alternative, we can actually divide out the common factors before multiplying the fractions. In our example, we can divide out the common factor of 3 and then multiply.

$$\frac{3}{4} \cdot \frac{5}{6} = \frac{3}{2 \cdot 2} \cdot \frac{5}{2 \cdot 3} = \frac{5}{8}$$

Note: Some people prefer to divide out the common factors without breaking down to primes like this:

$$\frac{\overset{1}{\cancel{3}}}{4} \cdot \frac{5}{\underset{2}{\cancel{6}}} = \frac{5}{8}$$

We can divide out common factors before or after multiplying because every factor in the original numerators becomes a factor in the product's numerator and every factor in the original denominators becomes a factor in the product's denominator. Now let's put this together with the sign rules for multiplication.

EXAMPLE 4 Multiply.

a. $-\frac{2}{3} \cdot \left(\frac{7}{8}\right) =$

Solution $-\frac{2}{3} \cdot \left(\frac{7}{8}\right) = -\frac{2}{3} \cdot \left(\frac{7}{2 \cdot 2 \cdot 2}\right)$ Divide out the common factor, 2.

$$= -\frac{7}{12}$$ Because we are multiplying two numbers that have different signs, the product is negative.

Calculator TIPS

To use a calculator for Example 4 (a), type:

b. $-\dfrac{4}{15} \cdot \left(-\dfrac{9}{16}\right) =$

Solution $-\dfrac{4}{15} \cdot \left(-\dfrac{9}{16}\right) = -\dfrac{2 \cdot 2}{3 \cdot 5} \cdot \left(-\dfrac{3 \cdot 3}{2 \cdot 2 \cdot 2 \cdot 2}\right)$ Divide out the common factors.

$= \dfrac{3}{20}$ Because we are multiplying two numbers that have the same sign, the product is positive.

CONNECTION ────────────────────────

Using the slash marks style, Example 4(b) looks like this:

$$-\dfrac{\overset{1}{\cancel{4}}}{\underset{5}{\cancel{15}}} \cdot \left(-\dfrac{\overset{3}{\cancel{9}}}{\underset{4}{\cancel{16}}}\right) = \dfrac{3}{20}$$

Notice that we are still dividing out a factor of 4 in the 4 and 16, and a common factor of 3 in the 9 and 15.

YOUR TURN   **Multiply.**

a. $\dfrac{5}{8} \cdot \left(-\dfrac{3}{10}\right) =$ **b.** $-\dfrac{4}{5} \cdot \left(\dfrac{1}{6}\right) \cdot \left(-\dfrac{3}{9}\right)$

Multiplying Decimal Numbers

Recall that decimal numbers name fractions with denominators that are powers of 10. In the problem $0.3 \cdot 0.7$, we could multiply by replacing the decimal factors with their fraction equivalents.

$$0.3 \cdot 0.7 = \dfrac{3}{10} \cdot \dfrac{7}{10} = \dfrac{21}{100} = 0.21$$

Because we multiply the denominators, if we multiply tenths times tenths, then the product will be hundredths. Notice that if we count the number of decimal places in the factors this corresponds to the total places in the product.

$$\begin{array}{rcl} 0.7 & \longrightarrow & 1 \text{ place} \\ \times\ 0.3 & \longrightarrow & +\ 1 \text{ place} \\ \hline 0.21 & \longrightarrow & 2 \text{ places} \end{array}$$

Likewise, hundredths (2 places) times tenths (1 place) would equal thousandths (3 places).

$$0.03 \cdot 0.7 = \dfrac{3}{100} \cdot \dfrac{7}{10} = \dfrac{21}{1000} = 0.021 \quad \text{OR} \quad \begin{array}{rcl} 0.7 & \longrightarrow & 1 \text{ place} \\ \times\ 0.03 & \longrightarrow & +\ 2 \text{ places} \\ \hline 0.021 & \longrightarrow & 3 \text{ places} \end{array}$$

ANSWERS

a. $-\dfrac{3}{16}$ **b.** $\dfrac{2}{45}$

PROCEDURE MULTIPLYING DECIMAL NUMBERS

To multiply decimal numbers:

1. Multiply as if they named whole numbers.
2. Place the decimal in the product so that it has the same number of places as the total number of places in the factors.

EXAMPLE 5 Multiply. $(-5.6)(0.03) =$

Solution First, we calculate the value and disregard the signs for now.

$$
\begin{array}{rl}
0.03 & \text{2 places} \\
\times \ 5.6 & \underline{+ \text{ 1 place}} \\
\hline
0\,1\,8 & \\
+ \ 0\,1\,5 & \\
\hline
0.1\,6\,8 & \longleftarrow \text{3 places}
\end{array}
$$

Answer: -0.168 ◄ **Note:** When we multiply two numbers with different signs, the product is negative.

YOUR TURN Multiply.

a. $(-0.07)(-2.65)$　　　　b. $(-1)(-0.9)(-24)(0.2)$

OBJECTIVE 4 **Find the multiplicative inverse of a number.** Two numbers that multiply to equal 1 are called **multiplicative inverses**.

DEFINITION *Multiplicative inverses:* Two numbers whose product is 1.

For example, $\dfrac{2}{3}$ and $\dfrac{3}{2}$ are multiplicative inverses because their product is 1.

$$\frac{2}{3} \cdot \frac{3}{2} = \frac{6}{6} = 1$$

Notice that to write a number's multiplicative inverse, we simply invert the numerator and denominator. Multiplicative inverses are also known as *reciprocals*.

EXAMPLE 6 Find the multiplicative inverse.

a. $\dfrac{5}{8}$

Answer: The multiplicative inverse of $\dfrac{5}{8}$ is $\dfrac{8}{5}$ because $\dfrac{5}{8} \cdot \dfrac{8}{5} = 1$.

b. $-\dfrac{1}{4}$

Answer: The multiplicative inverse of $-\dfrac{1}{4}$ is -4 because

$$-\dfrac{1}{4} \cdot (-4) = -\dfrac{1}{4} \cdot \left(-\dfrac{4}{1}\right) = 1.$$

c. -6

Answer: The multiplicative inverse of -6 is $-\dfrac{1}{6}$ because

$$-6 \cdot \left(-\dfrac{1}{6}\right) = -\dfrac{6}{1} \cdot \left(-\dfrac{1}{6}\right) = 1.$$

YOUR TURN Find the multiplicative inverse.

a. $\dfrac{4}{5}$ **b.** $\dfrac{1}{7}$ **c.** -9

OBJECTIVE 5 Divide rational numbers. Let's now turn our attention to division. The parts of a division statement are shown below.

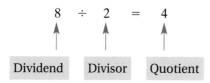

$$8 \;\;\div\;\; 2 \;\;=\;\; 4$$

Dividend Divisor Quotient

CONNECTION

Division can also be expressed in fraction form. For example, the division statement $8 \div 2 = 4$ can also be written as $\dfrac{8}{2} = 4$.

Sign Rules for Division

The sign rules for division are the same as for multiplication. We can see this by looking at the relationship between division and multiplication. We can use one operation to check the other. For example, if $8 \cdot 2 = 16$, then $16 \div 2 = 8$ or $16 \div 8 = 2$. In checking a multiplication, we divide the product by one of the factors to equal the other factor. If we place signs in the problem, we could say that if $8 \cdot (-2) = -16$, then $-16 \div (-2) = 8$ or $-16 \div 8 = -2$.

RULE **DIVIDING SIGNED NUMBERS**

When dividing two numbers that have the *same* sign, the quotient is positive.

When dividing two numbers that have *different* signs, the quotient is negative.

EXAMPLE 7 Divide.

a. $48 \div (-6) =$

Solution $48 \div (-6) = -8$

b. $-35 \div (-7) =$

Solution $-35 \div (-7) = 5$

YOUR TURN Divide.

a. $(-60) \div 12$ **b.** $-42 \div -6$

Division Involving 0

What if 0 is involved in division? First, consider when the *dividend* is 0 and the *divisor* is not 0, as in $0 \div 9$. The result should be $0 \div 9 = 0$. Remember that we can check division by multiplying the quotient and the divisor to equal the dividend, so we can verify that $0 \div 9 = 0$ by writing the following multiplication:

$$\text{Quotient} \cdot \text{Divisor} = \text{Dividend}$$
$$0 \quad \cdot \quad 9 \quad = \quad 0$$

Conclusion   If the dividend is 0 and the divisor is not zero, then the quotient is 0.

What if the divisor is 0 as in $8 \div 0$? Again, think about the check statement:

$$\text{Quotient} \cdot \text{Divisor} = \text{Dividend}$$
$$? \quad \cdot \quad 0 \quad = \quad 8$$

There is no number that can make $? \cdot 0 = 8$ true because any number multiplied by 0 would equal 0, not 8. So there is also no number that can make $8 \div 0 = ?$ true, and we say that dividing a nonzero number by 0 is *undefined*.

Conclusion   If the divisor is 0 with a nonzero dividend, the statement is undefined.

What if both the dividend and divisor are 0 as in $0 \div 0$? Again, think about the check statement.

$$\text{Quotient} \cdot \text{Divisor} = \text{Dividend}$$
$$? \quad \cdot \quad 0 \quad = \quad 0$$

Notice any number would work for this quotient because any number times 0 equals 0. Because we cannot determine a unique quotient here, we say that 0 ÷ 0 is *indeterminate*.

Conclusion   If both the dividend and divisor are 0, then the statement is indeterminate.

Following is a summary of these rules.

RULE **DIVISION INVOLVING 0**

$0 \div n = 0$ when $n \neq 0$.

$n \div 0$ is undefined when $n \neq 0$.

$0 \div 0$ is indeterminate.

Calculator
TIPS

Try dividing any number by 0 on your calculator. When you press the equal key the screen will display error, *which is your calculator's way of indicating undefined or indeterminate.*

Dividing Fractions

To determine how to divide when the divisor is a fraction, consider this problem:

How many quarters are in $5? A quarter is $\frac{1}{4}$ of a dollar, so this question translates to the division problem:

$$\text{Number of quarters } = 5 \div \frac{1}{4}$$

Now, since there are 4 quarters in each of those 5 dollars, then there must be 20 quarters in $5. Notice in analyzing the situation we translated the division problem into a multiplication problem. We can say:

$$\text{Number of quarters } = 5 \div \frac{1}{4} = 5 \cdot 4 = 20$$

Notice that $\frac{1}{4}$ and 4 are multiplicative inverses. We can write a division problem as an equivalent multiplication problem by changing the divisor to its multiplicative inverse.

RULE **DIVIDING FRACTIONS**

$\frac{a}{b} \div \frac{c}{d} = \frac{a}{b} \cdot \frac{d}{c}$, where $b \neq 0$, $c \neq 0$, and $d \neq 0$.

EXAMPLE 8 Divide $-\dfrac{7}{8} \div \dfrac{5}{6}$.

Solution $-\dfrac{7}{8} \div \dfrac{5}{6} = -\dfrac{7}{8} \cdot \dfrac{6}{5}$ Write an equivalent multiplication.

$$= -\dfrac{7}{2 \cdot 2 \cdot 2} \cdot \dfrac{2 \cdot 3}{5}$$ Divide out the common factor, 2.

$$= -\dfrac{21}{20}$$ Because we are dividing two numbers that have different signs, the result is negative.

CONNECTION

Using slash marks, Example 8 looks like this:

$$-\dfrac{7}{8} \div \dfrac{5}{6} = -\dfrac{7}{\overset{}{\underset{4}{8}}} \cdot \dfrac{\overset{3}{6}}{5} = -\dfrac{21}{20}$$

YOUR TURN   Divide.

a. $\dfrac{4}{9} \div \left(-\dfrac{8}{15}\right)$ **b.** $-\dfrac{8}{21} \div \left(-\dfrac{6}{7}\right)$

Dividing Decimal Numbers

Division with decimal numbers can be broken into two categories: one, if the divisor is an integer, and two, if the divisor contains a decimal point. The following procedure summarizes how to divide decimal numbers.

PROCEDURE   DIVIDING DECIMAL NUMBERS

To divide decimal numbers, set up a long division and consider the divisor.

Case 1: If the divisor is an integer, then divide as if the dividend were a whole number and place the decimal point in the quotient directly above its position in the dividend.

Case 2: If the divisor is a decimal number, then

1. Move the decimal point in the divisor to the right enough places to make the divisor an integer.
2. Move the decimal point in the dividend the same number of places.
3. Divide the divisor into the dividend as if both numbers were whole numbers. Be sure to align the digits in the quotient properly.
4. Write the decimal point in the quotient directly above its position in the dividend.

In either of the two cases, continue the division process until you get a remainder of 0 or a repeating digit or block of digits in the quotient.

ANSWERS

a. $-\dfrac{5}{6}$ **b.** $\dfrac{4}{9}$

EXAMPLE 9 Divide $-12.8 \div (-0.09)$.

Solution Because the divisor is a decimal number, we move the decimal point enough places to the right to create an integer—in this case two places. Then move the decimal point two places to the right in the dividend. Because we are dividing two numbers with the same sign, the result is positive.

$$
\begin{array}{r}
142.22 \\
9\overline{)1280.00} \\
-9 \\
\hline
38 \\
-36 \\
\hline
20 \\
-18 \\
\hline
20
\end{array}
$$

Note: Because the 2 digit repeats without end, we write a repeat bar over the 2 just after the decimal point.

Answer: $142.\overline{2}$

ANSWERS

a. -2.67 b. -2.05
c. $1.4\overline{6}$

YOUR TURN Divide.

a. $32.04 \div (-12)$ b. $-13.12 \div 6.4$ c. $-0.88 \div (-0.6)$

1.4 EXERCISES

For Extra Help

InterAct Math

Tutor Center — Addison-Wesley Math Tutor Center

Video 1/DVT 1

MyMathLab

MathXL — MathXL

For Exercises 1–4, state whether the result is positive or negative.

1. When multiplying or dividing two numbers that have the same sign, the sign of the result is _____ .

2. When multiplying or dividing two numbers that have different signs, the sign of the result is _____ .

3. When a multiplication problem contains an odd number of negative factors, the sign of the product is _____ .

4. When a multiplication problem contains an even number of negative factors, the sign of the product is _____ .

5. Why is $5 \div 0$ undefined?

6. Why is $0 \div 0$ indeterminate?

7. Why are 5 and $\frac{1}{5}$ multiplicative inverses?

8. Explain how to divide fractions.

9. Explain why $\frac{-a}{b} = \frac{a}{-b} = -\frac{a}{b}$, where a and b are any real numbers and $b \neq 0$.

10. Explain the difference between $6(-7)$ and $6 - 7$.

For Exercises 11–24, indicate whether the equation illustrates the multiplicative property of 0, the multiplicative identity, the commutative property of multiplication, the associative property of multiplication, or the distributive property.

11. $6(3 + 4) = 6 \cdot 3 + 6 \cdot 4$

12. $1 \cdot (-7) = -7$

13. $4(3 \cdot 2) = (4 \cdot 3)2$

14. $(3.4)(7.8) = (7.8)(3.4)$

15. $1 \cdot \left(-\frac{5}{6}\right) = -\frac{5}{6}$

16. $\frac{2}{3} \cdot 0 = 0$

17. $\frac{3}{5} \cdot \left(-\frac{1}{4}\right) = -\frac{1}{4} \cdot \frac{3}{5}$

18. $\frac{2}{3} \cdot 5 + \frac{2}{3} \cdot 4 = \frac{2}{3}(5 + 4)$

19. $0 \cdot (-6.3) = 0$

20. $(0.5)(7.4)(-2) = (0.5)(-2)(7.4)$

21. $-\frac{2}{3} \cdot \left(\frac{5}{5}\right) = -\frac{10}{15}$

22. $-(5 + 9) = -5 - 9$

23. $(-4.5)(2 + 0.6) = (2 + 0.6)(-4.5)$

24. $[(-6.2)(-4.3)]7.1 = (-6.2)[(-4.3)7.1]$

For Exercises 25–44, multiply.

25. $6(-3)$

26. $(-10)(5)$

27. $-4(5)$

28. $(-4)(-3)$

29. $6(-9)(-1)$

30. $12(-6)(-2)(-1)$

31. $(-1)(-6)(-40)(-3)$

32. $(-1)(-1)(4)(-5)(-3)$

33. $\dfrac{1}{2} \cdot \left(-\dfrac{1}{2}\right)$ **34.** $\dfrac{4}{5} \cdot \left(\dfrac{20}{3}\right)$ **35.** $\left(-\dfrac{2}{3}\right)\left(-\dfrac{3}{4}\right)$ **36.** $\left(-\dfrac{5}{6}\right)\left(-\dfrac{6}{5}\right)$

37. $\left(-\dfrac{5}{6}\right)\left(\dfrac{8}{15}\right)$ **38.** $\left(\dfrac{2}{9}\right)\left(-\dfrac{21}{26}\right)$ **39.** $3.8(-10)$ **40.** $8(-2.5)$

41. $-1.6(-4.2)$ **42.** $-7.1(0.5)$ **43.** $-4.3(1.52)$ **44.** $(-8.1)(-2.75)$

For Exercises 45–52, find the multiplicative inverse.

45. -5 **46.** $-\dfrac{6}{7}$ **47.** $\dfrac{2}{3}$ **48.** 17

49. $-\dfrac{5}{2}$ **50.** $\dfrac{20}{3}$ **51.** 0 **52.** -1

For Exercises 53–76, divide.

53. $-6 \div 2$ **54.** $-12 \div (-4)$ **55.** $42 \div (-7)$ **56.** $56 \div (-4)$

57. $\dfrac{-18}{-9}$ **58.** $\dfrac{75}{-3}$ **59.** $\dfrac{-8}{0}$ **60.** $\dfrac{51}{-17}$

61. $0 \div 0$ **62.** $-21 \div 0$ **63.** $6 \div \dfrac{2}{3}$ **64.** $-8 \div \dfrac{3}{4}$

65. $\dfrac{3}{5} \div \left(-\dfrac{9}{10}\right)$ **66.** $-\dfrac{1}{3} \div \left(-\dfrac{3}{2}\right)$ **67.** $-\dfrac{2}{7} \div \left(-\dfrac{8}{21}\right)$ **68.** $-\dfrac{4}{5} \div \dfrac{4}{5}$

69. $-\dfrac{4}{9} \div \dfrac{10}{21}$ **70.** $\dfrac{7}{15} \div \left(-\dfrac{35}{24}\right)$ **71.** $9.03 \div 4.3$ **72.** $-10.65 \div 7.1$

73. $-36.72 \div (-0.4)$ **74.** $8.1 \div 0.6$ **75.** $-14 \div 0.3$ **76.** $-19 \div (-0.06)$

For Exercises 77–82, solve.

77. In a poll where respondents can agree, disagree, or have no opinion, $\frac{3}{4}$ of the respondents said they agreed and $\frac{2}{3}$ of those that agreed were women. What fraction of all respondents were women who agreed with the statement in the poll?

78. On a standard size guitar, the length of the strings between the saddle and nut is $25\frac{1}{2}$ inches. Guitar makers must place the 12th fret at exactly half of the length of the string. How far from the saddle or nut should one measure to place the 12th fret?

Saddle

Nut

12th fret

79. A financial planner estimates that at the rate one of his clients is increasing debt she will have $3\frac{1}{2}$ times her current debt in five years. If her current debt is $-\$2480$, what will be her debt in five years?

80. Mario finds that $\frac{2}{3}$ of his credit card debt is from dining out. If his total credit card debt is $-\$858$, how much of the debt is from dining out?

81. In 1994, the Garret family's only debt was $-\$258.75$ on a credit card. In 2000, they have a mortgage, two car loans, and three credit cards with a total debt of $-\$158,572.85$. How many times greater is their debt in 2000 than it was in 1994?

82. A company's stock loses value by $-\$\frac{3}{8}$ each day for four days. What was the total loss in value at the close of the fourth day?

For Exercises 83–86, use the fact that an object's weight is a downward force due to gravity and is calculated by multiplying the object's mass by the acceleration due to gravity, which is a constant (force = mass × acceleration). The following table indicates the units.

	Force	=	Mass	×	Acceleration
American measurement	pounds (lb)		slugs (s)		-32.2 ft./sec.²
Metric measurement	newtons (N)		kilograms (kg)		-9.8 m/sec.²

83. Find the weight of an object with a mass of 12.5 slugs. What does the sign of the weight indicate?

84. Find the weight of a person with a mass of 70.4 kilograms.

85. The blue whale is the largest animal on Earth. The largest blue whale ever caught weighed 1,658,181.8 newtons, which means the downward force due to gravity was −1,658,181.8 newtons. What was its mass in kilograms?
(*Source: The Whale Watcher's Guide: Whale Watching Trips in North America*, Corrigan, 1999)

86. The Liberty Bell weighs 2080 pounds, which means the downward force due to gravity is −2080 pounds. What is its mass in slugs?
(*Source: Ring in the Jubilee: The Epic of America's Liberty Bell*, Boland, 1973)

For Exercises 87 and 88, use the fact that in an electrical circuit, voltage (V) is equal to the product of the current, measured in amperes (A), and the resistance of the circuit, measured in ohms (Ω) (voltage = current × resistance).

87. Suppose the current in a circuit is −6.4 amperes and the resistance is 8 ohms. Find the voltage.

88. An electrical technician measures the voltage in a circuit to be −15 volts and the current to be −8 amperes. What is the resistance of the circuit?

For Exercises 89 and 90, use the fact that in an electrical circuit, power, which is measured in watts (W), is the product of the voltage and the current (power = voltage × current).

89. An engineer measures the voltage in a circuit to be −120 volts and the current to be −30 amperes. What is the power?

90. An electrical technician determines that the power in a circuit is 400 watts. If the current is measured to be −6.5 amperes, what is the resistance in the circuit?

REVIEW EXERCISES

1. Determine whether π is a rational or irrational number.

2. Graph −7.2 on a number line.

3. Add: $-7 + (-15)$

4. Find the additive inverse of $\frac{2}{3}$.

5. Subtract: $-\frac{5}{8} - \left(-\frac{1}{6}\right)$

6. Simplify: $|-6.8|$

1.5 Exponents, Roots, and Order of Operations

OBJECTIVES

1. Evaluate numbers in exponential form.
2. Evaluate square roots.
3. Use the order-of-operations agreement to simplify numerical expressions.
4. Find the mean of a set of data.

OBJECTIVE 1 Evaluate numbers in exponential form. Sometimes problems involve repeatedly multiplying the same number. In such problems, we can use an **exponent** to indicate that a **base** number is repeatedly multiplied.

> **DEFINITIONS** *Exponent:* A symbol written to the upper right of a base number that indicates how many times to use the base as a factor.
> *Base:* The number that is repeatedly multiplied.

When we write a number with an exponent, we say the expression is in *exponential form.* The expression 2^4 is in exponential form, where the base is 2 and the exponent is 4, it is read "two to the fourth power," or simply "two to the fourth." To *evaluate* 2^4, write 2 as a factor 4 times, then multiply.

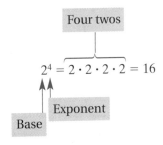

Four twos

$$2^4 = 2 \cdot 2 \cdot 2 \cdot 2 = 16$$

Exponent

Base

PROCEDURE EVALUATING AN EXPONENTIAL FORM

To evaluate an exponential form raised to a natural number exponent, write the base as a factor the number of times indicated by the exponent, then multiply.

EXAMPLE 1 Evaluate $(-7)^2$.

Solution The exponent 2 indicates we have two factors of -7. Because we multiply two negative numbers, the result is positive.

$$(-7)^2 = (-7)(-7) = 49$$

> **Note:** If a base is raised to the second power we can say it is *squared*. The expression $(-7)^2$ can be read as "negative seven *squared*."

EXAMPLE 2 Evaluate $\left(-\dfrac{2}{3}\right)^3$.

Solution The exponent 3 means we must multiply the base by itself three times.

$$\left(-\frac{2}{3}\right)^3 = \left(-\frac{2}{3}\right)\left(-\frac{2}{3}\right)\left(-\frac{2}{3}\right)$$

$$= -\frac{8}{27}$$

Note: If a base is raised to the third power we can say it is *cubed*. The expression $\left(-\dfrac{2}{3}\right)^3$ can be read as "negative two-thirds *cubed*."

Examples 1 and 2 suggest the following sign rules.

RULE   **EVALUATING EXPONENTIAL FORMS WITH NEGATIVE BASES**

If the base of an exponential form is a negative number and the exponent is even, then the product is positive.

If the base is a negative number and the exponent is odd, then the product is negative.

YOUR TURN **Evaluate.**

a. $\left(-\dfrac{1}{2}\right)^5$

b. $(-0.3)^4$

Calculator TIPS

Use the *key to indicate an exponent. To evaluate* $\left(-\dfrac{2}{3}\right)^3$, *type:*

(| (−) | 2 | ÷ | 3 |) | ^ | 3 | ENTER

The answer will be given as a decimal. To change the decimal answer to a fraction, press the MATH *key, select 1:Frac, then press* ENTER.

On some scientific calculators, you use a y^x *key. To evaluate* $\left(-\dfrac{2}{3}\right)^3$ *on one of these, type:*

2 | $a^{b/c}$ | 3 | +/− | y^x | 3 | =

ANSWERS

a. $-\dfrac{1}{32}$ **b.** 0.0081

OBJECTIVE 2 Evaluate square roots. Roots are inverses of exponents. More specifically, a square root is the inverse of a square, so a square root of a given number is a number that, when squared, equals the given number. For example, 5 is a square root of 25 because $5^2 = 25$. Notice -5 is also a square root of 25 because $(-5)^2 = 25$.

Conclusion For every real number there are *two* square roots, a positive root and a negative root.

For convenience, if asked to find all square roots of 25, we can write both the positive and negative answers in a compact expression like this: ± 5.

EXAMPLE 3 Find all square roots of the given number.

a. 36

Answer: ± 6

b. -25

Answer: Not a real number.

Note: We do not say that this situation is undefined or indeterminate, because we eventually will define the square root of a negative number. Such roots will be defined using imaginary numbers.

Explanation: Because of the sign rules for multiplication, there is no way to square a real number and get anything other than a positive result.

Our work so far suggests the following rules.

RULE   SQUARE ROOTS

Every positive number has two square roots, a positive root and a negative root. The square root of a negative number is not a real number.

YOUR TURN Find all square roots of the given number.

a. 144   **b.** -81

The Principal Square Root

A symbol, $\sqrt{}$, called the *radical*, is used to indicate finding only the *positive* (or *principal*) square root of a given number. The given number or expression inside the radical is called the *radicand*.

What about the square root of a fraction like $\sqrt{\dfrac{4}{9}}$? Since $\left(\dfrac{2}{3}\right)^2 = \dfrac{2}{3} \cdot \dfrac{2}{3} = \dfrac{4}{9}$ we can say that $\sqrt{\dfrac{4}{9}} = \dfrac{2}{3}$. Notice that we could evaluate $\sqrt{\dfrac{4}{9}}$ by finding the square root of the numerator and denominator separately, or $\sqrt{\dfrac{4}{9}} = \dfrac{\sqrt{4}}{\sqrt{9}} = \dfrac{2}{3}$.

We now have the following rules for square roots involving the radical sign.

SQUARE ROOTS INVOLVING THE RADICAL SIGN

The radical symbol $\sqrt{}$ denotes only the positive (principal) square root.

$$\sqrt{\frac{a}{b}} = \frac{\sqrt{a}}{\sqrt{b}}, \text{ where } a \geq 0 \text{ and } b > 0.$$

Calculator TIPS

To calculate a square root, use the $\sqrt{}$ function, which may require pressing [2nd] *and then the* [x²] *key. To find the square root of 144, type:*

[2nd] [x²] [1] [4] [4] [ENTER]

Some scientific calculators have a [√x] *key. To find the square root of 144 on one of these, type:*

[1] [4] [4] [√x]

EXAMPLE 4 Evaluate the square root.

a. $\sqrt{144} =$

b. $\sqrt{\dfrac{49}{81}}$

Solution $\sqrt{144} = 12$

Solution $\sqrt{\dfrac{49}{81}} = \dfrac{\sqrt{49}}{\sqrt{81}} = \dfrac{7}{9}$

c. $\sqrt{0.81}$

d. $\sqrt{-81} =$

Solution $\sqrt{0.81} = 0.9$

Solution No real-number solution.

YOUR TURN **Evaluate the square root.**

a. $\sqrt{121}$

b. $\sqrt{\dfrac{81}{169}}$

c. $\sqrt{-0.09}$

OBJECTIVE 3 Use the order-of-operations agreement to simplify numerical expressions. Some expressions are too complicated to evaluate using only the associative and commutative properties. If an expression contains a mixture of operations, then we must be careful about the order in which we perform the operations. Performing the operations in a different order can change the results. Consider the expression $2 + 5 \cdot 3$.

Adding first: $2 + 5 \cdot 3$
             $= 7 \cdot 3$
             $= 21$

Multiplying first: $2 + 5 \cdot 3$
                  $= 2 + 15$
                  $= 17$

ANSWERS

a. 11   **b.** $\dfrac{9}{13}$

c. No real number solution.

Changing the order of the operations changes the outcome. To ensure that everyone arrives at the same answer, mathematicians developed an order of operations that we all agree to follow.

PROCEDURE   **ORDER-OF-OPERATIONS AGREEMENT**

Perform operations in the following order:

1. Grouping symbols: parentheses ( ), brackets [ ], braces { }, absolute value | |, and radicals $\sqrt{}$.
2. Exponents/Roots from left to right, in order as they occur.
3. Multiplication/Division from left to right, in order as they occur.
4. Addition/Subtraction from left to right, in order as they occur.

If we follow this agreement to simplify $2 + 5 \cdot 3$, the correct order is to multiply first and then add so that the answer is 17.

Embedded Parentheses

When one set of grouping symbols contains another set of grouping symbols, we say they are nested or embedded. Work from the innermost set of grouping symbols outward.

EXAMPLE 5   Simplify $(-5)^2 - 8[4 - (6 + 3)] - \sqrt{36}$.

Solution $(-5)^2 - 8[4 - (6 + 3)] - \sqrt{36}$

$= (-5)^2 - 8[4 - 9] - \sqrt{36}$	Calculate within the innermost parentheses: $6 + 3 = 9$.
$= 25 - 8(-5) - 6$	Evaluate the exponential form, brackets, and square root.
$= 25 - (-40) - 6$	Multiply $8(-5) = -40$.
$= 25 + 40 - 6$	Write $25 - (-40)$ as an equivalent addition.
$= 65 - 6$	Add $25 + 40 = 65$.
$= 59$	Subtract $65 - 6 = 59$.

Warning: The expressions $(-5)^2$ and -5^2 are different. The expression $(-5)^2$ means the square of the number -5, whereas -5^2 means the additive inverse of the square of 5.

$$(-5)^2 = (-5)(-5) = 25 \qquad \text{whereas} \qquad -5^2 = -(5 \cdot 5) = -25$$

YOUR TURN   **Simplify.**

$15.8 - 0.2[7^2 - (12 + 8) \div 4]$

Radical Symbols

Radical symbols can be grouping symbols. Consider $\sqrt{36 \cdot 4}$. Following the order of operations, we multiply first, then find the root of the product. However, if a radical contains multiplication, we can find the roots of the factors and multiply those roots to get the same answer.

ANSWER

7

Multiplying first:		Multiplying the roots:

Multiplying first:

$$\sqrt{36 \cdot 4} = \sqrt{144}$$
$$= 12$$

or

Multiplying the roots:

$$\sqrt{36 \cdot 4} = 6 \cdot 2$$
$$= 12$$

Either approach yields the same result. The same holds for division under a radical.

$$\sqrt{36 \div 4} = \sqrt{9}$$
$$= 3$$

or

$$\sqrt{36 \div 4} = 6 \div 2$$
$$= 3$$

RULE **SQUARE ROOT OF A PRODUCT OR QUOTIENT**

If a square root contains multiplication or division, we can multiply or divide first, then find the square root of the result, or we can find the square roots of the individual numbers, then multiply or divide the square roots.

Note that this rule does *not* hold for addition or subtraction under a radical, such as in $\sqrt{9 + 16}$.

Adding first:

$$\sqrt{9 + 16} = \sqrt{25}$$
$$= 5$$

Adding the roots:

$$\sqrt{9 + 16} = 3 + 4$$
$$= 7$$

Because the results are different, we do not have a choice of how to approach the situation when the radical contains addition. The correct approach is to treat the radical as a grouping symbol and add first. Then find the root of the sum. Thus, the correct answer for $\sqrt{9 + 16}$ is 5. The same procedure applies when a radical contains subtraction.

RULE **SQUARE ROOT OF A SUM OR DIFFERENCE**

When a radical contains addition or subtraction, we must add or subtract first, then find the root of the sum or difference.

EXAMPLE 6 Simplify $12.4 \div 5(-3)^2 + 4\sqrt{169 - 25}$.

Solution $12.4 \div 5(-3)^2 + 4\sqrt{169 - 25}$

$= 12.4 \div 5(-3)^2 + 4\sqrt{144}$ Subtract within the radical: $169 - 25 = 144$.

$= 12.4 \div 5(9) + 4(12)$ Evaluate the exponential form and root.

$= 2.48(9) + 48$ Divide $12.4 \div 5 = 2.48$.

$= 22.32 + 48$ Multiply $2.48(9) = 22.32$.

$= 70.32$ Add $22.32 + 48 = 70.32$.

Note: We divide before multiplying here because the order is to multiply or divide from left to right in the order in which those operations occur.

EXAMPLE 7 Simplify $\left(-\dfrac{1}{4}\right)^3 - \left(\dfrac{1}{2} + \dfrac{3}{8}\right) \div \sqrt{\dfrac{48}{3}}$.

Solution $\left(-\dfrac{1}{4}\right)^3 - \left(\dfrac{1(4)}{2(4)} + \dfrac{3}{8}\right) \div \sqrt{\dfrac{48}{3}}$

$= \left(-\dfrac{1}{4}\right)^3 - \left(\dfrac{4}{8} + \dfrac{3}{8}\right) \div \sqrt{\dfrac{48}{3}}$ Write equivalent fractions with a common denominator in order to add within the parentheses.

$= \left(-\dfrac{1}{4}\right)^3 - \dfrac{7}{8} \div \sqrt{16}$ Add within parentheses and divide within the radical.

$= -\dfrac{1}{64} - \dfrac{7}{8} \div 4$ Evaluate the exponential form and square root.

$= -\dfrac{1}{64} - \dfrac{7}{8} \cdot \dfrac{1}{4}$ Write an equivalent multiplication using the reciprocal of the divisor.

$= -\dfrac{1}{64} - \dfrac{7(2)}{32(2)}$ Multiply $\dfrac{7}{8} \cdot \dfrac{1}{4}$ to get $\dfrac{7}{32}$ then write it as an equivalent fraction with the common denominator 64 in order to subtract.

$= -\dfrac{1}{64} - \dfrac{14}{64}$

$= -\dfrac{15}{64}$ Subtract.

YOUR TURN Simplify.

    a. $5 + (0.2)^3 - 3(8 - 14)$ **b.** $-\dfrac{3}{5} \div \dfrac{1}{10} \cdot 4 + \sqrt{64 + 36}$

Fraction Lines

Sometimes fraction lines are used as grouping symbols. When they are, we simplify the numerator and denominator separately, then divide the results.

CONNECTION

The expression in Example 8 is equivalent to the expression $[(-2)^3 + 5(-6)] \div [3(7) - 2]$.

EXAMPLE 8 Simplify $\dfrac{(-2)^3 + 5(-6)}{3(7) - 2}$.

Solution $\dfrac{(-2)^3 + 5(-6)}{3(7) - 2}$

$$= \frac{-8 + 5(-6)}{21 - 2}$$ Evaluate the exponential form in the numerator and multiply in the denominator.

$$= \frac{-8 + (-30)}{19}$$ Multiply in the numerator and subtract in the denominator.

$$= \frac{-38}{19}$$ Add in the denominator.

$$= -2$$ Divide.

EXAMPLE 9 Simplify $\dfrac{9(3) + 15}{4^2 + 2(-8)}$.

Solution $\dfrac{9(3) + 15}{4^2 + 2(-8)}$

$$= \frac{27 + 15}{16 + 2(-8)}$$ Multiply in the numerator and evaluate the exponential form in the denominator.

$$= \frac{42}{16 + (-16)}$$ Add in the numerator and multiply in the denominator.

$$= \frac{42}{0}$$ Add in the denominator.

Because the denominator or divisor is 0, the answer is undefined.

YOUR TURN Simplify.

a. $\dfrac{7(3 - 7) + 1}{(-2)^3 - 1}$

b. $\dfrac{2[9 - 4(3 + 5)]}{25 - (6 - 1)^2}$

OBJECTIVE 4 Find the mean of a set of data. Fraction line notation can be used to indicate the proper sequence of operations in calculating an arithmetic mean, or average, of a set of data.

PROCEDURE FINDING THE ARITHMETIC MEAN

To find the arithmetic mean, or average, of n numbers, divide the sum of the numbers by n.

$$\text{Arithmetic mean} = \frac{x_1 + x_2 + \cdots + x_n}{n}$$

EXAMPLE 10 Jacky has the following test scores: 86, 95, 78, 82, 84. Find the average of her test scores.

Solution $\dfrac{86 + 95 + 78 + 82 + 84}{5} = \dfrac{425}{5}$

$$= 85$$

This table has the daily rainfall accumulations for the month of April in a certain city. Find the average daily accumulation for April.

Date	Accumulation
April 2	0.5 in.
April 7	1.25 in.
April 9	1.0 in.
April 15	1.25 in.
April 21	0.25 in.
April 26	1.0 in.

ANSWER

0.875 in.

1.5 EXERCISES

For Extra Help

InterAct Math

Addison-Wesley Math Tutor Center

Video 2/DVT 1

MyMathLab

MathXL

1. What is another way to say "two to the third power"?

2. What is another way to say "three to the second power"?

3. Explain the difference between a number's square and its square root.

4. If the base of an exponential form is a negative number and the exponent is even, will the sign of the result be positive or negative?

5. If a radical contains multiplication or division of two numbers, what is the proper order of operations?

6. If a radical contains addition or subtraction of two numbers, what is the proper order of operations?

For Exercises 7–10, identify the base and the exponent, then translate the expression to words.

7. $(-5)^3$

8. 7^2

9. -2^7

10. -3^8

For Exercises 11–26, evaluate.

11. 3^4

12. 2^5

13. -8^2

14. $(-3)^2$

15. $(-2)^3$

16. -5^3

17. $(-1)^6$

18. $(-1)^5$

19. $\left(-\dfrac{1}{5}\right)^2$

20. $\left(\div\dfrac{2}{7}\right)^2$

21. $\left(-\dfrac{3}{4}\right)^3$

22. $\left(-\dfrac{1}{3}\right)^4$

23. $(0.2)^3$

24. $(0.3)^4$

25. $(-4.1)^3$

26. $(-0.2)^4$

For Exercises 27–34, find all square roots of the given number.

27. 121

28. 49

29. 81

30. 36

31. 196

32. 169

33. 256

34. 225

For Exercises 35–46, evaluate the square root.

35. $\sqrt{16}$

36. $\sqrt{36}$

37. $\sqrt{144}$

38. $\sqrt{289}$

39. $\sqrt{0.49}$

40. $\sqrt{0.01}$

41. $\sqrt{-64}$

42. $\sqrt{\dfrac{9}{100}}$

43. $\sqrt{\dfrac{64}{81}}$

44. $\sqrt{-25}$

45. $\sqrt{\dfrac{50}{2}}$

46. $\sqrt{\dfrac{48}{3}}$

For Exercises 47–74, simplify using the order of operations.

47. $5 \cdot 3 + 4$

48. $8 \div 2 + 4$

49. $5(9 - 6)$

50. $3^2 - 18 \div (6 - 3)$

51. $-8 + 2(-2)^2 + 7$

52. $-4 + 3(-1)^4 + 18 \div 3$

53. $(-2)^3 + 8 - 7(4 - 3)$

54. $18 \div (-6 + 3)(4 + 1)$

55. $13.02 \div (-3.1) + 6^2 - \sqrt{25}$

56. $-\dfrac{3}{4} \div \dfrac{1}{8} + \left(-\dfrac{2}{5}\right)(-3)(-4)$

57. $6 + 4^2 \div \sqrt{13 + 3}$

58. $\sqrt{100 - 64} + 18 \div (-3)(-2)$

59. $\dfrac{9}{8} \cdot \left(-\dfrac{2}{3}\right) + \left(\dfrac{1}{5} - \dfrac{2}{3}\right) \div \sqrt{\dfrac{125}{5}}$

60. $7 + (0.4)^2 - 5(6 - 9)$

61. $\dfrac{2}{5} \div \left(-\dfrac{1}{10}\right) \cdot (-3) + \sqrt{64 + 36}$

62. $(1.3)^3 - (-7)(2) + 5(1 - 7)$

63. $-2|6 - 4| \div (-2 - 2)$

64. $|-9| \div 3 \cdot 2$

65. $-16 \cdot \dfrac{3}{4} \div (-2) + |9 - 3(4 + 1)|$

66. $-\dfrac{15}{5} \div 3 + \sqrt{16 + 9}$

67. $18.2 - 3.4[6^2 - (5 + 2) \div 7]$

68. $-2|9 - 15| + 5^2 - (-3)^2$

69. $\dfrac{7 - |0 - 7|}{6^2 - 3(2 - 14)}$

70. $\dfrac{\sqrt{25} + 2^2 - 9}{3^4 - (6 + 3)}$

71. $\dfrac{4[5 - 8(2 + 1)]}{3 - 6 - (-4)^2}$

72. $\dfrac{6^2 - 3(4 + 2^5)}{5 + 20 - (2 + 3)^2}$

73. $\dfrac{4(5^2 - 10) + 3}{\sqrt{25 - 16} - 3}$

74. $\dfrac{5(4 - 9) + 1}{7 - 2^3 - \sqrt{49}}$

In Exercises 75–78, a property of arithmetic was used as an alternative to the order of operations. Determine what property of arithmetic was applied, and explain how it is different from the order-of-operations agreement.

75. $14 - 2 \cdot 6 \cdot 3 + 8^2$
$= 14 - 2 \cdot 18 + 64$
$= 14 - 36 + 64$
$= 42$

76. $2(3 + 8) - \sqrt{81}$
$= 6 + 16 - 9$
$= 13$

77. $-6[5 + 3^2] - \sqrt{14 + 11}$
$= -6[5 + 9] - \sqrt{25}$
$= -30 + (-54) - 5$
$= -89$

78. $(-4)^3 + 2[-10 + 8 + (-3)]$
$= -64 + 2[-13 + 8]$
$= -64 + 2[-5]$
$= -64 + (-10)$
$= -74$

For Exercises 79–82, explain the mistake. Then simplify correctly.

79. $24 \div 4 \cdot 2 - 11$
$= 24 \div 8 - 11$
$= 3 - 11$
$= -8$

80. $19 - 6(10 - 8)$
$= 19 - 6(2)$
$= 13(2)$
$= 26$

81.
$$40 \div 2 + \sqrt{25 - 9}$$
$$= 20 + \sqrt{25 - 9}$$
$$= 20 + 5 - 3$$
$$= 22$$

82.
$$-3^4 + 20 \div 5 - (16 - 24)$$
$$= 81 + 4 - (-8)$$
$$= 85 + 8$$
$$= 93$$

For Exercises 83–90, solve.

83. Tomeka has the following test scores in a history course: 82, 76, 64, 90, 74. What is the average of her test scores?

84. Will's math instructor gives quizzes worth 10 points each. Will has the following quiz scores: 9, 8, 4, 8, 7, 7, 6, 9, 8. If his instructor drops the lowest quiz score, what is Will's quiz average?

85. Michael can exempt his chemistry exam if he has a test average greater than or equal to 90 on the five tests in the course. His current test scores are 96, 88, 86, 84. Using trial and error, determine the minimum test score on the last test that will give him a test average of 90.

86. To get an A in her psychology course, Lisa must have a test average greater than or equal to 90 on four out of five tests (the lowest test score is dropped). Her test scores on the first four tests are 98, 68, 84, 86. What is the minimum score on the last test that will give her a test average of 90?

87. In May of 2001, the Lundberg survey reported the highest gas prices ever. Following are some cities from the report and their respective gas prices. What was the average price of gas for the listed cities? (*Source:* CNN.com, "U.S. Gas Prices Rise 9¢ a Gallon," May 6, 2001)

San Francisco: $2
Miami: $1.66
Dallas: $1.65
Hartford: $1.75
Detroit: $1.85
Baltimore: $1.67
Los Angeles: $1.90
Memphis: $1.57

Of Interest

Though the prices reported in 2001 were the highest to date, with inflation factored in, they still fell short of the 1981 gas prices. Using the May 2001 value of the dollar, the 1981 price would be equivalent to $2.76.

88. The following graphic shows the number of people 16 years of age or older who were employed in the United States for each month during 2000. Find the average number of people employed during 2000.

ALL IN A MONTH'S WORK

Year 2000	Employees (in thousands)
January	134,976
February	135,120
March	135,013
April	135,517
May	134,843
June	135,183
July	134,898
August	134,939
September	135,310
October	135,464
November	135,478
December	135,836

Source: Bureau of Labor Statistics, *Labor Force Statisitics from Current Population Survey*

89. The following bar graph shows the daily closing price of Wal-Mart stock over a one-month period. Find the average of the daily prices for the week ending April 27, 2001.

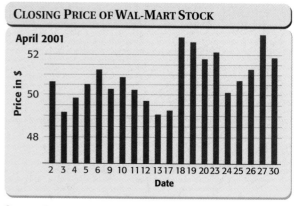

CLOSING PRICE OF WAL-MART STOCK

Source: Fool.com

90. The following bar graph shows the Dow Jones Industrial Average at the close of each of the days listed. What is the average of the DJIA for that week?

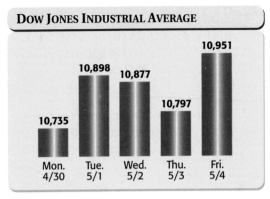

DOW JONES INDUSTRIAL AVERAGE

10,735   10,898   10,877   10,797   10,951

Mon. 4/30   Tue. 5/1   Wed. 5/2   Thu. 5/3   Fri. 5/4

Source: USA Today, May 6, 2001

Collaborative Exercises

Weather or Not

The average July temperatures for Savannah, Georgia, from 1964 to 1993, are reported below. The daily maximum and minimum temperatures for July 2001 are also listed below.

Average July Temperatures, Savannah, Georgia, 1964 to 1993

Year	Average Temperature (°F)
1964	79.1
1965	79.8
1966	80.9
1967	79.9
1968	81.6
1969	82.2
1970	81.6
1971	80.9
1972	81.0
1973	82.2
1974	79.2
1975	78.4
1976	81.7
1977	83.3
1978	82.2
1979	82.1
1980	84.4
1981	84.4
1982	81.3
1983	84.1
1984	80.6
1985	82.7
1986	85.7
1987	83.6
1988	82.7
1989	82.9
1990	84.4
1991	83.2
1992	84.3
1993	86.7

Source: Weather of U.S. Cities, 5th ed., Gale Research, 1996

Daily Maximum and Minimum Temperatures July 2001

Date	Maximum (°F)	Minimum (°F)
1	91	71
2	91	72
3	90	71
4	89	71
5	90	73
6	91	73
7	92	72
8	91	73
9	95	75
10	94	75
11	93	76
12	94	75
13	90	74
14	86	68
15	85	67
16	87	66
17	88	67
18	92	67
19	94	71
20	84	75
21	84	73
22	89	72
23	85	74
24	84	73
25	87	71
26	89	74
27	91	72
28	86	73
29	92	72
30	93	76
31	87	73

Source: National Weather Service.

continued

1. Study the average temperatures for July 1964 to 1993.
   a. Do the values in the table appear to follow any trend?
   b. What do you think is the cause of this trend (if any) that you described in step 1?

2. Use the data to calculate the average daily temperature for each day in July 2001. The average daily temperature for a particular day is the average of the minimum and the maximum temperatures for that day.

3. Find the average temperature for July 2001. The average temperature for a month is the average of the average daily temperatures (you found these in step 2).

4. How does the average temperature for July 2001 compare to the average July temperatures from 1964 to 1993?

5. Consider the trend that you described in step 1.
   a. Does the average temperature for July 2001 support your conclusion?
   b. If the trend continues, what would you expect to see occurring to average July temperatures over the next several years?

REVIEW EXERCISES

1. Is $5x - 4y$ an equation or an expression? Explain.

2. Write a set containing the whole numbers through 10.

3. Graph $-\dfrac{4}{5}$ on a number line.

4. Multiply: $-4(3)(-2)$

5. Divide: $\dfrac{-15}{-5}$

6. Find the multiplicative inverse of -9.

1.6 Translating Word Phrases to Expressions

OBJECTIVE **1** Translate word phrases to expressions.

An important technique in solving problems is to translate the words in the problem to math symbols. In this section, we focus on how to translate word phrases to expressions containing variables. The first step in translating is to identify the unknown amount. If a variable is not already given, then select a variable to represent that unknown amount.

Translating Basic Phrases

The following table contains some basic phrases and their translations.

Addition	Translation
The sum of x and 3	$x + 3$
h plus k	$h + k$
7 added to t	$t + 7$
3 more than a number	$n + 3$
y increased by 2	$y + 2$

Note: Since addition is a commutative operation, it does not matter in what order we write the translation.
For "the sum of x and 3" we can write $x + 3$ or $3 + x$.

Subtraction	Translation
The difference of x and 3	$x - 3$
h minus k	$h - k$
7 subtracted from t	$t - 7$
3 less than a number	$n - 3$
y decreased by 2	$y - 2$

Note: Subtraction is not a commutative operation; therefore, the way we write the translation matters. We must translate each key phrase exactly as it was presented above. Notice when we translate "less than" or "subtracted from," the translation is in reverse order from what we read.

Multiplication	Translation
The product of x and 3	$3x$
h times k	hk
twice a number	$2n$
triple the number	$3n$
$\frac{2}{3}$ of a number	$\frac{2}{3}n$

Note: Like addition, multiplication is a commutative operation. This means we can write the translation order any way we wish.

h times k can be hk or kh.

Division	Translation
The quotient of x and 3	$x \div 3$
h divided by k	$h \div k$
h divided into k	$k \div h$
the ratio of a to b	$a \div b$

Note: Division is like subtraction in that it is not a commutative operation; therefore, we must translate division phrases exactly as presented above. Notice how "divided into" is translated in reverse order of what we read.

Exponents	Translation
c squared	c^2
the square of b	b^2
k cubed	k^3
the cube of b	b^3
n to the fourth power	n^4
y raised to the fifth power	y^5

Roots	Translation
The square root of x	$\sqrt{x}$

The key words *sum, difference, product,* and *quotient* indicate the answer for their respective operations. Notice that they all involve the word *and.* In the translation, the word *and* separates the parts and can therefore be translated to the operation symbol indicated by the key word *sum, difference, product,* or *quotient.*

sum of *x* and 3

$x + 3$

difference of *x* and 3

$x - 3$

product of *x* and 3

$x \cdot 3$

quotient of *x* and 3

$x \div 3$

Combinations of Basic Phrases

Now that we have seen the basic phrases, let's translate phrases that involve combinations of the basic phrases.

EXAMPLE 1 Translate to an algebraic expression.

a. four more than three times a number

Translation: $4 + 3n$ or $3n + 4$

> **Note:** The commutative property of addition allows us to write the expression in either order.

b. six less than the square of a number

Translation: $n^2 - 6$

> **Note:** Because subtraction is not commutative, we must translate the subtraction in a specific order. *Less than* indicates that the 6 is the subtrahend and the *square of a number* is the minuend.

c. the sum of *h* raised to the fifth power and fifteen

Translation: $h^5 + 15$

d. the ratio of *m* times *n* to *r* cubed

Translation: $\dfrac{mn}{r^3}$

> **Note:** When coupled with the word *ratio,* the word *to* translates to the fraction line. The amount to the left of the word *to* goes in the numerator and the amount to the right of the word *to* goes in the denominator.

e. one-half of *v* divided by the square root of *t*

Translation: $\dfrac{1}{2}v \div \sqrt{t}$

> **Note:** When the word *of* is preceded by a fraction, it means multiply.

ANSWERS

a. $9n - \dfrac{2}{3}$

b. $-6.2 + 9.8n$

YOUR TURN Translate to an algebraic expression.

a. two-thirds subtracted from the product of nine and a number

b. −6.2 increased by 9.8 times a number

Translating Phrases Involving Parentheses

Sometimes the word phrases imply an order of operations that would require us to use parentheses in the translation. These situations arise when the phrase indicates that a sum or difference is to be calculated before performing a higher-order operation such as multiplication, division, exponent, or root.

EXAMPLE 2 Translate to an algebraic expression.

a. five times the sum of x and y

Translation: $5(x + y)$

> **Note:** Without the parentheses, the expression is $5x + y$, which indicates that we are to multiply 5 times x first, and then add y to the result. The parentheses indicate that we add x and y first, then multiply the resulting sum by 5.

b. the square root of the difference of the square of x and the square of y

Translation: $\sqrt{x^2 - y^2}$

> **Note:** *The square root of the difference* indicates that we are to calculate the difference before calculating the square root. We indicate this symbolically by placing the entire subtraction under the radical.

c. the product of x and y divided by the sum of x^2 and 5

Translation: $xy \div (x^2 + 5)$ or $\dfrac{xy}{x^2 + 5}$

YOUR TURN Translate to an algebraic expression.

a. a number minus twice the sum of the number and seven
b. the difference of m and n, all raised to the fourth power

ANSWERS

a. $n - 2(n + 7)$
b. $(m - n)^4$

EXAMPLE 3 Isaac Newton developed an expression that describes the gravitational effect between two objects. The relationship is the product of the masses of the two objects divided by the square of the distance between the objects. Translate the relationship to an expression.

Translation:

$Mm \div d^2$ or $\dfrac{Mm}{d^2}$

> **Note:** If the same letter is desired to describe two different quantities, uppercase and lowercase can be used to distinguish the quantities. Subscripts can also be used, as in $m_1 m_2 \div d^2$.

YOUR TURN

ANSWER

$d \div (v_1 + v_2)$ or $\dfrac{d}{v_1 + v_2}$

Two asteroids are on a collision course toward each other. We can calculate the time to impact by dividing the distance that separates them by the sum of their velocities. Translate the relationship to an expression.

1.6 EXERCISES

For Extra Help

InterAct Math

Addison-Wesley
Math Tutor Center

Video 2/DVT 1

MyMathLab

MathXL

1. List three key words that indicate addition.

2. List three key words that indicate subtraction.

3. List three key words that indicate multiplication.

4. List three key words that indicate division.

5. The phrase "9 subtracted from n" translates to $n - 9$, which is in reverse order of what we read. What other key words for subtraction translate in reverse order?

6. What key words for division translate in reverse order?

For Exercises 7–30, translate the phrase to an algebraic expression.

7. four times a number

8. the product of a number and four

9. the sum of four times x and sixteen

10. six less than T

11. five more than y

12. the ratio of seven to the square of m

13. a number to the fifth power

14. thirteen subtracted from twice a number

15. fourteen divided into m

16. the sum of b cubed and seven

17. one-fifth subtracted from the product of seven and a number

18. a number plus triple the sum of the number and four

19. the difference of two and l, all raised to the third power

20. the square root of n

21. five less than the product of m and n

22. the product of three and a number, increased by five

23. negative twenty-seven decreased by the sum of a and b

24. eighty-one hundredths increased by eight times a number

25. four-fifths of a number decreased by seven

26. the sum of a and b subtracted from the difference of c and d

27. the product of m and n subtracted from the square root of y

28. the quotient of x and y decreased by two-thirds

29. a number minus three times the difference of the number and six

30. six times the sum of h and k

For Exercises 31–34, explain the mistake. Then translate the phrase correctly.

31. nine times the sum of x and y

Translation: $9x + y$

32. seventeen less than three times t

Translation: $17 - 3t$

33. four subtracted from the square of m

Translation: $4 - m^2$

34. Nineteen divided into the product of h and k

Translation: $19 \div hk$

For Exercises 35–44, translate the indicated phrase.

35. The length of a rectangle is five more than the width. If the width is represented by w, then write an expression that describes the length.

36. The width of a rectangle is four less than the length. If the length is represented by l, then write an expression that describes the width.

37. The length of a rectangle is three times the width. If the width is represented by the variable w, then write an expression that describes the length.

38. The width of a rectangle is one-fourth of the length. If the length is represented by l, then write an expression that describes the width.

39. The radius of a circle is one-half of the diameter. If d represents the diameter, then write an expression for the radius.

40. The diameter of a circle is twice the radius. If r represents the radius, then write an expression for the diameter.

41. Lindsey has 42 coins in her change purse that are all either dimes or quarters. If n represents the number of quarters she has, then write an expression in terms of n that describes the number of dimes.

42. Sherice owns a total of 60 shares of stock in two companies. If n represents the number of the higher-priced stock, then write an expression in terms of n for the number of shares of the lower-priced stock.

43. Don passes mile marker 51 on the highway. One-fourth of an hour later, a state trooper traveling in the same direction passes the same mile marker. If t represents the amount of time it takes the trooper to catch up to Don, write an expression in terms of t that describes the amount of time Don has traveled since passing marker 51.

44. Berry is jogging along a trail and passes a sign. One-third of an hour later, Dedra, who is traveling in the same direction, passes the same sign. If t represents the amount of time it takes Dedra to catch up to Berry, write an expression in terms of t that describes the amount of time Berry has traveled since passing the sign.

Exercises 45–52 contain word descriptions of expressions from mathematics and physics. Translate the descriptions to symbolic form.

45. The perimeter of a rectangle is twice the width plus twice the length.

46. The area of a circle can be found by multiplying the square of the radius by π.

47. The volume of a cone is one-third of the product of π, the square of the radius, and the height of the cone.

48. The volume of a sphere is four-thirds of the product of π and the cube of the radius.

49. René Descartes developed an expression for the distance between two points in the coordinate plane. The distance between two points is the square root of the sum of the square of the difference of x_2 and x_1, and the square of the difference of y_2 and y_1.

50. The centripetal acceleration of an object traveling in a circular path is found by an expression that is the ratio of the square of the velocity, v, to the radius, r, of the orbital path.

51. Albert Einstein developed an expression that describes the energy of a particle at rest, which is the product of the mass, m, of the particle and the square of the speed of light, which is represented by c.

52. Albert Einstein's theory of relativity includes a mathematical expression that is the square root of the difference of one and the ratio of the square of the velocity, v, of an object in motion to the square of the speed of light, c.

Of Interest

Though Albert Einstein (1879–1955) is considered one of the greatest intellects of all time, he actually struggled academically. As a teen, he was expelled from school because his apathy was a bad influence on his classmates. He failed his first attempt at the technical college entrance exam. He later passed the exam after taking a year of remedial classes at a secondary school. His marginal academic performance did not deter him from pursuing on his own his love of math and science, and in 1905 while working as a clerk in the Swiss patent office, he submitted three papers to a German science journal that outlined quantum theory and his theory of relativity. The genius of these papers was quickly recognized and he was asked to lecture at several European universities. Ironically, he never left academia thereafter, accepting a position at Princeton University to escape persecution by the Nazis. He remained at Princeton until his death in 1955.

For Exercises 53–60, translate the expression to a word phrase.

53. The area of a triangle: $\frac{1}{2}bh$

54. The area of a trapezoid: $\frac{1}{2}h(a + b)$

55. The volume of a box: lwh

56. The volume of cylinder: $\pi r^2 h$

57. The surface area of a box: $2(lw + lh + wh)$

58. The surface area of a cylinder: $2\pi r(r + h)$

59. The slope of a line: $\dfrac{y_2 - y_1}{x_2 - x_1}$

60. A quadratic expression: $ax^2 + bx + c$

Consecutive integers follow one another in a pattern like 1, 2, 3, . . . Let n represent the first integer in a set of consecutive integers.

a. Write expressions for the next two consecutive integers.

b. Suppose n is odd. Write expressions for the next two consecutive odd integers.

c. Suppose n is even. Write expressions for the next two consecutive even integers.

REVIEW EXERCISES

1. Define rational numbers.

2. What property of addition is represented by $6 + 0 = 6$?

3. What property is represented by $3(4 + 2) = 12 + 6$?

For Exercises 4–6, simplify.

4. $-6^2 \div 3 + 4(5 - 9)$ **5.** $-4|3 - 8| + 2^5$ **6.** -3^2

1.7 **Evaluating and Rewriting Expressions**

OBJECTIVES

1 Evaluate an expression.

2 Determine all values that cause an expression to be undefined.

3 Rewrite an expression using the distributive property.

4 Rewrite an expression by combining like terms.

Because algebraic expressions form equations, we must learn how to manipulate those expressions. There are two actions we can perform with algebraic expressions. We can *evaluate* them or *rewrite* them.

OBJECTIVE **1** Evaluate an expression. First we learn how to evaluate an expression.

PROCEDURE **EVALUATING AN ALGEBRAIC EXPRESSION**

To evaluate an algebraic expression:

1. Replace the variables with their corresponding given numbers.
2. Calculate the numerical expression using the order of operations.

EXAMPLE **1** Evaluate $y^2 - 0.2xy + 5$ when $x = 6$ and $y = -4$.

Solution $y^2 - 0.2xy + 5$

$$= (-4)^2 - 0.2(6)(-4) + 5 \qquad \text{Replace } x \text{ with 6 and } y \text{ with } -4.$$
$$= 16 - 0.2(6)(-4) + 5 \qquad \text{Begin calculating by simplifying the exponential form.}$$
$$= 16 - (-4.8) + 5 \qquad \text{Multiply.}$$
$$= 16 + 4.8 + 5 \qquad \text{Write the subtraction as an equivalent addition.}$$
$$= 25.8 \qquad \text{Add from left to right.}$$

YOUR TURN

Evaluate $m^3 - 6n^2$ when $m = -2$ and $n = 3$.

ANSWER

-62

EXAMPLE 2 Evaluate $\dfrac{5r^2}{d-3}$ when $r = 2$ and $d = -5$.

Solution $\dfrac{5r^2}{d-3}$

$= \dfrac{5(2)^2}{-5-3}$ Replace r with 2 and d with -5.

$= \dfrac{5(4)}{-8}$ Calculate the top and bottom expressions separately.

$= \dfrac{20}{-8}$

$= -\dfrac{5}{2}$ or -2.5

OBJECTIVE 2 Determine all values that cause an expression to be undefined. In Example 2, what if $d = 3$? Notice that the denominator would become 0: $\dfrac{5r^2}{3-3} = \dfrac{5r^2}{0}$, which is undefined no matter what replaces r. When asked to evaluate a division expression in which the divisor or denominator contains a variable or variables, we must be careful about what values replace the variable(s). Often we'll need to know what values could replace the variable(s) and cause the expression to be undefined.

EXAMPLE 3 Determine all values that cause the expression to be undefined.

a. $\dfrac{5}{x+7}$

Answer: If $x = -7$, then the expression is undefined because the denominator is 0.

b. $\dfrac{-y}{(y+2)(y-6)}$

Solution Note that because $(y+2)(y-6)$ is a product, if either factor is 0, then the entire denominator is 0.

Notice if $y = -2$, we have: $\dfrac{-(-2)}{(-2+2)(-2-6)}$

$= \dfrac{2}{(0)(-8)}$

$= \dfrac{2}{0}$, which is undefined

Also, if $y = 6$, we have: $\dfrac{-6}{(6+2)(6-6)}$

$= \dfrac{-6}{(8)(0)}$

$= \dfrac{-6}{0}$, which is undefined

Answer: Both $y = -2$ and $y = 6$ cause the expression to be undefined.

CONNECTION ——
This principle, that if either factor is 0 then the entire product is 0, is known as the *zero-factor theorem*. We will use this theorem extensively in Chapter 6 when solving equations like $(y+2)(y-6) = 0$.

Determine all values that cause the expression to be undefined.

a. $\dfrac{9}{n + 6}$

b. $\dfrac{2m}{(m - 3)(m + 4)}$

OBJECTIVE ③ **Rewrite an expression using the distributive property.** Recall from Section 1.4 the distributive property of multiplication over addition.

RULE **THE DISTRIBUTIVE PROPERTY OF MULTIPLICATION OVER ADDITION**

$$a(b + c) = ab + ac$$

This property gives us an alternative to the order of operations. Look at the numerical expression $2(5 + 6)$ and compare using the order of operations versus using the distributive property.

Following order of operations:

$$2(5 + 6) = 2(11)$$
$$= 22$$

or

Using the distributive property:

$$2(5 + 6) = 2 \cdot 5 + 2 \cdot 6$$
$$= 10 + 12$$
$$= 22$$

Note: Using the order of operations, we add within the parentheses first, then multiply.

Note: Using the distributive property, we multiply each addend by the factor outside the parentheses, then add the products.

The result is the same either way, so the distributive property allows us to say that $2(5 + 6)$ and $2 \cdot 5 + 2 \cdot 6$ are equivalent expressions.

Conclusion: We can use the distributive property to rewrite an expression in another form that is equivalent to the original form.

EXAMPLE ④ Use the distributive property to write an equivalent expression.

a. $2(x + 5)$

Solution $2(x + 5) = 2 \cdot x + 2 \cdot 5$
$$= 2x + 10$$

b. $-6(n - 9)$

Solution $-6(n - 9) = -6 \cdot n - (-6) \cdot 9$
$$= -6n - (-54)$$
$$= -6n + 54$$

CONNECTION

We can apply the distributive property to $-6(n - 9)$ even though the parentheses contain subtraction instead of addition because subtraction can be expressed as addition like this:

$$-6(n - 9) = -6(n + (-9))$$

In fact, it is helpful to think of the expression in this way because when we multiply the -6 by the -9, we get positive 54.

ANSWERS

a. $n = -6$

b. Both $m = 3$ and $m = -4$

c. $\dfrac{3}{8}\left(2m + \dfrac{4}{5}\right)$

Solution $\dfrac{3}{8}\left(2m + \dfrac{4}{5}\right) = \dfrac{3}{8} \cdot 2m + \dfrac{3}{8} \cdot \dfrac{4}{5}$ Use the distributive property.

$$= \dfrac{3}{\overset{4}{\cancel{8}}} \cdot \dfrac{\overset{1}{\cancel{2}}m}{1} + \dfrac{3}{\overset{2}{\cancel{8}}} \cdot \dfrac{\overset{1}{\cancel{4}}}{5}$$ Divide out common factors.

$$= \dfrac{3}{4}m + \dfrac{3}{10}$$

> **YOUR TURN**   **Use the distributive property to write an equivalent expression.**
>
> **a.** $-9(7 - 3x)$ **b.** $\dfrac{4}{5}\left(\dfrac{1}{2}y - 10\right)$

OBJECTIVE **4** **Rewrite an expression by combining like terms.** Many expressions are a sum or difference of expressions called **terms**.

> **DEFINITION**
>
> **Term:** An expression that is separated by addition or subtraction.

For example, the terms in the expression $4x^2 - 6x + 8$ are as follows:

    First term: $4x^2$
    Second term: $-6x$ **Note:** Since the term 8 is a constant,
    Third term: 8 it is called a *constant* term.

The second term is negative because we can write the expression using addition like this: $4x^2 + (-6x) + 8$. Note that the sign to the left of the term is the sign of that term, so we actually don't have to write subtraction as addition to identify the terms.

The numerical factor in a term is called the numerical **coefficient**, or simply the coefficient of the term.

> **DEFINITION**
>
> **Coefficient:** The numerical factor in a term.

The coefficient of $4x^2$ is 4.
The coefficient of $-6x$ is -6.
The coefficient of 8 is 8.

Another way that we can rewrite an expression is to combine **like terms**.

> **DEFINITION**
>
> **Like terms:** Variable terms that have the same variable(s) raised to the same exponents, or constant terms.

Examples of like terms:

$3x$ and $5x$
$4y^2$ and $9y^2$
$7xy$ and $3xy$
6 and 15

Examples of unlike terms:

$2x$ and $8y$ (different variables)
$4t^2$ and $4t^3$ (different exponents)
x^2y and xy^2 (different exponents)
12 and $12x$ (different variables)

Combining Like Terms

Now, consider an expression that is a sum of like terms, such as $3x + 5x$. We can rewrite this expression in a more compact form. Multiplication of 3 times x means that x is repeatedly added three times. Likewise 5 times x means that x is repeatedly added five times. We can expand the expression out like this:

$$
\underbrace{3x}_{} \quad + \quad \underbrace{5x}_{}
$$
$$
= \underbrace{x + x + x \; + \; x + x + x + x + x}_{}
$$
$$
= \qquad\qquad 8x
$$

Note: After expanding, we see that there are a total of eight x's repeatedly added. We can write those eight x's as a single term, $8x$.

Notice that when we combine the like terms, we get an expression that is more compact. When we rewrite an expression in a more compact form, we say that we are *simplifying* the expression. Also notice that when combining like terms, we simply add the coefficients and keep the variable the same.

PROCEDURE   COMBINING LIKE TERMS

To combine like terms, add or subtract the coefficients and keep the variables and their exponents the same.

CONNECTION ——————————————————————————

We can use the distributive property in reverse to confirm our procedure for combining like terms. Consider the expression $3x + 5x$ again. Notice there is a common factor of x in both terms. In applying the distributive property in reverse we write this common factor outside the parentheses and write the remaining factors as addends within the parentheses. It looks like this:

$$3x + 5x = (3 + 5)x$$

Because the parentheses contain only the addition of two numbers, we can simplify by adding the numbers.

$$3x + 5x = (3 + 5)x$$
$$= 8x$$

EXAMPLE **5**   Combine like terms.

a. $9y + 7y$

Solution $9y + 7y = 16y$ **Note:** We think: Nine y's plus seven y's equals sixteen y's.

b. $7x - 2x$

Solution $7x - 2x = 5x$

> **Note:** We think: Seven x's minus two x's leaves five x's.

c. $14n^2 - n^2$

Solution $14n^2 - n^2 = 13n^2$

> **Note:** The coefficient of n^2 is 1, so we think fourteen n^2's minus one n^2 leaves thirteen n^2's.

YOUR TURN Combine like terms.

a. $6.5x + 2.3x$

b. $\dfrac{1}{3}n^3 - \dfrac{3}{4}n^3$

Collecting Like Terms

Sometimes expressions are more complex and contain different sets of like terms. In such cases, we combine the like terms and copy any unlike terms in the final expression. Many people like to use the commutative property of addition and rearrange the expression first so that the like terms are together, then combine them. This type of manipulation is called *collecting* the like terms.

EXAMPLE 6 Combine like terms in $4y^2 + 5 + 3y^2 - 9$.

Solution $4y^2 + 5 + 3y^2 - 9$

$= 4y^2 + 3y^2 + 5 - 9$ Collect the like terms.

$= 7y^2 - 4$ Combine like terms.

Collecting like terms is optional. Alternatively, many people mark through the terms as they combine them as a way of keeping track of what has been combined. Using this technique with the expression from Example 6 looks like this:

$$4y^2 + 5 + 3y^2 - 9$$
$$= 7y^2 - 4$$

EXAMPLE 7 Combine like terms in $15y + 8x - y - 8x$.

Solution $15y + 8x - y - 8x$.

> **Note:** The term $-y$ is equivalent to $-1y$.

$= 14y + 0$

$= 14y$

Sometimes collecting the like terms is beneficial, as in Example 8, where there are like terms with fraction coefficients. Some people find it easier to "see" the steps in writing the equivalent fractions when the fractions are closer together.

EXAMPLE 8 Combine like terms in $\frac{1}{3}n - 9m + 2 - m + \frac{1}{4}n$.

Solution $\frac{1}{3}n - 9m + 2 - m + \frac{1}{4}n$

$= \frac{1}{3}n + \frac{1}{4}n - 9m - m + 2$ Collect the like terms.

$= \frac{1(4)}{3(4)}n + \frac{1(3)}{4(3)}n - 9m - m + 2$ Write fraction coefficients as equivalent fractions with their LCD, 12.

$= \frac{4}{12}n + \frac{3}{12}n - 9m - m + 2$

$= \frac{7}{12}n - 10m + 2$ Combine like terms.

YOUR TURN Combine like terms.

a. $\frac{3}{8}y - x + 2 - \frac{3}{4}y + 5x$ **b.** $4.2x^2 - 6x + 9 - 0.7x - 12$

CONNECTION

One way to verify that two expressions are equivalent is to evaluate them both using the same value(s) for the variable(s). Consider the expression in Example 6. If we evaluate the original expression $4y^2 + 5 + 3y^2 - 9$ and the simplified expression $7y^2 - 4$ using the same chosen value for y, we should get the same answer. Let's choose the number 2 for y.

Original Expression:

$4y^2 + 5 + 3y^2 - 9$

$4(2)^2 + 5 + 3(2)^2 - 9$

$= 4(4) + 5 + 3(4) - 9$

$= 16 + 5 + 12 - 9$

$= 24$

Simplified expression:

$7y^2 - 4$

$7(2)^2 - 4$

$= 7(4) - 4$

$= 28 - 4$

$= 24$

Warning: Some expressions that are not equivalent may give the same answer using this method, so this "check" does not guarantee that the expressions are equivalent. It is, however, a good sign.

Notice the simplified expression gave the same answer, yet was easier to work with. This is one of the reasons we simplify expressions. If we can simplify an expression, then we have an easier time evaluating it.

ANSWERS

a. $-\frac{3}{8}y + 4x + 2$

b. $4.2x^2 - 6.7x - 3$

1.7 EXERCISES

For Extra Help

 InterAct Math

 Addison-Wesley Math Tutor Center

Video 2/DVT 1

 MyMathLab

Math$_{XP}$ MathXL

1. In your own words, explain how to evaluate an expression.

2. Explain the difference between evaluating and rewriting an expression.

3. What causes an expression to be undefined?

4. What are like terms?

5. What is the coefficient in a term?

6. Explain how to combine like terms.

For Exercises 7–18, evaluate the expression using the given values.

7. $3(a + 5) - 4b; a = 2, b = 3$

8. $8 - (m + 1); m = 5$

9. $-0.2(x + 3) + 4; x = 1$

10. $n^2 - 8.5n + 1; n = -1$

11. $2y^2 - 4y + 3; y = -\dfrac{1}{2}$

12. $-6 - 2(l - 5); l = -0.4$

13. $-|3x| + |y^3|; x = 5, y = -1$

14. $|m^2 + 2n|; m = 4, n = -5$

15. $\sqrt{c} + 4ab^2; a = -1, b = -2, c = 16$

16. $\sqrt{h} + \sqrt{k}; h = 16, k = 9$

17. $\dfrac{6x^3}{y - 8}; x = 2, y = 7$

18. $\dfrac{5 - 2a^2}{3\sqrt{x + y}}; a = 1, x = 64, y = 36$

19. The expression $b^2 - 4ac$ is called the discriminant and is used to determine the types of solutions for quadratic equations. Find the value of the discriminant given the following values.

    a. $a = 2, b = 1, c = -3$

    b. $a = 1, b = -2, c = 4$

20. The expression $ad - bc$ is used to calculate the determinant of a matrix. Find the determinant given the following values.

    a. $a = 1, b = 0.5, c = -4, d = 6$ **b.** $a = -3, b = \frac{4}{5}, c = 2, d = \frac{1}{2}$

21. The expression $\frac{y_2 - y_1}{x_2 - x_1}$ is used to calculate the slope of a line. Find the slope given each set of the following values.

    a. $x_1 = 2, y_1 = 1, x_2 = 5, y_2 = 7$ **b.** $x_1 = -1, y_1 = 2, x_2 = -7, y_2 = -2$

22. The expression $\sqrt{(x_2 - x_1)^2 + (y_2 - y_1)^2}$ is used to calculate the distance between two points in the coordinate plane. Evaluate the expression using the following values.

    a. $x_1 = 2, y_1 = 1, x_2 = 5, y_2 = 7$ **b.** $x_1 = -1, y_1 = 2, x_2 = -7, y_2 = -2$

For Exercises 23–30, determine all values that cause the expression to be undefined.

23. $\dfrac{-7}{5 + y}$ **24.** $\dfrac{8}{x + 3}$ **25.** $\dfrac{6m}{(m + 1)(m - 3)}$ **26.** $\dfrac{-5a}{(a - 4)(a - 2)}$

27. $\dfrac{6 + x^2}{x}$ **28.** $\dfrac{7 - y}{y}$ **29.** $\dfrac{x + 1}{3x - 2}$ **30.** $\dfrac{3y}{2y + 1}$

For Exercises 31–38, use the distributive property to write an equivalent expression.

31. $6(a + 2)$ **32.** $4(b - 5)$ **33.** $-8(4 - 3y)$ **34.** $-7(3 - 2m)$

35. $\dfrac{7}{8}\left(\dfrac{1}{2}c - 16\right)$ **36.** $\dfrac{4}{5}\left(-10h + \dfrac{2}{9}\right)$ **37.** $0.2(3n - 8)$ **38.** $-1.5(6x + 7)$

For Exercises 39–46, identify the coefficient of each term.

39. $-6x^3$ **40.** y^5 **41.** m^9 **42.** $-1.4y$

43. $-b$ **44.** $\dfrac{2}{3}c$ **45.** $\dfrac{y}{5}$ **46.** $-\dfrac{u}{3}$

For Exercises 47–66, simplify by combining like terms.

47. $3y + 5y$ **48.** $6m + 7m$ **49.** $\dfrac{1}{2}b^2 - \dfrac{5}{6}b^2$ **50.** $\dfrac{3}{4}z - \dfrac{7}{5}z$

51. $6.3n - 8.2n$ **52.** $5.1x^4 + 3.4x^4$ **53.** $-11c - 10c - 5c$ **54.** $-15w - 6w - 11w$

55. $6x + 7 - x - 8$ **56.** $5y^2 + 6 + 3y^2 - 8$ **57.** $-8x + 3y - 7 - 2x + y + 6$

58. $4a + 9b - a + 5 + 7c + 2b - 8 - 7c$ **59.** $-10m + 7n + 1 + m - 2n - 12 + y$

60. $-3h + 7k - 5 - 8h - 7k + 19$ **61.** $\frac{1}{6}c + 3d + \frac{2}{3}c + \frac{1}{7} - \frac{1}{2}d$ **62.** $\frac{5}{8}y + \frac{1}{2} - \frac{3}{4}x + \frac{2}{3} - \frac{1}{4}y$

63. $7a + \frac{4}{5}b^2 - 5 - \frac{3}{4}a - \frac{2}{7}b^2 + 9$ **64.** $\frac{1}{2}m - 3n + 14 - \frac{3}{8}m - \frac{9}{10}n - 5$

65. $1.5x + y - 2.8x + 0.3 - y - 0.7$ **66.** $0.4t^2 + t - 2.8 - t^2 + 0.9t - 4$

67. a. Translate to an algebraic expression: fourteen plus the difference of six times a number and eight times the same number.
   b. Simplify the expression.
   c. Evaluate the expression when the number is -3.

68. a. Translate to an algebraic expression: the sum of negative five times a number and eight minus two times the same number.
   b. Simplify the expression.
   c. Evaluate the expression when the number is 0.2.

PUZZLE PROBLEM

Each letter in the following addition problem represents a different whole number, 0–9. What number does each letter represent?

```
  F O R T Y
    T E N
+   T E N
  S I X T Y
```

REVIEW EXERCISES

For Exercises 1–4, perform the indicated operation(s).

1. $-8 + (-14)$ **2.** $1.2 - (-4.9)$ **3.** $\frac{1}{4}(-2)(-3)$ **4.** $0 \div (-9.8)$

5. Evaluate $6m^2 - 4a$ when $m = -2$ and $a = 5$.

6. Translate to an algebraic expression: sixteen more than three times a number.

Defined Terms

Section 1.1
Variable (p.2)
Constant (p.2)
Expression (p. 2)
Equation (p.2)
Inequality (p. 3)
Set (p. 3)
Rational number (p. 4)
Irrational number (p. 5)
Absolute value (p. 7)

Section 1.2
Fraction (p. 13)
Multiple (p. 15)
LCM (p. 15)
LCD (p. 15)
Factors (p. 15)
Prime number (p. 16)
Prime factorization
 (p. 16)
Lowest terms (p. 17)

Section 1.3
Additive inverses (p. 33)

Section 1.4
Multiplicative inverses
 (p. 49)

Section 1.5
Exponent (p. 59)
Base (p. 59)

Section 1.7
Term (p. 84)
Coefficient (p. 84)
Like terms (p. 84)

The Real Number System

Real Numbers

Irrational Numbers: Any real number that is not rational, such as: $-\sqrt{2}$, $-\sqrt{3}$, $\sqrt{0.8}$, and π.

Rational Numbers: Real numbers that can be expressed in the form $\frac{a}{b}$, where a and b are integers and $b \neq 0$, such as: $-4\frac{3}{4}$, $-\frac{2}{3}$, 0.018, $0.\overline{3}$, and $\frac{5}{8}$.

Integers: $\ldots, -3, -2, -1, 0, 1, 2, 3, \ldots$

Whole Numbers: $0, 1, 2, 3, \ldots$

Natural Numbers: $1, 2, 3, \ldots$

Arithmetic Summary Diagram

Each operation has an inverse operation. In the diagram the operations build from the top down. Addition leads to multiplication, which leads to exponents. Subtraction leads to division, which leads to roots.

Properties of Arithmetic

In each of the following a, b, and c represent real numbers.

Additive Identity
$$a + 0 = a$$

Commutative Property of Addition
$$a + b = b + a$$

Associative Property of Addition
$$(a + b) + c = a + (b + c)$$

Multiplicative Identity
$$a \cdot 1 = a$$

Multiplicative Property of 0
$$a \cdot 0 = 0$$

Commutative Property of Multiplication
$$ab = ba$$

Associative Property of Multiplication
$$(ab)c = a(bc)$$

Distributive Property
$$a(b + c) = ab + ac$$

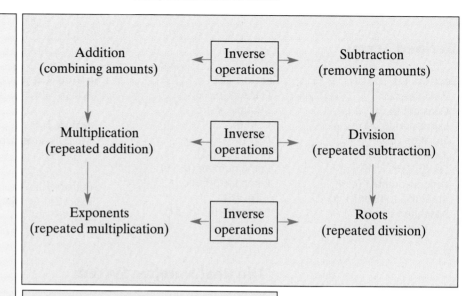

Order of Operations
1. Grouping symbols
2. Exponents and roots
3. Multiply or divide from left to right
4. Add or subtract from left to right

Procedures, Rules, and Key Examples

Procedures/Rules	Key Examples
Section 1.1 Number Sets and the Structure of Algebra To write a set, write the elements or members of the set within braces.	**Example 1:** Write the set containing the natural numbers divisible by 3. Answer: $\{3, 6, 9, 12, \ldots\}$
The absolute value of a number is positive or 0.	**Example 2:** Find the absolute value. **a.** $\|8\| = 8$ **b.** $\|-15\| = 15$ **c.** $\|0\| = 0$

continued

Procedures/Rules	Key Examples

Section 1.1 (continued)

For any two real numbers a and b, a is greater than b if a is to the right of b on a number line.

Example 3: Use $<$ or $>$ to make a true sentence.

a. $-8 \blacksquare -10$ Answers: **a.** $-8 > -10$

b. $-6 \blacksquare 9$ **b.** $-6 < 9$

Section 1.2 Fractions

For any fraction, we can write an equivalent fraction by multiplying or dividing both its numerator and denominator by the same number.

Example 1: Find the missing number in $\frac{5}{6} = \frac{?}{24}$ so that the fractions are equivalent.

Answer: $\frac{5}{6} = \frac{20}{24}$

To simplify a fraction to lowest terms:
1. Replace the numerator and denominator with their prime factorizations.
2. Eliminate (divide out) all prime factors common to the numerator and denominator.
3. Multiply the remaining factors.

Example 2: Reduce $\frac{54}{60}$ to lowest terms.

$$\frac{54}{60} = \frac{2 \cdot 3 \cdot 3 \cdot 3}{2 \cdot 2 \cdot 3 \cdot 5} = \frac{9}{10}$$

Section 1.3 Adding and Subtracting Real Numbers; Properties of Real Numbers

When adding two numbers that have the same sign, add their absolute values and keep the same sign.

When adding two numbers that have different signs, subtract their absolute values and keep the sign of the number with the greater absolute value.

To write a subtraction statement as an equivalent addition statement, change the operation symbol from a minus sign to a plus sign and change the subtrahend to its additive inverse.

Example 1: Add.

a. $4 + 9 = 13$

b. $-4 + (-9) = -13$

c. $-4 + 9 = 5$

d. $4 + (-9) = -5$

Example 2: Subtract.

a. $4 - 9 = 4 + (-9) = -5$

b. $-4 - 9 = -4 + (-9) = -13$

c. $4 - (-9) = 4 + 9 = 13$

d. $-4 - (-9) = = -4 + 9 = 5$

To add fractions:
1. Find the LCD.
2. Write each fraction as an equivalent fraction with the LCD.
3. Add the numerators and keep the LCD.
4. Simplify.

Example 3: Add $-\frac{3}{4} + \frac{5}{6}$.

$$-\frac{3(3)}{4(3)} + \frac{5(2)}{6(2)}$$

$$= -\frac{9}{12} + \frac{10}{12}$$

$$= \frac{1}{12}$$

continued

Procedures/Rules	Key Examples

Procedures/Rules

Section 1.4 Multiplying and Dividing Real Numbers; Properties of Real Numbers

The product of an even number of negative factors is positive, whereas the product of an odd number of negative factors is negative.

Rule for multiplying fractions: $\dfrac{a}{b} \cdot \dfrac{c}{d} = \dfrac{ac}{bd}$, where $b \neq 0$ and $d \neq 0$.

To multiply decimal numbers:
1. Multiply as if they named whole numbers.
2. Place the decimal in the product so that it has the same number places as the total number of places in the factors.

When dividing two numbers that have the same sign, the quotient is positive.

When dividing two numbers that have different signs, the quotient is negative.

Rules for division involving 0: $0 \div n = 0$, when $n \neq 0$

$n \div 0$ is undefined, when $n \neq 0$

$0 \div 0$ is indeterminate

Rule for dividing fractions: $\dfrac{a}{b} \div \dfrac{c}{d} = \dfrac{a}{b} \cdot \dfrac{d}{c}$, where $b \neq 0$, $c \neq 0$, and $d \neq 0$.

Key Examples

Example 1: Multiply.
a. $(5)(7) = 35$
b. $(-5)(-7) = 35$
c. $(-1)(-2)(-5)(-7) = 70$
d. $(-5)(7) = -35$
e. $(5)(-7) = -35$
f. $(-2)(-5)(-7) = -70$

Example 2: Multiply.

a. $\dfrac{3}{4} \cdot \dfrac{5}{7} = \dfrac{15}{28}$

b. $\dfrac{5}{6} \cdot \dfrac{4}{9} = \dfrac{5}{2 \cdot 3} \cdot \dfrac{2 \cdot 2}{9} = \dfrac{10}{27}$

Example 3: Multiply $(0.6)(2.4)$.

$$\begin{array}{r} 2.4 \\ \times\, 3.6 \\ \hline 144 \\ +\, 72 \\ \hline 8.64 \end{array}$$

Example 4: Divide.
a. $24 \div 8 = 3$
b. $-24 \div (-8) = 3$
c. $-24 \div 8 = = -3$
d. $24 \div (-8) = -3$

e. $0 \div 5 = 0$
f. $14 \div 0$ is undefined

Example 5: Divide $\dfrac{5}{8} \div \dfrac{3}{4}$.

$$\dfrac{5}{8} \div \dfrac{3}{4} = \dfrac{5}{\underset{2}{8}} \cdot \dfrac{\overset{1}{4}}{3} = \dfrac{5}{6}$$

continued

Procedures/Rules	Key Examples

Section 1.4 Multiplying and Dividing Real Numbers; Properties of Real Numbers (continued)

When dividing decimal numbers, set up a long division and consider the divisor.

Case 1: If the divisor is an integer, then divide as if the dividend were a whole number, and place the decimal point in the quotient directly above its position in the dividend.

Case 2: If the divisor is a decimal number, then:
1. Move the decimal point in the divisor to the right enough places to make the divisor an integer.
2. Move the decimal point in the dividend the same number of places.
3. Divide the divisor into the dividend as if both numbers were whole numbers. Be sure to align the digits in the quotient properly.
4. Write the decimal point in the quotient directly above its position in the dividend.

In either of the two cases, continue the division process until you get a remainder of 0 or a repeating digit or block of digits in the quotient.

Example 6: Divide $12.88 \div 0.06$.

$$
\begin{array}{r}
214.66 \\
6\overline{)1288.00} \\
-12 \\
\hline
08 \\
-6 \\
\hline
28 \\
-24 \\
\hline
40 \\
-36 \\
\hline
40 \\
\end{array}
$$

Answers: $214.\overline{6}$

Section 1.5 Exponents, Roots, and Order of Operations

To evaluate an exponential form raised to a natural number exponent, write the base as a factor the number of times indicated by the exponent, then multiply.

If the base of an exponential form is a negative number and the exponent is even, then the product is positive.

If the base is a negative number and the exponent is odd, then the product is negative.

Every positive number has two square roots, a positive root and a negative root.

The radical symbol $\sqrt{}$ indicates to find only the positive (principal) square root.

The square root of a negative number is not a real number.

$$\sqrt{\frac{a}{b}} = \frac{\sqrt{a}}{\sqrt{b}}, \text{ where } a \geq 0 \text{ and } b > 0.$$

Example 1: Evaluate.
a. $3^4 = 3 \cdot 3 \cdot 3 \cdot 3 = 81$

b. $(-2)^4 = (-2)(-2)(-2)(-2) = 16$

c. $(-2)^5 = (-2)(-2)(-2)(-2)(-2) = -32$

Example 2: Find all square roots of 49.

Answer: ± 7

Example 3: Simplify.
a. $\sqrt{81} = 9$

b. $\sqrt{-81}$ is not a real number.

c. $\sqrt{\dfrac{25}{64}} = \dfrac{\sqrt{25}}{\sqrt{64}} = \dfrac{5}{8}$

continued

Procedures/Rules	Key Examples

Section 1.5 Exponents, Roots, and Order of Operations (continued)

Order of Operations

Perform operations in the following order:

1. Grouping symbols.
2. Exponents/Roots from left to right, in order as they occur.
3. Multiplication/Division from left to right, in order as they occur.
4. Addition/Subtraction from left to right, in order as they occur.

To find the arithmetic mean, or average, of n numbers, divide the sum of the numbers by n.

$$\text{Arithmetic mean} = \frac{x_1 + x_2 + \ldots x_n}{n}$$

Example 4: Simplify.

$$35 - [29 - 2(12 + 4)] + 4^3$$
$$= 35 - [29 - 2(16)] + 4^3$$
$$= 35 - [29 - 32] + 4^3$$
$$= 35 - (-3) + 4^3$$
$$= 35 - (-3) + 64$$
$$= 35 + 3 + 64$$
$$= 102$$

Example 5: Carl has the following test scores: 68, 76, 82, 78. Find the average of his test scores.

$$\frac{68 + 76 + 82 + 78}{4} = \frac{304}{4} = 76$$

Section 1.6 Translating Word Phrases to Expressions

To translate a word phrase to an expression, identify the variables, constants, and key words, then write the corresponding symbolic form.

Example 1: Translate to an algebraic expression:

a. Five more than seven times a number
    Answer: $7n + 5$

b. Six times the difference of a number and nine
    Answer: $6(n - 9)$

c. Four divided by a number cubed
    Answer: $4 \div n^3$

Section 1.7 Evaluating and Rewriting Expressions

To evaluate an algebraic expression:

1. Replace the variables with the corresponding given numbers.
2. Calculate following the order of operations.

Example 1: Evaluate $4x^2 - 5x$ when $x = -3$.

$$4(-3)^2 - 5(-3)$$
$$= 4(9) - 5(-3)$$
$$= 36 - (-15)$$
$$= 36 + 15$$
$$= 51$$

The Distributive Property of Multiplication over Addition
$$a(b + c) = ab + ac$$

Example 2: Use the distributive property to write an equivalent expression.
$$3(x + 7) = 3 \cdot x + 3 \cdot 7$$
$$= 3x + 21$$

To combine like terms, add or subtract the coefficients and keep the variables and their exponents the same.

Example 3: Combine like terms.

$$15x^2 + 8x + 3x^2 + 7 - 9x$$
$$= 15x^2 + 3x^2 + 8x - 9x + 7$$
$$= 18x^2 - x + 7$$

For Exercises 1–6, answer true or false.

1. π is a rational number.

2. $-\sqrt{2}$ is a rational number.

3. $-6x$ and $2x^3$ are like terms.

4. $8xy$ and yx are like terms.

5. $-6^2 = -36$

6. $(-6)^2 = 36$

For Exercises 7–10, fill in the blank.

7. The absolute value of every number is either _____ or 0.

8. A natural number that has exactly two different factors, 1 and the number itself, is a(n) _____ number.

9. When adding two addends that have the same sign, add their absolute values and _____ the sign.

10. If there are a(n) _____ number of negative factors, then the product is positive.

For Exercises 11–14, write a set representing each description.

11. The months beginning with the letter M

12. The letters in the word *algebra*

13. The even natural numbers

14. The even integers

For Exercises 15–18, graph each on a number line.

15. $-3\dfrac{2}{5}$

16. $2\dfrac{1}{4}$

17. 8.2

18. -4.6

For Exercises 19–22, simplify.

19. $|-6.3|$

20. $|8.46|$

21. $\left|2\dfrac{1}{6}\right|$

22. $\left|-4\dfrac{3}{8}\right|$

For Exercises 23–26, use =, +, <, or > to write a true sentence.

23. $|-6.4|$ �" 6.4

24. -6.9 ▪ 0

25. -14 ▪ -9

26. $\left|5\dfrac{6}{7}\right|$ ▪ $\left|-5\dfrac{6}{7}\right|$

For Exercises 27 and 28, name the fraction represented by the shaded region.

27.

28.

For Exercises 29–32, find the missing number that makes the fractions equivalent.

29. $-\dfrac{6}{7} = -\dfrac{?}{14}$

30. $\dfrac{2}{5} = \dfrac{10}{?}$

31. $\dfrac{15}{30} = \dfrac{?}{10}$

32. $\dfrac{-5}{2} = \dfrac{?}{4}$

For Exercises 33–36, write the prime factorization for each number.

33. 51

34. 100

35. 108

36. 84

For Exercises 37–40, simplify to lowest terms.

37. $\dfrac{152}{200}$

38. $\dfrac{61}{122}$

39. $\dfrac{250}{360}$

40. $\dfrac{26}{39}$

For Exercises 41–46, indicate whether the expression illustrates the commutative property of addition, the associative property of addition, the commutative property of multiplication, the associative property of multiplication, or the distributive property of multiplication over addition.

41. $4 + (7 + 3) = (4 + 7) + 3$

42. $3(9 + 1) = 27 + 3$

43. $\dfrac{1}{3} \cdot \dfrac{5}{6} = \dfrac{5}{6} \cdot \dfrac{1}{3}$

44. $5.6 + (11.2 + 4.3) = 5.6 + (4.3 + 11.2)$

45. $(n - 8)(-3) = -3(n - 8)$

46. $(2 \cdot 3) \cdot 7 = 2 \cdot (3 \cdot 7)$

For Exercises 47–68, perform the indicated operation(s).

47. $-8 - (-4.2)$

48. $6 - (-9.1)$

49. $\dfrac{2}{5} + \dfrac{1}{3}$

50. $-\dfrac{1}{4} + \dfrac{2}{3}$

51. $-7(-13)$

52. $6(-8)$

53. $-25 \div 5$

54. $-30 \div (-3)$

55. -6^2

56. $(-3)^2$

57. $7(6 + 2) - 48 \div 12$

58. $6(9 - 4) \div 3 - 1$

59. $(-2)^3(4)(-5)$

60. $3^5 \div 3^2 \div 3^2 \div 3$

61. $-\dfrac{1}{3} - \dfrac{3}{2} \div \dfrac{1}{6}$

62. $-6.8 + (-4.1 + 2.3)$

63. $4|-2| \div (-8)$

64. $10 \div 5 + (-5)(-5)$

65. $6 + \{3(4 - 5) + 2[6 + (-4)]\}$

66. $-1 - 3[4^2 - 3(-6)]$

67. $\dfrac{-6 + 3^2(8 - 10)}{2^3 - 16}$

68. $\dfrac{3(4 - 5) - 4 \cdot 3 + (11 - 20)}{(3 - 4)^3}$

For Exercises 69–72, translate to an algebraic expression.

69. twice a number subtracted from fourteen

70. the ratio of y to 7

71. a number minus twice the sum of the number and 4

72. 7 times the difference of m and n

For Exercises 73–78, evaluate the expression using the given values.

73. $b^2 - 4ac$; $a = -3$, $b = -1$, $c = 5$

74. $b^2 - 4ac$; $a = 2$, $b = 6$, $c = -2$

75. $-|4x| + |y^2|$; $x = -3$, $y = -2$

76. $\sqrt{m} + \sqrt{n}$; $m = 16$, $n = 9$

77. $\sqrt{m + n}$; $m = 16$, $n = 9$

78. $\dfrac{3l^2}{4 - n}$; $l = -4$, $n = 16$

For Exercises 79 and 80, determine the value that could replace the variable and cause the expression to be undefined.

79. $\dfrac{n}{n + 6}$

80. $\dfrac{y}{(y - 4)(y + 3)}$

For Exercises 81–84, use the distributive property to write an equivalent expression.

81. $5(x + 6)$

82. $-3(5n - 8)$

83. $\dfrac{1}{4}(8y + 3)$

84. $0.6(4.5m - 2.1)$

For Exercises 85–90, simplify by combining like terms.

85. $6x + 3y - 9x - 6y - 15$

86. $5y^2 - 6y + 4y - y^2 + 9y$

87. $-6xy + 9xy - xy$

88. $8x^3 - 4x - 6x^2 - 10x^3 + 4x$

89. $-4m - 4m - 4n + 4n$

90. $14 - 6x + 4y - 8 - 10y - 8x$

91. The following credit card statement shows when each transaction was posted to the account. Find the new balance.

4-01-01	Beginning balance	−$685.92
4-03-01	Dillard's	−$45.80
4-05-01	Payment	$250.00
4-08-01	Applebee's	−$36.45
4-16-01	CVS Pharmacy	−$12.92
4-24-01	Target credit	$32.68
4-31-01	Finance charge	−$5.18

92. The following graph shows the closing price of Microsoft stock each day for the week beginning May 6, 2002.

   a. Find the difference between the closing price at the beginning of the week and the closing price at the end of the week.

   b. If a person bought at the close of 5/6 and sold at the close of 5/8, what would be that person's net profit or loss on each share?

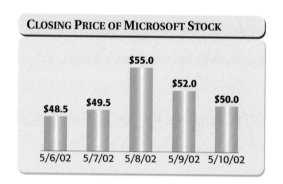

CLOSING PRICE OF MICROSOFT STOCK

93. Refer to the graph in Exercise 92.

   a. If Kaye bought 200 shares at the close of 5/6, how much did she spend?

   b. If she sold all 200 shares on 5/8, what was her profit?

94. The voltage in a circuit can be calculated by multiplying the current and the resistance. Suppose the current in a circuit is −9.5 amperes and the resistance is 16 ohms. Find the voltage.

95. In 1990, Larry has a loan balance of −$2405.80. If he makes no payment for 10 years, his balance will become −$14,896.08. How many times greater is his debt in 2000 than it was in 1990?

96. A company reports the following revenues for the first six months of 2001. Find the average revenue.

Month	Revenue
January	$45,320
February	$38,250
March	$61,400
April	$42,500
May	$74,680
June	$62,800

1. Graph $-3\frac{2}{7}$ on a number line.

2. $|-3.67| =$

3. Write the prime factorization of 100.

4. Simplify $\frac{17}{51}$.

5. Find the missing number: $-\frac{6}{5} = -\frac{?}{10}$.

For Exercises 6 and 7, indicate whether the expression illustrates the commutative property of addition, the associative property of addition, the commutative property of multiplication, the associative property of multiplication, or the distributive property of multiplication over addition.

6. $3(2 + 5) = 6 + 15$

7. $4(7 + 1) = 4(1 + 7)$

For Exercises 8–12, calculate.

8. $8 + (-4)$

9. $\frac{7}{8} - \left(-\frac{5}{6}\right)$

10. $(-1.5)(-0.4)$

11. $-\frac{5}{6} \div \frac{2}{3}$

12. $(-4)^3$

13. Find all square roots of 81.

For Exercises 14–19, simplify.

14. $-12 \div 3 \cdot 2$

15. $2^5 + 14 - 8(4 - 1)$

16. $\frac{6^2 + 14}{(2 - 7)^2}$

17. $-8 \div |4 - 2| + 3^2$

18. $(2 - 1)^4 - (2 - 3)^3$

19. $\sqrt{25 - 16} + [(14 + 2) - 3^2]$

20. After making a payment, Karen's balance on her credit card is −$854.80. If her balance prior to making the payment was −$1104.80, how much did she pay toward the balance?

21. A weather report has the following snowfall measurements during a 24-hour period for a region. Calculate the average amount of snowfall in a 24-hour period for this region.

November 24	6.0 in.
December 10	10.0 in.
December 11	4.0 in.
December 24	3.0 in.
January 5	2.5 in.
January 7	8.0 in.

22. Evaluate $-|3x^2 + 2y|$, when $x = -1$ and $y = 4$.

23. Evaluate $\sqrt{a - b}$, when $a = 64$ and $b = -36$.

24. Use the distributive property to write an equivalent expression: $-5(4y + 9)$

25. Simplify: $3.5x - 8 + 2.1x + 9.3$

Solving Linear Equations and Inequalities

In Chapter 1, we reviewed numbers, arithmetic, and expressions, which form a foundation for studying equations. Remember from Chapter 1 that our most basic components are variables and constants, which make up expressions. We saw that we can perform two actions with expressions: 1. evaluate or 2. rewrite. In this chapter we will build on that foundation by creating equations and inequalities out of those expressions, and then we will *solve* those equations and inequalities. This development will be repeated throughout the course. We will study and classify new types of expressions, evaluate and rewrite those expressions, and then solve equations and inequalities made from the new expressions.

Remember, as we build algebra concepts, they become more complex, just as a building becomes more complex during its construction. Try not to get discouraged if a new concept does not make sense right away. Learning new concepts is like the construction of a building: during the early stages, the building is not as solid as it will be upon completion. This chapter is like the frame of the walls of a house: the frame is not as solid as the walls will be after the drywall and roof are put in place. Similarly, as we revisit equations and inequalities throughout the course, we will reinforce and solidify the concepts you will learn in this chapter.

"The difficulties you meet will resolve themselves as you advance. Proceed, and light will dawn, and shine with increasing clearness on your path."

—*D'Alembert*
French mathematician (1717–1783)

2.1 Equations, Formulas, and the Problem-Solving Process

OBJECTIVES

1. Verify solutions to equations.
2. Determine whether an equation is an identity.
3. Use formulas to solve problems.

We now move to the top tier of the Algebra Pyramid and focus on equations. Recall from Section 1.1 the definition of an **equation**.

DEFINITION

Equation: Two expressions set equal.

For example, $4x + 5 = 9$ is an equation made from the expressions $4x + 5$ and 9.

CONNECTION

We can think of an equation as a complete sentence and the equal sign as the verb "is." We can read $4x + 5 = 9$ as "Four x plus five is equal to nine" or "Four x plus five is nine."

$$4x + 5 = 9$$

$$\updownarrow$$

"Four x plus five is 9."

THE ALGEBRA PYRAMID

EQUATIONS AND INEQUALITIES
$4x + 5 = 9$

EXPRESSIONS
$4x + 5, \ 9$

CONSTANTS AND VARIABLES
$4, 5, 9, x$

OBJECTIVE 1 Verify solutions to equations. An equation can be either true or false. For example, the equation $3 + 1 = 4$ is true, while the equation $3 + 1 = 5$ is false. When given an equation that contains a variable for a missing number, such as $x + 2 = 7$, our goal is to solve it, which means to find every number that can replace the variable and make it true. Such a number is called a **solution** to the equation.

DEFINITION

Solution: A number that makes an equation true when it replaces the variable in the equation.

For example, the number 5 is a solution to the equation $x + 2 = 7$ because when 5 replaces x, it makes the equation true.

$$x + 2 = 7$$
$$\downarrow$$
$$5 + 2 = 7 \quad \text{True.}$$

By showing that the equation $x + 2 = 7$ is a true statement when x is replaced with 5, we verify or *check* that 5 is the solution to the equation.

PROCEDURE CHECKING A POSSIBLE SOLUTION

To check a value as a solution to a given equation:

1. Replace the variable(s) in the equation with the value.
2. If the resulting equation is true, then the value is a solution.

EXAMPLE 1 Check to see if -2 is a solution to $4x + 9 = 1$.

Solution $4x + 9 = 1$

$4(-2) + 9 \stackrel{?}{=} 1$ Replace x with -2 and see if the equation is true.

$-8 + 9 \stackrel{?}{=} 1$

$1 = 1$

Note: The symbol $\stackrel{?}{=}$ indicates we are asking, "How does the left side compare with the right side? Are they equal or not equal?"

Since -2 makes the equation true, it is a solution to $4x + 9 = 1$.

EXAMPLE 2 Check to see if 0.4 is a solution to $n^2 + 0.8 = 3n - 0.1$.

Solution $n^2 + 0.8 = 3n - 0.1$

$(0.4)^2 + 0.8 \stackrel{?}{=} 3(0.4) - 0.1$ Replace n with 0.4 and see if the equation is true.

$0.16 + 0.8 \stackrel{?}{=} 1.2 - 0.1$

$0.96 \neq 1.1.$

Since 0.4 does not make the equation true, it is not a solution to $n^2 + 0.8 = 3n - 0.1$.

EXAMPLE 3 Check to see if $5\frac{1}{3}$ is a solution to $\frac{3}{4}y - 1 = \frac{y}{2} + \frac{1}{3}$.

Solution $\frac{3}{4}y - 1 = \frac{1}{2}y + \frac{1}{3}$

Note: It is convenient to write $\frac{y}{2}$ as $\frac{1}{2}y$, which we can do because $\frac{y}{2} = y \div 2 = y \cdot \frac{1}{2} = \frac{1}{2}y$.

$\frac{3}{4}\left(5\frac{1}{3}\right) - 1 \overset{?}{=} \frac{1}{2}\left(5\frac{1}{3}\right) + \frac{1}{3}$ Replace y with $5\frac{1}{3}$ and see if the equation is true.

$\overset{1}{\underset{1}{\frac{3}{4}}}\left(\overset{4}{\underset{1}{\frac{16}{3}}}\right) - 1 \overset{?}{=} \overset{1}{\underset{1}{\frac{1}{2}}}\left(\overset{8}{\underset{1}{\frac{16}{3}}}\right) + \frac{1}{3}$ Divide out common factors, then multiply.

$4 - 1 \overset{?}{=} \frac{8}{3} + \frac{1}{3}$

$3 = 3$ Therefore, $5\frac{1}{3}$ is a solution to $\frac{3}{4}y - 1 = \frac{y}{2} + \frac{1}{3}$.

YOUR TURN

a. Check to see if 6 is a solution to $5n - 8 = 3n + 4$.

b. Check to see if -1.5 is a solution to $4y - 3.5 = 2y + 3(y + 0.2)$.

c. Check to see if $\frac{1}{6}$ is a solution to $\frac{2}{3}x + \frac{1}{4} = \frac{1}{5} - x^2$.

OBJECTIVE 2 Determine whether an equation is an identity. Consider $3x + 9 = 3x + 9$. Because the expressions on each side of the equal sign are identical, they will each produce the same result no matter what number replaces the variable x. This means that every real number is a solution. Such an equation is called an **identity**.

DEFINITION

Identity: An equation that has every real number as a solution (excluding any numbers that cause an expression in the equation to be undefined).

The equation $5y - 4 = 7y - 3$ is not an identity because not every real number is a solution. Consider the number 2, for example. If we replace y with 2 in the equation, we see that it does not check:

$5(2) - 4 \overset{?}{=} 7(2) - 3$
$10 - 4 \overset{?}{=} 14 - 3$ The real number we picked, 2, is not a solution for
$6 \neq 11$ $5y - 4 = 7y - 3$, so the equation is not an identity.

Sometimes it is not obvious that an equation is an identity. Simplifying the expressions on each side of the equal sign can make the identity apparent.

PROCEDURE RECOGNIZING AN IDENTITY

To determine whether an equation is an identity, simplify the expressions on each side of the equal sign. If, after simplifying, the expressions are identical, then the equation is an identity.

EXAMPLE 4 Determine whether the equation is an identity.

a. $5x - 8x + 7 = 13 - 3x - 6$

Solution Simplify the expressions on each side of the equal sign. If the simplified expressions are identical, then the equation is an identity.

$$5x - 8x + 7 = 13 - 3x - 6$$
$$-3x + 7 = -3x + 7 \qquad \text{Combine like terms.}$$

The expressions on each side of the equal sign are identical. Therefore, the equation is an identity.

b. $2(3y - 4) - 10y = 15 - 4(y + 2)$

Solution $2(3y - 4) - 10y = 15 - 4(y + 2)$
$$6y - 8 - 10y = 15 - 4y - 8 \qquad \text{Distribute to clear parentheses.}$$
$$-4y - 8 = -4y + 7 \qquad \text{Combine like terms.}$$

The expressions on each side of the equal sign are not identical. Therefore, this equation is not an identity.

YOUR TURN Determine whether the equation is an identity.

a. $0.5(3n - 8) = n - 4 + 0.5n$ **b.** $10 + \dfrac{1}{4}t^2 - 9 = t^2 - 1 - 0.75t^2$

OBJECTIVE 3 **Use formulas to solve problems.** A primary purpose of studying mathematics is to develop and improve problem-solving skills. George Polya proposed the idea that all problem solving follows a four-step outline: 1. understand the problem, 2. devise a plan for solving the problem, 3. execute the plan, and 4. check the results. Following is an outline for problem solving based on Polya's four stages. We'll see Polya's four stages illustrated throughout the rest of the text in application problems.

PROCEDURE **PROBLEM-SOLVING OUTLINE**

1. **Understand** the problem.
   a. Read the question(s) (not the whole problem, just the question at the end) and write a note to yourself about what it is you are to find.
   b. Now read the whole problem.
   c. Underline key words.
   d. Draw a picture (if possible).
   e. Make a list or table.
   f. Simulate the situation.
   g. Search for a related example problem.
2. Devise a **plan**.
   a. Search for a formula.
   b. Translate the key words to an equation.
3. **Execute** the plan by solving the equation/formula.
4. **Answer** the question. Look back at the note you made to yourself about what it was you were to find and make sure you answer that question. Be sure to include appropriate units.

continued

5. Check results. Make sure the answer is reasonable.
   a. Repeat the solution in a different way.
   b. Reverse the process.
   c. Estimate the answer and make sure the estimate and actual answer are reasonably close.

As the course develops, we will explore the various strategies listed in the outline. In this section we focus on using **formulas** to solve problems. Later in this chapter, we will solve problems by translating key words to an equation.

DEFINITION

Formula: An equation that describes a mathematical relationship.

First, let's consider using formulas from geometry. The following table lists some common geometric formulas. Before we examine the formulas, let's review a few terms.

DEFINITIONS

Perimeter: The distance around a figure.
Area: The total number of square units that fill a figure.
Volume: The total number of cubic units that fill a space.
Circumference: The distance around a circle.
Radius: The distance from the center of a circle to any point on the circle.
Diameter: The distance across a circle through its center.

Geometric Formulas

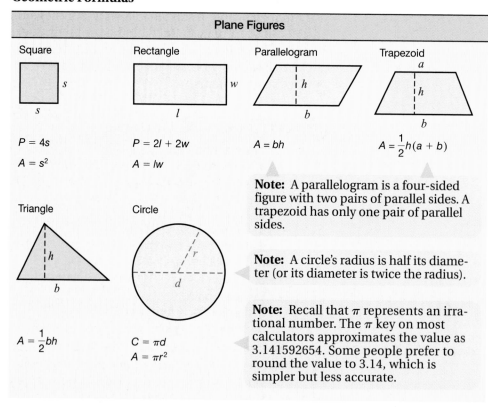

Plane Figures

Square
$P = 4s$
$A = s^2$

Rectangle
$P = 2l + 2w$
$A = lw$

Parallelogram
$A = bh$

Trapezoid
$A = \frac{1}{2}h(a + b)$

Triangle
$A = \frac{1}{2}bh$

Circle
$C = \pi d$
$A = \pi r^2$

Note: A parallelogram is a four-sided figure with two pairs of parallel sides. A trapezoid has only one pair of parallel sides.

Note: A circle's radius is half its diameter (or its diameter is twice the radius).

Note: Recall that π represents an irrational number. The π key on most calculators approximates the value as 3.141592654. Some people prefer to round the value to 3.14, which is simpler but less accurate.

Solids			

Box	Pyramid	Cylinder	Cone

$V = lwh$

$SA = 2lw + 2lh + 2wh$

$V = \dfrac{1}{3}lwh$

$V = \pi r^2 h$

$V = \dfrac{1}{3}\pi r^2 h$

Sphere

$V = \dfrac{4}{3}\pi r^3$

CONNECTION

Notice that the volume formulas for a box and pyramid are similar. This is because a pyramid has $\frac{1}{3}$ the volume of a box with the same-size base and same height. The same relationship is true for cylinders and cones with the same-size base and same height.

PROCEDURE USING A FORMULA

To use a formula:

1. Replace the variables with the corresponding given values.

2. Solve for the missing value.

Problems Involving a Single Formula

EXAMPLE 5 A mason is building a rectangular foundation wall that is to be 75 feet by 40 feet. What is the total distance around the wall?

Understand The "total distance around the wall" is the perimeter.

Plan Because the shape is a rectangle, we can use the formula $P = 2l + 2w$.

Execute Replace l with **75** feet and w with **40** feet, then calculate.

$$P = 2l + 2w$$
$$P = 2(75) + 2(40)$$
$$P = 150 + 80$$
$$P = 230 \text{ ft.}$$

Answer The mason must have enough supplies to build a wall that is 230 feet long.

Check In this case, the solution can be verified using an alternative method. One could add all four side lengths.

$$P = 75 + 75 + 40 + 40 = 230 \text{ ft.}$$

Problems Involving Combinations of Formulas

Sometimes a problem requires a combination of formulas. This occurs if we are finding the area of a figure that is composed of two or more simpler figures.

EXAMPLE 6 Below is a drawing of a room that is to be carpeted. Calculate the area of the room.

Understand This figure can be thought of as a trapezoid and a rectangle. The height of the trapezoid is found by subtracting 15 feet from 17 feet.

Plan Because the figure consists of a trapezoid and rectangle combined, we can add the areas of these shapes to find the total area of the figure. The area of a rectangle is found using the formula $A = lw$ and the area of a trapezoid is found using the formula $A = \frac{1}{2}h(a + b)$. The total area is:

$A =$ Area of the rectangle + area of the trapezoid

$A = \qquad lw \qquad + \qquad \frac{1}{2}h(a + b)$

Execute Replace the variables with the corresponding values and calculate.

$$A = lw + \frac{1}{2}h(a + b)$$

$$A = (15)(18) + \frac{1}{2}(2)(18 + 9)$$

$$A = (15)(18) + \frac{1}{2}(2)(27)$$

$$A = 270 + 27$$
$$A = 297 \text{ ft.}^2$$

Answer The total area is 297 square feet.

Check We can verify the reasonableness of the answer using estimation. Suppose the figure had been an 18 feet by 17 feet rectangle. We would expect the area of this rectangle to be slightly greater than the area of the actual figure. The area of an 18 feet by 17 feet rectangle is $A = (18)(17) = 306 \text{ ft.}^2$, which indicates that 297 square feet is reasonable.

Sometimes a figure may be placed within another figure and we must calculate the area of the region between the outside figure and inside figure.

 EXAMPLE 7 A portion of the front of a house is to be covered with siding (shaded in the figure). The window is 2.8 feet by 4.8 feet. Calculate the area to be covered.

Understand The shaded area is the entire triangle excluding the rectangular window.

Plan We can exclude the area of the window by calculating the area of the triangle and subtracting the area of the window.

A = Area of the triangle − area of the rectangle

$$A = \frac{1}{2}bh \quad - \quad lw$$

Execute $A = \dfrac{1}{2}(12.5)(8) - (2.8)(4.8)$

$A = 50 - 13.44$
$A = 36.56 \text{ ft.}^2$

Answer The area to be covered is 36.56 square feet.

Check Estimate the area by rounding the decimal numbers. Rounding the measurements of the triangle to the nearest ten, we have a base of 10 feet and a height of 10 feet which means an area of 50 square feet. Rounding the window dimensions to the nearest whole, we have a 3 feet by 5 feet window, which has an area of 15 square feet. Subtracting these areas, we have a final area of 35 square feet, which indicates our answer of 36.56 square feet is quite reasonable.

 YOUR TURN

A house is to be situated on a lot as shown. Once the house is built, the remaining area will be landscaped. Calculate the area to be landscaped.

Problems Involving Nongeometric Formulas

Following is a list of other formulas you may encounter.

The distance, d, an object travels given its rate, r, and the time of travel, t: $d = rt$

The average rate of travel, r, given the total distance, d, and total time, t: $r = \dfrac{d}{t}$

ANSWER

8425 ft.^{2}

The voltage, V, in a circuit with a current, i, in amperes (A), and a resistance, r, in ohms (Ω): $V = ir$

The temperature in degrees Celsius given degrees Fahrenheit:

$$C = \frac{5}{9}(F - 32)$$

The temperature in degrees Fahrenheit given degrees Celsius:

$$F = \frac{9}{5}C + 32$$

EXAMPLE 8 A truck driver begins a delivery at 9 A.M. and travels 150 miles. He then takes a 30-minute break. He then travels another 128 miles, arriving at his destination at 2:00 P.M. What was his average speed?

Understand We are given travel distances and times and we are to find the average speed, which is a rate.

Plan To find the average rate, we first need the total distance traveled and the total time spent traveling. We can then use the formula $r = \dfrac{d}{t}$.

Execute Total distance = 150 + 128 = 278 miles

Total time = 5 − 0.5 = 4.5 hours

Note: From 9 A.M. to 2:00 P.M. is 5 hours, but we must deduct the 0.5-hour (30 minute) break because he was not traveling during that time.

Now we can calculate the average rate:

$$r = \frac{278}{4.5}$$

$r = 61.\overline{7}$ mph (miles per hour)

Answer His average rate was $61.\overline{7}$ mph.

Check We can use $d = rt$ to verify that if he traveled an average rate of $61.\overline{7}$ miles per hour for 4.5 hours he would travel 278 miles.

$$d \approx 61.8(4.5)$$
$$d \approx 278.1$$

Note: We rounded $61.\overline{7}$ so the distance is an approximation.

YOUR TURN

A person hikes 10 miles then rests for 20 minutes, after which she hikes another 8 miles. If she began her hike at 10:30 A.M. and stopped at 4:30 P.M., what was her average rate?

ANSWER

$3\dfrac{3}{17}$ or ≈ 3.2 mph

2.1 EXERCISES

For Extra Help

 InterAct Math

 Addison-Wesley Math Tutor Center

Video 3/DVT 1

MyMathLab

MathXL

1. Explain the difference between an expression and an equation.

2. What is a solution for an equation?

3. Explain how to check a value to see if it is a solution for a given equation.

4. If an equation is also an identity, what does this indicate about its solutions?

For Exercises 5–20, determine if the given number is a solution.

5. $3x - 5 = 41; x = 2$

6. $4a + 7 = 51; a = 11$

7. $7y - 1 = y + 3; y = \dfrac{2}{3}$

8. $-8t - 3 = 2t - 15; t = -\dfrac{6}{5}$

9. $-4(x - 5) + 2 = 5(x - 1) + 3; x = -2$

10. $2(3m + 2) = 5m - 1; m = -3$

11. $-\dfrac{1}{2} + \dfrac{1}{3}y = \dfrac{3}{2}; y = 6$

12. $\dfrac{1}{2}p - 1 = \dfrac{2}{5}p + 3; p = 20$

13. $4.3z - 5.71 = 2.1z + 0.07; z = 2.2$

14. $12.7a + 12.6 = a + 5.4a; a = -2$

15. $b^2 - 4b = b^3 + 4; b = -1$

16. $-x^3 - 9 = 2x^2; x = -3$

17. $|x^2 - 9| = -x + 3; x = 3$

18. $-|2u - 3| = -3u + 8; u = 5$

19. $\dfrac{4x - 3}{x - 4} = \sqrt{x + 8}; x = 17$

20. $\dfrac{-y}{10 + y} = \dfrac{\sqrt{4 - y}}{3}; y = -5$

For Exercises 21–30, determine whether the equation is an identity.

21. $3(x + 7) = 3x + 21$

22. $-5x + 8x - 1 = 3x - 4 + 2$

23. $3(y + 1) - 2y = y + 3$

24. $2(2 + m) = 6m - 4m + 4$

25. $12 - 2(x + 8) = 3x + 4 - 2x$

26. $5n - 20 - 9n = 8 - 4(n + 5)$

27. $\frac{1}{3}(t + 6) - t = \frac{5}{6}t + 2 - \frac{1}{2}t$

28. $\frac{1}{2}x^2 + 3 - \frac{4}{5}x^2 = -3\left(\frac{1}{10}x^2 + 1\right)$

29. $0.27p + 3p = 7(p + 1) - (3.73 + 1)$

30. $0.4(n^2 - 6) - 3.6 = 1.9n^2 - 1.5(n + 4)$

For Exercises 31–50, solve using the geometric formulas. (Answers to exercises involving π were calculated using the π key on a calculator, which approximates the value as 3.141592654.)

31. Janet wants to put a wallpaper border around her den. The room is 13 feet by 20 feet.

   a. What is the total length of wallpaper border she needs?

   b. If the wallpaper border comes in packages of 12 feet, how many packages must she buy?

   c. If the packages are priced at $8.99 each, what will be the total cost of the wallpaper?

32. Jamal is planning to install crown molding where the walls and ceiling of his living room meet. The room is 16.5 feet by 22 feet.

   a. What is the total length of crown molding he needs?

   b. If the crown molding comes in strips of 8 feet, how many strips must he purchase?

   c. If the strips cost $9.99 each, what will be the cost of the crown molding?

33. Fermi National Accelerator Laboratory is a circular tunnel that is used to accelerate elementary particles. The radius of the tunnel is about 2 kilometers. If a particle travels one complete revolution through the tunnel, what distance does it travel? (*Source:* Cesare Emiliani, *The Scientific Companion,* 2nd ed., Wiley, 1995)

34. The Chicxulub crater, which is buried partly beneath the Yucatan peninsula and partly beneath the Gulf of Mexico, is about 180 kilometers in diameter. What is the circumference of this crater?

35. Tina needs to buy enough pine straw to cover a rectangular flower bed that measures 6 feet by 13 feet.

   a. What is the total area she must cover?

   b. A landscape consultant indicates that she should use 1 bale of pine straw for every 10 square feet to be covered. If she follows this advice, how many bales must Tina purchase?

   c. If Tina's local garden center charges $4.50 per bale, what will be the cost of covering her flower bed?

36. Juan is considering installing carpet in his new office space. The office is rectangular and measures 42 feet by 36 feet. Juan has a budget of $3000.

   a. What is the total area that will be carpeted?

   b. How many square yards is this area? (There are 9 square feet in 1 square yard.)

   c. If the contractor quotes Juan a price of $22.50 per square yard to install the carpet, how much will it cost to carpet the space? Is this feasible with Juan's budget?

Of Interest

The Chicxulub crater was created when Earth was hit by an asteroid about 65 million years ago. The devastation of the impact is thought to have caused the extinction of about 50 percent of the species on Earth, including the dinosaurs.

37. At one time, the Venetian Hotel and Casino in Las Vegas boasted having the largest hotel rooms in the world. If one of the rooms is 70 feet by 100 feet, what is the total square footage in the room?

38. Hardwood floors are to be installed in a bedroom that measures 14 feet by 15 feet. If the contractor charges $34.50 per square foot to install the floors, what will be the cost of this improvement?

39. A kitchen floor is to be re-covered with granite tiles. There is a fixed island in the center of the floor (see figure). If the contractor charges $28.80 per square foot to install the new floor, what will be the installation cost?

40. Conner and Ellyn purchase a new home. The area around a small storage building in the backyard is to be landscaped.

   a. What is the total area that is to be landscaped?

   b. If $\frac{2}{3}$ of this area is to be grass, what amount will be covered with grass?

   c. If each pallet of grass sod covers 504 square feet, how many pallets will be needed?

   d. Each pallet costs $106. What is the total cost of the sod?

41. The wall shown is to be painted. The door, which will not be painted, is 2.5 feet by 7.5 feet. What area will be painted?

42. The side of a house needs new siding. The window measures 3 feet by 4.5 feet. Find the area to be covered in siding.

43. A company cuts two circular pieces out of a sheet of metal and the rest of the sheet is recycled. The sheet is 4 feet by 8 feet and the diameter of each circle is 4 feet. Find the area of the sheet that is recycled.

44. A compact disc has a diameter of $5\frac{3}{4}$ inches. On a CD, no information is placed in a circle in the center of the CD with a diameter of $1\frac{3}{4}$ inches. Find the area on a CD that can contain information.

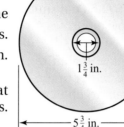

$1\frac{3}{4}$ in.

$5\frac{3}{4}$ in.

45. A small storage building is shaped like a box that measures 2 meters by 5 meters by 4.75 meters. What is the storage building's volume?

46. A dorm refrigerator measures 2 feet by $1\frac{1}{2}$ feet by 4 feet. What is its volume?

47. A cylindrical bottle has a diameter of 8 centimeters and a height of 20 centimeters. Find the volume of the bottle.

48. Earth is roughly a sphere with a radius of approximately 6370 kilometers. What is the approximate volume of Earth?

49. The base of the Great Pyramid at Giza, Egypt, is 745 feet by 745 feet and its height is 449 feet. Find the volume of the pyramid.

50. If a funnel is approximately a cone, approximate the volume of a funnel with a diameter of $8\frac{1}{2}$ inches and a height of 6 inches.

For Exercises 51–56, use the formulas relating distance, rate, and time.

51. A flight departs Atlanta at 8:30 A.M. EST to arrive in Las Vegas at 10:00 A.M. PST. If the plane flies at an average rate of $388\frac{1}{3}$ mph, what distance does it travel?

52. A flight departs at 7:30 A.M. EST from Philadelphia and travels to Dallas-Fort Worth, arriving at 10:10 A.M. CST. If the plane is averaging a speed of 368.2 miles per hour, what distance does it travel? (*Hint:* Express the time using fractions.)

53. A family began a trip of 516 miles at 8 A.M. They arrived at their final destination at 5:30 P.M. If they took two 15-minute breaks and took an hour for lunch, what was their average rate?

54. A family began a trip at 7:30 A.M. At the beginning of the trip, the car's odometer read 45,362.6. They arrived at their final destination at 6:00 P.M. and the odometer read 45,785.2. If they took three 15-minute breaks and took an hour for lunch, what was their average rate?

55. A truck driver is to make a delivery that requires him to drive 285 miles by 4:00 P.M. If he starts at 10:00 A.M. and must make three stops of 15 minutes each at weigh stations along the route, what must his average rate be to make the delivery on time?

56. In 2002, Lance Armstrong won the Tour de France, which covered a distance of 3300 kilometers, with a time of 82 hours, 5 minutes, and 12 seconds. What was his average rate in kilometers per hour for the race? (*Source:* VeloNews, *Journal of Competitive Cycling*)

For Exercises 57 and 58, use the formula relating voltage, current, and resistance.

57. A technician measures the current in a circuit to be −6.5 amperes and the resistance is 8 ohms. Find the voltage.

58. The current in a circuit is measured to be 4.2 amperes. If the resistance is 16 ohms, find the voltage.

For Exercises 59–64, use the formulas for converting degrees Fahrenheit to degrees Celsius or degrees Celsius to degrees Fahrenheit.

59. On a drive to work, Tim notices that the temperature shown on a bank's digital thermometer is 34°C. What is this temperature in degrees Fahrenheit?

60. Liquid oxygen boils at a temperature of −183°C. What is this temperature in degrees Fahrenheit?

61. Iron melts at a temperature of 1535°C. What is this temperature in degrees Fahrenheit?

62. The average temperature on Neptune is approximately −360°F. What is this temperature in degrees Celsius?

63. At the South Pole, temperatures can get down to −76°F. What is this temperature in degrees Celsius?

64. The average temperature on Venus is approximately 890°F. What is this temperature in degrees Celsius?

A contest has people guess how many marbles are inside a jar. The jar is a cylinder with a diameter of 9 inches and a height of 12 inches. Suppose the marbles have a diameter of 0.5 inches. Calculate the number of marbles that might fit inside the jar. Discuss the accuracy of your calculation.

REVIEW EXERCISES

For Exercises 1–2, combine like terms.

1. $5x - 14 + 3x + 9$

2. $6.2y + 7 - 1.5 - 0.8y$

For Exercises 3–4, multiply using the distributive property.

3. $\frac{1}{3}(x + 6)$

4. $-(4w - 6)$

For Exercises 5–6, simplify.

5. $14 + 9(7 - 2)$

6. $6(2 + 3^2) - 4(2 + 3)^2$

2.2 The Addition Principle

OBJECTIVES

1. Determine whether a given equation is linear.
2. Solve linear equations in one variable using the addition principle.
3. Solve equations with variables on both sides of the equal sign.
4. Solve application problems.

OBJECTIVE 1 Determine whether a given equation is linear. There are many types of equations. In this section, we introduce and begin solving **linear equations**.

> **DEFINITION**
>
> **Linear equation:** An equation in which each variable term contains a single variable raised to an exponent of 1.

Equations that are not linear are *nonlinear equations*.

EXAMPLE 1 Determine whether the equation is linear or nonlinear.

a. $3x + 5 = 12$

Answer: This equation is linear because the variable x has an exponent of 1.

b. $7x^3 + x = 1$

Answer: This equation is nonlinear because the variable in the $7x^3$ term has an exponent of 3.

YOUR TURN Determine whether the given equation is linear.

a. $3x - 4 = 11$ **b.** $-6x + y = 30$ **c.** $n^2 = 5m + 4$ **d.** $y = 3x^2 + 1$

In this chapter, we will focus on equations with the same variable throughout the equation. If a linear equation has the same variable throughout, then we say it is a **linear equation in one variable**.

> **DEFINITION**
>
> **Linear equation in one variable:** An equation that can be written in the form $ax + b = c$, where a, b, and c are real numbers and $a \neq 0$.

ANSWERS

a. Yes   **b.** Yes
c. No   **d.** No

$2x + 9 = 5x - 12$ is a linear equation in one variable.

$x^2 - 2x + 5 = 8$ is a nonlinear equation in one variable.

In Chapter 4, we will consider linear equations in two variables, such as $x + y = 3$.

OBJECTIVE **2** **Solve linear equations in one variable using the addition principle.** In Section 2.1, we discussed solutions to equations and how to check an equation's solution. We now consider a technique for solving an equation (which means to find its solution[s]). This technique is called the *balance technique*. Imagine an equation as scales with the equal sign as the fulcrum.

When weight is added or removed on one side of a balanced set of scales, the scales tip out of balance. In this case, we add 4 to the left side.

To maintain balance, the same weight must be added or removed on *both* sides of the scale. Adding 4 to the right side balances the scale.

In the language of mathematics, this balance technique is stated in two principles: the *addition principle of equality* and the *multiplication principle of equality*. In this section, we will consider the addition principle of equality. In Section 2.3 we will consider the multiplication principle of equality. The addition principle of equality says that we can add (or subtract) the same amount on both sides of an equation without affecting its solution(s).

RULE   **THE ADDITION PRINCIPLE OF EQUALITY**

If $a = b$ then $a + c = b + c$ is true for all real numbers a, b, and c.

In solving an equation, the goal is to write the equation in a simpler form $x = c$, where the variable is alone, or isolated, on one side of the equal sign and the solution appears on the other side. Returning to the original equation, $x + 5 = 12$, to isolate x, we need to clear, or undo, $+5$ from the left-hand side. To clear the $+5$, we add -5 (or subtract 5) because $+5$ and -5 are additive inverses, which means their sum is 0.

| To isolate x, we clear $+5$ by adding -5 (or subtracting 5) on the left side. | $\begin{aligned} x + 5 &= 12 \\ \underline{-5} \quad &\underline{-5} \\ x + 0 &= 7 \\ x &= 7 \end{aligned}$ | To keep the equation balanced, we add -5 (or subtract 5) on the right side as well. |

The statement $x = 7$ is the solution statement.

PROCEDURE   **USING THE ADDITION PRINCIPLE**

To use the addition principle:

1. Decide what term you want to clear.
2. Add the additive inverse of that term to both sides of the equation (that is, add or subtract appropriately so that the term you want to clear becomes 0).

EXAMPLE 2   Solve and check.

a. $x - 13 = -22$

Solution

| Add 13 to the left-hand side so that $-13 + 13 = 0$. | $\begin{aligned} x - 13 &= -22 \\ \underline{+13} \quad &\underline{+13} \\ x + 0 &= -9 \\ x &= -9 \end{aligned}$ | Since we added 13 on the left side, we *must* add 13 on the right side as well. |

Check   Recall from Section 2.1 that to check, we replace x in the original equation with -9 and verify that the equation is true.

$$x - 13 = -22$$
$$-9 - 13 \stackrel{?}{=} -22 \qquad \text{Replace } x \text{ with } -9.$$
$$-22 = -22$$

True, so -9 is the solution.

Learning Strategy

If you are a visual learner, to get comfortable with the addition principle, consider writing the amount added or subtracted on both sides of an equation in a different color than the rest of the equation. Once you are comfortable seeing the amount added or subtracted, you may not need the color any longer.

b. $\dfrac{3}{5} = y + \dfrac{1}{3}$

Note: Here, we illustrate another popular style of writing the addition principle, in which $-\dfrac{1}{3}$ is written beside the expressions on both sides of the equation instead of underneath the corresponding terms.

Solution $\dfrac{3}{5} - \dfrac{1}{3} = y + \dfrac{1}{3} - \dfrac{1}{3}$ Subtract $\dfrac{1}{3}$ from both sides to isolate y.

$\dfrac{9}{15} - \dfrac{5}{15} = y + 0$ Rewrite fractions with their LCD.

$\dfrac{4}{15} = y$

Check $\dfrac{3}{5} = y + \dfrac{1}{3}$

$\dfrac{3}{5} \overset{?}{=} \dfrac{4}{15} + \dfrac{1}{3}$ Replace y in the original equation with $\dfrac{4}{15}$ and verify that the equation is true.

$\dfrac{3}{5} \overset{?}{=} \dfrac{4}{15} + \dfrac{5}{15}$ Rewrite fractions with their LCD.

$\dfrac{3}{5} \overset{?}{=} \dfrac{9}{15}$ Add fractions, then simplify to lowest terms.

$\dfrac{3}{5} = \dfrac{3}{5}$

True, so $\dfrac{4}{15}$ is the solution.

YOUR TURN   Solve and check.

   a. $4.7 = x - 9.8$ **b.** $n + \dfrac{3}{4} = \dfrac{1}{6}$

Simplifying before Isolating the Variable

Some equations have expressions that can be simplified. If like terms are on the same side of the equation, we will combine the like terms before isolating the variable.

ANSWERS

  a. 14.5 **b.** $-\dfrac{7}{12}$

EXAMPLE 3 Solve and check $8y - 2.1 - 7y = -3.7 + 9$.

Solution Simplify the expressions, then isolate y.

$$8y - 2.1 - 7y = -3.7 + 9$$

Combine $8y - 7y$ to equal y. $\qquad y - 2.1 = 5.3$ $\qquad$ Combine $-3.7 + 9$ to equal 5.3.

$$\begin{array}{r} +2.1 \quad +2.1 \\ \hline y + 0 = 7.4 \\ y = 7.4 \end{array}$$ Add 2.1 on both sides to isolate y.

Check Replace y in the original equation with 7.4 and verify that the equation is true. We will leave this to the reader.

Using the Distributive Property

If the equation to be solved contains parentheses, we use the distributive property to clear the parentheses before isolating the variable.

EXAMPLE 4 Solve and check $6(n - 3) - 5n = -25 + 4$.

Solution Use the distributive property, simplify, and then isolate n.

$$6(n - 3) - 5n = -25 + 4$$ $\qquad$ Combine $-25 + 4$ to equal -21.

Distribute to clear parentheses. $\qquad 6n - 18 - 5n = -21$

Combine $6n - 5n$ to equal n. $\qquad\qquad n - 18 = -21$

$$\begin{array}{r} +18 \quad +18 \\ \hline n + 0 = \quad -3 \\ n = -3 \end{array}$$ Add 18 on both sides to isolate n.

Check Replace n with -3 in the original equation and verify that the equation is true. We will leave this to the reader.

YOUR TURN Solve and check.

    a. $9x + 4 - 8x = 1 - 6$ **b.** $6.2 - 0.4 = 1.5(m - 6) - 0.5m$

OBJECTIVE ③ **Solve equations with variables on both sides of the equal sign.** Some equations have variable terms on both sides of the equal sign. In order to isolate the variable, we must first get the variable terms together on the same side of the equal sign. To do this, we can use the addition principle.

ANSWERS

    a. -9 **b.** 14.8

EXAMPLE 5 Solve and check $5x - 8 = 4x - 13$.

Solution Use the addition principle to get the variable terms together on the same side of the equal sign. Then isolate the variable.

$$
\begin{aligned}
5x - 8 &= 4x - 13 \\
\underline{-4x \qquad\quad -4x} & \qquad \text{Subtract } 4x \text{ from both sides.}\\
x - 8 &= 0 - 13 \\
x - 8 &= -13 \\
\underline{+8 \qquad +8} & \qquad \text{Add 8 on both sides to isolate } x.\\
x + 0 &= -5 \\
x &= -5
\end{aligned}
$$

Note: By choosing to subtract $4x$ on both sides, we clear the $4x$ term from the right side and combine it with the $5x$ on the left side. It does not matter which term you move first (see explanation below).

Check
$$
\begin{aligned}
5x - 8 &= 4x - 13 \\
5(-5) - 8 &\overset{?}{=} 4(-5) - 13 \\
-25 - 8 &\overset{?}{=} -20 - 13 \\
-33 &= -33
\end{aligned}
$$

Replace x in the original equation with -5 and verify that the equation is true.

True, so -5 is the solution.

When variable terms appear on both sides of the equal sign, it does not matter which term you choose to clear first, although some choices make the process easier. Suppose in Example 5, we clear the $5x$ term first:

$$
\begin{aligned}
5x - 8 &= 4x - 13 \\
\underline{-5x \qquad\quad -5x} & \\
0 - 8 &= -x - 13 \\
-8 &= -x - 13 \\
\underline{+13 \qquad\qquad +13} & \\
5 &= -x + 0 \\
5 &= -x \\
-5 &= x
\end{aligned}
$$

Note: Clearing the $5x$ term first gives $-x$, whereas when we cleared the $4x$ term, we got x. Most people prefer working with positive coefficients.

Note: $5 = -x$ means we must find a number whose additive inverse is 5, so x must be equal to -5. In effect, we simply changed the signs of both sides. In Section 2.3, we will see another way to interpret $-x$.

Notice the term we chose to clear in the first step did not affect the solution, but it did affect our approach to the rest of the problem.

Conclusion In solving an equation that has variable terms on both sides of the equal sign, when we select a variable term to clear, we can avoid negative coefficients by clearing the term with the lesser coefficient.

ANSWERS

a. 13   b. −6.9

YOUR TURN Solve and check.

a. $7y - 3 = 6y + 10$

b. $2.6 + 3n = 9.5 + 4n$

EXAMPLE 6 Solve and check $y - (3y + 11) = 5(y - 2) - 8y$.

Solution Simplify both sides of the equation. Then isolate y.

Note: When we distribute a minus sign in an expression like $-(3y + 11)$, we can think of the minus sign as -1, so that we have:
$$-(3y + 11) = -1(3y + 11)$$
$$= -1(3y) + (-1)(11)$$
$$= -3y - 11$$

$$y - (3y + 11) = 5(y - 2) - 8y$$

Distribute the minus sign. $\quad y - 3y - 11 = 5y - 10 - 8y \quad$ Distribute 5.

Combine like terms. $\qquad -2y - 11 = -3y - 10 \qquad$ Combine like terms.

$$\underline{+3y \qquad\quad +3y} \qquad\qquad \text{Add } 3y \text{ on both sides.}$$

Note: The coefficient of $-3y$ is less than the coefficient of $-2y$, so we chose to clear $-3y$ by adding $3y$.

$$y - 11 = \quad 0 \ - 10$$
$$y - 11 = -10$$
$$\underline{+11 \quad +11} \qquad\qquad \begin{matrix}\text{Add 11 on both} \\ \text{sides to isolate } y.\end{matrix}$$
$$y + 0 = \quad 1$$
$$y = 1$$

Check Replace y in the original equation with 1 and verify that the equation is true. We will leave this to the reader.

Following is an outline for solving equations based on everything we've learned so far.

PROCEDURE SOLVING LINEAR EQUATIONS

To solve linear equations requiring the addition principle only:

1. Simplify both sides of the equation as needed.
   a. Distribute to clear parentheses.
   b. Combine like terms.
2. Use the addition principle so that all variable terms are on one side of the equation and all constants are on the other side. Then combine like terms.

Tip: Clear the variable term that has the lesser coefficient to avoid negative coefficients.

YOUR TURN Solve and check.

    a. $7n - (n - 3) = 4(n + 4) + n$ **b.** $11 + 3(5m - 2) = 7(2m - 1) + 9$

Number of Solutions

In general, a linear equation in one variable has only one real number solution. However, there are two special cases that we need to consider. First, if an equation is an identity, as in Example 7, every real number is a solution. Second, as we'll see in Example 8, some linear equations have no solution.

Equations with an Infinite Number of Solutions

EXAMPLE 7 Solve and check $5(2t + 1) - 8 = 4t - 3(1 - 2t)$.

Solution Simplify both sides of the equation. Then isolate the variable.

$$5(2t + 1) - 8 = 4t - 3(1 - 2t)$$

Distribute 5. $10t + 5 - 8 = 4t - 3 + 6t$ Distribute -3.

Combine like terms. $10t - 3 = 10t - 3$ Combine like terms.

$$\frac{-10t \qquad -10t}{0 - 3 = 0 - 3}$$

$$-3 = -3$$

Note: We can stop here because the equation is obviously an identity. However, if we continue and apply the addition principle, the equation is still an identity.

Because the equation is an identity, every real number is a solution.

Check Every real number is a solution for an identity, so any number we choose should check in the original equation. We will choose to test the number 1.

$$5(2t + 1) - 8 = 4t - 3(1 - 2t)$$
$$5(2(1) + 1) - 8 \stackrel{?}{=} 4(1) - 3(1 - 2(1))$$ Replace t with 1.
$$5(3) - 8 \stackrel{?}{=} 4 - 3(-1)$$
$$15 - 8 \stackrel{?}{=} 4 + 3$$
$$7 = 7$$

The equation is true, so 1 is a solution, which supports, but does not prove, our conclusion that the equation is an identity. To be more certain, we could choose another number to check in the equation (remember every number should work). We will leave this to the reader.

Conclusion If, after simplifying, the equation is an identity, then every real number is a solution to the equation.

Equations with No Solution

Some linear equations do not have a real number solution. Linear equations of this type can be recognized by the fact that after the expressions on each side of the equation have been simplified, the variable terms will match but the constant terms will not. For example, $2x - 5 = 2x - 3$ is an equation with no real number solution. Watch what happens when we try to solve the equation:

EXAMPLE 8 Solve and check $2x - 5 = 2x - 3$.

Solution $2x - 5 = 2x - 3$

$$\frac{-2x \qquad -2x}{0 - 5 = 0 - 3}$$ Subtract $2x$ from both sides.

$$-5 = -3$$ **Note:** This final statement is a contradiction.

Because the final statement is false, we can conclude that there is no solution.

Check Because the variable terms, $2x$, are identical on both sides of the equal sign, replacing x with any number will yield an identical product. Since we subtract 5 from that product on the left side and we subtract 3 from that product on the right side, the equation can never be true. Therefore, there is no real number solution.

Conclusion If, after simplifying, the variable terms clear out of the equation and the resulting equation is a contradiction, then there is no real number solution.

OBJECTIVE 4 Solve application problems. Now consider some problems that can be solved using the addition principle.

EXAMPLE 9 Laura wants to buy a car stereo that costs $275. She currently has $142. How much more is needed?

Understand We are given the total required and the amount she currently has, and we must find how much she needs.

Plan Let x represent the amount Laura needs. We will write an equation, then solve.

Execute

Current amount + needed amount = 275

$$142 \quad + \quad x \quad = 275$$

$$142 + x = 275$$

$$\underline{-142 \qquad -142} \qquad \text{Subtract 142 from both sides}$$
$$0 + x = 133 \qquad \text{to isolate } x.$$
$$x = 133$$

Answer Laura needs $133 to buy the stereo.

Check Does $142 plus the additional $133 equal $275?

$$142 + 133 \stackrel{?}{=} 275$$
$$275 = 275 \qquad \text{It checks.}$$

YOUR TURN

Daryl has a balance of −$568 on a credit card. How much must he pay to bring his balance to −$480?

ANSWER

$88

EXAMPLE 10 The following figure shows the front-view design of a bookcase. Find the missing length.

Understand In the drawing, the sum of the dimensions on the left side must be equal to 32 inches.

Plan Let d represent the unknown distance. We will write an equation and solve for d.

Note: As in Example 2(b), since we are subtracting a fraction form on both sides, we are writing the subtraction horizontally instead of vertically. Use whichever approach, horizontal or vertical, you prefer.

Execute

$$12\frac{1}{2} + 10\frac{3}{4} + d = 32$$

$$23\frac{1}{4} + d = 32 \qquad \text{Combine like terms.}$$

$$23\frac{1}{4} - 23\frac{1}{4} + d = 32 - 23\frac{1}{4} \qquad \text{Subtract } 23\frac{1}{4} \text{ from both sides to isolate } d.$$

$$d = 8\frac{3}{4}$$

Answer The unknown length is $8\frac{3}{4}$ inches.

Check Verify that the sum of the three lengths on the left side is equal to 32 inches. We will leave this to the reader.

YOUR TURN

In researching the grades given in a particular course during one semester, a department chair finds that 89 students received an A, 154 received a B, and 245 received a C. If the department does not give Ds and there were 645 students that initially enrolled in the course, how many students did not receive a passing grade in the course?

ANSWER

157 students

1. In your own words, explain what the addition principle says.

2. How do you use the addition principle to clear a positive term?

3. How do you use the addition principle to clear a negative term?

4. If an equation has like terms on the same side of the equation, what should you do first?

5. If an equation has variable terms on both sides of the equation, what should you do first?

6. How do you handle parentheses in an equation?

For Exercises 7–20, determine whether the given equation is linear.

7. $4y + 7 = 2y - 8$
8. $6x = 64$
9. $m^2 + 4 = 20$
10. $w^2 + 5w - 2 = 4w - 1$

11. $7u + u^2 = 16$
12. $6n - 9n = 4n + 6$
13. $4x - 8 = 2(x + 3)$
14. $5(t - 2) + 7 = 3t - 1$

15. $6x - y = 11$
16. $y = x^2 + 4$
17. $x^2 + y^2 = 1$
18. $y = 8$

19. $2c = 3d + 5$
20. $7 - a = 4(a + 2)$

For Exercises 21–56, solve and check.

21. $x - 7 = 2$
22. $a + 8 = 30$
23. $-16 = y + 9$
24. $-24 = n - 11$

25. $m + \dfrac{7}{8} = \dfrac{4}{5}$
26. $\dfrac{3}{4} = c - \dfrac{2}{3}$
27. $-\dfrac{5}{8} = y - \dfrac{1}{6}$
28. $k + \dfrac{5}{9} = -\dfrac{1}{3}$

29. $15.8 + y = 7.6$
30. $b + 4.8 - 9.2 = 5.4$
31. $-2.1 = n - 7.5 - 0.8$
32. $x + 0.4 - 1.6 = -12.5$

33. $18 = 7x - 3 - 6x$ **34.** $2z + 6 - z = 5 - 9$ **35.** $5m = 4m + 7$ **36.** $-6y = -7y - 8$

37. $7x - 9 = 6x + 4$ **38.** $12y + 22 = 11y - 3$ **39.** $-2y - 11 = -3y - 5$ **40.** $4t - 9 = 5t - 1$

41. $2.6 + 7a + 5 = 8a - 5.6$ **42.** $9c + 4.8 = 7.5 + 4.8 + 8c$

43. $7x - 2 + x = 10x - 1 - x$ **44.** $2.5y - 3.4 - 1.2y = 6.7 - 10.1 + 1.3y$

45. $8(h + 6) - 7h = 14 - 8$ **46.** $19 - 3(m + 4) + 4m = 42$

47. $5 - \dfrac{1}{2}(4x + 6) = -3x - 8$ **48.** $\dfrac{2}{3}(6b - 9) + 6 = 3b - 21$

49. $5y - (3y + 7) = y + 9$ **50.** $-15 - 2x = 16 - (3x - 9)$

51. $1.5(x + 8) - (6.2 + 0.5x) = -7 + 4$ **52.** $15 - (3c + 10) + 8c = -2(c + 8) + 6(c - 1)$

53. $20z + \dfrac{2}{3}(6z - 48) - 9z = 14z + 0.2(50z - 250) - 9z$

54. $6b - 1.5(8 + 2b) + 12 = 5b - \dfrac{1}{8}(24b + 48)$

55. $8(m - 4) + 20(m - 5) = 4m + 5(m - 20) - 32 + 19m$

56. $\dfrac{4}{5}(10x + 15) - 20 = 2x + \dfrac{3}{4}(8x + 4)$

For Exercises 57–68, translate to an equation, then solve.

57. Latonia is planning to buy a new car. The down payment is $2373. She has $1947 saved. How much more does she need?

58. Kent owes $12,412 on his Visa card. What payment should he make to get the balance to −$10,500?

59. Robert knows the distance from his home to work is 42 miles. Unfortunately, he gets a flat tire 16 miles from work. How far did Robert drive before his flat tire?

60. A patient must receive 350 cc of a medication in three injections. He has received two injections at 110 cc each. How much should the third injection be?

61. Susan is playing Yahtzee. The "chance" score is found by adding the value shown on five dice. If she has a total of 23 on four of the dice, what does the fifth die need to be so that her score will be 28?

62. On May 16, Nikki's balance in her checking account was $1741.62. Afterwards, she writes four checks and doesn't record the amount of the last check. The first is for $16.82, the second is for $150.88, and the third is for $192.71. On May 21 she finds her balance to be $1286.65. If the checks were the only transactions that could have cleared the bank during that time, what was the amount of the fourth check?

63. The perimeter of the trapezoid shown is 67.2 centimeters. Find the length of the missing side.

64. The perimeter of the triangle shown is $84\frac{1}{2}$ inches. Find the length of the missing side.

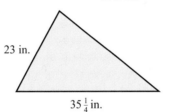

65. In the following blueprint, what is the distance x?

66. John sells scanning devices. The company quota is set at $8500 each month. The spreadsheet that follows shows John's sales as of the end of the second week of April. How much more does John need to sell to make the quota? Do you think John will make the quota? Why?

Date	Item no.	Quantity	Price per unit
4/2	32072	1	$900
4/9	17032	3	$1500
4/13	48013	2	$1245

67. In a survey, respondents were given a statement, and they could agree, disagree, or have no opinion regarding the statement. The results indicate that $\frac{1}{3}$ of the respondents agree with the statement while $\frac{2}{5}$ disagree. What fraction had no opinion?

68. In a *Time*/CNN poll, $\frac{19}{50}$ of the respondents said that they felt that the increase in the number of divorces was due more to changes in women's attitude toward marriage, while $\frac{9}{50}$ said the increase was due more to changes in men's attitude toward marriage. The rest felt the increase was a result of both men and women equally. What fraction of the respondents felt the increase was due to men and women equally?

PUZZLE PROBLEM

In a study of an experimental medication, half of the participants are given the medication and half are given a placebo. The results indicate that $\frac{1}{3}$ of the participants had improvement in their condition, $\frac{1}{8}$ had no improvement in their condition but experienced known side effects of the medication, and the rest showed no effect at all. What fraction of the group that received the actual medication showed no discernible effects from the medication?

REVIEW EXERCISES

For Exercises 1–4, simplify.

1. $4(-8.3)$

2. $27 \div (-3)$

3. $16 - (7 + 4) + 32$

4. $\dfrac{20 + 3(2 + 4)}{-6(2) + 10}$

5. Combine like terms: $3x^2 - 6x + 5x^2 + x - 9$

6. Use the distributive property to rewrite: $4(3x - 5)$

The Multiplication Principle

OBJECTIVES

1. Solve linear equations using the multiplication principle.
2. Solve linear equations using both the addition and the multiplication principles.
3. Use the multiplication principle to clear fractions and decimals from equations.
4. Solve application problems.

OBJECTIVE **1** **Solve equations using the multiplication principle.** In Section 2.2, when we isolated the variable, the coefficient of that variable was always 1. What if we ended up with a coefficient other than 1? Look at the equation $6n - 11 = 4n + 3$. Using the methods of Section 2.2, we have:

$$
\begin{array}{r}
6n - 11 = 4n + 3 \\
\underline{-4n \qquad -4n} \\
2n - 11 = 0 + 3 \\
\underline{+11 \qquad +11} \\
2n + 0 = \quad 14 \\
2n = \quad 14
\end{array}
$$

Note: The coefficient is not equal to 1.

How can we solve $2n = 14$? To isolate n we must clear the coefficient 2 so that we have $1n$, or simply n. Because coefficients multiply variables, we must undo that multiplication using its inverse operation, division. If we are going to divide, we must divide on both sides to keep the equation balanced. We write:

$$2n \div 2 = 14 \div 2 \text{ or the more popular form: } \frac{2n}{2} = \frac{14}{2}$$

Notice the coefficient 2 divides out.

The 2s divide out leaving $1n$, which can be simplified as n.

$$
\frac{\cancel{2}n}{\cancel{2}} = \frac{14}{2} \\
1n = 7 \\
n = 7
$$

You might remember in Chapter 1 we discussed that multiplicative inverses are numbers whose product is 1. Multiplying the coefficient by its multiplicative inverse would be another way to clear the coefficient. In the above case, if we multiply the coefficient 2 by its multiplicative inverse $\frac{1}{2}$, we get the desired $1n$.

$$\frac{1}{2} \cdot 2n = 14 \cdot \frac{1}{2}$$

$$\frac{1}{\cancel{2}_1} \cdot \frac{\cancel{2}^1 n}{1} = \frac{\cancel{14}^7}{1} \cdot \frac{1}{\cancel{2}_1}$$

$$1n = 7$$

$$n = 7$$

Note: Throughout the rest of the book, we will omit this step of writing the resulting 1 coefficient.

CONNECTION ───────

Remember that when we divide by a fraction, we write an equivalent multiplication using the multiplicative inverse. Multiplying by the multiplicative inverse is the same as dividing.

If we manipulate an equation using the addition principle and we still have a coefficient other than 1, we can clear that coefficient by dividing (or multiplying by its multiplicative inverse). This is the main purpose of the *multiplication principle*. The multiplication principle says that we can multiply (or divide) both sides of an equation by the same amount without affecting its solution(s).

RULE **THE MULTIPLICATION PRINCIPLE OF EQUALITY**

If $a = b$, then $ac = bc$ is true for all real numbers a, b, and c, where $c \neq 0$.

PROCEDURE **USING THE MULTIPLICATION PRINCIPLE**

To use the multiplication principle to clear a coefficient:

1. Determine the coefficient you want to clear.
2. Multiply both sides by the multiplicative inverse of that coefficient, or divide both sides by the coefficient.

EXAMPLE 1 Solve and check $-\frac{4}{5}m = \frac{6}{7}$.

Solution $-\frac{4}{5}m = \frac{6}{7}$

$$-\frac{\cancel{5}^1}{\cancel{4}_1} \cdot -\frac{\cancel{4}^1}{\cancel{5}_1}m = \frac{\cancel{6}^3}{7} \cdot -\frac{5}{\cancel{4}_2}$$

Clear the coefficient $-\frac{4}{5}$ by multiplying both sides by its multiplicative inverse, $-\frac{5}{4}$.

$$m = -\frac{15}{14}$$

Check $-\frac{4}{5}m = \frac{6}{7}$

$$-\frac{\cancel{4}^2}{\cancel{5}_1}\left(-\frac{\cancel{15}^3}{\cancel{14}_7}\right) \stackrel{?}{=} \frac{6}{7}$$

Replace m in the original equation with $-\frac{15}{14}$ and verify that the equation is true.

$$\frac{6}{7} = \frac{6}{7}$$

True, therefore $-\frac{15}{14}$ is correct.

a. $-56 = -7t$ **b.** $-\dfrac{3}{8}y = 9$ **c.** $0.4x = -0.92$

OBJECTIVE   **2** **Solve linear equations using both the addition and multiplication principles.**   Now let's put the multiplication principle together with the addition principle. We follow the same outline as in Section 2.2, with one new step: using the multiplication principle to clear any remaining coefficient at the end.

PROCEDURE SOLVING LINEAR EQUATIONS

To solve linear equations in one variable:

1. Simplify both sides of the equation as needed.
   a. Distribute to clear parentheses.
   b. Combine like terms.
2. Use the addition principle so that all variable terms are on one side of the equation and all constants are on the other side. (Clear the variable term with the lesser coefficient.) Then combine like terms.
3. Use the multiplication principle to clear the remaining coefficient.

EXAMPLE 2   Solve and check $-3x - 11 = 7$.

Solution Use the addition principle to separate the variable term and constant terms, then use the multiplication principle to clear the remaining coefficient.

$$
\begin{array}{rl}
-3x - 11 = & 7 \qquad \text{Add 11 on both sides to isolate the } -3x \text{ term.} \\
\underline{+11 \quad +11} & \\
-3x + 0 = & 18 \\
\dfrac{-3x}{-3} = & \dfrac{18}{-3} \qquad \text{Divide both sides by } -3 \text{ to clear the } -3 \text{ coefficient.} \\
x = & -6
\end{array}
$$

Note: Multiplying both sides by $-\dfrac{1}{3}$ would also work:

$$-\dfrac{1}{\overset{1}{3}} \cdot -\dfrac{\overset{1}{3x}}{1} = \dfrac{\overset{6}{18}}{1} \cdot -\dfrac{1}{\overset{1}{3}}$$

$$x = -6$$

Check
$$
\begin{array}{rl}
-3x - 11 = & 7 \\
-3(-6) - 11 \overset{?}{=} & 7 \qquad \text{Replace } x \text{ in the original equation with } -6 \text{ and} \\
18 - 11 \overset{?}{=} & 7 \qquad \text{verify that the equation is true.} \\
7 = & 7
\end{array}
$$

True, therefore -6 is correct.

YOUR TURN   Solve and check.

a. $6y - 19 = -22$ **b.** $-12 = 20 - 8t$

Solving Equations with Variable Terms on Both Sides

Recall that when variable terms appear on both sides of the equal sign, we use the addition principle to get the variable terms on one side of the equal sign and the constant terms on the other side.

EXAMPLE 3 Solve and check $8y - 5 = 2y - 29$.

Solution Use the addition principle to get the variable terms on one side of the equation and the constant terms on the other side, then use the multiplication principle to clear the remaining coefficient.

$$8y - 5 = 2y - 29$$

$8y - 5 = 2y - 29$ $\underline{-2y \qquad -2y}$ $6y - 5 = 0 - 29$	Subtract $2y$ from both sides (2 is the lesser coefficient).
$6y - 5 = -29$ $\underline{+5 \qquad +5}$ $6y + 0 = -24$	Add 5 to both sides to isolate the $6y$ term.
$\dfrac{6y}{6} = \dfrac{-24}{6}$ $y = -4$	Divide both sides by 6 to clear the 6 coefficient.

Check $8y - 5 = 2y - 29$

$8(-4) - 5 \stackrel{?}{=} 2(-4) - 29$ $-32 - 5 \stackrel{?}{=} -8 - 29$ $-37 = -37$	Replace y in the original equation with -4 and verify that the equation is true.

True, therefore -4 is correct.

Simplifying First

Recall that when an equation contains parentheses or like terms that appear on the same side of the equal sign, we first clear the parentheses and combine like terms. Then we use the addition principle to separate the variable terms and constant terms.

EXAMPLE 4 Solve and check $17 - (4n - 5) = 9n - 3(n + 7)$.

Solution

$17 - (4n - 5) = 9n - 3(n + 7)$	
$17 - 4n + 5 = 9n - 3n - 21$	Distribute to clear parentheses.
$22 - 4n = 6n - 21$	Combine like terms.
$22 - 4n = 6n - 21$ $\underline{+4n \quad +4n}$ $22 + 0 = 10n - 21$	Add $4n$ to both sides ($-4n$ is the lesser coefficient).
$22 = 10n - 21$ $\underline{+ 21 = \qquad + 21}$ $43 = 10n + 0$	Add 21 on both sides to isolate $10n$.
$\dfrac{43}{10} = \dfrac{10n}{10}$	Divide both sides by 10 to clear the 10 coefficient.
$4.3 = n$	

Check $17 - (4n - 5) = 9n - 3(n + 7)$
$17 - (4(4.3) - 5) \stackrel{?}{=} 9(4.3) - 3(4.3 + 7)$
$17 - (17.2 - 5) \stackrel{?}{=} 38.7 - 3(11.3)$
$17 - 12.2 \stackrel{?}{=} 38.7 - 33.9$
$4.8 = 4.8$

Replace *n* in the original equation with 4.3 and verify that the equation is true.

True, therefore 4.3 is correct.

YOUR TURN Solve and check.

a. $3t - 10 + 9t = 17 + 4t - 3$ **b.** $13 - 6(x + 2) = 3x - (11x + 5)$

OBJECTIVE ③ Use the multiplication principle to clear fractions and decimals from equations. To clear fractions and decimals from an equation, we can use the multiplication principle of equality. Though equations can be solved without clearing the fractions or decimals, most people find equations that contain only integers easier to solve.

Clearing Fractions in an Equation

If the equation contains fractions, we multiply both sides by a number that will clear all the denominators. We could multiply both sides by any multiple of the denominators, however, using the LCD (least common denominator) results in the simplest equations.

EXAMPLE 5 Solve and check $\frac{1}{3}x - \frac{3}{4} = \frac{5}{6}x + 2$.

Solution Clear the fractions by multiplying through by the LCD. The LCD of 3, 4, and 6 is 12.

Note: We have chosen to clear the fractions in our first step. Remember, you can clear the fractions at *any* step in solving the equation, or not at all.

$12\left(\frac{1}{3}x - \frac{3}{4}\right) = \left(\frac{5}{6}x + 2\right)12$

Clear the fractions by multiplying both sides by the LCD, 12.

$\frac{\overset{4}{\cancel{12}}}{1} \cdot \frac{1}{\underset{1}{\cancel{3}}}x - \frac{\overset{3}{\cancel{12}}}{1} \cdot \frac{3}{\underset{1}{\cancel{4}}} = \frac{\overset{2}{\cancel{12}}}{1} \cdot \frac{5}{\underset{1}{\cancel{6}}}x + 12 \cdot 2$

Distribute 12, then divide out the denominators.

$4x - 9 = 10x + 24$

$4x - 9 = 10x + 24$

$\underline{-4x \qquad\qquad -4x}$

Subtract 4x from both sides.

$0 - 9 = 6x + 24$

$-9 = 6x + 24$

$\underline{-24 \qquad\quad -24}$

Subtract 24 from both sides.

$-33 = 6x + 0$

$\dfrac{-33}{6} = \dfrac{6x}{6}$

Divide both sides by 6 to clear the coefficient.

$-\dfrac{11}{2} = x$

Simplify.

Check
$$\frac{1}{3}x - \frac{3}{4} = \frac{5}{6}x + 2$$

$$\frac{1}{3}\left(-\frac{11}{2}\right) - \frac{3}{4} \overset{?}{=} \frac{5}{6}\left(-\frac{11}{2}\right) + 2 \qquad$$ Replace x in the original equation with $-\frac{11}{2}$ and verify that the equation is true.

$$-\frac{11}{6} - \frac{3}{4} \overset{?}{=} -\frac{55}{12} + 2$$

$$-\frac{22}{12} - \frac{9}{12} \overset{?}{=} -\frac{55}{12} + \frac{24}{12} \qquad$$ Write equivalent fractions with their LCD.

$$-\frac{31}{12} = -\frac{31}{12}$$

True, therefore $-\frac{11}{2}$ is correct.

Clearing Decimals in an Equation

The multiplication principle can be used to clear decimals in a similar way. Because multiplying by a power of 10 will cause the decimal point to move to the right, we can clear decimals by multiplying both sides of the equation by an appropriate power of 10. The power of 10 we use depends on the decimal number with the most places.

EXAMPLE 6 Solve and check $0.2(y - 6) = 0.48y + 3$.

Solution Decimals can be cleared at any step in the process of solving an equation. In this case, we will distribute to clear the parentheses first, then clear the decimals.

$$0.2y - 1.2 = 0.48y + 3 \qquad$$ Distribute to clear parentheses.

$$100(0.2y - 1.2) = (0.48y + 3)100$$

$$20y - 120 = 48y + 300$$

$$20y - 120 = 48y + 300$$

$$\underline{-20y \qquad\qquad -20y}$$

$$0 - 120 = 28y + 300$$

$$-120 = 28y + 300$$

$$\underline{-300 \qquad\qquad -300}$$

$$-420 = 28y + 0$$

$$\frac{-420}{28} = \frac{28y}{28}$$

$$-15 = y$$

Note: The number 0.48 has two places, which is more places than any of the other decimal numbers, so we multiply both sides by 100 to clear the decimals.

CONNECTION

Clearing decimals and fractions uses the same process. Decimals represent fractions with denominators that are powers of 10. For example, $0.48 = \frac{48}{100}$ and $0.2 = \frac{2}{10}$. The LCD for these fractions is 100.

Check $\quad 0.2(y - 6) = 0.48y + 3$

$\qquad 0.2(-15 - 6) \overset{?}{=} 0.48(-15) + 3 \qquad$ Replace y in the original equation with -15 and verify that the equation is true.

$\qquad 0.2(-21) \overset{?}{=} -7.2 + 3$

$\qquad -4.2 = -4.2$

True, therefore -15 is correct.

We can amend the outline for solving equations to include clearing fractions and decimals. Remember that this process is optional.

PROCEDURE SOLVING LINEAR EQUATIONS

To solve linear equations in one variable:

1. Simplify both sides of the equation as needed.
   a. Distribute to clear parentheses.
   b. Clear fractions or decimals by multiplying through by the LCD. In the case of decimals, the LCD is a power of 10 determined by the decimal number with the most places. (Clearing fractions and decimals is optional.)
   c. Combine like terms.
2. Use the addition principle so that all variable terms are on one side of the equation and all constants are on the other side. (Clear the variable term with the lesser coefficient.) Then combine like terms.
3. Use the multiplication principle to clear the remaining coefficient.

YOUR TURN   **Solve and check.**

a. $\dfrac{1}{3}(y - 2) = \dfrac{3}{5}y + \dfrac{1}{6}$ **b.** $2.5n - 1.04 = 0.15n - 0.1$

OBJECTIVE **4** **Solve application problems.**   In Section 2.1, we began solving problems using formulas. Recall that to use a formula, we replace the variables with the corresponding given numbers, then solve for the unknown amount.

EXAMPLE **7** The perimeter of the figure below is 188 feet. Find the width and length.

Understand   The width is represented by w and the length is represented by $w + 7$. We are given the perimeter, so we can use the formula $P = 2l + 2w$.

Plan   In the perimeter formula, replace P with **188**, l with $w + 7$, and solve for w.

Execute

$$P = 2l + 2w$$
$$188 = 2(w + 7) + 2w$$

$$188 = 2w + 14 + 2w \qquad \text{Distribute.}$$
$$188 = 4w + 14 \qquad \text{Combine like terms.}$$
$$188 = 4w + 14$$
$$\underline{-14 \qquad\quad -14} \qquad \text{Subtract 14 from both sides.}$$
$$174 = 4w + 0$$
$$\frac{174}{4} = \frac{4w}{4} \qquad \text{Divide both sides by 4.}$$
$$43.5 = w$$

Answer The width is 43.5 feet. To find the length, we evaluate the expression that represents the length, $w + 7$ with $w = 43.5$.

$$\text{Length} = 43.5 + 7 = 50.5 \text{ feet}$$

Check First, verify that the length is 7 more than the width: $50.5 = 7 + 43.5$. Second, verify that a rectangle with a length of 50.5 feet and a width of 43.5 feet has a perimeter of 188 feet.

$$P = 2l + 2w$$
$$P = 2(50.5) + 2(43.5)$$
$$P = 101 + 87$$
$$P = 188$$

It checks.

EXAMPLE 8 The total material allotted for the construction of a metal box is 4754 square inches. The length is to be 30 inches and the width is to be 28 inches. Find the height.

Understand The material will be used to create the outer shell that is the box, which means that 4754 square inches is the surface area of the box. The formula for the surface area of a box is $SA = 2lw + 2lh + 2wh$.

Plan Replace SA with 4754, l with 30, w with 28, and solve for h.

Execute $SA = 2lw + 2lh + 2wh$

$$4754 = 2(30)(28) + 2(30)h + 2(28)h$$
$$4754 = 1680 + 60h + 56h \qquad \text{Simplify.}$$
$$4754 = 1680 + 116h \qquad \text{Combine like terms.}$$
$$4754 = 1680 + 116h$$
$$\underline{-1680 \quad -1680} \qquad \text{Subtract 1680 from both sides.}$$
$$3074 = \quad 0 + 116h$$
$$\frac{3074}{116} = \frac{116h}{116} \qquad \text{Divide both sides by 116.}$$
$$26.5 = h$$

Answer The height must be 26.5 inches.

Check Does a 30-inch by 28-inch by 26.5-inch box have a surface area of 4754 square inches?

$$4754 \stackrel{?}{=} 2(30)(28) + 2(30)(26.5) + 2(28)(26.5)$$
$$4754 \stackrel{?}{=} \quad 1680 \quad + \quad 1590 \quad + \quad 1484$$
$$4754 = 4754$$

It checks.

YOUR TURN

The desired surface area for a case is 98 square feet. If the length of the case is to be 4 feet and the height is to be 6 feet, what will the width be?

ANSWER

2.5 ft.

2.3 EXERCISES

For Extra Help

InterAct Math

Addison-Wesley Math Tutor Center

Video 3/DVT 1

MyMathLab

MathXL

1. In your own words, explain what the multiplication principle says.

2. How do you use the multiplication principle to clear a coefficient that is an integer?

3. How do you use the multiplication principle to clear a coefficient that is a fraction?

4. To clear a negative coefficient, do you multiply/divide by a positive or negative number? Why?

5. If an equation contains fractions, how do you transform the equation so that it contains only integers?

6. If an equation contains decimal numbers, how do you transform the equation so that it contains only integers?

For Exercises 7–32, solve and check.

7. $6x = -18$

8. $-5y = 20$

9. $5n - 2n = 21$

10. $7t - 3t = 20$

11. $-6a - a = 35$

12. $c + 2c + 3 + 4c = 24$

13. $5y - 7 = 2y + 13$

14. $6 - 12m = -20m + 22$

15. $\dfrac{5}{8}x - 2 = 8$

16. $7 = \dfrac{3}{4}x + 13$

17. $2(x - 3) = -6$

18. $4(5x + 7) = 13$

19. $-7b + 4(2b - 3) = 16$

20. $-3 + 6(x - 4) = 9$

21. $2(y - 33) + 3(y - 2) = 8$

22. $4m + 2 - 3(m - 5) = 0$

23. $\dfrac{3}{4}n - 2 = -\dfrac{7}{8}n + 4$

24. $-\dfrac{2}{5}t + 1 = \dfrac{3}{10}t - 3$

25. $0.3(a - 18) = 0.9a$

26. $4.2(5a + 1) - 7 = 1.4$

27. $\dfrac{1}{2}(y + 5) = \dfrac{3}{2} - y$

28. $\dfrac{1}{6}(m - 4) = \dfrac{2}{3}(m + 1) - \dfrac{1}{2}m$

29. $4 - (6x + 5) = 7 - 2(3x + 4)$

30. $-1 - (3z - 2) = -17 - 5z$

31. $8\left(p + \dfrac{1}{2}\right) = -12$

32. $-6\left(n - \dfrac{1}{3}\right) = 17$

For Exercises 33–36, the check indicates a mistake was made. Find and correct the mistake.

33.
$$
\begin{aligned}
6x - 11 &= 8x - 15 \\
-6x \quad\ \ &\quad\ -6x \\
\hline
11 &= 2x - 15 \\
+15 \quad\ \ &\quad\ +15 \\
\hline
26 &= 2x \\
\frac{26}{2} &= \frac{2x}{2} \\
13 &= x
\end{aligned}
$$

Check: $6x - 11 = 8x - 15$

$6(13) - 11 \overset{?}{=} 8(13) - 15$

$78 - 11 \overset{?}{=} 104 - 15$

$67 \neq 89$

34. $6n + 1 = -2n + 21$

$$\begin{array}{r} \underline{-1 \qquad\qquad -1} \\ 6n + 0 = -2n + 20 \\ 6n = -2n + 20 \\ \underline{+2n \qquad +2n} \\ 8n = 0 + 20 \\ \dfrac{8n}{8} = \dfrac{20}{8} \\ n = \dfrac{5}{2} \end{array}$$

Check: $6n + 1 = -2n + 21$

$$\dfrac{\overset{3}{\cancel{6}}}{1}\left(\dfrac{5}{\underset{1}{\cancel{2}}}\right) + 1 \;\overset{?}{=}\; -\dfrac{\overset{1}{\cancel{2}}}{1}\left(\dfrac{5}{\underset{1}{\cancel{2}}}\right) + 21$$

$$3 + 1 \overset{?}{=} -5 + 21$$

$$4 \neq 16$$

35. $4 - (x + 2) = 2x + 3(x - 8)$

$$\begin{array}{r} 4 - x + 2 = 2x + 3x - 24 \\ 6 - x = 5x - 24 \\ \underline{+x \qquad +x} \\ 6 + 0 = 6x - 24 \\ 6 = 6x - 24 \\ \underline{+24 \qquad\quad +24} \\ 30 = 6x + 0 \\ \dfrac{30}{6} = \dfrac{6x}{6} \\ 5 = x \end{array}$$

Check: $4 - (x + 2) = 5x + 3(x - 8)$

$$4 - (5 + 2) \overset{?}{=} 5(5) + 3(5 - 8)$$

$$4 - 7 \overset{?}{=} 25 + 3(-3)$$

$$-3 \overset{?}{=} 25 + (-9)$$

$$-3 \neq 16$$

36. $\dfrac{1}{2}n - 3 = \dfrac{2}{3}n + \dfrac{1}{4}$

$$\dfrac{\overset{6}{\cancel{12}}}{1}\cdot\dfrac{1}{\underset{1}{\cancel{2}}}n - 3 = \dfrac{\overset{4}{\cancel{12}}}{1}\cdot\dfrac{2}{\underset{1}{\cancel{3}}}n + \dfrac{\overset{3}{\cancel{12}}}{1}\cdot\dfrac{1}{\underset{1}{\cancel{4}}}$$

$$\begin{array}{r} 6n - 3 = 8n + 3 \\ \underline{-6n \qquad -6n} \\ 0 - 3 = 2n + 3 \\ \underline{-3 \qquad\quad -3} \\ -6 = 2n + 0 \\ \dfrac{-6}{2} = \dfrac{2n}{2} \\ -3 = n \end{array}$$

Check: $\dfrac{1}{2}n - 3 = \dfrac{2}{3}n + \dfrac{1}{4}$

$$\dfrac{1}{2}\left(-\dfrac{3}{1}\right) - 3 \;\overset{?}{=}\; \dfrac{2}{\underset{1}{\cancel{3}}}\left(-\dfrac{\overset{1}{\cancel{3}}}{1}\right) + \dfrac{1}{4}$$

$$-\dfrac{3}{2} - 3 \overset{?}{=} -2 + \dfrac{1}{4}$$

$$-\dfrac{3}{2} - \dfrac{6}{2} \overset{?}{=} -\dfrac{8}{4} + \dfrac{1}{4}$$

$$-\dfrac{9}{2} \neq -\dfrac{7}{4}$$

For Exercises 37–50, solve for the unknown amount.

37. The area of a deck is known to be 140 square feet. Find the length if the width must be 10 feet. (Use $A = lw$)

38. A crate manufacturer receives an order for a crate with a volume of 128 cubic feet. The crate must be 5 feet, 4 inches by 6 feet. What must be the crate's height? (Use $V = lwh$)

39. The area of the metal plate shown in the figure is 45.6 square centimeters. What is the length of the side labeled h? (Use $A = \frac{1}{2}bh$)

h

4.8 cm

40. The Worthy family is planning a 465-mile trip. If they travel at an average speed of 62 miles per hour, what will be their travel time? (Use $d = rt$)

41. A fence is to be installed around a rectangular field. The field's perimeter is 212 feet. Find the width if the length of the field is 54 feet. (Use $P = 2l + 2w$)

42. A building has a perimeter of 241 feet. The length is 16 feet more than the width as shown in the figure. Find the dimensions of the building.

w

$w + 16$

43. The area of the plot of land shown in the figure is 7500 square feet. What is the length of the side labeled b? (Use $A = \frac{1}{2}h(a + b)$)

90 ft.

75 ft.

b

44. The area of the L-shaped room shown is 321 square feet. Find all the missing dimensions.

w

? 12 ft.

15 ft.

$w + 2$

?

45. A small storage box is designed so that its surface area is 105 square inches. If the length is 7.5 inches and the width is 4 inches, find the height. (Use $SA = 2lw + 2lh + 2wh$)

46. Jerry has 1992 square inches of wood to build a case for storing musical equipment. He wants the length to be 22 inches and the width to be 18 inches. What will be the height? (Use $SA = 2lw + 2lh + 2wh$)

47. The circumference of Earth along the equator is approximately 40,053.84 kilometers. What is the equatorial radius? (Use $C = 2\pi r$)

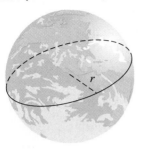

48. A giant sequoia is measured to have a circumference of 26.4 meters. What is the diameter? (Use $C = \pi d$)

49. An engineer is designing a chemical storage tank in the shape of a cylinder. The volume of the tank is to be 60 cubic meters and the radius is to be 2 meters. Find the height. (Use $V = \pi r^2 h$)

50. The liquid in a standard soda can occupies a volume of 355 cubic centimeters. If the diameter is 6 centimeters, find the height of the liquid inside the can. (Use $V = \pi r^2 h$)

The formula $V = ir$ describes the voltage, V, in a circuit when the current, i, passes through a resistance, r. Voltage is measured in volts (V), current is measured in amps (A), and resistance is measured in ohms (Ω). Use the formula for voltage to solve Exercises 51 and 52.

51. The voltage in a circuit is measured to be -76 V. If the resistance is 8 Ω, find the current.

52. A technician applies 15 V to a circuit. The current is measured to be 2.5 A. What is the resistance?

The formula $F = ma$ describes the force that an object with mass m experiences when accelerated an amount a. If acceleration is measured in meters per second squared (m/sec.²), and the mass is measured in kilograms (kg), then the force is in units of newtons (N). Use the force formula to solve Exercises 53–56.

53. A car with a mass of 1200 kg exerts a force of 4800 N. Find the acceleration.

54. An object weighs 44.1 N, that is, it exerts a downward force of -44.1 N. The object is dropped so that it accelerates at -9.8 m/sec.². Find the mass of the object.

55. The Liberty Bell weighs 9265.9 N, so that its downward force is −9265.9 N. If the acceleration due to gravity is about −9.8 m/sec.², what is the mass of the bell?

56. Jacob weighs 160 lb. (a downward force of −160 lb.). If the acceleration due to gravity is about −32.2 ft./sec.², what is his mass? (The mass will be in terms of the unit *slugs*.)

57. The formula $w = \dfrac{s}{f}$ describes the wavelength w of a musical note if f is the frequency and s is the speed of sound. A note is sounded with a wavelength of 0.94 m/cycle and a frequency of 440 Hz. Find the speed of sound.

58. The formula $C = 0.3m + 25$ describes the total cost of using a digital phone over 100 minutes, where m represents the number of minutes beyond 100 minutes. Pam's bill shows that she used the phone for 187 minutes. What is the total cost of her use of the phone?

59. A rental car agency rents an SUV for a weekend using the formula $C = 0.27m + 200$, where m represents the number of miles. Laura rents an SUV from the company over a weekend that the company is offering 100 free miles. If she returns her rental car and has driven 421 miles, what is the total cost of the rental?

60. The formula $d = v_i t + \dfrac{1}{2}at^2$ describes the distance d an object travels if the object has an initial velocity (speed) v_i, an acceleration of a, for a time t. An object travels 135 feet after being accelerated for 3 seconds with an acceleration of 20 ft./sec.². Find the object's initial velocity.

REVIEW EXERCISES

1. Simplify: $-6 + 3(5^2 + 2) - (6 - 1)$

2. Combine like terms: $-3.2x^2 + 0.4x + 7x^2 - 8.03x + 1.5$

3. Use the distributive property to rewrite: $-\dfrac{3}{4}(6x + 8)$

For Exercises 4–6, solve and check.

4. $6 + x = -14 + 2x$

5. $-\dfrac{2}{5}x = 10$

6. $2x + 3(x - 3) = 10x - (3x - 11)$

2.4 ～ Applying the Principles to Formulas

OBJECTIVE ① Isolate a variable in a formula using the addition and multiplication principles.

So far when using formulas we have substituted given numbers for all the variables except one. We then used the addition and multiplication principles of equality to isolate that variable. However, we can actually isolate a particular variable in a formula without being given numbers for the other variables. We follow the same steps as if we had been given numbers.

Consider the area formula, $A = lw$. If we were given values for A and l and asked to isolate w, or solve for w, we would divide both sides by the value of l. We can perform this manipulation symbolically.

$$\frac{A}{l} = \frac{\cancel{l}w}{\cancel{l}} \qquad \text{To isolate } w \text{, we clear } l \text{ by dividing both sides by } l.$$

$$\frac{A}{l} = w$$

This development suggests the following procedure.

PROCEDURE ISOLATING A VARIABLE IN A FORMULA

To isolate a particular variable in a formula:

1. Treat all other variables like constants.
2. Isolate the desired variable using the outline for solving equations.

Isolate a Variable Using the Addition Principle

Recall that the addition principle is used to clear terms that are added or subtracted.

EXAMPLE 1 Isolate R in the formula $P = R - C$.

Of Interest

The formula $P = R - C$ is used to calculate profit, given revenue, R, and cost, C.

Solution

$$
\begin{array}{rl}
P = R & - C \\
\underline{+C \qquad +C} & \\
P + C = R & + 0 \\
P + C = R &
\end{array}
$$

To isolate R we must clear C. Because C is subtracted from R, we add C on both sides.

Note: Because P and C are not like terms, they cannot be combined.

Isolate v_i in the formula that describes the velocity of an object after being accelerated from an initial velocity (v_i), $v = v_i + at$.

$v - at = v_i$

Isolate a Variable Using the Multiplication Principle

Now let's use the multiplication principle to isolate a variable in a formula.

EXAMPLE 2   **a.** Isolate w in the formula for the volume of a box, $V = lwh$.

Solution

$$\frac{V}{lh} = \frac{\cancel{l}\,w\,\cancel{h}}{\cancel{l}\,\cancel{h}}$$

To isolate w, we must clear l and h. Because l and h are multiplying w, we divide both sides by l and h.

$$\frac{V}{lh} = w$$

b. Isolate P in the formula $B = P(1 + rt)$

Solution

$$\frac{B}{1 + rt} = \frac{P\cancel{(1 + rt)}}{\cancel{1 + rt}}$$

To isolate P, we must clear $(1 + rt)$. Since $(1 + rt)$ is multiplying P, we divide both sides by $1 + rt$.

$$\frac{B}{1 + rt} = P$$

Of Interest

The formula $B = P(1 + rt)$ is used in finance to calculate the balance for a simple-interest investment or loan. The variable B represents the final balance, P represents the principle, r represents the interest rate, and t represents time.

a. Isolate b in the formula for the area of a parallelogram, $A = bh$.
b. Isolate i in the voltage formula for a series circuit with two resistors, $V = i(R + r)$.

a. $\dfrac{A}{h} = b$ **b.** $\dfrac{V}{R + r} = i$

Isolate a Variable Using Both Principles

Now consider formulas that require the use of both principles to isolate the desired variable.

EXAMPLE 3 **a.** Isolate l in the perimeter formula for rectangles, $P = 2l + 2w$.

Solution First, use the addition principle to isolate the term that contains l. Then use the multiplication principle to clear the coefficient.

$$P = 2l + 2w$$

$$\underline{-2w \qquad -2w}$$

$$P - 2w = 2l + \ 0 \qquad \text{Isolate } 2l \text{ by subtracting } 2w \text{ on both sides.}$$

$$\frac{P - 2w}{2} = \frac{2l}{2} \qquad \text{Isolate } l \text{ by dividing both sides by 2.}$$

$$\frac{P - 2w}{2} = l$$

b. Isolate d in the formula from physics, $v^2 = v_0^2 + \dfrac{2Fd}{m}$.

Solution First, use the addition principle to isolate the term that contains d. Then use the multiplication principle to clear the coefficient.

$$v^2 = v_0^2 + \frac{2Fd}{m} \qquad \text{Isolate } \frac{2Fd}{m} \text{ by subtracting } v_0^2 \text{ on both sides.}$$

$$\underline{-v_0^2 \quad -v_0^2}$$

$$v^2 - v_0^2 = 0 + \frac{2Fd}{m}$$

$$v^2 - v_0^2 = \frac{2Fd}{m}$$

$$\frac{m}{2F} \cdot \frac{v^2 - v_0^2}{1} = \frac{\cancel{m}}{\cancel{2F}} \cdot \frac{\cancel{2F}d}{\cancel{m}} \qquad \text{Clear the coefficient } \frac{2F}{m} \text{ by multiplying both sides by its reciprocal.}$$

$$\frac{m(v^2 - v_0^2)}{2F} = d$$

Of Interest

In this example, the 0 subscript is used to indicate that v_0 is an initial velocity. We have also seen v_i used to indicate initial velocity. Both forms are used in physics and engineering.

YOUR TURN

a. Isolate t in the formula that describes the velocity of an object after being accelerated from an initial velocity, $v = v_i + at$.

b. Isolate n in the formula $T = \dfrac{nt}{r} - R$.

ANSWERS

a. $\dfrac{v - v_i}{a} = t$

b. $\dfrac{r}{t}(T + R) = n$

or

$\dfrac{r(T + R)}{t} = n$

1. What would you do to isolate w in the formula $l = w - 2$? Why?

2. What would you do to isolate p in the formula $C = np$? Why?

3. To isolate y in $2x + 3y = 8$, what would you do first? Why?

4. How can you verify that your answer is correct after isolating a variable in a formula?

For Exercises 5–22, solve for the indicated variable.

5. $t - 4u = v;\ t$

6. $\dfrac{x}{2} = a + 3y;\ a$

7. $-5y = x;\ y$

8. $2n = a;\ n$

9. $2x - 3 = b;\ x$

10. $y = mx + b;\ m$

11. $ab + c = 0;\ b$

12. $3m + b = y;\ m$

13. $3x + 4y = 8;\ y$

14. $ax + by = 7;\ a$

15. $\dfrac{mn}{4} - Y = f;\ Y$

16. $5(2n + a) = bn - c;\ a$

17. $6(c + d) = 7;\ c$

18. $\dfrac{a}{4} + \dfrac{b}{4} = 10;\ a$

19. $\dfrac{3}{4} + 5a = \dfrac{n}{c};\ n$

20. $\dfrac{x}{6} + 5y = 4;\ x$

21. $\dfrac{c}{5} = \dfrac{3(x + 4)}{2};\ x$

22. $S = \dfrac{C}{1 - M};\ M$

For Exercises 23–36, solve the geometric formula for the indicated variable.

23. $P = a + b + c;\ a$

24. $a + b + c = 180;\ b$

25. $A = lw;\ l$

26. $C = 2\pi r;\ r$

27. $V = \pi r^2 h;\ r^2$

28. $V = \dfrac{4}{3}\pi r^3;\ r^3$

29. $\dfrac{C}{d} = \pi;\ d$

30. $V = \dfrac{1}{3}Bh;\ B$

31. $A = \dfrac{1}{2}bh$; b **32.** $S = \dfrac{1}{2}gt^2$; t^2 **33.** $P = 2l + 2w$; w **34.** $S = \pi r^2 - 2\pi rh$; h

35. $S = \dfrac{n}{2}(a + l)$; l **36.** $A = \dfrac{1}{2}h(a + b)$; b

For Exercises 37–42, solve the financial formula for the indicated variable.

37. $P = R - C$; C **38.** $P = R - C$; R **39.** $I = Prt$; r **40.** $I = Prt$; t

41. $A = P + Prt$; r **42.** $A = P + Prt$; t

For Exercises 43–54, solve the physics formula for the indicated variable.

43. $d = rt$; t **44.** $F = ma$; m **45.** $W = Fd$; d **46.** $K = \dfrac{1}{2}mv^2$; m

47. $v = -32t + v_0$; t **48.** $x = x_0 + vt$; v **49.** $F = \dfrac{9}{5}C + 32$; C **50.** $C = \dfrac{5}{9}(F - 32)$; F

51. $P = \dfrac{W}{t}$; t **52.** $a = \dfrac{v - v_0}{t}$; t **53.** $F = G\dfrac{Mm}{R^2}$; m **54.** $\dfrac{P_1 V_1}{T_1} = \dfrac{P_2 V_2}{T_2}$; V_2

For Exercises 55–58, find and explain the mistake in each solution.

55. $3n + 7t = 54$; isolate t

$$3n + 7t = 54$$
$$\underline{-3n \qquad\qquad -3n}$$
$$7t = 54 - 3n$$
$$7t = 54 - 3n$$
$$\underline{-7 \quad -7}$$
$$t = 47 - 3n$$

56. $\dfrac{1}{4}kt = r$; isolate t

$$\dfrac{1}{4}kt = r$$
$$\dfrac{4}{1} \cdot \dfrac{1}{4}kt = 4r$$
$$\dfrac{1}{k} \cdot kt = 4r\dfrac{k}{1}$$
$$t = 4kr$$

57. $\dfrac{3m - 2}{5} = nk$; isolate m

$$\frac{3m - 2}{5} = nk$$

$$\frac{5}{1} \cdot \frac{3m - 2}{5} = nk \cdot 5$$

$$3m - 10 = 5nk$$

$$3m - 10 = 5nk$$

$$\underline{+10 \quad +10}$$

$$3m = 5nk + 10$$

$$\frac{3m}{3} = \frac{5nk + 10}{3}$$

$$m = \frac{5nk + 10}{3}$$

58. $7(y - 5) = xv$;   isolate y

$$7(y - 5) = xv$$

$$7y - 5 = xv$$

$$7y - 5 = xv$$

$$\underline{+5 \quad +5}$$

$$\frac{7y}{7} = \frac{xv + 5}{7}$$

$$y = \frac{xv + 5}{7}$$

Collaborative Exercises

Where Does Speeding Get You?

1. The formula that relates distance, rate, and time is $d = rt$. Solve this formula for t.

2. Use the formula you found in Problem 1 to determine how long it would take you to travel 20 miles (an average daily commute) at a rate of 55 miles per hour. Note that your result will be in hours. Convert the time to minutes by multiplying by 60, and round to the nearest tenth of a minute.

3. Now use the formula from Problem 1 to determine how long it would take you to travel 20 miles at 65 miles per hour. Again, convert the time to minutes rounded to the nearest tenth of a minute.

4. How much time does speeding save you? To find this, subtract the time found in problem 3 from the time found in Problem 2.

5. Discuss in your group whether the additional time is worth the additional risk that comes with increasing the speed by 10 miles per hour.

continued

6. Complete the table, rounding all values to the nearest tenth. (You've already calculated the values for the 20-mile trip.)

Trip Distance	Time at 55 mph	Time at 65 mph	Time saved
20 miles			
30 miles			
40 miles			

7. Is the time saved on longer trips worth the additional risk of increasing speed by 10 miles per hour? Why or why not? Does the longer distance also increase risk? Why or why not?

REVIEW EXERCISES

1. Use $<$, $>$, or $=$ to make a true sentence:

$-7 \quad -9$

For Exercises 2–4, evaluate.

2. $(0.6)^2$ **3.** -4^2 **4.** $-|-6|$

For Exercises 5 and 6, translate the phrase to an expression.

5. four more than seven times a number

6. nine less than three times the sum of a number and two

2.5 Translating Word Sentences to Equations

OBJECTIVE 1 Translate sentences to equations using key words, then solve.

In previous sections, we have translated phrases to expressions (Section 1.6) and solved problems using formulas (Section 2.1). In this section, we will further develop problem-solving strategies by translating sentences to equations. Since we have not learned how to solve equations containing exponents and roots, we will explore translations that involve only addition, subtraction, multiplication, and division. See the following table, which we first saw in Section 1.6.

Key Words and Their Translations

Addition	Translation
The sum of x and 3	$x + 3$
h plus k	$h + k$
7 added to t	$t + 7$
3 more than a number	$n + 3$
y increased by 2	$y + 2$

Note: Since addition is a commutative operation, it does not matter in what order we write the translation.

For "the sum of x and 3" we can write $x + 3$ or $3 + x$.

Subtraction	Translation
The difference of x and 3	$x - 3$
h minus k	$h - k$
7 subtracted from t	$t - 7$
3 less than a number	$n - 3$
y decreased by 2	$y - 2$

Note: Subtraction is not a commutative operation. Therefore, the way we write the translation matters. We must translate each key phrase exactly as presented above. Particularly note how *less than* and *subtracted from* are translated.

Multiplication	Translation
The product of x and 3	$3x$
h times k	hk
twice a number	$2n$
triple the number	$3n$
$\frac{2}{3}$ of a number	$\frac{2}{3}n$

Note: Like addition, multiplication is a commutative operation. This means we can write the translation order any way we wish.

"h times k" can be hk or kh.

Division	Translation
The quotient of x and 3	$x \div 3$
h divided by k	$h \div k$
h divided into k	$k \div h$
the ratio of a to b	$a \div b$

Note: Division is like subtraction in that it is not a commutative operation. We must translate division phrases exactly as presented above. Particularly note how *divided into* is translated.

In order to translate a sentence to an equation, you must also know the key words that indicate an equal sign.

Key words for an equal sign:	is equal to
	is the same as
	is
	produces
	yields
	results in

To translate a word sentence to an equation, identify the variable(s), constants, and key words, then write the corresponding symbolic form.

Problems Involving Addition or Subtraction

EXAMPLE 1 The sum of twenty-five and a number is equal to fourteen. Translate to an equation, then solve for the number.

Understand The key word *sum* indicates addition, *is equal to* indicates an equal sign, and *a number* indicates a variable.

Plan Translate the key words to an equation, and then solve the equation. We will use n as the variable.

Execute Translate: The sum of twenty-five and a number is equal to fourteen.

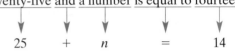

$$25 \quad + \quad n \quad = \quad 14$$

Solve: $25 + n = 14$
$$\underline{-25 \qquad -25}$$ Subtract 25 on both sides to isolate n.
$$0 + n = -11$$

Answer $n = -11$

Check In the original sentence, replace the unknown amount with -11 and verify that it makes the sentence true. Verify that the sum of twenty-five and a negative eleven is equal to fourteen.

$$25 + (-11) \stackrel{?}{=} 14$$
$$14 = 14$$

Yes, the sum of 25 and -11 is equal to 14.

EXAMPLE 2 Nineteen less than a number is seven. Translate to an equation and then solve.

Understand The key words *less than* indicate subtraction. The key word *is* means an equal sign.

Plan Translate the key words and then solve. We'll use n for the variable.

Execute Translate: <u>Nineteen</u> less than <u>a number</u> <u>is</u> <u>seven</u>.

$$n \quad - \quad 19 \qquad\qquad = \quad 7$$

> **Note:** The key words *less than* require careful translation. In the sentence, the word *nineteen* comes before the words *less than* and the words *a number* come after. In the translation, this order is reversed.

Solve: $n - 19 = 7$

$$\underline{+19 \quad +19}$$ **Add 19 to both sides.**

$$n + 0 = 26$$

Answer $n = 26$

Check Verify that 19 less than 26 is 7.

$$26 - 19 \stackrel{?}{=} 7$$
$$7 = 7$$

Yes, 19 less than 26 is 7.

  YOUR TURN   **Translate to an equation, and then solve.**

    a. Fifteen more than a number is negative seven.
    b. The difference of x and thirty-five is negative nine.

Problems Involving Multiplication

EXAMPLE 3   Three-fourths of a number is negative five-eighths. Translate to an equation and then solve.

Understand When *of* is preceded by a fraction, it means multiply. The word *is* means an equal sign.

Plan Translate the key words and then solve.

Execute Translate: <u>Three-fourths</u> <u>of</u> <u>a number</u> <u>is</u> <u>negative five-eighths</u>.

$$\frac{3}{4} \quad \cdot \quad n \qquad = \quad -\frac{5}{8}$$

Solve: $\dfrac{\overset{1}{\cancel{4}}}{\underset{1}{\cancel{3}}} \cdot \dfrac{\overset{1}{\cancel{3}}}{\underset{1}{\cancel{4}}} n = -\dfrac{\overset{1}{\cancel{5}}}{\underset{2}{\cancel{8}}} \cdot \dfrac{\overset{1}{\cancel{4}}}{3}$ **Clear the coefficient $\dfrac{3}{4}$ by multiplying both sides by its reciprocal $\dfrac{4}{3}$.**

Answer $n = -\dfrac{5}{6}$

Check   Verify that $\dfrac{3}{4}$ of $-\dfrac{5}{6}$ is $-\dfrac{5}{8}$.

$$\dfrac{\overset{1}{\cancel{3}}}{4} \cdot -\dfrac{5}{\underset{2}{\cancel{6}}} \overset{?}{=} -\dfrac{5}{8}$$

$$-\dfrac{5}{8} = -\dfrac{5}{8}$$

Yes, $\dfrac{3}{4}$ of $-\dfrac{5}{6}$ is $-\dfrac{5}{8}$, so $-\dfrac{5}{6}$ is correct.

YOUR TURN

Translate to an equation and then solve. The product of 0.6 and a number is -25.08.

Problems Involving More Than One Operation

EXAMPLE 4   Seven less than the product of four and a number is equal to negative five. Translate to an equation, and then solve.

Understand   *Less than* indicates subtraction in reverse order, *product* indicates multiplication, and *is equal to* indicates an equal sign.

Plan   Translate to an equation using the key words and then solve the equation.

Execute   Translate:

Seven less than the product of four and a number is equal to negative five.

$$4n \quad - \quad 7 \qquad\qquad\qquad\qquad = \qquad -5$$

Solve: $4n - 7 = -5$

$$\underline{ +7 \quad +7}$$ Add 7 to both sides to isolate $4n$.

$$4n + 0 = 2$$

$$\dfrac{4n}{4} = \dfrac{2}{4}$$ Divide both sides by 4 to clear the coefficient.

Answer $n = 0.5$

Check   Verify that 7 less than the product of 4 and 0.5 is equal to -5.

$$4(0.5) - 7 \overset{?}{=} -5$$
$$2 - 7 \overset{?}{=} -5$$
$$-5 = -5$$

Yes, 7 less than the product of 4 and 0.5 is equal to -5.

ANSWER

$0.6n = -25.08$; $n = -41.8$

Translations Requiring Parentheses

Sometimes a translation requires parentheses. This occurs when a sum or difference is multiplied or divided. When we multiply or divide a sum or difference, the addition or subtraction is to be calculated first. Because addition and subtraction follow multiplication and division in the order of operations, parentheses must be inserted around the addition or subtraction so that these operations are performed first.

EXAMPLE 5 Two-thirds of the difference of six and a number is the same as the number divided by four. Translate to an equation and then solve.

Understand *Of* means to multiply, and the word *difference* indicates subtraction. Because the difference is being multiplied, we write the difference expression in parentheses. *Divided by* indicates division.

Plan Translate the key words to an equation and then solve.

Execute Translate:

Two-thirds of the difference of six and a number is the same as the number divided by four.

$$\frac{2}{3}(6 - n) = \frac{n}{4}$$

Solve: $\dfrac{2}{3}(6 - n) = \dfrac{n}{4}$

$$\frac{2}{\cancel{3}_{1}} \cdot \frac{\cancel{6}^{2}}{1} - \frac{2}{3}n = \frac{n}{4}$$ Distribute to clear the parentheses.

$$4 - \frac{2}{3}n = \frac{n}{4}$$

$$12\left(4 - \frac{2}{3}n\right) = \left(\frac{n}{4}\right)12$$ Multiply both sides by 12 to clear the fractions.

$$12 \cdot 4 - \frac{\cancel{12}^{4}}{1} \cdot \frac{2}{\cancel{3}_{1}}n = \frac{n}{\cancel{4}_{1}} \cdot \frac{\cancel{12}^{3}}{1}$$ Distribute 12 and simplify.

$$\begin{array}{rcl} 48 - 8n &=& 3n \\ \underline{+8n} & & \underline{+8n} \\ 48 + 0 &=& 11n \end{array}$$ Add 8*n* to both sides to separate constant term and variable terms.

$$\frac{48}{11} = \frac{11n}{11}$$ Clear the coefficient by dividing both sides by 11.

Answer $\dfrac{48}{11} = n$

Check Verify that $\frac{2}{3}$ of the difference of 6 and $\frac{48}{11}$ is the same as $\frac{48}{11}$ divided by 4.

$$\frac{2}{3}\left(6 - \frac{48}{11}\right) \stackrel{?}{=} \frac{1}{4}\left(\frac{48}{11}\right)$$

Write 6 as a fraction with a denominator of 11.

$$\frac{2}{3}\left(\frac{66}{11} - \frac{48}{11}\right) \stackrel{?}{=} \frac{1}{\overset{}{\underset{1}{4}}}\left(\frac{\overset{12}{48}}{11}\right)$$ Divide out 4.

$$\frac{2}{3}\left(\frac{18}{11}\right) \stackrel{?}{=} \frac{12}{11}$$

Divide out 3.

$$\frac{2}{\underset{1}{3}}\left(\frac{\overset{6}{18}}{11}\right) \stackrel{?}{=} \frac{12}{11}$$

$$\frac{12}{11} = \frac{12}{11}$$

Yes, $\frac{2}{3}$ of the difference of 6 and $\frac{48}{11}$ is the same as $\frac{48}{11}$ divided by 4.

ANSWERS

a. $3(n - 5) = 7n - 12$;
   $\quad n = -\frac{3}{4}$

b. $\frac{1}{4}n + 6 = \frac{1}{5}(n + 10)$;
   $\quad n = -80$

YOUR TURN **Translate to an equation and then solve.**

a. Three times the difference of a number and five is equal to seven times the number, minus twelve.

b. The sum of one-fourth of a number and six is equal to one-fifth of the sum of the number and ten.

2.5 EXERCISES

For Extra Help

 InterAct Math

 Addison-Wesley Math Tutor Center

Video 4/DVT 1

 MyMathLab

 MathXL

1. List three key words that indicate addition.

2. List three key words that indicate subtraction.

3. List three key words that indicate multiplication.

4. List three key words that indicate division.

5. Why are "four minus n" and "four less than n" different?

6. Write a sentence that would require parentheses when translated and explain why your sentence requires parentheses.

For Exercises 7–42, translate each sentence to an equation and then solve.

7. Six decreased by the number *x* is equal to negative three.

8. Eleven times *m* is negative ninety-nine.

9. *c* divided by four is seven.

10. Eighteen subtracted from a number is equal to negative three.

11. Five less than *y* is negative one.

12. The product of negative ten and *y* is negative fifty.

13. Seven more than three times a number is equal to thirty-four.

14. Nineteen more than triple a number is one hundred.

15. Nine less than five times a number is seventy-six.

16. Twenty-five less than four times a number is eleven.

17. Eight times the sum of eight and a number is equal to one hundred sixty.

18. One-fourth of the sum of a number and five is equal to eight.

19. Eleven less than three times a number is equal to that number added to five.

20. Five times the sum of a number and three is the same as twice the number.

21. The product of three times the sum of a number and four added to twice the number yields negative three.

22. Seven times the sum of a number and six results in negative forty-two.

23. Sixty is equal to the ratio of the product of four and *y* to three.

24. Tripling the difference of *x* and five produces negative fifteen.

25. Four divided into *m* is 1.6.

26. Four-fifths of a number is five-eighths.

27. The quotient of a number and -6.5 is 4.2.

28. Nine less than the product of three and a number is equal to thirty-three.

29. Ten is the result when one is subtracted from the ratio of a number to four.

30. Two added to the quotient of a number and five is six.

31. The difference of a number and twelve is subtracted from the difference of twice the number and eight so that the result is eleven.

32. The sum of a number and six is added to the difference of the number tripled and four so that the result is fourteen.

33. One-third of a number is five-sixths.

34. Twice a number is the same as one-third.

35. One-third of the sum of a number and two is one.

36. Half of the difference of a number and two results in four.

37. Five times the sum of a number and one-third is equal to three times the number increased by two-thirds.

38. Two-thirds of the difference of three and a number is the same as twice the number.

39. A number increased by four is equal to triple the number added to sixteen.

40. One-third of a number is the same as half of the number increased by one.

41. The quotient of three less than a number and three is the same as the number divided by six.

42. Negative two times the difference of one and three times a number is equal to three subtracted from five times the difference of one and the number.

43. $4x + 3 = 7$

44. $-3y + 8 = 10$

45. $6(y + 4) = -10y$

46. $8(w - 2) = 3(w + 3)$

47. $\frac{1}{2}x - 3 = \frac{2}{3}(x - 8)$

48. $\frac{1}{2}(t + 1) = \frac{1}{3}(t - 5)$

49. $0.05m + 0.06(m - 11) = 22$

50. $0.05v + 0.03(v - 4.5) = 0.465$

51. $\frac{4}{3}(x + 1) = x + 4$

52. $\frac{1}{2}a + \frac{1}{3}a + \frac{1}{6}a = 5$

For Exercises 53–58, explain the mistake in the translation.

53. Ten less than a number is forty.
   Translation: $10 - n = 40$

54. Twelve divided into a number is negative eight.
   Translation: $12 \div n = -8$

55. Five times the difference of a number and six is equal to negative two.
   Translation: $5x - 6 = -2$

56. Nine times the sum of a number and eight is the same as three subtracted from the number.
   Translation: $9(y + 8) = 3 - y$

57. Twice a number minus the sum of the number and three is equal to negative six.
   Translation: $2t - t + 3 = -6$

58. Four times the sum of a number and two is equal to the number minus the difference of the number and seven.
   Translation: $4(y + 2) = y - (7 - y)$

For Exercises 59–66, translate to a formula, then use the formula to solve the problem.

59. The perimeter of a rectangle is equal to twice the sum of its length and width. Find the perimeter with the following length and width.

Length	Width
a. 24 ft.	18 ft.
b. 12.5 cm	18 cm
c. $8\frac{1}{4}$ in.	$10\frac{3}{8}$ in.

Width [rectangle diagram] Length

60. The surface area of a box is equal to twice the sum of its length times its width, its length times its height, and its width times its height. Find the surface area of a box with the following length, width, and height.

Length	Width	Height
a. 15 in.	6 in.	4 in.
b. 9.2 cm	12 cm	6.5 cm.
c. 8 ft.	$2\frac{1}{2}$ ft.	3 ft.

Height [box diagram] Length Width

61. An isosceles triangle has two sides that are equal in length. The perimeter of an isosceles triangle is the sum of its base and twice the length of one of the other sides. Find the perimeter of each listed isosceles triangle.

Base	Sides
a. 5 in.	11 in.
b. 1.5 m	0.6 m.
c. $15\frac{1}{4}$ in.	$9\frac{1}{2}$ in.

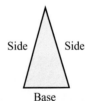

Side / Side
Base

62. The perimeter of a semicircle is the sum of the diameter and π times the radius. Find the perimeter of each semicircle listed.

a. Diameter = 18 in.
b. Radius = 4.5 m
c. Diameter = $2\frac{3}{4}$ ft.

Diameter

63. The simple interest earned after investing an amount of money, called principal, is equal to the product of the principal, the interest rate, and the time in years that the money remains invested. Use the formula to calculate the interest of each of the following investments.

	Principal	Rate	Time
a.	$4000	0.05	2 years
b.	$500	0.03	½ year
c.	$2000	0.06	¼ year

64. Investors often consider a stock's price-to-earnings ratio to help them decide whether a stock is worth investing in. The price-to-earnings ratio is equal to the stock's price divided by its company's earnings per share over the last 12 months. Use your formula to calculate each company's price-to-earnings ratio. (*Source:* David and Tom Gardner, *The Motley Fool Investment Guide,* Simon & Schuster, 1996)

	Company	Price	Earnings per share
a.	Microsoft	$47.98	$1.41
b.	Intel	$18.79	$0.29
c.	General Motors	$46.55	$4.01

65. Sky divers often use a formula to estimate free fall time. The amount of free fall time in seconds is approximately equal to the difference of the exit altitude and the parachute deployment altitude divided by 153.8. Use this formula to calculate free fall times if a sky diver deploys at each of the following altitudes after falling from an exit altitude of 12,500 feet.

a. 5000 feet

b. 4500 feet

c. 3000 feet

66. A person's body mass index (BMI) is a measure of the amount of body fat based on height, in inches, and weight, in pounds. A person's BMI is equal to the ratio of the product of his or her weight and 704.5 to his or her height multiplied by itself. Use your formula to find the BMI of each person listed below. (*Source:* National Institutes of Health)

	Height	Weight
a.	5'4"	138
b.	5'9"	155
c.	6'2"	220

Fill in each blank with a whole number so that the resulting rational number has the following properties:

The rational number is between 300 and 500.
The tenths placeholder is less than the tens placeholder.
The sum of the tens place and the tenths place is 8.
The product of all four placeholders is 42.

$$\underline{\quad}\ \underline{\quad}\ \underline{\quad}\ \cdot\ \underline{\quad}$$

REVIEW EXERCISES

1. Is $19 \geq 19$ true or false? Explain.

For Exercises 2–3, use $<, >$, or $=$ to make a true sentence.

2. $-|-6|\ \blacksquare\ -2(3)$

3. $5(4) - 23\ \blacksquare\ 2(3 - 8)$

For Exercises 4–5, solve and check.

4. $-6z = 72$

5. $\dfrac{1}{3}(y - 2) = \dfrac{3}{5}y + 6$

6. Solve: $-\dfrac{6}{8} = \dfrac{\blacksquare}{40}$

2.6 Solving Linear Inequalities

OBJECTIVES

1. Represent solutions to inequalities graphically and using set notation.
2. Solve linear inequalities.
3. Solve problems involving linear inequalities.

OBJECTIVE 1 Represent solutions to inequalities graphically and using set notation. Not all problems translate to equations. Sometimes a problem can have a range of values as solutions. In mathematics we can write inequalities to describe situations where a range of solutions is possible. Following are the inequality symbols and their meanings:

$$
\begin{array}{ll}
< & \text{is less than} \\
> & \text{is greater than} \\
\leq & \text{is less than or equal to} \\
\geq & \text{is greater than or equal to}
\end{array}
$$

Throughout the chapter, we have explored techniques for solving linear equations. In this section we will explore how to solve **linear inequalities**.

DEFINITION *Linear inequality:* An inequality containing expressions in which each variable term contains a single variable with an exponent of 1.

Following are some examples of linear inequalities:

$$x > 5 \qquad n + 2 < 6 \qquad 2(y - 3) \leq 5y - 9 \qquad 2x + 3y \geq 6$$

A solution for an inequality is any number that can replace the variable(s) in the inequality and make it true. Consider the inequality $x > 5$. Any number greater than 5 is a solution to the inequality. Notice that 5 itself is not a solution because the greater than symbol indicates values that are strictly greater than 5. Solutions for $x \geq 5$ would be any number greater than 5 and 5 itself.

Solutions to an inequality can be represented using a number line. In the case of $x > 5$, we represent every solution by shading to the right of 5 on a number line. Because 5 itself is not a solution we draw an open circle around the mark for 5.

$x > 5$ is represented by

Because $x \geq 5$ indicates numbers greater than 5 or 5 itself, we shade to the right of 5 and use a solid circle on the mark for 5.

$x \geq 5$ is represented by

Solutions to inequalities can also be represented using set notation. The inequality $x > 5$ can be written in set notation like this:

$$\{ \, x \mid x > 5 \, \}$$

In words: The set of all x such that x is greater than 5

PROCEDURE **GRAPHING INEQUALITIES**

To graph an inequality:

1. If the symbol allows for equality ($\leq$ or $\geq$), draw a solid circle on the number line at the indicated number. If the symbol does not allow for equality ($<$ or $>$), draw an open circle on the number line at the indicated number.
2. If the variable is greater than the indicated number, shade to the right of the indicated number. If the variable is less than the indicated number, shade to the left of the indicated number.

EXAMPLE 1   Graph and write in set notation.

a. $x \leq -6$

Graph: Because the inequality indicates that the solutions are less than or equal to -6, we shade to the left of -6 and draw a solid circle on the mark for -6.

Set notation: $\{x \mid x \leq -6\}$

b. $n < 0$

Graph: Because the inequality indicates that the solutions are strictly less than 0, we shade to the left of 0 and draw an open circle on the mark for 0.

Set notation: $\{n \mid n < 0\}$

CONNECTION

$\{n \mid n < 0\}$ is a way to express the set of all negative real numbers.

ANSWERS

a. Graph:

Set notation: $\{x \mid x < 2\}$

b. Graph:

Set notation: $\{t \mid t \geq -1\}$

YOUR TURN   **Graph and write in set notation.**

a. $x < 2$ **b.** $t \geq -1$

Compound Inequalities

Inequalities containing two inequality symbols are called *compound inequalities*. Compound inequalities are useful in writing a range of values between two numbers. For example, $1 < x < 6$ is a compound inequality meaning x can be any number that is greater than 1 and less than 6. In other words, a solution would be any number between 1 and 6. Again, we can represent this inequality graphically and using set notation.

Graph of $1 < x < 6$:

Set notation: $\{x \mid 1 < x < 6\}$

EXAMPLE 2 Graph $-5 \le x < 1$ and write in set notation.

Graph: A solution for $-5 \le x < 1$ is any number that is greater than or equal to -5 and less than 1. We shade in between -5 and 1 with a solid circle at -5 and an open circle at 1.

Set notation: $\{x \mid -5 \le x < 1\}$

ANSWERS

Graph:

Set notation:
$\{y \mid -4 < y \le -1\}$

YOUR TURN

Graph $-4 < y \le -1$ and write set notation.

OBJECTIVE 2 **Solve linear inequalities.** To solve inequalities like $n + 2 < 6$ and $2(x - 3) \le 5x - 9$, we will follow essentially the same process as for solving equations. The addition and multiplication principles of inequalities are similar to the principles of equality. First, let's examine the addition principle of inequality.

Consider the inequality $3 < 7$. According to the addition principle, if we add or subtract the same amount on both sides, then the inequality should still be true. Let's test this by choosing some numbers to add and subtract on both sides.

Add 2 on both sides:	Subtract 9 on both sides:
$3 < 7$	$3 < 7$
$\underline{+2\ +2}$	$\underline{-9\ -9}$
$5 < 9$ Still true.	$-6 < -2$ Still true.

RULE **THE ADDITION PRINCIPLE OF INEQUALITY**

If $a < b$, then $a + c < b + c$ is true for all real numbers a, b, and c.

We can add (or subtract) the same amount on both sides of an inequality without affecting its solution(s).

Note: Though the principle is written in terms of the $<$ symbol, the principle is true for any inequality symbol.

The multiplication principle, on the other hand, does not work as neatly as it did for equations. As we shall see, sometimes multiplying or dividing both sides of an inequality by the same number can turn a true inequality into a false inequality. Let's multiply both sides of $3 < 7$ by a positive number and then by a negative number.

Multiply both sides by 4:
$$4(3) < 4(7)$$
$$12 < 28 \quad \text{Still true.}$$

Multiply both sides by -2:
$$-2(3) < -2(7)$$
$$-6 < -14 \quad \text{Not true!}$$

Conclusion   An inequality remains true when we multiply both sides by a positive number, but not when we multiply both sides by a negative number.

RULE   **THE MULTIPLICATION PRINCIPLE OF INEQUALITY**

If a and b are real numbers, where $a < b$, then $ac < bc$ is true if c is a positive real number.

In other words, we can multiply (or divide) both sides of an inequality by the same positive amount without affecting its solution(s).

If a and b are real numbers, where $a < b$, then $ac > bc$ is true if c is a negative real number.

In other words, if we multiply (or divide) both sides of an inequality by the same negative amount, we must reverse the direction of the inequality symbol to maintain the truth of the inequality.

Note: Though we have written the principle in terms of the $<$ symbol, the principle is true for any inequality symbol.

EXAMPLE 3   Solve. Then graph the solutions and write the solution set.

a. $3x > -7$

Solution

Note: Dividing both sides by a positive number does not affect the inequality.

$$\frac{3x}{3} > \frac{-7}{3}$$

$$x > -\frac{7}{3}$$

To isolate x, we clear the 3 coefficient by dividing by 3 on both sides.

Graph:

Set notation: $\left\{ x \mid x > -\dfrac{7}{3} \right\}$

b. $-8x \geq 24$

Solution

> **Note:** Because we divided both sides by a negative number, we reversed the direction of the inequality symbol.

$$\frac{-8x}{-8} \leq \frac{24}{-8}$$
$$x \leq -3$$

To isolate x, we clear the -8 coefficient by dividing by -8 on both sides.

Graph:

$$\begin{array}{ccccccccc} -6 & -5 & -4 & -3 & -2 & -1 & 0 & 1 & 2 \end{array}$$

Set notation: $\{x \mid x \leq -3\}$

c. $\dfrac{5}{6} \leq -\dfrac{3}{4}n$

Solution

> **Note:** You may find it helpful to rewrite the inequality so that the variable is on the left side. The inequality $-\dfrac{10}{9} \geq n$ is the same as $n \leq -\dfrac{10}{9}$.

$$-\frac{\overset{2}{\cancel{4}}}{3}\left(\frac{5}{\underset{3}{\cancel{6}}}\right) \geq -\frac{\overset{1}{\cancel{4}}}{3}\left(-\frac{\overset{1}{\cancel{3}}}{\underset{1}{\cancel{4}}}n\right)$$
$$-\frac{10}{9} \geq n$$

To isolate n, we clear the $-\dfrac{3}{4}$ coefficient by multiplying by its reciprocal $-\dfrac{4}{3}$ on both sides. Multiplying by a negative number means we must reverse the direction of the inequality symbol.

Graph:

$$\overset{-\frac{10}{9}}{}$$

$$\begin{array}{cccc} -2 & -1 & 0 & 1 \end{array}$$

Set notation: $\left\{n \mid n \leq -\dfrac{10}{9}\right\}$

ANSWER

a. $t > -\dfrac{9}{4}$ or $-2\dfrac{1}{4}$

Graph:

$$\begin{array}{cccc} -4 & -3 & -\frac{9}{4}\,{-2} & -1 \end{array}$$

Set notation:
$$\left\{t \mid t > -\dfrac{9}{4}\right\}$$

YOUR TURN

Solve $-\dfrac{2}{5}t < \dfrac{9}{10}$. Then graph the solutions and write the solution set.

We can solve more complex linear inequalities using an outline similar to that for solving linear equations.

PROCEDURE   SOLVING LINEAR INEQUALITIES

To solve linear inequalities:

1. Simplify both sides of the inequality as needed.
   a. Distribute to clear parentheses.
   b. Clear fractions or decimals by multiplying through by the LCD just as we did for equations. (Clearing fractions and decimals is optional.)
   c. Combine like terms.
2. Use the addition principle so that all variable terms are on one side of the inequality and all constants are on the other side. Then combine like terms.
3. Use the multiplication principle to clear any remaining coefficient. If you multiply (or divide) both sides by a negative number, reverse the direction of the inequality symbol.

Notice that clearing the term with the smaller coefficient results in a positive coefficient, so you won't have to reverse the direction of the inequality.

EXAMPLE 4 Solve $7x + 9 > 4x - 6$. Then graph the solutions and write the solution set.

Solution Use the addition principle to separate the variable terms and constant terms, then clear the remaining coefficient.

Note: Clearing the $4x$ term results in a positive coefficient after combining like terms, so we won't have to reverse the inequality when we clear the 3 coefficient.

$$7x + 9 > 4x - 6$$
$$\underline{-4x \qquad -4x}$$ Subtract $4x$ from both sides.
$$3x + 9 > 0 - 6$$
$$3x + 9 > -6$$
$$\underline{-9 \qquad -9}$$ Subtract 9 from both sides.
$$3x + 0 > -15$$
$$\frac{3x}{3} > \frac{-15}{3}$$ Divide both sides by 3 to isolate x.
$$x > -5$$

Graph:
$$-7\ -6\ -5\ -4\ -3\ -2\ -1\ \ 0\ \ 1$$

Set notation: $\{x | x > -5\}$

YOUR TURN

Solve $7(x - 3) \leq 4x + 3$. Then graph the solutions and write the solution set.

OBJECTIVE 3 **Solve problems involving linear inequalities.** Problems requiring inequalities can be translated using key words much like those we used to translate sentences to equations. The following table lists common key words that indicate inequalities.

Less than:		Greater than:	
A number is less than 7.	$n < 7$	A number is greater than 2.	$n > 2$
A number must be smaller than 5.	$n < 5$	A number must be greater than 3.	$n > 3$
		A number must be more than -6.	$n > -6$

Less than or equal to:		Greater than or equal to:	
A number is at most 9.	$n \leq 9$	A number is at least 2.	$n \geq 2$
The maximum is 14.	$n \leq 14$	The minimum is 18.	$n \geq 18$

EXAMPLE 5 Three-fourths of a number is at least eighteen. Translate to an inequality, then solve.

Understand Because the word *of* is preceded by a fraction, it means multiplication. The key words *at least* indicate a greater-than or equal-to symbol.

Plan Translate the key words, then solve. We'll use n for the variable.

Execute Translate:

$$\frac{3}{4} \cdot n \qquad \geq \qquad 18$$

Solve: $\frac{3}{4} \cdot n \geq 18$

$$\frac{\cancel{4}^{1}}{\cancel{3}_{1}} \cdot \frac{\cancel{3}^{1}}{\cancel{4}_{1}} n \geq \frac{4}{\cancel{3}_{1}} \cdot \frac{\cancel{18}^{6}}{1} \qquad \text{Multiply both sides by } \frac{4}{3} \text{ to isolate } n.$$

Answer: $n \geq 24$

Check Verify that 24 and any number greater than 24 satisfy the original sentence.

Test 24:

Is $\frac{3}{4}$ of 24 at least 18?

$$\frac{3}{\cancel{4}_{1}} \cdot \frac{\cancel{24}^{6}}{1} \geq 18$$

$$18 \geq 18 \quad \text{True.}$$

Test a number greater than 24, like 25:

Is $\frac{3}{4}$ of 25 at least 18?

$$\frac{3}{4} \cdot \frac{25}{1} \geq 18$$

$$\frac{75}{4} \geq 18$$

$$18\frac{3}{4} \geq 18 \quad \text{True.}$$

YOUR TURN

0.2 times the sum of a number and eight is less than the difference of the number and five. Translate to an inequality, then solve.

EXAMPLE 6 Darwin is planning a garden area, which he will enclose with a fence. Cost restricts him to a total of 240 feet of fencing materials. He wants the garden to span the entire width of his lot, which is 80 feet. What is the maximum length of the garden?

Understand The fence surrounds the garden, so that 240 feet is the maximum perimeter for the garden. The width of the garden is to be 80 feet, and we must find the length.

ANSWER

$0.2(n + 8) < n - 5$;
$n > 8.25$

Plan   The formula for the perimeter of a rectangle is $P = 2l + 2w$. Because 240 feet is a *maximum* perimeter, we write the perimeter formula as an inequality so that the expression used to calculate perimeter $(2l + 2w)$ is less than or equal to 240. Because the width is to be 80 feet, we replace w in the formula with 80.

Execute   Translate:

The calculation for perimeter must be less than or equal to 240.

$$2l + 2(80) \qquad \leq \qquad 240$$

Solve: $2l + 160 \leq 240$ **Subtract 160 from both sides.**

$$\underline{-160 \quad -160}$$
$$2l + 0 \leq 80$$

$$\frac{2l}{2} \leq \frac{80}{2}$$ **Divide both sides by 2 to isolate *l*.**

$$l \leq 40$$

Answer   The length must be less than or equal to 40 feet, which means the maximum length is 40 feet.

Check   Verify that a garden with a length of 40 feet or less and a width of 80 feet has a perimeter that is 240 feet or less. This will be left to the reader.

YOUR TURN

It is desired that the surface area of a box be at least 678 square centimeters. If the length is to be 12 centimeters and the width is to be 9 centimeters, find the range of values for the height that satisfies the minimum surface area of 678 square centimeters.

ANSWER

$h \geq 11$ cm

2.6 EXERCISES

For Extra Help

InterAct Math

Tutor Center   Addison-Wesley Math Tutor Center

Video 4/DVT 1

MyMathLab

MathXP   MathXL

1. What is a solution for an inequality?

2. Explain the difference between $x < 8$ and $x \leq 8$.

3. In your own words, explain the meaning of $x \geq -5$.

4. What action causes the direction of an inequality symbol to change?

For Exercises 5–16, graph and write in set notation.

5. $x \geq -3$

6. $n \leq -5$

7. $h < 6$

8. $x > 0$

9. $n < -\dfrac{2}{3}$

10. $a \geq \dfrac{4}{5}$

11. $t \leq 2.4$

12. $p \leq -0.6$

13. $-3 < x < 6$

14. $4 < c \leq 10$

15. $0 \leq n \leq 5$

16. $-1 < a < 5$

For Exercises 17–42: **a.** *Solve.*
b. *Write the solution set.*
c. *Graph the solution set.*

17. $m - 5 < 20$

18. $x + 8 \geq -12$

19. $16y \geq 32$

20. $-4x < -16$

21. $5y + 1 > 16$

22. $3y - 2 > 10$

23. $1 - 6x < 25$

24. $-2c + 3 > 17$

25. $\dfrac{a}{2} + 1 < \dfrac{3}{2}$

26. $2 > \dfrac{2}{3} - \dfrac{l}{3}$

27. $1 - f > 2 + f$

28. $9 + 2n < 3 - 4n$

29. $3 - 6u \geq -5 - 2u$

30. $4 - 9k < -4k + 19$

31. $2(c - 3) < 3(c + 2)$

32. $6(n + 1) + 2(4n - 2) > 3(5n - 1)$

33. $10(9w + 13) - 5(3w - 1) \leq 8(11w + 12)$

34. $4(8u + 9) - 3(u + 6) - 2(9u - 2) \geq 0$

35. $\dfrac{1}{2}(x + 1) \leq \dfrac{1}{3}(x - 5)$

36. $\dfrac{1}{5}(6m - 7) > \dfrac{1}{2}(3m - 1)$

37. $\dfrac{1}{6}(5 + n) - \dfrac{1}{3}(10 - n) < 1$

38. $-\dfrac{1}{3}(2y - 5) + \dfrac{1}{6}(5y - 2) \geq 1$

39. $0.4l - 0.37 < 0.3(l + 2.1)$

40. $0.6t + 0.2(t - 8) \leq 0.5t - 0.6$

41. $2.1s + 3(1.4s - 1.5) \geq 0.09s + 20.34$

42. $0.05 + 0.03(v - 4.5) \geq 0.465$

For Exercises 43–50, translate to an inequality, then solve.

43. Four more than a number is greater than twenty-four.

44. Three-fifths of a number is greater than negative twenty-one.

45. Eight times a number less thirty-six must be more than sixty.

46. Three subtracted from a number times eleven is at least forty-one.

47. Five added to half of a number is at most two.

48. Triple a number is less than the number increased by eight.

49. The sum of two-fifths of a number and two will not exceed eight.

50. Negative five is greater than fifteen less than five times a number.

51. A school is planning a playground. The area of a rectangular playground may not exceed 170 square feet. If the width is 17 feet, what range of values can the length have?

52. Scotty needs to make a rectangular pool cover that is at least 128 square feet. If the pool length is 16 feet, what range of values must the width be?

53. The design of a storage box calls for a width of 27 inches and a length of 41 inches. If the surface area must be at least 4254 square inches, find the range of values for the height.

54. Andre builds tables. One particular circular table requires him to glue a strip of veneer around the circumference of the top. He is designing a new circular table and has a strip of veneer that is 150 inches long that he wishes to use. What range of values can the radius have?

55. Jon never exceeds 65 miles per hour when driving on the highway. If he is traveling 410 miles, what range of values can his time spent driving have?

56. A truck driver must make a delivery 298 miles away in 5 hours or less. What range of values can his average rate be so that he meets the schedule?

57. The design of a circuit specifies that the voltage cannot exceed 12 volts. If the resistance of the circuit is 8 ohms, find the range of values that the current can have.

58. A label in an elevator indicates that the maximum load is 9000 newtons, that is, the maximum force is -9000 newtons. If the acceleration due to gravity is -9.8 meters per second squared, find the range of values of mass that can be loaded onto the elevator.

59. The fourth-quarter profit goal for a software company is at least $850,000. If it is known that the cost for the quarter will be $625,000, find the revenue that must be generated to achieve the goal.

60. Silver is a solid up to a temperature of $960.8\,°C$. Write an inequality to describe this range of temperatures in degrees Fahrenheit.

61. To get an A in her math course, Tina needs the average of five tests to be at least 90. Her current test scores are 82, 91, 95, and 84. What range of scores on the fifth test would get her an A in the course?

62. In Aaron's English course, the final grade is determined by the average of five papers. The department requires any student whose average falls below 75 to repeat the course. Aaron's scores on the first four papers are 68, 78, 80, and 72. What range of scores on the fifth paper would cause him to have to repeat the course?

PUZZLE PROBLEM

Write the set of all values such that
$|x| = -x.$

Making the Grade

Students often want to know what they need to score on a test to achieve a certain average. Suppose that you want a final average grade of B in your algebra class. Suppose also that in this class you were to have five tests, T_1, T_2, T_3, T_4, and T_5.

1. Write an algebraic expression for the average of the five test scores.

2. Suppose the minimum average to get a B in a course is 80. Write an inequality indicating that the average of the five tests is at least 80.

3. Solve the inequality for T_5.

4. Select values for T_1, T_2, T_3, and T_4 between 90 and 100. Using those scores, find the range of values that T_5 can have in order to receive a B in the course.

5. Complete the following table, selecting scores in the indicated range for T_1, T_2, T_3, and T_4, and then solving for the corresponding range of scores for T_5.

Range for T_1, T_2, T_3, T_4	T_1	T_2	T_3	T_4	T_5
90 to 100					
80 to 90					
70 to 80					
60 to 70					

6. Is there a range of scores for T_1, T_2, T_3, and T_4 for which a B in the course is impossible? Which range?

7. What is the minimum total points that a person must acquire on all five tests to achieve a B average?

8. If a person were to receive a 100 on T_5, what must the total of T_1, T_2, T_3, and T_4 be so that the average of the five tests is exactly an 80?

9. Using the total you found for T_1, T_2, T_3, and T_4 in question 8, what is the average of the four tests?

10. Given any values for T_1, T_2, T_3, and T_4, if the average of those four scores is below the average you found in question 9, then a B is impossible to achieve with any score (even 100) on the fifth test. Translate the previous sentence to an inequality.

REVIEW EXERCISES

1. Simplify: $-6 \div 0$

2. Evaluate $b^2 - 4ac$ when $a = 1$, $b = -2$, and $c = -3$.

3. Evaluate $-|3x| + |y^3|$ when $x = 5$ and $y = -1$.

4. Write a set representing the natural numbers up to 10.

5. What property of addition is represented by $-6.5 + 3.4 = 3.4 + (-6.5)$?

6. Translate to an expression: six more than twice the difference of a number and five.

Defined Terms

Section 2.1
- Equation (p. 104)
- Solution (p. 104)
- Identity (p. 106)
- Formula (p. 108)
- Perimeter (p. 108)
- Area (p. 108)
- Volume (p. 108)
- Circumference (p. 108)
- Radius (p. 108)
- Diameter (p. 108)

Section 2.2
- Linear equation (p. 119)
- Linear equation in one variable (p. 119)

Section 2.6
- Linear inequality (p. 166)

Learning Strategy

The summaries in the textbook are like the study sheet suggested in the To the Student section (see page xvii). Remember that your study sheet is a list of the rules, procedures, and formulas that you need to know. If you are a tactile or visual learner, spend a lot of time reviewing and writing the rules or procedures. Try to get to the point where you can write the essence of each rule and procedure from memory. If you are an audio learner, record yourself saying each rule and procedure, then listen to the recording over and over. Consider developing clever rhymes or songs for each rule or procedure to help you remember them.

PROCEDURE PROBLEM-SOLVING OUTLINE

1. **Understand** the problem.
   a. Read the question(s) (not the whole problem, just the question at the end) and write a note to yourself about what it is you are to find.
   b. Now read the whole problem.
   c. Underline key words.
   d. Draw a picture (if possible).
   e. Make a list or table.
   f. Simulate the situation.
   g. Search for a related example problem.
2. Devise a **plan**.
   a. Search for a formula.
   b. Translate the key words to an equation.
3. **Execute** the plan by solving the equation/formula.
4. **Answer** the question. Look back at the note you made to yourself about what it was you were to find and make sure you answer that question. Be sure to include appropriate units.
5. **Check** results. Make sure the answer is reasonable.
   a. Repeat the solution in a different way.
   b. Reverse the process.
   c. Estimate the answer and make sure the estimate and actual answer are reasonably close.

Procedures, Rules, and Key Examples

Procedures/Rules	Key Examples

Section 2.1 Equations, Formulas, and the Problem-Solving Process

To check a value to see if it is a solution for a given equation:
1. Replace the variable(s) in the equation with the value.
2. If the resulting equation is true, the value is a solution.

Example 1: Check to see if the given number is a solution to the equation.

a. $7n - 13 = 3n + 11; n = 6$

$7(6) - 13 \stackrel{?}{=} 3(6) + 11$

$42 - 13 \stackrel{?}{=} 18 + 11$

$29 = 29$ Yes, 6 is a solution.

b. $2(x - 3) = 5x + 4; x = 1$

$2(1 - 3) \stackrel{?}{=} 5(1) + 4$

$2(-2) \stackrel{?}{=} 5 + 4$

$-4 \neq 9$ No, 1 is not a solution.

To determine whether an equation is an identity, simplify the expressions on each side of the equal sign. If, after simplifying, the expressions are identical, the equation is an identity.

Example 2: Is $5x - 9 - x = 4(x - 2) - 1$ an identity?

$4x - 9 = 4x - 8 - 1$

$4x - 9 = 4x - 9$ Yes, it is an identity.

To use a formula:
1. Replace the variables with the corresponding given values.
2. Solve for the missing value.

Example 3: Find the volume of a box with a length of 4 cm, width of 5 cm, and height of 6 cm.

$V = lwh$

$V = (4)(5)(6)$

$V = 120 \text{ cm}^3$

For composite figures:
1. To calculate the area of a figure composed of two or more figures that are next to each other, add the areas of the individual figures.
2. To calculate the area of a region defined by a smaller figure within a larger figure, subtract the area of the smaller figure from the area of the larger figure.

Example 4: Find the shaded area in the figure shown.

$A = \dfrac{1}{2}h(a + b) - lw$

$A = \dfrac{1}{2}(6)(12 + 8) - (3)(4)$

$A = \dfrac{1}{2}(6)(20) - 12$

$A = 60 - 12$

$A = 48 \text{ ft.}^2$

continued

Procedures/Rules	Key Examples

Section 2.2 The Addition Principle

To use the addition principle:
1. Decide what term you want to clear.
2. Add the additive inverse of that term to both sides of the equation. (That is, add or subtract appropriately so that the term you want to clear becomes 0.)

To solve equations requiring the addition principle only:
1. Simplify both sides of the equation as needed.
   a. Distribute to clear parentheses.
   b. Combine like terms.
2. Use the addition principle so that all variable terms are on one side of the equation and all constants are on the other side. Then combine like terms.

Tip: Clear the variable term that has the lesser coefficient to avoid negative coefficients.

Example 1: Solve.

a.
$$\begin{array}{r} x + 9 = 15 \\ \underline{-9 \quad -9} \\ x + 0 = 6 \\ x = 6 \end{array}$$

b.
$$\begin{array}{r} y - 5 = 7 \\ \underline{+5 \quad +5} \\ y + 0 = 12 \\ y = 12 \end{array}$$

c.
$$\begin{array}{r} 5(n - 2) + 3n = 7n - 4 \\ 5n - 10 + 3n = 7n - 4 \\ 8n - 10 = 7n - 4 \\ \underline{-7n \qquad -7n} \\ n - 10 = 0 - 4 \\ n - 10 = -4 \\ \underline{+10 \quad +10} \\ n + 0 = 6 \\ n = 6 \end{array}$$

Section 2.3 The Multiplication Principle

To use the multiplication principle to clear a coefficient:
1. Determine the coefficient you want to clear.
2. Multiply both sides by the multiplicative inverse of that coefficient or divide both sides by the coefficient.

To solve linear equations in one variable:
1. Simplify both sides of the equation as needed.
   a. Distribute to clear parentheses.
   b. Clear fractions or decimals by multiplying through by the LCD. In the case of decimals, the LCD is a power of 10 determined by the decimal number with the most places. (Clearing fractions and decimals is optional.)
   c. Combine like terms.
2. Use the addition principle so that all variable terms are on one side of the equation and all constants are on the other side. (Clear the variable term with the lesser coefficient.) Then combine like terms.
3. Use the multiplication principle to clear the remaining coefficient.

Example 1: Solve.

a.
$$\begin{array}{r} -5x = 15 \\ \dfrac{-5x}{-5} = \dfrac{15}{-5} \\ x = -3 \end{array}$$

b.
$$\frac{3}{4}n = \frac{5}{6}$$
$$\frac{\overset{1}{4}}{\underset{1}{3}} \cdot \frac{\overset{1}{3}}{\underset{1}{4}}n = \frac{5}{\underset{3}{6}} \cdot \frac{\overset{2}{4}}{3}$$
$$n = \frac{10}{9}$$

c.
$$\begin{array}{r} 7n - (n - 3) = 2n - 5 \\ 7n - n + 3 = 2n - 5 \\ 6n + 3 = 2n - 5 \\ \underline{-2n \qquad -2n} \\ 4n + 3 = 0 - 5 \\ 4n + 3 = -5 \\ \underline{-3 \quad -3} \\ 4n + 0 = -8 \\ \dfrac{4n}{4} = \dfrac{-8}{4} \\ n = -2 \end{array}$$

continued

Procedures/Rules	Key Examples

Section 2.4 Applying the Principles to Formulas

To isolate a particular variable in a formula:
1. Imagine all other variables to be constants.
2. Isolate the desired variable using the outline for solving equations.

Example 1: Isolate w in the formula
$$P = 2l + 2w.$$

$$P = 2l + 2w$$
$$\underline{-2l \quad -2l}$$
$$P - 2l = 0 + 2w$$
$$\frac{P - 2l}{2} = \frac{2w}{2}$$
$$\frac{P - 2l}{2} = w$$

Section 2.5 Translating Word Sentences to Equations

To translate a word sentence to an equation, identify the variable(s), constants, and key words, then write the corresponding symbolic form.

Example 1: Translate to an equation.
a. Five more than seven times a number is twelve.
 Answer: $7n + 5 = 12$
b. Six times the difference of a number and nine is equal to four times the sum of a number and five.
 Answer: $6(n - 9) = 4(n + 5)$

Section 2.6 Solving Linear Inequalities

To graph an inequality:
1. If the symbol allows for equality ($\leq$ or $\geq$), draw a solid circle on the number line at the indicated number. If the symbol does not allow for equality ($<$ or $>$), draw an open circle on the number line at the indicated number.
2. If the variable is greater than the indicated number, shade to the right of the indicated number. If the variable is less than the indicated number, shade to the left of the indicated number.

Example 1: Graph $x \leq 6$ on a number line.

Example 2: Graph $x > -2$ on a number line.

continued

Procedures/Rules	Key Examples

Section 2.6 Solving Linear Inequalities (continued)

To solve linear inequalities:

1. Simplify both sides of the inequality as needed.
   a. Distribute to clear parentheses.
   b. Clear fractions or decimals by multiplying through by the LCD just as we did for equations. (Clearing fractions and decimals is optional.)
   c. Combine like terms.
2. Use the addition principle so that all variable terms are on one side of the inequality and all constants are on the other side. Then combine like terms. (Remember, moving the term with the lesser coefficient results in a positive coefficient.)
3. Use the multiplication principle to clear any remaining coefficient. If you multiply (or divide) both sides by a negative number, reverse the direction of the inequality symbol.

Example 3: Solve.

a. $-6x > 24$

$$\frac{-6x}{-6} < \frac{24}{-6}$$

$$x < -4$$

b. $8y - 2 \geq 5y + 7$

$$8y - 2 \geq 5y + 7$$
$$\underline{-5y \qquad -5y}$$
$$3y - 2 \geq 0 + 7$$
$$3y - 2 \geq 7$$
$$\underline{+2 \quad +2}$$
$$3y + 0 \geq 9$$
$$\frac{3y}{3} \geq \frac{9}{3}$$
$$y \geq 3$$

Formulas

Perimeter of a rectangle: $P = 2l + 2w$

Circumference of a circle: $C = \pi d$ or $C = 2\pi r$

Area of a parallelogram: $A = bh$

Area of a triangle: $A = \frac{1}{2}bh$

Area of a trapezoid: $A = \frac{1}{2}h(a + b)$

Area of a circle: $A = \pi r^2$

Surface area of a box: $SA = 2lw + 2lh + 2wh$

Volume of a box: $V = lwh$

Volume of a pyramid: $V = \frac{1}{3}lwh$

Volume of a cylinder: $V = \pi r^2 h$

Volume of a cone: $V = \frac{1}{3}\pi r^2 h$

Volume of a sphere: $V = \frac{4}{3}\pi r^3$

Distance, d, an object travels given its rate, r, and the time of travel, t: $d = rt$

The temperature in degrees Celsius given degrees Fahrenheit: $C = \frac{5}{9}(F - 32)$

The temperature in degrees Fahrenheit given degrees Celsius: $F = \frac{9}{5}C + 32$

The profit, P, after cost, C, is deducted from revenue, R: $P = R - C$

For Exercises 1–6, answer true or false.

1. $6x + 3y = 7x^2 - 4y$ is an equation.

2. $5(x + 2) = 10 - 3x + 8x$ is an identity.

3. 1 is a solution for $b^2 - 4b = 4$.

4. "Five less than a number is fourteen" translates to $5 - n = 14$.

5. $5x - 3y = 2$ is a linear equation.

6. $3 \leq x < 5$ is a compound inequality.

For Exercises 7–9, fill in the blank.

7. If an equation is an _____, then every real number for which its expressions are defined is a solution.

8. An equation that is made of expressions in which each variable term contains a single variable raised to an exponent equal to 1 is a _____ equation.

9. To solve equations:

   1. _____ both sides of the equation as needed.

      a. _____ to clear parentheses.

      b. Clear fractions or decimals by _____ by their _____.

      c. _____ like terms.

   2. Use the _____ principle so that all variable terms are on one side of the equation and all constants are on the other side.

   3. Use the _____ principle to clear any remaining coefficient.

10. In your own words, explain how to check a potential solution.

For Exercises 11–16, check to see if the given number is a solution for the given equation.

11. $y + 6 + 3y - 4 = 14$; check $y = 3$

12. $\frac{1}{4}(n - 1) = \frac{1}{2}(n + 1)$; check $n = 5$

13. $x^2 + 6.1 = 3x - 0.1$; check $x = -0.1$

14. $\frac{5}{6}m - 1 = \frac{m}{3} + \frac{1}{2}$; check $m = 3$

15. $\frac{3}{5}(a + 10) = -2$; check $a = -20$

16. $6.1(7.2 + b) = 43.92$; check $b = 0$

For Exercises 17–28, solve and check.

17. $2(c + 4) - 5c = 17 - 8c$

18. $5(2a - 3) - 6(a + 2) = -7$

19. $\dfrac{5}{3} + 7x - 2x = \dfrac{1}{3}(9x + 2)$

20. $2 - \dfrac{2}{3}m = 6\left(m - \dfrac{1}{3}\right) - 4m + 2$

21. $\dfrac{1}{2}(-7y + 3) = y - 30$

22. $\dfrac{1}{6}x = \dfrac{1}{3}(x + 9)$

23. $6a + 1 - 2a = -5a + 3[-3a + 8 + 6a]$

24. $\dfrac{1}{12}(4x + 9) = \dfrac{2}{3}x$

25. $5(b - 1) - 3b = 7b - 5(1 + b)$

26. $1 - 2(3c - 5) = 10 - 6c$

27. $3(3 + 2d) = 23 - 3(5 - 2d)$

28. $4.1c = -11.07$

For Exercises 29–38, solve each for the indicated variable.

29. $A = bh$; for b

30. $C = \pi d$; for d

31. $a^2 + b^2 = c^2$; for a^2

32. $x + y = 1$; for x

33. $A = \dfrac{1}{2}bh$; for h

34. $F = \dfrac{mv^2}{r}$; for m

35. $E = mc^2$; for m

36. $V = \dfrac{1}{3}\pi r^2 h$; for h

37. $P = 2l + 2w$; for w

38. $A = \dfrac{1}{2}h(a + b)$; for h

For Exercises 39–44, *a. Solve.*
                                  b. Write the solution set.
                                  c. Graph the solution set.

39. $-6x > 18$

40. $-3x + 2x \leq 5$

41. $1 - 2z + 4 \leq -3(z + 1) + 7$

42. $\dfrac{1}{4}v < 3 + v$

43. $\dfrac{1}{2}(m - 1) > 5 - \dfrac{7}{2}$

44. $2c - (5c - 7) + 4c \geq 8 - 10$

For Exercises 45–52, translate to an equation or inequality and solve.

45. Six times a number is the same as negative eighteen.

46. Three subtracted from half of a number is equal to nine less than one-fourth of the number.

47. One less than two times the sum of a number and four results in one.

48. Twice the difference of a number and one added to triple the number yields twenty less than six times the number.

49. Thirteen is greater than three minus ten times a number.

50. Three increased by twice a number is less than three times the difference of a number and five.

51. Negative one-half of a number is greater than or equal to four.

52. Negative two is less than or equal to one minus one-fourth of a number.

For Exercises 53–62, solve.

53. Randy is snowskiing at Lake Tahoe. He knows the trail is 1.8 miles long and he is one-third of the way down. How far does he have left to ski?

54. Bonnie is making a shower curtain. The pattern says the curtain will use 5760 square inches. If the length is to be 72 inches, how wide will the curtain be?

55. Best Buy is selling an oak entertainment center. Elijah is concerned it will take up too much room in his den. Using the figure shown, determine the total area.

56. Find the area of the figure shown.

57. A long-distance company uses the formula $C = 3 + 0.05m$ to describe a calling plan where C is the total cost of a call lasting m minutes. If a customer uses 415 minutes, what is the total bill?

58. The surface area of a trunk is 2610 square inches. If the height of the trunk is 15 inches and the width is 36 inches, what is the length?

59. The area of the trapezoid shown is 58 square inches. Find the missing length.

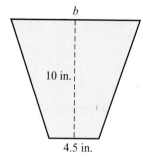

60. The cost of border material limits the circumference of a circular flower garden to 10 feet. What is the range of values for the diameter of the garden?

61. Terrel has to travel at least 350 miles each day to keep on his schedule. If he averages 60 miles per hour, what is the range of values for his travel time?

62. A company has a first quarter profit goal of at least $350,000. If the projected cost is $475,000, what range of values can revenue have to meet the profit goal?

1. Is $5m - 14 = 21$ an expression or an equation? Why?

2. Is $5y + 14 = y^2 - 9$ linear or nonlinear? Why?

3. Check to see if 3 is a solution for $3x^2 + x = 10x$.

4. Check to see if $\frac{2}{3}$ is a solution for $4 - 3(x + 2) = 0$.

Learning Strategy

Since your actual test most likely will not have the same problems as the Practice Test, it is important that you understand how to solve each type of problem. Before taking the Practice Test, write the definition, rule, procedure, or formula that applies to each problem. This will help you focus on how to solve each type. Once you've done this, take a break and then take the test. Repeat this process until every problem on the Practice Test seems easy to solve.

For Exercises 5–10, solve and check.

5. $6x + 9 = -9$

6. $-5.6 + 8a = 7a + 7.6$

7. $1 - 2(y + 8) - 6 + 6y = 8y + 15 - (3y + 10)$

8. $1 - 3(c + 4) + 8c = 1 - 42 + 5(c + 7)$

9. $\frac{1}{2}x + \frac{5}{2} = \frac{3}{2} - x$

10. $-\frac{1}{8}(6m - 32) = \frac{3}{4}m - 2$

For Exercises 11–14, solve for the indicated variable.

11. $2x + y = 7;\ x$

12. $I = Prt;\ r$

13. $A = \frac{1}{2}bh;\ h$

14. $C = \frac{5}{9}(F - 32);\ F$

For Exercises 15–20: a. Solve.
b. Write the solution set.
c. Graph the solution set.

15. $-5m + 3 > -12$

16. $-2(2 - x) \le -20$

17. $5(y + 3) + 5 - y \ge 7(8 + y)$

18. $6(p - 2) - 5 > 3 - 4(p + 6)$

19. $-0.3 < -0.7x + 0.7 + 0.8x$

20. $\frac{1}{5}(l + 10) \ge \frac{1}{2}(l + 5)$

For Exercises 21–24, translate and then solve.

21. Two-thirds of a number minus one-sixth is the same as twice the number.

22. The product of negative seven and a number added to twelve yields five.

23. Three subtracted from five times the difference of a number and two is equal to ten minus four times the difference of the number and one.

24. One minus a number is greater than twice the number.

For Exercises 25–30, solve.

25. A carpenter is to determine the amount of wood needed to trim a window. If the window is 4 feet by 6 feet, what is the perimeter of the window?

26. Find the area of the figure shown.

7 in.

14 in.

6.5 in.

27. The perimeter of a trapezoid shown is 63 centimeters. Find the length of the missing side.

20 cm

9 cm

27 cm

28. A rent-to-own store advertises a VCR for $10 down with $10 payments each week. The formula $c = 10 + 10w$ represents the cost, C, after making payments for w weeks. If the customer signs a contract for one year, what is the total cost of the VCR?

29. The area of a sandbox cannot exceed 9 square feet. If the width of the sandbox is to be 3 feet, what range of values can the length have?

30. In Dan's psychology class, if his average on five tests is at least 92, then he can be exempt from the final exam. His current scores are 90, 84, 89, and 96. What range of scores can he have on the last test that would allow him to be exempt from the final exam?

Problem Solving

In Chapter 2 we learned to solve linear equations and developed some techniques for solving problems, namely using formulas and translating key words. In this chapter we will expand the problem-solving techniques introduced in Chapter 2 and solve problems that are a little more complex. These problems will involve proportions, percents, and two unknown amounts.

Some of the exercises you encounter in this chapter will target a specific strategy such as translating key words or using a table to organize information. In the same way that an athlete uses exercise to target specific muscles, we use rather contrived problems to exercise specific problem-solving strategies and techniques. Most of the exercises an athlete performs in weight training are not sport-specific, meaning they are generic exercises designed to strengthen raw muscle tissue. Similarly, many of the problems in the chapter will not be what you would encounter in your job or everyday life. Rather, they are designed to exercise the strategies and techniques of problem solving.

> "Difficulty, my brethren, is the nurse of greatness—a harsh nurse, who roughly rocks her foster-children into strength and athletic proportion."
>
> —*WILLIAM CULLEN BRYANT (1794–1878)*
> U.S. poet, editor

> "It is one of man's curious idiosyncrasies to create difficulties for the pleasure of resolving them."
> —*JOSEPH DE MAISTRE (1753–1821)*
> French diplomat, philosopher

3.1 Ratios and Proportions

OBJECTIVES

1. Solve problems involving ratios.
2. Solve for a missing number in a proportion.
3. Solve proportion problems.
4. Use proportions to solve for missing lengths in figures that are similar.

OBJECTIVE 1 Solve problems involving ratios. In this section we will solve problems involving ratios.

DEFINITION

> *Ratio:* A comparison of two quantities using a quotient.

Ratios are usually expressed in fraction form. When a ratio is written in English, the word *to* separates the numerator and denominator quantities.

The ratio of **12** to **17** translates to $\dfrac{12}{17}$.

Numerator   Denominator

Note: The quantity preceding the word *to* is written in the numerator and the quantity following *to* is written in the denominator.

EXAMPLE 1 A small college has 2450 students and 140 faculty. Write the ratio of students to faculty in simplest form.

Solution The ratio of students to faculty $= \dfrac{2450}{140} = \dfrac{35}{2}$

The ratio in Example 1 indicates that there are 35 students for every 2 faculty members. Expressing a ratio as a **unit ratio** often makes the ratio easier to interpret.

DEFINITION *Unit ratio:* A ratio with a denominator of 1.

To express the ratio from Example 1 as a unit ratio, we divide.

$$\frac{2450}{140} = \frac{35}{2} = \frac{17.5}{1}$$

The unit ratio $\dfrac{17.5}{1}$ indicates that there are 17.5 students for every faculty member at the college.

YOUR TURN

The price of a 12.5-ounce box of cereal is $2.85. Write the unit ratio of price to weight.

OBJECTIVE **2** **Solve for a missing number in a proportion.** Some problems involve two equal ratios in a **proportion**.

DEFINITION **Proportion:** An equation in the form $\frac{a}{b} = \frac{c}{d}$, where $b \neq 0$ and $d \neq 0$.

In Section 1.2 we discussed equivalent fractions and saw that fractions like $\frac{3}{8}$ and $\frac{6}{16}$ are equivalent because $\frac{3}{8}$ can be rewritten as $\frac{6}{16}$ and $\frac{6}{16}$ can be reduced to $\frac{3}{8}$. We also saw that when fractions are equivalent, the cross products are equal.

$$16 \cdot 3 = 48 \qquad 8 \cdot 6 = 48$$
$$\frac{3}{8} \times \frac{6}{16}$$

Let's see why all proportions have equal cross products. We begin with the general form of a proportion from the definition: $\frac{a}{b} = \frac{c}{d}$, where $b \neq 0$ and $d \neq 0$.

We can use the multiplication principle of equality to clear the fractions by multiplying both sides by the LCD. In this case, the LCD is bd.

$$\frac{\overset{1}{\cancel{b}}d}{1} \cdot \frac{a}{\underset{1}{\cancel{b}}} = \frac{b\overset{1}{\cancel{d}}}{1} \cdot \frac{c}{\underset{1}{\cancel{d}}}$$

$$ad = bc$$

Note that ad and bc are the cross products.

$$ad \qquad\qquad bc$$
$$\frac{a}{b} \times \frac{c}{d}$$

Conclusion If two ratios are equal, then their cross products are equal.

RULE **PROPORTIONS AND THEIR CROSS PRODUCTS**

If $\frac{a}{b} = \frac{c}{d}$, where $b \neq 0$ and $d \neq 0$, then $ad = bc$.

ANSWER

$\dfrac{\$0.228}{1 \text{ oz.}}$

We can use cross products to solve for a missing number in a proportion.

EXAMPLE 2 Solve $\dfrac{7}{16} = \dfrac{x}{10}$.

Solution $10 \cdot 7 = 70$ $\qquad$ $16 \cdot x = 16x$ $\qquad$ Calculate the cross products.

$$70 = 16x$$

Set the cross products equal to one another.
Divide both sides by 16 to isolate x.

$$\frac{70}{16} = \frac{16x}{16}$$

Note: We could have expressed the answer as 4.375.

$$\frac{35}{8} = x$$

YOUR TURN

Solve $\dfrac{9}{y} = -\dfrac{5}{6}$.

Proportions can contain fractions or decimals.

EXAMPLE 3 Solve $\dfrac{3\frac{1}{5}}{12} = \dfrac{-9}{n}$.

Solution

$3\dfrac{1}{5} \cdot n = \dfrac{16}{5}n$ $\qquad$ $12 \cdot (-9) = -108$ $\qquad$ Calculate the cross products.

$$\frac{16}{5}n = -108$$

Set the cross products equal to one another.

$$\frac{\overset{1}{\cancel{5}}}{\cancel{16}} \cdot \frac{\overset{1}{\cancel{16}}}{\cancel{5}}n = \frac{-\overset{27}{\cancel{108}}}{1} \cdot \frac{5}{\cancel{16}}$$

Multiply both sides by $\dfrac{5}{16}$ to isolate n

$$n = -\frac{135}{4}$$

ANSWER

-10.8 or $-10\dfrac{4}{5}$

a. $\dfrac{\dfrac{2}{3}}{n} = \dfrac{-5}{4\dfrac{1}{2}}$

b. $-\dfrac{2.4}{8.1} = -\dfrac{t}{12}$

OBJECTIVE 3 Solve proportion problems. In a typical proportion problem, you are given a ratio and then asked to find an equivalent ratio. The key to translating problems to a proportion is to write the proportion so that the numerators and denominators correspond to each other in a logical way.

EXAMPLE 4 A company determines that it spends $254.68 every 5 business days on equipment failure. At this rate, how much would the company spend over 28 business days?

Understand We must calculate the amount spent on equipment failure over 28 business days. We are given an expense rate of $254.68 every 5 business days, which translates to the ratio $\dfrac{\$254.68}{5 \text{ days}}$. If we translate the rate with the unknown amount in the same way $\left(\dfrac{\$}{\text{days}}\right)$ we have:

$$\text{amount spent in 28 business days} = \dfrac{\$n}{28 \text{ days}}$$

Plan The words *at this rate* indicate that the ratios are equal, so we can set up a proportion and then solve for the unknown quantity.

Execute $\dfrac{\$254.68}{5 \text{ days}} = \dfrac{\$n}{28 \text{ days}}$

> **Note:** The units in the numerators and denominators correspond.
>
> *Tip:* One approach to setting up a proportion is to write the ratios so that the numerator units match and the denominator units match.

$28 \cdot 254.68 = 7131.04 \qquad 5 \cdot n = 5n$

$$\dfrac{254.68}{5} \diagup\!\!\!\!\diagdown \dfrac{n}{28}$$

$7131.04 = 5n$ Set the cross products equal.

$\dfrac{7131.04}{5} = \dfrac{5n}{5}$ Divide both sides by 5 to isolate *n*.

$1426.208 = n$

Answer Because the answer is in dollars, we round to the nearest hundredth. The company spends $1426.21 over a 28-day cycle.

Check Do a quick estimate to verify the reasonableness of the answer. The company spends $254.68 every 5 business days. Since 28 days is a little less than 6 times the number of days, the company must spend a little less than 6 times the $254.68 spent in the 28 business days. To estimate the calculation we round $254.68 to $250. We then have $6 \cdot \$250 = \$1500$, which is reasonably close to $1426.21.

Checking Proportion Problems

One of the nice things about proportion problems is that if you set the proportion up incorrectly, you will usually get an unreasonable answer. A quick estimate check can be a fast way to recognize when you have set up a proportion incorrectly. For example, suppose we had incorrectly set up Example 4 this way:

$$\frac{\$254.68}{5 \text{ days}} = \frac{28 \text{ days}}{\$n}$$

Warning: This setup is incorrect because the numerators and denominators do not correspond.

$254.68 \cdot n = 254.68n \qquad\qquad 5 \cdot 28 = 140$ **Cross multiply.**

$$\frac{254.68}{5} \diagup\!\!\!\!\diagdown \frac{28}{n}$$

$254.68n = 140$ **Set the cross products equal.**

$$\frac{254.68n}{254.68} = \frac{140}{254.68}$$ **Divide both sides by 254.68 to isolate n.**

$n \approx 0.55$ **Warning:** This answer indicates that the company spends only $0.55 over 28 business days based on spending $254.68 every 5 business days, which is clearly too low.

There are many correct ways to set up a proportion for a particular problem. For example, following are some other correct ways we could have set up the proportion for Example 4. Note that the proportions involve rates that correspond with each other.

$$\frac{5 \text{ days}}{\$254.68} = \frac{28 \text{ days}}{\$n} \qquad \frac{5 \text{ days}}{28 \text{ days}} = \frac{\$254.68}{\$n} \qquad \frac{28 \text{ days}}{5 \text{ days}} = \frac{\$n}{\$254.68}$$

Think:

In 5 days, $254.68 is spent.   In 28 days, $n is spent.

Think of this version as an analogy:

If 5 days means $254.68, then 28 days means $n.

This version is a rearrangement of the previous version.

PROCEDURE   **SOLVING PROPORTION APPLICATION PROBLEMS**

To solve proportion problems:

1. Set up the given ratio any way you wish.
2. Set the given ratio equal to the other ratio with the unknown so that the numerators and denominators correspond logically.
3. Solve using cross products.

YOUR TURN

Pam drove 351.6 miles using 16.4 gallons of gasoline. At this rate, how much gasoline would it take for her to drive 1000 miles?

OBJECTIVE **4** Use proportions to solve for missing lengths in figures that are similar. Consider triangles *ABC* and *DEF*.

 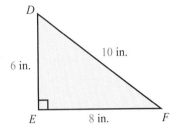

Though the triangles are different in size, their corresponding angle measurements are **congruent**.

DEFINITION *Congruent angles:* Angles that have the same measure. The symbol for congruent is $\cong$.

In symbols, we write: $\angle B \cong \angle E$, $\angle A \cong \angle D$, and $\angle C \cong \angle F$.

Besides having all angles congruent, notice that the sides of triangle *DEF* are twice as long as the corresponding sides of triangle *ABC*. Or, we could say that the sides of triangle *ABC* are all half as long as the corresponding sides of triangle *DEF*. In other words, the corresponding side lengths are proportional.

$$\begin{matrix} \text{Ratio of Triangle } DEF \\ \text{to} \\ \text{Triangle } ABC \end{matrix} \begin{matrix} \rightarrow \\ \\ \rightarrow \end{matrix} \frac{6}{3} = \frac{8}{4} = \frac{10}{5} = \frac{2}{1}$$

or

$$\begin{matrix} \text{Ratio of Triangle } ABC \\ \text{to} \\ \text{Triangle } DEF \end{matrix} \begin{matrix} \rightarrow \\ \\ \rightarrow \end{matrix} \frac{3}{6} = \frac{4}{8} = \frac{5}{10} = \frac{1}{2}$$

Because the corresponding angles in the two triangles are congruent and the side lengths are proportional, we say these are **similar figures**.

DEFINITION *Similar figures:* Figures with congruent angles and proportional side lengths.

ANSWER

≈ 46.6 gal.

EXAMPLE 5 The two trapezoids below are similar. Find the missing lengths.

Understand Since the figures are similar, the lengths of the corresponding sides are proportional. The corresponding sides can be shown as follows:

Ratio of larger trapezoid
to → $\dfrac{9.6}{4.2} = \dfrac{8}{a} = \dfrac{9.6}{b} = \dfrac{12.8}{c}$
Smaller trapezoid →

Plan Write a proportion to find each missing side length.

Execute It is not necessary to write a proportion to find b. Since two sides in the larger trapezoid measure 9.6 cm, the corresponding sides in the smaller trapezoid must be equal in length so we can conclude that $b = 4.2$ cm.

To find a:

$$\dfrac{9.6}{4.2} = \dfrac{8}{a}$$

$a \cdot 9.6 = 9.6a \qquad 4.2 \cdot 8 = 33.6$

$$\dfrac{9.6}{4.2} = \dfrac{8}{a}$$

$9.6a = 33.6$

$\dfrac{9.6a}{9.6} = \dfrac{33.6}{9.6}$

$a = 3.5$

To find c:

$$\dfrac{9.6}{4.2} = \dfrac{12.8}{c}$$

$c \cdot 9.6 = 9.6c \qquad 4.2 \cdot 12.8 = 53.76$

$$\dfrac{9.6}{4.2} = \dfrac{12.8}{c}$$

$9.6c = 53.76$

$\dfrac{9.6c}{9.6} = \dfrac{53.76}{9.6}$

$c = 5.6$

Answer The missing lengths are: $a = 3.5$ cm, $b = 4.2$ cm, and $c = 5.6$ cm.

Check We can verify the reasonableness of the answers by doing quick estimates of the missing side lengths. Comparing the smaller trapezoid to the larger, we see that the 4.2 cm side is a little less than half of the corresponding 9.6 cm side in the larger trapezoid. Each of the corresponding sides should follow this same rule. The 8 cm side of the larger trapezoid should correspond to a length of a little less than 4 for side a of the smaller trapezoid and the 12.8 cm side should correspond to a length of a little less than 6.4 for side c on the smaller trapezoid. These do check out.

YOUR TURN   **Find the missing lengths in the similar shapes.**

$a = 20\dfrac{3}{10}$ ft. or 20.3 ft.

$b = 14$ ft.

$c = 16\dfrac{4}{5}$ ft. or 16.8 ft.

3.1 EXERCISES

For Extra Help

InterAct Math

Addison-Wesley Math Tutor Center

Video 5/DVT 2

MyMathLab

MathXL

1. In the phrase "the ratio of the smallest side to shortest side," how do you determine the numerator and denominator of the ratio?

2. In your own words, define a proportion.

3. Explain how to determine whether two ratios are proportional.

4. Explain how to solve for a missing number in a proportion.

5. How do you set up and solve a word problem that translates to a proportion?

6. Why are $\dfrac{5}{8} \cdot \dfrac{x}{10}$ and $\dfrac{5}{8} = \dfrac{x}{10}$ different?

For Exercises 7–14, write the ratio in simplest form.

7. One molecule of lactose contains 12 carbon atoms, 22 hydrogen atoms, and 11 oxygen atoms.

   a. Write the ratio of carbon atoms to total atoms in the molecule.

   b. Write the ratio of carbon atoms to hydrogen atoms.

   c. Write the ratio of hydrogen atoms to oxygen atoms.

Of Interest

Lactose is a molecule found in milk. Its chemical formula is $C_{12}H_{22}O_{11}$. Notice that the subscripts indicate the number of atoms of each element in the molecule.

8. The roof of a house rises 0.75 feet for every 1.5 feet of horizontal length. Write the ratio of rise to horizontal length as a fraction in simplest form.

9. In a certain gear on a bicycle, the back wheel rotates $1\frac{1}{3}$ times for every 2 rotations of the pedals. Write the ratio of rotations of the back wheel to pedal rotations as a fraction in simplest form.

10. A recipe calls for $2\frac{1}{2}$ cups of flour and $1\frac{1}{3}$ cups of milk. Write the ratio of milk to flour as a fraction in simplest form.

11. In 2000, La-Z-Boy's stock earned $1.59 per share. The current price of La-Z-Boy stock is $19.50. Write a unit ratio that expresses the price to earnings. What does this ratio indicate?

12. The price of a 10.5-ounce can of soup is $0.79. Write the unit ratio that expresses the price to weight.

13. A 2-liter bottle of soft drink costs $1.09, and a 3-liter bottle of the same soft drink costs $1.49. Which is the better buy? Why?

14. The price of a 10.5-ounce box of corn flakes is $2.29, and a 14-ounce box of the same cereal is $3.19. Which is the better buy? Why?

For Exercises 15–16, use the following graph.

15. a. Write the unit ratio of convictions to total cases in 1985.

b. Write the unit ratio of convictions to total cases in 1995.

c. Compare the ratios for 1985 to 1995. What do these ratios indicate?

16. a. Write the unit ratio of imprisonment to convictions in 1985.

b. Write the unit ratio of imprisonment to convictions in 1995.

c. Compare the ratios for 1985 to 1995. What do these ratios indicate?

DEFENDANTS IN CASES CONCLUDED IN U.S. DISTRICT COURT

Source: Compendium of Federal Justice Statistics.

For Exercises 17 and 18, use the following table.

Median Annual Income of Year-Round Full-Time Workers 25 Years Old and Over, by Level of Education Completed and Sex: 1990 to 1998 (in constant 1998 dollars)

	Men				Women		
Year	All Education Levels	High School Graduate	Bachelor's Degree	Year	All Education Levels	High School Graduate	Bachelor's Degree
1990	38,328	33,240	48,935	1990	26,654	22,846	34,941
1992	37,244	31,694	48,046	1992	26,883	22,570	35,233
1994	36,779	30,837	48,023	1994	26,836	22,408	34,911
1996	37,007	31,903	47,628	1996	26,811	21,998	34,828
1998	37,906	31,477	51,405	1998	27,956	22,780	36,559

Source: U.S. Department of Education, National Center for Education Statistics, *Digest of Education Statistics,* 1999

17. a. In 1990, what was the unit ratio of women's median income to men's median income for all educational levels? What does this ratio indicate?

b. Calculate the same unit ratio for the year 1998.

c. Compare the ratio you found in 1990 to the ratio you found in 1998. What do these ratios indicate?

18. a. In 1992, what was the unit ratio of women's median income to men's median income for high school graduates? What does this ratio indicate?

b. Calculate the same unit ratio for the year 1996.

c. Compare the ratio you found in 1992 to the ratio you found in 1996. What do these ratios indicate?

For Exercises 19–26, determine whether the ratios are equal.

19. $\dfrac{15}{25} \overset{?}{=} \dfrac{3}{5}$

20. $\dfrac{7}{4} \overset{?}{=} \dfrac{15}{6.2}$

21. $\dfrac{10}{14} \overset{?}{=} \dfrac{2.5}{3.5}$

22. $\dfrac{15}{21} \overset{?}{=} \dfrac{130.2}{93}$

23. $\dfrac{15.2}{40.2} \overset{?}{=} \dfrac{60.8}{120.4}$

24. $\dfrac{6.5}{14} \overset{?}{=} \dfrac{10.5}{42}$

25. $\dfrac{\frac{2}{5}}{\frac{3}{10}} \overset{?}{=} \dfrac{15}{10\frac{1}{2}}$

26. $\dfrac{3\frac{1}{4}}{5\frac{1}{2}} \overset{?}{=} \dfrac{4\frac{2}{3}}{6\frac{1}{2}}$

For Exercises 27–38, solve for the missing number.

27. $\dfrac{2}{3} = \dfrac{7}{x}$

28. $\dfrac{7}{8} = \dfrac{y}{40}$

29. $\dfrac{-9}{2} = \dfrac{63}{n}$

30. $\dfrac{11}{m} = \dfrac{132}{-24}$

31. $\dfrac{21}{5} = \dfrac{h}{2.5}$

32. $\dfrac{29}{7} = \dfrac{k}{1.75}$

33. $\dfrac{-17}{b} = \dfrac{8.5}{4}$

34. $\dfrac{1.6}{4.96} = \dfrac{-4.8}{c}$

35. $\dfrac{2\frac{1}{4}}{6\frac{2}{3}} = \dfrac{d}{10\frac{2}{3}}$

36. $\dfrac{\frac{2}{5}}{4\frac{1}{2}} = \dfrac{7\frac{2}{3}}{j}$

37. $\dfrac{-1.5}{60} = \dfrac{-2\frac{7}{8}}{t}$

38. $\dfrac{u}{-4\frac{7}{16}} = \dfrac{-60}{1.5}$

For Exercises 39–50, solve.

39. LaTonia drives 110 miles in 3 hours. At this rate, how far will she go in 5 hours?

40. If 90 feet of wire weighs 18 pounds, what will 110 feet of the same type of wire weigh?

41. Carson is planning a trip to London. He knows that it takes 1.4608 U.S. dollars to equal a British pound. If he has $800 to take with him, how much will he have in British pounds?

42. Sydney is planning a trip to Mexico. It takes 9.7 Mexican pesos to equal 1 U.S. dollar. If she has $400 to take with her, how many pesos will she have?

43. Jodi pays taxes of $1040 on her house, which is valued at $154,000. She is planning to buy a house valued at $200,000. At the same rate, how much will her taxes be on the new house?

44. A dozen oranges costs $1.60. How much will 8 oranges cost?

45. Fletcher is building a model of an F-16 jet. If the scale is $1\frac{1}{2}$ inches to 1 foot, what length on the model represents 3.25 feet?

46. Mary Kennon learns in science class that baking powder can be made by mixing 21 parts of sodium bicarbonate with 47 parts of cream of tartar. She has already used 5 ounces of cream of tartar. How much sodium bicarbonate does she need to mix?

47. Ford estimates that its 2001 Expedition will travel 546 highway miles on one tank of gas. If the gas tank of the Expedition holds 26 gallons, how far can a driver expect to travel on 20 gallons?

48. Chevrolet estimates that its 2001 Tahoe will travel 442 highway miles on one tank of gas. If the gas tank of the Tahoe holds 26 gallons, how far can a driver expect to travel on 20 gallons?

49. The Hoover Dam has 17 generators and can supply all the electricity needed by a city of 750,000 people. If only 14 generators are running, how many people will be supplied with electricity?

50. Until the early 1950s the erosion rate for Niagara Falls was an average of 3 feet per year. The suggested rate of erosion may now be as low as 1 foot per 10 years for Niagara Falls. Given this rate, how much erosion may be expected over the next 50 years?

For Exercises 51–58, find the missing lengths in the similar figures.

51.

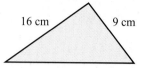

16 cm 9 cm

20 cm x

52.

$12\frac{1}{2}$ ft.

$6\frac{1}{3}$ ft.

15 ft.

w

53.

8 ft.

$9\frac{1}{2}$ ft. $9\frac{1}{2}$ ft.

$10\frac{2}{3}$ ft.

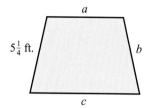

a

$5\frac{1}{4}$ ft.   b

c

54.

10.5 km

6.5 km

8 km

z

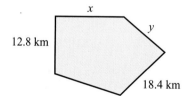

x

y

12.8 km

18.4 km

55. To estimate the height of the Eiffel Tower, a tourist uses the concept of similar triangles. The Eiffel Tower is found to have a shadow measuring 200 meters in length. The tourist has his own shadow measured at the same time. His shadow measures 1.2 meters. If he is 1.8 meters tall, how tall is the Eiffel Tower?

?

1.8 m

1.2 m

200 m

56. To estimate the height of the Great Pyramid in Egypt, the Greek mathematician Thales performed a procedure similar to that in Exercise 55, except that he supposedly used a staff instead of his body. Suppose the shadow from the staff was measured to be 6 feet and at the same time the shadow of the pyramid was measured to be $524\frac{2}{3}$ feet. If the staff was $5\frac{1}{2}$ feet tall, then what was the height of the pyramid?

$524\frac{2}{3}$ ft. $5\frac{1}{2}$ ft. 6 ft.

Not to scale.

Of Interest

The Greek mathematician Thales lived from 640 B.C. to 546 B.C. He is the first to have proposed the importance of proof and is considered to be the father of science and mathematics in the Greek culture. (*Source:* D. E. Smith, *History of Mathematics,* Dover, 1951)

57. To estimate the distance across a river, an engineer creates similar triangles. The idea is to create two similar right triangles as shown. Calculate the width of the river.

d 5 m 16.5 m 8 m

58. An engineer wants to estimate distance across the Amazon River at his current location. The engineer creates two similar right triangles as shown. Calculate the distance across the river.

d 5 m 340 m 24 m

Of Interest

At 6275 km long, the Amazon River is the second longest river in the world. Though the Nile River is slightly longer, the Amazon is the largest river in terms of watershed area, number of tributaries, and volume of water discharged.

A car is traveling up a hill that is 1 mile long at an average speed of 30 miles per hour. When the driver reaches the top of the hill, he starts down the other side. The downhill side is also 1 mile long. How fast must the driver go down the hill in order to average 60 miles per hour? (Hint: The answer is not 90 miles per hour.)

REVIEW EXERCISES

1. Define the set of integers.

For Exercises 2–5 solve and check.

2. $3x - 7 = x + 13$

3. $\dfrac{b}{5} = 15$

4. $25 + 2(x - 6) + 8x + 18 = 4(x + 5) + 53$

5. $\dfrac{2}{3}n + \dfrac{5}{4}n = \dfrac{23}{24}$

3.2 Percents

OBJECTIVES

1 Write a percent as a fraction or decimal number.

2 Write a fraction or decimal number as a percent.

3 Translate and solve percent sentences.

4 Solve problems involving percents.

OBJECTIVE 1 Write a percent as a fraction or decimal number.

The word *percent* is a compound word made from the prefix *per* and the suffix *cent*. The prefix *per* means "for each." The suffix *cent* comes from the Latin word *centum*, which means "100." Therefore, **percent** means "for each 100." In

DEFINITION **Percent:** A ratio representing some part out of 100.

other words, percent is a ratio with the denominator 100.

The symbol for percent is %. For example:

$$20\% = 20 \text{ out of } 100$$

Notice that the definition can be used to write percents as fractions or decimal numbers.

$$20\% = 20 \text{ out of } 100 = \frac{20}{100} = \frac{1}{5} = 0.2$$

PROCEDURE **REWRITING A PERCENT**

To write a percent as a fraction or decimal:

1. Write the percent as a ratio with 100 in the denominator.
2. Simplify to the desired form.

Note: When simplifying, remember that dividing a decimal number by 100 moves the decimal point two places to the left.

EXAMPLE 1 Write each percent as a decimal number and a fraction.

a. 20.5%

Solution Write 20.5 over 100, then simplify. In this case it might be helpful to write the decimal number first, then write the equivalent fraction from that decimal number.

$$20.5\% = \frac{20.5}{100} \qquad \text{Write as a ratio with 100 in the denominator.}$$

$$= 0.205 \qquad \text{Write the decimal form.}$$

$$= \frac{205}{1000} \qquad \text{Write the fraction form.}$$

$$= \frac{41}{200} \qquad \text{Reduce to lowest terms.}$$

b. $30\frac{1}{4}\%$

Solution Write $30\frac{1}{4}$ over 100, then simplify. In contrast to part *a,* you could find the fraction form first, and then find the decimal form.

$$30\frac{1}{4}\% = \frac{30\frac{1}{4}}{100} \qquad \text{Write as a ratio with 100 in the denominator.}$$

$$= 30\frac{1}{4} \div 100$$

$$= \frac{121}{4} \cdot \frac{1}{100} \qquad \text{Write an equivalent multiplication.}$$

$$= \frac{121}{400} \text{ or } 0.3025 \qquad \text{Multiply.}$$

> **YOUR TURN** Write each percent as a fraction in simplest form and as a decimal.
>
> **a.** 26% **b.** 9.25% **c.** $30\frac{2}{3}\%$

OBJECTIVE ② Write a fraction or decimal number as a percent. We have seen that $20\% = \frac{20}{100} = \frac{1}{5}$ or 0.2. To write a percent as a fraction or a decimal, we omit the % sign and divide by 100. To write a fraction or decimal as a percent, we should multiply by 100 and add the % sign.

ANSWERS

a. $\frac{13}{50}$ or 0.26

b. $\frac{37}{400}$ or 0.0925

c. $\frac{23}{75}$ or $0.30\overline{6}$

PROCEDURE   WRITING A FRACTION OR DECIMAL AS A PERCENT

To write a fraction or decimal number as a percent:
1. Multiply by 100%.
2. Simplify.

EXAMPLE 2 Write as a percent.

a. $\dfrac{5}{8}$

Solution $\dfrac{5}{8} = \dfrac{5}{\overset{}{\underset{2}{8}}} \cdot \dfrac{\overset{25}{\cancel{100}}}{1}\%$ *Multiply by 100%.*

$= \dfrac{125}{2}\%$ *Simplify.*

$= 62\dfrac{1}{2}\%$ or 62.5%

b. 0.267

Solution $0.267 = 0.267 \cdot 100\%$ *Multiply by 100%.*
$= 26.7\%$

CONNECTION

Multiplying an amount by 100% is a way of rewriting the amount because multiplying by 100% is equivalent to multiplying by 1. The percent symbol means $\dfrac{1}{100}$, so that 100% literally means:

$100 \cdot \% = 100 \cdot \dfrac{1}{100} = 1$

YOUR TURN Write as a percent.

a. $\dfrac{5}{9}$ **b.** 2.4

OBJECTIVE 3 Translate and solve percent sentences. To solve problems involving percents, it is often helpful to reduce the problem to the following simple percent sentence:

> A **percent** of a **whole** is a **part** of the whole.

Note that there are three pieces in the simple sentence: the *percent*, the *whole*, and the *part*. Any one of those three pieces could be unknown in a problem.

	A **percent** of	a **whole**	is a **part**.
Unknown part:	42% of	68	is what amount?
Unknown whole:	76% of	what number	is 63.84?
Unknown percent:	What percent of	72	is 63?

To solve for an unknown piece of a simple percent sentence, we can translate the sentence to an equation. We will look at two methods of translation: 1. translating word for word, and 2. translating to a proportion.

ANSWERS

a. $55\dfrac{5}{9}\%$ or $55.\overline{5}\%$

b. 240%

Method 1. Translate the sentence word for word:

1. Select a variable for the unknown.
2. Translate the word *is* to an equal sign.
3. If *of* is preceded by the percent then, it is translated to multiplication.
   If *of* is preceded by a whole number, translate it to division.

Method 2. Translate to a proportion by writing the following form:

$$\text{Percent} = \frac{\text{part}}{\text{whole}}$$

where the percent is expressed as a fraction with a denominator of 100.

First, let's consider the case where the part is unknown.

Simple Percent Sentence with an Unknown Part

EXAMPLE 3   42% of 68 is what number?

Note: This question can also be worded this way: "What is 42% of 68?"

Solution Notice the part is the unknown.

42%	of	68	is	what number?
↑		↑		↑
Percent	of	the whole	is	the part

Method 1. Word for word translation: Since *of* is preceded by the percent, it means multiply, and *is* means equals.

42% of 68 is what number?

Note: In order to calculate, we write the percent as a decimal (or fraction).

$$0.42 \cdot 68 = n$$
$$28.56 = n$$

Method 2. Proportion: The percent, which is a ratio, is equivalent to the ratio of the part to the whole, so we can write a proportion in the following form:

Note: This is the percent written as 42 to 100.

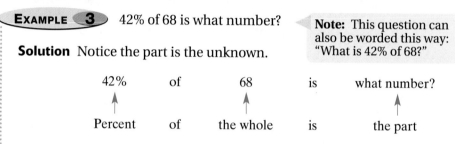

$$\text{Percent} = \frac{\text{part}}{\text{whole}}$$

$$\frac{42}{100} = \frac{n}{68} \quad \leftarrow \text{Part} \\ \leftarrow \text{Whole}$$

Note: The whole amount will always follow *of*.

2856 = 100*n*   Equate the cross products.

$$\frac{2856}{100} = \frac{100n}{100}$$ Divide both sides by 100 to isolate *n*.

28.56 = *n*

YOUR TURN

What number is 15% of 36?

ANSWER

5.4

Simple Percent Sentence with an Unknown Whole

Now consider the situation in which the whole amount is unknown.

EXAMPLE 4   76% of what number is 63.84?

Note: This question can be worded: "63.84 is 76% of what number?"

Solution   Note the three pieces:

76%	of	what number	is	63.84
↑		↑		↑
Percent	of	the whole	is	the part

Method 1. Word-for-word translation:

76% of what number is 63.84?

$$0.76 \cdot n = 63.84$$

$$0.76n = 63.84$$

$$\frac{0.76n}{0.76} = \frac{63.84}{0.76}$$

$$n = 84$$

Method 2. Proportion:

$$\text{Percent} = \frac{\text{part}}{\text{whole}}$$

Note: This is the percent written as 76 to 100.

$$\frac{76}{100} = \frac{63.84}{n}$$ ← Part
← Whole

$$76n = 6384$$ Equate the cross products.

$$\frac{76n}{76} = \frac{6384}{76}$$ Divide both sides by 76 to isolate n.

$$n = 84$$

YOUR TURN

48 is 120% of what number?

ANSWER

40

Simple Percent Sentence with an Unknown Percent

Finally, consider the case in which the percent is the unknown.

EXAMPLE 5 What percent of 72 is 63?

Note: This question can be worded: "63 is what percent of 72?"
or
"What percent is 63 out of 72?"
or
"What percent is 63 of 72?"

Note how 72, the whole amount, always follows *of*.

Solution Note the pieces.

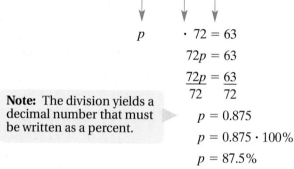

What percent   of   72   is   63?

Percent   of   the whole   is   the part

Method 1. Word-for-word translation:

What percent of 72 is 63?

$$p \cdot 72 = 63$$

$$72p = 63$$

$$\frac{72p}{72} = \frac{63}{72}$$

Note: The division yields a decimal number that must be written as a percent.

$$p = 0.875$$

$$p = 0.875 \cdot 100\%$$

$$p = 87.5\%$$

Method 2. Proportion:

Note: This is the unknown percent written as a ratio.

$$\frac{p}{100} = \frac{63}{72}$$

← Part

← Whole

$$72p = 6300$$ Equate the cross products.

$$\frac{72p}{72} = \frac{6300}{72}$$ Divide both sides by 72 to isolate p.

$$p = 87.5\%$$

Note: In calculating the cross products, we multiplied by 100, which yields an answer that is a percent.

As Example 5 shows, when the percent is unknown, the proportion method gives the answer as a percent, so you may want to consider using proportions in this case.

YOUR TURN

21.6 is what percent of 90?

ANSWER

24%

OBJECTIVE 4 **Solve problems involving percent.** If a problem involves a percent, it can be helpful to try to write the information in the form of a simple percent sentence.

> **PROCEDURE** SOLVING PERCENT APPLICATION PROBLEMS
>
> To solve problems involving percent:
> 1. Determine whether the percent, the whole, or the part is unknown.
> 2. Write the problem as a simple percent sentence.
> 3. Translate the sentence to an equation (word for word or proportion).
> 4. Solve for the unknown.

EXAMPLE 6 Twelve of the 28 students in an English class received an A. What percent of the class received an A?

Understand The unknown is the percent of the class that received an A. The number of students in the class is 28, which is the whole. The number of students that received an A, which is the part, is 12.

Plan Write a simple percent sentence, then solve. The percent is the unknown, so we will translate to a proportion because using a proportion yields an answer that is a percent.

Execute What percent of 28 is 12?

$$\frac{p}{100} = \frac{12}{28} \quad \begin{array}{l}\leftarrow \text{Part} \\ \leftarrow \text{Whole}\end{array}$$

$$28p = 1200 \qquad \text{Equate the cross products.}$$

$$\frac{28p}{28} = \frac{1200}{28} \qquad \text{Divide both sides by 28 to isolate } p.$$

$$p \approx 42.9\% \qquad \boxed{\textbf{Note:} \text{ We rounded to the nearest tenth of a percent.}}$$

Answer About 42.9% of the students in the class received an A on their first paper.

Check Verify that 42.9% of the 28 students is about 12 students.

$$0.429 \cdot 28 = 12.012$$

Since 12.012 is reasonably close to 12, we can conclude that 42.9% is correct.

YOUR TURN

ANSWER

1842

On May 18, 2001, *USA Today* reported that Envisionet plans to cut 700 jobs, which is 38% of its workforce. How many jobs does Envisionet have in all?

Percent Problems Involving Increase or Decrease

Sometimes a percent problem involves an increase or decrease of some initial amount. Sales tax, interest, and discount are situations in which an initial amount of money is increased or decreased by an amount of money that is a percent of the initial amount.

EXAMPLE 7 If the sales tax rate in a certain city is 6%, what is the total cost of:

Raisin Bran	$3.79
Milk	$1.29
Orange Juice	$1.49
Yogurt	$2.59

Understand The total cost is the purchase price plus the sales tax, which is a percent of the purchase price.

Plan We must first calculate the purchase price. We can then write a simple percent sentence and solve for the sales tax. Finally, we add the tax to the purchase price.

Execute Purchase price $= 3.79 + 1.29 + 1.49 + 2.59 = \$9.16$

Simple percent sentence: The sales tax is 6% of $9.16

$$x = 0.06 \cdot 9.16$$
$$x = 0.5496$$

Because the tax is a dollar amount, we round to the nearest hundredth. The sales tax is $0.55.

Answer The total cost is $9.16 + \$0.55 = \$9.71$.

Check Determine whether the total cost of $9.71 represents a 6% increase in the total sale of $9.16. To determine the percent of increase, we would need to calculate the amount of the increase by subtracting 9.16 from 9.71.

$$9.71 - 9.16 = 0.55 \qquad \text{This verifies that the final addition was correct.}$$

CONNECTION

The steps in this check will help us understand how to approach a type of problem in which the percent of increase or decrease is unknown. We will see this type of problem in Example 8.

Now, to determine the percent of the increase we can set up a proportion.

$$\frac{p}{100} = \frac{0.55}{9.16} \xleftarrow{} \text{Part} \atop \xleftarrow{} \text{Whole amount}$$

$9.16p = 55$ Equate the cross products.

$\dfrac{9.16p}{9.16} = \dfrac{55}{9.16}$ Divide both sides by 9.16 to isolate p.

$p \approx 6\%$ **Note:** This is rounded to the nearest percent. It is not exactly 6% because we rounded the sales tax amount.

YOUR TURN

An advertisement indicates that a coat is on sale at 30% off the marked price. If the marked price is $79.95, what will be the price after the discount?

The check in Example 7 is actually a common type of percent problem in which the percent of increase or decrease is unknown.

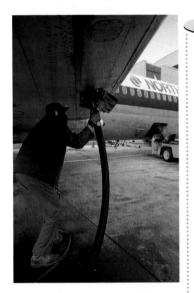

EXAMPLE 8 The Air Transport Association reports that airline fuel prices jumped from 72.1 cents in January of 2000 to 86.6 cents in January of 2001. What is the percent of the increase?

Understand We are to determine the percent of the increase given the initial price and the price after the increase.

Plan To determine the percent of the increase, we need the amount of the increase, which is found by subtracting the initial amount from the final amount.

Execute Amount of increase = $86.6 - 72.1 = 14.5$

Simple percent sentence: What percent of 72.1 is 14.5?

$$\frac{p}{100} = \frac{14.5}{72.1}$$

Note: We chose to translate to a proportion because the percent is the unknown.

$$72.1p = 1450$$ Equate the cross products.

$$\frac{72.1p}{72.1} = \frac{1450}{72.1}$$ Divide both sides by 72.1 to isolate p.

$$p \approx 20.1\%$$

Answer The percent of the increase in fuel price is about 20.1%.

Check Determine whether increasing the initial price of 72.1 cents by 20.1% results in a price of 72.1 cents.

$$0.201(72.1) = n$$
$$14.5 \approx n$$

Note: When rounded to the nearest tenth, this amount of the increase corresponds to our earlier calculation.

Now add the amount of the increase: $14.5 + 72.1 = 86.6$.

YOUR TURN

On May 29, 2001, Kmart's stock closed at a price of 11\frac{1}{2}$. On May 30, 2001, the stock closed at 11\frac{9}{32}$. Find the percent of the decrease in the price of Kmart's stock. (*Source:* Motleyfool.com)

ANSWER

$\approx 1.9\%$

1. Explain how to write a percent as a decimal or fraction.

2. Explain how to write a decimal or fraction as a percent.

3. When preceded by a percent, the word *of* indicates what operation?

4. Explain how to translate a simple percent sentence to a proportion.

5. When using word-for-word translation to find a percent, the calculation will not be a percent, whereas if you use the proportion method, the result will be a percent. Why?

6. When given an initial amount and a final amount after an increase, how do you find the percent of the increase?

For Exercises 7–18, write each percent as a decimal and as a fraction in simplest form.

7. 20%   8. 30%   9. 15%   10. 85%

11. 14.8%   12. 18.6%   13. 3.75%   14. 6.25%

15. $45\frac{1}{2}\%$ 16. $65\frac{1}{4}\%$ 17. $33\frac{1}{3}\%$ 18. $18\frac{1}{6}\%$

For Exercises 19–38, write as a percent rounded to the nearest hundredth.

19. $\frac{3}{5}$ 20. $\frac{1}{5}$ 21. $\frac{3}{8}$ 22. $\frac{5}{8}$

23. $\frac{5}{6}$ 24. $\frac{2}{3}$ 25. $\frac{4}{9}$ 26. $\frac{5}{11}$

27. 0.96   28. 0.42   29. 0.8   30. 0.7

31. 0.09   **32.** 0.01   **33.** 1.2   **34.** 3.58

35. 0.028   **36.** 0.065   **37.** 4.051   **38.** 0.007

For Exercises 39–50, translate word for word or to a proportion, then solve.

39. 2.5% of 12,450 is what number?

40. What number is $9\frac{1}{4}$% of 64?

41. 120% of 86 is what number?

42. What number is 250% of 62.8?

43. 15 is 30% of what number?

44. $45\frac{3}{5}$ is 5% of what number?

45. 7.8 is $12\frac{1}{2}$% of what number?

46. 13.5% of 954 is what number?

47. 8 is what percent of 24?

48. 24 is what percent of 8?

49. What percent is 19 out of 27?

50. 42 of 60 is what percent?

For Exercises 51–74, solve.

51. A county's annual property taxes are 4% of the market value of the property. How much is the tax on a property valued at $120,000? If the tax is paid in monthly installments, what amount must be paid each month?

52. How much is the tax on a house valued at $230,000 in the same county as that in Exercise 51? If the tax is paid in monthly installments, what amount must be paid each month?

53. To qualify for a loan, Alicia is told that her loan payment cannot exceed 20% of her gross monthly income. If her gross monthly income is $2314, what is the maximum that her payment can be to qualify?

54. In general, lenders do not want home buyers to spend more than 28% of their gross monthly income on a house payment. If a buyer has a monthly gross income of $3210, what is the most a lender will allow for a mortgage payment?

> ## Of Interest
>
> In finance, the ratio of the house payment to the gross monthly income is known as the *front-end ratio* and is one of the factors lenders use to determine whether someone qualifies for a loan.

55. In general, lenders do not want home buyers to spend more than 36% of gross income a month on debts. Using the buyer in Exercise 54, what is the most a lender will accept in total debt payments per month? If she makes the highest mortgage payment allowed each month, what does she have left to spend on other debt and still be within the lender's guidelines?

56. A TI-83 graphing calculator is normally sold at Circuit City for $99.99. If price is discounted $10, what is the percent of the discount?

57. A Best Buy ad has all major appliances on sale for 10% off the regular price. If the regular price of a refrigerator is $949, what is the sale price?

58. Dustin is planning to purchase a new Corvette. The sticker price for the car is $46,605. He plans to offer 12% less. How much is Dustin planning to offer for the car?

Of Interest

In finance, the ratio of the monthly debts to the gross monthly income is known as the *back-end ratio*. The front-end and back-end ratios, along with a person's credit history, are primary factors involved in determining loan qualification.

For Exercises 59 and 60, use the following circle graph. The graph shows expenditures for the city of Amarillo, Texas, in the 1997/1998 fiscal year.

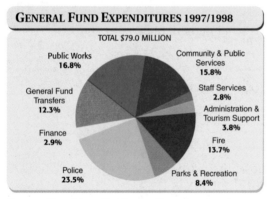

GENERAL FUND EXPENDITURES 1997/1998

TOTAL $79.0 MILLION

Public Works 16.8%
Community & Public Services 15.8%
General Fund Transfers 12.3%
Staff Services 2.8%
Administration & Tourism Support 3.8%
Finance 2.9%
Fire 13.7%
Police 23.5%
Parks & Recreation 8.4%

Source: City of Amarillo Finance Department

59. How much money was spent on police?

60. How much money was spent on parks and recreation?

61. A small company has the following costs in one quarter:

Category	Cost (in thousands)
Plant operations	124.2
Employee wages	248.6
Research and development	115.7

What percent of the company's total cost is spent on research and development?

62. A pharmaceutical company conducts a study on the side effects of a new drug. Below is a report of the data from the study.

	Number of Subjects
No side effects	1462
Headache	35
Rash	24
Fever	65

What percent of the subjects in the study experienced fever?

63. In the 2000–01 season, Allen Iverson made 762 field goals out of 1813 attempts. What percent of his field goal attempts did he score?

64. As of June 1, 2001, Rich Aurilia had the #1 batting average in the National League with 71 hits out of 192 at-bats. What percent of his at-bats are hits?

65. Margaret mixes 50 milliliters of a solution that is 10% HCl with a 100 milliliters of a solution that is 25% HCl. What percent of the resulting solution is HCl?

66. Leon invests $3400 in a stock that returns 12% in the first year and $2200 in stock that returns 15% during the same year. What is his total return as a percent of his total investment?

67. Video game sales in a department store increased from $2600 in November to $4550 in December. What was the amount of increase? What was the percent of increase?

68. After the Vallina family purchased a smaller car, they realized their monthly gasoline bill dropped from $162 per month to $124 per month. What is the percent of decrease in the monthly gasoline bill?

69. The price of a TI-82 calculator dropped from $120 to $79. What is the percent of decrease?

70. Fifteen years ago, James inherited a marble-top table. It was appraised at $420. The insurance company has asked that he have it reappraised. It is now worth $1200. What is the percent of increase?

71. At 9:30 A.M. on May 18, 2001, the Dow Jones Industrial Average (DJIA) was 11,216. At 4:00 P.M., the DJIA was at 11,249. What was the percent of the increase? (*Source: USA Today*, May 18, 2001)

72. Using the graph to the right, calculate the percent of the increase in the minimum wage from 1938 to 1997.

73. A report indicates that on May 29, 2001, the price of Microsoft's stock closed at $69.19, which was a change of −1.63% from the previous day's closing price. What was the price of Microsoft's stock at closing on May 28?

74. The U.S. Bureau of the Census reported a 61.8% increase in the number of single fathers from 1990 to 2000. If the number of single fathers in 2000 was 2,190,989, how many single fathers were there in 1990? (*Source: USA Today*, May 18, 2001)

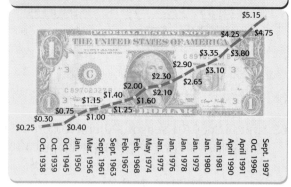

THE FEDERAL HOURLY MINIMUM WAGE SINCE ITS INCEPTION:

Collaborative Exercises

Occupation Growth

Complete the following table by calculating the amount of change and the percent of the increase for each occupation, then answer the questions. Round to the nearest tenth of a percent.

The 10 Fastest-Growing Occupations, 1998–2008 (numbers in thousands of jobs)

| Occupation | Employment | | Change | |
	1998	2008	Amount	Percent
Computer engineers	299	622		
Computer support specialists	429	869		
Systems analysts	617	1194		
Database administrators	87	155		
Desktop publishing specialists	26	44		
Paralegals and legal assistants	136	220		
Personal care and home health aides	746	1179		
Medical assistants	252	398		
Social and human service assistants	268	410		
Physician assistants	66	98		

Source: Bureau of Labor Statistics

1. In 1998, in which occupation were the greatest number of people employed?

2. By 2008, which occupation is projected to have the greatest number of people employed?

continued

3. Explain why the two occupations you listed in Exercises 1 and 2 are not at the top of the list.

4. Which is a better indicator of the demand for people in a particular occupation: the number of people employed in a particular year, the amount of change projected from 1998 to 2008, or the percent of increase in employment from 1998 to 2008?

5. Based on your conclusions in Exercise 4, what occupation will have the greatest demand? What college majors might have the greatest potential for employment in that occupation? Write your conclusions and present them to the class.

REVIEW EXERCISES

1. Multiply: $-3(x + 7)$

2. Combine like terms: $7xy + 4x - 8xy + 3y$

3. Simplify: -3^0

For Exercise 4, translate to an equation and then solve.

4. Three times the sum of a number and nine is equal to eleven less than the number.

5. Solve: $\dfrac{1}{5} = \dfrac{n}{10.5}$

6. The figures below are similar. Find the missing side length.

4 cm

6 cm

7 cm

?

3.3 Problems with Two or More Unknowns

OBJECTIVES

  1 Solve problems involving two unknowns.

  2 Use a table in solving problems with two unknowns.

OBJECTIVE **1** **Solve problems involving two unknowns.** Sometimes problems have two or more unknown amounts. In general, if there are two unknowns, then there will be two relationships. Our approach to solving these problems will be to use the relationships to write an equation that we can solve. The following outline gives a more specific approach.

PROCEDURE SOLVING PROBLEMS WITH TWO OR MORE UNKNOWNS

To solve problems with two or more unknowns:

1. Determine which unknown will be represented by a variable.
   Tip: Let the unknown that is acted on be represented by the variable.
2. Use one of the relationships to describe the other unknown(s) in terms of the variable.
3. Use the other relationship to write an equation.
4. Solve the equation.

First, let's look at problems in which we simply translate key words.

EXAMPLE **1** One number is four times another. The sum of the numbers is 45. What are the numbers?

Understand We must find the two unknown numbers. We are given two relationships that we can translate.

Plan Translate each relationship to an equation, and then solve.

Execute Use the first relationship to determine which unknown will be represented by a variable.

Relationship 1: One number is four times another.

one number $= 4 \cdot$ another number

one number $= 4 \cdot n$

> **Note:** In general, we choose the unknown amount that is acted on to be the variable. Because *another number* is multiplied by 4 and *one number* is isolated, we will let n represent *another number*.

Now we use the second relationship to write an equation that we can solve.

Relationship 2: The sum of the numbers is 45.

$$\text{one number} + \text{another number} = 45$$
$$4n \quad + \quad n \quad = 45$$
$$5n = 45 \qquad \text{Combine like terms.}$$
$$\frac{5n}{5} = \frac{45}{5} \qquad \text{Divide both sides by 5}$$
$$\text{to isolate } n.$$
$$n = 9$$

Answer   Use relationship 1 to answer the question. If $n = 9$ and one number $= 4 \cdot n$, then one number $= 4(9) = 36$, so the two numbers are 9 and 36.

Check   Verify that 9 and 36 satisfy both relationships described in the problem. 36 is four times 9, and the sum of 36 and 9 is 45.

YOUR TURN

One positive number is one-third of another positive number. The larger number minus the smaller number is equal to 15.

Geometry Problems

Sometimes problems do not give all the needed relationships in an obvious manner. This often occurs in geometry problems in which the definition of a geometry term provides the needed relationship. For example, we must know that *perimeter* means the sum of the lengths of the sides in order to solve a problem that involves perimeter.

EXAMPLE 2   A rectangular frame is to be built so that the length is 4 inches more than the width. The perimeter of the frame is to be 72 inches. Find the dimensions of the frame.

Understand   Draw a picture.

List the relationships:

Relationship 1: The length is 4 inches more than the width.

Relationship 2: The perimeter must be 72 inches.

Plan   Translate to an equation using the key words, and then solve the equation.

ANSWER

22.5 and 7.5

Execute   Relationship 1: The length is 4 inches more than the width.

length = 4 + width
length = 4 + w

Note: Since 4 is added to the width and the length is isolated, we let w represent width.

Now use the other given relationship to write an equation to solve.

Relationship 2: The perimeter is 72 inches. Recall that the perimeter is the sum of the lengths of all the sides.

Perimeter = 72

length + width + length + width = 72
4 + w + w + 4 + w + w = 72

$4w + 8 = 72$ **Combine like terms.**

$\underline{\quad -8 \quad -8 \quad}$ **Subtract 8 from both sides to isolate the**
$4w + 0 = 64$ **variable term.**

$\dfrac{4w}{4} = \dfrac{64}{4}$

Divide both sides by 4 to isolate w.

$w = 16$

Answer   The width, w, is 16 inches. To determine the length, use relationship 1. If the length = 4 + w, then the length = 4 + 16 = 20 inches.

Check   Verify that both conditions in the problem are satisfied. First, the length must be 4 inches more than the width, and 20 inches is 4 inches more than 16 inches. Second, the perimeter must be 72 inches, which is true because 20 + 16 + 20 + 16 = 72 inches.

YOUR TURN

Karen is constructing a rectangular flower garden with a wooden border. She wants the length to be twice the width and has 30 feet of border material. Find the dimensions of the garden.

In Example 2, the word *perimeter* was a key word that helped us write the equation that we solved. There are other terms in geometry that, by their definitions, give information to help us translate a problem into an equation. For example, a problem might involve angles that are **complementary** or **supplementary**.

ANSWER

5 ft. by 10 ft.

Complementary angles: Two angles are complementary if the sum of their measures is 90°.

In the following figure, $\angle ABD$ and $\angle DBC$ are complementary because $32° + 58° = 90°$.

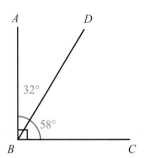

Note: The small square indicates a 90° angle.

Supplementary angles: Two angles are supplementary if the sum of their measures is 180°.

A straight line forms an angle that measures 180°. Any line that intersects a straight line will divide the 180° angle into two angles that are supplementary. In the following figure, line segment DB joins line segment AC, forming two angles, $\angle ABD$ and $\angle DBC$. These two angles are supplementary angles because $20° + 160° = 180°$.

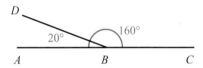

Now look at a problem containing one of these terms.

EXAMPLE 3 A steel beam is to be welded to a cross beam, creating two angles, an inner angle and an outer angle. If the outer angle is to be 15° less than twice the inner angle, what are the angle measurements?

Understand We must find the inner and outer angle measurements. The sketch for this situation is shown (notice it is like the preceding figure).

Next, list the relationships:

Relationship 1: The outer angle is 15° less than twice the inner angle.

Relationship 2: inner angle + outer angle = 180

Note: The second relationship comes from the sketch. It shows that the inner and outer angles are supplementary, which means the sum of their measures is 180°.

Plan Translate each relationship to an equation, and then solve.

Execute Relationship 1: The outer angle is 15° less than twice the inner angle.

$$\text{outer angle} = 2 \cdot \text{inner angle} - 15$$

$$\text{outer angle} = 2a - 15$$

Note: Since the measure of the inner angle is multiplied by 2, we should let the variable represent the inner angle. Let's use the letter a.

Now use relationship 2 to write an equation that we can solve.

$$\text{inner angle} + \text{outer angle} = 180$$
$$a \quad + \quad 2a - 15 \quad = 180$$
$$3a - 15 = 180 \qquad \text{Combine like terms.}$$
$$\underline{+15 \quad +15} \qquad \text{Add 15 on both sides.}$$
$$3a + 0 = 195$$
$$\frac{3a}{3} = \frac{195}{3} \qquad \text{Divide both sides by 3 to isolate } a.$$
$$a = 65$$

Answer The inner angle, represented by a, is 65°. According to relationship 1, the outer angle $= 2a - 15$, so the outer angle $= 2(65) - 15 = 115°$.

Check Verify that all the conditions in the problem are satisfied. The outer angle measuring 115° is 15° less than twice the inner angle measuring 65°. The sum of the inner and outer angles is 180°, which verifies that they are supplementary.

YOUR TURN

The wooden joists supporting a roof are connected as shown. The measure of $\angle DBC$ is 6° less than three times the measure of $\angle ABD$. Find the measure of $\angle DBC$ and $\angle ABD$.

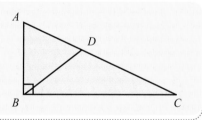

Problems Involving Consecutive Integers

Another group of problems in which the relationships are not obvious involve **consecutive integers**. The word *consecutive* means that the numbers are in sequence.

Consecutive integers: ... $-3, -2, -1, 0, 1, 2, 3, \ldots$

What relationship exists in the pattern of consecutive integers? Notice to get from one integer in the sequence to the next integer, we add 1. For the sake of finding the general relationship, suppose the first integer in a sequence of consecutive integers is 1. By noting how we get from the number 1 to its consecutive integer, we can find a general pattern when the first integer is an unknown number, n, as the following table illustrates.

	With 1 as the First Integer	With n as the First Integer
First integer:	1	n
Next integer:	$1 + 1 = 2$	$n + 1$
Next integer:	$1 + 2 = 3$	$n + 2$

What would be the pattern for consecutive odd integers? Suppose the first odd integer in a sequence of odd integers is 1, then we can write:

	With 1 as the First Odd Integer	With n as the First Odd Integer
First odd integer:	1	n
Next odd integer:	$1 + 2 = 3$	$n + 2$
Next odd integer:	$1 + 4 = 5$	$n + 4$

Warning: The pattern $n, n + 1, n + 3, \ldots$ is not correct for consecutive odd integers.

Notice that the pattern would be exactly the same for consecutive even integers. Suppose the first even integer in a sequence of even integers is 2, then we can write:

	With 2 as the First Even Integer	With n as the First Even Integer
First even integer:	2	n
Next even integer:	$2 + 2 = 4$	$n + 2$
Next even integer:	$2 + 4 = 6$	$n + 4$

EXAMPLE 4 The sum of three consecutive even integers is 78. What are the integers?

Understand The key word *sum* means to "add." If the smallest of the even integers is n, then the pattern for three consecutive even integers is:

smallest unknown even integer: n
next even integer: $n + 2$
third even integer: $n + 4$

Plan Translate to an equation, then solve.

Execute Translate:

The sum of three consecutive even integers is 78.

Smallest even integer	+	Next even integer	+	Third even integer	= 78
n	+	$n + 2$	+	$n + 4$	= 78

$$3n + 6 = 78 \quad \text{Combine like terms.}$$
$$\underline{-6 \qquad -6} \quad \text{Subtract 6 from both sides.}$$
$$3n + 0 = 72$$
$$3n = 72$$
$$\frac{3n}{3} = \frac{72}{3} \quad \text{Divide both sides by 3 to isolate } n.$$
$$n = 24$$

Answer The smallest of the three unknown integers, represented by n, is 24.

Next even integer: $n + 2 = 24 + 2 = 26$
Third even integer: $n + 4 = 24 + 4 = 28$

Check The numbers 24, 26, and 28 are three consecutive even integers and $24 + 26 + 28 = 78$.

YOUR TURN

The sum of three consecutive integers is 96. What are the integers?

OBJECTIVE **2** **Use a table in solving problems with two unknowns.** We have seen how using key words, drawing pictures, making lists, and looking for patterns have helped us understand a problem. Now we consider a group of problems in which a table is helpful in organizing information.

EXAMPLE **5** A home improvement store sells two sized cans of wood stain. A small can sells for $8.95 and a large can sells for $15.95. One day the store sold twice as many of the large cans as the small cans. If the total revenue that day for these cans of wood stain was $694.45, how many cans of each size were sold?

Understand We know that the total revenue is $694.45. From this we can say:

Revenue from the small cans	+	Revenue from the large cans	= 694.45

We can also describe the revenue from the sale of each size can. We know the price of each size. If we knew the number of cans sold, we could multiply that number by the price per can to calculate the revenue from that particular size.

Price per can	·	Number of cans sold	=	Revenue for that size can

ANSWER

31, 32, 33

With so much information, it is helpful to use a table to list it all. We use a four-column table with the labels: categories, price, number of cans, and revenue. The categories are the two sizes of wood stain.

Categories	Price per Can	Number of Cans	Revenue
Small	$ 8.95	n	$8.95n$
Large	$15.95	$2n$	$15.95(2n)$

We were given these.

We selected n to represent the number of small cans, then translated "twice as many of the large cans as the small cans."

We multiplied straight across because:

$$\frac{\text{Price}}{\text{per can}} \cdot \frac{\text{Number}}{\text{of cans}} = \frac{\text{Revenue from}}{\text{each size}}$$

The expressions in the last column in the table give the revenue from the sale of each size can.

Plan Translate the information to an equation, and then solve.

Execute Now we can use our initial relationship:

$$\frac{\text{Revenue from}}{\text{the small cans}} + \frac{\text{Revenue from}}{\text{the large cans}} = 694.45$$

$$8.95n + 15.95(2n) = 694.45$$
$$8.95n + 31.9n = 694.45 \qquad \text{Combine like terms.}$$
$$40.85n = 694.45$$
$$\frac{40.85n}{40.85} = \frac{694.45}{40.85} \qquad \text{Divide both sides by } 40.85 \text{ to isolate } n.$$
$$n = 17$$

Answer The number of small cans sold, n, is 17. The number of large cans sold, $2n$, is 34 (found by multiplying $2 \cdot 7$).

Check First, verify that twice the number of small cans sold, 17, is 34, which is true. Second, verify that the total revenue from the sale of wood stain is $694.45.

$$8.95(17) + 15.95(34) = 152.15 + 542.30 = 694.45 \qquad \text{It checks.}$$

The table used in Example 5 is a common type of table. In general, when a problem involves two or more categories, we can construct a table with a column labeled *Categories* and a row for each category in the problem. We then create a column for each parameter described in the problem. For example, parameters in a problem might be the *value* of each item, the *number* of items, and the total *amount* of money in each item, and the table would be:

Categories	Value	Number	Amount

Since *value · number = amount,* the column labeled *Amount,* contains the products of the expressions in the Value and Number columns.

YOUR TURN **Complete a table, write an equation, and then solve.**

A computer store sells two different ink cartridges for a particular printer. The black ink cartridge sells for $12.75 and the color cartridge sells for $24.95. A company orders 5 more black ink cartridges than color cartridges. If the total cost is $629.25, how many of each cartridge were purchased?

Now consider a problem in which a total number of items is given and you must determine how they are split up.

EXAMPLE 6 A company sells two versions of the same software product. The regular version sells for $29.95. The deluxe version sells for $39.95. Due to an inventory error, the number sold of each version was lost. However, the company knows that there were 458 units of sales of the two versions combined and the total revenue was $14,677.10. How many of each version was sold?

Understand We are given the combined income, $14,677.10, which indicates the following relationship:

$$\text{Revenue from the regular version} + \text{Revenue from the deluxe version} = 14{,}677.10$$

Because the problem involves two different items, their value, and the number of each item, we can use a four-column table to organize the information. We begin filling in the table with the categories and the given value of each version.

Categories	Selling Price (Value)	Number	Revenue (Amount)
Regular	29.95		
Deluxe	39.95		

Filling in the number column is tricky. We are told that a total of 458 units were sold, so we can say:

$$\text{Number of the regular version} + \text{Number of the deluxe version} = 458$$

Notice we can isolate one of the two unknowns using related subtraction statements. We will isolate the number of the regular version and let n represent the number of the deluxe version so that we have:

$$\text{Number of the regular version} = 458 - \text{Number of the deluxe version}$$

$$= 458 - n$$

Note: By letting the variable represent the number of the larger-valued item, which is the deluxe version, we will avoid negative coefficients in solving the equation.

ANSWER

20 black ink and 15 color

Now we can complete the table.

Categories	Selling price (Value)	Number	Revenue (Amount)
Regular	29.95	$458 - n$	$29.95(458 - n)$
Deluxe	39.95	n	$39.95n$

In general, if given a total number of items, select a variable to represent the number of one of the items. The number of the other item will be:

$$\text{total number} - \text{variable}$$

Since selling price · number = revenue, we multiply straight across the columns to get the expressions in the revenue column.

Plan   Translate the information to an equation, and then solve.

Execute   Use the initial relationship:

$$\boxed{\text{Revenue from the regular version}} + \boxed{\text{Revenue from the deluxe version}} = 14{,}677.10$$

$$
\begin{array}{ll}
29.95(458 - n) + \quad\quad 39.95n = 14{,}677.10 & \\
13{,}717.10 - 29.95n + 39.95n = 14{,}667.10 & \text{Distribute.} \\
13{,}717.10 + \quad 10n = 14{,}667.10 & \text{Combine like terms.} \\
\underline{-13{,}717.10 \quad\quad -13{,}717.10} & \text{Subtract 13,717.10 from} \\
0 + \quad 10n = \quad\quad 950.00 & \text{both sides.} \\
\dfrac{10n}{10} = \dfrac{950}{10} & \text{Divide both sides by 10 to isolate } n. \\
n = 95 &
\end{array}
$$

Answer   The number of the deluxe versions, n, is 95. The number of the regular versions, $458 - n$, is 363 (found by subtracting $458 - 95$).

Check   First, verify that the sum of 95 deluxe versions and 363 regular versions gives 458 copies of the software sold, which is true. Second, check the total revenue.

$$29.95(363) + 39.95(95) = 10{,}871.85 + 3795.25 = 14{,}677.10$$

It checks.

YOUR TURN   **Complete a table, write an equation, and then solve.**

A company offers two dental plans for its employees: individual coverage at $22.75 each month, or family coverage at $87.50 each month. The company has 26 employees with dental coverage paying a total of $1757 each month in premiums. How many employees have individual coverage and how many have family coverage?

ANSWER

8 have individual coverage, 18 have family coverage

3.3 EXERCISES

For Extra Help

InterAct Math

Addison-Wesley Math Tutor Center

Video 5/DVT 2

MyMathLab

MathXL

1. Suppose a problem gives the following information: "one number is five times a second number." Which is easier: letting the *one number* be represented by n or letting the *second number* be represented by n? What expression would describe the other unknown?

2. Suppose you are given that a rectangle has a width that is 5 centimeters less than the length. Which is easier: representing the length or the width with a variable? What expression would describe the other unknown?

3. In describing consecutive even integers or consecutive odd integers, we use the same pattern:

   Consecutive Odd

   first odd integer = n
   second odd integer = $n + 2$
   third odd integer = $n + 4$

   Consecutive Even

   first even integer = n
   second even integer = $n + 2$
   third even integer = $n + 4$

   Explain why the same pattern can be used to describe both odd and even consecutive integers.

4. Consider the following problem:
   Grant has a jar containing only quarters and half-dollars. If there are a total of 42 coins and their total value is $13.75, how many of each type of coin are there?
   a. Though we could let either unknown be represented by n, what is advantageous about letting n represent the number of half-dollars, which is the greater-valued coin?

   b. If n represents the number of half-dollars, what expression describes the number of quarters?

For Exercises 5–36, translate to an equation, then solve.

5. The larger of two numbers is 5 less than twice the smaller number. The sum of the two numbers is 28. What are the numbers?

6. The second of two numbers is three times the first. Twice the sum of 15 and the first number is equal to the 15 plus the second. What are the numbers?

7. The greater of two numbers is 2 more than the lesser. The sum of 25 times the greater number and 50 times the lesser number is 500. What are the numbers?

8. The difference between two numbers is 6. If three times the larger is subtracted from 52, the answer is the same as when 56 is subtracted from seven times the smaller number. Find the numbers.

9. One number is three times another. The sum of the numbers is 52. What are the numbers?

10. A positive number is one-fifth of another positive number. The larger number less the smaller number is 8. What are the numbers?

11. Marie is four times as old as her youngest sister Susan. In 8 years, Marie will be only twice as old as Susan. What are their ages now?

12. Jason is three times as old as his brother Michael. In 5 years, Jason will be twice as old as Michael. What are their ages now?

13. Arianna is 25 years older than Emma. In 15 years, Arianna will be twice as old as Emma. What will their ages be in 15 years?

14. Bob is now twice as old as his son. Sixteen years ago he was four times as old. How old is each now?

15. The length of a rectangular garden is 20 feet more than its width and the perimeter is 240 feet. What are the length and width?

16. The length of a rectangular hallway is three times its width. The perimeter is 32 feet. Find the length and width.

17. The perimeter of a rectangular frame is 36 centimeters. If the length is 3 less than twice the width, find the length and width.

18. The width a rectangular field is 12 meters less than its length. Find the length and width if the perimeter is 400 meters.

19. The width of a rectangular bathroom is $\frac{2}{3}$ of its length. What are its length and width if the perimeter is 25 feet?

20. An artist creates a rectangular sculpture with a width that is $\frac{3}{5}$ of the length. If the perimeter of the sculpture is 32 feet, find the length and width.

21. The base of an isosceles triangle is 9 meters shorter than the other two sides. If the perimeter is 69 meters, find the length of each side.

22. Two sides of an isosceles triangle are each 3 inches less than the base. If the perimeter is 30 inches, find the length of each side.

23. Two angles are supplementary. One of the angles is 20° more than twice the other angle. What are the measures of the two angles?

24. Two angles are supplementary. One of the angles is 30° less than the other angle. What are the measures of the two angles?

25. Two angles are complementary. One angle is 15° less than two times the other angle. What are the measures of the two angles?

26. Two angles are complementary. The smaller angle is $\frac{1}{4}$ of the measure of the larger angle. What are the measures of the two angles?

27. The sum of two consecutive integers is 147. What are the integers?

28. The sum of three consecutive integers is 279. What are the integers?

29. The sum of three consecutive even integers is 234. What are the integers?

30. The sum of three consecutive even integers is 42. What are the integers?

31. The sum of four consecutive odd integers is 48. What are the integers?

32. The sum of three consecutive odd integers is 339. What are the integers?

33. There are three consecutive integers such that the sum of the first and second integers decreased by the third integer will result in 68. What are the integers?

34. Twice the lesser of two consecutive odd integers decreased by the greater integer is 53. What are the integers?

35. Find three consecutive odd integers such that twice the smallest plus three times the largest will result in the middle integer being increased by 70. What are the integers?

36. Find three consecutive odd integers whose sum is 5 more than twice the next odd integer.

For Exercises 37–44, complete a table, write an equation, then solve.

37. The local high school is having a talent show. The tickets cost $3.50 for students and $5.00 for the general public. Although the organizers lost track of the ticket count, they know that 110 more student tickets had been sold than general public tickets and that the total sales were $980. How many of each type of ticket were sold?

38. A vending machine has 12-ounce and 16-ounce drinks. The 12-ounce drinks are $0.50 and the 16-ounce drinks are $1.00. If 3600 total drinks were sold in one month, and the total sales were $2225, how many of each size drink were sold?

39. Darryl has some $5 bills and $10 bills in his wallet. If he has a total of 16 bills worth a total of $110, then how many of each bill is in the wallet?

40. A tax preparation software is sold in the standard version for $14.95 and the deluxe version for $10 more. If a store sold 174 copies of the software for total sales of $3141.30, how many of each type were sold?

41. Ember is playing Monopoly and has two different types of money. The more valuable bill is worth 10 times the less valuable bill. She has 12 of the more valuable bills and 23 of the other bills for a total of $715. What is the value of each kind of bill?

42. Kelly is purchasing plants from a garden center where the prices are based on the size of the pot. She is purchasing 13 plants in one-gallon pots and 7 plants in half-gallon pots. A single one gallon and single half gallon have a combined price of $14.50. The total for all the plants she is buying is $155.50. What is the individual price of each size pot?

43. Rachel has two different stocks. She notices in the newspaper that one of her stocks is $5 more valuable than the other. She has 36 shares of the more valuable stock and 21 shares of the other stock. Her total assets in the stocks are $351. How much is each stock worth?

44. Aimee sells textbooks and instructor's materials. The combined price of a tutorial CD-ROM and a study guide is $24. One day she sold 6 CD-ROM and 13 study guides for a total of $207. Find the individual prices of the CD-ROM and study guide.

PUZZLE PROBLEM

The product of three integers is 84. Two of the integers are prime numbers and the third is not. If the sum of the composite number and one of the two prime numbers is 17, then what are the three numbers?

REVIEW EXERCISES

1. Check to see if $x = -2$ is a solution for $5x + 7 = x - 1$.

2. Is $3x - 8 = 12$ a linear equation?

For Exercises 3–5, solve and check.

3. $\dfrac{-3}{4}x = 12$ **4.** $x - 5 = -27$ **5.** $9x - 7(x + 2) = -3x + 1$

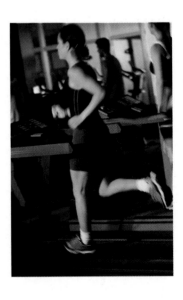

6. Carla knows from running on a treadmill that her average running speed is about 4.5 mph. If she maintains this rate running on roads in her neighborhood for 1.5 hours, what distance will she travel? (Use $d = rt$)

3.4 Rates

OBJECTIVES

1. Solve problems involving two objects traveling in opposite directions.
2. Solve problems involving two objects traveling in the same direction.

In Section 3.3, we developed the use of tables to organize information. In this section we will use a similar table to organize information in problems that involve distance, rate, and time. Recall that the general formula relating distance, rate, and time is $d = rt$.

OBJECTIVE 1 **Solve problems involving two objects traveling in opposite directions.** First, consider the situation in which two people or objects are traveling in opposite directions. Whether the two people or objects are moving in opposite directions toward or away from each other does not matter because the math involved will be the same.

EXAMPLE 1 In practicing maneuvers, two fighter jets fly toward each other. One flies east at 582 miles per hour and the other flies west at 625 miles per hour. If the two are 22 miles apart, how much time will it take for them to meet?

Understand Draw a picture of the situation:

Eastbound jet at 582 mph   Westbound jet at 625 mph

22 miles

Note: The westbound jet will travel the greater distance because it is going faster.

From the diagram we see that the sum of the individual distances traveled will be the total distance separating the two jets, which is 22 miles.

$$\boxed{\text{Distance traveled by the eastbound jet}} + \boxed{\text{Distance traveled by the westbound jet}} = 22$$

Use a table to find expressions for the individual distances.

Learning Strategy

If you are a tactile learner, you may find it helpful to simulate the situations in rate problems with toy cars or pencils.

Categories	Rate	Time	Distance
Eastbound	582 mph	t	$582t$
Westbound	625 mph	t	$625t$

Both jets start at the same time and meet at the same moment in time, so they traveled the same amount of time, t.

Multiplying the rate value by the time value gives the distance value.

Plan   Use the information to write an equation, then solve.

Execute

$$\boxed{\begin{array}{c}\text{Distance traveled by}\\\text{the eastbound jet}\end{array}} + \boxed{\begin{array}{c}\text{Distance traveled by}\\\text{the westbound jet}\end{array}} = 22$$

$$582t \quad + \quad 625t \quad = 22$$
$$1207t = 22$$
$$\frac{1207t}{1207} = \frac{22}{1207}$$
$$t \approx 0.018$$

CONNECTION

When we combined the coefficients in the like terms, we added the individual speeds of the jets, yielding the relative closing speed of the jets.

$$582 \text{ mph} + 625 \text{ mph} = 1207 \text{ mph}$$

The closing speed for these jets is 1207 mph, which is the same as if one of them remained still and the other traveled 1207 mph for the entire 22 miles.

The same reasoning applies if the two jets were going away from each other. They would separate at a rate of 1207 mph.

Note: If solving for time, the time unit will match the unit of time in the rate. In this problem, the rate is in miles per *hour*, so the time unit is hours.

Answer   The jets will meet in 0.018 hours, which is approximately 1.09 minutes, or about 1 minute and 6 seconds.

Check   Verify that in 0.018 hours, the jets will travel a combined distance of 22 miles.

Note: The westbound jet goes farther because it is traveling faster.

$$\text{Eastbound jet: } d = 582(0.018) = 10.476 \text{ mi.}$$
$$\text{Westbound jet: } d = 625(0.018) = 11.25 \text{ mi.}$$

The combined distance, in miles, is $10.476 + 11.25 = 21.726 \approx 22$. Because we rounded the time, 21.726 is reasonable.

We can summarize how to solve problems involving two people traveling in opposite directions as follows:

> **PROCEDURE** **TWO OBJECTS TRAVELING IN OPPOSITE DIRECTIONS**
>
> To solve for time when two objects are moving in opposite directions:
>
> 1. Use a table with columns for categories, rate, time, and distance. Use the fact that rate · time = distance.
> 2. Write an equation that is the sum of the individual distances equal to the total distance of separation.
> 3. Solve the equation.

YOUR TURN

Kari is running west at 5 miles per hour along a trail. Vernon is walking east on the same trail at $3\frac{1}{4}$ miles per hour. If they are $\frac{1}{2}$ mile apart, how long will it be until they meet?

OBJECTIVE **2** **Solve problems involving two objects traveling in the same direction.** If two objects are traveling in the same direction, we can use the same type of four-column table to organize the information. However, we will see that the equation is different.

EXAMPLE **2** Carol and Richard are traveling north in separate cars on the same highway. Carol is traveling at 65 miles per hour and Richard at 70 miles per hour. Carol passes Exit 102 at 1:30 P.M. Richard passes the same exit at 1:45 P.M. At what time will Richard catch up to Carol?

Understand To determine the time at which Richard catches up to Carol, we must calculate the amount of time it will take him to catch up to her. We can then add that amount of time to 1:45 P.M.

Richard passed the same exit 15 minutes after Carol did. In order to use $d = rt$, the time units must match the time units in the rate. Since 65 and 70 were in miles per hour (mph), the time units must be hours. What portion of an hour is 15 minutes? Since there are 60 minutes in an hour, we can write:

$$\frac{15}{60} = \frac{1}{4} = 0.25 \text{ hr.}$$

Now we can begin completing the table. We let t represent the amount of time it takes Richard to catch up. Carol's travel time is a bit tricky. Suppose Richard catches up at 2:30 P.M. Notice this is 1 hour after Carol passed Exit 102 and 45 minutes after Richard passed Exit 102. We see that at the time Richard catches up, Carol's travel time from Exit 102 is 15 minutes (0.25 hours) more than Richard's travel time from the same exit. This will be the case no matter what time he catches up, so Carol's travel time is $t + 0.25$.

ANSWERS

$\frac{2}{33}$ hr. or 3 min. and 38 sec.

Categories	Rate	Time	Distance
Carol	65	$t + 0.25$	$65(t + 0.25)$
Richard	70	t	$70t$

Since $d = rt$, we multiply the rate and time values to equal the distance values.

What can we conclude about the distances they each will have traveled when Richard catches up? Notice Exit 102 is the common reference point from which we describe the time for Richard to catch up. When Richard catches up, he and Carol will be the same distance from that exit, so their distances are equal.

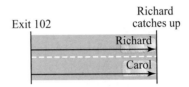

Plan Set the expressions of their individual distances equal, and solve for t.

Execute Richard's distance = Carol's distance (at the time he catches up)

CONNECTION

The coefficients of the like terms are the individual rates, so $70 - 65 = 5$ means that Richard's rate relative to Carol's is 5 mph. This means he must make up the 16.25 miles that separates them at a rate of 5 mph.

$$
\begin{aligned}
70t &= 65(t + 0.25) \\
70t &= 65t + 16.25 \\
\underline{-65t} \quad &\underline{-65t} \\
5t &= 0 + 16.25 \\
\frac{5t}{5} &= \frac{16.25}{5} \\
t &= 3.25
\end{aligned}
$$

CONNECTION

Since 65 is the rate and 0.25 is the additional amount of time, their product, 16.25 miles, is the distance that Carol is ahead of Richard. Richard must make up this distance to catch her.

CONNECTION

Dividing 16.25 miles by 5 mph calculates the time it takes Richard to make up the distance between Carol and him.

Answer It will take Richard 3.25 hours, which is 3 hours and 15 minutes, after passing Exit 102 to catch up to Carol. Since he passed Exit 102 at 1:45 P.M., he catches up to Carol at 5:00 P.M.

Check Verify that Richard and Carol are equal distances from Exit 102 after traveling for their respective times. If Richard travels for 3.25 hours after Exit 102, then Carol travels for $3.25 + 0.25 = 3.5$ hours after Exit 102. Using $d = rt$:

Richard: $d = (70)(3.25)$ Carol: $d = (65)(3.5)$
$d = 227.5$ mi. $d = 227.5$ mi.

Following is a summary of the process for problems involving two objects traveling in the same direction in which we are to calculate the time for one of the objects to catch up to the other.

PROCEDURE **TWO OBJECTS TRAVELING IN THE SAME DIRECTION**

To solve problems involving two objects traveling in the same direction in which the objective is to determine the time for one object to catch up to the other:

1. Use a table to organize the rates and times. Let t represent the time for the object to catch up, and add the time difference to t to represent the other object's time.
2. Set the expressions for the individual distances equal.
3. Solve the equation.

YOUR TURN

Juan and Angela are bicycling along the same trail. Juan passes a marker at 9:00 A.M. and Angela passes the same marker at 9:05 A.M. Juan is traveling at 8 miles per hour while Angela is traveling at 10 miles per hour. What time will Angela catch up to Juan?

ANSWER

9:25 A.M.

3.4 EXERCISES

1. Suppose two objects begin moving toward each other on a collision course. What can we conclude about the amount of time it takes them to meet?

2. Suppose two objects start "back to back" and travel in opposite directions. Object 1 travels $24t$ miles while object 2 travels $45t$ miles. In the following figure, which arrow represents which object? Explain.

?   start   ?

3. Suppose the two objects described in Exercise 2 are separated by 10 miles after traveling t hours. How do you write the equation?

4. A car passes under a bridge on a highway. Forty-five minutes later, a second car passes under the same bridge traveling in the same direction at a speed greater than the first car. Suppose the second car catches up to the first car in t hours. Write an expression for the first car's travel time (in hours) since passing under the bridge.

5. Draw a picture of the situation described in Exercise 4.

6. What can you say about the individual distances of the cars described in Exercise 4?

Exercises 7–18 involve people or objects moving in opposite directions. Complete a four-column table, write an equation, and then solve.

7. An Expedition and a Camaro leave a gas station and travel in opposite directions. The Camaro travels at 50 miles per hour and the Expedition travels at 45 miles per hour. In how many hours will the two cars be 380 miles apart?

8. Two planes leave Denver at the same time, one flying north and the other flying south. If the plane flying south is traveling at 653 miles per hour and the plane flying north is traveling 560 miles per hour, after how many hours will the planes be 2426 miles apart?

9. If a plane flying from Cincinnati travels west at 900 kilometers per hour and another leaving at the same time travels east at 1050 kilometers per hour, in how many hours will the planes be 6825 kilometers apart?

10. Monica and Chandler pass each other along a straight road traveling in opposite directions. Monica is driving at 40 miles per hour and Chandler at 50 miles per hour. In how many hours will the cars be 225 miles apart?

11. Two cars leave a restaurant at the same time with one heading north and the other heading south. If the first car travels at 55 miles per hour and the second travels at 60, how long will it take them to be 230 miles apart?

12. If two planes leave the Dallas airport at the same time with one flying west at 530 miles per hour and the other flying east at 620 miles per hour, how long will it take them to be 1725 miles apart?

13. Two trains started at the same time from Grand Central Station and traveled in opposite directions. One traveled at a rate of 45 miles per hour and the other at the rate of 60 miles per hour. In how many hours were they 210 miles apart?

14. Two trains start at the same time from Victoria Station and travel in opposite directions. One travels at a rate of 40 miles per hour and the other at the rate of 55 miles per hour. After how many hours will they be 190 miles apart?

15. Kim and Karmen begin walking toward each other from opposite directions at 8 A.M. If Kim is walking at 3.5 miles per hour and Karmen is walking at 4 miles per hour, what time will they meet if they are 12 miles apart?

16. Two trains traveling on parallel tracks are going toward each other from a distance of 378 miles. If the freight train is moving at 22 miles per hour and the passenger train is moving at 50 miles per hour, how long will it take for them to pass each other?

17. Two cars pass each other traveling in opposite directions on a highway. One car is going $1\frac{1}{2}$ times as fast as the other. At the end of $3\frac{1}{2}$ hours, they are $192\frac{1}{2}$ miles apart. Find each car's rate.

18. Dustin drives 5 miles per hour faster than Kelly. If he leaves town one hour after Kelly and they travel in opposite directions, they will have driven equal distances after Dustin has traveled seven hours. How far apart are they?

Exercises 19–30 involve people or objects moving in the same direction.
Complete a table, write an equation, and then solve.

19. At 9 A.M., a freight train leaves Seattle traveling at 45 miles per hour. At 10 A.M., a passenger train leaves the same station traveling in the same direction at 55 miles per hour. How long will it take the passenger train to overtake the freight train? How far will they be from the station at this time?

20. At 6 A.M., a freight train leaves Washington, D.C., traveling at 50 miles per hour. At 9 A.M., a passenger train leaves the same station traveling in the same direction at 75 miles per hour. How long will it take the passenger train to overtake the freight train? How far will they be from the station at this time?

21. Dan and Kathy are both traveling south for spring break. Dan is taking a bus traveling 60 miles per hour and Kathy is driving her own car at 70 miles per hour. If Dan's bus passes Jacksonville at 2:00 P.M. and Kathy passes through Jacksonville at 2:30 P.M., what time will Kathy catch up to Dan?

22. Two truckers are delivering materials from the same site. One leaves the site at 8 A.M. while the other leaves at 8:15 A.M. If the first trucker travels at 60 miles per hour and the second at 65 miles per hour, at what time will the second catch up to the first?

23. Daphne and Niles are both driving west on Route 66. At 4:00, Daphne is 5 miles west of Niles. A little later, Niles passes Daphne. If Niles is driving at 55 miles per hour and Daphne at 50 miles per hour, at what time does Niles pass Daphne?

24. Janet and Paul were both traveling west from Charleston, West Virginia, to the Rupp Arena in Lexington, Kentucky, for a concert. At 6:00 P.M. Janet, traveling at 45 miles per hour, was 10 miles west of Paul. A little later, Paul, traveling at 50 miles per hour, passed Janet. What time did Paul pass Janet?

25. Jane leaves Oklahoma City at 9 A.M. heading for Kansas City. Harrison leaves at 10 A.M. on the same highway, heading toward Kansas City. By driving 25 miles per hour faster, Harrison overtakes Jane at 1 P.M.

   a. How fast is Jane driving?

   b. How far have they traveled when Harrison catches up with Jane?

26. Morrison leaves Cincinnati at 4 P.M. heading for New York City. Barbara leaves at 4:30 P.M. on the same highway, heading toward New York City. By driving 10 miles per hour faster, Barbara overtakes Morrison at 7:30 P.M.

   a. How fast is Barbara driving?

   b. How far have they traveled when Barbara catches up with Morrison?

27. If two cars start driving in the same direction from the same place, while one is traveling 45 miles per hour and the other 50 miles per hour, how long will it take them to be 30 miles apart?

28. A car traveling 70 miles per hour passes a truck traveling 62 miles per hour in the same direction on the highway. If they maintain their speeds, how long will it take them to be 1 mile apart?

29. If two joggers leave the same point and run in the same direction, while one is jogging at 1.5 miles per hour and the other jogs at 2.5 miles per hour, how long will it take them to be 4 miles apart?

30. Two brothers start a bicycling race at the same time. If one brother rides at a rate of 15 miles per hour and the other brother rides at a rate of 12 miles per hour, how long will it take them to be 0.75 miles apart?

For Exercises 31–34, use a table and the techniques discussed in this section to solve.

31. On the way to a ski weekend in Vail, Carlita drove 2 hours in a snowstorm. When it finally quit snowing, she was able to increase her speed by 35 miles per hour and drove another 3 hours. If the entire trip was 325 miles, how fast did she drive when it finally quit snowing?

32. While driving from Los Angeles, Tori had a visibility problem due to the smog. After $1\frac{1}{2}$ hours, she hit the open highway outside of the city and was able to increase her speed by 40 miles per hour driving another $5\frac{1}{2}$ hours. If the entire trip was 514 miles, how fast was she traveling in the smog?

33. A trucker traveling along a highway encounters road construction so that traffic is down to a single lane. After driving 30 minutes, the road construction ends and the trucker is able to increase his speed by 64 miles per hour for 4.5 hours. If his entire trip is 303 miles, what was his speed in the road construction? What was his speed after the road construction?

34. An athlete is training for a triathlon. She is working on the running and cycling legs of the event. She runs for 45 minutes then bikes for 1.5 hours. If her combined distance running and cycling is 31.5 miles and her cycling speed is 3 times her speed running, what is her running speed?

REVIEW EXERCISES

1. Simplify: $|14 - 9 \cdot 3| + (4 - 9)^2$

2. Evaluate $-3x^2 - 4y^3$, where $x = 5$ and $y = -2$.

3. Simplify: $9x - 5y - 7 + 12y + 13 - x$

4. What property of arithmetic is illustrated by $3 \cdot (5 \cdot 7) = 3 \cdot (7 \cdot 5)$?

5. Solve: $0.15n + 0.4n = 0.25(200)$

6. 12% of what number is 84?

3.5 Investment and Mixture

OBJECTIVES

1 Use a table to solve problems involving two investments.

2 Use a table to solve problems involving mixtures.

In Sections 3.3 and 3.4 we began using tables to organize information in problems with two unknowns. In this section we continue the use of tables in problems involving investments and those involving chemical mixtures.

OBJECTIVE 1 **Use a table to solve problems involving two investments.** When money is invested in an account that earns a percentage of the money invested, the amount earned is called *interest* and is represented by I. The amount of money invested is called the *principal* and is represented by P. When dealing with interest, the *percentage rate*, r, is usually an annual percentage rate (APR), which is used to calculate the interest if the principal is invested for one year. The relationship for the interest after one year is $I = Pr$.

For example, if a person invests $500 at 8% for one year, the interest would be calculated as follows:

$$I = 0.08(500)$$
$$I = \$40$$

> **CONNECTION**
>
> The formula $I = Pr$ comes from the basic percent sentence:
>
> Percent   of a   whole   is a   part   of the whole.
>
> Percent   of the   principal   is the   interest.
>
> r · P = I

Let's explore problems in which an investor invests money in two different accounts.

EXAMPLE 1 Han invests a total of $8000 in two different accounts. The first account earns 5% while the second account earns 8%. If the total interest earned after 1 year is $565, what principal was invested in each account?

Understand Since Han invests a total of $8000, we can say:

Principal invested in first account + Principal invested in second account = 8000

Note that we can isolate one of the unknown amounts by writing a related subtraction statement. We will isolate the principal invested in the first account and let P represent the principal invested in the second account.

$$\begin{array}{l}\text{Principal invested in first account} = 8000 - \text{Principal invested in second account}\\ \qquad\qquad\qquad\qquad\qquad = 8000 - P\end{array}$$

CONNECTION ——

This is the same idea that we used in Example 6 of Section 3.3. By letting the variable represent the principal in the account with the larger APR, we will avoid negative coefficients in solving the equation.

We can use a table similar to those used in prior sections. Because $I = Pr$, we need columns for rate, principal, and interest, plus a column for the account names.

Accounts	Rate	Principal	Interest
First account	0.05	$8000 - P$	$0.05(8000 - P)$
Second account	0.08	P	$0.08P$

These rates were given. Don't forget to write the percents as decimals.

From our analysis above, if P represents the principal invested in the second account, then the principal invested in the first account is $8000 - P$.

Multiply each principal by the corresponding rate to get the expressions for interest.

rate · principal = interest

Plan   Write an equation. Then solve for P.

Execute   Since the total interest after 1 year is $565, we can say:

$$\begin{array}{l}\text{Interest from first account} + \text{Interest from second account} = \text{total interest}\end{array}$$

$$\begin{array}{ll}0.05(8000 - P) + \qquad 0.08P \quad= 565 & \\ 400 - 0.05P + 0.08P = 565 & \text{Distribute.}\\ 400 + 0.03P = 565 & \text{Combine like terms.}\\ \underline{-400 \qquad\qquad -400} & \text{Subtract 400 from both sides.}\\ 0 + 0.03P = 165 & \\ \dfrac{0.03P}{0.03} = \dfrac{165}{0.03} & \text{Divide both sides by 0.03.}\\ P = 5500 & \end{array}$$

Answer   The principal Han invested in the second account, P, is $5500. In the first account, he invested $8000 - P$, or $2500 (found by subtracting $8000 - 5500$).

Check   Verify that investing $5500 at 8% and 2500 at 5% results in a total of $565 in interest.

$$5500(0.08) + 2500(0.05) = 440 + 125 = 565 \qquad \text{It checks.}$$

The following procedure can be used as a general guide for solving the type of problems illustrated in Example 1.

> **PROCEDURE** SOLVING PROBLEMS INVOLVING INVESTMENT
>
> To solve problems involving two interest rates, in which the total interest is given:
>
> 1. Use a table to organize the interest rates and principals. Multiply the individual rates and principals to get expressions for the individual interests.
> 2. Write an equation that is the sum of the expressions for interest set equal to the given total interest.
> 3. Solve the equation.

YOUR TURN

Marvin invests $12,000 in two plans. Plan 1 is at an APR of 6% while plan 2 is at an APR of 9%. If the total interest after 1 year is $828, what principal was invested in each plan?

OBJECTIVE 2 **Use a table to solve problems involving mixtures.** Chemicals are often mixed to achieve a solution that has a particular concentration. Concentration refers to the portion of a solution that is pure. For example, an 80-milliliter solution may have a 5% concentration of hydrochloric acid (HCl), which means that 5% of that 80 milliliters is pure HCl and the rest is water. To calculate the actual number of milliliters that is pure HCl, we use the simple percent sentence:

$$\text{Percent of a whole is a part of the whole.}$$
$$0.05 \cdot 80 = 4$$

This means that 4 milliliters out of the 80 milliliters is pure HCl. The general relationship is:

$$\text{concentration} \cdot \text{whole solution volume} = \text{volume of the particular chemical}$$

Let's consider a problem in which two solutions are mixed to obtain a solution with a particular concentration.

EXAMPLE 2 Margaret has a bottle containing 50 milliliters of 10% HCl solution and a bottle of 25% HCl solution. She wants a 20% solution. How much of the 25% solution must be added to the 10% solution so that a 20% concentration is created?

Understand Because more than one solution is involved, a table is helpful in organizing the information. There are three solutions in this problem: the 10% solution, the 25% solution, and the new 20% solution that is created by mixing the 10% and 25% solutions. The table therefore has a row for each solution. We need columns for the percent of HCl, the solution volume, and the volume of HCl in each solution. Because we are to find the volume of the 25% solution, we choose a variable, *n* to represent this volume.

ANSWER

$3600 at 9% and $8400 at 6%

Solutions	Concentration of HCl	Volume of Solution	Volume of HCl
10% solution	0.10	50	0.10(50)
25% solution	0.25	n	0.25n
20% solution	0.20	50 + n	0.20(50 + n)

These are the concentration percents written in decimal form.

Because the 10% solution and 25% solution are combined to form the 20% solution, we add their volumes to get the volume of the 20% solution.

Multiply straight across the columns to generate the expressions for the volume of HCl in each solution.

$$\text{Concentration} \cdot \frac{\text{Volume}}{\text{of solution}} = \frac{\text{Volume}}{\text{of HCl}}$$

Plan   Write an equation that describes the mixture. Then solve for n.

Execute   The 10% solution and 25% solution combine to equal the 20% solution. We can say:

$$\frac{\text{Volume of HCl}}{\text{in the 10% solution}} + \frac{\text{Volume of HCl}}{\text{in the 25% solution}} = \frac{\text{Volume of HCl}}{\text{in the 20% solution}}$$

$$0.10(50) + 0.25n = 0.20(50 + n)$$

$5 + 0.25n = 10 + 0.20n$ **Multiply 0.10(50) and distribute.**

$$\begin{array}{rcl} 5 + 0.25n &=& 10 + 0.20n \\ -0.20n & & \quad -0.20n \\ \hline 5 + 0.05n &=& 10 \ + \ 0 \end{array}$$ **Subtract 0.20n from both sides.**

$$\begin{array}{rcl} -5 & & -5 \\ \hline 0 + 0.05n &=& 5 \end{array}$$ **Subtract 5 from both sides.**

$$\frac{0.05n}{0.05} = \frac{5}{0.05}$$ **Divide both sides by 0.05.**

$$n = 100$$

Answer   100 milliliters of the 25% solution must be added to 50 milliliters of 10% solution to create a solution that is 20% HCl.

Check   Verify that the volume of HCl in the two original solutions combined is equal to the volume of HCl in the combined solution.

$$0.10(50) + 0.25(100) \stackrel{?}{=} 0.20(150)$$
$$5 + 25 \stackrel{?}{=} 30$$
$$30 = 30 \quad \text{True.}$$

Following is a procedure that can be used as a general guide for solving problems of the type illustrated in Example 2.

YOUR TURN

ANSWERS

150 ml

Andre has 75 milliliters of a 5% sulfuric acid solution. How much of a 20% sulfuric acid solution must be added to create a solution that is 15% sulfuric acid?

3.5 EXERCISES

1. Suppose an investor has $10,000 to invest in two accounts. The following table lists some ways to split the principal that he is considering.

Account 1	Account 2
$9000	$1000
$8500	?
?	$6400
$3200	?

   a. Complete the table.

   b. How did you determine the unknown principals in the table?

   c. In general, if he invests P in one of the accounts, how do you describe the amount invested in the other account?

2. An investor always invests into two accounts according to the pattern shown in the following table.

Account 1	Account 2
$500	$1000
$700	$1400
$900	?
?	$2400

   a. Complete the table.

   b. In general, if he invests P in account 1, then how would you describe the amount invested in account 2?

   c. If he invests P in account 2, then how would you describe the amount invested in account 1?

3. Given the total interest from an investment into two plans, once you have the expressions for the two interest amounts, how do you write the equation?

4. When completing a table for mixture problems, what are the three categories?

5. The following table shows some possible combinations of two solutions.

a. Look for a pattern, then complete the table.

Solution 1	Solution 2	Combined Solution
25 ml	50 ml	75 ml
40 ml	80 ml	?
100 ml	?	?

b. How do you determine the amount of solution 2?

c. If solution 1 is represented by n, how would you describe solution 2?

d. If solution 1 is represented by n, how would you describe the combined solution?

6. a. If the volume of solution 1 from Exercise 5 is represented by n, and the solution is 20% HCl, write an expression for the volume of HCl in solution 1.

b. Refer to your answer to part c of Exercise 5. If 5% of solution 2 is HCl, write an expression for the volume of HCl in solution 2.

c. Refer to your answer to part d of Exercise 5. Suppose the combined solution is to be a 10% HCl solution. Write an expression for the volume of HCl in the combined solution.

Exercises 7–20 involve investment. Complete a table, write an equation, then solve.

7. Janice invests in a plan that has an APR of 7%. She invests three times as much in a plan that has an APR of 10%. If the total interest from the investments is $1702, how much was invested in each plan?

8. Gayle invests in a plan that has an APR of 5%. She invests twice as much in a plan that has an APR of 8%. If the total interest from the investments is $1029, how much was invested in each plan?

9. Boyd invests in a plan that has an APR of 8%. He invests in a 12% APR account $650 more than what he invested in the 8% account. If the total interest from the investments is $328, how much was invested in each plan?

10. Zelda invests a sum of money into two plans. She invests two-fifths of the sum at 8% and the rest at 7.6%. If the total interest from the investments is $620.80, how much was invested in each plan?

11. Agnes invests $4500 in two plans. Plan 1 is at an APR of 4%, and plan 2 is at an APR of 6%. If the total interest after 1 year is $234, what principal was invested in each plan?

12. Juan invests $12,800 in two plans. Plan 1 is at an APR of 7%, and plan 2 is at an APR of 5%. If the total interest after 1 year is $706, what principal was invested in each plan?

13. Annette invested $5000. She put some into an investment earning 5.5% and the rest earning 7%. If her income during this year was $335, how much did she invest at each rate?

14. Kennon invested $1600. She was able to put some of this into a CD at 5%, and the rest into a CD at only 4%. If she achieved an annual income of $73.40 on these investments, how much was invested in each plan?

15. Buddy inherited $12,000. He invested part of the money at 4% and the rest at $5\frac{1}{2}$%. If his annual income from these investments is $532.50, how much did he invest at each rate?

16. Shanequa has $4000. She put some of the money into savings that pays 6% and the rest in an account that pays 7%. If her total interest for the year is $264, how much did she invest at each rate?

17. Deloris invests in a plan that has an APR of 6%. She invests twice as much in a plan that ended up as a loss of 2% (that is, a −2% APR). If the net gain from the investments is $84, how much was invested in each plan?

18. Dominique invests $10,000 in two plans. Plan 1 is at an APR of 5% while plan 2 ended up returning a loss of 9% (that is, a −9% APR). If the net return on the investments after 1 year is a net loss of $312, then what principal was invested in each plan?

19. Cody has two investments totaling $10,000. Plan A is at an APR of 4% and plan B is at an APR of 8%. After one year, the interest earned from plan B is $320 more than the interest from plan A. How much did he invest in each plan?

20. Curtis has two investments totaling $12,000. Plan A is at an APR of 5% and plan B is at an APR of 7%. After one year, the interest earned from plan B is $360 more than the interest from plan A. How much did he invest in each plan?

Exercises 21–38 involve mixtures. Complete a table, write an equation, then solve.

21. A cough medicine contains 25% alcohol. How much liquid should a pharmacist add to 120 milliliters of the cough medicine so that it contains only 20% alcohol?

22. How many liters of a 40% solution of HCl are added to 2000 liters of a 20% solution to obtain a 35% solution?

23. A solution of gasoline and oil used for a gas-powered weed trimmer is 8% oil. How much gasoline must be added to 3 gallons of the solution to obtain a new solution that is 5% oil?

24. A 30-milliliter solution of alcohol and water is 10% alcohol. How much water must be added to yield a 5% solution?

25. How many quarts of pure antifreeze must be added to 6 quarts of a 40% antifreeze solution to obtain a 50% antifreeze solution?

26. A pharmacist has a 45% acid solution and a 35% acid solution. How many liters of each must be mixed to form 80 liters of a 40% acid solution?

27. The dairy is making a 30% buttermilk cream. If they mix a 26% buttermilk cream with a 35% buttermilk cream, how much of each do they need to use to produce 300 pounds of the 30% buttermilk cream?

28. Sharon has a bottle containing 45 milliliters of 15% HCl solution and a bottle of 35% HCl solution. She wants a 25% solution. How much of the 35% solution must be added to the 15% solution so that a 25% concentration is created?

29. Wilson has a bottle containing 35 milliliters of 15% saline solution and a bottle of 40% saline solution. She wants a 30% solution. How much of the 40% solution must be added to the 15% solution so that a 30% concentration is created?

30. Charles has 80 milliliters of a 15% acid solution. How much of a 20% acid solution must be added to create a solution that is 18% acid?

31. To form a 10% copper alloy weighing 75 grams, a 6% copper alloy is combined with an 18% copper alloy. How much of each type should be used?

32. A jeweler is mixing two silver alloys. How many ounces of an alloy containing 25% silver must be mixed with an alloy containing 20% silver to obtain 50 ounces of a 22% silver alloy?

33. A candy store is selling two types of candy, one at $1.20 per pound and the other at 90 cents per pound. To sell a mixture at $1.11 a pound, how many pounds of each must the store mix to make 80 pounds?

34. Pam has a mixture of peanuts and cashews that weighs 12 ounces. Of that, 25% is cashews. If she eats 2 ounces of peanuts, what percent of the remaining mixture is cashews?

35. The Market Street Sweets sells two types of pralines: a caramel version for $1.65 a pound and a chocolate version for $1.25 a pound. A mixture of these pralines can be sold for $1.49 a pound. How many pounds of each kind should be used to make 5 pounds of the $1.49 mixture?

36. Daly's Farm Feed sells two kinds of horse feed. One is 20% oats and the other is 28% oats. How many pounds of each should be mixed to form 50 pounds of a feed that is 25% oats?

37. Cromer's sells salted peanuts for $2.85 per pound and cashews for $7 per pound. How many pounds of each are needed to obtain a 10-pound mixture costing $4.50 per pound?

38. The Candy Shoppe wants to mix 115 pounds of candy to sell for $0.80 per pound. How many pounds of $0.60 candy must be mixed with a candy costing $1.20 per pound to make the desired mix?

PUZZLE PROBLEM

Herschal has $18,000 to invest. He invests $\frac{2}{3}$ in one account and the rest in a second account. The first account returned 2% more than the second account. The total return on his investment was $1050. What were the percentage rates of the two accounts?

REVIEW EXERCISES

1. Use $<$, $>$, or $=$ to make a true statement: $-|-14|$ ▮ 14

For Exercises 2–3, simplify.

2. $\dfrac{3}{4} - \dfrac{5}{6} \cdot \left(\dfrac{2}{3}\right)^2$

3. $(-4)^3$

4. Write an expression in simplest form that describes the perimeter of the rectangle shown.

For Exercises 5–6, solve.

5. $4(y - 3) + y = 7y - 8$

6. $\dfrac{3}{4}x - 5 = \dfrac{2}{3}x + 1$

Defined Terms

Section 3.1
Ratio (p. 194)
Unit ratio (p. 194)
Proportion (p. 195)
Congruent angles (p. 199)
Similar figures (p. 199)

Section 3.2
Percent (p. 209)

Section 3.3
Complementary angles (p. 227)
Supplementary angles (p. 227)
Consecutive integers (p. 229)

Procedures, Rules, and Key Examples

Procedures/Rules	Key Examples

Section 3.1 Ratios and Proportions

If $\dfrac{a}{b} = \dfrac{c}{d}$, where $b \neq 0$ and $d \neq 0$, then $ad = bc$.

Determine whether the ratios are proportional.

$$\frac{5}{6} \; ? \; \frac{15}{18}$$

$18 \cdot 5 = 90 \qquad 6 \cdot 15 = 90$

$$\frac{5}{6} \;{\diagup\!\!\!\!\diagdown}\; \frac{15}{18}$$

Since the cross products are equal, the ratios are proportional.

Solve for the missing number in the proportion.

$$\frac{5}{12} = \frac{x}{18}$$
$$90 = 12x$$
$$\frac{90}{12} = \frac{12x}{12}$$
$$7.5 = x$$

To solve a proportion using cross products:
1. Calculate the cross products.
2. Set the cross products equal to one another.
3. Use the multiplication principle of equality to isolate the variable.

To solve proportion problems:
1. Set up the given ratio any way you wish.
2. Set the given ratio equal to the other ratio with the unknown so that the numerators and denominators correspond logically.
3. Solve using cross products.

A car can travel about 350 miles on 16 gallons of gasoline. How many gallons would be needed to travel 600 miles?

$$\frac{350}{16} = \frac{600}{x}$$
$$350x = 9600$$
$$\frac{350x}{350} = \frac{9600}{350}$$
$$x \approx 27.4 \text{ gal.}$$

continued

Procedures/Rules	Key Examples

Section 3.2 Percents

To write a percent as a fraction or decimal:
1. Write the percent as a ratio with 100 in the denominator.
2. Simplify to the desired form.
Note: When simplifying, remember that dividing a decimal number by 100 moves the decimal point two places to the left.

Example 1: Write each percent as a fraction and as a decimal.

a. $42\% = \dfrac{42}{100} = \dfrac{21}{50} = 0.42$

b. $8.5\% = \dfrac{8.5}{100} = \dfrac{85}{1000} = \dfrac{17}{200} = 0.085$

c. $20\dfrac{1}{2}\% = \dfrac{20\frac{1}{2}}{100} = \dfrac{41}{200} = 0.205$

To write a fraction or decimal number as a percent:
1. Multiply by 100%.
2. Simplify.

Example 2: Write as a percent.

a. $0.453 = 0.453 \cdot 100\% = 45.3\%$

b. $\dfrac{5}{8} = \dfrac{5}{\overset{8}{\underset{2}{8}}} \cdot \dfrac{\overset{25}{100}}{1}\% = \dfrac{125}{2}\% = 62\dfrac{1}{2}\%$

Method 1. Translate the sentence word-for-word:
1. Select a variable for the unknown.
2. Translate the word *is* to an equal sign.
3. If *of* is preceded by the percent then it is translated to multiplication. If *of* is preceded by a whole number, translate it to division.

Example 3: 15% of what number is 33?
Method 1: Direct Translation.
$$0.15n = 33$$
$$\dfrac{0.15n}{0.15} = \dfrac{33}{0.15}$$
$$n = 220$$

Method 2. Translate to a proportion by writing the following form:

$$\text{Percent} = \dfrac{\text{part}}{\text{whole}}$$

where the percent is expressed as a fraction with a denominator of 100.

Method 2: Proportion.
$$\dfrac{15}{100} = \dfrac{33}{n}$$
$$15n = 3300$$
$$\dfrac{15n}{15} = \dfrac{3300}{15}$$
$$n = 220$$

To solve problems involving percent, try to write a basic sentence:
1. Determine whether the percent, whole, or part is unknown.
2. Write the problem as a simple percent sentence.
3. Translate the sentence to an equation (word-for-word or proportion).
4. Solve for the unknown.

Example 4: A student answered 24 questions out of 30 correctly. What percent of the questions did the student answer correctly?

Simple percent sentence: What percent of 30 is 24?

Direct Translation:
$$p \cdot 30 = 24$$
$$\dfrac{30p}{30} = \dfrac{24}{30}$$
$$p = 0.8$$
$$p = 0.8 \cdot 100\%$$
$$p = 80\%$$

Proportion:
$$\dfrac{p}{100} = \dfrac{24}{30}$$
$$30p = 2400$$
$$\dfrac{30p}{30} = \dfrac{2400}{30}$$
$$p = 80\%$$

continued

Procedures/Rules	Key Examples

Section 3.3 Problems with Two or More Unknowns

To solve problems with two or more unknowns:
1. Determine which unknown will be represented by a variable.
   Tip: Let the unknown that is acted on be represented by the variable.
2. Use one of the relationships to describe the other unknown(s) in terms of the variable.
3. Use the other relationship to write an equation.
4. Solve the equation.

Example 1: One number is fifteen more than twice another. The sum of the numbers is 39. Find the numbers.

Relationship 1: "One number is fifteen more than twice another."

Translation: $2x + 15$ where x represents the smaller number.

Relationship 2: "The sum of the numbers is 39."

Translation:
$$x + 2x + 15 = 39$$
$$3x + 15 = 39$$
$$\underline{-15 \quad -15}$$
$$3x + 0 = 24$$
$$\frac{3x}{3} = \frac{24}{3}$$
$$x = 8$$

The smaller number is 8 and the other number is $2(8) + 15 = 31$

Section 3.4 Rates

To solve for time when two objects are moving in opposite directions:
1. Use a table with columns for categories, rate, time, and distance. Use the fact that rate · time = distance. The travel times are equal.
2. Write an equation that is the sum of the individual distances equal to the total distance of separation.
3. Solve the equation.

Example 1: Mark and Latonia are 17 miles apart traveling toward each other. Mark is traveling at a rate of 45 miles per hour and Latonia is traveling at a rate of 40 miles per hour. How long will it be until they meet?

Categories	Rate	Time	Distance
Mark	45	t	$45t$
Latonia	40	t	$40t$

Equation:
$$45t + 40t = 17$$
$$85t = 17$$
$$\frac{85t}{85} = \frac{17}{85}$$
$$t = 0.2 \text{ hr.}$$

continued

| Procedures/Rules | Key Examples |

Procedures/Rules

Section 3.4 (continued)

To solve problems involving two people traveling in the same direction in which the objective is to determine the time for one person to catch up to the other:

1. Use a table to organize the rates and times. Let t represent the time for the person to catch up and add the time difference to t to represent the other person's time.
2. Set the expressions for the individual distances equal.
3. Solve the equation.

Key Examples

Example 2: Frank is traveling at a rate of 64 miles per hour and passes Exit 43 at 4:35 P.M. Juanita is traveling at a rate of 72 miles per hour in the same direction and passes the same exit at 4:50 P.M. At what time does she catch up with Frank?

Categories	Rate	Time	Distance
Frank	64	$t + 0.25$	$64(t + 0.25)$
Juanita	72	t	$72t$

Equation: $64(t + 0.25) = 72t$
$$64t + 16 = 72t$$
$$\underline{-64t \qquad\quad -64t}$$
$$16 = 8t$$
$$\frac{16}{8} = \frac{8t}{8}$$
$$2 = t$$

Answer: Juanita will catch up at 6:50 P.M.

Section 3.5 Investment and Mixture

To solve problems involving two interest rates, in which the total interest is given:

1. Use a table to organize the interest rates and principals. Multiply the individual rates and principals to get expressions for the individual interests.
2. Write an equation that is the sum of the expressions for interest set equal to the given total interest.
3. Solve the equation.

Example 1: Dan invests a total of $4000 in two different plans. The first plan returns 9% while the second plan returns 6%. If the total interest earned after 1 year is $285, what principal was invested in each plan?

Categories	Rate	Principal	Interest
Plan 1	0.09	p	$0.09p$
Plan 2	0.06	$4000 - p$	$0.06(4000 - p)$

Equation: $0.09p + 0.06(4000 - p) = 285$
$$0.09p + 240 - 0.06p = 285$$
$$0.03p + 240 = 285$$
$$\underline{-240 \quad -240}$$
$$0.03p + 0 = 45$$
$$\frac{0.03p}{0.03} = \frac{45}{0.03}$$
$$p = 1500$$

Answer: $1500 in plan 1 and
$4000 - $1500 = $2500 in plan 2

continued

Procedures/Rules	Key Examples

Section 3.5 (continued)

To solve problems involving mixing solutions:
1. Use a table to organize the concentrations and volumes. Multiply the individual concentrations and solution volumes to get expressions for the volume of the particular chemical.
2. Write an equation that is the sum of the volumes of the chemical in each solution set equal to the volume in the combined solution.
3. Solve the equation.

Example 2: 200 milliliter of a 15% alcohol solution is mixed with a 30% alcohol solution to make a solution that is 20% alcohol. How much of the 30% solution is needed?

Categories	Concentration	Volume of Solution	Volume of Alcohol
15% solution	0.15	200	0.15(200)
30% solution	0.30	n	0.30n
20% solution	0.20	200 + n	0.20(200 + n)

Equation:
$$0.15(200) + 0.30n = 0.20(200 + n)$$
$$30 + 0.30n = 40 + 0.20n$$
$$\underline{-0.20n \qquad\qquad -0.20n}$$
$$30 + 0.10n = 40 + 0$$
$$30 + 0.10n = 40$$
$$\underline{-30 \qquad\qquad\quad -30}$$
$$0.10n = 10$$
$$\frac{0.10n}{0.10} = \frac{10}{0.10}$$
$$n = 100$$

Answer: 100 ml of the 30% solution is needed

Formulas

Simple percent sentence: Percent of a whole is the part.

Distance, rate, and time: $d = rt$

Simple interest: $I = Pr$

For Exercises 1–6, answer true or false.

1. The cross products of a proportion are equal.

2. A percent is a ratio out of 100.

3. Consecutive odd integers can be represented as $x, x + 1, x + 3, \ldots.$

4. $7.8\% = 0.78$.

5. Complementary angles have a sum of 180°.

6. "One less than a number" can be represented as $1 - x$.

For Exercises 7–10, fill in the blank.

7. To write a percent as a fraction or decimal:
   1. Write the numeral over _____.
   2. Simplify to the desired form.

8. Angles that have the same measurement are _____.

9. Angles whose sum is 90° are _____.

10. An equation in the form $\dfrac{a}{b} = \dfrac{c}{d}$, where $b \neq 0$ and $d \neq 0$, is a _____.

11. A recipe calls for $\dfrac{3}{4}$ of a cup of sugar and 2 cups of flour. Write the ratio of sugar to flour as a fraction in simplest form.

12. The price of a 10.5-ounce can of vegetables is $0.89. Write the unit ratio of price to weight rounded to the nearest thousandth.

For Exercises 13–16, determine whether the ratios are equal.

13. $\dfrac{2}{5} \stackrel{?}{=} \dfrac{10}{20}$

14. $\dfrac{1}{3} \stackrel{?}{=} \dfrac{13}{39}$

15. $-\dfrac{2}{8} \stackrel{?}{=} -\dfrac{1}{4}$

16. $\dfrac{15}{21} \stackrel{?}{=} \dfrac{130.2}{93}$

For Exercises 17–22, solve for the missing number.

17. $\dfrac{3}{5} = \dfrac{15}{x}$

18. $\dfrac{-11}{12} = \dfrac{n}{-24}$

19. $\dfrac{\frac{2}{5}}{4\frac{1}{2}} = \dfrac{7}{p}$

20. $\dfrac{k}{15.3} = \dfrac{5}{8}$

21. $\dfrac{-60}{w} = \dfrac{20}{40}$

22. $\dfrac{c}{12.7} = \dfrac{5}{3}$

For Exercises 23–28, translate to a proportion and solve.

23. Tomasa is planning a trip to Spain. She knows that it takes 0.00554 U.S. dollars to equal a Spanish peseta. If she has $1200 to take with her, how much will she have in Spanish peseta?

24. Jason has driven 225 miles in 3 hours. If he averages the same speed after lunch, how far can he expect to travel in 4 hours?

25. Gutzon Borglum used a simple ratio of 1:12 (one inch to one foot) when he transferred the plaster models to the granite for Mt. Rushmore. Each of the sculptures is 60 feet tall. How tall was a plaster model?

Of Interest

Sculptor Gutzon Borglum began work on the series of 60-foot sculptures collectively called Mount Rushmore in 1927. The Washington head was formally dedicated in 1930, followed by Jefferson in 1936, Lincoln in 1937, and Teddy Roosevelt in 1939.

26. Tamika pays property taxes of $957 on her house, which is assessed at $125,500. She is planning to buy a house assessed at $175,000. At the same rate, how much will her taxes be on the new home?

27. It costs $3.20 to mail a 12-ounce package. How much will it cost to mail a 20-ounce package?

28. Ferrari estimates that the 2000 550 Maranello two-door coupe will travel 241 city miles on one tank of gas. If the gas tank of the Ferrari holds 30.1 gallons, how far can a driver expect to travel on 25 gallons?

For Exercises 29–32, find the missing lengths in the similar shapes.

29.

8 in.   10 in.   y   15 in.

30.

24 m   8.5 m   *l*   15.3 m

31.

8 ft. $4\frac{1}{2}$ ft.   4 ft.   *x* $6\frac{3}{4}$ ft.   *y*

32.

7 in. $2\frac{3}{4}$ in.   *b* $7\frac{7}{8}$ in.   12 in.   *a*   8 in.   *c*

For Exercises 33–38, write each percent as a decimal and as a fraction in simplest form.

33. 15%

34. 82.5%

35. $12\frac{1}{2}\%$

36. 2.45%

37. 10%

38. $33\frac{1}{3}\%$

For Exercises 39–44, write as a percent.

39. $\dfrac{2}{5}$

40. $\dfrac{1}{3}$

41. $\dfrac{4}{11}$

42. 0.35

43. 1.2

44. 2.016

For Exercises 45–68, solve.

45. 6 is what percent of 20?

46. 14.5% of what is 42?

47. 16% of 91 is what?

48. A county assesses annual property taxes at a rate of 4% of the appraised value of the property. Suppose a property is appraised for $325,000. What are the property taxes?

49. Tom is planning to purchase a 2001 Porsche Boxster convertible. The sticker price for the car is $50,200. He plans to offer 12% less. How much is Tom planning to offer for the car?

50. Summertime electricity usage rose 32% in Silicon Valley from 1994 to 2000. If a business had a July 1994 electric bill of $510, what can they expect to pay in 2000 if the bill increases at the same rate as the usage? (*Source:* CNN)

51. In a recent Gallup poll, 47% of those polled knew someone who lost a job during the past year. If 1016 Americans were polled, how many knew someone who lost a job? (*Source:* Gallup News Service 2/15/01)

52. The larger of two numbers is five more than three times the smaller number. The sum of the two numbers is 15. What are the numbers?

53. Find three consecutive integers such that twice the sum of the first two is six less than triple the third number.

54. The length of a rectangle is 15 feet more than twice its width, and the perimeter is 120 feet. What are the length and width?

55. Two sides of an isosceles triangle are 2 centimeters more than the other side. If the perimeter is 55 centimeters, find the length of each side.

56. Two angles are supplementary. One of the angles is 32° less than the measure of the other angle. What are the measures of the two angles?

57. Two angles are complementary. One of the angles is three less than twice the other. What are the measures of the two angles?

58. While buying pantyhose, Sophie notices that the regular fit cost $6.50 and control top cost $9.50. If she buys 5 pairs of hose and spends $38.50, how many of each type did she buy?

59. Candice owns stock in Coca-Cola and Pepsi. She notices that the Coca-Cola stock is worth about $12 more per share. If she has $3646 invested and owns 20 shares of the Pepsi and 47 shares of the Coca-Cola, how much is one share of each stock worth?

60. In Cincinnati, a Land Rover and an Escort enter I-71 at Exit 2 at the same time and travel in opposite directions. The Land Rover traveling south is caught in traffic and travels at 45 miles per hour. The Escort traveling north encounters light traffic and is able to travel at 65 miles per hour. In how many hours will the cars be 385 miles apart?

61. If a plane flying from O'Hare travels west at 550 miles per hour and another leaving at the same time travels east at 450 miles per hour, how many hours until the planes are 2600 miles apart?

62. Wayne and Libby are both driving south. At 3:00, Wayne was 5 miles south of Libby. A little later, Libby passed Wayne. If Libby is driving at 55 miles per hour and Wayne at 50 miles per hour, what time did Libby pass Wayne?

63. Two trains started from Victoria Station and traveled in the same direction on parallel tracks. The freight train leaves at 10 A.M. The passenger train leaves at 11 A.M., but traveling at 25 miles per hour faster, overtakes the freight train at 2 P.M. How fast is the freight train traveling?

64. Terrell invests in a retirement plan that has an APR of 8%. He invests twice as much in a second plan that has an APR of 10%. If the total interest from the investments is $5600, how much was invested in each plan?

65. Brian invests in a retirement plan that has an APR of 7%. He invests three times as much in a second plan that ended up as a loss of 2% (that is, at −2% APR). If the net gain from the investments is $102, how much was invested in each plan?

66. A solution that is 90% alcohol is to be mixed with one that is 40% alcohol. The chemist is preparing 100 liters of a 60% alcohol solution. How many liters of each must he use?

67. To make 100 pounds of concrete containing 25% cement, a builder must add how much pure cement to the 10% cement mixture he has?

68. How many ounces of a 20% saline solution must be added to 40 ounces of a 10% solution to produce a 16% solution?

1. Determine whether the ratios are equal: $\dfrac{11}{4} \overset{?}{=} \dfrac{130}{48}$

For Exercises 2 and 3, solve for the missing number.

2. $\dfrac{m}{8} = \dfrac{5}{12}$

3. $\dfrac{-32}{-9} = \dfrac{8}{b}$

For Exercises 4 and 5, write each percent as a decimal and as a fraction in simplest form.

4. 22%

5. $3\dfrac{1}{3}\%$

For Exercises 6 and 7, write as a percent.

6. 3.2

7. $\dfrac{2}{5}$

For Exercises 8–20, solve.

8. 12 is what percent of 60?

9. 20.1% of 70 is what number?

10. 70 is 40% of what number?

11. According to a survey conducted by the National Center for Health, 37.1% of Americans say they are in excellent health. Based on these findings, in a group of 500 Americans, how many would say they are in excellent health? (*Source:* National Center for Health Statistics)

12. Shaniqua pays taxes of $852 on her house, which is assessed at $89,000. She is planning to buy a house assessed at $75,000. What will the taxes be on her new home?

13. The sum of two consecutive even integers is 110. Find the integers.

14. If five times a number is reduced by 14, the result is 26. Find the number.

15. The length of a rectangle is one more than double the width. If the perimeter of the rectangle is 32 kilometers, what are the length and width?

16. Christe owns stock in both Wal-Mart and Kmart. She notices that the Wal-Mart stock is worth about $11 more per share. If she has 22 shares of Wal-Mart stock and 15 shares of Kmart stock for a total investment of $1759, what was the value of each stock?

17. Two runners begin running toward each other along the same path. One person runs at a speed of 4 miles per hour and the other person runs at 4.5 miles per hour. If they begin $4\frac{1}{4}$ miles apart, how long will it be until they meet?

18. An F-18 jet fighter passes a landmark at a speed of 450 miles per hour. A second F-18 traveling in the same direction at 570 miles per hour passes the same landmark 10 minutes later. How long will it take the second jet to catch up with the first jet?

19. Kari invests $4500 in two plans. After one year, she found that plan A earned 8% interest and plan B earned 6% interest. If she earned a total of $330, how much did she invest in each plan?

20. How many ounces of a 15% saline solution must be added to 30 ounces of a 25% solution to produce a 20% solution?

For Exercises 1–6, answer true or false.

1. A constant is a symbol that can vary in value.

2. $\frac{1}{2}x = 10$ is a linear equation.

3. $\sqrt{2}$ is an irrational number.

4. $-x = -1x$

5. When evaluating an expression using the order of operations, one must always do multiplication before division.

6. "Four less than a number" can be translated to $4 - n$.

For Exercises 7–10, fill in the blank.

7. A prime number is a natural number that has exactly two different factors, 1 and _____.

8. To clear fractions from an equation, we can use the multiplication principle of equality and multiply both sides by the _____.

9. A percent is a ratio representing some part out of _____.

10. The _____ products of equivalent fractions are equal.

11. Explain in your own words how to solve a proportion.

12. Write a set representing the natural numbers.

13. Graph $-3\frac{1}{4}$ on a number line.

14. Determine the values that make $\frac{4x}{x - 3}$ undefined.

15. Prove $4(x + 3) = 4x + 12$ is true.

16. Find the prime factorization of 120.

For Exercises 17–25, simplify.

17. $\left(-\dfrac{2}{3}\right)^2$

18. $\sqrt{\dfrac{4}{81}}$

19. $\sqrt{-36}$

20. $(-1)^2 - 8.5 + 1$

21. $\sqrt{16} + \sqrt{9}$

22. $\dfrac{5 - 2(-1)^2}{3\sqrt{64} + 36}$

23. $-|3 \cdot (-5)| + |(-1)^3|$

24. $2\left(-\dfrac{1}{2}\right)^2 - 4\left(\dfrac{3}{5}\right) + 3$

25. $-6 - 2(3 + 0.2)^2$

26. Evaluate $\dfrac{2x^3}{y - 3}$, where $x = -1$ and $y = 3$.

27. Use the distributive property to simplify $-3\left(\dfrac{1}{6}x - 9\right)$.

For Exercises 28–31, indicate whether the given equation illustrates the additive identity, commutative property of addition, associative property of addition, or additive inverse.

28. $-5 + (4 - 4) = -5 + 0$

29. $6 + (-3) = -3 + 6$

30. $4 + (5 + 6) = (4 + 5) + 6$

31. $-4 + 4 = 0$

For Exercises 32–37, solve and check.

32. $4(x + 1) = 2(x - 3)$

33. $3.23c - 8.75 = 1.41c + 7.63$

34. $2m + \dfrac{1}{2} = 4m - 3\dfrac{1}{2} + 8m$

35. $\dfrac{1}{8}w = \dfrac{4}{3} - \dfrac{2}{3}w$

36. $6x - 2 = 3\left(2x - \dfrac{2}{3}\right)$

37. $-7 = -6(p + 2) + 5(2p - 3)$

For Exercises 38–41, solve for the indicated variable.

38. $A = \dfrac{1}{2}bh$ for b

39. $A = P + Prt$ for r

40. $P = a + b + c$ for c

41. $d = rt$ for t

For Exercises 42–50, solve.

42. Twice a number is twelve less than three times the number. Find the number.

43. What is 15% of 42.3?

44. The sum of three consecutive odd integers is -3. What are the integers?

45. Jon made an 82 on his first test and a 70 on his second test. What range of scores does he need to make on his third test to have an average of at least 80?

46. How many liters of a 60% solution of boric acid should be added to 10 liters of a 30% solution to obtain a 50% solution?

47. Kathie has $6.50 in dimes and quarters. If there are 35 coins in all, how many of each coin does she have?

48. Archie and Veronica are both driving east on Riverdale Drive. At 3:00, Archie, driving at 50 miles per hour, was 5 miles east of Veronica. A little later, Veronica, driving at 55 miles per hour, passed Archie. At what time did Veronica pass Archie?

49. Cromers sells cashews for $7.00 per pound and peanuts for $3.00 per pound. If a 10-pound bag is mixed selling at $4.50 per pound, how many pounds of each type nut should be used?

50. If a car travels 120 miles on 4 gallons of gasoline, how far can it travel on 10 gallons?

Graphing Linear Equations and Inequalities

All equations and inequalities have corresponding pictures, or graphs. The graph of an equation or inequality offers information about the solutions to the equation or inequality. Graphs can be helpful in solving problems. For example, we may be given an equation or inequality that describes a particular problem. The graph of that equation may offer additional insight about possible solutions to the problem. Or, a problem may have some data that we can graph. We can often use such a graph to help us write an equation or inequality that describes the situation. Both approaches can be quite useful in solving problems.

There's a compelling reason to master information and news. Clearly there will be better job and financial opportunities. Other high stakes will be missed by people if they don't master and connect information.

—*EVERETTE DENNIS*
Media foundation executive and professor of media and entertainment

Where is the wisdom we have lost in knowledge? Where is the knowledge we have lost in information?

—*T. S. ELIOT*
American-born British critic and writer (1888–1965)

4.1 The Rectangular Coordinate System

OBJECTIVES

1. Determine the coordinates of a given point.
2. Plot points in the coordinate plane.
3. Determine the quadrant for a given coordinate.
4. Determine whether the graph of a set of data points is linear.

OBJECTIVE 1 **Determine the coordinates of a given point.** In 1619 René Descartes, the French philosopher and mathematician, recognized that positions of points in a plane could be described using two number lines that intersect at a right angle. Each number line is called an **axis**.

DEFINITION *Axis:* A number line used to locate a point in a plane.

Two perpendicular axes form the *rectangular,* or Cartesian, *coordinate system,* named in honor of René Descartes. Usually, we call the horizontal axis the *x-axis* and the vertical axis the *y-axis.*

Note: When letters are used to label the axes, they are usually placed so that the horizontal axis is the first letter in alphabetical order. For example, if the letters *t* and *u* are used, then the horizontal axis will be the *t*-axis and the vertical axis will be the *u*-axis.

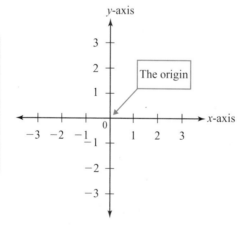

The point where the axes intersect is 0 for both the *x*-axis and the *y*-axis. This position is called the *origin.* The positive numbers are to the right and up from the origin, whereas negative numbers are to the left and down from the origin.

Any point in the plane can be described using two numbers, one number from each axis. To avoid confusion, these two numbers are written in a specific order. The number representing a point's *horizontal distance* from the origin is given *first* and the number representing a point's *vertical distance* from the origin is given *second.* Because the order in which we say or write these two numbers matters, we say that they form an *ordered pair.* Each number in an ordered pair is called a *coordinate* of the ordered pair. The notation for writing ordered pairs is: (horizontal coordinate, vertical coordinate).

Consider the point labeled *A* in the coordinate plane shown below. The point is drawn at the intersection of the 3rd line to the right of the origin and the 4th line up from the origin. The ordered pair that describes point *A* is (3, 4).

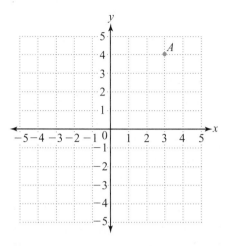

PROCEDURE **IDENTIFYING THE COORDINATES OF A POINT**

To determine the coordinates of a given point in the rectangular system:

1. Follow a vertical line from the point to the *x*-axis (horizontal axis). The number at this position on the *x*-axis is the first coordinate.
2. Follow a horizontal line from the point to the *y*-axis (vertical axis). The number at this position on the *y*-axis is the second coordinate.

EXAMPLE 1 Write the coordinates for each point shown.

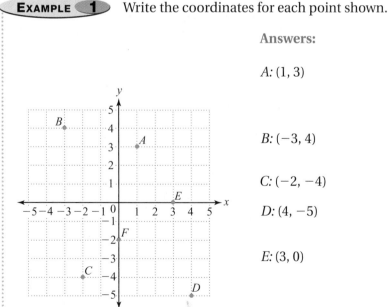

Answers:

A: (1, 3)

B: (−3, 4)

C: (−2, −4)

D: (4, −5)

E: (3, 0)

F: (0, −2)

Write the coordinates for each point shown.

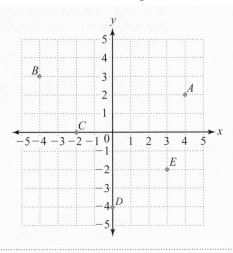

OBJECTIVE 2 Plot points in the coordinate plane. Any point can be plotted in the coordinate plane using its coordinates. Remember that the first coordinate in an ordered pair indicates the distance to move right or left, and the second coordinate indicates the distance to move up or down. To plot the ordered pair (3, 4), we would move to the right 3, then up 4 and draw a dot to indicate the point.

PROCEDURE PLOTTING A POINT

To graph or plot a point given its coordinates:

1. Beginning at the origin, (0, 0) move to the right or left along the *x*-axis the amount indicated by the first coordinate.
2. From that position on the *x*-axis, move up or down the amount indicated by the second coordinate.
3. Draw a dot to represent the point described by the coordinates.

EXAMPLE 2 Plot the point described by the coordinates.

a. $(3, -4)$ **b.** $(-4, 2)$ **c.** $(0, 5)$

Solution

Learning Strategy

If you are a tactile learner, when plotting a point, move your pencil along the *x*-axis first, then move up or down to the point location, just as we've done with the arrows.

Note: For $(-4, 2)$, we begin at the origin and move to the left 4, then up 2.

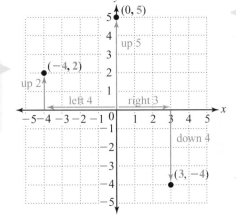

Note: For $(0, 5)$, because the first coordinate is 0, we do not move right or left. Since the second coordinate is 5, we move straight up 5, drawing the point on the *y*-axis.

Note: For $(3, -4)$, we begin at the origin and move to the right 3, then down 4.

ANSWERS

A: (4, 2) D: (0, −4)
B: (−4, 3) E: (3, −2)
C: (−2, 0)

Plot the point described by the coordinates.

A: $(-3, 2)$ B: $(1, 4)$ C: $(-2, -4)$ D: $(3, 0)$

OBJECTIVE **3** **Determine the quadrant for a given coordinate.** The two perpendicular axes divide the coordinate plane into four regions called **quadrants**.

DEFINITION

Quadrant: One of four regions created by the intersection of the axes in the coordinate plane.

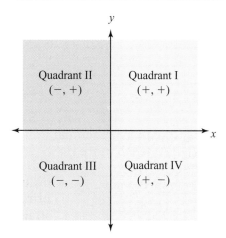

The quadrants are numbered using Roman numerals as shown to the left. Note the signs of the coordinates determine the quadrant in which a point lies. For points in quadrant I, both coordinates are positive. Every point in quadrant II has a negative first coordinate (horizontal) and a positive second coordinate (vertical). Points in quadrant III have both coordinates negative. Every point in quadrant IV has a positive first coordinate and a negative second coordinate.

PROCEDURE **IDENTIFYING QUADRANTS**

To determine the quadrant for a given ordered pair, consider the signs of the coordinates.

$(+, +)$ means the point is in quadrant I

$(-, +)$ means the point is in quadrant II

$(-, -)$ means the point is in quadrant III

$(+, -)$ means the point is in quadrant IV

ANSWER

EXAMPLE **3** State the quadrant in which each point is located.

a. $(-61, 23)$

Answer: Quadrant II (upper left), because the first coordinate is negative and the second coordinate is positive.

b. $\left(14, -37\frac{2}{3}\right)$

Answer: Quadrant IV (lower right), because the first coordinate is positive and the second coordinate is negative.

c. $(0, -12)$

Answer: Since the x-coordinate is 0, this point is on the y-axis and is not in a quadrant.

YOUR TURN   State the quadrant in which each point is located.

a. $(-42, -109)$ **b.** $(37, -15.9)$ **c.** $(4.7, 0)$ **d.** $\left(-16\frac{3}{4}, 124\right)$

OBJECTIVE 4 **Determine whether the graph of a set of data points is linear.**   In many problems data are listed as ordered pairs. For example, in a report on stock prices, we might have the following listing of the closing price of a particular stock for each day over four days.

Day	Closing price
1	$24\frac{1}{2}$
2	$25\frac{3}{4}$
3	$27
4	$28\frac{1}{4}$

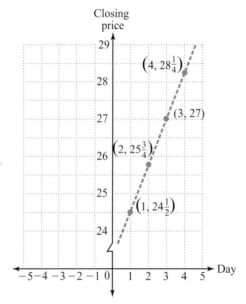

Note: Because the vertical axis has dollar values that involve fractions, there are two marks for every whole dollar. This allows halves to be plotted accurately and fourths to be estimated well.

If we plot each pair of data as an ordered pair in the form (day, closing price), we see that the points lie on a straight line.

Because the points can be connected to form a straight line they are said to be *linear*. Points that do not form a straight line are *nonlinear*.

Note: The broken line in the axis indicates that values between 0 and 24 were skipped to conserve space.

EXAMPLE 4   The following data points track the trajectory, or path, of an object over a period of time. Plot the points with the time along the horizontal axis and the height along the vertical axis. Then state whether the trajectory is linear or nonlinear.

Time (in seconds)	Height (in feet)
0	6.00
0.5	7.50
1.0	8.25
1.5	7.50
2.0	6.00
2.5	2.00

ANSWERS

a. III   **b.** IV
c. on the *x*-axis   **d.** II

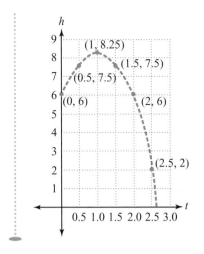

(1, 8.25)

(1.5, 7.5)

(0.5, 7.5)

(0, 6)   (2, 6)

(2.5, 2)

Solution Because the data points do not form a straight line when connected, the trajectory is nonlinear.

Note: It seems here that our axis labels violate the alphabetical convention. However, the problem states to plot the time along the horizontal axis. Also, in general, when a graph involves time, the time values are plotted along the horizontal axis.

NSWER

The points are linear.

(3, 3.5)

(1, 2.5)

(2, 3)

(0, 2)

YOUR TURN

The following data points show the velocity of an object as time passes. Plot the points with the time along the horizontal axis and the velocity along the vertical axis. Then state whether the points are linear or nonlinear.

Time (in seconds)	Height (in feet)
0	2.0
1	2.5
2	3.0
3	3.5

4.1 EXERCISES

1. When writing an ordered pair, which is written first, the horizontal-axis coordinate or the vertical-axis coordinate?

2. Describe how to use the coordinates of an ordered pair to plot a point in the rectangular coordinate system.

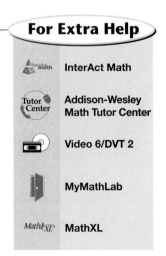

3. Draw the axes in the rectangular coordinate system, and then label the quadrants. Include the signs for coordinates in each quadrant.

4. Describe how to determine whether a set of data points is linear.

For Exercises 5–8, write the coordinates for each point.

5.

6.

7.

8.

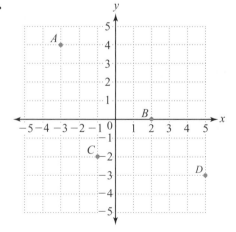

For Exercises 9–12, on a piece of graph paper, draw and label x- and y-axes. Then plot and label the points indicated by the coordinate pairs.

9. $(5, 4), (2, -3), (-1, -3), (2, 0)$

10. $(-3, 0), (0, 1), (-2, -2), (4, -1)$

11. $(-2, 0), (3, -3), (-1, -5), (2, 2)$

12. $(-1, -4), (0, -2), (4, -2), (-4, 2)$

For Exercises 13–24, state the quadrant in which the point is located.

13. $(-6.5, 1050)$

14. $\left(-42\frac{1}{3}, -500\right)$

15. $(620, 50)$

16. $(-42, 50)$

17. $(57, -82.71)$

18. $(16, 27)$

19. $\left(-41\frac{1}{2}, -82\right)$

20. $\left(47, 51\frac{5}{8}\right)$

21. $(0, -9)$

22. $(-16.8, -5.3)$

23. $(0.2, -0.1)$

24. $(8.2, 0)$

For Exercises 25–30, determine whether the set of points is linear or nonlinear.

25. (copy costs, number of copies): $(15, 300)$, $(18, 400)$, $(21, 500)$, $(24, 600)$

26. (number of hours, plumbing cost): $(0, 125)$, $(1, 200)$, $(2, 275)$, $(3, 350)$

27. (time, height): $(0, 8.0)$, $(0.5, 9.5)$, $(1.0, 11.0)$, $(1.5, 9.5)$, $(2.0, 8.0)$, $(2.5, 6.5)$

28. (days, stock closing price): $(1, 23)$, $(2, 21)$, $(3, 19)$, $(4, 17)$

29. (number of minutes, wireless phone cost): (300, 20), (310, 21), (320, 22)

30. (days, store profit): (1, 3012), (2, 2002), (3, 1145), (4, 5003)

31. List the coordinates of three points on the line shown. (Answers may vary.)

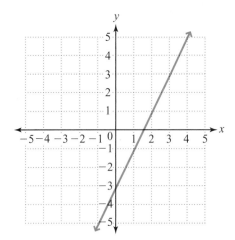

32. List the coordinates of each vertex of the figure shown.

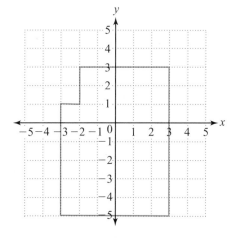

33. A parallelogram has been moved from its original position, shown with the dashed lines, to a new position, shown with solid lines.

   a. List the coordinates of each vertex of the original position.

   b. List the coordinates of each vertex of the new position.

   c. Write a rule that describes in mathematical terms how to move each point on the parallelogram in its original position to each point on the parallelogram in its new position.

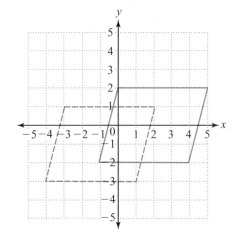

34. A trapezoid has been moved from its original position, shown with the dashed lines, to a new position, shown with solid lines.

   a. List the coordinates of each vertex of the original position.

   b. List the coordinates of each vertex of the new position.

   c. Write a rule that describes in mathematical terms how to move each point on the trapezoid in its original position to each point on the trapezoid in its new position.

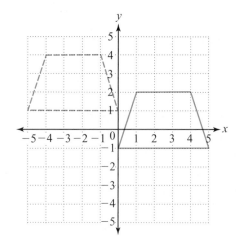

35. The ordered pairs $(-3, 1)$, $(-1, 1)$, $(-1, 3)$, $(2, 3)$, $(2, -2)$, and $(-3, -2)$ form the vertices of a figure.

  a. Plot the points and connect them to form the figure.

  b. Find the perimeter of the figure.

  c. Find the area of the figure.

36. The ordered pairs $(-4, 2)$, $(-2, 2)$, $(-2, 4)$, $(1, 4)$, $(1, 3)$, $(3, 3)$, $(3, -4)$, and $(-4, -4)$ form the vertices of a figure.

  a. Plot the points and connect them to form the figure.

  b. Find the perimeter of the figure.

  c. Find the area of the figure.

REVIEW EXERCISES

1. Graph -3 on a number line.

2. Evaluate $2x - 3y$, when $x = 4$ and $y = 2$.

3. Evaluate $-4x - 5y$, when $x = -3$ and $y = 0$

For Exercises 4–6, solve.

4. $2x - 5 = 10$

5. $-\dfrac{5}{6}x = \dfrac{3}{4}$

6. $7n - (2n - 9) = 3(4n + 1) - 8$

4.2 Graphing Linear Equations

OBJECTIVES

1. Determine whether a given pair of coordinates is a solution to a given equation with two unknowns.
2. Find solutions for an equation with two unknowns.
3. Graph linear equations.

OBJECTIVE 1 Determine whether a given pair of coordinates is a solution to a given equation with two unknowns. In Chapter 2 we considered linear equations with one variable. We now consider linear equations that have two variables such as $x + y = 4$ and $y = 2x - 3$.

A solution for an equation with two variables is a pair of numbers, one number for each variable, that can replace the corresponding variables and make the equation true. Since these solutions are ordered pairs, we can write them using coordinates. The ordered pair $(1, 3)$ is a solution for $x + y = 4$ because replacing x with 1 and y with 3 makes the equation true.

$$x + y = 4$$
$$\downarrow \quad \downarrow$$

Note: Recall from Section 4.1 that we assume coordinates are written in alphabetical order, so $(1, 3)$ means that $x = 1$ and $y = 3$.

$1 + 3 = 4$ This is true, so the ordered pair is a solution.

PROCEDURE CHECKING A POTENTIAL SOLUTION FOR AN EQUATION WITH TWO VARIABLES

To determine whether a given ordered pair is a solution for an equation with two variables:

1. Replace the variables in the equation with the corresponding coordinates.
2. Verify that the equation is true.

EXAMPLE 1 Determine whether the ordered pair is a solution for the equation.

a. $(-4, -10); y = 2x - 3$

Solution $-10 \overset{?}{=} 2(-4) - 3$ Replace x with -4 and y with -10 and see if the
$-10 \overset{?}{=} -8 - 3$ equation is true.

$-10 \neq -11$ Because the equation is not true, $(-4, -10)$ is not a solution for $y = 2x - 3$.

b. $(4, -1); 2m - n = 9$

Solution $2(4) - (-1) \overset{?}{=} 9$ Unless otherwise specified, ordered pairs are expressed in
$8 + 1 \overset{?}{=} 9$ alphabetical order; therefore, replace m with 4 and n with -1 and see if the equation is true.

$9 = 9$ This checks, which means that $(4, -1)$ is a solution for $2m - n = 9$.

c. $\left(-2, -\dfrac{1}{3}\right); y = \dfrac{2}{3}x + 1$

Solution $-\dfrac{1}{3} \overset{?}{=} \dfrac{2}{3}(-2) + 1$ Replace x with -2 and y with $-\dfrac{1}{3}$ and see if the equation is true.

$$-\dfrac{1}{3} \overset{?}{=} -\dfrac{4}{3} + \dfrac{3}{3}$$

$$-\dfrac{1}{3} = -\dfrac{1}{3}$$ This checks, which means that $\left(-2, -\dfrac{1}{3}\right)$ is a solution for $y = \dfrac{2}{3}x + 1$.

YOUR TURN Determine whether the ordered pair is a solution for the equation.

a. $(-8, -2); x - 3y = -2$ **b.** $(1, -6); q = -5p + 1$ **c.** $\left(\dfrac{3}{4}, \dfrac{9}{4}\right); y = \dfrac{x}{3} + 2$

OBJECTIVE **2** **Find solutions for an equation with two unknowns.** Consider the equation $x + y = 4$. We have already seen that $(1, 3)$ is a solution. But there are other solutions. We list some of those solutions in the following table.

x	y	Ordered Pair
0	4	(0, 4)
1	3	(1, 3)
2	2	(2, 2)
3	1	(3, 1)
4	0	(4, 0)
5	−1	(5, −1)

Note: Every solution for $x + y = 4$ is an ordered pair of numbers whose sum is 4.

In fact, $x + y = 4$ has an infinite number of solutions. For every x-value there is a corresponding y-value that will add to the x-value to equal 4 and vice versa, which gives a clue about how to find solutions. We can simply choose a value for either x or y and solve for the corresponding value of the other variable. (In this case, our equation was easy enough that we could solve for y mentally.)

PROCEDURE **FINDING SOLUTIONS TO EQUATIONS WITH TWO VARIABLES**

To find a solution to an equation in two variables:
1. Choose a value for one of the variables (any value).
2. Replace the corresponding variable with your chosen value.
3. Solve the equation for the value of the other variable.

EXAMPLE 2 Find three solutions for the equation $3x + y = 2$.

Solution To find a solution we replace one of the variables with a chosen value and then solve for the value of the other variable.

For the first solution, we will choose x to be 0.

$$3x + y = 2$$
$$3(0) + y = 2$$
$$y = 2$$

Solution $(0, 2)$

> **Note:** Choosing x (or y) to equal 0 usually makes the equation very easy to solve.

For the second solution, we will choose x to be 1.

$$3x + y = 2$$
$$3(1) + y = 2$$
$$3 + y = 2$$
$$y = -1$$

Subtract 3 on both sides.

Solution $(1, -1)$

For the third solution, we will choose x to be 2.

$$3x + y = 2$$
$$3(2) + y = 2$$
$$6 + y = 2$$
$$y = -4$$

Subtract 6 on both sides.

Solution $(2, -4)$

We can summarize the solutions in a table:

If we choose x to be 0, then y is equal to 2.
If we choose x to be 1, then y is equal to -1.
If we choose x to be 2, then y is equal to -4.

x	y	Ordered Pair
0	2	$(0, 2)$
1	-1	$(1, -1)$
2	-4	$(2, -4)$

Keep in mind that there are an infinite number of correct solutions for a given equation in two variables, so your solutions may be different from the solutions someone else finds.

EXAMPLE 3 Find three solutions for the equation $y = -\frac{1}{3}x + 2$.

Solution Notice this equation has y isolated. If we select values for x, we will not have to isolate y as we did in Example 2. We will simply calculate the y value. Also notice that the coefficient for x is a fraction. Since we can choose any value for x, let's choose values like 3 and 6 that will divide out nicely with the denominator of 3.

For the first solution, we will choose x to be 0.

$$y = -\frac{1}{3}x + 2$$

$$y = -\frac{1}{3}(0) + 2$$

$$y = 2$$

For the second solution, we will choose x to be 3.

$$y = -\frac{1}{3}x + 2$$

$$y = -\frac{1}{3}(3) + 2$$

$$y = -\frac{1}{\underset{1}{\cancel{3}}}\left(\frac{\overset{1}{\cancel{3}}}{1}\right) + 2$$

For the third solution, we will choose x to be 6.

$$y = -\frac{1}{3}x + 2$$

$$y = -\frac{1}{3}(6) + 2$$

$$y = -\frac{1}{\underset{1}{\cancel{3}}}\left(\frac{\overset{2}{\cancel{6}}}{1}\right) + 2$$

Solution $(0, 2)$ $y = -1 + 2$ $y = -2 + 2$

                          $y = 1$ $y = 0$

                     Solution $(3, 1)$ **Solution** $(6, 0)$

In summary,

x	y	Ordered Pair
0	2	(0, 2)
3	1	(3, 1)
6	0	(6, 0)

If we choose x to be 0, then y is equal to 2.
If we choose x to be 3, then y is equal to 1.
If we choose x to be 6, then y is equal to 0.

CONNECTION

When an equation is written with y isolated, as in the equation $y = -\dfrac{1}{3}x + 2$ from Example 3, we can think of x-values as inputs and the corresponding y-values as outputs. In Example 3, the x-input values were 0, 3, and 6 and the corresponding y-output values were 2, 1, and 0. This is the fundamental way of thinking about a mathematical concept called a *function*, which we will introduce in Section 4.7.

YOUR TURN   **Find three solutions for each equation. (Answers may vary.)**

   a. $x + 4y = 8$ **b.** $3x - 2y = 6$ **c.** $y = \dfrac{1}{2}x + 5$

OBJECTIVE 3 Graph linear equations.   We have learned that equations in two variables have an infinite number of solutions. Because of this fact, there is no way that all solutions to an equation can be found, much less listed. However, all the solutions can be represented using a graph. Consider some of the solutions for $x + y = 4$, which we listed in Table 4.1: $(0, 4)$ $(1, 3)$ $(2, 2)$ $(3, 1)$ $(4, 0)$ $(5, -1)$.

Note: The arrows on either end of the line indicate that the solutions continue beyond our "window" in both directions.

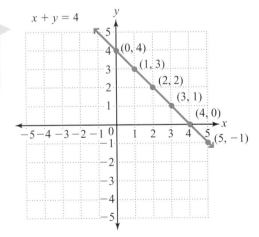

Notice that when we plot our ordered pair solutions for $x + y = 4$, the corresponding points lie in a straight line. In fact, all solutions for $x + y = 4$ are on that same line, which is why we say it is a *linear* equation. By connecting the points, we are graphically representing every possible solution for $x + y = 4$.

The graph of the solutions of every linear equation will be a straight line. Since two points determine a line in a plane, we need a minimum of two ordered pairs. This means we must find at least two solutions in order to graph the line. However, it is wise to find three solutions, using the third solution as a check. If we plot all three points and they cannot be connected with a straight line, then we know something is wrong.

PROCEDURE   **GRAPHING LINEAR EQUATIONS**

To graph a linear equation:

1. Find at least two solutions to the equation.
2. Plot the solutions as points in the rectangular coordinate system.
3. Connect the points to form a straight line.

EXAMPLE 4   Graph $3x + y = 2$.

Solution We found three solutions to this equation in Example 2. Recall those solutions:

x	y	Ordered Pair
0	2	(0, 2)
1	−1	(1, −1)
2	−4	(2, −4)

Now we plot each solution as a point in the rectangular coordinate system, and then connect the points to form a straight line.

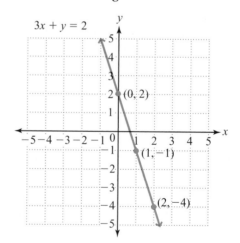

EXAMPLE 5   Graph $y = -\dfrac{1}{3}x + 2$.

Solution We found three solutions to this equation in Example 3. Recall those solutions:

x	y	Ordered Pair
0	2	(0, 2)
3	1	(3, 1)
6	0	(6, 0)

Now we plot each solution as a point in the rectangular coordinate system, and then connect the points to form a straight line.

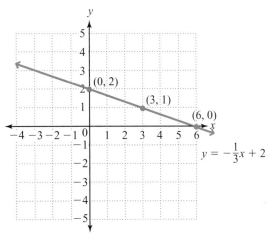

$y = -\frac{1}{3}x + 2$

YOUR TURN **Graph each equation.**

a. $x + 2y = 4$

b. $y = 2x - 1$

We have graphed linear equations in two variables. Now let's consider the graphs equations in one variable, such as $y = 3$ or $x = 4$, in which the variable is equal to a constant.

EXAMPLE 6 Graph $y = 3$.

Solution The equation $y = 3$ indicates that y is equal to a constant, 3. To establish ordered pairs we can rewrite the equation $y = 3$ as $0x + y = 3$. Because the coefficient of x is 0, y is always 3 no matter what we choose for x. We could complete a table of solutions like this:

If we choose x to be 0, then y equals 3.
If we choose x to be 2, then y is 3.
If we choose x to be 4, then y is still 3.

x	y	Ordered Pair
0	3	(0, 3)
2	3	(2, 3)
4	3	(4, 3)

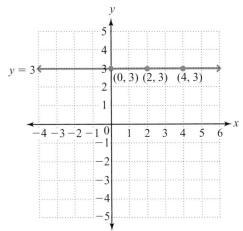

Notice that the graph of $y = 3$ is a horizontal line parallel to the x-axis that passes through the y-axis at a point with coordinates (0, 3).

ANSWERS

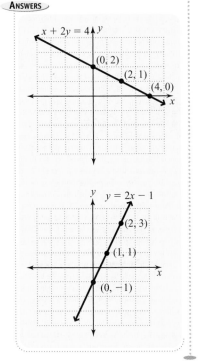

The graph of $y = c$, where c is a real number constant, is a horizontal line parallel to the x-axis that passes through the y-axis at a point with coordinates $(0, c)$.

EXAMPLE 7   Graph $x = 4$.

Solution The equation $x = 4$ indicates that x is equal to a constant, 4. If the y-variable is missing from an equation, it means that its coefficient is 0, so we can write $x = 4$ as $x + 0y = 4$. Note that x is always 4 no matter what we choose for y. We could complete a table of solutions like this:

If we choose y to be 0, then x equals 4.
If we choose y to be 1, then x is 4.
If we choose y to be 2, then x is still 4.

x	y	Ordered Pair
4	0	(4, 0)
4	1	(4, 1)
4	2	(4, 2)

a.

b.
$x = -3$
$(-3, 0)$

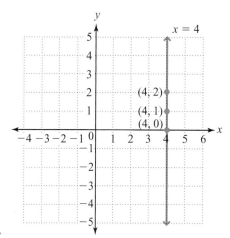

Notice that $x = 4$ is a vertical line parallel to the y-axis that passes through the x-axis at a point with coordinates $(4, 0)$.

RULE   VERTICAL LINES

The graph of $x = c$, where c is a real number constant, is a vertical line parallel to the y-axis that passes through the x-axis at a point with coordinates $(c, 0)$.

YOUR TURN   Graph.

  a. $y = -4$ **b.** $x = -3$

4.2 EXERCISES

For Extra Help

 InterAct Math

 Tutor Center   Addison-Wesley Math Tutor Center

 Video 6/DVT 2

MyMathLab

 MathXP MathXL

1. How do you determine if an ordered pair is a solution for a given equation?

2. If a given ordered pair is not a solution for a given equation, does that mean that there are no solutions for the equation? Explain.

3. How do you find a solution for a given equation with two unknowns?

4. What does the graph of an equation represent?

5. What is the minimum number of points needed to draw a straight line? Explain.

6. In your own words, explain the process of graphing a linear equation.

For Exercises 7–18, determine whether the given ordered pair is a solution for the given equation.

7. $(2, 3)$; $x + 2y = 8$

8. $(3, 1)$; $2x - y = 5$

9. $(-5, 2)$; $y - 4x = 3$

10. $(4, -6)$; $y = 3x - 1$

11. $(9, 0)$; $y = -2x + 18$

12. $(0, -1)$; $y = 2x + 1$

13. $(6, -2)$; $y = -\frac{2}{3}x$

14. $(0, 0)$; $y = \frac{2}{5}x$

15. $\left(-1\frac{2}{5}, 0\right)$; $y - 3x = 5$

16. $\left(\frac{2}{3}, -4\frac{5}{6}\right)$; $y = \frac{1}{4}x - 5$

17. $(2.2, -11.2)$; $y + 6x = -2$

18. $(-1.5, -1.3)$; $y = 0.2x - 1$

For Exercises 19–46, find three solutions for the given equation. Then graph.

19. $x - y = 8$

20. $x + y = -5$

21. $2x + y = 6$

22. $3x - y = 9$

23. $y = x$

24. $y = -x$

25. $y = 2x$

26. $y = -2x$

27. $y = -5x$

28. $y = 3x$

29. $y = x - 3$

30. $y = -x + 5$

31. $y = -2x + 4$

32. $y = 2x - 5$

33. $y = 3x + 2$

34. $y = -5x - 1$

35. $y = \dfrac{1}{2}x$

36. $y = -\dfrac{3}{4}x$

37. $y = -\dfrac{2}{3}x + 4$

38. $y = \dfrac{4}{5}x - 1$

39. $x - \dfrac{1}{4}y = 2$ **40.** $\dfrac{1}{3}x + y = -1$ **41.** $y = 0.4x - 2.5$ **42.** $1.2x + 0.5y = 6$

43. $y = -5$ **44.** $x = -6$ **45.** $x = 7$ **46.** $y = 4$

47. Compare the graphs of $y = x$, $y = 2x$, and $y = 3x$ from Exercises 23, 25, and 28. For an equation in the form $y = ax$, what effect does increasing a seem to have on the graph?

48. Compare the graphs of $y = x$ and $y = -x$ from Exercises 23 and 24, then compare the graphs of $y = 2x$ and $y = -2x$ from Exercises 25 and 26. For an equation in the form $y = ax$, what can you conclude about the graph when a is positive versus when a is negative?

49. Compare the graphs of $y = 3x$ and $y = 3x + 2$ from Exercises 28 and 33, then compare the graphs of $y = -2x$ and $y = -2x + 4$ from Exercises 26 and 31. For an equation in the form $y = ax + b$, what effect does adding b to $y = ax$ seem to have on the graph?

50. Compare the graphs of $y = x$ and $y = x - 3$ from Exercises 23 and 29, then compare the graphs of $y = -5x$ and $y = -5x - 1$ from Exercises 27 and 34. For an equation in the form $y = ax - b$, what effect does subtracting b from $y = ax$ seem to have on the graph?

For Exercises 51–56, solve.

51. A plumber charges $75 plus $25 per hour of labor. The equation $c = 25n + 75$ describes the total that she would charge for a service visit where n represents the number of hours of labor and c is the total cost.

   a. Find the total cost if labor is 2 hours.

   b. If a client's total charges are $175, for how many hours of labor was the client charged?

   c. Graph the equation with n along the horizontal axis and c along the vertical axis.

52. A copy center charges $15 for the first 300 copies plus $0.03 per copy after that. The equation $c = 0.03n + 15$ describes the total that would be charged for a copy service where n represents the number of copies above 300 and c is the total cost.

   a. Find the total cost for 400 copies.

   b. If a client's total charges are $21, for how many copies was the client charged?

   c. Graph the equation with n along the horizontal axis and c along the vertical axis.

53. A stock begins a 5-day decline in price at $23. The equation $c = -2n + 23$ describes the total that the stockbroker could sell the stock for where n represents the number of days into the decline and c is the total cost.

   a. Find the total cost if the stock is at day 4.

   b. If the stock is worth $19, how many days into the decline is the price?

   c. Graph the equation with n along the horizontal axis and c along the vertical axis.

54. A wireless company charges $20 per month for 300 anytime minutes plus $0.10 per minute above 300. The equation $c = 0.10n + 20$ describes the total that the company would charge for a monthly bill where n represents the number of minutes used above 300.

   a. Find the total cost if a customer uses 400 anytime minutes.

   b. If a client's total charges are $22, for how many additional minutes was the client charged?

   c. Graph the equation with n along the horizontal axis and c along the vertical axis.

55. A mechanic charges $125 plus $25 per hour of labor. The equation $c = 25n + 125$ describes the total that she would charge for a service visit where n represents the number of hours of labor and c is the total cost.

   a. Find the total cost if labor is 3.5 hours.

b. If a client's total charges are \$175, for how many hours of labor was the client charged?

c. Graph the equation with n along the horizontal axis and c along the vertical axis.

56. An academic tutor charges \$20 per month plus \$5 per hour of tutoring. The equation $c = 5n + 20$ describes the total that he would charge for tutoring where n represents the number of hours of tutoring and c is the total cost.

a. Find the total cost if the tutor works 3 hours.

b. If a client's total charges are \$45, for how many hours of labor was the client charged?

c. Graph the equation with n along the horizontal axis and c along the vertical axis.

PUZZLE PROBLEM

Without lifting your pencil from the paper, connect the dots using only four straight lines.

REVIEW EXERCISES

1. Evaluate $-32t + 70$ when $t = 4.2$.

2. Calculate the perimeter of the shape.

3. Calculate the area of the triangle with a base of 4.2 inches and a height of 2.7 inches.

10.4 m
8.2 m
9 m
6.1 m
6.3 m

4. Isolate x in the formula $v = mx + k$.

5. Find the value of x in the equation $2x + 3y = 6$ when $y = 0$.

4.3 Graphing Using Intercepts

OBJECTIVES

1 Given an equation, find the coordinates of the *x*- and *y*-intercepts.
2 Graph linear equations using intercepts.

OBJECTIVE **1** **Given an equation, find the coordinates of the *x*- and *y*-intercepts.** Look at the graph below. Notice the line intersects the *x*-axis at (4, 0) and the *y*-axis at (0, −2). These points are called *intercepts*.

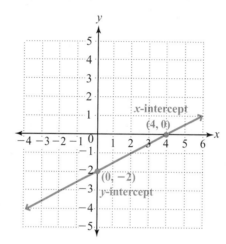

The point where a graph intersects the *x*-axis is called the **x-intercept**. The point where a graph intersects the *y*-axis is called the **y-intercept**.

DEFINITIONS
x-intercept: A point where a graph intersects the *x*-axis.
y-intercept: A point where a graph intersects the *y*-axis.

Note that the *y*-coordinate of any *x*-intercept will always be 0. Similarly, the *x*-coordinate of any *y*-intercept will always be 0. These facts about intercepts suggest the following procedures.

PROCEDURE **FINDING THE *x*- AND *y*-INTERCEPTS**
To find an *x*-intercept:

1. Replace *y* with 0 in the given equation.
2. Solve for *x*.

To find a *y*-intercept:

1. Replace *x* with 0 in the given equation.
2. Solve for *y*.

EXAMPLE 1 For the equation $5x - 2y = 10$, find the x- and y-intercepts.

Solution

For the x-intercept, replace y with 0 and solve for x.

$$5x - 2(0) = 10$$
$$5x = 10$$
$$x = 2 \quad \text{Divide both sides by 5.}$$

x-intercept: $(2, 0)$

For the y-intercept, replace x with 0 and solve for y.

$$5(0) - 2y = 10$$
$$-2y = 10$$
$$y = -5 \quad \text{Divide both sides by } -2.$$

y-intercept: $(0, -5)$

Note: We will no longer show the steps containing the addition or multiplication principles as we did in Chapters 2 and 3.

YOUR TURN

For the equation $3x - 4y = 8$, find x- and y-intercepts.

The equation in Example 1 is in the form $Ax + By = C$, where A, B, and C are real number constants with A and B not both equal to 0. A linear equation written in this form is said to be in *standard form*. One of the advantages of having a linear equation in standard form is that it is easy to find the intercepts. Throughout this section we introduce other forms for linear equations and explore some of the information that can be discerned from equations written in various forms.

EXAMPLE 2 For the equation $y = \dfrac{2}{3}x$, find the x- and y-intercepts.

Solution Replace x with 0 and solve for y. Then replace y with 0 and save for x.

y-intercept: $\quad y = \dfrac{2}{3}(0)$
$$y = 0$$

x-intercept: $\quad 0 = \dfrac{2}{3}x$
$$0 = x \quad \text{Multiply both sides by } \dfrac{3}{2}.$$

x- and y-intercept: $(0, 0)$

The equation in Example 2 is a linear equation in the form $y = mx$, where m is any real number. The graph of any equation in the form $y = mx$ is always a line that passes through the origin, which means that the x- and y-intercepts are both at $(0, 0)$.

ANSWERS

x-intercept: $\left(\dfrac{8}{3}, 0\right)$

y-intercept: $(0, -2)$

RULE   **INTERCEPTS FOR $y = mx$**

If an equation can be written in the form $y = mx$, where m is a real number other than 0, then the x- and y-intercepts are at the origin, $(0, 0)$.

EXAMPLE 3 For the equation $y = 2x + 1$, find the coordinates of the x- and y-intercepts.

Solution Replace x with 0 and solve for y. Then replace y with 0 and save for x.

x-intercept: $0 = 2x + 1$

$-1 = 2x$ Subtract 1 on both sides.

$-\dfrac{1}{2} = x$ Divide both sides by 2.

x-intercept: $\left(-\dfrac{1}{2}, 0\right)$

y-intercept: $y = 2(0) + 1$

$y = 1$

y-intercept: $(0, 1)$

Note: When we replace x with 0, we are left with the constant 1 for the y-intercept.

The equation in Example 3 is in the form $y = mx + b$, where m and b can be any real number. In this form, when we replace x with 0 to find the y-intercept, we are left with the constant b.

RULE THE y-INTERCEPT FOR $y = mx + b$

If an equation is in the form $y = mx + b$, where m and b are real numbers, then the y-intercept will be $(0, b)$.

YOUR TURN

For the equation $y = 3x - 5$, find the x- and y-intercepts.

EXAMPLE 4 For the equation $y = 3$, find the x- and y-intercepts.

Solution Remember from Section 4.2, Example 6, that the graph of $y = 3$ is a horizontal line parallel to the x-axis that passes through the y-axis at the point $(0, 3)$. Notice that this point is the y-intercept. Since the line is parallel to the x-axis, it will never intersect the x-axis. Therefore, there is no x-intercept.

Example 4 suggests the following rule.

RULE INTERCEPTS FOR $y = c$

The graph of an equation in the form $y = c$, where c is a real number constant, has no x-intercept and the y-intercept is $(0, c)$.

We can draw a similar conclusion about the graph of an equation in the form $x = c$.

RULE INTERCEPTS FOR $x = c$

The graph of an equation in the form $x = c$, where c is a real number constant, has no y-intercept and the x-intercept is $(c, 0)$.

a. $y = -5$ **b.** $x = -2$

OBJECTIVE 2 **Graph linear equations using intercepts.**   Because two points determine a line in a plane, we can use the *x*- and *y*-intercepts to draw the graph of a linear equation.

EXAMPLE 5   Graph $5x - 2y = 10$ using the *x*- and *y*-intercepts.

Solution In Example 1, the intercepts were found to be $(2, 0)$ and $(0, -5)$. We plot these intercepts and connect them to graph the line. It is helpful to find a third solution and verify that all three points can be connected to form a straight line. For this third point, we will choose $x = 3$ and solve for *y*.

Third solution: (check) If $x = 3$, then

$$5(3) - 2y = 10$$
$$15 - 2y = 10$$
$$-2y = -5 \qquad \text{Subtract 15 from both sides.}$$
$$y = \frac{5}{2} \qquad \text{Divide both sides by 22.} \qquad \textbf{Solution } \left(3, \frac{5}{2}\right)$$

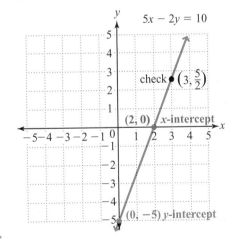

Table of Solutions

	x	y
x-intercept:	2	0
y-intercept:	0	-5
Third solution: (check)	3	$\frac{5}{2}$

ANSWERS

a. no *x*-intercept
   y-intercept: $(0, -5)$

b. *x*-intercept: $(-2, 0)$
   no *y*-intercept

ANSWER

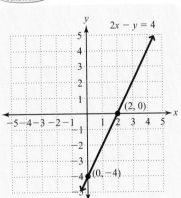

YOUR TURN

Graph $2x - y = 4$ using the *x*- and *y*-intercepts.

If the *x*- and *y*-intercepts are both at the origin, then we must find an additional point in order to graph the linear equation.

EXAMPLE 6 Graph $y = \frac{2}{3}x$ using the x- and y-intercepts.

Solution In Example 2 we found (0, 0) to be both the x- and the y-intercepts. Because this single point is not enough to determine the line in the plane, we must find at least one more solution. We will also find a third solution as a check.

Second solution:
Choose x to be 3

$$y = \frac{2}{3}(3)$$

$$y = \frac{2}{\cancel{3}}\left(\frac{\cancel{3}^1}{1}\right)$$

$$y = 2$$

Third solution (check):
Choose x to be -3

$$y = \frac{2}{3}(-3)$$

$$y = \frac{2}{\cancel{3}}\left(-\frac{\cancel{3}^1}{1}\right)$$

$$y = -2$$

Second solution (3, 2)   **Third solution** $(-3, -2)$

Now plot the solutions and connect the points to form the line.

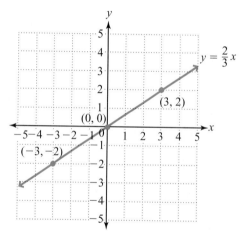

Table of Solutions

	x	y
x- and y-intercept:	0	0
Second solution:	3	2
Third solution: (check)	-3	-2

EXAMPLE 7 Graph $y = \frac{2}{3}x + 2$ using the x- and y-intercepts.

Solution Notice that this equation is in the form $y = mx + b$. Because $b = 2$ in this case, we can conclude that the y-intercept is (0, 2). For the x-intercept, let $y = 0$ and solve for x. Also, as a check, we will choose $x = 3$ to generate a third point.

x-intercept: $0 = \dfrac{2}{3}x + 2$

$-2 = \dfrac{2}{3}x$ **Subtract 2 from both sides.**

$\dfrac{3}{\cancelto{1}{2}} \cdot \dfrac{-\cancelto{1}{2}}{1} = \dfrac{\cancelto{1}{3}}{\cancelto{1}{2}} \cdot \dfrac{\cancelto{1}{2}}{\cancelto{1}{3}}x$ **Multiply both sides by $\dfrac{3}{2}$.**

$-3 = x$ ***x*-intercept** $(-3, 0)$

Third solution: (check) $y = \dfrac{2}{3}(3) + 2$

$y = \dfrac{2}{\cancelto{1}{3}} \cdot \dfrac{\cancelto{1}{3}}{1} + 2$

$y = 2 + 2$

$y = 4$ **Third solution** $(3, 4)$

Now plot the solutions and connect the points to form the line.

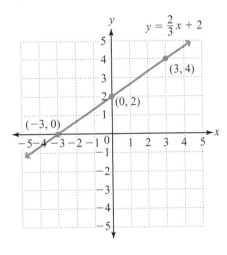

Table of Solutions

	x	y
x-intercept:	-3	0
y-intercept:	0	2
Third solution: (check)	3	4

Compare the graphs in Examples 6 and 7. If we place both lines on the same grid, we see that the graph of $y = \dfrac{2}{3}x + 2$ is a line parallel to the line $y = \dfrac{2}{3}x$. Further, each point on the graph of $y = \dfrac{2}{3}x + 2$ is 2 units higher than each point with the same x-coordinate on the graph of $y = \dfrac{2}{3}x$.

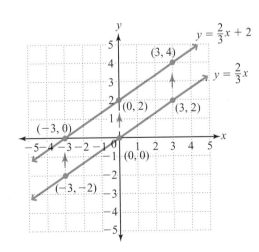

Table of Solutions

x	$y = \dfrac{2}{3}x$	$y = \dfrac{2}{3}x + 2$
-3	-2	0
0	0	2
3	2	4

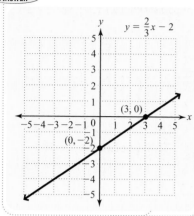

From the table we see that the difference in an equation in the form $y = mx$ and an equation in the form $y = mx + b$ is the value of the y-intercept. In other words, the line $y = mx$ shifts upwards or downwards by b units. The graph of $y = \frac{2}{3}x - 2$ is a line parallel to $y = \frac{2}{3}x$, only shifted down 2 units so that its y-intercept is $(0, -2)$. We will explore this further in the next section.

YOUR TURN

Graph $y = \frac{2}{3}x - 2$ using the x- and y-intercepts.

4.3 EXERCISES

For Extra Help

InterAct Math

Addison-Wesley
Math Tutor Center

Video 6/DVT 2

MyMathLab

MathXL

1. Explain what an x-intercept is.

2. Explain what a y-intercept is.

3. How do you find an x-intercept?

4. How do you find a y-intercept?

5. If an equation is in the form $y = mx$, where m is a real number constant, what are the coordinates of the x- and y-intercepts? Explain.

6. Compare the graphs of $y = 2x$ and $y = 2x + 5$.

For Exercises 7–20, find the x- and y-intercepts.

7. $2x + 3y = 6$ **8.** $y = 2x + 5$ **9.** $\frac{3}{2}x - y = 3$ **10.** $3x + 4y = 6$

11. $3x + 4y = -10$ **12.** $2x + 3y = 12$ **13.** $x - y = 4$ **14.** $x + y = 7$

15. $y = 2x$ **16.** $x + 3y = 0$ **17.** $x - 5 = 0$ **18.** $y = 7$

19. $y = \dfrac{2}{5}x$ **20.** $y = 3x - 1$

For Exercises 21–40, graph using the x- and y-intercepts.

21. $x - y = 7$ **22.** $x + y = 10$ **23.** $3x - y = 6$ **24.** $x - 5y = 15$

25. $2x + 3y = 6$ **26.** $3x + 4y = 12$ **27.** $5x + 2y = 10$ **28.** $3x - 2y = 18$

29. $y = \dfrac{2}{5}x$ **30.** $-\dfrac{1}{4}x = y$ **31.** $y = x - 4$ **32.** $y = 3 - 2x$

33. $-4x - y = 5$ **34.** $2x - y = -1$ **35.** $6x + 3y = 9$ **36.** $\dfrac{1}{2}x - y = 0$

37. $x = -4$ **38.** $y = 5$ **39.** $y - 2 = 0$ **40.** $x + 3 = 0$

41. Which of the following could be the graph of $1.4x - 0.9y = 2.7$? Explain.

a.   **b.**   **c.**   **d.**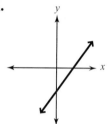

42. Which of the following could be the graph of $0.65x - 2.9y = -3.5$? Explain.

a.   **b.**   **c.**   **d.**

43. If a circle with a radius of 2.7 units were to be drawn in the coordinate plane with its center at the origin, what would be the coordinates of its x- and y-intercepts?

44. If a circle with a diameter of 6.8 units were to be drawn in the coordinate plane with its center at the origin, what would be the coordinates of its x- and y-intercepts?

45. A salesperson sells two different sizes of facial cleansing lotion. She sells the small size for $15 and the large size for $20.

    a. If x represents the number of units sold of the small size and y represents the number of units of the large size sold, then write an expression that describes her total sales.

    b. If her supervisor sets a goal of selling $2000 worth of the cleansing lotion, use the expression from part a to write an equation describing the sales goal.

    c. Find the x- and y-intercepts.

    d. What do the x- and y-intercepts mean in terms of sales?

e. Find a third combination of sales numbers that would equal the sales goal.

f. Graph the equation.

46. A comic book collector will pay $50 for any Superman comic published prior to 1970 if it is in mint condition and $30 if it is not in mint condition.

   a. If x represents the number of comic books in mint condition and y represents the number of comic books not in mint condition, then write an expression that describes the amount he would pay for a combination of comic books in mint and nonmint condition.

   b. Suppose the collector visits a convention for collectors and has a budget of $1800 to spend. Use the expression from part a to write an equation describing his expenditure on a combination of the comics if he spends all of his budget.

   c. Find the x- and y-intercepts.

   d. What do the x- and y-intercepts mean in terms of purchases?

   e. Find a third combination of comic books he could purchase to equal his budget amount.

   f. Graph the equation.

REVIEW EXERCISES

1. Solve for w in the equation $P = 2l + 2w$.

2. Translate to an equation, then solve. Four more than three times a number is equal to five times the sum of the number and six.

3. Translate to an equation, then solve. The sum of three consecutive odd integers is 141. Find the integers.

4. Two steel rods are welded to form a 90° angle. A third rod is welded to the other two as shown. If one of the angles is 15° more than the other, what are the two angles?

5. In what quadrant is (150, 200) located?

6. Determine whether $(-8, -2)$ is a solution for $x + 3y = -5$.

Slope-Intercept Form

OBJECTIVES

1. Compare lines with different slopes.
2. Graph equations in slope-intercept form.
3. Find the slope of a line given two points on the line.

In Section 4.3 we discussed various forms of linear equations, one of which was $y = mx + b$, where m and b are real number constants. Further, we discovered that the graph of an equation in this form is a line with its y-intercept at $(0, b)$. In this section we explore how the coefficient m affects the graph.

OBJECTIVE 1 **Compare lines with different slopes.** First, we consider graphs of equations in which $b = 0$, so that the equation is $y = mx$ and the graphs all have the origin $(0, 0)$ as the x- and y-intercepts.

EXAMPLE 1 Graph each of the following on the same grid.

$$y = x \qquad y = 2x \qquad y = 3x$$

Solution In the table below, the same choice for x has been substituted into each equation and the corresponding y-coordinate has been found.

If x is	and $y = x$, then y is	and $y = 2x$, then y is	and $y = 3x$, then y is
0	0	0	0
1	1	2	3
2	2	4	6

Now plot the ordered pairs and connect the points to graph the lines.

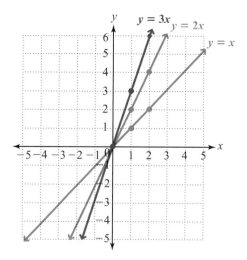

From Example 1 we see that for equations of the form $y = mx$, as the coefficient m increases, the graphs get steeper. Because the coefficient m affects how steep a line is, m is called the *slope* of the line.

For $y = x$, the slope is 1 ($m = 1$).

For $y = 2x$, the slope is 2 ($m = 2$).

For $y = 3x$, the slope is 3 ($m = 3$).

We have seen that if the slope is greater than 1, then the corresponding line has a steeper incline than the graph of $y = x$. What values for m would cause a line to be less inclined than $y = x$? The pattern in the graphs suggests we should explore values less than 1. First consider values less than 1 but greater than 0, such as $\frac{1}{2}$ and $\frac{1}{3}$.

EXAMPLE 2 Graph each of the following on the same grid.

$$y = x \qquad y = \frac{1}{2}x \qquad y = \frac{1}{3}x$$

Solution For consistency, we use the same three choices for x that we used in Example 1.

If x is	and $y = x$, then y is	and $y = \frac{1}{2}x$, then y is	and $y = \frac{1}{3}x$, then y is
0	0	0	0
1	1	$\frac{1}{2}$	$\frac{1}{3}$
2	2	1	$\frac{2}{3}$

Now plot the ordered pairs and connect the points to graph the lines.

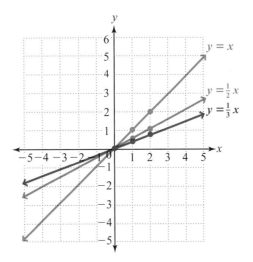

Notice that if the slope is a fraction between 0 and 1, then the smaller the fraction is, the less inclined or flatter the lines get. So far, we have considered only positive slope values. If we "travel" from left to right along lines with positive slopes, we travel uphill. What would a line with a negative slope look like?

EXAMPLE 3 Graph each of the following on the same grid.

$$y = -x \qquad y = -2x \qquad y = -\frac{1}{2}x$$

Solution Again, we make a table and graph the three equations.

If x is	and $y = -x$, then y is	and $y = -2x$, then y is	and $y = -\frac{1}{2}x$, then y is
0	0	0	0
1	−1	−2	$-\frac{1}{2}$
2	−2	−4	−1

Now plot the ordered pairs and connect the points to graph the lines.

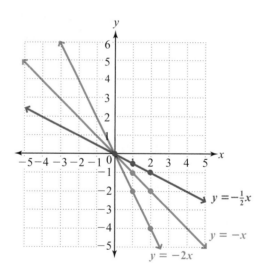

Note: When we speak of traveling along a graph, the convention is to travel from left to right, just as we do when we read English.

Notice that if we "travel" from left to right along lines with negative slopes, we travel downhill. As the slope becomes more negative, the downhill incline becomes steeper.

RULE **GRAPHS OF y = mx**

Given an equation of the form $y = mx$, the graph of the equation is a line passing through the origin and having the following characteristics.

If $m > 0$, then the graph is a line that slants uphill from left to right.

If $m < 0$, then the graph is a line that slants downhill from left to right.

The greater the absolute value of m, the steeper the line.

YOUR TURN

a. Is $y = 5x$ steeper or less inclined than $y = x$?

b. Is the graph of $y = -\dfrac{2}{3}x$ uphill or downhill from left to right?

OBJECTIVE **2** **Graph equations in slope-intercept form.** We have seen that slope determines the incline of a line. What more can we discover about slope? Notice if we isolate m in $y = mx$, we get $m = \dfrac{y}{x}$, which suggests that **slope** is a ratio of the amount of vertical change to the amount of horizontal change.

DEFINITION **Slope:** The ratio of the vertical change to the horizontal change of a line.

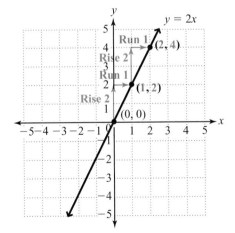

Look at the graph of $y = 2x$. The slope of this line is 2, which can be written as the ratio $\dfrac{2}{1}$. Given any point on the line as a starting point, we can get to another point on the line by moving vertically, or "rising" 2 units, and then moving horizontally, or "running" 1 unit, from the starting point. For example, if we rise 2 units and run 1 unit from (0, 0), we end up at (1, 2). Similarly, rising 2 units and running 1 unit from (1, 2) puts us at (2, 4).

Recall that in an equation of the form $y = mx + b$, the y-intercept is $(0, b)$. We can use the y-intercept as the starting point and then find other points on the line by using the slope.

ANSWERS

a. $y = 5x$ is steeper than $y = x$.

b. downhill

EXAMPLE 4 For the equation $y = -\frac{1}{3}x + 2$, determine the slope and the y-intercept. Then graph the equation.

Solution Because the equation is in the form $y = mx + b$, where $m = -\frac{1}{3}$ and $b = 2$, the slope of the line is $-\frac{1}{3}$ and the y-intercept is $(0, 2)$. To graph the line, we can plot the y-intercept and then use the slope to find other points. To interpret $-\frac{1}{3}$ in terms of "rise" and "run," we place the negative sign in either the numerator or denominator. We can "rise" -1 and "run" 3, which gives the point $(3, 1)$. Or we can "rise" 1 and "run" -3, which gives the point $(-3, 3)$. Either interpretation gives another point on the same line.

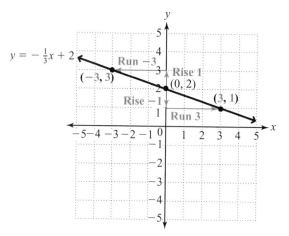

The points we found using the slope can be verified in the equation. We can check the ordered pair $(-3, 3)$ in the equation:

$$y = -\frac{1}{3}x + 2$$

$$3 \overset{?}{=} -\frac{1}{3}(-3) + 2$$

$$3 \overset{?}{=} -\frac{1}{\overset{1}{\cancel{3}}} \cdot -\frac{\overset{1}{\cancel{3}}}{1} + 2$$

$$3 \overset{?}{=} 1 + 2$$

$$3 = 3 \qquad \text{This checks.}$$

We will leave the check of $(3, 1)$ to the reader.

YOUR TURN

For the equation $y = \frac{2}{5}x - 2$, determine the slope and the y-intercept. Then graph the equation.

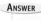

ANSWER

$m = \frac{2}{5}$, y-intercept: $(0, -2)$

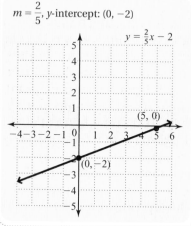

When an equation is in the form $y = mx + b$, we say it is in *slope-intercept* form because we can easily determine the slope, m, and y-intercept, $(0, b)$. If an equation is not in slope-intercept form and we need to determine the slope and y-intercept, then we can write the equation in slope-intercept form by isolating y.

EXAMPLE 5 For the equation $3x + 4y = 8$, determine the slope and the y-intercept. Then graph the equation.

Solution Write the equation in slope-intercept form by isolating y.

$$3x + 4y = 8$$

$$4y = -3x + 8 \qquad \text{Subtract } 3x \text{ from both sides to isolate } 4y.$$

$$y = -\frac{3}{4}x + \frac{8}{4} \qquad \text{Divide both sides throughout by 4 to isolate } y.$$

$$y = -\frac{3}{4}x + 2 \qquad \text{Simplify the fraction.}$$

ANSWER

$m = \frac{3}{2}$, y-intercept: $(0, -3)$

The slope is $-\frac{3}{4}$ and the y-intercept is $(0, 2)$. To graph the line we begin at $(0, 2)$ and then rise -3 and run 4, which gives the point $(4, -1)$. Or we could begin at $(0, 2)$ and then rise 3 and run -4, which gives the point $(-4, 5)$.

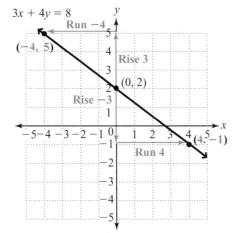

We can check the ordered pairs in the equation:

For $(-4, 5)$:

$$3x + 4y = 8$$
$$3(-4) + 4(5) \stackrel{?}{=} 8$$
$$-12 + 20 \stackrel{?}{=} 8$$
$$8 = 8$$

For $(4, -1)$:

$$3x + 4y = 8$$
$$3(4) + 4(-1) \stackrel{?}{=} 8$$
$$12 + (-4) \stackrel{?}{=} 8$$
$$8 = 8$$

Both ordered pairs check.

YOUR TURN

For the equation $3x - 2y = 6$, determine the slope and the y-intercept. Then graph the equation.

OBJECTIVE 3 **Find the slope of a line given two points on the line.** Given two points on a line, we can determine the slope of the line. Recall that slope is the ratio of the vertical change to the horizontal change. Consider points on the graph of $y = \frac{2}{3}x + 2$, which we developed in Example 7 of Section 4.3. Notice the slope of this line is $\frac{2}{3}$.

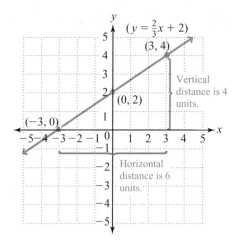

Table of Solutions

	x	y
x-intercept:	−3	0
y-intercept:	0	2
Third solution:	3	4

Given any two points on the graph of $y = \frac{2}{3}x + 2$, the ratio of their vertical distance to their horizontal distance simplifies to $\frac{2}{3}$. For example, the points $(-3, 0)$ and $(3, 4)$ are 4 vertical units apart and 6 horizontal units apart, so the ratio is:

$$\frac{\textbf{vertical distance}}{\textbf{horizontal distance}} = \frac{4}{6} = \frac{2}{3}$$

How could we calculate those distances without consulting the graph? Notice that the vertical distance is the difference between the y-coordinates and the horizontal distance is the difference between the x-coordinates. In the case of $(-3, 0)$ and $(3, 4)$

$$\frac{\text{vertical distance}}{\text{horizontal distance}} = \frac{4 - 0}{3 - (-3)} = \frac{4}{6} = \frac{2}{3}$$

We can write a formula for slope.

PROCEDURE THE SLOPE FORMULA

Given two points (x_1, y_1) and (x_2, y_2), where $x_2 \neq x_1$, the slope of the line connecting the two points is given by the formula $m = \dfrac{y_2 - y_1}{x_2 - x_1}$.

EXAMPLE 6 Find the slope of the line connecting the given points.

a. $(3, 5)$ and $(-1, 7)$

Solution Use the slope formula, $m = \dfrac{y_2 - y_1}{x_2 - x_1}$.

$m = \dfrac{7 - 5}{-1 - 3}$ $\quad$ Replace the variables with their corresponding values, then calculate.

$m = \dfrac{2}{-4}$ $\quad$ Simplify.

$m = -\dfrac{1}{2}$

Note: For this problem, the coordinate labels are:

$(-3, -4)$ $\quad$ $(1, -4)$
(x_1, y_1) $\quad$ (x_2, y_2)

b. $(-3, -4)$ and $(1, -4)$

Solution $m = \dfrac{-4 - (-4)}{1 - (-3)}$ $\quad$ Replace the variables with their corresponding values, then calculate.

$m = \dfrac{0}{4}$ $\quad$ Simplify.

$m = 0$

What does a line with a slope of 0 look like? Let's plot the points $(-3, -4)$ and $(1, -4)$ from Example 6, part b.

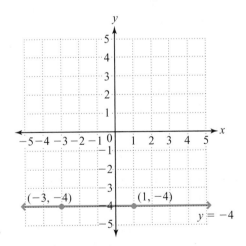

Because the two points have the same y-coordinate, they form a horizontal line parallel to the x-axis. The equation of this line is $y = -4$.

RULE **ZERO SLOPE**

Two points with different x-coordinates and the same y-coordinates, (x_1, c) and (x_2, c), will form the line with a slope 0 (a horizontal line) and equation $y = c$.

What is the slope of a vertical line? Two points on a vertical line would have the same x-coordinates but different y-coordinates.

EXAMPLE 7 Find the slope of the line connecting the points (2, 4) and (2, −3).

Solution Note that (2, 4) and (2, −3) connect to form a vertical line whose equation is $x = 2$. The slope of this line is found using the slope formula:

$$m = \frac{-3 - 4}{2 - 2}$$ Replace the variables in $m = \frac{y_2 - y_1}{x_2 - x_1}$ with their corresponding values.

$$m = \frac{-7}{0}$$ Simplify.

Because we never divide by 0, the slope is *undefined*.

RULE **UNDEFINED SLOPE**

Two points with the same x-coordinates and different y-coordinates, (c, y_1) and (c, y_2) will form a line with a slope that is undefined (a vertical line) and equation $x = c$.

YOUR TURN Find the slope of the line connecting the points with the given coordinates.

a. (2, 5) and (6, 1)   b. (−5, −2) and (7, −5)   c. (4, −1) and (−3, −1)

Calculator
TIPS

To graph an equation on a graphing calculator, you first enter the equation using the [Y=] *key. When you press the* [Y=] *key, you will see a list of Y's: Y1, Y2, Y3, etc. Each of the Ys can be used to enter an equation, which means you can graph multiple equations on the same graph if desired.*

To enter the equation $y = 3x + 2$, press [Y=] *and at Y1, type:*

Notice that the [X, T, θ, n] *key is used to type the variable x. To have the calculator draw the graph, press* [GRAPH] .

Sometimes you may need to adjust or change your viewing window. To change the range of marks on your axes, press [WINDOW] *. For example, if you want the x-axis to range from*

−10 to 10, then press [WINDOW] *and at Xmin, enter* [−][1][0] *and at Xmax, enter* [1][0] . *Xscl allows you to change the distance between each mark. If Xscl is 1, then you will see all ten marks from −10 to 0 and all ten marks from 0 to 10. If Xscl is 2, then you will only see five marks from −10 to 0 and five marks from 0 to 10, meaning it is only showing you marks at 2, 4, 6, 8, and 10. Similarly, Ymax and Ymin change the range on the y-axis, and Yscl affects how many marks are shown in that range.*

ANSWERS

a. −1   b. $-\frac{1}{4}$ c. 0

4.4 EXERCISES

For Extra Help

 InterAct Math

 Addison-Wesley Math Tutor Center

 Video 6/DVT 2

 MyMathLab

MathXL

1. In your own words, what is the slope of a line?

2. Does the graph of $y = -2x + 1$ go uphill from left to right or downhill? Why?

3. Does the graph of $3x - 5y = 10$ go uphill from left to right or downhill? Why?

4. The equations $y = 5x - 1$ and $y = 5x + 2$ have the same slope but different y-intercepts. What does this indicate about their graphs? Why?

5. Describe a line with a slope of 0. Explain why a slope of 0 produces this type of line.

6. Describe a line with a slope that is undefined. Explain why an undefined slope produces this type of line.

For Exercises 7–14, graph each set of equations on the same grid. For each set of equations, compare the slopes, y-intercepts, and their effects on the graphs.

7. $y = \dfrac{1}{2}x$

   $y = x$

   $y = 2x$

8. $y = \dfrac{1}{3}x$

   $y = x$

   $y = 3x$

9. $y = x$

   $y = \dfrac{1}{2}x$

   $y = \dfrac{1}{5}x$

10. $y = x$

    $y = \dfrac{1}{3}x$

    $y = \dfrac{1}{6}x$

11. $y = x$

    $y = -x$

    $y = -2x$

12. $y = x$

    $y = -x$

    $y = -3x$

13. $y = x$
   $y = x + 2$
   $y = x - 2$

14. $y = 2x$
   $y = 2x - 1$
   $y = 2x - 3$

For Exercises 15–32, determine the slope and the y-intercept. Then graph the equation.

15. $y = 2x + 3$

16. $y = -3x + 5$

17. $x - y = 6$

18. $x + 2y = -5$

19. $2x + 3y = 7$

20. $4x - 3y = 12$

21. $2x + y = 4$

22. $y = \dfrac{1}{2}x - 4$

23. $y = -\dfrac{2}{3}x + 8$

24. $y = 7 - \dfrac{4}{3}x$

25. $y + 3x = -1$

26. $2x - y = 5$

27. $5x + 12y = 2$

28. $3y = 2x - 4$

29. $5x - 2y = 2$

30. $3x - 2y + 7 = 0$

31. $2x - 3y + 8 = 0$

32. $x - y = 4$

For Exercises 33–44, find the slope of the line through the given points.

33. $(-3, 5), (4, 7)$

34. $(-4, -6), (2, -3)$

35. $(0, 5), (4, 0)$

36. $(-3, -1), (4, -2)$

37. $(4, 1), (8, 11)$

38. $(-4, 6), (6, -2)$

39. $(6, 1), (6, -8)$

40. $(-3, 2), (-3, 5)$

41. $(-3, 5), (4, 5)$

42. $(-2, -8), (4, -8)$

43. $(-3, 1), (-2, -2)$

44. $(6, -2), (5, 5)$

For Exercises 45–50, match the equation with the appropriate graph.

45. $y = 2x + 4$

46. $3x + 2y = 6$

47. $x = y$

48. $y = 3$

49. $x - 2 = 0$

50. $x - 2y = 8$

a.

b.

c.

d.

e.

f.
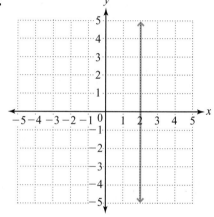

For Exercises 51–54, find the slope of the line, the y-intercept, and write the equation of the line in slope-intercept form.

51.

52.

53.

54.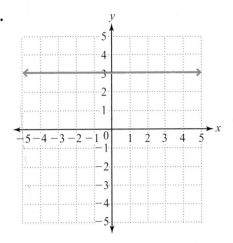

55. What is the equation of a horizontal line through $(0, -2)$?

56. What is the equation of a vertical line through $(-3.5, 0)$?

57. What equation describes the *x*-axis?

58. What equation describes the *y*-axis?

59. A parallelogram has vertices at $(1, -2)$, $(3, 4)$, $(-3, -2)$, and $(-1, 4)$.

    a. Plot the vertices in a coordinate plane, then connect them to form the parallelogram.

    b. Find the slope of each side of the parallelogram.

    c. What do you notice about the slopes of the parallel sides?

60. A right triangle has vertices at $(-2, 3)$, $(2, 1)$, and $(3, 3)$.

    a. Plot the vertices in a coordinate plane, then connect them to form the triangle.

    b. Find the slope of each side of the triangle.

    c. What do you notice about the slopes of the perpendicular sides?

61. In architecture, the slope of a roof is called its *pitch*. The roof of a house rises 3.5 feet for every 6 feet of horizontal distance. Find the pitch of the roof. Write the pitch as a decimal number and a fraction in simplest form.

62. The slope of a hill is referred to as the *grade* of the hill. Heartbreak Hill, which is part of the Boston Marathon, rises from an elevation of 150 feet to an elevation of 225 feet over a horizontal distance of about 5000 feet. Find the grade of Heartbreak Hill. (*Sources:* Boston Athletic Association)

63. The Great Pyramid at Giza rises 29.3 meters for every 23 meters of horizontal distance. Find the slope of the pyramid.

64. The steps of the Pyramid of the Sun at Teotihuacan, Mexico, rise 65 meters over a horizontal distance of 112.5 meters. Find the slope of the steps of the Pyramid of the Sun.

65. The graph to the right shows the change in percentage rate for Treasury bonds on July 17, 2001.

    a. Draw a straight line from the point on the graph at 9 A.M. to the point on the graph at 2 P.M.

    b. Find the slope of this line.

    c. If the same trend continued, predict the T-bond percentage at 4 P.M.

Treasury bond

66. The graph below shows a comparison of an average diver's consumption of gases using a standard open circuit tank versus a closed circuit breathing system.

    a. What is the slope of the line for the open circuit tank?

    b. What is the slope of the line for the closed circuit breathing system?

    c. What are some conclusions you could draw from the graphs?

Of Interest

Closed-circuit breathing systems operate by recycling the exhaled breath of the diver. A special filter absorbs carbon dioxide from the exhaled breath, then sensors detect how much oxygen remains, and a computer regulates the amount of new oxygen that is mixed in from a tank for the diver to inhale.

Source: http://ambientpressurediving.com/Graphs.html

Population Growth

The following data show the total U.S. population by decade.

Decade	Population (in millions)
1900	76.2
1910	92.2
1920	106.0
1930	123.2
1940	132.2
1950	151.3
1960	179.3
1970	203.3
1980	226.5
1990	248.7
2000	281.4

1. Plot each ordered pair as a point on a graph with the decades along the horizontal axis and the population along the vertical axis.

2. Do the points have a linear relationship?

3. Draw a straight line that connects the point at 1900 to the point at 1950. Find the slope of this line.

4. Draw a straight line that connects the point at 1950 to the point at 2000. Find the slope of this line.

5. Compare the slopes of the two lines you have drawn. Which has the greater slope? What does this indicate about population growth in the United States during the 20th century?

6. Use the line you drew from 1950 to 2000 to predict the population at 2110.

REVIEW EXERCISES

1. Simplify: $12x - 9y + 7 - x + 9y$

2. Isolate x in the equation $Ax + By = C$.

3. Multiply: $-2(x + 7)$

4. Solve and check: $(7x - 3) - (2x + 4) - (3x - 7) = 21$

5. Graph: $2x + 3y = 8$

6. Find the x- and y-intercepts for $3x - 5y = 7$.

4.5 Point-Slope Form

OBJECTIVES

1. Use slope-intercept form to write the equation of a line.
2. Use point-slope form to write the equation of a line.
3. Write the equation of a line parallel to a given line.
4. Write the equation of a line perpendicular to a given line.

In this section, we will explore how to write an equation of a line given information about the line, such as its slope and the coordinates of a point on the line.

OBJECTIVE 1 Use slope-intercept form to write the equation of a line. If we are given the y-intercept $(0, b)$ of a line, to write the equation of the line we will need either the slope or another point so that we can calculate the slope. First, consider a situation in which we are given the y-intercept and the slope of a line.

> **EXAMPLE 1** A line with a slope of 5 crosses the y-axis at the point $(0, 3)$. Write the equation of the line.
>
> **Solution** We use $y = mx + b$, the slope-intercept form of the equation, replacing m with the slope, 5, and b with the y-coordinate of the y-intercept, 3.
>
> $$y = 5x + 3$$

YOUR TURN

A line has a slope of -3. If the y-intercept is $(0, -4)$, write an equation of the line in slope-intercept form.

Now suppose we are given the y-intercept and a second point on the line. In order to write the equation in slope-intercept form we need the slope. So our first step is to calculate the slope using the formula from Section 4.4, $m = \dfrac{y_2 - y_1}{x_2 - x_1}$, where the two points, (x_1, y_1) and (x_2, y_2), are the y-intercept and the other given point.

> **EXAMPLE 2** The following graph below shows the relationship between Celsius and Fahrenheit temperatures. The table lists two ordered pairs of equivalent temperatures on the two scales.
>
> **a.** Write the equation for converting degrees Celsius to degrees Fahrenheit.
> **b.** Use the equation to convert 60°C to degrees Fahrenheit.

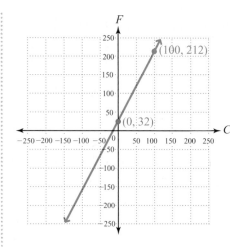

	Celsius (C)	Fahrenheit (F)
Freezing point of water:	0	32
Boiling point of water:	100	212

Solution

a. We are given (0, 32), which is like the *y*-intercept, and a second point, (100, 212). To use the slope-intercept equation, we must first find the slope using the slope formula:

Note: In this case, the *y*-coordinates are degrees Fahrenheit and the *x*-coordinates are degrees Celsius.

$$m = \frac{212 - 32}{100 - 0} \qquad \text{Use } m = \frac{y_2 - y_1}{x_2 - x_1}.$$

$$m = \frac{180}{100}$$

$$m = \frac{9}{5}$$

CONNECTION —————————

A slope of $\frac{9}{5}$ means that for every 9 degrees of change on the Fahrenheit scale, there is a 5-degree change on the Celsius scale.

Since the *F*-intercept is (0, 32) and the slope is $\frac{9}{5}$, we can write an equation in slope-intercept form. We let *C* (degrees Celsius) take the place of the *x*-variable, and *F* (degrees Fahrenheit) take the place of the *y*-variable.

$$F = \frac{9}{5}C + 32$$

b. Now we can use this equation to convert Celsius temperatures to degrees Fahrenheit. We are to convert 60°C to degrees Fahrenheit.

$$F = \frac{9}{5}(60) + 32 \qquad \text{Replace } C \text{ with 60.}$$

$$F = 108 + 32$$

$$F = 140°F$$

Examples 1 and 2 suggest the following procedure.

To write the equation of a line given its *y*-intercept, $(0, b)$, and its slope, *m*, use the slope-intercept form of the equation, $y = mx + b$. If given a second point and not the slope, we must first calculate the slope using $m = \dfrac{y_2 - y_1}{x_2 - x_1}$, then use $y = mx + b$.

YOUR TURN   **Write the equation of the line connecting the given points in slope-intercept form.**

$(0, -5)$, $(2, 4)$

OBJECTIVE   **2** **Use point-slope form to write the equation of a line.**   In our first objective we used the slope-intercept form of the equation of a line because we were given the *y*-intercept. However, we need a more general approach so that we can write the equation of a line given *any* two points on the line. For this more general approach, we can use the slope formula to derive a new form of the equation of a line called the *point-slope* form. Recall the slope formula:

CONNECTION

We can also view the formula for slope as a proportion and cross multiply to get the point-slope form:

$$\dfrac{m}{1} \diagdown \dfrac{y_2 - y_1}{x_2 - x_1}$$

$(x_2 - x_1)m = y_2 - y_1$

$$m = \frac{y_2 - y_1}{x_2 - x_1}$$

$$(x_2 - x_1) \cdot m = \left(\frac{y_2 - y_1}{x_2 - x_1}\right) \cdot \left(\frac{x_2 - x_1}{1}\right) \qquad \text{Multiply both sides by } x_2 - x_1 \text{ to isolate the } y\text{'s.}$$

$$(x_2 - x_1)m = y_2 - y_1 \qquad \text{Simplify.}$$

$$y_2 - y_1 = m(x_2 - x_1) \qquad \begin{array}{l}\text{Rewrite with } y\text{'s on the left side to} \\ \text{resemble slope-intercept form.}\end{array}$$

With this point-slope form, we can write the equation of a line given its slope and any point on the line, or any two points on the line.

To write the equation of a line given its slope and any point, (x_1, y_1), on the line, use the point-slope form of the equation of a line, $y - y_1 = m(x - x_1)$. If given a second point, (x_2, y_2), and not the slope, we first calculate the slope using $m = \dfrac{y_2 - y_1}{x_2 - x_1}$, then use $y - y_1 = m(x - x_1)$.

EXAMPLE   **3**   Write the equation of a line with a slope of 2 and passing through the point $(3, 5)$. Write the equation in slope-intercept form.

ANSWER

$y = \dfrac{9}{2}x - 5$

Solution   Because we are given the coordinates of a point and the slope of a line passing through the point, we begin with the point-slope formula. After replacing m, x_1, and y_1 with their corresponding values, we isolate y to get slope-intercept form.

$$y - y_1 = m(x - x_1)$$ Replace m with 2, x_1 with 3, and y_1 with 5.
$$y - 5 = 2(x - 3)$$
$$y - 5 = 2x - 6$$ Simplify.
$$y = 2x - 1$$ Add 5 to both sides to isolate y.
This is now slope-intercept form.

YOUR TURN **Write the equation of the line in slope-intercept form with the given slope passing through the given point.**

$m = 0.6$; $(-4, 2)$

ANSWER

$y = 0.6x + 4.4$

EXAMPLE 4 Write the equation of a line passing through the points $(3, 2)$ and $(-3, 6)$. Write the equation in slope-intercept form.

Solution Since we do not have the slope, we first calculate it.

$$m = \frac{6 - 2}{-3 - 3} \qquad \text{Use } m = \frac{y_2 - y_1}{x_2 - x_1}.$$

$$m = \frac{4}{-6}$$

$$m = -\frac{2}{3}$$

Now we can use the point-slope form, then isolate y to write slope-intercept form. Since we were given two points we can use either point for (x_1, y_1) in the point-slope equation. We will select $(3, 2)$ to be (x_1, y_1).

$$y - y_1 = m(x - x_1)$$

$$y - 2 = -\frac{2}{3}(x - 3) \qquad \text{Replace } m \text{ with } -\frac{2}{3}, x_1 \text{ with 3, and } y_1 \text{ with 2.}$$

$$y - 2 = -\frac{2}{3}x + 2 \qquad \text{Simplify.}$$

$$y = -\frac{2}{3}x + 4 \qquad \text{Add 2 to both sides to isolate } y.$$

ANSWER

$y = \frac{1}{2}x - 3$

YOUR TURN **Write the equation of the line connecting the given points in slope-intercept form.**

$(-2, -4)$, $(4, -1)$

Writing Linear Equations in Standard Form

Equations can also be written in standard form. Recall from Section 4.3 that standard form is $Ax + By = C$, where A, B, and C are real numbers. We will manipulate the point-slope form of the equation so that the x and y terms are on the same side of the equation and a constant appears on the other side of the equation.

EXAMPLE 5 The following data points relate the velocity of an object as time passes (You may recall this data from the last Your Turn problem in Section 4.1). The graph shows that the points are in a line. Write the equation of the line in standard form.

Time (in seconds)	Velocity (in ft./sec.)
0	2.0
1	2.5
2	3.0
3	3.5

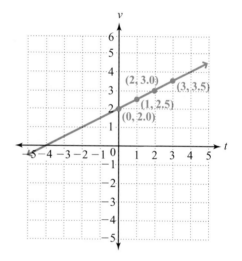

Solution First, we find the slope of the line using any two ordered pairs in the slope formula. We will use $(0, 2)$ and $(1, 2.5)$.

$$m = \frac{2.5 - 2}{1 - 0} \qquad \text{Use } m = \frac{y_2 - y_1}{x_2 - x_1}.$$

$$m = 0.5 \text{ or } \frac{1}{2}$$

Because $(0, 2)$ is the y-intercept, we can write the equation in slope-intercept form, then manipulate the equation so that x and y appear on the same side of the equal sign to get standard form.

$$y = \frac{1}{2}x + 2 \qquad \text{Slope-intercept form.}$$

$$y - \frac{1}{2}x = \frac{1}{2}x - \frac{1}{2}x + 2 \qquad \text{Subtract } \frac{1}{2}x \text{ on both sides to get } x \text{ and } y \text{ together.}$$

$$y - \frac{1}{2}x = 2$$

When we write an equation in standard form ($Ax + By = C$), it is customary to write the equation so that the x term is first with a positive coefficient and, if possible, so that A, B, and C are all integers. Let's manipulate $y - \dfrac{1}{2}x = 2$ to this more polished form.

$$-\dfrac{1}{2}x + y = 2$$

Use the commutative property of addition to write the x term first.

$$(-2)\left(-\dfrac{1}{2}x + y\right) = (-2)(2)$$

Multiply both sides by -2 so that the coefficient of x is a positive integer.

$$x - 2y = -4$$

Standard form with A, B, and C integers and $A > 0$.

EXAMPLE 6 A line connects the points $(1, 5)$ and $(-3, 2)$. Write the equation of the line in the form $Ax + By = C$, where A, B, and C are integers and $A > 0$.

Solution Find the slope, then write the equation of the line using point-slope form. To finish, rewrite the equation in standard form.

Find the slope: $m = \dfrac{2 - 5}{-3 - 1}$ Use $m = \dfrac{y_2 - y_1}{x_2 - x_1}$.

$$m = \dfrac{-3}{-4}$$ Simplify.

$$m = \dfrac{3}{4}$$

Because we were not given the y-intercept, we use the point-slope form of the linear equation:

$$y - 5 = \dfrac{3}{4}(x - 1)$$ Use $y - y_1 = (x - x_1)$.

Now manipulate the equation to get the form $Ax + By = C$, where A, B, and C are integers, and $A > 0$.

$$y - 5 = \dfrac{3}{4}x - \dfrac{3}{4}$$ Distribute $\dfrac{3}{4}$ to clear parentheses.

$$4(y - 5) = 4\left(\dfrac{3}{4}x - \dfrac{3}{4}\right)$$ Multiply both sides by the LCD, 4.

$$4y - 20 = 3x - 3$$

$$4y - 3x - 20 = -3$$ Subtract $3x$ on both sides to get x and y together.

$$4y - 3x = 17$$ Add 20 on both sides to get the constant terms on the right-hand side of the equation.

$$-3x + 4y = 17$$ Use the commutative property of addition to write the x term first.

$$(-1)(-3x + 4y) = (-1)(17)$$ Multiply both sides by -1 so that the coefficient of x is positive.

$$3x - 4y = -17$$ Standard form with A, B, and C integers and $A > 0$.

YOUR TURN Write the equation of the line through the given points in the form $Ax + By = C$, where A, B, and C are integers and $A > 0$.

$(-3, 6)$, $(1, -4)$

OBJECTIVE ③ **Write the equation of a line parallel to a given line.** Consider the graphs of $y = 2x - 3$ and $y = 2x + 1$. These two lines are parallel, which means that they will never intersect. Notice in the equations that their slopes are equal, which is why they are parallel.

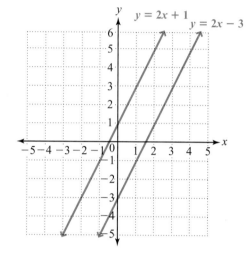

RULE **PARALLEL LINES**

The slopes of parallel lines are equal.

EXAMPLE ⑦ Write the equation of a line that passes through $(1, -4)$ and parallel to $y = -3x + 5$. Write the equation in slope-intercept form.

Solution In the given equation, we see that the slope of the line is -3, so the slope of the parallel line will also be -3. We can now use the point-slope form to write the equation of the parallel line.

$$y - y_1 = m(x - x_1)$$
$$y - (-4) = -3(x - 1)$$ Replace m with -3, x_1 with 1, and y_1 with -4.
$$y + 4 = -3x + 3$$ Simplify.
$$y = -3x - 1$$ Subtract 4 from both sides to isolate y.

YOUR TURN Write the equation of the line in slope-intercept form that passes through the given point and is parallel to the given line.

$(-4, -2)$; $y = \dfrac{3}{4}x - 7$

OBJECTIVE ④ **Write the equation of a line perpendicular to a given line.** Consider the graphs of $y = \dfrac{2}{3}x - 4$ and $y = -\dfrac{3}{2}x + 1$. These two lines are perpendicular, which means that they intersect at a 90° angle. Notice in the equations that their slopes are negative reciprocals. In other words $-\dfrac{3}{2}$ is the negative reciprocal of $\dfrac{2}{3}$.

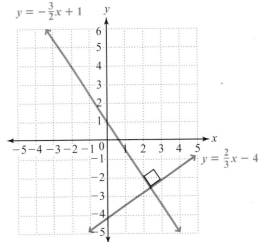

PERPENDICULAR LINES

The slope of a line perpendicular to a line with a slope of $\dfrac{a}{b}$ will be $-\dfrac{b}{a}$.

EXAMPLE 8 Write the equation of a line that passes through $(3, -1)$ and perpendicular to $3x + 4y = 8$. Write the equation in slope-intercept form.

Solution To write the equation of the perpendicular line, we need its slope. To determine its slope, we need the slope of the given line. Since the equation of the given line is in standard form, to determine its slope, we rearrange it in slope-intercept form.

$$3x + 4y = 8$$

$$4y = -3x + 8 \qquad \text{Subtract } 3x \text{ from both sides.}$$

$$y = -\frac{3}{4}x + 2 \qquad \text{Divide both sides by 4.}$$

We now see that the slope of the given line is $-\dfrac{3}{4}$, so the slope of the perpendicular line will be $\dfrac{4}{3}$. We can now use point-slope form to write the equation of the perpendicular line.

$$y - y_1 = m(x - x_1)$$

$$y - (-1) = \frac{4}{3}(x - 3) \qquad \text{Replace } m \text{ with } \frac{4}{3}, x_1 \text{ with 3, and } y_1 \text{ with } -1.$$

$$y + 1 = \frac{4}{3}x - 4 \qquad \text{Simplify.}$$

$$y = \frac{4}{3}x - 5 \qquad \text{Subtract 1 from both sides to isolate } y.$$

YOUR TURN

Write the equation of a line that passes through $(-10, 3)$ and perpendicular to $5x - 2y = 6$. Write the equation in slope-intercept form.

4.5 EXERCISES

1. If given the slope and the y-intercept and asked to write the equation of the line, which form of the equation would be easiest to use? Why?

2. If given two ordered pairs, neither of which is the y-intercept and asked to write the equation of the line, which form of the equation would be easiest to use? Why?

3. What is your first step when you are given two ordered pairs and asked to write the equation of the line connecting the two points?

4. When given two ordered pairs and asked to write an equation of a line, does it matter which ordered pair is used in the point-slope equation? Explain.

5. How do you rewrite an equation that is in standard form in slope-intercept form?

6. How do you rewrite an equation that is in point-slope form in standard form?

For Exercises 7–18, write the equation of the line in slope-intercept form given the *slope and the coordinate of the y-intercept.*

7. $m = 2$; $(0, 3)$

8. $m = 3$; $(0, -2)$

9. $m = -3$; $(0, -9)$

10. $m = \dfrac{4}{3}$; $(0, -5)$

11. $m = -\dfrac{2}{5}$; $\left(0, \dfrac{7}{8}\right)$

12. $m = \dfrac{3}{7}$; $(0, -1)$

13. $m = -9$; $\left(0, -\dfrac{3}{4}\right)$

14. $m = -0.75$; $(0, 2.5)$

15. $m = 0.8$; $(0, -5.1)$

16. $m = \dfrac{5}{2}$; $\left(0, \dfrac{3}{8}\right)$

17. $m = 4.1$; $(0, 5.85)$

18. $m = 1$; $(0, 0)$

For Exercises 19–30, write the equation of a line connecting the given points in slope-intercept form.

19. $(3, 1), (5, 6)$

20. $(-5, 3), (8, 1)$

21. $(-5, -2), (7, 5)$

22. $(0, 0), (-5, 1)$

23. $(4.2, 5), (3, 2.1)$

24. $(-1, -2), (-5, 4)$

25. $(0, 0), (-1, -7)$

26. $(5, 2), (5, -4)$

27. $(8, -1), (8, -10)$

28. $(-2, 5), (6, 5)$

29. $(-9, -1), (2, -1)$

30. $(3.9, -2.1), (4.1, -1)$

For Exercises 31–42, write the equation of a line in slope-intercept form with the given slope passing through the given point.

31. $m = 3; (5, 2)$

32. $m = 2; (3, -5)$

33. $m = \dfrac{2}{3}; (0, 3)$

34. $m = -\dfrac{3}{2}; (-2, 3)$

35. $m = \dfrac{3}{4}; (-1, -5)$

36. $m = 2; (3, -4)$

37. $m = -\dfrac{4}{3}; (-2, 4)$

38. $m = -5; (2, 0)$

39. $m = -1; (-1, -1)$

40. $m = -\dfrac{1}{8}; (3, 7)$

41. $m = \dfrac{2}{9}; (-1, -7)$

42. $m = 1; (0, 0)$

For Exercises 43–54, write the equation of a line through the given points in the form $Ax + By = C$, where A, B, and C are integers and $A > 0$.

43. $(2, 3), (4, 8)$

44. $(4, 1), (8, 2)$

45. $(2, 0), (0, -3)$

46. $(-4, -1), (7, -5)$

47. $(-6, 0), (0, 6)$

48. $(5, -2), (-3, -3)$

49. $(-1, 0), (-1, 8)$

50. $(4, 8), (-4, -8)$

51. $(7, -2), (-1, -5)$

52. $(2, -4), (3, -4)$

53. $(3, -1), (4, -1)$

54. $(-2, 5), (-2, -6)$

For Exercises 55–62, determine whether the given lines are parallel, perpendicular, or neither.

55. $y = 4x - 7$
$y = 4x + 2$

56. $y = 3x + 2$
$y = -3x - 2$

57. $y = \dfrac{1}{2}x$
$y = -2x + 4$

58. $y = -x + 3$
$y = x - 2$

59. $y = x - 2$
$y = x + 2$

60. $y = -\dfrac{2}{3}x + 1$
$y = -\dfrac{2}{3}x$

61. $y = 1$
$x = -3$

62. $3x - 2y = -5$
$y = \dfrac{3}{2}x + 3$

For Exercises 63–72, write the equation of a line that passes through the given point and is parallel to the given line. a. Write the equation in slope-intercept form. b. Write the equation in the form $Ax + By = C$, where A, B, and C are integers and $A > 0$.

63. $(4, 2); y = 4x + 1$

64. $(0, -1); y = 3x + 2$

65. $(-2, 1); y = -3x + 1$

66. $(-1, -3); y = -\frac{3}{4}x + 1$

67. $(5, 2); y = \frac{1}{3}x + 4$

68. $(-1, -1); 2x + 3y = 6$

69. $(2, -3); 2x + y = 5$

70. $(2, 0); x - y = -4$

71. $(5, 1); 6x + 3y = 8$

72. $(1, -3); y - 2x + 8 = 0$

For Exercises 73–82, write the equation of a line that passes through the given point and perpendicular to the given line. a. Write the equation in slope-intercept form. b. Write the equation in the form $Ax + By = C$, where A, B, and C are integers and $A > 0$.

73. $(-4, -1); y = -x + 3$

74. $(2, 3); y = \frac{1}{5}x - 4$

75. $(2, -4); y = -2x + 7$

76. $(-1, -1); y = -3x + 4$

77. $(3, -4); y = -\frac{2}{3}x + 4$

78. $(4, 2); y = \frac{1}{4}x - 11$

79. $(-2, 3); x - y = 5$

80. $(3, 7); x + 2y = 3$

81. $(4, 7); y - 3x = 6$

82. $(2, -5); 2x + 3y = 9$

83. The annual claim payment per person by the state of South Carolina's health insurance plan for its state employees went from \$855 in 1990 to \$1390 in 1995 and seems to follow a linear trend. Let 1990 be year 0, so that 1995 is year 5.

  a. Graph the line.
  b. Find the slope of the line.
  c. Write an equation of the line in slope-intercept form.
  d. If the insurance claims follow the same trend, predict the amount of claims paid per person in 2010.

84. Another report by the state of South Carolina shows the costs of prescription drugs for senior citizens (age 65 and older). In 1997 the amount was $27 million and in 1999 the amount was $41.8 million. Let 1997 be year 0, so that 1999 is year 2.

    a. Graph the line.

    b. Find the slope of the line.

    c. Write an equation of the line in slope-intercept form.

    d. If the prescription drug costs follow the same trend, predict the costs in 2008.

Collaborative Exercises

Pyramid Profile

In Exercise 63 in Section 4.4, you found the slope of the Great Pyramid at Giza.

1. Suppose we were to draw the side view of the pyramid in the coordinate plane with one corner at the origin and the top of the pyramid in the first quadrant. Write an equation in slope-intercept form that describes the line that will be the left side of the pyramid.

2. Suppose the top of the pyramid is located at a point directly over a point that is 115 meters to the right of the left corner. What is the height of the pyramid?

3. Write an equation that describes the line that will be the right side of the pyramid.

4. Draw the side view of the pyramid in the coordinate plane so that each unit represents 20 meters.

REVIEW EXERCISES

1. Solve: $\dfrac{3}{4}x - \dfrac{4}{5} = \dfrac{1}{2}x + 1$

2. What is 12% of 86?

3. Use $<$, $>$, or $=$ to write a true statement: $0.6 \; ?? \; \dfrac{3}{4}$

4. Solve $4x - 9 \geq 6x + 5$ and graph the solution set on a number line.

5. Solve $2(x + 1) - 5x \geq 3x - 6$ and graph the solution set on a number line.

6. Find the x- and y-intercepts for $2x - 7y = 14$.

4.6 Graphing Linear Inequalities

OBJECTIVES

1 Determine whether an ordered pair is a solution for a linear inequality with two variables.

2 Graph linear inequalities.

OBJECTIVE 1 **Determine whether an ordered pair is a solution for a linear inequality with two variables.** Now that we have seen how to graph linear equations, let's turn our attention to linear inequalities. Linear inequalities have the same form as linear equations, except that they contain an inequality symbol instead of an equal sign. Examples of linear inequalities are:

$$2x + 3y > 6$$
$$y \leq x - 3$$

Remember that a solution for a linear equation in two variables is an ordered pair that makes the equation true. Similarly, a solution to a linear inequality is an ordered pair that makes the inequality true. For example, the ordered pair $(2, 1)$ is a solution for $2x + 3y > 6$:

$$2x + 3y > 6$$
$$2(2) + 3(1) \overset{?}{>} 6 \qquad \text{Replace } x \text{ with 2 and } y \text{ with 1.}$$
$$4 + 3 \overset{?}{>} 6$$
$$7 > 6 \qquad \text{This is true, so } (2, 1) \text{ is a solution.}$$

PROCEDURE CHECKING AN ORDERED PAIR

To determine whether an ordered pair is a solution for an inequality, replace the variables with the corresponding coordinates and see if the resulting inequality is true. If so, the ordered pair is a solution.

EXAMPLE 1 Determine whether $(-4, 7)$ is a solution for $y \geq -2x + 3$.

Solution
$$y \geq -2x + 3$$
$$7 \overset{?}{\geq} -2(-4) + 3 \qquad \text{Replace } x \text{ with } -4 \text{ and } y \text{ with 7, and then}$$
$$7 \overset{?}{\geq} 8 + 3 \qquad \text{determine whether the inequality is true.}$$
$$7 \geq 11$$

This statement is false, so $(-4, 7)$ is not a solution.

OBJECTIVE 2 **Graph linear inequalities.**   Graphing linear inequalities is very much like graphing linear equations. Consider $2x + 3y \geq 6$. The greater than or equal to sign indicates that solutions to the inequality $2x + 3y \geq 6$ will be ordered pairs that satisfy $2x + 3y > 6$ and also $2x + 3y = 6$. The graph of $2x + 3y = 6$ is a line. Let's graph this line using intercepts.

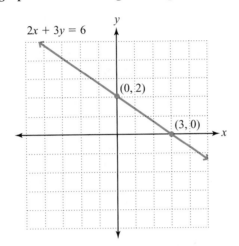

$2x + 3y = 6$

Find the x- and y-intercepts.

x-intercept:	y-intercept:
$2x + 3y = 6$	$2x + 3y = 6$
$2x + 3(0) = 6$	$2(0) + 3y = 6$
$2x = 6$	$3y = 6$
$x = 3$	$y = 2$

x-intercept coordinates: $(3, 0)$
y-intercept coordinates: $(0, 2)$

The "greater than" portion of the inequality consists of ordered pairs that satisfy $2x + 3y > 6$. To find one of these pairs, we select a point on one side of the line and see if its coordinates satisfy $2x + 3y > 6$. Let's select the point with coordinates $(1, 4)$.

$$2x + 3y > 6$$
$$2(1) + 3(4) \overset{?}{>} 6 \qquad \text{Replace } x \text{ with 1 and } y \text{ with 4.}$$
$$2 + 12 \overset{?}{>} 6$$
$$14 > 6 \qquad \text{This statement is true, therefore } (1, 4) \text{ is a solution.}$$

It can be shown that every ordered pair in the region containing the point $(1, 4)$, which is above the line, will satisfy $2x + 3y > 6$. It can also be shown that no ordered pair on the other side of the line will satisfy $2x + 3y > 6$. For example, consider the origin $(0, 0)$.

$$2x + 3y > 6 \qquad \text{Replace } x \text{ with 0 and } y \text{ with 0.}$$
$$2(0) + 3(0) \overset{?}{>} 6$$
$$0 > 6 \qquad \text{This statement is false, so } (0, 0) \text{ is not a solution.}$$

The line that is the graph of $2x + 3y = 6$ is called a *boundary*. The region on one side of this boundary line contains all the ordered pair solutions satisfying $2x + 3y > 6$. The region on the other side contains all ordered pairs that satisfy $2x + 3y < 6$. To indicate that every ordered pair above the boundary line is a solution to the inequality, we shade the region above the line.

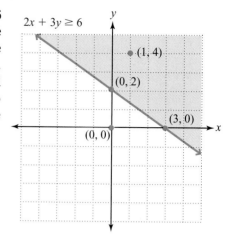

If the inequality had been $2x + 3y > 6$, the ordered pairs that satisfy $2x + 3y = 6$ would not be solutions. Because these ordered pairs form the boundary line, we draw a dashed line instead of a solid line to indicate that those ordered pairs are not solutions. The graph of $2x + 3y > 6$ looks like this:

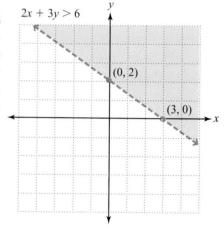

For the inequalities $2x + 3y \leq 6$ and $2x + 3y < 6$, we shade the region on the other side of the boundary line.

The graph of $2x + 3y \leq 6$:

The graph of $2x + 3y < 6$:

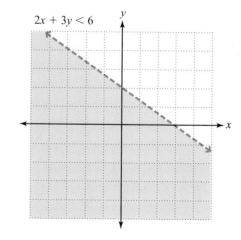

When graphing inequalities with the symbols $\leq$ or $\geq$ on a number line, we used a closed circle to indicate that equality is included. When the inequalities have $<$ or $>$, we used an open circle to indicate that equality is excluded. When graphing linear inequalities in the rectangular coordinate system, a solid line is like a closed circle and a dashed line is like an open circle.

PROCEDURE **GRAPHING LINEAR INEQUALITIES**

To graph a linear inequality:

1. Graph the related equation. The related equation has an equal sign in place of the inequality symbol. If the inequality symbol is $\leq$ or $\geq$, then draw a solid line. If the inequality symbol is $<$ or $>$, then draw a dashed line.
2. Choose an ordered pair on one side of the boundary line and test this ordered pair in the inequality. If the ordered pair satisfies the inequality, then shade the region that contains it. If the ordered pair does not satisfy the inequality, then shade the region on the other side of the boundary line.

EXAMPLE 2 Graph $y > 3x + 1$.

Solution First, graph the related equation $y = 3x + 1$. Two ordered pairs that satisfy $y = 3x + 1$ are $(0, 1)$ and $(1, 4)$.

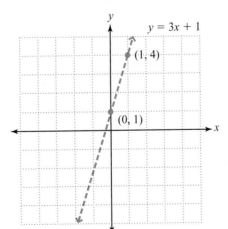

Because the inequality is strictly greater than, we draw a dashed line to indicate that ordered pairs along this boundary line are not solutions.

Now we choose an ordered pair on one side of the line and test this ordered pair in the inequality. We will choose (3, 0).

$$y > 3x + 1$$
$$0 \overset{?}{>} 3(3) + 1 \quad \text{Replace } x \text{ with 3 and } y \text{ with 0.}$$
$$0 > 10 \qquad \text{This statement is false, so (3, 0) is not a solution for the inequality.}$$

Since (3, 0) did not satisfy the inequality, we shade the region on the other side of the boundary line.

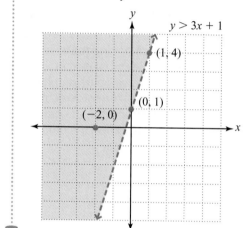

We can confirm that our shading is correct by choosing an ordered pair in this region. We will choose the ordered pair (−2, 0):

$$y > 3x + 1$$
$$0 \overset{?}{>} 3(-2) + 1 \quad \text{Replace } x \text{ with } -2 \text{ and } y \text{ with 0.}$$
$$0 > -5 \qquad \text{This statement is true, which confirms that we have shaded the correct side of the boundary line.}$$

EXAMPLE 3 Graph $x - 2y \leq 4$.

Solution First, graph the related equation $x - 2y = 4$. Two ordered pairs that satisfy $x - 2y = 4$ are (0, −2) and (4, 0).

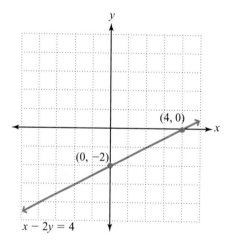

Because the inequality symbol is less than or equal to, we draw a solid line to indicate that ordered pairs along the boundary line are solutions.

Now we choose an ordered pair on one side of the line and test this ordered pair in the inequality. We will choose the origin (0, 0).

$$x - 2y \leq 4$$
$$0 - 2(0) \overset{?}{\leq} 4 \quad \text{Replace } x \text{ with 0 and } y \text{ with 0.}$$
$$0 \leq 4 \quad \text{This statement is true, therefore, } (0, 0) \text{ is a solution for the inequality.}$$

Since (0, 0) satisfies the inequality, we shade the region that contains it.

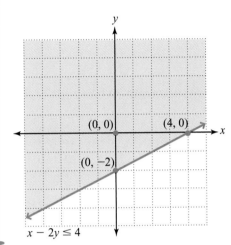

We can confirm that the region on the other side of the line should not be shaded by choosing a point in that region, such as (3, −3):

$$x - 2y \leq 4$$
$$3 - 2(-3) \overset{?}{\leq} 4 \quad \text{Replace } x \text{ with 3 and } y \text{ with } -3.$$
$$3 + 6 \overset{?}{\leq} 4$$
$$9 \leq 4$$

This statement is false, which confirms that the region containing the point with coordinates (3, −3) should not be shaded.

YOUR TURN   **Graph the linear inequality.**

$$y < -\frac{1}{3}x + 2$$

Graphing Calculator
TIPS

Shading can be added to a graph on a graphing calculator. Using the ⃞Y= key, enter the related equation for the given inequality. On the same line where you have entered the equation, use the arrow keys to move the cursor to the left of the variable y. Normally, in this position, you see a slash, which indicates that the graph will be a solid line. Pressing enter will change this slash to various other options for the graph. There are two options for shading:

   ◥ *means the graph will have the region* above *the line shaded.*
   ◤ *means the graph will have the region* below *the line shaded.*

Notice that it is up to you to determine which region should be shaded.

ANSWER

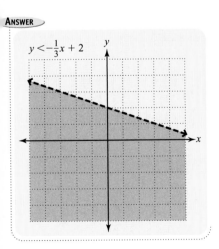

1. How do you determine if an ordered pair is a solution to an inequality?

2. How do you determine the related equation for a given inequality?

3. In graphing a linear inequality, when do you draw a dashed line and when do you draw a solid line?

4. When graphing a linear inequality, how do you determine which side of the boundary line to shade?

For Exercises 5–12, determine whether the ordered pair is a solution for the linear inequality.

5. $(5, 5); y > -x + 2$

6. $(0, 1); y > -x + 2$

7. $(2, 0); y \geq \dfrac{1}{2}x + 1$

8. $(0, 0); 2x + 3y < 7$

9. $(-1, 2); 3x - y \leq -8$

10. $(3, -1); y > x + 2$

11. $(-1, -2); 2y - x < 7$

12. $(0, -2); x + 4y \geq 8$

For Exercises 13–38, graph the linear inequality.

13. $y \leq -x + 1$

14. $y \leq x - 8$

15. $4y - 3x \leq 8$

16. $2x + y \geq -3$

17. $y > \dfrac{1}{3}x$

18. $y < \dfrac{2}{5}x$

19. $5x + 2y < 10$

20. $3x - 2y \geq 6$

21. $y > 2x$

22. $y < x$

23. $2y + 4x \le 3$

24. $3x + y \ge 8$

25. $y + 3x \ge 0$

26. $-3x - 2y < 4$

27. $\dfrac{1}{4}x + 2y > -8$

28. $-\dfrac{2}{3}x - 6y > 18$

29. $2x + 3y < 9$

30. $5x + y \ge -10$

31. $x + y > -2$

32. $x - y < 7$

33. $x > -4$

34. $x \le 6$

35. $y \le 2$

36. $y > -1$

37. $y > -x$

38. $x < -y$

39. A company sells two different sized bottles of hand lotion. The revenue for each unit of the large bottle is $2.00. The revenue for each unit of the small bottle is $1.50. For the company to break even during the first quarter, the company must generate $280,000 in revenue. The inequality $2x + 1.5y \geq 280,000$ describes the amount of revenue that must be generated from each size bottle in order to break even or turn a profit.

a. What do x and y represent?

b. Graph the inequality.

c. What does the boundary line represent? What does the shaded region represent?

d. List three combinations of sales numbers of each size that allow the company to break even.

e. List three combinations of sales numbers of each size that allow the company to turn a profit.

f. In reality, is every combination of units sold represented by the line and shaded region a possibility? Explain.

40. A company produces two versions of tax-preparation software. The regular version costs $12.50 per unit to produce, and the deluxe version costs $15.00 per unit to produce. Management plans for the cost of production to be a maximum of $300,000 for the first quarter. The inequality $12.50x + 15y \leq 300,000$ describes the total cost as prescribed by management.

a. What do x and y represent?

b. Graph the inequality.

c. What does the line represent?

d. What does the shaded region represent?

e. List three combinations of numbers of units produced of each version of software that yield a total cost of $300,000.

d. List three combinations of numbers of units produced of each version of software that yield a total cost less than $300,000.

f. In reality, is every combination of units produced represented by the line and shaded region a possibility? Explain.

41. Andrea visits a garden center to purchase some new plants for her garden. She has a gift certificate in the amount of $100. The plants that she is considering come in two sizes and are priced according to size. Plants in 1-gallon containers are priced at $5.50 each, and plants in 5-gallon containers are priced at $12.50 each.

a. Select a variable for the number of each plant size, and write an inequality that has Andrea's total cost within the amount of the gift certificate.

b. Graph the inequality.

c. What does the line represent?

d. What does the shaded region represent?

e. What are some possible combinations of containers she could purchase?

f. In reality, are there any combinations that she could purchase that would yield a total in the exact amount of the gift certificate? Explain.

42. Michelle is coordinating a fund-raiser that will sell two different cookbooks. One cookbook sells for $10 and the other for $12. The goal is to raise at least $30,000.

a. Select variables for each type of cookbook, and write an inequality in which the total sales is at least $30,000.

b. Graph the inequality.

c. What does the line represent?

d. What does the shaded region represent?

e. Find two combinations of book sales that yield exactly $30,000 in total sales.

f. Find two combinations of units sold that yield more than $30,000.

43. A building with a rectangular base is to be designed so that the maximum perimeter is 180 feet.

a. Select variables for the length and width, and write an inequality in which the maximum perimeter is 180 feet.

b. Graph the inequality.

c. What does the line represent?

d. What does the shaded region represent?

e. Find three combinations of length and width that yield a perimeter of exactly 180 feet.

f. Find three combinations of length and width that yield a perimeter of less than 180 feet.

44. Steel beams are to be welded together to form a frame that is an isosceles triangle. The perimeter of the frame cannot exceed 60 feet.

  a. Select a variable for the length of the base and a second variable for the sides of equal length. Then write an inequality in which the maximum perimeter is 60 feet.

  b. Graph the inequality.

  c. What does the line represent?

  d. What does the shaded region represent?

  e. Find sets of dimensions that yield a perimeter of exactly 60 feet.

  f. Find sets of dimensions that yield a perimeter of less than 60 feet.

PUZZLE PROBLEM

The sum of two different whole numbers is equal to 8 while the difference of the two numbers is greater than or equal to 1. Find all possible ordered pairs that satisfy both constraints.

REVIEW EXERCISES

1. Write a set containing all vowels in the alphabet.

2. Evaluate: 6^0

3. Solve and check: $3x - 4(x + 7) = -15$

4. The length of a rectangle is four more than the width. If the perimeter is 88 feet, find the dimensions.

5. Find the x and y intercepts for $3x - 2y = 8$.

6. Graph: $-x + y = 7$

4.7 Introduction to Functions and Function Notation

OBJECTIVES

1. Identify the domain and range of a relation.
2. Identify functions and their domains and ranges.
3. Find the value of a function.
4. Graph linear functions.

In this section, we study relations and functions, which develop naturally from our work with graphing because they involve ordered pairs.

OBJECTIVE **1** **Identify the domain and range of a relation.** In order to formally define a function, we will first discuss **relations**.

> **DEFINITION**
>
> **Relation:** A set of ordered pairs.

A linear equation like $y = 2x$ is a relation because it pairs a given x-value with a corresponding y-value. You might recall that with an equation in this form, we thought of the x-values as input values and the y-values as output values. In a relation, the set of all the input values (x-values) is called the **domain** and the set of all the output values (y-values) is called the **range**.

> **DEFINITIONS**
>
> **Domain:** The set of all input values for a relation.
> **Range:** The set of all output values for a relation.

The following table lists some values in the domain and range for $y = 2x$.

Domain (x-values)	Range (y-values)
0	0
1	2
2	4
3	6

EXAMPLE **1** Determine the domain and range of the relation $\{(2, -1), (3, 0), (4, 5), (5, 8)\}$.

Solution The domain is the set containing all the x-values $\{2, 3, 4, 5\}$, and the range is the set containing all the y-values $\{-1, 0, 5, 8\}$.

Domain and Range with a Graph

We can determine the domain and range of a relation from its graph. Look at the graph of $y = 2x$.

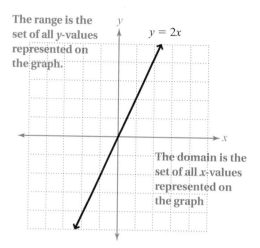

The range is the set of all y-values represented on the graph.

The domain is the set of all x-values represented on the graph

It is convenient to imagine traveling along a graph from left to right. As we travel along $y = 2x$ from left to right, every x-value along the x-axis has a corresponding y-value. Thus the domain is a set containing all real numbers. Similarly, every y-value is paired with an x-value, so the range also contains all real numbers.

PROCEDURE **DETERMINING DOMAIN AND RANGE**

To determine the domain of a relation given its graph, answer the question: what are all the x-values that have a corresponding y-value? To determine the range, answer the question: what are all the y-values that have a corresponding x-value?

EXAMPLE **2** Determine the domain and range of the relation.

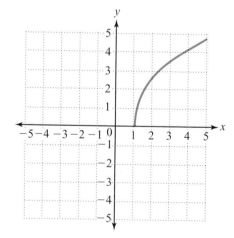

Solution From left to right, this graph "begins" at the point $(1, 0)$. Values along the x-axis begin at 1 and continue infinitely, so the domain is $\{x | x \geq 1\}$. For the range, the y-values begin at 0 and continue infinitely, so the range is $\{y | y \geq 0\}$.

Determine the domain and range of the relation.

a. $\{(-2, -4), (0, -1), (3, 6), (7, 5)\}$ **b.**

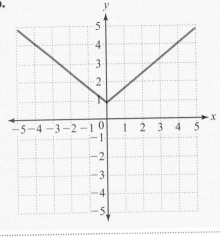

OBJECTIVE **2** **Identify functions and their domains and ranges.** Now we turn to a special type of relation called a function. Let's reconsider the trajectory data from Example 4 of Section 4.1 (pages 274–275), which we have listed here as a relation in the following table.

Time in Seconds (domain)	Height in Feet (range)
0	6.00
0.5	7.50
1.0	8.25
1.5	7.50
2.0	6.00
2.5	2.00

Note that each value for time is paired with one and only one height value. This is the key feature that makes a relation a **function**.

DEFINITION | ***Function:*** A relation in which every value in the domain is paired with exactly one value in the range.

From the data in the preceeding table, we can see that a function can have different values in the domain assigned to the same value in the range. For example, domain values 0.5 seconds and 1.5 seconds each correspond to the same range value 7.50 feet. Also, the domain values 0 seconds and 2.0 seconds each correspond to 6.00 feet in the range. Using arrows to match the values in the domain with their corresponding values in the range can be a helpful technique in determining whether a relation is a function. Such a map might look like this:

Domain: {0, 0.5, 1.0, 1.5, 2.0, 2.5}

Range: {2.00, 6.00, 7.50, 8.25}

Note: Each element in the domain has a single arrow pointing to an element in the range.

Note that every function is a relation, but not every relation is a function. For example, consider the following relation that assigns baseball outfield positions (the domain) to the 2002 New York Yankees outfielders (the range).

Domain: {left field, center field, right field}

Range: {Spencer, White, Williams, Mondesi, Rivera, Thames, Vander Wal}

Notice that left field and right field are assigned to more than one player. Since an element in the domain (left or right field) is assigned to more than one element in the range, this relation is not a function.

Conclusion   If any value in the domain is assigned to more than one value in the range, then the relation is not a function.

EXAMPLE 3   Determine whether the relation is a function.

a. The following table indicates the percent of people in the United States under 65 years of age with private health care coverage.

Year (domain)	Percent (range)
1984	76.8
1989	75.9
1993	71.5
1994	70.3
1995	71.6
1996	71.4
1997	70.7

Solution This relation is a function because each year value in the domain has only one percent value in the range assigned to it.

b. The following relation assigns a birthdate to the corresponding person.

Domain: {May 1, June 4, July 8, August 2}

Range: {Danielle, Gerard, Juan, René, Candice}

Solution This relation is not a function because an element in the domain, July 8, is assigned to two people in the range.

YOUR TURN Determine whether the relation is a function.

a.

Index Topic	Page Number(s)
(domain)	(range)
Celestial sphere	56
Celsius	36
Centigrade	36
Centimeter	6
Centripedal force	239, 305

b. The table shows the top five finishers in the 2000 Daytona 500 race.

Final Position	Car Number
(domain)	(range)
1st	88 (Dale Jarrett)
2nd	99 (Jeff Burton)
3rd	94 (Bill Elliott)
4th	2 (Rusty Wallace)
5th	6 (Mark Martin)

We can determine whether a relation is a function by its graph. Remember that in Example 4 of Section 4.1 (p. 275) we created a graph of the trajectory data (that data also appears in this section on p. 343) by plotting the ordered pairs as points in the coordinate plane.

Notice the domain (time in this case) is plotted along the horizontal axis and the range (height) is plotted along the vertical axis.

ANSWERS

a. not a function

b. function

To determine whether a relation is a function from its graph, we can perform a test called the *vertical line test*. The nature of the test is to draw or imagine a vertical line through every point in the domain.

Here, we've drawn lines in blue through each value in the domain. Note that each vertical line intersects the graph at one and only one point, which means that each value in the domain corresponds to exactly one value in the range. This relation is a function.

PROCEDURE   **THE VERTICAL LINE TEST**

To determine whether a relation is a function from its graph, perform a vertical line test:

1. Draw or imagine vertical lines through each point in the domain.
2. If each vertical line intersects the graph at only one point, then the graph is the graph of a function.
3. If any vertical line intersects the graph at two different points, then the graph is not the graph of a function.

EXAMPLE 4   For each graph, determine the domain and range. Then state whether the relation is a function.

a.

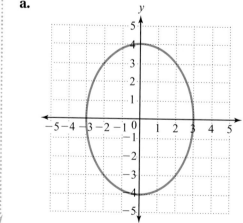

Solution Domain: $\{x \mid -3 \leq x \leq 3\}$
Range: $\{y \mid -4 \leq y \leq 4\}$

This relation is not a function because a vertical line can be drawn that intersects the graph at two different points.

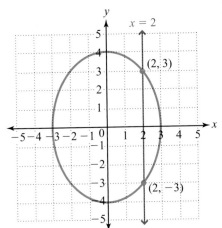

For example, the vertical line $x = 2$ (shown in blue) passes through two different points on the graph, $(2, 3)$ and $(2, -3)$.

b.

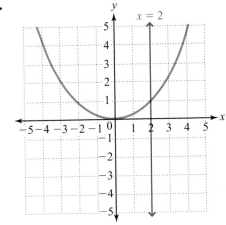

Solution Domain: all real numbers
Range: $\{y | y \geq 0\}$

This relation is a function. A vertical line through any value along the horizontal axis will intersect the graph at only one point. We have shown $x = 2$ to illustrate one such line.

c.

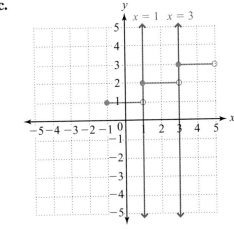

Solution Domain: $\{x | -1 \leq x < 5\}$
Range: $\{y | y = 1, 2, \text{ or } 3\}$

This relation is a function. It may seem that the vertical lines $x = 1$ and $x = 3$ (shown in blue) each intersect the graph at two different points. However, the open circles at $(1, 1)$, $(3, 2)$, and $(5, 3)$ indicate that those points are not part of the graph. The closed circles at $(-1, 1)$, $(1, 2)$, and $(3, 3)$ are part of the graph. The vertical line $x = 1$ "passes through" the open circle at $(1, 1)$ and intersects the graph at only one point; the closed circle at $(1, 2)$. Similarly $x = 3$ passes through the open circle at $(3, 2)$ and intersects the graph at only one point, $(3, 3)$.

d.

Source: USA Today, June 6, 2001

Solution Domain:
$$\{t | 9{:}30\,\text{A.M.} \le t \le 4{:}30\,\text{P.M.}\}$$
Range:
$$\{N | 2218 \le N \le 2264\}$$

This relation is a function. A vertical line through any value along the horizontal axis will intersect the graph at only one point. In the terminology from the graph, at each time during this business day, the Nasdaq had only one value.

YOUR TURN   **Determine whether the relation is a function.**

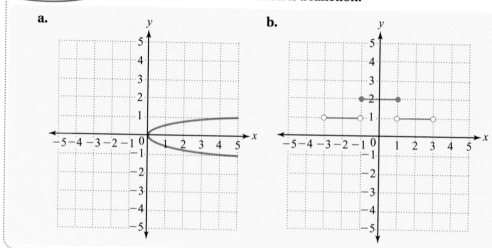

a.

b.

OBJECTIVE ❸ **Find the value of a function.**

Function Notation

When written as an equation, the notation for a function is a modification of an equation in two variables. Instead of $y = 2x + 5$, in function notation, we would write $f(x) = 2x + 5$ to mean the same equation. Notice that the notation $f(x)$, which is read "a function in terms of x," or "f of x," replaces the variable y. The following table shows how equations in two variables can be written using function notation.

Equation in two variables:	Function notation:
$y = 3x$	$f(x) = 3x$
$y = -4x + 1$	$f(x) = -4x + 1$
$y = \frac{2}{3}x - 5$	$f(x) = \frac{2}{3}x - 5$

Warning: The notation $f(x)$ does not mean multiply f and x.

Finding the Value of a Function

Remember that when we found solutions to equations with y isolated, we often spoke of the x-value as an input value and the corresponding y-value as the output value. Finding the value of a function is the same as finding a solution to an equation in two variables: we input a value for x, and calculate the output value, $f(x)$. The following tables list the same ordered pairs. The table on the left is for $y = 2x + 5$ and the table on the right is for $f(x) = 2x + 5$. Notice the only difference is in labeling the output values.

(input)	(output)
x	y
0	$2(0) + 5 = 5$
1	$2(1) + 5 = 7$
2	$2(2) + 5 = 9$

(input)	(output)
x	$f(x)$
0	$2(0) + 5 = 5$
1	$2(1) + 5 = 7$
2	$2(2) + 5 = 9$

Function notation offers a clever way to indicate that a specific x-value is to be used. For example, given a function $f(x) = 3x$, the notation $f(2)$ means to find the value of the function, where $x = 2$, so that $f(2) = 3(2) = 6$.

PROCEDURE   FINDING THE VALUE OF A FUNCTION

Given a function $f(x)$, to find $f(a)$, where a is a real number in the domain of f, replace x in the function with a and calculate the value.

EXAMPLE 5   For the function $f(x) = -4x + 1$, find the following.

a. $f(2)$

Solution $f(2) = -4(2) + 1$ Replace x in $-4x + 1$ with 2, then calculate.
$= -8 + 1$
$= -7$

b. $f(-3)$

Solution $f(-3) = -4(-3) + 1$ Replace x in $-4x + 1$ with -3, then calculate.
$= 12 + 1$
$= 13$

CONNECTION

We use similar procedures to find the value of a function and to evaluate an expression.

Function language:

For $f(x) = -4x + 1$, find $f(2)$.

Expression language:

Evaluate the expression $-4x + 1$ where $x = 2$.

YOUR TURN   For the function $f(x) = \dfrac{2}{3}x - 5$, find the following.

a. $f(-3)$

b. $f(6)$

Finding the Value of a Function Given Its Graph

We can also find the value of a function given its graph. For a given value in the domain (x-value), we look on the graph for the corresponding value in the range (y-value).

EXAMPLE 6 Using the graph, find the value of the function.

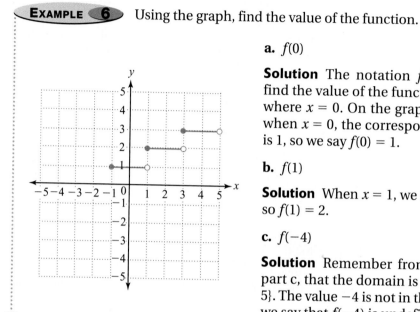

a. $f(0)$

Solution The notation $f(0)$ means to find the value of the function (y-value), where $x = 0$. On the graph we see that when $x = 0$, the corresponding y-value is 1, so we say $f(0) = 1$.

b. $f(1)$

Solution When $x = 1$, we see that $y = 2$, so $f(1) = 2$.

c. $f(-4)$

Solution Remember from Example 4, part c, that the domain is $\{x | -1 \leq x < 5\}$. The value -4 is not in the domain, so we say that $f(-4)$ is undefined.

YOUR TURN Use the graph to find the value of the function.

a. $f(-1)$
b. $f(0)$
c. $f(5)$

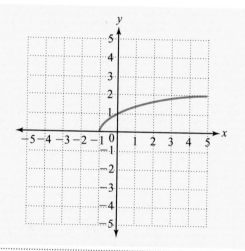

OBJECTIVE 4 **Graph linear functions.** We create the graph of a function the same way that we create the graph of an equation in two variables. In this chapter we will graph only linear functions. Linear functions have a form similar to linear equations in slope-intercept form.

Slope-intercept form: $y = mx + b$

Linear function: $f(x) = mx + b$

ANSWERS

a. 0 **b.** 1 **c.** 2

EXAMPLE 7 Graph $f(x) = -2x + 3$.

Solution Think of the function as the equation $y = -2x + 3$. We could make a table of ordered pairs, or use the fact that the slope is -2 and the y-intercept is 3.

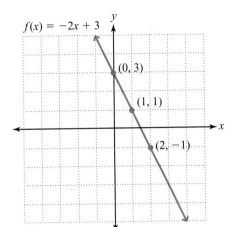

$f(x) = -2x + 3$

$(0, 3)$

$(1, 1)$

$(2, -1)$

Table of ordered pairs

x	f(x)
0	3
1	1
2	−1

ANSWER

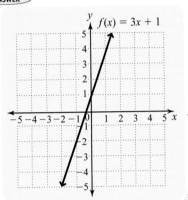

$f(x) = 3x + 1$

YOUR TURN

Graph $f(x) = 3x + 1$.

4.7 EXERCISES

1. Explain domain of a relation in your own words.

2. Explain range of a relation in your own words.

3. Given a graph of a relation, how do you determine the domain?

4. Given a graph of a relation, how do you determine the range?

5. When using the vertical line test on a graph, how do you determine if the graph is the graph of a function?

6. True or false: All relations are also functions. Explain.

For Extra Help

InterAct Math

Addison-Wesley
Math Tutor Center

Video 7/DVT 2

MyMathLab

MathXL

For Exercises 7–12, determine the domain and range of the relation.

7. $\{(2, 1), (3, -2), (1, 4), (-1, -1)\}$

8. $\{(-2, 1), (4, 3), (2, 5), (-3, -2)\}$

9. $\{(4, 1), (2, 1), (3, -5), (-4, 0)\}$

10. $\{(-2, 1), (8, 11), (-3, 5), (4, -4)\}$

11. $\{(0, 0), (-3, -3), (2, 9), (8, -16)\}$

12. $\{(5, 4), (-3, 8), (10, -2), (0, -5)\}$

For Exercises 13–24, determine if the relation is a function and explain your answer.

13. $\{(2, 7), (3, 4), (-1, 4), (0, 5)\}$

14. $\{(-1, 5), (2, 6), (2, 9), (-3, -2)\}$

15. $\{(8, 2), (4, -5), (3, 0), (8, -1)\}$

16. $\{(1, 2), (4, -7), (9, 5), (6, 14)\}$

17. The grades on a recent calculus test:

Student	Score
Jonathon	54
Nathan	79
Rebecca	84
Terri	97
Allen	97

18. Baby-sitter prices per hour around the country:

Rate	Location
$10.00	Chappaqua, New York
$7.00	Yellowstone National Park
	Menlo Park, California
$6.00	Coral Springs, Florida
	St. Louis, Missouri
	Hawi, Hawaii
$5.00	North Pole, Alaska
$2.00	Las Vegas, Nevada

Source: Good Housekeeping, October 2002

19. Number of bridge-construction applications for South Carolina coastal counties:

Applications	Year Applied
8	1990
	1998
7	1993
6	1991
5	1994
	1996
	2000
	2001
1	1999

Source: South Carolina Department of Health and Environmental Control, Coastal Division

20. The sources of distraction among inattentive drivers:

Distraction	Percentage of Drivers
Outside person, object, or event	29.4
Adjusting radio/cassette/CD	11.4
Other occupant	10.9
Using/dialing cell phone	1.5
Smoking related	0.9

Source: AAA Foundation for Traffic Safety/UNC Safety Research Center

21.

22.

23.

24.

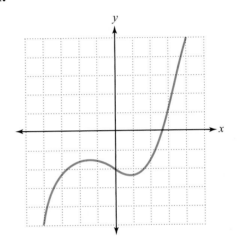

For Exercises 25–34, determine if the relation is a function. In each table, the domain is in the left column and the range is in the right column.

25. Average high temperature for Orlando, Florida:

Average High	Month
92	July
	August
91	June
89	September
88	May
84	October
83	April
77	March
	November
73	February
	December
72	January

26. Average rainfall for Seattle:

Avg. Rain (inches)	Month
6	January
	December
4	February
	March
3	October
2	April
	May
	September
1	June
	July
	August

27. Results of the 2001 Masters golf tournament:

Position	Name
1st	Tiger Woods
2nd	David Duval
3rd	Phil Mickelson
4th	Mark Calcavecchia, Toshi Izawa

Source: www.masters.org

28. AKC registered dogs in the United States:

Breed	Number of Dogs
Labrador Retriever	154,897
Golden Retriever	62,652
German Shepherd	57,256
Dachshund	50,772
Beagle	49,080

29. Average Home Sales Prices for Top 10 U.S. Metropolitan Areas, 1999:

Metropolitan Areas	Sales Price
San Francisco-Oakland-San Jose, CA	$342,700
Honolulu, HI	$278,700
San Diego, CA	$272,400
Los Angeles-Riverside-Orange Co., CA	$258,500
Washington-Baltimore, DC-MD-VA-WV	$246,100
New York-N. NJ-Long Island, NY-NJ-CT-PA	$240,300
Seattle-Tacoma-Bremerton, WA	$239,100
Denver-Boulder-Greeley, CO	$224,200
Boston-Worcester-Lawrence, MA-NH-ME-CT	$217,500
Sacramento-Yolo, CA	$213,900

Source: Federal Housing Finance Board

30. Average cost of college tuition in the United States for 2000–2001:

Sector	Costs
Two-year public	$1,705
Two-year private	$7,458
Four-year public	$3,510
Four-year private	$16,332

Source: Annual Survey of Colleges

31. Average low temperature for Charleston, South Carolina:

Average Low Temperature	Month
38	January
39	December
46	March, November
53	April
56	October
62	May
67	September
69	June
72	July, August

32. The top-5 money earners on the concert circuit during 2000:

Performer	Revenue in Millions of Dollars
Tina Turner	$80.2
*NSync	$76.4
Dave Matthews Band	$68.2
Kiss	$62.7
Tim McGraw/Faith Hill	$48.8

Source: Pollstar

33. Top selling cars in Hawaii for 2000:

Car	Number Sold
Toyota Camry	2228
Honda Accord	2096
Honda Civic	1972
Chevrolet Cavalier	1960
Toyota Corolla	1744

34. Top 5 records according to *Billboard* for 2000:

Artist	Record
Faith Hill	*Breathe*
Santana	*Maria Maria, Smooth*
Joe	*I Wanna Know*
Vertical Horizon	*Everything You Want*

Source: Billboard magazine

For Exercises 35–44, determine the domain and range. Then state whether the relation is a function.

35. This graph shows the number of Catholics in the United States from 1900 to 2001.

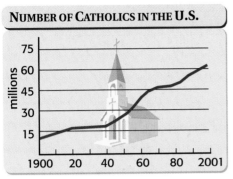

NUMBER OF CATHOLICS IN THE U.S.

Source: Newsweek, March 4, 2002

36.

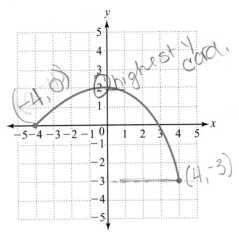

37. This graph represents the number of docks on the tax rolls in South Carolina.

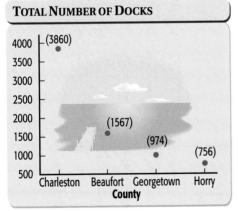

TOTAL NUMBER OF DOCKS

Source: South Carolina Office of Ocean and Coastal Resource Mgmt.

38. This graph shows the total number of people enrolled in a program each month after the program began.

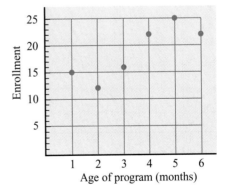

39. The line graph represents postage costs from 1991 to 2002.

Source: Postal Rate Commission

40.

41.

42.

43.

44.

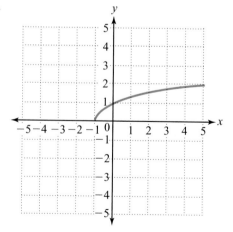

For Exercises 45–64, find the value of the function.

45. $f(x) = 2x - 5$
   a. $f(0)$
   b. $f(1)$
   c. $f(-2)$
   d. $f\left(\dfrac{1}{2}\right)$

46. $f(x) = 5x + 3$
   a. $f(1)$
   b. $f(0)$
   c. $f(-1)$
   d. $f(t)$

47. $f(x) = x^2 - 3x + 7$
   a. $f(0)$
   b. $f(1)$
   c. $f(-1)$
   d. $f(-2)$

48. $f(x) = 3x^2 + 2x - 1$
   a. $f(-2)$
   b. $f(0.2)$
   c. $f(a)$
   d. $f(3)$

49. $f(x) = x^2 + 4$
   a. $f(0)$
   b. $f(1)$
   c. $f\left(\dfrac{1}{4}\right)$
   d. $f(-1)$

50. $f(x) = \sqrt{x^2 + 2x}$
   a. $f(-2)$
   b. $f(-1)$
   c. $f(0)$
   d. $f(1)$

51. $f(x) = \dfrac{1}{x - 1}$
   a. $f(0)$
   b. $f(1)$
   c. $f(-1)$
   d. $f(-2)$

52. $f(x) = \dfrac{2}{x + 2}$
   a. $f(-2)$
   b. $f(-1)$
   c. $f(a)$
   d. $f(3)$

53. $f(x) = x^2 - 2x - 3$
   a. $f(0)$
   b. $f(0.1)$
   c. $f(a)$
   d. $f(a - 3)$

54. $f(x) = 2x^2 + x - 3$
   a. $f\left(\dfrac{3}{4}\right)$
   b. $f(-1)$
   c. $f(a)$
   d. $f(x + 2)$

55. $f(x) = \sqrt{x + 3}$
   a. $f(0)$
   b. $f(2)$
   c. $f(-1)$
   d. $f(-2)$

56. $f(x) = \sqrt{x - 5}$
   a. $f(-2)$
   b. $f(1)$
   c. $f(a)$
   d. $f(3)$

57. $f(x) = \dfrac{1}{3}x - 2$

    a. $f(0)$

    b. $f(1)$

    c. $f(-1)$

    d. $f(-3)$

58. $f(x) = \dfrac{2}{3}x^2 + 1$

    a. $f(-3)$

    b. $f(-1)$

    c. $f(t)$

    d. $f(3)$

59. $f(x) = \dfrac{1}{\sqrt{x + 2}}$

    a. $f(-2)$

    b. $f(1)$

    c. $f(-1)$

    d. $f(0)$

60. $f(x) = \dfrac{2}{\sqrt{2 - x}}$

    a. $f(-2)$

    b. $f(-1)$

    c. $f(a)$

    d. $f(3)$

61. $f(x) = \dfrac{x + 3}{x - 4}$

    a. $f(0)$

    b. $f(1)$

    c. $f(-2)$

    d. $f(3)$

62. $f(x) = \dfrac{9 - x}{x + 2}$

    a. $f(0)$

    b. $f(-2)$

    c. $f(3)$

    d. $f(-1)$

63. $f(x) = |x - 2|$

    a. $f(0)$

    b. $f(1.5)$

    c. $f(-1)$

    d. $f(-2)$

64. $f(x) = |3x^2 - 5|$

    a. $f(-2)$

    b. $f(-1)$

    c. $f(2)$

    d. $f(3)$

For Exercises 65–68, use the given graph to determine the value of the function.

65.

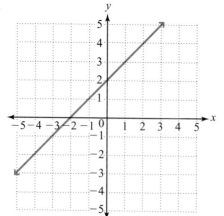

    a. $f(1)$

    b. $f(2)$

    c. $f(0)$

66.

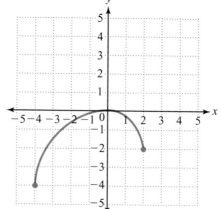

    a. $f(0)$

    b. $f(2)$

    c. $f(-4)$

67.

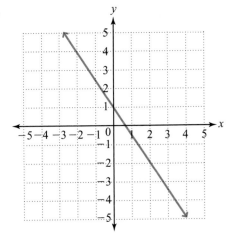

a. $f(4)$

b. $f(2)$

c. $f(0)$

68.

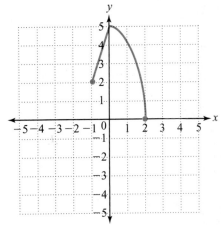

a. $f(0)$

b. $f(2)$

c. $f(-1)$

For Exercises 69–78, graph.

69. $f(x) = 2x + 1$

70. $f(x) = 3x - 5$

71. $f(x) = \dfrac{1}{3}x + 5$

72. $f(x) = \dfrac{1}{2}x - 3$

73. $f(x) = -4x + 1$

74. $f(x) = -x - 1$

75. $f(x) = -\dfrac{2}{3}x$

76. $f(x) = \dfrac{2}{5}x$

77. $f(x) = -\dfrac{1}{4}x - 2$

78. $f(x) = -3x + 5$

Function Quest

For this exercise, you will need to use the Internet or other media source.

1. Find a chart or graph that is an example of a function.

2. Explain why the chart or graph is a function.

3. Find a chart or graph that represents a relation but not a function.

4. Explain why the chart or graph is a relation but not a function.

REVIEW EXERCISES

For Exercises 1 and 2, simplify.

1. $\dfrac{2}{5} - \dfrac{5}{8} \div \dfrac{3}{4}$

2. $x + 3y - 5x + 7 - 3y - 9$

3. Which property is illustrated in the equation $4(x + 9) = 4(9 + x)$?

For Exercises 4 and 5, solve.

4. $3(x + 5) - x = 5x - (x + 1)$

5. $\dfrac{1}{2}x - \dfrac{4}{5} = \dfrac{1}{4}x + 1$

6. Angie graduated in the top 5% of her class. If there were 700 people in her class, how many people were in the top 5%?

Defined Terms

Section 4.1
Axis (p. 270)
Quadrant (p. 273)

Section 4.3
x-intercept (p. 292)
y-intercept (p. 292)

Section 4.4
Slope (p. 305)

Section 4.7
Relation (p. 341)
Domain (p. 341)
Range (p. 341)
Function (p. 343)

Procedures, Rules, and Key Examples

Procedures/Rules	Key Examples

Section 4.1 The Rectangular Coordinate System

To determine the coordinates of a given point in the rectangular system:
1. Follow a vertical line from the point to the x-axis (horizontal axis). The number at this position on the x-axis is the first coordinate.
2. Follow a horizontal line from the point to the y-axis (vertical axis). The number at this position on the y-axis is the second coordinate.

To graph or plot a point given its coordinates:
1. Beginning at the origin, move to the right or left along the x-axis the amount indicated by the first coordinate.
2. From that position on the x-axis, move up or down the amount indicated by the second coordinate.
3. Draw a dot to represent the point described by the coordinates.

To determine the quadrant for a given ordered pair, consider the signs of the coordinates.

$(+, +)$ means the point is in quadrant I
$(-, +)$ means the point is in quadrant II
$(-, -)$ means the point is in quadrant III
$(+, -)$ means the point is in quadrant IV

Note: If either coordinate is 0, then the point is on an axis and not in a quadrant.

Example 1: Plot each of the following. $(-2, 4), (5, 0), (0, -2), (-4, -3), (3, -4), (2, 3)$

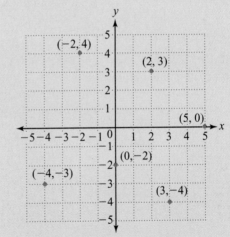

Example 2: State the quadrant in which each point is located.

$(19, 78)$: quadrant I
$(-67, 45)$: quadrant II
$(-107, -36)$: quadrant III
$(58, -92)$: quadrant IV
$(0, -16)$: Not in a quadrant. This point is on the y-axis.

continued

Procedures/Rules	Key Examples

Section 4.2 Graphing Linear Equations

To determine whether a given ordered pair is a solution for an equation with two variables:
1. Replace the variables in the equation with the corresponding coordinates.
2. Verify that the equation is true.

Example 1: Determine whether $(3, 5)$ is a solution for $4x - y = 7$.

Solution: $4(3) - 5 \overset{?}{=} 7$
$12 - 5 \overset{?}{=} 7$
$7 = 7$
$(3, 5)$ is a solution.

Example 2: Determine whether $(4, -9)$ is a solution for $4x - y = 7$.

Solution: $4(4) - (-9) \overset{?}{=} 7$
$16 + 9 \overset{?}{=} 7$
$25 \neq 7$
$(4, -9)$ is not a solution.

To find a solution to an equation in two variables:
1. Choose a value for one of the variables (any value).
2. Replace the corresponding variable with your chosen value.
3. Solve the equation for the value of the other variable.

Example 3: Find two solutions for the equation $y = 2x - 3$.

First solution: Second solution:
Let $x = 0$ Let $x = 1$

$y = 2(0) - 3$ $y = 2(1) - 3$
$y = -3$ $y = 2 - 3$
                $y = -1$

Solution: $(0, -3)$ Solution: $(1, -1)$

To graph a linear equation:
1. Find at least two solutions to the equation.
2. Plot the solutions as points in the rectangular coordinate system.
3. Connect the points to form a straight line.

Example 4: Graph $y = 2x - 3$.

We found two solutions above; $(0, -3)$ and $(1, -1)$

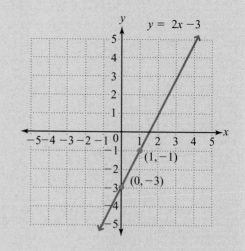

continued

Procedures/Rules	**Key Examples**

Procedures/Rules

The graph of $y = c$, where c is a real number constant, is a horizontal line parallel to the x-axis that passes through the y-axis at a point with coordinates $(0, c)$.

Key Examples

Example 5: Graph $y = 4$.

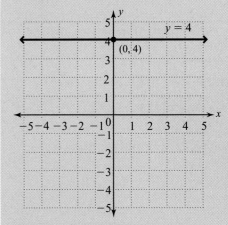

The graph of $x = c$, where c is a real number constant, is a vertical line parallel to the y-axis that passes through the x-axis at a point with coordinates $(c, 0)$.

Example 6: Graph $x = -3$.

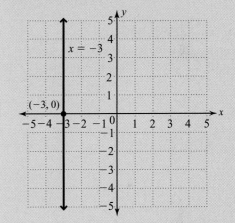

Section 4.3 Graphing Using Intercepts

To find an x-intercept:
1. Replace y with 0 in the given equation.
2. Solve for x.

To find a y-intercept:
1. Replace x with 0 in the given equation.
2. Solve for y.

Example 1: Find the x- and y-intercepts for $2x - 3y = 18$.

Solution:

x-intercept:	y-intercept:
$2x - 3(0) = 18$	$2(0) - 3y = 18$
$2x = 18$	$-3y = 18$
$x = 9$	$y = -6$
x-intercept: $(9, 0)$	y-intercept: $(0, -6)$

continued

Procedures/Rules	**Key Examples**

Procedures/Rules

Section 4.3 Graphing Using Intercepts (continued)

If an equation can be written in the form $y = mx$, where m is a real number constant other than 0, then the x- and y-intercepts are at the origin, $(0, 0)$.

If an equation is in the form $y = mx + b$, where m and b are nonzero real number constants, then the y-intercept will be $(0, b)$.

The graph of an equation in the form $y = c$, where c is a real number constant, has no x-intercept and the y-intercept is at $(0, c)$.

The graph of an equation in the form $x = c$, where c is a real number constant, has no y-intercept and the x-intercept is at $(c, 0)$.

Section 4.4 Slope-Intercept Form

Given an equation of the form $y = mx + b$,
if $m > 0$, then the graph is a line that slants uphill from left to right.
if $m < 0$, then the graph is a line that slants downhill from left to right.

Given two points (x_1, y_1) and (x_2, y_2), the slope of the line connecting the two points is given by the formula

$$m = \frac{y_2 - y_1}{x_2 - x_1}.$$

Key Examples

Example 2: The graph of the equation $y = 3x$ has $(0, 0)$ as both x- and y-intercepts.

Example 3: The y-intercept of the graph of $y = 3x - 2$ is $(0, -2)$.

Example 4: The graph of the equation $y = 4$ has no x-intercept and its y-intercept is at $(0, 4)$.

Example 5: The graph of the equation $x = -3$ has no y-intercept and its x-intercept is at $(-3, 0)$.

Example 1: Graph $y = 2x + 3$ and $y = -2x + 3$.

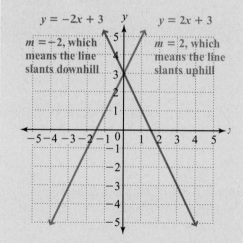

Example 2: Find the slope of a line passing through the points $(2, 7)$ and $(-8, 3)$.

Solution: $m = \dfrac{3 - 7}{-8 - 2} = \dfrac{-4}{-10} = \dfrac{2}{5}$

continued

Procedures/Rules	Key Examples
Two points with different x-coordinates and the same y-coordinates, (x_1, c) and (x_2, c), will form the line with a slope 0 (a horizontal line) and equation $y = c$.	**Example 3:** The slope of a line connecting the points $(2, 4)$ and $(1, 4)$ is 0 and the equation of the line is $y = 4$.
Two points with the same x-coordinates and different y-coordinates, (c, y_1) and (c, y_2), will form a line with a slope that is undefined (a vertical line) and equation $x = c$.	**Example 4:** The slope of a line connecting the points $(-3, 7)$ and $(-3, -2)$ is undefined and the equation of the line is $x = -3$.

Section 4.5 Point-Slope Form

Given the slope of a line passing through a point (x_1, y_1), the equation of the line can be found using the point-slope form of the equation of a line: $y - y_1 = m(x - x_1)$.

Given two points on a line, to write the equation of the line:

1. Find the slope using $m = \dfrac{y_2 - y_1}{x_2 - x_1}$.
2. If one of the points is the y-intercept, then write the equation of the line using slope-intercept form $y = mx + b$. If neither point is the y-intercept, then select one of the points to be (x_1, y_1) and write the equation of the line using the point-slope form $y - y_1 = m(x - x_1)$.
3. Manipulate the equation, if necessary, so that it is in the desired form.

Example 1: Write slope-intercept form of the equation for a line with slope -3 and passing through $(2, 5)$.

Solution:
$$y - y_1 = m(x - x_1)$$
$$y - 5 = -3(x - 2)$$
$$y - 5 = -3x + 6$$
$$y = -3x + 11$$

Example 2: Write an equation of a line connecting $(2, 3)$ and $(-3, -1)$ in the form $Ax + By = C$, where A, B, and C are integers and $A > 0$.

Solution: Find the slope:
$$m = \frac{-1 - 3}{-3 - 2} = \frac{-4}{-5} = \frac{4}{5}$$

Write the equation: $y - 3 = \dfrac{4}{5}(x - 2)$

$$5(y - 3) = 5 \cdot \frac{4}{5}(x - 2) \quad \left\{ \begin{array}{l} \text{Multiply} \\ \text{by 5 to} \\ \text{clear the} \\ \text{fraction.} \end{array} \right.$$

$$5y - 15 = 4(x - 2)$$

$$5y - 15 = 4x - 8 \quad \text{Distribute 4.}$$

$$5y - 4x - 15 = -8 \quad \begin{array}{l} \text{Subtract } 4x \text{ from} \\ \text{both sides.} \end{array}$$

$$5y - 4x = 7 \quad \begin{array}{l} \text{Add 15 on both} \\ \text{sides.} \end{array}$$

$$-4x + 5y = 7 \quad \begin{array}{l} \text{Rearrange so that} \\ \text{the } -4x \text{ term is first.} \end{array}$$

$$-1(-4x + 5y) = -1 \cdot 7 \quad \left\{ \begin{array}{l} \text{Multiply by } -1 \\ \text{so that the } x \\ \text{term is positive.} \end{array} \right.$$

$$4x - 5y = -7 \quad \text{Simplify.}$$

continued

Procedures/Rules	Key Examples

Section 4.5 Point-Slope Form (continued)

The slopes of parallel lines are equal.

The slope of a line perpendicular to a line with a slope of $\dfrac{a}{b}$ will be $-\dfrac{b}{a}$.

Example 3: Find the slope of a line parallel and perpendicular to $3x + 2y = 4$.

Solution: Write $3x + 2y = 4$ in slope-intercept form.

$2y = -3x + 4$ **Subtract $3x$ from both sides.**

$y = -\dfrac{3}{2}x + 2$ **Divide both sides by 2.**

The slope of $3x + 2y = 4$ is $-\dfrac{3}{2}$.

The slope of a line parallel to $3x + 2y = 4$ is $-\dfrac{3}{2}$.

The slope of line perpendicular to $3x + 2y = 4$ is $\dfrac{2}{3}$.

Section 4.6 Graphing Linear Inequalities

To determine whether an ordered pair is a solution for an inequality, replace the variables with the corresponding coordinates and see if the resulting inequality is true. If so, the ordered pair is a solution.

Example 1: Determine whether $(4, 3)$ is a solution for $2x - 4y \leq 8$.

$2x - 4y \leq 8$

$2(4) - 4(3) \overset{?}{\leq} 8$

$8 - 12 \overset{?}{\leq} 8$

$-4 \leq 8$ **This is true, so $(4, 3)$ is a solution.**

Example 2: Determine whether $(3, -1)$ is a solution for $2x - 4y \leq 8$.

$2x - 4y \leq 8$

$2(3) - 4(-1) \overset{?}{\leq} 8$

$6 + 4 \overset{?}{\leq} 8$

$10 \leq 8$ **This is false, so $(3, -1)$ is not a solution.**

continued

Procedures/Rules	Key Examples

Procedures/Rules

To graph a linear inequality:
1. Graph the related equation. The related equation has an equal sign in place of the inequality symbol. If the inequality symbol is $\leq$ or $\geq$, then draw a solid line. If the inequality symbol is $<$ or $>$, then draw a dashed line.
2. Choose an ordered pair on one side of the boundary line and test this ordered pair in the inequality. If the ordered pair satisfies the inequality, then shade the region that contains the chosen ordered pair. If the ordered pair does not satisfy the inequality, then shade the region on the other side of the boundary line.

Section 4.7 Introduction to Functions and Function Notation

To determine the domain of a graph, answer the question: what are all the x-values that have a corresponding y-value?

To determine the range, answer the question: what are all the y-values that have a corresponding x-value?

Key Examples

Example 3: Graph $2x - 4y \leq 8$.
Solution: We found $(4, 3)$ to be a solution, therefore shade the region containing $(4, 3)$.

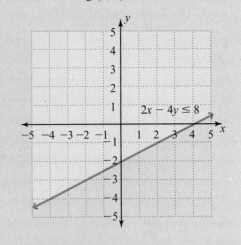

Example 1: Give the domain and range of the function shown.

Domain: all real numbers
Range: $\{y \mid y \geq 2\}$

continued

Procedures/Rules	Key Examples

Section 4.7 Introduction to Functions and Function Notation (continued)

To determine whether a graph is a graph of a function, perform a vertical line test:

1. Draw or imagine vertical lines through each point in the domain.
2. If each vertical line intersects the graph at only one point, then the graph is the graph of a function.
3. If any vertical line intersects the graph at two different points, then the graph is not the graph of a function.

Example 2: Determine whether the graph is the graph of a function.

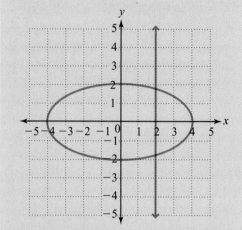

Solution: The relation in the graph is not a function because a vertical line can be drawn that intersects the graph at two points. This means that there is a domain value that corresponds to two values in the range.

Given a function $f(x)$, to find $f(a)$, where a is a real number in the domain of f, replace x in the function with a and calculate the value.

Example 3: For the function
$$f(x) = x^2 - 3x + 1, \text{ find}$$
$$f(-2).$$
$$f(-2) = (-2)^2 - 3(-2) + 1$$
$$= 4 + 6 + 1$$
$$= 11$$

Forms of a Linear Equation

Slope-intercept form: $y = mx + b$

Point-slope form: $y - y_1 = m(x - x_1)$

Standard form: $Ax + By = C$

For Exercises 1–6, answer true or false.

1. When writing coordinates, the horizontal coordinate is written first.

2. An *x*-intercept is where a graph crosses the horizontal axis.

3. The slope formula is $m = \dfrac{x_2 - x_1}{y_2 - y_1}$.

4. A linear inequality has just one solution.

5. Given any two points on a line, the equation of the line may be found.

6. Given the slope and any point on a line, the equation of the line may be found.

For Exercises 7–10, complete the rule.

7. To find an *x*-intercept:
   1. Replace _____ with 0 in the given equation.
   2. Solve for _____ .

8. To find a *y*-intercept:
   1. Replace _____ with 0 in the given equation.
   2. Solve for _____ .

9. If an equation is the form $y = mx + b$, where *m* and *b* are nonzero real number constants, then the _____ will be (0, *b*).

10. Given an equation of the form $y = mx + b$,
    if $m > 0$, then the graph will be a line that slants _____ from left to right.
    if $m < 0$, then the graph will be a line that slants _____ from left to right.

For Exercises 11 and 12, write the coordinates for each point.

11.

12.

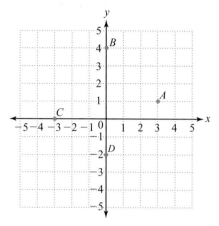

For Exercises 13 and 14, plot and label the points indicated by the coordinate pairs.

13. $(3, 4), (-2, 0), (0, 1), (-1, -6)$

14. $(-2, 2), (0, -3), (4, 1), (5, 0)$

For Exercises 15–18, state the quadrant in which the point is located.

15. $(-2.8, 1203)$

16. $\left(-\dfrac{1}{8}, -16\right)$

17. $(421, 5300)$

18. $\left(52\dfrac{2}{3}, -0.36\right)$

For Exercises 19–24, determine whether the given pair of coordinates is a solution for the given equation.

19. $(-1, 3); x + 3y = 8$

20. $(2, 1); 2x - y = 3$

21. $\left(\dfrac{3}{2}, -\dfrac{1}{2}\right); x + y = 2$

22. $(2.1, 3.2); y - 3x = 5$

23. $(0, 0); 6x = 2y + 1$

24. $(-2, -2); y = \dfrac{2}{3}x$

For Exercises 25–30, find three solutions for the given equation. Then graph.

25. $y = -2x$

26. $y + \dfrac{1}{3}x = 3$

27. $x = y$

28. $2x + 3y = 6$

29. $x + 4y = 5$

30. $y = -x + 7$

For Exercises 31–36, find the coordinates for the x- and y-intercepts.

31. $x + 2y = 12$

32. $y = 7$

33. $x = -2$

34. $2x - 3.1y = 6.2$

35. $y = \dfrac{2}{3}x + 5$

36. $x - y = 0$

For Exercises 37–42, graph using the x- and y-intercepts.

37. $3x + 2y = 6$

38. $y = -2x + 1$

39. $7 = y$

40. $2x = 5y + 10$

41. $y = -\dfrac{1}{5}x + 3$

42. $x = -1$

For Exercises 43 and 44, graph each of the following on the same grid.

43. $y = x$
$y = 2x$
$y = 4x$

44. $y = x$
$y = -x$
$y = -2x$

For Exercises 45–50, determine the slope and the coordinates of the y-intercept, then graph.

45. $y = -2x + 3$

46. $2x - 3y = 8$

47. $y = -\dfrac{1}{4}x + 2$

48. $x + y = 5$

49. $x - 3y = 7$

50. $y = 3x$

For Exercises 51–56, find the slope of the line through the given points.

51. $(2, 7), (-1, -2)$

52. $(0, -1), (3, 2)$

53. $(-3, 1), (3, -1)$

54. $(-6, 2), (-1, -1)$

55. $(2, 8), (-1, 8)$

56. $(-7, 3), (-7, -5)$

For Exercises 57–60, write the equation of the line in slope-intercept form given the slope and the coordinate of the y-intercept.

57. $m = -1; (0, 7)$

58. $m = -\dfrac{1}{5}; (0, -8)$

59. $m = 0.2; (0, 6)$

60. $m = 1; (0, 0)$

For Exercises 61–66, write the equation of a line in slope-intercept form with the slope passing through the given point.

61. $m = -2; (1, 7)$

62. $m = 0.4; (-1, -2)$

63. $m = \dfrac{1}{3}; (-5, 0)$

64. $m = 1; (3, -6)$

65. $m = 6.2; (-2, -3)$

66. $m = -\dfrac{2}{5}; (3, 3)$

For Exercises 67–72, write the equation of a line connecting the given points in slope-intercept form and in standard form.

67. $(3, 2), (5, 1)$

68. $(5, -3), (9, 2)$

69. $(7, -3), (2, 2)$

70. $(-1, -1), (2, -2)$

71. $(-3, -2), (-8, -1)$

72. $(5, 0), (0, 6)$

73. Find the equation of a line passing through $(0, -2)$ and parallel to the line $y = 2x + 7$.

74. Find the equation of a line passing through $(1, 5)$ and parallel to the line $2x + 3y = 6$.

75. Find the equation of a line passing through $(-1, -2)$ and perpendicular to the line
$$y = \frac{3}{5}x - 1.$$

76. Find the equation of a line passing through $(-2, 4)$ and perpendicular to the line
$$y = -\frac{2}{5}x - 6$$

For Exercises 77–80, determine whether the ordered pair is a solution for the linear inequality.

77. $(3, 2); x + 2y > 5$ **78.** $(-1, 0); y < x + 2$ **79.** $(-5, 8); 2x - 6y \leq 17$ **80.** $(0, 0); x + y \geq 5$

For Exercises 81–86, graph the linear inequality.

81. $y < 3x - 7$

82. $2x + 5y > 16$

83. $-3x - 5y \leq -15$

84. $y > -\frac{2}{3}x$

85. $x - y \geq 9$

86. $x \geq -1$

For Exercises 87–88, find the domain and range of the relation.

87. $\{(2, 3), (-1, 5), (3, 6), (-3, -4)\}$

88. Cars holding their value for resale:

Automobile	Percentage of Original Value
Volkswagen	52.2
Honda	49.7
Toyota	49.0
Subaru	47.8
Nissan	45.8

Source: Automotive Lease Guide

For Exercises 89–90, determine whether the relation is a function.

89. The following correspondence shows the courses taught by each instructor during one semester.

Domain (instructor name)	Range (courses taught)
Hames	Math 100
Carson	Math 100, Math 102
Pritchard	Math 035
Webb	Math 100, Math 110

90. The following correspondence shows the price of a particular brand of dog food based on the size of the bag.

Domain (size of bag)	Range (price)
5 lb.	$3.95
10 lb.	$7.90
20 lb.	$15.00
40 lb.	$29.95

For Exercises 91–93, give the domain and range. Then determine whether the graph is the graph of a function.

91.

92.

93.

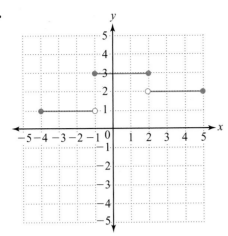

94. Use the graph in Exercise 93 to find the value of the function at the indicated values.
   a. $f(-3)$
   b. $f(-1)$
   c. $f(0)$
   d. $f(3)$

95. Find the value of the function $f(x) = x^3 - 5x + 2$
   a. $f(2)$
   b. $f(0)$
   c. $f(-3)$

96. For what values of x is the function $f(x) = \dfrac{x}{x^2 - 9}$ undefined?

1. Determine the coordinates for each point.

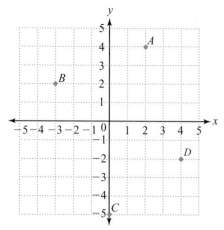

2. Plot and label the points indicated by the coordinate pairs.
$(0, -3), (4, -1), (-3, 2), (2, 0), (-1, -5)$

3. State the quadrant in which $\left(4\frac{2}{3}, 5115\right)$ is located.

4. Determine whether $(1, 7)$ is a solution for
$$y = -\frac{2}{5}x - 8.$$

For Exercises 5–10, find the coordinates for the x- and y-intercepts for each of the following, then graph.

5. $x - 2y = 4$

6. $9 = 2x + 3y$

7. $y = \frac{1}{2}x + 4$

8. $y = x$

9. $2x - y = 8$

10. $y = \frac{3}{4}x - 1$

For Exercises 11 and 12, determine the slope and the coordinates of the y-intercept.

11. $y = \dfrac{3}{4}x + 11$

12. $5x - 3y = 8$

For Exercises 13 and 14, determine the slope of the line through the given points.

13. $(-5, 6), (-2, -4)$

14. $(6, 2), (3, 2)$

For Exercises 15–18, write the equation of the line in slope-intercept form.

15. $m = \dfrac{3}{5}$; *y*-intercept $(0, 4)$

16. $m = -1$; *y*-intercept $(0, -5)$

17. through the points $(1, 7), (-1, -2)$

18. through the points $(8, -1), (-7, 5)$

19. Write the equation of a line in the form $Ax + By = C$ through the points $(2, 1)$ and $(5, 3)$.

20. Write the equation of a line in the form $Ax + By = C$ through the point $(6, 1)$ and parallel to the line $y = \dfrac{2}{3}x - 5$.

21. Write the equation of a line in the form $Ax + By = C$ through the point $(4, -2)$ and perpendicular to the line $y = -3x + 4$.

22. Determine whether $(2, 7)$ is a solution for the linear inequality $y \geq 2x - 9$.

For Exercises 23–25, graph the linear inequality.

23. $y \geq \dfrac{2}{5}x - 7$

24. $x - 3y < 8$

25. $8x + y > 11$

26. The following correspondence shows government spending (range) for each year (domain) from 1990 to 1999. Determine whether the correspondence is a function.

Year	Government Spending (in billions of dollars; 1990 prices)
1990	974.9
1991	978.99
1992	978.63
1993	971.17
1994	968.15
1995	973.13
1996	1014.39
1997	1013.07
1998	1027.75
1999	1052.16

27. Determine whether the graph is the graph of a function.

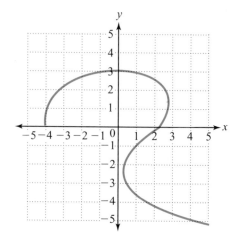

28. For what value(s) is the function $f(x) = \dfrac{x^2}{x - 4}$ undefined?

29. Find the indicated value of the function $f(x) = \dfrac{x^2}{x - 4}$.

a. $f(-2)$

b. $f(3)$

c. $f(0)$

30. a. Give the domain and range of the function.

b. Find $f(1)$.

CHAPTER 5

Polynomials

In Chapter 1, we reviewed real numbers and introduced algebraic expressions. We learned there are two actions we can perform on an expression: 1. we can evaluate an expression or 2. we can rewrite an expression. In this chapter, we will focus on polynomial expressions, exploring how to evaluate and rewrite these expressions. We will also see that we can add, subtract, multiply, and divide polynomial expressions in much the same way that we add, subtract, multiply, and divide numbers.

"The ability to simplify means to eliminate the unnecessary so that the necessary may speak."

—HANS HOFMANN
Early 20th-century teacher and painter

"Everything should be made as simple as possible, but not simpler."

—ALBERT EINSTEIN
Early 20th-century physicist

5.1 Exponents and Scientific Notation

OBJECTIVES

1 Evaluate exponential forms with integer exponents.

2 Write scientific notation in standard form.

3 Write standard form numbers in scientific notation.

OBJECTIVE **1** Evaluate exponential forms with integer exponents.

Positive Integer Exponents

We learned in Section 1.5 that exponents indicate repeated multiplication of a base number.

$$2^4 = \underbrace{2 \cdot 2 \cdot 2 \cdot 2}_{\text{four factors of two}} = 16 \qquad (-5)^3 = \underbrace{(-5)(-5)(-5)}_{\text{three factors of negative five}} = -125$$

We also discovered some rules for evaluating exponential forms with negative bases.

RULE | **EVALUATING EXPONENTIAL FORMS WITH NEGATIVE BASES**

If the base of an exponential form is a negative number and the exponent is even, then the product is positive.

If the base is a negative number and the exponent is odd, then the product is negative.

What if the base is a fraction, as in $\left(\dfrac{2}{3}\right)^4$?

$$\left(\frac{2}{3}\right)^4 = \frac{2}{3} \cdot \frac{2}{3} \cdot \frac{2}{3} \cdot \frac{2}{3} = \frac{2^4}{3^4} = \frac{16}{81}$$

This example suggests the following conclusion and rule.

Conclusion If a fraction is raised to a power, we can write its numerator and denominator to that power.

RULE | **RAISING A QUOTIENT TO A POWER**

If a and b are real numbers, where $b \neq 0$ and n is a natural number, then $\left(\dfrac{a}{b}\right)^n = \dfrac{a^n}{b^n}$.

EXAMPLE 1 Evaluate each exponential form.

a. $(-3)^5$

Solution $(-3)^5 = (-3)(-3)(-3)(-3)(-3) = -243$

b. -2^4

Solution $-2^4 = -[2 \cdot 2 \cdot 2 \cdot 2] = -16$

Warning: Notice -2^4 does not mean that four -2's are to be multiplied. Rather, it means find the additive inverse after multiplying four 2's.

c. $\left(\dfrac{4}{5}\right)^3$

Solution $\left(\dfrac{4}{5}\right)^3 = \dfrac{4^3}{5^3} = \dfrac{64}{125}$

Nonpositive Integer Exponents

So far, we have only considered evaluating exponential forms with positive exponents. However, exponents can be positive, zero, or negative. What does an exponent of 0 indicate? Consider the pattern below:

$$2^4 = 2 \cdot 2 \cdot 2 \cdot 2 = 16$$
$$2^3 = 2 \cdot 2 \cdot 2 \quad = 8$$
$$2^2 = 2 \cdot 2 \quad\quad = 4$$
$$2^1 = 2 \quad\quad\quad\quad = 2$$
$$2^0 = \quad\quad\quad\quad\quad = 1$$

Notice that as the exponent decreases, we can divide out a 2 to determine the next result; that is, since 2^4 is 16, 2^3 is $16 \div 2$, which is 8.

To continue the pattern, when we go from 2^1 to 2^0, we divide the result for 2^1 by 2 to determine the result for 2^0, and $2 \div 2 = 1$.

The pattern suggests the following conclusion and rule.

Conclusion   A base number raised to the 0 power is equal to 1. It can be shown that any nonzero number raised to the 0 power equals 1.

RULE   **ZERO AS AN EXPONENT**

If a is a real number and $a \neq 0$, then $a^0 = 1$.

If we continue the pattern, we can see what a negative exponent means.

$$2^0 = 1$$
$$2^{-1} = \frac{1}{2}$$

To continue the pattern, when we go from 2^0 to 2^{-1}, we divide the result for 2^0 by 2 to determine the result for 2^{-1}, and $1 \div 2 = \frac{1}{2}$.

$$2^{-2} = \frac{1}{2^2} = \frac{1}{4}$$

To continue the pattern, when we go from 2^{-1} to 2^{-2}, we divide the result for 2^{-1} by 2 to determine the result for 2^{-2}, and

$$2^{-3} = \frac{1}{2^3} = \frac{1}{8}$$

$$\frac{1}{2} \div 2 = \frac{1}{2} \cdot \frac{1}{2} = \frac{1}{2^2} = \frac{1}{4}$$

The pattern suggests the following conclusion:

Conclusion   An exponential form with a negative exponent is equal to its reciprocal with the exponent made positive.

It can be shown that our conclusion is true for all real number bases.

If a is a real number, where $a \neq 0$ and n is a natural number, then $a^{-n} = \dfrac{1}{a^n}$.

EXAMPLE 2 Evaluate the exponential form.

a. 10^{-3}

Solution $10^{-3} = \dfrac{1}{10^3} = \dfrac{1}{1000}$

b. $(-2)^{-4}$

Solution $(-2)^{-4} = \dfrac{1}{(-2)^4} = \dfrac{1}{(-2)(-2)(-2)(-2)} = \dfrac{1}{16}$

c. $(-5)^{-3}$

Solution $(-5)^{-3} = \dfrac{1}{(-5)^3} = \dfrac{1}{(-5)(-5)(-5)} = \dfrac{1}{-125} = -\dfrac{1}{125}$

Calculator TIPS

To evaluate exponential forms with negative exponents, like $(-2)^{-4}$ in part b of Example 2, using a calculator that has $\boxed{\wedge}$ for exponents and $\boxed{(-)}$ for negatives, type:

$\boxed{(}$ $\boxed{(-)}$ $\boxed{2}$ $\boxed{)}$ $\boxed{\wedge}$ $\boxed{(-)}$ $\boxed{4}$ $\boxed{\text{ENTER}}$

The result will be a decimal number, 0.0625. To see the fraction form, press $\boxed{\text{MATH}}$ and select 1: Frac from the menu, then press $\boxed{\text{ENTER}}$. Or, on a scientific calculator, use the F ◄► D function.

If the calculator has $\boxed{y^x}$ and $\boxed{+/-}$ keys, type:

$\boxed{(}$ $\boxed{2}$ $\boxed{+/-}$ $\boxed{)}$ $\boxed{y^x}$ $\boxed{4}$ $\boxed{+/-}$ $\boxed{=}$

What if a negative exponent is in the denominator of a fraction?

EXAMPLE 3 Evaluate $\dfrac{1}{2^{-3}}$.

Solution $\dfrac{1}{2^{-3}} = \dfrac{1}{\dfrac{1}{2^3}} = \dfrac{1}{\dfrac{1}{8}} = \dfrac{8}{1} = 8$ **Rewrite the denominator using $a^{-n} = \dfrac{1}{a^n}$.**

CONNECTION

Remember that $\dfrac{1}{\frac{1}{8}}$ means $1 \div \dfrac{1}{8}$ and $1 \div \dfrac{1}{8} = 1 \cdot \dfrac{8}{1} = \dfrac{8}{1} = 8$.

Example 3 suggests the following rule.

If a is a real number where $a \neq 0$ and n is a natural number, then $\dfrac{1}{a^{-n}} = a^n$.

What if a fraction is raised to a negative exponent?

EXAMPLE 4 Evaluate $\left(\dfrac{2}{5}\right)^{-2}$.

Solution $\left(\dfrac{2}{5}\right)^{-2} = \dfrac{1}{\left(\dfrac{2}{5}\right)^2} = \dfrac{1}{\dfrac{4}{25}} = \dfrac{25}{4}$

Example 4 suggests that when we evaluate a fraction raised to a negative exponent, we can write the reciprocal of the fraction and change the sign of the exponent.

RULE If a and b are real numbers, where $a \neq 0$ and $b \neq 0$ and n is a natural number, then $\left(\dfrac{a}{b}\right)^{-n} = \left(\dfrac{b}{a}\right)^n$.

YOUR TURN Evaluate the exponential form.

a. $\left(\dfrac{5}{8}\right)^2$ **b.** $(-7)^0$ **c.** 6^{-3} **d.** $\left(-\dfrac{2}{5}\right)^{-4}$

OBJECTIVE 2 Write scientific notation in standard form. Sometimes we use very large or very small numbers, such as when we describe the vast distances between stars and galaxies or the tiny size of bacteria and atomic structures. For example, the distance from the Sun to the next nearest star, Proxima Centauri, is about 24,700,000,000,000 miles; a single streptococcus bacterium (photo at left) is about 0.00000075 meters in diameter. The large number of zero digits in these numbers makes them tedious to write. **Scientific notation** gives us a shorthand way to write such numbers.

DEFINITION *Scientific notation:* A number expressed in the form $a \times 10^n$, where a is a decimal number with $1 \leq |a| < 10$ and n is an integer.

The number 3.58×10^4 is in scientific notation because 3.58 is greater than or equal to 1, but less than 10, and the exponent 4 is an integer. The number 35.8×10^3 is not in scientific notation because 35.8 is greater than 10. The number 0.358×10^5 is also not in scientific notation because 0.358 is less than 1.

ANSWERS

a. $\dfrac{25}{64}$ **b.** 1 **c.** $\dfrac{1}{216}$ **d.** $\dfrac{625}{16}$

What does 3.58×10^4 mean?

$$3.58 \times 10^4 = 3.58 \times 10 \times 10 \times 10 \times 10$$
$$= \quad 35.8 \times 10 \times 10 \times 10$$
$$= \quad 358 \times 10 \times 10$$
$$= \quad 3580 \times 10$$
$$= \quad 35{,}800$$

Note: 3.58×10^4, 35.8×10^3, and 0.358×10^5 all represent the same number, 35,800. Only 3.58×10^4 is in scientific notation.

Notice each multiplication by 10 moves the decimal point one place to the right. Since we multiplied by four factors of 10, the decimal point moved a total of four places to the right from where it started in 3.58.

$3.58 \times 10^4 = 3\,5\,8\,0\,0 = 35{,}800$ **The decimal point moves four places to the right.**

Our example suggests that the exponent of the power of 10 determines the number of places that the decimal point moves.

> **PROCEDURE** CHANGING SCIENTIFIC NOTATION (POSITIVE EXPONENT) TO STANDARD FORM
>
> To change from scientific notation with a positive integer exponent to standard notation, move the decimal point to the right the number of places indicated by the exponent.

EXAMPLE 5 Write each number in standard form.

a. 4.5×10^6

Solution Multiplying 4.5 by 10^6 means that the decimal point will move six places to the right.

$4.5 \times 10^6 = 4{,}500{,}000$

> **CONNECTION**
>
> Since 10^6 is another way to represent 1,000,000, it is popular to read 4.5×10^6 as 4.5 million.

b. 9×10^5

Solution Multiplying 9 by 10^5 means that the decimal point will move five places to the right.

$9 \times 10^5 = 900{,}000$

YOUR TURN Write each number in standard form.

a. 7.305×10^6 　　　　　　　　　　　　**b.** 8×10^7

What if the scientific notation contains a negative exponent?

EXAMPLE 6 Write 8.45×10^{-2} in standard form.

Solution To evaluate 10^{-2}, we write it with a positive exponent, $\dfrac{1}{10^2}$.

$$8.45 \times 10^{-2} = 8.45 \times \frac{1}{10^2}$$

$$= \frac{8.45}{10^2}$$

$$= \frac{8.45}{100}$$

Note: Dividing by 100 causes the decimal point to move two places to the *left*.

$$= 0.0845$$

0.0845

EXAMPLE 7 Write 4.2×10^{-5} in standard form.

Solution To evaluate 10^{-5}, we write it with a positive exponent, $\frac{1}{10^5}$.

$$4.2 \times 10^{-5} = 4.2 \times \frac{1}{10^5}$$

$$= \frac{4.2}{10^5}$$

$$= \frac{4.2}{100,000}$$

Note: Dividing by 100,000 causes the decimal point to move five places to the *left*.

$$= 0.000042$$

0.000042

Examples 6 and 7 suggest the following procedure.

PROCEDURE CHANGING SCIENTIFIC NOTATION (NEGATIVE EXPONENT) TO STANDARD FORM

To write a number expressed in scientific notation with a negative exponent in standard form, move the decimal point to the left the same number of places as the absolute value of the exponent.

YOUR TURN Write in standard form.

a. 7.24×10^{-6} **b.** 5.621×10^{-7}

Calculator TIPS

To enter scientific notation on a graphing or scientific calculator, we can use the EE *function. For example, to enter 4.2×10^{-5}, type:*

4 · 2 EE (−) 5 *(You may have to press* 2nd *to get to the* EE *function.)*

Notice that the base 10 is understood and is not entered when using the EE *function. We can also enter scientific notation without using* EE *like this:*

4 · 2 × 1 0 ^ (−) 5

After entering scientific notation, if you press ENTER *, the calculator will change the number to standard form. However, if the number has too many digits for the calculator to display, then it will leave it in scientific notation. Similarly, if the answer to a calculation is too large for the calculator to display, then it will show the result in scientific form.*

ANSWERS

a. 0.00000724
b. 0.0000005621

OBJECTIVE ③ **Write standard form numbers in scientific notation.** We mentioned that the distance from the Sun to the next nearest star, Proxima Centauri, is about 24,700,000,000,000 miles. Suppose we want to express 24,700,000,000,000 in scientific notation. Since scientific notation begins with a decimal number greater than or equal to 1 but less than 10, our first step will be to determine the position of the decimal point. The only place the decimal point could go to satisfy the criteria for scientific notation is between the 2 and 4 digits.

$$24{,}700{,}000{,}000{,}000$$

The decimal goes here to express a decimal number
greater than or equal to 1 but less than 10.

Notice that there are 13 place values to the right of the decimal point in its new position between the 2 and 4. To make the scientific notation equal to the original number, we must account for the 13 place values by expressing them as a power of 10.

$$24{,}700{,}000{,}000{,}000 = 2.4700000000000 \times 10^{13}$$

The 13 places to the right of the new decimal position are
expressed as a power of 10.

$$24{,}700{,}000{,}000{,}000 = 2.4700000000000 \times 10^{13} = 2.47 \times 10^{13}$$

These zeros to the right of the 7 digit are no longer needed.

PROCEDURE CHANGING STANDARD FORM TO SCIENTIFIC NOTATION

To write a number greater than 1 in scientific notation:

1. Move the decimal point so that the number is greater than or equal to 1, but less than 10. *(Tip: Place the decimal point to the right of the first nonzero digit.)*
2. Write the decimal number multiplied by 10^n, where n is the number of places between the new decimal position and the original decimal position.
3. Delete zeros to the right of the last nonzero digit.

EXAMPLE ⑧ Write 986,000 in scientific notation.

Solution Place the decimal to the right of the first nonzero digit, 9. Next we count the places to the right of this decimal position, which is five places. We write the 5 as the exponent of 10. Finally, we delete all the 0's to the right of the last nonzero digit, which is the 6 in this case.

$$986{,}000 = 9.86 \times 10^5$$

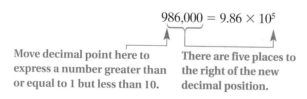

Move decimal point here to
express a number greater than
or equal to 1 but less than 10.

There are five places to
the right of the new
decimal position.

We learned that a number written in scientific notation that has a negative exponent, such as 5×10^{-2}, has an absolute value less than 1. Therefore, a positive number less than 1, such as 0.00000075, will have a negative exponent when written in scientific notation. To write 0.00000075 in scientific notation, we place the decimal between the 7 and 5 digits so that the number is greater than or equal to 1, but less than 10. Moving the decimal point to this position means that we have seven places to account for, which we will indicate with a -7 exponent.

$$0.00000075 = 7.5 \times 10^{-7}$$

This suggests the following procedure.

PROCEDURE   CHANGING STANDARD FORM TO SCIENTIFIC NOTATION

To write a positive decimal number that is less than 1 in scientific notation:

1. Move the decimal point so that the number is greater than or equal to 1 but less than 10. (*Tip: Place the decimal point to the right of the first nonzero digit.*)
2. Write the decimal number multiplied by 10^{n}, where n is a negative integer whose absolute value is the number of places between the new decimal position and the original decimal position.

EXAMPLE 9   Write 0.0000608 in scientific notation.

Solution: Place the decimal point between the 6 and 0 digits so that we have 6.08, which is a decimal number greater than 1 but less than 10. Since there are five decimal places between the original decimal position and the new position, the exponent is -5.

$$0.0000608 = 6.08 \times 10^{-5}$$

There are five decimal places in between the original position and the new position, so the exponent is -5.

Decimal point moves here to get a number whose absolute value is greater than or equal to 1 but less than 10.

1. Explain how to rewrite an expression with a negative exponent so that its exponent is positive.

2. After simplifying -5^2, is the result positive or negative? Explain.

3. After simplifying 2^{-3}, is the result positive or negative? Explain.

4. For what kinds of numbers is scientific notation a useful notation?

5. How do we write a number that is greater than 1 in scientific notation?

6. How do we write a number greater than 0 but less than 1 in scientific notation?

For Exercises 7–30, evaluate the exponential form.

7. 4^0
8. 3^0
9. -2^3
10. -3^2
11. $(-5)^3$
12. $(-2)^5$
13. -4^2

14. $\left(-\dfrac{1}{5}\right)^2$
15. $\left(-\dfrac{3}{4}\right)^3$
16. $\left(-\dfrac{5}{6}\right)^2$
17. $(0.2)^3$
18. $(-1.2)^2$
19. $(-2.3)^2$
20. $(-0.04)^3$

21. 2^{-3}
22. 5^{-1}
23. $\left(\dfrac{5}{7}\right)^{-2}$
24. $\left(\dfrac{1}{6}\right)^{-3}$
25. $(-4)^{-3}$
26. $(-1)^{-5}$
27. -2^{-4}

28. $-\dfrac{3^{-2}}{5}$
29. $(0.2)^{-3}$
30. $(-0.1)^{-4}$

For Exercises 31–36, use a scientific or graphing calculator to evaluate each exponential form.

31. $(-24)^{-5}$ **32.** $(-45)^{-4}$ **33.** -0.06^{-4} **34.** $(3.7)^{-6}$

35. $\left(-\dfrac{5}{8}\right)^{-3}$ **36.** $-\left(\dfrac{4}{9}\right)^{-4}$

For Exercises 37–40, explain the mistake, then work the problem correctly.

37. $-3^4 = (-3)(-3)(-3)(-3) = 81$ **38.** $(-5)^3 = -75$

39. $5^{-2} = -25$ **40.** $\left(\dfrac{3}{4}\right)^{-2} = \dfrac{9}{16}$

For Exercises 41–56, write the number in standard form.

41. Scientists speculate that the universe is about 1.65×10^{10} years old.

42. The solar system formed approximately 4.6×10^9 years ago.

43. In empty space, light travels at a constant speed of approximately 2.998×10^8 meters per second.

44. The equatorial radius of the Sun is approximately 6.96×10^8 meters.

45. The equatorial radius of Earth is approximately 6.378×10^6 meters.

46. A light-year is the distance that light travels in a year's time and is equal to about 5.76×10^{12} miles.

47. Earth is about 9.292×10^7 miles from the Sun.
(*Source: The Friendly Guide to the Universe.* Nancy Hathaway, 1994. Winokur/Boates)

48. The next nearest star to ours (the Sun) is Proxima Centauri and is about 2.47×10^{13} miles from the Sun.

49. Mitochondria are organelles that are about 2.5×10^{-6} meters across.

50. The diameter of human DNA is about 2×10^{-9} meters.

51. The rest mass of a proton is about 1.67×10^{-27} kilogram.

Of Interest

The light from the Sun takes about 8.2 minutes to reach Earth and about 4.26 years to reach Proxima Centauri. We would say that Earth is at a distance of about 8.2 light-minutes from the Sun and Proxima Centauri is at a distance of about 4.26 light-years.

52. Light with a wavelength of 7.6×10^{-11} meters is in the x-ray portion of the spectrum.

53. Light with a wavelength of 2.95×10^{-9} meters is in the ultraviolet portion of the spectrum.

54. Light with a wavelength of 4.5×10^{-7} meters is in the visible spectrum and is blue in color.

55. The mass of an alpha particle, which is emitted in the radioactive decay of plutonium 239, is 6.645×10^{-27} kilograms.

56. Protons have a positive electrical charge of 1.602×10^{-19} coulombs.

For Exercises 57–72, write the number in scientific notation.

57. There are approximately 25,000,000,000,000 red blood cells in the average person's bloodstream.

58. A single red blood cell contains about 250,000,000 molecules of hemoglobin.

59. Human DNA consists of about 5,300,000,000 nucleotide pairs.

60. The Andromeda galaxy is about 2,140,000 light-years from the Sun.

61. Uranium 236 is a radioactive isotope of uranium and has a half-life of 23,420,000 years.

62. The population of the United States is about 275,000,000.

63. The most populated country in the world is China, with a population of about 1,008,000,000.

64. In 1996, the population of Earth was about 5,772,000,000.

65. Light with a wavelength of 0.000000586 meters is yellow in color.

66. Light with a wavelength of 0.000000712 meters is red in color.

67. Protons have a mass of 0.00000000000000000000000000167 kilograms.

68. The size of the HIV virus that causes AIDS is about 0.0000001 meters.

69. The most common diameter for the core of a fiber-optic cable is 0.0000625 meters.

70. A human hair is about 0.00005 meters wide.

71. Celeron processors are 0.000025 meters wide.

72. It takes 0.00841 U.S. dollars to equal one Japanese yen.

For Exercises 73 and 74, write the numbers in order from smallest to largest.

73. $7.5 \times 10^6,$ $8.95 \times 10^5,$ $9 \times 10^7,$ $1.3 \times 10^8,$ 7.2×10^6

74. $2.8 \times 10^{-5},$ $3.7 \times 10^{-7},$ $6.2 \times 10^{-2},$ $9.8 \times 10^{-5},$ 6.1×10^{-2}

75. In Exercises 52 and 53, we saw that light with a wavelength of 7.6×10^{-11} meters is in the x-ray portion of the spectrum and light with a wavelength of 2.95×10^{-9} meters is in the ultraviolet portion of the spectrum. About how many times greater is ultraviolet light's wavelength compared to x-rays?

76. In Exercises 55 and 67, we saw that an alpha particle has a rest mass about 6.645×10^{-27} kilograms and a proton has a mass of 0.00000000000000000000000000167 kilograms. Which has the greater mass? About how many times greater is the mass of the larger of the two particles?

REVIEW EXERCISES

1. Use the distributive property to rewrite the expression $-4(y-3)$.

2. Find the slope and y-intercept for $2x - 3y = 6$.

3. 40 is 15% of what?

4. Solve: $\dfrac{x}{8} = \dfrac{5}{12}$

5. Find the slope through $(2, -1)$ and $(-3, -2)$.

6. Given the function $f(x) = 3x^2 - 5$, find $f(4)$.

5.2 Introduction to Polynomials

OBJECTIVES

1. Identify monomials.
2. Identify the coefficient and degree of a monomial.
3. Classify polynomials.
4. Identify the degree of a polynomial.
5. Evaluate polynomials.
6. Write polynomials in descending order of degree.
7. Combine like terms.

OBJECTIVE 1 Identify monomials. We have explored some simple algebraic expressions (Chapter 1) and linear equations and inequalities (Chapters 2 through 4). We will now move back to the expression level of our algebra pyramid and explore a particular class of expressions known as *polynomials*.

THE ALGEBRA PYRAMID

EQUATIONS AND INEQUALITIES
$2 + 5(7) = 37$ or
$x + 2y > 5$

EXPRESSIONS
$2 + 5(7)$ or $x + 2y$

CONSTANTS AND VARIABLES
$2, 5, 7, x, y$

First, we consider a special type of polynomial expression called a **monomial**.

DEFINITION

> *Monomial:* An expression that is a constant or a product of a constant and variables that are raised to whole number powers.

EXAMPLE 1 Is the given expression a monomial? Explain.

a. 7 ◀ **CONNECTION**

Though no variable factor appears with the constant 7, it can be expressed with a variable factor. Recall from Section 5.1 that any nonzero number raised to the 0 power simplifies to 1. Therefore, 7 can be written as $7x^0$ because $x^0 = 1$ (assuming $x \neq 0$). We will use this fact to help us understand the next objective concerning degree.

Answer: 7 is a monomial because it is a constant and by definition, a constant is a monomial.

b. $3x^2$

Answer: $3x^2$ is a monomial because it is a product of a constant, 3, and a variable, x, which has a whole number exponent of 2.

c. $9y$

Answer: Remember that a number or variable with no apparent exponent has an understood 1 exponent, so $9y = 9y^1$. Therefore, $9y$ is a monomial because it is a product of a constant, 9, and a variable, y, which has a whole number exponent of 1.

d. $-0.6xy^2$

Answer: $-0.6xy^2$ is a monomial because it is a product of a constant, -0.6, and variables, xy^2, which have whole number exponents. As we saw in Example 1c, we could write $-0.6xy^2$ as $-0.6x^1y^2$. Both of the variables' exponents, 1 and 2, are whole numbers.

e. y^4

Answer: It might seem that y^4 is not a monomial because no numerical factor is visible. However, y^4 is the same as $1y^4$. We don't need to write the 1 because 1 multiplied by an amount does not affect the amount.

Now consider some examples of expressions that are not monomials.

EXAMPLE 2 Explain why the given expressions are not monomials.

a. $\dfrac{5}{x^2}$

Answer: The expression $\dfrac{5}{x^2}$ can be written as a product by writing the divisor with a negative exponent, $5x^{-2}$. However, when written as a product, the variable's exponent is not a whole number, which means it is not a monomial.

b. $4y^2 + 9y - 3$

Answer: The expression is not a monomial because it is not a product of a constant and variables. Instead, addition and subtraction are involved.

YOUR TURN Is each of the given expressions a monomial? Explain.

a. x^3 **b.** $-y^2$ **c.** $-\dfrac{2}{3}tu$ **d.** $\dfrac{4m^2}{n}$ **e.** 0.56 **f.** $3y + 7$

In our introduction to expressions in Section 1.7, we defined a **coefficient** as the numerical or constant factor in a term. The same is true for monomials. Another piece of information about a monomial is its **degree**, which relates to the exponents of the variables.

DEFINITIONS

> *Coefficient:* The numerical factor in a monomial.
>
> *Degree:* The sum of the exponents of all variables in a monomial.

EXAMPLE **3** Identify the coefficient and degree of each monomial.

a. $-5t^2$

Answer: The coefficient is -5 because it is the numerical factor. The degree is 2 because it is the exponent for the variable t.

b. $-mn^4$

Answer: We can express $-mn^4$ as $-1m^1n^4$. In this form, we can see that the coefficient is -1 and that the exponents for the variables are 1 and 4. Since the degree is the sum of the variables' exponents, the degree is 5.

c. 7

Answer: Remember that $7 = 7x^0$, where x is any real number except 0. In this alternative form, we can see that 7 is the coefficient and 0 is the degree because it is the variable's exponent.

YOUR TURN Identify the coefficient and degree of each monomial.

a. $-4x^4$ **b.** $-0.9n^2p^5$ **c.** $\dfrac{5}{8}ab$ **d.** 13

OBJECTIVE **3** Classify polynomials. Now we are ready to formally define a **polynomial**.

DEFINITION

> *Polynomial:* A monomial or an expression that can be written as a sum of monomials.

Examples of polynomials: $4x$ $4x + 8$ $2x^2 + 5xy + 8y$ $2x^3 - 5x^2 + 3x - 9$

Note: Though it seems that the subtractions violate the word *sum* in the definition, the expression $2x^3 - 5x^2 + 3x - 9$ is a polynomial because it can be written as a sum of monomials by rewriting the subtractions as equivalent additions.

$$2x^3 - 5x^2 + 3x - 9 = 2x^3 + (-5x^3) + 3x + (-9)$$

ANSWERS

a. c: -4, d: 4
b. c: -0.9, d: 7
c. c: $\dfrac{5}{8}$, d: 2   **d.** c: 13, d: 0

Notice that the polynomial $2x^3 - 5x^2 + 3x - 9$ has the same variable x in each variable term whereas $2x^2 + 5xy + 8y$ has a mixture of two variables, x and y. We call a polynomial like $2x^3 - 5x^2 + 3x - 9$ a **polynomial in one variable**.

> **DEFINITION**
>
> **Polynomial in one variable:** A polynomial in which every variable term has the same variable.

$x^2 - 5x + 2$ is a polynomial in one variable, x.

$x^2 + 8y$ is a polynomial in two variables, x and y.

Special names have been given to some polynomials. We have already discussed monomials, like $4x$, which are single-term polynomials. Notice the prefix *mono* in the word *monomial* indicates "one" term. Continuing with the prefixes, a two-term polynomial, like $4x + 8$, is called a **binomial** and a three-term polynomial, like $2y^2 + 5y + 8$, is called a **trinomial**. No special names are given to polynomials with more than three terms.

> **DEFINITIONS**
>
> **Binomial:** A polynomial containing two terms.
> **Trinomial:** A polynomial containing three terms.

EXAMPLE 4 Determine whether the expression is a monomial, binomial, trinomial, or none of these.

a. $2.9xy^3z$

Answer: $2.9xy^3z$ is a monomial because it has a single term.

b. $-7a^2 + b$

Answer: $-7a^2 + b$ is a binomial because it contains two terms.

c. $9t^2 + 3.4t - 8.2$

Answer: $9t^2 + 3.4t - 8.2$ is a trinomial because it contains three terms.

d. $y^3 + 6y^2 - y + 15$

Answer: None of these, because $y^3 + 6y^2 - y + 15$ has four terms. Therefore, it has no special name.

YOUR TURN Determine whether the expression is a monomial, binomial, trinomial, or none of these.

a. $b + 2$ **b.** $7r^3 - 9r^2 + r - 6$ **c.** $5.7x^4y$ **d.** $x^2 - 7xy - y^2$

ANSWERS

a. binomial
b. none of these
c. monomial
d. trinomial

Previously, we defined the degree of a monomial to be the sum of the exponents of the variables. We can also talk about the **degree of a polynomial**.

> **DEFINITION**
>
> **Degree of a polynomial:** The greatest degree of any of the terms in the polynomial.

EXAMPLE **5** Identify the degree of each polynomial.

a. $x^3 + 12x^7 - 10x^2 + 9x - 14$

Answer: The degree is 7 because it is the greatest degree of all the terms.

b. $12x^6 - 3x^5y^3 - xy^3 + 4y^2 - 16$

Answer: To determine the degree of the monomial $-3x^5y^3$, we add the exponents of its variables, so its degree is 8. Likewise, we add the exponents of the variables in $-xy^3$ and find that its degree is 4. Comparing these degrees with the degrees of the other terms, we see that 8 is the greatest degree. Therefore, 8 is the degree of the polynomial.

YOUR TURN Identify the degree of each polynomial.

a. $y^7 - 3y^2 - 8y^9 + 4y + 5$　　　　**b.** $2x^5 - 9x^6y + x^3y^3 - 15y - 7$

OBJECTIVE **5** Evaluate polynomials. In Chapter 1, we learned that we can evaluate an expression. Let's now evaluate a polynomial expression. To evaluate an expression, we replace the variable(s) with given values and calculate a numerical result.

EXAMPLE **6** Evaluate $x^3 - 5x^2 + 7$ when $x = -3$.

Solution $x^3 - 5x^2 + 7$

$(-3)^3 - 5(-3)^2 + 7$　　**Replace each x with -3.**

$= -27 - 5(9) + 7$

$= -27 - 45 + 7$

$= -65$

EXAMPLE **7** If we neglect air resistance, the polynomial $-16t^2 + h_0$ describes the height of a falling object after falling from an initial height h_0 for t seconds. It is said that Galileo Galilei dropped cannonballs from the leaning tower of Pisa in order to study gravitational effects. If the tower is 180 feet tall, how high would a cannonball be after 3 seconds?

ANSWERS

a. 9　**b.** 7

Solution Evaluate $-16t^2 + h_0$, where $t = 3$ and $h_0 = 180$.

$$-16(3)^2 + 180 \qquad \text{Replace } t \text{ with 3 and } h_0 \text{ with 180.}$$

$$= -16(9) + 180$$

$$= -144 + 180$$

$$= 36 \text{ feet}$$

YOUR TURN

Use the polynomial in Example 7 to find the height of an object after it falls for 4 seconds from an initial height of 300 feet.

OBJECTIVE 6 **Write polynomials in descending order of degree.** Now, let's consider some ways that we can rewrite a polynomial. Recall from Chapter 1 that rewriting an expression means we write an equivalent expression by applying properties of arithmetic. For example, we could apply the commutative property of addition and rearrange the terms in $-8 + 4x^3 - 7x^2 + 2x$ to get an equivalent expression, $4x^3 - 7x^2 + 2x - 8$. Notice the signs move with the terms. Also, notice that the terms in $4x^3 - 7x^2 + 2x - 8$ are written in order from the greatest degree to the least degree.

A polynomial written in this way is said to be in *descending order of degree*.

PROCEDURE **WRITING A POLYNOMIAL IN DESCENDING ORDER**

To write a polynomial in descending order, place the highest degree term first, then the next highest degree, and so on.

CONNECTION

Because $4x^3 - 7x^2 + 2x - 8$ is a rearrangement of $-8 + 4x^3 - 7x^2 + 2x$, equating them forms the identity: $4x^3 - 7x^2 + 2x - 8 = -8 + 4x^3 - 7x^2 + 2x$. This means that we could evaluate both expressions using any real number and they will both give the same result.

Writing a polynomial in more than one variable in descending order can be complicated, so we will only write polynomials in one variable in descending order.

ANSWER

44 feet

EXAMPLE 8 Write each polynomial in descending order.

a. $9x^2 + 7x^4 - 2x^3 - 5 + 3x$

Solution Rearrange the terms so that the highest degree term is first, then the next highest degree, and so on.

$$7x^4 \qquad -2x^3 \qquad +9x^2 \qquad +3x \qquad -5$$

Degree 4   Degree 3   Degree 2   Degree 1   Degree 0

Answer: $7x^4 - 2x^3 + 9x^2 + 3x - 5$

b. $-9y^3 + 2y^6 + y - 3y^4 + 12 - 6y^2$

Answer: $2y^6 - 3y^4 - 9y^3 - 6y^2 + y + 12$

YOUR TURN Write each polynomial in descending order.

a. $2n^3 + n^5 - 9n^2 + 8 - 3n^4 + 4n$ **b.** $-4t^2 + 2t + 16 + 5t^3 - 13t^6$

OBJECTIVE 7 Combine like terms. Another way we can rewrite a polynomial expression is by combining like terms. We learned in Section 1.7 that combining like terms is one way to rewrite an expression in simplest form. We also learned that like terms have the same variables raised to the same powers.

EXAMPLE 9 Combine like terms and write the resulting polynomial in descending order.

$$12x^3 + 5x^2 - 2x^3 - 13 + 6x + x^2 + 2 - 6x$$

Solution $12x^3 + 5x^2 - 2x^3 - 13 + 6x + x^2 + 2 - 6x$

$= 12x^3 - 2x^3 + 5x^2 + x^2 + 6x - 6x - 13 + 2$ Collect like terms as needed.

$= \quad 10x^3 \quad + \quad 6x^2 \quad + \quad 0 \quad - 11$ Combine like terms.

$= 10x^3 + 6x^2 - 11$ Drop 0 and bring the terms together.

Warning: When combining like terms, it may be tempting to add the exponents. But we *add only the coefficients*, keeping the exponents the same.

Some people prefer to skip the collecting step and strike through like terms as they are combined to keep track of the terms that have been combined. It is helpful to combine the terms in descending order of degree so that the answer will be in descending order.

Alternative Solution $12x^3 + 5x^2 - 2x^3 - 13 + 6x + x^2 + 2 - 6x$

$= 10x^3 + 6x^2 + 0 - 11$

$= 10x^3 + 6x^2 - 11$

Note: We consider the terms in the following order.
Degree 3 terms: $12x^3 - 2x^3 = 10x^3$
Degree 2 terms: $5x^2 + x^2 = 6x^2$
Degree 1 terms: $6x - 6x = 0$
Degree 0 terms: $-13 + 2 = -11$

Combine like terms and write the resulting polynomial in descending order.

a. $10x^3 - 4x^2 - 3x^3 + 8 - 5x^2 - 9 + 6x$ **b.** $3y^4 - 12y - y^2 + y - 6 + y^2 + 5y^4$

ANSWERS

a. $7x^3 - 9x^2 + 6x - 1$
b. $8y^4 - 11y - 6$

Remember that polynomials may have multiple variables. Multivariable monomials are like terms if they have the same variables raised to the same exponents.

EXAMPLE 10 Combine like terms.

$$x^4 + 8x^2y - 7 - 13x^2y + 4y - 2xy^2 + 6y - x^4 + 2$$

Solution Collecting the like terms and then combining them gives us:

$$x^4 + 8x^2y - 7 - 13x^2y + 4y - 2xy^2 + 6y - x^4 + 2$$
$$= x^4 - x^4 + 8x^2y - 13x^2y - 2xy^2 + 4y + 6y + 2 - 7$$
$$= \quad 0 \quad\quad - 5x^2y \quad\quad - 2xy^2 \quad + 10y \quad\quad - 5$$
$$= -5x^2y - 2xy^2 + 10y - 5$$

Learning Strategy

If you are a tactile learner, collecting the like terms first will probably be more comfortable for you.

If you are a visual learner, striking through the terms as you combine them will probably be more comfortable.

Instead of first collecting the like terms, we could strike through like terms as they are combined:

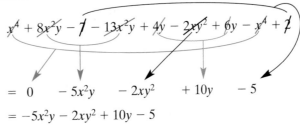

$$= \quad 0 \quad\quad - 5x^2y \quad\quad - 2xy^2 \quad + 10y \quad\quad - 5$$
$$= -5x^2y - 2xy^2 + 10y - 5$$

YOUR TURN **Combine like terms.**

a. $-x^5 + 6x^3y^2 + 2y - 9x^3y^2 + 3x^5 - 4 - 2y$
b. $-6t^3u + 12tu - 4t^2u + 9t^3u - u + 3t^2u - 12tu - u$

ANSWERS

a. $2x^5 - 3x^3y^2 - 4$
b. $3t^3u - t^2u - 2u$

5.2 EXERCISES

For Extra Help

InterAct Math

Addison-Wesley
Math Tutor Center

Video 8/DVT 3

MyMathLab

MathXL

1. How do we determine the degree of a monomial?

2. Explain the differences between a monomial, a binomial, and a trinomial.

3. How do we determine the degree of a polynomial?

4. What parts are alike in monomials that are like terms?

5. Given a polynomial in one variable, what does *descending order of degree* mean?

6. Explain how to combine like terms.

For Exercises 7–18, determine whether the expression is a monomial. Explain.

7. $-3x$ **8.** $5y^3$ **9.** $\dfrac{1}{5}$ **10.** $\dfrac{4}{7}y$ **11.** $\dfrac{6m}{4y^3}$ **12.** $\dfrac{-8t}{x}$

13. $4x^2 - 3x + 7$ **14.** y **15.** 0.76 **16.** $6u^3v^4w$ **17.** $3m^4n^2$ **18.** $p^2 - q^2$

For Exercises 19–30, identify the coefficient and degree of each monomial.

19. $-5m^3$ **20.** $8p^2$ **21.** $-xy^4$ **22.** $-m$ **23.** -9 **24.** 18

25. $4.2n^3p$ **26.** $-6.7uv^7$ **27.** $16abc$ **28.** $-8.1lkm$ **29.** y **30.** $\dfrac{2}{3}w$

For Exercises 31–42, determine whether the expression is a monomial, binomial, trinomial, or none of these.

31. $6x + 8y$ **32.** $17.3x^2 - 3x + 2.1$ **33.** $-7m^2n$ **34.** 18

35. $5.2x^3 - 3x^2 + 4.1x - 11$ **36.** $x^2 - 4x + 4$ **37.** $5u^3 - 16u$ **38.** $7k + 5k^3 - 19$

39. -21 **40.** $abc + a^2b^2c^2 - a^3b^3c^3$ **41.** $25 - x^2$ **42.** $14j^5 + 2j^4 - 21j^3 - 5j + 18$

For Exercises 43–48, identify the degree of each polynomial.

43. $19 - 7y^4 + 3y - 2y^3 - 7$ **44.** $6a^5 - 19a^3 + a^9 + 5a - 14$ **45.** $16z^4 + 7z^2 - z^5 - 4z^3 + z$

46. $11 + 4t^5 - 5t + 7t^3 - 18t^2 + t^8$ **47.** $2u^2 + 5u^3 - u^7 + 13u^4 - 3u$ **48.** $22j^3 + 5j^2 - 16j^4 + 21 - 14j$

For Exercises 49–54, evaluate the polynomial using the given values.

49. $-3xy^2$; $x = -5$, $y = 2$

50. $-2x^2y$; $x = -1$, $y = 4$

51. $x^2 - 6x + 1$; $x = 3$

52. $n^2 - 8n - 3$; $n = -4$

53. $a^3 + 0.5ab + 2.4b$; $a = -1$, $b = -2$

54. $m^3 - 0.2n^2$; $m = -3$, $n = 4$

55. If we neglect air resistance, the polynomial $-16t^2 + h_0$ describes the height of a falling object after falling from an initial height h_0 for t seconds. A marble is dropped from a tower at a height of 50 ft.

   a. What is its height after 0.5 seconds?

   b. What is its height after 1.2 seconds?

56. The polynomial $2lw + 2lh + 2wh$ describes the surface area of a box.

   a. Find the surface area of a box with a length of 8 inches, width of 6.5 inches, and height of 4 inches.

   b. Find the surface area of a box with a length of 4 inches, width of 3 inches, and height of 2.5 inches.

57. The polynomial $5p^2 - 6p + 3$ describes the number of units sold for every 1000 people in a particular region based on the price of the product, which is represented by p.

   a. Find the number of units sold if the price of each unit is $3.00.

   b. Find the number of units sold if the price of each unit is $3.50.

58. The polynomial $7r^2 - 2r + 6$ describes the voltage in a circuit where r represents the resistance in the circuit.

   a. Find the voltage if the resistance is 6 ohms.

   b. Find the voltage if the resistance is 8 ohms.

59. An engineer is designing a chemical-storage tank that is capsule shaped. The polynomial $\frac{4}{3}\pi r^3 + \pi r^2 h$ describes the volume of the tank.

   a. Find the volume of the tank if the radius is 4 feet and the height is 20 feet. Round the result to the nearest tenth.

   b. Find the volume of the tank if the radius is 3 feet and the height is 15 feet. Round the result to the nearest tenth.

60. The polynomial $lwh - \pi r^2 h$ describes the volume of metal remaining in a block after a cylinder has been bored into the block of metal.

   a. Find the volume of metal remaining if the length is 15 inches, the width is 7 inches, the height is 4 inches, and the radius of the cylinder is 3 inches. Round to the nearest tenth.

   b. Find the volume of metal remaining if the length is 15 inches, the width is 7 inches, the height is 4 inches, and the radius of the cylinder is 2 inches. Round to the nearest tenth.

For Exercises 61–66, write each polynomial in descending order.

61. $7x^4 + 5x^7 - 8x + 14 - 3x^5$

62. $-4y^3 + 7y - 8y^5 - 6y^2 + 9 + 2y^4$

63. $4r - 3r^2 + 18r^5 + 7r^3 - 8r^6$

64. $u^4 + 7u^2 - u^9 + 15u + 27 - 3u^5 - 8u^3$

65. $20 - w^3 + 5w + 11w^2 - 12w^4$

66. $a - 6 + 7a^5 + 3a^2 - 4a^3$

For Exercises 67–76, combine like terms and write the resulting polynomial in descending order of degree.

67. $2x^2 + 5x - 7x^2 + 8 - 5x$

68. $3m^5 + 7m^2 - 8m + 9m^3 - 7m - m^2$

69. $15 - 4y + 8y^2 - 4y - 3y^2 + 7 - y$

70. $6l - 5l^3 - 2l^4 + 7l^4 + 5l - l^3$

71. $9k - 5k^2 + 6k^3 + 2k^2 + k^4 - 3k$

72. $11p^5 - p^2 + 12p^3 + 5p^2 - 7p^3 + 20 - 4p$

73. $7a^2 - 5a^4 + 6a^3 - 5a^4 + 12 - 6a^3 + a^2$

74. $6c^9 + 7c^3 - 4c^5 + 8c - 14c^2 + 13c^5 - 7 - c^9$

75. $19v^2 - 20 - 14v^5 - 5v^3 + 16v^2 - 20v^5$

76. $b^3 - 13b^2 + b^4 + 6 - 2b^3 - 15b^4 - 17b^3 + 8$

For Exercises 77–86, combine like terms.

77. $2x^2 + 3xy - 4x^2 + 6xy - 5x^2 - 2x^2$

78. $3a - 5b + 2 - 4a + 7 - 5b + 6b - 3$

79. $6a^2b^2 + 3a^2b^3 + 3a^2b^2 - 4a^2b^2 - a^2b^3$

80. $x^4 + xy^3 - 6xy + xy^3 + 3y^4 - 4y^2 + 3xy - 8$

81. $-3w^2z - 9 + 6wz^2 + 3w^2 - w^2z + 8 + 9w^2$

82. $-m^3n - 8mn + 4 - 5m^2n^2 + 3m^3n - 7 + mn - 16m^2n^2$

83. $18k^4 + 3kl - 5k^2l^2 - 3k^3l + 14 + 9k^2l^2 - 9 - 8l^4$

84. $5u^3v - 4uv^2 + 6u^2v + 5uv^2 - 8 + u^3v - 7uv - 2u^2v + 5u^3v^2 + 16 - 5uv$

85. $-15xyz + 4x^2yz^3 - 4xy^2z + 18xyz + x^2yz^3 - 21 - 4x^2yz^3 + 15xyz$

86. $5a^2b^2 + 15a^2 - 20b^2 - 3abc^2 + 12 - 4ab + b^2 - 2a^2b^2 + 5ab - 2b^2 + 18abc^2$

PUZZLE PROBLEM

Rearrange the numbers in the figure shown so that no two consecutive numbers are next to each other horizontally, vertically, or diagonally.

REVIEW EXERCISES

1. Evaluate: -7^0

2. The Andromeda galaxy (the closest one to our Milky Way galaxy) contains at least 200,000,000,000 stars. Write this in scientific notation.

3. Solve and check: $\frac{1}{4}x = 3$

4. How many liters of a 40% solution must be added to 200 liters of a 20% solution to obtain a 35% solution?

5. Write 14.2% as a fraction in simplest form.

6. Simplify: $(-2)^3 - 9\,|12 - 14 \div (-2)|$

5.3 Adding and Subtracting Polynomials

OBJECTIVES

1 Add polynomials.

2 Subtract polynomials.

OBJECTIVE 1 Add polynomials.

Understanding Polynomial Addition

We can add and subtract polynomials in the same way that we add and subtract numbers. In fact, polynomials are like whole numbers that are in an expanded form. Consider the polynomial $4x^2 + 3x + 6$. If we replace the x's with the number 10, then we have the expanded form for the number 436.

$$4x^2 + \quad 3x \quad + 6$$
$$\downarrow \qquad\qquad \downarrow$$
$$4 \cdot 10^2 \ + \ 3 \cdot 10 \ + 6 \qquad \text{Replacing } x \text{ with 10.}$$
$$= \quad 400 \quad + \quad 30 \quad + 6$$
$$= \quad 436$$

In our base-ten number system each place value is a power of 10. We can think of polynomials as a variable-base number system. In other words, the place values are variables, where x^2 is like the hundreds place (10^2), and x is like the tens place (10^1). To add whole numbers, we add the digits in like place values. Polynomials are added in a similar way. However, instead of adding digits in like place values, we add like terms. Consider the following comparison:

Numeric addition:

$436 + 251 = 687$

Polynomial addition:

$(4x^2 + 3x + 6) + (2x^2 + 5x + 1) = 6x^2 + 8x + 7$

$$\begin{array}{r} 436 \\ + \ 251 \\ \hline 687 \end{array}$$
In numeric addition, like place values are added.

$$\begin{array}{r} 4x^2 + 3x + 6 \\ + \ 2x^2 + 5x + 1 \\ \hline 6x^2 + 8x + 7 \end{array}$$
In polynomial addition, like terms are added.

From this simple case, we see that polynomials are added by combining the like terms.

PROCEDURE ADDING POLYNOMIALS

To add polynomials, combine like terms.

Adding Polynomials

Although we stacked the polynomials above, we do not actually have to stack the polynomials in order to add them.

EXAMPLE 1 Add $(2x^3 + 5x^2 + 4x + 1) + (6x^3 + 3x + 8)$.

Solution Combine like terms. Notice combining in order of degree places the resulting polynomial in descending order of degree.

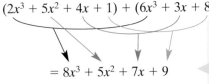

$(2x^3 + 5x^2 + 4x + 1) + (6x^3 + 3x + 8)$

$= 8x^3 + 5x^2 + 7x + 9$

Note: Our thinking flows in this order:
Degree 3 terms: $2x^3 + 6x^3 = 8x^3$
Degree 2 terms: $5x^2$ has no like term, so it is copied into the answer.
Degree 1 terms: $4x + 3x = 7x$
Degree 0 terms: $1 + 8 = 9$

If we had stacked the polynomials in Example 1, we would have had to leave a blank space under the $5x^2$ because there was no x^2 term in the second polynomial. It is the same as having a 0 in that place.

$$\begin{aligned} 2x^3 + 5x^2 + 4x + 1 \\ + 6x^3 + 0 + 3x + 8 \\ \hline 8x^3 + 5x^2 + 7x + 9 \end{aligned}$$

As the polynomials get more complex, these blanks become more prevalent. This is why the stacking method is not a preferred method for adding polynomials.

EXAMPLE 2 Add $(y^5 + 9y^3 - 3y - 7) + (2y^5 - 4y^3 + 3y + 5)$.

Solution $(y^5 + 9y^3 - 3y - 7) + (2y^5 - 4y^3 + 3y + 5)$ Combine like terms.

$= 3y^5 + 5y^3 + 0 - 2$

$= 3y^5 + 5y^3 - 2$

YOUR TURN   **Add.**

   a. $(x^4 + 3x^3 + 4x + 2) + (5x^3 + 2x + 7)$

   b. $(14n^5 - 6n^3 - 12n + 15) + (2n^5 - 4n^3 + 8n - 5)$

Sometimes the polynomials may contain terms that have several variables or coefficients that are fractions or decimals.

EXAMPLE 3 Add $\left(7t^4 + 2.9tu^2 - \dfrac{3}{4}tu + 11\right) + \left(-8t^4 - 7tu^2 + \dfrac{1}{6}tu - 5\right)$.

Solution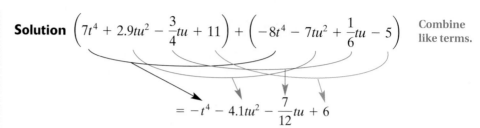

$$\left(7t^4 + 2.9tu^2 - \frac{3}{4}tu + 11\right) + \left(-8t^4 - 7tu^2 + \frac{1}{6}tu - 5\right)$$ Combine like terms.

$$= -t^4 - 4.1tu^2 - \frac{7}{12}tu + 6$$

Note: Our thinking flows like this:

The t^4 terms: $7t^4 + (-8t^4) = -1t^4 = -t^4$

The tu^2 terms: $2.9tu^2 + (-7tu^2) = -4.1tu^2$

The tu terms: $-\frac{3}{4}tu + \frac{1}{6}tu = -\frac{9}{12}tu + \frac{2}{12}tu = -\frac{7}{12}tu$

The constant terms: $11 + (-5) = 6$

YOUR TURN   Add.

a. $(12x^3 - 6.2xy^2 - 0.3xy - y^2) + (x^3 + 7xy^2 - 0.9xy + 0.4y^2)$

b. $\left(a^4 - \frac{4}{5}ab^3 - \frac{2}{3}ab + 9\right) + \left(a^4 + \frac{1}{4}ab^3 + ab - 5\right)$

EXAMPLE 4   Write an expression in simplest form for the perimeter of the rectangle shown.

$3x + 7$ [rectangle] $5x - 1$

Understand   Perimeter means the total distance around the shape. Therefore, we need to add the lengths of all the sides of the shape.

Plan   In this case, the lengths of the sides are represented by polynomials. Therefore, we add the polynomials to represent the perimeter.

Execute   Perimeter   =   length   +   width   +   length   +   width
= $(5x - 1)$ + $(3x + 7)$ + $(5x - 1)$ + $(3x + 7)$
= $5x + 3x + 5x + 3x - 1 + 7 - 1 + 7$
= $16x + 12$

Answer   The expression for the perimeter is $16x + 12$.

Check   To check, we (1) choose a value for x and evaluate the original expressions for length and width, (2) determine the corresponding numeric perimeter, and (3) evaluate the perimeter expression using the same value for x and verify that we get the same numeric perimeter. Let's choose $x = 2$.

Length: $5x - 1$ Width: $3x + 7$ Perimeter:
$5(2) - 1$ $3(2) + 7$ The perimeter of the rectangle with a
$= 10 - 1$ $= 6 + 7$ length of 9 and width of 13 is:
$= 9$ $= 13$ $9 + 13 + 9 + 13 = 44$

Now evaluate the perimeter expression where $x = 2$ and we should find that the result is 44.

$$\text{Perimeter expression:} \quad 16x + 12$$
$$16(2) + 12$$
$$= 32 + 12$$
$$= 44 \quad \text{This agrees with our}$$
$$\text{calculation on page 410.}$$

YOUR TURN Write an expression in simplest form for the perimeter of the following shape.

OBJECTIVE **2** Subtract polynomials.

Understanding Polynomial Subtraction

Subtracting polynomials is similar to subtracting signed numbers. When we subtract signed numbers it is often simpler to write the subtraction statement as an equivalent addition statement. Consider the following comparison of numeric subtraction and polynomial subtraction.

Numeric subtraction:

$975 - 621$

We can stack by place:

$$\begin{array}{r} 975 \\ -\ 621 \\ \hline 354 \end{array}$$

Polynomial subtraction:

$(9x^2 + 7x + 5) - (6x^2 + 2x + 1)$

We can stack like terms:

$$\begin{array}{r} 9x^2 + 7x + 5 \\ -\ (6x^2 + 2x + 1) \\ \hline 3x^2 + 5x + 4 \end{array}$$

In numeric subtraction, we subtract digits in the same place value, whereas in polynomial subtraction, we subtract like terms. Note that subtracting 1 from 5 is equivalent to combining 5 and -1, subtracting $2x$ from $7x$ is equivalent to combining $7x$ and $-2x$, and subtracting $6x^2$ from $9x^2$ is equivalent to combining $9x^2$ and $-6x^2$. This suggests that we can write polynomial subtraction as equivalent polynomial addition by changing the sign of each term in the second polynomial (the subtrahend) like this:

$$(9x^2 + 7x + 5) - (6x^2 + 2x + 1)$$
$$= (9x^2 + 7x + 5) + (-6x^2 - 2x - 1)$$
$$= 3x^2 + 5x + 4$$

Note: We are allowed to change the signs because of the distributive property. If we disregard the initial polynomial, we have:

$$-(6x^2 + 2x + 1) = -1(6x^2 + 2x + 1)$$

The distributive property tells us we can distribute the -1 (or minus sign) to each term inside the parentheses.

$$= -1 \cdot 6x^2 - 1 \cdot 2x - 1 \cdot 1$$
$$= -6x^2 - 2x - 1$$

ANSWER

$6x + 3$

Subtracting Polynomials

Our exploration suggests the following procedure for subtracting polynomials.

PROCEDURE SUBTRACTING POLYNOMIALS

To subtract polynomials:

1. Write the subtraction statement as an equivalent addition statement.
   a. Change the operation symbol from a minus sign to a plus sign.
   b. Change the subtrahend (second polynomial) to its additive inverse. To get the additive inverse we change the sign of each term in the polynomial.
2. Combine like terms.

CONNECTION ──────

The procedure for subtracting polynomials is the same as the procedure for subtracting signed numbers (Section 1.3). In both cases, we write subtraction as equivalent addition by changing the subtrahend to its additive inverse.

EXAMPLE 5 Subtract $(7x^3 + 8x^2 + 6x + 9) - (3x^3 + 6x^2 + 5x + 1)$.

Solution Write an equivalent addition statement, then combine like terms.

$$(7x^3 + 8x^2 + 6x + 9) - (\ 3x^3 + 6x^2 + 5x + 1)$$

Change the minus sign to plus sign   Change all signs in the subtrahend.

$$= (7x^3 + 8x^2 + 6x + 9) + (-3x^3 - 6x^2 - 5x - 1)$$
$$= 4x^3 + 2x^2 + x + 8 \qquad \text{Combine like terms.}$$

CONNECTION ──────────────────────

The preceding polynomial subtraction is equivalent to the following numeric subtraction:

$$\begin{array}{r} 7869 \\ -\ 3651 \\ \hline 4218 \end{array}$$

Notice the numeric result 4218 corresponds to the polynomial result $4x^3 + 2x^2 + x + 8$. We could use the stacking method for subtracting polynomials; however, as we saw with adding polynomials, it is not the best method for all cases because we often have missing terms or terms that are not like terms.

You may have noticed that the polynomials in Example 5 contained only plus signs. It is important to recognize that the polynomials may contain a mixture of signs and multiple variables. This is shown in Examples 6 and 7.

EXAMPLE 6 Subtract $(15y^3 + 3y^2 + y - 2) - (7y^3 - 6y^2 + 5y - 8)$.

Solution Write an equivalent addition statement, then combine like terms.

$$(15y^3 + 3y^2 + y - 2) - (\ 7y^3 - 6y^2 + 5y - 8)$$

Change the minus sign to plus sign.   Change all signs in the subtrahend.

$$= (15y^3 + 3y^2 + y - 2) + (-7y^3 + 6y^2 - 5y + 8)$$
$$= 8y^3 + 9y^2 - 4y + 6 \qquad \text{Combine like terms.}$$

EXAMPLE 7 Subtract $(9.7x^5 - 2x^2y + xy^2 - 14.6xy - 7y^2) - (x^5 - 5.8xy^2 + 10.3xy - 15y^2)$.

Solution We write an equivalent addition statement, then combine like terms.

$$(9.7x^5 - 2x^2y + xy^2 - 14.6xy - 7y^2) - (\;x^5 - 5.8xy^2 + 10.3xy - 15y^2)$$

Change the minus sign to plus sign.   Change all signs in the subtrahend.

$$= (9.7x^5 - 2x^2y + xy^2 - 14.6xy - 7y^2) + (-x^5 + 5.8xy^2 - 10.3xy + 15y^2)$$

$$= 8.7x^5 - 2x^2y + 6.8xy^2 - 24.9xy + 8y^2$$

Note: The $-2x^2y$ term had no like term, so we rewrote it in the final expression.

ANSWERS

a. $6t^4 + 8t + 3$

b. $-0.6x^4 - 24x^3 + 8x^2 + 5$

c. $5a^5 - 2a^3b^2 + 9a^2b^2 - ab^2 - 5b - 3$

YOUR TURN Subtract.

a. $(8t^4 + 5t^2 + 9t + 7) - (2t^4 + 5t^2 + t + 4)$

b. $(12.5x^4 - 15x^3 + 2x^2 - 9) - (13.1x^4 + 9x^3 - 6x^2 - 14)$

c. $(7a^5 - a^3b^2 + 9a^2b^2 - 2ab^2 + b - 18) - (2a^5 + a^3b^2 - ab^2 + 6b - 15)$

5.3 EXERCISES

For Extra Help

InterAct Math

Tutor Center — Addison-Wesley Math Tutor Center

Video 8/DVT 3

MyMathLab

MathXL

1. How do we add two polynomials?

2. What is the function of parentheses in the addition problem $(3x + 9) + (4x - 2)$?

3. How do we find the additive inverse of a polynomial?

4. How do we subtract two polynomials?

For Exercises 5–24, add and write the resulting polynomial in descending order.

5. $(3x + 2) + (5x - 1)$

6. $(5y + 4) + (3y + 1)$

7. $(8y + 7) + (2y + 5)$

8. $(3m - 4) + (5m + 1)$

9. $(2x + 3y) + (5x - 3y)$

10. $(4a - 3b) + (6a - 2b)$

11. $(2x + 5) + (3x^2 - 4x + 7)$

12. $(5p^2 + 3p - 1) + (4p + 8)$

13. $(z^2 - 3z + 7) + (5z^2 - 8z - 9)$

14. $(2w^2 - 5w - 1) + (4w^2 - 5w + 1)$

15. $(3r^2 - 2r + 10) + (2r^2 - 5r - 11)$

16. $(7m^2 + 8m - 1) + (-4m^2 - 3m - 2)$

17. $(4y^2 - 8y + 1) + (5y^2 + 3y + 2)$

18. $(3k^3 - 5k^2 - k - 1) + (4k^3 + 4k^2 + k + 7)$

19. $(9a^3 + 5a^2 - 3a - 1) + (-4a^3 - 2a^2 + 6a - 2)$

20. $(4u^3 - 6u^2 + u + 11) + (-5u^3 - 3u^2 + u - 5)$

21. $(7p^3 - 9p^2 + 5p - 1) + (-4p^3 + 8p^2 + 2p + 10)$

22. $(12r^4 - 5r^2 + 8r - 15) + (7r^4 + 3r^3 + 2r - 9)$

23. $(-5w^4 - 3w^3 - 8w^2 + w - 14) + (-3w^4 + 6w^3 + w^2 + 12w + 5)$

24. $(-r^4 + 3r^2 - 12r - 14) + (5r^4 - 2r^3 - 3r^2 + 10r - 5)$

For Exercises 25–30, add.

25. $(8x^5 - x^3 + 11x^2 - 12x + 1) + (-2x^4 + x^3 - 14x^2 + 5)$

26. $(x^2y^3 + 8xy^2 - 12x^2y - 4xy + 7y^2 - 9) + (-3x^2y^2 + 2xy^2 + 4x^2y - xy + 6y^2 + 4)$

27. $(8m^3n^4 - 6m^2n^2 - 13mn^2 + 14m^2n - 12n + 6) + (-4n^4m + 10m^2n^2 + mn^2 - 19n + 3)$

28. $(-13a^6 + a^3b^2 + 3ab^3 - 8b^3) + (11a^6 - 3a^2b^3 - 4ab^3 + 10ab^2 + 4b^3)$

29. $(-2u^4 + 6uv^3 - 8u^2v^3 + 7u^3v - u + v^2 + 9) + (-5u^4 - 6uv^3 + 9u^2v^3 + 8u - 14v^2 - 12)$

30. $(14abc - 2a^2b^2 - ab^2c + 4b - 9c) + (-15abc + 5a^2b^2 - 13ab^2c + 7a - 12b + 6)$

For Exercises 31–34, write an expression for the perimeter in simplest form.

31.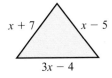

$x + 7$ $x - 5$

$3x - 4$

32.

$5x - 7$

$4x + 1$

33.

$a - 8$

$6a + 2$

34.

$m - 2$ $4m + 5$

$3m + 1$

For Exercises 35–52, subtract and write the resulting polynomial in descending order.

35. $(6x + 3) - (2x + 1)$

36. $(5m + 8) - (2m + 3)$

37. $(18a^2 + 3) - (5a^2 + 4)$

38. $(12b^2 + 4b) - (5b^2 + b)$

39. $(2x^2 - 3x + 4) - (6x + 2)$

40. $(7y^2 - 3y + 2) - (8y + 5)$

41. $(6z^2 - 3z + 1) - (4z^2 + 4z - 8)$

42. $(5y^2 - 3y - 10) - (5y^2 + 2y - 10)$

43. $(8a^2 + 11a - 12) - (-4a^2 + a + 3)$

44. $(y^2 + 3y + 6) - (-5y^2 + 3y - 8)$

45. $(6u^3 + 3u^2 - 8u + 5) - (7u^3 + 5u^2 - 2u - 3)$

46. $(-8t^3 + 3t^2 - 7t - 9) - (-10t^3 - 5t^2 + 3t - 11)$

47. $(-w^5 + 3w^4 + 8w^2 - w - 1) - (10w^5 + 3w^2 - w - 3)$

48. $(-s^4 + 7s^3 + 6s^2 - 13s - 11) - (10s^5 + 3s^4 - 3s^3 - 2s^2 - s + 5)$

49. $(-5p^4 + 8p^3 - 9p^2 + 3p + 1) - (6p^4 + 5p^3 + 4p^2 - 2p + 7)$

50. $(6r^5 - 9r^4 - 2r^2 + 8) - (5r^5 + 2r^4 + 7r^3 - 8r^2 - r - 9)$

51. $(16v^4 - 29v^3 + 21v^2 + 4v - 8) - (5v^4 - 3v^3 - 2v^2 + 9v + 1)$

52. $(12q^4 + 5q^3 - 2q + 15) - (18q^5 - 14q^4 - 3q^2 + 15q + 8)$

For Exercises 53–58, subtract.

53. $(x^2 + xy + y^2) - (x^2 - xy + y^2)$

54. $(a^2 - 4ab + b^2) - (a^2 + 4ab + b^2)$

55. $(4a^2 + 5ab - 6b^2) - (10ab - 6a^2 - 8b^2)$

56. $(4ab - 5bc + 6ac) - (10ab - bc - ac)$

57. $(-r^5 + 2r^3s^2 + 3r^2s^3 - 6rs + 10) - (4r^5 - 3r^3 + s^2 - 18rs - 5)$

58. $(15a^3y^4 + a^2y^2 + 5ay^3 - 7ay + 12y^2 - 11) - (-a^3y^4 + a^2y^2 + 2ay^3 - ay + 9)$

For Exercises 59–62, solve. (Hint: Profit = Revenue − Cost)

59. The polynomial $54.95x + 92.20y + 25.35z$ describes the revenue a company generates from the sale of three different portable CD players. The expression $22.85x + 56.75y + 19.34z$ describes the cost of producing each of the CD players. Write a polynomial in simplest form that describes the company's net profit.

60. The polynomial $27.50x + 14.70y + 42.38z$ describes the revenue a company generates from the sale of three printer cartridges. The expression $12.75x + 5.25y + 27.42z$ describes the cost of producing each of the cartridges. Write a polynomial in simplest form that describes the company's net profit.

61. The polynomial $6.50a + 3.38b + 25.00c$ describes the revenue a department store generates from the sale of three different candles. The expression $2.28a + 1.75b + 12.87c$ describes the cost the store pays to sell each of the candles. Write a polynomial in simplest form that describes the store's net profit.

62. The polynomial $10.55m + 14.75n + 27.50p$ describes the revenue a pet store generates from the sale of three different litter boxes. The expression $5.73m + 8.26n + 15.22p$ describes the cost the store pays to sell each of the products. Write a polynomial in simplest form that describes the store's net profit.

For Exercises 63–64, find and explain the mistake, then work the problem correctly.

63. $(3x^2 - 5x + 10) - (7x^2 - 3x + 5) = (3x^2 - 5x + 10) + (-7x^2 - 3x + 5)$
$$= 3x^2 - 5x + 10 - 7x^2 - 3x + 5$$
$$= -4x^2 - 8x + 15$$

64. $(7y^2 - 3y + 1) - (y^2 + 4y - 1) = (-7y^2 + 3y - 1) + (-y^2 - 4y + 1)$
$$= -7y^2 + 3y - 1 - y^2 - 4y + 1$$
$$= -8y^2 - y$$

Collaborative Exercises

Building Furniture and Profits

Suppose your group operates a furniture manufacturing company that produces two different sizes of coffee table. The larger coffee table sells for $350 each and the smaller one for $280 each.

1. Write a monomial expression for the revenue if x number of large tables are sold per month.

2. Write a monomial expression for the revenue if y number of small tables are sold per month.

3. Write a polynomial expression that describes the total revenue generated from the sale of the two tables per month.

Suppose that it costs $225 to produce each large table and $135 to produce each small table.

4. Write a monomial expression for the cost of producing x number of large tables per month.

5. Write a monomial expression for the cost of producing y number of small tables per month.

6. Your company also spends $18,500 per month to pay for the lease, utilities, and salaries. Write a polynomial expression that describes the total cost involved in production.

7. Using your polynomials for revenue and cost, write a polynomial in simplest form for the profit if x number of large tables are produced and sold and y number of small tables are produced and sold.

continued

8. What does each coefficient in the polynomial for profit indicate?

9. Suppose in one month, 200 large tables are produced and sold and 400 small tables are produced and sold. Find the revenue, cost, and profit for the month.

10. Suppose in one month, 56 large tables are produced and sold and 65 small tables are produced and sold. Find the revenue, cost, and profit for the month. Explain the meaning of your answer.

11. Suppose in one month, only large tables are produced and sold. How many would have to be produced and sold to break even (profit is 0)?

12. Suppose only small tables are produced and sold in one month. How many would have to be produced and sold to break even?

REVIEW EXERCISES

For Exercises 1–4, evaluate.

1. $2(-341)$ **2.** $(-2)^5$ **3.** 10^{-4} **4.** $(2^3)(2^4)$

5. Write 5.89×10^7 in standard form.

6. A swimming pool measures 20 meters by 15 meters. What is the surface area of the water?

5.4 Exponent Rules and Multiplying Monomials

5.4

OBJECTIVES

1. Multiply monomials.
2. Multiply numbers in scientific notation.
3. Simplify a monomial raised to a power.

In this section we will learn how to multiply monomials, multiply numbers in scientific notation, and simplify a monomial raised to a power. These simplifications require some new rules for exponents.

OBJECTIVE 1 Multiply monomials. Because monomials contain variables in exponential form, in order to multiply monomials, we must develop a rule for multiplying exponential forms.

Multiplying Exponential Forms

Consider $2^3 \cdot 2^4$, which is a product of exponential forms. To simplify $2^3 \cdot 2^4$ we could follow the order of operations and evaluate the exponential forms first, then multiply.

$$2^3 \cdot 2^4 = 8 \cdot 16 = 128$$

However, there is an alternative. We can write the result in exponential form by first writing 2^3 and 2^4 in their factored forms.

Since there are a total of seven 2's multiplied, we can express the product as 2^7.

Notice that $2^7 = 128$, which agrees with our earlier calculation. Also notice that in the alternative method the resulting exponent is the sum of the original exponents.

$$2^3 \cdot 2^4 = 2^{3+4} = 2^7$$

Conclusion To multiply exponential forms that have the same base, we can add the exponents and keep the same base.

This suggests the following rule.

RULE PRODUCT RULE FOR EXPONENTS

If a is a real number and m and n are integers, then $a^m \cdot a^n = a^{m+n}$.

We apply this rule when multiplying monomials.

> **EXAMPLE** **1** Multiply $(x^5)(x^3)$.
>
> **Solution** Because the bases are the same, we can add the exponents and keep the same base.
>
> $$(x^5)(x^3) = x^{5+3} = x^8$$

> **YOUR TURN** **Multiply.**
>
> **a.** $y^2 \cdot y^5$ **b.** $(c^7)(c^3)$ **c.** $t^5(t^3)$

Multiplying Monomials

Now let's consider multiplying monomials that have coefficients like $(4x^3)(2x^6)$. Because monomials are products by definition, we can use the commutative property of multiplication and write

$$
\begin{aligned}
(4x^3)(2x^6) &= 4 \cdot x^3 \cdot 2 \cdot x^6 \\
&= 4 \cdot 2 \cdot x^3 \cdot x^6 \\
&= 8 \cdot x^{3+6} \qquad \text{We multiply the coefficients and use the product} \\
&= 8x^9 \qquad\quad \text{rule for exponents to simplify } x^3 \cdot x^6.
\end{aligned}
$$

> **EXAMPLE** **2** Multiply $(9y^5)(3y^8)$.
>
> **Solution** Multiply the coefficients, and add the exponents of the like bases.
>
> $$
> \begin{aligned}
> (9y^5)(3y^8) &= 9 \cdot 3y^{5+8} \\
> &= 27y^{13}
> \end{aligned}
> $$

What if the monomial factors have some variables that are different, as in $(5x^2)(7xy)$? If we expand out the monomials in factored form and use the commutative property of multiplication, we have:

$$
\begin{aligned}
(5x^2)(7xy) &= 5 \cdot x^2 \cdot 7 \cdot x \cdot y \\
&= 5 \cdot 7 \cdot x^2 \cdot x \cdot y \\
&= 35x^3y
\end{aligned}
$$

Notice the unlike variable base, y, is simply written unchanged in the result. Our examples suggest the following procedure.

> **PROCEDURE** **MULTIPLYING MONOMIALS**
>
> To multiply monomials:
> 1. Multiply coefficients.
> 2. Add the exponents of the like bases.
> 3. Write any unlike variable bases unchanged in the product.

EXAMPLE 3 Multiply $\left(-\dfrac{5}{6}a^2bc^3\right)\left(\dfrac{2}{3}a^5b^4\right)$.

Solution Multiply the coefficients, add the exponents of the like bases, and write the unlike variable base c unchanged in the product.

$$\left(-\frac{5}{6}a^2bc^3\right)\left(\frac{2}{3}a^5b^4\right) = -\frac{5}{\underset{3}{\cancel{6}}}\cdot\frac{\overset{1}{\cancel{2}}}{3}a^{2+5}b^{1+4}c^3$$

Note: Remember that a variable or number with no apparent exponent has an understood exponent of 1.

$$-\frac{5}{6}a^2bc^3 = -\frac{5}{6}a^2b^1c^3$$

$$= -\frac{5}{9}a^7b^5c^3$$

YOUR TURN **Multiply.**

    a. $(9n^4)(5n^8)$ **b.** $-2.5x^2y^3\left(-6xy^2z^3\right)$

CONNECTION ————————————————————————

Multiplying distance measurements in area and volume is like multiplying monomials.

The area of the parallelogram above is:

$$A = bh$$
$$A = (8\text{ m})(6\text{ m})$$
$$A = 48\text{ m}^2$$

The area calculation above is similar to multiplying two monomials with m as the variable.

Multiply: $(8m)(6m) = 48m^2$

The volume of the box above is:

$$V = lwh$$
$$V = (5\text{ m})(3\text{ m})(4\text{ m})$$
$$V = 60\text{ m}^3$$

The volume calculation above is similar to multiplying three monomials with m as the variable.

Multiply: $(5m)(3m)(4m) = 60m^3$

EXAMPLE 4 Write an expression in simplest form for the volume of the box shown.

Understand We are given a box with side lengths that are monomial expressions.

Plan The volume of a box is found by multiplying the length, width, and height.

Execute $V = lwh$

$$V = (7x)(x)(2x)$$

$$V = 14x^3$$

Answer   The expression for volume is $14x^3$.

Check   Since $(7x)(x)(2x) = 14x^3$ is an identity, we can evaluate $(7x)$, (x), $(2x)$, and $14x^3$ using any chosen value for x and get the same result. Let's choose a value for x such as 3 and try it.

$$(7x)(x)(2x) = 14x^3$$

$$[7(3)](3)[2(3)] \stackrel{?}{=} 14(3)^3$$

$$(21)(3)(6) \stackrel{?}{=} 14(27)$$

$$378 = 378 \quad \text{This is true.}$$

Note: Choosing only one value does not guarantee that our simplification is correct, but it is a good indicator. Testing several values would give more conclusive assurance.

OBJECTIVE **2** **Multiply numbers in scientific notation.**   We multiply numbers in scientific notation using the same procedure we used to multiply monomials. If we replace a with 10 in $(4a^3)(2a^6)$, we have a product of two numbers in scientific notation $(4 \times 10^3)(2 \times 10^6)$. Following the same procedure for multiplying monomials, we multiply the $4 \cdot 2$ to get 8 and then add the exponents of the like bases.

Monomials:

$$(4a^3)(2a^6) = 4 \cdot 2a^{3+6}$$

$$= 8a^9$$

Scientific notation:

$$(4 \times 10^3)(2 \times 10^6) = 4 \times 2 \times 10^{3+6}$$

$$= 8 \times 10^9$$

EXAMPLE **5**   Multiply $(2.3 \times 10^5)(7.5 \times 10^6)$. Write the answer in scientific notation.

Solution   Multiply 2.3 and 7.5, then add the exponents for base 10's.

$$(2.3 \times 10^5)(7.5 \times 10^6) = 2.3 \times 7.5 \times 10^{5+6}$$

$$= 17.25 \times 10^{11}$$

$$= 1.725 \times 10^{12}$$

Note: The product 17.25×10^{11} is not in scientific notation because the absolute value of 17.25 is *not less than 10*. To fix this problem, we must move the decimal point one place to the left and account for this move by increasing the exponent by 1. If you are unconvinced that this is correct, consider the fact that 17.25×10^{11} is equal to 1,725,000,000,000,000. If we now write that standard form number in scientific notation, the proper position for the decimal point is between the 1 and the 7 digits, which means we must account for 12 places.

YOUR TURN

Multiply $(4.2 \times 10^5)(2.8 \times 10^8)$. Write the answer in scientific notation.

OBJECTIVE **3** **Simplify a monomial raised to a power.** We now discuss how to simplify an expression like $(3x^2)^4$, which is a monomial raised to a power. In order to simplify such an expression, we need to discuss two new rules of exponents: 1. raising a power to a power and 2. raising a product to a power.

Raising a Power to a Power

Consider a power raised to a power, like $(2^3)^2$. To calculate $(2^3)^2$, we could follow the order of operations and evaluate the exponential form 2^3 within the parentheses first, then square the result.

$$(2^3)^2 = (2 \cdot 2 \cdot 2)^2 = 8^2 = 64$$

However, there is an alternative. The outside exponent, 2, indicates to multiply the inside exponential form 2^3 by itself.

$$(2^3)^2 = 2^3 \cdot 2^3 \qquad \text{Since this is a multiplication of exponential}$$
$$= 2^{3+3} \qquad\qquad \text{forms that have the same base, we can add}$$
$$= 2^6 \qquad\qquad\;\; \text{the exponents.}$$

Notice that $2^6 = 64$. Also, in comparing $(2^3)^2$ with 2^6, notice that the exponent in 2^6 is the product of the original exponents.

$$(2^3)^2 = 2^{3 \cdot 2} = 2^6$$

Conclusion   If an exponential form is raised to a power, we can multiply the exponents and keep the same base.

RULE   **A POWER RAISED TO A POWER**

If a is a real number and m and n are integers, then $(a^m)^n = a^{mn}$.

Now let's consider a product raised to a power, like $(3 \cdot 4)^2$. We could write the factored form twice and then use the commutative property to rearrange the like bases.

$$(3 \cdot 4)^2 = (3 \cdot 4)(3 \cdot 4) \qquad \text{Since there are two factors of 3 and two}$$
$$= 3 \cdot 3 \cdot 4 \cdot 4 \qquad\quad \text{factors of 4, we can write each of these}$$
$$= 3^2 \cdot 4^2 \qquad\qquad\; \text{in exponential form.}$$

Conclusion   If a product is raised to a power, then we can evaluate the factors raised to that power.

RULE   **RAISING A PRODUCT TO A POWER**

If a and b are real numbers and n is an integer, then $(ab)^n = a^n b^n$.

Raising a Monomial to a Power

We apply these exponent rules to simplify an expression like $(3x^2)^4$, which is a monomial raised to a power. Because the monomial $3x^2$ expresses a product, we can use the rule for raising a product to a power and raise the factors 3 and x^2 to the 4th power.

$$(3x^2)^4 = (3^4)(x^2)^4 \qquad \text{To simplify } (x^2)^4, \text{ we use the rule for raising a}$$
$$= 81x^{2 \cdot 4} \qquad \text{power to a power.}$$
$$= 81x^8$$

Notice that the outside exponent is distributed to the coefficient and the variable(s) within the monomial. This suggests the following procedure.

> **PROCEDURE** **SIMPLIFYING A MONOMIAL RAISED TO A POWER**
>
> To simplify a monomial raised to a power:
> 1. Evaluate the coefficient raised to that power.
> 2. Multiply each variable's exponent by the power.

EXAMPLE 6 Simplify.

a. $(5a^6)^3$

Solution $(5a^6)^3 = (5)^3 a^{6 \cdot 3}$ Write the coefficient, 5, raised to the 3rd power, and
multiply the variable's exponent by 3.

$$= 125a^{18} \qquad \text{Simplify.}$$

b. $\left(-\dfrac{2}{3}xy^3z^5\right)^4$

Note: Recall that $x = x^1$.

Solution $\left(-\dfrac{2}{3}xy^3z^5\right)^4 = \left(-\dfrac{2}{3}\right)^4 x^{1 \cdot 4} y^{3 \cdot 4} z^{5 \cdot 4}$ Write the coefficient, $-\dfrac{2}{3}$, raised to the
4th power, and multiply each variable's
exponent by 4.

$$= \frac{16}{81}x^4 y^{12} z^{20} \qquad \text{Simplify.}$$

c. $(-0.4mn^6)^3$

Solution $(-0.4mn^6)^3 = (-0.4)^3 m^{1 \cdot 3} n^{6 \cdot 3}$
$$= -0.064m^3 n^{18}$$

YOUR TURN Simplify.

a. $(2y^4)^5$ **b.** $(-5t^2 u^6)^3$ **c.** $(-0.2a^4 b^5 c)^6$

Now, let's consider expressions that require us to use all of these exponent rules.

EXAMPLE 7 Simplify.

a. $(3x^2)(4x^5)^3$

Solution Since the order of operations is to simplify exponents before multiplying, we will simplify $(4x^5)^3$ first, then multiply the result by $3x^2$.

$$
\begin{aligned}
(3x^2)(4x^5)^3 &= (3x^2)(4^3 x^{5 \cdot 3}) \\
&= (3x^2)(64x^{15}) &&\text{Simplify.} \\
&= 3 \cdot 64 x^{2+15} &&\text{Multiply coefficients and add exponents of like variables.} \\
&= 192x^{17}
\end{aligned}
$$

b. $(-0.2a)^3(4ab)(1.5a^3c)^2$

Solution We follow the order of operations and simplify the monomials raised to a power first, then multiply the monomials.

$$
\begin{aligned}
(-0.2a)^3(4ab)(1.5a^3c)^2 &= ((-0.2)^3 a^{1 \cdot 3})(4ab)((1.5)^2 a^{3 \cdot 2} c^{1 \cdot 2}) \\
&= (-0.008a^3)(4ab)(2.25a^6c^2) &&\text{Simplify.} \\
&= -0.008 \cdot 4 \cdot 2.25 a^{3+1+6} bc^2 &&\text{Multiply coefficients and add exponents of like variables.} \\
&= -0.072a^{10}bc^2
\end{aligned}
$$

YOUR TURN Simplify.

a. $\left(\dfrac{1}{4}y^3\right)^2\left(\dfrac{8}{9}y^5\right)$

b. $(-5mn^4)(-2mp^2)^3(1.5m^2n)$

ANSWERS

a. $\dfrac{1}{18}y^{11}$ **b.** $60m^6n^5p^6$

5.4 EXERCISES

1. What rule can be applied when multiplying two exponential forms that have the same base?

2. Explain why $2^3 \cdot 5^2 \neq 10^5$.

3. Explain how to multiply monomials.

For Extra Help

InterAct Math

Addison-Wesley Math Tutor Center

Video 9/DVT 3

MyMathLab

MathXL

4. In multiplying the powers of 10 in $(8 \times 10^4)(6 \times 10^5)$, we add the exponents to get 10^9. However, the correct answer in scientific notation is 4.8×10^{10}. Why is the exponent 10 instead of 9?

5. How do we simplify $(x^3)^8$?

6. Explain how to simplify $(3x^4)^2$.

For Exercises 7–32, multiply.

7. $a \cdot a^4$

8. $b^2 \cdot b^3$

9. $2^3 \cdot 2^{10}$

10. $5^4 \cdot 5^3$

11. $a^3 \cdot a^2 b$

12. $3y^4 \cdot y$

13. $2x \cdot 3x^2$

14. $5ab^2 \cdot 2a^2 b$

15. $(-2mn)(4m^2 n)$

16. $(-3uv^2)(2u^3 v)$

17. $\left(\frac{5}{8}st\right)\left(-\frac{2}{7}s^2 t^7\right)$

18. $\left(\frac{3}{8}x^2 yz^2\right)\left(\frac{4}{9}xy^2 z\right)$

19. $(5ab^2 c)(0.2a^2 b^3 c^5)$

20. $(-3.1mn)(2.4m^2 n)$

21. $(r^2 s)(2r^2 s^3)(-3rs^6)$

22. $(-2jkl)(-5j^3 kl^4)(7jk^4 l^2)$

23. $xz^2(-x^3 tz^2)(tz^5)$

24. $-5mn^4(-m^3 n^9 p)(4m^3 np^5)$

25. $(5xy)(9x^2 y^4)(2x^3)$

26. $6hj^3 k \cdot 7h^5 k \cdot hkj$

27. $-5a^2\left(-\frac{1}{15}abc\right)\left(-\frac{3}{4}a^3 b^2 c\right)$

28. $\left(-\frac{1}{4}l^2 m\right)\left(\frac{5}{6}lm^3 n\right)\left(-\frac{2}{3}l^5\right)$

29. $(0.4wxy)(w^2 y^2)(2.5x^2 y^2)$

30. $(0.4u^2 vw)(-3.2u^2 v^3 w^3)(0.2uvw^9)$

31. $(5q^2 r^4 s^9)(-q^3 rs^2)(-r^5 s^{10})$

32. $(a^3 b^3 c^3)(-a^3 b^3 c^3)$

For Exercises 33–36, find and explain the mistake(s), then work the problem correctly.

33. $7x^3 \cdot 5x^4y = 35x^{12}y$

34. $-3m^2n \cdot 10m^5n = -30m^7n$

35. $(9x^3y^2z)(6xy^4) = 54x^3y^6z$

36. $(a^3b^2c)(ab^5) = a^4b^5$

For Exercises 37 and 38, write an expression in simplest form for the area of the shape.

37.

w

$5w$

38.

h

$7h$

For Exercises 39 and 40, write an expression in simplest form for volume.

39.

$0.5w$

$9w$

w

40.

h

$0.1h$

$0.2h$

41. The height of a trapezoid is half the base. The top side of the trapezoid is one-fourth the base.

    a. If b represents the length of the base, write expressions for the height and the length of the top side of the trapezoid.

    b. Using your expressions from part a, write an expression in simplest form for the area of the trapezoid.

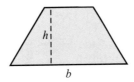

h

b

42. The height of a box is three times the width. The length is five times the width.

    a. If w represents the width, write expressions for the height and the length of the box.

    b. Using your expressions from part a, write an expression in simplest form for the volume of the box.

    c. Using your expressions from part a, write an expression in simplest form for the surface area of the box.

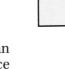

h

l

w

For Exercises 43–48, multiply and write your answer in scientific notation.

43. $(2 \times 10^5)(6 \times 10^3)$ **44.** $(9 \times 10^6)(5 \times 10^5)$

45. $(3.2 \times 10^5)(7.5 \times 10^8)$ **46.** $(4.7 \times 10^{10})(8.4 \times 10^7)$

47. $(9.5 \times 10^7)(2.63 \times 10^{11})$ **48.** $(8.52 \times 10^{12})(3.4 \times 10^9)$

49. Tennessee is shaped like a parallelogram with a base of approximately 350 miles and a height of approximately 120 miles. Calculate the area of Tennessee and write the answer in scientific notation. (Use $A = bh$.)

50. Idaho is shaped like a triangle with a base of approximately 320 miles and a height of approximately 520 miles. Calculate the area of Idaho and write the answer in scientific notation. $\left(\text{Use } A = \frac{1}{2}bh.\right)$

For Exercises 51 and 52, use the formula for the circumference of a circle: $C = 2\pi r$. (Use 3.14 to approximate π.)

51. The path of Earth's orbit around the Sun is approximately a circle with a radius of about 9.3×10^7 miles. What is the circumference of Earth's orbit?

Not to scale.

52. The radius of Earth is approximately 6.37×10^6 meters. Calculate the circumference of the equator.

For Exercises 53–80, simplify.

53. $(x^2)^3$ **54.** $(y^3)^5$ **55.** $(3^2)^4$ **56.** $(2^3)^2$

57. $(-h^2)^4$ **58.** $(-j^3)^2$ **59.** $(-y^2)^3$ **60.** $(-x^4)^5$

61. $(xy)^3$ **62.** $(abc)^3$ **63.** $\left(\frac{1}{4}m^2n\right)^3$ **64.** $\left(\frac{5}{6}u^3v\right)^2$

65. $(-0.2x^2y^5z)^3$

66. $(-0.3r^4st^3)^4$

67. $(-2p^4q^4)^6$

68. $(-6u^2v^4w^2)^3$

69. $(2x^3)(3x^2)^2$

70. $(-3y^2)(y^2)^3$

71. $(rs^2)(r^2s)^2$

72. $(m^2n)^3(mn)$

73. $(2ab)^2(5a^2b)^2$

74. $(4x)^2(2x^2)^2$

75. $\left(\frac{1}{4}a^2b^3c\right)^2\left(\frac{1}{4}a^2b^3c\right)^3$

76. $\left(-\frac{1}{2}s^2tu^4\right)^2\left(-\frac{1}{2}s^2tu^4\right)^5$

77. $(0.6u)^2(u^2v)(-3uv)^2$

78. $(-2w)^4(-w^2)^3(-w^4z)^2$

79. $(6u)^2(u^2v)(-3uv)^2$

80. $x^2(xy^3)(0.2x^3yz)^3$

For Exercises 81–84, find and explain the mistake(s), then work the problem correctly.

81. $(5x^3)^2 = 25x^5$

82. $(-4mn^5)^3 = -12m^3n^{15}$

83. $(2x^4y)^4 = 8x^8y$

84. $(-2ab^5)^3 = -8ab^{15}$

For Exercises 85 and 86, use the formula for the volume of a sphere: $V = \frac{4}{3}\pi r^3$. (Use 3.14 to approximate π.)

85. The radius of the Sun is approximately 6.96×10^8 m. Calculate the volume of the Sun, and write the answer using scientific notation.

86. The radius of Earth is approximately 6.37×10^6 m. Calculate the volume of Earth, and write the answer using scientific notation.

Earth   Sun

a. *If there are about 10^{11} stars in a galaxy and about 10^{11} galaxies in the visible universe, then about how many stars are in the visible universe?*

b. *Suppose one star out of every billion stars has a solar system with one planet that could support life. How many stars in the visible universe would have a solar system with one planet that could support life? (Hint: One out of every billion is one billionth, or 10^{-9}.)*

REVIEW EXERCISES

1. Add: $(3x + 4y - z) + (-7x - 4y - 3z)$

2. Evaluate: -2^4

3. Add: $(-4) + (-57)$

4. Find three consecutive integers whose sum is 15.

5. What is the degree of the monomial $-2r^5s$?

6. A business report indicates that revenues are $541,200 for the quarter, which is 12% less than the previous quarter. What was the revenue in the previous quarter?

5.5 Multiplying Polynomials; Special Products

OBJECTIVES

1. Multiply a polynomial by a monomial.
2. Multiply binomials.
3. Multiply polynomials.
4. Determine the product when given special polynomial factors.

In this section, we will use the rules for multiplying monomials to help us multiply multiple-term polynomials.

OBJECTIVE **1** **Multiply a polynomial by a monomial.** To multiply polynomials, we use the distributive property, which we introduced in Section 1.7 and used in solving linear equations and inequalities in Chapter 2. For example, we have applied the distributive property in situations like $2(3x + 4)$.

$$2(3x + 4) = 2 \cdot 3x + 2 \cdot 4$$
$$= 6x + 8$$

Notice that $2(3x + 4)$ is a product of a monomial 2 and a binomial $3x + 4$. This suggests the following procedure.

PROCEDURE **MULTIPLYING A POLYNOMIAL BY A MONOMIAL**

To multiply a polynomial by a monomial, use the distributive property to multiply each term in the polynomial by the monomial.

CONNECTION ————————————————————————

Multiplying a polynomial by a monomial is like multiplying a multidigit number by a single digit number. Consider $2 \cdot 34$:

$$
\begin{array}{r}
34 \\
\times\ 2 \\
\hline
68
\end{array}
$$

Notice that in the numerical multiplication, we distribute the 2 to each digit in the 34. In fact, if we expand 34 by expressing it as $30 + 4$, the problem takes the same form as $2(3x + 4)$ where $x = 10$:

$$2(30 + 4) = 2 \cdot 30 + 2 \cdot 4 \qquad\qquad 2(3x + 4) = 2 \cdot 3x + 2 \cdot 4$$
$$= 60 + 8 \qquad\qquad\qquad\qquad\qquad = 6x + 8$$
$$= 68$$

Now let's multiply a polynomial by a monomial that contains variables.

EXAMPLE 1 Multiply.

a. $2y(3y^2 + 4y + 1)$

Solution $2y \qquad (3y^2 + 4y + 1)$ Multiply each term in the polynomial by $2y$.

$2y \cdot 3y^2$
$2y \cdot 4y$
$2y \cdot 1$

$= 2y \cdot 3y^2 + 2y \cdot 4y + 2y \cdot 1$

$= 6y^3 + 8y^2 + 2y$

Note: When multiplying the individual terms by the monomial, we multiply the coefficients and add the exponents of the like bases.

b. $-7x^2(5x^2 + 3x - 4)$

Solution $-7x^2 \qquad (5x^2 + 3x - 4)$ Multiply each term in the polynomial by $-7x^2$.

$-7x^2 \cdot 5x^2$
$-7x^2 \cdot 3x$
$-7x^2 \cdot (-4)$

$= -7x^2 \cdot 5x^2 - 7x^2 \cdot 3x - 7x^2 \cdot (-4)$

$= -35x^4 - 21x^3 + 28x^2$

Note: When we multiply a polynomial by a negative monomial, the signs of the resulting polynomial will be the opposite of the signs in the original polynomial.

$-7x^2 (\ 5x^2\ +\ 3x\ -\ 4)$

$= -35x^4 - 21x^3 + 28x^2$

c. $-0.3t^2u(6t^3u + tu^2 - 3t^3 + 0.2uv)$

Solution $-0.3t^2u \quad (\quad 6t^3u\ +\ tu^2\ -\ 3t^3\ +\ 0.2uv)$ Multiply each term in the polynomial by $-0.3t^2u$.

$-0.3t^2u \cdot 6t^3u$
$-0.3t^2u \cdot tu^2$
$-0.3t^2u \cdot (-3t^3)$
$-0.3t^2u \cdot 0.2uv$

$= -0.3t^2u \cdot 6t^3u - 0.3t^2u \cdot tu^2 - 0.3t^2u \cdot (-3t^3) - 0.3t^2u \cdot 0.2uv$

$= \quad -1.8t^5u^2 \quad - \quad 0.3t^3u^3 \quad + \quad 0.9t^5u \quad - \quad 0.06t^2u^2v$

Learning Strategy

If you are a visual learner, try drawing lines connecting the terms that are to be multiplied.

Note: When multiplying multivariable terms, it is helpful to multiply the coefficients first, then the variables in alphabetical order.

For $-0.3t^2u \cdot 6t^3u$, think: $-0.3 \cdot 6 = -1.8$; $t^2 \cdot t^3 = t^5$; $u \cdot u = u^2$

Result: $-1.8t^5u^2$

YOUR TURN Multiply.

a. $4x^3(9x^2 - 7x + 3)$

b. $-3x^3yz\,(5x^2z - y^3z^2 + 8xz)$

ANSWERS

a. $36x^5 - 28x^4 + 12x^3$

b. $-15x^5yz^2 + 3x^3y^4z^3 - 24x^4yz^2$

Understanding Binomial Multiplication

Now that we have seen how to multiply a polynomial by a monomial, we are ready to multiply two binomials, which is like multiplying a pair of two-digit numbers. For example, consider $(12)(13)$ and compare this with $(x + 2)(x + 3)$.

$$
\begin{array}{r}
12 \\
\times\ 13 \\
\hline
36 \\
+\ 12\ \\
\hline
156
\end{array}
$$

Note: We think to ourselves "3 times 2 is 6, then 3 times 1 is 3," which creates the 36. We then move to the 1 in the tens place of the 13 and do the same thing. Since this 1 digit is in the tens place, it really means 10, so when we multiply this 10 times 12 it makes 120. We usually omit writing the 0 in the ones place and write 12 in the next two places.

Again, the distributive property is the governing principle. We multiply each digit in one number by each digit in the other number and shift underneath as we move to each new place. The same process applies to the binomials. We can stack them as we just did. Notice, however, that we will only stack this once in order to make the connection to numeric multiplication. As the polynomials get more complex, the stacking method becomes too tedious.

$$
\begin{array}{r}
x + 2 \\
x + 3 \\
\hline
3x + 6 \\
x^2 + 2x\ \ \ \ \ \\
\hline
x^2 + 5x + 6
\end{array}
$$

Note: We think "3 times 2 is 6, then 3 times x is $3x$." Now move to the x and think "x times 2 is $2x$, then x times x is x^2." Notice how we shifted so that the $2x$ and $3x$ line up. This is because they are like terms. It is the same as lining up the tens column when multiplying numbers. Note that the numeric result, 156, is basically the same as the algebraic result, $x^2 + 5x + 6$.

Notice that each term in $x + 2$ is multiplied by each term in $x + 3$. In general this is how we multiply two polynomials.

PROCEDURE **MULTIPLYING POLYNOMIALS**

To multiply two polynomials:

1. Multiply every term in the first polynomial by every term in the second polynomial.
2. Combine like terms.

Multiply Binomials

EXAMPLE **2** Multiply $(x + 4)(x + 3)$.

Solution Multiply each term in $x + 3$ by each term in $x + 4$.

$(x + 4)\ (x + 3)$

CONNECTION

If every term in both binomials is positive, the like terms are added and both signs in the resulting trinomial are plus signs. Noting these sign patterns will help us when factoring in Chapter 6.

$$= x \cdot x + x \cdot 3 + 4 \cdot x + 4 \cdot 3$$
$$= x^2 + \underbrace{3x + 4x} + 12$$
$$= x^2 + 7x + 12 \qquad \text{Combine like terms: } 3x + 4x = 7x.$$
$$= x^2 + 7x + 12$$

The word FOIL is a popular way to remember the process of multiplying two binomials. FOIL stands for **F**irst **O**uter **I**nner **L**ast. We will use Example 2 to demonstrate.

Warning: FOIL only helps keep track of products when multiplying two binomials. When multiplying larger polynomials, just remember to multiply every term in the second polynomial by every term in the first polynomial.

First terms: $x \cdot x = x^2$
$(x + 4)\ (x + 3)$

Outer terms: $x \cdot 3 = 3x$
$(x + 4)\ (x + 3)$

$(x + 4)\ (x + 3)$
Inner terms: $4 \cdot x = 4x$

$(x + 4)\ (x + 3)$
Last terms: $4 \cdot 3 = 12$

EXAMPLE 3 Multiply $(3x + 5)(x - 4)$.

Solution Multiply each term in $x - 4$ by each term in $3x + 5$ (think FOIL).

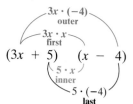

First Outer Inner Last
$$= 3x \cdot x + 3x \cdot (-4) + 5 \cdot x + 5 \cdot (-4)$$
$$= 3x^2 - 12x + 5x - 20$$
$$= 3x^2 - 7x - 20 \qquad \text{Combine like terms: } -12x + 5x = -7x.$$
$$= 3x^2 - 7x - 20$$

CONNECTION ——

Notice that $3x + 5$ has a plus sign whereas $x - 4$ has a minus sign. When multiplying binomials like these, the last term of the resulting trinomial will always be negative. The middle term could be positive or negative, depending on the coefficients of the like terms.

In Example 4, we will switch the signs in the binomials from Example 3. Look at how this changes the sign of the middle term in the resulting trinomial.

EXAMPLE 4 Multiply $(3x - 5)(x + 4)$.

CONNECTION ——

In Example 3, the result was $3x^2 - 7x - 20$, whereas in Example 4, the result is $3x^2 + 7x - 20$. Notice switching the signs like we did merely changed the sign of the middle term.

Solution Multiply each term in $x + 4$ by each term in $3x - 5$.

$$\text{First} \quad \text{Outer} \quad \text{Inner} \quad \text{Last}$$
$$= 3x \cdot x + 3x \cdot 4 + (-5) \cdot x + (-5) \cdot 4$$
$$= 3x^2 \qquad + 12x \qquad - 5x \qquad - 20$$
$$= 3x^2 \qquad\qquad + 7x \qquad\qquad - 20 \qquad \text{Combine like terms: } 12x - 5x = 7x.$$
$$= 3x^2 + 7x - 20$$

Now consider a case when the signs between terms in the binomials are both negative.

EXAMPLE 5 Multiply $(2x - 3)(4x - 5)$.

Solution Multiply each term in $4x - 5$ by each term in $2x - 3$ (think FOIL).

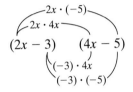

CONNECTION

When multiplying binomials like $2x - 3$ and $4x - 5$, notice the last term in the resulting trinomial is positive and the middle term is negative.

$$\text{First} \qquad \text{Outer} \qquad \text{Inner} \qquad \text{Last}$$
$$= 2x \cdot 4x + 2x \cdot (-5) + (-3) \cdot 4x + (-3) \cdot (-5)$$
$$= 8x^2 \qquad - 10x \qquad - 12x \qquad + 15$$
$$= 8x^2 \qquad\qquad - 22x \qquad\qquad + 15 \qquad \text{Combine like terms: } -10x - 12x = -22x.$$
$$= 8x^2 - 22x + 15$$

Keep in mind that Examples 2–5 do not illustrate all possible sign combinations that you could be given. The point of our discussion of signs is to note some patterns with the signs in the given binomial factors and see how they affect the resulting polynomial. Noticing these sign patterns now will help us when factoring in Chapter 6.

YOUR TURN Multiply.

    a. $(n + 6)(n - 2)$ **b.** $(2y - 5)(y - 3)$

CONNECTION

The product of two binomials can be shown in terms of geometry. Consider the rectangle below, with a length of $x + 7$ and a width of $x + 5$. We can describe its area in two ways: (1) the product of the length and width, $(x + 7)(x + 5)$; or (2) the sum of the areas of the four internal rectangles, $x^2 + 5x + 7x + 35$. Since these two approaches describe the same area, the two expressions must be equal.

	x	7
x	x^2	$7x$
5	$5x$	35

Length · width = sum of the areas of the four internal rectangles

$$(x + 7)(x + 5) = x^2 + 5x + 7x + 35$$
$$= x^2 + 12x + 35 \qquad \text{Combine like terms.}$$

ANSWERS

    a. $n^2 + 4n - 12$
    b. $2y^2 - 11y + 15$

OBJECTIVE ③ **Multiply polynomials.** Now consider an example of multiplication involving a larger polynomial, such as a trinomial. Remember that no matter how many terms are in the polynomials, the process is to multiply every term in the second polynomial by every term in the first polynomial.

EXAMPLE ⑥ Multiply $(x + 2)(3x^2 + 5x - 4)$.

Solution Multiply each term in $3x^2 + 5x - 4$ by each term in $x + 2$.

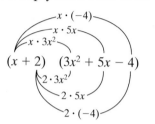

Warning: FOIL does not make sense here because there are too many terms. FOIL handles only the four terms from two binomials.

$$= x \cdot 3x^2 + x \cdot 5x + x \cdot (-4) + 2 \cdot 3x^2 + 2 \cdot 5x + 2 \cdot (-4)$$
$$= 3x^3 \quad + 5x^2 \quad - 4x \quad + 6x^2 \quad + 10x \quad - 8$$
$$= 3x^3 + 11x^2 + 6x - 8 \quad \text{Combine like terms.}$$

YOUR TURN

Multiply $(x - 3)(x^2 - 6x + 2)$.

OBJECTIVE ④ **Determine the product when given special polynomial factors.**

Multiplying Conjugates

In Objective 2 we discussed how the signs in the binomial factors affect the signs in the product in order to prepare for factoring in Chapter 6. To further prepare for factoring, it is helpful to note some patterns that occur when multiplying certain special polynomial factors. First, we will multiply **conjugates**.

DEFINITION | *Conjugates:* Binomials that differ only in the sign separating the terms.

The following binomial pairs are conjugates:

$$x + 9 \text{ and } x - 9 \qquad 2x + 3 \text{ and } 2x - 3 \qquad -6x + 5 \text{ and } -6x - 5$$

In Example 7 we will see something interesting about the product of conjugates.

ANSWER

$x^3 - 9x^2 + 20x - 6$

EXAMPLE 7 Multiply $(2x + 3)(2x - 3)$.

Solution

$$
\begin{array}{c}
\overset{\displaystyle 2x \cdot (-3)}{\overset{\displaystyle 2x \cdot 2x}{(2x + 3) \quad (2x - 3)}} \\
\underset{\displaystyle 3 \cdot (-3)}{\underset{\displaystyle 3 \cdot 2x}{}}
\end{array}
$$

$$
\begin{array}{cccc}
\text{First} & \text{Outer} & \text{Inner} & \text{Last} \\
= 2x \cdot 2x + 2x \cdot (-3) + 3 \cdot 2x + 3 \cdot (-3) \\
= \quad 4x^2 \quad\quad - 6x \quad\quad + 6x \quad\quad - 9 \\
= \quad 4x^2 \quad\quad\quad\quad + 0 \quad\quad\quad - 9 \\
= \quad 4x^2 - 9
\end{array}
$$

Combine like terms: $-6x + 6x = 0$.

Note: The sum of the like terms is 0, so the resulting binomial is the square of the first term minus the square of the last term.

When multiplying conjugates, the like terms will always be additive inverses, so their sum will always be 0. This suggests the following rule.

RULE **MULTIPLYING CONJUGATES**

If a and b are real numbers, variables, or expressions, then
$(a + b)(a - b) = a^2 - b^2$.

YOUR TURN

Multiply $(5y - 3)(5y + 3)$.

Squaring a Binomial

Now let's consider squaring a binomial, as in $(3x + 4)^2$. Remember that squaring a number or expression means to multiply that number or expression by itself, so $(3x + 4)^2 = (3x + 4)(3x + 4)$.

EXAMPLE 8 Multiply $(3x + 4)^2$.

Solution Rewrite $(3x + 4)^2$ as $(3x + 4)(3x + 4)$, then multiply.

$$
\begin{array}{c}
\overset{\displaystyle 3x \cdot 4}{\overset{\displaystyle 3x \cdot 3x}{(3x + 4)^2 = (3x + 4) \quad (3x + 4)}} \\
\underset{\displaystyle 4 \cdot 4}{\underset{\displaystyle 4 \cdot 3x}{}}
\end{array}
$$

$$
\begin{array}{cccc}
\text{First} & \text{Outer} & \text{Inner} & \text{Last} \\
= 3x \cdot 3x + 3x \cdot 4 + 4 \cdot 3x + 4 \cdot 4 \\
= \quad 9x^2 \quad\quad + 12x \quad + 12x \quad + 16 \\
= \quad 9x^2 \quad\quad\quad\quad + 24x \quad\quad + 16 \\
= \quad 9x^2 + 24x + 16
\end{array}
$$

Warning: Notice that $(3x + 4)^2$ does *not* equal $9x^2 + 16$.

Combine like terms: $12x + 12x = 24x$.

ANSWER

$25y^2 - 9$

5.5 Multiplying Polynomials; Special Products   **431**

From Example 8, we can see that there is a pattern to squaring a binomial.

$$(3x + 4)^2 = 9x^2 + 24x + 16$$

This first term is the square of the first term in the binomial: $(3x)^2 = 9x^2$

This last term is the square of the second term in the binomial: $(4)^2 = 16$

This middle term is twice the product of the two terms in the binomial: $2(3x)(4) = 24x$

Before we write a rule, let's consider a case with a minus sign separating the terms in a binomial that is to be squared.

EXAMPLE 9 Multiply $(3x - 4)^2$.

Solution Rewrite $(3x - 4)^2$ as $(3x - 4)(3x - 4)$, then multiply.

$$(3x - 4)^2 = (3x - 4)(3x - 4)$$

$3x \cdot (-4)$
$3x \cdot 3x$
$(-4) \cdot 3x$
$(-4) \cdot (-4)$

First   Outer   Inner   Last
$$= 3x \cdot 3x + 3x \cdot (-4) + (-4) \cdot 3x + (-4) \cdot (-4)$$
$$= \quad 9x^2 \quad - 12x \quad - 12x \quad + 16$$
$$= \quad 9x^2 \quad\quad - 24x \quad\quad + 16 \qquad \text{Combine like terms:}$$
$$= \quad 9x^2 - 24x + 16 \qquad\qquad -12x - 12x = -24x.$$

Warning: Notice that $(3x - 4)^2$ does not equal $9x^2 - 16$.

The pattern in Example 9 is similar to the pattern in Example 8.

$$(3x - 4)^2 = 9x^2 - 24x + 16$$

This first term is the square of the first term in the binomial: $(3x)^2 = 9x^2$

This last term is the square of the second term in the binomial: $(-4)^2 = 16$

This middle term is twice the product of the two terms in the binomial: $2(3x)(-4) = -24x$

The patterns in Examples 8 and 9 suggest the following rules.

RULES   **SQUARING A BINOMIAL**

$$(a + b)^2 = a^2 + 2ab + b^2$$
$$(a - b)^2 = a^2 - 2ab + b^2$$

ANSWERS

a. $9m^2 + 42m + 49$
b. $16t^2 - 40t + 25$

YOUR TURN   **Multiply.**

a. $(3m + 7)^2$ **b.** $(4t - 5)^2$

5.5 EXERCISES

For Extra Help

InterAct Math

Addison-Wesley
Math Tutor Center

Video 9/DVT 3

MyMathLab

MathXL

1. What is the mathematical property applied when multiplying polynomials?

2. Explain, in general, how to multiply two polynomials.

3. Explain how to recognize conjugates.

4. What happens when we multiply conjugates?

5. Explain why $(a + b)^2 \neq a^2 + b^2$.

6. Explain why $(a - b)^2 \neq a^2 - b^2$.

For Exercises 7–34, multiply the polynomial by the monomial.

7. $5(x + 3)$

8. $-2(y + 7)$

9. $-6(2x + 4)$

10. $4(-3y - 5)$

11. $4n(7n - 2)$

12. $-6n(3 - n)$

13. $9a(a - b)$

14. $3(y - z)$

15. $\frac{1}{8}m(2m - 5)$

16. $\frac{2}{3}k^2(2k - 3l)$

17. $-0.3p^2(p^3 - 2)$

18. $0.4b(b^2 - c^2)$

19. $-5x(4x^2 + 2x - 5)$

20. $-5a^2(3a^3 - 2a^2 + a + 6)$

21. $-x^2(7x^3 - 5x + 1)$

22. $4w(-w^4 - 3wx^2 + 5w^2x)$

23. $-r^2s(r^2s^3 + 3rs^2 - s)$

24. $-x^2y^3(-2xy^2 + 3x^2y^5 + 1 - y^3)$

25. $-2a^2b^2(2a - 6b + 3ab - a^2b^2)$

26. $-5hk^3(k^3 - 2h^2k - 3hk^4 + 4h)$

27. $abc(a^2b^2c^2 - abc + 1)$

28. $5xy^2z(2xy^3 - 4x^2z^2 + x^2yz)$

29. $4b\left(b^5 - \frac{1}{4}b^3 + \frac{1}{20}b^2 - \frac{1}{8}b + 4\right)$

30. $5r^2\left(r^4 - \frac{1}{20}r^3 - \frac{1}{5}r^2 + \frac{1}{10}r - 1\right)$

31. $-0.4rt^2(2.5r^3t - 8r^2t^2 + 4.1rt^3 - 2)$

32. $0.25suv^3(10s^2u - 2u^2v - 18sv^4 + 4)$

33. $-\frac{5}{6}x^2(30x^4y - 18x^3y^2 + 60xy^9)$

34. $\frac{2}{3}ab^4(-30a^2 + 18ab^3 - 12a^2b^2 + 6)$

5.5 Multiplying Polynomials: Special Products **433**

For Exercises 35 and 36, a larger rectangle is formed out of smaller rectangles:

   a. Write an expression in simplest form for the length (along the top).
   b. Write an expression in simplest form for the width (along the side).
   c. Write an expression that is the product of the length and width that you found in parts a and b.
   d. Write an expression in simplest form that is the sum of the areas of each of the smaller rectangles.
   e. Explain why the expressions in parts c and d are equivalent.

35.

36.

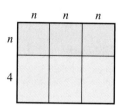

For Exercises 37–52, multiply the binomials (use FOIL).

37. $(x + 3)(x - 4)$ **38.** $(a + 4)(a - 9)$ **39.** $(x - 7)(x + 2)$ **40.** $(n + 7)(n - 5)$

41. $(2y - 5)(3y + 2)$ **42.** $(2x + 4)(x - 3)$ **43.** $(5m - 3)(3m + 4)$ **44.** $(2x + 3)(3x - 2)$

45. $(3t - 5)(4t - 2)$ **46.** $(2t + 7)(t - 1)$ **47.** $(5q - 3t)(3q + 4t)$ **48.** $(7k - 2j)(3k + j)$

49. $(7y + 3x)(2y + 4x)$ **50.** $(2a - 5b)(3a - 4b)$ **51.** $(a^2 + b)(a^2 - 2b)$ **52.** $(x^2 + 4)(x^2 - 5)$

For Exercises 53 and 54, a larger rectangle is formed out of smaller rectangles.

   a. Write an expression in simplest form for the length (along the top).
   b. Write an expression in simplest form for the width (along the side).
   c. Write an expression that is the product of the length and width that you found in parts a and b.
   d. Write an expression in simplest form that is the sum of the areas of each of the smaller rectangles.
   e. Explain why the expressions in parts c and d are equivalent.

53.

54.

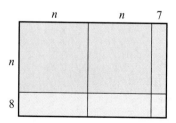

For Exercises 55–68, multiply the polynomials.

55. $(x - 1)(x^2 - 2x + 1)$ **56.** $(x + 1)(x^2 + 2x + 1)$ **57.** $(a - 1)(3a^2 - a - 2)$

58. $(x + 1)(x^2 - 2x + 1)$ **59.** $(c + 2)(2c^2 - c - 3)$ **60.** $(2x + 3)(4x^2 + 3x - 5)$

61. $(5p - 1)(3p^2 + 4p + 10)$ **62.** $(f + 3g)(2f^2 - 3fg - 9g^2)$ **63.** $(6y - 3z)(2y^2 - 3yz + 4z^2)$

64. $(2m - 3n)(4m^2 - 6mn + 8)$ **65.** $(x^2 + xy + y^2)(x^2 + 2xy - y^2)$ **66.** $(x^2 + 10x + 25)(x^2 + 4x + 4)$

67. $(u^2 - u - 1)(u^2 + 2u + 1)$ **68.** $(x^2 + 1)(x^6 - x^4 - x^2 + 1)$

For Exercises 69–76, state the conjugate of the given binomial.

69. $x - 3$ **70.** $y + 8$ **71.** $4x - 2y$ **72.** $2g + 3$

73. $4d - 3c$ **74.** $-5m - 2n$ **75.** $-3j - k$ **76.** $u^2 + v^2$

For Exercises 77–98, multiply using the rules for special products.

77. $(x - 5)(x + 5)$ **78.** $(y + 4)(y - 4)$ **79.** $(2m - 5)(2m + 5)$ **80.** $(3p + 2)(3p - 2)$

81. $(x + y)(x - y)$ **82.** $(a + b)(a - b)$ **83.** $(8r - 10s)(8r + 10s)$ **84.** $(4b - 5c)(4b + 5c)$

85. $(-2x - 3)(-2x + 3)$ **86.** $(-h + 2k)(-h - 2k)$ **87.** $(x - 3)^2$ **88.** $(y + 5)^2$

89. $(4t + 1)^2$ **90.** $(9k - 2)^2$ **91.** $(m + n)^2$ **92.** $(a - b)^2$

93. $(9 - 5y)^2$ **94.** $(6 - 7t)^2$ **95.** $(2u + 3v)^2$ **96.** $(3r + 7s)^2$

97. $(9w - 4z)^2$ **98.** $(2q + 11c)^2$

For Exercises 99–102, express the area.

99.

2yz

5x + y

100.

$3x - 2$

$2x - 7$

101.

$h + 7$

h

$h + 1$

102.

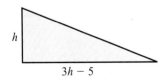

h

$3h - 5$

103. A rectangular room has a length that is 3 feet more than the width. Write an expression for the area of the room.

104. A circular metal plate for a machine has radius r. A smaller plate with radius $r - 2$ is to be used in another part of the machine. Write an expression in simplest form for the sum of the areas of the two circles.

For Exercises 105 and 106, express the volume.

105.

106.

107. A crate is to be designed with a length that is twice the width, and its height is to be 3 feet more than the length. Let w represent the width. Write an expression for the volume in terms of w.

108. A fish tank's length is 8 inches more than its width. The height of the tank is 5 inches greater than its width. Let w represent the width. Write an expression for the volume of the tank in terms of w.

109. A cylinder is to have a height that is 2 inches more than the radius. Write an expression for the volume of the cylinder in terms of the radius.

110. A cone is to have a radius that is 5 cm less than the height. Write an expression for the volume of the cone in terms of the height.

REVIEW EXERCISES

1. Simplify: $-42 \div 8$

2. Multiply: $x^2 \cdot x^5$

3. The larger of two numbers is four times the smaller number. The difference between the numbers is 27. What are the numbers?

4. Write 6.304×10^6 in standard form.

5. Simplify: -2^0

6. Isolate r in the formula $\dfrac{5 - d}{r} = m$.

5.6 Exponent Rules and Dividing Polynomials

OBJECTIVES

1. Divide exponential forms with the same base.
2. Divide numbers in scientific notation.
3. Divide monomials.
4. Divide a polynomial by a monomial.
5. Use long division to divide polynomials.
6. Simplify expressions using rules of exponents.

In Section 5.4 we began by learning the product rule of exponents, which we used in multiplying polynomials. In this section we follow a similar development and begin by learning the quotient rule of exponents so that we can divide polynomials.

OBJECTIVE 1 Divide exponential forms with the same base. In Section 5.4 we found that when we multiply exponential forms that have the same base, we can add the exponents and keep the same base. For example,

$$2^3 \cdot 2^4 = 2^{3+4} = 2^7$$

Now consider how to reverse this process and divide exponential forms with the same base. Remember that multiplication and division are inverse operations, which means they "undo" each other. A related division statement for the multiplication above is $2^7 \div 2^4 = 2^3$, which implies that, in the division problem, the exponents can be subtracted. To verify this result, it is helpful to write the division in fraction form and use the definition of exponents.

$$\frac{2^7}{2^4} = \frac{2 \cdot 2 \cdot 2 \cdot 2 \cdot 2 \cdot 2 \cdot 2}{2 \cdot 2 \cdot 2 \cdot 2}$$
$$= \frac{2 \cdot 2 \cdot 2}{1}$$
$$= 2^3$$

> **Note:** We can divide out four of the common factors. There are three 2's left in the top, which can be expressed in exponential form.

This result supports our thinking that we can subtract the exponents.

$$2^7 \div 2^4 = 2^{7-4} = 2^3$$

Notice that the divisor's exponent is subtracted from the dividend's exponent. This order is important because subtraction is not a commutative operation.

Conclusion To divide exponential forms that have the same base, we can subtract the divisor's exponent from the dividend's exponent and keep the same base.

QUOTIENT RULE FOR EXPONENTS

If m and n are integers and a is a real number, where $a \neq 0$, then $a^m \div a^n = a^{m-n}$.

Note: We cannot let a equal 0 because if a were replaced with 0, we would have $0^m \div 0^n$ and in Section 1.4 we concluded that $0 \div 0$ is indeterminate.

EXAMPLE 1 Divide $x^6 \div x^2$.

Note: Throughout the remainder of this section, assume that variables will not be replaced by values that cause denominators to be 0.

Solution Because the exponential forms have the same base, we can subtract the exponents and keep the same base.

$$x^6 \div x^2 = x^{6-2} = x^4$$

YOUR TURN Divide.

a. $5^9 \div 5^2$ 　　　 b. $t^8 \div t^2$ 　　　 c. $\dfrac{n^5}{n}$

ANSWERS

a. 5^7 　 b. t^6 　 c. n^4

We can apply the quotient rule for exponential forms when the exponents are negative. After simplifying, if the exponent in the result is negative, then we will rewrite the exponential form so that the exponent is positive. Recall the rule from Section 5.1: If n is any real number not equal to 0 and a is a natural number, then $a^{-n} = \dfrac{1}{a^n}$.

EXAMPLE 2 Divide and write the result with a positive exponent.

a. $x^6 \div x^{-2}$

Solution $x^6 \div x^{-2} = x^{6-(-2)}$ 　　Subtract exponents and keep the same base.

$= x^{6+2}$ 　　**Note:** To evaluate the subtraction, we write an equivalent addition.

$= x^8$

b. $n^{-5} \div n^{-3}$

Solution $n^{-5} \div n^{-3} = n^{-5-(-3)}$ 　　Subtract the exponents and keep the same base.

$= n^{-5+3}$ 　　Rewrite the subtraction as addition.

$= n^{-2}$ 　　Simplify.

$= \dfrac{1}{n^2}$ 　　Write with a positive exponent.

ANSWERS

a. $\dfrac{1}{y^4}$ 　 b. m^4 　 c. $\dfrac{1}{t^9}$

YOUR TURN Divide and write the result with a positive exponent.

a. $y^3 \div y^7$ 　　　 b. $m^{-2} \div m^{-6}$ 　　　 c. $\dfrac{t^{-4}}{t^5}$

OBJECTIVE ② **Divide numbers in scientific notation.** We can use the quotient rule of exponents to divide numbers in scientific notation.

EXAMPLE ③ Divide $\dfrac{1.08 \times 10^9}{2.4 \times 10^4}$ and write the result in scientific notation.

Solution The decimal factors and powers of 10 can be separated into a product of two fractions. This allows us to calculate the decimal division and divide the powers of 10 separately.

$$\frac{1.08 \times 10^9}{2.4 \times 10^4} = \frac{1.08}{2.4} \times \frac{10^9}{10^4} \quad \text{Separate the decimal numbers and powers of 10.}$$

$$= 0.45 \times 10^{9-4} \quad \text{Divide decimal numbers and subtract exponents.}$$

$$= 0.45 \times 10^5$$

$$= 4.5 \times 10^4$$

Note: This is not scientific notation because 0.45 is not a decimal number greater than 1. We must move the decimal point one place to the right. This means we have one less factor of 10 to account for with the exponent, so we subtract 1 from the exponent.

YOUR TURN

Divide $\dfrac{1.82 \times 10^{-9}}{6.5 \times 10^4}$ and write the result in scientific notation.

OBJECTIVE ③ **Divide monomials.** Dividing monomials is much like dividing numbers in scientific notation. Let's consider $\dfrac{24x^5}{6x^2}$. We can separate the coefficients and variables in the same way that we separated the decimal factors and powers of ten when dividing numbers in scientific notation:

$$\frac{24x^5}{6x^2} = \frac{24}{6} \cdot \frac{x^5}{x^2}$$

We can now divide the coefficients and subtract the exponents of the like bases.

$$\frac{24x^5}{6x^2} = \frac{24}{6} \cdot \frac{x^5}{x^2} = 4x^{5-2} = 4x^3$$

This suggests the following procedure.

PROCEDURE DIVIDING MONOMIALS

To divide monomials:

1. Divide the coefficients.
2. Use the quotient rule for the exponents with like bases.
3. Unlike bases are written unchanged in the quotient.
4. Write the final expression so that all exponents are positive.

ANSWER

2.8×10^{-14}

5.6 Exponent Rules and Dividing Polynomials **439**

EXAMPLE 4 Divide $\dfrac{16a^4b^7c}{20a^5b^2}$.

Solution $\dfrac{16a^4b^7c}{20a^5b^2} = \dfrac{16}{20} \cdot \dfrac{a^4}{a^5} \cdot \dfrac{b^7}{b^2} \cdot \dfrac{c}{1}$

$= \dfrac{16}{20} \cdot a^{4-5} \cdot b^{7-2} \cdot c$ Divide coefficients and subtract exponents of the like bases. The unlike base, c, will remain unchanged.

$= \dfrac{4}{5}a^{-1}b^5c$

$= \dfrac{4}{5} \cdot \dfrac{1}{a} \cdot \dfrac{b^5}{1} \cdot \dfrac{c}{1}$

$= \dfrac{4b^5c}{5a}$ **Note:** Since $a^{-1} = \dfrac{1}{a}$, the variable a is written in the denominator as a factor with the 5. Because the b^5 and c have positive exponents, they are expressed in the numerator as factors with the 4.

YOUR TURN Divide.

   a. $\dfrac{-32m^2n}{8mn}$ **b.** $\dfrac{56t^2u^6v}{16t^2u^4}$

OBJECTIVE 4 **Divide a polynomial by a monomial.** Let's now consider how to divide a polynomial by a monomial, as in $\dfrac{24x^5 + 18x^3}{6x^2}$. Recall that when fractions with a common denominator are added (or subtracted), the numerators are added and the denominator stays the same.

$$\frac{1}{7} + \frac{2}{7} = \frac{1+2}{7}$$

This process can be reversed so that a sum in the numerator of a fraction can be broken into fractions, with each addend over the same denominator.

$$\frac{1+2}{7} = \frac{1}{7} + \frac{2}{7}$$

RULE   If a, b, and c are real numbers, variables, or expressions with $c \neq 0$, then

$$\frac{a+b}{c} = \frac{a}{c} + \frac{b}{c}$$

We can apply this rule to the division problem $\dfrac{24x^5 + 18x^3}{6x^2}$.

$\dfrac{24x^5 + 18x^3}{6x^2} = \dfrac{24x^5}{6x^2} + \dfrac{18x^3}{6x^2}$ We now have a sum of monomial divisions, which we can simplify separately.

$= 4x^{5-2} + 3x^{3-2}$

$= 4x^3 + 3x$

Our illustration suggests the following procedure.

PROCEDURE **DIVIDING A POLYNOMIAL BY A MONOMIAL**

To divide a polynomial by a monomial, divide each term in the polynomial by the monomial.

EXAMPLE 5 Divide.

a. $\dfrac{32y^6 + 24y^4 - 48y^2}{8y^2}$

Solution $\dfrac{32y^6 + 24y^4 - 48y^2}{8y^2} = \dfrac{32y^6}{8y^2} + \dfrac{24y^4}{8y^2} - \dfrac{48y^2}{8y^2}$

$= 4y^4 + 3y^2 - 6y^0$ Divide each term in the polynomial by the monomial.

$= 4y^4 + 3y^2 - 6$

b. $\dfrac{42t^7u^2 - 18t^4u - 10t}{2t^3u}$

Solution $\dfrac{42t^7u^2 - 18t^4u - 10t}{2t^3u} = \dfrac{42t^7u^2}{2t^3u} - \dfrac{18t^4u}{2t^3u} - \dfrac{10t}{2t^3u}$

Note: When dividing t by t^3, because the denominator's exponent is greater than the numerator's exponent, the result is a negative exponent. Because u has no like base in the third term, it is written unchanged.

$= 21t^4u - 9t^1u^0 - \dfrac{5t^{-2}}{u}$

$= 21t^4u - 9t - \dfrac{5}{ut^2}$ Rewrite with positive exponents.

YOUR TURN Divide.

a. $\dfrac{45x^6 - 27x^5 + 9x^3}{9x^3}$

b. $\dfrac{28m^2n^6 + 36m^4n - 40m^2}{4m^2n}$

OBJECTIVE 5 **Use long division to divide polynomials.** To divide a polynomial by a polynomial, we can use long division. Consider the following numeric long division and think about the process.

Divide: $\dfrac{157}{12}$ ⟶ Quotient

$$
\begin{array}{r}
13 \\
12\overline{)157} \\
-12 \\
\hline
37 \\
-36 \\
\hline
1
\end{array}
$$

Divisor ⟶ $12\overline{)157}$ ⟵ Dividend

Remainder ⟶ 1

Because the answer has a remainder, we write the result as a mixed number, $13\dfrac{1}{12}$. Notice we could check the answer by multiplying the divisor by the quotient then adding the remainder. The result should be the dividend.

$$\text{Quotient} \cdot \text{Divisor} + \text{Remainder} = \text{Dividend}$$
$$13 \quad \cdot \quad 12 \quad + \quad 1 \quad = 157$$

Now let's consider polynomial division, which follows the same long division process.

EXAMPLE 6 Divide $\dfrac{x^2 + 5x + 7}{x + 2}$.

Solution Use long division. First, we determine what term will multiply by the first term in the divisor to equal the first term in the dividend. A clever way to determine this first term in the quotient is to divide the first term in the dividend by the first term in the divisor.

$$x + 2 \overline{)x^2 + 5x + 7}$$

Divide these first terms to
determine the first term in
the quotient: $x^2 \div x = x$.

Next, we multiply the divisor $x + 2$ by the x in the quotient.

multiply

$$x + 2 \overline{)x^2 + 5x + 7}$$
$$x^2 + 2x$$

Next we subtract. Note that to subtract the binomial, we change the signs of the terms and then combine like terms. After combining terms, we bring down the next term in the dividend, which is 7.

$$\begin{array}{r} x \\ x + 2 \overline{)x^2 + 5x + 7} \\ -(x^2 + 2x) \end{array}$$

Change signs.

$$\begin{array}{r} x \\ x + 2 \overline{)x^2 + 5x + 7} \\ -x^2 - 2x \\ \hline 3x + 7 \end{array}$$

Combine like terms
and bring down the
next term.

Now we repeat the process with $3x + 7$ as the dividend. The next term in the quotient is found by dividing the first term of $3x + 7$ by the first term in the divisor, $x + 2$. This gives $3x \div x = 3$. We then multiply the divisor by this 3.

multiply

$$\begin{array}{r} x + 3 \\ x + 2 \overline{)x^2 + 5x + 7} \\ -x^2 - 2x \\ \hline 3x + 7 \\ 3x + 6 \end{array}$$

Then we subtract. As before, we change the signs of the binomial to be subtracted.

$$\begin{array}{r} x + 3 \\ x + 2 \overline{)x^2 + 5x + 7} \\ \underline{-x^2 - 2x} \\ 3x + 7 \\ \underline{-(3x + 6)} \end{array}$$ Change signs. $$\begin{array}{r} x + 3 \\ x + 2 \overline{)x^2 + 5x + 7} \\ \underline{-x^2 - 2x} \\ 3x + 7 \\ \underline{-3x - 6} \\ 1 \end{array}$$ Combine like terms.

Notice that we have a remainder, 1. Recall that in the numeric version, we wrote the answer as a mixed number, $13\frac{1}{12}$, which means $13 + \frac{1}{12}$ and is in the form:

quotient $+ \dfrac{\text{remainder}}{\text{divisor}}$. With polynomials we write a similar expression:

$$\underbrace{x + 3}_{\text{quotient}} + \underbrace{\frac{1}{x + 2}}_{}$$ remainder / divisor

To check, we can multiply the quotient and the divisor, then add the remainder. The result should be the dividend.

$$\begin{aligned} (x + 3)(x + 2) + 1 &= x^2 + 2x + 3x + 6 + 1 \\ &= x^2 + 5x + 7 \qquad \text{It checks.} \end{aligned}$$

PROCEDURE **DIVIDING A POLYNOMIAL BY A POLYNOMIAL**

To divide a polynomial by a polynomial:

1. Use long division.

2. If there is a remainder, write the result in the form: quotient $+ \dfrac{\text{remainder}}{\text{divisor}}$.

In Example 6, the coefficient of the first term in the divisor and dividend were both 1. What if those coefficients are other than 1?

EXAMPLE 7 Divide $\dfrac{15x^2 - 22x + 14}{3x - 2}$.

Solution Begin by dividing the first term in the dividend by the first term in the divisor: $15x^2 \div 3x = 5x$.

$$\begin{array}{r} 5x \\ 3x - 2 \overline{)15x^2 - 22x + 14} \\ \underline{-(15x^2 - 10x)} \end{array}$$ Change signs. $$\begin{array}{r} 5x \\ 3x - 2 \overline{)15x^2 - 22x + 14} \\ \underline{-15x^2 + 10x} \\ -12x + 14 \end{array}$$ Combine like terms, then bring down the next term.

Determine the next part of the quotient by dividing $-12x$ by $3x$, which is -4, and repeat the multiplication and subtraction steps.

$$
\begin{array}{r}
5x - 4 \\
3x - 2\overline{)15x^2 - 22x + 14} \\
\underline{-15x^2 + 10x} \\
-12x + 14 \\
\underline{-(-12x + 8)}
\end{array}
$$

Change signs. →

$$
\begin{array}{r}
5x - 4 \\
3x - 2\overline{)15x^2 - 22x + 14} \\
\underline{-15x^2 + 10x} \\
-12x + 14 \\
\underline{+12x - 8} \\
6
\end{array}
$$

Combine like terms.

Answer: $5x - 4 + \dfrac{6}{3x - 2}$

YOUR TURN   **Divide.**

a. $\dfrac{x^2 + 6x + 8}{x + 2}$ **b.** $\dfrac{24x^2 + 14x + 4}{6x - 1}$

Using a Place Marker in Long Division

In polynomial division, it is important that we write the terms in descending order of degree. If there is a missing term, then we write that term with a 0 coefficient as a place marker.

EXAMPLE 8   Divide $\dfrac{-14x^2 - 9 + 8x^4}{2x - 1}$.

Solution First, write the dividend in descending order of degree. Also, because degree 3 and degree 1 terms are missing in the dividend, we write those terms with 0 coefficients as place holders.

$$
\begin{array}{r}
4x^3 \\
2x - 1\overline{)8x^4 + 0x^3 - 14x^2 + 0x - 9} \\
\underline{-(8x^4 - 4x^3)}
\end{array}
$$

Change signs. →

$$
\begin{array}{r}
4x^3 \\
2x - 1\overline{)8x^4 + 0x^3 - 14x^2 + 0x - 9} \\
\underline{-8x^4 + 4x^3} \\
4x^3 - 14x^2
\end{array}
$$

Combine like terms, then bring down the next term.

Determine the next part of the quotient by dividing $4x^3$ by $2x$, and repeat the multiplication and subtraction steps.

$$
\begin{array}{r}
4x^3 + 2x^2 \\
2x - 1\overline{)8x^4 + 0x^3 - 14x^2 + 0x - 9} \\
\underline{-8x^4 + 4x^3} \\
4x^3 - 14x^2 \\
\underline{-(4x^3 - 2x^2)}
\end{array}
$$

Change signs. →

$$
\begin{array}{r}
4x^3 + 2x^2 \\
2x - 1\overline{)8x^4 + 0x^3 - 14x^2 + 0x - 9} \\
\underline{-8x^4 + 4x^3} \\
4x^3 - 14x^2 \\
\underline{-4x^3 + 2x^2} \\
-12x^2 + 0x
\end{array}
$$

Combine like terms, then bring down the next term.

Determine the next part of the quotient by dividing $-12x^2$ by $2x$, and repeat the multiplication and subtraction steps.

$$
\begin{array}{r}
4x^3 + 2x^2 - 6x \\
2x - 1\overline{)8x^4 + 0x^3 - 14x^2 + 0x - 9} \\
\underline{-8x^4 + 4x^3} \\
4x^3 - 14x^2 \\
\underline{-4x^3 + 2x^2} \\
-12x^2 + 0x \\
\underline{-(-12x^2 + 6x)}
\end{array}
$$

Change signs. →

$$
\begin{array}{r}
4x^3 + 2x^2 - 6x \\
2x - 1\overline{)8x^4 + 0x^3 - 14x^2 + 0x - 9} \\
\underline{-8x^4 + 4x^3} \\
4x^3 - 14x^2 \\
\underline{-4x^3 + 2x^2} \\
-12x^2 + 0x \\
\underline{+12x^2 - 6x} \\
-6x - 9
\end{array}
$$

Combine like terms, then bring down the next term.

Determine the final term by dividing $-6x$ by $2x$, which is -3.

$$
\begin{array}{r}
4x^3 + 2x^2 - 6x - 3 \\
2x - 1\overline{)8x^4 + 0x^3 - 14x^2 + 0x - 9} \\
\underline{-8x^4 + 4x^3} \\
4x^3 - 14x^2 \\
\underline{-4x^3 + 2x^2} \\
-12x^2 + 0x \\
\underline{12x^2 - 6x} \\
-6x - 9 \\
\underline{-(-6x + 3)}
\end{array}
$$

Change signs. ⟶

$$
\begin{array}{r}
4x^3 + 2x^2 - 6x - 3 \\
2x - 1\overline{)8x^4 + 0x^3 - 14x^2 + 0x - 9} \\
\underline{-8x^4 + 4x^3} \\
4x^3 - 14x^2 \\
\underline{-4x^3 + 2x^2} \\
-12x^2 + 0x \\
\underline{12x^2 - 6x} \\
-6x - 9 \\
\underline{+6x - 3} \\
-12
\end{array}
$$

Combine like terms.

Answer: $4x^3 + 2x^2 - 6x - 3 + \dfrac{-12}{2x - 1}$ or $4x^3 + 2x^2 - 6x - 3 - \dfrac{12}{2x - 1}$

YOUR TURN

Divide $\dfrac{7 + 8x^3 - 18x}{4x - 2}$.

OBJECTIVE 6 **Simplify expressions using rules of exponents.** The quotient rule for exponents completes our rules of exponents in this chapter. So far the problems involving exponent rules have only required that we use one or two rules. Let's now consider problems that require us to use several of the exponent rules. Following is a summary of all of the rules of exponents.

RULE **EXPONENTS SUMMARY**

Assume that no denominators are 0, that a and b are real numbers, and that m and n are integers.

Zero as an exponent:	$a^0 = 1$, where $a \neq 0$
	0^0 is indeterminate
Negative exponents:	$a^{-n} = \dfrac{1}{a^n}$
	$\dfrac{1}{a^{-n}} = a^n$
	$\left(\dfrac{a}{b}\right)^{-n} = \left(\dfrac{b}{a}\right)^n$
Product rule for exponents:	$a^m \cdot a^n = a^{m+n}$
Quotient rule for exponents:	$a^m \div a^n = a^{m-n}$
Raising a power to a power:	$(a^m)^n = a^{mn}$
Raising a product to a power:	$(ab)^n = a^n b^n$
Raising a quotient to a power:	$\left(\dfrac{a}{b}\right)^n = \dfrac{a^n}{b^n}$

ANSWER

$2x^2 + x - 4 + \dfrac{-1}{4x - 2}$

5.6 Exponent Rules and Dividing Polynomials **445**

EXAMPLE 9 Simplify. Write all answers with positive exponents.

a. $\left(\dfrac{n^{-5}}{n^2}\right)^3$

Solution Following the order of operations, we simplify within the parentheses first, then consider the exponent outside the parentheses.

$$\left(\dfrac{n^{-5}}{n^2}\right)^3 = (n^{-5-2})^3 \qquad \text{Use the quotient rule for exponents.}$$
$$= (n^{-7})^3$$
$$= n^{-7\cdot 3} \qquad \text{Use the rule for raising a power to a power.}$$
$$= n^{-21}$$
$$= \dfrac{1}{n^{21}} \qquad \text{Write with a positive exponent.}$$

b. $\dfrac{(m^2)^{-3}}{m^4 \cdot m^{-5}}$

Solution $\dfrac{(m^2)^{-3}}{m^4 \cdot m^{-5}} = \dfrac{m^{2\cdot(-3)}}{m^{4+(-5)}}$ In the numerator, use the rule for raising a power to a power. In the denominator, use the product rule for exponents.

$$= \dfrac{m^{-6}}{m^{-1}}$$
$$= m^{-6-(-1)} \qquad \text{Use the quotient rule for exponents.}$$
$$= m^{-6+1} \qquad \text{Write the subtraction as an equivalent addition.}$$
$$= m^{-5}$$
$$= \dfrac{1}{m^5} \qquad \text{Write with a positive exponent.}$$

c. $\dfrac{(4y^3)^2}{(2y^4)^5}$

Solution $\dfrac{(4y^3)^2}{(2y^4)^5} = \dfrac{4^2 y^{3\cdot 2}}{2^5 y^{4\cdot 5}}$ Use the rule for raising a product to a power.

$$= \dfrac{16y^6}{32y^{20}}$$
$$= \dfrac{16}{32} \cdot \dfrac{y^6}{y^{20}} \qquad \text{Separate the coefficients and variables.}$$
$$= \dfrac{16}{32} \cdot y^{6-20} \qquad \text{Use the quotient rule for exponents.}$$
$$= \dfrac{1}{2} \cdot y^{-14}$$
$$= \dfrac{1}{2} \cdot \dfrac{1}{y^{14}} \qquad \text{Write with a positive exponent.}$$
$$= \dfrac{1}{2y^{14}} \qquad \text{Simplify.}$$

ANSWERS

a. $\dfrac{1}{x^{28}}$ **b.** y^{35} **c.** $\dfrac{y^6}{8x^8}$

YOUR TURN Simplify. Write all answers with a positive exponent.

a. $\left(\dfrac{x^2}{x^{-5}}\right)^{-4}$ **b.** $(y^{-2} \cdot y^{-3})^{-4} \div y^{-15}$ **c.** $\dfrac{(2x^4)^{-3}}{(x^2y^3)^{-2}}$

For Extra Help

InterAct Math

Tutor Center   Addison-Wesley
Math Tutor Center

Video 9/DVT 3

MyMathLab

MathXL   MathXL

5.6 EXERCISES

1. Explain how to divide exponential forms that have the same base.

2. How do we change the sign of the exponent in an exponential form that has a negative exponent?

3. How do we divide two numbers in scientific notation?

4. How do we divide two monomials?

5. Explain how to divide a polynomial by a monomial.

6. When do we use a 0 place holder in the process of dividing two polynomials by long division?

For Exercises 7–30, simplify. Write all answers with positive exponents.

7. a^{-3} **8.** n^{-7} **9.** 2^{-5} **10.** 3^{-4}

11. $\dfrac{3^4}{3^3}$ **12.** $\dfrac{6^{11}}{6^5}$ **13.** $\dfrac{4^3}{4^9}$ **14.** $\dfrac{5}{5^9}$

15. $\dfrac{x^3}{x^5}$ **16.** $\dfrac{n^4}{n}$ **17.** $\dfrac{y^8}{y^3}$ **18.** $\dfrac{m^4}{m^{10}}$

19. $a^8 \div a^6$ **20.** $u^5 \div u^3$ **21.** $w^7 \div w^{10}$ **22.** $j^6 \div j^{14}$

23. $\dfrac{a^{-2}}{a^5}$ **24.** $\dfrac{x^{-3}}{x^{-8}}$ **25.** $\dfrac{y^{-7}}{y^{-15}}$ **26.** $\dfrac{u^5}{u^{-12}}$

27. $\dfrac{t^{-4}}{t^{-4}}$ **28.** $\dfrac{p^{-1}}{p^5}$ **29.** $\dfrac{s^{-8}}{s}$ **30.** $\dfrac{m^{-5}}{m^5}$

For Exercises 31–38, divide and write your answers in scientific notation.

31. $\dfrac{1.2 \times 10^5}{3.2 \times 10^6}$ **32.** $\dfrac{1.2 \times 10^9}{3.2 \times 10^6}$ **33.** $\dfrac{1.4 \times 10^{-4}}{2.1 \times 10^3}$ **34.** $\dfrac{1.32 \times 10^{-5}}{6.6 \times 10^{15}}$

35. $\dfrac{9.088 \times 10^1}{1.28 \times 10^{-9}}$ **36.** $\dfrac{4.25 \times 10^{-4}}{1.88 \times 10^2}$ **37.** $\dfrac{8.75 \times 10^{-3}}{3.2 \times 10^{-7}}$ **38.** $\dfrac{7.9 \times 10^1}{2.2 \times 10^{-7}}$

For Exercises 39–42, solve.

39. If light travels at 3×10^8 meters per second, how long does it take light to travel the 1.5×10^{11} meters from the Sun to Earth?

40. In 1999, the population of the United States was approximately 2.7×10^8. Calculate the approximate number of people per square mile in the United States in 1999 if the estimated amount of land in the United States is 3.5×10^6 square miles. (*Source:* U.S. Bureau of the Census)

41. As of June 25, 2002, the national debt was about $\$6.108 \times 10^{12}$. If each of the 2.876×10^8 people in the United States contributed an equal share towards the debt, then what would each person have to contribute to pay off the debt? (*Source:* U.S. National Debt Clock)

42. There are about 2.3×10^6 stone blocks in the Great Pyramid. If the total weight of the pyramid is about 1.3×10^{10} pounds and if we assume each block is of equal size, then how much does each block weigh? (*Source:* Nova Online)

For Exercises 43–50, divide the monomials.

43. $\dfrac{3x^5}{3x^2}$ **44.** $\dfrac{14a^6}{2a^4}$ **45.** $\dfrac{9m^3n^5}{-3mn}$ **46.** $\dfrac{-42x^5y^3}{6xy^2}$

47. $\dfrac{-24p^3q^5}{15p^3q^2}$ **48.** $\dfrac{48t^4u^7}{-18t^4u^2}$ **49.** $\dfrac{12a^4bc^3}{9a^6bc^2}$ **50.** $\dfrac{28p^5q^4r}{42p^2q^7}$

For Exercises 51–62, divide the polynomial by the monomial.

51. $\dfrac{7a + 14b}{7}$ **52.** $\dfrac{4m + 16y}{4}$ **53.** $\dfrac{5x^2y^2 - 15xy^3}{5xy}$

54. $\dfrac{12k^2l^2 + 15k^2l^3}{3k^2l}$ **55.** $\dfrac{abc^2 - 4a^2b^2c}{-abc}$ **56.** $\dfrac{x^3yz^5 - 15xyz^2}{-xyz^2}$

57. $\dfrac{x^3 + x^2 - x}{x}$ **58.** $\dfrac{3x^2 + 6x - 3}{3x}$ **59.** $\dfrac{y^5 + y^7 - 3y^8 + y}{y^3}$

60. $\dfrac{12x^6 - 9x^4 + 18x^2 - x}{6x^2}$ **61.** $\dfrac{36u^3v^4 + 12uv^5 - 15u^2v^2}{3uv}$ **62.** $\dfrac{16hk^4 - 25hk - 4h^2k^2}{4hk}$

63. The area of the parallelogram shown is described by the monomial $35mn^2$. Find the height.

64. The area of the triangle shown is $18x^2y$. Find the base.

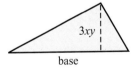

65. An engineer is designing a steel cover plate in the shape of a rectangle. The area of the plate is described by the polynomial $9x^2 - 15x + 12$ and the length must be $3x$.

   a. Find an expression for the width.

   b. Find the area, length, and width if $x = 2$.

66. The voltage in a circuit is described by the polynomial $36s^3 - 24s^2 + 42s$. The resistance is described by $6s$.

   a. Find an expression for the current. (Hint: Voltage = Current × Resistance)

   b. Find the voltage, current, and resistance if $s = 3$.

For Exercises 67–90, use long division to divide the polynomials.

67. $\dfrac{x^2 + 7x + 12}{x + 3}$

68. $\dfrac{y^2 + 4y + 4}{y + 2}$

69. $\dfrac{x^2 - 17x + 64}{x - 5}$

70. $\dfrac{x^2 + 6x + 10}{x + 3}$

71. $\dfrac{m^2 + 8m + 15}{m + 3}$

72. $\dfrac{u^2 - 6u + 10}{u - 3}$

73. $\dfrac{2x^3 - x^2 + 5}{x - 2}$

74. $\dfrac{2x^3 - 2x^2 + 5}{x + 3}$

75. $\dfrac{8p^3 + 125}{2p + 5}$

76. $\dfrac{t^3 - 48}{t + 2}$

77. $\dfrac{x^3 + 8}{x + 2}$

78. $\dfrac{21u^3 - 19u^2 + 14u - 6}{3u + 2}$

79. $\dfrac{14y^2 - 8y + 17}{7y + 3}$

80. $\dfrac{6a^2 + 2a - 28}{3a + 7}$

81. $\dfrac{14b + b^3 - 6b^2 - 12}{b - 3}$

82. $\dfrac{-10v + 5v^2 + 2v^3 - 3}{2v - 3}$

83. $\dfrac{4k^3 + 8k^2 - k}{k + 2}$

84. $\dfrac{12x^2 - 7x - 10}{4x - 5}$

85. $\dfrac{6x^2 + 7x + 5}{3x - 1}$

86. $\dfrac{y^3 - y^2 + 3 - 11y}{y + 2}$

87. $\dfrac{2w^3 - 7 - w^2 + 16w}{2w - 1}$

88. $\dfrac{x^5 - 3x^4 - 4x^2 + 10x}{x - 3}$

89. $\dfrac{y^4 + 2y^3 + 2y + 6}{y + 2}$

90. $\dfrac{2b^4 + 3b^3 - 4b - 6}{2b + 3}$

450 CHAPTER 5 Polynomials

91. The volume of a restaurant's walk-in refrigerator is described by $(6x^2 + 42x + 72)$ cubic feet. The width has to be 6 feet and the length is described by $(x + 3)$ feet.

   a. Find an expression for the height of the refrigerator.

   b. Find the length, height, and volume of the refrigerator if $x = 4$.

92. In an architectural project, the specifications for a rectangular room call for the area to be $(10x^2 + 3x - 18)$ square feet. It is decided that the length should be described by the binomial $(2x + 3)$ feet.

   a. Find an expression for the width.

   b. Find the length, width, and area of the room if $x = 5$.

For Exercises 93–112, simplify. Write all answers with positive exponents.

93. $-3x^{-2}$

94. $-2xy^{-3}$

95. $6x^{-4}$

96. $5y^{-2}$

97. $\left(\dfrac{-2}{3}\right)^{-3}$

98. $\left(\dfrac{1}{5}\right)^{-2}$

99. $\left(\dfrac{a^3}{b}\right)^{-4}$

100. $\left(\dfrac{m^5}{n^2}\right)^{-3}$

101. $\left(\dfrac{3x^{-1}y^2}{z^2}\right)^3$

102. $\left(\dfrac{2x^{-2}y}{z^{-1}}\right)^4$

103. $\dfrac{4xy^{-2}z^3}{x^{-3}y^3z^{-1}}$

104. $\dfrac{3m^6n^{-1}p^{-2}}{m^{-1}n^{-3}p}$

105. $\dfrac{(2ab)^3}{12a^4b^5}$

106. $\dfrac{(2xy^2)^2}{16x^2y^3}$

107. $(6m^2n^{-2})^2$

108. $(-4a^3b^{-2}c)^2$

109. $(4rs^3)^3(-0.5r^3s^{-1})^2$

110. $(-3abc^{-2})^{-1}(a^2b^{-3})^{-2}$

111. $\dfrac{(-3x^4)^3}{27x^2y^{-4}}$

112. $\dfrac{(-2y^{-1})^3}{-8x^{-4}y^{-3}}$

For Exercises 113–116 use Einstein's formula $E = mc^2$, which describes the rest energy contained within a mass m, where c represents the speed of light, which is a constant with a value of 3×10^8 meters per second. The units for energy are joules, which are equivalent to $\frac{kg \cdot m^2}{s^2}$.

113. Find the rest energy of a particle with a mass of 2.5×10^{-6} kilograms.

114. Find the rest energy of a particle with a mass of 1.4×10^{-9} kilograms.

115. The first atom bomb released an energy equivalent of about 8.4×10^{13} joules. What amount of mass was converted to energy?

116. The largest atom bomb detonated was a hydrogen bomb tested by the Soviet Union in 1961. The bomb released 2.4×10^{17} joules of energy. What amount of mass was converted to energy in that explosion?
(*Source:* Cesare Emiliani, *The Scientific Companion;* Wiley Popular Science, 1995)

PUZZLE PROBLEM

Fill the blanks below with the numbers 1, 2, 3, 4, 5, 6, 7, 8, and 9, using each number once so that the equation is true.

$$\boxed{}\boxed{} \div \boxed{} = \boxed{}\boxed{} \div \boxed{} = \boxed{}\boxed{} \div \boxed{}$$

Of Interest

The first atomic bomb, tested in 1945, used the process of nuclear *fission*, which involves the splitting of atoms of heavy elements, such as uranium or plutonium. The split atoms, in turn, split other atoms resulting in a chain reaction in which some of the atomic mass is converted into energy.

In contrast, the hydrogen bomb uses nuclear *fusion*, which is how the stars release energy. In a fusion reaction, two nuclei are fused together to form a new atom. For example, two hydrogen nuclei can be forced together to form a helium nucleus. During this reaction, some of the mass of the two hydrogen nuclei is converted to pure energy so that the resulting helium nucleus has slightly less mass than the two original hydrogen nuclei.

REVIEW EXERCISES

1. The mass of a proton is 0.00000000000000000000000167 grams. Write this in scientific notation.

2. The speed of light is approximately 2.998×10^{10} centimeters per second. Write this in standard form.

3. Translate to an algebraic expression: one number plus half of a second number.

4. Find the area of a playground that is 17 feet by 20 feet.

5. Solve and check: $3 - y = 27$

6. Does the following graph represent a function? Explain.

Defined Terms

Section 5.1
Scientific notation (p. 387)

Section 5.2
Monomial (p. 396)

Coefficient (p. 398)
Degree of a monomial (p. 398)
Degree of a polynomial (p. 400)
Polynomial (p. 398)

Polynomial in one variable (p. 399)
Binomial (p. 399)
Trinomial (p. 399)

Section 5.5
Conjugates (p. 436)

Procedures, Rules, and Key Examples

Procedures/Rules	Key Examples

Section 5.1 Exponents and Scientific Notation

If the base of an exponential form is a negative number and the exponent is even, then the product is positive.

If the base is a negative number and the exponent is odd, then the product is negative.

If a and b are real numbers with $b \neq 0$ and n is a natural number, then $\left(\dfrac{a}{b}\right)^n = \dfrac{a^n}{b^n}$.

If a is a real number and $a \neq 0$, then $a^0 = 1$.

If a is a real number, where $a \neq 0$ and n is a natural number, then $a^{-n} = \dfrac{1}{a^n}$.

If a is a real number, where $a \neq 0$ and n is a natural number, then $\dfrac{1}{a^{-n}} = a^n$.

If a and b are real numbers, where $a \neq 0$ and $b \neq 0$ and n is a natural number, then $\left(\dfrac{a}{b}\right)^{-n} = \left(\dfrac{b}{a}\right)^n$.

To change from scientific notation with a positive integer exponent to standard notation, move the decimal point to the right the number of places indicated by the exponent.

To write a number expressed in scientific notation with a negative exponent in standard form, move the decimal point to the left the same number of places as the absolute value of the exponent.

Example 1: Evaluate.

a. $(-2)^4 = (-2)(-2)(-2)(-2) = 16$

b. $(-2)^5 = (-2)(-2)(-2)(-2)(-2) = -32$

Example 2: Simplify.

a. $\left(\dfrac{3}{4}\right)^2 = \dfrac{3^2}{4^2} = \dfrac{9}{16}$

b. $7^0 = 1$

c. $2^{-4} = \dfrac{1}{2^4} = \dfrac{1}{16}$

d. $\dfrac{1}{3^{-2}} = 3^2 = 9$

e. $\left(\dfrac{2}{3}\right)^{-4} = \left(\dfrac{3}{2}\right)^4 = \dfrac{81}{16}$

Example 3: Write in standard form.

a. $3.4 \times 10^5 = 340{,}000$

b. $4.2 \times 10^{-5} = 0.000042$

continued

Procedures/Rules	Key Examples

Procedures/Rules

To write a number greater than 1 in scientific notation:
1. Move the decimal point so that the number is greater than or equal to 1, but less than 10. (Place the decimal point to the right of the first nonzero digit.)
2. Write the decimal number multiplied by 10^n, where n is the number of places between the new decimal position and the original decimal position.
3. Delete 0's to the right of the last nonzero digit.

To write a positive decimal number that is less than 1 in scientific notation:
1. Move the decimal point so that the number is greater than or equal to 1, but less than 10. (*Tip:* Place the decimal point to the right of the first nonzero digit.)
2. Write the decimal number multiplied by 10^n, where n is a negative integer whose absolute value is the number of places between the new decimal position and the original decimal position.

Key Examples

Example 4: Write in scientific notation.
$$7{,}230{,}000 = 7.23 \times 10^6$$

Example 5: Write in scientific notation.
$$0.00000056 = 5.6 \times 10^{-7}$$

Section 5.2 Introduction to Polynomials

Descending order agreement: To write a polynomial in descending order, place the highest degree term first, then the next highest degree, and so on.

Example 1: Write in descending order.
$$2x^2 + 4x^5 + 9 - 6x^3 = 4x^5 - 6x^3 + 2x^2 + 9$$

Section 5.3 Adding and Subtracting Polynomials

To add polynomials, combine like terms.

Example 1: Add.
$$(5x^3 + 12x^2 - 9x + 1) + (7x^2 - x - 13)$$
$$= 5x^3 + 12x^2 + 7x^2 - 9x - x + 1 - 13$$
$$= 5x^3 + 19x^2 - 10x - 12$$

To subtract polynomials:
1. Write the subtraction statement as an equivalent addition statement.
   a. Change the operation symbol from a minus sign to a plus sign.
   b. Change the subtrahend (second polynomial) to its additive inverse. To get the additive inverse, we change the sign of each term in the polynomial.
2. Combine like terms.

Example 2: Subtract.
$$(6y^3 - 5y + 18) - (y^3 - 5y + 7)$$
$$= (6y^3 - 5y + 18) + (-y^3 + 5y - 7)$$
$$= 6y^3 - y^3 - 5y + 5y + 18 - 7$$
$$= 5y^3 + 11$$

Section 5.4 Exponent Rules and Multiplying Monomials

If a is a real number and m and n are integers, then $a^m \cdot a^n = a^{m+n}$. (Product Rule for Exponents)

To multiply monomials:
1. Multiply coefficients.
2. Add the exponents of the like bases.
3. Write any unlike variable bases unchanged in the product.

If a is a real number and m and n are integers, then $(a^m)^n = a^{mn}$.

If a and b are real numbers and n is an integer, then $(ab)^n = a^n b^n$.

To simplify a monomial raised to a power:
1. Evaluate the coefficient raised to that power.
2. Multiply each variable's exponent by the power.

Example 1: Multiply.
a. $x^3 \cdot x^4 = x^{3+4} = x^7$

b. $-6x^4yz^2 \cdot 5x^5y^3 = -30x^{5+4}y^{1+3}z^2$
$$= -30x^9y^4z^2$$

Example 2: Simplify.
a. $(x^3)^4 = x^{3 \cdot 4} = x^{12}$
b. $(m^2n^3)^5 = m^{2 \cdot 5}n^{3 \cdot 5} = m^{10}n^{15}$
c. $(3x^2y^5)^4 = 3^4x^{2 \cdot 4}y^{5 \cdot 4} = 81x^8y^{20}$

continued

Procedures/Rules	Key Examples

Procedures/Rules

Section 5.5 Multiplying Polynomials; Special Products

To multiply a polynomial by a monomial, use the distributive property to multiply each term in the polynomial by the monomial.

To multiply two polynomials:
1. Multiply every term in the first polynomial by every term in the second polynomial.
2. Combine like terms.

Special Products:

Conjugates: $(a + b)(a - b) = a^2 - b^2$

Squaring a sum: $(a + b)^2 = a^2 + 2ab + b^2$

Squaring a difference: $(a - b)^2 = a^2 - 2ab + b^2$

Section 5.6 Exponent Rules and Dividing Polynomials

Quotient rule for exponents: If m and n are integers and a is a real number, where $a \neq 0$, then $a^m \div a^n = a^{m-n}$.

To divide monomials:
1. Divide the coefficients.
2. Use the quotient rule for the exponents with like bases.
3. Unlike bases are written unchanged in the quotient.
4. Write the final expression so that all exponents are positive.

If a, b, and c are real numbers, variables, or expressions with $c \neq 0$, then

$$\frac{a + b}{c} = \frac{a}{c} + \frac{b}{c}$$

To divide a polynomial by a monomial, divide each term in the polynomial by the monomial.

To divide a polynomial by a polynomial:
1. Use long division.
2. If there is a remainder, write the result in the form:

$$\text{quotient} + \frac{\text{remainder}}{\text{divisor}}$$

Key Examples

Example 1: Multiply.

a. $3x(x^4 - 7x^2y + 9y^2 - 2)$
$= 3x \cdot x^4 - 3x \cdot 7x^2y + 3x \cdot 9y^2 - 3x \cdot 2$
$= 3x^5 - 21x^3y + 27xy^2 - 6x$

b. $(3x + 5)(2x - 1)$
$= 3x \cdot 2x + 3x \cdot (-1) + 5 \cdot 2x + 5(-1)$
$= 6x^2 - 3x + 10x - 5$
$= 6x^2 + 7x - 5$

c. $(4x + 3)(4x - 3) = 16x^2 - 9$

d. $(3y + 2)^2 = 9y^2 + 12y + 4$

e. $(3y - 2)^2 = 9y^2 - 12y + 4$

Example 1: Divide.

a. $x^7 \div x^2 = x^{7-2} = x^5$

b. $-39a^6b^5c^2 \div 13a^4b^5c$
$= -3a^{6-4}b^{5-5}c^{2-1}$
$= -3a^2b^0c^1$
$= -3a^2(1)c$
$= -3a^2c$

c. $\dfrac{30y^6 + 45y^3}{5y^2} = \dfrac{30y^6}{5y^2} + \dfrac{45y^3}{5y^2}$
$= 6y^{6-2} + 9y^{3-2}$
$= 6y^4 + 9y$

d. $\dfrac{6x^2 + 7x - 23}{3x - 4}$

$$
\begin{array}{r}
2x + 5 \\
3x - 4 \overline{) 6x^2 + 7x - 23} \\
-6x^2 + 8x \\
\hline
15x - 23 \\
-15x + 20 \\
\hline
-3
\end{array}
$$

Answer: $2x + 5 + \dfrac{-3}{3x - 4}$ or

$2x + 5 - \dfrac{3}{3x - 4}$

For Exercises 1–6, answer true or false.

1. A negative exponent will always make the coefficient negative.

2. The degree of a binomial is two.

3. To raise an exponential form to a power, multiply the exponents.

4. Conjugates occur when both binomials contain subtraction.

5. FOIL can be used for all types of polynomial multiplication.

6. $(x - 4)^2 = x^2 + 16$

For Exercises 7–10, complete the rule.

7. To add polynomials, _____ like terms.

8. To multiply monomials:

   a. Multiply the _____.

   b. Add the _____ of the like bases.

   c. Rewrite any unlike bases in the product.

9. $(a^m)^n = a^?$, where a is a real number and m and n are integers.

10. $(a + b)^2 = a^2 +$ _____ $+ b^2$

For Exercises 11–14, evaluate the exponential form.

11. $\left(\dfrac{2}{5}\right)^3$

12. -4^2

13. 5^{-2}

14. 13^{-1}

For Exercises 15–18, write the number in standard form.

15. The radioactive half-life of uranium 238 is 4.5×10^9 years.

16. The temperature at the Sun's core is 1.38×10^7 degrees Celsius.

17. The mass of a hydrogen atom is 1.663×10^{-24} grams.

18. The mass of an electron is 2.006×10^{-30} pounds.

For Exercises 19–22, write the number in scientific notation.

19. Because atoms are so small, their weight is measured in atomic mass units (AMU). 1 atomic mass unit is 0.0000000000000000000001661 grams.

20. One molecule of NaCl, sodium chloride (table salt) is 0.00000000000000000000000963 grams.

21. The speed of light is about 300,000,000 meters per second.

22. The approximate radius of Earth is 6,370,000 meters.

For Exercises 23–26, determine whether the expression is a monomial.

23. $3xy^5$

24. $-\dfrac{1}{2}x$

25. $\dfrac{4}{ab^3}$

26. $2x^2 - 9$

For Exercises 27–30, identify the coefficient and degree of each monomial.

27. $6x^4$

28. 27

29. $-2.6xy^3$

30. $-m$

For Exercises 31–34, determine whether the expression is a monomial, binomial, trinomial, or none of these.

31. $4x^2 - 25$

32. $-st^5$

33. $2x^2 - 5x + 7$

34. $5x^3 - 6x^2 - 3x + 11$

For Exercises 35–38, identify the degree of each polynomial and write each polynomial in descending order.

35. $15 - 2x^9 - 3x + 21x^5 - 19x^3$

36. $22j^2 - 19 - j^4 + 5j$

37. $4u^5 - 18u^3 + 13u^4 - u - 21$

38. $16v^3 + 21v^4 - 2v + 6v^8 - 19 - v^2$

For Exercises 39–49, combine like terms. If possible, write the resulting polynomial in descending order.

39. $8y^5 - 3y^4 + 2y - 4y^5 - 2y + 8 + 7y^4$

40. $8l + 5l^4 - 2l^3 - l + 18l^5 + 6l^2 + 20$

41. $3m - 4 - m^2 - 2m^3 + 7 - 2m + 8m^2$

42. $-9 + 7y - 2y - 6 + 2y - 3y^2$

43. $a^2bc - 3ab^2c + 2ab^2c - 6$

44. $5xyz^2 - 4x^2yz - 7xyz^2 - 5x^2yz - 2x^2zy$

45. $6cd - 4cd^2 + 3cd^2 + 8c^3 + 5cd - 12c^3 + 2d + 8$

46. $4a^2bc + 5abc^3 - 8abc - 9a^5 - 2abc^3 - 4a^2bc + 8 + a^5$

47. $18jk - 2j^4 + 12jk^3 - j^4 - 8jk - 9jk^3 + 12 + k^2$

48. $4m^2 - 3mn + 2mn^2 - 8n^2 - 6mn - 4m^2 + 7 + mn^2$

49. $2a + 3b - 4ab - c - 18c - 20a + b - 4b + 8$

For Exercises 50–59, add or subtract and write the resulting polynomial in descending order.

50. $(5y - 2) + (6y - 8)$

51. $(8x^2 - 3) + (2x^2 + 3x - 1)$

52. $(5m - 1) - (2m + 7)$

53. $(5n + 8) - (2n^3 - 3n^2 + n + 5)$

54. $(2y - 4) + (4y + 8) - (6y - 5)$

55. $(3x^3 - 2x + 8) + (4x^2 - 3x - 1) - (4x^3 - 5x - 10)$

56. $(4p^3 + 2p^2 - 8p - 1) + (-5p^3 + 8p^2 - 2p + 10)$

57. $(2x^3 - 3x^2 - 2x - 1) - (18x^3 - 3x^2 + 2x + 1)$

58. $(5x^2 - 2xy + y^2) + (3x^2 + xy + 5y^2)$

59. $(m^2 + 5mn + 6) - (4m^2 - 3mn + 8)$

For Exercises 60–74, simplify. Write all answers with positive exponents.

60. $m \cdot m^4$

61. $2a \cdot 5a^8$

62. $(-3x^2y)(2x^4y^5)$

63. $(5x^2y)^2$

64. $(-6u^3)^2(2u^4)$

65. $(2x^2)(2x^2)^4$

66. $\dfrac{x^3}{x^{-5}}$

67. $\dfrac{u^{-1}}{u^{-8}}$

68. $\dfrac{s^{-5}}{s}$

69. $\left(\dfrac{1}{x^3}\right)^{-2}$

70. $(2a^3b^{-2})^{-3}$

71. $\dfrac{4hj^4k^{-2}}{(3hj^{-2})^2}$

72. $-5x^{-2}$

73. $8x^0 - (2x)^0$

74. $(-18x^4y^{-5})^0$

For Exercises 75–92, multiply.

75. $2(x - 21)$

76. $-8(2x - 1)$

77. $4a(a - 3)$

78. $-6b(2b^2 - 4b - 1)$

79. $4m^3(-2m^4 - 3m^2 + m + 5)$

80. $-4abc^2(3a - 4ab^3 + 2abc^2 + 8)$

81. $(x + 5)(x - 1)$

82. $(y - 3)(y - 8)$

83. $(2m + 5)(6m - 1)$

84. $(5a + 2b)(3a - 2b)$

85. $(2x + 1)(2x - 1)$

86. $(x - 3)(x^2 + 2x + 5)$

87. $(3y - 1)(y^2 - 2y + 4)$

88. $(a^2 + ab + b^2)(a^2 - 2ab - b^2)$

89. $(4 - x)(4 + x)$

90. $(x - 6)^2$

91. $(3r - 5)^2$

92. $(6s + 2r)(6s - 2r)$

93. a. Write an expression for the area of the figure shown.

b. Calculate the area if $x = 3$ feet.

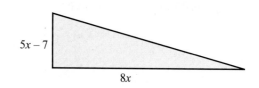
$5x - 7$
$8x$

94. A park ranger is studying the number of trees per square meter in a national forest. The area in the study is rectangular and measures 9×10^3 meters by 2.5×10^4 meters. Find the area of this region. Write your answer in scientific notation.

95. In a circuit, the current is measured to be 3.5×10^{-3} amperes. If the circuit has a resistance of 8.4×10^{-2} ohms, find the voltage. (Hint: Voltage = current × resistance)

For Exercises 96–101, divide.

96. $\dfrac{4x - 20}{4}$

97. $\dfrac{5y^2 - 10y + 15}{-5}$

98. $\dfrac{2st + 20s^2t^4 - 4st^5}{2st}$

99. $\dfrac{a^3bc^3 - abc^2 + 2a^2bc^4}{abc^3}$

100. $\dfrac{6a^2b^2 + 3a^2b^3}{3a^2b^2}$

101. $\dfrac{2xyz^2 - 5x^2y^2z^2 + 10xy^2z}{5xy^3z}$

102. a. The area of the parallelogram shown is described by the polynomial $54x^2 - 60x$. Find the height.

b. Find the base, height, and area of this parallelogram if $x = 4$ inches.

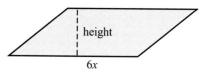
height
$6x$

103. Proxima Centauri is approximately 2.47×10^{13} miles from the Sun. If light travels 5.881×10^{12} miles per year, how long does the light from Proxima Centauri take to reach Earth? (Hint: Distance = rate × time)

For Exercises 104–111, use long division to divide the polynomials.

104. $\dfrac{z^2 + 8z + 15}{z + 3}$

105. $\dfrac{3m^2 - 17m + 11}{m - 5}$

106. $\dfrac{2y^2 - 7y - 8}{2y + 1}$

107. $\dfrac{6m^2 - 10m + 5}{3m - 2}$

108. $\dfrac{x^3 - 1}{x + 1}$

109. $\dfrac{8x^3 - 27}{2x - 3}$

110. $\dfrac{2s^3 - 2s - 3}{s - 1}$

111. $\dfrac{10y^4 - 5y^3 - 6y^2 + 7y - 2}{2y - 1}$

112. A storage room is to have an area described by $4x^2 + 23x - 35$. If the length is described by $x + 7$, then find an expression for the width.

For Exercises 1 and 2, evaluate the exponential form.

1. 2^{-3}

2. $\left(\dfrac{2}{3}\right)^{-2}$

3. Write 6.201×10^{-3} in standard form.

4. Write 275,000,000 in scientific notation.

For Exercises 5 and 6, identify the degree.

5. $-7x^2y$

6. $4x^2 - 9x^4 + 8x - 7$

7. Evaluate $-6mn - n^3$, where $m = 4$ and $n = -2$.

8. Combine like terms and write the resulting polynomial in descending order of degree.

$$-5x^4 + 7x^2 + 6x^2 - 5x^4 + 12 - 6x^3 + x^2$$

9. Add $(3x^2 + 4x - 2) + (5x^2 - 3x - 2)$.

10. Subtract $(7x^4 - 3x^2 + 4x + 1) - (2x^4 + 5x - 7)$.

For Exercises 11–18, multiply.

11. $(4x^2)(3x^5)$

12. $(2ab^3c^7)(-a^5b)$

13. $(4xy^3)^2$

14. $3x(x^2 - 4x + 5)$

15. $-6t^2u(4t^3 - 8tu^2)$

16. $(n - 1)(n + 4)$

17. $(2x - 3)^2$

18. $(x + 2)(x^2 - 4x + 3)$

19. Write an expression for the area of the shape shown.

$2n - 5$

$3n + 4$

For Exercises 20–25, simplify.

20. $x^9 \div x^4$

21. $\dfrac{(x^3)^{-2}}{x^4 \cdot x^{-5}}$

22. $\dfrac{(3y)^{-2}}{(x^3y^2)^{-3}}$

23. $\dfrac{24x^5 + 18x^3}{6x^2}$

24. $\dfrac{x^2 - x - 12}{x + 3}$

25. $\dfrac{15x^2 - 22x + 14}{3x - 2}$

Factoring

In Chapter 5 we learned about a specific class of expressions called *polynomials*. We learned that polynomials can be added, subtracted, multiplied, and divided in much the same way as numbers. In Chapter 6 we will explore polynomials further and in Sections 6.1 through 6.5, we will learn how to factor polynomial expressions. Factoring is another way to rewrite an expression and is one of the most important skills you will acquire in algebra. It is used extensively in simplifying the more complex expressions that we will study in future chapters. In Section 6.6 we will use factoring to solve certain equations that contain polynomial expressions. Finally, in Section 6.7, we will graph equations and functions containing polynomial expressions.

"Discipline is the keynote to learning. Discipline has been the great factor in my life."

—*ROSE HOFFMAN*
Elementary schoolteacher

"The higher processes are all simplification."

—*WILLA CATHER*
U.S. novelist (1873–1947)

6.1 **Greatest Common Factor and Factoring by Grouping**

OBJECTIVES

1 List all possible factors for a given number.

2 Find the greatest common factor of a set of numbers or monomials.

3 Write a polynomial as a product of a monomial GCF and a polynomial.

4 Factor by grouping.

Often in mathematics we need to consider the factors of a number or expression. A number or expression written as a product of factors is said to be in **factored form**.

DEFINITION | *Factored form:* A number or expression written as a product of factors.

Following are some examples of factored form:

An integer written in factored form with integer factors: $28 = 2 \cdot 14$

A monomial written in factored form: $8x^5 = 4x^2 \cdot 2x^3$

A polynomial written in factored form with a monomial factor and a polynomial factor: $2x + 8 = 2(x + 4)$

A polynomial written as a product of polynomial factors:
$x^2 + 5x + 6 = (x + 2)(x + 3)$

Notice we can check the factored form by multiplying the factors to equal the product. The process of writing factored form is called *factoring*.

OBJECTIVE **1** **List all possible factors for a given number.** We begin our exploration of factoring by listing natural number factors when given an integer.

EXAMPLE **1** List all natural number factors of 24.

Solution To list all the natural number factors, we can divide 24 by 1, 2, 3, and so on, writing each divisor and quotient pair as a product until we have all possible combinations.

> **Note:** The number 5 does not divide 24 evenly and 6 is already in the 4 · 6 pair. When you reach a natural number that you already have listed, you can stop.

$1 \cdot 24$
$2 \cdot 12$
$3 \cdot 8$
$4 \cdot 6$

The natural number factors of 24 are 1, 2, 3, 4, 6, 8, 12, 24.

ANSWERS
a. 1, 2, 3, 5, 6, 10, 15, 30
b. 1, 2, 3, 6, 7, 14, 21, 42

YOUR TURN List all natural number factors of each given number.

a. 30 **b.** 42

OBJECTIVE **2** **Find the greatest common factor of a set of numbers or monomials.** When factoring polynomials, the first step is to determine if there is a monomial factor that is common to all the terms in the polynomial. Further, we will want that monomial factor to be the **greatest common factor** of the terms. Let's first consider how to find the greatest common factor of numbers.

DEFINITION

Greatest common factor (GCF): The largest natural number that divides all given numbers evenly.

For example, the greatest common factor of 12 and 18 is 6 because 6 is the largest number that divides into both 12 and 18 evenly.

Using Listing to Find the GCF of Numbers

One way to find the GCF of a given set of numbers is by listing factors.

PROCEDURE LISTING METHOD FOR FINDING GCF

To find the GCF of a set of numbers by listing:
1. List all possible factors for each given number.
2. Search the lists for the largest factor common to all lists.

EXAMPLE **2** Find the GCF of 24 and 60.

Solution Factors of 24: 1, 2, 3, 4, 6, 8, 12, 24

Factors of 60: 1, 2, 3, 4, 5, 6, 10, 12, 15, 20, 30, 60

The GCF of 24 and 60 is 12.

Using Prime Factorization to Find the GCF

The listing method is a good method for smaller numbers, but for larger numbers, it is not the most efficient method to use. It turns out we can use prime factorization, which we explained in Section 1.2, to find the GCF of a given set of numbers. Consider the prime factorizations for the numbers from Example 2.

$$24 = 2 \cdot 2 \cdot 2 \cdot 3 = 2^3 \cdot 3$$
$$60 = 2 \cdot 2 \cdot 3 \cdot 5 = 2^2 \cdot 3 \cdot 5$$

In Example 2 we found the GCF to be 12. Let's look at the prime factorization of 12 to see if we can discover a rule for finding the GCF.

$$12 = 2 \cdot 2 \cdot 3 = 2^2 \cdot 3$$

Notice that the 12 contains two factors of 2 and one factor of 3, which are the factors that are common to 24's and 60's factorizations. This suggests that we only use primes that are common to all factorizations involved. The factor 5 is not common to both 24's and 60's factorizations, so it is not included in the GCF.

$$24 = 2 \cdot 2 \cdot 2 \cdot 3 = 2^3 \cdot 3$$
$$60 = 2 \cdot 2 \cdot 3 \cdot 5 = 2^2 \cdot 3 \cdot 5$$
$$GCF = 2 \cdot 2 \cdot 3 = 2^2 \cdot 3 = 12$$

Notice we could also compare exponents for a given common prime factor to determine how many of that factor should be included in the GCF. For example, 2^2 is included instead of 2^3 because 2^2 has the smaller exponent. This suggests the following procedure.

PROCEDURE PRIME FACTORIZATION METHOD FOR FINDING GCF

To find the greatest common factor of a given set of numbers:

1. Write the prime factorization of each given number in exponential form.
2. Create a factorization for the GCF that includes only those prime factors common to all the factorizations, each raised to its smallest exponent in the factorization.
3. Multiply.

Note: If there are no common prime factors, then the GCF is 1.

EXAMPLE 3 Find the GCF of 3024 and 2520.

Solution Write the prime factorization of 3024 and 2520 in factored form.

$$3024 = 2^4 \cdot 3^3 \cdot 7$$
$$2520 = 2^3 \cdot 3^2 \cdot 5 \cdot 7$$

The common prime factors are 2, 3, and 7. Now we compare the exponents of each of these common factors. The smallest exponent of the factor of 2 is 3. The smallest exponent of 3 is 2. The smallest exponent of 7 is 1. The GCF will be the product of 2^3, 3^2, and 7.

$$GCF = 2^3 \cdot 3^2 \cdot 7 = 8 \cdot 9 \cdot 7 = 504$$

GCF of Monomials

We use a similar approach to find the GCF of a set of monomials that have variables. The variables in the monomials are treated like prime factors.

EXAMPLE 4 Find the GCF of $24x^2y$ and $60x^3$.

Solution Write the prime factorization of each monomial, treating the variables like prime factors.

$$24x^2y = 2^3 \cdot 3 \cdot x^2 \cdot y$$
$$60x^3 = 2^2 \cdot 3 \cdot 5 \cdot x^3$$

The common prime factors are 2, 3, and x. The smallest exponent for the factor of 2 is 2. The smallest exponent of 3 is 1. The smallest exponent of the x is 2. The GCF will be the product of 2^2, 3, and x^2.

$$GCF = 2^2 \cdot 3 \cdot x^2 = 12x^2$$

YOUR TURN Find the GCF.

    a. 63, 84, and 105   **b.** $32a^2b$ and $40abc$ **c.** $35x^2$ and $18y$

OBJECTIVE 3 Write a polynomial as a product of a monomial GCF and a polynomial. Recall that when we multiply a monomial by a polynomial, we apply the distributive property.

$$2(x + 4) = 2 \cdot x + 2 \cdot 4$$
$$= 2x + 8$$

Now, let's reverse this process. Suppose we are given the product, $2x + 8$, and we want to write the factored form, $2(x + 4)$. Notice that terms in $2x + 8$ have 2 as a common factor. When writing factored form, we first determine the GCF of the terms. We then create a missing factor statement like this:

$$2x + 8 = 2\,(?)$$

Notice that the missing factor must be $x + 4$. To determine this missing factor, we divide the product $2x + 8$ by the known factor, 2.

$$2x + 8 = 2\,(?)$$
$$\frac{2x + 8}{2} = (?)$$
$$\frac{2x}{2} + \frac{8}{2} = (?)$$
$$x + 4 = (?)$$

CONNECTION

In Section 5.6, we learned that we divide a polynomial by a monomial by dividing each term in the polynomial by the monomial.

The factored form is written as $2x + 8 = 2(x + 4)$, which suggests the following procedure.

To factor a monomial GCF out of a given polynomial:

1. Find the GCF of the terms that make up the polynomial.
2. Rewrite the given polynomial as a product of the GCF and parentheses that contain the result of dividing the given polynomial by the GCF.

$$\text{given polynomial} = \text{GCF}\left(\frac{\text{given polynomial}}{\text{GCF}}\right)$$

EXAMPLE 5 Factor $12x^2 + 18x$.

Solution 1. Find the GCF of $12x^2$ and $18x$.

The GCF of $12x^2$ and $18x$ is $6x$.

Note: 6 is the largest number that divides both 12 and 18 evenly, and x has the smaller exponent of the x^2 and x.

2. Write the given polynomial as the product of the GCF and parentheses containing the quotient of the given polynomial and the GCF.

$$12x^2 + 18x = 6x\left(\frac{12x^2 + 18x}{6x}\right)$$

$$= 6x\left(\frac{12x^2}{6x} + \frac{18x}{6x}\right) \quad \text{Separate the terms.}$$

$$= 6x(2x + 3) \quad \text{Divide the terms by the GCF.}$$

Check We can check by multiplying the factored form using the distributive property.

$$6x(2x + 3) = 6x \cdot 2x + 6x \cdot 3 \quad \text{Distribute } 6x.$$
$$= 12x^2 + 18x \quad \text{The product is the original polynomial.}$$

CONNECTION

Keep in mind that when we factor, we are simply writing the original expression in a different form called *factored form*. When written equal to each other, the factored form and product make an identity, which means that every real number is a solution to the equation. Consider $12x^2 + 18x = 6x(2x + 3)$. Let's choose a value such as $x = 2$ and verify that it satisfies the equation. (Remember, we can select any real number.)

$$12x^2 + 18x = 6x(2x + 3)$$
$$12(2)^2 + 18(2) \stackrel{?}{=} 6(2)(2(2) + 3)$$
$$12(4) + 36 \stackrel{?}{=} 12(4 + 3)$$
$$48 + 36 \stackrel{?}{=} 12(7)$$
$$84 = 84 \quad \text{It checks.}$$

EXAMPLE 6 Factor $24x^2y - 60x^3$.

Solution 1. Find the GCF of $24x^2y$ and $60x^3$.

We found this GCF in Example 4 to be $12x^2$.

2. Write the given polynomial as the product of the GCF and the parentheses containing the quotient of the given polynomial and the GCF.

$$24x^2y - 60x^3 = 12x^2 \left(\frac{24x^2y - 60x^3}{12x^2} \right)$$

$$= 12x^2 \left(\frac{24x^2y}{12x^2} - \frac{60x^3}{12x^2} \right)$$

$$= 12x^2 (2y - 5x)$$

CONNECTION ————

Remember that when we divide exponential forms that have the same base, we subtract exponents.

Check Multiply the factored form using the distributive property.

$$12x^2 (2y - 5x) = 12x^2 \cdot 2y - 12x^2 \cdot 5x$$
$$= 24x^2y - 60x^3$$

Factoring When the First Term Is Negative

Consider factoring the expression $-6x + 10y$. Because the first term of the polynomial is negative, when we factor out the GCF, 2, the first term inside the parentheses is also negative, so that we have $2(-3x + 5y)$. However, it is considered undesirable to have the first term inside parentheses negative. We can avoid a negative first term in the parentheses by factoring the negative of the GCF, so in our example we would factor out -2. This will change the sign of each term inside the parentheses so that we have $-2(3x - 5y)$.

EXAMPLE 7 Factor $-18x^4y^3 + 9x^2y^2z - 12x^3y$.

Solution 1. Find the GCF of $-18x^4y^3$, $9x^2y^2z$, and $-12x^3y$.

Because the first term in the polynomial is negative, we will factor out the negative of the GCF to avoid a negative first term inside the parentheses. We will factor out $-3x^2y$.

2. Write the given polynomial as the product of the GCF and the parentheses containing the quotient of the given polynomial and the GCF.

$$-18x^4y^3 + 9x^2y^2z - 12x^3y = -3x^2y \left(\frac{-18x^4y^3 + 9x^2y^2z - 12x^3y}{-3x^2y} \right)$$

$$= -3x^2y \left(-\frac{18x^4y^3}{-3x^2y} + \frac{9x^2y^2z}{-3x^2y} - \frac{12x^3y}{-3x^2y} \right)$$

$$= -3x^2y (6x^2y^2 - 3yz + 4x)$$

Check Multiply the factored form using the distributive property.

$$-3x^2y\,(6x^2y^2 - 3yz + 4x) = -3x^2y \cdot 6x^2y^2 - 3x^2y \cdot (-3yz) - 3x^2y \cdot 4x$$
$$= -18x^4y^3 + 9x^2y^2z - 12x^3y$$

YOUR TURN   Factor.

    a. $30xy - 45x$ **b.** $-48a^4b^5 - 24a^3b^4 + 16ab^2c$

Before we discuss additional techniques of factoring, it is important to state that no matter what type of polynomial we are asked to factor, we will always first consider whether a monomial GCF (other than 1) can be factored out of the polynomial.

Factoring When the GCF Is a Polynomial

Sometimes, when factoring, the GCF is a polynomial.

EXAMPLE 8   Factor $y(x + 2) + 7(x + 2)$.

Solution Notice that this expression is a sum of two products, y and $(x + 2)$, and 7 and $(x + 2)$. Further, note that $(x + 2)$ is the GCF of the two products.

$$y(x + 2) + 7(x + 2) = (x + 2)\left(\frac{y(x + 2) + 7(x + 2)}{x + 2}\right)$$

$$= (x + 2)\left(\frac{y(x + 2)}{x + 2} + \frac{7(x + 2)}{x + 2}\right)$$

$$= (x + 2)(y + 7)$$

Note: The parentheses are filled in the same way, by dividing the original expression by the GCF.

YOUR TURN

Factor $4a(a - 3) - b(a - 3)$.

OBJECTIVE 4 **Factor by grouping.** The process of factoring out a polynomial GCF, as we did in Example 8, is an intermediate step in a process called *factoring by grouping*, which is a technique that we try when factoring a four-term polynomial such as $xy + 2y + 7x + 14$. The method is called *grouping* because we group pairs of terms together and look for a common factor within each group. We begin by pairing the first two terms together as one group and the last two terms as a second group.

$$xy + 2y + 7x + 14 = (xy + 2y) + (7x + 14)$$

Note that the first two terms have a common factor of y and the last two terms have 7 as a common factor. If we factor the y out of the first two terms and the 7 out of the last two terms, we have the same expression that we factored in Example 8:

$$
\begin{aligned}
xy + 2y + 7x + 14 &= (xy + 2y) + (7x + 14) \\
&= y(x + 2) + 7(x + 2) \\
&= (x + 2)(y + 7) \qquad \text{Factor out } (x + 2).
\end{aligned}
$$

PROCEDURE FACTORING BY GROUPING

To factor a four-term polynomial by grouping:

1. Factor out any monomial GCF (other than 1) that is common to all four terms.
2. Group together pairs of terms and factor the GCF out of each pair.
3. If there is a common binomial factor, then factor it out.
4. If there is no common binomial factor, then interchange the middle two terms and repeat the process. If there is still no common binomial factor, then the polynomial cannot be factored by grouping.

EXAMPLE 9 Factor $6x^3 - 8x^2 + 3xy - 4y$.

Solution First, we look for a monomial GCF (other than 1). This polynomial does not have one. Because the polynomial has four terms, we now try to factor by grouping.

$$
\begin{aligned}
6x^3 - 8x^2 + 3xy - 4y &= (6x^3 - 8x^2) + (3xy - 4y) && \text{Factor } 2x^2 \text{ out of } 6x^3 \text{ and} \\
& && 8x^2, \text{ then factor } y \text{ out of } 3xy \\
&= 2x^2(3x - 4) + y(3x - 4) && \text{and } 4y. \\
&= (3x - 4)(2x^2 + y) && \text{Factor out } 3x - 4.
\end{aligned}
$$

EXAMPLE 10 Factor $12mn^2 - 20mn - 24n^2 - 40n$.

Solution Note that in this case there is a monomial GCF, $4n$, so we first factor this GCF out of all four terms.

$$
12mn^2 - 20mn - 24n^2 + 40n = 4n(3mn - 5m - 6n + 10)
$$

Because the polynomial in the parentheses has four terms, we try to factor by grouping.

$$= 4n[(3mn - 5m) + (-6n + 10)]$$

Factor m out of $3mn$ and $5m$, then factor -2 out of $-6n$ and 10.

$$= 4n[m(3n - 5) - 2(3n - 5)]$$

$$= 4n(3n - 5)(m - 2)$$

Factor out $3n - 5$.

Note: Now that the expression is in factored form, we no longer need the brackets.

a. $(2x + 5z)(3x + y)$
b. $2a(3b - 4)(7a - 2)$

YOUR TURN   **Factor.**

a. $6x^2 + 15xz + 2xy + 5yz$

b. $42a^2b - 56a^2 - 12ab + 16a$

6.1 EXERCISES

InterAct Math

Tutor Center   Addison-Wesley Math Tutor Center

Video 10/DVT 3

MyMathLab

MathXL   MathXL

1. Explain in your own words how to list all possible factors of a number.

2. Define the GCF of a number in your own words.

3. Explain how to use prime factorization to determine the GCF of a given set of numbers.

4. Explain how to factor a monomial GCF out of the terms in a polynomial.

472   **CHAPTER 6** Factoring

5. What kinds of polynomials are factored by grouping?

6. Explain how to factor by grouping.

For Exercises 7–20, list all possible factors of the given number.

7. 60

8. 36

9. 18

10. 21

11. 45

12. 49

13. 9

14. 33

15. 44

16. 12

17. 90

18. 81

19. 56

20. 64

For Exercises 21–32, find the GCF.

21. 21, 30

22. 72, 80

23. 34, 51

24. 35, 60

25. 12, 42, 60

26. 10, 18, 36

27. $8x, 2xy$

28. $5x^4, 10x^3$

29. $4xy, xy$

30. $25h^2, 60h$

31. $6a^5b, 15a^4b^2$

32. $7m^6n, 21m^5n^4$

For Exercises 33–59, factor out the GCF.

33. $5c - 20$

34. $8y - 24$

35. $x^2 - x$

36. $3k^2 - 24k$

37. $6a^2b - 3ab^2$

38. $9m^2n - 2n^2$

39. $7uv^2 - 7uv$

40. $8x + 8y - 2xy$

41. $25xy - 50xz + 100x$

42. $x^2y + xy^2 + x^3y^3$

43. $w^3 + 3w^2v + 3wv^2$

44. $am^2 + am + 3a$

45. $18ab^3c - 36a^2b^2c$

46. $-20p^2q - 24pq + 16pq^2$

47. $4xy - 8x^2y + 15xy^2$

48. $6a^2c - 18abc + 12ac^2$

49. $24m - 16n + 24b$

50. $4axy^2 - 6axyz + 2ayz$

51. $3l^2mn^5 + 18lmn^3 - 6lmn$

52. $21t^4u^3 - 105t^5u^3v + 63t^6u^2v^4$

53. $105a^3b^2 - 63a^2b^3 + 84a^6b^4$

54. $10g + 40g^3 - 100g^4 + 120g^5$ **55.** $18x^4 - 9x^3 + 30x^2 - 12x$ **56.** $a(x + y) + b(x + y)$

57. $2(a - 3) + y(a - 3)$ **58.** $4m(2x - 3y) - 5(2x - 3y)$ **59.** $ay(4x - 3) - 2(4x - 3)$

For Exercises 60–81, factor by grouping.

60. $cx + cy + bx + by$ **61.** $bx + 2b + cx + 2c$ **62.** $x^3 + 2x^2 + 3x + 6$

63. $y^4 + 4y^3 - by - 4b$ **64.** $1 - m + m^2 - m^3$ **65.** $3xy + 5y + 6x + 10$

66. $xy + y + x + 1$ **67.** $ax^2 - 5y^2 + ay^2 - 5x^2$ **68.** $4b^2 - b + 4b - 1$

69. $u^2 - 6u - 4u + 24$ **70.** $3ax + 6ay + 8by + 4bx$ **71.** $6 + 2b - 3a - ab$

72. $x^3 - 4x^2 + x - 4$ **73.** $am - an - bm + bn$ **74.** $3x + 3y - x^2 - xy$

75. $x^2y - x^2s - ry + rs$ **76.** $8u + uv - 5v^2 - 40v$ **77.** $w^2 + 3wz + 5w + 15z$

78. $ac + 2ad + 2bc + 4bd$ **79.** $3st + 3ty - 2s - 2y$ **80.** $my + 3y - 5m - 15$

81. $4ab + 4x + abc + cx$

For Exercises 82–90, factor completely.

82. $10a^2b^2 - 10b^3 + 15a^2b - 15b^2$ **83.** $3m^2n - 12m^2 - 12m^3 + 3mn$

84. $2a^2 + 2ab + 6a + 6b$ **85.** $2x^3 + 6x^2y + 10x^2 + 30xy$

86. $10uv^2 - 5v^2 + 30uv - 15v$ **87.** $7gh^2 + 7h^3 + gh^3 + h^4$

88. $12pq + 8q + 30p + 20$ **89.** $3a^2y - 12a^2 + 9ay - 36a$

90. $4a^2b - 4a^2 - 4ab + 4a$

For Exercises 91–92, write an expression for the area of the shaded region, then factor completely.

91.

92.

93. The diagram below shows the floor plan of a room with the hearth of a fireplace shown. Write an expression in factored form of the area of the room excluding the hearth, which is 4 feet by 1 foot.

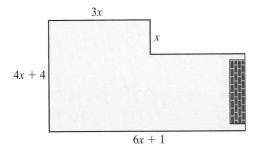

94. Write an expression in factored form for the area of the side of the house shown excluding the window.

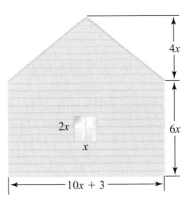

For Exercises 95–96, write an expression for the volume of the object shown, then factor completely.

95.

96.

On the Surface

Recall that surface area describes the area of the outside surfaces of an object. In this exercise, you will write an expression for the surface area of a cylinder.

1. Draw and label each surface of a cylinder. (*Hint:* Imagine separating the surfaces and laying them flat.)

2. Write an expression for the area of each surface.

3. Write an expression for the surface area of a cylinder.

4. Write the expression for the surface area of a cylinder in factored form.

5. Measure the radius and height of a soda can. Use these values to calculate the surface area of one soda can.

6. If a factory produces 5000 soda cans each day, how many square feet of aluminum are needed to produce that many cans?

REVIEW EXERCISES

1. Write 3.74×10^9 in standard form.

2. Write 45,600,000 in scientific notation.

For Exercises 3–6, multiply.

3. $(x + 2)(x + 5)$

4. $(x - 4)(x + 3)$

5. $(x - 5)(x - 3)$

6. $x(x - 7)$

6.2

Factoring Trinomials of the Form $x^2 + bx + c$

OBJECTIVES

1. Factor trinomials of the form $x^2 + bx + c$.
2. Factor out a monomial GCF, then factor the trinomial of the form $x^2 + bx + c$.

OBJECTIVE 1 **Factor trinomials of the form $x^2 + bx + c$.** In this section we consider trinomials of the form $x^2 + bx + c$. Notice that the coefficient of the squared term is 1. In Section 6.3, we will consider trinomials of this form in which the squared term has a coefficient other than 1. Following are some examples of trinomials of the form $x^2 + bx + c$.

$$x^2 + 5x + 6 \quad \text{or} \quad x^2 - 7x + 12 \quad \text{or} \quad x^2 + 2x - 15 \quad \text{or} \quad x^2 - 5x - 24$$

Trinomials of the form $x^2 + bx + c$ are the result of the product of two binomials. Consider $(x + 2)(x + 3)$. Recall from Section 5.5 that we use FOIL to multiply two binomials. In our use of FOIL to multiply $(x + 2)(x + 3)$, pay attention to how the last term and the middle term of the resulting trinomial are generated.

$$
\begin{array}{cccc}
& \text{F} & \text{O} \quad \text{I} & \text{L} \\
(x + 2)(x + 3) = & x^2 & + 3x + 2x & + 6 \\
= & x^2 & + 5x & + 6
\end{array}
$$

Note: These two numbers multiply to equal the last term, 6, and add to equal the coefficient of the middle term, 5.

sum of 2 and 3

product of 2 and 3

When we factor a trinomial of the form $x^2 + bx + c$, we reverse the FOIL process, using the fact that b is the sum of the last terms in the binomials and c is the product of the last terms in the binomials. For example, to factor $x^2 + 5x + 6$, we must find two binomials with last terms whose product is 6 and whose sum is 5.

$$x^2 + 5x + 6 = (x +)(x +)$$

The product of these numbers must be 6 and their sum must be 5.

If unsure of what those two numbers are, we could list all factor pairs of 6, and then find the pair whose sum is 5.

Factors Pairs of 6	Corresponding Sum
$1 \cdot 6$	$1 + 6 = 7$
$2 \cdot 3$	$2 + 3 = 5$ ← This is the correct pair.

Notice the 2 and 3 become the last terms in the binomials so that factored form for $x^2 + 5x + 6$ looks like this:

$$x^2 + 5x + 6 = (x + 2)(x + 3)$$

This suggests the following procedure.

PROCEDURE FACTORING $x^2 + bx + c$

To factor a trinomial of the form $x^2 + bx + c$:

1. Find two numbers with a product equal to c and a sum equal to b.
2. The factored trinomial will have the form:
   (x + first number)(x + second number).

Note: The signs in the binomial factors can be minus signs, depending on the signs of b and c.

In the following examples, we explore cases where minus signs occur. Pay attention to how we determine the signs in the binomial factors.

EXAMPLE 1 Factor $x^2 - 7x + 12$.

Solution We must find a pair of numbers whose product is 12 and whose sum is -7. If two numbers have a positive product and negative sum, they must both be negative. Following is a table listing the products and sums:

Product	Sum
$(-1)(-12) = 12$	$-1 + (-12) = -13$
$(-2)(-6) = 12$	$-2 + (-6) = 28$
$(-3)(-4) = 12$	$-3 + (-4) = -7$ ◄—

This is the correct combination so -3 and -4 are the second terms in each binomial of the factored form.

Answer: $x^2 - 7x + 12 = (x - 3)(x - 4)$

Note: We could have first written $(x + (-3))(x + (-4))$ to follow the procedure and then simplified.

Check We can check by multiplying the binomial factors to see if their product is the original polynomial.

$$(x - 3)(x - 4) = x^2 - 4x - 3x + 12$$
$$= x^2 - 7x + 12$$

Multiply the factors using FOIL.
The product is the original polynomial.

YOUR TURN

Factor $y^2 - 11y + 18$.

ANSWER

$(y - 2)(y - 9)$

EXAMPLE 2 Factor $x^2 + 2x - 15$.

Solution We must find a pair of numbers whose product is -15 and whose sum is 2. Since the product is negative, the two numbers must have different signs. Since the sum is positive, the number with the greater absolute value will be positive.

Product	Sum
$(-1)(15) = -15$	$-1 + 15 = 14$
$(-3)(5) = -15$	$-3 + 5 = 2$

This is the correct combination. So -3 and 5 are the second terms in each binomial of the factored form.

Answer: $x^2 + 2x - 15 = (x - 3)(x + 5)$

Check $(x - 3)(x + 5) = x^2 + 5x - 3x - 15$ Multiply the factors using FOIL.
$$= x^2 + 2x - 15$$ The product is the original polynomial.

YOUR TURN

Factor $t^2 + 3t - 28$.

ANSWER

$(t - 4)(t + 7)$

EXAMPLE 3 Factor $x^2 - 5x - 24$.

Solution We must find a pair of numbers whose product is -24 and whose sum is -5. Since the product is negative, the two numbers must have different signs. Since the sum is also negative, the number with the greater absolute value will be negative.

Product	Sum
$(1)(-24) = -24$	$1 + (-24) = -23$
$(2)(-12) = -24$	$2 + (-12) = -10$
$(3)(-8) = -24$	$3 + (-8) = -5$

This is the correct combination. So 3 and -8 are the second terms in each binomial of the factored form.

Note: In Examples 3–6, we leave the checks to the reader.

Answer: $x^2 - 5x - 24 = (x + 3)(x - 8)$

YOUR TURN

Factor $n^2 - n - 30$.

Now let's see how to factor a polynomial containing two variables.

EXAMPLE 4 Factor $x^2 - 10xy + 25y^2$.

Solution Since y is in the last term, it is helpful to rearrange the order of the variables in the middle term, writing the polynomial as $x^2 - 10yx + 25y^2$, so that we view the "coefficient" of $-10yx$ as $-10y$. We must find a pair of terms whose product is $25y^2$ and whose sum is $-10y$. These terms would have to be $-5y$ and $-5y$.

ANSWER

$(n + 5)(n - 6)$

Answer: $x^2 - 10xy + 25y^2 = (x - 5y)(x - 5y)$

EXAMPLE 5 Factor $a^2 + 2ab - 15b^2$.

Solution As in Example 4, we can view the coefficient of a in the middle term as $2b$. We must find a pair of terms whose product is $-15b^2$ and whose sum is $2b$. These terms would have to be $-3b$ and $5b$.

Answer: $a^2 + 2ab - 15b^2 = (a - 3b)(a + 5b)$

YOUR TURN Factor.

a. $m^2 + 7mn - 18n^2$ **b.** $x^2 - xy - 12y^2$

OBJECTIVE 2 **Factor out a monomial GCF, then factor the trinomial of the form $x^2 + bx + c$.** Sometimes there is a monomial GCF (other than 1) in the three terms of a trinomial. Consider $2x^3 + 10x^2 + 12x$. Notice the monomial $2x$ is the GCF of the terms. Factoring out this monomial, we have:

$$2x^3 + 10x^2 + 12x = 2x(x^2 + 5x + 6)$$

Now we try to factor the trinomial within the parentheses. We look for two numbers whose product is 6 and whose sum is 5. Note that 2 and 3 work, so we can write:

$$2x^3 + 10x^2 + 12x = 2x(x^2 + 5x + 6)$$
$$= 2x(x + 2)(x + 3) \quad \longleftarrow \boxed{\text{Factored form}}$$

Whenever factoring polynomials, always look for a monomial GCF among the terms.

EXAMPLE 6 Factor $3mn^3 + 12mn^2 - 96mn$.

Solution First, we look for a monomial GCF (other than 1). Notice that the GCF of the terms is $3mn$. Factoring out this monomial, we have:

$$3mn^3 + 12mn^2 - 96mn = 3mn(n^2 + 4n - 32)$$

Now try to factor the trinomial to two binomials. We must find a pair of numbers whose product is -32 and whose sum is 4.

Product	Sum	
$(-1)(32) = -32$	$-1 + 32 = 31$	This is the correct combination.
$(-2)(16) = -32$	$-2 + 16 = 14$	So -4 and 8 are the second terms
$(-4)(8) = -32$	$-4 + 8 = 4$	$\longleftarrow$ in each binomial.

Answer: $3mn^3 + 12mn^2 - 96mn = 3mn(n - 4)(n + 8)$

YOUR TURN Factor.

a. $y^4 - 10y^3 + 9y^2$ **b.** $4a^3b - 20a^2b - 56ab$

EXAMPLE 7 Factor $x^4 + 6x^3 + 14x^2$.

Solution First, we factor out the monomial GCF, x^2.

$$x^4 + 6x^3 + 14x^2 = x^2(x^2 + 6x + 14)$$

Now we try to factor the trinomial to two binomials. We must find a pair of numbers whose product is 14 and whose sum is 6.

Product	Sum
$(1)(14) = 14$	$1 + 14 = 15$
$(2)(7) = 14$	$2 + 7 = 9$
$(-1)(-14) = 14$	$-1 + (-14) = -15$
$(-2)(-7) = 14$	$-2 + (-7) = -9$

Note: We listed all factor pairs of 14 and found no combination whose sum is 6. This means that $x^2 + 6x + 14$ has no binomial factors with integer terms, so $x^2(x^2 + 6x + 14)$ is the final factored form.

Note, a polynomial like $x^2 + 6x + 14$ that cannot be factored is like a prime number in that its factors are only 1 and the expression itself. We say such an expression is *prime*.

ANSWER

$4x(y^2 - 2y + 3)$

YOUR TURN

Factor $4xy^2 - 8xy + 12x$.

6.2 EXERCISES

1. If given a trinomial in the form $x^2 + bx + c$, where $c > 0$ and $b > 0$, what can you conclude about the pair of numbers in the binomial factors?

2. If given a trinomial in the form $x^2 - bx + c$, where $c > 0$ and $b > 0$, what can you conclude about the pair of numbers in the binomial factors?

3. If given a trinomial in the form $x^2 + bx - c$, where $c > 0$ and $b > 0$, what can you conclude about the pair of numbers in the binomial factors?

4. If given a trinomial in the form $x^2 - bx - c$, where $c > 0$ and $b > 0$, what can you conclude about the pair of numbers in the binomial factors?

For Extra Help

InterAct Math

Addison-Wesley Math Tutor Center

Video 10/DVT 3

MyMathLab

MathXL

For Exercises 5–12, fill in the missing values in the factors.

5. $x^2 + 7x + 10 = (x + 2)(x +)$

6. $y^2 + 8y + 15 = (y + 5)(y +)$

7. $n^2 - 12n + 20 = (n - 10)(n -)$

8. $a^2 + a - 12 = (a + 4)(a -)$

9. $m^2 - 2m - 24 = (m -)(m + 4)$

10. $t^2 - 9t + 18 = (t -)(t - 3)$

11. $u^2 + 13u - 30 = (u -)(u + 15)$

12. $n^2 - 9n - 36 = (n -)(n + 3)$

For Exercises 13–40, factor.

13. $a^2 + 7a + 12$

14. $w^2 - w - 12$

15. $x^2 - 11x + 30$

16. $b^2 - 3b - 40$

17. $y^2 - 12y + 36$

18. $m^2 - 2m - 8$

19. $n^2 - 11n + 30$

20. $x^2 + 14x + 45$

21. $x^2 - x - 30$

22. $b^2 - 6b - 16$

23. $r^2 + 4r + 3$

24. $y^2 + 5y + 6$

25. $u^2 - 6u + 8$

26. $t^2 + 8t + 7$

27. $k^2 - 3k - 10$

28. $y^2 - 13y + 12$

29. $u^2 + u - 6$

30. $n^2 - 5n + 4$

31. $b^2 - 10b + 21$

32. $x^2 - 5x - 24$

33. $r^2 - 9r + 18$

34. $p^2 - 13p + 36$

35. $a^2 - 10a - 24$

36. $x^2 - 10x + 24$

37. $m^2 - 4m + 6$

38. $u^2 - 8u - 20$

39. $x^2 - 6x - 8$

40. $a^2 + 3a - 20$

For Exercises 41–48, factor the trinomials containing two variables.

41. $p^2 + 6pq + 9q^2$

42. $x^2 - 9xy + 14y^2$

43. $a^2 - 6ab - 27b^2$

44. $m^2 + mn - 16n^2$

45. $x^2 - 14xy + 24y^2$

46. $t^2 + 5tu - 24u^2$

47. $r^2 - rs - 30s^2$

48. $h^2 - 8hk - 20k^2$

For Exercises 49–66, factor completely.

49. $2m^3 - 14m^2 + 12m$

50. $2k^2y + 10ky - 28y$

51. $3a^2b^2 - 15ab^2 - 72b^2$

52. $3a^2y + 6ay - 72y$

53. $4x^2 - 24x + 36$

54. $6a^2b^2c - 36ab^2c + 48b^2c$

55. $n^4 + 5n^3 + 6n^2$

56. $5x^5 + 20x^4 + 15x^3$

57. $6a^2r^2 + 12ar^2 + 6r^2$

58. $a^2b^2c^2 + 6ab^2c^2 + 9b^2c^2$

59. $7u^4 + 28u^3 - 35u^2$

60. $3m^2 - 33m + 54$

61. $2x^3 + 16x^2 + 30x$

62. $y^5 - 18y^4 + 17y^3$

63. $3x^3 + 6x^2y + 3xy^2$

64. $5m^2 - 30mn + 40n^2$

65. $2a^2b - 6ab^2 - 36b^3$

66. $7h^2k + 14hk^2 - 168k^3$

For Exercises 67–70, find and correct the mistake.

67. $x^2 - 5x - 6 = (x - 3)(x - 2)$

68. $x^2 - 5x + 6 = (x - 6)(x + 1)$

69. $x^2 - 3x - 4 = (x - 1)(x + 4)$

70. $x^2 + x + 2 = (x + 2)(x + 1)$

For Exercises 71–74, find all natural number values of b that make the trinomial factorable.

71. $x^2 + bx + 12$

72. $x^2 + bx + 18$

73. $x^2 + bx - 21$

74. $x^2 + bx - 10$

For Exercises 75–78, find a natural number c that makes the trinomial factorable.

75. $x^2 + 10x + c$

76. $x^2 + 11x + c$

77. $x^2 - 7x + c$

78. $x^2 - 5x + c$

REVIEW EXERCISES

1. What property of arithmetic is illustrated by $(6x + 1)(5x - 2) = (5x - 2)(6x + 1)$?

For Exercises 2–4, multiply.

2. $(2x + 3)(5x + 1)$

3. $(3y - 4)(5y + 6)$

4. $4n(6n - 1)(3n - 2)$

5. List all factors of 48.

6. What is the GCF of $6x$ and $35y$?

6.3 Factoring Trinomials of the Form $ax^2 + bx + c$, where $a \neq 1$

OBJECTIVES

1 Factor trinomials of the form $ax^2 + bx + c$, where $a \neq 1$, by trial.

2 Factor trinomials of the form $ax^2 + bx + c$, where $a \neq 1$, by grouping.

OBJECTIVE 1 Factor trinomials of the form $ax^2 + bx + c$, where $a \neq 1$, by trial. In Section 6.2 we factored trinomials in which the coefficient of the squared term was 1. Now we focus on factoring trinomials in which the coefficient of the squared term is other than 1, such as the following:

$$3x^2 + 17x + 10 \qquad 8x^2 + 29x - 12$$

In general, like trinomials of the form $x^2 + bx + c$, trinomials of the form $ax^2 + bx + c$, where $a \neq 1$, also have two binomials factors. When a is not 1, we must consider the factors of the ax^2 term and the factors of the c term. First we will develop a trial-and-error method for finding the factored form. Consider $3x^2 + 17x + 10$. Since all the terms are positive, we know that the terms in the binomial factors will all be positive.

The *first* terms must multiply to equal $3x^2$. The logical factors are $3x$ and x.

$$3x^2 + 17x + 10 = (\ \square\ + \ \square\ )(\ \square\ + \ \square\ )$$

The *last* terms must multiply to equal 10. These factors could be 1 and 10, or 2 and 5.

We now try various combinations of these *first* and *last* terms in binomial factors and consider the resulting trinomial products. If the sum of the inner and outer terms matches the original trinomial, then we know we have the correct combination in the binomial factors.

$$(3x + 10)(x + 1) = 3x^2 + 3x + 10x + 10 = 3x^2 + 13x + 10$$
$$(3x + 1)(x + 10) = 3x^2 + 30x + x + 10 = 3x^2 + 31x + 10 \quad \left.\begin{array}{c} \\ \\ \end{array}\right\} \text{Incorrect combinations.}$$
$$(3x + 5)(x + 2) = 3x^2 + 6x + 5x + 10 = 3x^2 + 11x + 10$$
$$(3x + 2)(x + 5) = 3x^2 + 15x + 2x + 10 = 3x^2 + 17x + 10 \quad \text{Correct combination.}$$

You may find that drawing lines to connect each product is helpful.

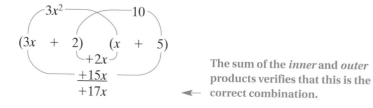

The sum of the *inner* and *outer* products verifies that this is the correct combination.

Our example suggests the following procedure.

PROCEDURE   FACTORING BY TRIAL AND ERROR

To factor a trinomial of the form $ax^2 + bx + c$, where $a \neq 1$, by trial and error:

1. Look for a monomial GCF in all the terms. If there is one, factor it out.
2. Write a pair of *first* terms whose product is ax^2.

$$(\Box + \ )(\Box + \ )$$

with ax^2 as the product of the first terms.

3. Write a pair of *last* terms whose product is c.

$$(\Box + \Box)(\Box + \Box)$$

with c as the product of the last terms.

4. Verify that the sum of the *inner* and *outer* products is bx (the middle term of the trinomial).

$$(\Box + \Box) \qquad (\Box + \Box)$$

$$\underbrace{\text{Inner}}$$
$$\underbrace{+ \text{ Outer}}$$
$$bx$$

If the sum of the inner and outer products is not bx, then try the following:

a. Exchange the first terms and repeat step 4.
b. Exchange the last terms and repeat step 4.
c. Repeat steps 2–4 with a different combination of first and last terms.

EXAMPLE 1   Factor $8x^2 + 29x - 12$.

Solution

The *first* terms must multiply to equal $8x^2$.
These could be x and $8x$, or $2x$ and $4x$.

$$8x^2 + 29x - 12 = (\quad + \quad)(\quad - \quad)$$

The *last* terms must multiply to equal -12. Because -12 is negative, the last terms in the binomials must have different signs. We have already written the appropriate signs, so these factor pairs could be 1 and 12, 2 and 6, or 3 and 4.

Now we multiply binomials with various combinations of these first and last terms until we find a combination whose inner and outer products combine to equal $29x$.

$$(8x + 1)(x - 12) = 8x^2 - 96x + x - 12 = 8x^2 - 95x - 12$$
$$(8x + 12)(x - 1) = 8x^2 - 8x + 12x - 12 = 8x^2 + 4x - 12$$
$$(8x + 2)(x - 6) = 8x^2 - 48x + 2x - 12 = 8x^2 - 46x - 12$$
$$(8x + 6)(x - 2) = 8x^2 - 16x + 6x - 12 = 8x^2 - 10x - 12$$
$$(8x + 3)(x - 4) = 8x^2 - 32x + 3x - 12 = 8x^2 - 29x - 12$$

Incorrect combinations.

Note: This middle term has the correct absolute value, but the incorrect sign (we are looking for $+29x$). Switching the $+$ and $-$ signs in the binomials will give us $+29x$.

$$(8x - 3)(x + 4) = 8x^2 + 32x - 3x - 12 = 8x^2 + 29x - 12$$ Correct combination.

Answer: $8x^2 + 29x - 12 = (8x - 3)(x + 4)$

YOUR TURN

Factor $6y^2 + 13y - 5$.

Factoring Out a Monomial GCF First

Remember that we should always look to see if a monomial GCF can be factored out of the terms in the trinomial.

EXAMPLE 2 Factor $40x^3 - 68x^2 + 12x$.

Solution First, we factor out the monomial GCF, $4x$.

$$40x^3 - 68x^2 + 12x = 4x(10x^2 - 17x + 3)$$

Now we factor the trinomial within the parentheses. The fact that 3 is positive and $-17x$ is negative indicates that the signs of the last terms in the binomial factors will both be minus signs.

The *first* terms must multiply to equal $10x^2$.
These could be x and $10x$, or $2x$ and $5x$.

$$4x(10x^2 - 17x + 3) = 4x(\quad - \quad)(\quad - \quad)$$

The *last* terms must multiply to equal 3. Because 3 is a prime number, its factors are 1 and 3.

Now multiply binomials with various combinations of these first and last terms until we find a combination whose inner and outer products combine to equal $-17x$.

$$4x(x - 1)(10x - 3) = 4x(10x^2 - 3x - 10x + 3) = 4x(10x^2 - 13x + 3)$$
$$4x(10x - 1)(x - 3) = 4x(10x^2 - 30x - x + 3) = 4x(10x^2 - 31x + 3)$$
$$4x(2x - 1)(5x - 3) = 4x(10x^2 - 6x - 5x + 3) = 4x(10x^2 - 11x + 3)$$

Incorrect combinations.

$$4x(5x - 1)(2x - 3) = 4x(10x^2 - 15x - 2x + 3) = 4x(10x^2 - 17x + 3)$$

Correct combination.

Answer: $40x^3 - 68x^2 + 12x = 4x(5x - 1)(2x - 3)$

ANSWER

$(2y + 5)(3y - 1)$

YOUR TURN

Factor $16y^3 - 102y^2 + 36y$.

EXAMPLE 3 Factor $8x^3 - 10x^2y - 18xy^2$.

Solution First, factor out the monomial GCF, $2x$.

$$8x^3 - 10x^2y - 18xy^2 = 2x(4x^2 - 5xy - 9y^2)$$

Now we factor the trinomial within the parentheses. The fact that the $-9y^2$ is negative and $-5xy$ is negative indicates that the signs of the last terms in the binomial factors will be different.

The *first* terms must multiply to equal $4x^2$.
These could be x and $4x$ or $2x$ and $2x$.

$$2x(4x^2 - 5xy - 9y^2) = 2x(\boxed{} + \boxed{})(\boxed{} - \boxed{})$$

The *last* terms must multiply to equal $9y^2$.
These terms could be y and $9y$ or $3y$ and $3y$.

Now we multiply binomials with various combinations of these first and last terms until we find a combination whose inner and outer products combine to equal $-5xy$.

$$\left.\begin{aligned} 2x(4x + y)(x - 9y) &= 2x(4x^2 - 36xy + xy - 9y^2) \\ &= 2x(4x^2 - 35xy - 9y^2) \\ 2x(4x + 9y)(x - y) &= 2x(4x^2 - 4xy + 9xy - 9y^2) \\ &= 2x(4x^2 + 5xy - 9y^2) \end{aligned}\right\} \begin{aligned} &\text{Incorrect} \\ &\text{combinations.} \end{aligned}$$

Note that the middle term has the *correct absolute value* but the *wrong sign*. As we saw in Example 1, we can change the signs of the second terms in the binomial factors to get the correct middle term.

$$\begin{aligned} 2x(4x - 9y)(x + y) &= 2x(4x^2 + 4xy - 9xy - 9y^2) \qquad \text{Correct combination.} \\ &= 2x(4x^2 - 5xy - 9y^2) \end{aligned}$$

Answer: $8x^3 - 10x^2y - 18xy^2 = 2x(4x - 9y)(x + y)$

ANSWER

$2y(y-6)(8y-3)$

YOUR TURN

Factor $36m^2n - 120mn^2 - 21n^3$.

ANSWER

$3n(6m + n)(2m - 7n)$

OBJECTIVE 2 **Factor trinomials of the form $ax^2 + bx + c$, where $a \neq 1$, by grouping.** The major drawback about using trial and error to factor is that it can be tedious to find the correct combination. An alternative method is to factor by grouping, which we introduced in Section 6.1. Recall that in the grouping method, we group pairs of terms in a four-term polynomial, then factor out the GCF from each pair of terms. Since a trinomial of the form $ax^2 + bx + c$ has only three terms, we will split the bx term into two like terms to create a four-term polynomial that we can factor by grouping. To determine how to split up the bx term, we use the fact that if $ax^2 + bx + c$ is factorable, then b will always equal the sum of a pair of factors of the product of a and c.

Consider again the trinomial $3x^2 + 17x + 10$, which we factored by trial and error at the beginning of this section. Notice that $a = 3$, $b = 17$, and $c = 10$, so the product of a and c is $(3)(10) = 30$. To split up the bx term, which is $17x$ in this case, we look for a pair of factors of 30 whose sum is 17. It is helpful to list the factor pairs and their corresponding sums in a table:

Factors of ac	Sum of Factors of ac
$(1)(30) = 30$	$1 + 30 = 31$
$(2)(15) = 30$	$2 + 15 = 17$
$(3)(10) = 30$	$3 + 10 = 13$
$(5)(6) = 30$	$5 + 6 = 11$

← Notice that 2 and 15 is the only factor pair of 30 whose sum is 17.

Now we can write $17x$ as $15x + 2x$, then factor by grouping.

$$3x^2 + 17x + 10 = 3x^2 + 15x + 2x + 10$$
$$= 3x(x + 5) + 2(x + 5)$$
$$= (x + 5)(3x + 2)$$

Note: We could have also written $17x$ as $2x + 15x$ to get the same result.

Every trinomial factorable by grouping has only one factor pair of ac whose sum is b, which suggests the following procedure.

PROCEDURE FACTORING $ax^2 + bx + c$, WHERE $a \neq 1$, BY GROUPING

To factor a trinomial of the form $ax^2 + bx + c$, where $a \neq 1$, by grouping:

1. Look for a monomial GCF in all the terms. If there is one, factor it out.
2. Multiply a and c.
3. Find two factors of this product whose sum is b.
4. Write a four-term polynomial in which bx is written as the sum of two like terms whose coefficients are the two factors you found in step 3.
5. Factor by grouping.

EXAMPLE 4 Factor by grouping $8x^2 - 22x + 5$.

Solution Notice that for this trinomial, $a = 8$, $b = -22$, and $c = 5$. We begin by multiplying a and c: $(8)(5) = 40$. Now we find two factors of 40 whose sum is -22. Notice that these two factors must both be negative. It is helpful to list the combinations in a table.

Factors of ac	Sum of Factors of ac	
$(-1)(-40) = 40$	$-1 + (-40) = -41$	
$(-2)(-20) = 40$	$-2 + (-20) = -22$	← Correct
$(-4)(-10) = 40$	$-4 + (-10) = -14$	
$(-5)(-8) = 40$	$-5 + (-8) = -13$	

You do not need to list all possible combinations as we did here. We listed them to illustrate that only one combination is correct. We now write $-22x$ as $-2x - 20x$ and then factor by grouping.

$$8x^2 - 22x + 5 = 8x^2 - 2x - 20x + 5 \qquad \text{Write } -22x \text{ as } -2x - 20x.$$

$$= 2x(4x - 1) - 5(4x - 1) \qquad \text{Factor } 2x \text{ out of } 8x^2 - 2x;$$
$$\text{factor } -5 \text{ out of } -20x + 5.$$

$$= (4x - 1)(2x - 5) \qquad \text{Factor out } (4x - 1).$$

EXAMPLE 5 Factor by grouping $18y^4 + 15y^3 - 12y^2$.

Solution Notice that there is a monomial GCF, $3y^2$, that we can factor out.

$$18y^4 + 15y^3 - 12y^2 = 3y^2(6y^2 + 5y - 4)$$

Now we factor the trinomial within the parentheses. For this trinomial, $a = 6$, $b = 5$, and $c = -4$. The product of ac is $(6)(-4) = -24$. Now find two factors of -24 whose sum is 5. Since the product is negative, the two factors will have different signs. Since the sum is positive, the factor with the greater absolute value must be positive.

Factors of ac	Sum of the Factors of ac	
$(-1)(24) = -24$	$-1 + 24 = 23$	
$(-2)(12) = -24$	$-2 + 12 = 10$	
$(-3)(8) = -24$	$-3 + 8 = 5$	← Correct

Now write $5y$ as $-3y + 8y$ and then factor by grouping.

$$3y^2(6y^2 + 5y - 4) = 3y^2(6y^2 - 3y + 8y - 4) \qquad \text{Write } 5y \text{ as } -3y + 8y.$$

$$= 3y^2[3y(2y - 1) + 4(2y - 1)] \qquad \text{Factor } 3x \text{ out of } 6y^2 - 3y;$$
$$\text{factor } 4 \text{ out of } 8y - 4.$$

$$= 3y^2(2y - 1)(3y + 4) \qquad \text{Factor out } (2y - 1).$$

YOUR TURN

Factor by grouping $8x^2y - 20xy - 72y$.

ANSWER

$4y(2x - 9)(x + 2)$

6.3 EXERCISES

For Extra Help

 InterAct Math

 Addison-Wesley Math Tutor Center

 Video 10/DVT 3

 MyMathLab

MathXL

1. In factoring a polynomial, your first step should be to look for a _____.

2. Given a trinomial of the form $ax^2 + bx + c$, where $a \neq 1$, how do you determine the first terms in its binomial factors?

3. Given a trinomial of the form $ax^2 + bx + c$, where $a \neq 1$, how do you determine the last terms in its binomial factors?

4. When factoring a trinomial of the form $ax^2 + bx + c$, assuming there is no monomial GCF, what is the first step?

For Exercises 5–10, fill in the missing values in the factors.

5. $3y^2 + 14y + 8 = (3y + \quad)(y + \quad)$

6. $6x^2 + 11x + 4 = (2x + \quad)(3x + \quad)$

7. $4t^2 + 19t - 30 = (\quad + 6)(\quad - 5)$

8. $8m^2 - 2m - 15 = (\quad - 3)(\quad + 5)$

9. $6x^2 - 25x + 14 = (2x - \quad)(\quad - 2)$

10. $9n^2 - 56n + 12 = (9n - \quad)(\quad - 6)$

For Exercises 11–38, factor completely.

11. $2y^2 - 3y - 5$

12. $2x^2 + 7x + 3$

13. $3m^2 - 10m + 8$

14. $4u^2 - 17u - 15$

15. $4a^2 - 19a + 12$

16. $4y^2 + 8y + 3$

17. $6a^2 + 13a + 7$

18. $3m^2 - 10m + 3$

19. $6p^2 + 2p + 1$

20. $8x^2 - 14x + 5$

21. $2k^2 + 7k - 30$

22. $6u^2 + 3u - 7$

23. $2j^2 + 5j + 2$

24. $3y^2 - 10y + 7$

25. $3p^2 - 13pq + 4q^2$

26. $2u^2 + 5uv + 3v^2$

27. $5k^2 - 7kh - 12h^2$

28. $9x^2 + 9xy - 10y^2$

29. $12a^2 - 40ab + 25b^2$

30. $8m^2 - 27mn + 9n^2$

31. $4w^2 + 7wv + 3v^2$

32. $6a^2 - 20a + 16$

33. $8x^2y - 4xy - y$

34. $20xy^2 - 50xy + 30x$

35. $6ab^2 - 20ab + 14a$

36. $12y^2 + 24y + 9$

37. $6w^2v^2 + 10wv^2 + 4v^2$

38. $6x^3 + 21x^2y - 90xy^2$

For Exercises 39–66, factor by grouping.

39. $3a^2 + 4a + 1$

40. $5m^2 + 11m + 2$

41. $2y^2 - y - 6$

42. $3l^2 + 2l - 5$

43. $8r^2 - 6r - 9$

44. $8m^2 - 10m + 3$

45. $2t^2 - 3t + 1$

46. $3a^2 - 10a + 8$

47. $4k^2 - 10k + 6$

48. $6x^2 + 5x - 6$

49. $20x^2 - 23x + 6$

50. $6a^2 + 17a + 10$

51. $3h^2 - 5h + 2$

52. $5p^2 + 13p - 6$

53. $6u^2 - 7u - 5$

54. $3x^2 + 5xy + 2y^2$

55. $5x^2 - 26xy + 5y^2$

56. $2t^2 - 3tu + u^2$

57. $6k^2 + 7jk + 2j^2$

58. $6u^2 + 5uv - 6v^2$

59. $10s^2 + st - 2t^2$

60. $12m^3 + 10m^2 - 12m$

61. $14u^2 - 6uv - 8v^2$

62. $24x^2 - 64xy - 24y^2$

63. $36x^2y - 78xy + 36y$

64. $24x^3 - 20x^2y - 24xy^2$

65. $28m^2n + 2mn - 8n$

66. $36x^3 - 93x^2y + 60xy^2$

For Exercises 67–70, explain the mistake, then factor the trinomial correctly.

67. $6x^2 + 13x + 5 = (6x + 1)(x + 5)$

68. $2x^2 + x - 3 = (2x - 3)(x + 1)$

69. $4n^2 + 12n + 8 = (4n + 8)(n + 1)$

70. $4h^2 - 3hk - 10k^2 = (h - 2)(4h + 5)$

For Exercises 71–72, given the area of the figure, find expressions for the missing side lengths.

71.

Area = $15x^2 - 11x + 2$

72.

Area = $2x^2 + 13x + 20$

6.3 Factoring Trinomials of the Form $ax^2 + bx + c$, where $a \neq 1$ **491**

For Exercises 73–78, find all natural numbers that can replace b and make the expression factorable.

73. $2x^2 + bx + 6$

74. $3x^2 + bx + 10$

75. $5x^2 + bx - 9$

76. $2x^2 + bx - 15$

77. $6x^2 + bx - 7$

78. $4x^2 + bx - 5$

REVIEW EXERCISES

For Exercises 1–4, multiply.

1. $(3x + 5)(3x - 5)$

2. $(x - 2)(x^2 + x + 4)$

3. $(2x - 5)^2$

4. $(6y + 1)^2$

For Exercises 5–6, simplify.

5. $(4x)^2$

6. $(2y)^3$

6.4 Factoring Special Products

1. Factor perfect square trinomials.
2. Factor a difference of squares.
3. Factor a difference of cubes.
4. Factor a sum of cubes.

In Section 5.5 we explored some special products found by multiplying conjugates and squaring binomials. In this section we see how to factor those special products.

OBJECTIVE 1 Factor perfect square trinomials. First, we consider the product that is a result of squaring a binomial. The trinomial product is a perfect square. Recall the rules for squaring a binomial that we developed in Section 5.5:

$$(a + b)^2 = a^2 + 2ab + b^2$$
$$(a - b)^2 = a^2 - 2ab + b^2$$

Consider $(2x + 3)^2$ and $(2x - 3)^2$. Using the above rules, we have

$$(2x + 3)^2 = (2x)^2 + 2(2x)(3) + (3)^2 \qquad (2x - 3)^2 = (2x)^2 - 2(2x)(3) + (3)^2$$
$$= 4x^2 + 12x + 9 \qquad\qquad\qquad = 4x^2 - 12x + 9$$

To use these rules when factoring, we look at the terms in the given trinomial to determine whether the trinomial fits the form of a perfect square. If we were asked to factor $4x^2 + 12x + 9$, we would note that the first and last terms are perfect squares and that twice the product of their roots equals the middle term.

$$4x^2 + 12x + 9$$

The square root of $4x^2$ is $2x$.   The square root of 9 is 3.

Twice the product of $2x$ and 3 is the middle term, $12x$.

Because $4x^2 + 12x + 9$ is a perfect square trinomial fitting the form $a^2 + 2ab + b^2$, where $a = 2x$ and $b = 3$, we can write the factored form as $(a + b)^2$, which is $(2x + 3)^2$.

RULE   FACTORING PERFECT SQUARE TRINOMIALS

$$a^2 + 2ab + b^2 = (a + b)^2$$
$$a^2 - 2ab + b^2 = (a - b)^2$$

EXAMPLE 1 Factor.

a. $9y^2 + 30y + 25$

Solution This trinomial is a perfect square because the first and last terms are perfect squares and twice the product of their roots is the middle term.

$$9y^2 + 30y + 25$$

The square root of $9y^2$ is $3y$.　　　The square root of 25 is 5.

Twice the product of $3y$ and 5 is $2(3y)(5) = 30y$, which is the middle term.

Answer:　$9y^2 + 30y + 25 = (3y + 5)^2$　Use $a^2 + 2ab + b^2 = (a + b)^2$, where $a = 3y$ and $b = 5$.

b. $36x^2 - 84x + 49$

Solution This trinomial is a perfect square.

$$36x^2 - 84x + 49$$

The square root of $36x^2$ is $6x$.　　　The square root of 49 is 7.

Twice the product of $6x$ and 7 is $2(6x)(7) = 84x$, which is the middle term.

Answer:　$36x^2 - 84x + 49 = (6x - 7)^2$　Use $a^2 - 2ab + b^2 = (a - b)^2$, where $a = 6x$ and $b = 7$.

YOUR TURN Factor.

　　a. $25n^2 + 20n + 4$ 　　　　**b.** $49x^2 - 14x + 1$

Note that in the perfect square trinomial forms, both a and b can have variables. Also, don't forget that you should always look for a monomial GCF in the terms as a first step.

EXAMPLE 2 Factor.

a. $16x^2 - 8xy + y^2$

Solution $16x^2 - 8xy + y^2 = (4x - y)^2$　Use $a^2 - 2ab + b^2 = (a - b)^2$, where $a = 4x$ and $b = y$.

b. $4x^3y - 36x^2y + 81xy$

Solution

$$4x^3y - 36x^2y + 81xy = xy(4x^2 - 36x + 81) \qquad \text{Factor out the monomial GCF, } xy.$$
$$= xy(2x - 9)^2 \qquad \text{Factor } 4x^2 - 36x + 81 \text{ using } a^2 - 2ab + b^2 = (a - b)^2, \text{ where } a = 2x \text{ and } b = 9.$$

YOUR TURN Factor.

a. $9m^2 + 48mn + 64n^2$ **b.** $2xy^4 + 24xy^3 + 72xy^2$

OBJECTIVE 2 **Factor a difference of squares.** Another special product that we considered in Section 5.5 was found by multiplying conjugates. Recall that conjugates are binomials that differ only in the sign separating the terms. For example, $3x + 2$ and $3x - 2$ are conjugates. Note that multiplying conjugates produces a product that is a difference of squares.

$$(3x + 2)(3x - 2) = 9x^2 - 4$$

This term is the square of $3x$.   This term is the square of 2.

The rule for multiplying conjugates that we developed in Section 5.5 is:

$$(a + b)(a - b) = a^2 - b^2$$

To turn this rule into a rule for factoring a difference of squares, we reverse it.

RULE **FACTORING A DIFFERENCE OF SQUARES**

$$a^2 - b^2 = (a + b)(a - b)$$

Warning: A *sum* of squares $a^2 + b^2$ is prime and cannot be factored.

EXAMPLE 3 Factor.

a. $25x^2 - 36y^2$

Solution The binomial is a difference of squares where $a = 5x$ and $b = 6y$.

$$25x^2 - 36y^2 = (5x + 6y)(5x - 6y) \qquad \text{Use } a^2 - b^2 = (a + b)(a - b).$$

b. $9x^4 - 16$

Solution It may seem like this binomial is not a difference of squares because of the $9x^4$ term, but $9x^4$ is the square of $3x^2$, so this binomial is a difference of squares where $a = 3x^2$ and $b = 4$.

$$9x^4 - 16 = (3x^2 + 4)(3x^2 - 4) \qquad \text{Use } a^2 - b^2 = (a + b)(a - b).$$

c. $25y^7 - 64y^9$

Solution

$$25y^7 - 64y^9 = y^7(25 - 64y^2)$$
$$= y^7(5 + 8y)(5 - 8y)$$

Factor out the monomial GCF, y^7.

Factor $25 - 64y^2$, which is a difference of squares, where $a = 5$ and $b = 8y$.

YOUR TURN   Factor.

a. $n^2 - 49$ **b.** $9t^6 - 49u^4$ **c.** $50x^3 - 18x$

Sometimes the factors themselves can be factored.

EXAMPLE 4   Factor $x^4 - 81$.

Solution This binomial is a difference of squares, where $a = x^2$ and $b = 9$.

$$x^4 - 81 = (x^2 + 9)(x^2 - 9)$$
$$= (x^2 + 9)(x + 3)(x - 3)$$

Use $a^2 - b^2 = (a + b)(a - b)$.

Factor $x^2 - 9$, which is a difference of squares, where $a = x$ and $b = 3$.

Note: The factor $x^2 + 9$ is a sum of squares, which is prime.

YOUR TURN

Factor $y^4 - 16$.

OBJECTIVE 3 **Factor a difference of cubes.**   Another common binomial form that we can factor is a difference of cubes. A difference of cubes is a result of a multiplication like the following:

$$(x - 2)(x^2 + 2x + 4) = x^3 + 2x^2 + 4x - 2x^2 - 4x - 8$$
$$= x^3 - 8 \leftarrow \text{Difference of cubes}$$

Writing factored form reverses this process so that we have:

$$x^3 - 8 = (x - 2)(x^2 + 2x + 4)$$

This suggests the following rule for factoring a difference of cubes.

RULE   **FACTORING A DIFFERENCE OF CUBES**

$$a^3 - b^3 = (a - b)(a^2 + ab + b^2)$$

EXAMPLE 5 Factor.

a. $64y^3 - 27$

Solution This binomial is a difference of cubes because $64y^3 - 27 = (4y)^3 - (3)^3$. To factor, we use the rule $a^3 - b^3 = (a - b)(a^2 + ab + b^2)$ with $a = 4y$ and $b = 3$.

$$a^3 - b^3 = (a - b) \ (a^2 + a \ b + b^2)$$
$$64y^3 - 27 = (4y - 3)((4y)^2 + (4y)(3) + (3)^2) \qquad \text{Replace } a \text{ with } 4y \text{ and } b \text{ with 3.}$$
$$= (4y - 3)(16y^2 + 12y + 9) \qquad \text{Simplify.}$$

Note: This trinomial may seem like a perfect square. However, to be a perfect square, the middle term should be $2ab$. In this trinomial, we only have ab, so it cannot be factored.

b. $27m^4 - 8mn^3$

Solution

$$27m^4 - 8mn^3 = m(27m^3 - 8n^3)$$

$$= m(3m - 2n)((3m)^2 + (3m)(2n) + (2n)^2)$$

$$= m(3m - 2n)(9m^2 + 6mn + 4n^2)$$

Factor out the monomial GCF, m.
Factor $27m^3 - 8n^3$, which is a difference of cubes with $a = 3m$ and $b = 2n$
Simplify.

YOUR TURN Factor.

a. $27 - u^3$ **b.** $xy^3 - 64x$

OBJECTIVE 4 Factor a sum of cubes. A sum of cubes can be factored using a pattern similar to the difference of cubes. A sum of cubes is a result of a multiplication like the following:

$$(x + 2)(x^2 - 2x + 4) = x^3 - 2x^2 + 4x + 2x^2 - 4x + 8$$
$$= x^3 + 8$$

Writing factored form reverses this process so that we have:

$$x^3 + 8 = (x + 2)(x^2 - 2x + 4)$$

In general, we can write the following rule for factoring a sum of cubes.

RULE   **FACTORING A SUM OF CUBES**

$$a^3 + b^3 = (a + b)(a^2 - ab + b^2)$$

ANSWERS
a. $(3 - u)(9 + 3u + u^2)$
b. $x(y - 4)(y^2 + 4y + 16)$

EXAMPLE 6 Factor.

a. $27x^3 + 8$

Solution This binomial is a sum of cubes because $27x^3 + 8 = (3x)^3 + (2)^3$. To factor, we use the rule $a^3 + b^3 = (a + b)(a^2 - ab + b^2)$ with $a = 3x$ and $b = 2$.

$$a^3 + b^3 = (\;a\;+\;b)(\;a^2\;-\;a\;\;b\;+\;b^2)$$

$$27x^3 + 8 = (3x + 2)((3x)^2 - (3x)(2) + (2)^2) \quad \text{Replace } a \text{ with } 3x \text{ and } b \text{ with 2, then simplify.}$$

$$= (3x + 2)(9x^2 - 6x + 4) \quad \blacktriangleleft \text{ Note: This trinomial cannot be factored.}$$

b. $5h + 40hk^3$

Solution

$$5h + 40hk^3 = 5h(1 + 8k^3) \qquad \text{Factor out the monomial GCF, } 5h.$$

$$= 5h(1 + 2k)((1)^2 - (1)(2k) + (2k)^2) \qquad \text{Factor } 1 + 8k^3, \text{ which is a sum of}$$

$$= 5h(1 + 2k)(1 - 2k + 4k^2) \qquad \text{cubes with } a = 1 \text{ and } b = 2k.$$

ANSWERS

a. $(4 + m)(16 - 4m + m^2)$

b. $x^2(5x - 3)(25x^2 + 15x + 9)$

YOUR TURN Factor.

a. $64 + m^3$

b. $125x^5 - 27x^2$

6.4 EXERCISES

For Extra Help

 InterAct Math

 Addison-Wesley Math Tutor Center

 Video 10/DVT 3

MyMathLab

 MathXL MathXL

1. Explain in your own words how to recognize a perfect square trinomial.

2. How do you determine the sign in the factored form of a perfect square trinomial?

3. The factors of a difference of squares are binomials known as _____.

4. There is only one minus sign in the factors of a difference of cubes. Where is that minus sign placed in the factors?

5. There is only one minus sign in the factors of a sum of cubes. Where is that minus sign placed in the factors?

6. How do you determine the a and b terms in a sum or difference of cubes?

For Exercises 7–24, factor the perfect square.

7. $x^2 + 14x + 49$

8. $y^2 + 6y + 9$

9. $b^2 - 8b + 16$

10. $m^2 + 6m + 9$

11. $25u^2 - 30u + 9$

12. $9m^2 - 24m + 16$

13. $n^2 + 24n + 144$

14. $4r^2 - 20r + 25$

15. $y^2 + 12y + 36$

16. $w^2 + 2wv + v^2$

17. $4p^2 - 28pq + 49q^2$

18. $25a^2 + 10a + 1$

19. $16g^2 + 24gh + 9h^2$

20. $4y^2 - 12by + 9b^2$

21. $9r^2 - 12rs + 4s^2$

22. $4t^2 + 20t + 25$

23. $100w^2 + 20w + 1$

24. $9q^2 - 12q + 4$

For Exercises 25–40, factor the difference of squares.

25. $x^2 - 4$

26. $m^2 - n^2$

27. $16 - y^2$

28. $a^2 - 121$

29. $25u^2 - 16$

30. $16x^2 - y^2$

31. $9x^2 - b^2$

32. $16x^2 - 49y^2$

33. $25x^6 - 9y^2$

34. $25r^8 - 4$

35. $u^4 - v^2$

36. $121p^2 - 9q^2$

37. $100w^2 - 1$

38. $81 - a^2b^2$

39. $49j^2 - k^2$

40. $64m^2 - 25n^2$

For Exercises 41–50, factor the difference of cubes.

41. $n^3 - 27$

42. $x^3 - 1$

43. $y^3 - 64$

44. $m^3 - 27$

45. $27k^3 - 8$

46. $a^3 - 8b^3$

47. $27x^3 - 64y^3$

48. $x^6 - y^6$

49. $1000a^3 - 27b^3$

50. $125m^3n^3 - 64p^3$

For Exercises 51–60, factor the sum of cubes.

51. $x^3 + 27$

52. $y^3 + 8$

53. $u^3 + 125$

54. $27k^3 + 8$

55. $125x^3 + 64y^3$

56. $27a^3 + 1000b^3$

57. $m^3 + n^3$

58. $u^6 + v^6$

59. $8p^3 + q^3z^3$

60. $x^3y^3 + 27y^3$

For Exercises 61–76, factor.

61. $16x^2 - \dfrac{25}{49}$

62. $25y^2 - \dfrac{1}{36}$

63. $50x^3 - 2x$

64. $27m^3 - 363mn^2$

65. $y^5 - 16y^3b^2$

66. $ab^3 - a^3b$

67. $2u^3 - 2u$

68. $4 - 16x^2$

69. $2x^2 - 50$

70. $2x^3 - 54$

71. $16c^4 + 2cd^3$

72. $16m^5 + 54m^2n^3$

73. $3y^3 - 24z^3$

74. $8d^4 - 64c^3d$

75. $c^3 - \dfrac{8}{27}$

76. $8y^3 - \dfrac{1}{125}$

For Exercises 77–84, use the rules for a difference of squares, sum of cubes, or difference of cubes to factor completely.

77. $(2a - b)^2 - c^2$

78. $(b - 3)^2 - 1$

79. $1 - 9(x - y)^2$

80. $25a^2 - (b + c)^2$

81. $x^3 + 27(y + z)^3$

82. $m^3 - 8(y + b)^3$

83. $64d^3 - (x + y)^3$

84. $(t + u)^3 + 64$

For Exercises 85–88, find a natural number b that makes the expression a perfect square trinomial.

85. $16x^2 + bx + 9$

86. $25x^2 + bx + 4$

87. $4x^2 - bx + 81$

88. $36x^2 - bx + 49$

For Exercises 89–92, find the natural number c that completes the perfect square trinomial.

89. $x^2 + 10x + c$

90. $x^2 + 8x + c$

91. $9x^2 - 24x + c$

92. $4x^2 - 28x + c$

For Exercises 93–94, write a polynomial for the area of the shaded region, then factor completely.

93.

94.

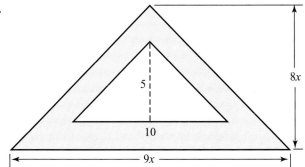

For Exercises 95–96, write a polynomial for the volume of the shaded region, then factor completely.

95.

96.

REVIEW EXERCISES

1. List all the factors of 36.

2. Find the prime factorization of 100.

3. Add $(4x^3 - 5x^2 + 9x - 11) + (7x^2 - 12x - 15)$.

4. Subtract $(10n^4 + n^3 - 15n - 21) - (19n^3 - 12n^2 + n - 15)$.

5. Multiply $(4x - 5)^2$.

6. Divide $\dfrac{5x^3 + 4x^2 - 5x + 17}{x + 2}$.

6.5 Strategies for Factoring

OBJECTIVE **1** Factor polynomials.

One of the difficulties in factoring expressions is determining which technique to use. In this section, we use a general outline for factoring. Following is the outline.

PROCEDURE FACTORING A POLYNOMIAL

To factor a polynomial, first factor out any monomial GCF, then consider the number of terms in the polynomial. If the polynomial has:

I. Four terms, then try to factor by grouping.

II. Three terms, then determine if the trinomial is a perfect square or not.

    A. If the trinomial is a perfect square, then consider its form.
      1. If in the form $a^2 + 2ab + b^2$, then the factored form is $(a + b)^2$.
      2. If in the form $a^2 - 2ab + b^2$, then the factored form is $(a - b)^2$.

    B. If the trinomial is not a perfect square, then consider its form.
      1. If in the form $x^2 + bx + c$, then find two factors of c whose sum is b, and write the factored form as $(x + \text{first number})(x + \text{second number})$.
      2. If in the form $ax^2 + bx + c$, where $a \neq 1$, then use trial and error. Or, find two factors of ac whose sum is b; write these factors as coefficients of two like terms that, when combined, equal bx; and then factor by grouping.

III. Two terms, then determine if the binomial is a difference of squares, sum of cubes, or difference of cubes.

    A. If given a binomial that is a difference of squares $a^2 - b^2$, then the factors are conjugates and the factored form is $(a + b)(a - b)$. Note that a sum of squares cannot be factored.

    B. If given a binomial that is a sum of cubes, $a^3 + b^3$, then the factored form is $(a + b)(a^2 - ab + b^2)$.

    C. If given a binomial that is a difference of cubes, $a^3 - b^3$, then the factored form is $(a - b)(a^2 + ab + b^2)$.

Note: Always look to see if any of the factors can be factored.

EXAMPLE **1**   Factor $12x^2 + 28x - 5$.

Solution There is no monomial GCF. Since this expression is a trinomial, we look to see if it is a perfect square. It is not a perfect square because the first and last terms are not perfect squares. It is a trinomial of the form $ax^2 + bx + c$, where $a \neq 1$. Therefore, we use trial and error or the alternative method that involves factoring by grouping. We will use trial and error here.

The *first* terms multiply to equal $12x^2$. The factors could be x and $12x$, $2x$ and $6x$, or $3x$ and $4x$.

$$12x^2 + 28x - 5 = (\boxed{} + \boxed{})(\boxed{} - \boxed{})$$

The *last* terms multiply to equal -5. We have already written in the appropriate signs, so these factors could be 1 and 5.

Now list possible combinations of these first and last factor pairs and check by multiplying to see if the sum of the products of the inner and outer terms equals $28x$.

$$\left.\begin{aligned}(x + 5)(12x - 1) &= 12x^2 - x + 60x - 5 = 12x^2 + 59x - 5 \\ (12x + 5)(x - 1) &= 12x^2 - 12x + 5x - 5 = 12x^2 - 7x - 5\end{aligned}\right\} \quad \begin{aligned}&\text{Incorrect} \\ &\text{combinations.}\end{aligned}$$

$$(2x + 5)(6x - 1) = 12x^2 - 2x + 30x - 5 = 12x^2 + 28x - 5 \qquad \text{Correct combination.}$$

Answer: $12x^2 + 28x - 5 = (2x + 5)(6x - 1)$

YOUR TURN

Factor $24y^2 - 53y - 7$.

EXAMPLE 2 Factor $7x^3 + 14x^2 - 105x$.

Solution

$$7x^3 + 14x^2 - 105x = 7x(x^2 + 2x - 15) \qquad \text{Factor out the monomial GCF, } 7x.$$

The trinomial in the parentheses is not a perfect square and is in the form $x^2 + bx + c$, so we look for two numbers whose product is -15 and whose sum is 2. Since the product is negative, the two factors must have different signs. Since the sum is positive, the factor with the greater absolute value will be the positive factor.

Product	Sum
$(-1)(15) = -15$	$-1 + 15 = 14$
$(-3)(5) = -15$	$-3 + 5 = 2$

Correct combination.

Answer: $7x^3 + 14x^2 - 105x = 7x(x - 3)(x + 5)$

ANSWER

$(3y - 7)(8y + 1)$

ANSWER

$2t(t - 6)(t - 3)$

YOUR TURN

Factor $2t^3 - 18t^2 + 36t$.

EXAMPLE 3 Factor $12m^4 - 48n^2$.

Solution

$$12m^4 - 48n^2 = 12(m^4 - 4n^2)$$ Factor out the monomial GCF, 12.

The binomial in the parentheses is a difference of squares in the form $a^2 - b^2$, where $a = m^2$ and $b = 2n$. Using $a^2 - b^2 = (a + b)(a - b)$ gives us:

$$= 12(m^2 + 2n)(m^2 - 2n)$$

Note: Neither of these binomial factors can be factored further.

EXAMPLE 4 Factor $10y^5 + 40y^3$.

Solution

$$10y^5 + 40y^3 = 10y^3(y^2 + 4)$$ Factor out the monomial GCF, $10y^3$.

The binomial in the parentheses is a sum of squares, which cannot be factored.

YOUR TURN

Factor $3m^2n^3 + 15m^2n^2 + 6m^2n$.

EXAMPLE 5 Factor $45x^3y^3 - 120x^2y^4 + 80xy^5$.

Solution

$$45x^3y^3 - 120x^2y^4 + 80xy^5 = 5xy^3(9x^2 - 24xy + 16y^2)$$ Factor out the monomial GCF, $5xy^3$.

The trinomial in the parentheses is a perfect square in the form $a^2 - 2ab + b^2$, where $a = 3x$ and $b = 4y$. Using $a^2 - 2ab + b^2 = (a - b)^2$ gives us:

$$= 5xy^3(3x - 4y)^2$$

EXAMPLE 6 Factor $x^5 - x^3 - 8x^2 + 8$.

Solution There is no monomial GCF common to all terms. Since this polynomial has four terms, we try to factor by grouping.

$$x^5 - x^3 - 8x^2 + 8 = x^3(x^2 - 1) - 8(x^2 - 1)$$
$$= (x^2 - 1)(x^3 - 8)$$

The two binomial factors can be factored further. The first binomial, $x^2 - 1$, is a difference of squares with $a = x$ and $b = 1$. The second binomial, $x^3 - 8$, is a difference of cubes with $a = x$ and $b = 2$.

$$= (x + 1)(x - 1)(x - 2)(x^2 + 2x + 4)$$

ANSWER

$3m^2n(n^2 + 5n + 2)$

ANSWER

$(2x + 3)(2x - 3)(3x - 2)$

YOUR TURN

Factor $12x^3 - 8x^2 - 27x + 18$.

6.5 EXERCISES

For Extra Help

InterAct Math

Addison-Wesley
Math Tutor Center

Video 11/DVT 3

MyMathLab

MathXL

1. What should you look for first when factoring?

2. What are three different types of factorable binomials?

3. Describe one type of binomial that cannot be factored.

4. What are the different types of trinomials that you might encounter?

5. When do you factor by grouping?

6. How might you check a factored form to see if it is correct?

For Exercises 7–76, factor completely.

7. $3xy^2 + 6x^2y$

8. $(u + v)r + (u + v)s$

9. $2x^2 - 32$

10. $b^3 + 125$

11. $ax + ay + bx + by$

12. $12a^3b^2c + 3a^2b^2c^2 + 5abc^3$

13. $x^2 + 8x + 15$

14. $x^4 - 16$

15. $6b^2 + b - 2$

16. $ax - xy - ay + y^2$

17. $x^2 + 25$

18. $2a^2 - 5a - 12$

19. $3y^3 - 24$

20. $ax^2 + 4ax + 4a$

21. $6ab - 36ab^2$

22. $x^2 + x + 2$

23. $x^2 - 49$

24. $p^2 + p - 30$

25. $8x^3 - 27y^3$

26. $2x^2 - 2y^2$

27. $15x^2 + 7x - 2$

28. $u^3 - u$

29. $2b^2 + 14b + 24$

30. $5m^2 - 45$

31. $xy^2 - xy - 20x$

32. $6r^2 - 15r^3$

33. $7u^2 - 14u - 105$

34. $2a + ab$

35. $t^4 - 81$

36. $h^4 + h^3 + h^2$

37. $y^2 - 6y - 40$

38. $3r^2 - 108$

39. $5p^2 - 80$

40. $1 - 25x^2$

41. $q^4 - 6q^2 - 16$

42. $3w^2 - w - 2$

43. $12v^2 + 23v + 10$

44. $9a^2x + 12ax + 4x$

45. $2 - 50x^2$

46. $x^3 - 4xy^2$

47. $a - 81ax^4$

48. $80k^2 - 20k^2l^2$

49. $2j^2 + j - 3$

50. $15u^2 - 2u^2v^2 + 2uv^3$

51. $50 - 20t + 2t^2$

52. $3p^2 - 13pq + 4q^2$

53. $5ax - 5a^2x$

54. $7a(x + y) - (x + y)$

55. $3ax - 6ay - 8by + 4bx$

56. $x^3 - x^2 - x + 1$

57. $3m^2 - 8m + 5$

58. $4x^2 - 28x + 49$

59. $2x^4 - 162$

60. $4x^4 - 4x$

61. $5x^2 - 10x$

62. $a^2 + 2ab + ab + 2b^2$

63. $6x^2 - 5x - 6$

64. $9 - 4m^2$

65. $x^5 - 4xy^2$

66. $4m^2 + 20m + 25$

67. $3x^3 - 12x$

68. $3k^2 - 2k - 1$

69. $2ax + 6bx - ay - 3by$

70. $x^2 + 3x + 4$

71. $54x - 2xy^3$

72. $x^3 + a^3$

73. $3m^2 + 9m + 27$

74. $20x^2 + 3xy + 2y^2$

75. $y^4 - 6y^3$

76. $8m^3 + 125b^3$

For Exercises 77–78, given an expression for the area of each rectangle, find the length and width.

77.
Area $= 6x^2 - 11x - 7$

78.
Area $= 25n^2 - 49$

For Exercises 79–80, given an expression for the volume of each box, find the length, width, and height.

79.

Volume $= 18x^3 + 60x^2 + 50x$

80.

Volume $= 4x^3 + 8x^2 - 60x$

REVIEW EXERCISES

1. Find three consecutive integers whose sum is 54.

2. Find the area of a rectangular garden that is 14 feet by 16 feet.

3. Find the product of 6 and negative 7.

4. Divide $\dfrac{2x^2 - 7x - 8}{2x + 1}$.

5. Find the GCF of $12x^2$ and $-6x^3y^2$.

6.6 Solving Quadratic Equations by Factoring

OBJECTIVES

1 Use the zero-factor theorem to solve equations containing expressions in factored form.

2 Solve quadratic equations by factoring.

3 Solve problems involving quadratic equations.

4 Use the Pythagorean theorem to solve problems.

OBJECTIVE **1** **Use the zero-factor theorem to solve equations containing expressions in factored form.**
So far in Chapters 5 and 6, we have rewritten polynomial expressions. We are now ready to work with polynomial equations, which we will solve using factoring. Notice we are moving to the equation level of our Algebra Pyramid, which is built upon the foundation of expressions, variables, and constants.

THE ALGEBRA PYRAMID

Note: Inequalities are also in the top level of the Pyramid, but we will not be exploring polynomial inequalities at this point.

EQUATIONS AND INEQUALITIES
$x^2 + 2x + 3 = 0$

EXPRESSIONS
$x^2 + 2x + 3$

CONSTANTS AND VARIABLES
$2, 3, x$

CONNECTION ——
Throughout this book, after developing new types of expressions and ways of rewriting those expressions, we then put those skills into the context of solving equations.

When solving polynomial equations by factoring, we make use of the **zero-factor theorem**, which states that if the product of two factors is 0, then one or the other of the two factors, or both factors, are equal to 0.

RULE **ZERO-FACTOR THEOREM**

If a and b are real numbers and $ab = 0$, then $a = 0$ or $b = 0$.

EXAMPLE **1** Solve $(x + 2)(x - 3) = 0$.

Solution In this equation, the product of two factors, $x + 2$ and $x - 3$, is equal to 0. According to the zero-factor theorem, one of the two factors, or both factors, must equal 0. In solving, we determine what values for x cause each individual factor to equal 0.

$$x + 2 = 0 \quad \text{or} \quad x - 3 = 0$$

Now solve each of the linear equations to get two solutions.

$$x = -2 \qquad x = 3$$

Check Verify that -2 and 3 satisfy the original equation, $(x + 2)(x - 3) = 0$.

For $x = -2$: $(-2 + 2)(-2 - 3) \overset{?}{=} 0$ $\qquad$ For $x = 3$: $(3 + 2)(3 - 3) \overset{?}{=} 0$

$\qquad\qquad\qquad\qquad (0)(-5) = 0$ $\qquad\qquad\qquad\qquad\qquad (5)(0) = 0$

Both -2 and 3 check, therefore, they are both solutions.

YOUR TURN

Solve $(n - 6)(n + 1) = 0$.

Example 1 suggests the following procedure.

PROCEDURE SOLVING EQUATIONS WITH TWO OR MORE FACTORS EQUAL TO 0

To solve an equation in which two or more factors are equal to 0, use the zero-factor theorem:

1. Set each factor equal to zero.
2. Solve each of those equations.

EXAMPLE 2 Solve $y(3y + 4) = 0$.

Solution In this equation, the product of two factors, y and $3y + 4$, is equal to 0. We set each factor equal to 0, and then solve each of those equations.

$$y = 0 \qquad \text{or} \qquad 3y + 4 = 0$$

$\qquad\qquad\qquad\qquad\qquad\qquad 3y = -4$ $\qquad$ Subtract 4 on both sides to isolate $3y$.

Note: This equation is already solved.

$\qquad\qquad\qquad\qquad\qquad\qquad y = -\dfrac{4}{3}$ $\qquad$ Divide both sides by 3 to isolate y.

Check Verify that 0 and $-\dfrac{4}{3}$ satisfy the original equation, $y(3y + 4) = 0$.

For $y = 0$: $(0)(3(0) + 4) \overset{?}{=} 0$ $\qquad$ For $y = -\dfrac{4}{3}$: $\left(-\dfrac{4}{3}\right)\left[\dfrac{\overset{1}{\cancel{3}}}{1}\left(-\dfrac{4}{\cancel{3}}\right) + 4\right] \overset{?}{=} 0$

$\qquad\qquad\qquad\qquad (0)(4) = 0$ $\qquad\qquad\qquad\qquad\qquad\qquad \left(-\dfrac{4}{3}\right)[-4 + 4] \overset{?}{=} 0$

$\qquad\qquad\qquad\qquad\qquad\qquad\qquad\qquad\qquad\qquad\qquad \left(-\dfrac{4}{3}\right)(0) = 0$

ANSWER

$n = 6 \text{ or } -1$

Both 0 and $-\dfrac{4}{3}$ check, therefore, they are both solutions.

ANSWER

$x = 0 \text{ or } \dfrac{2}{3}$

YOUR TURN

Solve $5x(3x - 2) = 0$.

EXAMPLE 3 Solve $(x - 4)^2 = 0$.

Solution Note that $(x - 4)^2 = 0$ can be written as $(x - 4)(x - 4) = 0$. Because the two factors are identical, there is no need to write two separate equations. We set $x - 4$ equal to 0, then solve.

$$x - 4 = 0$$
$$x = 4 \qquad \text{Add 4 on both sides to isolate } x.$$

Check Verify that 4 satisfies the original equation, $(x - 4)^2 = 0$.

$$(4 - 4)^2 \overset{?}{=} 0$$
$$(0)^2 = 0 \qquad \text{4 checks, therefore it is a solution.}$$

EXAMPLE 4 Solve $n(n + 6)(5n - 3) = 0$.

Solution In this equation, the product of three factors, n, $n + 6$, and $5n - 3$, is 0. We set each factor equal to 0, and then solve each of those equations.

$$n = 0 \quad \text{or} \quad n + 6 = 0 \quad \text{or} \quad 5n - 3 = 0$$
$$n = -6 \qquad\qquad 5n = 3$$
$$n = \dfrac{3}{5}$$

Check To check, we verify that 0, -6, and $\dfrac{3}{5}$ satisfy the original equation. We leave this check to the reader.

ANSWER

$t = 0, \dfrac{3}{4}, \text{ or } -1$

YOUR TURN

Solve $t(4t - 3)(t + 1) = 0$.

OBJECTIVE 2 Solve quadratic equations by factoring. Now that we have seen how to use the zero-factor theorem to solve equations containing expressions that are in factored form, we are ready to consider equations that have an expression that is not in factored form. More specifically, we will use factoring to solve equations called **quadratic equations**.

DEFINITION

> **Quadratic equation:** An equation that can be written in the form $ax^2 + bx + c = 0$, where a, b, and c are all real numbers and $a \neq 0$.

The fact that a cannot equal zero means that the ax^2 term must be present to be a quadratic equation. A quadratic equation written in the form $ax^2 + bx + c = 0$ is said to be in *standard form*. Following are some examples of quadratic equations.

$$x^2 + 5x + 6 = 0 \qquad 3x^2 - 48 = 0 \qquad x^2 = 10 - 11x$$

Note: The bx term is missing, which means that $b = 0$.

Note: Though not in standard form, the equation is still quadratic. Written in standard form, the equation would be: $x^2 + 11x - 10 = 0$.

Consider the equation $x^2 - x + 6 = 0$. To use the zero-factor theorem, we need to have an expression in factored form equal to 0, which means we need to factor $x^2 - x + 6$.

$$x^2 - x - 6 = 0$$
$$(x + 2)(x - 3) = 0$$

After factoring $x^2 - x - 6$, we have the same equation that we solved in Example 1. This suggests the following procedure.

PROCEDURE SOLVING QUADRATIC EQUATIONS USING FACTORING

To solve a quadratic equation:

1. Manipulate the equation as needed so that one side of the equation is 0.
2. Write the variable expression in factored form.
3. Use the zero-factor theorem to solve.

EXAMPLE 5 Solve $2x^2 + 9x - 5 = 0$.

Solution This equation is in standard form, so we can simply factor the variable expression.

$$(2x - 1)(x + 5) = 0 \qquad \text{Factor } 2x^2 + 9x - 5.$$

Now use the zero-factor theorem to solve.

$$2x - 1 = 0 \qquad \text{or} \qquad x + 5 = 0$$
$$2x = 1 \qquad\qquad\qquad x = -5$$
$$x = \frac{1}{2}$$

Check To check, we verify that $\frac{1}{2}$ and -5 satisfy the original equation. We leave this check to the reader.

ANSWER

$y = -\dfrac{2}{3}$ or $-\dfrac{1}{4}$

YOUR TURN

Solve $12y^2 + 11y + 2 = 0$.

Sometimes we may have to factor out a monomial GCF.

EXAMPLE **6** Solve $x^3 - 3x^2 - 10x = 0$.

Note: Though $x^3 - 3x^2 - 10x = 0$ is not a quadratic, after we factor out the x, the parentheses contain a quadratic form: $x^2 - 3x - 10$.

Solution

$$x(x^2 - 3x - 10) = 0 \qquad \text{Factor out the monomial GCF, } x.$$
$$x(x - 5)(x + 2) = 0 \qquad \text{Factor } x^2 - 3x - 10.$$

Now use the zero-factor theorem to solve.

$$x = 0 \quad \text{or} \quad x - 5 = 0 \quad \text{or} \quad x + 2 = 0$$
$$x = 5 \qquad\qquad x = -2$$

Check Verify that 0, 5, and -2 each satisfy the original equation. We leave the check to the reader.

YOUR TURN

Solve $12x^3 + 20x^2 - 8x = 0$.

Rewrite Quadratic Equations in Standard Form

If a quadratic equation is not in standard form ($ax^2 + bx + c = 0$), we first put it in standard form, then factor and use the zero-factor theorem.

EXAMPLE **7** Solve.

a. $6y^2 - 5y = 4 - 28y$

Solution We first need to manipulate the equation so that it is in standard form.

$$6y^2 + 23y = 4 \qquad\qquad \text{Add } 28y \text{ to both sides.}$$
$$6y^2 + 23y - 4 = 0 \qquad\qquad \text{Subtract 4 from both sides to get standard form.}$$
$$(6y - 1)(y + 4) = 0 \qquad\qquad \text{Factor using trial and error.}$$
$$6y - 1 = 0 \quad \text{or} \quad y + 4 = 0 \qquad \text{Use the zero-factor theorem to solve.}$$
$$6y = 1 \qquad\qquad y = -4$$
$$y = \frac{1}{6}$$

ANSWER

$x = 0, \dfrac{1}{3}, \text{ or } -2$

b. $x(x - 5) = -6$

Solution

Note: We could have multiplied $x(x - 5)$ first and then moved the 6.

$$x(x - 5) + 6 = 0$$ Add 6 to both sides so that the right-hand side is 0.

$$x^2 - 5x + 6 = 0$$ Multiply $x(x - 5)$ to get standard form.

$$(x - 2)(x - 3) = 0$$ Factor by looking for two numbers whose product is 6 and sum is -5.

$$x - 2 = 0 \quad \text{or} \quad x - 3 = 0$$ Use the zero-factor theorem to solve.

$$x = 2 \qquad\qquad x = 3$$

YOUR TURN

Solve $15n^2 + 20n = -8 - 6n$.

OBJECTIVE ③ **Solve problems involving quadratic equations.** Many problems and applications are solved using quadratic equations.

EXAMPLE 8 The product of two consecutive odd natural numbers is 195. Find the numbers.

Understand Odd natural numbers are 1, 3, 5, . . . Note that adding 2 to a given odd natural number gives a consecutive odd natural number, which suggests the pattern:

First odd natural number: x
Consecutive odd natural number: $x + 2$

The word *product* means that two numbers are multiplied to equal 195.

CONNECTION

In Chapter 3 we studied problems that dealt with the sum or difference of consecutive integers. The pattern for consecutive odd natural numbers is the same as for consecutive odd integers because every natural number is an integer.

Plan Translate to an equation, then solve.

Execute $x(x + 2) = 195$ The equation has the product of x and $x + 2$ set equal to 195.

$$x(x + 2) - 195 = 0$$ Subtract 195 on both sides.

$$x^2 + 2x - 195 = 0$$ Multiply $x(x + 2)$ to get the form $ax^2 + bx + c = 0$.

$$(x - 13)(x + 15) = 0$$ Factor by finding a pair of numbers whose product is -195 and sum is 2.

$$x - 13 = 0 \text{ or } x + 15 = 0$$ Use the zero-factor theorem to solve.

$$x = 13 \qquad x = -15$$

Answer Because -15 is not a natural number and 13 is a natural number, the first number is 13. This means that the consecutive odd natural number is 15.

ANSWER

$n = -\dfrac{2}{5}$ or $-\dfrac{4}{3}$

Check 13 and 15 are consecutive odd natural numbers and their product is 195, so the answer is correct.

ANSWER

21 and 22

YOUR TURN

The product of two consecutive natural numbers is 462. Find the numbers.

EXAMPLE 9 An architect is designing an addition to a house. The addition will be in the shape of a rectangle and have an area of 270 square feet. The width of the room is to be 3 feet less than the length. Find the dimensions of the room.

Understand The area of a rectangle is found by multiplying the length and width. We are given a relationship about the width and length, which we can translate:

"the width is to be 3 feet less than the length"

Width = $l - 3$, where l represents length.

Plan Translate to an equation, then solve.

Execute $l(l - 3) = 270$ The equation has the product of the length, l, and width, $l - 3$, set equal to 270.

$l(l - 3) - 270 = 0$ Subtract 270 on both sides.

$l^2 - 3l - 270 = 0$ Multiply $l(l - 3)$ to get standard form.

$(l - 18)(l + 15) = 0$ Factor

$l - 18 = 0$ or $l + 15 = 0$ Use the zero-factor theorem to solve.

$l = 18$ $l = -15$

Answer Because the problem involves room dimensions, only the positive number makes sense as an answer. If the length is 18 feet, and the width is 3 feet less than that, then the width is 15 feet.

Check The width, 15 feet, is 3 feet less than the length, 18 feet. Also, the area of a 15 foot by 18 foot rectangle is 270 square feet.

YOUR TURN

Two discs are designed so that the smaller of the two has an area of 113.04 square centimeters and a radius that is 8 centimeters less than the larger disc's radius. Find the radius of the larger disc. (Use 3.14 to approximate π.)

Area = 113.04 cm^2

Note: This radius is 8 less than the other disc's radius. ▶

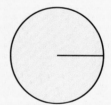

EXAMPLE 10 The equation $h = -16t^2 + h_0$ describes the height, h in feet of an object, after falling from an initial height h_0 for a time, t seconds. An object is dropped from a height of 600 feet. When the object reaches 344 feet, how much time has passed?

Understand We are given a formula and we are to find the time it takes an object to drop from an initial height of 600 ft. to a height of 344 ft.

ANSWER

14 cm

Plan Replace the variables in the formula with the given values to get an equation in terms of t, then solve for t.

Execute

$344 = -16t^2 + 600$	Replace h with 344 and h_0 with 600.
$-256 = -16t^2 + 0$	Subtract 600 on both sides.
$16t^2 - 256 = 0$	Add $16t^2$ to both sides.
$16(t^2 - 16) = 0$	Factor out the monomial GCF, 16.
$t^2 - 16 = 0$	We can divide out the constant factor, 16.
$(t - 4)(t + 4) = 0$	Factor
$t - 4 = 0$ or $t + 4 = 0$	Use the zero-factor theorem to solve.
$t = 4$ $t = -4$	

Answer Because the problem involves amount of time elapsed after an object was dropped, only the positive value, 4, makes sense. This means it took the object 4 seconds to fall from 600 feet to 344 feet.

Check Verify that after 4 seconds, an object dropped from an initial height of 600 feet will descend to a height of 344 feet.

$$h = -16(4)^2 + 600$$
$$h = -256 + 600$$
$$h = 344$$

YOUR TURN

An object is dropped from a height of 225 feet above the ground. When the object is 81 feet from the ground, how much time has passed?

ANSWER

3 s

OBJECTIVE **4** **Use the Pythagorean theorem to solve problems.** One of the most popular theorems in mathematics is the Pythagorean theorem, named after the Greek mathematician, Pythagoras. The theorem relates the side lengths of all right triangles. Recall that in a right triangle, the two sides that form the 90° angle are *legs* and the side directly across from the 90° angle is the *hypotenuse.*

RULE **THE PYTHAGOREAN THEOREM**

Given a right triangle, where a and b represent the lengths of the legs and c represents the length of the hypotenuse, then $a^2 + b^2 = c^2$.

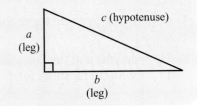

Of Interest

Pythagoras was a Greek mathematician who is believed to have been born around 569 B.C. and died in 500 B.C. The Pythagorean theorem is named after Pythagoras not because he discovered the relationship, as commonly thought, but because he and his followers are the first to have proved that the relationship is true for all right triangles. The relationship was known and used by other cultures prior to the Greeks; however, Pythagoras and others in the Greek culture believed it important to prove mathematical relationships.

CONNECTION

In Chapter 9 we will revisit the Pythagorean theorem. There we will use another technique involving square roots to solve equations like $c^2 = 25$.

We can use the Pythagorean theorem to find a missing length in a right triangle if we know the other two lengths.

EXAMPLE 11 Find the length of the missing side.

Solution Use the Pythagorean theorem, $a^2 + b^2 = c^2$. The hypotenuse, c, is the missing side length.

$3^2 + 4^2 = c^2$	Substitute $a = 3$ and $b = 4$.
$9 + 16 = c^2$	Simplify exponential forms.
$25 = c^2$	Add
$0 = c^2 - 25$	Subtract 25 from both sides to get standard form.
$0 = (c - 5)(c + 5)$	Factor
$c - 5 = 0$ or $c + 5 = 0$	Use the zero-factor theorem.
$c = 5$ $c = -5$	

Because we are dealing with lengths, only the positive solution is sensible, so the missing length is 5.

YOUR TURN

In constructing a roof, three beams are used to form a right triangle frame. A 12-foot beam and a 5-foot beam are brought together to form a 90° angle. How long must the third beam be?

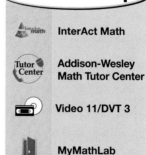

5 ft.   ?   12 ft.

ANSWER

13 ft.

6.6 EXERCISES

For Extra Help

InterAct Math

Tutor Center   Addison-Wesley Math Tutor Center

Video 11/DVT 3

MyMathLab

MathXP   MathXL

1. Explain the zero-factor theorem in your own words.

2. How do you use the zero-factor theorem to solve an equation that has an expression in factored form set equal to 0?

3. What is the first step in solving a quadratic equation?

4. Explain why we cannot write $(x + 2)(x - 2) = 32$ as $x + 2 = 32$ and $x - 2 = 32$.

For Exercises 5–36, solve.

5. $(x + 5)(x + 2) = 0$

6. $(x - 7)(x + 3) = 0$

7. $(a + 3)(a - 4) = 0$

8. $(b - 7)(b + 2) = 0$

9. $(3x - 4)(2x + 3) = 0$

10. $(3a + 2)(a - 5) = 0$

11. $x(x - 7) = 0$

12. $y(y + 3) = 0$

13. $m(m - 1)(m + 2) = 0$

14. $r(r - 2)(r + 3) = 0$

15. $b(b - 2)^2 = 0$

16. $(m - 4)^2 = 0$

17. $x^2 - 4 = 0$

18. $y^2 - 25 = 0$

19. $x^2 + x - 6 = 0$

20. $b^2 - 5b + 6 = 0$

21. $n^2 + 6n - 27 = 0$

22. $p^2 - 16p + 64 = 0$

23. $x^2 - 4x = 21$

24. $u^2 + 6u = 27$

25. $p^2 = 3p - 2$

26. $m^2 = 16$

27. $v^2 = 9$

28. $6k^2 - 7k = -1$

29. $4x^2 = 60 - x$

30. $4n^2 = 5 - 19n$

31. $x^2 + 3x = 18$

32. $-9x = -20 - x^2$

33. $b(b - 5) = 14$

34. $m(m + 6) = -9$

35. $4x(x + 7) = -49$

36. $c(2c - 11) = -5$

For Exercises 37–48, translate to an equation, then solve.

37. Find every number such that the square of the number added to 55 is the same as 16 times the number.

38. Find every number such that triple the square of the number is equal to four times that number.

39. One number is four times another. The product of the two numbers is 676. Find both numbers.

40. The product of two consecutive natural numbers is 306. Find the numbers.

41. The sum of the squares of two consecutive natural numbers is 365. Find the integers.

42. The difference of the squares of two consecutive even natural numbers is 60.

43. The length of a rectangular garden is 9 meters more than the width. If the area is 252 square meters, find the dimensions of the garden.

44. A design on the front of a marketing brochure calls for a triangle with a base that is 6 centimeters less than the height. If the area of the triangle is to be 216 square centimeters, what are the lengths of the base and height?

45. The design of a small building calls for a rectangular shape with dimensions of 22 feet by 28 feet. The architect decides to change the shape of the building to a circle, but wants it to have the same area. If we use $\dfrac{22}{7}$ to approximate π, what would be the radius of the circular building?

46. The design of a rectangular room calls for the perimeter to be 40 feet and the area to be 91 square feet. Find the dimensions of the room.

47. A steel plate is in the shape of a trapezoid and has an area of 85.5 square inches. The dimensions are shown. Note that the length of the base is equal to the height. Calculate the height.

10 in.

h

h

48. The front elevation of one wing of a house is shown. Because of budget constraints, the total area of the front of this wing must be 352 square feet. The height of the triangular portion is 14 feet less than the base. Find the base length.

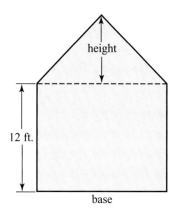

height

12 ft.

base

For Exercises 49 and 50, use the formula $h = -16t^2 + v_0 t + h_0$, where h is the final height in feet, t is the time of travel in seconds, v_0 is the initial velocity in feet per second, and h_0 is the initial height in feet of an object traveling upwards.

49. A ball is thrown upward at 4 feet per second from a building 29 feet high. When will the ball be 9 feet above the ground?

50. A toy rocket is fired from the ground with an upward velocity of 200 feet per second. How many seconds will it take to return to the ground?

For Exercises 51 and 52, use the formula $B = P(1 + rt)^n$, which is used to calculate the final balance of an investment or a loan after being compounded. Following is a list of what each variable represents:

B represents the final balance
P represents the principal, which is the amount invested
r represents the annual percentage rate
t represents the portion of the year in which the principal is compounded
n represents the number of times the principal is compounded

51. Carlita invests \$4000 in an account that is compounded annually. If, after two years, her balance is \$4840, what was the interest rate of the account? (*Hint:* Once you have the equation in the form $ax^2 + bx + c = 0$, multiply through by an appropriate power of 10 to clear the decimal numbers.)

52. Donovan invests $1000 in an account that is compounded semiannually (every six months). If, after one year, his balance is $1210, what is the interest rate of the account?

For Exercises 53–56, find length of the hypotenuse.

53.

9

?

12

54.

8

?

6

55.

?

24

7

56.

?

15

8

For Exercises 57–60, solve.

57. An artist welds a metal rod measuring 12 centimeters to a second rod measuring 35 centimeters to form a 90° angle. She wants a third rod to be welded to the ends of the first two to form a right triangle. What length must the third rod be?

58. A rectangular screen measures 27 inches by 36 inches. Find the diagonal distance between two opposing corners.

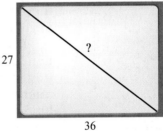

27

?

36

59. A person drives south 18 blocks then turns east and drives 24 blocks. Assuming all the city blocks are equal size and form a gird of streets which interesect at 90°, find the number of blocks that the person is from her original location.

60. To support a power pole, a wire is to be attached 24 feet from the ground then staked 10 feet from the base of the pole. How long must the wire be?

24 ft.

|←10 ft.→|

REVIEW EXERCISES

1. Factor $ax + bx + ay + by$.

2. Factor $x^2 - a^2$.

3. Solve and check $-4x + 8 = 20$.

4. Simplify $-4x^0$.

5. Write 0.0004203 in scientific notation.

6. Find the x- and y-intercepts for $3x - 4y = 24$.

6.7 Graphs of Quadratic Equations and Functions

OBJECTIVES

1. Graph quadratic equations in the form $y = ax^2 + bx + c$.
2. Graph quadratic functions.

In Chapter 4 we graphed linear equations in two variables and developed the concept of a function. In this section, we build on that work and graph quadratic equations and functions.

OBJECTIVE 1 Graph quadratic equations in the form $y = ax^2 + bx + c$. A quadratic equation in two variables will have the form $y = ax^2 + bx + c$, where $a \neq 0$. Following are some examples of quadratic equations in two variables.

$$y = x^2 \qquad y = 3x^2 + 2 \qquad y = x^2 - 4x + 1$$

We learned in Section 2.2 that in a linear equation, every term will be either degree 1 or degree 0. Since quadratic equations always have a degree 2 term, they are nonlinear, which means their graphs will not be straight lines. What might their graphs look like?

Recall that one way to graph an equation is to plot solutions to the equation in the coordinate plane. In Section 4.1 we said that a solution to an equation in two variables is an ordered pair (x, y) that satisfies the equation. For example, $(3, 9)$ is a solution to $y = x^2$ because $9 = (3)^2$. Let's create a table of solutions for $y = x^2$, plot those solutions, and see what the graph looks like if we connect the points.

EXAMPLE 1 Graph $y = x^2$.

Solution We complete a table of solutions, plot the solutions in the coordinate plane, then connect the points to form the graph.

Note: Since we know quadratic equations are nonlinear, we find many solutions to get a good sense of the shape of the graph.

x	y
−3	9
−2	4
−1	1
0	0
1	1
2	4
3	9

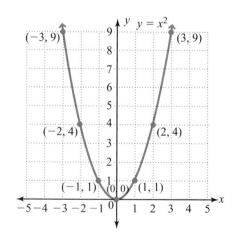

The graph of $y = x^2$ is a *parabola*. In fact, the graph of any quadratic equation in two variables is a parabola. Look closely at the graph and table of solutions in Example 1. In the table of solutions, notice that for each x-value on either side of 0, the y-values are the same. On the graph, notice that the left side of the y-axis is the mirror image of the right side of the y-axis. In mathematics, we say that this graph is *symmetrical* about the line $x = 0$ (the y-axis). All parabolas will have symmetry about a line called the **axis of symmetry**. Also, the axis of symmetry will always pass through a point called the **vertex**.

In the case of $y = x^2$, the axis of symmetry is the line $x = 0$ (y-axis) and the vertex is $(0, 0)$. In Section 10.5 we will learn more about the vertex and axis of symmetry. For now, we will graph by finding enough solutions to see where the parabola turns.

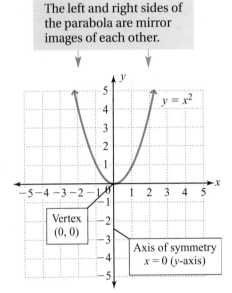

The left and right sides of the parabola are mirror images of each other.

PROCEDURE **GRAPHING QUADRATIC EQUATIONS**

To graph a quadratic equation:

1. Find ordered-pair solutions, and plot them in the coordinate plane. Continue finding and plotting solutions until the shape of the parabola can be clearly seen.
2. Connect the points to form a parabola.

EXAMPLE 2 Graph $y = 3x^2 + 2$.

Solution We complete a table of solutions, plot the solutions in the coordinate plane, and then connect the points to form the graph.

Note: The constant 2 in $y = 3x^2 + 2$ indicates the y-intercept because replacing x with 0 leaves us with $y = 3(0)^2 + 2 = 2$.

x	y
−2	14
−1	5
0	2
1	5
2	14

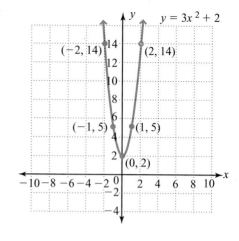

YOUR TURN

Graph $y = x^2 - 5$.

EXAMPLE 3 Graph $y = x^2 + 5x - 3$.

Solution We complete a table of solutions, plot the solutions in the coordinate plane, then connect the points to form the graph.

Note: Since −3 and −2 have the same y-values, the x-coordinate of the vertex will be halfway between −3 and −2 at −2.5.

x	y
−5	−3
−4	−7
−3	−9
−2	−9
−1	−7
0	−3

$y = x^2 + 5x - 3$

Graph $y = 2x^2 - x + 1$.

OBJECTIVE **2** **Graph quadratic functions.**   In Section 4.7, we learned about functions, function notation, and how to graph linear functions. Now we can write and graph quadratic functions. Where a quadratic equation in two variables has the form $y = ax^2 + bx + c$, a quadratic function will have the form $f(x) = ax^2 + bx + c$ (notice that, in essence, $f(x)$ replaces y).

Quadratic equation in two variables	Quadratic function
$y = -2x^2 + 1$	$f(x) = -2x^2 + 1$

Also recall from Section 4.7 that for a given function such as $f(x) = -2x^2 + 1$, the notation $f(1)$ means to find the value of the function, where $x = 1$.

$$f(1) = -2(1)^2 + 1$$
$$= -2 + 1$$
$$= -1$$

CONNECTION

Evaluating a function is like finding the y-value in an equation in two variables having been given a particular x-value.

We can create a table of ordered pairs and graph quadratic functions using the same technique that we used for quadratic equations in two variables. For the function $f(x) = -2x^2 + 1$, we have:

x	$f(x)$
-2	-7
-1	-1
0	1
1	-1
2	-7

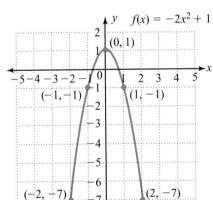

CONNECTION

This parabola opens downward. Given a quadratic function in the form $f(x) = ax^2 + bx + c$ or a quadratic equation in the form $y = ax^2 + bx + c$, if $a < 0$, then the parabola opens downward. We will formalize this rule in Section 10.5.

$y = 2x^2 - x + 1$

EXAMPLE **4** Graph $f(x) = x^2 + x - 2$.

Solution We complete a table of solutions, plot the solutions in the coordinate plane, then connect the points to form the graph.

Note: The y-values match for 0 and -1, which indicates that the x-coordinate of the vertex is halfway between 0 and -1 at $-\dfrac{1}{2}$.

x	y
−3	4
−2	0
−1	−2
0	−2
1	0
2	4

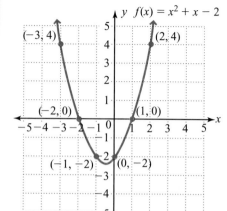

CONNECTION

This parabola opens upward. Given a quadratic function in the form $f(x) = ax^2 + bx + c$ or a quadratic equation in the form $y = ax^2 + bx + c$, if $a > 0$, then the parabola opens upward. We will formalize this rule in Section 10.5.

YOUR TURN

Graph $f(x) = -3x^2 + 2x + 1$.

ANSWER

$f(x) = -3x^2 + 2x + 1$

6.7 EXERCISES

1. In your own words, describe what is meant by *symmetry* in a parabola.

2. Given an equation in the form $y = ax^2 + bx + c$, what indicates whether the graph opens up or down?

3. Given an equation in the form $y = ax^2 + bx + c$, what indicates the y-coordinate of the y-intercept?

4. Given that the graph of $y = ax^2 + bx + c$, where $a \neq 0$, is a parabola that opens upward or downward, explain why $f(x) = ax^2 + bx + c$ is a function. (*Hint:* Refer to the definition of a function and the vertical line test in Section 4.7.)

For Exercises 5–10, use the graph of the parabola to determine the coordinates of the vertex.

5.

6.

7.

8.

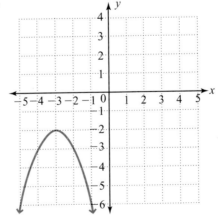

6.7 Graphs of Quadratic Equations and Functions   525

9.

10.

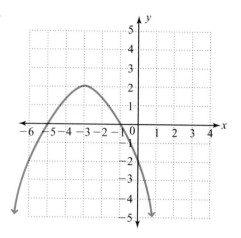

For Exercises 11–16, complete the table of solutions, then graph.

11. $y = -x^2$

x	y
−2	
−1	
0	
1	
2	

12. $y = x^2 + 3$

x	y
−2	
−1	
0	
1	
2	

13. $y = x^2 + 2x - 3$

x	y
−4	
−3	
−2	
−1	
0	
1	
2	

14. $y = x^2 - 4x - 5$

x	y
−1	
0	
1	
2	
3	
4	
5	

15. $y = 2x^2 - 4x + 1$

x	y
−1	
0	
1	
2	
3	

16. $y = 3x^2 + 6x - 4$

x	y
−3	
−2	
−1	
0	
1	

For Exercises 17–32, graph.

17. $y = x^2 + 1$

18. $y = x^2 - 4$

19. $y = -x^2 + 3$

20. $y = -x^2 + 2$

21. $y = x^2 + x - 2$ **22.** $y = x^2 + x - 6$ **23.** $y = 3x^2 + x - 2$ **24.** $y = 2x^2 + x - 3$

25. $y = -x^2 + 6x - 5$ **26.** $y = -x^2 + 4x - 5$ **27.** $y = x^2 - 6x + 3$ **28.** $y = 4x^2 - 8x - 1$

29. $y = x^2 - 3x + 2$ **30.** $y = 3x^2 + 4x + 1$ **31.** $y = 2x^2 - x - 6$ **32.** $y = 3x^2 - 5x + 2$

For Exercises 33–42, graph the function.

33. $f(x) = -x^2 - 4x - 3$ **34.** $f(x) = x^2 - 2x - 2$ **35.** $f(x) = x^2 + 4x - 1$

36. $f(x) = x^2 - 6x + 5$

37. $f(x) = x^2 + 3x - 3$

38. $f(x) = -x^2 + 5x - 8$

39. $f(x) = 2x^2 - 3x + 1$

40. $f(x) = 4x^2 - 10x + 3$

41. $f(x) = 2x^2 + 7x + 5$

42. $f(x) = -2x^2 + x + 5$

For Exercises 43–46, state whether the graph is the graph of a function.

43.

44.

45.

46.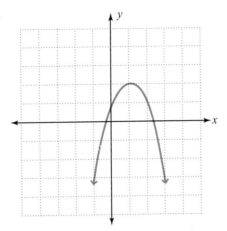

For Exercises 47–50, use a graphing calculator.

47. a. Enter the equation $y = x^2$, then graph. What are the coordinates of the y-intercept?

  b. Enter a second equation $y = x^2 + 2$, then graph. What are the coordinates of the y-intercept?

  c. On the same grid, graph $y = x^2 + 4$ and compare all three graphs. What do your explorations suggest about how c affects the graph of an equation in the form $y = x^2 + c$?

  d. On the same grid, graph $y = x^2 - 3$. Comparing this graph to the first three graphs, how did changing c to a negative value affect the graph?

48. a. Enter the equation $y = x^2$, then graph. What are the coordinates of the vertex?

  b. Enter a second equation: $y = 2x^2$. Compare the graph of $y = 2x^2$ and the graph of $y = x^2$, then discuss their similarities and differences.

  c. On the same grid, graph $y = 3x^2$ and compare all three graphs. What does this exploration suggest about how the size of a in $y = ax^2$ affects the graph?

  d. In $y = ax^2$, if $a = \dfrac{1}{2}$, how do you think the graph will be different from the graphs in parts a–c? Graph $y = \dfrac{1}{2}x^2$.

  e. What do these explorations suggest about how the size of a affects the graph of an equation in the form $y = ax^2$?

49. a. Graph $y = x^2$ and $y = -x^2$ on the same grid. What is different about the two graphs? What does this suggest about the sign of a in an equation of the form $y = ax^2$?

  b. Clearing the equations from part a, graph $y = 2x^2$ and $y = -2x^2$ together. Do these graphs confirm your conclusion from part a?

c. On a new grid, graph $y = \frac{1}{2}x^2$ and $y = -\frac{1}{2}x^2$. What is different about these graphs compared to those in part a and b?

d. What do these explorations suggest about how the absolute value of a affects the graph of an equation in the form $y = ax^2$?

50. a. Enter the equation $y = x^2 + x + 1$, then graph. What are the coordinates of the y-intercept?

b. On the same grid, graph $y = x^2 + 2x + 1$. Compare the graph of $y = x^2 + x + 1$ with the graph of $y = x^2 + 2x + 1$, then discuss their similarities and differences.

c. Using the same grid, graph $y = x^2 + 3x + 1$. Compare this graph with the graphs of $y = x^2 + x + 1$ and $y = x^2 + 2x + 1$, then discuss similarities and differences.

d. On the same grid, graph $y = x^2 - x + 1$, $y = x^2 - 2x + 1$, and $y = x^2 - 3x + 1$. Compare these graphs with your graphs from parts a–c. What do these explorations suggest about how b affects the graph of an equation in the form $y = ax^2 + bx + c$?

Collaborative Exercise

What Goes Up, Must Come Down

There are many sets of variables that a have a relationship that is roughly modeled by a parabolic shape. The data below give the high temperature and peak power load (in megawatts) for 31 days throughout the calendar year.

Temp	Peak Load	Temp	Peak Load
69	94	95	141
71	95	46	142
59	100	43	147
65	100	98	150
67	102	96	151
60	104	97	155
74	105	100	158
79	105	43	165
56	112	38	173
84	113	102	176
52	114	36	193
88	115	103	199
50	120	105	204
90	123	106	226
92	132	34	228
92	132		

1. On a sheet of graph paper, draw and label axes with the temperature along the horizontal axis and the peak load on the vertical axis. Then plot the points.

2. Draw a smooth parabola through the points that might best fit the data.

3. At what temperature does the minimum peak load seem to occur?

4. In your own words, describe what happens to the peak load as the temperature increases and decreases and explain why.

5. Use the model that you drew in part two to estimate the peak load on a day when the maximum temperature is 85.

6. Give another example of a situation that might be best modeled by a parabola.

REVIEW EXERCISES

For Exercises 1 and 2, multiply.

1. $3mn(m^2 - 4mn + 5)$

2. $7x(x - 3)(2x + 1)$

For Exercises 3 and 4, factor.

3. $x^2 - 25$

4. $3n^2 - 17n + 20$

For Exercises 5 and 6, solve.

5. $h(h - 3) = 0$

6. $y^2 + 3y - 18 = 0$

Defined Terms

Section 6.1
Factored Form (p. 472)
Greatest Common
Factor (GCF) (p. 473)

Section 6.6
Quadratic Equation
(p. 518)

Section 6.7
Axis of Symmetry
(p. 529)
Vertex (p. 529)

Procedures, Rules, and Key Examples

Procedures/Rules	Key Examples

Section 6.1 Greatest Common Factor and Factoring by Grouping

To find the GCF by listing:
1. List all possible factors for each given number.
2. Search the lists for the largest factor common to all lists.

Example 1: Find the GCF of 54 and 180 by listing.

Factors of 54: 1, 2, 3, 6, 9, 18, 27, 54
Factors of 180: 1, 2, 3, 4, 5, 6, 9, 10, 12, 15, 18, 20, 30, 36, 45, 60, 90, 180
$$GCF = 18$$

To find the greatest common factor of a given set of numbers:
1. Write the prime factorization of each given number in exponential form.
2. Create a factorization for the GCF that includes only those primes common to all factorizations, each raised to its smallest exponent in the factorizations.
3. Multiply.

Note: If there are no common prime factors, then the GCF is 1.

Example 2: Find the GCF of 54 and 180 using prime factorization.

$$54 = 2 \cdot 3^3$$
$$180 = 2^2 \cdot 3^2 \cdot 5$$
$$GCF = 2 \cdot 3^2 = 18$$

To factor a monomial out of a given polynomial:
1. Find the GCF of the terms that make up the polynomial.
2. Rewrite the given polynomial as a product of the GCF and parentheses that contain the result of dividing the given polynomial by the GCF.

$$\text{given polynomial} = GCF \left(\frac{\text{given polynomial}}{GCF} \right)$$

Example 3: Factor $54x^3y - 180x^2yz$.

$$54x^3y - 180x^2yz = 18x^2y\,(3x - 10z)$$

To factor a four-term polynomial by grouping:
1. Factor out any monomial GCF (other than 1) that is common to all four terms.
2. Group together pairs of terms and factor the GCF out of each pair.
3. If there is a common binomial factor, then factor it out.
4. If there is no common binomial factor, then interchange the middle two terms and repeat the process. If there is still no common binomial factor, then the polynomial cannot be factored by grouping.

Example 4: Factor by grouping
$$8x^2 - 20x - 6xy + 15y.$$

$$8x^2 - 20x - 6xy + 15y$$
$$= (8x^2 - 20x) - (6xy - 15y)$$
$$= 4x(2x - 5) - 3y(2x - 5)$$
$$= (2x - 5)(4x - 3y)$$

continued

Procedures/Rules	Key Examples

Procedures/Rules

Section 6.2 Factoring Trinomials of the form $x^2 + bx + c$

To factor a trinomial of the form $x^2 + bx + c$:
1. Find two numbers with a product equal to c and a sum equal to b.
2. The factored trinomial will have the form:
   $(x + \text{first number})(x + \text{second number})$.

Note: The signs in the binomial factors can be minus signs, depending on the signs of b and c.

Key Examples

Example 1: Factor $x^2 + 9x + 20$.
Find two numbers whose product is 20 and whose sum is 9.

$$
\begin{array}{ll}
\text{Product:} & \text{Sum:} \\
(1)(20) = 20 & 1 + 20 = 21 \\
(2)(10) = 20 & 2 + 10 = 12 \\
(4)(5) = 20 & 4 + 5 = 9 \quad \text{Correct} \\
& \qquad\qquad\qquad \text{combination.}
\end{array}
$$
$$x^2 + 9x + 20 = (x + 4)(x + 5)$$

Example 2: Factor $x^2 + 2x - 15$.

$$
\begin{array}{ll}
\text{Product:} & \text{Sum:} \\
(-1)(15) = -15 & -1 + 15 = 14 \\
(-3)(5) = -15 & -3 + 5 = 2 \quad \text{Correct} \\
& \qquad\qquad\qquad \text{combination.}
\end{array}
$$
$$x^2 + 2x - 15 = (x - 3)(x + 5)$$

Section 6.3 Factoring Trinomials of the form $ax^2 + bx + c$, where $a \neq 1$.

To factor a trinomial of the form $ax^2 + bx + c$, where $a \neq 1$, by trial and error:
1. Look for a monomial GCF in all the terms. If there is one, factor it out.
2. Write a pair of *first* terms whose product is ax^2.
3. Write a pair of *last* terms whose product is c.
4. Verify that the sum of the *inner* and *outer* products is bx (the middle term of the trinomial). If the sum of the inner and outer products is not bx, then try the following:
   a. Exchange the first terms and repeat step 4.
   b. Exchange the last terms and repeat step 4.
   c. Repeat steps 2–4 with a different combination of first and last terms.

To factor a trinomial of the form $ax^2 + bx + c$, where $a \neq 1$, by grouping:
1. Look for a monomial GCF in all the terms. If there is one, factor it out.
2. Multiply a and c.
3. Find two factors of this product whose sum is b.
4. Write a four-term polynomial in which bx is written as the sum of two like terms whose coefficients are the two numbers you found in step 3.
5. Factor by grouping.

Example 1: Factor $4x^2 - 9x + 5$.

Factors of 4 are 4 and 1, or 2 and 2.

Factors of 5 are 1 and 5.

Because the middle term $-9x$ is negative and the last term 5 is positive, we know that the second terms in each binomial factor will be negative.

$$(x - 1)(4x - 5) = 4x^2 - 9x + 5$$

Example 2: Factor $18y^3 + 12y^2 - 48y$.
Factor out the monomial GCF, $6y$.

$$18y^3 + 12y^2 - 48y = 6y(3y^2 + 2y - 8)$$

To factor the trinomial, multiply $3(-8) = -24$, then find two factors of -24 whose sum is 2. Note that -4 and 6 work. Write the middle term, $2y$, as $-4y + 6y$, then factor by grouping.

$$
\begin{aligned}
&= 6y(3y^2 - 4y + 6y - 8) \\
&= 6y[\, y(3y - 4) + 2(3y - 4)\,] \\
&= 6y(3y - 4)(y + 2)
\end{aligned}
$$

continued

Procedures/Rules	Key Examples

Section 6.4 Factoring Special Products
Rules for factoring special products.

Perfect square trinomials:
$$a^2 + 2ab + b^2 = (a + b)^2$$
$$a^2 - 2ab + b^2 = (a - b)^2$$

Difference of squares:
$$a^2 - b^2 = (a + b)(a - b)$$

Note: A sum of squares cannot be factored.

Difference of cubes:
$$a^3 - b^3 = (a - b)(a^2 + ab + b^2)$$

Sum of cubes:
$$a^3 + b^3 = (a + b)(a^2 - ab + b^2)$$

Example 1: Factor.
a. $9x^2 + 30x + 25 = (3x + 5)^2$
b. $36m^2 - 60mn + 25n^2 = (6m - 5n)^2$

c. $4y^2 - 81 = (2y + 9)(2y - 9)$

d. $8x^3 - 27 = (2x - 3)(4x^2 + 6x + 9)$

e. $n^3 + 64 = (n + 4)(n^2 - 4n + 16)$

Section 6.5 Strategies for Factoring
To factor a polynomial, first factor out any monomial GCF, then consider the number of terms in the polynomial. If the polynomial has:

I. Four terms, then try to factor by grouping.

II. Three terms, then determine if the trinomial is a perfect square or not.
  A. If the trinomial is a perfect square, then consider its form.
   1. If in the form $a^2 + 2ab + b^2$, then the factored form is $(a + b)^2$.
   2. If in the form $a^2 - 2ab + b^2$, then the factored form is $(a - b)^2$.
  B. If the trinomial is not a perfect square, then consider its form.
   1. If in the form $x^2 + bx + c$, then find two factors of c whose sum is b and write the factored form as $(x + \text{first number})(x + \text{second number})$.
   2. If in the form $ax^2 + bx + c$, where $a \neq 1$, then use trial and error. Or, find two factors of ac whose sum is b, write these factors as coefficients of two like terms that, when combined, equal bx, then factor by grouping.

III. Two terms, then determine if the binomial is a difference of squares, sum of cubes, or difference of cubes.
  A. If given a binomial that is a difference of squares $a^2 - b^2$, then the factors are conjugates and the factored form is $(a + b)(a - b)$. Note that a sum of squares cannot be factored.
  B. If given a binomial that is a sum of cubes, $a^3 + b^3$, then the factored form is $(a + b)(a^2 - ab + b^2)$.
  C. If given a binomial that is a difference of cubes, $a^3 - b^3$, then the factored form is $(a - b)(a^2 + ab + b^2)$.

Note: Always look to see if any of the factors can be factored.

The examples shown do not represent every possible type of problem. Each example will contain a monomial GCF other than 1:

Example 1: Factoring by grouping:
$$8x^3 + 24x^2 - 2x^2y - 6xy$$
$$= 2x(4x^2 + 12x - xy - 3y)$$
$$= 2x[4x(x + 3) - y(x + 3)]$$
$$= 2x(x + 3)(4x - y)$$

Example 2: Factoring a perfect square trinomial:
$$75x^4 + 60x^3y + 12x^2y^2$$
$$= 3x^2(25x^2 + 20xy + 4y^2)$$
$$= 3x^2(5x + 2y)^2$$

Example 3: Factoring a difference of squares:
$$81y^5 - 16y = y(81y^4 - 16)$$
$$= y(9y^2 + 4)(9y^2 - 4)$$
$$= y(9y^2 + 4)(3y + 2)(3y - 2)$$

Example 4: Factoring a sum of cubes:
$$54n^4 + 16nm^3 = 2n(27n^3 + 8m^3)$$
$$= 2n(3n + 2m)(9n^2 - 6mn + 4m^2)$$

continued

Procedures/Rules	Key Examples

Procedures/Rules

Section 6.6 Solving Quadratic Equations by Factoring

Zero-Factor Theorem: If a and b are real numbers and $ab = 0$, then
$$a = 0, \text{ or } b = 0.$$

To solve an equation in which two or more factors are equal to 0, use the zero-factor theorem:

1. Set each factor equal to zero.
2. Solve each of those equations.

To solve a quadratic equation,

1. Manipulate the equation as needed so that one side of the equation is 0.
2. Write the variable expression in factored form.
3. Use the zero-factor theorem to solve.

The Pythagorean Theorem

Given a right triangle where a and b represent the lengths of the legs and c represents the length of the hypotenuse, then $a^2 + b^2 = c^2$.

Key Examples

Example 1: Solve $(2x - 5)(x + 4) = 0$.

$2x - 5 = 0 \quad$ or $\quad x + 4 = 0$

$\qquad 2x = 5 \qquad\qquad\quad x = -4$

$\qquad x = \dfrac{5}{2}$

Example 2: Solve $2x^2 = 3 - x$.

$\qquad\qquad 2x^2 = 3 - x$

$\qquad\quad 2x^2 + x = 3 \qquad$ Add x on both sides.

$2x^2 + x - 3 = 0 \qquad$ Subtract 3 from both sides.

$(2x + 3)(x - 1) = 0 \qquad$ Factor

$2x + 3 = 0$ or $x - 1 = 0 \qquad$ Use the zero-factor theorem.

$\qquad 2x = -3 \qquad\quad x = 1$

$\qquad x = -\dfrac{3}{2}$

Example 3: The length of a small rectangular building is to be 15 ft. more than the width. The area is to be 700 ft.^{2} What must the dimensions of the building be?

Let w represent width. The length will be $w + 15$. Writing an equation for area, we have:

$\qquad\quad w(w + 15) = 700$

$\qquad\quad w^2 + 15w = 700 \qquad$ Distribute w.

$w^2 + 15w - 700 = 0 \qquad$ Subtract 700 from both sides.

$\qquad (w - 20)(w + 35) = 0 \qquad$ Factor

$w - 20 = 0$ or $w + 35 = 0 \qquad$ Use the zero-factor theorem.

$\qquad w = 20 \qquad\qquad w = -35$

Answer: The width is 20 ft. and the length is 35 ft.

Example 4: Find the unknown length in the following right triangle.

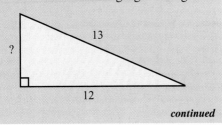

continued

Procedures/Rules	Key Examples

Section 6.6 Solving Quadratic Equations by Factoring (continued)

Solution: Use the Pythagorean theorem, $a^2 + b^2 = c^2$.

$$a^2 + 12^2 = 13^2$$
$$a^2 + 144 = 169$$
$$a^2 - 25 = 0 \quad \text{Subtract 169 from both sides.}$$
$$(a - 5)(a + 5) = 0 \quad \text{Factor}$$
$$a - 5 = 0 \ or \ a + 5 = 0 \quad \text{Use the zero-factor theorem.}$$
$$a = 5 \qquad a = -5$$

Answer: The unknown length is 5.

Section 6.7 Graphs of Quadratic Equations and Functions

Procedure: To graph a quadratic equation:
1. Find ordered-pair solutions, and plot them in the coordinate plane. Continue finding and plotting solutions until the shape of the parabola can be clearly seen.
2. Connect the points to form a parabola.

Example 1: Graph $y = x^2 - 3$.

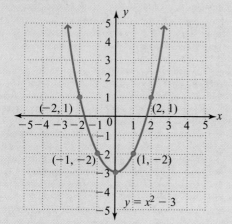

For Exercises 1–5, answer true or false.

1. The GCF must be in every term.

2. $x^2 + 16 = (x + 4)(x + 4)$

3. To factor a binomial, we use grouping.

4. If a and b are real numbers and $ab = 0$, then $a = 0$ or $b = 0$.

5. $x^2 + 8x + 16 = (x + 4)^2$

For Exercises 6–10, complete the rule.

6. To factor a monomial out of a given polynomial:

  1. Find the _____ of the terms that make up the polynomial.

  2. Rewrite the given polynomial as a product of the GCF and parentheses that contain _____.

7. $a^2 - b^2 = ($ $)($ $)$

8. $a^3 - b^3 = ($ $)($ $)$

9. $a^2 - 2ab + b^2 = ($ $)^2$

10. Procedure: To factor a trinomial of the form $x^2 + bx + c$:

  1. Find two numbers with a product equal to _____ and a sum equal to _____.

  2. The factored trinomial will have the form:
  $(x + \text{first number})(x + \text{second number})$

For Exercises 11–18, list all of the factors.

11. 60

12. 9

13. 81

14. 44

15. 45

16. 49

17. 30

18. 33

For Exercises 19–26, find the GCF.

19. $21, 30$ **20.** $34, 51$ **21.** $12, 42, 60$ **22.** $10, 15, 45$

23. $10y^4, 25y^2$ **24.** $4x, 20xy$ **25.** $15m^2n, 3mn^5$ **26.** $8k, 22k^9$

For Exercises 27–34, factor out the GCF.

27. $4x - 2$ **28.** $5m - 35m^3$ **29.** $y^3 - y$

30. $abc + a^2b^2c^2 - a^3b^3c^3$ **31.** $x^2y + xy^2 + x^3y^3$ **32.** $105a^3b^2 - 63a^2b^3 + 84a^6b^4$

33. $100k^4 + 120k^5 - 10k + 40k^3$ **34.** $18ab^3c - 36a^2b^2c$

For Exercises 35–42, factor by grouping.

35. $ax + ay + bx + by$ **36.** $ax + 2a + bx + 2b$ **37.** $y^3 + 2y^2 + 3y + 6$ **38.** $y^4 + 4y^3 - by - 4b$

39. $xy + y + x + 1$ **40.** $ax^2 - 5y^2 + ay^2 - 5x^2$ **41.** $2b^3 - 2b^2 + b - 1$ **42.** $u^2 - 3u + 4uv - 12v$

For Exercises 43–50, factor.

43. $x^2 - x - 12$ **44.** $x^2 + 14x + 45$ **45.** $n^2 - 6n + 8$ **46.** $a^2 + 3a - 20$

47. $h^2 + 51h + 144$ **48.** $y^2 - 10y - 24$ **49.** $4x^2 - 24x + 36$ **50.** $3m^2 - 33m + 54$

For Exercises 51–58, factor completely.

51. $6x^2 + 3x - 7$ **52.** $2u^2 + 5u + 2$ **53.** $3m^2 - 10m + 3$ **54.** $5k^2 - 7kh - 12h^2$

55. $6a^2 - 20a + 16$ **56.** $8x^2y - 4xy - y$ **57.** $3p^2 - 13pq + 4q^2$ **58.** $24x^2 - 64xy - 24y^2$

For Exercises 59–66, factor.

59. $v^2 - 8v + 16$ **60.** $u^2 + 6u + 9$ **61.** $4x^2 + 20x + 25$ **62.** $9y^2 - 12y + 4$

63. $x^2 - 4$ **64.** $25 - y^2$ **65.** $x^3 - 1$ **66.** $x^3 + 27$

For Exercises 67–76, factor completely.

67. $6b^2 + b - 2$ **68.** $4ab - 24ab^2$ **69.** $y^2 + 25$ **70.** $3x^2 - 3y^2$

71. $x^4 - 81$

72. $7u^2 - 14u - 105$

73. $8x^3 - 27y^3$

74. $2 - 50y^2$

75. $3am - 6an - 8bn + 4bm$

76. $3m^2 + 9m + 27$

For Exercises 77–84, solve.

77. $(m + 3)(m - 4) = 0$ **78.** $y^2 - 4 = 0$

79. $x^2 - 5x + 6 = 0$

80. $x^2 - 4x = 21$

81. $m^2 = 9$

82. $x^2 + 3x = 18$

83. $y(y + 6) = -9$

84. $c(c - 5) = 14$

For Exercises 85–90, translate to an equation, then solve.

85. Find every number such that five times the square of the number is equal to twice the number.

86. The sum of the squares of two consecutive natural numbers is 61. Find the numbers.

87. The product of a number and 4 times that same number is 100. Find all numbers.

88. Find the dimensions of a rectangle whose length is 5 more than its width and whose area is 91 square inches.

89. The product of two consecutive natural numbers is 110. Find the numbers.

90. Find the length of the missing side.

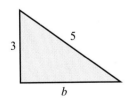

For Exercises 91–94, graph.

91. $y = -5x^2$

92. $y = x^2 - 3$

93. $f(x) = x^2 + 2$

94. $f(x) = -3x^2 + 6x + 2$

1. List all possible factors of 80.

2. Find the GCF of $25m^2$ and $40m$.

Factor completely.

3. $5y - 30$ **4.** $6x^2y - 2y^2$ **5.** $ax - ay - bx + by$ **6.** $m^2 + 7m + 12$

7. $c^2 - 25$ **8.** $ax^2 - 5ax - 24a$ **9.** $m^3 + 125$ **10.** $r^2 - 8r + 16$

11. $2 - 50u^2$ **12.** $3q^2 - 10q + 8$ **13.** $x^2 + 4$

14. $-4x^2 + 16$ **15.** $x^3 - 8$

Solve.

16. $a(a + 3) = 0$ **17.** $x^2 + 6x = 27$

18. The width of a rectangle is 5 feet less than the length. If the area is 36 square feet, find the length and width.

Graph.

19. $y = x^2 + 4$ **20.** $f(x) = x^2 - 6x + 9$

For Exercises 1–6, answer true or false.

1. $x^3 \cdot x^4 = x^{12}$

2. The GCF of 12 and 5 is 1.

3. The commutative property can be used for both addition and subtraction.

4. $2x - y^2 = 1$ is a linear equation.

5. The expression $5 + x$ can be simplified to equal $5x$.

6. $4x + 12 = 2(2x + 6)$ is factored completely.

For Exercises 7–10, fill in the blank.

7. To factor a trinomial of the form $ax^2 + bx + c$, where $a \neq 1$, by grouping:

   a. Look for a monomial _____ in all the terms. If there is one, factor it out.

   b. Multiply a and c.

   c. Find two factors of this product whose sum is b.

   d. Write a four-term polynomial in which bx is written as the sum of two like terms whose coefficients are the two numbers you found in step c.

   e. Factor by _____.

8. To write a polynomial in descending order, place the _____ degree term first, then the next highest degree, and so on.

9. Explain in your own words what it means to completely factor a polynomial.

10. To solve a quadratic equation:

   a. Manipulate the equation as needed so that one side is an expression and the other side is _____.

   b. Write the expression in _____ form.

   c. Use the zero-factor theorem to solve.

11. What are the x- and y-intercepts on a graph?

12. What is a binomial?

For Exercises 13–16, simplify.

13. $14 - 2 \cdot 6 \cdot 3 + 8^2$

14. $2(3 + 4) - \sqrt{16}$

15. $12 \div 3 \cdot 4$

16. $-6|5 + 3^2| - \sqrt{14 + 11}$

For Exercises 17–20, solve and check.

17. $2x - 9 = 11$

18. $\dfrac{3}{4} = x - \dfrac{2}{3}$

19. $7x - 9 = 6x + 4$

20. $2.6 + 7a + 5 = 8a - 5.6$

21. Solve for x in the equation $2x - 3y = 7$.

22. Determine whether $(-1, 3)$ is a solution for $2x + y = 9$.

23. Find the slope between the points $(6, 2)$ and $(5, 2)$.

24. Find the x- and y-intercepts for $y = 2x - 6$.

For Exercises 25 and 26, graph.

25. $y = 3x + 2$

26. $2x - y = 6$

For Exercises 27–36, simplify.

27. $(5x^2 + 6x - 1) - (2x - 3)$

28. $(7m^2 - 8) + (5m^2 - 6m + 1)$

29. $(6x^2)(3x^2y)$

30. $(3x)(4x^3)^2$

31. $(y - 8)(2y + 1)$

32. $(2x + 5)^2$

33. $u^3 \div u^2$

34. $\dfrac{16h^3 + 4h^2 - 8h}{4h^2}$

35. $(x^2 + 7x + 12) \div (x + 3)$ **36.** $(x^3 - 48) \div (x + 2)$

For Exercises 37–44, factor completely.

37. $x^2 - 121$

38. $10x^2 + 40$

39. $x^2 - 8x + 16$

40. $2x^2 + 5x - 12$

41. $x^3 - 5x^2 + 5x - 25$

42. $7m(n - 2) + 6(n - 2)$

43. $8x^3 - 27$

44. $3x^2 + 8x + 4$

For Exercises 45–46, solve.

45. $x^2 - 100 = 0$

46. $x^2 - 3x = 28$

For Exercises 47–50, solve.

47. Find three consecutive even integers whose sum is 24.

48. Janice invests a total of $8000 in two different accounts. The first account returns 5% while the second account returns 8%. If the total interest earned after 1 year is $565, what principal was invested in each account?

49. How much of a 35% HCl solution must be added to 50 milliliters of a 10% HCl solution to obtain a 20% solution?

50. Using the following poll results, if 1204 husbands were polled, how many said that they often or always told their wives that they were beautiful?

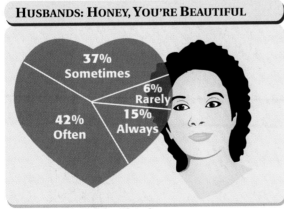

HUSBANDS: HONEY, YOU'RE BEAUTIFUL

37% Sometimes

6% Rarely

42% Often

15% Always

Source: Harris Poll for Clairol

Rational Expressions and Equations

Throughout this book, topics are developed in a specific pattern. We first define a new type of expression, then learn how to perform arithmetic operations on those expressions, and finally solve equations that contain those expressions. For example, in Chapter 5 we defined polynomials, then learned how to perform operations of arithmetic with the polynomials. In Chapter 6 we solved equations containing the polynomial expressions. We will now follow a similar pattern of development in Chapter 7. We first define rational expressions, then learn to perform arithmetic operations on these rational expressions, and, finally, solve equations that contain rational expressions.

"What is the most rigorous law of our being? Growth. No smallest atom of our moral, mental, or physical structure can stand still a year. It grows, it must grow . . . nothing can prevent it."

—MARK TWAIN
American author and humorist (1835–1910)

7.1 Simplifying Rational Expressions

OBJECTIVES

1. Evaluate rational expressions.
2. Find numbers that cause a rational expression to be undefined.
3. Simplify rational expressions containing only monomials.
4. Simplify rational expressions containing multiterm polynomials.

OBJECTIVE 1 Evaluate rational expressions. Now that we have explored polynomials, we are ready to develop a new class of expressions called **rational expressions**.

> **DEFINITION**
>
> **Rational expression:** An expression that can be written in the form $\dfrac{P}{Q}$, where P and Q are polynomials and $Q \neq 0$.

Some rational expressions are:

$$\frac{3x^5}{18x^2} \qquad \frac{7x}{x^2 - 9} \qquad \frac{x^2 + 2x - 15}{4x + 20}$$

> **CONNECTION**
>
> Notice how the definition of a rational expression is like the definition of a rational number. Recall from Chapter 1 that a rational number is a number that can be expressed in the form $\dfrac{a}{b}$, where a and b are integers and $b \neq 0$.

As with any expression, we can evaluate or rewrite rational expressions. First, let's look at evaluating rational expressions. Remember that to evaluate an algebraic expression, we replace the variables with given values and then simplify the resulting numerical expression.

EXAMPLE 1 Evaluate the expression $\dfrac{3x - 5}{x + 1}$, when

a. $x = 2$

Solution $\dfrac{3x - 5}{x + 1} = \dfrac{3(2) - 5}{2 + 1}$ Replace x with 2.

$$= \frac{6 - 5}{3}$$

$$= \frac{1}{3}$$

b. $x = -0.4$

Solution $\dfrac{3x-5}{x+1} = \dfrac{3(-0.4)-5}{-0.4+1}$ Replace x with -0.4.

$$= \dfrac{-1.2-5}{0.6}$$

$$= \dfrac{-6.2}{0.6}$$

$$= -10.\overline{3} \quad \text{or} \quad -10\dfrac{1}{3}$$

c. $x = -1$

Solution $\dfrac{3x-5}{x+1} = \dfrac{3(-1)-5}{-1+1}$ Replace x with -1.

$$= \dfrac{-3-5}{0}$$

$$= \dfrac{-8}{0}, \text{ which is undefined.}$$

> **YOUR TURN**   Evaluate the expression $\dfrac{x^2-9}{x-2}$, when
>
> **a.** 4   **b.** -3 **c.** 2

OBJECTIVE **2** **Find numbers that cause a rational expression to be undefined.** Note that in Example 1c, the expression $\dfrac{3x-5}{x+1}$ is undefined, when $x = -1$, because replacing x with -1 causes the denominator to be 0. We need to avoid values that cause a rational expression to be undefined, which means we need to identify and avoid values that make their denominators 0. This suggests the following procedure.

PROCEDURE **FINDING VALUES THAT MAKE A RATIONAL EXPRESSION UNDEFINED**

To determine the value(s) that make a rational expression undefined,

1. Write an equation that has the denominator set equal to zero.
2. Solve the equation.

ANSWERS

a. $\dfrac{7}{2}$ **b.** 0   **c.** undefined

EXAMPLE **2**   Find every value for the variable that makes the expression undefined.

a. $\dfrac{5y}{y-2}$

Solution

> **Note:** We do not consider the numerator because the rational expression is undefined only if the denominator is 0.

$$y - 2 = 0 \qquad \text{Set the denominator, } y - 2,$$
$$\text{equal to 0.}$$
$$y = 2 \qquad \text{Add 2 on both sides.}$$

The expression $\dfrac{5y}{y - 2}$ is undefined if y is replaced with 2 because the denominator would equal 0.

b. $\dfrac{7}{(x + 4)(x - 3)}$

Solution $\quad (x + 4)(x - 3) = 0 \qquad\qquad\qquad$ Set the denominator equal to 0.

$$x + 4 = 0 \quad \text{or} \quad x - 3 = 0 \qquad \text{Use the zero-factor theorem to solve.}$$
$$x = -4 \qquad\qquad x = 3$$

The original expression is undefined if x is replaced by -4 or 3.

c. $\dfrac{n - 6}{3n^3 + n^2 - 4n}$

Solution $\quad 3n^3 + n^2 - 4n = 0 \qquad$ Set the denominator equal to 0.

$$n(3n^2 + n - 4) = 0 \qquad \text{Factor out the monomial GCF, } n.$$
$$n(3n + 4)(n - 1) = 0 \qquad \text{Factor } 3n^2 + n - 4 \text{ by trial.}$$
$$n = 0 \quad \text{or} \quad 3n + 4 = 0 \quad \text{or} \quad n - 1 = 0 \qquad \text{Use the zero-factor theorem.}$$
$$3n = -4 \qquad\qquad n = 1$$
$$n = -\frac{4}{3}$$

The original expression is undefined if n is replaced by 0, $-\dfrac{4}{3}$, or 1.

YOUR TURN   **Find every value that can replace the variable in the expression and cause the expression to be undefined.**

a. $\dfrac{y + 4}{3y - 1}$ **b.** $\dfrac{2x}{x^2 + 5x}$ **c.** $\dfrac{m}{5m^2 + 28m - 12}$

OBJECTIVE **3** **Simplify rational expressions containing only monomials.**   Now that we have defined rational expressions, we are ready to begin simplifying them. Recall from Section 1.2 that a fraction is in lowest terms when the only common factor of its numerator and denominator is 1. Also in Section 1.2, we used the rule $\dfrac{an}{bn} = \dfrac{a \cdot 1}{b \cdot 1} = \dfrac{a}{b}$, where $b \neq 0$ and $n \neq 0$, to reduce fractions. We can rewrite this rule so that it applies to rational expressions.

$$\frac{PR}{QR} = \frac{P \cdot 1}{Q \cdot 1} = \frac{P}{Q}, \text{ where } P, Q, \text{ and } R \text{ are polynomials and } Q \text{ and } R \text{ are not } 0.$$

The rule indicates that a factor common to both the numerator and denominator can be divided out of a rational expression. Consider the following comparison between simplifying a fraction and simplifying a similar rational expression:

$$\frac{10}{15} = \frac{2 \cdot 5}{3 \cdot 5} = \frac{2 \cdot 1}{3 \cdot 1} = \frac{2}{3} \qquad \frac{2x}{3x} = \frac{2 \cdot x}{3 \cdot x} = \frac{2 \cdot 1}{3 \cdot 1} = \frac{2}{3}$$

5 is the common factor here.

x is the common factor here.

Note: When working with rational expressions, we will assume that the variables will not be replaced by any value that would cause the expression to be undefined or indeterminate.

This suggests the following procedure for simplifying rational expressions.

PROCEDURE SIMPLIFYING RATIONAL EXPRESSIONS

To simplify a rational expression to lowest terms:

1. Write the numerator and denominator in factored form.
2. Divide out all common factors in the numerator and denominator.
3. Multiply the remaining factors in the numerator and the remaining factors in the denominator.

We will first simplify rational expressions that are a monomial over a monomial.

EXAMPLE 3 Simplify $\dfrac{3x^5}{15x^2}$.

Solution Write the numerator and denominator in factored form, then divide out all common factors.

Note: From here on, we will leave out this step with the 1's and simply highlight the common factors that are eliminated.

$$\frac{3x^5}{15x^2} = \frac{3 \cdot x \cdot x \cdot x \cdot x \cdot x}{3 \cdot 5 \cdot x \cdot x}$$

$$= \frac{1 \cdot x \cdot x \cdot x \cdot 1}{1 \cdot 5 \cdot 1}$$

$$= \frac{x^3}{5}$$

Note: The common factors are a single 3 and two x's. These form the GCF that we divide out, which is $3x^2$.

There are different styles for showing the process of dividing out common factors. For example, we could have used cancellation marks to show that the two x's and one 3 divide out in Example 3.

$$\frac{3x^5}{15x^2} = \frac{\overset{1}{\cancel{3}} \cdot \overset{1}{\cancel{x}} \cdot \overset{1}{\cancel{x}} \cdot x \cdot x \cdot x}{\underset{1}{\cancel{3}} \cdot 5 \cdot \underset{1}{\cancel{x}} \cdot \underset{1}{\cancel{x}}} = \frac{x^3}{5}$$

Develop a style that works best for you, keeping in mind that no matter the style, the bottom line is that you will be dividing out the GCF of the numerator and denominator to get a rational expression in lowest terms.

EXAMPLE 4 Simplify.

a. $-\dfrac{4x^2}{28x^3}$

Solution $-\dfrac{4x^2}{28x^3} = -\dfrac{2 \cdot 2 \cdot x \cdot x}{2 \cdot 2 \cdot 7 \cdot x \cdot x \cdot x}$

Write the numerator and denominator in factored form, then eliminate the common factors, which are two 2's and two x's. Multiply the remaining factors.

Note: Since all of the factors in the numerator divided out, the numerator is 1.

$= -\dfrac{1}{7x}$

CONNECTION

Using cancellation marks:

$$-\frac{4x^2}{28x^3} = -\frac{\overset{1}{\cancel{2}} \cdot \overset{1}{\cancel{2}} \cdot \overset{1}{\cancel{x}} \cdot \overset{1}{\cancel{x}}}{\underset{1}{\cancel{2}} \cdot \underset{1}{\cancel{2}} \cdot 7 \cdot \underset{1}{\cancel{x}} \cdot \underset{1}{\cancel{x}} \cdot x} = -\frac{1}{7x}$$

b. $\dfrac{36ab^3c}{54a^2b^3}$

Solution $\dfrac{36ab^3c}{54a^2b^3} = \dfrac{2 \cdot 2 \cdot 3 \cdot 3 \cdot a \cdot b \cdot b \cdot b \cdot c}{2 \cdot 3 \cdot 3 \cdot 3 \cdot a \cdot a \cdot b \cdot b \cdot b}$

Write the numerator and denominator in factored form, then divide out the common factors, which are one 2, two 3's, one a, and three b's. Multiply the remaining factors.

$= \dfrac{2c}{3a}$

YOUR TURN Simplify.

a. $-\dfrac{12t^6}{30t^2}$ **b.** $\dfrac{9n}{18n^3}$ **c.** $-\dfrac{72x^2y}{60x^2y^3z}$

ANSWERS

a. $-\dfrac{2t^4}{5}$ **b.** $\dfrac{1}{2n^2}$ **c.** $-\dfrac{6}{5y^2z}$

OBJECTIVE 4 Simplify rational expressions containing multiterm polynomials. The rational expressions we have considered so far have had monomials in the both the numerator and denominator. We now consider rational expressions that contain multiterm polynomials. Remember that we can only divide out common factors, so we must first write the polynomials in factored form.

EXAMPLE 5 Simplify $\dfrac{9ab}{3a + 6}$.

Solution $\dfrac{9ab}{3a + 6} = \dfrac{3 \cdot 3 \cdot a \cdot b}{3 \cdot (a + 2)}$ Write the numerator and denominator in factored form, then divide out the common factor, which is 3. Multiply the remaining factors.

$$= \frac{3ab}{a + 2}$$

Warning: It may be tempting to try to divide out the a, however $\dfrac{3ab}{a + 2} \neq \dfrac{3b}{2}$. Though a is a factor in the numerator, it is not a factor in the denominator, so it cannot be divided out.

YOUR TURN Simplify.

a. $\dfrac{6xy}{3x + 12}$ b. $\dfrac{2a + 8}{10a^2}$

ANSWERS

a. $\dfrac{2xy}{x + 4}$ b. $\dfrac{a + 4}{5a^2}$

We now consider rational expressions with multiterm polynomials in both the numerator and denominator. Again, since we can only divide out common factors, we must first factor the polynomials.

EXAMPLE 6 Simplify.

a. $\dfrac{2x^2 + 8x}{3x^2 + 12x}$

Solution $\dfrac{2x^2 + 8x}{3x^2 + 12x} = \dfrac{2 \cdot x \cdot (x + 4)}{3 \cdot x \cdot (x + 4)}$ Factor the numerator and denominator, then divide out the common factors, x and $x + 4$.

$= \dfrac{2}{3}$

b. $\dfrac{x^2 + 2x - 15}{x^2 + x - 12}$

Solution $\dfrac{x^2 + 2x - 15}{x^2 + x - 12} = \dfrac{(x - 3)(x + 5)}{(x - 3)(x + 4)}$ Factor the numerator and denominator, then divide out the common factor, $x - 3$.

$= \dfrac{x + 5}{x + 4}$

YOUR TURN Simplify.

a. $\dfrac{5x + 10}{7x + 14}$ b. $\dfrac{3n^2 - 23n - 8}{3n^2 + 13n + 4}$

Recall from Chapter 6 that to factor multiterm polynomials, we must sometimes factor the factors. In Example 7, the factors themselves can be factored.

EXAMPLE 7 Simplify.

a. $\dfrac{6y^4 + 2y^3 - 4y^2}{36y^3 - 42y^2 + 12y}$

Solution $\dfrac{6y^4 + 2y^3 - 4y^2}{36y^3 - 42y^2 + 12y} = \dfrac{2y^2(3y^2 + y - 2)}{6y(6y^2 - 7y + 2)}$

$= \dfrac{2 \cdot y \cdot y \cdot (y + 1) \cdot (3y - 2)}{2 \cdot 3 \cdot y \cdot (2y - 1) \cdot (3y - 2)}$ Factor out the monomial GCFs, then factor the polynomial factors.

$= \dfrac{y(y + 1)}{3(2y - 1)}$ or $\dfrac{y^2 + y}{6y - 3}$ Divide out the common factors 2, y, and $3y - 2$.

b. $\dfrac{3x^2 + 6x - 24}{x^4 - 16}$

Solution Write the numerator and denominator in factored form, then divide out all common factors.

$$\frac{3x^2 + 6x - 24}{x^4 - 16} = \frac{3(x^2 + 2x - 8)}{(x^2 + 4)(x^2 - 4)}$$

$$= \frac{3 \cdot (x + 4) \cdot (x - 2)}{(x^2 + 4) \cdot (x + 2) \cdot (x - 2)}$$

Factor the numerator and denominator completely. Notice that $x^2 - 4$ is a difference of squares.

$$= \frac{3(x + 4)}{(x^2 + 4)(x + 2)} \text{ or } \frac{3x + 12}{x^3 + 2x^2 + 4x + 8}$$

Divide out the common factor, $x - 2$.

YOUR TURN

Simplify $\dfrac{14x^3 + 70x^2 + 84x}{7x^4 - 14x^3 - 105x^2}$.

Binomial factors that are additive inverses can be tricky:

EXAMPLE 8 Simplify $\dfrac{3x^2 - 7x - 20}{8 - 2x}$.

Solution Write the numerator and denominator in factored form, then divide out all common factors.

$$\frac{3x^2 - 7x - 20}{8 - 2x} = \frac{(3x + 5)(x - 4)}{2(4 - x)}$$

It appears that there are no common factors. However, $x - 4$ and $4 - x$ are additive inverses. We can get matching binomial factors by factoring -1 out of either $x - 4$ or $4 - x$.

$$x - 4 = -1(4 - x) \qquad \text{or} \qquad 4 - x = -1(x - 4)$$

We can use either form. We will use $x - 4 = -1(4 - x)$.

Note: We could have factored the -1 out of the denominator:

$$\frac{(3x + 5)(x - 4)}{2(-1)(x - 4)} = \frac{3x + 5}{-2}$$

$$= \frac{(3x + 5)(-1)(4 - x)}{2(4 - x)}$$

Divide out the common factor, $4 - x$, then multiply the remaining factors.

$$= \frac{-3x - 5}{2} \text{ or } -\frac{3x + 5}{2}$$

ANSWER

$\dfrac{2(x + 2)}{x(x - 5)}$ or $\dfrac{2x + 4}{x^2 - 5x}$

Example 8 also illustrates the rule of sign placement in a fraction or rational expression. Remember that with a negative fraction (or rational expression), the minus sign can be placed in the numerator, denominator, or aligned with the fraction line. Note, however, that it is generally considered unsightly to write the minus sign in the denominator.

RULE

$$-\frac{P}{Q} = \frac{-P}{Q} = \frac{P}{-Q}, \text{ where } P \text{ and } Q \text{ are polynomials and } Q \neq 0.$$

ANSWER

$$-\frac{x}{x+6}$$

YOUR TURN

Simplify $\dfrac{2x - x^2}{x^2 + 4x - 12}$.

7.1 EXERCISES

For Extra Help

InterAct Math

Tutor Center   Addison-Wesley Math Tutor Center

Video 12/DVT 4

MyMathLab

MathXP   MathXL

1. Explain how to evaluate an expression.

2. In general, what causes a rational expression to be undefined?

3. How do you determine what values cause a rational expression to be undefined?

4. When simplifying a rational expression, we divide out all common _____.

5. Explain why you cannot divide out the y's in $\dfrac{x + y}{2y}$.

6. What is the first step when simplifying rational expressions?

For Exercises 7–14, evaluate the rational expression.

7. $\dfrac{4x^2}{9y}$ **a.** when $x = -2, y = 5$ **b.** when $x = 3, y = -4$ **c.** when $x = 0, y = 7$

8. $-\dfrac{x}{y^3}$ **a.** when $x = 1$, $y = 2$ **b.** when $x = -2$, $y = 3$ **c.** when $x = -4$, $y = -2$

9. $\dfrac{2x + 5}{3x - 1}$ **a.** when $x = 3$ **b.** when $x = -1.3$ **c.** when $x = -2$

10. $\dfrac{4x}{x - 3}$ **a.** when $x = 3$ **b.** when $x = -3$ **c.** when $x = 1.2$

11. $\dfrac{x^2 - 4}{x + 3}$ **a.** when $x = -1$ **b.** when $x = -4.2$ **c.** when $x = 0$

12. $\dfrac{x^2 + 3}{2x + 1}$ **a.** when $x = 2.1$ **b.** when $x = -2$ **c.** when $x = 1$

13. $\dfrac{x}{5x^2 + 1}$ **a.** when $x = 1$ **b.** when $x = -2$ **c.** when $x = 3.1$

14. $\dfrac{3x^2}{x - 5}$ **a.** when $x = 3$ **b.** when $x = -2$ **c.** when $x = -1.1$

For Exercises 15–24, find every value for the variable that makes the expression undefined.

15. $\dfrac{2x}{x - 3}$ **16.** $\dfrac{3y}{y + 5}$ **17.** $\dfrac{3}{x^2 - 9}$ **18.** $\dfrac{4r}{r^2 - 4}$

19. $\dfrac{6}{x^2 - 14x + 45}$ **20.** $\dfrac{2x + 3}{x^2 + 5x + 6}$ **21.** $\dfrac{2m}{4m^2 - 17m - 15}$ **22.** $\dfrac{2p + 3}{p^2 + 6p}$

23. $\dfrac{4a}{2a^2 - 3a - 5}$ **24.** $\dfrac{t}{2t^2 + 7t + 3}$

For Exercises 25–72, simplify.

25. $\dfrac{9x^2}{12xy}$ **26.** $\dfrac{14h^3k}{21h}$ **27.** $-\dfrac{m^3n}{mn^2}$ **28.** $-\dfrac{5a^3b}{a^2b^4}$

29. $\dfrac{54x^4y}{36x^6yz^2}$

30. $-\dfrac{48t^5uv}{32tv^4}$

31. $-\dfrac{15m^2n}{5m + 10n}$

32. $\dfrac{-7}{7r + 14t}$

33. $\dfrac{7(x + 5)}{14}$

34. $\dfrac{9}{18(a - 1)}$

35. $\dfrac{3(m - 2)}{5(m - 2)}$

36. $\dfrac{6(k + 2)}{15(k + 2)}$

37. $\dfrac{10x - 20}{15x - 30}$

38. $\dfrac{12y + 2}{18y + 3}$

39. $\dfrac{30a + 15b}{18a + 9b}$

40. $\dfrac{8m - 24n}{12m - 36n}$

41. $\dfrac{x^2 + 5x}{4x + 20}$

42. $\dfrac{ab - b^2}{2a - 2b}$

43. $\dfrac{5y^2 - 10y - 15}{xy^2 - 2xy - 3x}$

44. $\dfrac{3m^3n - 2m^2n + mn}{6m^2n - 4mn + 2n}$

45. $\dfrac{x^2 - y^2}{x^2 + 2xy + y^2}$

46. $\dfrac{x^2 - 16}{x^2 + 6x + 8}$

47. $\dfrac{x^2 - 5x}{x^2 - 7x + 10}$

48. $\dfrac{x^3 - 4x}{x^3 - 4x^2 + 4x}$

49. $\dfrac{3x + 6}{x^4 - 16}$

50. $\dfrac{3x^2 + 16x - 35}{5x^2 + 33x - 14}$

51. $\dfrac{3x^3 + 3y^3}{2x^3 - 2x^2y + 2xy^2}$

52. $\dfrac{r^2 - s^2}{r^3 + s^3}$

53. $\dfrac{x^2 + 5x}{x + 5}$

54. $\dfrac{x^3 - 8}{x^2 + 2x - 8}$

55. $\dfrac{a^2 - b^2}{a^2 + 2ab + b^2}$

56. $\dfrac{m^3 - 4m}{m^3 - 4m^2 + 4m}$

57. $\dfrac{t^2 - 9}{t^2 + 5t + 6}$

58. $\dfrac{c^2 + d^2}{c + d}$

59. $\dfrac{k^2 - k}{k^2 + k - 2}$

60. $\dfrac{by - ay + bx - ax}{x^2 + ax + xy + ay}$

61. $\dfrac{5m - 20m^3}{5m + 40m^4}$

62. $\dfrac{5y^2 - 45y + 90}{180 - 5y^2}$

63. $\dfrac{6u^3 - 4u^2 - 3u + 2}{3u^2 + u - 2}$

64. $\dfrac{8n^3 + 20n^2 - 2n - 5}{4n^2 + 12n + 5}$

65. $\dfrac{4x - 4}{1 - x}$

66. $\dfrac{2y^2 - 8}{2 - y}$

67. $\dfrac{3 - m}{m^3 - 27}$

68. $\dfrac{a^2 - b^2}{b - a}$

69. $\dfrac{3 - w}{w^2 - 2w - 3}$

70. $\dfrac{12 - 4m}{m - 3}$

71. $\dfrac{20 - 5k}{2k - 8}$

72. $\dfrac{1 - u}{u^2 - 3u + 2}$

For Exercises 73–78, find and explain the mistake, then correct the mistake.

73. $-\dfrac{5x^2y}{10x^3y} = -\dfrac{x}{2}$

74. $\dfrac{8x^2y}{4x^3} = 2xy$

75. $\dfrac{x - \overset{3}{\cancel{6}}}{\underset{1}{\cancel{2}}y} = \dfrac{x-3}{y}$

76. $\dfrac{y - \overset{3}{\cancel{12}}}{\underset{1}{\cancel{4}}y - 3} = \dfrac{y-3}{y-3} = 1$

77. $\dfrac{yx - x^2}{x - y} = \dfrac{x(y-x)}{x-y} = x$

78. $\dfrac{x^2 - x - 6}{x^3 + 2x^2 - 9x - 18} = \dfrac{(x+2)(x-3)}{(x+2)(x^2-9)} = \dfrac{x-3}{x^2-9}$

79. A person's body mass index (BMI) is a measure of the amount of body fat based on their height, h, in inches, and weight, w, in pounds. The formula for BMI is

$$\text{BMI} = \frac{704.5w}{h^2}$$

a. Use the formula to find the BMI of each person listed below.

Height	Weight	BMI
5' 4"	138	
5' 9"	155	
6' 2"	220	

b. Use the following table from the National Institutes of Health to classify each person in part (a) as underweight, normal weight, overweight, or obese.

Underweight:	BMI ≤ 18.5
Normal Weight:	BMI = 18.5 − 24.9
Overweight:	BMI = 25 − 29.9
Obesity:	BMI ≥ 30

> **CONNECTION**
> In Section 2.4, Exercise 66, you translated the BMI formula from words to symbols.

80. Recall the formula for slope from Chapter 4,

$$m = \frac{y_2 - y_1}{x_2 - x_1}$$

> **CONNECTION**
> Though we hadn't formally defined rational expressions in Chapter 4, we were evaluating the rational expression, $\dfrac{y_2 - y_1}{x_2 - x_1}$, to find slope.

a. Find the slope of a line passing through the points $(3, 5)$ and $(-1, -4)$.

b. Find the slope of a line passing through the points $(-4, 7)$ and $(-2, -1)$.

81. Given the diameter in inches, d, and the number of revolutions per minute, N, of a revolving tool such as a circular saw or drill, the cutting speed, C, in feet per minute, of the tool can be found by the formula

$$C = \frac{\pi dN}{12}$$

(*Source*: Robert D. Smith, *Mathematics for Machine Technology*, 4th ed., Delmar Publishers, 1999)

Use the formula to complete the following table for a table saw.

Blade Diameter	rpm	Cutting Speed
8 inches	800 rpm	
8 inches	1000 rpm	
10 inches	800 rpm	
10 inches	1000 rpm	

82. The cutting time, T, in minutes, that it takes a drill to cut through something depends on its speed of revolution, N, in rpm, the thickness of the material, L, in inches, and the tool feed, F, in inches per revolution. The formula for calculating cutting time is

$$T = \frac{L}{FN}$$

(*Source*: Robert D. Smith, *Mathematics for Machine Technology*, 4th ed., Delmar Publishers, 1999)

Use the formula to complete the following table for a drill operating at a speed of 480 rpm and a feed of 0.04 inches per revolution.

Material Thickness (inches)	Cut Time (minutes)
0.25	
0.5	
0.75	
1	

83. The formula for the volume of a cone is $V = \dfrac{1}{3}\pi r^2 h$.

  a. Solve this formula for h, so that we have a formula for finding the height of a cone given its volume and radius. (The formula for height of a cone will contain a rational expression.)

  b. Suppose a cone has a volume of approximately 150.8 cm³ with a radius of 4 cm. Find the height of the cone.

84. The formula for the volume of a pyramid is $V = \dfrac{1}{3} lwh$.

  a. Solve this formula for h, so that we have a formula for finding the height of a pyramid given its volume, length, and width.

  b. Suppose a decorative onyx pyramid has a volume of 126 in.³, a width of 3.5 in., and a length of 4.5 in. Find the height of the pyramid.

85. In analyzing circuits, electrical engineers often have to simplify rational expressions. Suppose the following rational expression describes an electrical circuit:

$$\frac{s^2 - 9}{s^2 + 7s + 12}$$

Simplify the expression to lowest terms.

86. An engineer derives the following rational expression, which describes a certain circuit.

$$\frac{4s^2 - 6s}{2s^2 + 7s - 15}$$

Simplify the expression to lowest terms.

Collaborative Exercise

Graphs of Rational Expressions

Use a graphing utility.

1. Graph $f(x) = \dfrac{x}{x - 3}$. The vertical line you see is called an *asymptote*.
   a. What is the equation of the asymptote?
   b. What do you think the asymptote indicates?

2. Graph $f(x) = \dfrac{5}{(x + 2)(x - 4)}$.
   a. Why are there two asymptotes?

   b. What are the equations of the two asymptotes?

3. Graph $f(x) = \dfrac{x^2 - x - 6}{x^2 + 3x + 2}$.
   a. Are there any asymptotes? If so, write the equation for each.

   b. Rewrite $f(x) = \dfrac{x^2 - x - 6}{x^2 + 3x + 2}$ so that the numerator and denominator are in factored form.

   c. Find $f(-2)$.

continued

d. Use the zoom function (or resize the window) so that you can clearly see the graph of $f(x) = \dfrac{x^2 - x - 6}{x^2 + 3x + 2}$ in the 2nd quadrant. The blank space you see in the graph where $x = -2$ is called a hole. How is a hole different from an asymptote?

4. Writing $\dfrac{x^2 - x - 6}{x^2 + 3x - 2}$ in simplest form we have $\dfrac{x - 3}{x + 1}$. Graph $f(x) = \dfrac{x - 3}{x + 1}$.

   a. Compare the graph of $f(x) = \dfrac{x - 3}{x + 1}$ with the graph of $f(x) = \dfrac{x^2 - x - 6}{x^2 + 3x + 2}$. What do you notice about the graphs?

5. Given a rational function, how can you identify its asymptotes and holes?

REVIEW EXERCISES

For Exercises 1 and 2, multiply.

1. $\dfrac{2}{3} \cdot \dfrac{6}{4}$

2. $-\dfrac{15}{16} \cdot \dfrac{4}{5}$

For Exercises 3 and 4, divide.

3. $\dfrac{3}{8} \div \dfrac{5}{6}$

4. $-\dfrac{7}{9} \div \left(-\dfrac{5}{12}\right)$

5. Factor completely: $ax + bx - ay - by$

6. Simplify: $\dfrac{a - b}{b - a}$

7.2 Multiplying and Dividing Rational Expressions

OBJECTIVES

1. Multiply rational expressions.
2. Divide rational expressions.
3. Convert units of measurement using dimensional analysis.

OBJECTIVE **1** Multiply rational expressions. In Section 1.4, we used the rule $\dfrac{a}{b} \cdot \dfrac{c}{d} = \dfrac{ac}{bd}$, where $b \neq 0$ and $d \neq 0$, to multiply fractions. We can rewrite this rule so that it applies to multiplying rational expressions.

RULE $\dfrac{P}{Q} \cdot \dfrac{R}{S} = \dfrac{PR}{QS}$, where P, Q, R, and S are polynomials where $Q \neq 0$ and $S \neq 0$.

Also, remember that when multiplying fractions, we can simplify after multiplying, or we can simplify before multiplying by dividing out factors common to both the numerator and denominator. For example,

Multiply, then eliminate common factors: $\dfrac{8}{15} \cdot \dfrac{3}{4} = \dfrac{24}{60} = \dfrac{2 \cdot 2 \cdot 2 \cdot 3}{2 \cdot 2 \cdot 3 \cdot 5} = \dfrac{2}{5}$

or

Eliminate common factors, then multiply. $\dfrac{8}{15} \cdot \dfrac{3}{4} = \dfrac{2 \cdot 2 \cdot 2}{3 \cdot 5} \cdot \dfrac{3}{2 \cdot 2} = \dfrac{2}{5}$

Multiplying rational expressions works the same way.

PROCEDURE **MULTIPLYING RATIONAL EXPRESSIONS**

To multiply rational expressions:

1. Write each numerator and denominator in factored form.
2. Divide out any numerator factor with any matching denominator factor.
3. Multiply numerator by numerator and denominator by denominator.
4. Simplify as needed.

EXAMPLE 1 Multiply $\dfrac{9x}{5y} \cdot \dfrac{15xy^2}{12}$.

Solution $\dfrac{9x}{5y} \cdot \dfrac{15xy^2}{12} = \dfrac{3 \cdot 3 \cdot x}{5 \cdot y} \cdot \dfrac{3 \cdot 5 \cdot x \cdot y \cdot y}{2 \cdot 2 \cdot 3}$

Write the numerators and denominators in factored form.

$= \dfrac{3 \cdot 3 \cdot x}{1} \cdot \dfrac{x \cdot y}{2 \cdot 2}$

Divide out the common factors, which are 3, 5, and y.

$= \dfrac{9x^2 y}{4}$

Multiply the remaining numerator factors and denominator factors.

YOUR TURN **Multiply.**

a. $-\dfrac{36m}{8n^4} \cdot \dfrac{14mn^2}{28n}$

b. $\dfrac{-9}{4xy^2} \cdot \dfrac{-20xz}{15y}$

Now consider rational expressions that contain multiterm polynomials.

EXAMPLE 2 Multiply.

a. $-\dfrac{2x}{3x - 9} \cdot \dfrac{5x - 15}{20x^3}$

Solution $-\dfrac{2x}{3x - 9} \cdot \dfrac{5x - 15}{20x^3}$

$= -\dfrac{2 \cdot x}{3 \cdot (x - 3)} \cdot \dfrac{5 \cdot (x - 3)}{2 \cdot 2 \cdot 5 \cdot x \cdot x \cdot x}$

Write the numerators and denominators in factored form.

$= -\dfrac{1}{3} \cdot \dfrac{1}{2 \cdot x \cdot x}$

Divide out the common factors, which are 2, 5, x, and $(x - 3)$.

$= -\dfrac{1}{6x^2}$

Multiply the remaining numerator factors and denominator factors.

b. $\dfrac{12 - 4x}{15x} \cdot \dfrac{5x^2}{2x - 6}$

Solution $\dfrac{12 - 4x}{15x} \cdot \dfrac{5x^2}{2x - 6} = \dfrac{4 \cdot (3 - x)}{3 \cdot 5 \cdot x} \cdot \dfrac{5 \cdot x \cdot x}{2 \cdot (x - 3)}$

Note: We can factor 4 further. Also, since $3 - x$ and $x - 3$ are additive inverses, we can factor -1 out of one of them.

$= \dfrac{2 \cdot 2 \cdot (-1)(x - 3)}{3 \cdot 5 \cdot x} \cdot \dfrac{5 \cdot x \cdot x}{2 \cdot (x - 3)}$

$= \dfrac{2 \cdot (-1)}{3} \cdot \dfrac{x}{1}$

Divide out the common factors, which are 2, 5, x, and $(x - 3)$.

$= \dfrac{-2x}{3}$

Multiply the remaining numerator factors and denominator factors.

c. $\dfrac{x^2 + x - 6}{9x^3} \cdot \dfrac{12x^2 - 6x}{2x^2 - 5x + 2}$

Solution $\dfrac{x^2 + x - 6}{9x^3} \cdot \dfrac{12x^2 - 6x}{2x^2 - 5x + 2}$

$= \dfrac{(x - 2)(x + 3)}{3 \cdot 3 \cdot x \cdot x \cdot x} \cdot \dfrac{2 \cdot 3 \cdot x \cdot (2x - 1)}{(x - 2)(2x - 1)}$ Write the numerators and denominators in factored form.

$= \dfrac{x + 3}{3 \cdot x \cdot x} \cdot \dfrac{2}{1}$ Divide out the common factors, which are $(x - 2)$, 3, x, and $(2x - 1)$.

$= \dfrac{2(x + 3)}{3x^2} \; or \; \dfrac{2x + 6}{3x^2}$ Multiply the remaining numerator factors and denominator factors.

YOUR TURN   **Multiply.**

a. $\dfrac{-8h^2}{2h + 10} \cdot -\dfrac{6h + 30}{18h^4}$ **b.** $-\dfrac{9y - 3}{y^2 + 3y} \cdot \dfrac{2y + 6}{2 - 6y}$ **c.** $\dfrac{m^2 - 25}{20m} \cdot \dfrac{30m + 10}{3m^2 + 16m + 5}$

OBJECTIVE   **2**   **Divide rational expressions.**   In Section 1.4, we saw how to divide fractions using the rule $\dfrac{a}{b} \div \dfrac{c}{d} = \dfrac{a}{b} \cdot \dfrac{d}{c}$, where b, c, and d are not 0, which indicates to multiply by the reciprocal of the divisor. We can rewrite this rule so that it applies to dividing rational expressions.

RULE $\dfrac{P}{Q} \div \dfrac{R}{S} = \dfrac{P}{Q} \cdot \dfrac{S}{R}$, where P, Q, R, and S are polynomials where $Q \neq 0$, $R \neq 0$, and $S \neq 0$.

Following is a procedure for dividing rational expressions.

PROCEDURE   **DIVIDING RATIONAL EXPRESSIONS**

To divide rational expressions:
1. Write an equivalent multiplication statement with the reciprocal of the divisor.
2. Write each numerator and denominator in factored form. (Steps 1 and 2 are interchangeable.)
3. Divide out any numerator factor with any matching denominator factor.
4. Multiply numerator by numerator and denominator by denominator.
5. Simplify as needed.

ANSWERS

a. $\dfrac{4}{3h^2}$ **b.** $\dfrac{3}{y}$ **c.** $\dfrac{m - 5}{2m}$

EXAMPLE **3** Divide $-\dfrac{9x^2y}{28z^3} \div \dfrac{12y}{7z}$.

Solution

$$-\frac{9x^2y}{28z^3} \div \frac{12y}{7z} = -\frac{9x^2y}{28z^3} \cdot \frac{7z}{12y}$$

Write an equivalent multiplication statement by changing the division sign to multiplication and changing the divisor to its reciprocal.

$$= -\frac{3 \cdot 3 \cdot x \cdot x \cdot y}{2 \cdot 2 \cdot 7 \cdot z \cdot z \cdot z} \cdot \frac{7 \cdot z}{2 \cdot 2 \cdot 3 \cdot y}$$

Factor the numerators and denominators.

$$= -\frac{3 \cdot x \cdot x}{2 \cdot 2 \cdot z \cdot z} \cdot \frac{1}{2 \cdot 2}$$

Divide out the common factors, which are 3, 7, y, and z.

$$= -\frac{3x^2}{16z^2}$$

Multiply the remaining factors.

YOUR TURN **Divide.**

a. $\dfrac{15a^3}{28b^2} \div \dfrac{10ab}{7}$

b. $-\dfrac{2t}{33u^3v} \div -\dfrac{8t^3v}{21u^4}$

Now consider dividing rational expressions that contain multiterm polynomials.

EXAMPLE **4** Divide.

a. $\dfrac{2x - 10}{15x^3} \div \dfrac{x^2 - 5x}{12x}$

Solution

$$\frac{2x - 10}{15x^3} \div \frac{x^2 - 5x}{12x} = \frac{2x - 10}{15x^3} \cdot \frac{12x}{x^2 - 5x}$$

Write an equivalent multiplication statement by changing the division sign to multiplication and changing the divisor to its reciprocal.

$$= \frac{2 \cdot (x - 5)}{3 \cdot 5 \cdot x \cdot x \cdot x} \cdot \frac{2 \cdot 2 \cdot 3 \cdot x}{x \cdot (x - 5)}$$

Factor the numerators and denominators.

$$= \frac{2}{5 \cdot x \cdot x} \cdot \frac{2 \cdot 2}{x}$$

Divide out the common factors, which are 3, x, and $(x - 5)$.

$$= \frac{8}{5x^3}$$

Multiply the remaining factors.

b. $\dfrac{4x^2 - 7x - 2}{2x^2 + 3x - 9} \div \dfrac{4x^2 + x}{6x^3 + 18x^2}$

ANSWERS

a. $\dfrac{3a^2}{8b^3}$ **b.** $\dfrac{7u}{44t^2v^2}$

Solution

$$\frac{4x^2 - 7x - 2}{2x^2 + 3x - 9} \div \frac{4x^2 + x}{6x^3 + 18x^2}$$

$$= \frac{4x^2 - 7x - 2}{2x^2 + 3x - 9} \cdot \frac{6x^3 + 18x^2}{4x^2 + x}$$

$$= \frac{(4x + 1)(x - 2)}{(2x - 3)(x + 3)} \cdot \frac{6x^2(x + 3)}{x(4x + 1)}$$

$$= \frac{(4x + 1) \cdot (x - 2)}{(2x - 3) \cdot (x + 3)} \cdot \frac{2 \cdot 3 \cdot x \cdot x \cdot (x + 3)}{x \cdot (4x + 1)}$$

$$= \frac{x - 2}{2x - 3} \cdot \frac{2 \cdot 3 \cdot x}{1}$$

$$= \frac{6x(x - 2)}{2x - 3} \text{ or } \frac{6x^2 - 12x}{2x - 3}$$

Write an equivalent multiplication statement by changing the division sign to multiplication and changing the divisor to its reciprocal.

Factor the numerators and denominators.

Divide out the common factors, which are x, $(4x + 1)$, and $(x + 3)$.

Multiply the remaining factors.

c. $\dfrac{8x^2 - 6x^3}{6xy} \div \dfrac{9x^2 - 16}{5y^2 - 10y}$

Solution

$$\frac{8x^2 - 6x^3}{6xy} \div \frac{9x^2 - 16}{5y^2 - 10y} = \frac{8x^2 - 6x^3}{6xy} \cdot \frac{5y^2 - 10y}{9x^2 - 16}$$

Write an equivalent multiplication statement by changing the division sign to multiplication and changing the divisor to its reciprocal.

$$= \frac{2x^2(4 - 3x)}{6xy} \cdot \frac{5y(y - 2)}{(3x + 4)(3x - 4)}$$

Factor the numerators and denominators.

Note: $4 - 3x$ and $3x - 4$ are additive inverses, so we factored -1 out of $4 - 3x$.

$$= \frac{2 \cdot x \cdot x \cdot (-1) \cdot (3x - 4)}{2 \cdot 3 \cdot x \cdot y} \cdot \frac{5 \cdot y \cdot (y - 2)}{(3x + 4) \cdot (3x - 4)}$$

$$= \frac{x \cdot (-1)}{3} \cdot \frac{5 \cdot (y - 2)}{(3x + 4)}$$

Divide out the common factors, which are 2, x, y, and $(3x - 4)$.

$$= \frac{-5x(y - 2)}{3(3x + 4)} \text{ or } -\frac{5xy - 10x}{9x + 12} \text{ or } \frac{10x - 5xy}{9x + 12}$$

Multiply the remaining factors.

YOUR TURN   Divide.

a. $\dfrac{6yz^3}{2y - y^2} \div \dfrac{8yz}{6 - 3y}$

b. $\dfrac{mn - 3m}{14m^4} \div \dfrac{n}{7n + 7}$

c. $\dfrac{4y^2 - 25}{9y^2 + 18y} \div \dfrac{2y^2 + 3y - 5}{3y^4 - 3y^3}$

d. $\dfrac{3x^2 + 10x + 8}{18 - 6x} \div \dfrac{10 + 5x}{4x^3 - 12x^2}$

OBJECTIVE 3 **Convert units of measurement using dimensional analysis.** We can convert from one unit of a measurement to another using a method called *dimensional analysis,* which will involve writing measurement facts as conversion factors. A conversion factor is a ratio of equivalent measures.

Measurement fact:	Conversion factor:	

1 foot = 12 inches

$$\frac{1\text{ foot}}{12\text{ inches}} \quad \text{or} \quad \frac{12\text{ inches}}{1\text{ foot}}$$

1 pound = 16 ounces

$$\frac{1\text{ pound}}{16\text{ ounces}} \quad \text{or} \quad \frac{16\text{ ounces}}{1\text{ pound}}$$

Note: Because a conversion factor is a ratio of equivalent measures, its ratio will always equal 1.

To use dimensional analysis, we multiply a measurement by appropriate conversion factors so that the undesired units divide out, leaving only the desired units of measurement. For example, if we wish to convert 6 feet to inches, we multiply by the conversion factor $\dfrac{12\text{ inches}}{1\text{ foot}}$ so that the units of feet divide out, leaving inches.

$$6\text{ ft.} = \frac{6\text{ ft.}}{1} \cdot \frac{12\text{ in.}}{1\text{ ft.}} = 72\text{ in.}$$

CONNECTION

The ratio of the conversion factor $\dfrac{12\text{ in.}}{1\text{ ft.}}$ is 1, so multiplying 6 ft. by this ratio means the result, 72 in., is equal to 6 ft.

PROCEDURE USING DIMENSIONAL ANALYSIS TO CONVERT BETWEEN UNITS OF MEASUREMENT

To convert units using dimensional analysis, multiply the given measurement by conversion factors so that the undesired units divide out, leaving the desired units.

EXAMPLE 5 Convert.

a. 500 ounces to pounds

Solution There are 16 ounces to 1 pound. We write this fact as a unit ratio and multiply so that ounces divides out, leaving pounds.

$$500\text{ oz.} = \frac{500\text{ oz.}}{1} \cdot \frac{1\text{ lb.}}{16\text{ oz.}} = \frac{500}{16}\text{ lb.} = 31.25\text{ lb.}$$

CONNECTION

Dividing out units in dimensional analysis is like dividing out common factors when multiplying rational expressions.

b. 6 miles to yards

Solution There are 5280 feet to 1 mile and 3 feet to 1 yard. So, we write these facts as unit fractions so that miles and feet divide out, leaving yards.

$$6\text{ mi.} = \frac{6\text{ mi.}}{1} \cdot \frac{5280\text{ ft.}}{1\text{ mi.}} \cdot \frac{1\text{ yd.}}{3\text{ ft.}} = \frac{31{,}680}{3}\text{ yd.} = 10{,}560\text{ yd.}$$

c. 40 miles per hour to feet per second

Solution First, 40 miles per hour means $\dfrac{40\text{ miles}}{1\text{ hour}}$. Notice we must convert miles to feet and hours to seconds. For the distance conversion, we use the fact that there are 5280 feet to 1 mile. For the time conversion, there are 60 minutes in 1 hour and 60 seconds in 1 minute.

$$\frac{40\text{ miles}}{1\text{ hour}} = \frac{40\text{ mi.}}{1\text{ hr.}} \cdot \frac{5280\text{ ft.}}{1\text{ mi.}} \cdot \frac{1\text{ hr.}}{60\text{ min.}} \cdot \frac{1\text{ min.}}{60\text{ sec.}} = \frac{211{,}200\text{ ft.}}{3600\text{ sec.}} = 58.\overline{6}\text{ ft./sec.}$$

1. When multiplying rational expressions, why is it important to write the numerators and denominators in factored form?

2. If, after factoring, you determine that there are two binomial factors that are additive inverses, how can you get them to be common binomial factors?

3. Why are $\dfrac{x}{y}$ and $\dfrac{y}{x}$ multiplicative inverses (or reciprocals)?

4. Explain how to divide rational expressions.

5. What is a unit fraction? Give two examples.

6. Explain how to convert one unit of measurement to another using dimensional analysis.

For Exercises 7–28, multiply.

7. $\dfrac{m}{n} \cdot \dfrac{1}{n}$

8. $\dfrac{x}{y} \cdot \dfrac{3x}{2y}$

9. $\dfrac{x}{y} \cdot \dfrac{my}{nx}$

10. $\dfrac{ab}{xy} \cdot \dfrac{x}{a^3 b^2}$

11. $\dfrac{3}{y} \cdot \dfrac{y^3}{3}$

12. $\dfrac{7mn^2}{8m^2 n} \cdot \dfrac{16mz^2}{49n^2 z}$

13. $\dfrac{ab^2}{2} \cdot \dfrac{p}{abc}$

14. $\dfrac{a}{b^2} \cdot \dfrac{3b^2}{4a} \cdot \dfrac{ab}{9}$

15. $\dfrac{8n}{5} \cdot \dfrac{n}{2} \cdot \dfrac{3}{n^2}$

16. $\dfrac{9}{2a + 4} \cdot \dfrac{3a + 6}{15}$

17. $\dfrac{9 - r^2}{r + 3} \cdot \dfrac{r}{3 - r}$

18. $\dfrac{m^2 + 5m}{m^2 - 16} \cdot \dfrac{m^2 - 4m}{m^2 - 25}$

19. $\dfrac{w^2 - 6w + 5}{w^2 - 1} \cdot \dfrac{w - 1}{w^2 - 10w + 25}$

20. $\dfrac{y^2 - 2y - 24}{y^2 - 16} \cdot \dfrac{(y + 5)^2}{y^2 - y - 30}$

21. $\dfrac{n^2 - 6n + 9}{n^2 - 9} \cdot \dfrac{n^2 + 4n + 3}{n^2 - 3n}$

22. $\dfrac{4u^2 + 4u + 1}{2u^2 + u} \cdot \dfrac{u}{2u^2 - u - 1}$

23. $\dfrac{12q}{q^2 - 2q + 1} \cdot \dfrac{q^2 - 1}{24q^2}$

24. $\dfrac{b^2 - 5b}{2b^2} \cdot \dfrac{b^2 - b - 12}{5b}$

25. $\dfrac{r^2 - s^2}{3r^2 - 21r + 30} \cdot \dfrac{3r^2 - 75}{r^2 - 2rs + s^2}$

26. $\dfrac{4u^2 + 4u + 1}{u + 2u^2} \cdot \dfrac{u}{2u^2 - u - 1}$

27. $\dfrac{x^2 + 7xy + 10y^2}{x^2 + 6xy + 5y^2} \cdot \dfrac{x + 2y}{y} \cdot \dfrac{x + y}{x^2 + 4xy + 4y^2}$

28. $\dfrac{2m - 3n}{m^2 + 4mn + 4n^2} \cdot \dfrac{5m^2 + 10mn}{3mn - 3n^2} \cdot \dfrac{4m^2 - 4n^2}{4m^2 - 9n^2}$

For Exercises 29–50, divide.

29. $\dfrac{a}{b} \div \dfrac{c}{d}$

30. $\dfrac{x}{2} \div \dfrac{x}{4}$

31. $\dfrac{xy}{bc} \div \dfrac{x}{y}$

32. $\dfrac{5x^2}{8xy} \div 15x$

33. $\dfrac{2m}{3n^2} \div \dfrac{8m^3}{15n}$

34. $\dfrac{7a^2b}{2c^2} \div \dfrac{7a^2}{b}$

35. $\dfrac{12w^2y}{5z^2} \div \dfrac{12w^2}{y}$

36. $\dfrac{25r^2}{9s^2t^2} \div \dfrac{100r^3}{18st^3}$

37. $\dfrac{14y^2}{10a^2} \div \dfrac{21y^2}{15a^2}$

38. $\dfrac{a^2 - b^2}{x^2 - y^2} \div \dfrac{a + b}{x - y}$

39. $\dfrac{a^3 - 6a^2 + 8a}{5b} \div \dfrac{2a - 4}{10a + 40}$

40. $\dfrac{5r^2 - 5rs}{rs + s^2} \div \dfrac{5r^2 - 5s^2}{s}$

41. $\dfrac{v + 1}{3} \div \dfrac{2v + 2}{9}$

42. $\dfrac{3a + 6}{5} \div \dfrac{4a + 8}{10a}$

43. $\dfrac{4k + 4}{k + 3} \div \dfrac{4k + 8}{2k + 6}$

44. $\dfrac{t^2 - 2t}{t + 1} \div \dfrac{t^2 + 3t}{t^2 + 4t + 3}$

45. $\dfrac{a^2 - b^2}{a^2 + 2ab + b^2} \div \dfrac{a^2 - 3ab + 2b^2}{a^2 + 3ab + 2b^2}$

46. $\dfrac{3w^2 - 7w - 6}{w^2 - 9} \div \dfrac{9w^2 - 4}{3w^2 + 7w - 6}$

47. $\dfrac{3k + 2}{5k^2 - k} \div \dfrac{6k^2 + k - 2}{10k^2 + 3k - 1}$

48. $\dfrac{5y - 3}{4y^2 - y - 3} \div \dfrac{5y^3 - 3y^2}{4y^2 + 3y}$

49. $\dfrac{x^2 + 4x + 4}{x + 1} \div (3x^2 + 5x - 2)$

50. $\dfrac{u^2 - 2u - 8}{u^2 + 3u + 2} \div (u^2 - 3u - 4)$

For Exercises 51–56, perform the indicated operations. (Remember that the order of operations is to multiply or divide from left to right.)

51. $\dfrac{15 - 3y}{y^2 - 4} \cdot \dfrac{y^2 - 4y + 4}{y - 5}$

52. $\dfrac{2m^2 - 7m - 15}{4m^2 - 9} \cdot \dfrac{2m + 1}{1 - 2m} \div \dfrac{5 - m}{2m - 1}$

53. $\dfrac{y^2 + y - 6}{2y^2 - 3y - 2} \cdot \dfrac{2y^2 + 9y + 4}{y^2 + 7y + 12} \div \dfrac{4y + y^2}{7 + y}$

54. $\dfrac{12h^2 + 11h - 5}{h^4 - 16} \div \dfrac{h - 3h^2}{h^3 + 4h - 2h^2 - 8} \div \dfrac{4h + 5}{h^3}$

55. $\dfrac{6x^2}{x^2 - 9} \cdot \dfrac{3x^2 + x - 24}{20y - 6xy} \cdot \dfrac{3 - x}{3x - 8} \div \dfrac{3x^2 + 9x}{3x^2 - x - 30}$

56. $\dfrac{t^2 - 2t}{2t} \cdot \dfrac{2}{t} \cdot \dfrac{2t}{t^2 - 4t + 4} \div \dfrac{t + 2}{t - 2}$

For Exercises 57–60, explain the mistake, then correct the mistake.

57. $\dfrac{5y}{9} \div \dfrac{3y}{10} = \dfrac{y}{3} \cdot \dfrac{y}{2} = \dfrac{y^2}{6}$

58. $\dfrac{12x^3y}{5} \cdot \dfrac{3y}{6x^2} = \dfrac{2xy}{5} \cdot \dfrac{3y}{1} = \dfrac{6y}{5}$

59. $\dfrac{x^2 - 9}{15y} \cdot \dfrac{y - 2}{x + 3} = \dfrac{(x - 3)(x + 3)}{15y} \cdot \dfrac{y - 2}{x + 3}$

$= \dfrac{x - 3}{15} \cdot \dfrac{-2}{x + 3}$

$= \dfrac{-2x + 6}{15x + 45}$

60. $\dfrac{x^2 + 3x + 2}{x^2 - 1} \cdot \dfrac{x - 1}{x + 1} = \dfrac{(x + 1)(x + 2)}{1} \cdot \dfrac{1}{x + 1}$

For Exercises 61–64, use dimensional analysis and the following facts to convert units of length.

$$1 \text{ foot} = 12 \text{ inches}$$
$$1 \text{ yard} = 3 \text{ feet}$$
$$1 \text{ mile} = 5280 \text{ feet}$$

61. 18.5 feet to inches   **62.** 24.2 miles to feet   **63.** $6\frac{1}{4}$ miles to yards   **64.** 2400 yards to miles

For Exercises 65–68, use dimensional analysis and the following values to convert units of weight.

$$1 \text{ pound} = 16 \text{ ounces}$$
$$1 \text{ ton} = 2000 \text{ pounds}$$

65. 5.5 pounds to ounces

66. 36 ounces to pounds

67. 25,400 pounds to tons

68. 2.5 tons to pounds

For Exercises 69–74, use dimensional analysis.

69. The speed of sound is 741.8 miles per hour at 32°F at sea level. How many feet per second is this?

70. If a car is traveling at 65 miles per hour, what is its speed in feet per second?

71. A sprinter's best time in the 40-yard dash is 4.3 seconds. What is that speed in miles per hour?

72. A runner knows that she can run 400 yards in 87.2 seconds. If she can maintain that pace, how many minutes will it take her to run a mile?

73. One light-year is the distance that light travels in a year. If light travels at a speed of approximately 186,000 miles per second, how many miles does light travel in one year?

74. The light from the Sun takes about 8 minutes to reach Earth. Using the fact that light travels at a speed of about 186,000 miles per second, determine Earth's distance from the Sun in miles.

> ## Of Interest
>
> The nearest star to the Sun is Proxima Centauri, which is about 4.2 light-years away. This means that when we look at Proxima Centauri, we are seeing an image of that star that is 4.2 years old. Some objects we can see in the night sky are much farther. For example, the Andromeda galaxy is about 2,140,000 light-years from Earth, which means that when we look at it, we are seeing it as it was 2,140,000 years ago.

For Exercises 75–80, use dimensional analysis and the following table of exchange rates to convert.

	USD($)	GBP(£)	CAD($)	EUR(€)
USD($)	1	1.49819	0.652784	0.964599
GBP(£)	0.667468	1	0.435712	0.643839
CAD($)	1.5319	2.29509	1	1.47766
EUR(€)	1.0367	1.55318	0.676741	1

Source: Federal Reserve Bank of New York

Note: Each number is the ratio of the row's unit to 1 of the column's unit. For example, 1.47766 means there are 1.47766 Can$ to 1€.

75. Convert $500 into £ (Great Britain pound).

76. Convert $250 into Can$ (Canadian dollars).

77. Convert 450 € (euros) into U.S. dollars.

78. Convert 700 £ into U.S. dollars.

79. Convert 650 £ to €.

80. Convert 400 Can$ to €.

REVIEW EXERCISES

For Exercises 1 and 2, find the LCM.

1. 12, 15, and 24

2. $2x^2y$ and $4x^3y^2$

For Exercises 3–6, simplify.

3. $\dfrac{1}{3} + \dfrac{2}{3}$

4. $\dfrac{1}{4} - \dfrac{3}{4}$

5. $\dfrac{1}{2}x + \dfrac{3}{2}x$

6. $\dfrac{3}{4}y - 5 - \dfrac{2}{3}y + 1$

7.3 Adding and Subtracting Rational Expressions with the Same Denominator

OBJECTIVE **1** Add or subtract rational expressions with the same denominator.

Recall that when adding or subtracting fractions with the same denominator, we add or subtract the numerators and keep the same denominator.

$$\frac{3}{7} + \frac{1}{7} = \frac{3+1}{7} = \frac{4}{7} \qquad \text{or} \qquad \frac{7}{9} - \frac{2}{9} = \frac{7-2}{9} = \frac{5}{9}$$

We add and subtract rational expressions the same way.

$$\frac{3x}{7} + \frac{x}{7} = \frac{3x+x}{7} = \frac{4x}{7} \qquad \text{or} \qquad \frac{7x}{9} - \frac{2x}{9} = \frac{7x-2x}{9} = \frac{5x}{9}$$

PROCEDURE ADDING OR SUBTRACTING RATIONAL EXPRESSIONS (SAME DENOMINATOR)

To add or subtract rational expressions that have the same denominator:

1. Add or subtract the numerators and keep the same denominator.
2. Simplify to lowest terms (remember to write the numerators and denominators in factored form in order to simplify).

EXAMPLE 1 Add $\dfrac{5x}{14} + \dfrac{x}{14}$.

Solution $\dfrac{5x}{14} + \dfrac{x}{14} = \dfrac{5x+x}{14}$ Since the rational expressions have the same denominator, we add numerators and keep the same denominator.

$= \dfrac{6x}{14}$ Combine like terms.

$= \dfrac{2 \cdot 3 \cdot x}{2 \cdot 7}$ Factor

$= \dfrac{3x}{7}$ Divide out the common factor, 2.

ANSWER

$-\dfrac{a^2}{3}$

YOUR TURN

Subtract $\dfrac{2a^2}{15} - \dfrac{7a^2}{15}$.

Numerators with No Like Terms

In Example 1, the numerators were like terms. If the numerators have no like terms, we write a multiterm polynomial expression in the numerator of the sum or difference.

EXAMPLE 2 Subtract.

a. $\dfrac{y}{y + 5} - \dfrac{3}{y + 5}$

Solution $\dfrac{y}{y + 5} - \dfrac{3}{y + 5} = \dfrac{y - 3}{y + 5}$ Subtract numerators and keep the same denominator. Since y and 3 are not like terms, we express their difference as a polynomial.

▲ **Warning:** Though it may be tempting, we cannot divide out the y's because they are terms, not factors.

b. $\dfrac{n^2}{n + 3} - \dfrac{9}{n + 3}$

Solution $\dfrac{n^2}{n + 3} - \dfrac{9}{n + 3} = \dfrac{n^2 - 9}{n + 3}$ **Note:** The numerator can be factored, so we may be able to simplify.

$= \dfrac{(n - 3)(n + 3)}{n + 3}$ Factor the numerator.

$= n - 3$ Divide out the common factor, $n + 3$.

c. $\dfrac{4m}{15} - \left(-\dfrac{8}{15}\right)$

Solution $\dfrac{4m}{15} - \left(-\dfrac{8}{15}\right) = \dfrac{4m - (-8)}{15}$

$= \dfrac{4m + 8}{15}$

Note: $4m + 8$ can be factored.

$$4m + 8 = 4(m + 2)$$

However, since there are no common factors in the denominator, we cannot simplify further.

YOUR TURN Add or subtract.

a. $\dfrac{x}{x - 4} + \dfrac{6}{x - 4}$ **b.** $\dfrac{16n^2}{4n + 5} - \dfrac{25}{4n + 5}$ **c.** $\dfrac{5x^2}{7} - \left(-\dfrac{2x}{7}\right)$

Numerators That Are Multiterm Polynomials

So far, the numerators in the given rational expressions were monomials. Now look at some cases where the numerators in the given rational expressions are multiterm polynomials. Remember from Chapter 5 that to add polynomials, we combine like terms.

EXAMPLE 3 Add.

a. $\dfrac{7y - 13}{5} + \dfrac{y + 8}{5}$

Solution $\dfrac{7y - 13}{5} + \dfrac{y + 8}{5} = \dfrac{(7y - 13) + (y + 8)}{5}$ Combine $7y$ with y to get $8y$ and combine -13 with 8 to get -5.

$= \dfrac{8y - 5}{5}$ ◀ **Warning:** It may be tempting to divide out the 5, but the 5 in $8y - 5$ is not a factor. Because $8y - 5$ cannot be factored, we cannot simplify this rational expression.

b. $\dfrac{x^2 - 5x - 13}{x - 6} + \dfrac{3x - 11}{x - 6}$

Solution $\dfrac{x^2 - 5x - 13}{x - 6} + \dfrac{3x - 11}{x - 6} = \dfrac{(x^2 - 5x - 13) + (3x - 11)}{x - 6}$

Note: Since the numerator can be factored, we may be able to simplify.

$= \dfrac{x^2 - 2x - 24}{x - 6}$ Combine like terms in the numerator.

$= \dfrac{(x - 6)(x + 4)}{x - 6}$ Factor the numerator.

$= x + 4$ Divide out the common factor, $x - 6$.

YOUR TURN Add.

a. $\dfrac{n + 7}{11n} + \dfrac{2 - 5n}{11n}$

b. $\dfrac{3t^2 - 5t}{t - 3} + \dfrac{t^2 - 8t + 3}{t - 3}$

Simplifying When the Denominator Is Factorable

Sometimes you may also be able to factor the denominator. If the denominator can be factored, you will need to write it in factored form in order to simplify.

EXAMPLE 4 Add $\dfrac{4x^2 + 9x + 1}{2x^3 - 18x} + \dfrac{6x^2 + 23x + 5}{2x^3 - 18x}$.

Solution $\dfrac{4x^2 + 9x + 1}{2x^3 - 18x} + \dfrac{6x^2 + 23x + 5}{2x^3 - 18x} = \dfrac{(4x^2 + 9x + 1) + (6x^2 + 23x + 5)}{2x^3 - 18x}$

$= \dfrac{10x^2 + 32x + 6}{2x^3 - 18x}$ Combine like terms in the numerator.

$= \dfrac{2(5x^2 + 16x + 3)}{2x(x^2 - 9)}$ Factor 2 out of the numerator and $2x$ out of the denominator.

$= \dfrac{2(5x + 1)(x + 3)}{2x(x - 3)(x + 3)}$ Factor the trinomial in the numerator and the difference of squares in the denominator.

$= \dfrac{5x + 1}{x(x - 3)}$ or $\dfrac{5x + 1}{x^2 - 3x}$ Divide out the common factors, 2 and $x + 3$.

ANSWERS

a. $\dfrac{9 - 4n}{11n}$ **b.** $4t - 1$

Subtracting with Numerators That Are Multiterm Polynomials

Now let's look at subtraction of rational expressions that have the same denominator and multiterm polynomials in the numerators. Recall from Chapter 5 that when subtracting polynomials, we write an equivalent addition with the additive inverse of the subtrahend.

EXAMPLE 5 Subtract.

a. $\dfrac{7t^2 - t + 5}{t + 2} - \dfrac{3t^2 - 6t - 2}{t + 2}$

Solution $\dfrac{7t^2 - t + 5}{t + 2} - \dfrac{3t^2 - 6t - 2}{t + 2} = \dfrac{(7t^2 - t + 5) - (3t^2 - 6t - 2)}{t + 2}$

Note: To write an equivalent addition, change the operation symbol from a minus sign to a plus sign and change all the signs in the subtrahend (second) polynomial.

$$= \dfrac{(7t^2 - t + 5) + (-3t^2 + 6t + 2)}{t + 2}$$

$$= \dfrac{4t^2 + 5t + 7}{t + 2} \qquad \text{Combine like terms.}$$

b. $\dfrac{x^3 + 4x - 3}{5x^2 - 10x} - \dfrac{4x + 5}{5x^2 - 10x}$

Solution $\dfrac{x^3 + 4x - 3}{5x^2 - 10x} - \dfrac{4x + 5}{5x^2 - 10x} = \dfrac{(x^3 + 4x - 3) - (4x + 5)}{5x^2 - 10x}$

$$= \dfrac{(x^3 + 4x - 3) + (-4x - 5)}{5x^2 - 10x} \qquad \text{Write the equivalent addition.}$$

Note: The numerator is a difference of cubes.

$$= \dfrac{x^3 - 8}{5x^2 - 10x} \qquad \text{Combine like terms.}$$

$$= \dfrac{(x - 2)(x^2 + 2x + 4)}{5x(x - 2)} \qquad \begin{array}{l}\text{Factor the numerator and}\\ \text{denominator. Divide out the}\\ \text{common factor, } x - 2.\end{array}$$

$$= \dfrac{x^2 + 2x + 4}{5x}$$

ANSWER

$\dfrac{x - 1}{3x(x - 2)}$ or $\dfrac{x - 1}{3x^2 - 6x}$

ANSWER

$\dfrac{y^2 + 3y + 9}{3y + 2}$

YOUR TURN

Subtract $\dfrac{y^3 + y - 20}{3y^2 - 7y - 6} - \dfrac{y + 7}{3y^2 - 7y - 6}.$

7.3 EXERCISES

For Extra Help

  InterAct Math

  Addison-Wesley
Math Tutor Center

Video 12/DVT 4

MyMathLab

*Math*XL MathXL

1. Two rational expressions with monomial numerators are added to form a rational expression whose numerator is a multiterm polynomial. Why would this happen?

2. After adding or subtracting two rational expressions, what is the first step in simplifying the resulting rational expression?

3. Two rational expressions that have the same denominator and numerators that are multiterm polynomials are to be subtracted. Explain the process.

4. After adding or subtracting two rational expressions, the resulting rational expression contains a multiterm polynomial in the numerator that can be factored. Does this mean that the rational expression can be simplified? Explain.

For Exercises 5–40, add or subtract. Simplify your answers to lowest terms.

5. $\dfrac{3x}{10} + \dfrac{2x}{10}$

6. $\dfrac{2x}{9} + \dfrac{x}{9}$

7. $\dfrac{4a^2}{9} - \dfrac{a^2}{9}$

8. $\dfrac{3}{w^2} + \dfrac{2}{w^2}$

9. $\dfrac{3}{a^2 + a} + \dfrac{2}{a^2 + a}$

10. $\dfrac{3}{a^2 + 2ab} + \dfrac{2}{a^2 + 2ab}$

11. $\dfrac{2 - 2a}{2a - 3} + \dfrac{8a - 11}{2a - 3}$

12. $\dfrac{16x + y}{x - y} + \dfrac{10x - 15y}{x - y}$

13. $\dfrac{2r}{r + 2} + \dfrac{r}{r + 2}$

14. $\dfrac{2x + 5}{x + 1} - \dfrac{x + 4}{x + 1}$

15. $\dfrac{m + 3}{m^2 - 1} - \dfrac{4}{m^2 - 1}$

16. $\dfrac{r - 5s}{r^2 - s^2} - \dfrac{2r - 6s}{r^2 - s^2}$

17. $\dfrac{2b}{b + c} - \dfrac{b - c}{b + c}$

18. $\dfrac{1 - 2q}{3q^2} + \dfrac{5q + 2}{3q^2}$

19. $\dfrac{x^2 + 2}{x + 1} - \dfrac{6 - x^2}{x + 1}$

20. $\dfrac{8y + 5}{3y} - \dfrac{2y + 5}{3y}$

21. $\dfrac{1 - 3t}{2t - 3} - \dfrac{8t - 2}{2t - 3}$

22. $\dfrac{2 - 5a}{a^2b} + \dfrac{5a - 2}{a^2b}$

23. $\dfrac{w^2 + 5}{w + 1} + \dfrac{6 - w^2}{w + 1}$

24. $\dfrac{u^2 + 5u}{u + 1} + \dfrac{2 - 2u}{u + 1}$

25. $\dfrac{9x^2 - 24x + 16}{2x + 1} + \dfrac{2x^2 + 3x - 9}{2x + 1}$

26. $\dfrac{z^2 + 2z}{z - 7} - \dfrac{12z - 21}{z - 7}$

27. $\dfrac{5t + 5}{u} - \dfrac{4t - 6}{u}$

28. $\dfrac{7p + 3z}{p - 1} + \dfrac{2p - 5z}{p - 1}$

29. $\dfrac{g}{g^2 - 9} - \dfrac{3}{g^2 - 9}$

30. $\dfrac{s^2 + 2s}{s^2 + 4s + 4} - \dfrac{s^2 + s - 2}{s^2 + 4s + 4}$

31. $\dfrac{y^2}{y^2 + 3y} - \dfrac{9}{y^2 + 3y}$

32. $\dfrac{v - 8}{v^2 - 16} + \dfrac{4}{v^2 - 16}$

33. $\dfrac{2w + 3}{5} + \dfrac{5w - 2}{5} - \dfrac{w + 6}{5}$

34. $\dfrac{6q + 21}{q + 1} - \dfrac{28 - 14q}{q + 1} + \dfrac{6q - 5}{q + 1}$

35. $\dfrac{h^3 - 3}{h + 3} - \dfrac{h}{h + 3} + \dfrac{3h^2}{h + 3}$

36. $\dfrac{2m^2 - 6m}{m^2 - 5m + 6} + \dfrac{2m - 4}{m^2 - 5m + 6} + \dfrac{4}{m^2 - 5m + 6}$

37. $\dfrac{x^2 + 1}{x^3 - 2x^2 + x - 2} - \dfrac{x^2 - 2x}{x^3 - 2x^2 + x - 2} - \dfrac{x + 3}{x^3 - 2x^2 + x - 2}$

38. $\dfrac{3x - 3}{x^2 - 1} + \dfrac{1}{x^2 - 1} + \dfrac{x}{x^2 - 1}$

39. $\dfrac{x - 5}{x^2 - 25} + \dfrac{2x + 10}{x^2 - 25} + \dfrac{4}{x^2 - 25}$

40. $\dfrac{p^2 - 1}{2p^2 - 2} - \dfrac{12p - 12}{2p^2 - 2} + \dfrac{2p^2 + 2p}{2p^2 - 2}$

For Exercises 41–44, explain the mistake, then find the correct sum or difference.

41. $\dfrac{4y}{5} + \dfrac{1}{5} = \dfrac{5y}{5} = y$

42. $\dfrac{7}{m} + \dfrac{n}{m} = \dfrac{7 + n}{2m}$

43. $\dfrac{x+1}{3} - \dfrac{x-5}{3} = \dfrac{(x+1)-(x-5)}{3}$

$\qquad\qquad\qquad = \dfrac{(x+1)+(-x-5)}{3}$

$\qquad\qquad\qquad = -\dfrac{4}{3}$

44. $\dfrac{x}{x-2} + \dfrac{x-4}{x-2} = \dfrac{x^2-4}{x-2}$

$\qquad\qquad\qquad = \dfrac{(x+2)(x-2)}{x-2}$

$\qquad\qquad\qquad = x+2$

For Exercises 45–48, find the perimeter of the figure shown.

45.

46.

47.

48.

REVIEW EXERCISES

1. Find the least common multiple of 24 and 18.

For Exercises 2–4, add or subtract.

2. $\dfrac{1}{2} + \dfrac{3}{5}$

3. $\dfrac{5}{6} - \dfrac{3}{8}$

4. $\dfrac{4}{7} - \dfrac{-3}{5}$

5. Combine like terms: $\dfrac{1}{3}x - \dfrac{1}{6} + \dfrac{2}{5}x + \dfrac{3}{4}$

6. Solve: $\dfrac{5}{6}x - \dfrac{3}{4} = \dfrac{2}{3}x + \dfrac{1}{2}$

7.4 Adding and Subtracting Rational Expressions with Different Denominators

OBJECTIVES

1 Find the LCD of two or more rational expressions.

2 Given two rational expressions, write equivalent rational expressions with their LCD.

3 Add or subtract rational expressions with different denominators.

Adding or subtracting rational expressions with different denominators follows three main stages: 1. finding the LCD, 2. rewriting the expressions with the LCD, and 3. adding or subtracting the numerators. We will explore the first two stages in objectives 1 and 2, then put them back into their proper context in Objective 3.

OBJECTIVE 1 **Find the LCD of two or more rational expressions.** Remember that when adding or subtracting fractions with different denominators, we must first find a common denominator. It is helpful to use the least common denominator (LCD), which is the smallest number that is evenly divisible by all the denominators.

For example, given the fractions $\frac{1}{8}$ and $\frac{7}{12}$, the LCD is 24.

To help us find the LCD when denominators are more complex, as in rational expressions, let's consider a method involving prime factorizations. Consider the prime factorizations of the denominators, 8 and 12, compared to the prime factorization of the LCD, 24, in order to determine a procedure.

$$8 = 2^3 \qquad 12 = 2^2 \cdot 3$$

$$LCD = 2^3 \cdot 3 = 24$$

Both 8 and 12 have factors of 2, but 8 has the greatest number of 2's, which is why 2^3 is included in the factorization for the LCD instead of 2^2. Similarly, 12 has the greatest number of factors of 3 (a single factor, whereas 8 has no factors of 3), so a single factor of 3 is included in the factorization of the LCD. This suggests the following procedure.

PROCEDURE FINDING THE LCD

To find the LCD of two or more rational expressions:

1. Factor each denominator.

2. For each unique factor, compare the number of times it appears in each factorization. Write a factored form that includes each factor the greatest number of times it appears in the denominator factorizations.

First, let's consider cases where the denominators are monomials.

EXAMPLE 1 Find the LCD of $\dfrac{3}{4x^2}$ and $\dfrac{5}{6x^3}$.

Solution We first factor the denominators $4x^2$ and $6x^3$ by writing their prime factorizations.

$$4x^2 = 2^2 \cdot x^2$$
$$6x^3 = 2 \cdot 3 \cdot x^3$$

The unique factors are 2, 3, and x. To generate the LCD, include 2, 3 and x the greatest number of times each appears in any of the factorizations.

The greatest number of times that 2 appears is twice (in $2^2 \cdot x^2$).
The greatest number of times that 3 appears is once (in $2 \cdot 3 \cdot x^3$).
The greatest number of times that x appears is three times (in $2 \cdot 3 \cdot x^3$).

$$\text{LCD} = 2^2 \cdot 3 \cdot x^3 = 12x^3$$

Note: We can compare exponents in the prime factorizations to create the LCD. If two factorizations have the same prime factor, we write that prime factor in the LCD with the greater of the two exponents.

YOUR TURN Find the LCD.

a. $\dfrac{y}{12x^5}$ and $\dfrac{5}{9x^2}$

b. $\dfrac{p}{10mn}$ and $\dfrac{7}{15n^2}$

Now consider denominators that are multiterm polynomials.

EXAMPLE 2 Find the LCD.

a. $\dfrac{7}{5x - 15}$ and $\dfrac{5}{2x - 6}$

Solution Factor the denominators $5x - 15$ and $2x - 6$.

$$5x - 15 = 5(x - 3)$$
$$2x - 6 = 2(x - 3)$$

The unique factors are 2, 5, and $x - 3$.
The greatest number of times that $x - 3$ appears in any of the factorizations is once.
The greatest number of times that 2 appears is once (in $2(x - 3)$).
The greatest number of times that 5 appears is once (in $5(x - 3)$).

$$\text{LCD} = 2 \cdot 5 \cdot (x - 3) = 10(x - 3) \text{ or } 10x - 30$$

ANSWERS

a. $36x^5$ **b.** $30mn^2$

b. $\dfrac{7}{x^2 + 10x + 25}$ and $\dfrac{17}{6x^4 + 30x^3}$

Solution Factor $x^2 + 10x + 25$ and $6x^4 + 30x^3$.

$$x^2 + 10x + 25 = (x + 5)^2$$
$$6x^4 + 30x^3 = 6x^3(x + 5) = 2 \cdot 3 \cdot x^3(x + 5)$$

The factors are 2, 3, x, and $x + 5$.

The greatest number of times that 2 appears is once (in $2 \cdot 3 \cdot x^3(x + 5)$).

The greatest number of times that 3 appears is once (in $2 \cdot 3 \cdot x^3(x + 5)$).

The greatest number of times that $x + 5$ appears is twice (in $(x + 5)^2$).

The greatest number of times that x appears is three times as x^3 (in $2 \cdot 3 \cdot x^3(x + 5)$).

$$\text{LCD} = 2 \cdot 3 \cdot x^3 \cdot (x + 5)^2 = 6x^3(x + 5)^2$$

YOUR TURN Find the LCD.

a. $\dfrac{y}{3y + 6}$ and $\dfrac{5}{4y + 8}$

b. $\dfrac{9}{x^2 - 2x + 1}$ and $\dfrac{17}{2x^3 - 2x}$

OBJECTIVE 2 Given two rational expressions, write equivalent rational expressions with their LCD.

Writing equivalent rational expressions is much like writing equivalent fractions, which we reviewed in Section 1.2. Remember that to write an equivalent fraction we multiply the numerator and denominator by the same nonzero number.

EXAMPLE 3 Write $\dfrac{3}{4x^2}$ and $\dfrac{5}{6x^3}$ as equivalent rational expressions with their LCD.

Solution In Example 1, we found the LCD to be $12x^3$. For each rational expression, we multiply both the numerator and denominator by an appropriate factor so that the denominator becomes $12x^3$.

$$\frac{3}{4x^2} = \frac{3 \cdot 3x}{4x^2 \cdot 3x} = \frac{9x}{12x^3}$$

$$\frac{5}{6x^3} = \frac{5 \cdot 2}{6x^3 \cdot 2} = \frac{10}{12x^3}$$

Note: We can determine the appropriate factor by dividing the LCD by the original denominator. In this case, $12x^3 \div 4x^2 = 3x$.

Now consider rational expressions with denominators that are multiterm polynomials. Rewriting these rational expressions with their LCD is usually easier if we leave their denominators in factored form.

EXAMPLE 4 Write $\dfrac{7}{x^2 + 10x + 25}$ and $\dfrac{17}{6x^4 + 30x^3}$ as equivalent rational expressions with their LCD.

Solution In Example 2, we found the LCD to be $6x^3(x+5)^2$.

$$\frac{7}{x^2 + 10x + 25} = \frac{7}{(x+5)^2}$$

Write the denominator in factored form.

$$= \frac{7 \cdot 6x^3}{(x+5)^2 \cdot 6x^3}$$

Multiply the numerator and denominator by the same factor, $6x^3$, to get the LCD, $6x^3(x+5)^2$.

Note: Another way to determine the appropriate factor is to think of it as the factor in the LCD that is missing from the original denominator. In this case, $6x^3$ is in the LCD, but not in the original denominator.

$$= \frac{42x^3}{6x^3(x+5)^2}$$

Note: Though we could multiply out $6x^3(x+5)^2$, it is simpler to leave it in this form.

$$\frac{17}{6x^4 + 30x^3} = \frac{17}{6x^3(x+5)}$$

Write the denominator in factored form.

Note: There are two factors of $x + 5$ in the LCD, whereas only one factor of $x + 5$ appears in the original denominator.

$$= \frac{17 \cdot (x+5)}{6x^3(x+5) \cdot (x+5)}$$

Multiply the numerator and denominator by the same factor, $x + 5$, to get the LCD, $6x^3(x+5)^2$.

$$= \frac{17(x+5)}{6x^3(x+5)^2}$$

YOUR TURN Write equivalent rational expressions with their LCD.

a. $\dfrac{p}{10mn}$ and $\dfrac{7}{15n^2}$ **b.** $\dfrac{9}{x^2 - 2x + 1}$ and $\dfrac{17}{2x^3 - 2x}$

OBJECTIVE 3 **Add and subtract rational expressions with different denominators.** To add or subtract rational expressions with different denominators, we use the same process that we use to add or subtract fractions.

PROCEDURE **ADDING OR SUBTRACTING RATIONAL EXPRESSIONS WITH DIFFERENT DENOMINATORS**

To add or subtract rational expressions with different denominators:

1. Find the LCD.
2. Write each rational expression as an equivalent expression with the LCD.
3. Add or subtract the numerators and keep the LCD.
4. Simplify.

ANSWERS

a. $\dfrac{3np}{30mn^2}$ and $\dfrac{14m}{30mn^2}$

b. $\dfrac{18x(x+1)}{2x(x-1)^2(x+1)}$ and

$\dfrac{17(x-1)}{2x(x-1)^2(x+1)}$

EXAMPLE 5 Add or subtract as indicated.

a. $\dfrac{3}{4x^2} + \dfrac{5}{6x^3}$

Solution We found the LCD for these expressions in Example 1. We then wrote equivalent expressions in Example 3. Now we add the numerators and keep the LCD.

$$\frac{3}{4x^2} + \frac{5}{6x^3} = \frac{9x}{12x^3} + \frac{10}{12x^3} \qquad \text{Rewrite with the LCD.}$$

$$= \frac{9x + 10}{12x^3} \qquad \text{Add numerators.}$$

Note: Because $9x$ and 10 are not like terms, we must indicate the addition as a polynomial expression. Also, since $9x + 10$ cannot be factored, we cannot simplify this rational expression.

b. $\dfrac{7}{x^2 + 10x + 25} - \dfrac{17}{6x^4 + 30x^3}$

Solution We found the LCD for these expressions in Example 2 and wrote equivalent expressions in Example 4. Now we combine the numerators and keep the LCD.

$$\frac{7}{x^2 + 10x + 25} - \frac{17}{6x^4 + 30x^3} = \frac{42x^3}{6x^3(x+5)^2} - \frac{17(x+5)}{6x^3(x+5)^2} \qquad \text{Rewrite with the LCD.}$$

$$= \frac{42x^3 - 17(x+5)}{6x^3(x+5)^2} \qquad \text{Combine numerators.}$$

$$= \frac{42x^3 - 17x - 85}{6x^3(x+5)^2} \qquad \begin{array}{l}\text{Simplify in the} \\ \text{numerator.}\end{array}$$

Note: Leaving the LCD in factored form helps us determine whether the final expression can be reduced. Since $42x^3 - 17x - 85$ cannot be factored, we cannot reduce this rational expression.

Now that we have seen all the pieces to the process spread out over several examples, let's put them all together using new rational expressions.

EXAMPLE 6 Add or subtract as indicated.

a. $\dfrac{2x}{9y^3} + \dfrac{3}{4xy}$

Solution First, find the LCD.

$$4xy = 2^2 \cdot x \cdot y \qquad 9y^3 = 3^2 \cdot y^3 \qquad \text{LCD} = 2^2 \cdot 3^2 \cdot x \cdot y^3 = 36xy^3$$

$$\frac{2x}{9y^3} + \frac{3}{4xy} = \frac{2x(4x)}{9y^3(4x)} + \frac{3(9y^2)}{4xy(9y^2)} \qquad \begin{array}{l}\text{Write equivalent rational expressions with} \\ \text{the LCD, } 36xy^3.\end{array}$$

$$= \frac{8x^2}{36xy^3} + \frac{27y^2}{36xy^3}$$

$$= \frac{8x^2 + 27y^2}{36xy^3} \qquad \text{Add numerators.}$$

Note: The numerator cannot be factored, so we cannot reduce.

7.4 Adding and Subtracting Rational Expressions with Different Denominators   **583**

b. $\dfrac{x + 5}{2x} + \dfrac{4x - 1}{3x^2}$

Solution The LCD is $6x^2$.

$$\dfrac{x + 5}{2x} + \dfrac{4x - 1}{3x^2} = \dfrac{(x + 5)(3x)}{(2x)(3x)} + \dfrac{(4x - 1)(2)}{(3x^2)(2)}$$

Write equivalent rational expressions with the LCD, $6x^2$.

$$= \dfrac{3x^2 + 15x}{6x^2} + \dfrac{8x - 2}{6x^2}$$

$$= \dfrac{3x^2 + 15x + 8x - 2}{6x^2}$$

Add numerators.

$$= \dfrac{3x^2 + 23x - 2}{6x^2}$$

Note: Remember that to add polynomials, we combine like terms.

c. $\dfrac{x}{x + 2} - \dfrac{10}{x^2 - x - 6}$

Solution First, find the LCD by factoring $x + 2$ and $x^2 - x - 6$.

Factored forms: $x + 2$ cannot be factored; $x^2 - x - 6 = (x + 2)(x - 3)$
LCD $= (x + 2)(x - 3)$

$$\dfrac{x}{x + 2} - \dfrac{10}{x^2 - x - 6} = \dfrac{x}{x + 2} - \dfrac{10}{(x + 2)(x - 3)}$$

Factor the denominators.

$$= \dfrac{(x)(x - 3)}{(x + 2)(x - 3)} - \dfrac{10}{(x + 2)(x - 3)}$$

Write equivalent rational expressions with the LCD, $(x + 2)(x - 3)$.

$$= \dfrac{x^2 - 3x}{(x + 2)(x - 3)} - \dfrac{10}{(x + 2)(x - 3)}$$

Distribute

$$= \dfrac{x^2 - 3x - 10}{(x + 2)(x - 3)}$$

Combine numerators.

$$= \dfrac{(x + 2)(x - 5)}{(x + 2)(x - 3)}$$

Factor the numerator.

$$= \dfrac{x - 5}{x - 3}$$

Divide out the common factor, $x + 2$.

d. $\dfrac{x - 1}{x^2 + 4x} - \dfrac{3x + 1}{x^2 - 16}$

Solution To determine the LCD, we write the factored forms of $x^2 + 4x$ and $x^2 - 16$.

Factored forms: $x^2 + 4x = x(x + 4)$; $x^2 - 16 = (x + 4)(x - 4)$
LCD $= x(x + 4)(x - 4)$

$$\dfrac{x - 1}{x^2 + 4x} - \dfrac{3x + 1}{x^2 - 16} = \dfrac{x - 1}{x(x + 4)} - \dfrac{3x + 1}{(x + 4)(x - 4)}$$

Factor the denominators.

$$= \dfrac{(x - 1)(x - 4)}{x(x + 4)(x - 4)} - \dfrac{(3x + 1)(x)}{(x + 4)(x - 4)(x)}$$

Write equivalent rational expressions with the LCD, $x(x + 4)(x - 4)$.

$$= \frac{x^2 - 5x + 4}{x(x + 4)(x - 4)} - \frac{3x^2 + x}{x(x + 4)(x - 4)}$$ Multiply in the numerator.

$$= \frac{(x^2 - 5x + 4) - (3x^2 + x)}{x(x + 4)(x - 4)}$$ Combine numerators.

$$= \frac{(x^2 - 5x + 4) + (-3x^2 - x)}{x(x + 4)(x - 4)}$$

$$= \frac{-2x^2 - 6x + 4}{x(x + 4)(x - 4)}$$

Note: Remember that to subtract polynomials, we write an equivalent addition.

Note: Though the numerator can be factored to $-2(x^2 + 3x - 2)$, none of those factors matches any factor in the denominator, so the result is in lowest terms.

ANSWERS

a. $\dfrac{20y^2 + 9x}{24x^2y^3}$

b. $\dfrac{-x^2 + 6x - 2}{x(x + 1)(x - 1)}$

YOUR TURN   Add or subtract as indicated.

a. $\dfrac{5}{6x^2y} + \dfrac{3}{8xy^3}$

b. $\dfrac{2}{x^2 + x} - \dfrac{x - 4}{x^2 - 1}$

If denominators are additive inverses, we can multiply the numerator and denominator of either rational expression by -1 to obtain the LCD.

EXAMPLE 7   Add $\dfrac{5x}{x - 3} + \dfrac{2}{3 - x}$.

Solution $\dfrac{5x}{x - 3} + \dfrac{2}{3 - x} = \dfrac{5x}{x - 3} + \dfrac{2(-1)}{(3 - x)(-1)}$

$$= \dfrac{5x}{x - 3} + \dfrac{-2}{x - 3}$$

$$= \dfrac{5x - 2}{x - 3}$$

Since $x - 3$ and $3 - x$ are additive inverses, we obtain the LCD by multiplying the numerator and denominator of one of the rational expressions by -1. We chose the second rational expression.

YOUR TURN   Add or subtract as indicated.

a. $\dfrac{3x}{x - 2} + \dfrac{6}{2 - x}$

b. $\dfrac{7y}{2y - 3} - \dfrac{4}{3 - 2y}$

ANSWERS

a. 3   b. $\dfrac{7y + 4}{2y - 3}$

7.4 EXERCISES

For Extra Help

InterAct Math

Addison-Wesley Math Tutor Center

Video 12/DVT 4

MyMathLab

MathXL

1. Explain how to find the LCD of two rational expressions.

2. Suppose the LCD of two fractions is $48x^3yz$. If one of the fraction's denominators is $8x^2y$, what factor do you multiply by when writing it as an equivalent fraction with the LCD? Explain how you determine this factor.

3. In general, if you are given rational expressions that contain multiterm polynomials in the denominator(s), what is your first step toward finding the LCD?

4. Discuss two ways that you could find a common denominator for $\dfrac{3}{2x - y} + \dfrac{5}{y - 2x}$. Which way works better for you? Why?

5. Suppose you have found the LCD of two rational expressions and have written the equivalent rational expressions. If the problem is now at the stage: $\dfrac{(7x + 21) - (2x - 10)}{(x - 5)(x + 3)}$, how do you proceed?

6. Suppose you have found the LCD of two rational expressions and have written the equivalent rational expressions. If the problem is now at the stage: $\dfrac{(7x + 21) - (2x - 10)}{(x - 5)(x + 3)}$, what might the original problem have been? Explain.

For Exercises 7–18, find the least common denominator for the rational expressions and write equivalent rational expressions with the LCD.

7. $\dfrac{2}{3a}, \dfrac{3}{5a^2}$

8. $\dfrac{5}{ab^2}, \dfrac{3}{a^3b}$

9. $\dfrac{3}{y + 2}, \dfrac{4y}{y - 2}$

10. $\dfrac{3 + x}{x - 5}, \dfrac{2}{x + 1}$

11. $\dfrac{2}{p^2 - 4}, \dfrac{3 + p}{p + 2}$

12. $\dfrac{3}{4w - 8}, \dfrac{5}{w^2 - 2w}$

13. $\dfrac{4u}{(u + 2)^2}, \dfrac{2}{3(u + 2)}$

14. $\dfrac{2y}{y^2 - xy}, \dfrac{7}{y}$

15. $\dfrac{c}{c^3 + 8}, \dfrac{2}{c + 2}$

16. $\dfrac{8t^2 - 1}{t^2 - 36}, \dfrac{t}{t^2 - 7t + 6}$

17. $\dfrac{2}{x - 1}, \dfrac{3}{x^2 - 3x + 2}, \dfrac{5x}{x - 2}$

18. $\dfrac{4a + 2}{a^2 - 9}, \dfrac{2a - 1}{a^2 - a - 6}$

For Exercises 19–60, add or subtract as indicated.

19. $\dfrac{3a - 5}{9} - \dfrac{2a + 7}{6}$

20. $\dfrac{3x - y}{6} - \dfrac{3x - 2y}{4}$

21. $\dfrac{5}{2c} + \dfrac{1}{6c}$

22. $\dfrac{5a}{b^2} - \dfrac{3a}{b}$

23. $\dfrac{5}{x} - \dfrac{2}{x^3}$

24. $\dfrac{3}{x - 2} - \dfrac{2}{x + 2}$

25. $\dfrac{2a - b}{a^2} - \dfrac{5a + 2b}{a}$

26. $\dfrac{5m}{m^2 - mn} + \dfrac{3}{m}$

27. $\dfrac{3}{y + 2} + \dfrac{5}{y - 4}$

28. $\dfrac{2}{c + 4} + \dfrac{3}{c + 3}$

29. $\dfrac{t + 2}{t + 4} - \dfrac{t - 1}{t + 6}$

30. $\dfrac{5}{r^2 - 9} + \dfrac{4}{r + 3}$

31. $\dfrac{6m}{(m - 2)^2} - \dfrac{3m}{2m - 4}$

32. $\dfrac{a + 6}{a^2 + 8a + 15} - \dfrac{a - 3}{a + 3}$

33. $\dfrac{x}{x - y} - \dfrac{y}{y - x}$

34. $\dfrac{u}{u-3} + \dfrac{1}{3-u}$

35. $\dfrac{3}{t-3} + \dfrac{6}{3-t}$

36. $\dfrac{10}{2r-1} - \dfrac{6}{1-2r}$

37. $\dfrac{5y+6}{y^2+3y} + \dfrac{y}{y+3}$

38. $\dfrac{x}{x-7} - \dfrac{x+3}{x^2-4x-21}$

39. $\dfrac{w}{6w+6} - \dfrac{1}{w+1}$

40. $\dfrac{2p}{p^2-2p+1} + \dfrac{8}{p-1}$

41. $\dfrac{3}{2c+10} + \dfrac{15}{c^2-25}$

42. $\dfrac{k+3}{k^2+6k+9} - \dfrac{7}{2k+6}$

43. $\dfrac{z}{2z-14} - \dfrac{5z}{z^2-4z-21}$

44. $\dfrac{u}{u-1} + \dfrac{2u}{u^2-2u+1}$

45. $\dfrac{2m-3n}{m+n} + \dfrac{m+n}{m-n}$

46. $\dfrac{y-8}{y+5} + \dfrac{y-9}{y-6}$

47. $\dfrac{ab}{a^2-b^2} + \dfrac{a}{a-b}$

48. $\dfrac{3s+1}{s-2} - \dfrac{4s+1}{s-3}$

49. $\dfrac{w-7}{w^2+4w-5} - \dfrac{w-9}{w^2+3w-10}$

50. $\dfrac{x+1}{x^2-4x+4} + \dfrac{4}{x^2+3x-10}$

51. $\dfrac{4}{2x-1} + \dfrac{x}{1-2x}$

52. $\dfrac{x}{6x-2} - \dfrac{3x}{1-3x}$

53. $\dfrac{y}{3y-6} - \dfrac{2}{2-y}$

54. $\dfrac{4}{t-4} + \dfrac{t}{4-t}$

55. $\dfrac{3r}{r^2-1} - \dfrac{2r+1}{r^2-2r+1}$

56. $\dfrac{m^2}{m^2-m+1} - \dfrac{m+1}{m}$

57. $\dfrac{q+5}{q^2-9} - \dfrac{1}{q^2+3q}$

58. $\dfrac{h+2}{h-4} - \dfrac{8h+4}{h^2-2h-8}$

59. $\dfrac{1}{d^3+d^2} + \dfrac{2}{d^2+d}$

60. $\dfrac{7u}{3} - \dfrac{14u}{3u-6}$

For Exercises 61–66, explain the mistake, then find the correct sum or difference.

61. $\dfrac{3}{x} - \dfrac{x}{2} = \dfrac{3-x}{x-2}$

62. $\dfrac{\cancel{5}}{2x} + \dfrac{3}{\cancel{10}x} = \dfrac{1}{2x} + \dfrac{3}{2x}$

$\qquad = \dfrac{4}{2x}$

$\qquad = \dfrac{2}{x}$

63. $\dfrac{2w}{x} + \dfrac{5w}{x} = \dfrac{7w}{2x}$

64. $\dfrac{3}{x} + \dfrac{2}{x+y} = \dfrac{3(+y)}{x(+y)} + \dfrac{2}{x+y}$

$\qquad = \dfrac{3+y+2}{x+y}$

$\qquad = \dfrac{5+y}{x+y}$

65. $\dfrac{3x}{x+2} - \dfrac{4x+7}{x+2} = \dfrac{3x-(4x+7)}{x+2}$

$\qquad = \dfrac{3x+(-4x+7)}{x+2}$

$\qquad = \dfrac{-x+7}{x+2}$

66. $\dfrac{2}{x} + \dfrac{3}{y} = \dfrac{6}{xy}$

For Exercises 67–70, solve.

67. Adam and Candace build and install cabinets in newly constructed homes. When working together, the expression $\dfrac{t}{2}$ represents the portion of the job completed by Adam and $\dfrac{t}{3}$ represents the portion completed by Candace. Write a rational expression that describes their combined work.

68. Two construction teams are combined in an effort to speed construction on a building. The expression $\dfrac{t}{5}$ describes the portion of the job that team 1 can complete and $\dfrac{t}{6}$ describes the portion of the job that team 2 can complete working independently. Write a rational expression that describes their combined work.

69. An engineer uses the expression $\dfrac{1}{x+3}$ to describes the width of a rectangular steel plate she is designing for a machine. The expression $\dfrac{1}{x}$ describes the length. Write a rational expression for the perimeter of the steel plate.

70. An electrical engineer is trying to determine the effects of using a thinner layer of a chemical on a circuit board. Currently, the thickness of the chemical layer is described by $\dfrac{1}{n-4}$. If the thinner layer is described by $\dfrac{1}{n-2}$, write a rational expression that describes the difference in the coating thickness?

For Exercises 71–80, use the order-of-operations agreement to simplify.

71. $\dfrac{x}{4x^2-1} + \dfrac{2}{4x-2} - \dfrac{3x+1}{1-2x}$

72. $\dfrac{c}{(c-d)^2} - \dfrac{2}{c+d} - \dfrac{c-3}{c^2-d^2}$

73. $\left(\dfrac{a+b}{a-b} + \dfrac{a-b}{a+b}\right) \div \dfrac{a+b}{a-b}$

74. $(x+y) \div \left(\dfrac{1}{x} + \dfrac{1}{y}\right)$

75. $\left(\dfrac{x}{x-4} - \dfrac{3}{x+3}\right) \div \left(\dfrac{1}{x} + \dfrac{2}{x-4}\right)$

76. $\left(\dfrac{25-9j^2}{2j+3}\right) \div (5j+3j^2)$

77. $\dfrac{1}{m}\left(m + \dfrac{1}{m}\right) \div m^2$

78. $\left(\dfrac{1}{x+2} - \dfrac{3}{x^2-4}\right) \div \dfrac{3}{x-2}$

79. $\dfrac{12}{ax} \cdot \dfrac{4a}{3x} \div \dfrac{6a}{12}$

80. $\dfrac{3}{x} + \dfrac{1}{x^2+3x} \cdot \dfrac{2x}{3}$

REVIEW EXERCISES

For Exercises 1 and 2, divide.

1. $\dfrac{1}{4} \div \dfrac{2}{3}$

2. $\dfrac{m}{6} \div \dfrac{18}{m}$

For Exercises 3 and 4, factor completely.

3. $4x^2 - 16$

4. $6y^3 - 27y^2 - 15y$

For Exercises 5 and 6, simplify.

5. $\dfrac{x^2-7x+6}{x^2-4x-12}$

6. $\dfrac{x+3}{x^3+27}$

7.5 Complex Rational Expressions

Sometimes a rational expression may contain rational expressions in its numerator or denominator. Such an expression is called a **complex rational expression** or a **complex fraction**.

> **DEFINITION**
>
> *Complex rational expression:* A rational expression that contains rational expressions in the numerator or denominator.

Here are some complex rational expressions:

$$\dfrac{\frac{3}{4}}{\frac{5}{6}} \qquad \dfrac{x + \frac{x}{3}}{\frac{y}{2}} \qquad \dfrac{7}{\frac{t+2}{t}} \qquad \dfrac{\frac{2}{n} + \frac{1}{4}}{\frac{1}{n} - \frac{n}{3}}$$

There are two common methods for simplifying complex rational expressions. One method is to rewrite the complex rational expression as a horizontal division problem. For example, the complex fraction $\dfrac{\frac{3}{4}}{\frac{5}{6}}$ can be expressed as the division problem $\dfrac{3}{4} \div \dfrac{5}{6}$, then simplified.

$$\dfrac{\frac{3}{4}}{\frac{5}{6}} = \frac{3}{4} \div \frac{5}{6} = \frac{3}{4} \cdot \frac{\overset{3}{\cancel{6}}}{5} = \frac{9}{10}$$

Another method is to multiply the numerator and denominator of the complex fraction by the LCD of the numerator and denominator fractions. In this case, the denominators are 4 and 6, so the LCD is 12.

> **CONNECTION**
>
> Notice that the essence of this second method is writing an equivalent fraction by multiplying both the numerator and denominator by the same factor.

$$\dfrac{\frac{3}{4} \cdot 12}{\frac{5}{6} \cdot 12} = \dfrac{\frac{3}{\cancel{4}} \cdot \frac{\overset{3}{\cancel{12}}}{1}}{\frac{5}{\cancel{6}} \cdot \frac{\overset{2}{\cancel{12}}}{1}} = \dfrac{\frac{9}{1}}{\frac{10}{1}} = \frac{9}{10}$$

EXAMPLE 1 Simplify $\dfrac{\dfrac{1}{4}+\dfrac{2}{3}}{\dfrac{3}{8}-\dfrac{1}{6}}$.

Method 1 Simplify the numerator and denominator, then divide.

$$\frac{\dfrac{1}{4}+\dfrac{2}{3}}{\dfrac{3}{8}-\dfrac{1}{6}} = \frac{\dfrac{1(3)}{4(3)}+\dfrac{2(4)}{3(4)}}{\dfrac{3(3)}{8(3)}-\dfrac{1(4)}{6(4)}}$$

Write the numerator fractions as equivalent fractions with their LCD, 12, and write the denominator fractions with their LCD, 24.

$$= \frac{\dfrac{3}{12}+\dfrac{8}{12}}{\dfrac{9}{24}-\dfrac{4}{24}}$$

$$= \frac{\dfrac{11}{12}}{\dfrac{5}{24}}$$

Add in the numerator and subtract in the denominator.

$$= \frac{11}{12} \div \frac{5}{24}$$

Write the complex fraction as a division problem.

$$= \frac{11}{\overset{}{\underset{1}{\cancel{12}}}} \cdot \frac{\overset{2}{\cancel{24}}}{5}$$

Write an equivalent multiplication problem, divide out the common factors, and then multiply.

$$= \frac{22}{5}$$

Method 2 Multiply the numerator and denominator of the rational expression by the LCD of all the rational expressions involved. In this case, that LCD is 24.

$$\frac{\dfrac{1}{4}+\dfrac{2}{3}}{\dfrac{3}{8}-\dfrac{1}{6}} = \frac{\left(\dfrac{1}{4}+\dfrac{2}{3}\right)\cdot 24}{\left(\dfrac{3}{8}-\dfrac{1}{6}\right)\cdot 24}$$

Multiply the numerator and denominator by 24.

$$= \cfrac{\dfrac{1}{\cancelto{1}{4}} \cdot \dfrac{\cancelto{6}{24}}{1} + \dfrac{2}{\cancelto{1}{3}} \cdot \dfrac{\cancelto{8}{24}}{1}}{\dfrac{3}{\cancelto{1}{8}} \cdot \dfrac{\cancelto{3}{24}}{1} - \dfrac{1}{\cancelto{1}{6}} \cdot \dfrac{\cancelto{4}{24}}{1}}$$

Distribute the 24 to each fraction, and divide out the common factors.

$$= \frac{6 + 16}{9 - 4}$$

Simplify the numerator and denominator.

$$= \frac{22}{5}$$

YOUR TURN

Simplify $\dfrac{2 - \dfrac{3}{4}}{\dfrac{4}{5} + \dfrac{1}{2}}$.

EXAMPLE 2 Simplify $\dfrac{5 + \dfrac{x}{3}}{\dfrac{4}{x} - \dfrac{1}{2x}}$.

Method 1 Simplify the numerator and denominator, then divide.

$$\frac{5 + \dfrac{x}{3}}{\dfrac{4}{x} - \dfrac{1}{2x}} = \frac{\dfrac{5(3)}{1(3)} + \dfrac{x(1)}{3(1)}}{\dfrac{4(2)}{x(2)} - \dfrac{1(1)}{2x(1)}}$$

Write the numerator fractions as equivalent fractions with their LCD, 3, and write the denominator fractions with their LCD, 2x.

$$= \frac{\dfrac{15}{3} + \dfrac{x}{3}}{\dfrac{8}{2x} - \dfrac{1}{2x}}$$

$$= \frac{\dfrac{15 + x}{3}}{\dfrac{7}{2x}}$$

Add in the numerator, and subtract in the denominator.

$$= \frac{15 + x}{3} \div \frac{7}{2x}$$

Write the complex fraction as a division problem.

$$= \frac{15 + x}{3} \cdot \frac{2x}{7}$$

Write an equivalent multiplication problem and then multiply.

$$= \frac{30x + 2x^2}{21}$$

ANSWER

$\dfrac{25}{26}$

Method 2 Multiply the numerator and denominator of the rational expression by the LCD of all the rational expressions involved. In this case, that LCD is $6x$.

$$\frac{5 + \dfrac{x}{3}}{\dfrac{4}{x} - \dfrac{1}{2x}} = \frac{\left(5 + \dfrac{x}{3}\right) \cdot 6x}{\left(\dfrac{4}{x} - \dfrac{1}{2x}\right) \cdot 6x}$$

Multiply the numerator and denominator by $6x$.

$$= \frac{\dfrac{5}{1} \cdot \dfrac{6x}{1} + \dfrac{x}{\overset{}{\underset{1}{3}}} \cdot \dfrac{\overset{2}{6x}}{1}}{\dfrac{4}{\underset{1}{x}} \cdot \dfrac{\overset{1}{6x}}{1} - \dfrac{1}{\underset{1}{2x}} \cdot \dfrac{\overset{3}{6x}}{1}}$$

Distribute the $6x$ to each fraction, and divide out the common factors.

$$= \frac{30x + 2x^2}{24 - 3}$$

Simplify the numerator and denominator.

$$= \frac{30x + 2x^2}{21}$$

YOUR TURN

Simplify $\dfrac{2 - \dfrac{1}{y}}{y - \dfrac{1}{3}}$.

EXAMPLE 3 Simplify $\dfrac{\dfrac{y}{4} - \dfrac{1}{y}}{\dfrac{1}{2} - \dfrac{1}{y}}$.

Method 1 Simplify the numerator and denominator, then divide.

$$\frac{\dfrac{y}{4} - \dfrac{1}{y}}{\dfrac{1}{2} - \dfrac{1}{y}} = \frac{\dfrac{y(y)}{4(y)} - \dfrac{1(4)}{y(4)}}{\dfrac{1(y)}{2(y)} - \dfrac{1(2)}{y(2)}}$$

Write the numerator fractions as equivalent fractions with their LCD, $4y$, and write the denominator fractions with their LCD, $2y$.

$$= \frac{\dfrac{y^2}{4y} - \dfrac{4}{4y}}{\dfrac{y}{2y} - \dfrac{2}{2y}}$$

$$= \frac{\dfrac{y^2 - 4}{4y}}{\dfrac{y - 2}{2y}}$$

Subtract in the numerator and in the denominator.

ANSWER

$$\frac{6y - 3}{3y^2 - y}$$

$$= \frac{y^2 - 4}{4y} \div \frac{y - 2}{2y}$$ Write the complex fraction as a division problem.

$$= \frac{y^2 - 4}{4y} \cdot \frac{2y}{y - 2}$$ Write an equivalent multiplication problem.

$$= \frac{(y + 2)\overset{1}{\cancel{(y - 2)}}}{\underset{2}{\cancel{4y}}} \cdot \frac{\overset{1}{\cancel{2y}}}{\underset{1}{\cancel{y - 2}}}$$ Factor, divide out the common factors, and then multiply the remaining factors.

$$= \frac{y + 2}{2}$$

Method 2 Multiply the numerator and denominator of the rational expression by the LCD of all the rational expressions involved. In this case, that LCD is $4y$.

$$\frac{\dfrac{y}{4} - \dfrac{1}{y}}{\dfrac{1}{2} - \dfrac{1}{y}} = \frac{\left(\dfrac{y}{4} - \dfrac{1}{y}\right) \cdot 4y}{\left(\dfrac{1}{2} - \dfrac{1}{y}\right) \cdot 4y}$$ Multiply the numerator and denominator by $4y$.

$$= \frac{\dfrac{y}{\underset{1}{\cancel{4}}} \cdot \dfrac{\overset{1}{\cancel{4y}}}{1} - \dfrac{1}{\underset{1}{\cancel{y}}} \cdot \dfrac{\overset{1}{\cancel{4y}}}{1}}{\dfrac{1}{\underset{1}{\cancel{2}}} \cdot \dfrac{\overset{2}{\cancel{4y}}}{1} - \dfrac{1}{\underset{1}{\cancel{y}}} \cdot \dfrac{\overset{1}{\cancel{4y}}}{1}}$$ Distribute the $4y$ to each fraction and then divide out the common factors.

$$= \frac{y^2 - 4}{2y - 4}$$ Simplify the numerator and denominator.

$$= \frac{(y + 2)\cancel{(y - 2)}}{2\cancel{(y - 2)}}$$ Factor and then divide out the common factors.

$$= \frac{y + 2}{2}$$

YOUR TURN

Simplify $\dfrac{\dfrac{3}{x} - \dfrac{1}{x^2}}{\dfrac{1}{4} + \dfrac{1}{x}}$.

ANSWER

$$\frac{12x - 4}{x^2 + 4x}$$

1. Identify the numerator and denominator of the complex rational expression

$$\frac{\dfrac{4}{x+3}}{\dfrac{3}{x}}.$$

2. Given the complex rational expression $\dfrac{\dfrac{2x}{x-3}}{4}$, is the numerator $2x$ or

$\dfrac{2x}{x-3}$? How could the expression be written so that the numerator and denominator are easily identified?

3. Write $\dfrac{2x}{x-3} \div \dfrac{4-x}{2x-6}$ as a complex rational expression.

4. Write $\dfrac{5}{x-1} \div 7x$ as a complex rational expression.

5. Which of the two methods discussed in this section would you use to

simplify $\dfrac{\dfrac{x}{6}-\dfrac{3}{4}}{\dfrac{2}{3}+x}$? Why?

6. Which of the two methods discussed in this section would you use to

simplify $\dfrac{\dfrac{x}{7}}{\dfrac{x}{x+1}}$? Why?

For Exercises 7–44, simplify.

7. $\dfrac{\dfrac{3}{4}}{\dfrac{3}{2}}$

8. $\dfrac{\dfrac{2}{3}}{\dfrac{3}{2}}$

9. $\dfrac{\dfrac{m}{n}}{\dfrac{q}{p}}$

10. $\dfrac{\dfrac{a}{b}}{\dfrac{c}{d}}$

11. $\dfrac{\dfrac{1}{2}+\dfrac{1}{3}}{\dfrac{1}{2}-\dfrac{1}{3}}$

12. $\dfrac{\dfrac{2}{3}+\dfrac{1}{4}}{1+\dfrac{1}{2}}$

13. $\dfrac{1+\dfrac{1}{2}}{1-\dfrac{1}{5}}$

14. $\dfrac{4-\dfrac{2}{3}}{\dfrac{1}{2}+\dfrac{3}{7}}$

15. $\dfrac{\dfrac{a}{a^2-b^2}}{\dfrac{b}{a+b}}$

16. $\dfrac{\dfrac{5}{2x-1}}{\dfrac{x}{x+1}}$

17. $\dfrac{\dfrac{x}{4x^2-1}}{\dfrac{5}{2x-1}}$

18. $\dfrac{\dfrac{x^2+4x}{y^2}}{\dfrac{x^2-16}{3y}}$

19. $\dfrac{x+2}{\dfrac{x^2-4}{x}}$

20. $\dfrac{\dfrac{a}{b}-1}{a^2-b^2}$

21. $\dfrac{x+y}{\dfrac{1}{x}+\dfrac{1}{y}}$

22. $\dfrac{4+\dfrac{1}{x}}{3+\dfrac{5}{x}}$

23. $\dfrac{\dfrac{1}{y}-1}{\dfrac{1}{y}+1}$

24. $\dfrac{x-\dfrac{1}{x}}{1+\dfrac{1}{x}}$

25. $\dfrac{2+\dfrac{2}{u-1}}{\dfrac{2}{u-1}}$

26. $\dfrac{1-\dfrac{2}{p}}{1-\dfrac{4}{p^2}}$

27. $\dfrac{r-\dfrac{3}{4}}{\dfrac{1}{2}-r}$

28. $\dfrac{\dfrac{k+2}{k^2-3k}}{\dfrac{k^2-4}{k}}$

29. $\dfrac{\dfrac{a}{a-1}-1}{1+\dfrac{a}{a-1}}$

30. $\dfrac{q-\dfrac{q^2-1}{q}}{1-\dfrac{q-1}{q}}$

31. $\dfrac{\dfrac{1}{w+1} - 1}{\dfrac{1}{w+1}}$

32. $\dfrac{\dfrac{1}{s^2} + \dfrac{1}{t^2}}{\dfrac{2}{st}}$

33. $\dfrac{\dfrac{1}{x-1} - \dfrac{1}{x+1}}{\dfrac{1}{x^2-1}}$

34. $\dfrac{\dfrac{1}{f+2} - \dfrac{1}{f-3}}{1 + \dfrac{1}{f^2-f-6}}$

35. $\dfrac{2 - \dfrac{1}{2y+1}}{\dfrac{1}{2y^2-5y-3} - \dfrac{1}{y-3}}$

36. $\dfrac{1 - \dfrac{3}{x}}{1 - \dfrac{2}{x} - \dfrac{3}{x^2}}$

37. $\dfrac{1 - \dfrac{1}{n}}{n - 2 + \dfrac{1}{n}}$

38. $\dfrac{\dfrac{v^2+v-2}{v^2+4v}}{\dfrac{2v^2-8}{v^2+2v-8}}$

39. $\dfrac{\dfrac{2}{x+h} - \dfrac{2}{x}}{h}$

40. $\dfrac{\dfrac{x}{x-1} - 1}{1 + \dfrac{x}{1-x}}$

41. $\dfrac{\dfrac{m^2+n^2}{m-n} + n}{\dfrac{m^2-n^2}{m+n} + n}$

42. $\dfrac{\dfrac{x^2-2x+xy-2y}{x+y}}{\dfrac{x^2-4x+4}{2}}$

43. $\dfrac{1 + \dfrac{1}{y} - \dfrac{2}{y^2}}{\dfrac{1}{y} + \dfrac{2}{y^2}}$

44. $\dfrac{1 - \dfrac{1}{u^2}}{1 + \dfrac{2}{u} + \dfrac{1}{u^2}}$

For Exercises 45–52, explain the mistake, then find the correct answer.

45. $\dfrac{x - \dfrac{1}{2}}{y - \dfrac{1}{2}} = \dfrac{x}{y}$

46. $\dfrac{\dfrac{2}{3}}{\dfrac{4}{6}} = \dfrac{\overset{1}{\cancel{2}}}{3} \cdot \dfrac{4}{\underset{3}{\cancel{6}}}$

$= \dfrac{4}{9}$

47. $\dfrac{\dfrac{m}{6}}{\dfrac{3}{m}} = \dfrac{\overset{1}{\cancel{m}}}{\underset{2}{\cancel{6}}} \div \dfrac{3}{\underset{1}{\cancel{m}}}$

$= \dfrac{1}{2}$

48. $\dfrac{\dfrac{1}{x} + \dfrac{1}{y}}{x^2 - y^2} = \dfrac{\dfrac{x}{1} \cdot \dfrac{1}{x} + \dfrac{y}{1} \cdot \dfrac{1}{y}}{x \cdot x^2 - y \cdot y^2}$

$= \dfrac{1 + 1}{x^3 - y^3}$

$= \dfrac{2}{x^3 - y^3}$

49. $\dfrac{c + \dfrac{1}{c}}{\dfrac{1}{c}} = \dfrac{c + \dfrac{c}{1} \cdot \dfrac{1}{c}}{\dfrac{c}{1} \cdot \dfrac{1}{c}}$

$= \dfrac{c + 1}{1}$

$= c + 1$

50. $\dfrac{1 + \dfrac{2}{r}}{1 - \dfrac{4}{r^2}} = \dfrac{r \cdot 1 + \dfrac{r}{1} \cdot \dfrac{2}{r}}{r \cdot 1 - \dfrac{r}{1} \cdot \dfrac{4}{r^2}}$

$= \dfrac{r + 2}{r - 4r}$

$= \dfrac{r + 2}{-3r}$

51. $\dfrac{\dfrac{3}{x}}{\dfrac{x}{6}} = \dfrac{3}{x} \div \dfrac{x}{6}$

$= \dfrac{3}{x} \cdot \dfrac{6}{x}$

$= \dfrac{18}{2x}$

$= \dfrac{9}{x}$

52. $\dfrac{\dfrac{1}{x} + \dfrac{1}{y}}{\dfrac{1}{x} - \dfrac{1}{y}} = \dfrac{\dfrac{xy}{1} \cdot \dfrac{1}{x} + \dfrac{xy}{1} \cdot \dfrac{1}{y}}{\dfrac{xy}{1} \cdot \dfrac{1}{x} - \dfrac{xy}{1} \cdot \dfrac{1}{y}}$

$= \dfrac{y + x}{y - x}$

$= -1$

For Exercises 53 and 54, average rate can be found using the formula:

$$\textbf{Average rate} = \frac{\textbf{Total distance}}{\textbf{Total time}}$$

53. Suppose a person travels 10 miles in $\dfrac{1}{3}$ of an hour, then returns by traveling that same 10 miles in $\dfrac{1}{4}$ of an hour. What is that person's average rate?

54. Juan travels 50 miles in $\dfrac{3}{4}$ of an hour and then encounters a traffic jam so that the next 15 miles takes $\dfrac{1}{3}$ of an hour. The last 40 miles of his trip takes him $\dfrac{2}{3}$ of an hour. What was his average rate for the trip?

55. Given the area and length of a rectangle, the width can be found using the formula $w = \dfrac{A}{l}$. Suppose the area of a rectangle is described by $\dfrac{5n - 2}{8}$ square feet and the length is described by $\dfrac{n - 1}{6}$ feet. Find an expression for the width.

56. Given the area and base of a triangle, the height can be found using the formula $h = \dfrac{2A}{b}$. Suppose the area of a triangle is described by $\dfrac{n + 1}{6}$ and its base is described by $\dfrac{5n - 2}{9}$. Find an expression for the height.

57. In electrical circuits, if two resistors R_1 and R_2 are wired in parallel, the resistance of the circuit is found using the complex rational expression

$$\dfrac{1}{\dfrac{1}{R_1} + \dfrac{1}{R_2}}.$$

a. Simplify this complex rational expression.

b. If a 100-ohm and 10-ohm resistor are wired in parallel, find the resistance of the circuit.

c. Two 16-ohm speakers are wired in parallel, find the resistance of the circuit.

58. Suppose three resistors R_1, R_2, and R_3 are wired in parallel. The resistance of the circuit is found using the complex rational expression

$$\dfrac{1}{\dfrac{1}{R_1} + \dfrac{1}{R_2} + \dfrac{1}{R_3}}.$$

a. Simplify this complex rational expression.

b. A 20-ohm, a 10-ohm, and a 40-ohm resistor are wired in parallel. Find the resistance of the circuit.

c. A 300-ohm, a 500-ohm, and a 200-ohm resistor are wired in parallel. Find the resistance of the circuit.

REVIEW EXERCISES

For Exercises 1–4, solve.

1. $-6x = 42$

2. $9t - 8 = 19$

3. $\dfrac{x}{4} = \dfrac{5}{8}$

4. $\dfrac{3}{4}y - \dfrac{1}{3} = \dfrac{1}{6}y + 2$

5. A runner travels 1.2 miles in 15 minutes. Find the runner's average rate in miles per hour.

6. An eastbound car and westbound car pass each other on a highway. The eastbound car is traveling 40 miles per hour and the westbound car is traveling 60 miles per hour. How much time elapses from the time they pass each other until they are 9 miles apart?

7.6 Solving Equations Containing Rational Expressions

OBJECTIVE 1 Solve equations containing rational expressions.

In Section 2.3 we learned that we can eliminate fractions from an equation by multiplying both sides of the equation by their LCD. For example, if given the equation $\frac{1}{2}x - \frac{3}{4} = \frac{2}{3}$, we can eliminate the fractions by multiplying both sides by 12, which is the LCD of $\frac{1}{2}, \frac{3}{4}$, and $\frac{2}{3}$:

$$\overset{6}{\cancel{12}} \cdot \frac{1}{\underset{1}{\cancel{2}}}x - \overset{3}{\cancel{12}} \cdot \frac{3}{\underset{1}{\cancel{4}}} = \overset{4}{\cancel{12}} \cdot \frac{2}{\underset{1}{\cancel{3}}}$$

$$6x - 9 = 8$$

Since the rewritten equation contains only integers, it is easier to solve. Similarly, we can use the LCD to eliminate rational expressions in an equation.

EXAMPLE 1 Solve $\frac{x}{5} = \frac{x}{2} - \frac{1}{4}$.

Solution Eliminate the rational expressions by multiplying both sides by the LCD, 20.

$$20\left(\frac{x}{5}\right) = 20\left(\frac{x}{2} - \frac{1}{4}\right) \qquad \text{Multiply both sides by 20.}$$

$$\overset{4}{\cancel{20}} \cdot \frac{x}{\underset{1}{\cancel{5}}} = \overset{10}{\cancel{20}} \cdot \frac{x}{\underset{1}{\cancel{2}}} - \overset{5}{\cancel{20}} \cdot \frac{1}{\underset{1}{\cancel{4}}} \qquad \begin{array}{l}\text{Use the distributive property to multiply every term} \\ \text{by 20 and then divide out the common factors.}\end{array}$$

$$4x = 10x - 5$$

$$-6x = -5 \qquad \text{Subtract } 10x \text{ on both sides.}$$

$$\frac{-6x}{-6} = \frac{-5}{-6} \qquad \text{Divide both sides by } -6.$$

$$x = \frac{5}{6}$$

Check Using the original equation, we replace x with $\frac{5}{6}$ and see if the equation is true. It is helpful to write $\frac{x}{5}$ as $\frac{1}{5}x$ and $\frac{x}{2}$ as $\frac{1}{2}x$.

$$\frac{1}{5}x = \frac{1}{2}x - \frac{1}{4}$$

$$\frac{1}{\cancel{5}_1}\left(\frac{\cancel{5}^1}{6}\right) \stackrel{?}{=} \frac{1}{2}\left(\frac{5}{6}\right) - \frac{1}{4} \qquad \text{Replace } x \text{ with } \frac{5}{6} \text{ and then simplify.}$$

$$\frac{1}{6} \stackrel{?}{=} \frac{5}{12} - \frac{3}{12}$$

$$\frac{1}{6} = \frac{2}{12} \qquad \text{True, therefore, } \frac{5}{6} \text{ is a solution.}$$

YOUR TURN

Solve $\dfrac{x}{2} = 4 - \dfrac{x}{3}$.

Now consider equations that contain rational expressions with variables in the denominator.

EXAMPLE 2 Solve $\dfrac{1}{2} = \dfrac{3}{x} - \dfrac{1}{4}$.

Solution Eliminate the rational expressions by multiplying both sides by the LCD, $4x$.

$$4x\left(\frac{1}{2}\right) = 4x\left(\frac{3}{x} - \frac{1}{4}\right) \qquad \text{Multiply both sides by } 4x.$$

$$\cancel{4}^2 x \cdot \frac{1}{\cancel{2}_1} = 4\cancel{x}^1 \cdot \frac{3}{\cancel{x}_1} - \cancel{4}^1 x \cdot \frac{1}{\cancel{4}_1} \qquad \begin{array}{l}\text{Use the distributive property to multiply every term}\\\text{by } 4x \text{ and then divide out the common factors.}\end{array}$$

$$2x = 12 - x$$

$$3x = 12 \qquad \text{Add } x \text{ to both sides.}$$

$$x = 4 \qquad \text{Divide both sides by 3.}$$

ANSWER

$$\frac{24}{5}$$

Check We will leave the check to the reader.

ANSWER

$$\frac{24}{11}$$

YOUR TURN

Solve $\dfrac{1}{x} + \dfrac{3}{8} = \dfrac{5}{6}$.

Extraneous Solutions

If an equation contains rational expressions with any variables in the denominators, we must make sure that its solutions do not cause any of the expressions to be undefined. In Example 2, if the solution had turned out to be 0, then the expression $\frac{3}{x}$ would be undefined. By inspecting the denominators of each rational expression, you can determine the values that would cause the expressions to be undefined before solving the equation. If by solving an equation, you obtain a number that causes an expression in the equation to be undefined, then we say that number is an *extraneous solution* and we discard it.

PROCEDURE SOLVING EQUATIONS CONTAINING RATIONAL EXPRESSIONS

To solve an equation that contains rational expressions:

1. Eliminate the rational expressions by multiplying both sides of the equation by their LCD.
2. Solve the equation using the methods we learned in Chapters 2 (linear equations) and 6 (quadratic equations).
3. Check your solution(s) in the original equation. Discard any extraneous solutions.

EXAMPLE 3 Solve $\dfrac{x}{x-3} = \dfrac{2}{5}$.

Solution Notice that if $x = 3$, then $\dfrac{x}{x-3}$ is undefined, so the solution cannot be 3. To solve, we first multiply both sides by the LCD, $5(x-3)$.

$$5(\overset{1}{\cancel{x-3}}) \cdot \frac{x}{\underset{1}{x-3}} = \overset{1}{\cancel{5}}(x-3) \cdot \frac{2}{\underset{1}{\cancel{5}}}$$ Divide out the common factors.

$$5x = (x-3)2$$

$$5x = 2x - 6$$ Distribute 2 to clear the parentheses.

$$3x = -6$$ Subtract $2x$ on both sides.

$$x = -2$$ Divide both sides by 3.

Check We will leave the check to the reader.

CONNECTION

The equation in Example 3 is a proportion. We could have solved the equation by cross multiplying, which leads to the same equation as in the second step of the solution in Example 3.

$$5x = (x-3)2$$

Warning: You can use this method only if the equation is a proportion.

In Examples 1 through 3, after eliminating the rational expressions, we were left with linear equations. Now let's look at some equations that transform to quadratic equations after we use the LCD to clear the rational expressions. Remember that when solving quadratic equations, we manipulate the equation so that one side is 0, then factor and use the zero-factor theorem.

EXAMPLE 4 Solve $\dfrac{3}{x} - 5 = x - 3$.

Solution Notice that because of the x in the denominator, the solution cannot be 0.

$$x\left(\frac{3}{x} - 5\right) = x(x - 3) \qquad \text{Multiply both sides by the LCD, } x.$$

$$\overset{1}{\cancel{x}} \cdot \frac{3}{\underset{1}{\cancel{x}}} - x \cdot 5 = x \cdot x - x \cdot 3 \qquad \text{Distribute } x, \text{ then divide out common factors.}$$

$$3 - 5x = x^2 - 3x \qquad \begin{array}{l}\text{Since this equation is quadratic, we will set it} \\ \text{equal to 0 by subtracting 3 and adding } 5x \text{ to} \\ \text{both sides.}\end{array}$$

$$0 = x^2 - 3x + 5x - 3$$
$$0 = x^2 + 2x - 3 \qquad \text{Combine like terms.}$$
$$0 = (x + 3)(x - 1) \qquad \text{Factor}$$

$$x + 3 = 0 \quad \text{or} \quad x - 1 = 0 \qquad \text{Use the zero-factor theorem.}$$
$$x = -3 \qquad\qquad x = 1$$

Check Verify that both -3 and 1 make the original equation true.

$$\frac{3}{-3} - 5 \overset{?}{=} -3 - 3 \qquad\qquad \frac{3}{1} - 5 \overset{?}{=} 1 - 3$$

$$-1 - 5 \overset{?}{=} -6 \qquad\qquad\qquad 3 - 5 \overset{?}{=} -2$$

$$-6 = -6 \quad \text{True} \qquad\qquad\quad -2 = -2 \quad \text{True}$$

Both -3 and 1 are solutions.

EXAMPLE 5 Solve $\dfrac{x^2}{x+2} - 3 = \dfrac{x+6}{x+2} - 4$.

Solution By inspecting the denominators, notice that the solution cannot be -2.

$$(x+2)\left(\dfrac{x^2}{x+2} - 3\right) = (x+2)\left(\dfrac{x+6}{x+2} - 4\right)$$

Multiply both sides by the LCD, $x+2$.

$$(\cancel{x+2})^1 \cdot \dfrac{x^2}{\cancel{x+2}_1} - (x+2) \cdot 3 = (\cancel{x+2})^1 \cdot \dfrac{x+6}{\cancel{x+2}_1} - (x+2) \cdot 4$$

Distribute $x+2$ and then divide out the common factors.

$$x^2 - 3x - 6 = x + 6 - 4x - 8$$
$$x^2 - 3x - 6 = -3x - 2$$

Combine like terms.

$$x^2 - 4 = 0$$

Since the equation is quadratic, we set it equal to 0 by adding $3x$ and 2 to both sides.

$$(x+2)(x-2) = 0$$

Factor

$$x + 2 = 0 \quad \text{or} \quad x - 2 = 0$$

Use the zero-factor theorem.

$$x = -2 \qquad\qquad x = 2$$

Check We have already noted that -2 cannot be a solution because it causes an expression in the equation to be undefined, so -2 is extraneous. Therefore, we only need to check 2.

$$\dfrac{2^2}{2+2} - 3 \overset{?}{=} \dfrac{2+6}{2+2} - 4$$

In the original equation, replace x with 2 and then simplify.

$$\dfrac{4}{4} - 3 \overset{?}{=} \dfrac{8}{4} - 4$$

$$1 - 3 \overset{?}{=} 2 - 4$$

$$-2 = -2$$

True, so 2 is a solution and the only solution.

YOUR TURN

Solve $\dfrac{3x+4}{x-4} = \dfrac{x^2}{x-4} + 3$.

EXAMPLE 6 Solve $\dfrac{5}{x-3} = \dfrac{x}{x-2} + \dfrac{x}{x^2 - 5x + 6}$.

Solution To determine the LCD of all the denominators, we first factor quadratic form $x^2 - 5x + 6$.

Note: Inspecting the denominators after factoring, we see that neither 2 nor 3 can be a solution.

$$\dfrac{5}{x-3} = \dfrac{x}{x-2} + \dfrac{x}{(x-2)(x-3)}$$

Factor the denominator $x^2 - 5x + 6$.

$$(x-2)(x-3)\left(\dfrac{5}{x-3}\right) = (x-2)(x-3)\left(\dfrac{x}{x-2} + \dfrac{x}{(x-2)(x-3)}\right)$$

Multiply both sides by the LCD, $(x-2)(x-3)$.

ANSWER

-4

$$(x - 2)(x - 3) \cdot \frac{5}{(x - 3)} = (x - 2)(x - 3) \cdot \frac{x}{(x - 2)}$$

$$+ (x - 2)(x - 3) \cdot \frac{x}{(x - 2)(x - 3)}$$

Distribute $(x - 2)(x - 3)$, then divide out the common factors.

$$(x - 2)5 = (x - 3)x + x$$

$$5x - 10 = x^2 - 3x + x \qquad \text{Distribute to clear the parentheses.}$$

$$5x - 10 = x^2 - 2x \qquad \text{Combine like terms.}$$

$$0 = x^2 - 7x + 10$$

Since the equation is quadratic, we set it equal to 0 by subtracting $5x$ and adding 10 on both sides.

$$0 = (x - 2)(x - 5) \qquad \text{Factor}$$

$$x - 2 = 0 \quad \text{or} \quad x - 5 = 0 \qquad \text{Use the zero-factor theorem.}$$
$$x = 2 \qquad\qquad x = 5$$

Check In the original equation, note that $\dfrac{x}{x - 2}$ is undefined when $x = 2$, so 2 is an extraneous solution. We will leave the check for 5 to the reader.

YOUR TURN

Solve $\dfrac{2}{x + 5} - \dfrac{6}{x^3 + 5x^2} = \dfrac{1}{x}$.

Warning: Be sure you understand the difference between performing operations with rational expressions and solving equations containing rational expressions. For example, $\dfrac{5}{x} + \dfrac{1}{3}$ is an *expression* that means two rational expressions are to be added. Remember, we cannot solve expressions, we can only evaluate or rewrite them. To rewrite this expression, we begin by writing each rational expression with the LCD, $3x$.

$$\frac{5}{x} + \frac{1}{3} = \frac{5(3)}{x(3)} + \frac{1(x)}{3(x)}$$

$$= \frac{15}{3x} + \frac{x}{3x}$$

$$= \frac{15 + x}{3x}$$

To solve an *equation* like $\dfrac{3}{x} + \dfrac{1}{4} = \dfrac{1}{2}$, we first eliminate the rational expressions by multiplying both sides of the equation by the LCD, $4x$.

$$4x\left(\frac{3}{x} + \frac{1}{4}\right) = 4x\left(\frac{1}{2}\right),$$

$$12 + x = 2x$$

$$12 = x \qquad \text{Subtract } x \text{ on both sides.}$$

1. Explain how to find the LCD of an equation that contains rational expressions.

2. What is the LCD for the expressions in $\dfrac{x}{x^2 + 4x - 5} = \dfrac{5}{x} - 2x$?

3. What is accomplished by multiplying both sides of an equation by the LCD of the rational expressions in the equation?

4. What is an extraneous solution? Why might an equation with rational expressions lead to one or more extraneous solutions?

5. What values might be extraneous solutions in the equation $\dfrac{3}{x - 4} = \dfrac{5x}{x + 2}$? Why?

6. What values might be extraneous solutions in the equation $\dfrac{x}{x^2 + 4x - 5} = \dfrac{5}{x} - 2x$? Why?

For Exercises 7–12, check the given values to see if they are solutions to the equation.

7. $\dfrac{3}{x} + \dfrac{2}{3} = \dfrac{5}{6};\ x = 9$

8. $\dfrac{3}{4} - \dfrac{1}{x} = \dfrac{2}{3};\ x = 12$

9. $\dfrac{4}{m} - \dfrac{2}{5} = \dfrac{3}{4m};\ m = \dfrac{65}{8}$

10. $\dfrac{5}{6n} + \dfrac{1}{4} = \dfrac{2}{3n};\ n = -\dfrac{2}{3}$

11. $\dfrac{1}{t + 2} + \dfrac{1}{4} = \dfrac{t}{16};\ t = -4$

12. $\dfrac{1}{y - 3} - \dfrac{5}{6} = -\dfrac{y}{12};\ y = -7$

For Exercises 13–60, solve and check.

13. $\dfrac{2n}{3} + \dfrac{3n}{2} = \dfrac{13}{3}$

14. $\dfrac{3x}{5} - \dfrac{x}{2} = \dfrac{19}{5}$

15. $3y - \dfrac{4y}{5} = 22$

16. $\dfrac{3t}{4} - 2 = \dfrac{t}{4}$

17. $\dfrac{r - 7}{2} - 1 = \dfrac{r + 9}{9} + \dfrac{1}{3}$

18. $\dfrac{1}{3}(d + 5) - 4 = \dfrac{d - 8}{4} - \dfrac{1}{2}$

19. $\dfrac{5}{6w} + 3 = \dfrac{1}{2w}$

20. $\dfrac{3}{2u} - \dfrac{1}{3} = \dfrac{5}{6u}$

21. $\dfrac{4}{x} + \dfrac{3}{2x} = \dfrac{11}{6}$

22. $\dfrac{2x + 1}{2x} = \dfrac{7}{12} + \dfrac{3x + 2}{3x}$

23. $\dfrac{4}{2t} + \dfrac{3}{5} - 3 = -\dfrac{2}{5t}$

24. $\dfrac{y - 2}{y + 3} = \dfrac{3}{8}$

25. $\dfrac{x}{x - 3} = \dfrac{6}{3}$

26. $\dfrac{1}{2} = \dfrac{-3}{q + 5}$

27. $\dfrac{5}{20} = \dfrac{x - 7}{x + 2}$

28. $\dfrac{4}{2d + 9} = \dfrac{3}{4d - 8}$

29. $\dfrac{6}{y - 2} = \dfrac{5}{y - 3}$

30. $\dfrac{2x^2}{3x - 4} + \dfrac{x}{3} = \dfrac{9x - 2}{9}$

31. $\dfrac{6f - 5}{6} = \dfrac{2f - 1}{2} - \dfrac{f + 2}{2f + 5}$

32. $\dfrac{m + 5}{m - 3} - 5 = \dfrac{4}{m - 3}$

33. $3 + \dfrac{9 - 2x}{8x} = \dfrac{5}{2x}$

34. $\dfrac{3}{y^2 - 4} = \dfrac{-2}{5y + 10}$

35. $\dfrac{2}{x^2 + x - 2} - \dfrac{1}{x + 2} = \dfrac{3}{x - 1}$

36. $\dfrac{-1}{3x^2 + 9x} = \dfrac{1}{x + 3} - \dfrac{x - 3}{3x}$

37. $\dfrac{5}{2x} + \dfrac{15}{2x + 16} = \dfrac{6}{x + 8}$

38. $\dfrac{5}{t^2 - 9} = \dfrac{3}{t + 3} - \dfrac{2}{t - 3}$

39. $\dfrac{3m + 1}{m^2 - 9} - \dfrac{m + 3}{m - 3} = \dfrac{1 - 5m}{m + 3}$

40. $\dfrac{2y^2 - 4y}{y + 5} - \dfrac{y}{y + 5} = \dfrac{5y - 10}{6}$

41. $\dfrac{1}{u - 4} + \dfrac{2}{u^2 - 16} = \dfrac{3}{u + 4}$

42. $\dfrac{m + 2}{m - 2} - \dfrac{2}{m + 2} = -\dfrac{14}{6}$

43. $\dfrac{5w + 3}{6w^2 + 3w} = \dfrac{w - 1}{2w + 1}$

44. $\dfrac{3}{r - 2} - \dfrac{4}{r - 3} = -\dfrac{6}{r^2 - 5r + 6}$

45. $\dfrac{2k}{k + 7} - 1 = \dfrac{1}{k^2 + 10k + 21} + \dfrac{k}{k + 3}$

46. $\dfrac{p - 5}{p + 5} - 2 + \dfrac{p + 15}{p - 5} = \dfrac{-25}{p^2 - 25}$

47. $\dfrac{1}{n^2 - 1} - \dfrac{n - 1}{n + 1} + \dfrac{n + 1}{n - 1} = 0$

48. $\dfrac{3a}{a^2 - 2a - 15} - \dfrac{a}{a + 3} = \dfrac{2a}{a - 5}$

49. $\dfrac{5}{5c^2 - c - 20} = \dfrac{3}{3c^2 - 3c - 28}$

50. $\dfrac{u^2 + 2}{u^2 + u - 2} - \dfrac{3u}{u + 2} = \dfrac{-2u - 1}{u - 1}$

51. $\dfrac{7n}{n^2 - 2n - 8} - \dfrac{4n}{n^2 - 5n + 4} = \dfrac{3}{n - 4}$

52. $1 - \dfrac{3}{x - 2} = -\dfrac{12}{x^2 - 4}$

53. $\dfrac{5}{h + 5} - \dfrac{2}{h^2 + 2h - 15} = \dfrac{2}{h - 3}$

54. $\dfrac{2z}{2z - 3} - \dfrac{3z}{2z + 3} = \dfrac{15 - 32z^2}{4z^2 - 9}$

55. $\dfrac{1}{a - 3} - \dfrac{6}{a^2 - 1} = \dfrac{12}{a^3 - 3a^2 - a + 3}$

56. $\dfrac{5}{2s^2 + s} - \dfrac{2}{2s + 1} = 1$

57. $\dfrac{9r}{r^2 - 4r} - \dfrac{r - 7}{2r^2 - 7r - 4} = \dfrac{3r}{2r^2 + r}$

58. $\dfrac{y + 2}{y^2 - 5y + 6} + \dfrac{2}{y^2 - 2y - 3} = \dfrac{-2y}{y^2 - y - 2}$

59. $\dfrac{1}{d^2 - 5d + 6} + \dfrac{d - 2}{d^2 - d - 6} - \dfrac{d}{d^2 - 4} = 0$

60. $\dfrac{6}{b + 2} - \dfrac{3}{b^2 - 1} - \dfrac{6}{b^3 + 2b^2 - b - 2} = 0$

For Exercises 61–62, explain the mistake, then find the correct solution(s).

61.
$$\frac{y}{3} + \frac{y}{y-3} = \frac{3}{y-3}$$

$$3(y-3) \cdot \frac{y}{3} + 3(y-3) \cdot \frac{y}{y-3} = 3(y-3) \cdot \frac{3}{y-3}$$

$$y^2 - 3y + 3y = 9$$

$$y^2 - 9 = 0$$

$$(y+3)(y-3) = 0$$

$$y = -3, 3$$

62.
$$\frac{x}{x+3} + \frac{3}{x+3} = 0$$

$$(x+3) \cdot \frac{x}{x+3} + (x+3) \cdot \frac{3}{x+3} = (x+3) \cdot 0$$

$$x + 3 = 0$$

$$x = -3$$

63. Explain how the LCD is used differently in $\dfrac{3x}{4} - \dfrac{5}{6}$ and in $\dfrac{3x}{4} - \dfrac{5}{6} = \dfrac{x}{2}$.

64. Explain how the LCD is used differently in $\dfrac{5x}{8} + \dfrac{1}{3}$ and in $\dfrac{5x}{8} + \dfrac{1}{3} = \dfrac{x}{6}$.

For Exercises 65–68, use the formula for the total resistance R in an electrical circuit with resistors R_1, R_2, R_3, …, R_n that are wired in parallel.

$$\frac{1}{R} = \frac{1}{R_1} + \frac{1}{R_2} + \frac{1}{R_3} + \cdots \frac{1}{R_n}$$

65. Two resistors are wired in parallel, one of which is 100 ohms. If the total resistance is to be 80 ohms, what must be the value of the other resistor?

66. Three resistors are wired in parallel, two of which are equal in value. If the third resistor is 200 ohms and the total resistance is to be 20 ohms, what are the values of the other two resistors?

67. Two resistors are wired in parallel. One resistor is to have a value that is 10 ohms more than the other resistor and the total resistance is to be 12 ohms. Find the value of both resistors.

68. Two resistors are wired in parallel. One resistor is to have a value that is 20 ohms less than the other resistor and the total resistance is to be 24 ohms. Find the value of both resistors.

In optics, a lens can be used to bend light from an object through a focal point to reproduce an inverted image that is larger (or smaller) than the original object. The following formula describes how the image of an object is affected by a lens where o represents the object's distance from the lens, i represents the image's distance from the lens, and f represents the focal length of the lens.

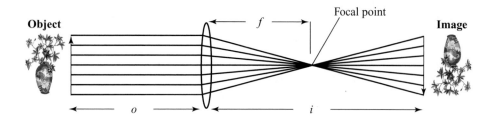

Formula: $\dfrac{1}{o} + \dfrac{1}{i} = \dfrac{1}{f}$

69. An object is 40 feet from a lens with a focal length of 2 feet. What will be the image's distance from the lens?

70. An object is placed 50 centimeters from your eye. Your lens focuses the image of the object onto your retina, which is about 2.5 centimeters from your lens. What is the focal length of your lens?

71. A lens is being designed so that the focal length will be 5 millimeters less than the the image length. Suppose an object is placed 60 millimeters from the lens. What are the focal length and image length?

72. A lens is being designed so that the image length is 9 millimeters more than the the focal length. Suppose an object is placed 40 millimeters from the lens. What are the focal length and image length?

REVIEW EXERCISES

1. Juanita has some quarters and four more nickels. If she has a total of $4.10, how many of each coin does she have?

2. Katlin is seven years younger than her brother is. If the sum of their ages is 39, how old are they?

3. A rectangular eraser has a perimeter of 36 millimeters. If the length is three less than twice the width, find the length and width.

4. The sum of two numbers is 75. Eight times the smaller number is the same as twice the larger number. Find the numbers.

5. If the sum of two number is 23 and their difference is 9, find the numbers.

6. Two numbers add to 116, and one of the number exceeds the other by 36. What are the numbers?

7.7 Applications with Rational Expressions

OBJECTIVES

1. Use tables to solve problems with two unknowns involving rational expressions.
2. Solve problems involving direct variation.
3. Solve problems involving inverse variation.

OBJECTIVE 1 **Use tables to solve problems with two unknowns involving rational expressions.** In Chapter 3 we used tables to organize information involving two unknown amounts. Let's now look at some similar problems that lead to equations containing rational expressions.

Problems Involving Work

Tables are helpful in problems involving two or more people (or machines) working together to complete a task. In these problems, we would be given each person's rate of work and asked to find the time for them to complete the task if they work together. For each person involved, the rate of work, time at work, and amount of the task completed are related as follows:

$$\boxed{\text{Person's rate of work}} \cdot \boxed{\text{Time at work}} = \boxed{\text{Amount of the task completed by that person}}$$

Since the people are working together, the sum of their individual amounts of the task completed equals the whole task.

$$\boxed{\text{Amount completed by one person}} + \boxed{\text{Amount completed by the other person}} = \boxed{\text{Whole task}}$$

EXAMPLE 1 Karen and Jeff own a cleaning business. Karen can clean an average-size house in about 2 hours. Jeff can clean a similar house in 3 hours. How long would it take them to clean an average-size house working together?

Understand Karen cleans at a rate of 1 house in 2 hours, or $\frac{1}{2}$ of a house per hour. Jeff cleans at a rate of 1 house in 3 hours, or $\frac{1}{3}$ of a house per hour.

Categories	Rate of Work	Time at Work	Amount of Task Completed
Karen	$\dfrac{1}{2}$	t	$\dfrac{1}{2}t$ or $\dfrac{t}{2}$
Jeff	$\dfrac{1}{3}$	t	$\dfrac{1}{3}t$ or $\dfrac{t}{3}$

Because they are working together on the same job for the same amount of time, we let t represent that amount of time.

Multiplying the rate of work and the time at work gives an expression of the amount of work completed. For example, if Karen works for 4 hours at a rate of $\frac{1}{2}$ of a house every hour, she can clean 2 houses.

The total job in this case is 1 house, so we can write an equation that combines their individual expressions for the task completed, and set this sum equal to 1 house.

Plan and Execute   Karen's amount completed + Jeff's amount completed = 1 house

$$\frac{t}{2} + \frac{t}{3} = 1$$

$$6\left(\frac{t}{2} + \frac{t}{3}\right) = 6(1) \qquad \text{Multiply both sides by the LCD, 6.}$$

$$\overset{3}{\cancel{6}} \cdot \frac{t}{\underset{1}{\cancel{2}}} + \overset{2}{\cancel{6}} \cdot \frac{t}{\underset{1}{\cancel{3}}} = 6 \qquad \text{Distribute and then divide out common factors.}$$

$$3t + 2t = 6$$

$$5t = 6 \qquad \text{Combine like terms.}$$

$$t = \frac{6}{5} \qquad \text{Divide both sides by 5.}$$

Answer   Working together, it takes Karen and Jeff $\dfrac{6}{5}$ or $1\dfrac{1}{5}$ hours to clean an average-size house.

Check   Karen cleans $\dfrac{1}{2}$ of a house per hour, so if she works alone $1\dfrac{1}{5}$ hours, she cleans $\dfrac{1}{2} \cdot \dfrac{6}{5} = \dfrac{3}{5}$ of a house. Jeff cleans $\dfrac{1}{3}$ of a house per hour, so in $1\dfrac{1}{5}$ hours, he cleans $\dfrac{1}{3} \cdot \dfrac{6}{5} = \dfrac{2}{5}$ of a house. Combining their individual amounts, we see that in $1\dfrac{1}{5}$ hours, they clean $\dfrac{3}{5} + \dfrac{2}{5} = \dfrac{5}{5} = 1$ house.

YOUR TURN

Terrell owns a landscaping company. Every Monday he services the same group of clients, and it takes him 5 hours to complete the service. On one occasion he hired a student, Jarod, to service those same clients, and it took him 6 hours. How much time would it take them to service those clients working together?

Motion Problems

Recall that the formula for calculating distance, given the rate of travel and time of travel, is $d = rt$. If we isolate r, we have $r = \dfrac{d}{t}$. If we isolate t, we have $t = \dfrac{d}{r}$. These equations suggest that we will use rational expressions when describing rate or time.

EXAMPLE 2 Warren runs 3 miles away from the gym, turns around and runs an average of 2 miles per hour faster on the return trip. If the total time of his run is $1\dfrac{1}{4}$ hours, what is his speed in the outbound leg and the inbound leg of his run?

Understand We are to find the rates for each leg of Warren's run. Because this situation involves two different rates, we will use a table to organize the distance, rate, and time of each leg of the run.

Categories	Distance	Rate	Time
Outbound	3 miles	r	$\dfrac{3}{r}$
Inbound	3 miles	$r + 2$	$\dfrac{3}{r + 2}$

Warren runs 2 mph faster during the inbound leg of his run, so we let r represent the outbound rate and add 2 mph to r for the inbound rate.

Since $d = rt$, to describe time, we divide the distance by the rate: $t = \dfrac{d}{r}$

ANSWER

$2\dfrac{8}{11}$ hours

Because the total time of the trip is $1\dfrac{1}{4}$ (or $\dfrac{5}{4}$) hours, we can write an equation that is the sum of the outbound and inbound times.

Plan and Execute   Outbound time + Inbound time = $\dfrac{5}{4}$ hours

$$\frac{3}{r} + \frac{3}{r+2} = \frac{5}{4}$$

$$4r(r+2)\left(\frac{3}{r} + \frac{3}{r+2}\right) = 4r(r+2)\left(\frac{5}{4}\right)$$

Multiply both sides by the LCD, $4r(r+2)$.

$$4r(r+2) \cdot \frac{3}{r} + 4r(r+2) \cdot \frac{3}{r+2} = 4r(r+2) \cdot \frac{5}{4}$$

Distribute, then divide out common factors.

$$12(r+2) + 12r = 5r(r+2)$$

$$12r + 24 + 12r = 5r^2 + 10r$$

Distribute

$$24r + 24 = 5r^2 + 10r$$

Combine like terms.

$$0 = 5r^2 - 14r - 24$$

Subtract $24r$ and 24 from both sides.

$$0 = (5r + 6)(r - 4)$$

Factor

$$5r + 6 = 0 \quad \text{or} \quad r - 4 = 0$$

Use the zero-factor theorem.

$$5r = -6 \qquad\qquad r = 4$$

$$r = -\frac{6}{5}$$

Answer   Though $-\dfrac{6}{5}$ is a solution to the equation, a negative rate does not make sense in this situation, so we do not consider it as an answer. Therefore the rate of the outbound leg is 4 mph. Since the rate of the inbound leg is 2 mph faster, then it must be 6 mph.

Check   Verify that traveling 3 miles at 4 mph, then 3 miles at 6 mph takes a total of $1\dfrac{1}{4}$ hours.

$$\text{Outbound time} = \frac{3}{4} \text{ hour} \qquad \text{Inbound time} = \frac{3}{6} = \frac{1}{2} \text{ hour}$$

$$\text{Total time} = \frac{3}{4} + \frac{1}{2} = \frac{3}{4} + \frac{2}{4} = \frac{5}{4} = 1\frac{1}{4} \text{ hour}$$

YOUR TURN

A plane travels 1200 miles against the jet stream, causing its airspeed to be decreased by 20 miles per hour. On the return flight, the plane travels with the jet stream so that its airspeed is increased by 20 miles per hour. If the total flight time of the round trip is $6\dfrac{1}{3}$ hours, what would be the plane's rate in still air?

OBJECTIVE ② **Solve problems involving direct variation.**   Suppose a vehicle travels at a constant rate of 30 miles per hour. Using the formula $d = rt$, with r replaced by 30, we have $d = 30t$. In the table on page 626, we use $d = 30t$ to determine distances for various values of time.

ANSWER

380 mph

Time t	Distance Traveled $d = 30t$
1 hour	30 miles
2 hours	60 miles
3 hours	90 miles

From the table above we see that as the time of travel increases, so does the distance traveled. Or, more formally, as values of t increase, so do values of d. In $d = 30t$, the two variables, d and t, are said to be in **direct variation** or are *directly proportional*.

> **DEFINITION**
>
> **Direct variation:** Two variables y and x are in direct variation if $y = kx$, where k is a constant.

In words, direct variation is written as "y varies directly as x" or "y is directly proportional to x" and these phrases translate to $y = kx$. Often the constant, k, is not given in problems involving variation, so the first objective is to find its value.

EXAMPLE 3 Suppose y varies directly as x. When $y = 9$, $x = 4$. Find y when $x = 7$.

Solution Translating "y varies directly as x," we have $y = kx$. We need to find the value of the constant k. We replace y with 9 and x with 4 in $y = kx$, then solve for k.

$$9 = k \cdot 4 \qquad \text{Replace } y \text{ with 9 and } x \text{ with 4.}$$
$$2.25 = k \qquad \text{Divide both sides by 4.}$$

Now, we can replace k with 2.25 in $y = kx$ so that we have $y = 2.25x$. We can now use this equation to find y when $x = 7$.

$$y = 2.25(7) = 15.75$$

EXAMPLE 4 The distance a vehicle can travel varies directly with the amount of fuel it carries. A family traveled 207 miles in their van using 9 gallons of fuel. How many gallons are required to travel 368 miles?

Understand Translating "the distance a vehicle travels varies directly with the amount of fuel," we write $d = kf$, where d represents distance and f represents the amount of fuel.

Plan Use $d = kf$, replacing d with 207 miles and f with 9 gallons, in order to solve for the value of k. Then use that value in $d = kf$ to solve for the number of gallons required to travel 368 miles.

Execute $207 = k \cdot 9$

$\qquad$ $23 = k$ $\qquad$ **Divide both sides by 9.**

◀ **CONNECTION** ─────

Consider the units of measurement. Because miles $= k \cdot$ gallons, when we isolate k, we have:

$k = \dfrac{\text{miles}}{\text{gallon}}$. The constant k represents the miles per gallon of this van, which is 23 miles per gallon. The number of miles a vehicle can travel per gallon of fuel is also called its *fuel economy*.

Replacing k with 23 in $d = kf$, we have $d = 23f$. We use $d = 23f$ to solve for f when d is 368 miles.

$$368 = 23f$$
$$16 = f \qquad \text{**Divide both sides by 23.**}$$

Answer   To travel 368 miles, the van needs 16 gallons of fuel.

Check   If the van's fuel economy is 23 miles per gallon, then with 16 gallons, the van would travel $23(16) = 368$ miles.

YOUR TURN

At a meat market, the price per pound, or unit price, of hamburger is constant. The amount you pay is directly proportional to quantity purchased. Juan notes that a 2.5-pound package is priced at \$3.58. If he wishes to buy 8 pounds of hamburger, how much will it cost?

OBJECTIVE ③ **Solve problems involving inverse variation.**   Suppose we know that the distance to a particular place is 60 miles. Given a rate, we can calculate the travel time to that place using the formula $t = \dfrac{d}{r}$. For example, if we travel at a rate of 10 miles per hour, it will take $\dfrac{60 \text{ mi}}{10 \text{ mph}} = 6$ hours to drive the 60 miles. In the following table, we use $t = \dfrac{60}{r}$ to see the relationship between rate and travel time for a fixed distance of 60 miles.

ANSWER

\$11.46

Rate	Travel time
r	$t = \dfrac{60}{r}$
10 mph	6 hours
20 mph	3 hours
30 mph	2 hours

From the table on page 627, we see that as we increase speed, the travel time decreases. More formally, as values of r increase, values of t decrease. In $t = \dfrac{60}{r}$, the two variables, t and r, are said to be in **inverse variation** or are *inversely proportional.*

DEFINITION

Inverse variation: Two variables, y and x, are in inverse variation if $y = \dfrac{k}{x}$, where k is a constant.

In words, inverse variation is written as "y varies inversely as x" or "y is inversely proportional to x," and these phrases translate to $y = \dfrac{k}{x}$.

EXAMPLE 5 A technician applies a constant voltage to a circuit that has a variable resistor. Varying the resistance causes the current to vary inversely. At a resistance of 4 ohms, the current is measured to be 8 amperes. Find the current when the resistance is 12 ohms.

Understand Because the current and resistance vary inversely, we can write $I = \dfrac{k}{R}$, where I represents current, R represents the resistance, and k represents the constant voltage.

Plan Use the fact that when the resistance is 4 ohms, the current is 8 amperes to determine the value of the constant k. Then use the constant to find the current when the resistance is 12 ohms.

Execute
$$8 = \frac{k}{4}$$

$$4 \cdot 8 = \overset{1}{\cancel{4}} \cdot \frac{k}{\underset{1}{\cancel{4}}} \qquad \text{Multiply both sides by 4 to eliminate the 4 in the denominator.}$$

$$32 = k \qquad \boxed{\textbf{Note:} \text{ This means that the voltage in the circuit is 32 volts.}}$$

Replacing k with 32 in $I = \dfrac{k}{R}$, we have: $I = \dfrac{32}{R}$. We use $I = \dfrac{32}{R}$ to solve for I when R is 12 ohms.

$$I = \frac{32}{12}$$

$$I = 2\frac{2}{3} \text{ or } 2.\overline{6}$$

Answer With a resistance of 12 Ω, the current is $2\dfrac{2}{3}$ or $2.\overline{6}$ A.

Check Recall from the exercises in Section 2.3 (page 145) that the formula for voltage is $V = ir$. Using this formula we can verify that a current of $2\frac{2}{3}$ A with a resistance of 12 Ω yields a voltage of 32 V.

$$V = \left(2\frac{2}{3}\right)12 = 32$$

YOUR TURN

If the wavelength of a wave remains constant, then the velocity, v, of a wave is inversely proportional to its period, T. In an experiment, waves are created in a fluid so that the period is 5 seconds and the velocity is 6 centimeters per second. If the period is increased to 8 seconds, what is the velocity?

ANSWER

3.75 cm/sec.

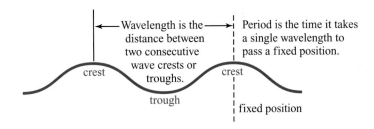

Wavelength is the distance between two consecutive wave crests or troughs.

Period is the time it takes a single wavelength to pass a fixed position.

crest   crest

trough

fixed position

7.7 EXERCISES

For Extra Help

InterAct Math

Tutor Center   Addison-Wesley Math Tutor Center

Video 13/DVT 4

MyMathLab

MathXL   MathXL

1. If a person can complete a task in x hours, write an expression that describes the portion of the task she completes in 1 hour.

2. If x represents the amount of time one person takes to complete a task, and y represents the amount of time for a second person to complete the same task, write an expression that describes the portion of the task completed-while working together.

3. If a vehicle travels a distance of 100 miles at a particular rate, *r*, write an expression that describes the amount of time it takes the vehicle to travel that 100 miles.

4. A cyclist travels 10 miles at an average rate, *r*, then travels 5 miles per hour faster during the return trip. Write an expression that describes the total time of the trip.

5. If two quantities vary directly, then as one of the quantities increases, what happens to the other?

6. If two quantities are inversely proportional, then as one quantity increases, what happens to the other?

For Exercises 7–22, use a table to organize the information, then solve.

7. Juan can paint a 3600-square-foot house in 12 days, and Carissa can do the same job in only 8 days. How long will it take them to paint a 3600-square-foot house if they work together?

8. After a large snowfall, Franklin can shovel the snow from the sidewalk in 6 minutes and Shawnna can do it 8 minutes. How long will it take them to clear the sidewalk if they work together?

9. The Smith children are responsible for mowing the lawn. If it takes Anna 20 minutes and Christina 30 minutes, assuming they have two lawn mowers, how long will it take them to mow the lawn working together?

10. A large field is being cleared to begin construction on a new library. To remove the dirt, two bulldozers are being used. The larger one could do the job in 6 days, while the smaller one would take 9 days. Working together, how long should the clearing take?

11. When filling a new swimming pool, the builder knows that it will take 6 hours using the smaller pipe. The larger pipe can do the job in only 4 hours. How long will it take the builder to fill the swimming pool if both pipes are used?

12. Karmin is planting vincas for her summer garden. Last summer, it took her 4 hours to complete the job. Her sister Kim planted the fall garden in only 3 hours. If the sisters work together, how long will the planting take?

13. To earn money for college, Peter and Susan address wedding invitations using calligraphy. Peter can complete a job of 150 invitations in 6 hours and Susan can complete the same number in 5 hours. Working together, how long will it take them to address 150 invitations?

14. Normally it takes Professor Gunther 3 hours to grade a class of final exam essays. If his graduate assistant grades the essays, it takes her 5 hours. How long will it take them to grade the essays together?

15. During an afternoon practice, two track team members took the same amount of time to run 18 miles and 12 miles, respectively. If one of the track team members ran 2 miles per hour faster than the other, find the rate of each team member.

16. If an airplane travels 1050 miles in the same amount of time that an automobile travels 150 miles, and the speed of the airplane is 50 mph more than six times the speed of the car, how fast is each moving?

17. A jet flies 90 miles with a 20 mph tailwind in the same amount of time as it takes to fly against the wind 80 miles. What would the jet's speed be in still air?

18. The river current is 8 miles per hour. A jet ski can travel 34 miles downstream in the same time that it takes to go 18 miles upstream. What is the speed of the jet ski with no current?

19. A car can travel 75 miles in the same amount of time that it takes a moped to travel 25 miles. If the moped is traveling at 40 mph less than the car, what is the rate of each?

20. The Quad Cities offer tours on the Mississippi River. If the *Quad City Queen* riverboat travels upstream 120 miles and then returns, and the river current is measured at 4 miles per hour, what is the speed of the *Quad City Queen* if the total travel time is 22.5 hours?

21. After training for a year, a cyclist discovers that she has increased her average rate by 4 miles per hour and can travel 15 miles in the same time that it used to take her to travel 12 miles. What was her old average rate? What is her new average rate?

22. During target practice, it is determined that a heavily armed bomber flying from base to a target 600 miles away required the same flight time as it did to fly 720 miles at a rate 30 miles per hour faster. What was the rate for each trip?

23. If y varies directly as x and $y = 45$ when $x = 9$, what is the value of y when $x = 6$?

24. If a varies directly as b and $a = 15$ when $b = 27$, what is a when $b = 54$?

25. If x varies directly as y and $x = 4$ when $y = 1$, what is the value of y when $x = 18$?

26. If m varies directly as n and $m = 18$ when $n = 4$, what is the value of n when $m = 20$?

27. At the Piggly Wiggly, the cost of London broil is constant. The purchase price increases with the quantity purchased. Darlene notes a 1.5-pound package is priced at $4.29. If she wished to buy 6 pounds of London broil, how much will it cost?

28. At Kroger, the cost of corn is constant. The purchase price increases with the quantity purchased. Uviv notes that 5 ears of corn cost $0.99. If she wished to buy 8 ears of corn, how much will it cost?

29. The distance a car can travel varies directly with the amount of gas it carries. On the highway, a 2001 Jeep Wrangler travels 112 miles using 7 gallons of fuel. How many gallons are required to travel 240 miles?

30. The distance a car can travel varies directly with the amount of gas it carries. On the highway, a 2001 BMW M3 convertible travels 64 miles using 4 gallons of fuel. How many gallons are required to travel 144 miles?

31. The weight of an object is directly proportional to its mass. A person who weighs 161 pounds has a mass of 5 slugs. Find the weight of a person with a mass of 4.5 slugs.

32. The weight of an object is directly proportional to its mass. An object with a mass of 55 kilograms has a weight of 539 newtons. Find the weight of a person with a mass of 60 kilograms.

33. On a stringed instrument, such as a guitar, the frequency of the vibrating string is inversely proportional to the string's length. Suppose the length of a string on a guitar is 25.5 inches with a frequency of 440 vibrations per second (vps). Now, if we shorten the string to a length of 17 inches by placing a finger on the fret board, what will be the frequency of the string?

34. The frequency of the vibrating string on a stringed instrument is directly proportional to the tension of the string. Suppose the tension of a string is 300 pounds with a frequency of 440 vibrations per second (vps). Now, if we increase the string tension to 340 pounds by turning one of the tuning keys, what will be the frequency of the string?

35. The volume of a gas is directly proportional to temperature. Suppose a balloon containing air has a volume of 525 cubic inches at a temperature of 75°F. If the temperature drops to 60°F, what will be the volume of the balloon?

36. In physics, the energy, E, of a photon of light is directly proportional to the frequency of the light, represented by ν (pronounced "nu"). A "hard" X-ray with a frequency of 2×10^{19} hertz has energy of 1.325×10^{-14} joules. What energy would a gamma ray with a frequency of 4×10^{22} hertz have?

37. If y varies inversely as x and $y = 4$ when $x = 10$, what is y when x is 8?

38. If m varies inversely as n and m is 6 when n is 1.5, what is m when n is 9?

39. If a varies inversely as b and b is 4 when a is 7, what is a when b is 2?

40. If x varies inversely as y and $y = 5$ when $x = 2$, what is y when x is 5?

41. The pressure that a gas exerts against the walls of a container is inversely proportional to the volume of the container. A gas is inside a cylinder with a piston at one end that can vary the volume of the cylinder. When the volume of the cylinder is 60 cubic inches, the pressure inside is 20 pounds per square inch (psi). Find the pressure of the gas if the piston compresses the gas to a volume of 40 cubic inches.

20 psi at
60 cubic inches

Compressed to
40 cubic inches

42. Find the pressure in the cylinder from Exercise 41 if the piston compresses the gas to a volume of 20 cubic inches.

43. A constant voltage is applied to a circuit that has a variable resistor. Varying the resistance causes the current to vary inversely. At a resistance of 6 ohms, the current is measured to be 10 amperes. Find the current when the resistance is 15 ohms.

44. Find the current of the circuit in Exercise 43 when the resistance is 20 ohms.

45. The weight of an object is inversely proportional to the square of the distance of the object from the center of Earth. At the surface of Earth, the space shuttle weighs 4.5 million pounds. It is then 4000 miles from the center of Earth. How much will the space shuttle weigh when it is 50 miles from the surface of Earth?

46. The intensity of light is inversely proportional to the square of the distance from the source. A light meter 6 feet from a lightbulb measures the intensity to be 16 foot-candles. What is the intensity at 20 feet from the bulb?

47. If the wavelength of a wave remains constant, then the velocity, v, of a wave is inversely proportional to its period, T. In an experiment, waves are created in a fluid so that the period is 7 seconds and the velocity is 4 centimeters per second. If the period is increased to 12 seconds, what is the velocity?

48. In quantum physics, the de Broglie wavelength of an object, represented by λ (pronounced "lambda"), is inversely proportional to the product of mass and velocity (momentum) of the object. A particle with a mass of 10^{-7} grams is moving at a velocity of 10^6 centimeters per second and has a de Broglie wavelength of 6.6×10^{-26} centimeters. Find the de Broglie wavelength of a particle with a mass of 2.5×10^{-6} grams moving at a velocity of 5×10^7 centimeters per second.

49. The following table lists data from a business of the number of units it supplied to the market and the corresponding price per unit.

Supply (in thousands)	Price per Unit
150	$1.25
175	$1.50
200	$2.00
250	$2.75

a. Plot the ordered pairs of data as points in the coordinate plane. Connect the points to form a graph.

b. Are supply and price directly proportional or inversely proportional? Explain.

c. Does the data represent a function? Explain.

50. The following table lists data for a business of the demand for its product and the price per unit.

Demand (in thousands)	Price per Unit
150	$2.25
175	$1.50
200	$1.00
250	$0.75

a. Plot the ordered pairs of data as points in the coordinate plane. Connect the points to form a graph.

b. Are demand and price directly proportional or inversely proportional? Explain.

c. Does the data represent a function? Explain.

PUZZLE PROBLEM

Assume that a plane travels 800 kilometers with no wind resistance. If it takes the same amount of time to travel 720 kilometers against the wind as it does to fly 880 kilometers with the wind, what is the wind speed?

Collaborative Exercise

Discovering π

This exercise requires three different sized circular objects. You can use various sized cans, a CD, drinking cups, and so on. You will also need a tape measure.

1. Measure the circumference and diameter of each object. Record your measurements in a table.
   Tip: The diameter will be the greatest distance across the circle. To measure the diameter, hold one end of the tape measure on one edge of the circle, then slowly sweep the other end across the other edge. As you sweep the distance will increase to a maximum value, which is the diameter, then decrease.

2. The circumference of a circle is directly proportional to its diameter. Find k for each of your circles and record your values in your table. Your table should look like this:

Object	Circumference	Diameter	k

continued

3. What do you notice about k? What does this indicate about circles?

4. Find the circumference of a circle with a diameter of 50 feet.

REVIEW EXERCISES

1. Determine if $(3, -1)$ is a solution to the equation $-2x + 3y = -1$.

2. Find the x- and y-intercepts of $y - 4x = 6$.

3. Graph: $y = -3x - 5$

4. If $f(x) = 4x^2 - 3x + 2$, find $f(2 - x)$.

5. If $f(x) = x^2 - 2x + 1$, find $f\left(-\dfrac{3}{4}\right)$.

6. Find two consecutive integers so that twice the first plus four times the second is equal to 34.

Defined Terms

Section 7.1
Rational expression (p. 554)

Section 7.5
Complex rational expression
(p. 599)

Section 7.7
Direct variation (p. 626)
Inverse variation (p. 628)

Procedures, Rules, and Key Examples

Procedures/Rules	Key Example(s)

Section 7.1 Simplifying Rational Expressions

To determine the value(s) that make a rational expression undefined, find the value(s) that cause the denominator to be 0.
1. Write an equation that has the denominator set equal to zero.
2. Solve the equation.

Example 1: Find every value for the variable that makes the expression undefined.

a. $\dfrac{5x}{x-6}$

Solution: $x - 6 = 0$
$$x = 6$$

b. $\dfrac{n}{n^2 - 4n - 5}$

Solution: $n^2 - 4n - 5 = 0$
$$(n-5)(n+1) = 0$$
$$n - 5 = 0 \quad \text{or} \quad n + 1 = 0$$
$$n = 5 \qquad\qquad n = -1$$

To simplify a rational expression to lowest terms:
1. Write the numerator and denominator in factored form.
2. Divide out all common factors in the numerator and denominator.
3. Multiply the remaining factors in the numerator and the remaining factors in the denominator.

Example 2: Simplify.

a. $\dfrac{18x^3y}{24x^2yz} = \dfrac{2 \cdot 3 \cdot 3 \cdot x \cdot x \cdot x \cdot y}{2 \cdot 2 \cdot 2 \cdot 3 \cdot x \cdot x \cdot y \cdot z}$

$$= \dfrac{3 \cdot x}{2 \cdot 2 \cdot z}$$

$$= \dfrac{3x}{4z}$$

b. $\dfrac{5m^2 - 15m}{m^2 - 9} = \dfrac{5 \cdot m \cdot (m-3)}{(m+3)(m-3)}$

$$= \dfrac{5m}{m+3}$$

Procedures/Rules	Key Example(s)

Section 7.2 Multiplying and Dividing Rational Expressions

To multiply rational expressions:
1. Write each numerator and denominator in factored form.
2. Divide out any numerator factor with any matching denominator factor.
3. Multiply numerator by numerator and denominator by denominator.
4. Simplify as needed.

Example 1: Multiply.

a.
$$-\frac{8u^3}{3t} \cdot \frac{9t^2}{10u^3} =$$

$$-\frac{2 \cdot 2 \cdot 2 \cdot u \cdot u \cdot u}{3 \cdot t} \cdot \frac{3 \cdot 3 \cdot t \cdot t}{2 \cdot 5 \cdot u \cdot u \cdot u}$$

$$= -\frac{2 \cdot 2}{1} \cdot \frac{3 \cdot t}{5}$$

$$= -\frac{12t}{5}$$

b.
$$\frac{x^2 + 4x}{x^2 - 3x - 10} \cdot \frac{3x + 6}{12x}$$

$$= \frac{x \cdot (x + 4)}{(x + 2) \cdot (x - 5)} \cdot \frac{3 \cdot (x + 2)}{2 \cdot 2 \cdot 3 \cdot x}$$

$$= \frac{(x + 4)}{(x - 5)} \cdot \frac{1}{2 \cdot 2}$$

$$= \frac{x + 4}{4(x - 5)} \text{ or } \frac{x + 4}{4x - 20}$$

To divide rational expressions:
1. Write an equivalent multiplication statement with the reciprocal of the divisor.
2. Write each numerator and denominator in factored form. (Steps 1 and 2 are interchangeable.)
3. Divide out any numerator factor with any matching denominator factor.
4. Multiply numerator by numerator and denominator by denominator.
5. Simplify as needed.

Example 2: Divide.

a.
$$\frac{3n^3}{14m} \div \frac{12mn^2}{7m}$$

$$= \frac{3n^3}{14m} \cdot \frac{7m}{12mn^2}$$

$$= \frac{3 \cdot n \cdot n \cdot n}{2 \cdot 7 \cdot m} \cdot \frac{7 \cdot m}{2 \cdot 2 \cdot 3 \cdot m \cdot n \cdot n}$$

$$= \frac{n}{2} \cdot \frac{1}{2 \cdot 2 \cdot m}$$

$$= \frac{n}{8m}$$

b.
$$\frac{r - 4}{3r^2 - 17r + 10} \div \frac{r^2 - 3r - 4}{3r^2 + r - 2}$$

$$= \frac{r - 4}{3r^2 - 17r + 10} \cdot \frac{3r^2 + r - 2}{r^2 - 3r - 4}$$

$$= \frac{r - 4}{(3r - 2) \cdot (r - 5)} \cdot \frac{(3r - 2) \cdot (r + 1)}{(r - 4) \cdot (r + 1)}$$

$$= \frac{1}{r - 5}$$

To convert units using dimensional analysis, multiply the given measurement by conversion factors so that the undesired units divide out, leaving the desired units.

Example 3: Convert.

a. 20 feet to inches

$$20 \text{ ft.} = \frac{20 \text{ ft.}}{1} \cdot \frac{12 \text{ in.}}{1 \text{ ft.}} = 240 \text{ in.}$$

continued

Procedures/Rules	Key Example(s)

Key Example(s)

b. 6 miles to yards

$$6 \text{ mi.} = \frac{6 \text{ mi.}}{1} \cdot \frac{5280 \text{ ft.}}{1 \text{ mi.}} \cdot \frac{1 \text{ yd.}}{3 \text{ ft.}} = 10{,}560 \text{ yd.}$$

Section 7.3 Adding and Subtracting Rational Expressions with the Same Denominator

To add or subtract rational expressions that have the same denominator:
1. Add or subtract the numerators and keep the same denominator.
2. Simplify to lowest terms (remember to write the numerators and denominators in factored form in order to simplify).

Example 1: Add or subtract.

a.
$$\frac{7y}{12} + \frac{y}{12} = \frac{7y + y}{12}$$
$$= \frac{8y}{12}$$
$$= \frac{2y}{3}$$

b.
$$\frac{u}{u + 2} - \frac{3}{u + 2} = \frac{u - 3}{u + 2}$$

c.
$$\frac{7x^2 + 5x}{x + 1} - \frac{x + 3}{x + 1}$$
$$= \frac{(7x^2 + 5x) - (x + 3)}{x + 1}$$
$$= \frac{(7x^2 + 5x) + (-x - 3)}{x + 1}$$
$$= \frac{7x^2 + 4x - 3}{x + 1}$$
$$= \frac{(7x - 3)(x + 1)}{x + 1}$$
$$= 7x - 3$$

Section 7.4 Adding and Subtracting Rational Expressions with Different Denominators

To find the LCD of two or more rational expressions:
1. Factor each denominator.
2. For each unique factor, compare the number of times it appears in each factorization. Write a factored form that includes each factor the greatest number of times it appears in the denominator factorizations.

Example 1: Find the LCD.

a.
$$\frac{5}{6t} \text{ and } \frac{1}{8t^2}$$
$$6t = 2 \cdot 3 \cdot t$$
$$8t^2 = 2^3 \cdot t^2$$
$$\text{LCD} = 24t^2$$

b.
$$\frac{2}{5t - 15} \text{ and } \frac{t}{t^2 - 9}$$
$$5t - 15 = 5(t - 3)$$
$$t^2 - 9 = (t - 3)(t + 3)$$
$$\text{LCD} = 5(t - 3)(t + 3)$$

To add or subtract rational expressions with different denominators:
1. Find the LCD.
2. Write each rational expression as an equivalent expression with the LCD.
3. Add or subtract the numerators and keep the LCD.
4. Simplify.

Example 2: Add or subtract.

a.
$$\frac{5}{6t} + \frac{1}{8t^2} = \frac{5(4t)}{6t(4t)} + \frac{1(3)}{8t^2(3)}$$
$$= \frac{20t}{24t^2} + \frac{3}{24t^2}$$
$$= \frac{20t + 3}{24t^2}$$

continued

Procedures/Rules	Key Example(s)

Section 7.4 (continued)

b. $\dfrac{2}{5t-15} - \dfrac{t}{t^2-9}$

$= \dfrac{2}{5(t-3)} - \dfrac{t}{(t-3)(t+3)}$

$= \dfrac{2(t+3)}{5(t-3)(t+3)} - \dfrac{t(5)}{(t-3)(t+3)(5)}$

$= \dfrac{2t+6}{5(t-3)(t+3)} - \dfrac{5t}{5(t-3)(t+3)}$

$= \dfrac{2t+6-5t}{5(t-3)(t+3)}$

$= \dfrac{-3t+6}{5(t-3)(t+3)}$

Section 7.5 Complex Rational Expressions

To simplify a complex rational expression, use one of the following methods:

Method 1
1. Simplify the numerator and denominator if needed.
2. Rewrite as a horizontal division problem.

Example 1: Simplify $\dfrac{\dfrac{5t}{t+1}}{\dfrac{10}{t^2-1}}$.

$\dfrac{\dfrac{5t}{t+1}}{\dfrac{10}{t^2-1}} = \dfrac{5t}{t+1} \div \dfrac{10}{t^2-1}$

$= \dfrac{5t}{t+1} \cdot \dfrac{t^2-1}{10}$

$= \dfrac{5 \cdot t}{t+1} \cdot \dfrac{(t+1)(t-1)}{2 \cdot 5}$

$= \dfrac{t}{1} \cdot \dfrac{t-1}{2}$

$= \dfrac{t^2-t}{2}$

Method 2
1. Multiply the numerator and denominator of the complex rational expression by their LCD.
2. Simplify.

Example 2: Simplify $\dfrac{\dfrac{5}{6} - \dfrac{t}{2}}{\dfrac{t}{4} + \dfrac{1}{3}}$.

$\dfrac{\dfrac{5}{6} - \dfrac{t}{2}}{\dfrac{t}{4} + \dfrac{1}{3}} = \dfrac{12\left(\dfrac{5}{6} - \dfrac{t}{2}\right)}{12\left(\dfrac{t}{4} + \dfrac{1}{3}\right)}$

$= \dfrac{\dfrac{\overset{2}{\cancel{12}}}{1} \cdot \dfrac{5}{\underset{1}{\cancel{6}}} - \dfrac{\overset{6}{\cancel{12}}}{1} \cdot \dfrac{t}{\underset{1}{\cancel{2}}}}{\dfrac{\overset{3}{\cancel{12}}}{1} \cdot \dfrac{t}{\underset{1}{\cancel{4}}} + \dfrac{\overset{4}{\cancel{12}}}{1} \cdot \dfrac{1}{\underset{1}{\cancel{3}}}}$

$= \dfrac{10 - 6t}{3t + 4}$

continued

Procedures/Rules	Key Example(s)

Section 7.6 Solving Equations Containing Rational Expressions

To solve an equation that contains rational expressions:

1. Eliminate the rational expressions by multiplying both sides of the equation by their LCD.
2. Solve the equation using the methods we learned in Chapters 2 (linear equations) and 6 (quadratic equations).
3. Check your solution(s) in the original equation. Discard any extraneous solutions.

Example 1: Solve:

$$\frac{x^2 - 6}{x - 3} + 2 = \frac{x^2 - 2x}{x - 3} - x.$$

Note that x cannot be 3 because it would cause the denominators to be 0, making those expressions undefined.

$$(x - 3)\left(\frac{x^2 - 6}{x - 3} + 2\right)$$

$$= (x - 3)\left(\frac{x^2 - 2x}{x - 3} - x\right)$$

$$(x - 3) \cdot \frac{x^2 - 6}{x - 3} + (x - 3) \cdot 2$$

$$= (x - 3) \cdot \frac{x^2 - 2x}{x - 3} - (x - 3) \cdot x$$

$$x^2 - 6 + 2x - 6 = x^2 - 2x - x^2 + 3x$$

$$x^2 + 2x - 12 = x$$

$$x^2 + x - 12 = 0$$

$$(x - 3)(x + 4) = 0$$

$$x - 3 = 0 \quad \text{or} \quad x + 4 = 0$$

$$x = 3 \qquad\qquad x = -4$$

3 is extraneous. Check –4 by substituting into the original equation:

$$\frac{(-4)^2 - 6}{(-4) - 3} + 2 \overset{?}{=} \frac{(-4)^2 - 2(-4)}{(-4) - 3} - (-4)$$

$$\frac{16 - 6}{-7} + 2 \overset{?}{=} \frac{16 + 8}{-7} + 4$$

$$-\frac{10}{7} + 2 \overset{?}{=} -\frac{24}{7} + 4$$

$$-\frac{10}{7} + \frac{14}{7} \overset{?}{=} -\frac{24}{7} + \frac{28}{7}$$

$$\frac{4}{7} = \frac{4}{7}$$

True, therefore -4 is the solution.

continued

Procedures/Rules	Key Example(s)

Section 7.7 Applications with Rational Expressions

If y varies directly as x, then we translate to an equation $y = kx$, where k is a constant.

Example 1: The price of photocopies is directly proportional to the number of copies produced. If 20 copies costs \$1.40, then how much will 150 copies cost?

Direct variation means $y = kx$. In this case the price of photocopies is y and the number of copies is x. We can use the fact that 20 copies costs \$1.40 in $y = kx$ to find k.

$$1.40 = k(20)$$
$$\frac{1.40}{20} = \frac{20k}{20}$$
$$0.07 = k$$

Now use $y = kx$ again with $k = 0.07$ to solve for y when $x = 150$ copies.

$$y = (0.07)(150)$$
$$y = \$10.50$$

If y varies inversely as x, then we translate to an equation $y = \dfrac{k}{x}$, where k is a constant.

Example 2: Given that $y = 20$ when $x = 8$, if y varies inversely as x, then find x when $y = 9$.

Inverse variation means $y = \dfrac{k}{x}$. We can use the fact that $y = 20$ and $x = 8$ to find k.

$$20 = \frac{k}{8}$$
$$8(20) = 8\left(\frac{k}{8}\right)$$
$$160 = k$$

Now use $y = \dfrac{k}{x}$ again with $k = 160$ to solve for x when $y = 9$.

$$9 = \frac{160}{x}$$
$$x(9) = x\left(\frac{160}{x}\right)$$
$$\frac{9x}{9} = \frac{160}{9}$$
$$x = 17\frac{7}{9}$$

For Exercises 1–5, answer true or false.

1. The rational expression $\dfrac{2x - 1}{x}$ is undefined when $x = \dfrac{1}{2}$.

2. $\dfrac{2x}{3} + \dfrac{x}{3} = \dfrac{3x}{3} = x$

3. A fraction is in lowest terms when the GCF of the numerator and denominator is 1.

4. $\dfrac{2 - x}{x - 2} = -1$

5. $-x + 5 = -1(x - 5)$

For Exercises 6–10, complete the rule.

6. To divide rational expressions:

   1. Write an equivalent _____ statement with the reciprocal of the divisor.

   2. Write each numerator and denominator in factored form.

      (Steps 1 and 2 are interchangeable.)

   3. _____ out any numerator factor with any matching denominator factor.

   4. Multiply numerator by numerator and denominator by denominator.

   5. Simplify as needed.

7. To solve an equation that contains rational expressions:

   1. Eliminate the rational expressions by multiplying through by their _____.

   2. Solve the equation using the methods we learned in Chapters 2 (linear equations) and 6 (quadratic equations).

   3. Check your solution(s) in the original equation.

8. To add or subtract rational expressions that have the same denominator:

   1. Add or subtract the _____ and keep the same _____.

2. Simplify to lowest terms (remember to write the numerators and denominators in factored form in order to reduce).

9. A _____ rational expression is a rational expression that contains rational expressions in the numerator or denominator.

10. One method of simplifying a complex rational expression is the following:

1. Simplify the _____ and _____ of the complex rational expression.

2. Divide.

For Exercises 11–12, evaluate the rational expression.

11. $\dfrac{4x + 5}{2x}$ when   **a.** $x = 3$ **b.** $x = -1.3$ **c.** $x = -2$

12. $\dfrac{5x}{x + 2}$ when   **a.** $x = 3$ **b.** $x = -3$ **c.** $x = 1.2$

For Exercises 13–16, find every value that can replace the variable in the expression and cause the expression to be undefined.

13. $\dfrac{9y}{5 - y}$ **14.** $\dfrac{2x}{x^2 - 4}$ **15.** $\dfrac{6}{x^2 + 5x - 6}$ **16.** $\dfrac{2x + 3}{x^2 + 6x}$

For Exercises 17–24, simplify to lowest terms.

17. $\dfrac{3(x + 5)}{9}$ **18.** $\dfrac{2y + 12}{3y + 18}$ **19.** $\dfrac{x^2 + 6x}{4x + 24}$ **20.** $\dfrac{xy - y^2}{3x - 3y}$

21. $\dfrac{a^2 - b^2}{a^2 - 2ab + b^2}$ **22.** $\dfrac{3x + 9}{x^4 - 81}$ **23.** $\dfrac{x^2 - y^2}{y - x}$ **24.** $\dfrac{1 - w}{w^2 + 2w - 3}$

For Exercises 25–32, multiply.

25. $\dfrac{x}{y} \cdot \dfrac{7x}{4y}$ **26.** $\dfrac{2n^2}{8m^2n} \cdot \dfrac{24mp}{10n^2p^5}$ **27.** $\dfrac{6}{2m + 4} \cdot \dfrac{3m + 6}{15}$ **28.** $\dfrac{x^2 + 3x}{x^2 - 16} \cdot \dfrac{x^2 - 4x}{x^2 - 9}$

29. $\dfrac{n^2 - 6n + 9}{n^2 - 9} \cdot \dfrac{n^2 + 4n + 3}{n^2 - 3n}$

30. $\dfrac{4r^2 + 4r + 1}{r + 2r^2} \cdot \dfrac{2r}{2r^2 - r - 1}$

31. $\dfrac{4m}{m^2 - 2m + 1} \cdot \dfrac{m^2 - 1}{16m^2}$

32. $\dfrac{w^2 - 2w - 24}{w^2 - 16} \cdot \dfrac{(w + 5)(w + 4)}{w^2 - w - 30}$

For Exercises 33–40, divide.

33. $\dfrac{y^2}{ab} \div \dfrac{x}{y}$

34. $\dfrac{7y^2}{10b^2} \div \dfrac{21y^2}{25b^2}$

35. $\dfrac{x^2 - y^2}{c^2 - d^2} \div \dfrac{x - y}{c + d}$

36. $\dfrac{b^3 - 6b^2 + 8b}{6b} \div \dfrac{2b - 4}{10b + 40}$

37. $\dfrac{v + 1}{3} \div \dfrac{3v + 3}{18}$

38. $\dfrac{3j + 2}{5j^2 - j} \div \dfrac{6j^2 + j - 2}{10j^2 + 3j - 1}$

39. $\dfrac{u^2 - 2u - 8}{u^2 + 3u + 2} \div (u^2 - 3u - 4)$ **40.** $\dfrac{2a + 4}{5} \div \dfrac{4a + 8}{25a}$

For Exercises 41–48, add or subtract. Simplify all answers.

41. $\dfrac{5x}{18} + \dfrac{x}{18}$

42. $\dfrac{2r}{r + 3} + \dfrac{r}{r + 3}$

43. $\dfrac{2x + 5}{x + 1} - \dfrac{x + 4}{x + 1}$

44. $\dfrac{2x}{x + y} - \dfrac{x - y}{x + y}$

45. $\dfrac{9x^2 - 24x + 16}{2x + 1} + \dfrac{2x^2 + 3x - 9}{2x + 1}$

46. $\dfrac{q^2 + 2q}{q - 7} - \dfrac{12q - 21}{q - 7}$

47. $\dfrac{2r - 5s}{r^2 - s^2} - \dfrac{2r - 6s}{r^2 - s^2}$

48. $\dfrac{8m + 5}{6m} - \dfrac{2m + 5}{6m}$

For Exercises 49–54, find the least common denominator and write equivalent rational expressions with the LCD.

49. $\dfrac{2}{4c}, \dfrac{3}{3c^3}$

50. $\dfrac{5}{xy^3}, \dfrac{3}{x^2y^2}$

51. $\dfrac{4}{m - 1}, \dfrac{4y}{m + 1}$

52. $\dfrac{x}{x^2 - 1}, \dfrac{2 + x}{x + 1}$

53. $\dfrac{2}{p^2 - 4}, \dfrac{p}{p^2 + 4p + 4}$

54. $\dfrac{3}{4h - 8}, \dfrac{5}{h^2 - 2h}$

For Exercises 55–62, add or subtract as indicated.

55. $\dfrac{3a - 4}{18} - \dfrac{2a + 5}{12}$

56. $\dfrac{4}{x + 3} - \dfrac{5}{x - 3}$

57. $\dfrac{3}{y - 2} + \dfrac{5}{y - 4}$

58. $\dfrac{5}{x^2 - 4} + \dfrac{4}{x + 2}$

59. $\dfrac{6t}{(t - 3)^2} - \dfrac{3t}{2t - 6}$

60. $\dfrac{a + 6}{a^2 + 7a + 12} - \dfrac{a - 3}{a + 3}$

61. $\dfrac{10}{2y - 1} - \dfrac{5}{1 - 2y}$

62. $\dfrac{3}{3x - 12} + \dfrac{15}{x^2 - 16}$

For Exercises 63–70, simplify.

63. $\dfrac{\dfrac{a}{b}}{\dfrac{x}{y}}$

64. $\dfrac{\dfrac{1}{3} - \dfrac{1}{2}}{\dfrac{1}{2} - \dfrac{1}{3}}$

65. $\dfrac{\dfrac{5}{3x + 5}}{\dfrac{x}{x - 2}}$

66. $\dfrac{\dfrac{x + 3}{x^2 - 9}}{x}$

67. $\dfrac{\dfrac{x}{4x^2 - 1}}{\dfrac{5}{2x + 1}}$

68. $\dfrac{r - \dfrac{r^2 - 1}{r}}{1 - \dfrac{r - 1}{r}}$

69. $\dfrac{\dfrac{1}{x^2} + \dfrac{1}{y^2}}{\dfrac{7}{xy}}$

70. $\dfrac{\dfrac{1}{h + 1} - 1}{\dfrac{1}{h + 1}}$

For Exercises 71–78, solve and check.

71. $\dfrac{3x}{2} - \dfrac{x}{5} = \dfrac{19}{5}$

72. $\dfrac{3t}{6} - 2 = \dfrac{t}{6}$

73. $\dfrac{2x + 1}{4x} = \dfrac{5}{12} + \dfrac{3x - 2}{3x}$

74. $\dfrac{y + 5}{y - 3} - 5 = \dfrac{4}{y - 3}$

75. $\dfrac{5}{4} = \dfrac{x + 1}{x - 2}$

76. $\dfrac{6g - 5}{6} = \dfrac{2g - 1}{2} - \dfrac{g + 2}{2g + 5}$

77. $\dfrac{5}{x^2 - 16} = \dfrac{3}{x + 4} - \dfrac{2}{x - 4}$

78. $\dfrac{3}{r - 2} - \dfrac{4}{r + 3} = -\dfrac{6}{r^2 + r - 6}$

For Exercises 79–84, solve.

79. After giving a math test, Yolonda can grade 20 tests in 60 minutes and Shawn can do it 68 minutes. How long will it take them to grade the tests if they work together?

80. During an afternoon practice, two football team members took the same amount of time to run sprints of 40 yards and 50 yards, respectively. If one of the football team members ran 2 yards per second faster than the other, find the rate of each team member.

81. If a varies directly as b and $a = 15$ when $b = 27$, what is a when $b = 54$?

82. If x varies inversely as y and $y = 5$ when $x = 2$, what is y when x is 5?

83. The distance a car can travel varies directly with the amount of gas it carries. On the highway, a car travels 80 miles using 4 gallons of fuel. How many gallons are required to travel 200 miles?

84. A constant voltage is applied to a circuit that has a variable resister. Varying the resistance causes the current to vary inversely. At a resistance of 8 ohms, the current is measured to be 12 amperes. Find the current when the resistance is 20 ohms.

For Exercises 1–2, find the value(s) that can replace the variable in the expression and cause the expression to be undefined.

1. $\dfrac{2x}{x-7}$

2. $\dfrac{8-m}{m^2-16}$

For Exercises 3–4, simplify to lowest terms.

3. $\dfrac{12-3x}{x^2-8x+16}$

4. $\dfrac{x-y}{x^2-y^2}$

For Exercises 5–6, find the least common denominator and write equivalent rational expressions with the LCD.

5. $\dfrac{6}{fg^3}, \dfrac{2}{f^2}$

6. $\dfrac{5x}{x^2-9}, \dfrac{2}{x^2+6x+9}$

For Exercises 7–18, perform the indicated operation.

7. $\dfrac{4a^2b^3}{15x^3y} \cdot \dfrac{25x^5y}{16ab}$

8. $\dfrac{9x^3y^4}{16ab^2} \div \dfrac{45x^5y^2}{14a^7b^9}$

9. $\dfrac{4ab-8b}{x^2} \div \dfrac{2b-ab}{x^3}$

10. $\dfrac{12x^2-6x}{x^2+6x+5} \cdot \dfrac{2x^2+10x}{4x^2-1}$

11. $\dfrac{2x}{2x+3} + \dfrac{5x}{2x+3}$

12. $\dfrac{x}{x^2-9} - \dfrac{3}{x^2-9}$

13. $\dfrac{6}{x^2-4} + \dfrac{x-3}{x^2-2x}$

14. $\dfrac{4}{x-1} + \dfrac{5}{x+2}$

15. $\dfrac{5}{x-4} - \dfrac{2}{x+1}$

16. $\dfrac{x}{2x+4} - \dfrac{2}{x^2+2x}$

17. $\dfrac{2+\dfrac{1}{y}}{3-\dfrac{1}{y}}$

18. $\dfrac{\dfrac{x-y}{2}}{\dfrac{x^2-y^2}{4}}$

For Exercises 19–22, solve the equation.

19. $\dfrac{3x - 1}{4} - \dfrac{7}{6} = \dfrac{2}{3}$

20. $\dfrac{4}{5m - 1} = \dfrac{2}{2m - 1}$

21. $\dfrac{y}{y + 2} - \dfrac{y + 6}{y^2 - 4} + \dfrac{2}{y - 2} = 0$

22. $\dfrac{g}{g - 1} = \dfrac{8}{g + 2}$

For Exercises 23–25, solve.

23. If Erin can paint a room in 3 hours and her husband can paint the same room in 2, how long will it take them to paint the room together?

24. Jake runs 3 miles away from the gym, turns around, and runs an average of 2 miles per hour slower on the return trip. If the total time of his run is $1\frac{1}{4}$ hours, what was his speed in each the outbound leg and the inbound leg of his run?

25. The weight of an object is directly proportional to its mass. Suppose an object with a mass of 40 kilograms weighs 392 newtons. How much would an object with a mass of 54 kilograms weigh?

CHAPTER 8

More on Inequalities, Absolute Value, and Functions

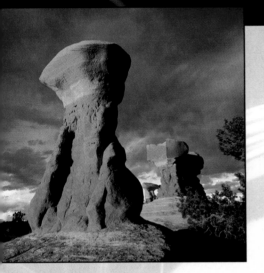

In this chapter we will build upon what we have learned about inequalities, absolute value, and functions. We have solved linear inequalities in which only a single inequality symbol is involved. Now we will learn to solve compound inequalities, which involve two inequality symbols. We will then learn to solve absolute value equations and inequalities. We will see that compound inequalities are used in solving absolute value inequalities. Finally, after a quick review of functions and graphing, we will learn to perform function operations.

"Inequality is a fact. Equality is a value."

—*MASON COOLEY*
U.S. aphorist

"Thinking is the function. Living is the functionary."

—*RALPH WALDO EMERSON*
U.S. essayist, poet, and philosopher (1803–1882)

8.1 Compound Inequalities

OBJECTIVES

1. Solve compound inequalities involving "and."
2. Solve compound inequalities involving "or."

In Section 2.6, we learned to solve linear inequalities and represent their solutions sets with set notation and graphs. For example, a solution set such as $x > 3$ is written in set notation as $\{x \mid x > 3\}$, which is read "the set of all numbers x such that x is a number greater than 3." The graph of this solution set is

Note: The open circle indicates that 3 is not included in the solution set.

In this section we will build on this foundation and explore **compound inequalities**, which involve two inequality symbols.

THE ALGEBRA PYRAMID

Note: In this chapter, we explore inequalities and equations, so we are at the top of our Algebra Pyramid.

EQUATIONS AND INEQUALITIES
$2 + 5(7) = 37$ or
$x + 2y > 5$

EXPRESSIONS
$2 + 5(7)$ or $x + 2y$

CONSTANTS AND VARIABLES
$2, 5, 7, x, y$

> **DEFINITION** **Compound inequality:** An inequality containing two inequality symbols.

Compound inequalities involve the words "and" and "or" as in

$$x > 3 \quad \text{and} \quad x \le 8 \qquad -2 \ge x \quad \text{or} \quad x > 4$$

Interpreting Compound Inequalities with "and"

First, let's consider inequalities involving "and," such as $x > 3$ *and* $x \leq 8$. The word "and" indicates that the solution set contains only values that satisfy *both* inequalities. Therefore, the solution set is the **intersection**, or overlap, of the two inequalities' solution sets.

> **DEFINITION**
>
> **Intersection:** For two sets A and B, the intersection of A and B, symbolized by $A \cap B$, is a set containing only elements that are in both A and B.

For example, if $A = \{1, 2, 3, 4, 5\}$ and $B = \{3, 4, 5, 6, 7\}$, then $A \cap B = \{3, 4, 5\}$ because 3, 4, and 5 are the only elements in both A and B. Let's look at our compound inequality $x > 3$ and $x \leq 8$. First we will graph the two inequalities separately, then consider their intersection:

Note: We use the blue field to indicate the intersection and the dashed lines to indicate the intersection boundaries.

Note: The intersection of the two inequalities contains elements that are only in *both* of their solution sets. Notice that 3 is excluded in the intersection because it was excluded in one of the individual graphs.

So $x > 3$ and $x \leq 8$ means that x is any number greater than 3 *and* less than or equal to 8. Since this inequality indicates x is between two values, we can write it without the word "and": $3 < x \leq 8$. In set-builder notation, we write $\{x \mid 3 < x \leq 8\}$.

Another popular notation is called *interval notation*. If an end value *is* included in the solution set, we use brackets: [or]. If an end value is *not* included in the solution set, we use parentheses: (or). To write $x > 3$ and $x \leq 8$ using interval notation we write $(3, 8]$. Many people use the parentheses and brackets from this notation when graphing on a number line.

Learning Strategy

With interval notation, imagine moving along the number line from left to right indicating all values in the solution set between two extremes.

Note: Parentheses are like open circles and brackets are like solid circles.

EXAMPLE 1 For the compound inequality $x > -2$ and $x < 3$, graph the solution set and write the compound inequality without "and," if possible. Then write in set-builder notation and in interval notation.

Solution The solution set is the region of intersection.

$x > -2$

$x < 3$

Solution set: $x > -2$ and $x < 3$

or

Without "and": $-2 < x < 3$

Set-builder notation: $\{x \mid -2 < x < 3\}$

Interval notation: $(-2, 3)$

Warning: Be careful not to confuse the interval notation $(-2, 3)$ with an ordered pair.

YOUR TURN For each compound inequality, graph the solution set and write the compound inequality without "and," if possible. Then write in set-builder notation and in interval notation.

    a. $x > 4$ and $x < 7$ **b.** $x \geq -8$ and $x \leq -2$

Your instructor may prefer to use parentheses and brackets when graphing inequalities. For simplicity, from here on we will use only the open and closed circles.

Solving Compound Inequalities with "and"

Now let's consider solution sets for more complex inequalities involving "and."

EXAMPLE 2 For the inequality $3x - 1 > 2$ and $2x + 4 \leq 14$, graph the solution set. Then write the solution set in set-builder notation and in interval notation.

Solution First, we solve each inequality in the compound inequality.

$3x - 1 > 2$	and	$2x + 4 \leq 14$
$3x > 3$	and	$2x \leq 10$ **Use the addition principle to isolate the x term.**
$x > 1$	and	$x \leq 5$ **Divide out each coefficient.**

The solution set is the intersection of the two individual solution sets.

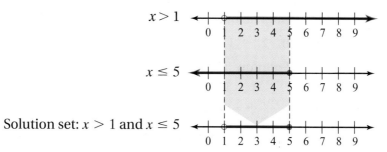

Solution set: $x > 1$ and $x \leq 5$

Set-builder notation: $\{x \mid 1 < x \leq 5\}$

Interval notation: $(1, 5]$

Example 2 suggests the following procedure.

PROCEDURE **SOLVING COMPOUND INEQUALITIES INVOLVING "AND"**

To solve a compound inequality involving "and,"

1. Solve each inequality in the compound inequality.
2. The solution set will be the intersection of the individual solution sets.

Finally, we consider some special situations with compound inequalities involving "and."

EXAMPLE 3 For the compound inequality, graph the solution set. Then write the solution set in set-builder notation and in interval notation.

a. $-2 \leq -3x + 7 \leq 1$

Note: Remember, when we multiply or divide both sides of an inequality by a negative number, we must change the direction of the inequality.

Solution Note that $-2 \leq -3x + 7 \leq 1$ means $-3x + 7 \geq -2$ and $-3x + 7 \leq 1$. Solving the two inequalities gives us

$-3x + 7 \geq -2$	and	$-3x + 7 \leq 1$
$-3x \geq -9$	and	$-3x \leq -6$ **Subtract 7 from both sides of each inequality.**
$x \leq 3$	and	$x \geq 2$ **Divide both sides of each inequality by -3.**

Remember, we can write $x \leq 3$ and $x \geq 2$ as $2 \leq x \leq 3$. Graphing this solution set we have

We could have also solved the compound inequality in its original form.

$$-2 \le -3x + 7 \le 1$$
$$-9 \le -3x \le -6 \qquad \text{Subtract 7 from all three parts of the compound inequality.}$$
$$3 \ge x \ge 2 \qquad \text{Divide all three parts of the compound inequality by } -3.$$

Note that $3 \ge x \ge 2$ is the same as $2 \le x \le 3$.

Set-builder notation: $\{x \mid 2 \le x \le 3\}$

Interval notation: $[2, 3]$

b. $2x + 11 \ge 5$ and $2x + 11 > 3$

Solution To graph the solution set, we first solve each inequality.

$$2x + 11 \ge 5 \qquad \text{and} \qquad 2x + 11 > 3$$
$$2x \ge -6 \qquad \text{and} \qquad 2x > -8 \qquad \text{Subtract 11 from both sides of each inequality.}$$
$$x \ge -3 \qquad \text{and} \qquad x > -4 \qquad \text{Divide both sides of each inequality by 2.}$$

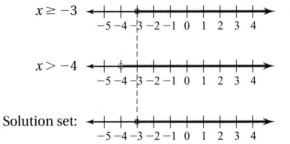

Note: The region of intersection for these graphs is from -3 to ∞.

Set-builder notation: $\{x \mid x \ge -3\}$

Interval notation: $[-3, \infty)$

Note: We cannot include $-\infty$ or ∞ as a boundary in interval notation, so we always use a parenthesis.

c. $-4x - 3 > 1$ and $-4x - 3 < -11$

Solution First, we solve each inequality.

$$-4x - 3 > 1 \qquad \text{and} \qquad -4x - 3 < -11$$
$$-4x > 4 \qquad \text{and} \qquad -4x < -8 \qquad \text{Add 3 to both sides of each inequality.}$$
$$x < -1 \qquad \text{and} \qquad x > 2 \qquad \text{Divide both sides of each inequality by } -4.$$

Warning: Do not write the symbol for empty set within braces, as in $\{\varnothing\}$. This no longer indicates an empty set because the braces now contain something, so they are not truly empty.

Note: There is no region of intersection for these graphs, so the solution set contains no values and is said to be empty.

To graph an empty set, we draw a number line with no shading.

Set-builder notation: The empty set is written as $\{\,\}$. Or we can use a symbol that means an empty set, $\varnothing$.

Interval notation: We cannot write interval notation because there are no values in the solution set.

OBJECTIVE 2 Solve compound inequalities involving "or."

Interpreting Compound Inequalities with "or"

In a compound inequality such as $x > 2$ or $x \leq -1$, the word "or" indicates that the solution set contains values that satisfy *either* inequality. Therefore, the solution set is the **union**, or joining, of the two inequalities' solution sets.

DEFINITION *Union:* For two sets A and B, the union of A and B, symbolized by $A \cup B$, is a set containing every element in A or in B.

Learning Strategy
If you are a visual learner, imagine placing one graph on top of the other to form their union.

For example, if $A = \{1, 2, 3, 4, 5\}$ and $B = \{3, 4, 5, 6, 7\}$, then $A \cup B = \{1, 2, 3, 4, 5, 6, 7\}$. Consider our compound inequality $x > 2$ or $x \leq -1$. Let's graph the two inequalities separately, then consider their union.

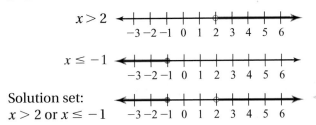

Note: This graph is the union of the two individual solution sets because it contains every element in *both* of the individual solution sets.

 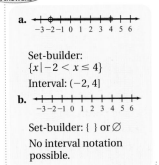
Using set-builder notation, we write the solution set as $\{x \mid x > 2 \text{ or } x \leq -1\}$. Interval notation is a little more challenging for compound inequalities like $x > 2$ or $x \leq -1$ because the solution set is a combination of two intervals. As we scan from left to right, the first interval is $(-\infty, -1]$. The second interval is $(2, \infty)$. To indicate the union of these two intervals, we write $(-\infty, -1] \cup (2, \infty)$.

Solving Compound Inequalities with "or"

Now let's solve more complex inequalities involving "or." The process is similar to the process we used for solving compound inequalities involving "and."

PROCEDURE SOLVING COMPOUND INEQUALITIES INVOLVING "OR"

To solve a compound inequality involving "or,"

1. Solve each inequality in the compound inequality.
2. The solution set will be the union of the individual solution sets.

EXAMPLE 4 For each compound inequality, graph the solution set. Then write the solution set in set-builder notation and in interval notation.

a. $\frac{2}{3}x - 5 \le -7$ or $\frac{2}{3}x - 5 \ge -1$

Solution First, we solve each inequality in the compound inequality.

$$\frac{2}{3}x - 5 \le -7 \quad \text{or} \quad \frac{2}{3}x - 5 \ge -1$$

$$\frac{2}{3}x \le -2 \quad \text{or} \quad \frac{2}{3}x \ge 4 \qquad \text{Add 5 to both sides of each inequality.}$$

$$x \le -3 \quad \text{or} \quad x \ge 6 \qquad \text{Multiply both sides by } \frac{3}{2} \text{ in each inequality.}$$

The solution set is the union of the two individual solution sets.

$x \le -3$

$x \ge 6$

Solution set:
$x \le -3$ or $x \ge 6$

Set-builder notation: $\{x \mid x \le -3 \text{ or } x \ge 6\}$
Interval notation: $(-\infty, -3] \cup [6, \infty)$

b. $-3x + 8 < -1$ or $-3x + 8 \le 20$

Solution
$$-3x + 8 < -1 \quad \text{or} \quad -3x + 8 \le 20$$
$$-3x < -9 \quad \text{or} \quad -3x \le 12 \qquad \text{Subtract 8 from both sides of each inequality.}$$
$$x > 3 \quad \text{or} \quad x \ge -4 \qquad \text{Divide both sides of each inequality by } -3.$$

Now we graph the individual solution sets and consider their union.

$x > 3$

$x \ge -4$

Solution set:
$x > 3$ or $x \ge -4$

Note: When we join the two graphs, the graph of $x \ge -4$ covers all of the graph of $x > 3$, so the union of the two graphs is actually $x \ge -4$.

Set-builder notation: $\{x \mid x \ge -4\}$
Interval notation: $[-4, \infty)$

c. $9x - 16 > -34$ or $9x - 16 \leq 29$

Solution $\quad 9x - 16 > -34 \qquad$ or $\qquad 9x - 16 \leq 29$

$\qquad\qquad\qquad 9x > -18 \qquad$ or $\qquad\qquad 9x \leq 45 \qquad$ Add 16 to both sides of each inequality.

$\qquad\qquad\qquad\quad x > -2 \qquad$ or $\qquad\qquad\quad x \leq 5 \qquad$ Divide both sides of each inequality by 9.

Now we graph the individual solution sets and consider their union.

$x > -2$ [number line from -5 to 8, open circle at -2, shaded right]

$x \leq 5$ [number line from -5 to 8, closed at 5, shaded left]

Note: When we join the two graphs, the entire number line is covered, so the union of the two graphs is the set of real numbers.

Solution set:
$x > -2$ or $x \leq 5$ [number line from -5 to 8, entirely shaded]

Set-builder notation: $\{x \mid x \text{ is a real number}\}$ or we can write the set of real numbers using the symbol $\Re$ or $\mathbb{R}$.

Interval notation: $(-\infty, \infty)$

ANSWERS

a. [number line from -6 to 3, open circle at -4, open circle at -1]

Set-builder:
$\{x \mid x < -4 \text{ or } x \geq -1\}$
Interval:
$(-\infty, -4) \cup [-1, \infty)$

b. [number line from -3 to 6, open circle at 4]

Set-builder: $\{x \mid x < 4\}$
Interval: $(-\infty, 4)$

YOUR TURN   For each compound inequality, graph the solution set. Then write the solution set in set-builder notation and in interval notation.

a. $-5x - 8 > 12$ or $-5x - 8 \leq -3$ **b.** $\dfrac{3}{4}x + 5 < -1$ or $\dfrac{3}{4}x + 5 < 8$

8.1 EXERCISES

1. What is a compound inequality?

2. Compound inequalities involve what two key words?

3. Describe the intersection of two sets.

4. Describe the union of two sets.

5. In general, describe how to graph a compound inequality involving "and."

6. In general, describe how to graph a compound inequality involving "or."

For Extra Help

InterAct Math

Addison-Wesley
Math Tutor Center

Video 14/DVT 4

MyMathLab

MathXL

7. $A = \{1,3,5\}$
  $B = \{1,3,5,7,9\}$

8. $A = \{2,4,6\}$
  $B = \{2,4,6,8,10\}$

9. $A = \{5, 6, 7\}$
  $B = \{7, 8, 9\}$

10. $A = \{8, 10, 12, 14\}$
  $B = \{14, 16, 18\}$

11. $A = \{c, a, t\}$
  $B = \{d, o, g\}$

12. $A = \{l, o, v, e\}$
  $B = \{m, a, t, h\}$

13. $A = \{x, y, z\}$
  $B = \{w, x, y, z\}$

14. $A = \{m, n, o, p, q\}$
  $B = \{a, b, c, \ldots x, y, z\}$

For Exercises 15–22, write each inequality without "and."

15. $x > -4$ and $x < 5$

16. $n \geq 3$ and $n < 7$

17. $y > -2$ and $y \leq 0$

18. $m \geq 0$ and $m < 3$

19. $w > -7$ and $w < 3$

20. $r > -1$ and $r \leq 1$

21. $u \geq 0$ and $u \leq 2$

22. $t > 7$ and $t < 15$

For Exercises 23–30, graph the compound inequality.

23. $x > 2$ and $x < 7$

24. $x > 3$ and $x < 9$

25. $x > -1$ and $x \leq 5$

26. $x \geq -4$ and $x < 0$

27. $1 \leq x \leq 10$

28. $-2 < x < -1$

29. $-3 \leq x < 4$

30. $0 < x < 8$

For Exercises 31–50: ***a.*** *Graph the solution set.*
                  b. *Write the solution set in set-builder notation.*
                  c. *Write the solution set in interval notation.*

31. $x > -3$ and $x < -1$

32. $x > 4$ and $x < 7$

33. $x + 2 > 5$ and $x - 4 \leq 2$

34. $x - 3 \geq 1$ and $x + 2 < 10$

35. $-x > 1$ and $-2x \leq -10$

36. $-4x \geq 8$ and $3x \leq 15$

37. $2x + 6 \geq -4$ and $3x - 1 < 8$

38. $3x + 5 \geq -1$ and $5x - 2 < 13$

39. $-3x - 8 < 1$ and $-4x + 5 \leq -3$

40. $-6x + 4 < -14$ and $-x - 3 \geq -2$

41. $-3 < x + 4 < -1$

42. $-2 \leq x - 2 \leq 2$

43. $-2 \leq 5x + 3 \leq 13$

44. $7 < 2x - 1 < 11$

45. $0 \leq 2 + 3x < 8$

46. $4 \leq -2 + x \leq 13$

47. $-6 < -3x + 3 \leq 3$

48. $-3 < 5 - 2x < 1$

49. $3 \leq 6 - x \leq 6$

50. $-2 < -6x + 10 < 16$

For Exercises 51–58, graph each inequality.

51. $x < -2$ or $x > 6$

52. $n \leq 3$ or $n > 4$

53. $y < -3$ or $y \geq 0$

54. $m \leq 2$ or $m > 8$

55. $w < -6$ or $w > 2$

56. $r > -3$ or $r \geq -1$

57. $u \leq 0$ or $u \geq 2$

58. $t < 6$ or $t > 13$

For Exercises 59–76: a. Graph the solution set.
b. Write the solution set in set-builder notation.
c. Write the solution set in interval notation.

59. $y + 2 < -7$ or $y + 2 > 7$

60. $a - 4 \leq -2$ or $a - 4 \geq 2$

61. $5x + 2 \leq -1$ or $5x + 2 \geq 3$

62. $2m - 1 < 0$ or $2m - 1 > 5$

63. $3r + 1 < -4$ or $3r + 1 > 0$

64. $-2x - 3 \leq -1$ or $-2x - 3 \geq 5$

65. $-w + 2 \leq -5$ or $-w + 2 \geq 3$

66. $4n - 3 < -8$ or $4n - 3 > 3$

67. $-2c + 3 < -4$ or $-2c + 3 < -6$

68. $5d + 3 < 8$ or $5d + 3 < 18$

69. $-3x + 2 \leq -1$ or $-3x + 2 \geq -2$

70. $4k - 7 \leq 1$ or $4k - 7 \geq -5$

71. $2x + 9 \leq 1$ or $2x + 9 \leq -3$

72. $5 + y \geq 4$ or $5 + y \geq 6$

73. $6 - 3m \leq -2$ or $6 - 3m \geq 3$

74. $4t + 5 \leq -7$ or $4t + 5 \geq -3$

75. $7 - 3k \leq -2$ or $7 - 3k \geq 1$

76. $8 - 2q \leq 2$ or $8 - 2q \geq 4$

For Exercises 77–88: a. Graph the solution set.
b. Write the solution set in set-builder notation.
c. Write the solution set in interval notation.

77. $4 < -2x < 6$

78. $9 \leq -3x \leq 15$

79. $-3 < x + 4$ and $x + 4 < 7$

80. $0 < x - 1$ and $x - 1 \leq 3$

81. $2x + 3 < -1$ or $2x + 3 > 7$

82. $3x - 2 < -5$ or $3x - 2 > 7$

83. $1 \leq 2x - 5 \leq 7$

84. $3x - 1 < 8$ or $3x - 1 \geq 10$

85. $-5 < 1 - 2x < -1$

86. $4 \leq -3x + 1 < 7$

87. $x - 3 < 2x + 1 < 3x$

88. $2x - 2 \leq x + 1 \leq 2x + 5$

For Exercises 89–98, solve and graph the solution set.

89. If Andrea's current long-distance bill is between $30 and $45 in a month, then she can save money by switching to another company. Her current rate is $0.09 per minute. What is the range of minutes that Andrea must use long distance to justify switching to the other company?

90. A mail-order music club offers one bonus CD if the total of an order is between $75 and $100. Dayle decides to buy the lowest priced CDs, which are $12.50 each. In what range would the number of CDs he orders have to be in order to receive a bonus CD?

91. Juan has taken four of the five tests in his history course. To get a B in the course, his average needs to be between 80 an 90. If his scores on the first four tests are 95, 80, 82, 88, in what range of values can his score on the fifth test be so that he has a B average?

92. Students in a chemistry course receive a C if the average of four tests is between 70 and 80. Suppose a student has the following scores: 60, 72, and 70. What range of scores on the fourth test would cause the student to receive a C?

93. To conserve energy, it is recommended that a home's thermostat be set at 5 degrees above 73°F during summer or 5 degrees below 73° during winter. If the heat pump does not run when the temperature is between those values, what range of temperatures is the heat pump off?

94. The house thermostat is set so that the heat pump/air conditioner comes on if the temperature is 2 degrees above or below 70°. What range of temperatures is the heat pump off?

95. To maintain a saltwater aquarium, the temperature should be 4 degrees above or below 76°. What range of temperatures would be acceptable for saltwater fish? (*Source: The Conscientious Marine Aquarist* by Robert M. Fenner © 2001.)

96. When driving in the state of Mississippi, motorists risk a ticket if they drive 15 mph more or less than 55 mph on a state highway. In what range of speeds would a motorist not risk receiving a ticket? (*Source:* Mississippi State Code 63-3-509.)

97. A building is to be designed in the shape of a trapezoid as shown. Find the range of values for the length of the back side of the building so that the square footage remains between 6000 and 8000 square feet. (Use $A = \frac{1}{2}h(a + b)$.)

60 ft.

front

50 ft.

back

98. At 32°F and below, water changes to a solid (ice). At 212°F and above, water changes to a gas (steam). Use a compound inequality to describe the range of values in degrees Celsius for water in its liquid state. (Use $C = \frac{5}{9}(F - 32)$.)

For Exercises 99–104, use the following graph and write each solution set using set-builder notation and interval notation.

99. If x represents time, write a solution set that describes the years in which drug defendants $\geq 20,000$ and public order defendants $\geq 20,000$

100. If x represents time, write a solution set that describes the years in which drug defendants $\geq 10,000$ and drug defendants $\leq 20,000$.

101. If x represents time, write a solution set that describes the years in which drug defendants $<$ property defendants and drug defendants $<$ public order defendants.

102. If x represents time, write a solution set that describes the years in which drug defendants $<$ property defendants or drug defendants $<$ public order defendants.

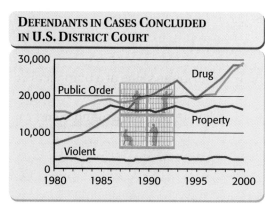

Source: Compendium of Federal Justice Statistics.

103. If x represents time, write a solution set that describes the years in which public order defendants $\geq$ property defendants and public order defendants $<$ drug defendants.

104. If x represents time, write a solution set that describes the years in which drug defendants $<$ 10,000 or property defendants $<$ drug defendants.

PUZZLE PROBLEM

A cyclist is involved in a multiple-day race. She feels she needs to complete today's 40-kilometers part in somewhere between 1 hour and 45 minutes and 2 hours. Find the range of values her rate can be to complete this leg of the race in her desired time frame.

REVIEW EXERCISES

For Exercises 1–2, simplify.

1. $-|2 - 3^2| - |-4|$

2. $-|-|-16||$

For Exercises 3–4, solve.

3. $3x - 14x = 17 - 12x - 11$

4. $\frac{3}{5}(25 - 5x) = 15 - \frac{3}{5}$

5. Is the absolute value of a number always positive? Explain.

6. One-half the sum of a number and 2 is zero. Find the number.

8.2 Equations Involving Absolute Value

OBJECTIVE 1 Solve equations involving absolute value.

We learned in Section 1.1 that the absolute value of a number is its distance from zero on a number line. For example, $|4| = 4$ because 4 is 4 units from 0 on a number line. Also, $|-4| = 4$ because -4 is *also* 4 units from 0 on a number line.

Now we will solve equations in which a variable appears within the absolute value symbols, as in $|x| = 7$. A solution to $|x| = 7$ will be any number whose absolute value is 7, so the solutions are 7 and -7. This suggests the following rule.

RULE   ABSOLUTE VALUE PROPERTY

If $|x| = a$, where x is a variable or an expression and $a \geq 0$, then $x = a$ or $x = -a$.

EXAMPLE 1 Solve.

a. $|2x - 5| = 9$

Solution Using the absolute value property, we separate the equation into a positive case and a negative case, then solve the two cases separately.

$$
\begin{array}{lll}
2x - 5 = 9 & \text{or} & 2x - 5 = -9 \\
2x = 14 & \text{or} & 2x = -4 \\
x = 7 & \text{or} & x = -2
\end{array}
$$

The solutions are 7 and -2.

b. $|4x + 1| = -11$

Solution This equation has the absolute value equal to a negative number, -11. Because the absolute value of every real number is a positive number, this equation has no solution.

Warning: An absolute value equation in the form $|n| = -a$ where $a > 0$ has no solution.

YOUR TURN Solve.

a. $|3x + 4| = 8$ **b.** $|2x - 1| = 0$ **c.** $|5x - 2| = -3$

To use the absolute value property, we need the absolute value isolated.

EXAMPLE 2 Solve $|4x - 3| + 9 = 16$.

Solution $|4x - 3| = 7$ Subtract 9 from both sides to isolate the absolute value.

$$4x - 3 = 7 \quad \text{or} \quad 4x - 3 = -7 \quad \text{Use the absolute value property.}$$
$$4x = 10 \quad \text{or} \quad 4x = -4$$
$$x = \frac{5}{2} \quad \text{or} \quad x = -1$$

The solutions are $\frac{5}{2}$ and -1.

Examples 1 and 2 suggest the following procedure.

PROCEDURE SOLVING EQUATIONS CONTAINING A SINGLE ABSOLUTE VALUE

To solve an equation containing a single absolute value,

1. Isolate the absolute value so that the equation is in the form $|ax + b| = c$.
2. Separate the absolute value into two equations, $ax + b = c$ and $ax + b = -c$.
3. Solve both equations.

YOUR TURN

Solve $|6x - 3| - 5 = 4$.

More Than One Absolute Value

Some equations contain more than one absolute value expression, such as $|x + 3| = |2x - 8|$. If two absolute values are equal, they must contain expressions that are *equal* or *opposites*. Some simple examples follow:

Equal		Opposites
$|7| = |7|$	or	$|7| = |-7|$
$|-5| = |-5|$	or	$|-5| = |5|$

For the equation $|x + 3| = |2x - 8|$, therefore, the expressions $x + 3$ and $2x - 8$ could be equal or opposites.

Equal		Opposites
$x + 3 = 2x - 8$	or	$x + 3 = -(2x - 8)$

This suggests the following procedure.

PROCEDURE SOLVING EQUATIONS IN THE FORM $|ax + b| = |cx + d|$

To solve an equation in the form $|ax + b| = |cx + d|$,

1. Separate the absolute value equation into two equations: $ax + b = cx + d$ and $ax + b = -(cx + d)$.
2. Solve both equations.

EXAMPLE 3 Solve $|3x + 5| = |x - 9|$.

Solution Separate the absolute value equation into two equations.

Equal		Opposites
$3x + 5 = x - 9$	or	$3x + 5 = -(x - 9)$
$2x + 5 = -9$	or	$3x + 5 = -x + 9$
$2x = -14$	or	$4x + 5 = 9$
$x = -7$	or	$4x = 4$
		$x = 1$

Note: Think of $-(x - 9)$ as $-1(x - 9)$.

The solutions are -7 and 1.

Sometimes when we separate an equation containing two absolute values into its two equivalent equations, we find that only one of them yields a solution.

EXAMPLE 4 Solve $|3x - 7| = |5 - 3x|$.

Solution

Equal		Opposites
$3x - 7 = 5 - 3x$	or	$3x - 7 = -(5 - 3x)$
$6x - 7 = 5$	or	$3x - 7 = -5 + 3x$
$6x = 12$	or	$-7 = -5$
$x = 2$		

Separate into two equations.

Note: Subtracting $3x$ from both sides gives us a false equation, so the "opposites" equation has no solution.

This absolute value equation has only one solution, 2.

YOUR TURN

ANSWER

-1

Solve $|9 + 4x| = |1 - 4x|$.

8.2 EXERCISES

1. What does absolute value mean in terms of a number line?

2. How do you interpret $|x| = 5$?

3. State the absolute value property in your own words.

4. If given an equation in the form $|ax + b| = c$ and $c < 0$, what can you conclude about its solution(s)?

5. Explain the first step in solving an equation containing two absolute values, such as $|2x + 1| = |3x - 5|$.

6. How can you tell that an absolute value equation containing two absolute values has only one solution?

For Exercises 7–22, solve using the absolute value property.

7. $|x| = 2$ **8.** $|y| = 5$ **9.** $|a| = -4$ **10.** $|r| = -1$

11. $|x + 3| = 8$ **12.** $|w - 1| = 4$ **13.** $|2m - 5| = 1$ **14.** $|3s + 7| = 10$

15. $|6 - 5x| = 1$ **16.** $|4 - 3w| = 6$ **17.** $|3x - 2| = 5$ **18.** $|2x - 1| = 2$

19. $|4m - 2| = -5$ **20.** $|6p + 5| = -1$ **21.** $|4w - 3| = 0$ **22.** $|6 - 3m| = 0$

For Exercises 23–38, isolate the absolute value, then use the absolute value property.

23. $|y - 1| + 2 = 4$ **24.** $|x + 3| - 1 = 5$ **25.** $|b + 4| - 6 = 2$ **26.** $|v - 2| + 2 = 4$

27. $|2y| - 3 = 5$ **28.** $|3r| + 7 = 10$ **29.** $3 + |5x - 1| = 7$ **30.** $2 + |3t - 2| = 5$

31. $1 - |2k + 3| = -4$ **32.** $-3 = 1 - |4u - 2|$ **33.** $4 - 3|z - 2| = -8$ **34.** $3 - 2|x - 5| = -7$

35. $6 - 2|3 - 2w| = -18$ **36.** $15 - 2|5 - 2x| = -5$ **37.** $|3x - 2(x + 5)| = 10$ **38.** $|4y + 2(7 - y)| = 5$

For Exercises 39–48, solve by separating the two absolute values into equal and opposite cases.

39. $|x + 3| = |2x - 4|$ **40.** $|p - 1| = |2p + 8|$ **41.** $|3v + 4| = |1 - 2v|$ **42.** $|3 - 6c| = |2c - 3|$

43. $|2x + 1| = |x + 5|$ **44.** $|2p - 5| = |3p + 1|$ **45.** $|n - 3| = |3 - n|$ **46.** $|2r - 1| = |1 - 2r|$

47. $|2k + 1| = |2k - 5|$ **48.** $|5 + 2q| = |2q + 8|$

For Exercises 49–56, solve.

49. $\left|\dfrac{4-3x}{2}\right| = \dfrac{3}{4}$

50. $\left|\dfrac{6-5w}{6}\right| = \dfrac{2}{3}$

51. $\left|\dfrac{b}{2}-1\right| = 4$

52. $\left|\dfrac{u}{3}+2\right| = 3$

53. $\left|2y+\dfrac{3}{2}\right| - \dfrac{1}{2} = 5$

54. $\left|p+\dfrac{2}{3}\right| - 1 = 7$

55. $|10-(5+h)| = 8$

56. $|7-(2-n)| = 17$

REVIEW EXERCISES

For Exercises 1 and 2, solve.

1. $6 - 4x > -5x - 1$

2. $-4x \le 12$

3. Graph the compound inequality: $-1 < x \le 0$

4. Solve the compound inequality: $-1 < \dfrac{3x+2}{4} < 4$

5. Use interval notation to represent the graph shown.

6. Translate to a linear inequality and solve: The sum of a number and -2 is less than 5.

8.3 Inequalities Involving Absolute Value

1. Solve absolute value inequalities involving less than.
2. Solve absolute value inequalities involving greater than.

OBJECTIVE 1 **Solve absolute value inequalities involving less than.** Now let's solve inequalities that contain absolute value. Think about $|x| \leq 3$. A solution for this inequality is any number whose absolute value is less than or equal to 3. This means the solutions are a distance of 3 units or less from 0 on a number line.

Note: Solutions for the *equal to* part of $|x| \leq 3$ are 3 and -3. The *less than* part means values like -2, -1, 0, 1, and 2 because their absolute values are less than 3.

Notice the solution region corresponds to the compound inequality $x \geq -3$ and $x \leq 3$, which we can write as $-3 \leq x \leq 3$. In set-builder notation the solution is $\{x | -3 \leq x \leq 3\}$ and in interval notation, we write $[-3, 3]$.

Our examples suggest the following procedure.

PROCEDURE SOLVING INEQUALITIES IN THE FORM $|x| < a$, WHERE $a > 0$

To solve an inequality in the form $|x| < a$, where $a > 0$,

1. Rewrite as a compound inequality involving "and": $x > -a$ and $x < a$. (We can also use $-a < x < a$.)
2. Solve the compound inequality.

Similarly, to solve $|x| \leq a$, we would write $x \geq -a$ and $x \leq a$ (or $-a \leq x \leq a$).

EXAMPLE 1 For each inequality, solve, graph the solution set, and write the solution set in both set-builder and interval notation.

a. $|x - 4| \leq 3$

Solution $x - 4 \geq -3$ and $x - 4 \leq 3$ Rewrite as a compound inequality.

$\qquad\qquad x \geq 1$ and $x \leq 7$ Add 4 to both sides of each inequality.

Recall that $x \geq 1$ and $x \leq 7$ means $1 \leq x \leq 7$, so our graph is as follows:

Note: The solution set for $|x - 4| \leq 3$ contains every number whose distance from 4 is 3 units or less.

Set-builder notation: $\{x | 1 \leq x \leq 7\}$

Interval notation: $[1, 7]$

b. $|3x + 1| < 5$

Solution Instead of $3x + 1 > -5$ and $3x + 1 < 5$, we will use the more compact form.

$$-5 < 3x + 1 < 5 \qquad \text{Rewrite as a compound inequality.}$$
$$-6 < 3x < 4 \qquad \text{Subtract 1 from both sides of each inequality.}$$
$$-2 < x < \frac{4}{3} \qquad \text{Divide both sides of each inequality by 3 to isolate } x.$$

Set-builder notation: $\left\{ x \mid -2 < x < \dfrac{4}{3} \right\}$

Interval notation: $\left(-2, \dfrac{4}{3} \right)$

c. $|-0.5x + 1| - 2 < 1$

Solution Notice that the equation is not in the form $|x| < a$, so our first step is to isolate the absolute value.

$$|-0.5x + 1| < 3 \qquad \text{Add 2 to both sides to isolate the absolute value.}$$
$$-3 < -0.5x + 1 < 3 \qquad \text{Rewrite as a compound inequality.}$$
$$-4 < -0.5x < 2 \qquad \text{Subtract 1 from both sides of each inequality.}$$
$$8 > x > -4 \qquad \text{Divide both sides of each inequality by } -0.5 \text{ to isolate } x.$$

Note: Because we divided by a negative, we changed the direction of the inequalities.

Set-builder notation: $\{ x \mid -4 < x < 8 \}$

Interval notation: $(-4, 8)$

d. $\left| \dfrac{1}{3}x - 4 \right| + 6 < -1$

Solution $\left| \dfrac{1}{3}x - 4 \right| < -7 \qquad \text{Subtract 6 from both sides to isolate the absolute value.}$

The absolute value is less than a negative number, but absolute values cannot be negative, so this inequality has no solution.

ANSWERS

a.
Set notation:
$\{ x \mid -7 < x < -3 \}$
Interval notation: $(-7, -3)$

b.
Set notation:
$\{ x \mid -2 \le x \le 5 \}$
Interval notation: $[-2, 5]$

c.
Set notation:
$\{ x \mid -6 \le x \le 12 \}$
Interval notation: $[-6, 12]$

d. No solution.

YOUR TURN For each inequality, solve, graph the solution set, and write the solution set in both set-builder notation and interval notation.

a. $|x + 5| < 2$ **b.** $|2x - 3| \le 7$ **c.** $\left| -\dfrac{1}{3}x + 1 \right| - 2 \le 1$ **d.** $\left| \dfrac{1}{2}x - 1 \right| + 3 < -2$

OBJECTIVE ② **Solve absolute value inequalities involving greater than.** Now we consider inequalities with greater than, such as $|x| \geq 5$. A solution for $|x| \geq 5$ is any number whose absolute value is greater than or equal to 5. As the graph shows, the solution set contains all values that are a distance of 5 units or more from 0.

> **Note:** Solutions for the *equal to* part of $|x| \geq 5$ are 5 and -5. The *greater than* part means values like 6 and -6, or 7 and -7, because their absolute values are greater than 5.

Notice that the solutions are equivalent to $x \leq -5$ or $x \geq 5$. An inequality such as $|x| > 2$ uses open circles:

The solutions are equivalent to $x < -2$ or $x > 2$. Our examples suggest that we can split these inequalities into a compound inequality involving "or."

PROCEDURE **SOLVING INEQUALITIES IN THE FORM** $|x| > a$ **WHERE** $a > 0$

To solve an inequality in the form $|x| > a$ where $a > 0$,

1. Rewrite as a compound inequality involving "or": $x < -a$ or $x > a$.
2. Solve the compound inequality.

Similarly, to solve $|x| \geq a$, we would write $x \leq -a$ or $x \geq a$.

EXAMPLE 2 For each inequality, solve, graph the solution set, and write the solution set in both set-builder notation and interval notation.

a. $|x - 2| \geq 5$

Solution
$$x - 2 \leq -5 \quad \text{or} \quad x - 2 \geq 5 \qquad \text{Rewrite as a compound inequality.}$$
$$x \leq -3 \quad \text{or} \quad x \geq 7 \qquad \text{Add 2 to both sides of each inequality.}$$

Set-builder notation: $\{x \mid x \leq -3 \text{ or } x \geq 7\}$
Interval notation: $(-\infty, -3] \cup [7, \infty)$

b. $|-3x - 4| > 5$

Solution
$$-3x - 4 < -5 \quad \text{or} \quad -3x - 4 > 5 \qquad \text{Rewrite as a compound inequality.}$$
$$-3x < -1 \quad \text{or} \quad -3x > 9 \qquad \text{Add 4 to both sides of each inequality.}$$
$$x > \frac{1}{3} \quad \text{or} \quad x < -3 \qquad \text{Divide both sides of each inequality by } -3.$$

Set-builder notation: $\left\{ x \mid x < -3 \text{ or } x > \dfrac{1}{3} \right\}$

Interval notation: $(-\infty, -3) \cup \left(\dfrac{1}{3}, \infty \right)$

c. $\left| \dfrac{3}{4}x - 2 \right| + 1 \geq 6$

Solution $\left| \dfrac{3}{4}x - 2 \right| \geq 5$ Subtract 1 from both sides to isolate the absolute value.

$$\dfrac{3}{4}x - 2 \leq -5 \quad \text{or} \quad \dfrac{3}{4}x - 2 \geq 5 \quad \text{Rewrite as a compound inequality.}$$

$$\dfrac{3}{4}x \leq -3 \quad \text{or} \quad \dfrac{3}{4}x \geq 7 \quad \text{Add 2 to both sides of each inequality.}$$

$$x \leq -4 \quad \text{or} \quad x \geq \dfrac{28}{3} \quad \text{Multiply both sides of each inequality by } \dfrac{4}{3}.$$

Set-builder notation: $\left\{ x \mid x \leq -4 \text{ or } x \geq \dfrac{28}{3} \right\}$

Interval notation: $(-\infty, -4] \cup \left[\dfrac{28}{3}, \infty \right)$

d. $|0.2x + 1| - 3 > -7$

Solution $|0.2x + 1| > -4$ Add 3 to both sides to isolate the absolute value.

This equation indicates that the absolute value is greater than a negative number. Since the absolute value of every real number is either positive or 0, the solution set is $\mathbb{R}$. The graph is the entire number line.

Set-builder notation: $\{ x \mid x \text{ is a real number} \}$

Interval notation: $(-\infty, \infty)$

YOUR TURN For each inequality, solve, graph the solution set, and write the solution set in both set-builder notation and interval notation.

a. $|x - 3| > 2$ **b.** $|4x + 1| \geq 9$ **c.** $|-0.4x + 1| + 2 \geq 3$ **d.** $\left| \dfrac{1}{3}x - 4 \right| - 2 > -9$

ANSWERS

a.

Set notation:
$\{ x \mid x < 1 \text{ or } x > 5 \}$

Interval notation:
$(-\infty, 1) \cup (5, \infty)$

b.

Set notation:
$\left\{ x \mid x \leq -\dfrac{5}{2} \text{ or } x \geq 2 \right\}$

Interval notation:
$\left(-\infty, -\dfrac{5}{2} \right] \cup [2, \infty)$

c.

Set-builder notation:
$\{ x \mid x \leq 0 \text{ or } x \geq 5 \}$

Interval notation:
$(-\infty, 0] \cup [5, \infty)$

d.

Set-builder notation:
$\{ x \mid x \text{ is a real number} \}$

Interval notation:
$(-\infty, \infty)$

For Extra Help

InterAct Math

Tutor Center   Addison-Wesley
Math Tutor Center

Video 14/DVT 4

MyMathLab

MathXL

1. What compound inequality is related to $|x| \le a$?

2. What compound inequality is related to $|x| \ge a$?

3. How would you characterize the graph of $|x| < a$?

4. How would you characterize the graph of $|x| > a$?

5. Under what conditions does $|x| < a$ have no solution?

6. Under what conditions does the solution set for $|x| > a$ contain all real numbers?

Note: For Exercises 1–4, we assume a is a positive number.

For Exercises 7–20, solve the inequality, then a. Graph the solution set.
b. Write the solution set in set-builder notation.
c. Write the solution set in interval notation.

7. $|x| < 5$

8. $|y| \le 3$

9. $|x + 3| \le 7$

10. $|m - 5| < 2$

11. $|s + 3| + 2 < 5$

12. $|p - 3| + 4 \le 8$

13. $|2m - 5| + 1 < 10$

14. $|4x + 7| + 2 < 5$

15. $|-3k + 5| + 7 \leq 8$

16. $|-5h - 1| - 6 < 8$

17. $2|x| - 7 \leq 3$

18. $3|u| + 2 < 8$

19. $2|w - 3| + 4 < 10$

20. $4|n + 2| - 3 \leq 9$

For Exercises 21–34, solve the inequality, then a. Graph the solution set.
                                                      b. Write the solution set in set-builder notation.
                                                      c. Write the solution set in interval notation.

21. $|c| > 12$

22. $|h| \geq 6$

23. $|y + 2| \geq 7$

24. $|a - 4| > 3$

25. $|p - 6| + 2 > 10$

26. $|x + 5| - 3 \geq 6$

27. $|3x + 2| - 4 \geq 9$

28. $|2x - 5| - 1 \geq 10$

29. $|-4n - 5| + 3 > 8$

30. $|-6h + 1| - 7 > 4$

31. $4|v| - 3 \geq 1$

32. $5|m| + 2 > 12$

33. $4|y + 2| - 1 > 3$

34. $3|x - 2| + 5 \geq 11$

For Exercises 35–54, solve the inequality, then a. Graph the solution set.
                                                      b. Write the solution set in set-builder notation.
                                                      c. Write the solution set in interval notation.

35. $|4m + 3| - 2 > 7$

36. $|2x + 4| - 5 \geq 10$

37. $|-3x - 1| < 5$

38. $|-2y + 3| \leq 3$

39. $|2r - 3| > -3$

40. $|3b + 7| < -2$

41. $2 - |x + 3| > 1$

42. $5 - 2|u + 4| \geq 1$

43. $6|2x - 1| + 3 < 9$

44. $4|3p + 5| - 2 < 10$

45. $5 - |w + 4| > -3$

46. $4 - |5 + k| > 7$

47. $\left| 2 - \dfrac{3}{2}k \right| \le 5$

48. $\left| 3 - \dfrac{1}{2}x \right| \ge 7$

49. $|0.25x - 3| + 2 > 4$

50. $|0.5y - 3| + 4 < 5$

51. $\left| 2.4 - \dfrac{3}{4}y \right| \le 7.2$

52. $\left| 5.3 - \dfrac{2}{3}w \right| \ge 5.3$

53. $|2b - 8| + 5 < 1$

54. $|8p + 7| + 4 > 3$

For Exercises 55–62, write an inequality involving absolute value that describes the graph shown.

55.

56.

57.

58.

59.

60.

61.

62.

PUZZLE PROBLEM

The compound inequality $x \le -1$ or $x \ge 0$ can be written as an inequality in the form $|ax + b| + c \ge d$, where a, b, c, and d are consecutive integers such that $a + b + c + d = 10$ and $b < a < c < d$. Find a, b, c, and d, then write the absolute value inequality.

REVIEW EXERCISES

For Exercises 1–2, graph the given equation.

1. $2x + y = 7$

2. $y = 2x^2$

For Exercises 3–4, find the intercepts.

3. $x + y = 6$

4. $y = x^2 + 14x + 45$

For Exercises 5–6, evaluate the expression when $x = 2$ and $y = -4$.

5. $3x - xy^2$

6. $-3x^2 - 2y^2$

8.4 Functions and Graphing

OBJECTIVES

1. Identify the domain and range of a relation and determine if the relation is a function.
2. Find the value of a function.
3. Graph functions.

In this section, we review and expand on what we have learned about functions and graphing.

OBJECTIVE 1 Identify the domain and range of a relation and determine if the relation is a function. Let's review **relations**, **domain**, **range**, and **functions**, which we introduced in Section 4.7.

> **DEFINITIONS**
>
> **Relation:** A set of ordered pairs.
> **Domain:** The set of all input values for a relation.
> **Range:** The set of all output values for a relation.
> **Function:** A relation in which every value in the domain is paired with exactly one value in the range.

For example, the linear equation $y = 2x$ is a relation because it pairs a given x-value with a corresponding y-value. We list here some values in the domain and range for $y = 2x$ and draw its graph.

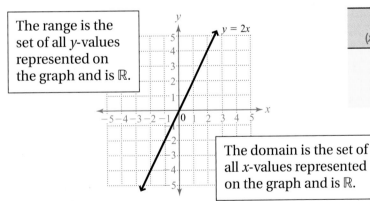

The range is the set of all y-values represented on the graph and is $\mathbb{R}$.

The domain is the set of all x-values represented on the graph and is $\mathbb{R}$.

Domain (x-values)	Range (y-values)
−1	−2
0	0
1	2
2	4

We can also conclude that $y = 2x$ is a function because every value in its domain is paired with only one value in its range. In Section 4.7, we developed the vertical line test to determine whether a graph is a graph of a function. With a function, a vertical line drawn through any point of its graph intersects the graph only once.

EXAMPLE 1 Identify the domain and range of each relation and determine whether it is a function.

a. The following relation pairs players on the Philadelphia Eagles football team with their position(s).

Player	Position
David Akers	Kicker
Lee Johnson	Punter
Koy Detmer	Holder
Mike Bartrum	Long Snapper
Brian Mitchell	Punt Return and Kick Return

Solution Domain: {David Akers, Lee Johnson, Koy Detmer, Mike Bartrum, Brian Mitchell}

Range: {Kicker, Punter, Holder, Long Snapper, Punt Return, Kick Return}

It is not a function because an element in the domain, Brian Mitchell, is paired with two elements in the range.

b. { (−2, −4), (−1, −1), (0, 0), (1, −1), (2, −4) }

Solution Domain: {−2, −1, 0, 1, 2}
Range: {−4, −1, 0, −1, −4}

It is a function because every value in the domain is paired with only one value in the range.

c.

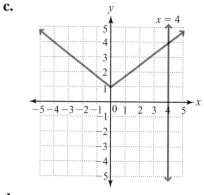

Solution Domain: All real numbers.
Range: $\{y \mid y \geq 1\}$

It is a function because it passes the vertical line test: A vertical line drawn through any point of the graph intersects it only once. We have drawn one such line, $x = 4$, in blue.

d.

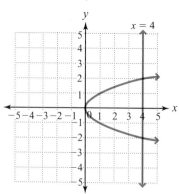

Solution Domain: $\{x \mid x \geq 0\}$
Range: All real numbers.

It is a not a function because there are values in the domain paired with more than one value in the range, as the vertical line test confirms. For example, the line $x = 4$ (in blue) intersects the graph twice so the value 4 in the domain is paired with two values in the range, 2 and −2.

YOUR TURN   Identify the domain and range of each relation and determine whether it is a function.

a. The following relation shows the forecasted high temperatures for Yellowstone National Park from March 25 to 29, 2003.

Day	High Temperature
Tuesday	32°
Wednesday	32°
Thursday	20°
Friday	30°
Saturday	36°

b. { (0, 2), (0, −2), (3, 0), (−3, 0) }

c.

d.

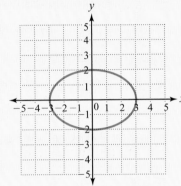

ANSWERS

a. Domain: {Tuesday, Wednesday, Thursday, Friday, Saturday}

Range: {20, 30, 32, 36}

It is a function.

b. Domain: {−3, 0, 3}

Range: {−2, 0, 2}

It is not a function.

c. Domain: $\{x \mid -3 \leq x \leq 3\}$

Range: $\{y \mid -2 \leq y \leq 2\}$

It is not a function.

d. Domain: $\{x \mid -3 \leq x < 5\}$

Range: {1, 2, 3}

It is a function.

OBJECTIVE ② **Find the value of a function.**   In Section 4.7, we introduced the function notation $f(x)$, which is read "a function in terms of x" or "f of x." Using function notation, we can write the equation $y = 3x$ as $f(x) = 3x$. Recall that we evaluate functions by replacing x with the given value. For example, the notation $f(2)$ means the value of f when $x = 2$, so if $f(x) = 3x$, then $f(2) = 3(2) = 6$.

8.4 Functions and Graphing   **675**

EXAMPLE 2 For the function $f(x) = -5x + 3$, find

a. $f(4)$

Solution $f(4) = -5(4) + 3$ Replace x with 4, then calculate.

$= -20 + 3$

$= -17$

b. $f(-2)$

Solution $f(-2) = -5(-2) + 3$ Replace x with -2, then calculate.

$= 10 + 3$

$= 13$

ANSWERS

a. -5

b. -1.5

YOUR TURN For the function $f(x) = 0.5x - 2$, find

a. $f(-6)$ **b.** $f(1)$

OBJECTIVE 3 **Graph functions.** Now let's review how to graph functions. In general, we can graph a function by finding enough ordered pairs to determine its shape.

Linear Functions

Given a linear function in the form $f(x) = mx + b$, where m is the slope and $(0, b)$ is the y-intercept, we need to find only two ordered pairs. Note that $(0, b)$ can be one of those two ordered pairs.

EXAMPLE 3 Graph $f(x) = -\frac{3}{4}x + 2$.

Solution We can use $(0, 2)$ as one ordered pair, then use the slope, $-\frac{3}{4}$, to find a second ordered pair. Recall that slope indicates the "rise" and "run" from any point on the line to another point on the line. So, $-\frac{3}{4}$ indicates we rise -3 and run 4 from the point $(0, 2)$ to reach a second point, $(4, -1)$.

ANSWER

YOUR TURN

Graph $f(x) = 2x - 3$.

Quadratic Functions

In Section 6.7, we introduced quadratic functions, which have the form $f(x) = ax^2 + bx + c$ and we saw their graphs are parabolas. To graph a parabola, we have to find many ordered pairs.

EXAMPLE 4 Graph $f(x) = -2x^2 + 3$.

Solution We create a table of ordered pairs, plot the points, and connect them with a smooth curve.

x	f(x)
−2	−5
−1	1
0	3
1	1
2	−5

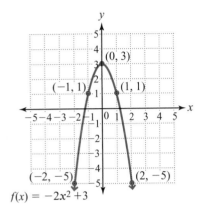

$f(x) = -2x^2 + 3$

CONNECTION

Recall from Section 6.7 that the sign of a in $f(x) = ax^2 + bx + c$ indicates whether the parabola opens up or down. If a is *positive*, the parabola opens *up*, and if a is *negative*, the parabola opens *down*.

ANSWER

YOUR TURN

Graph $f(x) = 3x^2 - 4$.

Absolute Value Functions

Now let's consider how to graph functions containing absolute value. In general, absolute value functions have a V shape.

EXAMPLE 5 Graph each function.

a. $f(x) = |x|$

Solution We find ordered pair solutions, plot those ordered pairs, and connect the points.

x	f(x)
−4	4
−2	2
0	0
2	2
4	4

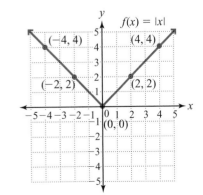

Note: $f(x) = |x|$ is equivalent to $f(x) = -x$ when $x < 0$ and $f(x) = x$ when $x \geq 0$. Notice $f(x) = -x$ and $f(x) = x$ are linear functions with opposite slopes, which explains the V shape of the graph.

b. $f(x) = |x| + 1$

Solution

x	f(x)
−4	5
−2	3
0	1
2	3
4	5

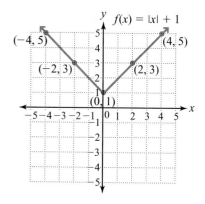

Note: Compare the graphs in Examples 5a and 5b. Notice that every point on $f(x) = |x| + 1$ is one higher than every point on $f(x) = |x|$.

c. $f(x) = |x - 2| + 1$

Solution

x	f(x)
−4	7
−2	5
0	3
2	1
4	3

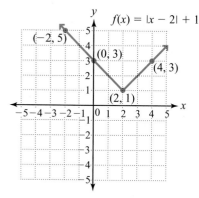

Note: As we saw in Example 5b, the $+ 1$ shifts the graph up 1. Notice that the 2 in the absolute value signs shifts the graph right 2.

d. $f(x) = -|x - 2| + 1$

Solution

x	f(x)
−4	−5
−2	−3
0	−1
2	1
4	−1

Note: Notice that the minus sign in front of the absolute value causes the graph to invert so that the V is "upside down."

ANSWERS

a.

b.

YOUR TURN Graph each function.

a. $f(x) = |x| - 2$

b. $f(x) = -2|x - 1| + 3$

8.4 EXERCISES

1. Explain what makes a relation a function.

2. How do you determine if a graph represents a function?

3. In your own words, define *domain* and *range.*

4. What is the shape of every linear function? How do you graph a linear function?

5. What is the shape of every quadratic function? How do you graph a quadratic function?

6. What is the shape of every absolute value function? How do you graph an absolute value function?

For Exercises 7–14, identify the domain and range of each relation and determine whether it is a function.

7. Favorite vacation states.

Rank	State
1	Hawaii
2	Florida
3	Maine
4	Alaska
5	California

Source: NFO Plog 2002 American Travel Survey according to Knight Ridder Tribune.

8. Best-mannered cities in the United States for 2002 according to etiquette expert Marjabelle Young Stewart.

Rank	City
1	Charleston, SC
2	San Diego, CA
3	New York, NY
4	Philadelphia, PA
5	San Francisco, CA

9. Percent of the workforce leaving for work commute.

Percent (%)	Time of departure
2	Work at home
3	Midnight to 4:59 A.M. *and* 10 A.M. to 11:59 A.M.
5	9 A.M. to 9:59 A.M.
14	Noon to 11:59 P.M.
15	8 A.M. to 8:59 A.M.
26	5 A.M. to 6:59 A.M.
33	7 A.M. to 7:59 A.M.

Source: U.S. Census 2000.

10. The percent of resale value to job cost for home repair/remodeling projects according to *Remodeling* magazine's 2001 Cost vs. Value Report.

Cost recouped (%)	Project
128	Bathroom remodel
87	Exterior paint
82	Attic bedroom
69	Minor kitchen remodel, Reroofing
59	Deck addition
58	Major kitchen remodel
37	Home office

11. $\{(2, 1), (-3, 2), (0, 5), (1, 8)\}$

12. $\{(1, 1), (2, 2), (-3, -3), (4, 4)\}$

13. $\{(-3, 2), (4, 1), (-3, 5), (2, 8)\}$

14. $\{(-1, 0), (-3, 0), (4, 5), (2, 6)\}$

For Exercises 15–22, identify the domain and range of each relation and determine whether it is a function.

15.

WOMEN ARE WAITING

The average age of American mothers giving birth to their first child: 21.4 21.8 22.7 23.7 24.2 24.5 24.9

'70 '75 '80 '85 '90 '95 '00

Source: Centers for Disease Control and Prevention.

16.

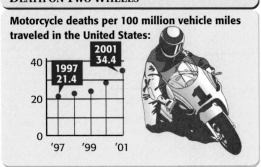

DEATH ON TWO WHEELS

Motorcycle deaths per 100 million vehicle miles traveled in the United States:

2001
34.4

1997
21.4

17.

18.

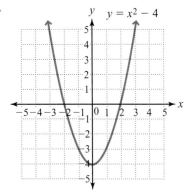

$y = x^2 - 4$

19.

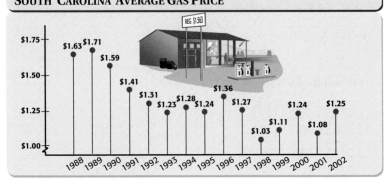

SOUTH CAROLINA AVERAGE GAS PRICE

REG $1.563

$1.75
$1.63 $1.71
$1.59
$1.50
$1.41
$1.31
$1.28
$1.23 $1.24 $1.36
$1.27 $1.24 $1.25
$1.25
$1.11
$1.03 $1.08

$1.00

1988 1989 1990 1991 1992 1993 1994 1995 1996 1997 1998 1999 2000 2001 2002

Source: S.C. Energy Office; AAA Carolinas.

20.

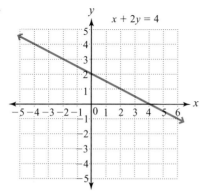

$x + 2y = 4$

21.

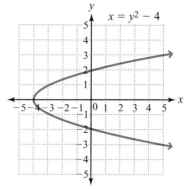

$x = y^2 - 4$

22.

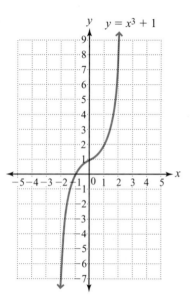

$y = x^3 + 1$

For Exercises 23–26, given $f(x) = 2x^2 - 3x + 1$, *find each of the following.*

23. $f(-2)$

24. $f(1.5)$

25. $f\left(\dfrac{1}{3}\right)$

26. $f(0)$

For Exercises 27–30, given $f(x) = |x - 4| + 2$, *find each of the following.*

27. $f(1)$

28. $f(0.5)$

29. $f\left(\dfrac{2}{5}\right)$

30. $f(-3)$

For Exercises 31–34, given $f(x) = x^3 - 2x^2 + 3$, *find each of the following.*

31. $f(0)$

32. $f(0.2)$

33. $f\left(\dfrac{1}{4}\right)$

34. $f(-2)$

For Exercises 35–38, given $f(x) = \dfrac{2x}{x-1}$, *find each of the following.*

35. $f(1)$

36. $f(-2)$

37. $f\left(\dfrac{2}{3}\right)$

38. $f(1.5)$

For Exercises 39–44, use the graph to determine the value of the function.

39. a. $f(2)$
   b. $f(-1)$
   c. $f(0)$

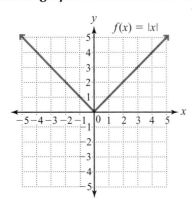

$f(x) = |x|$

40. a. $f(-2)$
   b. $f(2)$
   c. $f(-1)$

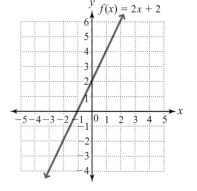

$f(x) = 2x + 2$

41. a. $f(3)$
   b. $f(-1)$
   c. $f(4)$

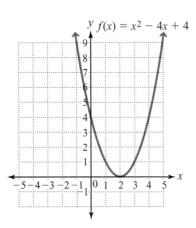

$f(x) = x^2 - 4x + 4$

42. a. $f(1)$
   b. $f(0)$
   c. $f(-1)$

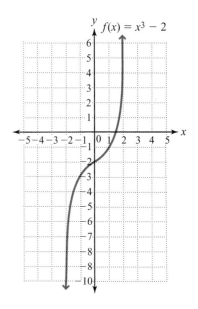

$f(x) = x^3 - 2$

43. a. $f(-1)$
  b. $f(-2)$
  c. $f(3)$

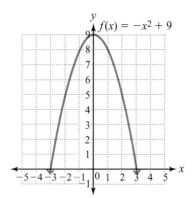

44. a. $f(3)$
  b. $f(0)$
  c. $f(4)$

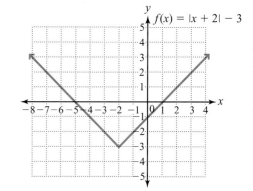

For Exercises 45–70, graph.

45. $f(x) = 2x + 3$

46. $f(x) = 3x - 1$

47. $f(x) = -x + 4$

48. $f(x) = -2x + 5$

49. $f(x) = \dfrac{1}{4}x - 2$

50. $f(x) = -\dfrac{2}{5}x + 1$

51. $f(x) = x^2 + 4$

52. $f(x) = x^2 - 3$

53. $f(x) = -x^2 + 1$

54. $f(x) = -x^2 + 9$

55. $f(x) = \dfrac{1}{4}x^2 - 2$

56. $f(x) = -\dfrac{1}{3}x^2 + 3$

57. $f(x) = x^2 - 12x + 20$

58. $f(x) = x^2 + 5x + 6$

59. $f(x) = |x| - 4$

60. $f(x) = |x| + 1$

61. $f(x) = |x + 3| - 2$

62. $f(x) = |x - 1| + 3$

63. $f(x) = -|x + 4|$

64. $f(x) = -|x - 3|$

65. $f(x) = 2|x - 1| + 3$

66. $f(x) = 3|x + 1| - 2$

67. $f(x) = -3|x + 2| - 1$

68. $f(x) = -2|x - 1| + 4$

69. $f(x) = -\dfrac{1}{2}|x + 1| + 1$

70. $f(x) = -\dfrac{1}{4}|x - 2| + 2$

Collaborative Exercises

In the following exercises, you will use a graphing utility to explore and discuss how numbers can **shift, reflect,** *or* **stretch** *a basic function. We will use the graph of* $f(x) = |x|$ *as our basic function.*

1. Graph $f(x) = |x + 2|$ and compare it with the graph of $f(x) = |x|$. What effect does adding 2 to x within the absolute value have on the graph?

2. Graph $f(x) = |x - 3|$ and compare it with the graph of $f(x) = |x|$. What effect does subtracting 3 from x within the absolute value have on the graph?

3. Graph $f(x) = |x| + 4$ and compare it with the graph of $f(x) = |x|$. What effect does adding 4 to $|x|$ have on the graph?

4. Graph $f(x) = |x| - 1$ and compare it with the graph of $f(x) = |x|$. What effect does subtracting 1 from $|x|$ have on the graph?

continued

5. Notice that adding or subtracting in a function *shifts* the position of the basic function. Without graphing, how would you expect the graph of $f(x) = |x + 1| - 3$ to be shifted as compared with the basic function $f(x) = |x|$?

6. Graph $f(x) = -|x|$. How is the graph of $f(x) = -|x|$ different from that of $f(x) = |x|$?

7. Recall that $f(x) = -|x|$ is the same as $f(x) = -1 \cdot |x|$. Multiplying the basic function by a negative number causes a *reflection* of the function about the x-axis. Without graphing, how would you describe the differences between the graph of $f(x) = -|x - 2| + 1$ and the basic function $f(x) = |x|$?

8. Graph $f(x) = 2|x|$. How is the graph of $f(x) = 2|x|$ different from that of $f(x) = |x|$?

9. Graph $f(x) = \frac{1}{2}|x|$. How is the graph of $f(x) = \frac{1}{2}|x|$ different from that of $f(x) = |x|$?

10. Notice that multiplying the basic function by a number other than 1 causes the graph to be *stretched*, so that it is wider or narrower than the basic function's graph. Without graphing, describe the differences in each of the following compared to the graph of $f(x) = |x|$.

    a. $f(x) = 3|x| + 1$

    b. $f(x) = -2|x - 4|$

    c. $f(x) = \frac{1}{3}|x + 4| - 2$

For further investigation, change the basic function to $f(x) = x^2$ and work through Exercises 1–10 again. For example, Exercise 1 would read: Graph $f(x) = (x + 2)^2$ and compare it with the graph of $f(x) = x^2$ What effect does adding 2 to x have on the graph? Exercise 3 would read: Graph $f(x) = x^2 + 4$ and compare it with the graph of $f(x) = x^2$. What effect does adding 4 to x^2 have on the graph?

REVIEW EXERCISES

For Exercises 1–6, perform the indicated operation.

1. $(2x^2 + 3x - 1) + (3x^2 - 3x + 7)$

2. $(5x^2 - 3x + 1) - (-2x^2 + 8)$

3. $(2x - 4)^2$

4. $(3x - 4)(2x + 1)$

5. $(12x^4 + 18x^3 - 30x^2) \div 6x^2$

6. $(x^4 - 81) \div (x - 3)$

8.5 Function Operations

OBJECTIVES

1 Add or subtract functions.
2 Multiply functions.
3 Divide functions.

Note: Letters other than f and g can be used to name functions, such as h, p, or q.

So far, we have seen how to evaluate and graph functions. Now we will consider how to add, subtract, multiply, and divide functions. Because we will be considering two different functions in the same problem we will write one function as f and the second as g.

OBJECTIVE 1 **Add or subtract functions.** Given two functions f and g, their sum or difference is found using the following rules.

RULE   **ADDING OR SUBTRACTING FUNCTIONS**

The sum of two functions, $f + g$, is found by $(f + g)(x) = f(x) + g(x)$.

The difference of two functions, $f - g$, is found by $(f - g)(x) = f(x) - g(x)$.

EXAMPLE 1   Given $f(x) = 2x + 1$ and $g(x) = 7x + 3$, find

a. $f + g$
b. $f - g$
c. $(f - g)(-4)$

Solution

a. $(f + g)(x) = f(x) + g(x)$

$\quad\quad\quad\quad\quad = (2x + 1) + (7x + 3)$ Write a sum of the two functions.

$\quad\quad\quad\quad\quad = 9x + 4$ Combine like terms.

b. $(f - g)(x) = f(x) - g(x)$

$\quad\quad\quad\quad\quad = (2x + 1) - (7x + 3)$ Write a difference of the two functions.

$\quad\quad\quad\quad\quad = (2x + 1) + (-7x - 3)$ Write as an equivalent addition.

$\quad\quad\quad\quad\quad = -5x - 2$ Combine like terms.

c. In part b, we found $(f - g)(x) = -5x - 2$, so to find $(f - g)(-4)$, we replace x with -4.

$\quad\quad (f - g)(-4) = -5(-4) - 2$ Replace x with -4.

$\quad\quad\quad\quad\quad\quad = 20 - 2$

$\quad\quad\quad\quad\quad\quad = 18$

ANSWERS

a. $2x^2 - 5x - 4$
b. $-x + 14$
c. 14

YOUR TURN Given $h(x) = x^2 - 3x + 5$ and $k(x) = x^2 - 2x - 9$, find

a. $h + k$ **b.** $h - k$ **c.** $(h + k)(-2)$

OBJECTIVE **2** **Multiply functions.** Given two functions f and g, their product is found using the following rule.

RULE **MULTIPLYING FUNCTIONS**

The product of two functions, $f \cdot g$, is found by $(f \cdot g)(x) = f(x)\, g(x)$.

EXAMPLE **2** Given $f(x) = 3x - 4$ and $g(x) = x + 5$, find $f \cdot g$.

Solution $(f \cdot g)(x) = f(x)\, g(x)$

$$= (3x - 4)(x + 5) \qquad \text{Write a product of the functions.}$$
$$= 3x^2 + 15x - 4x - 20 \qquad \text{Use FOIL to multiply the binomials.}$$
$$= 3x^2 + 11x - 20 \qquad \text{Combine like terms.}$$

ANSWERS

a. $35x^2 - 17x + 2$
b. 108

YOUR TURN Given $p(x) = 7x - 2$ and $q(x) = 5x - 1$, find

a. $p \cdot q$ **b.** $(p \cdot q)(2)$

OBJECTIVE **3** **Divide functions.** Given two functions f and g, their quotient is found using the following rule.

RULE **DIVIDING FUNCTIONS**

The quotient of two functions, f / g, is found by $(f/g)(x) = \dfrac{f(x)}{g(x)}$, where $g(x) \neq 0$.

EXAMPLE **3** Given $f(x) = 9x^3 - 12x^2 + 6x$ and $g(x) = 3x$, find f / g.

Note: Because we divided by $3x$, we are assuming that $x \neq 0$ in $(f/g)(x)$.

Solution $(f/g)(x) = \dfrac{f(x)}{g(x)} = \dfrac{9x^3 - 12x^2 + 6x}{3x}$ Write a quotient of the functions.

$$= \frac{9x^3}{3x} - \frac{12x^2}{3x} + \frac{6x}{3x} \qquad \text{Since the divisor is a monomial, we separate the terms.}$$
$$= 3x^2 - 4x + 2 \qquad \text{Divide.}$$

EXAMPLE 4 Given $f(x) = 3x^2 - 13x - 6$ and $g(x) = x - 5$, find f/g.

Solution $(f/g)(x) = \dfrac{f(x)}{g(x)} = \dfrac{3x^2 - 13x - 6}{x - 5}$ Write a quotient of the functions.

To find the quotient, we use long division.

$$\begin{array}{r} 3x \\ x - 5 \overline{)3x^2 - 13x - 6} \\ \underline{-(3x^2 - 15x)} \end{array}$$
Change signs. $\longrightarrow$
$$\begin{array}{r} 3x \\ x - 5 \overline{)3x^2 - 13x - 6} \\ \underline{-3x^2 + 15x} \\ 2x - 6 \end{array}$$
Combine like terms and bring down the next term.

$$\begin{array}{r} 3x + 2 \\ x - 5 \overline{)3x^2 - 13x - 6} \\ -3x^2 + 15x \\ 2x - 6 \\ \underline{-(2x - 10)} \end{array}$$
Change signs. $\longrightarrow$
$$\begin{array}{r} 3x + 2 \\ x - 5 \overline{)3x^2 - 13x - 6} \\ -3x^2 + 15x \\ 2x - 6 \\ \underline{-2x + 10} \\ 4 \end{array}$$
Combine like terms.

Note: The domain of the quotient function contains all real numbers except 5, so we would write the domain as $\{x \mid x \neq 5\}$ or, using interval notation, $(-\infty, 5) \cup (5, \infty)$.

Answer: $3x + 2 + \dfrac{4}{x - 5}$

ANSWERS

a. $f/g = 4x^2 + 12x - 7$;
$(f/g)(-2) = -15$

b. $m/n = 6x - 5 + \dfrac{-1}{x + 4}$

or $6x - 5 - \dfrac{1}{x + 4}$;

$(m/n)(1) = \dfrac{4}{5}$

YOUR TURN

a. Given $f(x) = 12x^4 + 36x^3 - 21x^2$ and $g(x) = 3x^2$, find f/g and $(f/g)(-2)$.

b. Given $m(x) = 6x^2 + 19x - 21$ and $n(x) = x + 4$, find m/n and $(m/n)(1)$.

8.5 EXERCISES

For Extra Help

InterAct Math

Addison-Wesley Math Tutor Center

Video 14/DVT 4

MyMathLab

MathXL

1. If $p(x)$ and $q(x)$ are both polynomial functions, how do you find $p + q$?

2. If $p(x)$ and $q(x)$ are both polynomial functions, how do you find $p - q$?

3. If $p(x)$ and $q(x)$ are both binomial functions, how do you find $p \cdot q$?

4. If $p(x)$ is a trinomial function and $q(x)$ is a monomial function, how do you find p/q?

For Exercises 5–14, find $f + g$ and $f - g$.

5. $f(x) = 2x, g(x) = 3x - 1$

6. $f(x) = 3x, g(x) = 4x + 2$

7. $f(x) = x - 5, g(x) = 2x - 3$

8. $f(x) = 5x + 1, g(x) = x - 8$

9. $f(x) = -4x + 7, g(x) = x + 1$

10. $f(x) = x^2 - 3x + 1, g(x) = -x - 2$

11. $f(x) = 2x^2 - 5x - 3, g(x) = x + 5$

12. $f(x) = -x^2 - 8x + 2, g(x) = x^2 + 2$

13. $f(x) = -5x^2 + 8, g(x) = 3x^2 - 2x - 1$

14. $f(x) = -2x^3 - x^2, g(x) = 2x - 9$

For Exercises 15–24, find $f \cdot g$.

15. $f(x) = x, g(x) = x - 5$

16. $f(x) = -2x, g(x) = x + 4$

17. $f(x) = x + 1, g(x) = x - 2$

18. $f(x) = x - 5, g(x) = x + 4$

19. $f(x) = 3x + 2, g(x) = x + 2$

20. $f(x) = 2x - 5, g(x) = 3x - 7$

21. $f(x) = x^2 - 3x + 2, g(x) = x^2 + 2x - 7$

22. $f(x) = x^2 - 2x - 1, g(x) = x^2 + 3x + 4$

23. $f(x) = 2x^2 - 3x + 1, g(x) = 3x^2 - x - 1$

24. $f(x) = 3x^2 - 2x - 1, g(x) = 2x^2 - 2x - 3$

For Exercises 25–34, find f / g.

25. $f(x) = 2x^2 - 6x, g(x) = 2x$

26. $f(x) = 3x^2 - 9x, g(x) = 3x$

27. $f(x) = 6x^2 - 3x + 6, g(x) = 3x$

28. $f(x) = -4x^2 + 4x - 8, g(x) = 2x$

29. $f(x) = x^2 - 14x + 45, g(x) = x - 9$

30. $f(x) = x^2 - x - 20, g(x) = x + 4$

31. $f(x) = 2x^2 - 3x - 5, g(x) = 2x - 5$

32. $f(x) = 3x^2 - 4x - 7, g(x) = 3x - 7$

33. $f(x) = 2x^2 - 5x - 7, g(x) = x + 5$

34. $f(x) = 6x^2 - 5x + 4, g(x) = 2x - 1$

For Exercises 35–38, find: a. $p + q$
b. $p - q$
c. $p \cdot q$
d. p / q

35. $p(x) = 5x - 1, q(x) = 2x + 3$

36. $p(x) = 3x - 4, q(x) = 7x + 2$

37. $p(x) = 2x^2 - 3x + 5, q(x) = x - 1$

38. $p(x) = 4x^2 - 4x - 1, q(x) = 2x + 3$

39. Using the functions in Exercise 35, find
   a. $(p + q)(-2)$ **b.** $(p - q)(5)$ **c.** $(p \cdot q)(-1)$ **d.** $(p / q)(7)$

40. Using the functions in Exercise 36, find
   a. $(p + q)(-1)$ **b.** $(p - q)(3)$ **c.** $(p \cdot q)(-2)$ **d.** $(p / q)(2)$

41. Using the functions in Exercise 37, find
   a. $(p + q)(3)$ **b.** $(p - q)(-4)$ **c.** $(p \cdot q)(2)$ **d.** $(p / q)(-3)$

42. Using the functions in Exercise 38, find
   a. $(p + q)(-2)$ **b.** $(p - q)(3)$ **c.** $(p \cdot q)(-2)$ **d.** $(p / q)(4)$

For Exercises 43–44: a. Find $f + g$.
b. Complete the table shown.
c. Graph f, g, and $f + g$ on the same graph.
d. What do the table and graphs suggest
about the sum of two functions?

43. $f(x) = 3x + 1, g(x) = x + 5$

x	f(x)	g(x)	(f + g)(x)
−4			
−2			
0			
2			
4			

44. $f(x) = -x + 3, g(x) = 4x + 3$

For Exercises 45–46: a. Find h − k.
      b. Complete the table shown.
      c. Graph h, k, and h − k on the same graph.
      d. What do the table and graphs suggest
       about the difference of two functions?

x	h(x)	k(x)	(h − k)(x)
−4			
−2			
0			
2			
4			

45. $h(x) = 2x - 9, k(x) = 3x + 5$

46. $h(x) = 4x + 1, k(x) = 2x - 7$

For Exercises 47 and 48, the function w(x) describes a company's monthly costs in wages and p(x) describes the cost of production, where x represents the number of units produced.

47. $w(x) = 3x + 2$, $p(x) = x + 2$

    a. Find the function, $t(x)$, that describes the total cost.

    b. Calculate the total cost if $x = 200$.

    c. Graph all three functions.

48. $w(x) = 2x + 1$, $p(x) = x - 1$

    a. Find the function, $t(x)$, that describes the total cost.

    b. Calculate the total cost if $x = 175$.

    c. Graph all three functions.

For Exercises 49 and 50, the function r(x) describes a company's monthly revenue and c(x) describes the monthly costs where x represents the number of units produced and then sold. Use the formula Net Profit = Revenue − Cost.

49. $r(x) = x^2 - 3x - 4$, $c(x) = x + 8$

    a. Find the function, $p(x)$, that describes the net profit or loss.

    b. Calculate the net profit or loss if $x = 100$. Is this value a profit or a loss?

    c. Graph all three functions.

    d. How many units must be sold for the company to break even (which means the net = 0)?

50. $r(x) = x^2 - 4x - 5$, $c(x) = x + 1$

    a. Find the function, $p(x)$, that describes the net profit or loss.

    b. Calculate the net profit or loss if $x = 300$. Is this value a profit or a loss?

    c. Graph all three functions.

    d. How many units must be sold for the company to break even (which means the net = 0)?

For Exercises 51 and 52, the function l(x) describes the length of a box and h(x) describes the height, where x represents the width.

51. $l(x) = 3x + 2$, $h(x) = 2x$

    a. Find the function that describes the area of the front face of the box.

    b. Calculate the area of the front face if $x = 3$ ft.

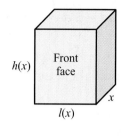

52. $l(x) = 5x - 1$, $h(x) = 3x$

    a. Find the function that describes the area of the front face of the box.

    b. Calculate the area of the front face if $x = 2$ m.

For Exercises 53 and 54, the function V(x) describes the volume of a box, l(x) describes its length, and w(x) describes its width.

53. $V(x) = 4x^3 - 7x^2 - 14x - 3$, $l(x) = x - 3$, $w(x) = x + 1$

    a. Find the function, $h(x)$, that describes the height of the box.

    b. Calculate the height of the box if $x = 4$ cm.

54. $V(x) = 6x^3 + 11x^2 - 47x + 20$, $l(x) = x + 4$, $w(x) = 2x - 1$

    a. Find the function, $h(x)$, that describes the height of the box.

    b. Calculate the height of the box if $x = 5$ in.

REVIEW EXERCISES

For Exercises 1 and 2, solve.

    1. $x^2 - 16 = 0$ **2.** $|x + 5| = -8$

For Exercises 3 and 4, graph.

    3. $y = 2x - 3$ **4.** $y = x^2 - 9$

For Exercises 5 and 6, factor completely.

    5. $3x - 3y - xy + y^2$ **6.** $-3x^2 + 27$

Defined Terms

Section 8.1
Compound inequality (p. 644)
Intersection (p. 645)
Union (p. 649)

Section 8.4
Relation (p. 673)
Domain (p. 673)
Range (p. 673)
Function (p. 673)

Procedures, Rules, and Key Examples

Procedures/Rules	Key Examples
Section 8.1 Compound Inequalities To solve a compound inequality involving "and," 1. Solve each inequality in the compound inequality. 2. The solution set will be the intersection of the individual solution sets.	For each example, solve the compound inequality, then **a.** Graph the solution set. **b.** Write the solution set in set-builder notation. **c.** Write the solution set in interval notation.

For each example, solve the compound inequality, then
a. Graph the solution set.
b. Write the solution set in set-builder notation.
c. Write the solution set in interval notation.

Example 1: $-3 < -2x - 5 \le 1$
Solution:
a.
$$-2x - 5 > -3 \quad \text{and} \quad -2x - 5 \le 1$$
$$-2x > 2 \quad \text{and} \quad -2x \le 6$$
$$x < -1 \quad \text{and} \quad x \ge -3$$

Graph:
$$-6\ -5\ -4\ -3\ -2\ -1\ \ 0\ \ 1$$

b. Set-builder notation: $\{x \mid -3 \le x < -1\}$

c. Interval notation: $[-3, -1)$

Example 2: $5x - 1 \le -16$ or $5x - 1 > 9$
Solution:
a.
$$5x - 1 \le -16 \quad \text{or} \quad 5x - 1 > 9$$
$$5x \le -15 \quad \text{or} \quad 5x > 10$$
$$x \le -3 \quad \text{or} \quad x > 2$$

Graph:
$$-5\ -4\ -3\ -2\ -1\ \ 0\ \ 1\ \ 2\ \ 3\ \ 4$$

b. Set-builder notation:
$\{x \mid x \le -3 \text{ or } x > 2\}$

c. Interval notation: $(-\infty, -3] \cup (2, \infty)$

To solve a compound inequality involving "or,"
1. Solve each inequality in the compound inequality.
2. The solution set will be the union of the individual solution sets.

continued

Procedures/Rules	Key Examples

Procedures/Rules

Section 8.2 Equations Involving Absolute Value

If $|x| = a$, where x is a variable or an expression and $a \geq 0$, then $x = a$ or $x = -a$.

To solve an equation containing a single absolute value,
1. Isolate the absolute value so that the equation is in the form $|ax + b| = c$.
2. Separate the absolute value into two equations, $ax + b = c$ and $ax + b = -c$.
3. Solve both equations.

To solve an equation in the form $|ax + b| = |cx + d|$,
1. Separate the absolute value equation into two equations:
   $ax + b = cx + d$ and $ax + b = -(cx + d)$.
2. Solve both equations.

Section 8.3 Inequalities Involving Absolute Value

To solve an inequality in the form $|x| < a$ where $a > 0$,
1. Rewrite as a compound inequality involving "and": $x > -a$ and $x < a$.
2. Solve the compound inequality.

   Similarly, to solve $|x| \leq a$, we would write $x \geq -a$ and $x \leq a$.

Key Examples

Example 1: Solve $|4x + 1| - 3 = 2$.
Solution: $|4x + 1| = 5$ Isolate the absolute value.

$$4x + 1 = 5 \quad \text{or} \quad 4x + 1 = -5$$
$$4x = 4 \quad \text{or} \quad 4x = -6 \quad \text{Use the}$$
$$x = 1 \quad \text{or} \quad x = -\frac{3}{2} \quad \begin{array}{l}\text{absolute} \\ \text{value} \\ \text{property.}\end{array}$$

Example 2: Solve $|5x - 7| = |3x + 1|$.
Solution: Separate into two equations.

Equal		Opposites
$5x - 7 = 3x + 1$	or	$5x - 7 = -(3x + 1)$
$2x - 7 = 1$	or	$5x - 7 = -3x - 1$
$2x = 8$	or	$8x = 6$
$x = 4$	or	$x = \dfrac{3}{4}$

For each example, solve the inequality, then
a. Graph the solution set.
b. Write the solution set in set-builder notation.
c. Write the solution set in interval notation.

Example 1: $|2x - 3| - 1 < 4$
Solution:
a. $|2x - 3| < 5$ Isolate the absolute value.

$$2x - 3 > -5 \quad \text{and} \quad 2x - 3 < 5 \quad \text{Separate}$$
$$2x > -2 \quad \text{and} \quad 2x < 8 \quad \text{using}$$
$$x > -1 \quad \text{and} \quad x < 4 \quad \text{"and."}$$

b. Set-builder notation: $\{x | -1 < x < 4\}$

c. Interval notation: $(-1, 4)$

continued

Procedures/Rules	Key Examples

Procedures/Rules

Section 8.3 Inequalities Involving Absolute Value (continued)

To solve an inequality in the form $|x| > a$,
1. Rewrite as a compound inequality involving "or": $x < -a$ or $x > a$.
2. Solve the compound inequality.

Similarly, to solve $|x| \geq a$, we would write $x \leq -a$ or $x \geq a$.

Section 8.4 Functions and Graphing

Relation: A set of ordered pairs.
Domain: The set of all input values for a relation.
Range: The set of all output values for a relation.
Function: A relation in which every value in the domain is paired with exactly one value in the range.

To find the value of a function replace x with the given value.

Key Examples

Example 2: $|3x - 5| - 7 \geq 4$
Solution:
a. $|3x - 5| \geq 11$ Isolate the absolute value.

$3x - 5 \leq -11$ or $3x - 5 \geq 11$ Separate using "or".

$\quad 3x \leq -6$ or $\quad\quad 3x \geq 16$

$\quad\quad x \leq -2$ or $\quad\quad x \geq \dfrac{16}{3}$

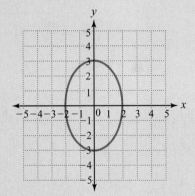

b. Set-builder notation:
$$\left\{ x \,\middle|\, x \leq -2 \text{ or } x \geq \frac{16}{3} \right\}$$

c. Interval notation: $(-\infty, -2] \cup \left[\dfrac{16}{3}, \infty\right)$

Example 1: For the relation shown, determine the domain, range, and if it is a function.

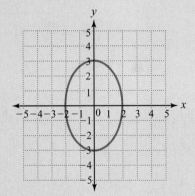

Domain: $\{x \mid -2 \leq x \leq 2\}$
Range: $\{y \mid -3 \leq y \leq 3\}$
It is not a function because it does not pass the vertical line test.

Example 2: Given $f(x) = 3x^2 - 4$, find $f(2)$.

$$f(2) = 3(2)^2 - 4$$
$$= 12 - 4$$
$$= 8$$

continued

Procedures/Rules	Key Examples

Section 8.4 Functions and Graphing (continued)

The graph of a linear function is a line.

A linear function has the form $f(x) = mx + b$, where m is the slope and $(0, b)$ is the y-intercept. To graph, we need to find only two ordered pairs.

Example 3: Graph $f(x) = -2x + 3$.

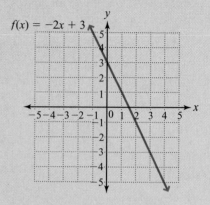

The graph of a quadratic function is a parabola.

Quadratic functions have the form $f(x) = ax^2 + bx + c$ and we saw their graphs are parabolas. To graph a parabola we have to find many ordered pairs.

Example 4: Graph $f(x) = x^2 - 2$.

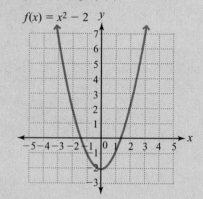

The graph of an absolute value function is a V shape.

To graph, find enough ordered pairs to generate the V shape.

Example 5: Graph $f(x) = |x + 2|$.

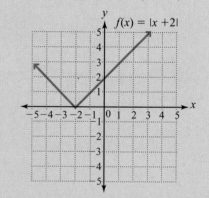

continued

Procedures/Rules	Key Examples
Section 8.5 Function Operations	

Section 8.5 Function Operations

The sum of two functions, $f + g$, is found by $(f + g)(x) = f(x) + g(x)$.

The difference of two functions, $f - g$, is found by $(f - g)(x) = f(x) - g(x)$.

The product of two functions, $f \cdot g$, is found by $(f \cdot g)(x) = f(x)\, g(x)$

The quotient of two functions, f / g, is found by $(f / g)(x) = \dfrac{f(x)}{g(x)}$, where $g(x) \neq 0$.

Key Examples

Example 1: Given $f(x) = 2x - 5$ and $g(x) = 3x + 4$, find
   a. $f + g$ **b.** $f - g$ **c.** $f \cdot g$

a. $(f + g)(x) = (2x - 5) + (3x + 4)$
$$= 5x - 1$$

b. $(f - g)(x) = (2x - 5) - (3x + 4)$
$$= (2x - 5) + (-3x - 4)$$
$$= -x - 9$$

c. $(f \cdot g)(x) = (2x - 5)(3x + 4)$
$$= 6x^2 + 8x - 15x - 20$$
$$= 6x^2 - 7x - 20$$

Example 2: Given $f(x) = 12x^3 - 20x^2 + 8x$ and $g(x) = 4x$, find f / g.

Solution:
$$(f / g)(x) = \frac{f(x)}{g(x)} = \frac{12x^3 - 20x^2 + 8x}{4x}$$
$$= \frac{12x^3}{4x} - \frac{20x^2}{4x} + \frac{8x}{4x}$$
$$= 3x^2 - 5x + 2,\ x \neq 0$$

For Exercises 1–6, answer true or false.

1. The notation $(2, 5)$ can represent an ordered pair or interval notation.

2. To solve absolute value equations, the absolute value must be isolated.

3. The notation $f(x)$ means $f \cdot x$.

4. All functions are also relations.

5. The domain of a function $f(x)$ is a set containing all values of x for which the function is defined.

6. Functions always pass the vertical line test.

For Exercises 7–10, fill in the blank.

7. For two sets A and B, the intersection of A and B, symbolized by _____, is a set containing elements that are only in both A and B.

8. To solve an equation containing a single absolute value,
   1. Isolate the absolute value so that the equation is in the form $|ax + b| = c$.
   2. Separate the absolute value into two equations, _____ and _____.
   3. Solve both equations.

9. The set of all output values for a relation is _____.

10. The sum of two functions $f + g$ is shown by $(f + g)(x) =$ _____.

For Exercises 11–12, find the intersection and union of the given sets.

11. $A = \{1, 5, 9\}$
    $B = \{1, 5, 7, 9\}$

12. $A = \{1, 2, 3\}$
    $B = \{4, 5, 6, 7\}$

For Exercises 13–24, solve the compound inequality, then a. Graph the solution set.

b. *Write the solution set in set-builder notation.*

c. *Write the solution set in interval notation.*

13. $4 < -2x < 6$

14. $9 \leq -3x \leq 15$

15. $-3 < x + 4 < 7$

16. $0 < x - 1 \leq 3$

17. $-1 \leq 2x + 3 < 3$

18. $-5 < 3x - 2 < 7$

19. $w + 4 \leq -2$ or $w + 4 \geq 2$

20. $3x + 2 \leq -2$ or $3x + 2 \geq 8$

21. $2m - 5 < 0$ or $2m - 5 > 5$

22. $4w - 3 < 1$ or $4w - 3 > 0$

23. $-x - 6 \leq -2$ or $-x - 6 \geq 3$

24. $-4w + 1 \leq -3$ or $-4w + 1 \geq 5$

For Exercises 25–34, solve.

25. $|x| = 4$

26. $|x - 4| = 7$

27. $|2w - 1| = 3$

28. $|3u - 4| = 2$

29. $|5r + 8| = -3$

30. $|q - 4| - 3 = 8$

31. $|5w| - 2 = 13$

32. $2|3x - 4| = 8$

33. $1 - |2x + 3| = -3$

34. $4 - 2|r - 5| = -8$

For Exercises 35–44, solve the inequality, then a. Graph the solution set.
                                              b. Write the solution set in set-builder notation.
                                              c. Write the solution set in interval notation.

35. $|x| < 5$

36. $|p| \geq 4$

37. $|x - 3| > 7$

38. $|2m + 6| < 4$

39. $|3s - 1| \leq -2$

40. $5|b| - 2 > 3$

41. $7|m + 3| \leq 21$

42. $-2|t - 5| < -10$

43. $5 - 2|2k - 3| \leq -15$

44. $3 - 7|2p + 4| \leq 24$

For Exercises 45–50, identify the domain and range of the relation, then determine whether it is a function.

45. Top-Selling Comic Books

Rank	Title
1	Batman, No. 613
2	Ultimate Spider-Man, No. 38
3	Ultimate Spider-Man, No. 37
4	Amazing Spider-Man, No. 51
5	New X-Men, No. 138
6	Ultimate X-Men, No. 31

Source: Diamond Comics Distributors. Rankings are based on orders placed by retailers and reflect total comics ordered as of 4/8/03.

46. **Top Business Schools for Graduate Studies as Ranked by** *U.S. News and World Report* **2004**

Rank	Business School
1	Harvard University (MA)
2	Stanford University (CA), University of Pennsylvania (Wharton)
3	Massachusetts Institute of Technology (Sloan), Northwestern University (Kellogg) (IL)
4	Columbia University (NY)

Source: U.S. News and World Report

47. $\{(2, 1), (4, -3), (2, 4), (3, -3)\}$

48. $\{(0, 3), (-2, 3), (1, 4)\}$

49.

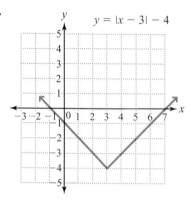

$y = |x - 3| - 4$

50.

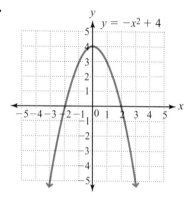

$y = -x^2 + 4$

51. Find $f(3)$ given $f(x) = 3x^2 - 4x + 2$.

52. Find $f(-2)$ given $f(x) = -|2x - 5|$.

For Exercises 53–56, graph.

53. $f(x) = -3x + 2$

54. $f(x) = 2x^2 - 3$

55. $f(x) = |x - 2|$

56. $f(x) = -2|x + 1| + 3$

For Exercises 57–60, find a. f + g
          b. f − g
          c. f · g

57. $f(x) = -3x + 9, g(x) = 4x + 1$

58. $f(x) = -x - 2, g(x) = 8x + 2$

59. $f(x) = 3x^2 - x + 5, g(x) = x + 2$

60. $f(x) = x^2 - 2x - 1, g(x) = 3x + 1$

For Exercises, 61–62, find f / g.

61. $f(x) = 6x^4 - 15x^3 + 12x^2, g(x) = 3x^2$

62. $f(x) = 2x^3 - 17x^2 + 36x - 17, g(x) = 2x - 5$

For Exercises 63–66, solve.

63. Students in a biology course receive a B if the average of four tests is between 80 and 90. Suppose a student has the following scores: 80, 89, and 83. What range of scores on the fourth test would cause the student to receive a B?

64. The width of a rectangular building is to be 80 feet. What range of values can the length have so that the area is between 12,000 and 16,000 square feet?

65. The function $w(x) = 3x + 5$ describes a company's monthly expenditure on wages. The function $p(x) = x - 1$ describes the total cost of production. Note that x represents the number of units produced.

   a. Find the function, $c(x)$, that describes the total cost.

   b. If $x = 225$, then calculate the total cost.

   c. Graph all three functions.

66. The function $V(x) = 6x^3 + 41x^2 + 26x - 24$ describes the volume of a box, $l(x) = 3x + 4$ describes its length, and $w(x) = x + 6$ describes its width.

   a. Find the function, $h(x)$, that describes the height of the box.

   b. Calculate the height of the box if $x = 9$ in.

1. Find the intersection and union of the given sets. $A = \{h, o, m, e\}$
$$B = \{h, o, u, s, e\}$$

For 2–5, solve.

2. $|x + 3| = 5$

3. $3 - |2x - 3| = -6$

4. $|2x + 3| = |x - 5|$

5. $|5x - 4| = -3$

For 6–15, solve and then a. Graph the solution set.
          b. Write the solution set in set-builder notation.
          c. Write the solution set in interval notation.

6. $-3 < x + 4 \leq 7$

7. $4 < -2x \leq 6$

8. $5x + 2 \leq -3$ or $5x + 2 \geq 12$

9. $6 - 2n < 2$ or $6 - 2n < 4$

10. $|x + 4| < 9$

11. $2|x - 1| > 4$

12. $3 - 2|x + 4| > -3$

13. $|3y - 2| < -2$

14. $|8t + 4| \geq -12$

15. $2|3x - 4| \leq 10$

16. Identify the domain and range of the relation, then determine whether it is a function.

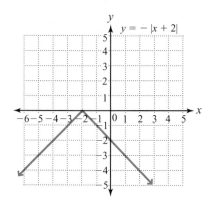

17. Given $f(x) = x^2 - 5x + 4$, find $f(-2)$.

For Exercises 18–20, graph.

18. $f(x) = -\dfrac{3}{4}x + 1$

19. $f(x) = x^2 - 3$

20. $f(x) = |x + 1| - 2$

For Exercises 21 and 22, find a. $f + g$
b. $f - g$
c. $f \cdot g$

21. $f(x) = 6x - 9, g(x) = x + 2$

22. $f(x) = x^2, g(x) = 3x^2 - 2$

23. Given $f(x) = 20x^4 + 36x^3 - 28x^2$ and $g(x) = 4x^2$, find f / g.

24. A mail-order music club offers one bonus book if the total of an order is between \$40 and \$50. Johnson decides to buy the lowest priced books, which are \$5 each. What range would the number of books he orders have to be in order to receive a bonus book?

25. The function $A(x) = 36x^2 - x - 2$ describes the area of the front face of a box and $l(x) = 4x - 1$ describes its length, where x represents its width.

a. Find the function, $h(x)$, that describes the height of the box.

b. Calculate the height of the box if $x = 15$ cm.

Systems of Equations

In Chapter 3, we encountered problems involving two unknown amounts, such as finding two consecutive integers whose sum is 63. At that time, we only knew how to solve equations with a *single* variable, so we had to let one of the unknown amounts be represented by a variable, then describe the other unknown amount in terms of that variable. For example, in the consecutive integer problem just mentioned, we would let x represent the first integer and then describe the consecutive integer as $x + 1$.

In this chapter we will again solve problems involving two or more unknowns; however, we will now assign a different variable to each unknown and translate the problem to several equations, called a *system of equations*. As we will see, it is easier to translate a problem involving two or more unknowns to a system of equations than trying to translate the problem to a single equation in terms of a single variable. The challenge with systems of equations is in solving them. We will learn several techniques for solving a system of equations: graphing (Section 9.1), substitution (Section 9.2), elimination (Section 9.3), matrices (Section 9.5), and Cramer's Rule (Section 9.6). Finally, in Section 9.7, we will solve systems of linear inequalities.

"Everybody's a mad scientist, and life is their lab. We're all trying to experiment to find a way to live, to solve problems, to fend off madness and chaos."

—*David Cronenberg*

"The problem is that we attempt to solve the simplest questions cleverly, thereby rendering them unusually complex. One should seek the simple solution."

—*Anton Chekhov*

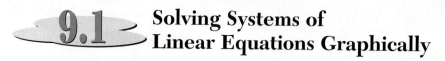

9.1 Solving Systems of Linear Equations Graphically

OBJECTIVES

① Determine if an ordered pair is a solution for a system of equations.

② Determine whether a system of equations is consistent, inconsistent, or dependent.

③ Solve a system of linear equations graphically.

OBJECTIVE ① **Determine if an ordered pair is a solution for a system of equations.** Problems involving multiple unknowns can be represented by a group of equations called a **system of equations**.

> **DEFINITION**
>
> **System of equations:** A group of two or more equations.

For example, look at the following problem:

The sum of two numbers is 3.
Twice the first number plus three times the second number is 8.

What are the two numbers? If x represents the first number and y represents the second number, then we can translate each sentence to an equation. The two equations together form a system of equations that describes the problem.

$$\text{System of equations} \quad \begin{cases} x + y = 3 & \text{(Equation 1)} \\ 2x + 3y = 8 & \text{(Equation 2)} \end{cases}$$

Our goal will be to find the **solution for the system of equations**.

> **DEFINITION**
>
> **Solution for a system of equations:** An ordered set of numbers that makes all equations in the system true.

For example, in the above system of equations, $x = 1$ and $y = 2$ make both equations true, so they form a solution to the system of equations. We can write the solution as an ordered pair: $(1, 2)$. We can check by substituting the values for the corresponding variables.

> **PROCEDURE** CHECKING A SOLUTION TO A SYSTEM OF EQUATIONS
>
> To verify or check a solution to a system of equations,
> 1. Replace each variable in each equation with its corresponding value.
> 2. Verify that each equation is true.

EXAMPLE 1 Determine whether each ordered pair is a solution to the system of equations.

$$\begin{cases} x - 2y = 3 & \text{(Equation 1)} \\ y = 3x + 1 & \text{(Equation 2)} \end{cases}$$

a. $(5, 1)$

Solution

Equation 1:
$$x - 2y = 3$$
$$5 - 2(1) \stackrel{?}{=} 3$$
$$5 - 2 \stackrel{?}{=} 3$$
$$3 = 3 \quad \text{True.}$$

Equation 2:
$$y = 3x + 1$$
$$1 \stackrel{?}{=} 3(5) + 1$$
$$1 \stackrel{?}{=} 15 + 1$$
$$1 = 16 \quad \text{False.}$$

In both equations, replace x with 5 and y with 1.

Because $(5, 1)$ does not satisfy both equations, it is not a solution for the system of equations.

b. $(-1, -2)$

Solution

Equation 1:
$$x - 2y = 3$$
$$-1 - 2(-2) \stackrel{?}{=} 3$$
$$-1 + 4 \stackrel{?}{=} 3$$
$$3 = 3 \quad \text{True.}$$

Equation 2:
$$y = 3x + 1$$
$$-2 \stackrel{?}{=} 3(-1) + 1$$
$$-2 \stackrel{?}{=} -3 + 1$$
$$-2 = -2 \quad \text{True.}$$

In both equations, replace x with -1 and y with -2.

Because $(-1, -2)$ satisfies both equations, it is a solution for the system of equations.

YOUR TURN Determine whether each ordered pair is a solution to the system of equations.

$$\begin{cases} 3x + y = 1 & \text{(Equation 1)} \\ y = 1 - 2x & \text{(Equation 2)} \end{cases}$$

a. $(2, -5)$ **b.** $(0, 1)$

ANSWERS
a. Not a solution.
b. A solution.

OBJECTIVE 2 Determine whether a system of equations is consistent, inconsistent, or dependent.

Consistent Systems: Systems with Solutions

In Example 1, we found that the system of equations $x - 2y = 3$ and $y = 3x + 1$ has a solution, $(-1, -2)$. If a system *has a solution*, then it is said to be **consistent**.

DEFINITION *Consistent system of equations:* A system of equations that has at least one solution.

We can use graphs to help visualize what it means for a system of equations to be consistent.

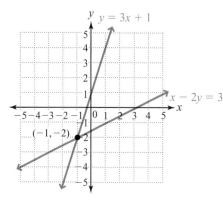

A consistent system

Notice that the graphs of $x - 2y = 3$ and $y = 3x + 1$ intersect at the point $(-1, -2)$, which is the solution for the system. If two linear graphs intersect at a single point, then the system is consistent with a single solution at the point of intersection. Also, notice from the graphs that the lines have different slopes, which is always the case when a consistent system of linear equations has a single solution.

Inconsistent Systems: Systems with No Solution

Some systems of equations have *no solution at all.* Such a system is said to be **inconsistent**.

> **DEFINITION**
>
> **Inconsistent system of equations:** A system of equations that does not have a solution.

For a system of equations to be inconsistent, the graphs of the equations must not intersect, which means they must be parallel lines.

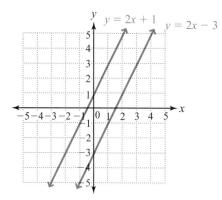

An inconsistent system

For example, the system $y = 2x - 3$ and $y = 2x + 1$ is inconsistent because the graphs of the equations are parallel lines; they have no point of intersection. We know that the graphs of $y = 2x - 3$ and $y = 2x + 1$ are parallel lines because they have the same slope, 2, but different y-intercepts.

Dependent and Independent Equations

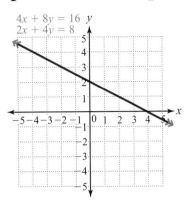

$4x + 8y = 16$
$2x + 4y = 8$

A consistent system with
dependent equations

Not all consistent systems have just one solution. Consider the system $4x + 8y = 16$ and $2x + 4y = 8$. Notice that their graphs are identical. Consequently, this system has an infinite number of solutions, which are the coordinates of any point on that line. We say that these equations are **dependent**.

DEFINITION | ***Dependent equations:*** Two equations with identical solution sets.

Note: Because two dependent equations have identical solution sets, their graphs will be identical.

We can see without graphing that the two equations describe the same line by writing them in slope-intercept form.

$$4x + 8y = 2 \qquad\qquad 2x + 4y = 1$$

Subtract $4x$ from both sides.	$8y = -4x + 2$	$4y = -2x + 1$	Subtract $2x$ from both sides.
Divide both sides by 8 to isolate y.	$y = -\dfrac{4}{8}x + \dfrac{2}{8}$	$y = -\dfrac{2}{4}x + \dfrac{1}{4}$	Divide both sides by 4 to isolate y.
Simplify the fractions.	$y = -\dfrac{1}{2}x + \dfrac{1}{4}$	$y = -\dfrac{1}{2}x + \dfrac{1}{4}$	Simplify the fractions.

Both equations are identical with the same slope, $-\dfrac{1}{2}$, and the same y-intercept, $\dfrac{1}{4}$. This will always be the case with dependent linear equations with two variables.

Equations that have different solution sets are said to be **independent**.

DEFINITION | ***Independent equations:*** Two equations with different solution sets.

Note: Because their solution sets are different, two independent linear equations in two variables will either intersect at a single point, or will be parallel.

Classifying Systems of Equations

We now summarize our discussion of consistent and inconsistent systems, and dependent and independent equations with this fact: Every system of two linear equations in two variables can be categorized in one of three ways:

| **Consistent system:** The system has a single solution at the point of intersection. | **Consistent system:** The system has an infinite number of solutions. | **Inconsistent system:** The system has no solution. |
| **Independent equations:** The graphs are different and intersect at one point. They have different slopes. | **Dependent equations:** The graphs are identical. They have the same slope and same y-intercept. | **Independent equations:** The graphs are different and are parallel. Though they have the same slope, they have different y-intercepts. |

Note that by observing the slope and y-intercept of the equations in a system of equations, we can tell which of the three cases we are dealing with.

PROCEDURE CLASSIFYING SYSTEMS OF EQUATIONS

To classify a system of equations, write the equations in slope-intercept form and compare the slopes and y-intercepts.

1. If the slopes are different, then the system is **consistent with independent equations** and has a single solution.
2. If the slopes are equal and the y-intercepts are also equal, then the system is **consistent with dependent equations** and has an infinite number of solutions.
3. If the slopes are equal with different y-intercepts, then the system is **inconsistent with independent equations** and has no solution.

EXAMPLE 2 Consider the system: $\begin{cases} 4x - 6y = 3 \\ 3y - 2x = 3 \end{cases}$

a. Is the system consistent or inconsistent?
b. Are the equations dependent or independent.
c. How many solutions does the system have?

Solution Write the equations in slope-intercept form, then compare the slopes and y-intercepts.

$$4x - 6y = 3 \qquad\qquad 3y - 2x = 3$$
$$-6y = -4x + 3 \qquad\qquad 3y = 2x + 3$$
$$y = \frac{-4}{-6}x + \frac{3}{-6} \qquad\qquad y = \frac{2}{3}x + \frac{3}{3}$$
$$y = \frac{2}{3}x - \frac{1}{2} \qquad\qquad y = \frac{2}{3}x + 1$$

a. The slopes are the same and the y-intercepts are different, so the graphs are parallel lines. Therefore, the system of equations is inconsistent.
b. The graphs are different, so the equations are independent.
c. The system has no solution.

Consider the system: $\begin{cases} x - 2y = 1 \\ x - 4y = 2 \end{cases}$

a. Determine whether the system is consistent with independent equations, consistent with dependent equations, or inconsistent with independent equations.

b. How man solutions does the system have?

OBJECTIVE ③ **Solve a system of linear equations graphically.** Our work in Objective 2 suggests the following graphical method for solving systems of equations.

PROCEDURE SOLVING SYSTEMS OF LINEAR EQUATIONS GRAPHICALLY

To solve a system of linear equations graphically,

1. Graph each equation.
   a. If the lines intersect at a single point, then the coordinates of that point form the solution. (A consistent system with independent equations.)
   b. If the lines are parallel, then there is no solution. (An inconsistent system with independent equations.)
   c. If the lines are identical, then there are an infinite number of solutions, which are coordinates of all the points on that line. (A consistent system with dependent equations.)

2. In any case, check your solution.

EXAMPLE ③ Solve the system of equations graphically.

a. $\begin{cases} y = -2x + 3 \\ x - 2y = 4 \end{cases}$

Solution Graph each equation.

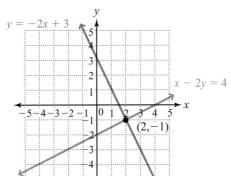

The lines intersect at a single point, which appears to be $(2, -1)$. We can verify that it is the solution by substituting into the equations.

$$y = -2x + 3 \qquad\qquad x - 2y = 4$$
$$-1 \overset{?}{=} -2(2) + 3 \qquad 2 - 2(-1) \overset{?}{=} 4$$
$$-1 \overset{?}{=} -4 + 3 \qquad\qquad 2 + 2 \overset{?}{=} 4$$
$$-1 = -1 \text{ True.} \qquad\qquad 4 = 4 \text{ True.}$$

Answer: Because $(2, -1)$ makes both equations true, it is the solution.

Warning: Graphing by hand can be imprecise and not all solutions are integers. Always check your solutions by substituting them back into the original equations.

b. $\begin{cases} 3x + 4y = 8 \\ y = -\dfrac{3}{4}x - 3 \end{cases}$

Solution Graph each equation.

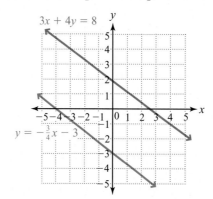

The lines appear to be parallel, which we can verify by comparing the slopes. The slope of $y = -\dfrac{3}{4}x - 3$ is $-\dfrac{3}{4}$. To determine the slope of $3x + 4y = 8$, we rewrite it in slope-intercept form.

$3x + 4y = 8$

$\qquad 4y = -3x + 8$ Subtract $3x$ from both sides.

$\qquad\quad y = -\dfrac{3}{4}x + 2$ Divide both sides by 4 to isolate y.

Answer: Since the slopes are equal and the y-intercepts are different, the lines are parallel, which means the system of equations is inconsistent and has no solution.

c. $\begin{cases} 6x + 2y = 4 \\ y = -3x + 2 \end{cases}$

Solution Graph each equation.

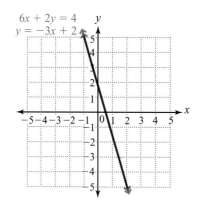

The lines appear to be identical, which we can verify by rewriting $6x + 2y = 4$ in slope-intercept form:

$6x + 2y = 4$

$\qquad 2y = -6x + 4$ Subtract $6x$ from both sides.

$\qquad\quad y = -3x + 2$ Divide both sides by 2 to isolate y.

The equations are the same, so they are dependent. There are an infinite number of solutions, which are the coordinates of all the points on the line.

ANSWER

Solution: $(-2, 4)$

YOUR TURN   **Solve the system of equations graphically.**

$$\begin{cases} 3x + y = -2 \\ x + 2y = 6 \end{cases}$$

To solve a system of equations using a graphing calculator, begin by graphing each equation. Remember that you must write the equations in slope-intercept form in order to enter them in the calculator. Using the [Y=] *key, enter one equation in Y1 and the second equation in Y2, then press* [GRAPH] *to see both graphs.*

Plot 1 Plot 2   Plot 3
$\backslash Y1 = -2X + 3$
$\backslash Y2 = X - 3$
$\backslash Y3 =$
$\backslash Y4 =$
$\backslash Y5 =$
$\backslash Y6 =$
$\backslash Y7 =$

You can find the coordinates of the point of intersection by using the CALC *feature. To use the* CALC *feature, press* [2nd] [TRACE] *, which will open a menu. Select* intersect*. The calculator will then prompt you to indicate or verify which lines you want it to find the intersection of, so press* [ENTER] *for each line. It will then prompt you to indicate which point of intersection. Because these straight lines will have only one point of intersection, you can simply press* [ENTER] *. The coordinates of the point of intersection will appear at the bottom of the screen.*

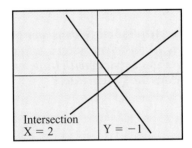

9.1 EXERCISES

1. How do you check a solution for a system of equations?

2. If a system of linear equations has at least one solution (consistent), what does this indicate about the graphs of the equations?

3. If a system of linear equations has no solution (inconsistent), what does this indicate about the graphs of the equations?

4. If two equations are dependent, what does this mean about their graphs?

5. Can a system of linear equations have more than one solution? Explain.

6. What are some weaknesses in using graphing to find a solution to a system of equations?

For Exercises 7–16, determine whether the given ordered pair is a solution to the given system of equations.

7. $(3, 2)$; $\begin{cases} x - y = 1 \\ -x + y = -1 \end{cases}$

8. $(4, 3)$; $\begin{cases} 3x + 4y = 24 \\ 3x + 2y = 18 \end{cases}$

9. $(1, 1)$; $\begin{cases} x = y \\ 9y - 13 = -4x \end{cases}$

10. $(3, -2)$; $\begin{cases} 2x = -3y \\ x - 2y = 7 \end{cases}$

11. $(2, -3)$; $\begin{cases} 2x - y = 9 \\ x - 4y = 8 \end{cases}$

12. $\left(\dfrac{3}{2}, -\dfrac{3}{2}\right)$; $\begin{cases} x - y = 3 \\ x + y = 0 \end{cases}$

13. $\left(\dfrac{2}{3}, -1\right)$; $\begin{cases} 3x + y = 1 \\ 3x + y = 3 \end{cases}$

14. $(1, 5)$; $\begin{cases} 0.25x + 0.75y = 4 \\ x + y = -7 \end{cases}$

15. $(5, -2)$; $\begin{cases} x + y = 3 \\ \dfrac{1}{5}x - \dfrac{1}{2}y = 2 \end{cases}$

16. $(-1, -2)$; $\begin{cases} \dfrac{2}{3}x - y = 7 \\ 0.25x + y = -1 \end{cases}$

For Exercises 17–22: a. Determine whether the graph shows a consistent system with independent equations, a consistent system with dependent equations, or an inconsistent system with independent equations.
b. How many solutions does the system have?

17.

18.

19.

20.

21.

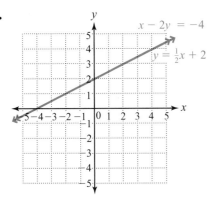

$x - 2y = -4$

$y = \frac{1}{2}x + 2$

22.

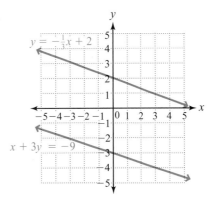

$y = -\frac{1}{3}x + 2$

$x + 3y = -9$

For Exercises 23–30, determine whether the system of equations is consistent with independent equations, consistent with dependent equations, or inconsistent with independent equations.

23. $\begin{cases} x - 4y = 9 \\ 2x - 3y = 8 \end{cases}$

24. $\begin{cases} -4y = 2x - 12 \\ x - 2y = -4 \end{cases}$

25. $\begin{cases} 3x + 2y = 5 \\ -6x - 4y = 1 \end{cases}$

26. $\begin{cases} x - y = 5 \\ x + 5y = 11 \end{cases}$

27. $\begin{cases} x - y = 1 \\ 2x - 2y = 2 \end{cases}$

28. $\begin{cases} 2x - y = 1 \\ -4x + 2y = -2 \end{cases}$

29. $\begin{cases} 2x - y = 1 \\ x + y = 4 \end{cases}$

30. $\begin{cases} 3x + 4y = 15 \\ 9x + 12y = 8 \end{cases}$

For Exercises 31–52, solve the system graphically.

31. $\begin{cases} x + y = 5 \\ 2x - y = 7 \end{cases}$

32. $\begin{cases} x - y = 4 \\ x + y = 8 \end{cases}$

33. $\begin{cases} y = x - 1 \\ 3x + y = 11 \end{cases}$

34. $\begin{cases} 3y = x \\ x - y = 8 \end{cases}$

35. $\begin{cases} y - x = 3 \\ x + y = 11 \end{cases}$

36. $\begin{cases} 2x + y = 1 \\ 2x + 3y = 3 \end{cases}$

37. $\begin{cases} 2x + 2y = -2 \\ 3x - 2y = 12 \end{cases}$

38. $\begin{cases} 2x - 4y = 2 \\ 3x - 2y = -1 \end{cases}$

39. $\begin{cases} x + y = 1 \\ y = 7 + x \end{cases}$

40. $\begin{cases} x + y = 4 \\ y - x = 6 \end{cases}$

41. $\begin{cases} x = y - 4 \\ x + y = 2 \end{cases}$

42. $\begin{cases} x = y + 4 \\ x + 4y = -1 \end{cases}$

43. $\begin{cases} 4x + y = 8 \\ 5x + 3y = 3 \end{cases}$

44. $\begin{cases} 2x + 3y = 6 \\ x + 4 = 2y \end{cases}$

45. $\begin{cases} y = 3 \\ x = -2 \end{cases}$

46. $\begin{cases} x = 1 \\ y = -4 \end{cases}$

47. $\begin{cases} y = -\dfrac{2}{5}x \\ x - y = 7 \end{cases}$

48. $\begin{cases} y = \dfrac{1}{4}x \\ 3x + 2y = 14 \end{cases}$

49. $\begin{cases} 3y = -2x + 6 \\ 4x + 6y = 18 \end{cases}$

50. $\begin{cases} x + y = 4 \\ -2x - 2y = -8 \end{cases}$

51. $\begin{cases} y = 3x \\ 6x - 2y = 0 \end{cases}$

52. $\begin{cases} 2x + y = 0 \\ 2y = 3 - 4x \end{cases}$

53. A business breaks even when its cost and revenue are equal. To the right is a graph of the cost of a product based on the number of units produced and the revenue based on the number of units sold.

   a. How many units must be produced and sold in order to break even?

   b. What amount of revenue is needed for the business to break even?

   c. Write an inequality that describes the number of units that must be sold for the business to make a profit.

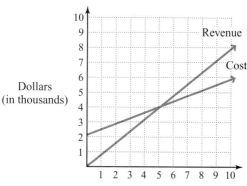

54. When choosing a cell phone plan, it is important to evaluate both the monthly cost of the phone and the number of minutes of expected use. The graph shows the relationship between cost and number of minutes for two plans. Plan 1 allows unlimited local minutes for $60. Plan 2 costs a flat rate of $30 per month, plus $.10 per minute.

   a. How many minutes must be used in order for the two plans to cost the same per month?

   b. Which plan is cheaper if the customer plans to use no more than 200 minutes per month?

   c. If the phone is only for emergencies, which plan is the less-expensive choice?

55. The graph shows two AT&T long-distance plans. The first plan costs $3.95 per month plus $0.10 per minute. This plan can be used to call anyone in the country. The second plan costs a flat $19.95 per month, but you may call only other AT&T customers at that rate.

   a. How many minutes would a person have to talk in one month using the first plan to equal the monthly cost of the second plan? (Assume all customers are AT&T customers.)

   b. If a person considering the first plan were to use 150 minutes calling only AT&T customers, is the first plan the best plan?

   c. What other factors need to be considered in choosing one of the two plans?

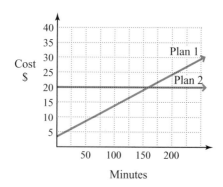

56. Deciding whether to purchase or rent a home is a difficult decision based on many factors. The graph shows the monthly mortgage payment ($850 per month) compared to the monthly rent payment (initially $600 per month and raised $50 every six months) for a 1200 square foot house.

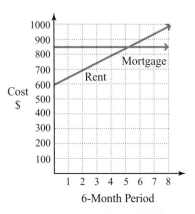

a. After how many months will the monthly rent be the same as the monthly mortgage payment?

b. If a customer plans to stay only two years (ignoring any tax benefits), which costs less per month, renting or buying?

c. In the fourth year, which costs less per month?

d. What other factors should a person consider when deciding whether to purchase or rent a home?

REVIEW EXERCISES

1. Given $f(x) = x + x^2$, find $f(x + 4)$.

2. One number exceeds another number by 62. If five times the smaller is subtracted from two times the larger, the difference is 91. What are the numbers?

For Exercises 3 and 4, factor.

3. $3w^2 - 6w + w - 2$

4. $4x^3 - 8x + 12$

5. Solve for y: $\quad x + 3y = 10$

6. Solve for x: $\quad \dfrac{1}{3}x - 2y = 10$

9.2 Solving Systems of Linear Equations by Substitution

OBJECTIVES

1 Solve systems of linear equations using substitution.

2 Solve applications involving two unknowns using a system of equations.

OBJECTIVE 1 **Solve systems of linear equations using substitution.** If the solution to a system of equations contains fractions or decimal numbers, it could be difficult to determine those values using the graphing method (Section 9.1). A more practical method is called the *substitution* method. Consider the system

$$\begin{cases} x + 3y = 10 & \text{(Equation 1)} \\ y = x + 4 & \text{(Equation 2)} \end{cases}$$

Remember that in a system's solution, the same x- and y-values satisfy both equations. In the preceding system, notice equation 2 indicates that y is equal to $x + 4$, so y must be equal to $x + 4$ in equation 1 as well. Therefore, we can substitute $x + 4$ for y in equation 1.

$$x + 3y = 10$$
$$x + 3\,\overbrace{(x + 4)} = 10 \quad \text{Substitute } x + 4 \text{ for } y.$$

Now we have an equation in terms of a single variable, x, which allows us to solve for the value of x.

$$\begin{aligned} x + 3(x + 4) &= 10 \\ x + 3x + 12 &= 10 \quad &&\text{Distribute 3.} \\ 4x + 12 &= 10 \quad &&\text{Combine like terms.} \\ 4x &= -2 \quad &&\text{Subtract 12 from both sides.} \\ x &= -\frac{1}{2} \quad &&\text{Divide both sides by 4 and simplify.} \end{aligned}$$

We can now find the y-value by substituting $-\dfrac{1}{2}$ for x in either of the original equations. Equation 2 is easier because y is isolated.

$$\begin{aligned} y &= x + 4 \\ y &= -\frac{1}{2} + 4 \quad \text{Substitute } -\frac{1}{2} \text{ for } x. \\ y &= 3\frac{1}{2} \end{aligned}$$

The solution to the system of equations is $\left(-\dfrac{1}{2}, 3\dfrac{1}{2} \right)$.

PROCEDURE SOLVING SYSTEMS OF EQUATIONS USING SUBSTITUTION

To find the solution of a system of two linear equations using the substitution method,

1. Isolate one of the variables in one of the equations.
2. In the other equation, substitute the expression you found in step 1 for that variable.
3. Solve this new equation. (It will now have only one variable.)
4. Using one of the equations containing both variables, substitute the value you found in step 3 for that variable and solve for the value of the other variable.
5. Check the solution in the original equations.

EXAMPLE 1 Solve the system of equations using substitution.

$$\begin{cases} x = 3 - y \\ 5x + 3y = 5 \end{cases}$$

Solution

Step 1: We isolate a variable in one of the equations. Since x is already isolated in the first equation, we proceed to step 2.

Step 2: Substitute the expression $3 - y$ for x in the second equation.

$$5x + 3y = 5$$
$$5(\overbrace{3 - y}) + 3y = 5 \qquad \text{Substitute } 3 - y \text{ for } x.$$

Step 3: Now that we have a linear equation in terms of a single variable, y, we solve for the value of y.

$$\begin{aligned} 15 - 5y + 3y &= 5 & \text{Distribute 5.} \\ 15 - 2y &= 5 & \text{Combine like terms.} \\ -2y &= -10 & \text{Subtract 15 on both sides.} \\ y &= 5 & \text{Divide both sides by } -2 \text{ to isolate } y. \end{aligned}$$

CONNECTION

If we were to solve the system in Example 1 graphically, the lines would intersect at the point whose coordinates are $(-2, 5)$.

Step 4: Now we solve for x by substituting 5 for y in one of the equations. We will use $x = 3 - y$.

$$\begin{aligned} x &= 3 - y \\ x &= 3 - 5 & \text{Substitute 5 for } y. \\ x &= -2 \end{aligned}$$

Note: We chose $x = 3 - y$ because x is isolated.

The solution is $(-2, 5)$. We will let the reader check that $(-2, 5)$ makes both equations true.

YOUR TURN Solve the system of equations using substitution.

$$\begin{cases} x - 3y = 11 \\ y = 2x + 3 \end{cases}$$

ANSWER

$(-4, -5)$

In step 1, if no variable is isolated in any of the given equations, then we select an equation and isolate one of its variables. It is easiest to isolate a variable that has a coefficient of 1, as this will avoid fractions.

EXAMPLE 2 Solve the system of equations using substitution.

$$\begin{cases} x + y = 6 \\ 5x + 2y = 8 \end{cases}$$

Solution

Step 1: Isolate a variable in one of the equations. Since both x and y have a coefficient of 1 in $x + y = 6$, isolating either variable is easy. We will isolate y.

$$x + y = 6$$
$$y = 6 - x \qquad \text{Subtract } x \text{ on both sides to isolate } y.$$

Step 2: Substitute $6 - x$ for y in the second equation.

$$5x + 2y = 8$$
$$5x + 2(\overbrace{6 - x}) = 8 \qquad \text{Substitute } 6 - x \text{ for } y.$$

Step 3: Solve for x using the equation we found in step 2.

$$5x + 12 - 2x = 8 \qquad \text{Distribute 2.}$$
$$3x + 12 = 8 \qquad \text{Combine like terms.}$$
$$3x = -4 \qquad \text{Subtract 12 on both sides.}$$
$$x = -\frac{4}{3} \qquad \text{Divide both sides by 3 to isolate } x.$$

Step 4: Find the value of y by substituting $-\frac{4}{3}$ for x in one of the equations containing both variables. Since we isolated y in $y = 6 - x$, we will use this equation.

$$y = 6 - x$$
$$y = 6 - \left(-\frac{4}{3}\right) \qquad \text{Substitute } -\frac{4}{3} \text{ for } x.$$
$$y = \frac{18}{3} + \frac{4}{3} \qquad \begin{array}{l}\text{Rewrite as addition of equivalent fractions}\\\text{with a common denominator.}\end{array}$$
$$y = \frac{22}{3}$$

The solution is $\left(-\frac{4}{3}, \frac{22}{3}\right)$.

Step 5: We will leave the check to the reader.

YOUR TURN Solve the system of equations by substitution.

$$\begin{cases} x - y = 1 \\ 3x + y = 11 \end{cases}$$

If none of the variables in the system of equations has a coefficient of 1, then for step 1, select the equation that seems easiest to work with and the variable in that equation that seems easiest to isolate. When selecting that variable, recognize that you'll eventually divide out its coefficient, so choose the variable whose coefficient will divide evenly into most, if not all, of the other numbers in its equation.

EXAMPLE 3 Solve the system of equations using substitution.

$$\begin{cases} 3x + 4y = 6 \\ 2x + 6y = -1 \end{cases}$$

Solution

Step 1: We will isolate x in $3x + 4y = 6$.

$3x + 4y = 6$

$3x = 6 - 4y$ — Subtract $4y$ on both sides.

$x = 2 - \dfrac{4}{3}y$ — Divide both sides by 3 to isolate x.

> **Note:** We chose to isolate x because its coefficient, 3, goes evenly into one of the other numbers, 6, in the equation, whereas y's coefficient, 4, does not go evenly into any of the other numbers.

Step 2: Substitute $2 - \dfrac{4}{3}y$ for x in $2x + 6y = -1$.

$$2x + 6y = -1$$

$$2\left(2 - \dfrac{4}{3}y\right) + 6y = -1 \quad \text{Substitute } 2 - \dfrac{4}{3}y \text{ for } x.$$

Step 3: Solve for y using the equation we found in step 2.

$4 - \dfrac{8}{3}y + 6y = -1$ — Distribute to clear the parentheses.

$3 \cdot 4 - 3 \cdot \dfrac{8}{3}y + 3 \cdot 6y = 3 \cdot (-1)$ — Multiply both sides by 3 to clear the fraction.

$12 - 8y + 18y = -3$

$12 + 10y = -3$ — Combine like terms.

$10y = -15$ — Subtract 12 from both sides.

$y = -\dfrac{15}{10}$ — Divide both sides by 10 to isolate y.

$y = -\dfrac{3}{2}$ — Simplify.

Step 4: Find the value of x by substituting $-\dfrac{3}{2}$ for y in one of the equations containing both variables. Since x is isolated in $x = 2 - \dfrac{4}{3}y$, we will use that equation.

$x = 2 - \dfrac{4}{3}y$

$x = 2 - \dfrac{\overset{2}{\cancel{4}}}{\cancel{3}} \cdot \left(-\dfrac{\overset{1}{\cancel{3}}}{\cancel{2}}\right)$ — Substitute $-\dfrac{3}{2}$ for y and simplify.

$x = 2 + 2$

$x = 4$

The solution is $\left(4, -\dfrac{3}{2}\right)$.

Step 5: We will leave the check to the reader.

YOUR TURN   **Solve the system of equations using substitution.**

$$\begin{cases} 5x - 2y = 10 \\ 3x - 6y = 2 \end{cases}$$

Inconsistent Systems of Equations

In Section 9.1 we learned that inconsistent systems of linear equations have no solution because the graphs of the equations are parallel lines. Let's see what happens when we solve an inconsistent system of equations using the substitution method.

EXAMPLE 4   Solve the system of equations using substitution.

$$\begin{cases} x + 2y = 4 \\ y = -\dfrac{1}{2}x + 1 \end{cases}$$

CONNECTION

If we were to solve the system in Example 4 using graphing, we would see that the lines are parallel.

Solution Since y is isolated in the second equation, we substitute $-\dfrac{1}{2}x + 1$ for y in the first equation.

$$x + 2y = 4$$

$$x + 2\left(-\dfrac{1}{2}x + 1\right) = 4 \qquad \text{Substitute } -\dfrac{1}{2}x + 1 \text{ for } y.$$

Now, solve for x.

$$x - x + 2 = 4 \qquad \text{Distribute to clear the parentheses.}$$
$$2 = 4 \qquad \text{Combine like terms.}$$

Notice that the $2 = 4$ no longer has a variable and is false. This false equation with no variable indicates that the system is inconsistent and has no solution.

Consistent Systems with Dependent Equations

Recall from Section 9.1 that if the equations in a system of equations are dependent, then there are an infinite number of solutions. Let's see what happens when we use the substitution method to solve a system of dependent equations.

EXAMPLE 5   Solve the system of equations using substitution.

$$\begin{cases} x = 3y + 4 \\ 2x - 6y = 8 \end{cases}$$

Solution Since x is isolated in the first equation, we substitute $3y + 4$ for x in the second equation.

CONNECTION

If we were to solve the system in Example 5 using graphing, we would see that the lines are identical.

$$2x - 6y = 8$$

$$2(\overbrace{3y + 4}) - 6y = 8 \qquad \text{Substitute } 3y + 4 \text{ for } x.$$

Now, solve for x.

$$6y + 8 - 6y = 8 \qquad \text{Distribute to clear the parentheses.}$$
$$8 = 8 \qquad \text{Combine like terms.}$$

Notice that $8 = 8$ no longer has a variable and is true. This true equation with no variable indicates that the equations in the system are dependent, so there are an infinite number of solutions, which are the coordinates of all the points on $2x - 6y = 8$ (or $x = 3y + 4$).

OBJECTIVE 2 Solve applications involving two unknowns using a system of equations. We will now use systems of equations to solve applications like some we solved in Chapter 3. We will select two variables to represent two unknowns, and then write two equations in terms of those variables. We can then solve the system of equations by substitution (or graphing).

PROCEDURE SOLVING PROBLEMS USING A SYSTEM OF EQUATIONS

To solve a problem with two unknowns using a system of equations,

1. Select a variable for each of the unknowns.
2. Translate each relationship to an equation.
3. Solve the system.

EXAMPLE 6 Aaron is designing a frame so that the length is three times the width. The perimeter is to be 96 inches. Find the length and width of the frame.

Understand We are given two relationships about this frame, and we are to find the length and width.

Plan and Execute We select a variable, l, for the length and another variable, w, for the width. Then we translate the relationships to a system of equations, and solve the system.

Relationship 1: The length is three times the width.

Translation: $l = 3w$

Relationship 2: The perimeter is 96 inches.

Translation: $2l + 2w = 96$

Our system: $\begin{cases} l = 3w \\ 2l + 2w = 96 \end{cases}$

Since the first equation has l isolated, we will substitute $3w$ for l in the second equation.

$$2l + 2w = 96$$
$$2(\overbrace{3w}) + 2w = 96 \qquad \text{Substitute } 3w \text{ for } l.$$
$$6w + 2w = 96$$
$$8w = 96 \qquad \text{Combine like terms.}$$
$$w = 12 \qquad \text{Divide both sides by 8 to isolate } w.$$

Now we can find the value of l using $l = 3w$ and substituting 12 for w.

$$l = 3(12) = 36$$

Answer   The length should be 36 inches and the width 12 inches.

Check   Verify that the length is three times the width: $3(12) = 36$. Also verify that the perimeter is 96: $2(36) + 2(12) = 72 + 24 = 96$.

YOUR TURN

Two steel beams are welded together forming two angles as shown. The greater of the two angles is four times the smaller. Find the two angles.

Angle 2

Angle 1

EXAMPLE 7   A computer salesperson's goal for May is to sell twice as many more-expensive computers as lower-priced computers. The more-expensive computer sells for $2500; the lower-priced computer sells for $1800. If her total sales revenue for May needs to be $27,200, how many of each computer must she sell to meet her goal?

Understand   The two unknowns are the number sold of each type of computer. There are also two relationships, one involving the number sold, and the other involving the total revenue.

Plan and Execute   Let x represent the number of lower-priced computers sold, and let y represent the number of more-expensive computers sold. We can use a table to organize the information.

Categories	Selling price	Number	Revenue
Lower priced	$1800	x	1800x
Higher priced	$2500	y	2500y

Note: Sales revenue is found by multiplying an item's price by the number sold.

Relationship 1: Sell twice as many of the more-expensive model as the lower-priced model.

Translation: $y = 2x$

Relationship 2: The total revenue needs to be $27,200.

Translation: $1800x + 2500y = 27{,}200$

Our system: $\begin{cases} y = 2x \\ 1800x + 2500y = 27{,}200 \end{cases}$

To solve, we replace y in the second equation with $2x$ from the first equation:

$$1800x + 2500y = 27{,}200$$

$$1800x + 2500(\mathbf{2x}) = 27{,}200 \quad \text{Substitute } 2x \text{ for } y.$$

$$1800x + 5000x = 27{,}200$$

$$6800x = 27{,}200 \quad \text{Combine like terms.}$$

$$x = 4 \quad \text{Divide both sides by 6800 to isolate } x.$$

Now we can find the value of y using $y = 2x$ and substituting 4 for x.

$$y = 2(4) = 8$$

Answer   Selling 4 of the lower-priced computers and 8 of the more-expensive computers would meet her goal.

Check   Verify both given relationships. The number of more-expensive computers sold, 8, is indeed twice as many as the number of lower-priced computers, 4. The total revenue, $1800(4) + 2500(8) = 27{,}200$, is also correct.

YOUR TURN

A cookware consultant sells two sizes of rectangular baking dish. The small size sells for \$24 and the large size sells for \$35. In one month she sold 6 more of the small-size dish than the large-size dish. If she made a total of \$439 from the sale of both dishes combined, then how many of each size baking dish did she sell?

ANSWER

11 small and 5 large

EXAMPLE 8   Jasmine and Darius are traveling east in separate cars on the same highway. Jasmine is traveling at 60 miles per hour and Darius at 70 miles per hour. Jasmine passes Exit 82 at 3:15 P.M. Darius passes the same exit at 3:30 P.M. At what time will Darius catch up to Jasmine?

Understand   To determine the time at which Darius catches up to Jasmine, we must calculate the amount of time it will take him to catch up to her. We can then add that amount of time to 3:30 P.M.

Plan and Execute   We will let x represent Jasmine's travel time after passing Exit 82 and y represent Darius' travel time after passing Exit 82.

Categories	Rate	Time	Distance
Jasmine	60	x	$60x$
Darius	70	y	$70y$

Note: Distance is the product of rate and time.

Relationship 1:   Darius passes the exit 15 minutes after Jasmine, so when he catches up to her, she will have traveled 15 minutes longer than he has.

Translation: $x = y + \dfrac{1}{4}$

Note: We use $\dfrac{1}{4}$ of an hour instead of 15 minutes because the rates are in miles per hour and the time units must be consistent.

Relationship 2: When Darius catches up, they will have traveled the same distance from Exit 82, so we set the expressions for their individual distances equal to each other.

Translation: $60x = 70y$

Our system: $\begin{cases} x = y + \dfrac{1}{4} \\ 60x = 70y \end{cases}$

To solve, we replace x in the second equation with $y + \dfrac{1}{4}$ from the first equation.

$$60x = 70y$$

$$60\left(y + \frac{1}{4}\right) = 70y \qquad \text{Substitute } y + \frac{1}{4} \text{ for } x.$$

$$60y + 15 = 70y \qquad \text{Distribute 60.}$$

$$15 = 10y \qquad \text{Subtract } 60y \text{ from both sides.}$$

$$1.5 = y \qquad \text{Divide both sides by 10 to isolate } y.$$

Answer   It will take Darius 1.5 hours to catch up to Jasmine, so he catches up at 5:00 P.M.

Check   At 60 miles per hour, in $\dfrac{1}{4}$ of an hour Jasmine travels 15 miles. Since Darius is traveling 70 miles per hour, he is gaining on her at a rate of 10 miles per hour (the difference of their rates). So he must make up the 15 miles that separate them at a rate of 10 miles per hour, which will indeed take 1.5 hours.

YOUR TURN

At 10:30 A.M., Celeste begins walking along a trail at an average rate of 3 miles per hour. At 10:40 A.M., Chris begins bicycling along the same trail at 8 miles per hour. At what time will Chris catch up to Celeste?

ANSWER

10:46 A.M.

1. What advantages does the method of substitution have over graphing for solving systems of equations?

2. Suppose you are given the system of equations: $\begin{cases} y = x - 6 \\ 4x + 3y = 1 \end{cases}$

   a. Which variable would you replace in the substitution process?

   b. What expression would replace that variable?

3. Suppose you are given the system of equations: $\begin{cases} 5x + y = 4 \\ 3x - 2y = 9 \end{cases}$

   Which variable in which equation would you isolate first? Why?

4. Suppose you are given the system of equations: $\begin{cases} 3x + 4y = 1 \\ 2x - 5y = 12 \end{cases}$

   Which variable would you isolate first? Why?

5. How do you know a system has no solution (inconsistent system) when using substitution?

6. How do you know a system has an infinite number of solutions (dependent equations) when using substitution?

For Exercises 7–30, solve the system of equations using substitution. Note that some systems may be inconsistent or consistent with dependent equations.

7. $\begin{cases} x + y = 12 \\ x = 3y \end{cases}$
    8. $\begin{cases} x - y = 8 \\ y = 2x \end{cases}$
    9. $\begin{cases} 2x - y = 4 \\ y = 3x + 6 \end{cases}$
    10. $\begin{cases} x = 6 - y \\ 2x - 3y = 2 \end{cases}$

11. $\begin{cases} 2x + y = -4 \\ 3x + 2y = -5 \end{cases}$
    12. $\begin{cases} 5x + y = 0 \\ 3x = 2y \end{cases}$
    13. $\begin{cases} 5x - y = 1 \\ 3x + 2y = 24 \end{cases}$
    14. $\begin{cases} x + 3y = 5 \\ 2y - 3x = 18 \end{cases}$

15. $\begin{cases} 5x - 3y = 4 \\ y - x = -2 \end{cases}$
    16. $\begin{cases} 2x + y = 41 \\ 14 + 3x = 74y \end{cases}$
    17. $\begin{cases} x = 3y \\ 2x - 5y = 4 \end{cases}$
    18. $\begin{cases} x + 2y = 9 \\ y = -2x \end{cases}$

19. $\begin{cases} 2x = 10 - y \\ 3x - y = 5 \end{cases}$

20. $\begin{cases} x + 3y = 7 \\ 2x - 3y = -4 \end{cases}$

21. $\begin{cases} x + 2y = 3 \\ 3x - 2y = 1 \end{cases}$

22. $\begin{cases} 2x + 3y = -1 \\ 5x - y = 6 \end{cases}$

23. $\begin{cases} y = -2x \\ 4x + 2y = 0 \end{cases}$

24. $\begin{cases} y = 2x + 1 \\ 6x - 3y = 3 \end{cases}$

25. $\begin{cases} 4x + 6y = 3 \\ 8x - 3y = -4 \end{cases}$

26. $\begin{cases} 2x + 3y = 6 \\ 8x - 3y = -1 \end{cases}$

27. $\begin{cases} 5x + 4y = 12 \\ 7x - 6y = 40 \end{cases}$

28. $\begin{cases} 4x - 3y = 2 \\ 3x + 5y = 16 \end{cases}$

29. $\begin{cases} 8x - 12y = 5 \\ 10x + 6y = -1 \end{cases}$

30. $\begin{cases} 3y - 12x = 4 \\ 2x - 9y = 3 \end{cases}$

For Exercises 31–34, solve using substitution then verify the solution by graphing the equations on a graphing utility.

31. $\begin{cases} y = 2x + 1 \\ 3x + y = 11 \end{cases}$

32. $\begin{cases} y = -4x + 3 \\ y - 2x = -3 \end{cases}$

33. $\begin{cases} 3x + 4y = 11 \\ x + 2y = 5 \end{cases}$

34. $\begin{cases} 4x - y = 12 \\ x - 3y = 14 \end{cases}$

For Exercises 35 and 36, find the mistake.

35. $\begin{cases} 3x + 2y = 5 \\ x - 3y = 6 \end{cases}$

Rewrite: $x = 6 + 3y$

Substitute: $3(6 + 3y) + 2y = 5$
$$18 + 3y + 2y = 5$$
$$18 + 5y = 5$$
$$5y = -13$$
$$y = -\frac{13}{5}$$

$x = 6 + 3y$

$x = 6 + 3\left(-\dfrac{13}{5}\right)$

$x = 6 - \dfrac{39}{5}$

$x = -\dfrac{9}{5}$

Solution: $\left(-\dfrac{9}{5}, -\dfrac{13}{5}\right)$

36. $\begin{cases} 2x + 3y = 8 \\ x + 2y = 6 \end{cases}$

Rewrite: $x = 6 - 2y$

Substitute: $\qquad\qquad x + 2y = 6$
$$6 - 2y + 2y = 6$$
$$6 - 6 - 2y + 2y = 6 - 6$$
$$0 = 0$$

Solution: $(0, 0)$

37. One number is 62 more than another. If twice the larger number is subtracted from five times the smaller number, the result is 155. What are the two numbers?

38. Find two numbers whose sum is 14 if one number is three times as large as the other number.

39. The larger of two integers is 10 more than twice the smaller. The sum of the two integers is -8. Find the integers.

40. The sum of the two integers is -13. The greater of the two integers is 23 more than three times the smaller. Find the integers.

41. Jon is 13 years older than Tony. The sum of their ages is 27. How old are they?

42. Patrick is twice as old as his sister. The sum of their ages is 12. How old are they?

43. During Super Bowl XXXVI, Patriots kicker Adam Vinatieri kicked a 48-yard field goal as time expired, and New England won by 3 points over the Rams. If the total points scored during the game was 37, how many points did each team score?

44. Jack Nicklaus has the most victories ever won by a golfer in the Masters Tournament. He exceeds the number of Arnold Palmer's victories by 2 wins. If they have won a total of 10 tournaments, how many times has each man won?

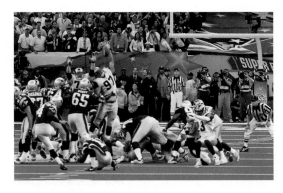

45. The length of a rectangle is 1 meter more than twice the width. What are the dimensions of the rectangle if the perimeter is 32 meters?

46. The length of a rectangle is 1 meter more than triple the width. The length is also five less than five times the width. Find the length and width of the rectangle.

47. What are the length and width of a rectangle if the length exceeds the width by 14 inches and the perimeter is 60 inches?

48. What are the length and width of a rectangle if the width is 27 feet less than its length and the perimeter is 398 feet?

49. A rectangular garden is 6 feet longer than it is wide. If the perimeter is 132 feet, what are the dimensions of the garden?

50. The width of a rectangular field is 4 feet less than the length. The perimeter of the field is 48 feet. Find the length and width.

51. The perimeter of the playing lines for a doubles tennis match is 228 feet. If the width is 42 feet less than the length, what are the dimensions?

52. The perimeter of a college basketball court is 288 feet. If the length of the court is 44 feet longer than it is wide, what are the dimensions?

53. The perimeter of the base of the Castillo at Chichen Itza (a pyramid in Mexico), is 300 feet. If the length is equal to the width, what are the dimensions?

54. The perimeter of the colonnade in the Lincoln Memorial is 612 feet. It is 70 feet longer than it is wide. What are the dimensions of the colonnade?

55. A support beam is attached to a wall and to the bottom of a ceiling truss. The angle made with the truss on one side of the support beam is three times the angle on the other side of the support beam. Find the two angles.

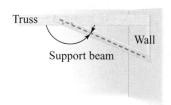

56. A laser beam is aimed at a flat photocell at an angle. The angle on one side of the beam is 25° more than the angle on the other side of the beam. Find the two angles.

57. Two angles are complementary. One of the angles is twice the other. Find the measure of the two angles.

58. Two angles are complementary. The larger angle is six less than twice the smaller angle. Find the measure of the two angles.

59. A security camera is placed in the corner of a room near the ceiling. The camera is aimed so that the angle made with the wall on one side of the camera is 15° less than the angle made with the wall on the other side of the camera. Find the angle made with the wall on each side of the camera.

60. Used as a security device, a laser is placed a few inches above the floor on one wall of a room and aimed at a photoreceptor in the corner of the room. At the receptor, the angle made by the beam with the wall on one side of the receptor is 12° more than the angle made on the other side of the receptor. Find the two angles.

61. At the University of South Carolina, a difference between a professor's salary and an associate professor's salary is $22,292. The sum of the salaries is $143,426. Find the salary of a USC associate professor. (*Source:* The University of South Carolina as reported in the *State* newspaper)

62. The sum of tuition and fees for both in-state and out-of-state students at the University of Kentucky is $14,022. The difference between the two types of tuition is $6286. What does it cost an out-of-state student to attend the University of Kentucky? (*Source:* The University of Kentucky)

63. A college recycled twice as many tons of paper as cardboard during the 2000–2001 school year. If the total amount recycled was 375 tons, how much paper and how much cardboard was recycled? (*Source:* The *State* newspaper, 8/18/2001)

64. A study by the National Athletic Trainers' Association lists general trauma as the most common injury in high school football. General trauma occurs nearly five times more than a fracture injury. If the local high school has 41 football players and 18 of them are injured during the season, how many players should be expected to experience general trauma? How many players should be expected to experience fractures? (*Source:* National Athletic Trainers' Association)

65. Studies show that 78% of college freshman have credit cards. Sue is one of them. She has two credit cards with a total balance of $2700. If she owes three times more on one of the cards than the other, how much does she owe on each card? *(Source The Dallas Morning News)*

66. During the close 2000 presidential election, the country watched as Miami-Dade County recounted the votes. The final number of certified votes was 618,341. Although George W. Bush ultimately won the election, he lost Miami-Dade County to Al Gore by 39,275 votes. How many votes did each man receive?

67. John and Karen are traveling south in separate cars on the same interstate. Karen is traveling at 65 miles per hour and John at 70 miles per hour. Karen passes Exit 38 at 1:15 P.M. John passes the same exit at 1:30 P.M. At what time will John catch up to Karen?

68. Diane and Curtis are both running in the Cooper River Bridge Run. Diane passes the first checkpoint at 10:30 A.M. and Curtis passes at 11:00 A.M. While Curtis is pushing himself to run at 5 miles per hour, Diane is pacing herself at 4 miles per hour. When will Curtis catch up with Diane?

69. Sharon and Rae are riding bicycles on the same trail and in the same direction. Sharon crosses a bridge at 12:18 P.M. Rae crosses the same bridge at 12:33 P.M. If Sharon is going 8 miles per hour and Rae is going 10 miles per hour, when will Rae catch up with Sharon?

70. Holly and Rebecca are jogging on the same road and in the same direction. Holly passes a McDonald's at 8:10 A.M. Rebecca passes the same McDonald's at 8:20 A.M. If Holly is going 4 miles per hour and Rebecca is going 6 miles per hour, when will Rebecca catch up with Holly?

REVIEW EXERCISES

For Exercises 1 and 2, add or subtract.

1. $(6x + 2y) + (3x - 2y)$

2. $(5x + 3y) - (5x - 10y)$

For Exercises 3 and 4, multiply.

3. $\frac{1}{4}(20x - 16y)$

4. $-3(x - 7y)$

For Exercises 5 and 6, solve.

5. $\frac{1}{4}x = 20$

6. $-8y = 12$

Solving Systems of Linear Equations by Elimination

OBJECTIVES

1. Solve systems of linear equations using elimination.
2. Solve applications using a system of equations.

The substitution method can be tedious when no coefficients are 1. For this reason we turn to a third method of solving a system of equations, the *elimination* method.

OBJECTIVE 1 Solve systems of linear equations using elimination. The elimination method uses the addition principle of equality to add equations so that a new equation emerges with one of the variables eliminated. We will work through some examples to get a sense of the method before stating it formally.

EXAMPLE 1 Solve the system of equations.

$$\begin{cases} x + y = 5 & \text{(Equation 1)} \\ 2x - y = 7 & \text{(Equation 2)} \end{cases}$$

Solution In Chapter 2 we learned that the addition principle of equality says that adding the same amount to both sides of an equation will not affect its solution(s). Since equation 1 indicates that $x + y$ and 5 are the same amount, if we add $x + y$ to the left side of equation 2 and 5 to the right side of equation 2, then we are applying the addition principle of equality. Most people prefer to stack the equations and combine like terms vertically like this:

$$\begin{array}{ll} x + y = 5 & \text{(Equation 1)} \\ \underline{2x - y = 7} & \text{(Equation 2)} \\ 3x + 0 = 12 & \text{Add equation 1 to equation 2} \end{array}$$

> **CONNECTION**
>
> The terms y and $-y$ are additive inverses, or opposites, which means their sum is 0.

Notice that y is eliminated in this new equation, so we can easily solve for the value of x.

$$\begin{array}{ll} 3x = 12 \\ x = 4 & \text{Divide both sides by 3 to isolate } x. \end{array}$$

Now that we have the value of x, we can find y by substituting 4 for x in one of the original equations. We will use $x + y = 5$:

$$\begin{array}{ll} x + y = 5 \\ 4 + y = 5 & \text{Substitute 4 for } y \\ y = 1 & \text{Subtract 4 from both sides to isolate } y. \end{array}$$

The solution is (4, 1). We can check by verifying that (4, 1) makes both of the original equations true. We will leave this to the reader.

$$\begin{cases} 4x + 3y = 8 \\ x - 3y = 7 \end{cases}$$

CONNECTION

In Chapter 2 we learned that the multiplication principle of equality says that we can multiply both sides of an equation by the same amount without affecting its solution(s).

Multiplying One Equation by a Number to Create Additive Inverses

You may have noted that the expressions $x + y$ and $2x - y$ conveniently contained the additive inverses y and $-y$, which eliminated the y's when the expressions were combined. If no such pairs of additive inverses appear in a system of equations, we will use the multiplication principle of equality to multiply both sides of one equation by a number in order to create additive inverse pairs.

EXAMPLE 2 ⟩ Solve the system of equations.

$$\begin{cases} x + y = 6 & \text{(Equation 1)} \\ 2x - 5y = -16 & \text{(Equation 2)} \end{cases}$$

Solution We will rewrite one of the equations so that it has a term that is the additive inverse of one of the terms in the other equation. We will multiply both sides of equation 1 by 5.

$$x + y = 6$$
$$5 \cdot x + 5 \cdot y = 5 \cdot 6 \qquad \text{Multiply both sides by 5.}$$
$$5x + 5y = 30 \qquad \text{(Equation 1 rewritten)}$$

Note: We could have multiplied both sides of equation 1 by -2. We would get a $-2x$ term in equation 1 that is the opposite of the $2x$ term in equation 2.

Now we add the rewritten equation 1 to equation 2 to eliminate a variable, just as we did in Example 1.

Note: Multiplying equation 1 by 5 made the elimination of the y terms possible.

$$\begin{array}{ll} 5x + 5y = 30 & \text{(Equation 1 rewritten)} \\ \underline{2x - 5y = -16} & \text{(Equation 2)} \\ 7x + 0 \ = 14 & \text{Add rewritten equation 1 to} \\ & \text{equation 2 to eliminate } y. \end{array}$$

We can now solve for x.

$$7x = 14$$
$$x = 2 \qquad \text{Divide both sides by 7 to isolate } x.$$

To finish, we substitute 2 for x in an equation containing both variables. We will use $x + y = 6$.

$$x + y = 6$$
$$2 + y = 6 \qquad \text{Substitute 2 for } x.$$
$$y = 4 \qquad \text{Subtract 2 from both sides to isolate } y.$$

The solution is $(2, 4)$. We will leave the check to the reader.

$$\begin{cases} -3x - 5y = 6 \\ x + 2y = -1 \end{cases}$$

Multiplying Each Equation by a Number to Create Additive Inverses

If every coefficient in a system of equations is other than 1, then we may have to multiply each equation by a number to generate a pair of additive inverses.

EXAMPLE 3 Solve the system of equations.

$$\begin{cases} 4x - 3y = -2 & \text{(Equation 1)} \\ 6x - 7y = 7 & \text{(Equation 2)} \end{cases}$$

Solution We will choose to eliminate x, so we will multiply equation 1 by 3 and equation 2 by -2, which makes the x terms additive inverses.

$$
\begin{array}{l}
4x - 3y = -2 \quad \xrightarrow{\text{Multiply by 3}} \quad 12x - 9y = -6 \\
6x - 7y = 7 \quad \xrightarrow{\text{Multiply by } -2} \quad -12x + 14y = -14
\end{array}
$$

Note: The x terms are now additive inverses, $12x$ and $-12x$.

Now we can add the rewritten equations to eliminate the x term.

$$
\begin{array}{r}
12x - 9y = -6 \\
-12x + 14y = -14 \\
\hline
0 + 5y = -20 \\
y = -4
\end{array}
$$

Add the rewritten equations to eliminate x.
Divide both sides by 5 to isolate y.

To finish, we substitute -4 for y in one of the equations and solve for x. We will use $4x - 3y = -2$.

Note: It does not matter which equation we use because x should be the same value in any equation in the system.

$$
\begin{aligned}
4x - 3y &= -2 \\
4x - 3(-4) &= -2 \qquad \text{Substitute } -4 \text{ for } y. \\
4x + 12 &= -2 \\
4x &= -14 \qquad \text{Subtract 12 from both sides.} \\
x &= -\frac{14}{4} \qquad \text{Divide both sides by 4 to isolate } y. \\
x &= -\frac{7}{2} \qquad \text{Simplify.}
\end{aligned}
$$

The solution is $\left(-\dfrac{7}{2}, -4 \right)$. We will leave the check to the reader.

Fractions or Decimals in a System

If any of the equations in a system of equations contains fractions or decimals, it is helpful to use the multiplication principle of equality to clear those fractions or decimals so that the equations contain only integers.

EXAMPLE 4 Solve the system of equations.

$$\begin{cases} \dfrac{1}{2}x - y = \dfrac{3}{4} & \text{(Equation 1)} \\ 0.4x - 0.3y = 1 & \text{(Equation 2)} \end{cases}$$

Solution To clear the fractions in equation 1, we multiply both sides by the LCD, which is 4. To clear the decimals in equation 2, we can multiply both sides by 10.

$$\frac{1}{2}x - y = \frac{3}{4} \xrightarrow{\text{Multiply by 4}} 2x - 4y = 3$$
$$0.4x - 0.3y = 1 \xrightarrow{\text{Multiply by 10}} 4x - 3y = 10$$

Now that both equations contain only integers, it will be easier to solve the system. We will choose to eliminate x, so we will multiply the first equation by -2, then combine the equations.

$$\begin{array}{l} 2x - 4y = 3 \\ 4x - 3y = 10 \end{array} \xrightarrow{\text{Multiply by } -2} \begin{array}{l} -4x + 8y = -6 \\ \underline{4x - 3y = 10} \\ 0 + 5y = 4 \end{array}$$

Add the rewritten equations to eliminate x.

$$y = \frac{4}{5} \text{ or } 0.8 \quad \text{Divide both sides by } 5 \text{ to isolate } y.$$

To finish, we let $y = 0.8$ in one of the equations and solve for x.

$0.4x - 0.3y = 1$	We will choose equation 2.
$0.4x - 0.3(0.8) = 1$	Substitute 0.8 for y.
$0.4x - 0.24 = 1$	Simplify.
$0.4x = 1.24$	Add 0.24 to both sides.
$x = 3.1$	Divide both sides by 0.4.

The solution is $(3.1, 0.8)$. We will leave the check to the reader.

YOUR TURN Solve the system of equations using elimination.

$$\begin{cases} 0.6x + y = 3 \\ \dfrac{2}{5}x + \dfrac{1}{4}y = 1 \end{cases}$$

Rewriting the Equations in the form $Ax + By = C$

Notice in Examples 1 through 4, all of the equations were in standard form, which we learned in Chapter 4 to be $Ax + By = C$. To use the elimination method, the equations need to be written in standard form. Consider the following system:

$$\begin{cases} y = 2 - x & \text{(Equation 1)} \\ 3x - y = 1 & \text{(Equation 2)} \end{cases}$$

We need to rewrite equation 1 in standard form before using elimination.

$$\begin{cases} x + y = 2 & \text{(Equation 1)} \\ 3x - y = 1 & \text{(Equation 2)} \end{cases}$$

We can now summarize the elimination method with the following procedure.

PROCEDURE SOLVING SYSTEMS OF EQUATIONS USING ELIMINATION

To solve a system of two linear equations using the elimination method,

1. Write the equations in standard form ($Ax + By = C$).
2. Use the multiplication principle to clear fractions or decimals.
3. Multiply one or both equations by a number (or numbers) so that they have a pair of terms that are additive inverses.
4. Add the equations. The result should be an equation in terms of one variable.
5. Solve the equation from step 4 for the value of that variable.
6. Using an equation containing both variables, substitute the value you found in step 5 for the corresponding variable and solve for the value of the other variable.
7. Check your solution in the original equations.

Inconsistent Systems and Dependent Equations

How would we recognize an inconsistent system or a system with dependent equations using the elimination method?

EXAMPLE 5 Solve the system of equations.

$$\begin{cases} 2x - y = 1 \\ 2x - y = -3 \end{cases}$$

Solution Notice that the left sides of the equations match. Multiplying one of the equations by -1, then adding the equations, will eliminate both variables.

CONNECTION

If we graph the equations in Example 5, the lines would be parallel, indicating there is no solution.

$$\begin{array}{c} 2x - y = 1 \\ 2x - y = -3 \end{array} \xrightarrow{\text{Multiply by } -1} \begin{array}{c} -2x + y = -1 \\ \underline{2x - y = -3} \\ 0 = -4 \end{array}$$

Add the rewritten equations.

Both variables have been eliminated and the resulting equation $0 = -4$ is false, therefore there is no solution. This system of equations is inconsistent.

EXAMPLE 6 Solve the system of equations.

$$\begin{cases} 3x + 4y = 5 \\ 9x + 12y = 15 \end{cases}$$

Solution To eliminate x, we could multiply the first equation by -3, then combine the equations.

$$
\begin{array}{ll}
3x + 4y = 5 \\
9x + 12y = 15
\end{array}
\quad \xrightarrow{\text{Multiply by } -3} \quad
\begin{array}{ll}
-9x - 12y = -15 \\
\underline{\;\;9x + 12y = 15} \\
\qquad\qquad 0 = 0
\end{array}
\quad
\begin{array}{l}
\text{Add the rewritten} \\
\text{equations.}
\end{array}
$$

Both variables have been eliminated and the resulting equation $0 = 0$ is true. This means the equations are dependent, so there are an infinite number of solutions which are the coordinates of all the points on $3x + 4y = 5$ (or $9x + 12y = 15$). Any solution for one of the equations will also be a solution for the other.

YOUR TURN **Solve the system of equations.**

a. $\begin{cases} x - 4y = 2 \\ 5x - 20y = 10 \end{cases}$ **b.** $\begin{cases} x + 2y = 3 \\ x + 2y = 1 \end{cases}$

OBJECTIVE 2 Solve applications using a system of equations. Now we focus on applications in which elimination is advantageous. These problems translate to a system of linear equations in which every equation is in standard form ($Ax + By = C$).

EXAMPLE 7 In constructing a roof frame, a support beam is connected to a horizontal joist forming two angles. If the larger angle minus twice the smaller angle equals 6°, what are the measures of the two angles?

Understand We are to find two angle measurements, but it appears that we are given only one relationship. The second relationship is that the horizontal joist represents an angle of 180°, therefore the two angles are supplementary.

Plan and Execute We will let x represent the measure of angle 1 and y represent the measure of angle 2. Now we translate the two relationships to a system of two equations.

Relationship 1: The larger angle minus twice the smaller angle equals 6°

Translation: $y - 2x = 6$

Relationship 2: From the picture we see that the angles are supplementary.

Translation: $x + y = 180$

We can now solve the system using elimination. We will eliminate y.

Note: We re-arranged the terms to align the like terms vertically.

$$\begin{cases} -2x + y = 6 \\ x + y = 180 \end{cases} \xrightarrow{\text{Multiply by } -1.} \begin{array}{l} 2x - y = -6 \\ \underline{x + y = 180} \\ 3x + 0 = 174 \end{array}$$

Add the rewritten

eliminate y.

$x = 58$ Divide both sides by 3 to isolate x.

Now we can find the value of y using $x + y = 180$.

$58 + y = 180$ Substitute 58 for x.

$y = 122$ Subtract 58 from both sides to isolate y.

Answer   Angle 1 measures 58° and angle 2 measures 122°.

Check   Verify that the measure of the larger angle minus twice the measure of the smaller angle equals 6°: $122 - 2(58) = 122 - 116 = 6$. Also, verify that the angles are supplementary: $58 + 122 = 180$.

YOUR TURN

The security system at a museum uses a laser placed in the corner just above the floor with its beam projecting across the room parallel to the floor. The beam makes two angles in the corner of the room. The difference of the measures of the two angles is 22°. Find the angle measurements.

EXAMPLE 8   Aaron sells concessions at a sporting event. In 1 hour, he sells 68 drinks. The drink sizes are small, which sells for $2 each, and large, which sells for $3. If his total sales revenue was $160, how many of each size did he sell?

Understand   The two unknowns are the number of each size drink. One relationship involves the number of drinks (68 total), and the other relationship involves the total money in sales ($160).

Plan and Execute   Let x represent the number of small drinks sold and y represent the number of large drinks sold. We can use a table to organize the information.

Categories	Selling price	Number	Revenue
Small	$2	x	$2x$
Large	$3	y	$3y$

Relationship 1: The total number of drinks sold is 68.

$$\text{Translation: } x + y = 68$$

Relationship 2: The total sales revenue is $160.

$$\text{Translation: } 2x + 3y = 160$$

We will solve the system by eliminating x.

$$\begin{cases} x + y = 68 \\ 2x + 3y = 160 \end{cases} \xrightarrow{\text{Multiply by } -2} \begin{array}{r} -2x - 2y = -136 \\ \underline{2x + 3y = 160} \\ y = 24 \end{array}$$

Now we can find the value of x by substituting 24 in place of y in one of the equations. We will use $x + y = 68$.

$$x + 24 = 68 \qquad \text{Substitute 24 for } y.$$
$$x = 44 \qquad \text{Subtract 24 from both sides to isolate } x.$$

Answer Aaron sold 44 small drinks and 24 large drinks.

Check Verify both given relationships. The total number of drinks sold: $24 + 44 = 68$.

The total sales: $2(44) + 3(24) = \$88 + \$72 = \$160$.

YOUR TURN

A college bookstore receives a shipment of 480 copies of a textbook. Some of the books have tutorial software shrink-wrapped with the book while others do not have the software. The bookstore pays $54 for each book with the tutorial software and $48 for each copy without the software. If the total invoice for the shipment is $24,000, how many books in the shipment had software and how many did not have software?

EXAMPLE 9 Airplanes traveling west travel against the jet stream, which decreases their average speed. When traveling east, on the other hand, the jet stream increases their average speed. Suppose a large jet is traveling from Atlanta to Los Angeles, a distance of approximately 2000 miles, and the flight lasts 5 hours. On the return trip, the plane takes only 4 hours. Find the plane's speed in still air and how much the jet stream changes the speed of the plane.

Understand We assume that without the jet stream, the plane would average the same speed going west as east. We also assume that the amount the jet stream adds to the plane's speed is the same as the amount subtracted, depending on which way the plane is traveling.

Plan and Execute   Let x represent the plane's speed in still air. Let y represent the amount that the jet stream increases or decreases the plane's speed, depending on which way it is traveling. We can use a table to organize the information.

Categories	Rate	Time	Distance
West	$x - y$	5	$5(x - y)$
East	$x + y$	4	$4(x + y)$

Relationship 1: The distance from Atlanta to Los Angeles is 2000 miles.

Translation: $5(x - y) = 2000$

$$\frac{5(x - y)}{5} = \frac{2000}{5} \qquad \text{Divide both sides by 5 to streamline the equation.}$$

$$x - y = 400$$

Relationship 2: The distance from Los Angeles to Atlanta is 2000 miles.

Translation: $4(x + y) = 2000$

$$\frac{4(x + y)}{4} = \frac{2000}{4} \qquad \text{Divide both sides by 4 to streamline the equation.}$$

$$x + y = 500$$

Solve the system:

$$x - y = 400$$
$$\underline{x + y = 500}$$
$$2x + 0 = 900 \qquad \text{Add the equations to eliminate } y.$$
$$x = 450 \qquad \text{Divide both sides by 2 to isolate } x.$$

Now we can find the value of y by substituting 450 in place of x in one of the equations. We will use $x + y = 500$.

$$450 + y = 500 \qquad \text{Substitute 450 for } x.$$
$$y = 50 \qquad \text{Subtract 450 from both sides.}$$

Answer   The plane travels 450 mph in still air. The jet stream changes the plane's speed by 50 mph.

Check   Verify the time of travel for each part of the round trip.

Rate going west: $450 - 50 = 400$ miles per hour

$$\text{Time: } t = \frac{d}{r} = \frac{2000}{400} = 5 \text{ hours}$$

Rate going east: $450 + 50 = 500$ miles per hour

$$\text{Time: } t = \frac{d}{r} = \frac{2000}{500} = 4 \text{ hours}$$

YOUR TURN

ANSWER

His rate in still water is 1.4 mph. The current increases or decreases his rate by 0.6 mph.

To train for the swimming portion of a triathlon, an athlete swims in a river to gauge how the current will affect him. When swimming against the current, he can swim a mile in 1.25 hours. When swimming with the current, he can swim a mile in 0.5 hours. What is his rate in still water? How much did the current increase or decrease his rate in still water?

EXAMPLE 10 Jeff invests $5000 in two different accounts. The first account has an annual percentage rate (APR) of 4%. The second account has an APR of 6%. If the total interest earned after 1 year is $252, what principal was invested in each account?

Understand The two unknowns are the amounts invested in each account. One relationship involves the total principal ($5000), and the other relationship involves the total interest earned ($252).

Plan and Execute Let x and y represent the two amounts invested. We can use a table to organize the information.

Note: Recall from Chapter 3 that we write the percentage rates as decimal numbers.

Categories	APR	Principal	Interest
Account 1	0.04	x	0.04x
Account 2	0.06	y	0.06y

Relationship 1: The total principal is $5000.

Translation: $x + y = 5000$

Relationship 2: The total interest is $252.

Translation: $0.04x + 0.06y = 252$

We will eliminate x by multiplying both sides of the first equation by -0.04.

Note: We could have first cleared the decimal numbers, but we chose not to.

$$
\begin{aligned}
x + y &= 5000 \quad \xrightarrow{\text{Multiply by } -0.04} \quad -0.04x - 0.04y = -200 \\
0.04x + 0.06y &= 252 \qquad\qquad\qquad\qquad\quad \underline{0.04x + 0.06y = 252} \\
&\qquad\qquad\qquad\qquad\qquad\qquad\qquad\quad 0.02y = 52 \\
&\qquad\qquad\qquad\qquad\qquad\qquad\qquad\qquad y = 2600
\end{aligned}
$$

Divide both sides by 0.02 to isolate y.

Now we can find x by substituting 2600 for y in one of the equations. We will use $x + y = 5000$.

$$x + 2600 = 5000 \qquad \text{Substitute 2600 for } y.$$
$$x = 2400 \qquad \text{Subtract 2600 from both sides to isolate } x.$$

Answer Jeff invested $2400 in the 4% account and $2600 in the 6% account.

Check The total investment is $2400 + $2600 = $5000 and the total interest is $0.04(2400) + 0.06(2600) = $96 + $156 = $252.

YOUR TURN

Monique invests a total of $8000 in two different accounts. The first account returns 5% while the second account returns 8%. If the total interest earned after 1 year is $532, what principal was invested in each account?

ANSWER

Monique invested $3600 at 5% and $4400 at 8%.

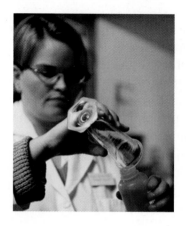

EXAMPLE 11 How much 10% HCl solution and 30% HCl solution must be mixed together to make 500 ml of 15% HCl solution?

Understand The two unknowns are the volumes of 10% solution and 30% solution that are mixed. One relationship involves concentrations of each solution in the mixture and the other relationship involves the total capacity of the final mixture (500 ml).

Plan and Execute Let x and y represent the two amounts to be mixed. We can use a table to organize the information.

Solution	Concentration	Volume	Amount of HCl
10% solution	0.10	x	0.10x
30% solution	0.30	y	0.30y
15% solution	0.15	$x + y$	0.15(500)

Relationship 1: The total volume is 500 ml.

$$\text{Translation: } x + y = 500$$

Relationship 2: The combined volumes of HCl in the two mixed solutions is to be 15% of the total capacity of the mixture.

$$\text{Translation: } 0.10x + 0.30y = 0.15(500)$$

$$0.10x + 0.30y = 75 \qquad \text{Simplify the right-hand side.}$$

We will eliminate x by multiplying both sides of the first equation by -0.10.

$$
\begin{array}{l}
x + y = 500 \quad\xrightarrow{\text{Multiply by } -0.10}\quad -0.10x - 0.10y = -50 \\
0.10x + 0.30y = 75 \qquad\qquad\qquad \underline{0.10x + 0.30y = 75} \\
\qquad\qquad\qquad\qquad\qquad\qquad\qquad 0.20y = 25 \\
\qquad\qquad\qquad\qquad\qquad\qquad\qquad\quad y = 125 \quad \text{Divide both sides by 0.20 to isolate } y.
\end{array}
$$

Now we can find x by substituting 125 for y in one of the equations. We will use $x + y = 500$.

$$x + 125 = 500 \qquad \text{Substitute 125 for } y.$$
$$x = 375 \qquad \text{Subtract 125 on both sides to isolate } x.$$

Answer Mixing 375 ml of 10% solution with 125 ml of 30% solution gives 500 ml of 15% solution.

Check The mixture volume is $125 + 375 = 500$. The amount of HCl is $0.10(375) + 0.30(125) = 37.5 + 37.5 = 75$. We can verify that 75 ml is 15% of 500 ml by multiplying $0.15(500) = 75$.

ANSWER

450 ml of 40% solution with 150 ml of 20% solution.

YOUR TURN

How much 20% solution and 40% solution must be mixed together to make 600 ml of 35% HCl solution.

9.3 EXERCISES

For Extra Help

 InterAct Math

Tutor Center Addison-Wesley Math Tutor Center

Video 15/DVT 5

 MyMathLab

MathXL MathXL

1. What advantages does the method of elimination have over graphing and substitution?

2. Suppose you are given the system of equations: $\begin{cases} x - y = 2 \\ 3x + y = 5 \end{cases}$

   Which variable would you eliminate? Why?

3. Suppose you are given the system of equations: $\begin{cases} 6x - 2y = 1 \\ 3x + 5y = 7 \end{cases}$

   a. Which variable is easier to eliminate? Why?

   b. How would you eliminate the variable you chose?

4. Suppose you are given the system of equations: $\begin{cases} 5x + 4y = 8 \\ 3x - 6y = 2 \end{cases}$

   a. Which variable is easier to eliminate? Why?

   b. How would you eliminate the variable you chose?

5. When using elimination, how do you know if a system of equations has no solution (inconsistent system)?

6. When using elimination, how do you know if a system of equations has an infinite number of solutions (dependent equations)?

For Exercises 7–34, solve the system of equations using the elimination method.

7. $\begin{cases} x - y = 14 \\ x + y = -2 \end{cases}$

8. $\begin{cases} 2x - 3y = 8 \\ 4x + 3y = 16 \end{cases}$

9. $\begin{cases} 3x + y = 9 \\ 2x - y = 1 \end{cases}$

10. $\begin{cases} 3x + y = 10 \\ -2x - y = -7 \end{cases}$

11. $\begin{cases} 3x + 2y = -7 \\ 5x - 2y = -1 \end{cases}$

12. $\begin{cases} x + y = 9 \\ 5x - y = 3 \end{cases}$

13. $\begin{cases} 4x + 3y = 17 \\ 2x + 3y = 13 \end{cases}$

14. $\begin{cases} x - 3y = 7 \\ x - 5y = 13 \end{cases}$

15. $\begin{cases} 5x + y = 14 \\ 2x + y = 5 \end{cases}$

16. $\begin{cases} 2x - 3y = -7 \\ 4x - 5y = -9 \end{cases}$

17. $\begin{cases} 11x - 3y = -34 \\ 8x + 10y = 24 \end{cases}$

18. $\begin{cases} 8x + 6y = 5 \\ 2x + y = 1 \end{cases}$

19. $\begin{cases} 5x + 6y = 11 \\ 2x - 4y = -2 \end{cases}$

20. $\begin{cases} -6x + 7y = -2 \\ 9x - 5y = -8 \end{cases}$

21. $\begin{cases} 10x + 5y = 3.5 \\ 3x - 4y = -0.6 \end{cases}$

22. $\begin{cases} 2x - y = 1.5 \\ 3x - 2y = -0.5 \end{cases}$

23. $\begin{cases} \dfrac{3}{2}x - \dfrac{4}{3}y = \dfrac{11}{3} \\ \dfrac{1}{4}x - \dfrac{2}{3}y = -\dfrac{7}{6} \end{cases}$

24. $\begin{cases} \dfrac{2}{3}x - \dfrac{3}{2}y = -1 \\ \dfrac{3}{2}x - \dfrac{7}{3}y = -1 \end{cases}$

25. $\begin{cases} 0.7x - \dfrac{1}{2}y = 9.4 \\ 0.9x + \dfrac{7}{10}y = 0 \end{cases}$

26. $\begin{cases} \dfrac{1}{3}x + 0.2y = -0.2 \\ \dfrac{2}{3}x - 0.75y = -5 \end{cases}$

27. $\begin{cases} 2x + y = -2 \\ 8x + 4y = -8 \end{cases}$

28. $\begin{cases} 5x + 5y = 50 \\ x + y = 2.5 \end{cases}$

29. $\begin{cases} y = 2x - 5 \\ x - y = 9 \end{cases}$

30. $\begin{cases} x + 3y = 8 \\ 2y = x + 6 \end{cases}$

31. $\begin{cases} y = \dfrac{3}{4}x + 1 \\ 4x = 2y + 3 \end{cases}$

32. $\begin{cases} y = -\dfrac{2}{3}x - 5 \\ 3y = 2x - 27 \end{cases}$

33. $\begin{cases} 0.2x - y = -3.2 \\ y = \dfrac{3}{4}x + 1 \end{cases}$

34. $\begin{cases} y = \dfrac{1}{5}x - 2 \\ 6y = 0.3x - 1.2 \end{cases}$

For Exercises 35–36, find the mistake.

35. $\begin{cases} x - y = 1 \\ 9x + 8y = 77 \end{cases}$

$\begin{aligned} -9(x - y) &= 1 \\ 9x + 8y &= 77 \end{aligned} \quad \longrightarrow \quad \begin{aligned} -9x + 9y &= 1 \\ \underline{9x + 8y} &= \underline{77} \\ 17y &= 78 \\ y &= \dfrac{78}{17} \end{aligned}$

$x - y = 1$

$x - \left(\dfrac{78}{17}\right) = 1$

$x = \dfrac{95}{17}$

Solution: $\left(\dfrac{95}{17}, \dfrac{78}{17}\right)$

36. $\begin{cases} x + y = 1 \\ x + 2y = 2 \end{cases}$

$\begin{aligned} x + y &= 1 \\ -1(x + 2y &= 2) \end{aligned} \quad \longrightarrow \quad \begin{aligned} x + y &= 1 \\ \underline{-x - 2y} &= \underline{2} \\ -y &= 3 \end{aligned}$

$y = -3$

$x + y = 1$

$x + (-3) = 1$

$x = 4$

Solution: $(4, -3)$

For Exercises 37–78, translate the problem to a system of equations, then solve using the elimination method.

37. Find two numbers whose sum is 18 and whose difference is 4.

38. Find two numbers whose difference is 6 and whose sum is 40.

39. The sum of Janet and her younger brother's ages is 15 and the difference of their ages is 4. What are their ages?

40. Morgan is older than Angela. Their ages add up to 56 and their age difference is 14. What are their ages?

41. The sum of three times a number and twice a smaller number is 46. The difference of the larger number and twice the smaller number is 2. Find the numbers.

42. The difference between two numbers is 4. Three times the larger is 2 more than five times the smaller. Find the numbers.

43. Two angles are complementary. The difference of their measures is 22°. Find the measures of the two angles.

44. Two angles are supplementary. The difference of their measures is 44°. Find the measures of the two angles.

45. A piece of steel is connected to a bridge for support as shown. The measure of the greater angle minus twice the measure of the smaller angle is 18°. Find the measures of the two angles.

Bridge

46. A laser is positioned in the corner of a room so that the beam forms two angles as shown. Three times the greater angle minus the other angle is 118°. What are the measures of the two angles?

47. During the 2002 Winter Olympic Games in Salt Lake City, Germany won the most medals, followed closely by the United States. The total medals won by both countries was 69 and the difference between the numbers of medals won by each country was 1. Find the total number of medals won by each country.

48. During the same Olympics, the difference between the number of gold medals won by Germany and the number of gold medals won by Canada was 6. If the total gold medals awarded to each country was 18, find the number of gold medals awarded to each.

49. If David has 19 bills in his wallet worth $125, all fives and tens, how many of each bill does he have?

50. During a game of Monopoly, Kari tells her opponent that she has 15 bills that total $3900, all $500 and $100 bills. How many of each bill does she have?

Of Interest

When the modern Olympics were revived in 1896, silver medals were given to first-place winners as gold was considered inferior. It was not until the 1904 games that gold medals were used to award first place. (*Source: Atlanta Journal-Constitution,* July 13, 1996.)

51. A shipment of books contains both algebra books for $80 and calculus books for $140. The packing slip has been lost, but the book buyer knows that there are 20 books for a total of $2080. How many of each book are there?

52. A bookstore sells TI-83 and TI-86 graphing calculators. The TI-83 sells for $89 and the TI-86 sells for $106. If 79 graphing calculators are sold in one day with a total revenue of $7541, how many of each type were sold?

53. Some of the tickets for Madonna's *Drowned World Tour* at the MGM Grand cost $350 and $250. Ulie won big at the slot machines and purchased 10 tickets for his friends, costing a total of $3300. How many of each ticket did he purchase?

54. The most expensive tickets for Aerosmith's *Just Push Play* concert at the New Orleans Arena were $54 and $74. Suppose the computer system went down and the cashier had to tally the number of tickets she sold during the first 20 minutes. Suppose she sold 234 tickets for a total of $15,036. How many of each ticket did she sell?

55. A movie theater charges $7.00 for adults and $4.25 for children. During a recent showing of *Pirates of the Caribbean*, 114 tickets were sold for a total of $583.50. How many of each ticket were sold?

56. When using a laser printer, Kinko's charges $1.40 per page for black-and-white documents and $3.20 per page for color documents. If a presentation containing 42 pages costs $69.60, how many color pages does it contain?

57. The cashier's office at the college has collected $7020 from students for a total of 42 part-time credit hours. If each credit hour costs $45, how many 3-hour classes and how many 5-hour classes were paid for?

58. During the 2000–2001 school year at the University of South Carolina, professors' salaries averaged $82,859 and associate professors' salaries averaged $60,567. If a department has a total of eight professors and associate professors with an annual salary budget of $529,120, how many of each type of professor work in the department? (*Source:* The University of South Carolina as reported in the *State* newspaper.)

59. A boat travels 24 miles upstream in 3 hours. Going downstream it can travel 36 miles in the same amount of time. Find the speed of the current and the speed of the boat in still water.

60. A boat travels 12 miles downstream in $1\frac{1}{2}$ hours. Going upstream it takes 6 hours to return. Find the speed of the current and the speed of the boat in still water.

61. If a plane can travel 500 miles per hour with the wind and only 440 miles per hour against the wind, find the speed of the wind and the speed of the plane in still air.

62. A plane can travel 200 miles per hour with the wind and only 120 miles per hour against the wind, find the speed of the wind and the speed of the plane in still air.

63. Adam inherited $15,500. He invested part of the money at an annual percentage rate (APR) of 4% and put the rest into a savings account at an APR of 3%. If his annual income from these investments is $585, how much did he invest at each rate?

64. Petunia invests some money into a plan that pays a 5% APR and twice as much in a plan that pays an 8% APR. If the total interest from the two plans is $1029, how much was invested in each?

65. Ellyn owes $13,250 on two student loans. One loan has an APR of 5% and the other, an APR of 8%. If the total interest after one year is $727, what is the amount of each loan?

66. Enrique has two credit cards. One is a Visa at an APR of 14% and the other is a department store card with an APR of 18%. If the total amount he owes is $1536 and the interest for one year is $245.36, what is the amount owed on each card?

67. Rhonda and Mike have two car loans totaling $16,250. One loan has an APR of 2.9% and the other has an APR of 6%. If the total interest they paid after one year is $657.16, what is the amount owed on each car?

68. Corey has borrowed a total of $250,000 in two loans. One loan is a mortgage at an APR of 8% and the other is a personal loan at an APR of 18%. If he paid $23,000 in interest during the first year, how much was loaned at each rate?

69. Jason invested $20,000 in two accounts, one paying 7% and the other paying 9% in annual interest. If the total interest he earned for the year was $1550, how much did he invest in each?

70. George invests some money into a bank account with an 8% APR, and then Martha invests twice as much in a movie project that pays 21% annually. If the total interest they earn is $6000, how much did they invest in each?

71. A pharmacist is preparing a 300 ml solution of 25% alcohol. If he has 15% and 45% alcohol solutions in stock, how much of each must he use?

72. During chemistry class, Tamika plans to mix a bottle containing 50 ml of 10% HCl solution with a bottle containing 25% HCl solution. If she wants a 20% solution, how much 25% HCl must be added?

73. How much of a 20% acid solution must be added to 80 ml of a 15% acid solution to obtain an 18% acid solution?

74. How many gallons of milk containing 1% butterfat must be added to 50 gallons of milk containing 6% butterfat to achieve a milk containing 2% butterfat?

75. A coffee shop is considering a new mixture of coffee beans. It will be created from Italian Roast beans costing $8.40 per pound and the Gold Coast Blend beans costing $11.60 per pound. The two types will be mixed to create 20 pounds of coffee to be sold for $9.76 per pound. How much of each type bean should be used?

76. A store is selling two types of jelly beans. One is $1.20 per pound and the other is $0.90 per pound. If the merchant wishes to sell a mixture at $1.11 per pound, how many pounds of each must be used to make 30 pounds?

77. Cromers P-Nuts sells several types of peanuts and cashews. The Virginia Fancy Peanuts cost $2.60 per pound and the Salted Cashews cost $12.50 per pound. If a caterer is planning to serve mixed nuts, how much of each type would be needed to make 9 pounds of a mixture that costs $7.00 per pound?

78. Liz invested part of her money at 5% and the rest at 4%. Her yearly income from these two investments was $35. If she had reversed the investments, the income would have been $37. How much did she invest in each?

A store sells meat in 3-pound and 5-pound packages. The 3-pound package sells for $3.80 and the 5-pound package sells for $6.20. The cost per pound for packaging the meat is the same for both packages, and the profit the store makes on each package is the same. What is the profit on each package? One day the store sells 27 pounds of meat and at least one of each package is sold. What is the total profit?

Collaborative Exercise

Tying It All Together—It's All Connected at the Center

Using the concepts and methods discussed in this chapter, we can determine the coordinates of the center of an equilateral octagon.

1. On a piece of graph paper, plot the following points: (9, 16), (14, 16), (18, 13), (18, 8), (14, 5), (9, 5), (5, 8), and (5, 13).

2. The points you plotted in step 1 are the vertices of an octagon. Draw line segments connecting these vertices to form the octagon.

3. Choose two vertices that are opposite each other on the octagon. Use a straightedge to draw a straight line that passes through these two points.

4. Use the coordinates of the vertices to determine the slope and the equation of the line that you've drawn. Remember that the formula for slope is $m = \dfrac{y_2 - y_1}{x_2 - x_1}$ and the point-slope equation of the line is $y - y_1 = m(x - x_1)$.

5. Now select another pair of opposing vertices and draw a straight line through the those two vertices.

6. Use the coordinates of the vertices to determine the slope and the equation of this second line.

7. If you've drawn your lines correctly, they will intersect at the center of the octagon. Notice that the coordinates of the center of the octagon satisfy both equations. Use the two equations that you found to form a system of equations. Solve this system of equations to determine the coordinates of the center point. (You might find it helpful to clear any fractions by multiplying both sides of the equation by the least common denominator.)

REVIEW EXERCISES

1. The total value of two bank accounts is $1100. The annual interest earned on the checking account is 5% while the CD pays 6%. The total interest earned is $61. How much was deposited into each account?

2. Monique has $4.05 in her purse. If she has 24 coins, all dimes and quarters, how many of each does she have?

3. Whitney's coin collection has seven more dimes than nickels. The total dollar amount is $3.25. How many of each coin does Whitney have?

4. Jay and Lisa are traveling west in separate cars on the same highway. Jay is traveling at 60 miles per hour and Lisa at 70 miles per hour. Jay passes Exit 54 at 2:30 P.M. Lisa passes the same exit at 2:36 P.M. At what time will Lisa catch up to Jay?

5. Use graphing to solve the system.
$$\begin{cases} x + y = -5 \\ 3x - y = 9 \end{cases}$$

6. Use substitution to solve the system.
$$\begin{cases} y = -2x \\ 2x + 5y = 10 \end{cases}$$

Solving Systems of Linear Equations in Three Variables

OBJECTIVES

1 Determine if an ordered triple is a solution for a system of equations.

2 Understand the graphs of systems of three equations.

3 Solve a system of three linear equations using the elimination method.

4 Solve applications involving three unknowns using a system of equations.

OBJECTIVE **1** **Determine if an ordered triple is a solution for a system of equations.** In this section, we will solve systems of three linear equations with three unknowns. Solutions of these systems will be *ordered triples* with the form (x, y, z). We check solutions for these systems just as we did in Section 9.1: Replace each variable with its corresponding value, and verify that each equation is true.

EXAMPLE 1 Determine whether each ordered triple is a solution to the system of equations.

$$\begin{cases} x + y + z = 3 & \text{(Equation 1)} \\ 2x + 3y + 2z = 7 & \text{(Equation 2)} \\ 3x - 4y + z = 4 & \text{(Equation 3)} \end{cases}$$

a. $(2, 1, 0)$

Solution In all three equations, replace x with 2, y with 1, and z with 0.

Equation 1:
$x + y + z = 3$
$2 + 1 + 0 \stackrel{?}{=} 3$
$3 = 3$ True.

Equation 2:
$2x + 3y + 2z = 7$
$2(2) + 3(1) + 2(0) \stackrel{?}{=} 7$
$4 + 3 + 0 \stackrel{?}{=} 7$
$7 = 7$ True.

Equation 3:
$3x - 4y + z = 4$
$3(2) - 4(1) + 0 \stackrel{?}{=} 4$
$6 - 4 + 0 \stackrel{?}{=} 4$
$2 = 4$ False.

Because $(2, 1, 0)$ does not solve all three equations in the system, it is not a solution for the system.

b. $(3, 1, -1)$

Solution

Equation 1:
$x + y + z = 3$
$3 + 1 + (-1) \stackrel{?}{=} 3$
$3 = 3$ True.

Equation 2:
$2x + 3y + 2z = 7$
$2(3) + 3(1) + 2(-1) \stackrel{?}{=} 7$
$6 + 3 - 2 \stackrel{?}{=} 7$
$7 = 7$ True.

Equation 3:
$3x - 4y + z = 4$
$3(3) - 4(1) + (-1) \stackrel{?}{=} 4$
$9 - 4 - 1 \stackrel{?}{=} 4$
$4 = 4$ True.

Because $(3, 1, -1)$ solves all three equations in the system, it is a solution for the system.

YOUR TURN Determine whether the following ordered triples are solutions to the system of equations.

a. $(4, 1, 1)$ **b.** $(-1, 5, 2)$

$$\begin{cases} x + y + z = 6 & \text{(Equation 1)} \\ 3x - 2y + 3z = -7 & \text{(Equation 2)} \\ 4x - 2y + z = -12 & \text{(Equation 3)} \end{cases}$$

OBJECTIVE **2** **Understand the graphs of systems of three equations.** Recall that we plot an ordered pair using two axes: x and y. Similarly, we plot an ordered triple like $(3, 4, 5)$ using three axes: x, y, and z, each of which is perpendicular to the other two, as shown.

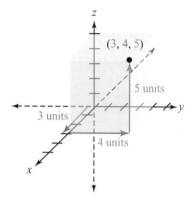

The graph of a linear equation in three variables is a *plane*, which is a flat surface much like a sheet of paper with infinite length and width. We draw a plane as a parallelogram (shown to the right), but remember that the length and width are infinite.

In Section 9.1 we found that two lines could intersect in one point (consistent system), no points (inconsistent system), or an infinite number of points (consistent system with dependent equations). Similarly, the intersection of the planes of a three-variable system can tell us about the number of solutions for the system.

A Single Solution

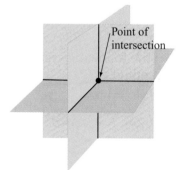

If the planes intersect at a single point, then that ordered triple is the solution to the system.

Infinite Number of Solutions

If the three planes intersect along a line, then the system has an infinite number of solutions, which are the coordinates of any point along that line.

If all three graphs are the same plane, then the system has an infinite number of solutions, which are the coordinates of any point in the plane.

No Solution

If all the planes are parallel, then the system has no solution.

Pairs of planes could also intersect, as shown. However, since all three planes do not have a common intersection, the system has no solution.

Note: There are other ways the planes could be configured so that they do not have a common intersection.

OBJECTIVE ③ **Solve a system of three linear equations using the elimination method.** Since graphing systems of three linear equations is difficult, we will first solve them using elimination. The process is much like what we learned in Section 9.3.

EXAMPLE 2 Solve the following systems of equations using the elimination method.

$$\begin{cases} x + y + z = 2 & \text{(Equation 1)} \\ 2x + y + 2z = 1 & \text{(Equation 2)} \\ 3x + 2y + z = 1 & \text{(Equation 3)} \end{cases}$$

Solution First we choose a variable and eliminate that variable from two pairs of equations. Let's eliminate y using equations 1 and 2. We multiply equation 1 by -1, then add the equations.

Note: Our initial goal is to generate two equations with two variables that are the same.

$$
\begin{array}{ll}
\text{(Eq. 1)} & x + y + z = 2 \quad\xrightarrow{\text{Multiply by } -1}\quad -x - y - z = -2 \\
\text{(Eq. 2)} & 2x + y + 2z = 1 \qquad\qquad\qquad\quad \underline{2x + y + 2z = 1} \\
& \qquad\qquad\qquad\qquad\qquad\qquad\qquad\quad x \quad\; + \; z = -1
\end{array}
$$

Add the equations. (Equation 4)

Now we need to eliminate y from another pair of equations. We will choose equations 1 and 3.

$$
\begin{array}{ll}
\text{(Eq. 1)} & x + y + z = 2 \quad\xrightarrow{\text{Multiply by } -2}\quad -2x - 2y - 2z = -4 \\
\text{(Eq. 3)} & 3x + 2y + z = 1 \qquad\qquad\qquad\quad \underline{3x + 2y + \; z = 1} \\
& \qquad\qquad\qquad\qquad\qquad\qquad\qquad\quad x \qquad\; - \; z = -3
\end{array}
$$

Add the equations. (Equation 5)

Equations 4 and 5 form a system of equations with variables x and z. We can eliminate z by adding the equations together.

$$
\begin{array}{ll}
\text{(Eq. 4)} & x + z = -1 \\
\text{(Eq. 5)} & \underline{x - z = -3} \qquad \text{Add the equations.} \\
& 2x \quad\;\; = -4 \\
& \qquad x = -2 \qquad \text{Divide both sides by 2 to isolate } x.
\end{array}
$$

To find z, we substitute -2 for x in equation 4 or 5. We will use equation 4.

$$
\begin{array}{ll}
x + z = -1 \\
-2 + z = -1 & \text{Substitute } -2 \text{ for } x. \\
\quad\;\; z = 1 & \text{Add 2 to both sides.}
\end{array}
$$

To find y, substitute -2 for x and 1 for z in any of the original equations. We will use equation 1.

$$
\begin{array}{ll}
x + y + z = 2 \\
-2 + y + 1 = 2 & \text{Substitute } -2 \text{ for } x \text{ and 1 for } z. \\
\quad\; y - 1 = 2 & \text{Simplify the left side.} \\
\quad\qquad y = 3 & \text{Add 1 to both sides to isolate } y.
\end{array}
$$

The solution is $(-2, 3, 1)$. We can check the solution by verifying that the ordered triple satisfies each of the three original equations. We will leave the check to the reader.

In some systems, one or more variables may be missing from one or more of the equations, which can simplify solving the system.

EXAMPLE 3 Solve the following systems of equations using the elimination method.

$$\begin{cases} 2x + 3y & = 3 & \text{(Equation 1)} \\ 2y - 3z = -8 & \text{(Equation 2)} \\ 4x \quad - \; z = 10 & \text{(Equation 3)} \end{cases}$$

Solution Notice that a variable is missing in each of the equations. We will eliminate z from equations 2 and 3.

(Eq. 2) $2y - 3z = -8$ Multiply by -3 $2y - 3z = -8$
(Eq. 3) $4x \quad - \; z = 10$ ⟶ $-12x \qquad + 3z = -30$ Add the equations.
$$-12x + 2y \qquad = -38 \qquad \text{(Equation 4)}$$

Because equations 1 and 4 do not have z terms, we do not need to eliminate z from another pair of equations. We will now use equations 1 and 4 to eliminate x.

                                    Multiply by 6
(Eq. 1) $2x + 3y = 3$ ⟶ $12x + 18y = 18$
(Eq. 4) $-12x + 2y = -38$ $-12x + \; 2y = -38$ Add the equations.
$$20y = -20$$
$$y = -1 \qquad \text{Divide both sides by 20 to isolate } y.$$

To find x, substitute -1 for y in equation 1 or 4. We will use equation 1.

$$2x + 3(-1) = 3 \qquad \text{Substitute } -1 \text{ for } y \text{ in equation 1.}$$
$$2x - 3 = 3 \qquad \text{Simplify.}$$
$$2x = 6 \qquad \text{Add 3 to both sides.}$$
$$x = 3 \qquad \text{Divide both sides by 2 to isolate } x.$$

To find z, substitute in either equation 2 or 3 (equation 1 has no z term). We will use equation 2.

$$2(-1) - 3z = -8 \qquad \text{Substitute } -1 \text{ for } y \text{ in equation 2.}$$
$$-2 - 3z = -8 \qquad \text{Simplify.}$$
$$-3z = -6 \qquad \text{Add 2 to both sides.}$$
$$z = 2 \qquad \text{Divide both sides by } -3 \text{ to isolate } z.$$

The solution is $(3, -1, 2)$. We will leave the check to the reader.

Following is a summary of the process of solving a system of three linear equations.

Let's see what happens when a system of three linear equations has no solution (inconsistent).

EXAMPLE 4 Solve the following system using elimination.

$$\begin{cases} 2x + y + 2z = -1 & \text{(Equation 1)} \\ -3x + 2y + 3z = -13 & \text{(Equation 2)} \\ 4x + 2y + 4z = 5 & \text{(Equation 3)} \end{cases}$$

Solution We need to eliminate a variable from two pairs of equations. Let's eliminate y using equations 1 and 2.

Multiply by -2

(Eq. 1) $2x + y + 2z = -1$ $\longrightarrow$ $-4x - 2y - 4z = 2$

(Eq. 2) $-3x + 2y + 3z = -13$ $\underline{-3x + 2y + 3z = -13}$ Add the equations.

$-7x \qquad - z = -11$ (Equation 4)

Now we eliminate y from equations 2 and 3.

Multiply by -1

(Eq. 2) $-3x + 2y + 3z = -13$ $\longrightarrow$ $3x - 2y - 3z = 13$

(Eq. 3) $4x + 2y + 4z = 5$ $\underline{4x + 2y + 4z = 5}$ Add the equations.

$7x \qquad + z = 18$ (Equation 5)

Equations 4 and 5 form a system in x and z. Add the equations together.

(Eq. 4) $-7x - z = -11$

(Eq. 5) $\underline{7x + z = 18}$

$0 = 7$

All variables are eliminated and the resulting equation is false, which means this system has no solution; it is inconsistent.

Note: We mentioned on p. 756 that we can identify systems of dependent equations when using the elimination method. A system of three equations is dependent if during the solution process we add two equations and get $0 = 0$ (we saw this with two equations in Example 6 of Section 9.3). For example, if equation 4 of Example 4 were $-7x - z = -18$, the sum of equations 4 and 5 would be $0 = 0$.

eq. 4 $-7x - z = -18$

eq. 5 $\underline{7x + z = 18}$

$0 = 0$

In such a case, we say the equations in the system are dependent.

ANSWERS

  a. $(-2, 4, -2)$

  b. No solution.

  c. Infinite solutions in
    $x + 3y - 6z = 9$
    (dependent equations)

YOUR TURN Solve the following systems of equations.

a. $\begin{cases} x + 3y + 2z = 6 \\ 2x - 3y + z = -18 \\ -3x + 2y + z = 12 \end{cases}$ **b.** $\begin{cases} 2x + y + 2z = 1 \\ x + y + z = 2 \\ 4x + 2y + 4z = 6 \end{cases}$ **c.** $\begin{cases} x + 3y - 6z = 9 \\ -7y + 6z = -3 \\ -2x + y + 6z = -15 \end{cases}$

4 Solve applications involving three unknowns using a system of equations. We follow the same procedure to solve applications involving three unknowns as we did in Sections 9.2 and 9.3 when solving applications with two unknowns.

EXAMPLE 5 At a movie theater, John buys one popcorn, one soft drink, and one candy bar, all for $7. Fred buys two popcorns, three soft drinks, and two candy bars for $16. Carla buys one popcorn, two soft drinks, and three candy bars for $12. Find the price of one popcorn, one soft drink, and one candy bar.

Understand We have three unknowns, three relationships, and we are to find the cost of each.

Plan Select a variable for each unknown, translate the relationships to a system of three equations, and then solve the system.

Execute Let x represent the cost of one popcorn, y represent the cost of one soft drink, and z represent the cost of one candy bar.

Relationship 1: One popcorn, one soft drink, and one candy bar cost $7.

Translation: $x + y + z = 7$

Relationship 2: Two popcorns, three soft drinks, and two candy bars cost $16.

Translation: $2x + 3y + 2z = 16$

Relationship 3: One popcorn, two soft drinks, and three candy bars cost $12.

Translation: $x + 2y + 3z = 12$

Our system: $\begin{cases} x + y + z = 7 & \text{(Equation 1)} \\ 2x + 3y + 2z = 16 & \text{(Equation 2)} \\ x + 2y + 3z = 12 & \text{(Equation 3)} \end{cases}$

We will choose to eliminate z from two pairs of equations. We will start with equations 1 and 2.

(Eq. 1) $\quad x + \ y + \ z = 7 \quad \xrightarrow{\text{Multiply by } -2.} \quad -2x - 2y - 2z = -14$

(Eq. 2) $\quad 2x + 3y + 2z = 16 \quad\quad\quad\quad\quad \underline{2x + 3y + 2z = 16}$ Add the equations.

$\quad\quad\quad\quad\quad\quad\quad\quad\quad\quad\quad\quad\quad\quad\quad\quad\quad y \quad\quad = 2 \quad$ (Equation 4)

Note: Although we were trying to eliminate z, we ended up eliminating both x and z.

Equation 4 gives us the value of y, indicating that a soft drink costs $2.00. Now, we choose another pair of equations and eliminate z again. We will use equations 1 and 3.

(Eq. 1) $\quad x + \ y + \ z = 7 \quad \xrightarrow{\text{Multiply by } -3.} \quad -3x - 3y - 3z = -21$

(Eq. 3) $\quad x + 2y + 3z = 12 \quad\quad\quad\quad\quad \underline{x + 2y + 3z = 12}$ Add the equations.

$\quad\quad\quad\quad\quad\quad\quad\quad\quad\quad\quad\quad\quad\quad\quad -2x - \ y \quad\quad = -9 \quad$ (Equation 5)

Since we already know $y = 2$, we can substitute for y in $-2x - y = -9$.

$$-2x - 2 = -9 \qquad \text{Substitute 2 for } y.$$
$$-2x = -7 \qquad \text{Add 2 to both sides.}$$
$$x = 3.50 \qquad \text{Divide both sides by } -2 \text{ to isolate } x.$$

Since x represents the cost of one popcorn, a popcorn costs \$3.50. To find z, substitute for x and y into one of the original equations. We will use equation 1.

$$3.5 + 2 + z = 7 \qquad \text{Substitute 3.5 for } x \text{ and 2 for } y.$$
$$5.5 + z = 7 \qquad \text{Simplify.}$$
$$z = 1.5 \qquad \text{Subtract 5.5 from both sides to isolate } z.$$

Since z represents the cost of one candy bar, a candy bar costs \$1.50.

Answer Popcorn costs \$3.50, a soft drink costs \$2.00, and a candy bar costs \$1.50.

Check Verify that at these prices, the amounts of money spent by John, Fred, and Carla are correct. We will leave the check to the reader.

YOUR TURN

ANSWER

Length = 20 in.,
width = 12 in.,
height = 14 in.

A small aquarium is in the shape of a rectangular solid. The sum of the length, width, and height is 46 inches. The sum of twice the length, three times the width, and the height is 90 inches. The sum of the length, twice the width, and three times the height is 86 inches. Find the dimensions of the aquarium.

9.4 EXERCISES

For Extra Help

InterAct Math

Addison-Wesley
Math Tutor Center

Video 15/DVT 5

MyMathLab

MathXL

1. When solving a system of three equations with three variables, you have to select two pairs of equations and eliminate the same variable in each pair. Does it matter which equations you choose? Why or why not?

2. When solving a system of three equations with three variables, you have to select two pairs of equations and eliminate the same variable in each pair. Does it matter which variable you eliminate? Why or why not?

3. Suppose you are given the system of equations $\begin{cases} x + y + z = 0 & \text{(Equation 1)} \\ 2x + 4y + 3z = 5 & \text{(Equation 2)} \\ 4x - 2y + 3z = -13 & \text{(Equation 3)} \end{cases}$

Which variable would you choose to eliminate, using which pairs of equations? Why?

4. When using elimination to solve a system of three equations with three variables, how do you know if it has no solution?

5. Two drawings were given in the text of inconsistent systems of equations with three unknowns. Make another drawing of an inconsistent system other than the ones given.

6. Where would a point be located if it solves two equations of a system of three equations in three variables but does not solve the third equation?

For Exercises 7–12, determine if the given point is a solution of the system.

7. $(3, -1, 1)$

$$\begin{cases} x + y + z = 3 \\ 2x - 2y - z = 4 \\ 2x + y - 2z = 3 \end{cases}$$

8. $(2, -2, 1)$

$$\begin{cases} 3x + 2y + z = 3 \\ 2x - 3y - 2z = 8 \\ -2x + 4y + 3z = -9 \end{cases}$$

9. $(1, 0, 2)$

$$\begin{cases} 2x + 3y - 3z = -4 \\ -2x + 4y - z = -4 \\ 3x - 4y + 2z = 5 \end{cases}$$

10. $(3, 4, 0)$

$$\begin{cases} x + 2y + 5z = 11 \\ 3x - 2y - 4z = 1 \\ 2x + 2y + 3z = 12 \end{cases}$$

11. $(2, -2, 4)$

$$\begin{cases} x + 2y - z = -6 \\ 2x - 3y + 4z = 26 \\ -x + 2y - 3z = -18 \end{cases}$$

12. $(0, -2, 4)$

$$\begin{cases} 3x + 2y - 3z = -16 \\ 2x - 4y - z = 4 \\ -3x + 4y - 2z = -16 \end{cases}$$

For Exercises 13–30, solve the systems of equations.

13. $\begin{cases} x + y + z = 5 \\ 2x + y - 2z = -5 \\ x - 2y + z = 8 \end{cases}$

14. $\begin{cases} x + y + z = 4 \\ 3x + y - z = -4 \\ 2x - 2y + 3z = 3 \end{cases}$

15. $\begin{cases} x + y + z = 2 \\ 4x - 3y + 2z = 2 \\ 2x + 3y - 2z = -8 \end{cases}$

16. $\begin{cases} x + y - z = 7 \\ -2x + 2y - z = -5 \\ 3x - 3y + 2z = 6 \end{cases}$

17. $\begin{cases} 2x + y + 2z = 5 \\ 3x - 2y + 3z = 4 \\ -2x + 3y + z = 8 \end{cases}$

18. $\begin{cases} 2x + 3y - 2z = -4 \\ 4x - 3y + z = 25 \\ x + 2y - 4z = -12 \end{cases}$

19. $\begin{cases} x + 2y - z = 1 \\ 2x + 4y - 2z = -8 \\ -0.5x - y + 0.5z = -2 \end{cases}$

20. $\begin{cases} -8x + 4y + 6z = -18 \\ 2x - y - 1.5z = 4.5 \\ 4x - 2y - 3z = 9 \end{cases}$

21. $\begin{cases} y = 2x + z - 8 \\ x = 2y - 3z + 11 \\ 2x + 3y - z = -6 \end{cases}$

22. $\begin{cases} z = -3x + y - 10 \\ 2x + 3y - 2z = 5 \\ x = 3y - 3z - 14 \end{cases}$

23. $\begin{cases} x = 4y - z + 1 \\ 3x + 2y - z = -8 \\ x + 6y + 2z = -3 \end{cases}$

24. $\begin{cases} 2x - 3y - 4z = 3 \\ y = -4x + 8z - 1 \\ x = -5y - 2z - 4 \end{cases}$

25. $\begin{cases} 4x + 2y + 3z = 9 \\ 2x - 4y - z = 7 \\ 3x - 2z = 4 \end{cases}$

26. $\begin{cases} 4x + 2y - 3z = 20 \\ 2x + 5z = -4 \\ 3x - 3y + 2z = 2 \end{cases}$

27. $\begin{cases} 3x + 2y = -2 \\ 2x - 3z = 1 \\ 0.4y - 0.5z = -2.1 \end{cases}$

28. $\begin{cases} 0.2x - 0.3z = -1.8 \\ 3x + 2y = -5 \\ 3y + 2z = 14 \end{cases}$

29. $\begin{cases} \dfrac{3}{2}x + y - z = 0 \\ 4y - 3z = -22 \\ -0.2x + 0.3y = -2 \end{cases}$

30. $\begin{cases} -0.2x - 0.3y + 0.1z = -0.3 \\ -3x + 2y = 13 \\ \dfrac{1}{4}y - \dfrac{1}{2}z = 2 \end{cases}$

For Exercises 31–62, translate the problem to a system of equations, then solve.

31. The sum of three numbers is 16. One of the numbers is 2 more than twice a second number, and 2 less than the third number. Find the numbers.

32. The sum of three numbers is 18. Twice the first number is 1 less than the third number, and twice the second number is 1 more than the third number. Find the numbers.

33. The sum of the measures of the angles in every triangle is 180°. Suppose the measure of one angle of a triangle is three times that of a second angle and the measure of the second angle is 5° less than the measure of the third. Find the measure of each angle of the triangle.

34. The measure of one angle of a triangle is 25° less than the measure of a second angle, and twice the measure of the second angle is 65° more than the measure of the third angle. Find the measure of each angle of the triangle.

35. The perimeter of a triangle is 33 inches. The sum of the length of the longest side and twice the length of the shortest side is 31 inches. Twice the length of the longest side minus both the other side lengths is 12 inches. Find the side lengths.

36. The perimeter of the figure shown is 118 centimeters. Twice x minus y is 1 and y minus z is 10 centimeters. Find the side lengths.

37. At a fast-food restaurant, one burger, one order of fries, and one drink cost $5, three burgers, two orders of fries, and two drinks cost $12.50, and two burgers, four order of fries, and three drinks cost $14. Find the individual cost of one burger, one order of fries, and one drink.

38. James went to the college bookstore and purchased two pens, two erasers, and one pack of paper for $6. Tamika purchased four pens, three erasers, and two packs of paper for $11.50. Jermaine purchased three pens, one eraser, and three packs of paper for $9.50. Find the cost of one pen, one eraser, and one pack of paper.

39. Tickets for a high school band concert were $3 for children, $5 for students, and $8 for adults. There was a total of 500 tickets sold and the total money received from the sale of the tickets was $2500. There were 150 fewer adult tickets sold than student tickets. Find the number of each type of ticket sold.

40. Tickets to a play were $2 for children, $4 for students, and $6 for adults. There was a total of 300 tickets sold and the total money received from the sale of the tickets was $1250. There were 35 more student tickets sold than children's tickets. Find the number of each type of ticket sold.

41. In a recent basketball game, Tracy McGrady of the Orlando Magic scored a total of 28 points. He made twice as many 2 point field goals as he did 3 point field goals, and he made one more free throw (worth 1 point each) than he did 2 point field goals. Find how many 3 point field goals, 2 point field goals, and free throws he made in this game. (*Source: Orlando Sentinel*, December 16, 2002)

42. At a track meet, 10 points are awarded for each first place finish, 5 points for each second, and 1 point for each third. Suppose a track team scored a total of 71 points and had two more firsts than seconds and one less first than third. Find the number of first-, second-, and third-place finishes for the team.

43. Tanisha has 25 coins consisting of nickels, dimes, and quarters with a total value of $3.60. If the sum of the number of nickels and dimes is five more than the number of quarters, how many of each type of coin does she have?

44. A vending machine accepts nickels, dimes, and quarters. When the owner checks the machine, there is a total of 56 coins worth $7.55. If the sum of the number of nickels and dimes is 2 more than twice the number of quarters, find the number of each type of coin.

45. Tommy empties his piggy bank and finds that he has 25 coins consisting of pennies, nickels, and dimes that are worth $1.36. If there are twice as many nickels as there are pennies, find the number of each type of coin.

46. At the end of each day, Josh puts all the change in his pockets into a jar. After one week he has 25 coins consisting of pennies, dimes, and quarters that are worth $2.17. If there are 3 more dimes than quarters, find the number of each type of coin.

47. Bronze is an alloy that is made of zinc, tin, and copper in a specified proportion. The amount of tin in bronze is three times the amount of zinc, and the amount of copper is 20 pounds more than 15 times the amount of tin. Suppose an order is placed for 1000 pounds of bronze. Find the number of pounds of zinc, tin, and copper. (*Source: World Almanac and Book of Facts* 2002)

48. The Nutty Bavarian sells nuts in bulk quantities. When bought in bulk, peanuts sell for $1.25 per pound, almonds for $2.10 per pound, and cashews for $3.75 per pound. Suppose a local specialty shop wants a mixture of 120 pounds that will cost $1.95 per pound. Find the number of pounds of each type of nut if the sum of the number of pounds of peanuts and cashews is twice the number of pounds of almonds. (*Source:* The Nutty Bavarian, in Sanford, Florida)

49. The Wunderbar delicatessen sells Black Forest ham for $11.96 per pound, breast of turkey for $8.76 per pound, and roast beef for $9.16 per pound. Suppose they make a 10 pound party tray of these three meats such that the average cost is $9.80 per pound. Find the number of pounds of each if the number of pounds of breast of turkey is equal to the sum of the number of pounds of Black Forest ham and roast beef. (*Source:* The Wunderbar, Altamonte Springs)

50. Barney's coffee shop sells Jamaican Blue Mountain coffee for $45.99 per pound, Hawaiian Kona for $36.99 per pound, and Sulawesi Kalossi for $12.99 per pound. A 25-pound mixture of these three coffees sells for $30.27 per pound. Find the number of pounds of each if the sum of the number of pounds of Jamaica Blue Mountain and Hawaiian Kona is 5 more than the number of pounds of Sulawesi Kalossi. (*Source: Barney's, Orlando, Florida*)

51. A total of $8000 is invested in three stocks, which paid 4%, 6%, and 7% dividends, respectively, in one year. The amount invested in the stock that returned 7% dividends is $500 less than twice the amount that returned 4% dividends. If the total dividends in one year were $475, find the amount invested in each stock.

52. A total of $5000 is invested in three funds. The money market fund pays 5% annually, the income fund pays 6% annually, and the growth fund pays 8% annually. The total earnings for one year from the three funds are $340. If the amount invested in the growth fund is $500 less than twice the amount invested in the money market fund, find the amount in each of the funds.

53. Nina inherited $100,000, which she invested in stocks, bonds, and certificates of deposit. After the first year, she had lost 6% in the stocks, earned 5% on the bonds, and earned 4% on the certificates of deposit. The sum of twice the principal in certificates of deposit and the principal in stocks is $20,000 more than the principal in bonds. Find the principal in each investment if the total income for the year was $1500.

54. An executive received a $30,000 end-of-the-year bonus. He invests the bonus in stocks, municipal bonds, and certificates of deposit. At the end of the first year, he had lost 5% in the stocks, earned 5% on the bonds, and earned 4% on the certificates of deposit, which resulted in a net return of $220. Find the principal in each of the investments if the principal in stocks is $6000 less than the sum of the principals in bonds and certificates.

55. Eggs are classified by the weight of a dozen. The three largest sizes are jumbo, extra large, and large. The total weight of one dozen of each size is 81 ounces. Three dozen jumbo eggs weigh 39 ounces more than the sum of the weight of one dozen extra large and one dozen large. The sum of the weights of two dozen extra large and one dozen large is 18 ounces more than the weight of two dozen jumbo. Find the weight of one dozen of each type of egg. (*Source: Numbers: How Many, How Long, How Far, How Much*)

56. The intensity of sound is measured in decibels. The decibel reading of a rock concert is 60 less than three times the reading of normal speaking and 10 less than that of a jet at takeoff. The decibel reading of a jet takeoff is 10 more than twice that of normal speaking. Find the decibel reading of each. (*Source: Numbers: How Many, How Long, How Far, How Much*)

57. The number of calories burned per hour bicycling is 120 more than the number burned from brisk walking. The number of calories burned per hour in climbing stairs is 180 more than from bicycling and is twice that from brisk walking. Find the number of calories burned per hour for each of the three activities. (*Source: Numbers: How Many, How Long, How Far, How Much*)

58. The human body is composed of many elements. About 98% of the human body is made up of carbon, oxygen, and hydrogen. The percent of the human body that is hydrogen is the sum of twice the percentage of oxygen and the percentage of carbon. The percent of oxygen is 6% more than twice the percentage of carbon. Find the percent of the human body that is composed of each of these three elements. (*Source: Numbers: How Many, How Long, How Far, How Much*)

59. The three countries in the world with the highest death rates per 1000 are Angola, Burundi, and Afghanistan. The sum of the death rates of all three countries is 58 per 1000. The sum of the death rate per 1000 of Angola and Burundi is 5 less than twice the death rate of Afghanistan, and the death rate per 1000 for Afghanistan is 2 more than the rate for Angola. Find the death rate per 1000 for each country. (*Source: Webster's New World Book of Facts*)

60. The three countries in the world with the highest birth rates per 1000 are Angola, Ethiopia, and Afghanistan. The sum of the birth rates is 149 per 1000. The sum of the birth rates per 1000 of Angola and Ethiopia is 7 less than twice the birth rate of Afghanistan. The birth rate per 1000 of Ethiopia is 4 less than the birth rate of Afghanistan. Find the birth rate per 1000 for each of these countries. (*Source: Webster's New World Book of Facts*)

61. An object is thrown upward with an initial velocity of v_0 from an initial height of h. The height of the object, h, is described by an equation of the form $h = at^2 + v_0 t + h_0$, where h and h_0 are in feet and t is in seconds. The following table shows the height for different values of t. Find the values of a, v_0, and h_0 and write the equation for h.

t	h
1	234
3	306
6	174

62. Use the same general equation from Exercise 61, $h = at^2 + v_0 t + h_0$, where h and h_0 are in feet and t is in seconds. The table at the right shows the distance above the ground for different values of t. Find the values of a, v_0, and h_0 and write the equation for h.

t	h
2	136
3	106
4	44

PUZZLE PROBLEM

The sum of four positive integers is 228. The greatest number subtracted from twice the smallest number gives 18. If twice the sum of the third number and fourth numbers is subtracted from the product of 6 and the second number, the result is 10. If the smallest two numbers are consecutive odd integers, find all numbers.

REVIEW EXERCISES

For Exercises 1 and 2, simplify.

1. $\frac{1}{2}(2x + 3y) - (x + 4y)$

2. $-2(x + y + 2z) + (2x + y - 2z)$

3. Given $2x - 3y = 6$, find x if $y = 4$.

4. Given $x + y + 2z = 7$, find x if $y = -1$ and $z = 3$.

For Exercises 5 and 6, solve.

5. $\begin{cases} 2x + 5y = 1 \\ -3x + 2y = 11 \end{cases}$

6. $\begin{cases} 3x + 4y = 6 \\ x = -2 \end{cases}$

9.5 Solving Systems of Linear Equations Using Matrices

OBJECTIVES

1 Write a system of equations as an augmented matrix.

2 Solve a system of linear equations by using the augmented matrix.

OBJECTIVE 1 **Write a system of equations as an augmented matrix.** Although the elimination method that we learned in Section 9.3 is effective for solving systems of linear equations, we can streamline the method by manipulating just the coefficients and constants in a **matrix**.

> **DEFINITION**
>
> **Matrix:** A rectangular array of numbers.

Following are some examples of matrices (plural of matrix).

$$\begin{bmatrix} 1 & -2 \\ -3 & 4 \end{bmatrix} \quad \begin{bmatrix} 1 & -5 & 6 \\ -2 & 4 & 0 \end{bmatrix} \quad \begin{bmatrix} -3 & 0 & 9 \\ -2 & 4 & 7 \\ 9 & -2 & 0 \end{bmatrix}$$

Matrices are made up of horizontal rows and vertical columns.

Column 1 Column 2 Column 3

$$\text{Row 1} \longrightarrow \begin{bmatrix} 1 & 4 & -2 \\ 5 & -1 & 3 \end{bmatrix} \text{Row 2}$$

Column 1 Column 2

$$\begin{array}{l} \text{Row 1} \longrightarrow \\ \text{Row 2} \longrightarrow \\ \text{Row 3} \longrightarrow \end{array} \begin{bmatrix} 3 & -4 \\ -1 & 2 \\ 5 & -3 \end{bmatrix}$$

Note: One way to remember that columns are vertical is to imagine the columns on a building, which are vertical.

We call a matrix like $\begin{bmatrix} 2 & 5 & -1 \\ 5 & 4 & 3 \end{bmatrix}$ a 2×3 (read "2 by 3") matrix because it

has two rows and three columns; $\begin{bmatrix} 2 & 5 \\ -4 & 6 \\ 3 & -2 \end{bmatrix}$ is a 3×2 matrix because it has

three rows and two columns. Each number in a matrix is called an *element*. To solve a system of equations using matrices, we first rewrite the system as an **augmented matrix**.

> **DEFINITION**
>
> **Augmented matrix:** A matrix made up of the coefficients and the constant terms of a system. The constant terms are separated from the coefficients by a dashed vertical line.

For example, to write the system $\begin{cases} 3x - 2y = 7 \\ 4x - 3y = 10 \end{cases}$ as an augmented matrix, we omit the variables and write $\begin{bmatrix} 3 & -2 & | & 7 \\ 4 & -3 & | & 10 \end{bmatrix}$.

EXAMPLE 1 Write $\begin{cases} 2x + 3y - 4z = 1 \\ 3x - 5y + z = -4 \\ 2x - 6y + 3z = 2 \end{cases}$ as an augmented matrix.

Solution $\begin{bmatrix} 2 & 3 & -4 & | & 1 \\ 3 & -5 & 1 & | & -4 \\ 2 & -6 & 3 & | & 2 \end{bmatrix}$

Note: It is helpful to think of the dashes as the equal sign.

YOUR TURN

ANSWER

$\begin{bmatrix} 1 & 3 & | & -2 \\ -4 & -5 & | & 7 \end{bmatrix}$

Write the augmented matrix for $\begin{cases} x + 3y = -2 \\ -4x - 5y = 7 \end{cases}$

OBJECTIVE 2 **Solve a system of linear equations by using the augmented matrix.** Recall that in the elimination method, we could interchange equations, add equations, or multiply equations by a number and then add the rewritten equations. Since each row of an augmented matrix contains the constants and coefficients of each equation in the system, we can interchange rows, add rows, or multiply rows by a number and then add the rewritten rows.

RULE **ROW OPERATIONS**

1. Any two rows may be interchanged.
2. The elements of any row may be multiplied (or divided) by any nonzero real number.
3. Any row may be replaced by a row resulting from adding the elements of that row (or multiples of that row) to a multiple of the elements of any other row.

We use these row operations to solve a system of linear equations by transforming the augmented matrix of the system into an equivalent matrix that is in **echelon form**.

DEFINITION

Echelon form: An augmented matrix whose coefficient portion has 1's on the diagonal from upper left to lower right and 0's below the 1's.

For example, $\begin{bmatrix} 1 & 3 & | & 3 \\ 0 & 1 & | & 4 \end{bmatrix}$ and $\begin{bmatrix} 1 & 2 & -4 & | & 5 \\ 0 & 1 & -5 & | & 7 \\ 0 & 0 & 1 & | & 5 \end{bmatrix}$ are in echelon form.

In the following examples we will let R_1 mean row 1, R_2 mean row 2, and so on.

EXAMPLE 2 Solve the following linear system by transforming its augmented matrix into echelon form.

$$\begin{cases} x - 2y = -4 & (\text{Equation 1}) \\ -2x + 5y = 9 & (\text{Equation 2}) \end{cases}$$

Solution First, we write the augmented matrix: $\begin{bmatrix} 1 & -2 & | & -4 \\ -2 & 5 & | & 9 \end{bmatrix}$

Now we perform row operations to get the matrix in echelon form. The element in the first row, first column is already 1, which is what we want. So we need to rewrite the matrix so that -2 in the second row, first column is 0. To do this, we multiply the first row by 2 and add it to the second row.

$$2R_1 + R_2 \longrightarrow \begin{bmatrix} 1 & -2 & | & -4 \\ 0 & 1 & | & 1 \end{bmatrix}.$$

The resulting matrix represents the system $\begin{cases} x - 2y = -4 \\ y = 1 \end{cases}$. Since $y = 1$, we can solve for x using substitution.

$$x - 2(1) = -4 \qquad \text{Substitute 1 for } y \text{ in } x - 2y = -4.$$
$$x - 2 = -4 \qquad \text{Simplify.}$$
$$x = -2 \qquad \text{Add 2 to both sides.}$$

The solution is $(-2, 1)$.

EXAMPLE 3 Solve the following linear system by transforming its augmented matrix into echelon form.

$$\begin{cases} 2x + 6y = 1 & (\text{Equation 1}) \\ x + 8y = 3 & (\text{Equation 2}) \end{cases}$$

Solution First, we write the augmented matrix: $\begin{bmatrix} 2 & 6 & | & 1 \\ 1 & 8 & | & 3 \end{bmatrix}$

Now we perform row operations to get echelon form.

$$\frac{1}{2} \cdot R_1 \longrightarrow \begin{bmatrix} 1 & 3 & | & \frac{1}{2} \\ 1 & 8 & | & 3 \end{bmatrix}$$
We need the 2 in row 1, column 1 to be a 1, so multiply each number in row 1 by $\frac{1}{2}$.

$$-1 \cdot R_1 + R_2 \longrightarrow \begin{bmatrix} 1 & 3 & | & \frac{1}{2} \\ 0 & 5 & | & \frac{5}{2} \end{bmatrix}$$
We need the 1 in row 2, column 1 to be 0, so multiply row 1 by -1 and add it to row 2.

$$\frac{1}{5} \cdot R_2 \longrightarrow \begin{bmatrix} 1 & 3 & | & \frac{1}{2} \\ 0 & 1 & | & \frac{1}{2} \end{bmatrix}$$
We need the 5 in row 2, column 2 to be 1, so multiply row 2 by $\frac{1}{5}$.

CONNECTION

Row operations correspond to equation manipulations in the elimination method. For example, $-1 \cdot R_1 + R_2$ correponds to multiplying $x + 3y = \frac{1}{2}$ by -1 and then adding the resulting equation to $x + 8y = 3$ so that we get $5y = \frac{5}{2}$.

This matrix represents the system $\begin{cases} x + 3y = \dfrac{1}{2} \\ y = \dfrac{1}{2} \end{cases}$. Since $y = \dfrac{1}{2}$, we can solve for x using substitution.

$$x + 3\left(\frac{1}{2}\right) = \frac{1}{2} \qquad \text{Substitute } \frac{1}{2} \text{ for } y \text{ in } x + 3y = \frac{1}{2}.$$

$$x + \frac{3}{2} = \frac{1}{2} \qquad \text{Simplify.}$$

$$x = -1 \qquad \text{Subtract } \frac{3}{2} \text{ from both sides.}$$

The solution is $\left(-1, \dfrac{1}{2}\right)$.

YOUR TURN   Solve the following system by transforming its augmented matrix to echelon from.

$$\begin{cases} 3x + 4y = 6 \\ x - 3y = -11 \end{cases}$$

Now let's use row operations to solve systems of three equations.

EXAMPLE 4   Use the echelon method to solve

$$\begin{cases} 2x + y - 2z = -3 & \text{(Equation 1)} \\ x + y + 2z = 7 & \text{(Equation 2)} \\ 4y + 3z = 5 & \text{(Equation 3)} \end{cases}$$

Solution Write the augmented matrix: $\begin{bmatrix} 2 & 1 & -2 & | & -3 \\ 1 & 1 & 2 & | & 7 \\ 0 & 4 & 3 & | & 5 \end{bmatrix}$

Now we perform row operations to get echelon form.

$$\begin{matrix} R_2 \longrightarrow \\ R_1 \longrightarrow \\ {} \end{matrix} \begin{bmatrix} 1 & 1 & 2 & | & 7 \\ 2 & 1 & -2 & | & -3 \\ 0 & 4 & 3 & | & 5 \end{bmatrix}$$

Because we need a 1 in row 1, column 1, we will interchange rows 1 and 2.

$$-2 \cdot R_1 + R_2 \longrightarrow \begin{bmatrix} 1 & 1 & 2 & | & 7 \\ 0 & -1 & -6 & | & -17 \\ 0 & 4 & 3 & | & 5 \end{bmatrix}$$

To get a 0 in row 2, column 1, we multiply the new row 1 by -2 and add it to row 2.

$$4 \cdot R_2 + R_3 \longrightarrow \begin{bmatrix} 1 & 1 & 2 & | & 7 \\ 0 & -1 & -6 & | & -17 \\ 0 & 0 & -21 & | & -63 \end{bmatrix}$$

Row 3 only needs a 0 in the second column, so we multiply row 2 by 4 and add it to row 3.

$$\begin{matrix} -1 \cdot R_2 \longrightarrow \\ -\dfrac{1}{21} \cdot R_3 \longrightarrow \end{matrix} \begin{bmatrix} 1 & 1 & 2 & | & 7 \\ 0 & 1 & 6 & | & 17 \\ 0 & 0 & 1 & | & 3 \end{bmatrix}$$

To get a 1 in row 2, column 2, we multiply row 2 by -1.

To get a 1 in row 3, column 3, we multiply row 3 by $-\dfrac{1}{21}$.

ANSWER

$(-2, 3)$

The resulting matrix represents the system $\begin{cases} x + y + 2z = 7 \\ \quad\ y + 6z = 17. \\ \qquad\quad z = 3 \end{cases}$

To find y, substitute 3 for z in $y + 6z = 17$.

$$y + 6(3) = 17 \qquad \text{Substitute 3 for } z.$$
$$y + 18 = 17 \qquad \text{Multiply 6 and 3.}$$
$$y = -1 \qquad \text{Subtract 18 from both sides.}$$

To find x, substitute -1 for y and 3 for z in $x + y + 2z = 7$.

$$x + (-1) + 2(3) = 7 \qquad \text{Substitute } -1 \text{ for } y \text{ and 3 for } z.$$
$$x + 5 = 7 \qquad \text{Simplify the left side of the equation.}$$
$$x = 2 \qquad \text{Subtract 5 from both sides of the equation.}$$

The solution is $(2, -1, 3)$. We can check by verifying that the ordered triple satisfies all three of the original equations.

YOUR TURN

Solve the following linear system by transforming its augmented matrix into echelon form.

$$\begin{cases} 3x + y + z = 2 \\ x + 2y - z = -4 \\ 2x - 2y + 3z = 9 \end{cases}$$

Calculator TIPS

To use a graphing calculator to solve the system in Example 4, press the MATRX *key then use the arrow keys to highlight **EDIT**. You will see a list of matrices and their current sizes. Press 1 to select the first matrix, named [A]. We need our matrix to be 3×4, so press* 3 ENTER*, then* 4 ENTER*. The cursor will now be in row 1, column 1 of a 3×4 matrix. You can now enter each number. Use the arrow keys to move the cursor to any position in the matrix. Note that the cursor position's row number, column number, and current value are displayed at the bottom of the screen. After entering each number for Example 4, you should see*

MATRX[A] 3×4
$\begin{bmatrix} 2 & 1 & -2 & -3 \\ 1 & 1 & 2 & 7 \\ 0 & 4 & 3 & 5 \end{bmatrix}$

Now, to put it into row echelon form, press MATRX *again and move the cursor to highlight* MATH *at the top of the screen. From the menu, select **A: ref (** and press* ENTER*. You will be prompted to enter the name of the desired matrix. Press* MATRX *again and select*

continued

ANSWER

$(1, -2, 1)$

1: [A] 3 × 4. You will now see ref ([A]. Use the ☐ key to close the parentheses and press ENTER . You now have the row echelon form and your screen should look like this:

ref([A])

$$\begin{bmatrix} 1 & .5 & -1 & -1.5 \\ 0 & 1 & .75 & 1.25 \\ 0 & 0 & 1 & 3 \end{bmatrix}$$

Note: The function **ref** stands for *row echelon form.*

Notice the last row indicates that z = 3 and we can use substitution to find the other values.

*Your calculator can also show the complete solution. Instead of selecting **A: ref (**, select **B: rref (** from the math menu. Your screen will look like this:*

rref([A])

$$\begin{bmatrix} 1 & 0 & 0 & 2 \\ 0 & 1 & 0 & -1 \\ 0 & 0 & 1 & 3 \end{bmatrix}$$

Note: This form is called *reduced row echelon form,* which is why the function is **rref**.

Note that the first row of this form indicates that x = 2, the second row indicates that y = −1, and the third row indicates that z = 3, which is the solution for our system.

9.5 EXERCISES

For Extra Help

 InterAct Math

Addison-Wesley Math Tutor Center

 Video 15/DVT 5

MyMathLab

 MathXL

1. How many rows and columns are in a 4 × 2 matrix?

2. How do you write a system of equations as an augmented matrix?

3. What does the dashed line in an augmented matrix correspond to in a system of equations?

4. Explain how solving a system of equations using row operations is similar to solving the system using elimination.

5. What is echelon form?

6. Once an augmented matrix is in echelon form, how do you find the values of the variables?

For Exercises 7–14, write the augmented matrix for the system of equations.

7. $\begin{cases} 14x + 7y = 6 \\ 7x + 6y = 8 \end{cases}$

8. $\begin{cases} 5x + 6y = 2 \\ 10x + 3y = -2 \end{cases}$

9. $\begin{cases} 7x - 6y = 1 \\ 8x - 12y = 6 \end{cases}$

10. $\begin{cases} 3x - y = 6 \\ 9x + 2y = -2 \end{cases}$

11. $\begin{cases} x - 3y + z = 4 \\ 2x - 4y + 2z = -4 \\ 6x - 2y + 5z = -4 \end{cases}$

12. $\begin{cases} 3x + 2y - 3z = 2 \\ 2x - 4y + 5z = -10 \\ 5x - 4y + z = 0 \end{cases}$

13. $\begin{cases} 4x + 6y - 2z = -1 \\ 8x + 3y = -12 \\ -y + 2z = 4 \end{cases}$

14. $\begin{cases} x - 3y = 3 \\ -3x + 4y + 9z = 3 \\ 2x + 7z = -9 \end{cases}$

For Exercises 15–18, given the matrices in echelon form, find the solution for the system.

15. $\begin{bmatrix} 1 & 5 & \vdots & -6 \\ 0 & 1 & \vdots & -2 \end{bmatrix}$

16. $\begin{bmatrix} 1 & -3 & \vdots & 7 \\ 0 & 1 & \vdots & -5 \end{bmatrix}$

17. $\begin{bmatrix} 1 & -4 & -8 & \vdots & 6 \\ 0 & 1 & -2 & \vdots & -7 \\ 0 & 0 & 1 & \vdots & 1 \end{bmatrix}$

18. $\begin{bmatrix} 1 & -2 & 6 & \vdots & 16 \\ 0 & 1 & -4 & \vdots & -13 \\ 0 & 0 & 1 & \vdots & 3 \end{bmatrix}$

For Exercises 19–24, complete the indicated row operation.

19. Perform $2R_1 + R_2$ on $\begin{bmatrix} 1 & 3 & \vdots & -1 \\ -2 & 5 & \vdots & 6 \end{bmatrix}$.

20. Perform $-3R_1 + R_2$ on $\begin{bmatrix} 1 & 2 & \vdots & -2 \\ 3 & 8 & \vdots & -4 \end{bmatrix}$.

21. Perform $-4R_2 + R_3$ on $\begin{bmatrix} 1 & -2 & 4 & \vdots & 6 \\ 0 & 2 & -1 & \vdots & -5 \\ 0 & 8 & -6 & \vdots & -3 \end{bmatrix}$.

22. Perform $3R_1 + R_2$ on $\begin{bmatrix} 1 & 5 & -3 & \vdots & 8 \\ -3 & 2 & -4 & \vdots & -6 \\ 0 & 1 & -2 & \vdots & 9 \end{bmatrix}$.

23. Perform $\frac{1}{4}R_1$ on $\begin{bmatrix} 4 & 8 & \vdots & -10 \\ -1 & 3 & \vdots & 2 \end{bmatrix}$.

24. Perform $-\frac{1}{6}R_3$ on $\begin{bmatrix} 1 & 3 & 4 & \vdots & 8 \\ 0 & 1 & -2 & \vdots & -6 \\ 0 & 0 & -6 & \vdots & 24 \end{bmatrix}$.

For Exercises 25–28, describe the row operation that should be performed to get the matrix closer to echelon form.

25. $\begin{bmatrix} 1 & 2 & \vdots & -1 \\ -3 & 4 & \vdots & 5 \end{bmatrix}$

26. $\begin{bmatrix} 1 & -3 & \vdots & 2 \\ 5 & 6 & \vdots & -4 \end{bmatrix}$

27. $\begin{bmatrix} 1 & -4 & 3 & \vdots & 6 \\ 0 & 1 & -2 & \vdots & 4 \\ 0 & 2 & -5 & \vdots & 1 \end{bmatrix}$

28. $\begin{bmatrix} 1 & -2 & 4 & \vdots & 7 \\ 0 & 1 & -2 & \vdots & 3 \\ 0 & -5 & 4 & \vdots & -9 \end{bmatrix}$

For Exercises 29–52, solve by transforming the augmented matrix into echelon form.

29. $\begin{cases} x - y = 2 \\ x + y = 2 \end{cases}$

30. $\begin{cases} x + y = 2 \\ x - y = -8 \end{cases}$

31. $\begin{cases} x + y = 3 \\ 3x - y = 1 \end{cases}$

32. $\begin{cases} x + 3y = -9 \\ -x + 2y = -11 \end{cases}$

33. $\begin{cases} x + 2y = -7 \\ 2x - 4y = 2 \end{cases}$

34. $\begin{cases} x - 3y = -13 \\ 3x + 2y = 5 \end{cases}$

35. $\begin{cases} -2x + 5y = 9 \\ x - 2y = -4 \end{cases}$

36. $\begin{cases} 2x + y = 12 \\ x - 3y = 6 \end{cases}$

37. $\begin{cases} 4x - 3y = -2 \\ 2x - 3y = -10 \end{cases}$

38. $\begin{cases} 6x - 3y = 0 \\ 2x + 4y = 0 \end{cases}$

39. $\begin{cases} 5x + 2y = 12 \\ 2x + 3y = -4 \end{cases}$

40. $\begin{cases} 4x + 3y = 14 \\ 3x - 2y = 2 \end{cases}$

41. $\begin{cases} x + y + z = 6 \\ 2x + y - 3z = -3 \\ 2x + 2y + z = 6 \end{cases}$

42. $\begin{cases} x + 2y + z = 2 \\ 3x + y - z = 3 \\ 2x + y + 2z = 7 \end{cases}$

43. $\begin{cases} x - y + z = -1 \\ 2x - 2y + z = 0 \\ x + 3y + 2z = 1 \end{cases}$

44. $\begin{cases} x + 3y - 2z = 4 \\ 2x - 3y + 2z = -7 \\ 3x + 2y - 2z = -1 \end{cases}$

45. $\begin{cases} 2x - y + z = 8 \\ x - 2y + 3z = 11 \\ 2x + 3y - z = -6 \end{cases}$

46. $\begin{cases} 2x + 3y - 2z = -21 \\ 2x - 4y + 3z = 15 \\ 3x + 2y - 3z = -24 \end{cases}$

47. $\begin{cases} 3x + 2y - 3z = 1 \\ -2x + 3y - 4z = 7 \\ 5x - 2y + z = -5 \end{cases}$

48. $\begin{cases} 2x - 2y + z = 6 \\ -2x + 4y + 3z = 4 \\ 4x - 3y - 2z = -7 \end{cases}$

49. $\begin{cases} 3x - 6y + z = -10 \\ 7y - z = 2 \\ 2x + 4z = 14 \end{cases}$

50. $\begin{cases} 6x - y + 5z = -28 \\ 4x + 2y = 10 \\ 5x + 6z = -23 \end{cases}$

51. $\begin{cases} 2x + 5y - z = 10 \\ x - y - z = 14 \\ x - 6y = 20 \end{cases}$

52. $\begin{cases} 2x + 4y - 5z = 16 \\ x - 2y - 11z = 1 \\ 4x + 5y = -6 \end{cases}$

For Exercises 53–60, solve using a calculator.

53. $\begin{cases} 2x + 5y = -14 \\ 6x + 7y = -10 \end{cases}$

54. $\begin{cases} 4x - 3y = 11 \\ 5x + 4y = 68 \end{cases}$

55. $\begin{cases} 4x - 7y = -80 \\ 9x + 5y = -14 \end{cases}$

56. $\begin{cases} 7x - 3y = 23 \\ 4x - 9y = -67 \end{cases}$

57. $\begin{cases} 3x + 2y - 5z = -19 \\ 4x - 7y + 6z = 67 \\ 5x - 6y - 4z = 18 \end{cases}$

58. $\begin{cases} 6x - 7y + 2z = -52 \\ 8x + 5y - 6z = 132 \\ 12x + 4y - 7z = 150 \end{cases}$

59. $\begin{cases} 8x - 7y + 2z = 108 \\ 6x + 11y - 10z = -74 \\ 9x - 2y + 13z = 140 \end{cases}$

60. $\begin{cases} 8x - 11y + 14z = 40 \\ 5x + 15y - 5z = 35 \\ 17x - 9y + 2z = -133 \end{cases}$

For Exercises 61–62, explain the mistake, then find the correct solution.

61. $\begin{cases} x + 3y = 13 \\ -4x - y = -26 \end{cases}$

$$\begin{bmatrix} 1 & 3 & | & 13 \\ -4 & -1 & | & -26 \end{bmatrix}$$

$4 \cdot R_1 + R_2 \longrightarrow \begin{bmatrix} 1 & 3 & | & 13 \\ 0 & 11 & | & -13 \end{bmatrix}$

$\dfrac{1}{11} \cdot R_2 \longrightarrow \begin{bmatrix} 1 & 3 & | & 13 \\ 0 & 1 & | & -\dfrac{13}{11} \end{bmatrix}$

Solution: $\left(-\dfrac{182}{11}, -\dfrac{13}{11} \right)$

62. $\begin{cases} x - y + z = 8 \\ 3x - z = -9 \\ 4y + z = -6 \end{cases}$

$$\begin{bmatrix} 1 & -1 & 1 & | & 8 \\ 3 & 0 & -1 & | & -9 \\ 4 & 0 & 1 & | & -6 \end{bmatrix}$$

$-3 \cdot R_1 + R_2 \longrightarrow \begin{bmatrix} 1 & -1 & 1 & | & 8 \\ 0 & 3 & -4 & | & -33 \\ 4 & 0 & 1 & | & -6 \end{bmatrix}$

$\begin{matrix} \dfrac{1}{3} \cdot R_2 \longrightarrow \\ -4 \cdot R_1 + R_3 \longrightarrow \end{matrix} \begin{bmatrix} 1 & -1 & 1 & | & 8 \\ 0 & 1 & -\dfrac{4}{3} & | & -11 \\ 0 & 4 & -3 & | & -38 \end{bmatrix}$

$-4 \cdot R_2 + R_3 \longrightarrow \begin{bmatrix} 1 & -1 & 1 & | & 8 \\ 0 & 1 & -\dfrac{4}{3} & | & -11 \\ 0 & 0 & \dfrac{7}{3} & | & 6 \end{bmatrix}$

$\dfrac{3}{7} \cdot R_3 \longrightarrow \begin{bmatrix} 1 & -1 & 1 & | & 8 \\ 0 & 1 & -\dfrac{4}{3} & | & -11 \\ 0 & 0 & 1 & | & \dfrac{18}{7} \end{bmatrix}$

Solution: $\left(-\dfrac{15}{7}, -\dfrac{53}{7}, \dfrac{18}{7} \right)$

For Exercises 63–74, translate the problem to a system of equations, then solve using matrices.

63. Brad purchased three grilled chicken sandwiches and two drinks for $9.90 and Angel purchased seven grilled chicken sandwiches and four drinks for $22.30. Find the price of one grilled chicken sandwich and one drink.

64. Sharika purchased three general admission tickets and two student tickets to a college play for $55 and Yo Chen purchased two general admission tickets and four student tickets for $50. Find the cost of one general admission and one student ticket.

65. A right triangle has one angle whose measure is 90°. One of the remaining angles is 6° less than twice the measure of the other. Find the measure of each of the remaining angles.

66. The measure of the largest angle of a triangle is 16° less than the sum of the measures of the other two. Twice the measure of the middle angle is 40° more than the sum of the largest and smallest. Find the measure of each of the angles.

67. The two longest rivers in the world, the Nile and the Amazon, have a combined length of 8050 miles. The Nile is 250 miles longer than the Amazon. Find the length of both rivers. *(Source: MS Bookshelf '98)*

68. The longest vehicular tunnel in the world is the Saint Gotthard Tunnel in Switzerland and the second longest is the Arberg Tunnel in Austria. The total length of the two tunnels is 18.8 miles and the Saint Gotthard tunnel is 1.4 miles longer than the Arberg. Find the length of each tunnel. *(Source: Webster's New World Book of Facts)*

69. Nikita invested a total of $10,000 in certificates of deposit that pay 5% annually and in a money market account that pays 6% annually. If the total interest earned in one year from the two investments is $536, find the principal that was invested in each.

70. An athelete received a $60,000 signing bonus, which he invested in three funds—a money market fund paying 5% annually, an income fund paying 5.5% annually, and a growth fund paying 7% annually. The sum of the principals invested in the money market and the growth fund equals the principal in the income fund. The total annual interest received from the three investments is $3400. Find the principal invested in each fund.

71. John bought two CDs, four tapes, and three DVDs for $164 and Tanelle bought five CDs, two tapes, and two DVDs for $160. If the sum of the cost of one CD and one tape equals the cost of one DVD, find the cost of one of each.

72. Angel bought 3 pounds of salmon, 2 pounds of tuna, and 5 pounds of cod for $63. Sara bought 6 pounds of salmon, 3 pounds of tuna, and 5 pounds of cod for $94. The cost of 1 pound of salmon and 1 pound of tuna is the same as the cost of 3 pounds of cod. Find the cost of 1 pound of each.

73. In Tampa Bay's 48–21 victory over Oakland in Super Bowl XXXVII, Tampa Bay scored touchdowns (6 points), extra points (1 point), and field goals (3 points). The number of touchdowns equaled the number of extra points and the number of touchdowns was two less than four times the number of field goals. Find how many of each type of score Tampa Bay had.

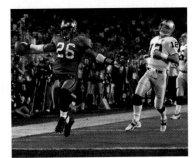

74. In a January 30, 2003, game against the Dallas Mavericks, the Minnesota Timberwolves guard Troy Vincent scored fourteen times for a total of 31 points using a combination of 3 point field goals, 2 point field goals, and free throws (1 point). If twice the number of 3 point field goals is 1 more than the sum of the number of 2 point field goals and free throws, find the number of each type. (*Source: Orlando Sentinel,* January 31, 2003)

75. An electrical circuit has three points of connection, with different voltage measurements at each of the three connections. An engineer has written the following equations to describe the voltages. Find each voltage.
$$4v_1 - v_2 = 30$$
$$-2v_1 + 5v_2 - v_3 = 10$$
$$-v_2 + 5v_3 = 4$$

76. An engineer has written the following system of equations to describe the forces in pounds acting on a steel structure. Find the forces.
$$6F_1 - F_2 = 350$$
$$9F_1 - 2F_2 - F_3 = -100$$
$$3F_2 - F_3 = 250$$

REVIEW EXERCISES

For Exercises 1–6, simplify.

1. $(-3)(4) - (-4)(5)$

2. $2(2 - 8) + 3[-1 - (-6)] - 4(4 - 6)$

3. $\dfrac{(-5)(-1) - (9)(3)}{(2)(-1) - (3)(3)}$

4. $\dfrac{2\left(\dfrac{5}{4}\right) - 8\left(\dfrac{1}{4}\right)}{\left(\dfrac{1}{2}\right)\left(\dfrac{5}{4}\right) - \left(\dfrac{3}{2}\right)\left(\dfrac{1}{4}\right)}$

5. $\dfrac{(1.1)(-0.2) - (1.7)(-0.5)}{(0.2)(-0.2) - (0.5)(-0.5)}$

6. $\dfrac{1[-45 - (-9)] + 7(-9 - 2) - 2(-27 - 30)}{1(6 - 3) - 2(-9 - 2) - 2[9 - (-4)]}$

Cramer's Rule

OBJECTIVES

1. Evaluate determinants of 2 × 2 matrices.
2. Evaluate determinants of 3 × 3 matrices.
3. Solve systems of equations using Cramer's Rule.

OBJECTIVE 1 Evaluate determinants of a 2 × 2 matrices. In this section, we will explore another way of solving systems of linear equations using a special type of matrix called a **square matrix**.

DEFINITION **Square matrix:** A matrix that has the same number of rows and columns.

For example, $\begin{bmatrix} 2 & 3 \\ 1 & -4 \end{bmatrix}$ and $\begin{bmatrix} 1 & -3 & 5 \\ -7 & 2 & 9 \\ 0 & 4 & -6 \end{bmatrix}$ are square matrices.

Every square matrix has a **determinant**. We write the determinant of a matrix, A, as $\det(A)$ or $|A|$. The method used to find the determinant of a matrix depends upon its size.

RULE **DETERMINANT OF A 2 × 2 MATRIX**

If $A = \begin{bmatrix} a_1 & b_1 \\ a_2 & b_2 \end{bmatrix}$, then $\det(A) = \begin{vmatrix} a_1 & b_1 \\ a_2 & b_2 \end{vmatrix} = a_1b_2 - a_2b_1$.

Notice that the determinant contains diagonal, or cross products as illustrated by the following.

$\begin{vmatrix} a_1 & b_1 \\ a_2 & b_2 \end{vmatrix} = a_1b_2 - a_2b_1$

Note: Since subtraction is not commutative, be sure to note the order of the two products.

EXAMPLE 1 Find the determinants of the following matrices.

a. $A = \begin{bmatrix} 3 & -2 \\ 2 & 4 \end{bmatrix}$

Solution $\det(A) = \begin{vmatrix} 3 & -2 \\ 2 & 4 \end{vmatrix} = (3)(4) - (2)(-2) = 12 + 4 = 16$

Learning Strategy

The products in the determinant are arranged with the downward cross product subtracting the upward cross product. A clever way to remember this order is that you have to "fall down before you can get up."

b. $B = \begin{bmatrix} -3 & 5 \\ 4 & -2 \end{bmatrix}$

Solution $\det(B) = (-3)(-2) - (4)(5) = 6 - 20 = -14$

c. $M = \begin{bmatrix} 1 & 3 \\ 3 & 9 \end{bmatrix}$

Solution $\det(M) = (1)(9) - (3)(3) = 9 - 9 = 0$

YOUR TURN

Find the determinant of $\begin{bmatrix} 1 & -3 \\ -4 & 2 \end{bmatrix}$.

OBJECTIVE 2 **Evaluate determinants of a 3 × 3 matrices.** There are various methods of evaluating the determinant of a 3×3 matrix. One of the most common methods is *expanding by minors*. Each element of a square matrix has a number called the **minor** for that element.

DEFINITION

Minor: The determinant of the remaining matrix when the row and column in which the element is located are ignored.

EXAMPLE 2 Find the minor of 2 in $\begin{bmatrix} 2 & -3 & -6 \\ -1 & 5 & -2 \\ 3 & -4 & 1 \end{bmatrix}$.

Solution To find the minor of 2, we ignore its row and column (shown in blue) and evaluate the determinant of the remaining matrix (shown in red).

$$\begin{bmatrix} 2 & -3 & -6 \\ -1 & 5 & -2 \\ 3 & -4 & 1 \end{bmatrix}$$

$$\begin{vmatrix} 5 & -2 \\ -4 & 1 \end{vmatrix} = (5)(1) - (-4)(-2) = 5 - 8 = -3$$

YOUR TURN

Find the minor of 6 in $\begin{bmatrix} 1 & -3 & 6 \\ -2 & 2 & 0 \\ 4 & -1 & 5 \end{bmatrix}$.

Note: We can actually expand by minors along *any* row or column to find the determinant of a 3×3 matrix. For simplicity, we have chosen to show expanding by minors only along the first column. You may learn how to expand by minors along other rows or columns in future courses.

RULE

EVALUATING THE DETERMINANT OF A 3 × 3 MATRIX

$$\begin{vmatrix} a_1 & b_1 & c_1 \\ a_2 & b_2 & c_2 \\ a_3 & b_3 & c_3 \end{vmatrix} = a_1 \begin{pmatrix} \text{minor} \\ \text{of } a_1 \end{pmatrix} - a_2 \begin{pmatrix} \text{minor} \\ \text{of } a_2 \end{pmatrix} + a_3 \begin{pmatrix} \text{minor} \\ \text{of } a_3 \end{pmatrix}$$

$$= a_1 \begin{vmatrix} b_2 & c_2 \\ b_3 & c_3 \end{vmatrix} - a_2 \begin{vmatrix} b_1 & c_1 \\ b_3 & c_3 \end{vmatrix} + a_3 \begin{vmatrix} b_1 & c_1 \\ b_2 & c_2 \end{vmatrix}$$

EXAMPLE 3 Find the determinant of $\begin{bmatrix} 2 & -3 & -4 \\ -1 & 2 & -2 \\ 3 & -4 & 1 \end{bmatrix}$.

Solution Using the rule for expanding by minors along the first column, we have:

$$\begin{vmatrix} 2 & -3 & -4 \\ -1 & 2 & -2 \\ 3 & -4 & 1 \end{vmatrix} = 2 \begin{pmatrix} \text{minor} \\ \text{of } 2 \end{pmatrix} - (-1) \begin{pmatrix} \text{minor} \\ \text{of } -1 \end{pmatrix} + 3 \begin{pmatrix} \text{minor} \\ \text{of } 3 \end{pmatrix}$$

$$= 2 \begin{vmatrix} 2 & -2 \\ -4 & 1 \end{vmatrix} - (-1) \begin{vmatrix} -3 & -4 \\ -4 & 1 \end{vmatrix} + 3 \begin{vmatrix} -3 & -4 \\ 2 & -2 \end{vmatrix}$$

$$= 2(2 - 8) + 1(-3 - 16) + 3(6 + 8)$$

$$= -12 + (-19) + 42$$

$$= 11$$

YOUR TURN

Find the determinant of $\begin{bmatrix} 3 & -2 & 4 \\ -2 & 3 & 1 \\ 2 & -4 & 2 \end{bmatrix}$.

ANSWER

26

Calculator TIPS

*The determinant of a matrix can be found using a graphing calculator. After entering the matrix (see p. 775), we use the **det(** function. Press the* **MATRX** *key and select the **det(** function from the **MATH** menu. Now indicate the matrix you want to find the determinant of by pressing* **MATRX** *and selecting the desired matrix from your list. After indicating the desired matrix, press* **ENTER**

DEFINITION

Cramer's Rule

The solution to the system of linear equations $\begin{cases} a_1x + b_1y = c_1 \\ a_2x + b_2y = c_2 \end{cases}$ is

$$x = \frac{\begin{vmatrix} c_1 & b_1 \\ c_2 & b_2 \end{vmatrix}}{\begin{vmatrix} a_1 & b_1 \\ a_2 & b_2 \end{vmatrix}} = \frac{D_x}{D} \quad \text{and} \quad y = \frac{\begin{vmatrix} a_1 & c_1 \\ a_2 & c_2 \end{vmatrix}}{\begin{vmatrix} a_1 & b_1 \\ a_2 & b_2 \end{vmatrix}} = \frac{D_y}{D}$$

The solution to the system of linear equations $\begin{cases} a_1x + b_1y + c_1z = d_1 \\ a_2x + b_2y + c_2z = d_2 \\ a_3x + b_3y + c_3z = d_3 \end{cases}$ is

$$x = \frac{\begin{vmatrix} d_1 & b_1 & c_1 \\ d_2 & b_2 & c_2 \\ d_3 & b_3 & c_3 \end{vmatrix}}{\begin{vmatrix} a_1 & b_1 & c_1 \\ a_2 & b_2 & c_2 \\ a_3 & b_3 & c_3 \end{vmatrix}} = \frac{D_x}{D}, \quad y = \frac{\begin{vmatrix} a_1 & d_1 & c_1 \\ a_2 & d_2 & c_2 \\ a_3 & d_3 & c_3 \end{vmatrix}}{\begin{vmatrix} a_1 & b_1 & c_1 \\ a_2 & b_2 & c_2 \\ a_3 & b_3 & c_3 \end{vmatrix}} = \frac{D_y}{D}, \quad \text{and} \quad z = \frac{\begin{vmatrix} a_1 & b_1 & d_1 \\ a_2 & b_2 & d_2 \\ a_3 & b_3 & d_3 \end{vmatrix}}{\begin{vmatrix} a_1 & b_1 & c_1 \\ a_2 & b_2 & c_2 \\ a_3 & b_3 & c_3 \end{vmatrix}} = \frac{D_z}{D}$$

Note: Each denominator, D, is the determinant of a matrix containing only the coefficients in the system. To find D_x, we replace the column of x-coefficients in the coefficient matrix with the constants from the system. To find D_y, we replace the column of y-coefficients in the coefficient matrix with the constant terms, and do likewise to find D_z.

EXAMPLE 4 Use Cramer's Rule to solve $\begin{cases} 2x + 3y = -5 \\ 3x - y = 9 \end{cases}$.

Solution First, we find D, D_x, and D_y.

$$D = \begin{vmatrix} 2 & 3 \\ 3 & -1 \end{vmatrix} = (2)(-1) - (3)(3) = -2 - 9 = -11$$

$$D_x = \begin{vmatrix} -5 & 3 \\ 9 & -1 \end{vmatrix} = (-5)(-1) - (9)(3) = 5 - 27 = -22$$

$$D_y = \begin{vmatrix} 2 & -5 \\ 3 & 9 \end{vmatrix} = (2)(9) - (3)(-5) = 18 + 15 = 33$$

Now we can find x and y.

$$x = \frac{D_x}{D} = \frac{-22}{-11} = 2 \qquad y = \frac{D_y}{D} = \frac{33}{-11} = -3$$

The solution is $(2, -3)$, which we can check by verifying that it satisfies both equations in the system. We will leave the check to the reader.

YOUR TURN

Use Cramer's Rule to solve $\begin{cases} 3x - 2y = -16 \\ x + 3y = 2 \end{cases}$.

EXAMPLE 5 Use Cramer's Rule to solve $\begin{cases} x + 2y - 2z = -7 \\ 3x - 2y + z = 15 \\ 2x + 3y - 3z = -9 \end{cases}$.

Solution We need to find D, D_x, D_y, and D_z.

Note: Recall that we find the determinant of a 3×3 matrix by expanding by minors along the first column.

$$D = \begin{vmatrix} 1 & 2 & -2 \\ 3 & -2 & 1 \\ 2 & 3 & -3 \end{vmatrix} = (1)\begin{vmatrix} -2 & 1 \\ 3 & -3 \end{vmatrix} - (3)\begin{vmatrix} 2 & -2 \\ 3 & -3 \end{vmatrix} + (2)\begin{vmatrix} 2 & -2 \\ -2 & 1 \end{vmatrix}$$

$$= 1(6 - 3) - 3(-6 + 6) + 2(2 - 4)$$
$$= 3 - 0 + (-4)$$
$$= -1$$

$$D_x = \begin{vmatrix} -7 & 2 & -2 \\ 15 & -2 & 1 \\ -9 & 3 & -3 \end{vmatrix} = (-7)\begin{vmatrix} -2 & 1 \\ 3 & -3 \end{vmatrix} - (15)\begin{vmatrix} 2 & -2 \\ 3 & -3 \end{vmatrix} + (-9)\begin{vmatrix} 2 & -2 \\ -2 & 1 \end{vmatrix}$$

$$= -7(6 - 3) - 15(-6 + 6) + (-9)(2 - 4)$$
$$= -21 - 0 + 18$$
$$= -3$$

$$D_y = \begin{vmatrix} 1 & -7 & -2 \\ 3 & 15 & 1 \\ 2 & -9 & -3 \end{vmatrix} = (1)\begin{vmatrix} 15 & 1 \\ -9 & -3 \end{vmatrix} - (3)\begin{vmatrix} -7 & -2 \\ -9 & -3 \end{vmatrix} + (2)\begin{vmatrix} -7 & -2 \\ 15 & 1 \end{vmatrix}$$

$$= 1(-45 + 9) - 3(21 - 18) + 2(-7 + 30)$$

$$= -36 - 9 + 46$$

$$= 1$$

$$D_z = \begin{vmatrix} 1 & 2 & -7 \\ 3 & -2 & 15 \\ 2 & 3 & -9 \end{vmatrix} = (1)\begin{vmatrix} -2 & 15 \\ 3 & -9 \end{vmatrix} - (3)\begin{vmatrix} 2 & -7 \\ 3 & -9 \end{vmatrix} + (2)\begin{vmatrix} 2 & -7 \\ -2 & 15 \end{vmatrix}$$

$$= 1(18 - 45) - 3(-18 + 21) + 2(30 - 14)$$

$$= -27 - 9 + 32$$

$$= -4$$

$$x = \frac{D_x}{D} = \frac{-3}{-1} = 3 \qquad y = \frac{D_y}{D} = \frac{1}{-1} = -1 \qquad z = \frac{D_z}{D} = \frac{-4}{-1} = 4$$

Note: After finding the values of two of the variables you could find the value of the third variable by substituting these values back into any one of the original equations.

The solution is $(3, -1, 4)$. We will leave the check to the reader.

YOUR TURN

Use Cramer's Rule to solve $\begin{cases} 2x + 3y - 2z = 11 \\ 3x + y + 4z = -8 \\ x - 3y - 2z = 4 \end{cases}$.

ANSWER

$(1, 1, -3)$

9.6 EXERCISES

For Extra Help

InterAct Math

Addison-Wesley Math Tutor Center

Video 15/DVT 5

MyMathLab

  MathXL

1. Is it possible to find the determinant of $\begin{bmatrix} 1 & 2 & 5 \\ 3 & 6 & 2 \end{bmatrix}$? Why or why not?

2. Explain the difference between a matrix and a determinant.

3. How do you find the minor for an element of a 3×3 matrix?

4. How do you find D_y when solving a system of equations using Cramer's Rule?

For Exercises 5–32, find the determinant.

5. $\begin{bmatrix} 3 & 2 \\ 1 & 5 \end{bmatrix}$

6. $\begin{bmatrix} 4 & 1 \\ 3 & 7 \end{bmatrix}$

7. $\begin{bmatrix} -3 & 5 \\ 2 & 4 \end{bmatrix}$

8. $\begin{bmatrix} -5 & 4 \\ 2 & 6 \end{bmatrix}$

9. $\begin{bmatrix} 3 & 6 \\ -2 & 4 \end{bmatrix}$

10. $\begin{bmatrix} 2 & 8 \\ -3 & 2 \end{bmatrix}$

11. $\begin{bmatrix} -3 & -4 \\ 2 & 5 \end{bmatrix}$

12. $\begin{bmatrix} -3 & -5 \\ 4 & 6 \end{bmatrix}$

13. $\begin{bmatrix} -2 & -3 \\ -4 & 5 \end{bmatrix}$

14. $\begin{bmatrix} -6 & -2 \\ 3 & -4 \end{bmatrix}$

15. $\begin{bmatrix} 0 & 3 \\ -5 & 7 \end{bmatrix}$

16. $\begin{bmatrix} -5 & 0 \\ -4 & 2 \end{bmatrix}$

17. $\begin{bmatrix} 1 & 2 & 1 \\ 3 & 1 & 4 \\ 2 & 3 & 2 \end{bmatrix}$

18. $\begin{bmatrix} 3 & 1 & 4 \\ 2 & 2 & 3 \\ 1 & 4 & 3 \end{bmatrix}$

19. $\begin{bmatrix} -1 & 2 & 0 \\ -3 & 2 & 4 \\ -4 & 2 & 3 \end{bmatrix}$

20. $\begin{bmatrix} -2 & 0 & -1 \\ 3 & -2 & 4 \\ -3 & 2 & 1 \end{bmatrix}$

21. $\begin{bmatrix} 2 & 1 & -3 \\ 0 & -3 & 2 \\ 4 & 1 & -3 \end{bmatrix}$

22. $\begin{bmatrix} 3 & -2 & 4 \\ 3 & 0 & 2 \\ -4 & -2 & 2 \end{bmatrix}$

23. $\begin{bmatrix} 1 & 4 & -2 \\ 3 & 2 & 0 \\ -1 & 4 & 3 \end{bmatrix}$

24. $\begin{bmatrix} 3 & -5 & 0 \\ 2 & -4 & 1 \\ -2 & -1 & -3 \end{bmatrix}$

25. $\begin{bmatrix} \frac{1}{2} & -\frac{1}{3} \\ \frac{2}{5} & \frac{3}{5} \end{bmatrix}$

26. $\begin{bmatrix} -\frac{3}{4} & -\frac{3}{5} \\ \frac{3}{2} & \frac{2}{5} \end{bmatrix}$

27. $\begin{bmatrix} 0.3 & -0.5 \\ 1.3 & -0.6 \end{bmatrix}$

28. $\begin{bmatrix} -0.4 & 1.6 \\ -4.7 & 3.1 \end{bmatrix}$

29. $\begin{bmatrix} \frac{1}{2} & -\frac{3}{4} & \frac{2}{5} \\ \frac{1}{3} & \frac{1}{5} & -\frac{3}{2} \\ -\frac{3}{4} & \frac{1}{2} & \frac{3}{5} \end{bmatrix}$

30. $\begin{bmatrix} -\frac{1}{4} & -\frac{3}{2} & \frac{4}{3} \\ \frac{1}{5} & -\frac{5}{4} & \frac{1}{2} \\ -\frac{5}{3} & \frac{1}{4} & -\frac{4}{5} \end{bmatrix}$

31. $\begin{bmatrix} -0.4 & 0.7 & -1.2 \\ 3.1 & 1.5 & -3.2 \\ 1.6 & -2.2 & -1.5 \end{bmatrix}$

32. $\begin{bmatrix} 1.7 & -3.2 & 4.1 \\ 5.3 & -6.2 & -1.1 \\ -1.3 & 2.3 & -4.5 \end{bmatrix}$

For Exercises 33 and 34, expand by minors about the first column.

33. $\begin{bmatrix} x & y & 1 \\ 2 & -1 & 3 \\ -2 & 0 & 1 \end{bmatrix}$

34. $\begin{bmatrix} x & y & 1 \\ -3 & -2 & 4 \\ 3 & -2 & 2 \end{bmatrix}$

For Exercises 35–48, solve using Cramer's Rule.

35. $\begin{cases} x + y = -5 \\ x - 2y = -2 \end{cases}$

36. $\begin{cases} x + 3y = 1 \\ x + y = -3 \end{cases}$

37. $\begin{cases} 2x - 3y = -6 \\ x - y = -1 \end{cases}$

38. $\begin{cases} 2x + 5y = -7 \\ x - y = -7 \end{cases}$

39. $\begin{cases} -x + 2y = -12 \\ 2x - 3y = 20 \end{cases}$

40. $\begin{cases} -x + 2y = -9 \\ 4x + 5y = -3 \end{cases}$

41. $\begin{cases} 2x - y = -4 \\ -x + 3y = -3 \end{cases}$

42. $\begin{cases} x - 3y = -19 \\ 2x + y = -3 \end{cases}$

43. $\begin{cases} 8x - 3y = 10 \\ 4x + 3y = 14 \end{cases}$

44. $\begin{cases} 2x - y = -4 \\ 4x - 5y = -51 \end{cases}$

45. $\begin{cases} \dfrac{1}{2}x - \dfrac{1}{4}y = 0 \\[2mm] \dfrac{3}{4}x + \dfrac{5}{2}y = \dfrac{23}{2} \end{cases}$

46. $\begin{cases} \dfrac{2}{3}x + \dfrac{1}{4}y = \dfrac{1}{2} \\[2mm] \dfrac{3}{4}x + \dfrac{4}{3}y = -\dfrac{23}{4} \end{cases}$

47. $\begin{cases} 0.2x + 0.5y = 3.4 \\ 0.7x - 0.3y = -0.4 \end{cases}$

48. $\begin{cases} 1.2x - 0.6y = -2.4 \\ 3.1x + 1.3y = -11.9 \end{cases}$

For Exercises 49–60, solve using Cramer's Rule.

49. $\begin{cases} x + y + z = 6 \\ 2x - 4y + 2z = 6 \\ 3x + 2y + z = 11 \end{cases}$

50. $\begin{cases} 2x + y - 3z = -1 \\ x + 2y - 2z = -3 \\ -3x - 4y + z = -3 \end{cases}$

51. $\begin{cases} 3x + y - z = -4 \\ 2x - y + 2z = -7 \\ x - 3y + z = -6 \end{cases}$

52. $\begin{cases} x - y + 3z = -10 \\ 5x + 4y - z = -7 \\ 2x + y - z = -4 \end{cases}$

53. $\begin{cases} 4x + 2y + 3z = 9 \\ 2x - 4y - z = 7 \\ 3x - 2z = 4 \end{cases}$

54. $\begin{cases} 2x - y = -1 \\ 5x - 3y + 2z = 0 \\ 3x + 2y - 3z = -8 \end{cases}$

55. $\begin{cases} 3x + 2y = -12 \\ 3y + 10z = -16 \\ 6x - 2z = 3 \end{cases}$

56. $\begin{cases} 4x + 2z = 7 \\ 8x - 2y = -7 \\ 10y - 3z = -5 \end{cases}$

57. $\begin{cases} \dfrac{1}{2}x + \dfrac{1}{3}y + \dfrac{3}{4}z = \dfrac{25}{12} \\[2mm] \dfrac{2}{3}x + \dfrac{3}{2}y - \dfrac{1}{4}z = -\dfrac{37}{12} \\[2mm] \dfrac{3}{4}x + \dfrac{1}{4}y - \dfrac{2}{3}z = -\dfrac{7}{4} \end{cases}$

58. $\begin{cases} \dfrac{3}{4}x - \dfrac{1}{2}y + \dfrac{1}{2}z = \dfrac{1}{4} \\[2mm] \dfrac{2}{5}x + \dfrac{5}{2}y - \dfrac{4}{3}z = -\dfrac{51}{5} \\[2mm] \dfrac{7}{4}x + \dfrac{5}{3}y - \dfrac{3}{5}z = -\dfrac{623}{60} \end{cases}$

59. $\begin{cases} 0.3x + 0.4y - 0.6z = 2.6 \\ 0.5x - 0.2y + 0.7z = -0.8 \\ 1.4x + 1.3y - 2.2z = 9.8 \end{cases}$

60. $\begin{cases} 1.2x + 2.1y - 0.5z = 7.3 \\ 3.2x - 2.4y + 1.3z = 6.1 \\ 2.5x + 1.3y - 1.7z = 8.4 \end{cases}$

For Exercises 61–64, find x.

61. $\begin{vmatrix} 9 & x \\ -6 & 5 \end{vmatrix} = 21$

62. $\begin{vmatrix} 12 & 2 \\ x & -3 \end{vmatrix} = -22$

63. $\begin{vmatrix} 2 & -1 & 0 \\ 0 & x & 1 \\ -2 & 0 & 4 \end{vmatrix} = -38$

64. $\begin{vmatrix} 1 & -2 & 4 \\ -1 & 0 & 2 \\ 0 & x & 5 \end{vmatrix} = -28$

For Exercises 65–70, use the following. If a triangle has vertices of (x_1, y_1), (x_2, y_2),

and (x_3, y_3) then the area of the triangle is given by $A = \dfrac{1}{2}\left| \det\begin{bmatrix} x_1 & y_1 & 1 \\ x_2 & y_2 & 1 \\ x_3 & y_3 & 1 \end{bmatrix}\right|$.

Note: This notation indicates half of the absolute value of the determinant.

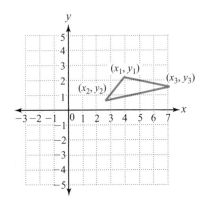

For example, the area of the triangle whose vertices are $(1, 2)$, $(3, -2)$, and $(-2, 5)$

is $A = \dfrac{1}{2}\left| \det\begin{bmatrix} 1 & 2 & 1 \\ 3 & -2 & 1 \\ -2 & 5 & 1 \end{bmatrix}\right|$. Expanding about the first column we have:

$$A = \frac{1}{2}\left| 1\det\begin{bmatrix} -2 & 1 \\ 5 & 1 \end{bmatrix} - 3\det\begin{bmatrix} 2 & 1 \\ 5 & 1 \end{bmatrix} - 2\det\begin{bmatrix} 2 & 1 \\ -2 & 1 \end{bmatrix}\right|$$

$$= \frac{1}{2}\left|(-2 - 5) - 3(2 - 5) - 2(2 + 2)\right|$$

$$= \frac{1}{2}\left|-7 + 9 - 8\right| = \frac{1}{2}\left|-6\right| = 3 \text{ square units.}$$

Find the area of the triangles with vertices at the given points.

65. $(2, 4)\ (4, 0)\ (6, 5)$

66. $(0, 2)\ (3\ {-2})\ (5, 5)$

67. $(-3, 1)\ (2, -3)\ (4, 4)$

68. $(-3, -2)\ (-1, 4)\ (3, -4)$

69. $(-4, -1)\ (1, 3)\ (3, -3)$

70. $(-3, -3)\ (2, 2)\ (4, -1)$

For Exercises 71–76, translate the problem to a system of equations, then solve using Cramer's Rule.

71. The two heaviest known meteorites to be found on Earth's surface are the Hoba West, which was found in Namibia, and the Ahnighito, which was found in Greenland. The total weight of the two meteorites is 90 tons. The Hoba West is twice as heavy as the Ahnighito. Find the weight of each. (*Source: Webster's New World Book of Facts*)

(Ahnighito meteorite)

72. The two largest expenses for the average American family are federal taxes and housing, including household expenses. Together these two items total 43.3% of the average family's income. The amount spent on taxes is 12.1% more than the amount spent on housing. Find the percent spent on each. (*Source: Numbers: How Many, How Long, How Far, How Much*)

73. A restaurant makes a soup that includes garbanzo and black turtle beans. The manager purchased 10 pounds of beans at a cost of $8.80. If the garbanzo beans cost $1.00 per pound and the black turtle beans cost $.70 per pound, how many pounds of each did he purchase?

74. The perimeter of a triangle is 21 inches and two sides are of equal length. The length of the third side is 3 inches less than the length of the two equal sides. Find the length of each side of the triangle.

75. Coinage bronze is made up of zinc, tin, and copper. The percent of tin is four times the percent of zinc. The percent of copper is nineteen times the sum of the percents of zinc and tin. Find the percent of zinc, tin, and copper in coinage bronze. (*Source: Webster's New World Book of Facts*)

76. On January 30, 2003, the Dallas Mavericks defeated the Houston Rockets 104–81. Dallas scored a total of 53 times. They scored six more 2 point field goals than 3 point field goals and two more free throws (1 point each) than 3 point field goals. How many of each did they score? (*Source: Orlando Sentinel*, January 31, 2003)

REVIEW EXERCISES

1. Find two consecutive integers whose sum is 39.

For Exercises 2 and 3, solve and graph the solution set.

2. $x > 2x - 3$

3. $2x + 3 \geq 10 - 5x$

For Exercises 4 and 5, graph.

4. $y > 3x + 2$

5. $x + y \leq -2$

6. Use substitution to solve $\begin{cases} x = 3 - y \\ 5x + 3y = 5 \end{cases}$.

9.7 Solving Systems of Linear Inequalities

1 Graph the solution set of a system of linear inequalities.

2 Solve applications involving systems of linear inequalities.

OBJECTIVE 1 Graph the solution set of a system of linear inequalities. In Section 2.6 we learned to solve linear inequalities in terms of one variable. In Section 4.6 we learned how to graph linear inequalities. In this section we will develop a graphical approach to solving *systems of linear inequalities.*

Consider the system of linear inequalities: $\begin{cases} x + 2y < 6 \\ 2x - y \geq 2 \end{cases}$

First, let's graph each inequality separately.

In the graph of $x + 2y < 6$, the shaded region contains all ordered pairs that make $x + 2y < 6$ true.

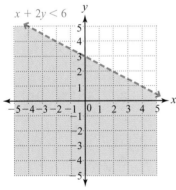

Note: Recall from Section 4.6 that we use a dashed line with < or >, and ordered pairs on a dashed line are not in the solution set.

Note: Recall from Section 4.6 that we determine which side of the line to shade by choosing an ordered pair and checking to see if it makes the inequality true. If it does, we shade on the side of the line containing that ordered pair. If it does not, we shade on the other side of the line.

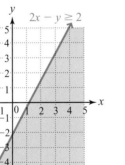

Note: Recall from Section 4.6 that we use a solid line with ≤ or ≥, and ordered pairs on a solid line are part of the solution set.

In the graph of $2x - y \geq 2$, ordered pairs in the shaded region and on the line itself make $2x - y \geq 2$ true.

Now we put the two graphs together on the same grid to determine the solution set for the system. A solution for a system of inequalities is an ordered pair that makes every inequality in the system true.

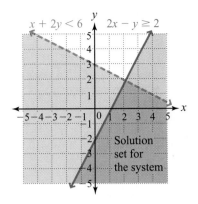

In our graph containing both $x + 2y < 6$ and $2x - y \geq 2$, the region where the shading overlaps contains ordered pairs that make both inequalities true, so all ordered pairs in this region are in the solution set for the system.

Also, ordered pairs on the solid line for $2x - y \geq 2$ where it touches the region of overlap are in the solution set for the system, whereas ordered pairs on the dashed line for $x + 2y < 6$ are not. This implies the following procedure.

PROCEDURE SOLVING A SYSTEM OF LINEAR INEQUALITIES

To solve a system of two linear inequalities,

1. Graph each inequality
2. The region of overlap for the two graphs is the solution set for the system of inequalities.

   Note: Solid lines touching the solution region are also part of the solution set.

EXAMPLE 1 Graph the solution set for the system of inequalities.

a. $\begin{cases} x + 3y > 6 \\ y \leq 2x - 1 \end{cases}$

Solution Graph each inequality. The region of overlap is the solution set for the system of inequalities.

Note: Ordered pairs on a dashed line are *not* part of the solution region.

b. $\begin{cases} y < 2 \\ 5x - y < 4 \end{cases}$

Solution Graph each inequality. The region of overlap is the solution set for the system of inequalities.

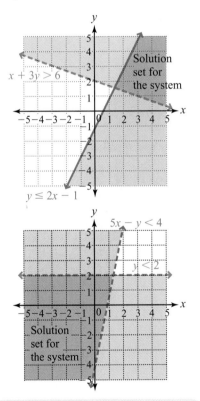

CONNECTION

Remember that for a system of two linear equations, a solution is a point of intersection of the graphs of the two equations. Similarly, for a system of linear inequalities, the solution set is a region of intersection of the graphs of the two inequalities.

a.

Solution set for the system

b.

Graph the solution set for the system of inequalities.

a. $\begin{cases} x + y > -1 \\ 4x - y \le 3 \end{cases}$ **b.** $\begin{cases} x > -3 \\ 3x - 4y \le -8 \end{cases}$

Inconsistent Systems

Some systems of linear inequalities have no solution. We say these systems are inconsistent.

EXAMPLE 2 Graph the solution set of the system of inequalities.

$$\begin{cases} x - 4y \ge 8 \\ y \ge \dfrac{1}{4}x + 3 \end{cases}$$

Solution Graph each inequality. The region of overlap is the solution set for the system of inequalities.

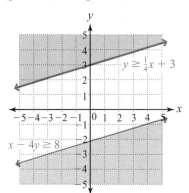

The lines appear to be parallel. We can verify by writing $x - 4y \ge 8$ in slope-intercept form and comparing the slopes.

$x - 4y \ge 8$

$\quad -4y \ge -x + 8$ Subtract x from both sides.

$\quad y \le \dfrac{1}{4}x - 2$ Divide both sides by -4, which changes the direction of the inequality.

The slopes are equal so the lines are in fact parallel. Since they are parallel and the shaded regions do not overlap, there is no solution region for this system. The system is inconsistent.

CONNECTION

Notice that the lines in Example 2 are parallel. Recall that in a system of linear equations, if the lines are parallel the system is inconsistent because there is no point of intersection. With linear inequalities, the lines must be parallel *and* the shaded regions must not overlap for the system to be inconsistent.

EXAMPLE 3 A home-interiors store stocks two different size prints from an artist. The manager wants to purchase at least 15 prints for the store. The artist sells the smaller print for $20 and the larger print for $30. The manager cannot spend more than $800 for the prints. Write a system of inequalities that describes the manager's order, then solve the system by graphing.

Understand We must translate to a system of inequalities, then solve the system.

Plan and Execute Let x represent the number of small prints and y represent the number of large prints ordered.

Relationship 1: The manager wants to purchase at least 15 prints.

The words *at least* indicate that the combined number of prints is to be greater than or equal to 15, so $x + y \geq 15$.

Relationship 2: The manager cannot spend more than $800.

Since the small prints cost $20 each, $20x$ describes the amount spent on small prints. Similarly, $30y$ describes the amount spent on large prints. The total cannot exceed $800, so $20x + 30y \leq 800$.

Our system: $\begin{cases} x + y \geq 15 \\ 20x + 30y \leq 800 \end{cases}$

Answer See graph.

Check Any ordered pair in the solution region should satisfy both relationships.

Note: Since the manager cannot order a negative numbers of prints, the solution region is confined to quadrant I.

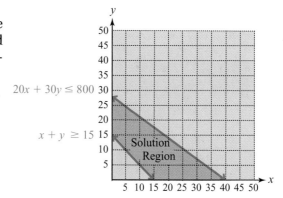

YOUR TURN

In designing a building with a rectangular base, the architect decides that the perimeter cannot exceed 200 feet. Also, it is decided that the length will be at least 10 feet more than the width. Write a system of inequalities that describes the design requirements, then solve the system graphically.

Note: Since length and width must be positive numbers, the solution region is confined to quadrant I.

ANSWERS

System: $\begin{aligned} 2l + 2w &\leq 200 \\ l &\geq w + 10 \end{aligned}$

9.7 EXERCISES

For Extra Help

InterAct Math

Addison-Wesley
Math Tutor Center

Video 15/DVT 5

MyMathLab

MathXL

1. How would you check a possible solution to a system of inequalities?

2. How do you determine the region that is the solution set for a system of linear inequalities?

3. What circumstances would cause a system of inequalities to have no solution?

4. Write a system of linear inequalities whose solution set is the entire first quadrant.

5. Write a system of linear inequalities whose solution set is the entire third quadrant.

6. Write a system of linear inequalities whose solution set is the entire fourth quadrant.

For Exercises 7–32, graph the solution set for the system of inequalities.

7. $\begin{cases} x + 2y < 10 \\ x - y < 4 \end{cases}$

8. $\begin{cases} 2x + y > 9 \\ x < 4y \end{cases}$

9. $\begin{cases} 3x + 4y \le -9 \\ 3x > y \end{cases}$

10. $\begin{cases} 9x + 4y < 14 \\ x + y \le 26 \end{cases}$

11. $\begin{cases} 4x - 3y \ge 2 \\ 3x + 5y \le 16 \end{cases}$

12. $\begin{cases} 3x - y \ge 5 \\ -x - 3y < 5 \end{cases}$

13. $\begin{cases} y < x \\ y > -x + 1 \end{cases}$

14. $\begin{cases} y < x \\ y < 2x - 3 \end{cases}$

15. $\begin{cases} x \geq -3y \\ y \geq 2x \end{cases}$

16. $\begin{cases} 3 + y > 1 \\ 3x + y < 4 \end{cases}$

17. $\begin{cases} 3y \leq 6 - 2x \\ 8x - 3y > -1 \end{cases}$

18. $\begin{cases} 5x - y \leq 25 \\ -5x - 2y \leq -10 \end{cases}$

19. $\begin{cases} 3x > -4y \\ 3x + 4y \leq -8 \end{cases}$

20. $\begin{cases} 4y - x < 12 \\ x - 4y > -16 \end{cases}$

21. $\begin{cases} x > 2 \\ 3x - 2y > 6 \end{cases}$

22. $\begin{cases} y \leq -1 \\ x + 2y > 3 \end{cases}$

23. $\begin{cases} 3x + 2y < -3 \\ 2x + 3y \leq -2 \end{cases}$

24. $\begin{cases} x + y > 8 \\ x - y \geq 4 \end{cases}$

25. $\begin{cases} 3x - 4y \geq -2 \\ 5x + 3y \geq 16 \end{cases}$

26. $\begin{cases} x + 2y \leq 2 \\ 6y \leq -2x - 2 \end{cases}$

27. $\begin{cases} y < 3x + 1 \\ 3x - y \leq 4 \end{cases}$

28. $\begin{cases} 2y - x \geq 6 \\ 2x - 4y \geq 5 \end{cases}$

29. $\begin{cases} 3x + y \geq 1 \\ -6x - y \geq -3 \end{cases}$

30. $\begin{cases} x > 2y \\ x + y > 6 \end{cases}$

31. $\begin{cases} y > 2 \\ x < -1 \end{cases}$

32. $\begin{cases} y \geq -1 \\ x < 2 \end{cases}$

For Exercises 33–36, explain the mistake.

33. $\begin{cases} x + y > 3 \\ x - y \leq 2 \end{cases}$

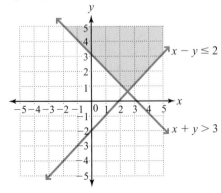

34. $\begin{cases} y < \frac{1}{2}x \\ y \leq -3x + 1 \end{cases}$

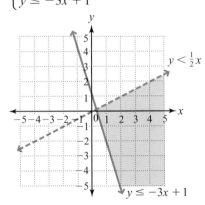

35. $\begin{cases} x < 3 \\ y > 4 \end{cases}$

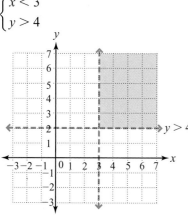

36. $\begin{cases} x > 3y \\ x + y \geq 2 \end{cases}$

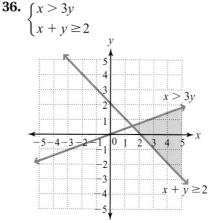

37. To be admitted to a certain college, prospective students must have a combined verbal and math score of at least 1100 on the SAT. To be admitted to the engineering program at this college, students must also have a math score of at least 500.

a. Write a system of inequalities that describes the requirements to be admitted to the college of engineering.

b. The maximum score on each part of the SAT is 800. Write two inequalities describing these maximum scores.

c. Solve the system by graphing.

d. Give two different combinations of verbal and math scores that satisfy the admission requirements to the college of engineering.

38. An architect is designing a rectangular platform for an auditorium. The client wants the length of the platform to be at least twice the width and the perimeter must not exceed 200 feet.

a. Write a system of inequalities that describes the specifications for the platform.

b. Solve the system by graphing.

c. Give two different combinations of length and width that satisfy the requirements of the client.

39. A company sells two versions of software: the regular version and the deluxe version. The regular version sells for $15.95 and the deluxe sells for $20.95. The company gives a bonus to any salesperson who sells at least 100 units and has at least $1700 in total sales.

a. Write a system of inequalities that describes the requirements a salesperson must meet in order to receive the bonus.

b. Solve the system by graphing.

c. Give two different combinations of number of units sold that meet the requirements for receiving the bonus.

40. An investor is trying to decide how much to invest in two different stocks. She has decided that she will invest at most $5000. The "safe" stock is very stable and is projected to return about 5% over the next year. The other stock is more risky, but could return about 9% if the company grows as expected. She wishes her total dividend at the end of the year to be at least $300.

a. Write a system of inequalities that describes the amount invested in the two stocks and the return on that investment.

b. Solve the system by graphing.

c. Give two different combinations of investment amounts that return at least $300.

Maximizing the Profit

An aspiring-artists group is preparing for a fund-raising sale. They design and construct two types of antique-reproduction tables: a semicircular foyer table and a side table. The materials cost $100 for each foyer table and $200 for each side table. Both types of table require 2 hours of cutting and carving. The foyer table requires 3.5 hours to assemble, sand, and finish; and the side table requires 1.75 hours to assemble, sand, and finish. The group has a total of $3300 to purchase all materials; 40 hours available for cutting and carving; and 63 hours available for assembling, sanding, and finishing. For each foyer table that they sell, their profit will be $350. For each side table, their profit will be $215. How many of each should they make to maximize their profit?

Linear programming is an area of mathematics that solves problems like the one described above. One of the fundamental elements of a linear programming model is the constraint. A constraint is simply an inequality that describes some limited resource required in the problem. For example, suppose you make two types of bicycles, style A and style B. Also, suppose that style A requires 40 minutes of welding for each bicycle and style B requires 25 minutes of welding. If you only have 600 minutes of welding time available, then the welding-time constraint would be $40A + 25B \leq 600$, where A represents the number of style A produced and B represents the number of style B produced.

1. For the artists fund-raising problem there are three basic constraints: (1) cost of materials, (2) cutting and carving, and (3) assembling-sanding-finishing. For each of these constraints, write an inequality similar to the one in the bicycle example. Let x represent the number of foyer tables and y the number of side tables.

2. Since x and y represent numbers of tables, they cannot represent negative values. Write inequalities for these two additional constraints.

(continued)

3. Now graph the system of inequalities described by these five constraints.

4. The region that is the solution of the system of inequalities is called the *feasible region* and includes all points that satisfy the system. The goal is to find the optimal solution, which, according to linear programming, is one of the corner points of the feasible region. There should be five corner points: one at the origin, one on each axis, and two more in quadrant I. Determine the coordinates of each corner point. Note that a corner point lying on an axis is the x- or y-intercept for the line passing through that point. Since two lines intersect to form a corner point that is not on an axis, make a system out of their two equations, then solve the system to find their point of intersection.

5. Next develop an algebraic expression describing the profit gained for selling x of the foyer tables and y of the side tables. This is called the *objective function*.

6. Using the objective function, test each of the points found in step 4 to see which one yields the maximum profit.

7. Using your solution, answer the following questions.
   a. How many of each type of table should the group produce?

   b. How many hours will be spent in each phase of the production?

   c. How much money will they need to purchase the materials?

   d. What amount of profit will the group receive?

REVIEW EXERCISES

For Exercises 1 and 2, evaluate.

1. 3^2

2. -10^2

3. Combine like terms: $3x - 2(2x + 7)$

4. Multiply: $4a^2b \cdot (-2ab^4)$

5. Divide: $\dfrac{-25a^3b^5}{5a^9}$

6. Solve: $x^2 - 5x + 6 = 0$

Defined Terms

Section 9.1
System of equations
(p. 714)
Solution for a system of
equations (p. 714)
Consistent system (p. 715)

Inconsistent system
(p. 716)
Dependent equations
(p. 717)
Independent equations
(p. 717)

Section 9.5
Matrix (p. 771)
Augmented matrix
(p. 771)
Echelon form (p. 772)

Section 9.6
Square matrix (p. 782)
Minor (p. 783)
Cramer's Rule (p. 785)

Procedures, Rules, and Key Examples

Procedures/Rules	Key Example(s)

Section 9.1 Solving Systems of Linear Equations Graphically
To verify or check a solution to a system of equations:
1. Replace each variable in each equation with its corresponding value.
2. Verify that each equation is true.

Example 1: Determine whether the ordered pair is a solution to the system.
$$\begin{cases} 2x + 3y = 8 \\ y = 5x - 3 \end{cases}$$

a. $(-2, 4)$

$$2x + 3y = 8 \qquad y = 5x - 3$$
$$2(-2) + 3(4) \stackrel{?}{=} 8 \qquad 4 \stackrel{?}{=} 5(-2) - 3$$
$$8 = 8 \qquad 4 \neq -13$$

$(-2, 4)$ is not a solution to the system.

b. $(1, 2)$

$$2x + 3y = 8 \qquad y = 5x - 3$$
$$2(1) + 3(2) \stackrel{?}{=} 8 \qquad 2 \stackrel{?}{=} 5(1) - 3$$
$$8 = 8 \qquad 2 = 2$$

$(1, 2)$ is a solution to the system.

To classify a system of equations, write the equations in slope-intercept form and compare the slopes and y-intercepts.
1. If the slopes are different, then the system is **consistent with independent equations** and has a single solution.
2. If the slopes are equal and the y-intercepts are also equal, then the system is **consistent with dependent equations** and has an infinite number of solutions.

Example 2: Classify $\begin{cases} x + 3y = 6 \\ 2x + 6y = -5 \end{cases}$ and discuss the number of solutions.

Solution: Write the equations in slope-intercept form.

$$x + 3y = 6 \quad \text{becomes} \quad y = -\frac{1}{3}x + 2$$

$$2x + 6y = -5 \quad \text{becomes} \quad y = -\frac{1}{3}x - \frac{5}{6}$$

continued

Procedures/Rules	Key Example(s)

Section 9.1 (continued)

3. If the slopes are equal with different y-intercepts, then the system is **inconsistent with independent equations** and has no solution.

Since the slopes are the same (both $-\frac{1}{3}$), but the y-intercepts are different, this system is inconsistent with independent equations. This means that there is no solution to the system.

To solve a system of linear equations graphically:

1. Graph each equation.
  a. If the lines intersect at a single point, then the coordinates of that point form the solution. (Consistent system with independent equations)
  b. If the lines are parallel, then there is no solution. (Inconsistent system with independent equations)
  c. If the lines are identical, then there are an infinite number of solutions, which are the coordinates of all the points on that line. (Consistent system with dependent equations)
2. In any case, check your solution.

Example 3: Solve $\begin{cases} 2x + 3y = 8 \\ y = 5x - 3 \end{cases}$ graphically.

Solution:

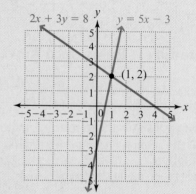

Solution: $(1, 2)$ is the solution.

Section 9.2 Solving Systems of Linear Equations by Substitution

To find the solution of a system of two linear equations using the substitution method:

1. Isolate one of the variables in one of the equations.
2. In the other equation, substitute the expression you found in step 1 for that variable.
3. Solve this new equation. (It will now have only one variable.)
4. Using one of the equations containing both variables, substitute the value you found in step 3 for that variable and solve for the value of the other variable.
5. Check the solution in the original equations.

Example 1: Solve $\begin{cases} y - 4x = -1 \\ 5x - y = 2 \end{cases}$ using substitution.

Solution: We will isolate y in the first equation.

$y = 4x - 1$ Add $4x$ on both sides.

Now we substitute $4x - 1$ in place of y in the second equation.

$$5x - y = 2$$
$$5x - (4x - 1) = 2$$
$$5x - 4x + 1 = 2 \quad \text{Distribute.}$$
$$x + 1 = 2 \quad \text{Combine like terms.}$$
$$x = 1 \quad \text{Subtract 1 on both sides.}$$

Now substitute 1 for x in one of the equations. We will use $y = 4x - 1$.

$$y = 4x - 1$$
$$y = 4(1) - 1$$
$$y = 3$$

Solution: $(1, 3)$ *continued*

Procedures/Rules

Section 9.3 Solving Systems of Linear Equations by Elimination
To solve a system of two linear equations using the elimination method:
1. Write the equations in standard form ($Ax + By = C$).
2. Use the multiplication principle of equality to clear fractions or decimals.
3. Multiply one or both equations by a number (or numbers) so that they have a pair of terms that are additive inverses.
4. Add the equations. The result should be an equation in terms of one variable.
5. Solve the equation from step 4 for the value of that variable.
6. Using an equation containing both variables, substitute the value you found in step 4 for the corresponding variable and solve for the value of the other variable.
7. Check your solution in the original equations.

Section 9.4 Solving Systems of Linear Equations in Three Variables
To solve a system of three equations with three unknowns:
1. If necessary, write each equation in the form $Ax + By + Cz = D$.
2. Choose a variable and choose two equations and eliminate that variable.
3. Choose two other equations and eliminate the same variable as was eliminated in step 2.
4. The two equations resulting from steps 2 and 3 form a system of two equations with two unknowns which can be solved using the methods of Section 9.3.
5. To find the third variable, substitute the values of the variables found in step 4 back into any of the three original equations that contain that variable.
6. Check the ordered triple in all three original equations.

Key Example(s)

Example 1: Solve $\begin{cases} x + 2y = -4 \\ 4x + y = 5 \end{cases}$ using elimination.

Solution: We will eliminate y.

$$x + 2y = -4 \qquad\qquad x + 2y = -4$$
$$4x + y = 5 \xrightarrow{\text{Multiply by } -2.} \underline{-8x - 2y = -10}$$
$$-7x + 0 = -14$$

Add equations.
Divide both sides by -7. $x = 2$

Now solve for the value of y.

$$4(2) + y = 5 \qquad \text{Substitute 2 in place of } x.$$
$$y = -3$$

Solution: $(2, -3)$

Example 1: Solve the following system of equations.

$$\begin{cases} x + 2y - z = -6 & \text{(Equation 1)} \\ 2x - 3y + 4z = 26 & \text{(Equation 2)} \\ -x + 2y - 3z = -18 & \text{(Equation 3)} \end{cases}$$

Eliminate x by adding equation 1 and equation 3.

$$x + 2y - z = -6$$
$$\underline{-x + 2y - 3z = -18}$$
$$4y - 4z = -24 \quad \text{Divide both sides by 4.}$$
$$y - z = -6 \quad \text{(Eq. 4)}$$

Eliminate x again.

$$2x - 3y + 4z = 26 \qquad\qquad 2x - 3y + 4z = 26$$
$$-x + 2y - 3z = -18 \xrightarrow{\text{Multiply by 2.}} \underline{-2x + 4y - 6z = -36}$$
$$y - 2z = -10 \quad \text{(Eq.5)}$$

Use equations 4 and 5 to solve for z.

$$y - z = -6 \qquad\qquad y - z = -6$$
$$y - 2z = -10 \xrightarrow{\text{Multiply by } -1.} \underline{-y + 2z = 10}$$
$$z = 4$$

Substitute 4 for z in equation 4 and solve for y.

$$y - 4 = -6$$
$$y = -2$$

continued

Procedures/Rules	Key Example(s)

Section 9.4 (continued)

Substitute -2 for y and 4 for z in equation 1 and solve for x.

$$x + 2(-2) - 4 = -6$$
$$x - 4 - 4 = -6$$
$$x = 2$$

Solution: $(2, -2, 4)$

Section 9.5 Solving Systems of Linear Equations Using Matrices

Elementary row operations:
1. Any two rows may be interchanged.
2. The elements of any row may be multiplied (or divided) by any nonzero real number.
3. Any row may be replaced by a row resulting from adding the elements of that row (or multiples of that row) to a multiple of the elements of any other row.

Elementary row transformations can be used to solve a system of linear equations by transforming the augmented matrix of the system into a matrix that is in echelon form. A matrix is in echelon form if the coefficient portion of the augmented matrix has 1's on the diagonal from upper left to lower right and 0's below the 1's.

Note: Systems of three equations with three unknowns are solved in a similar manner.

Example 1: Solve $\begin{cases} 2x + 3y = -5 \\ x + 2y = -4 \end{cases}$ using row operations.

The augmented matrix is $\begin{bmatrix} 2 & 3 & | & -5 \\ 1 & 2 & | & -4 \end{bmatrix}$

$\begin{bmatrix} 1 & 2 & | & -4 \\ 2 & 3 & | & -5 \end{bmatrix}$ We need row 1, column 1 to be a 1, so interchange the two rows.

$-2 \cdot R_1 + R_2 \longrightarrow \begin{bmatrix} 1 & 2 & | & -4 \\ 0 & -1 & | & 3 \end{bmatrix}$ We need row 2, column 1 to be 0, so multiply row 1 by -2 and add it to row 2.

$-1 \cdot R_2 \longrightarrow \begin{bmatrix} 1 & 2 & | & -4 \\ 0 & 1 & | & -3 \end{bmatrix}$ Multiply row 2 by -1 to get echelon form.

Row 2 means $y = -3$ and row 1 means $x + 2y = -4$.

$$x + 2(-3) = -4 \quad \text{Substitute } -3 \text{ for } y$$
$$x = 2 \quad \text{and solve for } x$$

Solution: $(2, -3)$

Section 9.6 Cramer's Rule

If $A = \begin{bmatrix} a_1 & b_1 \\ a_2 & b_2 \end{bmatrix}$, then $\det(A) = \begin{vmatrix} a_1 & b_1 \\ a_2 & b_2 \end{vmatrix} = a_1b_2 - a_2b_1$.

Example 1: Find the determinant of $\begin{bmatrix} 3 & -4 \\ 2 & -5 \end{bmatrix}$.

Solution:

$$\begin{vmatrix} 3 & -4 \\ 2 & -5 \end{vmatrix} = 3(-5) - 2(-4) = -15 + 8 = -7$$

continued

| Procedures/Rules | Key Example(s) |

Procedures/Rules

The determinant of a 3×3 matrix is found by

$$\begin{vmatrix} a_1 & b_1 & c_1 \\ a_2 & b_2 & c_2 \\ a_3 & b_3 & c_3 \end{vmatrix} = a_1 \begin{pmatrix} \text{minor} \\ \text{of } a_1 \end{pmatrix} - a_2 \begin{pmatrix} \text{minor} \\ \text{of } a_2 \end{pmatrix} + a_3 \begin{pmatrix} \text{minor} \\ \text{of } a_3 \end{pmatrix}$$

$$= a_1 \begin{vmatrix} b_2 & c_2 \\ b_3 & c_3 \end{vmatrix} - a_2 \begin{vmatrix} b_1 & c_1 \\ b_3 & c_3 \end{vmatrix} + a_3 \begin{vmatrix} b_1 & c_1 \\ b_2 & c_2 \end{vmatrix}$$

Cramer's Rule

The solution to $\begin{cases} a_1 x + b_1 y = c_1 \\ a_2 x + b_2 y = c_2 \end{cases}$ is

$$x = \frac{\begin{vmatrix} c_1 & b_1 \\ c_2 & b_2 \end{vmatrix}}{\begin{vmatrix} a_1 & b_1 \\ a_2 & b_2 \end{vmatrix}} = \frac{D_x}{D} \quad \text{and} \quad y = \frac{\begin{vmatrix} a_1 & c_1 \\ a_2 & c_2 \end{vmatrix}}{\begin{vmatrix} a_1 & b_1 \\ a_2 & b_2 \end{vmatrix}} = \frac{D_y}{D}$$

The solution to $\begin{cases} a_1 x + b_1 y + c_1 z = d_1 \\ a_2 x + b_2 y + c_2 z = d_2 \\ a_3 x + b_3 y + c_3 z = d_3 \end{cases}$ is

$$x = \frac{\begin{vmatrix} d_1 & b_1 & c_1 \\ d_2 & b_2 & c_2 \\ d_3 & b_3 & c_3 \end{vmatrix}}{\begin{vmatrix} a_1 & b_1 & c_1 \\ a_2 & b_2 & c_2 \\ a_3 & b_3 & c_3 \end{vmatrix}} = \frac{D_x}{D}, y = \frac{\begin{vmatrix} a_1 & d_1 & c_1 \\ a_2 & d_2 & c_2 \\ a_3 & d_3 & c_3 \end{vmatrix}}{\begin{vmatrix} a_1 & b_1 & c_1 \\ a_2 & b_2 & c_2 \\ a_3 & b_3 & c_3 \end{vmatrix}} = \frac{D_y}{D},$$

$$\text{and} \quad z = \frac{\begin{vmatrix} a_1 & b_1 & d_1 \\ a_2 & b_2 & d_2 \\ a_3 & b_3 & d_3 \end{vmatrix}}{\begin{vmatrix} a_1 & b_1 & c_1 \\ a_2 & b_2 & c_2 \\ a_3 & b_3 & c_3 \end{vmatrix}} = \frac{D_z}{D}$$

Key Example(s)

Example 2: Find the determinant of
$$\begin{bmatrix} -1 & -2 & 3 \\ 1 & -3 & 4 \\ -2 & 3 & -1 \end{bmatrix}.$$

Solution:
$$\begin{vmatrix} -1 & -2 & 3 \\ 1 & -3 & 4 \\ -2 & 3 & -1 \end{vmatrix} = -1 \begin{vmatrix} -3 & 4 \\ 3 & -1 \end{vmatrix} - 1 \begin{vmatrix} -2 & 3 \\ 3 & -1 \end{vmatrix} + (-2) \begin{vmatrix} -2 & 3 \\ -3 & 4 \end{vmatrix}$$

$$= -1(3 - 12) - 1(2 - 9) + (-2)(-8 + 9)$$
$$= 9 + 7 - 2$$
$$= 14$$

Example 3: Solve $\begin{cases} 3x + 2y = -2. \\ 5x + 2y = 2 \end{cases}$

Solution: $D = \begin{vmatrix} 3 & 2 \\ 5 & 2 \end{vmatrix} = -4$

$$x = \frac{\begin{vmatrix} -2 & 2 \\ 2 & 2 \end{vmatrix}}{-4} = \frac{-4 - 4}{6 - 10} = \frac{-8}{-4} = 2$$

$$y = \frac{\begin{vmatrix} 3 & -2 \\ 5 & 2 \end{vmatrix}}{-4} = \frac{6 + 10}{-4} = \frac{16}{-4} = -4$$

The solution is $(2, -4)$

Example 4: Solve $\begin{cases} x + 2y - z = -4 \\ 2x - 3y - 2z = 13. \\ 2x - 2y + 3z = 20 \end{cases}$

Solution: $D = \begin{vmatrix} 1 & 2 & -1 \\ 2 & -3 & -2 \\ 2 & -2 & 3 \end{vmatrix} = -35$

$$x = \frac{\begin{vmatrix} -4 & 2 & -1 \\ 13 & -3 & -2 \\ 20 & -2 & 3 \end{vmatrix}}{-35} = \frac{-140}{-35} = 4$$

$$y = \frac{\begin{vmatrix} 1 & -4 & -1 \\ 2 & 13 & -2 \\ 2 & 20 & 3 \end{vmatrix}}{-35} = \frac{105}{-35} = -3$$

$$z = \frac{\begin{vmatrix} 1 & 2 & -4 \\ 2 & -3 & 13 \\ 2 & -2 & 20 \end{vmatrix}}{-35} = \frac{-70}{-35} = 2$$

The solution is $(4, -3, 2)$

continued

Procedures/Rules	Key Example(s)

Procedures/Rules

Section 9.7 Solving Systems of Linear Inequalities

To solve a system of two linear inequalities:

1. Graph each inequality.
2. The region of overlap for the two graphs is the solution set for the system of inequalities.

Note: Solid lines touching the solution region are also part of the solution set.

Key Example(s)

Example 1: Solve $\begin{cases} x + y > -2 \\ y \le 3x - 1 \end{cases}$.

Solution:

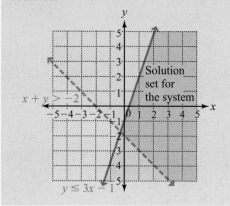

For Exercises 1–5, answer true or false.

1. The solution for a system of linear equations is always an ordered pair that satisfies all the equations of the system.

2. Given a system of two equations with two variables, if there is no solution then the graphs are parallel lines.

3. Any system of linear equations that can be solved, can be solved using the elimination method.

4. Solving a system of linear equations using the echelon method is like solving a system using the elimination method except the echelon method uses the coefficients and constant terms only.

5. The determinant of a matrix is another matrix.

For Exercises 6–10, complete the rule.

6. To verify or check a solution to a system of equations:
   1. _____ each variable in each equation with its corresponding value.
   2. Verify that each equation is true.

7. An inconsistent system has no solution. The graphs have the same _____ but different _____.

8. To solve a system of two linear inequalities:
   1. Graph each inequality.
   2. The solution of the system is _____.

9. A matrix is in echelon form if _____ _____.

10. When solving a system of linear equations using Cramer's Rule, the denominator is the _____ of the matrix containing only the coefficients of the system.

For Exercises 11–12, determine whether the given ordered pair is a solution to the given system of equations.

11. $(4, 3)$; $x - y = 1$
    $-x + y = -1$

12. $(1, 1)$; $2x - y = 7$
    $x + y = 8$

For Exercises 13–14: **a. Determine whether the system of equations is consistent with independent equations, inconsistent with independent equations, or consistent with dependent equations.**
b. How many solutions does the system have?

13.

14.

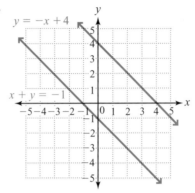

For Exercises 15–16: **a. Determine whether the system of equations is consistent with independent equations, inconsistent with independent equations, or consistent with dependent equations.**
b. How many solutions does the system have?

15. $\begin{cases} x - 3y = 10 \\ 2x + 3y = 5 \end{cases}$

16. $\begin{cases} x - 5y = 10 \\ 2x - 10y = 20 \end{cases}$

For Exercises 17–20, solve the system graphically.

17. $\begin{cases} 4x = y \\ 3x + y = -7 \end{cases}$

18. $\begin{cases} x - y = 4 \\ 2x + 3y = 3 \end{cases}$

19. $\begin{cases} y = -2x - 3 \\ 4x + 2y = 6 \end{cases}$

20. $\begin{cases} 3x - 2y = 6 \\ y = \dfrac{3}{2}x - 3 \end{cases}$

For Exercises 21–24, solve the system of equations using substitution. Note that some systems may be inconsistent or consistent with dependent equations.

21. $\begin{cases} 3x + 10y = 2 \\ x - 2y = 6 \end{cases}$

22. $\begin{cases} 2x + 5y = 8 \\ x - 10y = 9 \end{cases}$

23. $\begin{cases} 3y - 4x = 6 \\ y = \dfrac{4}{3}x + 2 \end{cases}$

24. $\begin{cases} 8x - 2y = 12 \\ y - 4x = 3 \end{cases}$

For Exercises 25–28, solve the system of equations using elimination.

25. $\begin{cases} x + y = 4 \\ x - y = -2 \end{cases}$

26. $\begin{cases} 3x + 2y = 4 \\ 2x - 3y = 7 \end{cases}$

27. $\begin{cases} 0.25x + 0.75y = 4 \\ x - y = -4 \end{cases}$

28. $\begin{cases} \dfrac{1}{5}x - \dfrac{1}{3}y = 2 \\ x + y = 2 \end{cases}$

For Exercises 29–32, solve the systems using the elimination method.

29. $\begin{cases} x + y + z = -2 \\ 2x + 3y + 4z = -10 \\ 3x - 2y - 3z = 12 \end{cases}$

30. $\begin{cases} x + y + z = 0 \\ 2x - 4y + 3z = -12 \\ 3x - 3y + 4z = 2 \end{cases}$

31. $\begin{cases} 3x + 4y = -3 \\ -2y + 3z = 12 \\ 4x - 3z = 6 \end{cases}$

32. $\begin{cases} 2x + 2y - 3z = 3 \\ x = -3y + 4z - 3 \\ z = -x - 2y + 6 \end{cases}$

For Exercises 33–36, solve the system using the echelon method.

33. $\begin{cases} x - 4y = 8 \\ x + 2y = 2 \end{cases}$

34. $\begin{cases} 2x - y = -4 \\ 2x - 3y = 0 \end{cases}$

35. $\begin{cases} x + y + z = 2 \\ 2x + y + 2z = 1 \\ 3x + 2y + z = 1 \end{cases}$

36. $\begin{cases} x + 2y + z = -1 \\ 2x - 2y + z = -10 \\ -x + 3y - 2z = 15 \end{cases}$

For Exercises 37–40, solve the system of equations using Cramer's Rule.

37. $\begin{cases} 9x + 4y = 18 \\ 5x - 2y = -28 \end{cases}$

38. $\begin{cases} 5x - y = 3 \\ 10x - 2y = 5 \end{cases}$

39. $\begin{cases} x + y + z = 5 \\ 3x - 2y + z = -3 \\ x + 3y + 4z = 10 \end{cases}$

40. $\begin{cases} x + 3y + 2z = 6 \\ 2x - 3y + z = -18 \\ -3x + 2y + z = 12 \end{cases}$

For Exercises 41–44, graph the solution set for the system of inequalities.

41. $\begin{cases} x + y > -5 \\ x - y \geq -1 \end{cases}$

42. $\begin{cases} 2x + y \geq -4 \\ -3x + y \geq -1 \end{cases}$

43. $\begin{cases} 2x + y > -1 \\ 3x + y \leq 4 \end{cases}$

44. $\begin{cases} -3x + 4y < 12 \\ 2x - y \leq -3 \end{cases}$

For Exercises 45–56, solve.

45. A movie theater charges $7.00 for adults and $4.25 for children. During a recent showing of *The Matrix Reloaded*, 139 tickets were sold for a total of $720. How many of each ticket were sold?

46. At the University of Kentucky, in 2001 there were 23,221 students enrolled in both undergraduate and graduate degree programs. There are about three times as many undergraduate students as graduate students. About how many graduate students are there? (*Source:* University of Kentucky)

47. At a local automobile dealership, 409 of the cars are either black or silver. There are 35 more black cars than silver cars. How many cars of each color are there?

48. The sum of two numbers is 16. Their difference is 4. What are the two numbers?

49. If two angles are complementary, find the measures of each if the smaller angle is 10 degrees less than one-third of the larger angle.

50. Duke is beginning to invest for his son's college fund. He put part of a $5000 account in a fund that yields 7% and the rest was placed in a fund that yields 9%. If he earned $380 in interest during the year, how much was invested in each fund?

51. Yolanda and Dee are traveling west in separate cars on the same highway. Yolanda is traveling at 60 miles per hour and Dee at 70 miles per hour. Yolanda passes the I-95 exit at 4:00 P.M. Dee passes the same exit at 4:15 P.M. At what time will Dee catch up to Yolanda?

52. Suppose a small plane is traveling with the jet stream and can travel 450 miles in 3 hours. When the plane returns, it takes 5 hours to travel the same distance against the jet stream. What is the speed of the plane in still air?

53. How much of a 10% saline solution is needed to combine with a 60% saline solution to obtain 50 milliliters of a 30% saline solution?

54. John went to the drugstore and bought some vitamin supplements that cost $8 each, some film that costs $4 per roll and some bags of candy that cost $3 per bag. He bought a total of eight items that cost $32. The number of rolls of film was one less than the number of bags of candy. Find the number of each type of item that he bought.

55. At a swim meet 5 points are awarded for each first place finish, 3 points for each second, and 1 point for each third. Shawnee Mission South High School scored a total of 38 points. The number of first-place finishes was one more than the number of second-place finishes. The number of third-place finishes was three times the number of second-place finishes. Find the number of first-, second-, and third-place finishes for the school.

56. A real estate course is offered in a continuing education program at a two-year college. The course is limited to 16 students. Also, to run the course, the college must receive at least $2750 in tuition. The college charges $250 per student if they live in the same county as the college and $400 if they live in any other county (or out-of-state).

   a. Write a system of inequalities that describes the number of in-county students enrolled in the course versus the number of out-of-county students and how much they paid.

   b. Solve the system.

For Exercises 1–2, determine whether the given ordered pair is a solution to the given system of equations.

1. $(-1, 3); \begin{cases} 3x + 2y = 3 \\ 4x - y = -7 \end{cases}$

2. $(2, 2, 3); \begin{cases} 2x - 3y + z = 1 \\ -2x + y - 3z = 11 \\ 3x + y + 3z = 14 \end{cases}$

3. Solve $\begin{cases} x + 3y = 1 \\ -2x + y = 5 \end{cases}$ by graphing.

For Exercises 4–9, solve the system of equations using substitution or elimination. Note that some systems may be inconsistent or consistent with dependent equations.

4. $\begin{cases} 2x + y = 15 \\ y = 7 - x \end{cases}$

5. $\begin{cases} x - 2y = 1 \\ 3x - 5y = 4 \end{cases}$

6. $\begin{cases} 3x - 2y = -8 \\ 2x + 3y = -14 \end{cases}$

7. $\begin{cases} 4x + 6y = 2 \\ 6x + 9y = 3 \end{cases}$

8. $\begin{cases} x + 2y + z = 2 \\ x + 4y - z = 12 \\ 3x - 3y - 2z = -11 \end{cases}$

9. $\begin{cases} x + 2y - 3z = -9 \\ 3x - y + 2z = -8 \\ 4x - 3y + 3z = -13 \end{cases}$

For Exercises 10–11, solve the equations using the echelon method.

10. $\begin{cases} x + 2y = -6 \\ 3x + 4y = -10 \end{cases}$

11. $\begin{cases} x + y + z = 6 \\ 3x - 2y + 3z = -7 \\ 4x - 2y + z = -12 \end{cases}$

For Exercises 12–13, solve the system of equations using Cramer's Rule.

12. $\begin{cases} 3x + 4y = 14 \\ 2x - 3y = -19 \end{cases}$

13. $\begin{cases} x + y + z = -1 \\ 3x - 2y + 4z = 0 \\ 2x + 5y - z = -11 \end{cases}$

For Exercises 14, graph the solution set for the system of inequalities.

14. $\begin{cases} 2x - 3y < 1 \\ x + 2y \le -2 \end{cases}$

For Exercises 15–20, solve.

15. Excedrin surveyed workers in various professions who get headaches at least once a year on the job. There were 9 more accountants than there were waiters/waitresses in the survey who got a headache. If the combined number of accountants and waiters/waitresses that got a headache was 163, how many people in each of those professions got a headache? (*Source:* The *State* newspaper)

16. When asked what the Internet most resembled, three times as many people said a library as opposed to a highway. If 240 people were polled, how may people considered the Internet to be a library? (*Source: bLINK* magazine, winter 2001)

17. A boat traveling with the current took 3 hours to go 30 miles. The same boat went 12 miles in 3 hours against the current. What is the rate of the boat in still water?

18. Janice invested $12,000 into two funds. One of the funds returned 6% interest and the other returned 8% interest after one year. If the total interest for the year was $880, how much did she invest in each fund?

19. Tickets for the senior play at Apopka High School cost $3 for children, $5 for students, and $8 for adults. There were 800 tickets sold for a total of $4750. The number of adult tickets sold was 50 more than two times the number of children's tickets. Find the number of each type of ticket.

20. A landscaper wants to plan a bordered rectangular garden area in a yard. Since she currently has 200 feet of border materials, the perimeter needs to be at most 200 feet. She thinks it would look best if the length is at least 10 feet more than the width.

   a. Write a system of inequalities to describe the situation.

   b. Solve the system of inequalities.

For Exercises 1–6, answer true or false.

1. $y = -3x + 5$ has a slope of $-3x$

2. $x^2 + 16$ is prime

3. Every linear inequality has a solution set.

4. $(1, 1)$ is a solution for the system of equations $\begin{cases} x = y \\ 9y - 13 = -4x \end{cases}$

5. A solution to a system of equations that has three variables will be an ordered triple.

6. To use the elimination method to solve a system of equations, all equations must be in standard form.

For Exercises 7–10, fill in the blank.

7. To find a solution to an equation in two variables:
   a. Choose a value for one of the variables.
   b. Replace the corresponding _____ with your chosen value.
   c. Solve the equation for the values of the other variable.

8. Complete using the distributive property: $4(2x + 3) =$ _____.

9. A proportion is two equal _____.

10. The degree of a monomial is the _____ of the exponents of all the variables.

For Exercises 11–13, simplify.

11. $-(6 - 2^2)^2 + 15$

12. $\sqrt{36} - 6^2 \div 0$

13. $-|2 - 8| + 3(2)$

For exercises 14–16, simplify.

14. $(-3x - 4) - (3x + 2)$

15. $(x - 3)(4x^2 - 2x + 1)$

16. $(x^3 + 5x^2 + 10x) \div 5x$

17. $(15m^2 - 22m + 14) \div (3m - 2)$

For Exercises 18–20, factor completely.

18. $m^3 + 8$

19. $25 + 10x + x^2$

20. $x^4 - 81$

For Exercises 21–34, simplify.

21. $\dfrac{4u^2 + 4u + 1}{u + 2u^2} \cdot \dfrac{u}{2u^2 - u - 1}$

22. $\dfrac{a^2 - b^2}{x^2 - y^2} \div \dfrac{a + b}{x - y}$

23. $\dfrac{\dfrac{x}{4x^2 - 1}}{\dfrac{5}{2x - 1}}$

24. $\dfrac{3}{y - 2} + \dfrac{y}{y + 3}$

25. $\dfrac{g}{g^2 - 9} - \dfrac{3}{g^2 - 9}$

26. $\dfrac{3}{x - 2} - \dfrac{2}{x + 2}$

For Exercises 27–30, solve.

27. $6x - 8 = 2(3x - 4)$

28. $-2 < x + 4 < 5$

29. $x^2 - 36 = 0$

30. $\dfrac{5}{x - 2} - \dfrac{3}{x} = \dfrac{11}{3x}$

31. $|3x + 4| + 3 = 7$

32. Solve for m in the formula $E = \dfrac{1}{2}mv^2$.

33. Determine if the following represents a function. List the domain and range.

The computer and related equipment that the University of Oregon recycled from May 2001 to October 2001. (*Source: The Chronicle of Higher Education, 2/14/03*)

CPUs	1163
Monitors	775
Keyboards	706
Printers	280
Peripherals	81
Macintoshes	67
Typewriters	62
Laptops	25
Televisions	21
Other	228

For Exercises 32–34, graph.

34. $y = -2x$

35. $f(x) = x^2 - 3$

36. $f(x) = |x + 2|$

37. Write the equation of a line in standard form that passes through the point $(-2, 4)$ and is perpendicular to the line $3x - 2y = 6$.

For Exercises 38–41, solve the systems.

38. $\begin{cases} 4x - 3y = -2 \\ 6x - 7y = 7 \end{cases}$

39. $\begin{cases} x + y + z = 5 \\ 2x + y - 2z = -5 \\ x - 2y + z = 8 \end{cases}$

40. Use Cramer's Rule to solve the system:
$\begin{cases} 2x + 3y = -5 \\ 3x - y = 9 \end{cases}$

41. Solve and graph the solution set: $\begin{cases} x + y < 3 \\ x - y \geq 4 \end{cases}$

For Exercises 42–50, solve.

42. The average size of a single-family home in 1970 was 1500 square feet. In 2000, the average size was 2,266. What was the percent of increase in the average size of a single-family home? (*Source: The State 2/1/03*)

43. In 2002, the median salary for all men was $154 more than the median salary for all women. If the combined median salary was $1216, what was the median salary for men and for women? (*Source: Parade Magazine 3/2/03*)

44. A chemist has a bottle containing 50 ml of 15% saline solution and a bottle of 40% saline solution. She wants a 30% solution. How much of the 40% solution must be added to the 15% solution so that a 30% concentration is created?

45. An event planner is trying to determine how many events she must plan in a year to make a profit of at least $50,000. She estimates that based on her current prices, she receives a profit of $250 for each small event and $500 for each large event that she plans.

   a. Select a variable for each type of event and write an inequality that describes a total profit of at least $50,000.

   b. Graph the solution region for this inequality.

   c. Give one combination of types of events that would give her a profit of exactly $50,000.

   d. Give one combination of types of events that would give her a profit of more than $50,000.

46. If light travels about 3×10^8 meters per second, how long does it take the light from the sun to travel to Pluto, which is a distance of about 5.9×10^{12} meters. Convert your answer to hours.

47. If the wavelength of a wave remains constant, then the velocity, v, of a wave is inversely proportional to its period, T. In an experiment, waves are created in a pool of water so that the period is 8 seconds and the velocity is 6 centimeters per second. If the period is increased to 15 seconds, what is the velocity?

48. LaTilla takes 3 hours to mow and edge the yard. Her husband can mow and edge in 2 hours. How long will it take them to if they work on the yard together?

49. A jet can fly 600 mph in calm air. If it travels with the wind, it can fly 2100 miles in the same amount of time that it takes to fly 1500 miles against the wind. Find the rate of the wind.

50. Joanna purchased 2 pens, 2 erasers, and 1 ream of paper for $6. Willa purchased 4 pens, 3 erasers, and 2 reams of paper for $11.50. On the same day, Scott purchased 3 pens, 1 eraser and 3 reams of paper for $9.50. Find the cost of each.

CHAPTER 10

Rational Exponents, Radicals, and Complex Numbers

In Section 1.5, we reviewed the basic rules of square roots. In this chapter, we will explore square roots and radicals in much more detail. We will begin in Section 10.1 with evaluating and simplifying radical expressions, much like how we started with evaluating and simplifying polynomial expressions in the first part of Chapter 5. Then, in Section 10.2, we will see that exponents can be fractions and that those rational exponents mean roots. In Sections 10.3–10.5 we move to rewriting expressions containing radicals. Once we have explored how to rewrite radical expressions, we will learn to solve equations that contain radicals in Section 10.6. Finally, in Section 10.7, we will discuss a new set of numbers called *complex numbers*, which come from square roots of negative numbers.

"Radical simply means 'grasping things at the root.'"

—*Angela Davis*
U.S. political activist, address June 25, 1987, to Spellman College.

Radical Expressions and Functions

OBJECTIVES

1. Find the nth root of a number.
2. Approximate roots using a calculator.
3. Simplify radical expressions.
4. Evaluate radical functions.
5. Find the domain of radical functions.
6. Solve applications involving radical functions.

In this section through Section 10.5, we focus on the expression portion of our algebra pyramid and explore square root and radical expressions.

THE ALGEBRA PYRAMID

RADICAL EQUATIONS
Examples: $\sqrt{x} = 5$ or
$\sqrt{x+3} = 8$ or
$\sqrt{x+5} = \sqrt{x-7}$

Note: We will explore radical equations in Section 10.6.

RADICAL EXPRESSIONS
Examples: $\sqrt{50}$ or $\sqrt{x}$

CONSTANTS AND VARIABLES
Examples: $2, 5, 7, x, y$

OBJECTIVE 1 **Find the nth root of a number.** In Section 1.5, we learned that a square root of a given number is a number that, when squared, equals the given number. For example, a square root of 16 is 4 since $4^2 = 16$. However, another number can be squared to equal 16: $(-4)^2 = 16$, so -4 is also a square root of 16. Consequently, 16 has two square roots: 4 and -4. We can write them more compactly as ± 4.

Similarly, a *cube root* of a given number is a number that, when cubed, equals the given number. For example, the cube root of 8 is 2 because $2^3 = 8$.

The *fourth root* of a number is a number that, when raised to the fourth power, equals the given number. For example, 3 is a fourth root of 81 because $3^4 = 81$. However, $(-3)^4 = 81$ also, so the fourth roots of 81 are ± 3. Our examples suggest the following definition of nth root.

DEFINITION $\quad$ **n^{th} root:** The number b is an nth root of a number a if $b^n = a$.

Recall from Section 1.5 that we used the symbol $\sqrt{}$, called a *radical sign*, to denote the square root of a number. To indicate roots other than square roots we use the symbol $\sqrt[n]{a}$, read "the nth root of a," where n is called the *root index* and indicates which root we are to find. If no root index is given, we assume it is 2, the square root. The number a is called the *radicand* and is the number or expression whose root we are to find. The entire expression is called a *radical* and any expression containing a radical is called a *radical expression.*

We have seen some numbers that have more than one root, such as 16, which has two square roots, 4 and -4. To keep things simple, the symbol $\sqrt[n]{a}$ means to find a single root. If n is even, then $\sqrt[n]{a}$ denotes the nonnegative root only and is called the *principal root*. For example, the principle square root of 16 is 4. Further, for $\sqrt[n]{a}$ to exist as a real number when n is even, a must be nonnegative because there is no real number that can be raised to an even power to equal a negative number. For example, $\sqrt{-4}$ does not exist as a real number because there is no real number whose square is -4.

If the root index is odd, the radicand can be positive or negative since a positive number raised to an odd power is positive and a negative number raised to an odd power is negative.

RULE   **EVALUATING nTH ROOTS**

When evaluating a radical expression $\sqrt[n]{a}$, the sign of a and the index n will determine possible outcomes.

If a is nonnegative, then $\sqrt[n]{a} = b$, where $b \geq 0$ and $b^n = a$.

If a is negative and n is even, then there is no real-number root.

If a is negative and n is odd, then $\sqrt[n]{a} = b$, where b is negative and $b^n = a$.

EXAMPLE 1   Evaluate the following roots.

a. $\sqrt{225}$

Solution $\sqrt{225} = 15$ Because $15^2 = 225$.

b. $-\sqrt{0.81}$

Note: It can be helpful to think of $-\sqrt{0.81}$ as $-1 \cdot \sqrt{0.81} = -1 \cdot 0.9 = -0.9$.

Solution $-\sqrt{0.81} = -0.9$ Because $(-0.9)^2 = 0.81$.

c. $\sqrt{-9}$

Solution No real number solution.

Note: In Section 10.7, we will define the square root of a negative number using a new set of numbers called the *imaginary numbers.*

d. $\pm\sqrt{121}$

Solution $\pm\sqrt{121} = \pm 11$ Because $(11)^2 = 121$ and $(-11)^2 = 121$.

e. $\sqrt{\dfrac{4}{9}}$

Solution $\sqrt{\dfrac{4}{9}} = \dfrac{2}{3}$ Because $\left(\dfrac{2}{3}\right)^2 = \dfrac{4}{9}$.

f. $\sqrt[3]{8}$

Solution $\sqrt[3]{8} = 2$ Because $2^3 = 8$.

g. $\sqrt[3]{-8}$

Solution $\sqrt[3]{-8} = -2$ Because $(-2)^3 = -8$.

h. $\sqrt[4]{16}$

Solution $\sqrt[4]{16} = 2$ Because $2^4 = 16$.

ANSWERS

a. 0.6   **b.** -2
c. No real number solution.
d. ±13 **e.** $\dfrac{5}{6}$
f. 4   **g.** -3
h. -3

YOUR TURN   **Evaluate the following expressions.**

a. $\sqrt{0.36}$ **b.** $-\sqrt{4}$ **c.** $\sqrt{-64}$ **d.** $\pm\sqrt{169}$

e. $\sqrt{\dfrac{25}{36}}$ **f.** $\sqrt[3]{64}$ **g.** $\sqrt[3]{-27}$ **h.** $-\sqrt[4]{81}$

OBJECTIVE 2   **Approximate roots using a calculator.** All the roots that we have considered so far have been rational. Some roots, like $\sqrt{3}$, are called irrational. As we learned in Section 1.1, we cannot express the exact value of an irrational number using rational numbers. In fact, writing $\sqrt{3}$ with the radical sign is the only way we can express the exact value. However, we may approximate $\sqrt{3}$ with a calculator or a table, such as the one in the end papers of this text.

For example, a calculator shows that $\sqrt{2} \approx 1.414213562$, which we can round to various decimal places:

Note: Recall that the symbol $\approx$ means "approximately equal to."

Approximating to two decimal places: $\sqrt{2} \approx 1.41$
Approximating to three decimal places: $\sqrt{2} \approx 1.414$

EXAMPLE 2   Approximate the roots using a calculator or Appendix A. Round to three decimal places.

a. $\sqrt{12}$

Answer: $\sqrt{12} \approx 3.464$

b. $-\sqrt{38}$

Answer: $-\sqrt{38} \approx -6.164$

c. $\sqrt[3]{45}$

Answer: $\sqrt[3]{45} \approx 3.557$

Calculator
TIPS

To evaluate roots higher than a square root on a calculator, use the $\boxed{\sqrt[x]{}}$ *key. First type the index, then press* $\boxed{\sqrt[x]{}}$*, then the radicand. For example, to evaluate* $\sqrt[3]{45}$*, type:* $\boxed{3}$ $\boxed{\sqrt[x]{}}$ $\boxed{4}$ $\boxed{5}$ $\boxed{\text{ENTER}}$.

On some calculators $\sqrt[x]{}$ *is a function in a menu accessed by pressing the* $\boxed{\text{MATH}}$ *key.*

ANSWERS

a. 4.359
b. -9.644
c. 3.979

YOUR TURN   **Approximate the following irrational numbers. Round to three decimal places.**

a. $\sqrt{19}$ **b.** $-\sqrt{93}$ **c.** $\sqrt[3]{63}$

The definition of a root can also be used to find roots with variable radicands. Recall that with an even index, the principal root is nonnegative, so at first we will assume all variables represent nonnegative values. We will use $(a^m)^n = a^{mn}$ to verify the roots.

Note: Remember that with an even index, the principal root is nonnegative. Since we are assuming that the variables are nonnegative, our result accurately indicates the principal square root.

CONNECTION

To raise a power to another power using $(a^m)^n = a^{mn}$, we multiply the powers. To find a root of a power, we divide the index into the exponent.

EXAMPLE **3** Find the root. Assume all variables represent nonnegative values.

a. $\sqrt{x^2}$

Solution $\sqrt{x^2} = x$ Because $(x)^2 = x^2$.

b. $\sqrt{a^6}$

Solution $\sqrt{a^6} = a^3$ Because $(a^3)^2 = a^6$.

c. $\sqrt{16x^8}$

Solution $\sqrt{16x^8} = 4x^4$ Because $(4x^4)^2 = 16x^8$.

d. $\sqrt{\dfrac{25x^8}{49y^2}}$

Solution $\sqrt{\dfrac{25x^8}{49y^2}} = \dfrac{5x^4}{7y}$ Because $\left(\dfrac{5x^4}{7y}\right)^2 = \dfrac{25x^8}{49y^2}$.

e. $\sqrt[3]{y^6}$

Solution $\sqrt[3]{y^6} = y^2$ Because $(y^2)^3 = y^6$.

f. $\sqrt[4]{16x^{12}}$

Solution $\sqrt[4]{16x^{12}} = 2x^3$ Because $(2x^3)^4 = 16x^{12}$.

CONNECTION

In parts c and f, notice the similarity between finding the root of a product and raising a product to a power. To raise a product to a power, we raise each factor to the power. To find a root of a product, we find the root of each factor.

YOUR TURN Find the root. Assume variables represent nonnegative values.

a. $\sqrt{x^4}$

b. $\sqrt{9x^{10}}$

c. $\sqrt{36a^{12}}$

d. $\sqrt{\dfrac{100x^4}{81y^6}}$

e. $\sqrt[3]{27y^9}$

f. $\sqrt[4]{b^8}$

If the variables can represent *any* real number and the index is even, then we must be careful to ensure that the principal root is nonnegative by using absolute value symbols. If the index is odd, however, we do not need absolute value signs because the root can be either positive or negative depending on the radicand's sign.

ANSWERS

a. x^2 **b.** $3x^5$

c. $6a^6$ **d.** $\dfrac{10x^2}{9y^3}$

e. $3y^3$ **f.** b^2

Note: To illustrate why $\sqrt{x^2} = |x|$, suppose we were to evaluate $\sqrt{x^2}$ when $x = -3$. We would have: $\sqrt{(-3)^2} = \sqrt{9} = 3$. Notice that the root, 3, is, in fact, the absolute value of -3. If we had incorrectly stated that $\sqrt{x^2} = x$, then $\sqrt{(-3)^2}$ would have to equal -3, which is not true.

EXAMPLE 4 Find the root. Assume variables represent any real number.

a. $\sqrt{x^2}$

Solution $\sqrt{x^2} = |x|$

b. $\sqrt{a^6}$

Solution $\sqrt{a^6} = |a^3|$

c. $\sqrt{(n+1)^2}$

Solution $\sqrt{(n+1)^2} = |n+1|$

d. $\sqrt{25y^8}$

Solution $\sqrt{25y^8} = 5y^4$

e. $\sqrt[3]{8n^3}$

Solution $\sqrt[3]{8n^3} = 2n$

f. $\sqrt[3]{(t-2)^3}$

Solution $\sqrt[3]{(t-2)^3} = t-2$

Note: In part d, we do not need absolute value because y^4, with its even exponent, will always be nonnegative. In parts e and f, we do not use absolute value because the indexes are odd.

YOUR TURN Find the root. Assume variables represent any real number.

a. $\sqrt{x^4}$

b. $\sqrt{9x^{10}}$

c. $\sqrt{36a^{12}}$

d. $\sqrt{1.21u^6t^2}$

e. $\sqrt{\dfrac{100x^4}{81y^6}}$

f. $\sqrt[3]{27y^9}$

g. $\sqrt[4]{b^8}$

ANSWERS

a. x^2 **b.** $3|x^5|$
c. $6a^6$ **d.** $1.1|u^3t|$
e. $\dfrac{10x^2}{9|y^3|}$ **f.** $3y^3$
g. b^2

OBJECTIVE 4 Evaluate radical functions. Many problems in math, science, and engineering can be solved using formulas that contain radicals that are in the form of **radical functions**.

DEFINITION

Radical function: A function of the form $f(x) = \sqrt[n]{P}$, where P is a polynomial.

EXAMPLE 5

a. Given $f(x) = \sqrt{3x - 2}$, find $f(3)$.

Solution To find $f(3)$, substitute 3 for x and simplify.
$f(3) = \sqrt{3(3) - 2} = \sqrt{9 - 2} = \sqrt{7}$.

b. Given $f(x) = \sqrt{2x - 6}$, find $f(0)$.

Solution To find $f(0)$, substitute 0 for x and simplify. $f(0) = \sqrt{2(0) - 6} = \sqrt{-6}$, which is not a real number.

ANSWER

$f(-1) = \sqrt{3}$

YOUR TURN

Given $f(x) = \sqrt{2x + 5}$, find $f(-1)$.

OBJECTIVE **5** **Find the domain of radical functions.** In Example 5b, we found that $f(0)$ did not exist because $\sqrt{-6}$ is not a real number. What does this suggest about the domains of functions involving radicals? Consider the graphs of $f(x) = \sqrt{x}$ and $f(x) = \sqrt[3]{x}$, which we can generate by choosing values for x.

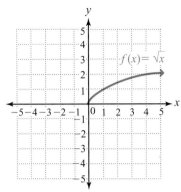

x	$f(x) = \sqrt{x}$
-1	Not real
0	0
1	1
4	2

Notice that $\sqrt{x}$ is not a real number if x is negative, which means the domain for $f(x) = \sqrt{x}$ is $\{x \mid x \geq 0\}$. Alternatively, $\sqrt[3]{x}$, with its odd index, is a real number when x is negative, so its domain is all real numbers. Our graphs suggest the following conclusion.

x	$f(x) = \sqrt[3]{x}$
-8	-2
-1	-1
0	0
1	1
8	2

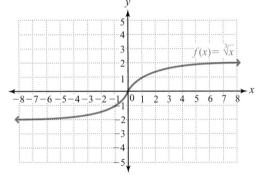

Conclusion: The domain of a radical function with an even index must contain values that keep its radicand nonnegative.

EXAMPLE **6** Find the domain of each of the following.

a. $f(x) = \sqrt{x - 4}$

Solution Since the index is even, the radicand must be nonnegative.

$$x - 4 \geq 0$$
$$x \geq 4 \qquad \text{Add 4 to both sides.}$$

Domain: $\{x \mid x \geq 4\}$, or $[4, \infty)$.

b. $f(x) = \sqrt{-2x + 6}$

Solution The radicand must be nonnegative.

$$-2x + 6 \geq 0$$
$$-2x \geq -6 \qquad \text{Subtract 6 from both sides.}$$
$$x \leq 3 \qquad \text{Divide both sides by } -2.$$

Domain: $\{x \mid x \leq 3\}$, or $(-\infty, 3]$

CONNECTION

Recall that when we divide *both* sides of an inequality by a negative number, we reverse the direction of the inequality.

ANSWERS

a. $\{x \mid x \geq 2\}$, or $[2, \infty)$
b. $\{x \mid x \leq -3\}$, or $(-\infty, -3]$

YOUR TURN Find the domain of each of the following.

a. $f(x) = \sqrt{2x - 4}$

b. $f(x) = \sqrt{-3x - 9}$

OBJECTIVE **6** **Solve applications involving radical functions.** Often, radical functions appear in real-world situations where one variable is a function of another.

EXAMPLE **7** The velocity of a free-falling object is a function of the distance that it has fallen. Ignoring air resistance, the velocity of an object v, in meters per second, can be found after falling h meters using the formula $v = -\sqrt{19.6h}$. Find the velocity of a stone that has fallen 30 meters after being dropped from a cliff.

Understand We are to find the velocity of an object after it falls 30 meters.

Plan Use the formula $v = -\sqrt{19.6h}$, replacing h with 30.

Execute $v = -\sqrt{19.6(30)}$ Replace h with 30.

$\qquad\quad v = -\sqrt{588}$ Multiply within the radical.

$\qquad\quad v \approx -24.2$ Evaluate the square root.

Note: A negative velocity indicates that the object is traveling downward.

Answer After falling 30 meters, the stone is traveling at a velocity of -24.2 meters per second.

Check We can verify the calculations, which we will leave to the reader.

YOUR TURN

ANSWER

-44.3 m/sec.

A sky diver jumps from a plane and puts her body into a dive position so that air resistance is minimized. Find her velocity after she falls 100 meters.

EXAMPLE **8** The period of a pendulum is the amount of time it takes the pendulum to swing from the point of release to the opposite extreme then back to the point of release. The period is a function of the length. The period, T, measured in seconds, can be found using the formula $T = 2\pi\sqrt{\dfrac{L}{9.8}}$, where L represents the length of the pendulum in meters. Find the period of a pendulum that is 0.5 meters long.

Understand We are to find the period of a pendulum that is 0.5 meters long.

Plan Use the formula $T = 2\pi\sqrt{\dfrac{L}{9.8}}$, approximating π with 3.14 and replacing L with 0.5.

Execute $T \approx 2(3.14)\sqrt{\dfrac{0.5}{9.8}}$ Simplify.

$\qquad\quad T \approx 6.28\sqrt{0.051}$

$\qquad\quad T \approx 6.28(0.226)$ Approximate the square root.

$\qquad\quad T \approx 1.42$

Point of release   Opposite extreme

Answer The period of a pendulum that is 0.5 meters long is 1.42 seconds.

Check We can verify the calculations, which we will leave to the reader.

YOUR TURN

Find the period of a pendulum that is 0.2 meters long. Round to the nearest thousandth.

ANSWER

≈ 0.897 sec.

10.1 EXERCISES

For Extra Help

 InterAct Math

 Addison-Wesley Math Tutor Center

Video 16/DVT 5

 MyMathLab

Math XL MathXL

1. Give an example of a number with a rational square root and an example of a number with an irrational square root. What makes the second number irrational?

2. How can we express the exact value of a root that is irrational?

3. Why are there two square roots for every positive real number?

4. If the index is even, what is the principal root of a number?

5. Why is an even root of any negative number not a real number?

6. If x is any real number, explain why $\sqrt{x^2} = |x|$ instead of x (with no absolute value).

For Exercises 7–14, find all square roots of the number given.

7. 36

8. 64

9. 121

10. 81

11. 196

12. 400

13. 225

14. 289

For Exercises 15–50, evaluate the roots, if possible.

15. $\sqrt{25}$

16. $\sqrt{-25}$

17. $-\sqrt{25}$

18. $-\sqrt{100}$

19. $\pm\sqrt{25}$

20. $\pm\sqrt{100}$

21. $\sqrt{1.44}$

22. $\sqrt{0.36}$

23. $\sqrt{-10.64}$

24. $\sqrt{2.25}$

25. $-\sqrt{0.0121}$

26. $-\sqrt{0.0009}$

27. $\sqrt{\dfrac{49}{81}}$

28. $\sqrt{\dfrac{64}{121}}$

29. $-\sqrt{\dfrac{144}{169}}$

30. $\sqrt{-\dfrac{25}{4}}$

31. $\sqrt[3]{27}$

32. $-\sqrt[3]{27}$

33. $\sqrt[3]{-27}$

34. $\sqrt[3]{-216}$

35. $\sqrt[3]{216}$

36. $-\sqrt[3]{-216}$

37. $\sqrt[4]{625}$

38. $\sqrt[4]{-625}$

39. $-\sqrt[4]{81}$

40. $\sqrt[4]{-81}$

41. $\sqrt[5]{32}$

42. $\sqrt[5]{-32}$

43. $-\sqrt[5]{-32}$

44. $\sqrt[5]{-243}$

45. $\sqrt[6]{64}$

46. $\sqrt[6]{-64}$

47. $\sqrt[3]{\dfrac{8}{27}}$

48. $\sqrt[3]{-\dfrac{64}{125}}$

49. $\sqrt[4]{\dfrac{16}{81}}$

50. $\sqrt[5]{\dfrac{1}{32}}$

For Exercises 51–66, use a calculator to approximate each root to the nearest thousandth.

51. $\sqrt{5}$

52. $\sqrt{12}$

53. $-\sqrt{11}$

54. $-\sqrt{41}$

55. $\sqrt[3]{50}$

56. $\sqrt[3]{21}$

57. $\sqrt[3]{-53}$

58. $\sqrt[3]{-83}$

59. $\sqrt[4]{65}$

60. $\sqrt[4]{123}$

61. $-\sqrt[4]{85}$

62. $-\sqrt[4]{77}$

63. $\sqrt[5]{89}$

64. $\sqrt[5]{-62}$

65. $\sqrt[6]{146}$

66. $\sqrt[6]{98}$

For Exercises 67–90, simplify. Assume variables represent nonnegative values.

67. $\sqrt{b^4}$

68. $\sqrt{r^8}$

69. $\sqrt{16x^2}$

70. $\sqrt{81t^6}$

71. $\sqrt{100r^2s^6}$

72. $\sqrt{121x^8y^{10}}$

73. $\sqrt{0.25a^6b^{12}}$

74. $\sqrt{0.36r^4s^{14}}$

75. $\sqrt[3]{m^3}$

76. $\sqrt[3]{n^6}$

77. $\sqrt[3]{27a^9b^6}$

78. $\sqrt[3]{64u^{12}t^9}$

79. $\sqrt[3]{-64a^3b^{12}}$

80. $\sqrt[3]{-27r^{15}s^3}$

81. $\sqrt[3]{0.008x^{18}}$

82. $\sqrt[3]{-0.125r^6}$

83. $\sqrt[4]{a^4}$

84. $\sqrt[4]{x^{12}}$

85. $\sqrt[4]{16x^{16}}$

86. $\sqrt[4]{81t^8}$

87. $\sqrt[5]{32x^{10}}$

88. $\sqrt[5]{243x^{15}}$

89. $\sqrt[6]{x^{12}y^6}$

90. $\sqrt[7]{s^7t^{21}}$

For Exercises 91–102, simplify. Assume variables represent any real number.

91. $\sqrt{36m^2}$

92. $\sqrt{9t^6}$

93. $\sqrt{(r-1)^2}$

94. $\sqrt{(k+3)^2}$

95. $\sqrt[4]{256y^{12}}$

96. $\sqrt[4]{16x^4}$

97. $\sqrt[3]{27y^3}$

98. $\sqrt[3]{125x^6}$

99. $\sqrt[4]{81x^8}$

100. $\sqrt{(x+2)^4}$

101. $\sqrt[3]{(y-4)^3}$

102. $\sqrt[4]{(n+5)^8}$

For Exercises 103–106, find the indicated value of the function.

103. $f(x) = \sqrt{2x+4}$, find $f(0)$

104. $f(x) = \sqrt{3x+2}$, find $f(3)$

105. $f(x) = \sqrt{4x+3}$, find $f(3)$

106. $f(x) = \sqrt{-2x+3}$, find $f(-2)$

For Exercises 107–110, find the domain.

107. $f(x) = \sqrt{2x - 8}$

108. $f(x) = \sqrt{3x + 12}$

109. $f(x) = \sqrt{-4x + 16}$

110. $f(x) = \sqrt{-2x + 6}$

For Exercises 111–114: ***a. Use a graphing calculator to graph the function.***
b. Find the domain of the function.

111. $f(x) = \sqrt{x - 2}$

112. $f(x) = \sqrt{x + 3}$

113. $f(x) = \sqrt[3]{x + 1}$

114. $f(x) = \sqrt[4]{x - 3}$

For Exercises 115 and 116, use the following. Neglecting air resistance, the velocity, v, in meters per second, of an object after falling h meters can be found using the formula $v = -\sqrt{19.6h}$.

Of Interest

One of the many topics Galileo Galilei studied was pendulums. His interest in them was piqued when he noticed a swinging light fixture in the cathedral of his hometown of Pisa and timed its period using his pulse.

115. Find the velocity of a rock that has been dropped from a cliff after it falls 16 meters.

116. Find the velocity of a ball that has been dropped from a roof after it falls 5 meters.

For Exercises 117 and 118, use the following. The period, T, of a pendulum in seconds can be found using the formula $2\pi\sqrt{\dfrac{L}{9.8}}$, *where L represents the length of the pendulum in meters.*

117. Find the period of a pendulum that is 3 meters long.

118. Find the period of a pendulum that is 2.5 meters long.

For Exercises 119 and 120, use the following. The formula $S = \dfrac{7}{2}\sqrt{2D}$ can be used to approximate the speed, S, in miles per hour, that a car was traveling prior to braking and skidding a distance, D, in feet, on asphalt. (Source: Harris Technical Services Traffic Accident Reconstructionists.)

119. Find the speed of a car if the skid distance is 15 feet.

120. Find the speed of a car if the skid distance is 40 feet.

For Exercises 121 and 122, use the following. Given the lengths a and b of the sides of a right triangle, we can find the hypotenuse, c, using the formula $c = \sqrt{a^2 + b^2}$.

121. Three pieces of lumber are to be connected to form a right triangle that will be part of the frame for a roof. If the horizontal piece is to be 12 feet and the vertical piece is to be 5 feet, how long must the connecting piece be?

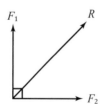

5 ft.

?

12 ft.

122. A counselor decides to create a ropes course with a zip line. She wants the line to connect from a 40-foot-tall tower to the ground at a point 100 feet from the base of the tower. Assuming the tower and ground form a right angle, find the length of the zip line.

40 ft.

?

100 ft.

CONNECTION ───────

In Section 6.6, we used the Pythagorean theorem, $c^2 = a^2 + b^2$, to find missing side lengths of a right triangle. The formula $c = \sqrt{a^2 + b^2}$ comes from isolating c in that theorem.

For Exercises 123 and 124, use the following. Two forces, F_1 and F_2, acting on an object at a 90° angle will pull the object with a resultant force, R, at an angle in between F_1 and F_2 (see the figure). The value of the resultant force can be found by the formula $R = \sqrt{F_1^2 + F_2^2}$.

F_1 R

F_2

123. Find the resultant force if $F_1 = 9$ N and $F_2 = 12$ N.

Note: The unit of force in the metric system is the newton (abbreviated N).

124. Find the resultant force if $F_1 = 6.2$ N and $F_2 = 4.5$ N.

REVIEW EXERCISES

For Exercises 1–4, multiply.

1. $x^5 \cdot x^3$

2. $(-9m^3n)(5mn^2)$

3. $4x^2(3x^2 - 5x + 1)$

4. $(7y - 4)(3y + 5)$

5. Write an expression for the area of the figure shown.

$x - 9$

$2x + 1$

6. Follow the order of operations to simplify.
   a. $\sqrt{16 \cdot 9}$

   b. $\sqrt{16} \cdot \sqrt{9}$

10.2 Rational Exponents

OBJECTIVES

1. Evaluate rational exponents.
2. Write radicals as expressions raised to rational exponents.
3. Simplify expressions with rational number exponents using the rules of exponents.
4. Use rational exponents to simplify radical expressions.

OBJECTIVE 1 Evaluate rational exponents.

Rational Exponents with a Numerator of 1

We now consider **rational exponents**, which are exponents that are fractions, as in $3^{1/2}$. To discover what these exponents mean, consider the following:

$$3^{1/2} \cdot 3^{1/2} = 3^{1/2+1/2} = 3^1 = 3 \qquad \text{Use } a^m \cdot a^n = a^{m+n}, \text{which we learned in Section 5.4.}$$

Notice that multiplying $\sqrt{3}$ by itself gives the same result:

$$\sqrt{3} \cdot \sqrt{3} = 3$$

These two results suggest that a base raised to the 1/2 power and the square root of that same base are equivalent expressions, or $a^{1/2} = \sqrt{a}$. Note that the denominator, 2, of the rational exponent is the index of the root (remember that square roots have an invisible index of 2). This relationship holds for other roots.

RULE **RATIONAL EXPONENTS WITH A NUMERATOR OF 1**

$a^{1/n} = \sqrt[n]{a}$, where n is a natural number other than 1.

> **Note:** If a is negative and n is odd, then the root will be negative. If a is negative and n is even, then there is no real number root.

EXAMPLE 1 Rewrite using radicals, then simplify, if possible. Assume all variables represent nonnegative values.

a. $36^{1/2}$

Solution $36^{1/2} = \sqrt{36} = 6$

b. $81^{1/4}$

Solution $81^{1/4} = \sqrt[4]{81} = 3$

We use the $\boxed{\wedge}$ key to evaluate a number raised to a rational exponent. For example, to evaluate $81^{1/4}$, we type:

On some scientific calculators, we use the $\boxed{y^x}$ key. For example, to evaluate the above example, we type:

$\boxed{8}\,\boxed{1}\,\boxed{y^x}\,\boxed{1}\,\boxed{a^{b/c}}\,\boxed{4}\,\boxed{=}$

Note: In $(-64)^{1/4}$, the parentheses group the minus sign with the radicand, whereas in $-36^{1/2}$, the minus sign is *not* part of the radicand.

c. $(-125)^{1/3}$
Solution $(-125)^{1/3} = \sqrt[3]{-125} = -5$

d. $(-64)^{1/4}$
Solution $(-64)^{1/4} = \sqrt[4]{-64}$ — There is no real number solution.

e. $-36^{1/2}$
Solution $-36^{1/2} = -\sqrt{36} = -6$

f. $x^{1/5}$
Solution $x^{1/5} = \sqrt[5]{x}$

Note: Only the z is raised to the one-fourth power.

g. $(64x^6)^{1/2}$
Solution $(64x^6)^{1/2} = \sqrt{64x^6} = 8x^3$

h. $12z^{1/4}$
Solution $12\sqrt[4]{z}$

i. $\left(\dfrac{a^6}{36}\right)^{1/2}$

Solution $\left(\dfrac{a^6}{36}\right)^{1/2} = \sqrt{\dfrac{a^6}{36}} = \dfrac{a^3}{6}$

ANSWERS

a. $\sqrt{49} = 7$
b. $\sqrt[4]{625} = 5$
c. $\sqrt[3]{-64} = -4$
d. $-\sqrt{81} = -9$
e. $\sqrt[4]{y}$ **f.** $\sqrt{25a^8} = 5a^4$
g. $18\sqrt[5]{n}$ **h.** $\sqrt{\dfrac{49}{x^{10}}} = \dfrac{7}{x^5}$

YOUR TURN   Rewrite each of the following using radical notation and evaluate, when possible. Assume all variables represent nonnegative values.

a. $49^{1/2}$ **b.** $625^{1/4}$

c. $(-64)^{1/3}$ **d.** $-81^{1/2}$

e. $y^{1/4}$ **f.** $(25a^8)^{1/2}$

g. $18n^{1/5}$ **h.** $\left(\dfrac{49}{x^{10}}\right)^{1/2}$

Rational Exponents with a Numerator Other Than 1

What if the numerator of a rational exponent is other than 1, as in $8^{2/3}$? Let's explore how we might rewrite. Note that we could write the fraction 2/3 as a product:

$$8^{2/3} = 8^{2 \cdot 1/3} \qquad \text{or} \qquad 8^{2/3} = 8^{1/3 \cdot 2}$$

Using the rule of exponents $(n^a)^b = n^{ab}$ in reverse, we can write:

$$8^{2/3} = 8^{2 \cdot 1/3} = (8^2)^{1/3} \qquad \text{or} \qquad 8^{2/3} = 8^{1/3 \cdot 2} = (8^{1/3})^2$$

Applying the rule $a^{1/n} = \sqrt[n]{a}$, we can write:

$$8^{2/3} = 8^{2 \cdot 1/3} = (8^2)^{1/3} = \sqrt[3]{8^2} \qquad \text{or} \qquad 8^{2/3} = 8^{1/3 \cdot 2} = (8^{1/3})^2 = (\sqrt[3]{8})^2$$

Notice that the denominator of the rational exponent becomes the index of the radical. The numerator of the rational exponent can be written either as the exponent of the radicand or as an exponent for the entire radical.

RULE   **GENERAL RULE FOR RATIONAL EXPONENTS**

$a^{m/n} = \sqrt[n]{a^m} = (\sqrt[n]{a})^m$, where $a \geq 0$ and m and n are natural numbers other than 1.

EXAMPLE 2   Rewrite using radicals, then simplify, if possible. Assume all variables represent nonnegative values.

a. $8^{2/3}$

Solution $8^{2/3} = \sqrt[3]{8^2}$ Rewrite.   or $8^{2/3} = (\sqrt[3]{8})^2$ Rewrite.

$\qquad\qquad\quad = \sqrt[3]{64}$ Square the radicand. $= (2)^2$ Evaluate the root.

$\qquad\qquad\quad = 4$ Evaluate the root. $= 4$ Evaluate the exponential form.

b. $625^{3/4}$

Solution   We could rewrite $625^{3/4}$ as $\sqrt[4]{625^3}$, but calculating 625^3 is tedious without a calculator, so we will use the other form of the rule.

$$625^{3/4} = (\sqrt[4]{625})^3 \qquad \text{Rewrite.}$$
$$= (5)^3 \qquad \text{Evaluate the root.}$$
$$= 125 \qquad \text{Evaluate the exponential form.}$$

Warning: In the expression $-9^{5/2}$, the negative sign tells us to find the opposite of the value of the exponential form.

$$-9^{5/2} = -(9^{5/2}) = -(\sqrt{9})^5$$

Or, some people find it helpful to think of $-9^{5/2}$ as $-1 \cdot 9^{5/2}$:

$$-9^{5/2} = -1 \cdot 9^{5/2}$$
$$= -1 \cdot (\sqrt{9})^5$$

c. $-9^{5/2}$

Solution $-9^{5/2} = -(\sqrt{9})^5 = -(3)^5 = -243$

d. $\left(\dfrac{1}{9}\right)^{5/2}$

Solution $\left(\dfrac{1}{9}\right)^{5/2} = \left(\sqrt{\dfrac{1}{9}}\right)^5 = \left(\dfrac{1}{3}\right)^5 = \dfrac{1}{243}$

e. $x^{2/3}$

Solution $x^{2/3} = \sqrt[3]{x^2}$ **Note:** Since the base is a variable, we put the exponent beneath the radical sign.

f. $(3x - 5)^{3/5}$

Solution $(3x - 5)^{3/5} = \sqrt[5]{(3x - 5)^3}$ **Note:** The simplification is usually easier if we write $a^{m/n}$ as $(\sqrt[n]{a})^m$ if a is a constant, and as $\sqrt[n]{a^m}$ if a is a variable.

ANSWERS

a. 8   **b.** -343

c. $\dfrac{1}{64}$ **d.** $\sqrt[6]{y^5}$

e. $\sqrt[5]{(3a + 4)^4}$

YOUR TURN   Simplify.

     a. $32^{3/5}$ **b.** $-49^{3/2}$ **c.** $\left(\dfrac{1}{16}\right)^{3/2}$ **d.** $y^{5/6}$ **e.** $(3a + 4)^{4/5}$

Negative Rational Exponents

Recall from Section 5.1 that $a^{-b} = \dfrac{1}{a^b}$. For example, $2^{-3} = \dfrac{1}{2^3}$. This same rule applies to negative rational exponents.

RULE **NEGATIVE RATIONAL EXPONENTS**

$a^{-m/n} = \dfrac{1}{a^{m/n}}$, where $a \neq 0$ and m and n are natural numbers with $n \neq 1$.

EXAMPLE 3 Rewrite using radicals, then simplify, if possible.

a. $81^{-1/2}$

Solution $81^{-1/2} = \dfrac{1}{81^{1/2}}$ Rewrite the exponential form with a positive exponent by inverting the base and changing the sign of the exponent.

$= \dfrac{1}{\sqrt{81}}$ Write the rational exponent in radical form.

$= \dfrac{1}{9}$ Evaluate the square root.

b. $16^{-3/4}$

Solution $16^{-3/4} = \dfrac{1}{16^{3/4}}$ Rewrite the exponential form with a positive exponent by inverting the base and changing the sign of the exponent.

$= \dfrac{1}{(\sqrt[4]{16})^3}$ Write the rational exponent in radical form.

$= \dfrac{1}{(2)^3}$ Evaluate the radical.

$= \dfrac{1}{8}$ Simplify the exponential form.

c. $\left(\dfrac{16}{25}\right)^{-1/2}$

CONNECTION ─────────

Though we used $a^{-m/n} = \dfrac{1}{a^{m/n}}$ in part c, we could have used $\left(\dfrac{a}{b}\right)^{-n} = \left(\dfrac{b}{a}\right)^{n}$, which we learned in Chapter 5.

Solution $\left(\dfrac{16}{25}\right)^{-1/2} = \dfrac{1}{\left(\dfrac{16}{25}\right)^{1/2}}$ Rewrite the exponential form with a positive exponent by inverting the base and changing the sign of the exponent.

$= \dfrac{1}{\sqrt{\dfrac{16}{25}}}$ Write the rational exponent in radical form.

$= \dfrac{1}{\dfrac{4}{5}}$ Evaluate the square root.

$= \dfrac{5}{4}$ Simplify the complex fraction.

d. $(-64)^{-2/3}$

Solution $(-64)^{-2/3} = \dfrac{1}{(-64)^{2/3}}$

$$= \dfrac{1}{(\sqrt[3]{-64})^2}$$

$$= \dfrac{1}{(-4)^2}$$

$$= \dfrac{1}{16}$$

YOUR TURN Simplify.

a. $49^{-1/2}$ **b.** $-27^{-2/3}$ **c.** $\left(\dfrac{16}{81}\right)^{-3/4}$ **d.** $(-8)^{-5/3}$

Of Interest

John Wallis (1616–1703) was one of the first mathematicians to explain rational and negative exponents; Wallis wrote extensively about physics and mathematics but, unfortunately, his work was overshadowed by the work of his countryman Isaac Newton. Newton added to what Wallis began with exponents and roots and popularized the notation that we use today. (*Source:* D. E. Smith, *History of Mathematics*, Dover, 1953, p. 415.)

OBJECTIVE **2** **Write radicals as expressions raised to rational exponents.** In upper-level math courses, it is often necessary to change radical expressions into exponential form. To do so, we use the facts that $\sqrt[n]{a^m} = a^{m/n}$ and $(\sqrt[n]{a})^m = a^{m/n}$.

EXAMPLE **4** Write each of the following in exponential form.

a. $\sqrt[4]{x^3}$

b. $\dfrac{1}{\sqrt[3]{x^2}}$

Solution $\sqrt[4]{x^3} = x^{3/4}$

Solution $\dfrac{1}{\sqrt[3]{x^2}} = \dfrac{1}{x^{2/3}} = x^{-2/3}$

c. $(\sqrt[5]{z})^3$

d. $\sqrt[6]{(3x-5)^5}$

Solution $(\sqrt[5]{z})^3 = z^{3/5}$

Solution $\sqrt[6]{(3x-5)^5} = (3x-5)^{5/6}$

YOUR TURN Write each of the following in exponential form.

a. $\sqrt[5]{y^2}$ **b.** $\dfrac{1}{\sqrt[4]{x^3}}$ **c.** $(\sqrt[7]{a})^3$ **d.** $\sqrt[3]{(3y-2)^5}$

Rational exponents follow the same rules that we established in Chapter 5 for integer exponents, which we review here:

RULE | **RULES OF EXPONENTS SUMMARY**

(Assume that no denominators are 0, that a and b are real numbers, and that m and n are integers.)

Zero as an exponent:	$a^0 = 1$, where $a \neq 0$
	0^0 is indeterminate.
Negative exponents:	$a^{-n} = \dfrac{1}{a^n}$
	$\dfrac{1}{a^{-n}} = a^n$
	$\left(\dfrac{a}{b}\right)^{-n} = \left(\dfrac{b}{a}\right)^n$
Product rule for exponents:	$a^m \cdot a^n = a^{m+n}$
Quotient rule for exponents:	$a^m \div a^n = a^{m-n}$
Raising a power to a power:	$(a^m)^n = a^{mn}$
Raising a product to a power:	$(ab)^n = a^n b^n$
Raising a quotient to a power:	$\left(\dfrac{a}{b}\right)^n = \dfrac{a^n}{b^n}$

EXAMPLE 5 Use the rules of exponents to simplify the following.

a. $x^{1/5} \cdot x^{3/5}$

Solution $x^{1/5} \cdot x^{3/5} = x^{1/5+3/5}$ Use $a^m \cdot a^n = a^{m+n}$.

$\qquad\qquad\qquad\quad = x^{4/5}$ Add the exponents.

b. $(2a^{1/2})(4a^{1/3})$

Solution $(2a^{1/2})(4a^{1/3}) = 8a^{1/2+1/3}$ Use $a^m \cdot a^n = a^{m+n}$.

$\qquad\qquad\qquad\qquad\quad = 8a^{3/6+2/6}$ Rewrite the exponents with a common denominator of 6.

$\qquad\qquad\qquad\qquad\quad = 8a^{5/6}$ Add the exponents.

c. $\dfrac{5^{2/7}}{5^{6/7}}$

Solution $\dfrac{5^{2/7}}{5^{6/7}} = 5^{2/7-6/7}$ Use $\dfrac{a^m}{a^n} = a^{m-n}$.

$\qquad\qquad\quad = 5^{-4/7}$ Subtract the exponents.

$\qquad\qquad\quad = \dfrac{1}{5^{4/7}}$ Rewrite with a positive exponent.

d. $(-4y^{-3/5})(5y^{4/5})$

Solution $(-4y^{-3/5})(5y^{4/5}) = -20y^{-3/5+4/5}$ Use $a^m \cdot a^n = a^{m+n}$.

$\qquad\qquad\qquad\qquad\qquad\; = -20y^{1/5}$ Add the exponents.

e. $(7w^{3/4})^2$

Solution $(7w^{3/4})^2 = 7^2 \cdot (w^{3/4})^2$ Use $(ab)^n = a^n \cdot b^n$.

$\qquad\qquad\qquad = 49 \cdot w^{3/4 \cdot 2}$ Use $(a^m)^n = a^{mn}$.

$\qquad\qquad\qquad = 49w^{3/2}$ Multiply the exponents.

f. $(2a^{2/3}b^{3/5})^3$

Solution $(2a^{2/3}b^{3/5})^3 = 2^3(a^{2/3})^3(b^{3/5})^3$ Use $(ab)^n = a^n \cdot b^n$.

$\qquad\qquad\qquad\qquad = 8a^{2/3 \cdot 3}\, b^{3/5 \cdot 3}$ Use $(a^m)^n = a^{mn}$.

$\qquad\qquad\qquad\qquad = 8a^2\, b^{9/5}$ Multiply the exponents.

g. $\dfrac{(3x^{4/3})^3}{x^3}$

Solution $\dfrac{(3x^{4/3})^3}{x^3} = \dfrac{3^3(x^{4/3})^3}{x^3}$ Use $(ab)^n = a^n \cdot b^n$ in the numerator.

$\qquad\qquad\qquad = \dfrac{27x^{4/3 \cdot 3}}{x^3}$ Use $(a^m)^n = a^{mn}$.

$\qquad\qquad\qquad = \dfrac{27x^4}{x^3}$ Simplify.

$\qquad\qquad\qquad = 27x^{4-3}$ Use $\dfrac{a^m}{a^n} = a^{m-n}$.

$\qquad\qquad\qquad = 27x$ Simplify.

YOUR TURN   Use the rules of exponents to simplify the following.

a. $(3x^{2/3})(-2x^{1/2})$ **b.** $\dfrac{12y^{2/7}}{3y^{5/7}}$ **c.** $(4b^{3/4})^2$ **d.** $(2x^{3/2}y^{2/3})^4$ **e.** $\dfrac{(2z^{5/3})^3}{z^2}$

OBJECTIVE **4** **Use rational exponents to simplify radical expressions.** Often, radical expressions can be simplified by rewriting them with rational exponents, simplifying, and then rewriting as a radical expression.

EXAMPLE **6** Rewrite as a radical with a smaller root index. Assume all variables represent nonnegative values.

a. $\sqrt[4]{49}$

Solution $\sqrt[4]{49} = 49^{1/4}$ Rewrite in exponential form.

$\qquad\qquad = (7^2)^{1/4}$ Write 49 as 7^2.

$\qquad\qquad = 7^{2 \cdot 1/4}$ Use $(a^m)^n = a^{mn}$.

$\qquad\qquad = 7^{1/2}$ Simplify.

$\qquad\qquad = \sqrt{7}$ Write in radical form.

ANSWERS

a. $-6x^{7/6}$ **b.** $\dfrac{4}{y^{3/7}}$

c. $16b^{3/2}$ **d.** $16x^6y^{8/3}$

e. $8z^3$

b. $\sqrt[6]{x^4}$

Solution $\sqrt[6]{x^4} = x^{4/6}$ Rewrite in exponential form.

$= x^{2/3}$ Simplify to lowest terms.

$= \sqrt[3]{x^2}$ Write in radical form.

c. $\sqrt[6]{a^4 b^2}$

Solution $\sqrt[6]{a^4 b^2} = (a^4 b^2)^{1/6}$ Rewrite in exponential form.

$= (a^4)^{1/6}(b^2)^{1/6}$ Use $(ab)^n = a^n b^n$.

$= a^{4 \cdot 1/6} b^{2 \cdot 1/6}$ Use $(a^m)^n = a^{mn}$.

$= a^{2/3} b^{1/3}$ Simplify.

$= (a^2 b)^{1/3}$ Use $(ab)^n = a^n b^n$.

$= \sqrt[3]{a^2 b}$ Write in radical form.

> **YOUR TURN**   **Rewrite as a radical with a smaller root index. Assume all variables represent nonnegative values.**
>
>   **a.** $\sqrt[4]{36}$ **b.** $\sqrt[8]{x^6}$ **c.** $\sqrt[8]{x^6 y^2}$

Writing radicals in exponential form also allows us to multiply and divide radical expressions with different root indices.

> **EXAMPLE 7**   Perform the indicated operations. Write the result using a radical.

a. $\sqrt{x} \cdot \sqrt[3]{x^2}$

Solution $\sqrt{x} \cdot \sqrt[3]{x^2} = x^{1/2} \cdot x^{2/3}$ Write in exponential form.

$= x^{1/2 + 2/3}$ Use $a^m \cdot a^n = a^{m+n}$.

$= x^{3/6 + 4/6}$ Rewrite the exponents with their LCD.

$= x^{7/6}$ Simplify.

$= \sqrt[6]{x^7}$ Write in radical form.

> **Note:** In Section 10.3, we will learn to simplify expressions like $\sqrt[6]{x^7}$ further.

b. $\dfrac{\sqrt[4]{x^3}}{\sqrt[3]{x}}$

Solution $\dfrac{\sqrt[4]{x^3}}{\sqrt[3]{x}} = \dfrac{x^{3/4}}{x^{1/3}}$ Write in exponential form.

$= x^{3/4 - 1/3}$ Use $\dfrac{a^m}{a^n} = a^{m-n}$.

$= x^{9/12 - 4/12}$ Rewrite the exponents with their LCD.

$= x^{5/12}$ Simplify.

$= \sqrt[12]{x^5}$ Write in radical form.

> **ANSWERS**
>
> **a.** $\sqrt{6}$
> **b.** $\sqrt[4]{x^3}$
> **c.** $\sqrt[4]{x^3 y}$

c. $\sqrt{3} \cdot \sqrt[3]{2}$

Solution $\sqrt{3} \cdot \sqrt[3]{2} = 3^{1/2} \cdot 2^{1/3}$ **Write in exponential form.**

$= 3^{3/6} \cdot 2^{2/6}$ **Write exponents with their LCD.**

$= (3^3 \cdot 2^2)^{1/6}$ **Use $(ab)^n = a^n b^n$.**

$= (27 \cdot 4)^{1/6}$ **Evaluate 3^3 and 2^2.**

$= 108^{1/6}$ **Multiply.**

$= \sqrt[6]{108}$ **Rewrite as a radical.**

ANSWERS

a. $\sqrt[12]{x^{11}}$
b. $\sqrt[12]{a}$
c. $\sqrt[4]{50}$

YOUR TURN   **Perform the indicated operations.**

a. $\sqrt[3]{x^2} \cdot \sqrt[4]{x}$ **b.** $\dfrac{\sqrt[4]{a^3}}{\sqrt[3]{a^2}}$ **c.** $\sqrt[4]{2} \cdot \sqrt{5}$

By writing radical expressions using rational exponents, we can also find the root of a root.

EXAMPLE 8   Write $\sqrt[3]{\sqrt{x}}$ with a single radical.

Solution $\sqrt[3]{\sqrt{x}} = \left(x^{1/2}\right)^{1/3}$ **Write in exponential form.**

$= x^{1/2 \cdot 1/3}$ **Apply $(a^m)^n = a^{mn}$**

$= x^{1/6}$ **Simplify.**

$= \sqrt[6]{x}$ **Write as a radical.**

ANSWER

$\sqrt[12]{y}$

YOUR TURN

Write $\sqrt[4]{\sqrt[3]{y}}$ with a single radical.

10.2 EXERCISES

1. If $4^{1/3}$ were written as a radical expression, what would be the radicand?

2. If $4^{3/5}$ were written as a radical expression, what is the root index?

3. If $a^{m/n}$, with n even and m/n reduced, were written as a radical expression, what restrictions would be placed on a? Why?

4. If $a^{m/n}$, with n odd, were written as a radical expression, what restrictions would be placed on a? Why?

5. Does $\sqrt[4]{100} = \sqrt{10}$? Why or why not?

6. Does $\sqrt[3]{3} \cdot \sqrt[5]{2} = \sqrt[15]{1944}$? Why or why not?

For Exercises 7–40, rewrite each of the following using radical notation and evaluate, when possible. Assume all variables represent nonnegative values.

7. $25^{1/2}$

8. $64^{1/2}$

9. $-100^{1/2}$

10. $-64^{1/2}$

11. $27^{1/3}$

12. $125^{1/3}$

13. $(-64)^{1/3}$

14. $(-125)^{1/3}$

15. $y^{1/4}$

16. $w^{1/8}$

17. $(144x^8)^{1/2}$

18. $(9z^{10})^{1/2}$

19. $18r^{1/2}$

20. $22a^{1/2}$

21. $\left(\dfrac{x^4}{121}\right)^{1/2}$

22. $\left(\dfrac{n^8}{36}\right)^{1/2}$

23. $8^{2/3}$

24. $64^{1/3}$

25. $-81^{3/4}$

26. $-16^{5/4}$

27. $(-8)^{4/3}$

28. $(-27)^{5/3}$

29. $16^{-3/2}$

30. $49^{-1/2}$

31. $x^{4/5}$

32. $m^{5/4}$

33. $8n^{2/3}$

34. $6a^{5/6}$

35. $(-32)^{-2/5}$

36. $(-216)^{-1/3}$

37. $\left(\dfrac{1}{25}\right)^{3/2}$

38. $\left(\dfrac{1}{32}\right)^{3/5}$

39. $(2a+4)^{5/6}$

40. $(5r-2)^{5/7}$

For Exercises 41–56, write each of the following in exponential form.

41. $\sqrt[4]{25}$

42. $\sqrt[8]{33}$

43. $\sqrt[6]{z^5}$

44. $\sqrt[7]{r^5}$

45. $\dfrac{1}{\sqrt[6]{5^5}}$

46. $\dfrac{1}{\sqrt[7]{6^2}}$

47. $\dfrac{5}{\sqrt[5]{x^4}}$

48. $\dfrac{8}{\sqrt[7]{n^3}}$

49. $(\sqrt[3]{5})^7$

50. $(\sqrt[5]{8})^3$

51. $(\sqrt[7]{x})^2$

52. $(\sqrt[3]{m})^8$

53. $\sqrt[4]{(4a-5)^7}$

54. $\sqrt[7]{(5w+3)^5}$

55. $(\sqrt[5]{2r-5})^8$

56. $(\sqrt[5]{3r-6})^9$

For Exercises 57–96, use the rules of exponents to simplify.

57. $x^{1/5} \cdot x^{3/5}$

58. $n^{1/4} \cdot n^{2/4}$

59. $x^{3/2} \cdot x^{-1/3}$

60. $n^{2/3} \cdot n^{-1/2}$

61. $a^{2/3} \cdot a^{3/4}$

62. $r^{5/2} \cdot r^{5/4}$

63. $(3w^{1/5})(4w^{2/5})$

64. $(5p^{1/9})(3p^{4/9})$

65. $(-3a^{2/3})(4a^{3/4})$

66. $(8c^{4/5})(-4c^{3/2})$

67. $\dfrac{7^{6/5}}{7^{3/5}}$

68. $\dfrac{3^{7/9}}{3^{2/9}}$

69. $\dfrac{x^{2/5}}{x^{4/5}}$

70. $\dfrac{y^{2/7}}{y^{5/7}}$

71. $\dfrac{x^{3/4}}{x^{1/2}}$

72. $\dfrac{x^{5/8}}{x^{1/2}}$

73. $\dfrac{r^{3/4}}{r^{2/3}}$

74. $\dfrac{m^{3/5}}{m^{1/2}}$

75. $\dfrac{x^{-3/7}}{x^{2/7}}$

76. $\dfrac{v^{-4/5}}{v^{2/5}}$

77. $\dfrac{a^{3/4}}{a^{-3/2}}$

78. $\dfrac{b^{4/3}}{b^{-2/3}}$

79. $(5s^{-2/7})(4s^{5/7})$

80. $(6u^{8/9})(-6u^{-5/9})$

81. $(-6b^{-5/4})(4b^{3/2})$

82. $(-6y^{-5/6})(-7y^{5/3})$

83. $(x^{2/3})^3$

84. $(r^{3/4})^4$

85. $(a^{5/6})^2$

86. $(n^{3/8})^4$

87. $(b^{2/3})^{3/5}$

88. $(m^{3/5})^{2/5}$

89. $(2x^{2/3}y^{1/2})^6$

90. $(3a^{1/4}b^{3/2})^4$

91. $(8q^{3/2}t^{3/4})^{1/3}$

92. $(16x^{2/3}y^{1/3})^{3/4}$

93. $\dfrac{(3a^{3/4})^4}{a^2}$

94. $\dfrac{(5v^{5/2})^2}{v^3}$

95. $\dfrac{(9z^{7/3})^{1/2}}{z^{5/6}}$

96. $\dfrac{(36x^{3/4})^{1/2}}{x^{1/8}}$

For Exercises 97–108, represent each of the following as a radical with a smaller root index. Assume all variables represent nonnegative values.

97. $\sqrt[4]{4}$

98. $\sqrt[4]{36}$

99. $\sqrt[6]{25}$

100. $\sqrt[6]{27}$

101. $\sqrt[4]{x^2}$

102. $\sqrt[6]{y^3}$

103. $\sqrt[8]{r^6}$

104. $\sqrt[10]{n^6}$

105. $\sqrt[8]{x^6y^2}$

106. $\sqrt[6]{y^2z^4}$

107. $\sqrt[10]{m^4n^6}$

108. $\sqrt[10]{a^2b^8}$

For Exercises 109–120, perform the indicated operations. Write the result using a radical.

109. $\sqrt[3]{x} \cdot \sqrt{x}$

110. $\sqrt[4]{y} \cdot \sqrt[3]{x^2}$

111. $\sqrt[4]{y^2} \cdot \sqrt[3]{y^2}$

112. $\sqrt[5]{x^4} \cdot \sqrt[3]{x^2}$

113. $\dfrac{\sqrt[3]{x^4}}{\sqrt{x}}$

114. $\dfrac{\sqrt[4]{y^3}}{\sqrt{y}}$

115. $\dfrac{\sqrt[5]{n^4}}{\sqrt[3]{n^2}}$

116. $\dfrac{\sqrt[6]{z^4}}{\sqrt{4z}}$

117. $\sqrt{5} \cdot \sqrt[3]{3}$

118. $\sqrt[3]{4} \cdot \sqrt{5}$

119. $\sqrt[4]{6} \cdot \sqrt[3]{2}$

120. $\sqrt[4]{4} \cdot \sqrt[3]{2}$

For Exercises 121–124, write each as a single radical.

121. $\sqrt[3]{\sqrt[3]{x}}$

122. $\sqrt[3]{\sqrt[5]{m}}$

123. $\sqrt{\sqrt[3]{n}}$

124. $\sqrt[4]{\sqrt[5]{z}}$

REVIEW EXERCISES

For Exercises 1 and 2, simplify using the order of operations agreement.

1. $\sqrt{16} \cdot \sqrt{9}$

2. $\sqrt[3]{27} \cdot \sqrt[3]{125}$

3. Write the prime factorization of 240.

For Exercises 4 and 5, simplify.

4. $(2.5 \times 10^6)(3.2 \times 10^5)$

5. $\left(\dfrac{3}{4}x^3y\right)\left(-\dfrac{5}{6}xyz^2\right)$

6. Use long division to find the quotient: $\dfrac{2x^3 - 2x^2 - 19x + 18}{x + 3}$

10.3 Multiplying, Dividing, and Simplifying Radicals

OBJECTIVES

1. Multiply radical expressions.
2. Divide radical expressions
3. Use the product rule to simplify radical expressions.

In this section, we explore some ways to simplify expressions that involve multiplication or division of radicals.

OBJECTIVE 1 **Multiply radical expressions.** Consider the expression $\sqrt{9} \cdot \sqrt{16}$. The usual approach would be to multiply the roots. However, we can also multiply the radicands and then find the roots of the product.

Find the roots first:	Multiply the radicands first:
$\sqrt{9} \cdot \sqrt{16} = 3 \cdot 4 = 12$	$\sqrt{9} \cdot \sqrt{16} = \sqrt{9 \cdot 16} = \sqrt{144} = 12$

Both approaches give the same result, suggesting the following rule:

RULE **PRODUCT RULE FOR RADICALS**

If $\sqrt[n]{a}$ and $\sqrt[n]{b}$ are both real numbers, then $\sqrt[n]{a} \cdot \sqrt[n]{b} = \sqrt[n]{a \cdot b}$.

EXAMPLE 1 Find the product and write the answer in simplest form. Assume all variables represent nonnegative values.

a. $\sqrt{3} \cdot \sqrt{27}$

Solution $\sqrt{3} \cdot \sqrt{27} = \sqrt{3 \cdot 27} = \sqrt{81} = 9$

> **CONNECTION**
> Example 1a illustrates that the product of two irrational numbers can be a rational number.

b. $\sqrt{11} \cdot \sqrt{x}$

Solution $\sqrt{11} \cdot \sqrt{x} = \sqrt{11x}$

c. $\sqrt[3]{4} \cdot \sqrt[3]{2}$

Solution $\sqrt[3]{4} \cdot \sqrt[3]{2} = \sqrt[3]{4 \cdot 2} = \sqrt[3]{8} = 2$

d. $\sqrt[3]{3x} \cdot \sqrt[3]{4x}$

Solution $\sqrt[3]{3x} \cdot \sqrt[3]{4x} = \sqrt[3]{3x \cdot 4x} = \sqrt[3]{12x^2}$

e. $\sqrt[4]{5} \cdot \sqrt[4]{7x^2}$

Solution $\sqrt[4]{5} \cdot \sqrt[4]{7x^2} = \sqrt[4]{5 \cdot 7x^2} = \sqrt[4]{35x^2}$

f. $\sqrt{\dfrac{5}{x}} \cdot \sqrt{\dfrac{y}{2}}$

Solution $\sqrt{\dfrac{5}{x}} \cdot \sqrt{\dfrac{y}{2}} = \sqrt{\dfrac{5}{x} \cdot \dfrac{y}{2}} = \sqrt{\dfrac{5y}{2x}}$

g. $\sqrt{x} \cdot \sqrt{x}$

Solution $\sqrt{x} \cdot \sqrt{x} = \sqrt{x \cdot x} = \sqrt{x^2} = x$

CONNECTION ———————

Expressions that have a fraction in a radical, like $\sqrt{\dfrac{5y}{2x}}$, are not considered to be in simplest form. We will learn how to simplify them in Section 10.5.

ANSWERS

a. 8   **b.** $\sqrt{5a}$

c. $\sqrt[3]{10xy}$ **d.** $\sqrt{\dfrac{2b}{7a}}$

YOUR TURN   **Find the product and write the answer in simplest form.**

a. $\sqrt{2} \cdot \sqrt{32}$ **b.** $\sqrt{5} \cdot \sqrt{a}$ **c.** $\sqrt[3]{5x} \cdot \sqrt[3]{2y}$ **d.** $\sqrt{\dfrac{2}{a}} \cdot \sqrt{\dfrac{b}{7}}$

Notice that in Example 1g, $\sqrt{x} \cdot \sqrt{x} = x$. It is also true that $\sqrt{x} \cdot \sqrt{x} = (\sqrt{x})^2$. Therefore, $(\sqrt{x})^2 = x$. Similarly, $\sqrt[3]{x} \cdot \sqrt[3]{x} \cdot \sqrt[3]{x} = \sqrt[3]{x \cdot x \cdot x} = \sqrt[3]{x^3} = x$. It is also true that $\sqrt[3]{x} \cdot \sqrt[3]{x} \cdot \sqrt[3]{x} = (\sqrt[3]{x})^3$, so $(\sqrt[3]{x})^3 = x$. These examples suggest the following rule.

RULE   **RAISING AN nTH ROOT TO THE nTH POWER**

For any nonnegative real number a, $(\sqrt[n]{a})^n = a$.

This means that $(\sqrt[3]{4})^3 = 4$, $(\sqrt[5]{19})^5 = 19$, and $(\sqrt[4]{3x^2})^4 = 3x^2$.

OBJECTIVE 2 Divide radical expressions.   Earlier, we developed the product rule for radicals. Now, we develop a similar rule for quotients like $\dfrac{\sqrt{100}}{\sqrt{25}}$. We can follow the order of operations and divide the roots, or we can divide the radicands, then find the square root of the quotient.

Find the roots first:

$$\dfrac{\sqrt{100}}{\sqrt{25}} = \dfrac{10}{5} = 2$$

Divide the radicands first:

$$\dfrac{\sqrt{100}}{\sqrt{25}} = \sqrt{\dfrac{100}{25}} = \sqrt{4} = 2$$

Both approaches give the same result, which suggests the following rule.

RULE   **QUOTIENT RULE FOR RADICALS**

If $\sqrt[n]{a}$ and $\sqrt[n]{b}$ are both real numbers, then $\dfrac{\sqrt[n]{a}}{\sqrt[n]{b}} = \sqrt[n]{\dfrac{a}{b}}$, where $b \neq 0$.

As with all equations, this rule can be used going from left to right or right to left.

EXAMPLE 2 Simplify.

a. $\sqrt{\dfrac{7}{36}}$

Solution $\sqrt{\dfrac{7}{36}} = \dfrac{\sqrt{7}}{\sqrt{36}} = \dfrac{\sqrt{7}}{6}$

b. $\dfrac{\sqrt{108}}{\sqrt{3}}$

Solution $\dfrac{\sqrt{108}}{\sqrt{3}} = \sqrt{\dfrac{108}{3}} = \sqrt{36} = 6$

c. $\sqrt[3]{\dfrac{9}{x^3}}$

Solution $\sqrt[3]{\dfrac{9}{x^3}} = \dfrac{\sqrt[3]{9}}{\sqrt[3]{x^3}} = \dfrac{\sqrt[3]{9}}{x}$

d. $\dfrac{\sqrt[3]{15}}{\sqrt[3]{5}}$

Solution $\dfrac{\sqrt[3]{15}}{\sqrt[3]{5}} = \sqrt[3]{\dfrac{15}{5}} = \sqrt[3]{3}$

e. $\sqrt[4]{\dfrac{y}{81}}$

Solution $\sqrt[4]{\dfrac{y}{81}} = \dfrac{\sqrt[4]{y}}{\sqrt[4]{81}} = \dfrac{\sqrt[4]{y}}{3}$

YOUR TURN   Simplify. Assume variables represent nonnegative values.

a. $\sqrt{\dfrac{x}{49}}$ b. $\dfrac{\sqrt{75}}{\sqrt{5}}$ c. $\sqrt[4]{\dfrac{6}{x^4}}$ d. $\dfrac{\sqrt[3]{32}}{\sqrt[3]{4}}$

OBJECTIVE 3 **Use the product rule to simplify radical expressions.** Several conditions exist in which a radical is not considered to be in simplest form. One such condition is if a radicand has a factor that can be written to a power greater than or equal to the index. For example, $\sqrt[3]{81}$ is not in simplest form because the perfect cube 27 is a factor of 81. Our first step in simplifying $\sqrt[3]{81}$ would be to rewrite it as $\sqrt[3]{27 \cdot 3}$ so that we can then use the product rule for radicals.

Note: We will explore other conditions that require simplification in future sections.

PROCEDURE SIMPLIFYING nTH ROOTS

1. Write the radicand as a product of the greatest possible perfect n^{th} power and a number that has no perfect n^{th} power factors.
2. Use the product rule $\sqrt[n]{ab} = \sqrt[n]{a} \cdot \sqrt[n]{b}$, where a is a perfect n^{th} power.
3. Find the nth root of the perfect n^{th} power radicand.

EXAMPLE 3 Simplify.

a. $\sqrt{18}$

Solution $\sqrt{18} = \sqrt{9 \cdot 2}$ The greatest perfect square factor of 18 is 9, so we write 18 as $9 \cdot 2$.

$\quad\quad\quad\quad = \sqrt{9} \cdot \sqrt{2}$ Use the product rule of roots to separate the factors into two radicals.

$\quad\quad\quad\quad = 3\sqrt{2}$ Simplify the square root of 9.

b. $5\sqrt{72}$

Solution $5\sqrt{72} = 5 \cdot \sqrt{36 \cdot 2}$ The greatest perfect square factor of 72 is 36, so we write 72 as $36 \cdot 2$.

$\quad\quad\quad\quad = 5 \cdot \sqrt{36} \cdot \sqrt{2}$ Use the product rule of roots to separate the factors into two radicals.

$\quad\quad\quad\quad = 5 \cdot 6 \cdot \sqrt{2}$ Simplify the square root of 36.

$\quad\quad\quad\quad = 30\sqrt{2}$ Multiply $5 \cdot 6$.

Note: We can use perfect n^{th} power factors other than the greatest perfect n^{th} power factor. For example, in simplifying $\sqrt{72}$, we would write:

$$\sqrt{72} = \sqrt{36 \cdot 2} = 6\sqrt{2} \quad \text{or}$$
$$\sqrt{72} = \sqrt{4 \cdot 18} = 2\sqrt{18} = 2\sqrt{9 \cdot 2} = 2 \cdot 3\sqrt{2} = 6\sqrt{2}$$

Notice that using the greatest perfect n^{th} factor saves steps.

c. $\sqrt[3]{40}$

Solution $\sqrt[3]{40} = \sqrt[3]{8 \cdot 5}$ The greatest perfect cube factor of 40 is 8.

$\quad\quad\quad\quad = \sqrt[3]{8} \cdot \sqrt[3]{5}$ Use the product rule of roots.

$\quad\quad\quad\quad = 2\sqrt[3]{5}$ Simplify the cube root of 8.

d. $\sqrt[4]{162}$

Solution $\sqrt[4]{162} = \sqrt[4]{81 \cdot 2}$ The greatest perfect fourth power factor of 162 is 81.

$\quad\quad\quad\quad = \sqrt[4]{81} \cdot \sqrt[4]{2}$ Use the product rule of roots.

$\quad\quad\quad\quad = 3\sqrt[4]{2}$ Simplify the fourth root of 81.

ANSWERS

a. $5\sqrt{6}$ **b.** $24\sqrt{5}$
c. $3\sqrt[3]{4}$ **d.** $2\sqrt[4]{5}$

YOUR TURN Simplify the following radical expressions.

a. $\sqrt{150}$ **b.** $6\sqrt{80}$ **c.** $\sqrt[3]{108}$ **d.** $\sqrt[4]{80}$

Using Prime Factorization

If the greatest perfect nth power of a particular radicand is not obvious, then try using the prime factorization of the radicand. Each prime factor that appears twice will have a square root equal to one of the two factors, each prime factor that appears three times will have a cube root that is one of the three factors, and so on. The remaining factors stay in the radical sign.

EXAMPLE 4 Simplify the following radicals using prime factorizations.

a. $\sqrt{375}$

Solution
$$\sqrt{375} = \sqrt{5 \cdot 5 \cdot 5 \cdot 3}$$ Write 375 as the product of its prime factors.
$$= 5\sqrt{3 \cdot 5}$$ The square root of the pair of 5's is 5.
$$= 5\sqrt{15}$$ Multiply the prime factors in the radicand.

b. $\sqrt[3]{324}$

Solution
$$\sqrt[3]{324} = \sqrt[3]{3 \cdot 3 \cdot 3 \cdot 3 \cdot 2 \cdot 2}$$ Write 324 as the product of its prime factors.
$$= 3\sqrt[3]{3 \cdot 2 \cdot 2}$$ The cube root of the three 3's is 3.
$$= 3\sqrt[3]{12}$$ Multiply the prime factors in the radicand.

c. $\sqrt[4]{240}$

Solution
$$\sqrt[4]{240} = \sqrt[4]{2 \cdot 2 \cdot 2 \cdot 2 \cdot 3 \cdot 5}$$ Write 240 as the product of its prime factors.
$$= 2\sqrt[4]{3 \cdot 5}$$ The fourth root of the four 2's is 2.
$$= 2\sqrt[4]{15}$$ Multiply the prime factors in the radicand.

YOUR TURN Simplify the following radicals using prime factorizations.

a. $\sqrt{294}$ **b.** $\sqrt[3]{324}$ **c.** $\sqrt[4]{486}$

Simplifying Radicals with Variables

We can use either procedure to simplify radicals whose radicands contain variables. Since $\sqrt[n]{a^m} = a^{m/n}$, we can find an exact root if n divides into m evenly. For example, $\sqrt{x^4} = x^{4/2} = x^2$ and $\sqrt[3]{x^{12}} = x^{12/3} = x^4$. Therefore, to find the nth root we rewrite the radicand as a product in which one factor has the greatest possible exponent divisible by the index n.

ANSWERS

a. $7\sqrt{6}$ **b.** $3\sqrt[3]{12}$
c. $3\sqrt[4]{6}$

EXAMPLE 5 Simplify. Assume all variables represent nonnegative values.

a. $\sqrt{x^7}$

Solution $\sqrt{x^7} = \sqrt{x^6 \cdot x}$ The greatest number smaller than 7 that is divisible by 2 is 6, so write x^7 as $x^6 \cdot x$.

$= \sqrt{x^6} \cdot \sqrt{x}$ Use the product rule of roots.

$= x^3\sqrt{x}$ Simplify: $\sqrt{x^6} = x^3$.

b. $3\sqrt{24a^5 \cdot b^9}$

Solution $3\sqrt{24a^5 \cdot b^9} = 3\sqrt{4 \cdot 6 \cdot a^4 \cdot a \cdot b^8 \cdot b}$ Write 24 as $4 \cdot 6$, a^5 as $a^4 \cdot a$, and b^9 as $b^8 \cdot b$.

$= 3\sqrt{4a^4 \cdot b^8 \cdot 6a \cdot b}$ Rearrange the factors so the perfect squares are together.

$= 3\sqrt{4a^4b^8} \cdot \sqrt{6ab}$ Use the product rule of roots.

$= 3 \cdot 2a^2b^4\sqrt{6ab}$ Simplify: $\sqrt{4} = 2$, $\sqrt{a^4} = a^2$, and $\sqrt{b^8} = b^4$.

$= 6a^2b^4\sqrt{6ab}$ Multiply $3 \cdot 2 = 6$.

c. $y^2\sqrt[3]{y^8}$

Solution $y^2\sqrt[3]{y^8} = y^2\sqrt[3]{y^6 \cdot y^2}$ The greatest number smaller than 8 that is divisible by 3 is 6, so write y^8 as $y^6 \cdot y^2$.

$= y^2\sqrt[3]{y^6} \cdot \sqrt[3]{y^2}$ Use the product rule of roots.

$= y^2 \cdot y^2\sqrt[3]{y^2}$ Simplify: $\sqrt[3]{y^6} = y^2$.

$= y^4\sqrt[3]{y^2}$ Multiply: $y^2 \cdot y^2 = y^4$.

d. $\sqrt[5]{64x^9y^{12}}$

Solution $\sqrt[5]{64x^9y^{12}} = \sqrt[5]{32 \cdot 2 \cdot x^5 \cdot x^4 \cdot y^{10} \cdot y^2}$ Write 64 as $32 \cdot 2$, x^9 as $x^5 \cdot x^4$, and y^{12} as $y^{10} \cdot y^2$.

$= \sqrt[5]{32x^5y^{10} \cdot 2x^4y^2}$ Regroup the factors.

$= \sqrt[5]{32x^5y^{10}} \cdot \sqrt[5]{2x^4y^2}$ Use the product rule of roots.

$= 2xy^2\sqrt[5]{2x^4y^2}$ Simplify: $\sqrt[5]{32} = 2$, $\sqrt[5]{x^5} = x$, and $\sqrt[5]{y^{10}} = y^2$.

YOUR TURN Simplify. Assume all variables represent nonnegative values.

a. $\sqrt{n^{11}}$ **b.** $\sqrt{45r^7s^3}$ **c.** $\sqrt[4]{m^{13}}$ **d.** $\sqrt[3]{a^8b^{10}}$

After using the product or quotient rules, it is often necessary to simplify the results.

ANSWERS

a. $n^5\sqrt{n}$ **b.** $3r^3s\sqrt{5rs}$
c. $m^3\sqrt[4]{m}$ **d.** $a^2b^3\sqrt[3]{a^2b}$

EXAMPLE 6 Find the product or quotient and simplify the results. Assume variables represent nonnegative values.

a. $\sqrt{3} \cdot \sqrt{6}$

Solution $\sqrt{3} \cdot \sqrt{6} = \sqrt{18}$ Use the product rule of roots to multiply.

$\phantom{\sqrt{3} \cdot \sqrt{6}} = \sqrt{9 \cdot 2}$ Write 18 as $9 \cdot 2$.

$\phantom{\sqrt{3} \cdot \sqrt{6}} = 3\sqrt{2}$ Simplify: $\sqrt{9} = 3$.

b. $5\sqrt{3x^3} \cdot 3\sqrt{15x^2}$

Solution $5\sqrt{3x^3} \cdot 3\sqrt{15x^2} = 5 \cdot 3\sqrt{3x^3} \cdot \sqrt{15x^2}$ Regroup the factors.

$\phantom{5\sqrt{3x^3} \cdot 3\sqrt{15x^2}} = 15\sqrt{45x^5}$ Multiply.

$\phantom{5\sqrt{3x^3} \cdot 3\sqrt{15x^2}} = 15\sqrt{9 \cdot 5 \cdot x^4 \cdot x}$ Write 45 as $9 \cdot 5$ and x^5 as $x^4 \cdot x$.

$\phantom{5\sqrt{3x^3} \cdot 3\sqrt{15x^2}} = 15 \cdot 3x^2\sqrt{5x}$ Simplify: $\sqrt{9} = 3$ and $\sqrt{x^4} = x^2$.

$\phantom{5\sqrt{3x^3} \cdot 3\sqrt{15x^2}} = 45x^2\sqrt{5x}$ Multiply.

c. $\dfrac{\sqrt{288}}{\sqrt{6}}$

Solution $\dfrac{\sqrt{288}}{\sqrt{6}} = \sqrt{\dfrac{288}{6}}$ Use the quotient rule of roots.

$\phantom{\dfrac{\sqrt{288}}{\sqrt{6}}} = \sqrt{48}$ Divide the radicand.

$\phantom{\dfrac{\sqrt{288}}{\sqrt{6}}} = \sqrt{16 \cdot 3}$ Write 48 as $16 \cdot 3$.

$\phantom{\dfrac{\sqrt{288}}{\sqrt{6}}} = 4\sqrt{3}$ Simplify: $\sqrt{16} = 3$.

d. $\dfrac{8\sqrt{756a^8b^5}}{2\sqrt{7a^4b^2}}$

Solution $\dfrac{8\sqrt{756a^8b^5}}{2\sqrt{7a^4b^2}} = 4\sqrt{\dfrac{756a^8b^5}{7a^4b^2}}$ Divide coefficients and use the quotient rule of radicals

$\phantom{\dfrac{8\sqrt{756a^8b^5}}{2\sqrt{7a^4b^2}}} = 4\sqrt{108a^4b^3}$ Divide the radicand.

$\phantom{\dfrac{8\sqrt{756a^8b^5}}{2\sqrt{7a^4b^2}}} = 4\sqrt{36a^4b^2 \cdot 3b}$ Rewrite the radicand with a perfect square factor.

$\phantom{\dfrac{8\sqrt{756a^8b^5}}{2\sqrt{7a^4b^2}}} = 4 \cdot 6a^2b\sqrt{3b}$ Find the square roots.

$\phantom{\dfrac{8\sqrt{756a^8b^5}}{2\sqrt{7a^4b^2}}} = 24a^2b\sqrt{3b}$ Multiply.

YOUR TURN Find the product or quotient and simplify the results. Assume variables represent nonnegative values.

a. $\sqrt{6} \cdot \sqrt{15}$ **b.** $4\sqrt{14x^3} \cdot 2\sqrt{6x^4}$ **c.** $\dfrac{\sqrt{1296}}{\sqrt{12}}$ **d.** $\dfrac{14\sqrt{315x^{11}y^8}}{2\sqrt{5x^6y^5}}$

ANSWERS

a. $3\sqrt{10}$ **b.** $16x^3\sqrt{21x}$

c. $6\sqrt{3}$ **d.** $21x^2y\sqrt{7xy}$

1. For $\sqrt{8} \cdot \sqrt{18}$, explain the difference between using the product rule for radicals and multiplying the approximate roots of 8 and 18.

2. Explain why $\sqrt{28}$ is not in simplest form.

3. Describe in your own words how to simplify a cube root containing a radicand with a perfect cube factor.

4. Explain why the expression $3x^2\sqrt[3]{x^5}$ is not in simplest form.

For Exercises 5–32, find the product and write the answer in simplest form. Assume variables represent nonnegative values.

5. $\sqrt{2} \cdot \sqrt{32}$

6. $\sqrt{3} \cdot \sqrt{12}$

7. $\sqrt{3x} \cdot \sqrt{27x^5}$

8. $\sqrt{8y^3} \cdot \sqrt{2y}$

9. $\sqrt{6xy^3} \cdot \sqrt{24xy}$

10. $\sqrt{50u^3v^2} \cdot \sqrt{uv^4}$

11. $\sqrt{2} \cdot \sqrt{7}$

12. $\sqrt{6} \cdot \sqrt{11}$

13. $\sqrt{15} \cdot \sqrt{x}$

14. $\sqrt{17} \cdot \sqrt{y}$

15. $\sqrt[3]{3} \cdot \sqrt[3]{9}$

16. $\sqrt[3]{4} \cdot \sqrt[3]{16}$

17. $\sqrt[3]{5y} \cdot \sqrt[3]{2y}$

18. $\sqrt[3]{6m} \cdot \sqrt[3]{2m}$

19. $\sqrt[4]{3} \cdot \sqrt[4]{7}$

20. $\sqrt[4]{8} \cdot \sqrt[4]{5}$

21. $\sqrt[4]{12w^3} \cdot \sqrt[4]{6w}$

22. $\sqrt[4]{21r^3} \cdot \sqrt[4]{7r^2}$

23. $\sqrt[4]{3x^2y} \cdot \sqrt[4]{5xy^2}$

24. $\sqrt[4]{ab^2} \cdot \sqrt[4]{ab}$

25. $\sqrt[5]{6x^3} \cdot \sqrt[5]{5x}$

26. $\sqrt[5]{m^2} \cdot \sqrt[5]{m^2}$

27. $\sqrt[6]{4x^2y^3} \cdot \sqrt[6]{2x^3y}$

28. $\sqrt[6]{ab^3} \cdot \sqrt[6]{7a^3b^2}$

29. $\sqrt{\dfrac{7}{2}} \cdot \sqrt{\dfrac{3}{5}}$

30. $\sqrt{\dfrac{5}{2}} \cdot \sqrt{\dfrac{11}{3}}$

31. $\sqrt{\dfrac{6}{x}} \cdot \sqrt{\dfrac{y}{5}}$

32. $\sqrt{\dfrac{a}{3}} \cdot \sqrt{\dfrac{7}{b}}$

For Exercises 33–48, use the quotient rule to simplify. Assume variables represent nonnegative values.

33. $\sqrt{\dfrac{25}{36}}$

34. $\sqrt{\dfrac{49}{64}}$

35. $\sqrt{\dfrac{10}{9}}$

36. $\sqrt{\dfrac{15}{81}}$

37. $\dfrac{\sqrt{196}}{\sqrt{4}}$

38. $\dfrac{\sqrt{81}}{\sqrt{3}}$

39. $\dfrac{\sqrt{15}}{\sqrt{5}}$

40. $\dfrac{\sqrt{21}}{\sqrt{3}}$

41. $\sqrt[3]{\dfrac{4}{w^6}}$

42. $\sqrt[3]{\dfrac{7}{v^3}}$

43. $\sqrt[3]{\dfrac{5y^2}{27x^9}}$

44. $\sqrt[3]{\dfrac{5a}{8r^6}}$

45. $\dfrac{\sqrt[3]{320}}{\sqrt[3]{5}}$

46. $\dfrac{\sqrt[3]{162}}{\sqrt[3]{6}}$

47. $\sqrt[4]{\dfrac{3u^3}{16x^8}}$

48. $\sqrt[4]{\dfrac{3x^2}{81y^4}}$

For Exercises 49–88, simplify. Assume variables represent nonnegative values.

49. $\sqrt{98}$

50. $\sqrt{48}$

51. $\sqrt{128}$

52. $\sqrt{180}$

53. $6\sqrt{80}$

54. $4\sqrt{50}$

55. $5\sqrt{112}$

56. $3\sqrt{245}$

57. $\sqrt{a^3}$

58. $\sqrt{d^5}$

59. $\sqrt{x^2y^4}$

60. $\sqrt{a^6b^2}$

61. $\sqrt{x^6y^8z^{10}}$

62. $\sqrt{p^4q^8r^8}$

63. $rs^2\sqrt{r^9s^5}$

64. $a^2b^3\sqrt{a^{11}b^3}$

65. $3\sqrt{72x^5}$

66. $6\sqrt{75c^5d^3}$

67. $\sqrt[3]{32}$

68. $\sqrt[3]{54}$

69. $\sqrt[3]{x^7}$

70. $\sqrt[3]{b^{11}}$

71. $\sqrt[3]{x^6y^5}$

72. $\sqrt[3]{m^{13}n^9}$

73. $\sqrt[3]{128z^8}$

74. $\sqrt[3]{48h^{14}}$

75. $2\sqrt[3]{24}$

76. $4\sqrt[3]{250}$

77. $\sqrt[4]{80}$

78. $\sqrt[4]{162}$

79. $3x^2\sqrt[4]{135x^9}$

80. $3a^4\sqrt[4]{48a^7}$

81. $\sqrt[5]{486x^{16}}$

82. $\sqrt[5]{160n^{18}}$

83. $\sqrt[6]{x^8y^{14}z^{11}}$

84. $\sqrt[7]{a^{16}b^9c^{12}}$

For Exercises 85–94, find the product and write the answer in simplest form. Assume variables represent nonnegative values.

85. $\sqrt{3} \cdot \sqrt{21}$

86. $\sqrt{5} \cdot \sqrt{15}$

87. $5\sqrt{10} \cdot 3\sqrt{14}$

88. $2\sqrt{6} \cdot 5\sqrt{21}$

89. $\sqrt{y^3} \cdot \sqrt{y^2}$

90. $\sqrt{m^7} \cdot \sqrt{m^4}$

91. $\sqrt{x^2y^3} \cdot \sqrt{x^4y^3}$

92. $x\sqrt{x^5y^2} \cdot y\sqrt{xy^3}$

93. $4\sqrt{6c^3} \cdot 3\sqrt{10c^5}$

94. $6\sqrt{15c^2} \cdot 2\sqrt{10c^5}$

For Exercises 95 and 96, write an expression in simplest form for the area of the figure.

95.

$4\sqrt{3}$

$5\sqrt{6}$

96.

$3\sqrt{5}$

$6\sqrt{10}$

For Exercises 97–108, find the quotient and write the answer in simplest form. Assume variables represent nonnegative values.

97. $\dfrac{\sqrt{48}}{\sqrt{6}}$

98. $\dfrac{\sqrt{54}}{\sqrt{3}}$

99. $\dfrac{9\sqrt{160}}{3\sqrt{8}}$

100. $\dfrac{10\sqrt{280}}{2\sqrt{10}}$

101. $\dfrac{\sqrt{c^5 d^4}}{\sqrt{cd^3}}$

102. $\dfrac{\sqrt{m^6 n^5}}{\sqrt{m^3 n^3}}$

103. $\dfrac{8\sqrt{45a^5}}{2\sqrt{5a}}$

104. $\dfrac{6\sqrt{48n^7}}{3\sqrt{3n^3}}$

105. $\dfrac{12\sqrt{72c^5}}{4\sqrt{6c^2}}$

106. $\dfrac{15\sqrt{48a^7}}{5\sqrt{2a^2}}$

107. $\dfrac{36\sqrt{96x^6 y^{11}}}{4\sqrt{3x^2 y^4}}$

108. $\dfrac{54\sqrt{240r^9 s^{10}}}{9\sqrt{5r^6 s^4}}$

For Exercises 109–114, find the products and write the answers in simplest form. Assume variables represent nonnegative values.

109. $\sqrt{\dfrac{3}{7}} \cdot \sqrt{\dfrac{8}{7}}$

110. $\sqrt{\dfrac{8}{5}} \cdot \sqrt{\dfrac{6}{5}}$

111. $\sqrt{\dfrac{a^3}{2}} \cdot \sqrt{\dfrac{a^5}{2}}$

112. $\sqrt{\dfrac{c^2}{6}} \cdot \sqrt{\dfrac{c^5}{6}}$

113. $\sqrt{\dfrac{3x^5}{2}} \cdot \sqrt{\dfrac{15x^5}{8}}$

114. $\sqrt{\dfrac{5y}{3}} \cdot \sqrt{\dfrac{10}{27}}$

For Exercises 115 and 116, write an expression in simplest form for the area of the figure.

115.

$\sqrt{\dfrac{3}{8}}$

$\sqrt{\dfrac{15}{2}}$

116.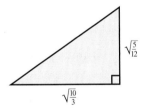

$\sqrt{\dfrac{5}{12}}$

$\sqrt{\dfrac{10}{3}}$

REVIEW EXERCISES

For Exercises 1–2, use $a = 6$ and $b = 8$.

1. Does $\sqrt{a^2 + b^2} = \sqrt{a^2} + \sqrt{b^2} = a + b$?

2. Does $\sqrt{(a + b)^2} = a + b$?

For Exercises 3–6, simplify.

3. $6x^2 - 4x - 3x + 2x^2$

4. $(2a - 3b)(4a + 3b)$

5. $(3m + 5n)(3m - 5n)$

6. $(2x - 3y)^2$

Adding, Subtracting, and Multiplying Radical Expressions

OBJECTIVES

1 Add or subtract like radicals.

2 Use the distributive property in expressions containing radicals.

3 Simplify radical expressions that contain mixed operations.

OBJECTIVE **1** **Add or subtract like radicals.** Recall that like terms, such as $-8a^2$ and $3a^2$, have identical variables and identical exponents. Similarly, **like radicals** have identical radicands and indexes.

DEFINITION

Like radicals: Radical expressions with identical radicands and identical indexes.

The radicals $3\sqrt{2}$ and $5\sqrt{2}$ are like. The radicals $3\sqrt{2}$ and $5\sqrt{3}$ are unlike because their radicands are different. The radicals $3\sqrt{2}$ and $5\sqrt[3]{2}$ are unlike because their indexes are different.

Adding or subtracting radicals is essentially the same as combining like terms. Remember that the distributive property is at work when we combine like terms because we factor out the common variable, which leaves us a sum or difference of the coefficients.

Like terms:
$$3x + 4x = (3 + 4)x$$
$$= 7x$$

Like radicals:
$$3\sqrt{5} + 4\sqrt{5} = (3 + 4)\sqrt{5}$$
$$= 7\sqrt{5}$$

Our example suggests the following procedure for adding like radical expressions.

PROCEDURE ADDING LIKE RADICALS

To add or subtract like radicals, add or subtract the coefficients and keep the radicals the same.

EXAMPLE **1** Simplify.

a. $7\sqrt{5} + 2\sqrt{5}$

Solution $7\sqrt{5} + 2\sqrt{5} = (7 + 2)\sqrt{5}$
$$= 9\sqrt{5}$$

b. $6\sqrt[3]{4} + \sqrt[3]{4}$

Solution $6\sqrt[3]{4} + \sqrt[3]{4} = (6 + 1)\sqrt[3]{4}$
$$= 7\sqrt[3]{4}$$

c. $6x\sqrt[4]{3x} - 2x\sqrt[4]{3x}$

Solution $6x\sqrt[4]{3x} - 2x\sqrt[4]{3x} = (6x - 2x)\sqrt[4]{3x}$
$$= 4x\sqrt[4]{3x}$$

d. $8\sqrt{3} + 6\sqrt{2} - 5\sqrt{3} + 3\sqrt{2}$

Solution $8\sqrt{3} + 6\sqrt{2} - 5\sqrt{3} + 3\sqrt{2} = 8\sqrt{3} - 5\sqrt{3} + 6\sqrt{2} + 3\sqrt{2}$ Regroup the terms.
$$= (8 - 5)\sqrt{3} + (6 + 3)\sqrt{2}$$
$$= 3\sqrt{3} + 9\sqrt{2}$$

ANSWERS

a. $7\sqrt{6}$
b. $6\sqrt[4]{5x^2}$
c. $-5y\sqrt[3]{7}$
d. $3x\sqrt{5x} + 5y\sqrt{3y}$

YOUR TURN   **Simplify.**

a. $5\sqrt{6} + 2\sqrt{6}$ **b.** $5\sqrt[4]{5x^2} + \sqrt[4]{5x^2}$

c. $7y\sqrt[3]{7} - 12y\sqrt[3]{7}$ **d.** $9x\sqrt{5x} - 2y\sqrt{3y} - 6x\sqrt{5x} + 7y\sqrt{3y}$

In a problem involving addition or subtraction of radicals, if the radicals are not like, it may be possible to simplify one or more of the radicals so that they are like.

EXAMPLE 2   Simplify. Assume variables represent nonnegative quantities.

a. $7\sqrt{3} + \sqrt{12}$

Solution $7\sqrt{3} + \sqrt{12} = 7\sqrt{3} + \sqrt{4 \cdot 3}$ Factor out the perfect square factor, 4.
$$= 7\sqrt{3} + \sqrt{4} \cdot \sqrt{3}$$ Use the product rule to separate the radicals.
$$= 7\sqrt{3} + 2\sqrt{3}$$ Simplify.
$$= 9\sqrt{3}$$ Combine like radicals.

b. $3\sqrt[3]{24} - 2\sqrt[3]{3}$

Solution $3\sqrt[3]{24} - 2\sqrt[3]{3} = 3\sqrt[3]{8 \cdot 3} - 2\sqrt[3]{3}$ Rewrite 24 as $8 \cdot 3$
$$= 3\sqrt[3]{8} \cdot \sqrt[3]{3} - 2\sqrt[3]{3}$$ Use the product rule.
$$= 3 \cdot 2\sqrt[3]{3} - 2\sqrt[3]{3}$$ Simplify $\sqrt[3]{8}$.
$$= 6\sqrt[3]{3} - 2\sqrt[3]{3}$$ Multiply.
$$= 4\sqrt[3]{3}$$ Combine like radicals.

c. $\sqrt{48x^3} + \sqrt{12x^3}$

Solution $\sqrt{48x^3} + \sqrt{12x^3} = \sqrt{16x^2 \cdot 3x} + \sqrt{4x^2 \cdot 3x}$ Rewrite $48x^3$ as $16x^2 \cdot 3x$ and $12x^3$ as $4x^2 \cdot 3x$.
$$= \sqrt{16x^2} \cdot \sqrt{3x} + \sqrt{4x^2} \cdot \sqrt{3x}$$ Use the product rule.
$$= 4x\sqrt{3x} + 2x\sqrt{3x}$$ Find $\sqrt{16x^2}$ and $\sqrt{4x^2}$.
$$= 6x\sqrt{3x}$$ Combine like radicals.

10.4   Adding, Subtracting, and Multiplying Radical Expressions   **853**

d. $3\sqrt[4]{32x^5} + \sqrt[4]{162x^9}$

Solution $3\sqrt[4]{32x^5} + \sqrt[4]{162x^9} = 3\sqrt[4]{16 \cdot 2 \cdot x^4 \cdot x} + \sqrt[4]{81 \cdot 2 \cdot x^8 \cdot x}$ Write 32 as $16 \cdot 2$, x^5 as $x^4 \cdot x$, 162 as $81 \cdot 2$ and x^9 as $x^8 \cdot x$.

$$= 3 \cdot 2x\sqrt[4]{2x} + 3x^2\sqrt[4]{2x} \qquad \text{Find the roots.}$$
$$= 6x\sqrt[4]{2x} + 3x^2\sqrt[4]{2x} \qquad \text{Multiply.}$$
$$= (6x + 3x^2)\sqrt[4]{2x} \qquad \text{Combine like radicals.}$$

Note: Although the radicals are like, we cannot add the coefficients because they are not like.

YOUR TURN   **Simplify. Assume variables represent nonnegative quantities.**

   a. $4\sqrt{24} - 6\sqrt{54}$ **b.** $4\sqrt{50x^5} - 2\sqrt{18x^5}$ **c.** $6a\sqrt[3]{54a^4} - 2\sqrt[3]{128a^7}$

OBJECTIVE 2 Use the distributive property in expressions containing radicals. Products involving sums and differences of radicals are found in much the same way as products of polynomials. We will use the distributive property, multiply binomials using FOIL, and square a binomial (all from Section 5.5).

EXAMPLE 3   Find the product. Assume variables represent nonnegative values.

a. $\sqrt{3}(\sqrt{3} + \sqrt{15})$

Solution $\sqrt{3}(\sqrt{3} + \sqrt{15}) = \sqrt{3} \cdot \sqrt{3} + \sqrt{3} \cdot \sqrt{15}$ Use the distributive property.
$$= \sqrt{3 \cdot 3} + \sqrt{3 \cdot 15} \qquad \text{Use the product rule.}$$
$$= \sqrt{9} + \sqrt{45} \qquad \text{Multiply.}$$
$$= 3 + 3\sqrt{5} \qquad \text{Find } \sqrt{9} \text{ and simplify } \sqrt{45}.$$

b. $2\sqrt{6}(3 + 5\sqrt{5})$

Solution $2\sqrt{6}(3 + 5\sqrt{5}) = 2\sqrt{6} \cdot 3 + 2\sqrt{6} \cdot 5\sqrt{5}$ Use the distributive property.
$$= 2 \cdot 3\sqrt{6} + 2 \cdot 5\sqrt{6 \cdot 5} \qquad \text{Use the product rule.}$$
$$= 6\sqrt{6} + 10\sqrt{30} \qquad \text{Find the products.}$$

c. $(2 + \sqrt{3})(\sqrt{5} - \sqrt{6})$

Solution $(2 + \sqrt{3})(\sqrt{5} - \sqrt{6}) = 2\sqrt{5} - 2\sqrt{6} + \sqrt{3} \cdot \sqrt{5} - \sqrt{3} \cdot \sqrt{6}$ Use FOIL.
$$= 2\sqrt{5} - 2\sqrt{6} + \sqrt{15} - \sqrt{18} \qquad \text{Use the product rule.}$$
$$= 2\sqrt{5} - 2\sqrt{6} + \sqrt{15} - 3\sqrt{2} \qquad \text{Simplify } \sqrt{18}.$$

d. $(3\sqrt{x} + \sqrt{y})(2\sqrt{x} - 5\sqrt{y})$

Solution $(3\sqrt{x} + \sqrt{y})(2\sqrt{x} - 5\sqrt{y}) =$
$$3\sqrt{x} \cdot 2\sqrt{x} - 3\sqrt{x} \cdot 5\sqrt{y} + \sqrt{y} \cdot 2\sqrt{x} - \sqrt{y} \cdot 5\sqrt{y} \quad \text{Use FOIL.}$$
$$= 6x - 15\sqrt{xy} + 2\sqrt{xy} - 5y \quad \text{Use the product rule.}$$
$$= 6x - 13\sqrt{xy} - 5y \quad \text{Combine like radicals.}$$

e. $(5 + \sqrt{3})^2$

Solution $(5 + \sqrt{3})^2 = 5^2 + 2 \cdot 5\sqrt{3} + (\sqrt{3})^2 \quad \text{Use } (a + b)^2 = a^2 + 2ab + b^2.$
$$= 25 + 10\sqrt{3} + 3 \quad \text{Simplify.}$$
$$= 28 + 10\sqrt{3} \quad \text{Add 25 and 3.}$$

YOUR TURN Find the following products. Assume variables represent nonnegative quantities.

a. $2\sqrt{11}(3 - 3\sqrt{6})$ **b.** $(2\sqrt{a} + 3\sqrt{b})(\sqrt{a} - 3\sqrt{b})$ **c.** $(3 + \sqrt{5})^2$

Radicals in Conjugates

Radical expressions can be conjugates. Like binomial conjugates, conjugates involving radicals differ only in the sign separating the terms. For example, $7 - \sqrt{3}$ and $7 + \sqrt{3}$ are conjugates. Let's explore what happens when we multiply conjugates containing radicals.

EXAMPLE 4 Multiply.

a. $(2 + \sqrt{3})(2 - \sqrt{3})$

Solution $(2 + \sqrt{3})(2 - \sqrt{3}) = 2^2 - (\sqrt{3})^2 \quad \text{Use } (a + b)(a - b) = a^2 - b^2 \text{ (from Section 5.5).}$
$$= 4 - 3 \quad \text{Simplify.}$$
$$= 1$$

b. $(\sqrt{5} - 3\sqrt{3})(\sqrt{5} + 3\sqrt{3})$

Note: The product of conjugates *always* results in a rational number. This will be useful in the Section 10.5 when rationalizing denominators.

Solution $(\sqrt{5} - 3\sqrt{3})(\sqrt{5} + 3\sqrt{3}) = (\sqrt{5})^2 - (3\sqrt{3})^2 \quad \text{Use } (a + b)(a - b) = a^2 - b^2.$
$$= 5 - 9 \cdot 3 \quad \text{Simplify.}$$
$$= 5 - 27$$
$$= -22$$

YOUR TURN Find the following products.

a. $(4 + \sqrt{10})(4 - \sqrt{10})$ **b.** $(3\sqrt{5} + 2\sqrt{6})(3\sqrt{5} - 2\sqrt{6})$

OBJECTIVE **3** **Simplify radical expressions that contain mixed operations.** Now let's use the order of operations to simplify radical expressions that have more than one operation.

EXAMPLE 5 Perform the following operations.

a. $\sqrt{2} \cdot \sqrt{10} + \sqrt{3} \cdot \sqrt{15}$

Solution $\sqrt{2} \cdot \sqrt{10} + \sqrt{3} \cdot \sqrt{15} = \sqrt{2 \cdot 10} + \sqrt{3 \cdot 15}$ Use the product rule.

$= \sqrt{20} + \sqrt{45}$ Multiply.

$= \sqrt{4 \cdot 5} + \sqrt{9 \cdot 5}$ Rewrite 20 as $4 \cdot 5$ and 45 as $9 \cdot 5$.

$= \sqrt{4} \cdot \sqrt{5} + \sqrt{9} \cdot \sqrt{5}$ Use the product rule.

$= 2\sqrt{5} + 3\sqrt{5}$ Find $\sqrt{9}$ and $\sqrt{4}$.

$= 5\sqrt{5}$ Combine like radicals.

b. $\dfrac{\sqrt{54}}{\sqrt{3}} + \sqrt{32}$

Solution $\dfrac{\sqrt{54}}{\sqrt{3}} + \sqrt{32} = \sqrt{\dfrac{54}{3}} + \sqrt{16 \cdot 2}$ Use the quotient rule and rewrite 32 as $16 \cdot 2$.

$= \sqrt{18} + \sqrt{16} \cdot \sqrt{2}$ Divide and use the product rule.

$= \sqrt{9 \cdot 2} + 4\sqrt{2}$ Rewrite 18 as $9 \cdot 2$ and find $\sqrt{16}$.

$= \sqrt{9} \cdot \sqrt{2} + 4\sqrt{2}$ Use the product rule.

$= 3\sqrt{2} + 4\sqrt{2}$ Find $\sqrt{9}$.

$= 7\sqrt{2}$ Combine like radicals.

Calculator
TIPS

We can use a calculator to simplify radical expressions. For Example 5b, we type:

[2nd] [x^2] [5] [4] [)] [÷] [2nd] [x^2] [3] [)] [+] [2nd] [x^2] [3] [2] [)] [ENTER]

The answer, rounded to the nearest thousandth, is 9.899.

YOUR TURN Perform the following operations.

a. $2\sqrt{3} \cdot \sqrt{6} + 4\sqrt{7} \cdot \sqrt{14}$

b. $\sqrt{63} + \dfrac{\sqrt{140}}{\sqrt{5}}$

ANSWERS

a. $34\sqrt{2}$

b. $5\sqrt{7}$

10.4 EXERCISES

For Extra Help

InterAct Math

Tutor Center — Addison-Wesley Math Tutor Center

Video 16/DVT 5

MyMathLab

Math XP MathXL

1. What must be identical in like radicals? What can be different?

2. Add $3x + 2x$ and then add $3\sqrt{2} + 2\sqrt{2}$. Discuss the similarities in the process.

3. Multiply $(x + 3)(x + 2)$ and then multiply $(\sqrt{5} + 3)(\sqrt{5} + 2)$. Discuss the similarities in the process.

4. Why is $\sqrt{a} + \sqrt{b} \neq \sqrt{a + b}$? Use examples if necessary.

For Exercises 5–18, simplify. Assume variables represent nonnegative values.

5. $9\sqrt{6} - 15\sqrt{6}$

6. $2\sqrt{10} - 13\sqrt{10}$

7. $7\sqrt{a} + 2\sqrt{a}$

8. $5\sqrt{y} + 7\sqrt{y}$

9. $4\sqrt{5} - 2\sqrt{6} + 8\sqrt{5} - 6\sqrt{6}$

10. $-6\sqrt{2} + 5\sqrt{7} + 3\sqrt{2} - 2\sqrt{7}$

11. $3a\sqrt{5a} - 4b\sqrt{7b} + 8a\sqrt{5a} + 2b\sqrt{7b}$

12. $12n\sqrt{2n} - 14m\sqrt{5m} - 8n\sqrt{2n} + 18m\sqrt{5m}$

13. $6x\sqrt[3]{9} - 3x\sqrt[3]{9}$

14. $4y\sqrt[3]{3} - y\sqrt[3]{3}$

15. $6x^2\sqrt[4]{5x} + 12x^2\sqrt[4]{5x}$

16. $3y^3\sqrt[4]{8y} - 9y^3\sqrt[4]{8y}$

17. $3x\sqrt{5x} + 4x\sqrt[3]{5x}$

18. $4z\sqrt[4]{2z} - 7z\sqrt{2z}$

For Exercises 19–38, simplify the radicals and then find the sum. Assume variables represent nonnegative values.

19. $\sqrt{48} - \sqrt{75}$

20. $\sqrt{80} - \sqrt{20}$

21. $\sqrt{80y} - \sqrt{125y}$

22. $\sqrt{27x} + \sqrt{75x}$

23. $\sqrt{80} - 4\sqrt{45}$

24. $\sqrt{20} - 2\sqrt{180}$

25. $3\sqrt{96} - 2\sqrt{54}$

26. $5\sqrt{63} + 2\sqrt{28}$

27. $6\sqrt{48a^3} - 2\sqrt{75a^3}$

28. $4\sqrt{98y^5} - 7\sqrt{128y^5}$

29. $\sqrt{150} - \sqrt{54} + \sqrt{24}$

30. $\sqrt{20} + \sqrt{125} - \sqrt{80}$

31. $2\sqrt{8} - 3\sqrt{48} + 2\sqrt{98} - \sqrt{75}$

32. $3\sqrt{216} - \sqrt{147} - 4\sqrt{96} - \sqrt{108}$

33. $\sqrt[3]{16} + \sqrt[3]{54}$

34. $\sqrt[3]{24} + \sqrt[3]{81}$

35. $4\sqrt[3]{135x^5} - 6x\sqrt[3]{320x^2}$

36. $3a^2\sqrt[3]{500a^4} + 6a\sqrt[3]{108a^7}$

37. $-4\sqrt[4]{x^9} + 2\sqrt[4]{16x^7}$

38. $6y\sqrt[4]{81y^5} - 2y\sqrt[4]{16y^5}$

For Exercises 39–46 use the distributive property. Assume variables represent nonnegative values.

39. $\sqrt{2}(3 + \sqrt{2})$

40. $\sqrt{5}(4 - \sqrt{5})$

41. $\sqrt{3}(\sqrt{3} - \sqrt{15})$

42. $\sqrt{6}(\sqrt{6} + \sqrt{2})$

43. $\sqrt{5}(\sqrt{3} + 2\sqrt{15})$

44. $\sqrt{7}(\sqrt{5} - 3\sqrt{14})$

45. $4\sqrt{3x}(2\sqrt{3x} - 4\sqrt{6x})$

46. $6\sqrt{2y}(3\sqrt{2y} + 2\sqrt{10y})$

For Exercises 47–66, multiply (use FOIL). Assume variables represent nonnegative values.

47. $(3 + \sqrt{5})(4 - \sqrt{2})$

48. $(3 + \sqrt{7})(7 - \sqrt{3})$

49. $(3 + \sqrt{x})(2 + \sqrt{x})$

50. $(5 - \sqrt{a})(2 + \sqrt{a})$

51. $(2 + 3\sqrt{3})(3 + 5\sqrt{2})$

52. $(7 - 3\sqrt{5})(2 - 2\sqrt{10})$

53. $(\sqrt{2} + \sqrt{3})(\sqrt{3} + \sqrt{5})$

54. $(\sqrt{5} + \sqrt{2})(\sqrt{2} + \sqrt{7})$

55. $(\sqrt{x} + \sqrt{y})(\sqrt{x} - 2\sqrt{y})$

56. $(2\sqrt{a} + \sqrt{b})(\sqrt{a} + \sqrt{b})$

57. $(4\sqrt{2} + 2\sqrt{5})(3\sqrt{7} - 3\sqrt{3})$

58. $(8\sqrt{2} - 2\sqrt{3})(2\sqrt{5} + 3\sqrt{10})$

59. $(2\sqrt{a} + 3\sqrt{b})(4\sqrt{a} - \sqrt{b})$

60. $(\sqrt{m} - 4\sqrt{n})(2\sqrt{m} - 3\sqrt{n})$

61. $(\sqrt[3]{4} + 5)(\sqrt[3]{4} - 8)$

62. $(\sqrt[3]{6} + 5)(\sqrt[3]{6} - 2)$

63. $(\sqrt[3]{9} + \sqrt[3]{4})(\sqrt[3]{3} - \sqrt[3]{2})$

64. $(\sqrt[3]{5} + \sqrt[3]{9})(\sqrt[3]{25} - \sqrt[3]{3})$

65. $(\sqrt[3]{x} + 2)(\sqrt[3]{x^2} - 2\sqrt[3]{x} + 4)$

66. $(\sqrt[3]{r} - 6)(\sqrt[3]{r^2} - 5\sqrt[3]{r} - 7)$

For Exercises 67–74, find the product.

67. $(4 + \sqrt{6})^2$

68. $(1 - \sqrt{2})^2$

69. $(6 + \sqrt{3})^2$

70. $(5 + \sqrt{2})^2$

71. $(2 + 2\sqrt{3})^2$

72. $(3 + 2\sqrt{5})^2$

73. $(2\sqrt{3} + 3\sqrt{2})^2$

74. $(4\sqrt{2} - 5\sqrt{6})^2$

For Exercises 75–88, multiply the conjugates. Assume all variables have nonnegative values.

75. $(2 + \sqrt{3})(2 - \sqrt{3})$

76. $(3 + \sqrt{5})(3 - \sqrt{5})$

77. $(\sqrt{2} + 4)(\sqrt{2} - 4)$

78. $(\sqrt{7} - 6)(\sqrt{7} + 6)$

79. $(6 + \sqrt{x})(6 - \sqrt{x})$

80. $(5 + \sqrt{y})(5 - \sqrt{y})$

81. $(\sqrt{3} + \sqrt{2})(\sqrt{3} - \sqrt{2})$

82. $(\sqrt{5} - \sqrt{3})(\sqrt{5} + \sqrt{3})$

83. $(\sqrt{x} + \sqrt{y})(\sqrt{x} - \sqrt{y})$

84. $(\sqrt{a} - \sqrt{b})(\sqrt{a} + \sqrt{b})$

85. $(4 + 2\sqrt{3})(4 - 2\sqrt{3})$

86. $(5 + 3\sqrt{3})(5 - 3\sqrt{3})$

87. $(3\sqrt{7} + \sqrt{13})(3\sqrt{7} - \sqrt{13})$

88. $(4\sqrt{5} - 3\sqrt{2})(4\sqrt{5} + 3\sqrt{2})$

For Exercises 89–96, simplify.

89. $\sqrt{3} \cdot \sqrt{15} + \sqrt{8} \cdot \sqrt{10}$

90. $\sqrt{6} \cdot \sqrt{8} + \sqrt{5} \cdot \sqrt{15}$

91. $3\sqrt{3} \cdot \sqrt{18} - 4\sqrt{18} \cdot \sqrt{12}$

92. $6\sqrt{80} - 5\sqrt{12} \cdot \sqrt{15}$

93. $\dfrac{\sqrt{40}}{\sqrt{5}} + \sqrt{50}$

94. $\dfrac{\sqrt{60}}{\sqrt{5}} + \sqrt{48}$

95. $\dfrac{\sqrt{540}}{\sqrt{3}} - 4\sqrt{125}$

96. $\dfrac{\sqrt{288}}{\sqrt{6}} - 6\sqrt{108}$

For Exercises 97 and 98, find the perimeter of the shape.

97.

$5\sqrt{3}$ $4\sqrt{12}$ $4\sqrt{27}$

98.

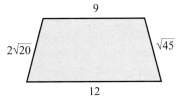

9 $2\sqrt{20}$ $\sqrt{45}$ 12

99. Crown molding, which is placed at the top of a wall, is to be installed around the perimeter of the room shown.

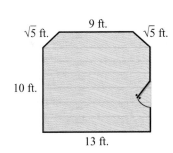

√5 ft. 9 ft. √5 ft. 10 ft. 13 ft.

a. Write an expression in simplest form for the perimeter of the room.

b. Use a calculator to approximate the perimeter, rounded to the nearest tenth.

c. If crown molding costs $1.89 per foot length, how much will the crown molding cost for this room?

100. A tabletop is to be fitted with veneer strips along the sides.

    a. Write an expression in simplest form for the perimeter of the tabletop.

    b. Use a calculator to approximate the perimeter, rounded to the nearest tenth.

$2\sqrt{2}$ in.   70 in. $2\sqrt{2}$ in.

46 in.   46 in.

$2\sqrt{2}$ in.   70 in. $2\sqrt{2}$ in.

REVIEW EXERCISES

1. For the equation $5y - 2x = 10$, find the slope and the y-intercept.

2. Graph: $y = -\dfrac{1}{3}x + 4$.

3. What is the conjugate of $2x - 5$?

4. Multiply: $(4x + 3)(4x - 3)$.

5. What factor could multiply $\sqrt{8}$ in order to equal $\sqrt{16}$?

6. What factor could multiply $\sqrt[3]{2}$ in order to equal $\sqrt[3]{8}$?

10.5 Rationalizing Numerators and Denominators of Radical Expressions

1 Rationalize denominators containing square roots.

2 Rationalize denominators containing higher-order roots.

3 Rationalize denominators that have a sum or difference with a square root term.

4 Rationalize numerators.

We are now ready to formalize the conditions for a radical expression that is in simplest form. A radical expression is in simplest form if:

1. All rational roots have been found.
2. There are no perfect n^{th} factors of radicands of the form $\sqrt[n]{a^n}$.
3. All possible products, quotients, sums, and differences have been found.
4. There are no radicals in the denominator of a fraction.

In this section, we explore how to simplify expressions that have a radical in the denominator of a fraction, as in $\dfrac{1}{\sqrt{2}}$.

OBJECTIVE 1 Rationalize denominators containing square roots. If the denominator of a fraction contains a radical, our goal will be to *rationalize* the denominator, which means to rewrite the expression so that it has a rational number in the denominator. In general, we multiply the fraction by a well-chosen 1 so that the radical is eliminated. For a square root, multiplying the root by itself eliminates the square root. Consider $\dfrac{1}{\sqrt{2}}$:

$$\frac{1}{\sqrt{2}} = \frac{1}{\sqrt{2}} \cdot \frac{\sqrt{2}}{\sqrt{2}} = \frac{\sqrt{2}}{\sqrt{4}} = \frac{\sqrt{2}}{2}$$

Note: We are not changing the value of $\dfrac{1}{\sqrt{2}}$ because we are multiplying it by 1 in the form of $\dfrac{\sqrt{2}}{\sqrt{2}}$.

PROCEDURE TO RATIONALIZE A DENOMINATOR WITH A SINGLE SQUARE ROOT,

1. Multiply both the numerator and denominator by the same square root as appears in the denominator.
2. Simplify.

EXAMPLE 1 Simplify. Assume variables represent nonnegative values.

a. $\dfrac{2}{\sqrt{5}}$

Solution $\dfrac{2}{\sqrt{5}} = \dfrac{2}{\sqrt{5}} \cdot \dfrac{\sqrt{5}}{\sqrt{5}}$ Multiply the numerator and denominator by $\sqrt{5}$.

$= \dfrac{2\sqrt{5}}{\sqrt{25}}$ Simplify.

$= \dfrac{2\sqrt{5}}{5}$

b. $\sqrt{\dfrac{5}{6}}$

Solution $\sqrt{\dfrac{5}{6}} = \dfrac{\sqrt{5}}{\sqrt{6}}$ Use the quotient rule of square roots to separate the numerator and denominator into two radicals.

$= \dfrac{\sqrt{5}}{\sqrt{6}} \cdot \dfrac{\sqrt{6}}{\sqrt{6}}$ Multiply the numerator and denominator by $\sqrt{6}$.

$= \dfrac{\sqrt{30}}{\sqrt{36}}$ Simplify.   **Warning:** Though it may be tempting, we cannot divide out the 6 and 30 because 30 is a radicand, whereas 6 is not. We never divide out factors common to a radicand and a number not under a radical.

$= \dfrac{\sqrt{30}}{6}$

c. $\dfrac{3}{\sqrt{2x}}$

Solution $\dfrac{3}{\sqrt{2x}} = \dfrac{3}{\sqrt{2x}} \cdot \dfrac{\sqrt{2x}}{\sqrt{2x}}$ Multiply numerator and denominator by $\sqrt{2x}$.

$= \dfrac{3\sqrt{2x}}{\sqrt{4x^2}}$ Simplify.

$= \dfrac{3\sqrt{2x}}{2x}$

ANSWERS

a. $\dfrac{\sqrt{7}}{7}$

b. $\dfrac{\sqrt{6}}{3}$

c. $\dfrac{3\sqrt{10x}}{10x}$

YOUR TURN Simplify. Assume variables represent nonnegative values.

a. $\dfrac{1}{\sqrt{7}}$ **b.** $\sqrt{\dfrac{2}{3}}$ **c.** $\dfrac{3}{\sqrt{10x}}$

OBJECTIVE 2 **Rationalize denominators containing higher-order roots.** If the denominator contains a higher-order root, such as a cube root, then we multiply appropriately to get a perfect cube radicand in the denominator so that we can eliminate the radical.

For example, $\dfrac{2}{\sqrt[3]{5}} = \dfrac{2}{\sqrt[3]{5}} \cdot \dfrac{\sqrt[3]{25}}{\sqrt[3]{25}} = \dfrac{2\sqrt[3]{25}}{\sqrt[3]{125}} = \dfrac{2\sqrt[3]{25}}{5}$. We summarize as follows.

PROCEDURE **Rationalizing a Denominator Containing a Higher-Order Radical**

To rationalize a denominator containing a single nth root, multiply the numerator and denominator by an expression that will make the radicand of the denominator a perfect nth power.

EXAMPLE 2   Rationalize the denominator. Assume variables represent non-negative values.

a. $\dfrac{3}{\sqrt[3]{2}}$

Solution $\dfrac{3}{\sqrt[3]{2}} = \dfrac{3}{\sqrt[3]{2}} \cdot \dfrac{\sqrt[3]{4}}{\sqrt[3]{4}}$ Since $\sqrt[3]{2} \cdot \sqrt[3]{4} = \sqrt[3]{8} = 2$, we multiply the fraction by $\dfrac{\sqrt[3]{4}}{\sqrt[3]{4}}$.

$\qquad\qquad = \dfrac{3\sqrt[3]{4}}{\sqrt[3]{8}}$ Simplify.

$\qquad\qquad = \dfrac{3\sqrt[3]{4}}{2}$

b. $\dfrac{\sqrt[3]{a}}{\sqrt[3]{b}}$

Solution $\dfrac{\sqrt[3]{a}}{\sqrt[3]{b}} = \dfrac{\sqrt[3]{a}}{\sqrt[3]{b}} \cdot \dfrac{\sqrt[3]{b^2}}{\sqrt[3]{b^2}}$ Since $\sqrt[3]{b} \cdot \sqrt[3]{b^2} = \sqrt[3]{b^3} = b$, we multiply the fraction by $\dfrac{\sqrt[3]{b^2}}{\sqrt[3]{b^2}}$.

$\qquad\qquad = \dfrac{\sqrt[3]{ab^2}}{\sqrt[3]{b^3}}$ Simplify.

$\qquad\qquad = \dfrac{\sqrt[3]{ab^2}}{b}$

c. $\sqrt[3]{\dfrac{5}{9a^2}}$

Solution $\sqrt[3]{\dfrac{5}{9a^2}} = \dfrac{\sqrt[3]{5}}{\sqrt[3]{9a^2}}$ Use the quotient rule to separate the numerator and denominator.

$\qquad\qquad = \dfrac{\sqrt[3]{5}}{\sqrt[3]{9a^2}} \cdot \dfrac{\sqrt[3]{3a}}{\sqrt[3]{3a}}$ Since $\sqrt[3]{9a^2} \cdot \sqrt[3]{3a} = \sqrt[3]{27a^3} = 3a$, multiply the fraction by $\dfrac{\sqrt[3]{3a}}{\sqrt[3]{3a}}$.

$\qquad\qquad = \dfrac{\sqrt[3]{15a}}{\sqrt[3]{27a^3}}$ Simplify.

$\qquad\qquad = \dfrac{\sqrt[3]{15a}}{3a}$

d. $\dfrac{5}{\sqrt[4]{3}}$

Solution $\dfrac{5}{\sqrt[4]{3}} = \dfrac{5}{\sqrt[4]{3}} \cdot \dfrac{\sqrt[4]{27}}{\sqrt[4]{27}}$ Since $\sqrt[4]{3} \cdot \sqrt[4]{27} = \sqrt[4]{81} = 3$, multiply the fraction by $\dfrac{\sqrt[4]{27}}{\sqrt[4]{27}}$.

$$= \dfrac{5\sqrt[4]{27}}{\sqrt[4]{81}} \qquad \text{Simplify.}$$

$$= \dfrac{5\sqrt[4]{27}}{3}$$

ANSWERS

a. $2\sqrt[3]{9}$ **b.** $\dfrac{\sqrt[3]{3x}}{x}$

c. $\dfrac{2\sqrt[3]{2y^2}}{y}$ **d.** $\dfrac{7\sqrt[4]{8}}{2}$

YOUR TURN   Rationalize the denominator of the following. Assume variables represent nonnegative values.

a. $\dfrac{6}{\sqrt[3]{3}}$ **b.** $\sqrt[3]{\dfrac{3}{x^2}}$ **c.** $\dfrac{4}{\sqrt[3]{4y}}$ **d.** $\dfrac{7}{\sqrt[4]{2}}$

OBJECTIVE 3   **Rationalize denominators that have a sum or difference with a square root term.** In Example 4 of Section 10.4, we saw that the product of conjugates containing square roots is an expression that does not contain any radicals. We can use this fact to rationalize a denominator that is a sum or difference containing a square root term.

Consider $\dfrac{5}{7 - \sqrt{3}}$. Multiplying the numerator and denominator by the conjugate of the denominator clears the radical from the denominator, thus, rationalizing it.

$$\dfrac{5}{7 - \sqrt{3}} = \dfrac{5}{7 - \sqrt{3}} \cdot \dfrac{7 + \sqrt{3}}{7 + \sqrt{3}} = \dfrac{5(7 + \sqrt{3})}{(7)^2 - (\sqrt{3})^2} = \dfrac{35 + 5\sqrt{3}}{49 - 3} = \dfrac{35 + 5\sqrt{3}}{46}$$

PROCEDURE   **RATIONALIZING A DENOMINATOR CONTAINING A SUM OR DIFFERENCE**

To rationalize a denominator containing a sum or difference with at least one square root term, multiply the numerator and denominator by the conjugate of the denominator.

EXAMPLE 3   Rationalize the denominator and simplify. Assume variables represent nonnegative values.

a. $\dfrac{9}{\sqrt{2} + 7}$

Solution $\dfrac{9}{\sqrt{2} + 7} = \dfrac{9}{\sqrt{2} + 7} \cdot \dfrac{\sqrt{2} - 7}{\sqrt{2} - 7}$ **Multiply the numerator and denominator by the conjugate of the denominator.**

$$= \dfrac{9(\sqrt{2} - 7)}{(\sqrt{2})^2 - (7)^2}$$

$$= \dfrac{9\sqrt{2} - 63}{2 - 49}$$

$$= \frac{9\sqrt{2} - 63}{-47}$$

We can simplify the negative denominator by factoring out -1 in the numerator and denominator.

$$= \frac{-1(63 - 9\sqrt{2})}{-1(47)}$$

After factoring out the -1, the signs of the terms change. Since 63 is now positive, we write it first.

$$= \frac{63 - 9\sqrt{2}}{47}$$

Divide out the common factor -1.

b. $\dfrac{2\sqrt{3}}{\sqrt{6} - \sqrt{2}}$

Solution $\dfrac{2\sqrt{3}}{\sqrt{6} - \sqrt{2}} = \dfrac{2\sqrt{3}}{\sqrt{6} - \sqrt{2}} \cdot \dfrac{\sqrt{6} + \sqrt{2}}{\sqrt{6} + \sqrt{2}}$

Multiply the numerator and denominator by the conjugate of the denominator.

$$= \frac{2\sqrt{3}(\sqrt{6} + \sqrt{2})}{(\sqrt{6})^2 - (\sqrt{2})^2}$$

$$= \frac{2\sqrt{18} + 2\sqrt{6}}{6 - 2}$$

Multiply in the numerator and evaluate the exponents in the denominator.

$$= \frac{2\sqrt{9 \cdot 2} + 2\sqrt{6}}{4}$$

Simplify $\sqrt{18}$ by factoring out a perfect square factor in 18.

$$= \frac{2 \cdot 3\sqrt{2} + 2\sqrt{6}}{4}$$

Simplify $\sqrt{9 \cdot 2}$ by finding the square root of 9.

Note: We cannot add $3\sqrt{2}$ and $\sqrt{6}$ because their radicands do not match.

$$= \frac{2(3\sqrt{2} + \sqrt{6})}{4}$$

Factor out the common 2 factor in the numerator.

$$= \frac{3\sqrt{2} + \sqrt{6}}{2}$$

Divide out the common factor 2.

c. $\dfrac{6}{\sqrt{x} - 5}$

Solution $\dfrac{6}{\sqrt{x} - 5} = \dfrac{6}{\sqrt{x} - 5} \cdot \dfrac{\sqrt{x} + 5}{\sqrt{x} + 5}$

Multiply the numerator and denominator by the conjugate of the denominator.

$$= \frac{6(\sqrt{x} + 5)}{(\sqrt{x})^2 - (5)^2}$$

$$= \frac{6\sqrt{x} + 30}{x - 25}$$

Multiply in the numerator and evaluate the exponents in the denominator.

ANSWERS

a. $9\sqrt{5} - 18$

b. $\dfrac{\sqrt{10} + \sqrt{6}}{2}$

c. $\dfrac{3\sqrt{x} - 12}{x - 16}$

YOUR TURN Rationalize the denominator and simplify. Assume variables represent nonnegative values.

a. $\dfrac{9}{\sqrt{5} + 2}$ **b.** $\dfrac{\sqrt{2}}{\sqrt{5} - \sqrt{3}}$ **c.** $\dfrac{3}{\sqrt{x} + 4}$

In later mathematics courses, you may need to rationalize the numerator. We use the same procedure that we use in rationalizing denominators.

EXAMPLE 4 Rationalize the numerator. Assume variables represent non-negative values.

a. $\dfrac{\sqrt{5x}}{4}$

Solution $\dfrac{\sqrt{5x}}{4} = \dfrac{\sqrt{5x}}{4} \cdot \dfrac{\sqrt{5x}}{\sqrt{5x}}$ Multiply the numerator and denominator by $\sqrt{5x}$.

$= \dfrac{\sqrt{25x^2}}{4\sqrt{5x}}$ Simplify.

$= \dfrac{5x}{4\sqrt{5x}}$

b. $\dfrac{3 + \sqrt{2x}}{4}$

Solution $\dfrac{3 + \sqrt{2x}}{4} = \dfrac{3 + \sqrt{2x}}{4} \cdot \dfrac{3 - \sqrt{2x}}{3 - \sqrt{2x}}$ Multiply the numerator and denominator by the conjugate of the numerator.

$= \dfrac{3^2 - (\sqrt{2x})^2}{4(3 - \sqrt{2x})}$ Simplify.

$= \dfrac{9 - 2x}{12 - 4\sqrt{2x}}$

ANSWERS

a. $\dfrac{3a}{7\sqrt{3a}}$

b. $\dfrac{25 - 3a}{10 + 2\sqrt{3a}}$

YOUR TURN Rationalize the numerators. Assume variables represent nonnegative values.

a. $\dfrac{\sqrt{3a}}{7}$

b. $\dfrac{5 - \sqrt{3a}}{2}$

10.5 EXERCISES

For Extra Help

InterAct Math

Addison-Wesley
Math Tutor Center

Video 16/DVT 5

MyMathLab

MathXL

1. Explain why each of the following expressions is not in simplest form.

   a. $\sqrt{\dfrac{3}{16}}$

   b. $\dfrac{5}{\sqrt{3}}$

2. Although $\dfrac{1}{\sqrt{3}}$ and $\dfrac{\sqrt{3}}{3}$ are equal, explain why $\dfrac{\sqrt{3}}{3}$ is considered simplest form.

3. Explain in your own words how to rationalize a denominator that is the square root of a number $\left(\text{like } \dfrac{2}{\sqrt{3}} \text{ or } \dfrac{\sqrt{5}}{\sqrt{7}}\right).$

4. Explain in your own words how to rationalize a denominator that is a sum or difference with a square root term $\left(\text{like } \dfrac{3}{5 + \sqrt{2}} \text{ or } \dfrac{2}{\sqrt{x} - \sqrt{y}}\right).$

For Exercises 5–28, rationalize the denominator. Assume variables represent non-negative values.

5. $\dfrac{1}{\sqrt{3}}$

6. $\dfrac{1}{\sqrt{5}}$

7. $\dfrac{2}{\sqrt{3}}$

8. $\dfrac{5}{\sqrt{12}}$

9. $\sqrt{\dfrac{36}{7}}$

10. $\sqrt{\dfrac{81}{5}}$

11. $\dfrac{\sqrt{7x^2}}{\sqrt{50}}$

12. $\dfrac{\sqrt{2x^2}}{\sqrt{3}}$

13. $\dfrac{\sqrt{8}}{\sqrt{56}}$

14. $\sqrt{\dfrac{2}{7}}$

15. $\dfrac{\sqrt{10}}{\sqrt{20}}$

16. $\sqrt{\dfrac{15}{2}}$

17. $\dfrac{5}{\sqrt{3a}}$

18. $\dfrac{11}{\sqrt{7b}}$

19. $\sqrt{\dfrac{3m}{11n}}$

20. $\sqrt{\dfrac{5r}{6s}}$

21. $\dfrac{10}{\sqrt{5x}}$

22. $\dfrac{20}{\sqrt{10a}}$

23. $\dfrac{\sqrt{6x}}{\sqrt{32x}}$

24. $\dfrac{\sqrt{10a}}{\sqrt{18a}}$

25. $\dfrac{3}{\sqrt{x^3}}$

26. $\dfrac{5}{\sqrt{b^5}}$

27. $\dfrac{8x^2}{\sqrt{2x}}$

28. $\dfrac{14a^3}{\sqrt{7a}}$

For Exercises 29 and 30, explain the mistake, then simplify correctly.

29. $\dfrac{\sqrt{3}}{\sqrt{2}} = \dfrac{\sqrt{3}}{\sqrt{2}} \cdot \dfrac{2}{2} = \dfrac{2\sqrt{3}}{2}$

30. $\sqrt{\dfrac{7}{3}} = \dfrac{\sqrt{7}}{\sqrt{3}} \cdot \dfrac{\sqrt{3}}{\sqrt{3}} = \dfrac{\sqrt{21}}{9}$

For Exercises 31–50, rationalize the denominators. Assume variables represent nonnegative values.

31. $\dfrac{5}{\sqrt[3]{3}}$

32. $\dfrac{7}{\sqrt[3]{5}}$

33. $\sqrt[3]{\dfrac{5}{2}}$

34. $\sqrt[3]{\dfrac{6}{5}}$

35. $\dfrac{6}{\sqrt[3]{4}}$

36. $\dfrac{9}{\sqrt[3]{9}}$

37. $\dfrac{m}{\sqrt[3]{n}}$

38. $\dfrac{p}{\sqrt[3]{q}}$

39. $\sqrt[3]{\dfrac{a}{b^2}}$

40. $\sqrt[3]{\dfrac{m}{n^2}}$

41. $\dfrac{4}{\sqrt[3]{2x}}$

42. $\dfrac{9}{\sqrt[3]{3a}}$

43. $\sqrt[3]{\dfrac{6}{25a^2}}$

44. $\sqrt[3]{\dfrac{5}{9b^2}}$

45. $\dfrac{5}{\sqrt[4]{2}}$

46. $\dfrac{7}{\sqrt[4]{3}}$

47. $\sqrt[4]{\dfrac{3}{x^2}}$

48. $\sqrt[4]{\dfrac{5}{y^3}}$

49. $\dfrac{9}{\sqrt[4]{3x^3}}$

50. $\dfrac{12}{\sqrt[4]{4x^2}}$

For Exercises 51–72, rationalize the denominator and simplify. Assume variables represent nonnegative values.

51. $\dfrac{\sqrt{3}}{\sqrt{2}+1}$

52. $\dfrac{3}{\sqrt{5}+2}$

53. $\dfrac{4}{2-\sqrt{3}}$

54. $\dfrac{\sqrt{2}}{6-\sqrt{7}}$

55. $\dfrac{5}{\sqrt{2}+\sqrt{3}}$

56. $\dfrac{7}{\sqrt{6}+\sqrt{2}}$

57. $\dfrac{4}{1-\sqrt{5}}$

58. $\dfrac{6}{2+\sqrt{3}}$

59. $\dfrac{\sqrt{3}}{\sqrt{3}-1}$

60. $\dfrac{\sqrt{5}}{1-\sqrt{5}}$

61. $\dfrac{2\sqrt{3}}{\sqrt{3}-4}$

62. $\dfrac{2\sqrt{5}}{1+\sqrt{2}}$

63. $\dfrac{4\sqrt{3}}{\sqrt{7}+\sqrt{2}}$

64. $\dfrac{8\sqrt{2}}{4\sqrt{2}-\sqrt{3}}$

65. $\dfrac{\sqrt{8}}{\sqrt{4}+\sqrt{3}}$

66. $\dfrac{5\sqrt{3}}{\sqrt{2}+4\sqrt{6}}$

67. $\dfrac{6\sqrt{y}}{\sqrt{y}+1}$

68. $\dfrac{\sqrt{x}}{\sqrt{x}+1}$

69. $\dfrac{3\sqrt{t}}{\sqrt{t}+2\sqrt{u}}$

70. $\dfrac{\sqrt{m}}{\sqrt{n}-\sqrt{m}}$

71. $\dfrac{\sqrt{2y}}{\sqrt{x}-\sqrt{6y}}$

72. $\dfrac{\sqrt{14h}}{\sqrt{2h}+\sqrt{k}}$

For Exercises 73–84, rationalize the numerator. Assume variables represent non-negative values.

73. $\dfrac{\sqrt{3}}{2}$

74. $\dfrac{\sqrt{5}}{8}$

75. $\dfrac{\sqrt{2x}}{5}$

76. $\dfrac{\sqrt{7y}}{3}$

77. $\dfrac{\sqrt{8n}}{6}$

78. $\dfrac{\sqrt{20t}}{8}$

79. $\dfrac{2 + \sqrt{3}}{5}$

80. $\dfrac{5 - \sqrt{2}}{6}$

81. $\dfrac{\sqrt{5x} - 6}{9}$

82. $\dfrac{\sqrt{2x} + 7}{3}$

83. $\dfrac{5\sqrt{n} + \sqrt{6n}}{2n}$

84. $\dfrac{4\sqrt{k} - \sqrt{10k}}{5k}$

85. Given $f(x) = \dfrac{5\sqrt{2}}{x}$, find each of the following. Express your answer in simplest form.

   a. $f(\sqrt{6})$ **b.** $f(\sqrt{10})$ **c.** $f(\sqrt{22})$

86. Given $g(x) = \dfrac{\sqrt{2}}{x - 1}$, find each of the following. Express your answer in simplest form.

   a. $g(\sqrt{5})$ **b.** $g(3\sqrt{2})$ **c.** $g(2\sqrt{6})$

 87. Graph $f(x) = \dfrac{1}{\sqrt{x}}$, then graph $g(x) = \dfrac{\sqrt{x}}{x}$.

   a. What do you notice about the two graphs? What does this indicate about the two functions?

   b. Simplify $f(x)$ by rationalizing the denominator. What do you notice?

88. Graph $f(x) = -\dfrac{1}{\sqrt{x}}$, then graph $g(x) = -\dfrac{\sqrt{x}}{x}$.

   a. What do you notice about the two graphs? What does this indicate about the two functions?

   b. Simplify $f(x)$ by rationalizing the denominator. What do you notice?

89. In Section 10.1, we used the formula $T = 2\pi\sqrt{\dfrac{L}{9.8}}$ to determine the period of a pendulum where T is the period in seconds and L is the length in meters.

   a. Rewrite the formula so that the denominator is rationalized.

   b. Rewrite the formula so that the numerator is rationalized.

90. The formula $t = \sqrt{\dfrac{h}{16}}$ can be used to find the time, t, in seconds for an object to fall a distance of h feet.

a. Rewrite the formula so that the denominator is rationalized.

b. Rewrite the formula so that the numerator is rationalized.

Of Interest

91. The formula $s = \sqrt{\dfrac{3V}{h}}$ can be used to find the length, s, of each side of the base of a pyramid having a square base, volume V in cubic feet, and height h in feet.

a. Rationalize the denominator in the formula.

b. The volume of the Great Pyramid at Giza in Egypt is approximately 83,068,742 cubic feet and its height is 449 feet. Find the length of each side of its base.

92. The formula $s = 2\sqrt{\dfrac{A}{6\sqrt{3}}}$ can be used to find the length, s, of each side of a regular hexagon having an area A.

a. Rationalize the denominator in the formula.

b. If A is 100 square meters, write an expression in simplest form for the side length.

c. Use a calculator to approximate the side length rounded s to three decimal places.

93. In AC circuits, voltage is often expressed as a *root-mean-square*, or *rms*, value. The formula for calculating the rms voltage, V_{rms}, given the maximum voltage, V_m, value is $V_{rms} = \dfrac{V_m}{\sqrt{2}}$.

a. Rationalize the denominator in the formula.

b. Given a maximum voltage of 163 V, write an expression for the rms voltage.

c. Use a calculator to approximate the rms voltage rounded to the nearest tenth.

94. The velocity, in meters per second, of a particle can be determined by the formula $v = \sqrt{\dfrac{2E}{m}}$, where E represents the kinetic energy, in joules, of the particle and m represents the mass, in kilograms, of the particle.

    a. Rationalize the denominator in the formula.

    b. A particle with a mass of 1×10^{-6} kilograms has 2.4×10^7 joules of kinetic energy. Write an expression for its velocity.

    c. Use a calculator to approximate the velocity rounded to the nearest tenth.

95. The resistance in a circuit is found to be $\dfrac{5\sqrt{2}}{3 + \sqrt{6}}\,\Omega$. Rationalize the denominator.

96. Two charged particles, q_1 and q_2, are separated by a distance of 8 centimeters. The values of the charges are $q_1 = 3 \times 10^{-6}$ coulombs and $q_2 = 1 \times 10^{-6}$ coulombs. The charged particles each exert an electrical field. At a point in between the two particles x centimeters away from q_1, the electric fields will cancel each other so that the value of the fields at the point x is 0.

$$\underset{\text{8 cm}}{\overset{\overset{x}{\longleftrightarrow}}{\underset{q_1 \,\bigcirc \hspace{4cm} \bigcirc\, q_2}{}}}$$

    a. Use the formula $x = \dfrac{l}{1 + \sqrt{\dfrac{q_2}{q_1}}}$ to find the distance from q_1 at which the electric field is canceled, where l is the distance separating the particles. Write the distance with a rationalized denominator.

    b. Use a calculator to approximate the distance rounded to the nearest tenth.

REVIEW EXERCISES

For Exercises 1 and 2, solve and check.

  1. $x^2 - 36 = 0$ **2.** $x^2 - 5x + 6 = 0$ **3.** Factor $x^2 - 6x + 9$.   **4.** Simplify $\pm\sqrt{28}$.

For Exercises 5 and 6, solve.

  5. $2x - 3 = 5$ **6.** $2x - 3 = -5$

10.6 Radical Equations and Problem Solving

> **OBJECTIVE** **1** Use the power rule to solve radical equations.

We now explore how to solve **radical equations**.

> **DEFINITION** **Radical equation:** An equation containing at least one radical expression whose radicand has a variable.

THE ALGEBRA PYRAMID

RADICAL EQUATIONS
Examples: $\sqrt{x} = 5$ or
$\sqrt{x+3} = 8$ or
$\sqrt{x+5} = \sqrt{x-7}$

RADICAL EXPRESSIONS
Examples: $\sqrt{50}$ or $\sqrt{x}$

CONSTANTS AND VARIABLES
Examples: $2, 5, 7, x, y$

Note: We move upward on the Algebra Pyramid, making equations out of the expressions we've studied.

> **OBJECTIVE** **1** **Use the power rule to solve radical equations.** To solve radical equations we will use a new principle of equality called the *power rule*.

> **RULE** **POWER RULE**
>
> If $a = b$, then $a^n = b^n$.

Isolated Radicals

First, we consider equations such as $\sqrt{x} = 9$ in which the radical is isolated. When we use the power rule, we raise both sides of the equation to the same power as the root index, then use the principle $(\sqrt[n]{x})^n = x$, which eliminates the radical, leaving us with its radicand.

EXAMPLE 1 Solve.

a. $\sqrt{x} = 9$

Solution $(\sqrt{x})^2 = (9)^2$ Since the root index is 2, we square both sides.

$$x = 81$$

Check $\sqrt{81} \stackrel{?}{=} 9$

$$9 = 9 \quad \text{True.}$$

b. $\sqrt[3]{y} = -2$

Solution $(\sqrt[3]{y})^3 = (-2)^3$ Since the root index is 3, we cube both sides.

$$y = -8$$

Check $\sqrt[3]{-8} \stackrel{?}{=} -2$

$$-2 = -2 \quad \text{True.}$$

In Section 7.6, we learned that some equations have *extraneous solutions*, which are apparent solutions that actually do not make the original equation true. The power rule can sometimes lead to extraneous solutions, so it is important to check solutions.

EXAMPLE 2 Solve.

$$\sqrt{x} = -9$$

Solution We can see there is no real-number solution for this equation because the square root of a real number cannot be a negative number. Applying the power rule, however, incorrectly suggests that there is a real-number solution.

$(\sqrt{x})^2 = (-9)^2$ Square both sides.   **Warning:** From the check, we will see that this result

$$x = 81$$ is *not* a solution.

Check   By checking 81 in the original equation we see that it is extraneous.

$$\sqrt{81} \stackrel{?}{=} -9$$

$$9 = -9 \quad \text{False. There is no real-number solution for this equation.}$$

Now consider some radical equations that are a little more complicated.

EXAMPLE 3 Solve.

$$\sqrt{x - 7} = 8$$

Solution $(\sqrt{x - 7})^2 = (8)^2$ Square both sides.

$$x - 7 = 64$$

$$x = 71 \quad \text{Add 7 to both sides.}$$

Check $\sqrt{71 - 7} \stackrel{?}{=} 8$

$$\sqrt{64} \stackrel{?}{=} 8$$

$$8 = 8 \quad \text{True.}$$

YOUR TURN   Solve.

a. $\sqrt{x - 5} = 6$ **b.** $\sqrt[3]{x + 2} = 3$ **c.** $\sqrt{x + 3} = -7$

As we will see in Example 4, the power rule can be used to solve equations with radicals on both sides of the equal sign.

EXAMPLE 4   Solve.

a. $\sqrt{6x - 1} = \sqrt{x + 2}$

Solution $(\sqrt{6x - 1})^2 = (\sqrt{x + 2})^2$ Square both sides.

$$6x - 1 = x + 2$$ Subtract x from both sides and add 1 to both sides.

$$5x = 3$$ Divide both sides by 5.

$$x = \frac{3}{5}$$

Check $\sqrt{6\left(\dfrac{3}{5}\right) - 1} \stackrel{?}{=} \sqrt{\dfrac{3}{5} + 2}$

$$\sqrt{\frac{18}{5} - \frac{5}{5}} \stackrel{?}{=} \sqrt{\frac{3}{5} + \frac{10}{5}}$$

$$\sqrt{\frac{13}{5}} = \sqrt{\frac{13}{5}}$$ True.

b. $\sqrt[3]{5x - 2} = \sqrt[3]{3x + 2}$

Solution $(\sqrt[3]{5x - 2})^3 = (\sqrt[3]{3x + 2})^3$ Cube both sides.

$$5x - 2 = 3x + 2$$

$$2x = 4$$ Subtract $3x$ from both sides and add 2 to both sides.

$$x = 2$$ Divide both sides by 2.

Check $\sqrt[3]{5(2) - 2} \stackrel{?}{=} \sqrt[3]{3(2) + 2}$

$$\sqrt[3]{10 - 2} \stackrel{?}{=} \sqrt[3]{6 + 2}$$

$$\sqrt[3]{8} \stackrel{?}{=} \sqrt[3]{8}$$

$$2 = 2$$ True.

YOUR TURN   Solve.

a. $\sqrt{8x + 5} = \sqrt{2x + 7}$ **b.** $\sqrt[3]{6x + 9} = \sqrt[3]{10x - 3}$

Radical equations may have multiple solutions if, after using the power rule, we are left with a quadratic form.

EXAMPLE 5 Solve.

$$x + 2 = \sqrt{5x + 16}$$

Solution $(x + 2)^2 = (\sqrt{5x + 16})^2$ Square both sides.

$$x^2 + 4x + 4 = 5x + 16$$ Use FOIL on the left-hand side.

$$x^2 - x - 12 = 0$$ Since the equation is quadratic, we set it equal to 0 by subtracting $5x$ and 16 from both sides.

$$(x + 3)(x - 4) = 0$$ Factor.

$x + 3 = 0$ or $x - 4 = 0$ Use the zero factor theorem.

   $x = -3$ $x = 4$

Checks $-3 + 2 \stackrel{?}{=} \sqrt{5(-3) + 16}$ $4 + 2 \stackrel{?}{=} \sqrt{5(4) + 16}$

              $-1 \stackrel{?}{=} \sqrt{-15 + 16}$ $6 \stackrel{?}{=} \sqrt{20 + 16}$

                    $-1 \stackrel{?}{=} \sqrt{1}$ $6 \stackrel{?}{=} \sqrt{36}$

                       $-1 = 1$ False. $6 = 6$ True.

Because -3 does not check, it is an extraneous solution. The only solution is 4.

ANSWER

2 and 1

YOUR TURN   Solve.

$$\sqrt{9x + 7} = x + 3$$

Radicals Not Isolated

Now we consider radical equations in which the radical term is not isolated. In such equations, we must first isolate the radical term.

EXAMPLE 6   Solve.

a. $\sqrt{x + 1} - 2x = x + 1$

Solution $\sqrt{x + 1} = 3x + 1$ Add $2x$ to both sides to isolate the radical term.

$$(\sqrt{x + 1})^2 = (3x + 1)^2$$ Square both sides.

$$x + 1 = 9x^2 + 6x + 1$$ Use FOIL on the right-hand side.

$$0 = 9x^2 + 5x$$ Since the equation is quadratic, we set it equal to 0 by subtracting x and 1 from both sides.

$$0 = x(9x + 5)$$ Factor.

$x = 0$ or $9x + 5 = 0$ Use the zero factor theorem.

                     $9x = -5$

                       $x = -\dfrac{5}{9}$

Checks $\sqrt{0 + 1} - 2(0) \overset{?}{=} 0 + 1$ $\qquad$ $\sqrt{-\dfrac{5}{9} + 1} - 2\left(-\dfrac{5}{9}\right) \overset{?}{=} -\dfrac{5}{9} + 1$

$\qquad\qquad\qquad\quad \sqrt{1} - 0 \overset{?}{=} 1$ $\qquad\qquad\quad \sqrt{-\dfrac{5}{9} + \dfrac{9}{9}} - 2\left(-\dfrac{5}{9}\right) \overset{?}{=} -\dfrac{5}{9} + \dfrac{9}{9}$

$\qquad\qquad\qquad\qquad\quad 1 = 1 \qquad$ **True.** $\qquad\qquad\qquad\quad \sqrt{\dfrac{4}{9}} + \dfrac{10}{9} \overset{?}{=} \dfrac{4}{9}$

$\qquad\qquad\qquad\qquad\qquad\qquad\qquad\qquad\qquad\qquad\qquad \dfrac{2}{3} + \dfrac{10}{9} \overset{?}{=} \dfrac{4}{9}$

$\qquad\qquad\qquad\qquad\qquad\qquad\qquad\qquad\qquad\qquad\qquad \dfrac{6}{9} + \dfrac{10}{9} \overset{?}{=} \dfrac{4}{9}$

> **Note:** This false equation indicates that $-\dfrac{5}{9}$ is an extraneous solution.

$\qquad\qquad\qquad\qquad\qquad\qquad\qquad\qquad\qquad\qquad\qquad \dfrac{16}{9} = \dfrac{4}{9} \qquad$ **False.**

The solution is 0.

b. $\sqrt[4]{3x + 4} + 5 = 7$

Solution $\sqrt[4]{3x + 4} = 2 \qquad$ **Subtract 5 from both sides to isolate the radical term.**

$\qquad\quad (\sqrt[4]{3x + 4})^4 = 2^4 \qquad$ **Raise both sides to the fourth power.**

$\qquad\qquad\qquad 3x + 4 = 16 \qquad$ **Simplify both sides.**

$\qquad\qquad\qquad\quad 3x = 12 \qquad$ **Subtract 4 from both sides.**

$\qquad\qquad\qquad\qquad x = 4 \qquad$ **Divide both sides by 3.**

Check $\sqrt[4]{3(4) + 4} + 5 \overset{?}{=} 7$

$\qquad\qquad \sqrt[4]{12 + 4} + 5 \overset{?}{=} 7$

$\qquad\qquad\quad \sqrt[4]{16} + 5 \overset{?}{=} 7$

$\qquad\qquad\qquad 2 + 5 \overset{?}{=} 7$

$\qquad\qquad\qquad\qquad 7 = 7 \qquad$ **True. The solution is 4.**

YOUR TURN **Solve.**

a. $\sqrt{5x^2 + 6x - 7} + 3x = 5x + 1$ $\qquad\qquad$ **b.** $\sqrt[4]{3x + 6} - 7 = -4$

Using the Power Rule Twice

Some equations may require that we use the power rule twice in order to eliminate all radicals.

EXAMPLE 7 **Solve.**

$$\sqrt{x + 21} = \sqrt{x} + 3$$

Solution $(\sqrt{x + 21})^2 = (\sqrt{x} + 3)^2 \qquad$ Since one of the radicals is isolated, we square both sides.

$\qquad\qquad\quad x + 21 = (\sqrt{x} + 3)(\sqrt{x} + 3)$

$\qquad\qquad\quad x + 21 = x + 3\sqrt{x} + 3\sqrt{x} + 9 \qquad$ Use FOIL on the right-hand side.

$\qquad\qquad\quad x + 21 = x + 6\sqrt{x} + 9 \qquad$ Combine like terms.

$\qquad\qquad\qquad\qquad\qquad\qquad\qquad\qquad\qquad\qquad$ Subtract x and 9 from both sides

$\qquad\qquad\qquad 12 = 6\sqrt{x} \qquad$ to isolate the remaining radical expression.

$$2 = \sqrt{x}$$
$$(2)^2 = (\sqrt{x})^2$$
$$4 = x$$

Divide both sides by 6.

Square both sides.

Check $\sqrt{4 + 21} \stackrel{?}{=} \sqrt{4} + 3$

$\sqrt{25} \stackrel{?}{=} 2 + 3$

$5 = 5$ True. The solution is 4.

Our examples suggest the following procedure.

PROCEDURE   SOLVING RADICAL EQUATIONS

To solve a radical equation:

1. Isolate the radical. (If there is more than one radical term, then isolate one of the radical terms.)
2. Raise both sides of the equation to the same power as the root index.
3. If all radicals have been eliminated, then solve. If a radical term remains, then isolate that radical term and raise both sides to the same power as its root index.
4. Check each solution. Any apparent solution that does not check is an extraneous solution.

YOUR TURN   Solve.

ANSWER

0 and 4

$$\sqrt{2x + 1} = \sqrt{x} + 1$$

10.6 EXERCISES

For Extra Help

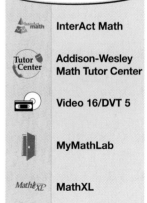

InterAct Math

Tutor Center — Addison-Wesley Math Tutor Center

Video 16/DVT 5

MyMathLab

MathXL — MathXL

1. Explain why we must check all potential solutions to radical equations.

2. What is an extraneous solution?

3. Explain why there is no real-number solution for the radical equation $\sqrt{x} = -6$.

4. Show why $(\sqrt{a})^2 = a$, assuming $a \geq 0$.

5. Given the radical equation $\sqrt{x + 2} + 3x = 4x - 1$, what would be the first step in solving the equation? Why?

6. Give an example of a radical equation that would require you to use the power rule twice in solving the equation. Explain why the principle would have to be used twice.

For Exercises 7–28, solve.

7. $\sqrt{x} = 2$ **8.** $\sqrt{y} = 5$ **9.** $\sqrt{k} = -4$ **10.** $\sqrt{x} = -1$

11. $\sqrt[3]{y} = 3$ **12.** $\sqrt[3]{m} = 4$ **13.** $\sqrt[3]{z} = -2$ **14.** $\sqrt[3]{p} = -5$

15. $\sqrt{n - 1} = 4$ **16.** $\sqrt{x + 3} = 7$ **17.** $\sqrt{t - 7} = 2$ **18.** $\sqrt{m + 8} = 1$

19. $\sqrt{3x - 2} = 4$ **20.** $\sqrt{2x + 5} = 3$ **21.** $\sqrt{2x + 17} = 4$ **22.** $\sqrt{3y - 2} = 5$

23. $\sqrt{2n - 8} = -3$ **24.** $\sqrt{5x - 1} = -6$ **25.** $\sqrt[3]{x - 3} = 2$ **26.** $\sqrt[3]{k + 2} = 4$

27. $\sqrt[3]{3y - 2} = -2$ **28.** $\sqrt[3]{2x + 4} = -3$

For Exercises 29–38 solve. First isolate the radical term.

29. $\sqrt{u - 3} - 10 = 1$ **30.** $\sqrt{y + 1} - 4 = 2$

31. $\sqrt{y - 6} + 2 = 9$ **32.** $\sqrt{r - 5} + 6 = 10$

33. $\sqrt{x - 5} - 2 = 3$ **34.** $\sqrt{x - 3} - 2 = 0$

35. $\sqrt[3]{n + 3} - 2 = -4$ **36.** $\sqrt[3]{x - 4} + 2 = 3$

37. $\sqrt[4]{x - 2} - 2 = -4$ **38.** $\sqrt[4]{m + 3} + 2 = 5$

For Exercises 39–64, solve.

39. $\sqrt{2 - 3x} = \sqrt{5}$ **40.** $\sqrt{m + 2} = \sqrt{2m - 3}$ **41.** $\sqrt{4x - 5} = \sqrt{6x + 5}$

42. $\sqrt{x + 1} = \sqrt{2x - 3}$ **43.** $\sqrt[3]{2r + 2} = \sqrt[3]{3r - 1}$ **44.** $\sqrt[3]{3h - 4} = \sqrt[3]{h + 4}$

45. $\sqrt[4]{4x + 4} = \sqrt[4]{5x + 1}$ **46.** $\sqrt[4]{2x + 4} = \sqrt[4]{3x - 2}$ **47.** $\sqrt{k + 4} = k - 8$

48. $\sqrt{5n - 1} = 4 - 2n$ **49.** $y - 1 = \sqrt{2y - 2}$ **50.** $3 + x = \sqrt{7 + 3x}$

51. $\sqrt{3x + 10} - 4 = x$ **52.** $\sqrt{4y + 1} + 5 = y$ **53.** $\sqrt{10n + 4} - 3n = n + 1$

54. $\sqrt{3 - 13h} + 2h = 5h - 1$ **55.** $\sqrt[3]{5x + 2} + 2 = 5$ **56.** $\sqrt[3]{4x - 1} - 4 = -1$

57. $\sqrt[3]{n^2 - 2n + 5} = 2$ **58.** $\sqrt[3]{y^2 + y + 7} = 3$ **59.** $1 + \sqrt{x} = \sqrt{2x + 1}$

60. $\sqrt{t + 2} - 1 = \sqrt{t}$ **61.** $\sqrt{3x + 1} + \sqrt{3x} = 2$ **62.** $\sqrt{x + 5} - 1 = \sqrt{x}$

63. $\sqrt{x + 8} - 2 = \sqrt{x}$ **64.** $\sqrt{5x - 1} - 1 = \sqrt{x + 2}$

For Exercises 65–68, explain the mistake, then solve correctly.

65. $\sqrt{x} = -9$
$(\sqrt{x})^2 = (-9)^2$
$x = 81$

66. $\sqrt{x} = -2$
$(\sqrt{x})^2 = (-2)^2$
$x = 4$

67. $\sqrt{x+3} = x - 3$
$x + 3 = x^2 - 9$
$0 = x^2 - x - 12$
$0 = (x+3)(x-4)$
$x = -3, 4$

68. $\sqrt{x+3} = 4$
$(\sqrt{x+3})^2 = 4^2$
$x + 9 = 16$
$x + 9 - 9 = 16 - 9$
$x = 7$

For Exercises 69–72, given the period of a pendulum, find the length of the pendulum. Use the formula $T = 2\pi\sqrt{\dfrac{L}{9.8}}$, where T is the period in seconds and L is the length in meters.

69. 2π seconds

70. 6π seconds

71. π seconds

72. $\dfrac{\pi}{3}$ seconds

For Exercises 73–76, find the distance an object has fallen. Use the formula $t = \sqrt{\dfrac{h}{16}}$, where t is the time in seconds and h is the distance that the object has fallen in feet.

73. 0.3 seconds

74. 0.5 seconds

75. 3 seconds

76. $\dfrac{1}{4}$ seconds

For Exercises 77–80, find the skid distance after braking hard at a given speed. Use the formula $S = \dfrac{7}{2}\sqrt{2D}$, where D represents the skid distance in feet on asphalt and S represents the speed of the car in miles per hour. (Source: Harris Technical Services, Traffic Accident Reconstructionists.)

77. 30 mph

78. 60 mph

79. 45 mph

80. 75 mph

For Exercises 81–84, use the following. Two forces, F_1 and F_2, acting on an object at a 90° angle will pull the object with a resultant force, R, at an angle in between F_1 and F_2 (see the figure). The value of the resultant force can be found by the formula $R = \sqrt{F_1^2 + F_2^2}$.

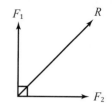

81. Find F_1 if $R = 5$ N and $F_2 = 3$ N.

82. Find F_1 if $R = 10$ N and $F_2 = 8$ N.

83. Find F_2 if $R = 3\sqrt{5}$ N and $F_1 = 3$ N.

84. Find F_2 if $R = 2\sqrt{13}$ N and $F_1 = 6$ N.

85. a. Complete the table of ordered pairs for the equation $y = \sqrt{x}$.

x	0	1	4	9	16	?
y	0	1	2	?	?	5

b. Graph the ordered pairs in the coordinate plane, then connect the points to make a curve.

c. Does the graph extend below or to the left of the origin? Explain.

d. Does the curve represent a function? Explain.

86. a. Complete the table of ordered pairs for the equation $y = \sqrt{x} + 2$.

x	0	1	4	9	16	?
y	2	3	4	?	?	7

b. Graph the ordered pairs in the coordinate plane, then connect the points to make a curve.

c. Does the graph extend below or to the left of the origin? Explain.

d. Does the curve represent a function? Explain.

87. Use a graphing calculator to graph $y = \sqrt{x}$, $y = 2\sqrt{x}$, and $y = 3\sqrt{x}$. Based on your observations, as you increase the size of the coefficient, what happens to the graph?

88. Use a graphing calculator to compare the graph of $y = \sqrt{x}$ to the graph of $y = -\sqrt{x}$. What effect does the negative sign have on the graph?

89. Use a graphing calculator to graph $y = \sqrt{x}$, $y = \sqrt{x} + 2$, and $y = \sqrt{x} - 2$. Based on your observations, what effect does adding or subtracting a constant have on the graph of $y = \sqrt{x}$?

90. Based on your conclusions from Exercises 87–89, without graphing, describe the graph of $y = -2\sqrt{x} + 3$. Use a graphing calculator to graph the equation to confirm your description.

Collaborative Exercises

Building Time

You are constructing a grandfather clock that operates using a pendulum. The period of the pendulum is the time required to complete one full swing. The formula giving the relationship between the length, L (in meters), and the period, T (in seconds), is $T = 2\pi\sqrt{\dfrac{L}{9.8}}$.

1. Find the length of the pendulum if the period is 1 second.

2. Suppose that you want the pendulum to complete one period in 2 seconds. Would you need to increase or decrease the length of the pendulum found in part a? Use the same formula to determine the length of the pendulum so that the period is 2 seconds.

3. In general, what can you conclude about the required length of the pendulum as the period increases?

REVIEW EXERCISES

For Exercises 1–3, evaluate.

1. 3^4

2. $(-0.2)^3$

3. $\left(\dfrac{2}{5}\right)^{-4}$

For Exercises 4–6, use the rules of exponents to simplify.

4. $(x^3)(x^5)$

5. $\dfrac{y^7}{y^3}$

6. $(n^4)^6$

10.7 Complex Numbers

OBJECTIVES

1. Write imaginary numbers using *i*.
2. Perform arithmetic operations with complex numbers.
3. Raise *i* to powers.

We have said that the square root of a negative number is not a real number. In this section we will develop another number system, called the *imaginary number system*, in which square roots of negative numbers will be expressed using a notation involving the letter *i*.

OBJECTIVE **1** **Write imaginary numbers using *i*.** Using the product rule of square roots, any square root of a negative number can be rewritten as a product of a real number and an **imaginary unit**, which we express as *i*.

DEFINITION | *Imaginary unit:* The number represented by *i*, where $i = \sqrt{-1}$ and $i^2 = -1$.

A number that can be expressed as a product of a real number and the imaginary unit is called an **imaginary number**.

DEFINITION | *Imaginary number:* A number that can be expressed in the form *bi*, where *b* is a real number and *i* is the imaginary unit.

Of Interest

When the idea of finding square roots of negative numbers was first introduced, members of the established mathematics community said this type of number existed, but only in the imagination of those finding them. From then on they've been called "imaginary numbers."

Note: In a product with the imaginary unit, it is customary to write integer factors first, then *i*, then square root factors.

EXAMPLE **1** Write each imaginary number as a product of a real number and *i*.

a. $\sqrt{-9}$

Solution $\sqrt{-9} = \sqrt{-1 \cdot 9}$ Factor out −1 in the radicand.

$\quad\quad\quad = \sqrt{-1} \cdot \sqrt{9}$ Use the product rule of square roots.

$\quad\quad\quad = i \cdot 3$

$\quad\quad\quad = 3i$

b. $\sqrt{-24}$

Solution $\sqrt{-24} = \sqrt{-1 \cdot 24}$ Factor out −1 in the radicand.

$\quad\quad\quad = \sqrt{-1} \cdot \sqrt{24}$ Use the product rule of square roots.

$\quad\quad\quad = i\sqrt{4 \cdot 6}$ Use the product rule again to simplify further.

$\quad\quad\quad = 2i\sqrt{6}$

Example 1 suggests the following procedure.

ANSWERS

a. $6i$
b. $i\sqrt{22}$
c. $5i\sqrt{2}$

YOUR TURN Write each imaginary number as a product of a real number and i.

a. $\sqrt{-36}$

b. $\sqrt{-22}$

c. $\sqrt{-50}$

OBJECTIVE **2** **Perform arithmetic operations with complex numbers.** We now have two distinct sets of numbers, the set of real numbers and the set of imaginary numbers. There is yet another set of numbers, called the set of **complex numbers**, that contains both the real and the imaginary numbers.

DEFINITION

Complex number: A number that can be expressed in the form $a + bi$, where a and b are real numbers and i is the imaginary unit.

When written in the form $a + bi$, a complex number is said to be in *standard form.* Following are some examples of complex numbers written in standard form:

$$2 + 3i \qquad\qquad 5 - 9i \qquad\qquad -7.9 - 3i\sqrt{5}$$

Note that if $a = 0$, then the complex number is purely an imaginary number, such as:

$$-7i \qquad 5i\sqrt{2}$$

If $b = 0$, then the complex number is a real number.

The following Venn Diagram shows how the set of complex numbers contains both the real numbers and imaginary numbers.

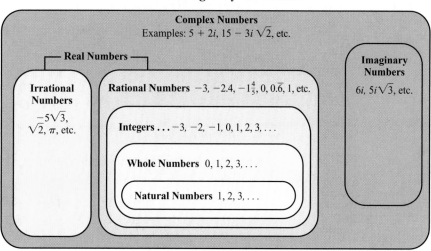

We can perform arithmetic operations with complex numbers. In general, we treat the complex numbers just like polynomials where i is like a variable.

Adding and Subtracting Complex Numbers

EXAMPLE 2 Add or subtract.

a. $(-9 + 2i) + (4 - 15i)$

Solution We add complex numbers just like we add polynomials by combining like terms.
$$(-9 + 2i) + (4 - 15i) = -5 - 13i$$

b. $(-1 + 8i) - (6 - 7i)$

Solution We subtract complex numbers just like we subtract polynomials by writing an equivalent addition and changing the signs in the second complex number.
$$(-1 + 8i) - (6 - 7i) = (-1 + 8i) + (-6 + 7i)$$
$$= -7 + 15i$$

Multiplying Complex Numbers

We multiply complex numbers in the same way that we multiply monomials and binomials. However, we must be careful when simplifying because these products may contain i^2, which is equal to -1.

EXAMPLE 3 Multiply.

a. $(6i)(-7i)$

Solution Multiply in the same way that we multiply monomials.
$$\begin{aligned} (6i)(-7i) &= -42i^2 \\ &= -42(-1) \quad \text{Replace } i^2 \text{ with } -1. \\ &= 42 \end{aligned}$$

b. $(4i)(9 - 5i)$

Solution Multiply in the same way that we multiply a binomial by a monomial.
$$\begin{aligned} (4i)(9 - 5i) &= 36i - 20i^2 & \text{Distribute.} \\ &= 36i - 20(-1) & \text{Replace } i^2 \text{ with } -1. \\ &= 36i + 20 \\ &= 20 + 36i & \text{Write in standard form.} \end{aligned}$$

c. $(8 - 3i)(2 + i)$

Solution Multiply in the same way that we multiply binomials.
$$\begin{aligned} (8 - 3i)(2 + i) &= 16 + 8i - 6i - 3i^2 & \text{Use FOIL.} \\ &= 16 + 2i - 3(-1) & \text{Combine like terms and replace } i^2 \text{ with } -1. \\ &= 16 + 2i + 3 & \text{Simplify.} \\ &= 19 + 2i & \text{Write in standard form.} \end{aligned}$$

d. $(3 - i)(3 + i)$

Solution Note that these complex numbers are conjugates.

$$(3 - i)(3 + i) = 9 + 3i - 3i - i^2 \quad \text{Use FOIL.}$$
$$= 9 - (\mathbf{-1}) \quad \text{Combine like terms and replace } i^2 \text{ with } -1.$$
$$= 9 + 1$$
$$= 10 \quad \boxed{\text{Note: The product of these two complex numbers is a real number.}}$$

The complex numbers that we multiplied in Example 3d are called **complex conjugates**.

> **DEFINITION**
>
> ***Complex conjugates:*** The complex conjugate of a complex number $a + bi$ is $a - bi$.

Some other examples of complex conjugates follow:

$$5 + 6i \quad \text{and} \quad 5 - 6i$$
$$9 - 7i \quad \text{and} \quad 9 + 7i$$

Example 3d also illustrates the fact that the product of complex conjugates will always be a real number.

ANSWERS

a. -24 **b.** $-3 - 6i$
c. $83 - i$ **d.** 73

YOUR TURN   **Multiply.**

a. $(-4i)(-6i)$ **b.** $(-3i)(2 - i)$ **c.** $(5 - 9i)(4 + 7i)$ **d.** $(8 + 3i)(8 - 3i)$

Dividing Complex Numbers

We have established that the product of complex conjugates is a real number. We can use that fact when the divisor is a complex number. The process is similar to rationalizing denominators.

EXAMPLE 4   Write in standard form.

a. $\dfrac{6 + 5i}{7 - 2i}$

CONNECTION

Recall that we rationalize denominators in order to clear undesired square root expressions from a denominator. The imaginary unit i represents a square root expression, $\sqrt{-1}$, which is why we rationalize denominators that contain i.

Solution $\dfrac{6 + 5i}{7 - 2i} = \dfrac{6 + 5i}{7 - 2i} \cdot \dfrac{7 + 2i}{7 + 2i}$ Multiply numerator and denominator by the complex conjugate of the denominator, which is $7 + 2i$.

$$= \frac{42 + 12i + 35i + 10i^2}{49 - 4i^2}$$

$$= \frac{42 + 47i + 10(-1)}{49 - 4(-1)} \quad \text{Simplify.}$$

$$= \frac{42 + 47i - 10}{49 + 4}$$

$$= \frac{32 + 47i}{53}$$

$$= \frac{32}{53} + \frac{47}{53}i \quad \text{Write in standard form.}$$

b. $\dfrac{9}{2i}$

Solution Think of $2i$ as $0 + 2i$ so the conjugate is $0 - 2i$ which is $-2i$.

$$\frac{9}{2i} = \frac{9}{2i} \cdot \frac{-2i}{-2i} \qquad \text{Multiply numerator and denominator by } -2i.$$

$$= \frac{-18i}{-4i^2}$$

$$= \frac{-18i}{-4(-1)} \qquad \text{Replace } i^2 \text{ with } -1.$$

$$= \frac{-18i}{4}$$

$$= -\frac{9}{2}i$$

Note: Remember the goal is to eliminate i from the denominator. Multiplying by $\dfrac{i}{i}$ would have worked, too.

$$\frac{9}{2i} = \frac{9}{2i} \cdot \frac{i}{i} = \frac{9i}{2i^2} = \frac{9i}{2(-1)} = -\frac{9}{2}i$$

YOUR TURN Write in standard form.

a. $\dfrac{4 + i}{5 - 2i}$

b. $\dfrac{8}{3i}$

OBJECTIVE 3 **Raise i to powers.** We have defined i as $\sqrt{-1}$ and seen that $i^2 = -1$. Raising i to other powers leads to an interesting pattern.

$$i^1 = i \qquad\qquad i^5 = i^4 \cdot i = 1 \cdot i = i$$
$$i^2 = -1 \qquad\qquad i^6 = i^4 \cdot i^2 = 1 \cdot (-1) = -1$$
$$i^3 = i^2 \cdot i = -1 \cdot i = -i \qquad\qquad i^7 = i^4 \cdot i^3 = 1 \cdot (-i) = -i$$
$$i^4 = (i^2)^2 = (-1)^2 = 1 \qquad\qquad i^8 = (i^4)^2 = 1^2 = 1$$

Note: If i's exponent is an even number not divisible by 4, then the result is -1. If i's exponent is divisible by 4, then the result is 1. If the exponent is an odd number that precedes a number divisible by 4, then the result is $-i$.

If we continue the pattern, we get:

$$i^1 = i \qquad\qquad i^5 = i \qquad\qquad i^9 = i$$
$$i^2 = -1 \qquad\qquad i^6 = -1 \qquad\qquad i^{10} = -1$$
$$i^3 = -i \qquad\qquad i^7 = -i \qquad\qquad i^{11} = -i$$
$$i^4 = 1 \qquad\qquad i^8 = 1 \qquad\qquad i^{12} = 1$$

Notice that i to any integer power is i, -1, $-i$, or 1. This pattern allows us to find i to any integer power. We will use the fact that since $i^4 = 1$, then $(i^4)^n = 1$ for any integer value of n.

EXAMPLE 5 Simplify.

a. i^{25}

Solution $i^{25} = i^{24} \cdot i$ Write i^{25} as $i^{24} \cdot i$ since 24 is the largest multiple of 4 that is smaller than 25.

$$= (i^4)^6 \cdot i \qquad \text{Write } i^{24} \text{ as } (i^4)^6 \text{ since } i^4 = 1.$$
$$= 1^6 \cdot i \qquad \text{Replace } i^4 \text{ with 1.}$$
$$= 1 \cdot i$$
$$= i$$

b. i^{-14}

Solution $i^{-14} = \dfrac{1}{i^{14}}$ Write i^{-14} with a positive exponent.

$\qquad\qquad = \dfrac{1}{i^{12} \cdot i^2}$ Write i^{14} as $i^{12} \cdot i^2$ since 12 is the largest multiple of 4 that is smaller than 14.

$\qquad\qquad = \dfrac{1}{(i^4)^3 \cdot (-1)}$ Write i^{12} as $(i^4)^3$ and replace i^2 with -1.

$\qquad\qquad = \dfrac{1}{1(-1)}$ $(i^4)^3 = 1^3 = 1$.

$\qquad\qquad = -1$

ANSWERS

a. $-i$ **b.** -1

YOUR TURN Simplify.

a. i^{43} **b.** i^{-18}

10.7 EXERCISES

For Extra Help

InterAct Math

Addison-Wesley Math Tutor Center

Video 16/DVT 5

MyMathLab

MathXL

1. As an imaginary number, what does i represent?

2. Is every real number a complex number? Explain.

3. Is every complex number an imaginary number? Explain.

4. Explain how to add complex numbers.

5. Explain how to subtract complex numbers.

6. Is the expression $\dfrac{5 - 4i}{3}$ in standard form for a complex number? Explain.

For Exercises 7–22, write the imaginary number using i.

7. $\sqrt{-16}$ 8. $\sqrt{-49}$ 9. $\sqrt{-20}$ 10. $\sqrt{-24}$

11. $\sqrt{-27}$ 12. $\sqrt{-75}$ 13. $\sqrt{-125}$ 14. $\sqrt{-12}$

15. $\sqrt{-63}$ 16. $\sqrt{-8}$ 17. $\sqrt{-45}$ 18. $\sqrt{-80}$

19. $\sqrt{-147}$ 20. $\sqrt{-3}$ 21. $\sqrt{-2}$ 22. $\sqrt{-360}$

For Exercises 23–38, add or subtract.

23. $(4 + 3i) + (-3 + 2i)$

24. $(2 - 3i) + (4 - 5i)$

25. $(3 + 2i) + (5 - 6i)$

26. $(7 + i) + (5 - 2i)$

27. $(-4 + 6i) - (2 + i)$

28. $(5 + i) - (-2 - 2i)$

29. $(10 - 3i) - (-5i)$

30. $(2i) - (4 + 3i)$

31. $(2 + 3i) + (-15 - 5i)$

32. $(6 - i) - (-4 + 2i)$

33. $(-5 - 9i) - (-5 - 9i)$

34. $(9 - 7i) + (-8 - i)$

35. $(4 + i) - (2 - 3i) + (6 - 8i)$

36. $(-2i) - (4 - 3i) - (8 + 4i)$

37. $(5 - 2i) - (3 - 4i) + (6i)$

38. $(-4 + 2i) + (6 + i) + (9 + 3i)$

For Exercises 39–54, multiply.

39. $(2i)(3i)$

40. $(4i)(-3i)$

41. $(-8i)(2i)$

42. $(6i)(i)$

43. $2i(5 - 6i)$

44. $3i(7 + i)$

45. $-3i(2 + 4i)$

46. $-2i(5 - 3i)$

47. $(2 + i)(3 - i)$

48. $(5 - 2i)(1 + i)$

49. $(6 + 5i)(5 - 3i)$

50. $(4 - 3i)(2 + 8i)$

51. $(3 + i)(3 - i)$

52. $(2 + 7i)(2 - 7i)$

53. $(3 + i)^2$

54. $(4 - 2i)^2$

For Exercises 55–74, write in standard form.

55. $\dfrac{2}{i}$

56. $\dfrac{6}{-i}$

57. $\dfrac{4}{5i}$

58. $\dfrac{6}{7i}$

59. $\dfrac{6}{2i}$

60. $\dfrac{8}{4i}$

61. $\dfrac{2 - i}{2i}$

62. $\dfrac{3 + i}{3i}$

63. $\dfrac{4 + 2i}{4i}$

64. $\dfrac{5 - 3i}{4i}$

65. $\dfrac{3 - 7i}{1 - i}$

66. $\dfrac{3 + i}{2 - i}$

67. $\dfrac{1 + 2i}{4 + 3i}$

68. $\dfrac{1 + i}{1 - i}$

69. $\dfrac{3}{2 + i}$

70. $\dfrac{4}{5 + i}$

71. $\dfrac{2i}{3 - 4i}$

72. $\dfrac{-4i}{2 + i}$

73. $\dfrac{3 + 2i}{4 - i}$

74. $\dfrac{5 - 6i}{3 + 2i}$

For Exercises 75–90, find the powers of i.

75. i^{19} **76.** i^{25} **77.** i^{46} **78.** i^{51}

79. i^{38} **80.** i^{41} **81.** i^{60} **82.** i^{52}

83. i^{-20} **84.** i^{-32} **85.** i^{-26} **86.** i^{-38}

87. i^{-21} **88.** i^{-45} **89.** i^{-35} **90.** i^{-45}

REVIEW EXERCISES

1. Factor: $4x^2 - 12x + 9$

2. Simplify: $\pm\sqrt{48}$

For Exercises 3 and 4, solve.

3. $3x - 2 = 4$

4. $2x^2 - x - 6 = 0$

5. Find the x- and y-intercepts for $2x + 3y = 6$.

6. Graph: $y = x^2 - 3$

Defined Terms

Section 10.1
n^{th} root (p. 818)
Radical function (p. 822)

Section 10.2
Rational exponent (p. 830)

Section 10. 4
like radicals (p. 852)

Section 10.6
Radical equation (p. 872)

Section 10.7
Imaginary unit (p. 882)
Imaginary number (p. 882)
Complex number (p. 883)
Complex conjugate (p. 885)

Procedures, Rules, and Key Examples

Procedures/Rules	Key Example(s)
Section 10.1 Radical Expressions and Functions If a is positive, then $\sqrt[n]{a} = b$, where $b \geq 0$ and $b^n = a$. If a is negative and n is even, then there is no real-number root. If a is negative and n is odd, then $\sqrt[n]{a} = b$, where b is negative and $b^n = a$.	**Example 1:** Simplify. **a.** $\sqrt{36} = 6$ **b.** $\sqrt{-25}$ is not a real number **c.** $-\sqrt{121} = -11$ **d.** $\sqrt{\dfrac{49}{100}} = \dfrac{7}{10}$ **e.** $\sqrt{16x^6} = 4|x^3|$ **f.** $\sqrt[3]{-27} = -3$ **Note:** If we assume that the variables represent nonnegative numbers, then the solution for part e would be $\sqrt{16x^6} = 4x^3$.
To evaluate a radical function, replace the variable with the indicated value and simplify.	**Example 2:** If $f(x) = \sqrt{4x + 3}$, find $f(3)$. Solution: $f(3) = \sqrt{4(3) + 3} = \sqrt{12 + 3} = \sqrt{15}$
For radical functions with an even index, the radicand must be nonnegative.	**Example 3:** Find the domain of $f(x) = \sqrt{2x - 4}$. Solution: Since $2x - 4$ must be nonnegative, we solve: $$2x - 4 \geq 0$$ $$2x \geq 4$$ $$x \geq 2$$

Procedures/Rules	Key Example(s)

Section 10.2 Rational Exponents

$a^{1/n} = \sqrt[n]{a}$, where n is a natural number other than 1.

Note: If a is negative and n is odd, then the root will be negative. If a is negative and n is even, then there is no real-number solution.

$a^{m/n} = \sqrt[n]{a^m} = (\sqrt[n]{a})^m$, where $a \geq 0$ and m and n are natural numbers other than 1.

$a^{-m/n} = \dfrac{1}{a^{m/n}}$, where $a \neq 0$ and m and n are natural numbers with $n \neq 1$.

Example 1: Evaluate.

a. $25^{1/2} = \sqrt{25} = 5$

b. $(-27)^{1/3} = \sqrt[3]{-27} = -3$

c. $(-16)^{1/4} = \sqrt[4]{-16}$, which is not a real number.

Example 2: Write the following in exponential form.

a. $\sqrt[3]{x^2} = x^{2/3}$

b. $(\sqrt[6]{x})^5 = x^{5/6}$

Example 3: Evaluate.

$$16^{-3/4} = \frac{1}{16^{3/4}} = \frac{1}{(\sqrt[4]{16})^3} = \frac{1}{2^3} = \frac{1}{8}$$

Example 4: Write as a radical expression with a smaller root index.

a. $\sqrt[4]{25} = 25^{1/4} = (5^2)^{1/4}$
$$= 5^{2 \cdot 1/4} = 5^{1/2} = \sqrt{5}$$

b. $\sqrt[8]{x^6} = x^{6/8} = x^{3/4} = \sqrt[4]{x^3}$

Example 5: Simplify.

a. $\sqrt[4]{x} \cdot \sqrt{x} = x^{1/4} \cdot x^{1/2} = x^{1/4+1/2}$
$$= x^{1/4+2/4}$$
$$= x^{3/4} = \sqrt[4]{x^3}$$

b. $\dfrac{\sqrt[3]{x^2}}{\sqrt[4]{x}} = \dfrac{x^{2/3}}{x^{1/4}} = x^{(2/3)-(1/4)}$
$$= x^{(8/12)-(3/12)} = x^{5/12} = \sqrt[12]{x^5}$$

c. $\sqrt{5} \cdot \sqrt[3]{2} = 5^{1/2} \cdot 2^{1/3} = 5^{3/6} \cdot 2^{2/6}$
$$= (5^3 \cdot 2^2)^{1/6} = (125 \cdot 4)^{1/6}$$
$$= 500^{1/6} = \sqrt[6]{500}$$

Example 6: Simplify $\sqrt[3]{\sqrt[5]{x}}$ to a single radical.

Solution: $\sqrt[3]{\sqrt[5]{x}} = (x^{1/5})^{1/3} = x^{(1/5) \cdot (1/3)}$
$$= x^{1/15} = \sqrt[15]{x}$$

Procedures/Rules	Key Example(s)

Section 10.3 Multiplying, Dividing, and Simplifying Radicals

Product rule for radicals:

If $\sqrt[n]{a}$ and $\sqrt[n]{b}$ are both real numbers, then $\sqrt[n]{a} \cdot \sqrt[n]{b} = \sqrt[n]{a \cdot b}$.

Example 1: Simplify. Assume variables represent nonnegative values.

a. $\sqrt{2} \cdot \sqrt{50} = \sqrt{2 \cdot 50} = \sqrt{100} = 10$

b. $\sqrt{3x} \cdot \sqrt{12x^3} = \sqrt{3x \cdot 12x^3}$
$\qquad = \sqrt{36x^4} = 6x^2$

c. $\sqrt[3]{5x} \cdot \sqrt[3]{3x} = \sqrt[3]{15x^2}$

d. $\sqrt[4]{27x^3} \cdot \sqrt[4]{3x} = \sqrt[4]{27x^3 \cdot 3x}$
$\qquad = \sqrt[4]{81x^4} = 3x$

Raising an nth root to the nth power:

For any nonnegative real number a, $(\sqrt[n]{a})^n = a$.

Example 2: Simplify $(\sqrt[4]{x^3})^4$.

Solution: $(\sqrt[4]{x^3})^4 = x^3$

Example 3: Simplify.

Quotient rule for radicals:

If $\sqrt[n]{a}$ and $\sqrt[n]{b}$ are both real numbers, then $\dfrac{\sqrt[n]{a}}{\sqrt[n]{b}} = \sqrt[n]{\dfrac{a}{b}}$, where $b \neq 0$.

a. $\dfrac{\sqrt{45}}{\sqrt{5}} = \sqrt{\dfrac{45}{5}} = \sqrt{9} = 3$

b. $\dfrac{\sqrt[3]{128x^7}}{\sqrt[3]{2x}} = \sqrt[3]{\dfrac{128x^7}{2x}} = \sqrt[3]{64x^6} = 4x^2$

c. $\sqrt{\dfrac{18x^5}{8x^3}} = \sqrt{\dfrac{9x^2}{4}} = \dfrac{\sqrt{9x^2}}{\sqrt{4}} = \dfrac{3x}{2}$

d. $\sqrt[4]{\dfrac{8}{x^4}} = \dfrac{\sqrt[4]{8}}{\sqrt[4]{x^4}} = \dfrac{\sqrt[4]{8}}{x}$

Simplifying an nth root.

1. Write the radicand as the product of the greatest possible n^{th} power and a number that has no perfect n^{th} power factors.
2. Use the product rule $\sqrt[n]{ab} = \sqrt[n]{a} \cdot \sqrt[n]{b}$, where a is the perfect n^{th} power.
3. Find the nth root of the perfect n^{th} power radicand.

Example 4: Simplify.

a. $\sqrt{50} = \sqrt{25 \cdot 2} = \sqrt{25} \cdot \sqrt{2} = 5\sqrt{2}$

b. $\sqrt{48x^3} = \sqrt{16x^2 \cdot 3x}$
$\qquad = \sqrt{16x^2} \cdot \sqrt{3x} = 4x\sqrt{3x}$

c. $\sqrt[4]{48a^9} = \sqrt[4]{16a^8 \cdot 3a}$
$\qquad = \sqrt[4]{16a^8} \cdot \sqrt[4]{3a} = 2a^2\sqrt[4]{3a}$

Example 5: Perform the operation and simplify.

$$\sqrt{6} \cdot \sqrt{10} = \sqrt{60} = \sqrt{4 \cdot 15}$$
$$= \sqrt{4} \cdot \sqrt{15} = 2\sqrt{15}$$

Procedures/Rules	Key Example(s)
Section 10.4 Adding, Subtracting and Multiplying Radical Expressions To add or subtract like radicals, add or subtract the coefficients and keep the radicals the same.	**Example 1:** Simplify. **a.** $2\sqrt{x} - 7\sqrt{x} = (2-7)\sqrt{x} = -5\sqrt{x}$ **b.** $4\sqrt[3]{5} + 2\sqrt[3]{5} = (4+2)\sqrt[3]{5} = 6\sqrt[3]{5}$

Example 2: Simplify the radicals, then combine like radicals.

$$9\sqrt{3} - \sqrt{12} + 6\sqrt{75}$$
$$= 9\sqrt{3} - \sqrt{4 \cdot 3} + 6\sqrt{25 \cdot 3}$$
$$= 9\sqrt{3} - 2\sqrt{3} + 6 \cdot 5\sqrt{3}$$
$$= 9\sqrt{3} - 2\sqrt{3} + 30\sqrt{3}$$
$$= 37\sqrt{3}$$

Example 3: Multiply then simplify.

$$(\sqrt{3} - 4)(\sqrt{3} + 6)$$
$$= \sqrt{3} \cdot \sqrt{3} + \sqrt{3} \cdot 6 - 4 \cdot \sqrt{3} - 4 \cdot 6 \qquad \textbf{Use FOIL.}$$
$$= \sqrt{9} + 6\sqrt{3} - 4\sqrt{3} - 24 \qquad \textbf{Multiply.}$$
$$= 3 + 6\sqrt{3} - 4\sqrt{3} - 24 \qquad \textbf{Simplify } \sqrt{9}.$$
$$= -21 + 2\sqrt{3} \qquad \textbf{Combine like radicals.}$$

Example 4: Perform the following operations.

a.
$$\sqrt{6} \cdot \sqrt{8} + \sqrt{5} \cdot \sqrt{15}$$
$$= \sqrt{48} + \sqrt{75}$$
$$= \sqrt{16 \cdot 3} + \sqrt{25 \cdot 3}$$
$$= 4\sqrt{3} + 5\sqrt{3} = 9\sqrt{3}$$

b.
$$\frac{\sqrt{60}}{\sqrt{5}} + \sqrt{48} = \sqrt{\frac{60}{5}} + \sqrt{48}$$
$$= \sqrt{12} + \sqrt{48}$$
$$= \sqrt{4 \cdot 3} + \sqrt{16 \cdot 3}$$
$$= 2\sqrt{3} + 4\sqrt{3} = 6\sqrt{3}$$

Procedures/Rules	Key Example(s)

Section 10.5 Rationalizing Numerators and Denominators of Radical Expressions

To rationalize a denominator containing a single square root:
1. Multiply both the numerator and denominator of the fraction by the same square root as appears in the denominator.
2. Simplify.

To rationalize a denominator containing a single n^{th} root, multiply the numerator and denominator by an expression that will make the radicand of the denominator a perfect n^{th} power.

To rationalize a denominator containing a sum or difference with at least one square root term, multiply the numerator and denominator by the conjugate of the denominator.

Summary of Simplest Form
A radical expression is in simplest form if:
1. All rational roots have been found.
2. There are no perfect n^{th} factors of radicands of the form $\sqrt[n]{a^n}$.
3. All possible products, quotients, sums, and differences have been found.
4. There are no radicals in the denominator of a fraction.

The numerator of a rational expression is rationalized exactly the same way a denominator is rationalized.

Example 1: Rationalize the denominator.

a. $\dfrac{7}{\sqrt{3}} = \dfrac{7}{\sqrt{3}} \cdot \dfrac{\sqrt{3}}{\sqrt{3}} = \dfrac{7\sqrt{3}}{3}$

b. $\dfrac{4}{\sqrt[3]{3}} = \dfrac{4}{\sqrt[3]{3}} \cdot \dfrac{\sqrt[3]{9}}{\sqrt[3]{9}} = \dfrac{4\sqrt[3]{9}}{\sqrt[3]{27}} = \dfrac{4\sqrt[3]{9}}{3}$

c. $\dfrac{6}{4 - \sqrt{5}}$

$\dfrac{6}{4 - \sqrt{5}} = \dfrac{6}{4 - \sqrt{5}} \cdot \dfrac{4 + \sqrt{5}}{4 + \sqrt{5}}$

$= \dfrac{6 \cdot 4 + 6 \cdot \sqrt{5}}{16 - 5}$

$= \dfrac{24 + 6\sqrt{5}}{11}$

Example 2: Rationalize the numerator.

a. $\dfrac{\sqrt{6}}{3} = \dfrac{\sqrt{6}}{3} \cdot \dfrac{\sqrt{6}}{\sqrt{6}} = \dfrac{\sqrt{36}}{3\sqrt{6}}$

$= \dfrac{6}{3\sqrt{6}} = \dfrac{2}{\sqrt{6}}$

b. $\dfrac{2 - \sqrt{3}}{5} = \dfrac{2 - \sqrt{3}}{5} \cdot \dfrac{2 + \sqrt{3}}{2 + \sqrt{3}}$

$= \dfrac{4 - 3}{5(2 + \sqrt{3})} = \dfrac{1}{10 + 5\sqrt{3}}$

Section 10.6 Radical Equations and Problem Solving

Power Rule: If $a = b$, then $a^n = b^n$.

To solve a radical equation:
1. Isolate the radical. (If there is more than one radical term, then isolate one of the radical terms.)
2. Raise both sides of the equation to the same power as the root index.
3. If all radicals have been eliminated, then solve. If a radical term remains, then isolate that radical term and raise both sides to the same power as its root index.
4. Check each solution. Any apparent solution that does not check is an extraneous solution.

Example 1: Solve $\sqrt{x - 5} = 7$.

$(\sqrt{x - 5})^2 = 7^2$ **Square both sides.**

$x - 5 = 49$

$x = 54$ **Add 5 on both sides.**

Check: $\sqrt{54 - 5} = 7$

$\sqrt{49} = 7$ **True.**

Procedures/Rules	Key Example(s)

Section 10.6 Radical Equations and Problem Solving (continued)

Example 2: Solve $\sqrt{n + 14} = n + 2$.

$(\sqrt{n + 14})^2 = (n + 2)^2$ Square both sides.

$n + 14 = n^2 + 4n + 4$

$0 = n^2 + 3n - 10$ Subtract n and 14 on both sides.

$0 = (n + 5)(n - 2)$ Factor.

$n + 5 = 0$ or $n - 2 = 0$ Use the zero factor theorem.

$n = -5 \qquad n = 2$

Check:

$\sqrt{-5 + 14} = -5 + 2 \quad \sqrt{2 + 14} = 2 + 2$

$\sqrt{9} = -3$ False. $\sqrt{16} = 4$ True.

The only solution is 2 (-5 is an extraneous solution).

Example 3: Solve $3 + \sqrt{t} = \sqrt{t + 21}$

$(3 + \sqrt{t})^2 = (\sqrt{t + 21})^2$ Square both sides.

$9 + 6\sqrt{t} + t = t + 21$

$6\sqrt{t} = 12$ Subtract 9 and t on both sides.

$\sqrt{t} = 2$ Divide both sides by 6.

$(\sqrt{t})^2 = 2^2$ Square both sides.

$t = 4$

We will leave the check to the reader.

Section 10.7 Complex Numbers

To write an imaginary number $\sqrt{-n}$ in terms of the imaginary unit i:
1. Separate the radical into two factors $\sqrt{-1} \cdot \sqrt{n}$.
2. Replace $\sqrt{-1}$ with i.
3. Simplify $\sqrt{n}$.

Example 1: Write using the imaginary unit.

a. $\sqrt{-36} = \sqrt{-1} \cdot \sqrt{36} = 6i$

b. $\sqrt{-32} = \sqrt{-1} \cdot \sqrt{32}$

$= i \cdot 4\sqrt{2} = 4i\sqrt{2}$

To add complex numbers, combine like terms.

Example 2: Add: $(5 - 6i) + (9 + 2i)$.

$(5 - 6i) + (9 + 2i) = 14 - 4i$

To subtract complex numbers, change the signs of the second complex numbers, then combine like terms.

Example 3: Subtract: $(7 - i) - (3 + 5i)$.

$(7 - i) - (3 + 5i) = (7 - i) + (-3 - 5i)$

$= 4 - 6i$

To multiply complex numbers, follow the same procedures as for multiplying monomials or binomials (FOIL). Remember that $i^2 = -1$.

Example 4: Multiply: $(6 + 5i)(2 - 3i)$.

$(6 + 5i)(2 - 3i) = 12 - 18i + 10i - 15i^2$

$= 12 - 8i - 15(-1)$

$= 12 - 8i + 15$

$= 27 - 8i$

continued

Procedures/Rules	Key Example(s)

Section 10.7 Complex Numbers (continued)

To divide complex numbers, rationalize the denominator using the complex conjugate.

Example 5: Write $\dfrac{2-3i}{4+5i}$ in standard form.

$$\frac{2-3i}{4+5i} = \frac{2-3i}{4+5i} \cdot \frac{4-5i}{4-5i}$$

$$= \frac{8 - 10i - 12i + 15i^2}{16 - 25i^2}$$

$$= \frac{8 - 22i + 15(-1)}{16 - 25(-1)}$$

$$= \frac{-7 - 22i}{41}$$

$$= -\frac{7}{41} - \frac{22}{41}i$$

Powers of i.

$i^1 = i$

$i^2 = -1$

$i^3 = -i$

$i^4 = 1$

Example 6: Simplify.

a. $i^{19} = i^{16} \cdot i^3 = (i^4)^4 \cdot (-i)$

$$= 1^4(-i) = 1(-i) = -i$$

b. $i^{-21} = \dfrac{1}{i^{21}} = \dfrac{1}{i^{20} \cdot i} = \dfrac{1}{(i^4)^5 \cdot i} = \dfrac{1}{1^5 \cdot i}$

$$= \frac{1}{i} = \frac{1}{i} \cdot \frac{-i}{-i} = \frac{-i}{-i^2}$$

$$= \frac{-i}{-(-1)} = \frac{-i}{1} = -i$$

For Exercises 1–5, answer true or false.

1. Every positive number has two square roots: a positive root and a negative root.

2. $\sqrt{x^2} = x$, where x is any real number.

3. The expression $\dfrac{3}{\sqrt{2}}$ is in simplest form.

4. Every radical equation has a single solution.

5. It is necessary to check all potential solutions to radical equations.

For Exercises 6–10, complete the rule.

6. Like radicals are two radical expression with identical _____ and indexes.

7. To add or subtract like radicals, add or subtract the _____ and keep the radicals the same.

8. To rationalize a denominator that is a single square root,
   1. _____ both the numerator and denominator of the fraction by the same square root as appears in the denominator.
   2. Simplify.

9. To rationalize a denominator containing a sum or difference with a square root term, multiply the numerator and denominator by the _____ of the denominator.

10. An apparent solution to a radical equation that does not solve the equation is considered _____ .

For Exercises 11 and 12, find all square roots of the given number.

11. 121

12. 49

For Exercises 13–16, evaluate the square root.

13. $\sqrt{169}$

14. $-\sqrt{49}$

15. $\sqrt{-36}$

16. $\sqrt{\dfrac{1}{25}}$

For Exercises 17 and 18, use a calculator to approximate each root to the nearest thousandth.

17. $\sqrt{7}$

18. $\sqrt{90}$

For Exercises 19–20, simplify. Assume variables can represent any real number.

19. $\sqrt{81x^2}$

20. $\sqrt[4]{(x-1)^8}$

For Exercises 21–36, simplify. Assume all variables represent nonnegative numbers.

21. $\sqrt{\dfrac{25}{16}a^4b^{10}}$

22. $\sqrt{\dfrac{81x^4y^2}{64a^8}}$

23. $\sqrt[3]{3} \cdot \sqrt[3]{9}$

24. $\sqrt[4]{8xy^3} \cdot \sqrt[4]{2xy}$

25. $\sqrt{2m} \cdot \sqrt{18m^3}$

26. $\sqrt{x^2yz^3} \cdot \sqrt{100yz}$

27. $\sqrt{18}$

28. $\sqrt{120}$

29. $\sqrt{6x^4b^3}$

30. $\sqrt[3]{-64a^3b^{12}}$

31. $\sqrt[3]{-0.008x^{18}}$

32. $\sqrt{\dfrac{1}{16m^4}}$

33. $\sqrt{\dfrac{5h^5}{20h}}$

34. $\sqrt[3]{-\dfrac{64}{125}}$

35. $\sqrt[4]{\dfrac{16}{81}}$

36. $\dfrac{\sqrt{51a^3b^4}}{\sqrt{3ab^4c^2}}$

For Exercises 37–40, rewrite using radical notation and evaluate. Assume all variables represent nonnegative quantities.

37. $(-64)^{1/3}$

38. $(24a^4)^{1/2}$

39. $\left(\dfrac{1}{32}\right)^{3/5}$

40. $(5r-2)^{5/7}$

For Exercises 41–46, write in exponential form.

41. $\sqrt[8]{33}$

42. $\dfrac{8}{\sqrt[7]{n^3}}$

43. $(\sqrt[5]{8})^3$

44. $(\sqrt[3]{m})^8$

45. $(\sqrt[4]{3xw})^3$

46. $\sqrt[3]{(a+b)^4}$

For Exercises 47–52, use the rules of exponents to simplify.

47. $x^{2/3} \cdot x^{4/3}$

48. $(4m^{1/4})(8m^{5/4})$

49. $\dfrac{y^{3/5}}{y^{4/5}}$

50. $\dfrac{b^{2/5}}{b^{-3/5}}$

51. $(k^{2/3})^{3/4}$

52. $(2xy^{1/5})^{3/4}$

For Exercises 53–62, find the products.

53. $\sqrt{5}(\sqrt{3} + \sqrt{2})$

54. $\sqrt[3]{7}(\sqrt[3]{3} + 2\sqrt[3]{7})$

55. $3\sqrt{6}(2 - 3\sqrt{6})$

56. $(\sqrt{2} - \sqrt{3})(\sqrt{5} + \sqrt{7})$

57. $(\sqrt[3]{2} - 4)(\sqrt[3]{4} + 2)$

58. $(\sqrt[4]{6x} + 2)(\sqrt[4]{2x} - 1)$

59. $(\sqrt{5a} + \sqrt{3b})(\sqrt{5a} - \sqrt{3b})$

60. $(2\sqrt{3} - \sqrt{5})(2\sqrt{3} + \sqrt{5})$

61. $(\sqrt{2} + 1)^2$

62. $(\sqrt{2} - \sqrt{5})^2$

For Exercises 63–72, rationalize the denominator and simplify.

63. $\dfrac{1}{\sqrt{2}}$

64. $\dfrac{3}{\sqrt[3]{3}}$

65. $\sqrt{\dfrac{4}{7}}$

66. $\dfrac{\sqrt[4]{5x^2}}{\sqrt[4]{2}}$

67. $\sqrt[3]{\dfrac{17}{3y^2}}$

68. $\dfrac{\sqrt[4]{9}}{\sqrt[4]{27}}$

69. $\dfrac{4}{\sqrt{2} - \sqrt{3}}$

70. $\dfrac{1}{4 + \sqrt{3}}$

71. $\dfrac{1}{2 - \sqrt{n}}$

72. $\dfrac{2\sqrt{3}}{3\sqrt{2} - 2\sqrt{3}}$

For Exercises 73–76, rationalize the numerator.

73. $\dfrac{\sqrt{10}}{6}$

74. $\dfrac{\sqrt{3x}}{5}$

75. $\dfrac{2 - \sqrt{3}}{8}$

76. $\dfrac{2\sqrt{t} + \sqrt{3t}}{5t}$

For Exercises 77–88, solve.

77. $\sqrt{x} = 9$

78. $\sqrt{y} = -3$

79. $\sqrt{w - 1} = 3$

80. $\sqrt[3]{3x - 2} = -2$

81. $\sqrt[4]{x - 2} - 3 = -1$

82. $\sqrt{y + 1} = \sqrt{2y - 4}$

83. $\sqrt{x - 6} = x + 2$

84. $\sqrt[3]{3x + 10} - 4 = 5$

85. $\sqrt[4]{x + 8} = \sqrt[4]{2x + 1}$

86. $\sqrt{5n - 1} = 4 - 2n$

87. $1 + \sqrt{x} = \sqrt{2x + 1}$

88. $\sqrt{3x + 1} = 2 - \sqrt{3x}$

For Exercises 89 and 90, write the imaginary number using i.

89. $\sqrt{-9}$

90. $\sqrt{-20}$

For Exercises 91 and 92, add or subtract.

91. $(3 + 2i) + (5 - 8i)$

92. $(7 - 3i) - (-2 + 4i)$

For Exercises 93–96, multiply and simplify.

93. $(3i)(4i)$

94. $2i(4 - i)$

95. $(6 + 2i)(4 - i)$

96. $(5 - i)^2$

For Exercises 97–102, write in standard form.

97. $\dfrac{5}{i}$

98. $\dfrac{3}{-i}$

99. $\dfrac{4}{3i}$

100. $\dfrac{7 + i}{5i}$

101. $\dfrac{3}{2 + i}$

102. $\dfrac{5 + i}{2 - 3i}$

For Exercises 103 and 104, simplify.

103. i^{20}

104. i^{15}

105. Write an expression in simplest form for the area of the figure.

For Exercises 106 and 107, use the following. The speed of a car can be determined by the length of the skid marks using the formula $S = 2\sqrt{2L}$, where L represents the length of the skid mark in feet and S represents the speed of the car in miles per hour.

106. **a.** Write an expression for the exact speed of a car if the length of the skid marks measures 40 feet long.

    b. Approximate the speed to the nearest tenth.

107. Find the length of skid marks if a driver brakes hard at a speed of 50 miles per hour.

For Exercises 108 and 109, use the formula $T = 2\pi\sqrt{\dfrac{L}{9.8}}$, where T is the period of a pendulum in seconds and L is the length of the pendulum in meters.

108. Find the period of a pendulum with a length of 2.45 meters.

109. Suppose the period of the pendulum is $\dfrac{\pi}{3}$ seconds. Find the length.

For Exercises 110–111, use the formula $t = \sqrt{\dfrac{h}{16}}$, where t is the time in seconds and h is the distance that the object has fallen in feet.

110. a. Write an expression in simplest form of the time that an object takes to fall 40 feet.

   b. Approximate the time to the nearest hundredth.

111. If an object falls for 0.3 seconds, how much distance does it fall?

112. Three pieces of lumber are to be connected to form a right triangle that will be part of the roof frame for a small storage building. If the horizontal piece is to be 4 feet and the vertical piece is to be 3 feet, how long must the connecting piece be?

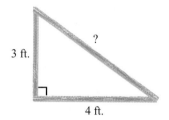

For Exercises 1 and 2, evaluate the square root.

1. $\sqrt{36}$

2. $\sqrt{-49}$

For Exercises 3–12, simplify. Assume variables represent nonnegative values.

3. $\sqrt{81x^2y^5}$

4. $\sqrt[3]{54}$

5. $\sqrt[4]{4x} \cdot \sqrt[4]{4x^5}$

6. $-\sqrt[3]{-27r^{15}}$

7. $\dfrac{\sqrt{5}}{\sqrt{45}}$

8. $\dfrac{\sqrt[4]{1}}{\sqrt[4]{81}}$

9. $6\sqrt{7} - \sqrt{7}$

10. $(\sqrt{3} - 1)^2$

11. $x^{2/3} \cdot x^{-4/3}$

12. $(\sqrt[3]{2} - 4)(\sqrt[3]{4} + 2)$

For Exercises 13 and 14, write in exponential form.

13. $\sqrt[5]{8x^3}$

14. $(\sqrt[3]{2xy})^2$

For Exercises 15–16, rationalize the denominator and simplify. Assume variables represent nonnegative values.

15. $\dfrac{1}{\sqrt[3]{4}}$

16. $\dfrac{\sqrt{x}}{\sqrt{x} + \sqrt{y}}$

For Exercises 17 and 18, solve the equation.

17. $\sqrt{3x - 2} = 8$

18. $\sqrt[4]{x + 8} = \sqrt[4]{2x + 1}$

For Exercises 19–22, simplify and write the answer in standard form ($a + bi$).

19. $(2 - i) - (4 + 3i)$

20. $(2i)(-i)$

21. $(4 - i)(4 + i)$

22. $\dfrac{2}{4 - 3i}$

23. Write an expression in simplest form for the area of the figure.

24. The formula $t = \sqrt{\dfrac{h}{16}}$ describes the amount of time t, in seconds, that an object falls a distance of h feet.

$2\sqrt{3}$

$5\sqrt{12}$

   a. Write an expression in simplest form for the exact amount of time an object falls a distance of 12 feet.

   b. Find the distance an object falls in 2 seconds.

25. The formula $S = \dfrac{7}{4}\sqrt{D}$ can be used to approximate the speed, S, in miles per hour, that a car was traveling prior to braking and skidding a distance D, in feet, on ice. (*Source:* Harris Technical Services, Traffic Accident Reconstructionists.)

   a. Find the speed of a car if the skid length measures 40 feet long.

   b. Find the length of the skid marks that a car traveling 30 miles per hour would make if it brakes hard and skids to a halt.

Quadratic Equations

In Section 6.6, we introduced quadratic equations and solved them by factoring. However, not all quadratic equations can be solved by factoring, so we need more powerful methods. In this chapter, we explore three other methods for solving quadratic equations, each of which will use square roots. In the first method, we will solve quadratic equations using a new principle of equality, called the *square root principle*. Like factoring, this first method has some limitations, so we will develop a more powerful method called *completing the square*. Although we can complete the square to solve any quadratic equation, it is rather tedious. So we will develop a third method that uses a general formula, called the *quadratic formula*, to solve any quadratic equation. We will then use these methods to solve equations that are quadratic in form and that are quadratic inequalities. Finally, we will add to what we learned in Section 6.7 and learn more about the graphs of quadratic functions.

"All our progress is unfolding like the vegetable bud."

—*RALPH WALDO EMERSON*
(1803—1882), U.S. essayist, poet, philosopher

"For beautiful variety no crop can be compared with this. Here is not merely the plain yellow of the grains, but nearly all the colors that we know . . ."
—*HENRY DAVID THOREAU*

Completing the Square

1 Use the square root principle to solve quadratic equations.

2 Solve quadratic equations by completing the square.

OBJECTIVE **1** Use the square root principle to solve quadratic equations. In Section 6.6, we solved quadratic equations such as $x^2 = 25$ by subtracting 25 from both sides, factoring, and then using the zero-factor theorem. Let's recall that process:

$$x^2 - 25 = 0$$
$$(x - 5)(x + 5) = 0 \qquad \text{Factor.}$$
$$x - 5 = 0 \quad \text{or} \quad x + 5 = 0 \qquad \text{Use the zero-factor theorem.}$$
$$x = 5 \qquad\qquad x = -5 \qquad \text{Isolate } x \text{ in each equation.}$$

Another approach to solving $x^2 = 25$ involves square roots. Notice that the solutions to this equation must be numbers that can be squared to equal 25. These numbers are the square roots of 25, which are 5 and -5. This suggests a new rule called the *square root principle*.

RULE   **THE SQUARE ROOT PRINCIPLE**

Note: The expression $\pm\sqrt{a}$ is read "plus or minus the square root of a."

If $x^2 = a$ and $a \geq 0$, then $x = \sqrt{a}$ or $x = -\sqrt{a}$.

It is common to indicate the positive and negative solutions by writing $\pm\sqrt{a}$.

For example, if $x^2 = 49$, then $x = \sqrt{49} = 7$ or $x = -\sqrt{49} = -7$. Or, we could simply write $x = \pm\sqrt{49} = \pm 7$.

Solve Equations in the Form $x^2 = a$

The square root principle is especially useful for solving equations in the form $x^2 = a$ when a is not a perfect square.

EXAMPLE 1 Solve $x^2 = 40$.

CONNECTION

We learned how to simplify square roots of numbers that have perfect square factors in Section 10.3.

Solution $x^2 = 40$
$$x = \pm\sqrt{40} \qquad \text{Use the square root principle.}$$
$$x = \pm\sqrt{4 \cdot 10} \qquad \text{Simplify by factoring out a perfect square.}$$
$$x = \pm 2\sqrt{10}$$

Note: Remember the $\pm$ symbol means that the two solutions are $2\sqrt{10}$ and $-2\sqrt{10}$.

We can check the two solutions using the original equation:

Check $2\sqrt{10}$: $(2\sqrt{10})^2 \overset{?}{=} 40$ *Check* $-2\sqrt{10}$: $(-2\sqrt{10})^2 \overset{?}{=} 40$

$4 \cdot 10 = 40$ True. $4 \cdot 10 = 40$ True.

For the remaining examples, the checks will be left to the reader.

ANSWERS

a. $x = \pm 6$
b. $x = \pm 3\sqrt{5}$

YOUR TURN Solve.

a. $x^2 = 36$

b. $x^2 = 45$

Solve Equations in the Form $ax^2 + b = c$

If an equation is in the form $ax^2 + b = c$, we use the addition and multiplication principles of equality to isolate x^2 and then use the square root principle.

EXAMPLE 2 Solve.

a. $x^2 + 2 = 100$

Solution $x^2 + 2 = 100$

$$x^2 = 98 \qquad \text{Subtract 2 on both sides to isolate } x^2.$$
$$x = \pm\sqrt{98} \qquad \text{Use the square root principle.}$$
$$x = \pm\sqrt{49 \cdot 2} \qquad \text{Simplify by factoring out a perfect square.}$$
$$x = \pm 7\sqrt{2}$$

b. $9x^2 = 27$

Solution $9x^2 = 27$

$$x^2 = 3 \qquad \text{Divide both sides by 9 to isolate } x^2.$$
$$x = \pm\sqrt{3} \qquad \text{Use the square root principle.}$$

c. $5x^2 + 2 = 62$

Solution $5x^2 + 2 = 62$

$$5x^2 = 60 \qquad \text{Subtract 2 from both sides.}$$
$$x^2 = 12 \qquad \text{Divide both sides by 5.}$$
$$x = \pm\sqrt{12} \qquad \text{Use the square root principle.}$$
$$x = \pm\sqrt{4 \cdot 3} \qquad \text{Simplify by factoring out a perfect square.}$$
$$x = \pm 2\sqrt{3}$$

ANSWERS

a. $\pm 4\sqrt{3}$
b. $\pm 2\sqrt{5}$
c. $\pm 3\sqrt{3}$

YOUR TURN Solve.

a. $x^2 - 6 = 42$

b. $3x^2 = 60$

c. $2x^2 + 11 = 65$

Solve Equations in the Form $(ax + b)^2 = c$

In an equation in the form $(ax + b)^2 = c$, notice the expression $ax + b$ is squared. We can use the square root principle to eliminate the square.

EXAMPLE 3 Solve.

a. $(x + 3)^2 = 16$

Solution $(x + 3)^2 = 16$

$x + 3 = \pm\sqrt{16}$ Use the square root principle.

$x + 3 = \pm 4$

$x = -3 \pm 4$ Subtract 3 from both sides.

$x = -3 + 4$ or $x = -3 - 4$ Simplify by separating the two solutions.

$x = 1$ $x = -7$

b. $(2x - 5)^2 = 6$

Solution $(2x - 5)^2 = 6$

$2x - 5 = \pm\sqrt{6}$ Use the square root principle.

$2x = 5 \pm \sqrt{6}$ Add 5 to both sides to isolate $2x$.

$x = \dfrac{5 \pm \sqrt{6}}{2}$ Divide both sides by 2.

c. $(6x - 7)^2 = -20$

Solution $(6x - 7)^2 = -20$

$6x - 7 = \pm\sqrt{-20}$ Use the square root principle.

$6x = 7 \pm \sqrt{-4 \cdot 5}$ Add 7 to both sides and simplify the square root.

$6x = 7 \pm 2i\sqrt{5}$ Rewrite the imaginary number using i notation.

$x = \dfrac{7 \pm 2i\sqrt{5}}{6}$ Divide both sides by 6.

$x = \dfrac{7}{6} \pm \dfrac{\sqrt{5}}{3}i$ Write the complex number in standard form.

> **Note:** The square root of a negative number is an imaginary number, which we can rewrite using i.

ANSWERS

a. 8 or 2

b. $\dfrac{-1 \pm \sqrt{5}}{3}$

c. $\dfrac{5}{4} \pm \dfrac{\sqrt{3}}{2}i$

YOUR TURN Solve.

a. $(x - 5)^2 = 9$ **b.** $(3x + 1)^2 = 5$ **c.** $(4x - 5)^2 = -12$

OBJECTIVE 2 **Solve quadratic equations by completing the square.** To make use of the square root principle, we need one side of the equation to be a perfect square and the other side to be a constant, as in $(x + 2)^2 = 5$. But, suppose we are given an equation like $x^2 + 6x = 2$ whose left-hand side is an "incomplete" square. We can use the addition principle of equality to add an appropriate number (9 in this case) to both sides so that the left-hand side becomes a perfect square. We call this process *completing the square*.

$$x^2 + 6x + 9 = 2 + 9 \quad \text{Adding 9 to both sides completes the square on the left side.}$$

We can now factor the left-hand side of the equation, then use the square root principle to finish solving the equation.

$(x + 3)^2 = 11$ Write the left-hand side in factored form.

$x + 3 = \pm\sqrt{11}$ Use the square root principle to eliminate the square.

$x = -3 \pm \sqrt{11}$ Subtract 3 from both sides to isolate x.

How do we know what number to add to complete the square?

Notice that squaring half of the coefficient of x gives the constant term in the completed square.

$$x^2 + 6x + 9$$

> Half of 6 is 3, which when squared is 9.

Also notice that half of the coefficient of x is the constant in the factored form.

$$x^2 + 6x + 9 = (x + 3)^2$$

> Half of 6 is 3, which is the constant in the factored form.

CONNECTION

Every perfect square in the form $(x + b)^2$ multiplies to equal the trinomial form $x^2 + 2bx + b^2$. Notice that half of x's coefficient, $2b$, is b, which is then squared to equal the last term.

Equations in the Form $x^2 + bx = c$

To solve a quadratic equation by completing the square, we need the equation in the form $x^2 + bx = c$, so that the coefficient of x^2 is 1. Once the equation is in the form $x^2 + bx = c$, we complete the square and then use the square root principle.

EXAMPLE 4 Solve by completing the square.

a. $x^2 + 8x + 5 = 0$

Solution We first write the equation in the form $x^2 + bx = c$.

Note: We found 16 by squaring half of 8.
$$\left(\frac{8}{2}\right)^2 = 4^2 = 16$$

$$x^2 + 8x = -5 \qquad \text{Subtract 5 from both sides to get the form } x^2 + bx = c.$$
$$x^2 + 8x + 16 = -5 + 16 \qquad \text{Complete the square by adding 16 on both sides.}$$
$$(x + 4)^2 = 11 \qquad \text{Factor.}$$
$$x + 4 = \pm\sqrt{11} \qquad \text{Use the square root principle.}$$
$$x = -4 \pm \sqrt{11} \qquad \text{Subtract 4 from both sides to isolate } x.$$

b. $x^2 - 5x + 3 = 4$

Solution

$$x^2 - 5x = 1 \qquad \text{Subtract 3 from both sides to get the form } x^2 + bx = c.$$
$$x^2 - 5x + \frac{25}{4} = 1 + \frac{25}{4} \qquad \text{Complete the square by adding } \frac{25}{4} \text{ on both sides.}$$
$$\left(x - \frac{5}{2}\right)^2 = \frac{29}{4} \qquad \text{Factor the left side and simplify the right.}$$
$$x - \frac{5}{2} = \pm\sqrt{\frac{29}{4}} \qquad \text{Use the square root principle.}$$
$$x = \frac{5}{2} \pm \sqrt{\frac{29}{4}} \qquad \text{Add } \frac{5}{2} \text{ to both sides to isolate } x.$$
$$x = \frac{5}{2} \pm \frac{\sqrt{29}}{2} \qquad \text{Simplify the square root.}$$
$$x = \frac{5 \pm \sqrt{29}}{2} \qquad \text{Combine the fractions.}$$

YOUR TURN   **Solve by completing the square.**

a. $x^2 + 10x + 1 = 0$ **b.** $x^2 - 3x - 1 = 2$

Equations in the Form $ax^2 + bx = c$, where $a \neq 1$

So far our equations have been in the form $x^2 + bx = c$, where the coefficient of the x^2 term is 1. To solve an equation such as $3x^2 + 6x = 8$, we will need to divide both sides of the equation by 3 $\left(\text{or multiply both sides by } \dfrac{1}{3}\right)$ so that the coefficient of x^2 is 1.

$$\frac{3x^2 + 6x}{3} = \frac{8}{3} \qquad \text{Divide both sides by 3.}$$

$$x^2 + 2x = \frac{8}{3}$$

We can now solve by completing the square.

$$x^2 + 2x + 1 = \frac{8}{3} + 1 \qquad \text{Add 1 to both sides to complete the square.}$$

$$(x + 1)^2 = \frac{11}{3} \qquad \text{Factor the left side and simplify the right.}$$

$$x + 1 = \pm\sqrt{\frac{11}{3}} \qquad \text{Use the square root principle.}$$

$$x = -1 \pm \sqrt{\frac{11}{3}} \qquad \text{Subtract 1 from both sides to isolate } x.$$

$$x = -1 \pm \frac{\sqrt{11}}{\sqrt{3}} \cdot \frac{\sqrt{3}}{\sqrt{3}} \qquad \text{Rationalize the denominator.}$$

$$x = -1 \pm \frac{\sqrt{33}}{3}$$

We can now write a procedure for solving any quadratic equation by completing the square.

PROCEDURE   SOLVING QUADRATIC EQUATIONS BY COMPLETING THE SQUARE

To solve a quadratic equation by completing the square:

1. Write the equation in the form $x^2 + bx = c$.
2. Complete the square by adding $\left(\dfrac{b}{2}\right)^2$ to both sides.
3. Write the completed square in factored form.
4. Use the square root principle to eliminate the square.
5. Isolate the variable.
6. Simplify as needed.

EXAMPLE 5 Solve by completing the square.

a. $4x^2 - 20x = 6$

Solution

$$\frac{4x^2 - 20x}{4} = \frac{6}{4}$$ Divide both sides by 4.

$$x^2 - 5x = \frac{3}{2}$$ Simplify.

$$x^2 - 5x + \frac{25}{4} = \frac{3}{2} + \frac{25}{4}$$ Add $\frac{25}{4}$ to both sides to complete the square.

$$\left(x - \frac{5}{2}\right)^2 = \frac{31}{4}$$ Factor the left side and simplify the right.

$$x - \frac{5}{2} = \pm\sqrt{\frac{31}{4}}$$ Use the square root principle.

$$x = \frac{5}{2} \pm \frac{\sqrt{31}}{2}$$ Add $\frac{5}{2}$ to both sides and simplify the square root.

$$x = \frac{5 \pm \sqrt{31}}{2}$$ Combine the fractions.

> **Note:** To complete the square, we square half of 5:
> $$\left(\frac{5}{2}\right)^2 = \frac{25}{4}$$

b. $3x^2 + 7x + 10 = 4$

Solution $3x^2 + 7x + 10 = 4$

$$3x^2 + 7x = -6$$ Subtract 10 from both sides.

$$\frac{3x^2 + 7x}{3} = \frac{-6}{3}$$ Divide both sides by 3.

$$x^2 + \frac{7}{3}x = -2$$ Simplify.

$$x^2 + \frac{7}{3}x + \frac{49}{36} = -2 + \frac{49}{36}$$ Add $\frac{49}{36}$ to both sides to complete the square.

$$\left(x + \frac{7}{6}\right)^2 = -\frac{23}{36}$$ Factor the left side and simplify the right.

$$x + \frac{7}{6} = \pm\sqrt{-\frac{23}{36}}$$ Use the square root principle.

$$x = -\frac{7}{6} \pm \frac{\sqrt{23}}{6}i$$ Subtract $-\frac{7}{6}$ from both sides and write the complex number in standard form.

> **Note:** To complete the square, we square half of $\frac{7}{3}$:
> $$\left(\frac{1}{2} \cdot \frac{7}{3}\right)^2 = \left(\frac{7}{6}\right)^2 = \frac{49}{36}$$

ANSWERS

a. $\dfrac{-5 \pm 3\sqrt{5}}{5}$ or

     $-1 \pm \dfrac{3\sqrt{5}}{5}$

b. $4 \pm 3i$

YOUR TURN Solve by completing the square.

a. $5x^2 + 10x = 4$ **b.** $x^2 - 8x + 11 = -14$

11.1 EXERCISES

For Extra Help

 InterAct Math

 Addison-Wesley Math Tutor Center

 Video 17/DVT 6

 MyMathLab

MathXL

1. Explain why there are two solutions to an equation in the form $x^2 = a$.

2. Write a formula for the solutions of $ax^2 - b = c$ by solving for x.

3. Write a formula for the solutions of $(ax - b)^2 = c$ by solving for x.

4. Given an equation in the form $x^2 + bx = c$, explain how to complete the square.

5. Consider the equation $x^2 - 6x + 8 = 0$. Which is a better method for solving the equation, factoring and then using the zero-factor theorem or completing the square? Explain.

6. Consider the equation $x^2 + 5x - 2 = 0$. Which is a better method for solving the equation, factoring and then using the zero-factor theorem or completing the square? Explain.

For Exercises 7–16, solve and check.

7. $x^2 = 16$

8. $x^2 = 36$

9. $y^2 = \dfrac{4}{9}$

10. $t^2 = \dfrac{1}{25}$

11. $n^2 = 0.81$

12. $p^2 = 1.21$

13. $t^2 = 50$

14. $r^2 = 27$

15. $w^2 = -25$

16. $c^2 = -49$

For Exercises 17–36, solve and check. Begin by using the addition or multiplication principles of equality to isolate the squared term.

17. $n^2 - 7 = 42$

18. $y^2 - 5 = 59$

19. $y^2 - 7 = 29$

20. $k^2 + 5 = 30$

21. $4n^2 = 36$

22. $5y^2 = 125$

23. $25t^2 = 9$

24. $16d^2 = 49$

25. $4h^2 = -16$

26. $-3k^2 = 15$

27. $\dfrac{5}{6}x^2 = \dfrac{3}{8}$

28. $-\dfrac{2}{3}m^2 = -\dfrac{4}{5}$

29. $2x^2 + 1 = 3$

30. $4x^2 + 5 = 21$

31. $6y^2 - 1 = -13$

32. $5n^2 + 24 = 9$

33. $\dfrac{3}{4}y^2 - 5 = 3$

34. $\dfrac{6}{5}m^2 - 5 = 15$

35. $0.5t^2 - 0.4 = 0.32$

36. $0.4p^2 + 0.6 = 0.744$

For Exercises 37–52, solve and check. Use the square root principle to eliminate the square.

37. $(x + 3)^2 = 16$

38. $(y + 5)^2 = 4$

39. $(2n + 3)^2 = 36$

40. $(3h - 1)^2 = 49$

41. $(m - 7)^2 = -12$

42. $(t - 5)^2 = -28$

43. $(4k - 1)^2 = 40$

44. $(3x - 7)^2 = 50$

45. $(m - 8)^2 = -1$

46. $(t - 2)^2 = -5$

47. $\left(y - \dfrac{2}{3}\right)^2 = \dfrac{4}{9}$

48. $\left(x + \dfrac{3}{4}\right)^2 = \dfrac{25}{16}$

49. $\left(\dfrac{2}{3}d - \dfrac{1}{4}\right)^2 = \dfrac{1}{9}$

50. $\left(\dfrac{3}{4}h - \dfrac{5}{6}\right)^2 = \dfrac{1}{16}$

51. $(0.3x + 2.2)^2 = 0.49$

52. $(0.2n - 5.1)^2 = 1.69$

For Exercises 53–56, explain the mistake, then find the correct solutions.

53. $x^2 - 3 = 6$
$$x^2 = 9$$
$$x = \sqrt{9}$$
$$x = 3$$

54. $x^2 = 20$
$$x = \sqrt{20}$$
$$x = 2\sqrt{5}$$

55. $(x - 3)^2 = -2$
$$x - 3 = \pm\sqrt{2}$$
$$x = 3 \pm \sqrt{2}$$

56. $(x - 1)^2 = -12$
$$x - 1 = \pm\sqrt{-12}$$
$$x = 1 \pm 2\sqrt{3}$$

For Exercises 57–62, solve, then use a calculator to approximate the irrational solutions rounded to three places.

57. $p^2 = 90$

58. $r^2 = 45$

59. $y^2 - 5 = 10$

60. $x^2 - 12 = 0$

61. $(n - 2)^2 = 3$

62. $(m - 1)^2 = 5$

For Exercises 63–74: a. Add a term to the expression to make it a perfect square.
                       b. Factor the perfect square.

63. $x^2 + 14x$

64. $c^2 + 8c$

65. $n^2 - 10n$

66. $a^2 - 4a$

67. $y^2 - 7y$

68. $m^2 - 11m$

69. $s^2 - \dfrac{2}{3}s$

70. $u^2 + \dfrac{1}{2}u$

71. $b^2 + \dfrac{3}{2}b$

72. $v^2 + \dfrac{1}{5}v$

73. $p^2 + 9p$

74. $z^2 - 15z$

For Exercises 75–90, solve the equation by completing the square.

75. $w^2 + 2w = 15$

76. $p^2 + 8p = 9$

77. $y^2 + 10y = -16$

78. $x^2 + 10x = -24$

79. $r^2 - 2r + 50 = 0$

80. $c^2 - 6c + 27 = 0$

81. $k^2 = 9k - 18$

82. $a^2 = 6a + 7$

83. $n^2 + 10n - 20 = 4$

84. $u^2 - 5u - 9 = -13$

85. $b^2 - 2b - 11 = 4$

86. $z^2 + 2z = 6$

87. $h^2 - 6h + 3 = -26$

88. $j^2 - 4j + 25 = -3$

89. $u^2 + \dfrac{1}{2}u = \dfrac{3}{2}$

90. $y^2 + \dfrac{1}{3}y = \dfrac{2}{3}$

For Exercises 91–104, solve the equation by completing the square. Begin by writing the equation in the form $x^2 + bx = c$.

91. $4x^2 + 12x = 7$ **92.** $3m^2 + 6m = 8$ **93.** $2h^2 - 6h = 0$ **94.** $3s^2 - 4s = 2$

95. $5k^2 + k - 2 = 0$ **96.** $2t^2 - 5 = -3t$ **97.** $2g^2 + g - 11 = -5$ **98.** $4l^2 + l - 30 = 30$

99. $6w^2 - 6 = -5w$ **100.** $3a^2 + 2 = 6a$ **101.** $2n^2 - n - 3 = 0$ **102.** $f^2 - 12f - 45 = 0$

103. $6d^2 - 13d - 20 = 8$ **104.** $3z^2 + 5z - 2 = 0$

For Exercises 105–106, explain the mistake.

105.
$$3x^2 + 4x = 2$$
$$3x^2 + 4x + 4 = 2 + 4$$
$$(3x + 2)^2 = 6$$
$$3x + 2 = \pm\sqrt{6}$$
$$3x = -2 \pm \sqrt{6}$$
$$x = \frac{-2 \pm \sqrt{6}}{3}$$

106.
$$x^2 + 4x = 5$$
$$x^2 + 4x + 4 = 5 + 4$$
$$(x + 2)^2 = 9$$
$$x + 2 = \sqrt{9}$$
$$x + 2 = 3$$
$$x = -2 + 3$$
$$x = 1$$

For Exercises 107–122, solve.

107. A square sheet of metal has an area of 196 square inches. What is the length of each side?

108. A severe thunderstorm warning is issued by the National Weather Service for a square area covering 14,400 square miles. What is the length of each side of the square area?

109. A tank is to be made for an aquarium so that the length is 8 feet, the width is twice the height, and the volume is 144 cubic feet. Find the height and width of the tank.

8 feet

110. The length of a swimming pool is to be three times the width. If the depth is to be a constant 4 feet and the volume is to be 4800 cubic feet, find the length and width.

111. The Arecibo radio telescope is like a giant satellite dish that covers a circular area of approximately $23{,}256.25\pi$ square meters. Find the diameter of the dish.

112. A field is planted in a circular pattern. A watering device is to be constructed with pipe in a line extending from the center of the field to the edge of the field. The pipe is set upon wheels so that it can rotate around the field and cover the entire field with water. If the area of the field is 7225π square feet, how long will the watering device be?

7225π ft.^{2}

113. The corner of a sheet of wood has been cut out as shown. If the area of the piece that was removed was 256 square inches, find x.

Removed piece

256 in.^{2}

2 in.

2 in.

x

x

2 in.

114. The area of the hole in the washer shown is 25π. Find r, the radius of the washer.

4 mm

r

115. Susan's den is 2 feet longer than it is wide. If the area is 168 square feet, what are the dimensions of the room?

116. A rectangular playground is designed so that the length is 20 feet more than the width. If the area is to be 2400 square feet, what are the dimensions of the playground?

length

width

117. A ramp forms a right triangle. If the base of the ramp is 7 feet longer than the height and the hypotenuse is 13 feet, what are the dimensions of the base and height?

118. Two identical right triangles are placed together to form the frame of a roof. If the height of each triangle is 4 feet less than its base and its hypotenuse is 20 feet, what are the dimensions of the base and height?

20 ft.

height

base

119. The metal panel shown is to have a total area of 825 square inches. Find the length x.

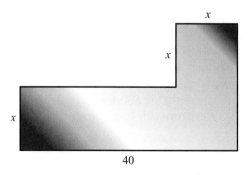

x

x

x

40

120. A side view of a concrete bridge support footing is shown. The side of the footing is to have a total area of 80 square feet. Find the length of x.

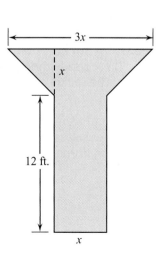

$3x$

x

12 ft.

x

121. A plastic panel is to have a rectangular hole cut as shown.

    a. Find w so that the area remaining after the hole is cut is 520 square inches.

    b. Find the length and width of the plastic panel.

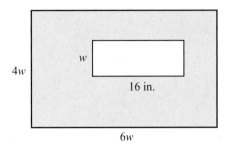

122. An 8-centimeter-wide groove with a height of h is to be cut into a steel block as shown.

    a. Find the height and width of the groove so that the volume remaining in the block after the groove is cut is 400 cubic centimeters.

    b. Find the height and width of the block.

For Exercises 123–126, if an object is dropped, the formula $d = 16t^2$ describes the distance d in feet that the object falls in t seconds.

123. Suppose a cover for a ceiling light falls from a 9-foot ceiling. How long does it take for the cover to hit the floor?

124. A construction worker tosses a scrap piece of lumber from the roof of a house. How long does it take the piece of lumber to reach the ground 25 feet below?

125. A toy rocket is launched straight up and reaches a height of 180 feet in 1.5 seconds then plummets back to the ground. Determine the total time of the rocket's round trip.

126. A lead weight is dropped from a tower at a height of 150 feet. How much time passes for the weight to be halfway to the ground? Write an exact answer, then approximate to the nearest tenth of a second.

For Exercises 127 and 128, use the following information. In physics, if an object is in motion it has kinetic energy. The formula $E = \dfrac{1}{2}mv^2$ is used to calculate the kinetic energy E of an object with a mass m and velocity v. If the mass is measured in kilograms and the velocity is in meters per second, then the kinetic energy will be in units called joules (J).

127. Suppose an object with a mass of 50 kilograms has 400 joules of kinetic energy. Find its velocity.

128. In a crash test, a vehicle with a mass of 1200 kilograms is found to have kinetic energy of 117,600 joules just before impact. Find the velocity of the vehicle just before impact.

PUZZLE PROBLEM

Without using a calculator, which of the following four numbers is a perfect square? (Hint: Write a list of smaller perfect squares and look for a pattern)

$$9{,}456{,}804{,}219{,}745{,}618$$
$$2{,}512{,}339{,}789{,}576{,}516$$
$$7{,}602{,}985{,}471{,}286{,}543$$
$$4{,}682{,}715{,}204{,}643{,}182$$

Collaborative Exercises

Isolating Squared Variables

1. The formula for the area of a circle is $A = \pi r^2$.

   a. Isolate r.

   b. The square root principle indicates there is one positive solution and one negative solution. Are both solutions reasonable in this situation? Explain.

   c. Use the formula you have written for r to find the radius of a circle with an area of 144π square inches.

   d. Find the radius of a circle with an area of 40π square feet. Write an exact answer in simplest form.

   e. Use a calculator to approximate your answer to part d to the nearest hundredth.

continued

2. The formula for the volume of a cone is $V = \dfrac{1}{3}\pi r^2 h$.

   a. Isolate r.

   b. The square root principle indicates there is one positive solution and one negative solution. Are both solutions reasonable in this situation? Explain.

   c. Use the formula you have written for r to find the radius of a cone with a height of 8 centimeters and a volume of 96π cubic centimeters.

   d. Find the radius of a cone with a height of 1.5 feet and a volume of 24π cubic feet. Write an exact answer in simplest form.

   e. Use a calculator to approximate your answer to part (d) to the nearest hundredth.

REVIEW EXERCISES

For Exercises 1 and 2, factor.

1. $x^2 - 10x + 25$

2. $x^2 + 6x + 9$

For Exercises 3–6, simplify. Assume variables represent nonnegative values.

3. $\sqrt{36x^2}$

4. $\sqrt{(2x + 3)^2}$

5. $\dfrac{-8 + \sqrt{8^2 - 4(2)(5)}}{2(2)}$

6. $\dfrac{-6 - \sqrt{6^2 - 4(5)(2)}}{2(5)}$

11.2 Solving Quadratic Equations Using the Quadratic Formula

OBJECTIVES

1. Solve quadratic equations using the quadratic formula.
2. Use the discriminant to determine the number of real solutions that a quadratic equation will have.
3. Find the x- and y-intercepts of a quadratic function.
4. Solve applications using the quadratic formula.

In this section, we will solve quadratic equations using a formula called the *quadratic formula*. This formula is much easier to use than completing the square.

OBJECTIVE 1 **Solve quadratic equations using the quadratic formula.** To derive the quadratic formula, we begin with the general form of the quadratic equation, $ax^2 + bx + c = 0$, and assume that $a \neq 0$. We will follow the procedure for solving a quadratic equation by completing the square.

$$ax^2 + bx + c = 0$$

$$ax^2 + bx = -c$$ Subtract c from both sides.

$$\frac{ax^2 + bx}{a} = \frac{-c}{a}$$ Divide both sides by a so that the coefficient of x^2 will be 1.

$$x^2 + \frac{b}{a}x = -\frac{c}{a}$$ Simplify.

> **Note:** To complete the square, we square half of $\frac{b}{a}$.
> $$\left(\frac{1}{2} \cdot \frac{b}{a}\right)^2 = \left(\frac{b}{2a}\right)^2$$
> $$= \frac{b^2}{4a^2}$$

$$x^2 + \frac{b}{a}x + \frac{b^2}{4a^2} = -\frac{c}{a} + \frac{b^2}{4a^2}$$ Complete the square.

$$\left(x + \frac{b}{2a}\right)^2 = -\frac{4ac}{4a^2} + \frac{b^2}{4a^2}$$ On the left side, write factored form. On the right side, rewrite $-\frac{c}{a}$ with the common denominator $4a^2$ so that we can combine the rational expressions.

$$\left(x + \frac{b}{2a}\right)^2 = \frac{b^2 - 4ac}{4a^2}$$ Combine the rational expressions. For simplicity, we rearrange the order of the b^2 and $-4ac$ terms in the numerator.

$$x + \frac{b}{2a} = \pm\sqrt{\frac{b^2 - 4ac}{4a^2}}$$ Use the square root principle to eliminate the square.

$$x + \frac{b}{2a} = \pm\frac{\sqrt{b^2 - 4ac}}{2a}$$ Simplify the square root in the denominator.

$$x = -\frac{b}{2a} \pm \frac{\sqrt{b^2 - 4ac}}{2a}$$ Subtract $\frac{b}{2a}$ from both sides to isolate x.

$$x = \frac{-b \pm \sqrt{b^2 - 4ac}}{2a}$$ Combine rational expressions.

This final equation is the *quadratic formula* and can be used to solve any quadratic equation simply by replacing *a*, *b*, and *c* with the corresponding numbers from the given equation.

PROCEDURE USING THE QUADRATIC FORMULA

To solve a quadratic equation in the form $ax^2 + bx + c = 0$, where $a \neq 0$, use the quadratic formula:

$$x = \frac{-b \pm \sqrt{b^2 - 4ac}}{2a}$$

CONNECTION

In Example 5 of Section 6.6, we solved the equation $2x^2 + 9x - 5 = 0$ by factoring.

EXAMPLE 1 Solve.

a. $2x^2 + 9x - 5 = 0$

Solution This equation is in the form $ax^2 + bx + c = 0$, where $a = 2$, $b = 9$, and $c = -5$, so we can use the quadratic formula, $x = \dfrac{-b \pm \sqrt{b^2 - 4ac}}{2a}$.

$x = \dfrac{-9 \pm \sqrt{9^2 - 4(2)(-5)}}{2(2)}$ Replace *a* with 2, *b* with 9, and *c* with −5.

$x = \dfrac{-9 \pm \sqrt{121}}{4}$ Simplify in the radical and the denominator.

$x = \dfrac{-9 \pm 11}{4}$

$x = \dfrac{-9 + 11}{4}$ or $x = \dfrac{-9 - 11}{4}$ Now we split up the ± to calculate the two solutions.

$x = \dfrac{2}{4}$ $x = \dfrac{-20}{4}$

$x = \dfrac{1}{2}$ $x = -5$

CONNECTION

Notice that the radicand 121 is a perfect square, which causes the solutions to be two rational numbers. As we consider more examples in this section, note how the radicand determines the type of solutions.

b. $x^2 + 12x + 7 = 11$

Solution $x^2 + 12x - 4 = 0$ Subtract 11 from both sides to get the form $ax^2 + bx + c = 0$.

$x = \dfrac{-12 \pm \sqrt{12^2 - 4(1)(-4)}}{2(1)}$ Use the quadratic formula replacing *a* with 1, *b* with 12, and *c* with −4.

$x = \dfrac{-12 \pm \sqrt{160}}{2}$ Simplify in the radical and the denominator.

$x = \dfrac{-12 \pm \sqrt{16 \cdot 10}}{2}$ Use the product rule of radicals.

$x = \dfrac{-12 \pm 4\sqrt{10}}{2}$ Simplify the radical.

$x = \dfrac{-12}{2} \pm \dfrac{4\sqrt{10}}{2}$ Separate into two rational expressions in order to simplify.

$x = -6 \pm 2\sqrt{10}$ Simplify by dividing out the 2.

CONNECTION

Notice that the radicand 160 is not a perfect square, which causes the solutions to be irrational numbers.

c. $12x + 9 = -4x^2$

Solution $4x^2 + 12x + 9 = 0$

Add $4x^2$ to both sides to get the form $ax^2 + bx + c = 0$.

$$x = \frac{-12 \pm \sqrt{12^2 - 4(4)(9)}}{2(4)}$$

Use the quadratic formula, replacing a with 4, b with 12, and c with 9

$$x = \frac{-12 \pm \sqrt{0}}{8}$$

Simplify in the radical and the denominator.

$$x = -\frac{3}{2}$$

Simplify the radical and write the fraction in lowest terms.

CONNECTION

Notice that the radicand is 0, which causes this equation to have only one solution.

d. $3x^2 + 9 = -8x + 2$

Solution $3x^2 + 8x + 7 = 0$

Add $8x$ and subtract 2 on both sides to get the form $ax^2 + bx + c = 0$.

CONNECTION

Notice that the negative radicand causes the solutions to be complex numbers.

$$x = \frac{-8 \pm \sqrt{8^2 - 4(3)(7)}}{2(3)}$$

Use the quadratic formula, replacing a with 3, b with 8, and c with 7.

$$x = \frac{-8 \pm \sqrt{-20}}{6}$$

Simplify in the radical and the denominator.

$$x = \frac{-8 \pm 2i\sqrt{5}}{6}$$

Simplify the radical.

$$x = \frac{-4 \pm i\sqrt{5}}{3}$$

Simplify to lowest terms.

Note: In standard form, the two complex solutions would be

$$-\frac{4}{3} + \frac{\sqrt{5}}{3}i \text{ and } -\frac{4}{3} - \frac{\sqrt{5}}{3}i.$$

ANSWERS

a. $\frac{5}{3}$ or -4

b. $\frac{-4 \pm \sqrt{6}}{5}$

c. 4

d. $-2 \pm i\sqrt{2}$

YOUR TURN Solve using the quadratic formula.

a. $3x^2 + 7x - 20 = 0$

b. $5x^2 + 2 = -8x$

c. $x^2 - 8x + 16 = 0$

d. $x^2 + 8 = 2 - 4x$

Choosing a Method for Solving Quadratic Equations

We have learned several methods for solving quadratic equations. The following table summarizes the methods and the conditions that would make each method the best choice.

Methods for Solving Quadratic Equations

Method	When the method is beneficial
1. Factoring (Section 6.6)	Use when the quadratic equation can be easily factored.
2. Square root principle (Section 11.1)	Use when the quadratic equation can be easily written in the form $ax^2 = c$ or $(ax + b)^2 = c$.
3. Completing the square (Section 11.1)	Rarely the best method, but important for future topics.
4. Quadratic formula (Section 11.2)	Use when factoring is not easy.

OBJECTIVE **2** Use the discriminant to determine the number of real solutions that a quadratic equation will have. In the Connection boxes for Example 1, we pointed out how the radicand in the quadratic formula affects the solutions to a given quadratic equation. The expression $b^2 - 4ac$, which is the radicand, is called the **discriminant**.

Discriminant: The radicand $b^2 - 4ac$ in the quadratic formula.

We use the discriminant to determine the number and type of solutions to a quadratic equation.

> **PROCEDURE** **USING THE DISCRIMINANT**

Note: When the discriminant is 0, the solution is $-\dfrac{b \pm \sqrt{0}}{2a} = -\dfrac{b}{2a}$. This real number solution is a rational number if $-\dfrac{b}{2a}$ is rational.

Given a quadratic equation in the form $ax^2 + bx + c = 0$, where $a \neq 0$, to determine the number and type of solutions a quadratic equation will have, evaluate the discriminant $b^2 - 4ac$.

If the **discriminant is positive**, then the equation has two real-number solutions. They will be rational if the discriminant is a perfect square and irrational otherwise.

If the **discriminant is 0**, then the equation has one real solution.

If the **discriminant is negative**, then the equation has two nonreal complex solutions.

> **EXAMPLE 2** Use the discriminant to determine the number and type of solutions.

a. $2x^2 - 9x = 6$

Solution $2x^2 - 9x - 6 = 0$ Write the equation in the form $ax^2 + bx + c = 0$.

Evaluate the discriminant, $b^2 - 4ac$.

$(-9)^2 - 4(2)(-6)$ Replace a with 2, b with -9, and c with -6.

$= 81 + 48$ **Warning:** 129 is the value of the discriminant,

$= 129$ not a solution for the equation $2x^2 - 9x - 6 = 0$.

Because the discriminant is positive, this equation has two real-number solutions. Since 129 is not a perfect square, the solutions are irrational.

b. $x^2 = \dfrac{4}{5}x - \dfrac{4}{25}$

Solution $x^2 - \dfrac{4}{5}x + \dfrac{4}{25} = 0$ Write the equation in the form $ax^2 + bx + c = 0$.

It is easier to evaluate the discriminant if the numbers are integers, so we multiply the equation through by the LCD, which is 25.

$$25 \cdot x^2 - \frac{25}{1} \cdot \frac{4}{5}x + \frac{25}{1} \cdot \frac{4}{25} = 25 \cdot 0$$
$$25x^2 - 20x + 4 = 0$$

Note: Since the discriminant is 0, the solution is $-\dfrac{b}{2a} = -\dfrac{(-20)}{2(25)} = \dfrac{20}{50} = \dfrac{2}{5}$

which is a rational number.

Now evaluate the discriminant, $b^2 - 4ac$.

$(-20)^2 - 4(25)(4)$ Replace a with 25, b with -20, and c with 4.

$= 400 - 400$

$= 0$

Because the discriminant is zero, this equation has only one real solution.

Note: We could avoid calculations with decimal numbers by multiplying the original equation through by 10 so that it becomes: $3x^2 - 7x + 8 = 0$.

c. $0.3x^2 - 0.7x + 0.8 = 0$

Solution $(-0.7)^2 - 4(0.3)(0.8)$ In $b^2 - 4ac$, replace a with 0.3, b with -0.7, and c with 0.8.

$= 0.49 - 0.96$

$= -0.47$

Because the discriminant is negative, there are two nonreal complex solutions.

YOUR TURN Use the discriminant to determine the number and type of solutions for the equation. If the solutions are real, state whether they are rational or irrational.

a. $5x^2 + 8x - 9 = 0$ **b.** $\dfrac{5}{2}x^2 + \dfrac{5}{6} = \dfrac{4}{3}x$ **c.** $x^2 - 0.12x = -0.0036$

OBJECTIVE ❸ **Find the x- and y-intercepts of a quadratic function.** In Chapter 4, we learned that x-intercepts are points where a graph intersects the x-axis. Since an x-intercept is on the x-axis, the y-coordinate of the point will be 0. In Section 6.7, we graphed quadratic functions, which have the form $f(x) = ax^2 + bx + c$ (or $y = ax^2 + bx + c$) and learned their graphs are parabolas. Notice in $y = ax^2 + bx + c$ when we replace y with 0 to find the x-intercepts, we have the quadratic equation $ax^2 + bx + c = 0$, which has two, one, or no real-number solutions. As a result, quadratic functions in the form $y = ax^2 + bx + c$ have two, one, or no x-intercepts. The following graphs illustrate the possibilities.

Note: In Section 6.7, we noted that if $a > 0$, the parabola opens up, and if $a < 0$, the parabola opens down.

Two x-intercepts: If $0 = ax^2 + bx + c$ has two real-number solutions, which occurs if $b^2 - 4ac > 0$, then the graph of $y = ax^2 + bx + c$ has two x-intercepts and looks like one of the following graphs.

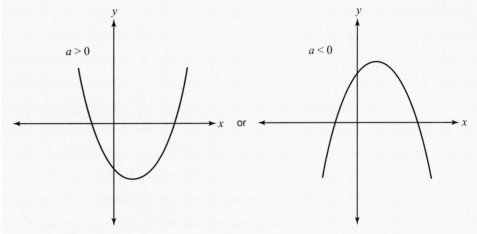

One x-intercept: If $0 = ax^2 + bx + c$ has one real-number solution, which occurs if $b^2 - 4ac = 0$, then the graph of $y = ax^2 + bx + c$ has one x-intercept and looks like one of the following graphs.

continued

11.2 Solving Quadratic Equations Using the Quadratic Formula **923**

No x-intercepts: If $0 = ax^2 + bx + c$ has no real-number solution, which occurs if $b^2 - 4ac < 0$, then the graph of $y = ax^2 + bx + c$ has no x-intercepts and looks like one of the following graphs.

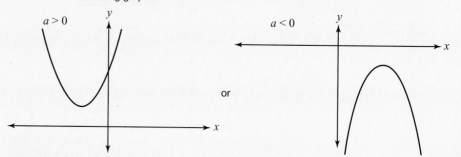

or

We have also learned that a y-intercept is where a graph intersects the y-axis. To find y-intercepts, we let $x = 0$ and solve for y. Notice when we replace x with 0 in $y = ax^2 + bx + c$, we have $y = a(0)^2 + b(0) + c = c$, so the y-intercept is $(0, c)$.

EXAMPLE 3 Find the x- and y-intercepts of $y = x^2 + 2x - 8$, then graph.

Solution For the x-intercepts, letting $y = 0$ gives the equation $0 = x^2 + 2x - 8$, which we will solve using the quadratic formula.

$$x = \frac{-2 \pm \sqrt{2^2 - 4(1)(-8)}}{2(1)} = \frac{-2 \pm \sqrt{36}}{2} = -1 \pm 3 = 2 \quad \text{or} \quad -4$$

x-intercepts: $(2, 0)$ and $(-4, 0)$

Since $c = -8$, the y-intercept is $(0, -8)$.

Note: We could also have calculated the y-intercept:
$$y = (0)^2 + 2(0) - 8 = -8$$

Now we can graph. We know the graph opens up because a is positive.

Note: Knowing the intercepts and whether the parabola opens up or down gives us a pretty good sense of its position. However, if we knew the exact location of the vertex, we could get a more accurate graph. We will learn how to find the coordinates of the vertex in Section 11.4.

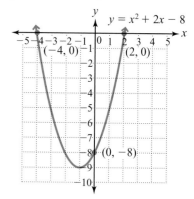

Calculator
TIPS

The ZERO function on a graphing calculator can be used to find the coordinates of x-intercepts. After graphing the parabola on the calculator, select ZERO from the CALC menu. You will then be prompted to enter the left and right bound of one of the intercepts. Move the cursor along the parabola until it is to the left of one of the intercepts and press ENTER. *Then move the cursor along the parabola to the right of the same intercept and press* ENTER *again. When prompted for the guess, press* ENTER *once more. The coordinates of the intercept will appear at the bottom of the screen.*

YOUR TURN Find the *x*- and *y*-intercepts. Verify on a graphing calculator.

a. $y = 3x^2 - 5x + 1$ b. $y = x^2 - 8x + 16$ c. $y = 2x^2 - 3x + 5$

OBJECTIVE ④ Solve applications using the quadratic formula. In physics, the general formula for describing the height of an object after it has been thrown upwards is $h = \dfrac{1}{2}gt^2 + v_0 t + h_0$, where *g* represents the acceleration due to gravity, *t* is the time in flight, v_0 is the initial velocity, and h_0 is the initial height. For Earth, the acceleration due to gravity is -32.2 ft./sec.^{2} or -9.8 m/sec.^{2}

EXAMPLE ④ In an extreme-games competition, a motorcyclist jumps with an initial velocity of 70 feet per second from a ramp height of 25 feet, landing on a ramp with a height of 15 feet. Find the time the motorcyclist is in the air.

Understand   We are given the initial velocity, initial height, and final height and we are to find the time the bike is in the air.

Plan   Use the formula $h = \dfrac{1}{2}gt^2 + v_0 t + h_0$. Since the units are in feet, we will use -32.2 ft./sec.^{2} for *g*.

Execute $15 = \dfrac{1}{2}(-32.2)t^2 + 70t + 25$ Substitute the values.

$15 = -16.1t^2 + 70t + 25$

$0 = -16.1t^2 + 70t + 10$ Subtract 15 from both sides to get the form $ax^2 + bx + c = 0$.

$x = \dfrac{-70 \pm \sqrt{70^2 - 4(-16.1)(10)}}{2(-16.1)}$ Use the quadratic formula, replacing *a* with -16.1, *b* with 70, and *c* with 10.

$x = \dfrac{-70 \pm \sqrt{5544}}{-32.2}$ Simplify in the radical and the denominator.

$x \approx -0.138$ or 4.486 Approximate the two irrational solutions.

Answer   Since the time cannot be negative, the motorcycle is in the air approximately 4.486 seconds.

Check   We can check by evaluating the original formula using $t = 4.486$ seconds to see if the motorcycle indeed lands at a height of 15 feet. We will leave this check to the reader.

YOUR TURN

A ball is thrown from an initial height of 1.5 meters with an initial velocity of 8 meters per second. Find the time for the ball to land on the ground ($h = 0$). Approximate the time to the nearest thousandth.

11.2 EXERCISES

For Extra Help

InterAct Math

Addison-Wesley Math Tutor Center

Video 17/DVT 6

MyMathLab

MathXL

1. Explain the general plan for deriving the quadratic formula.

2. Discuss the advantages and disadvantages of using the quadratic formula to solve quadratic equations.

3. Are there quadratic equations that cannot be solved using factoring? Explain.

4. What part of the quadratic formula is the discriminant?

5. Under what conditions will a quadratic equation have only one solution? Modify the quadratic formula to describe this single solution.

6. If x-intercepts for a quadratic function are imaginary numbers, what does that indicate about its graph?

For Exercises 7–14, rewrite each quadratic equation in the form $ax^2 + bx + c = 0$, then identify a, b, and c.

7. $x^2 - 4x + 9 = 0$

8. $x^2 + 5x - 12 = 0$

9. $2x^2 - 7x = 5$

10. $4x^2 + 9x = -16$

11. $0.5x^2 = x - 2.4$

12. $x = 0.6x^2 + 10.5$

13. $\dfrac{2}{3}x = -\dfrac{1}{2}x^2 + 8$

14. $\dfrac{5}{6}x^2 = \dfrac{1}{4}x - 3$

For Exercises 15–38, solve using the quadratic formula.

15. $x^2 + 9x + 18 = 0$

16. $x^2 - 4x - 21 = 0$

17. $x^2 - x - 1 = 0$

18. $x^2 + 3x - 5 = 0$

19. $x^2 + 6x = 16$

20. $2x^2 - x - 3 = 0$

21. $x^2 - 3x = 0$

22. $x^2 - 2x - 8 = 0$

23. $-x^2 + 8x = 16$

24. $x^2 + 6x = -9$

25. $x^2 + 2 = 2x$

26. $2x^2 = 6x - 9$

27. $3x^2 - 2x = 1$

28. $2x^2 + x - 5 = 0$

29. $3x^2 + 10x + 5 = 0$

30. $3x^2 - x - 3 = 0$

31. $-4x^2 + 6x = 5$

32. $-5x^2 = 3 - 4x$

33. $3x^2 + 8x + 4 = 0$

34. $10x^2 - 12 = -7x$

35. $5x^2 - 6x = -9$ **36.** $-3x^2 + 6x = 8$ **37.** $6x^2 - 3x = 4$ **38.** $6x^2 - 6 - 13x = 0$

For Exercises 39–50, solve using the quadratic formula. If the solutions are irrational, give an exact answer and an approximate answer rounded to the nearest thousandth. (Hint: You might first clear the fractions or decimals by multiplying both sides by an appropriately chosen number.)

39. $2x^2 + 0.1x = 0.03$

40. $3x^2 - 12.9x - 7.2 = 0$

41. $x^2 + \dfrac{1}{2}x - 3 = 0$

42. $x^2 + \dfrac{1}{3}x - \dfrac{2}{3} = 0$

43. $x^2 - \dfrac{9}{4} = 0$

44. $x^2 - 0.0144 = 0$

45. $\dfrac{1}{2}x^2 + \dfrac{3}{2} = x$

46. $\dfrac{1}{5}x^2 + 2 = \dfrac{1}{2}x$

47. $x^2 - 0.5 = -0.06x$

48. $\dfrac{1}{3}x^2 - \dfrac{1}{4}x - \dfrac{1}{24} = 0$

49. $1.2x^2 - 0.6x = -0.5$

50. $2.4x^2 + 4.5 = 6.3x$

For Exercises 51–54, explain the mistake, then solve correctly.

51. Solve $3x^2 - 7x + 1 = 0$ using the quadratic formula.
$$\dfrac{-7 \pm \sqrt{(-7)^2 - (4)(3)(1)}}{2(3)} = \dfrac{-7 \pm \sqrt{49 - 12}}{6}$$
$$= \dfrac{-7 \pm \sqrt{37}}{6}$$

52. Solve $x^2 - 4x + 2 = 0$ using the quadratic formula.
$$\dfrac{4 \pm \sqrt{(-4)^2 - (4)(1)(2)}}{2(1)} = \dfrac{4 \pm \sqrt{16 - 8}}{2}$$
$$= \dfrac{4 \pm \sqrt{8}}{2}$$

53. Solve $x^2 - 2 + 3x = 0$ using the quadratic formula.
$$\dfrac{-(-2) \pm \sqrt{(-2)^2 - (4)(1)(3)}}{2(1)} = \dfrac{2 \pm \sqrt{4 - 12}}{2}$$
$$= \dfrac{2 \pm \sqrt{-8}}{2}$$

54. Solve $x^2 - 4 = 0$ using the quadratic formula.

$$\frac{-(-4) \pm \sqrt{(-4)^2 - (4)(1)(0)}}{2(1)} = \frac{4 \pm \sqrt{16 - 0}}{2}$$

$$= \frac{4 \pm \sqrt{16}}{2}$$

$$= \frac{4 \pm 4}{2}$$

$$= 4, 0$$

For Exercises 55–64, use the discriminant to determine the number and type of solutions for the equation. If the solution(s) are real, state whether they are rational or irrational.

55. $x^2 - x + 3 = 0$

56. $2x^2 - 8x + 8 = 0$

57. $\frac{1}{2}x^2 - 2x = -2$

58. $\frac{1}{3}x^2 = -2x - 3$

59. $x^2 + 4x + 9 = 0$

60. $2x^2 - 3x - 2 = 0$

61. $x^2 + 10x = -25$

62. $x^2 + 4x + 5 = 0$

63. $x^2 - 6x + 6 = 0$

64. $3x^2 - 13x = -8$

For Exercises 65–74, discuss which of the following methods is the best choice for solving the given equation: factoring, using the principle of square roots, or using the quadratic formula.

65. $x^2 - 16 = 0$

66. $x^2 = 20$

67. $3x^2 - 36 = 0$

68. $2x^2 = 5x - 2$

69. $x^2 + x + 1 = 0$

70. $(x + 5)^2 = 16$

71. $x^2 + 6x = 0$

72. $x^2 + 14x + 45 = 0$

73. $5x + 8 = x^2$

74. $x^2 + 7x = 2.5$

For Exercises 75–82, find the x- and y-intercepts.

75. $y = x^2 - x - 2$

76. $y = x^2 - 3x + 2$

77. $y = -x^2 - 2x + 8$

78. $y = x^2 - x - 30$

79. $y = 2x^2 + 15x - 8$

80. $y = -15x^2 + x + 6$

81. $y = -2x^2 + 3x - 6$

82. $y = -3x^2 - 2x - 5$

For Exercises 83–90, translate to a quadratic equation then solve using the quadratic formula.

83. A positive integer squared plus five times its consecutive integer is equal to 71. Find the integers.

84. The square of a positive integer minus half of its consecutive integer is equal to 162. Find the integers.

85. A right triangle has side lengths that are three consecutive integers. Use the Pythagorean theorem to find the lengths of those sides. (Remember that the hypotenuse in a right triangle is always the longest side.)

86. A right triangle has side lengths that are consecutive even integers. Use the Pythagorean theorem to find the lengths of those sides.

87. The length of a rectangular air filter is 3 inches less than twice the width. Find the length and width of the filter if the area is 350 square inches.

88. A small access door for an attic storage area is designed so that the width is 4 feet less than the length. Find the length and width if the area is 5.25 square feet.

89. An architect is experimenting with two different shapes of a room, as shown.

    a. Find x so that the rooms have the same area.

    b. Complete the dimensions for the L-shaped room.

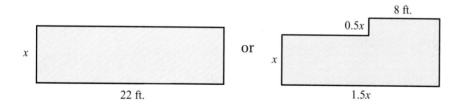

90. A cylinder is to be made so that its volume is equal to that of a sphere with a radius of 9 inches. If a cylinder is to have a height of 4 inches, find its radius.

For Exercises 91–94, use the formula $h = \frac{1}{2}gt^2 + v_0t + h_0$, where g represents the acceleration due to gravity, t is the time in flight, v_0 is the initial velocity, and h_0 is the initial height. For Earth, the acceleration due to gravity is -32.2 ft./sec.^{2} or -9.8 m/sec.^{2} Approximate irrational answers to the nearest hundredth.

91. In 2001, Robby Knieval jumped the Grand Canyon on a motorcycle. Suppose his motorcycle was traveling at 48 feet per second and at the end of the launch ramp his altitude was 485 feet above sea level. If he landed on a ramp that was 460 feet above sea level, how long was he in flight?

92. At the 2003 extreme-games competition, Brian Deegan successfully performed the first 360-degree flip ever attempted in a competition. Suppose his motorcycle was traveling at 12.8 meters per second from the end of a ramp that was 8 meters high and he landed on a ramp at a point 4 meters above the ground. How long was he in the air?

93. A platform diver dives from a platform that is 10 meters above the water.

  a. Write the equation that describes his height during the dive. (Assume his initial velocity is 0.)

  b. Find the time it takes for the diver to be at 5 meters above the water.

  c. Find the time for the diver to enter the water.

94. In a cliff diving championship in Acapulco, a diver dives from a cliff at a height of 70 feet.

  a. Write the equation that describes his height during the dive. (Assume his initial velocity is 0.)

  b. Find the time it takes for the diver to be at 50 feet above the water.

  c. Find the time for the diver to enter the water.

95. The expression $0.5n^2 + 2.5n$ describes the gross income from the sale of a particular software product, where n is the number of units sold in thousands. The expression $4.5n + 16$ describes the cost of producing each unit. Find the number of units that must be produced and sold for the company to break even. (To *break even* means that the gross income and cost are the same.)

96. An economist and marketing manager discover that the expression $2n^2 + 5n$ models the price of a CD based on the demand for it, where n is the number of units (in millions) that the market demands. The expression $-0.5n^2 + 17.1$ describes the price of the CD based on the number of units (also in millions) supplied to the market.

  a. Find the number of units that need to be demanded and supplied so that the price based on demand is equal to the price based on supply.

When the number of units demanded by the market is the same as the number of units supplied to the market, the product is said to be at equilibrium.

b. What is the price of the CD at equilibrium?

97. For the equation $2x^2 - 5x + c = 0$,

    a. Find c so that the equation has only one rational-number solution.

    b. Find the range of values of c for which the equation has two real-number solutions.

    c. Find the range of values of c for which the equation has no real-number solution.

98. For the equation $4x^2 + 6x + c = 0$,

    a. Find c so that the equation has only one rational-number solution.

    b. Find the range of values of c for which the equation has two real-number solutions.

    c. Find the range of values of c for which the equation has no real-number solution.

99. For the equation $ax^2 + 6x + 1 = 0$,

    a. Find a so that the equation has only one rational-number solution.

    b. Find the range of values of a for which the equation has two real-number solutions.

    c. Find the range of values of a for which the equation has no real-number solution.

100. For the equation $3x^2 + bx + 8 = 0$,

    a. Find b so that the equation has only one rational-number solution.

    b. Find all positive values of b for which the equation has two real-number solutions.

    c. Find all positive values of b for which the equation has no real-number solution.

In the equation shown, A, B, C, D, and E are 5 consecutive positive integers where A < B < C < D < E. What are they?

$$A^2 + B^2 + C^2 = D^2 + E^2$$

REVIEW EXERCISES

For Exercises 1–3, simplify.

1. $\sqrt{20}$

2. $(3x + 2) + (-4x + 5)$

3. $(-3x - 1) - (-2x + 5)$

For Exercises 4 and 5, multiply.

4. $-3x(2x - 1)$

5. $(y - 7)(y + 7)$

6. Divide: $\dfrac{2\sqrt{3}}{\sqrt{3} + 1}$

11.3 Solving Equations That Are Quadratic in Form

OBJECTIVES

1 Solve equations by rewriting them in quadratic form.

2 Solve equations that are quadratic in form by using substitution.

3 Solve applications problems using equations that are quadratic in form.

Many equations that are not quadratic equations are **quadratic in form** and can be solved using the methods for solving quadratic equations.

DEFINITION An equation is **quadratic in form** if it can be rewritten as a quadratic equation $au^2 + bu + c = 0$, where $a \neq 0$ and u is a variable or an expression.

OBJECTIVE 1 Solve equations by rewriting them in quadratic form.

Equations with Rational Expressions

In Section 7.6 we solved equations containing rational expressions by multiplying both sides of the equation by the LCD. Those rewritten equations are often quadratic.

CONNECTION

Recall that a solution for an equation with rational expressions is extraneous if it causes one or more of the denominators to equal 0.

EXAMPLE 1 Solve $\dfrac{3}{x + 1} = 1 - \dfrac{3}{x(x + 1)}$.

Solution $x(x + 1) \cdot \dfrac{3}{x + 1} = x(x + 1)\left(1 - \dfrac{3}{x(x + 1)}\right)$ Multiply both sides by the LCD $x(x + 1)$.

$x(x + 1) \cdot \dfrac{3}{x + 1} = x(x + 1) \cdot 1 - x(x + 1) \cdot \dfrac{3}{x(x + 1)}$ Distribute $x(x + 1)$.

$3x = x^2 + x - 3$ Simplify both sides.

$0 = x^2 - 2x - 3$ Subtract $3x$ from both sides to get the quadratic form $ax^2 + bx + c = 0$.

$0 = (x - 3)(x + 1)$ Factor.

$x - 3 = 0$ or $x + 1 = 0$ Use the zero-factor theorem.

$x = 3$ $x = -1$ Solve each equation.

Checks: $x = 3$ $x = -1$

$\dfrac{3}{3 + 1} = 1 - \dfrac{3}{3(3 + 1)}$ $\dfrac{3}{-1 + 1} = 1 - \dfrac{3}{-1(-1 + 1)}$

$\dfrac{3}{4} = 1 - \dfrac{3}{12}$ $\dfrac{3}{0} = 1 - \dfrac{3}{0}$

$\dfrac{3}{4} = \dfrac{3}{4}$ True.

Note: The expression $\dfrac{3}{0}$ is undefined, so -1 is extraneous.

The only solution is 3.

Solve $\dfrac{x}{x-2} = \dfrac{6}{x} + \dfrac{4}{x(x-2)}$.

Equations Containing Radicals

In Section 10.6, we found that after using the power rule on equations containing radicals the result was often a quadratic equation. Remember that these radical equations often have extraneous solutions.

EXAMPLE 2 Solve.

a. $\sqrt{x} + x = 6$

Solution

$\sqrt{x} = 6 - x$	Subtract x from both sides to isolate the radical.
$(\sqrt{x})^2 = (6 - x)^2$	Square both sides.
$x = 36 - 12x + x^2$	Simplify both sides.
$0 = x^2 - 13x + 36$	Subtract x from both sides to write in quadratic form.
$0 = (x - 4)(x - 9)$	Factor.
$x - 4 = 0 \quad$ or $\quad x - 9 = 0$	Use the zero-factor theorem.
$x = 4 \qquad\qquad x = 9$	Solve each equation.

Checks:

$$x = 4 \qquad\qquad x = 9$$
$$\sqrt{4} + 4 \overset{?}{=} 6 \qquad \sqrt{9} + 9 \overset{?}{=} 6$$
$$2 + 4 = 6 \quad \text{True.} \qquad 3 + 9 = 6 \quad \text{False.}$$

Because 9 does not check, it is an extraneous solution. The only solution is 4.

b. $\sqrt{x - 1} = 2x - 1$

Solution

$(\sqrt{x - 1})^2 = (2x - 1)^2$	Square both sides of the equation.
$x - 1 = 4x^2 - 4x + 1$	Simplify both sides.
$0 = 4x^2 - 5x + 2$	Subtract x and add 1 on both sides to write in quadratic form.

We cannot factor $4x^2 - 5x + 2$, so we will use the quadratic formula.

$$x = \frac{-(-5) \pm \sqrt{(-5)^2 - 4(4)(2)}}{2(4)} = \frac{5 \pm \sqrt{-7}}{8} = \frac{5 \pm i\sqrt{7}}{8}$$

Check: Because the solutions are complex, we will not check them.

Note: Written in standard form, these solutions are:

$$\frac{5}{8} + \frac{\sqrt{7}}{8}i \text{ and } \frac{5}{8} - \frac{\sqrt{7}}{8}i.$$

Solve.

a. $x - \sqrt{x - 2} = 0$

b. $\sqrt{6x - 11} = 3x - 1$

Recall that an equation that is quadratic in form can be written as $au^2 + bu + c = 0$, where $a \neq 0$ and u can be an *expression*. We now explore a method for solving these equations where we substitute u for an expression. To use substitution, it is important to note a pattern with the exponents of the terms in a quadratic equation. Notice that the degree of the first term, au^2, is 2, the degree of the middle term, bu, is 1, and the third term is a constant. If a trinomial has one term with an expression raised to the second power, a second term with that same expression raised to the first power, and a third term that is a constant, then the equation is quadratic in form and we can use substitution to solve it.

Consider the equation $x^4 - 13x^2 + 36 = 0$. Notice we can rewrite the equation as $(x^2)^2 - 13(x^2) + 36 = 0$, so it is quadratic in form. By substituting u for each x^2, we have a "friendlier" form of quadratic equation, which we can then solve by factoring, completing the square, or the quadratic formula.

> **Note:** Given an equation containing a trinomial, if the degree of the first term is twice the degree of the middle term, and the third term is a constant, then try using substitution.

$$(x^2)^2 - 13(x^2) + 36 = 0$$
$$u^2 - 13u + 36 = 0 \qquad \text{Substitute } u \text{ for } x^2.$$
$$(u - 9)(u - 4) = 0 \qquad \text{Factor.}$$
$$u - 9 = 0 \quad \text{or} \quad u - 4 = 0 \qquad \text{Use the zero-factor theorem.}$$
$$u = 9 \qquad\qquad u = 4$$

Note that these solutions are for u, not x. We must substitute x^2 back in place of u to finish solving for x.

> **Note:** The equation $x^4 - 13x^2 + 36 = 0$ has four solutions: 3, −3, 2, and −2.

$$x^2 = 9 \qquad x^2 = 4 \qquad \text{Substitute } x^2 \text{ for } u.$$
$$x = \pm\sqrt{9} \qquad x = \pm\sqrt{4}$$
$$x = \pm 3 \qquad x = \pm 2$$

Our example suggests the following procedure.

PROCEDURE **USING SUBSTITUTION TO SOLVE EQUATIONS THAT ARE QUADRATIC IN FORM**

To solve equations that are quadratic in form using substitution,

1. Rewrite the equation so that it is in the form $au^2 + bu + c = 0$.
2. Solve the quadratic equation for u.
3. Substitute for u and solve.
4. Check the solutions.

EXAMPLE 3 Solve.

a. $(x + 2)^2 - 2(x + 2) - 8 = 0$

Solution If we substitute u for $x + 2$, we have an equation in the form $au^2 + bu + c = 0$.

$$(x + 2)^2 - 2(x + 2) - 8 = 0$$
$$u^2 - 2u - 8 = 0 \qquad \text{Substitute } u \text{ for } x + 2.$$
$$(u - 4)(u + 2) = 0 \qquad \text{Factor.}$$
$$u - 4 = 0 \quad \text{or} \quad u + 2 = 0 \qquad \text{Use the zero-factor theorem.}$$
$$u = 4 \qquad\qquad u = -2 \qquad \text{Solve each equation for } u.$$
$$x + 2 = 4 \qquad x + 2 = -2 \qquad \text{Substitute } x + 2 \text{ for } u.$$
$$x = 2 \qquad\qquad x = -4 \qquad \text{Solve each equation for } x.$$

Check: Verify that 2 and −4 make $(x + 2)^2 - 2(x + 2) - 8 = 0$ true. We will leave this check to the reader.

b. $x^{2/3} - x^{1/3} - 6 = 0$

Solution Since $x^{2/3} - x^{1/3} - 6 = 0$ can be written as $(x^{1/3})^2 - x^{1/3} - 6 = 0$, it is quadratic in form. We will substitute u for $x^{1/3}$.

$(x^{1/3})^2 - x^{1/3} - 6 = 0$	Rewrite in quadratic form.
$u^2 - u - 6 = 0$	Substitute u for $x^{1/3}$.
$(u - 3)(u + 2) = 0$	Factor.
$u - 3 = 0$ or $u + 2 = 0$	Use the zero-factor theorem.
$u = 3$ $u = -2$	Solve each equation for u.
$x^{1/3} = 3$ $x^{1/3} = -2$	Substitute $x^{1/3}$ for u.
$(x^{1/3})^3 = 3^3$ $(x^{1/3})^3 = (-2)^3$	Cube both sides of the equations.
$x = 27$ $x = -8$	Simplify.

Checks: $x = 27$ $x = -8$

$27^{2/3} - 27^{1/3} - 6 \overset{?}{=} 0$ $(-8)^{2/3} - (-8)^{1/3} - 6 \overset{?}{=} 0$

$(\sqrt[3]{27})^2 - \sqrt[3]{27} - 6 \overset{?}{=} 0$ $(\sqrt[3]{-8})^2 - \sqrt[3]{-8} - 6 \overset{?}{=} 0$

$3^2 - 3 - 6 \overset{?}{=} 0$ $(-2)^2 - (-2) - 6 \overset{?}{=} 0$

$9 - 3 - 6 \overset{?}{=} 0$ $4 + 2 - 6 \overset{?}{=} 0$

$0 = 0$ True $0 = 0$ True

ANSWERS

a. $\pm 3, \pm 1$
b. ± 4
c. $8, -1$

YOUR TURN   Solve.

a. $x^4 - 10x^2 + 9 = 0$ **b.** $(n - 2)^2 + 4(n - 2) - 12 = 0$ **c.** $x^{2/3} - x^{1/3} - 2 = 0$

OBJECTIVE   **3**   Solve applications problems using equations that are quadratic in form.

EXAMPLE 4   The average speed of a car is 10 miles per hour more than the average speed of a bus. The bus takes 1 hour longer than the car to travel 200 miles. Find how long it takes the car to travel 200 miles.

Understand   Both the car and the bus travel 200 miles, but it takes the bus 1 hour longer than the car. The rate of the car is 10 miles per hour more than the rate of the bus.

Plan   We will use a table to organize the information, then write an equation, which we can solve.

Execute   We will let x represent the time for the car to travel 200 miles. Since the bus travels 1 more hour, $x + 1$ describes the time for the bus to travel 200 miles.

Vehicle	d	t	r
bus	200	$x + 1$	$\dfrac{200}{x + 1}$
car	200	x	$\dfrac{200}{x}$

We use $r = \dfrac{d}{t}$ to describe each rate.

Since the rate of the car is 10 miles per hour faster than the rate of the bus, we can say (the rate of the car) = (the rate of the bus) + 10.

$$\frac{200}{x} = \frac{200}{x+1} + 10$$

$$x(x+1)\left(\frac{200}{x}\right) = x(x+1)\left(\frac{200}{x+1} + 10\right)$$ **Multiply both sides by the LCD $x(x+1)$.**

$$(x+1)(200) = x(x+1)\left(\frac{200}{x+1}\right) + x(x+1)(10)$$ **Simplify both sides.**

$$200x + 200 = 200x + 10x^2 + 10x$$ **Simplify.**

$$0 = 10x^2 + 10x - 200$$ **Subtract 200x and 200 from both sides to get quadratic form.**

$$0 = x^2 + x - 20$$ **Divide both sides by 10.**

$$0 = (x+5)(x-4)$$ **Factor.**

$$x + 5 = 0 \quad \text{or} \quad x - 4 = 0$$ **Use the zero-factor theorem.**

$$x = -5 \qquad\qquad x = 4$$ **Solve each equation.**

Answer   Since time cannot be negative, it takes the car 4 hours to travel 200 miles.

Check   The rate of the car is $\dfrac{200}{x} = \dfrac{200}{4} = 50$ miles per hour, so in 4 hours the car travels $50(4) = 200$ miles. The bus travels 200 miles in $4 + 1 = 5$ hours at a rate of $\dfrac{200}{4+1} = \dfrac{200}{5} = 40$ miles per hour, so in 5 hours the bus travels $5(40) = 200$ miles.

Notice also that the car's rate is 10 miles per hour more than the bus's.

EXAMPLE 5   Bobby and Pam manufacture and install blinds. Working together, they can install blinds in every window of an average-sized house in $1\frac{1}{3}$ hours. Working alone, Bobby takes 2 hours longer than Pam to do the same installation. How long does each take to install blinds in an average-sized house working alone?

Understand   We are given the time it will take Bobby and Pam working together and are asked to find how long it would take each individually. We also know that it will take Bobby 2 hours longer than it would take Pam.

Plan   We will use a table to organize the information, write an equation, and then solve.

Execute   We will let x represent the number of hours for Pam to install working alone. Since Bobby takes 2 more hours, his time working alone is represented by $x + 2$.

Worker	Time to Complete the Job Alone	Rate of Work	Time at Work	Portion of Job Completed
Pam	x	$\dfrac{1}{x}$	$\dfrac{4}{3}$	$\dfrac{4}{3x}$
Bobby	$x + 2$	$\dfrac{1}{x + 2}$	$\dfrac{4}{3}$	$\dfrac{4}{3(x + 2)}$

Multiplying the rate of work and the time at work gives an expression of the amount of the job completed.

The total job in this case is 1 average-sized house, so we can write an equation that combines their individual expressions for work completed and set this sum equal to 1.

(portion Pam does) + (portion Bobby does) = 1 (the entire job)

$$\frac{4}{3x} + \frac{4}{3(x + 2)} = 1$$

$$3x(x + 2)\left(\frac{4}{3x} + \frac{4}{3(x + 2)}\right) = 3x(x + 2)(1) \qquad \text{Multiply both sides by the LCD } 3x(x + 2).$$

$$3x(x + 2)\left(\frac{4}{3x}\right) + 3x(x + 2)\left(\frac{4}{3(x + 2)}\right) = 3x^2 + 6x \qquad \text{Simplify both sides.}$$

$$4(x + 2) + 4x = 3x^2 + 6x \qquad \text{Continue simplifying.}$$

$$4x + 8 + 4x = 3x^2 + 6x \qquad \text{Distribute.}$$

$$8x + 8 = 3x^2 + 6x \qquad \text{Combine like terms.}$$

$$0 = 3x^2 - 2x - 8 \qquad \text{Subtract } 8x \text{ and 8 from both sides.}$$

$$0 = (3x + 4)(x - 2) \qquad \text{Factor.}$$

$$3x + 4 = 0 \quad \text{or} \quad x - 2 = 0 \qquad \text{Use the zero-factor theorem.}$$

$$3x = -4 \qquad\qquad x = 2 \qquad \text{Solve each equation.}$$

$$x = -\frac{4}{3}$$

Answer   Since negative time makes no sense in the context of this problem, it takes Pam 2 hours to install working alone and Bobby $x + 2 = 2 + 2 = 4$ hours working alone.

Check   Since Pam can install in 2 hours, she can install $\dfrac{1}{2}$ of an average-sized house in 1 hour.

Since Bobby takes 4 hours to do the same work, he can install $\dfrac{1}{4}$ of an average-sized house in 1 hour. Together, they can install $\dfrac{1}{2} + \dfrac{1}{4} = \dfrac{2}{4} + \dfrac{1}{4} = \dfrac{3}{4}$ of an average-sized house in 1 hour, so in $1\dfrac{1}{3}$ hours they can install blinds in $1\dfrac{1}{3} \cdot \dfrac{3}{4} = \dfrac{4}{3} \cdot \dfrac{3}{4} = 1$ average-sized house.

a. The average speed of a passenger train is 25 miles per hour more than the average speed of a car. The time required for the car to travel 300 miles is 2 hours more than the time required for the train. Find the average speed of the car.

b. Terri and Tommy run a flea market. It takes Terri 2 hours longer to put the merchandise out than it does Tommy. If they can put the merchandise out together in $1\frac{7}{8}$ hours, how long would it take each working alone?

11.3 EXERCISES

1. What does it mean to say that an equation is quadratic in form?

2. Is $x^{3/4} - x^{1/4} - 6 = 0$ quadratic in form? Why or why not?

3. If you solved $3\left(\dfrac{x+2}{3}\right)^2 + 13\left(\dfrac{x+2}{3}\right) - 10 = 0$ using substitution, what would your substitution be?

4. If you solved $x^4 - 7x^2 + 12 = 0$ using substitution, what would your substitution be?

5. If Alisha can clean her house in x hours, what part of her house can she clean in 1 hour?

6. If a car travels 400 miles in x hours, how many miles does it travel in 1 hour?

For Exercises 7–18, solve the equations with rational expressions.

7. $\dfrac{1}{x} + \dfrac{1}{x+2} = \dfrac{3}{4}$

8. $\dfrac{1}{x} + \dfrac{2}{x-3} = \dfrac{5}{6}$

9. $\dfrac{60}{x+2} = \dfrac{60}{x} - 5$

10. $\dfrac{120}{x+2} = \dfrac{120}{x} - 5$

11. $\dfrac{1}{p-4} + \dfrac{1}{4} = \dfrac{8}{p^2-16}$

12. $\dfrac{1}{y-5} - \dfrac{10}{y^2-25} = -\dfrac{1}{5}$

13. $\dfrac{6}{2y+5} = \dfrac{2}{y+5} + \dfrac{1}{5}$

14. $\dfrac{6}{2x+3} = \dfrac{2}{x-6} + \dfrac{4}{3}$

15. $1 + 2x^{-1} - 8x^{-2} = 0$

16. $1 + 5x^{-1} + 6x^{-2} = 0$

17. $3 + 13x^{-1} - 10x^{-2} = 0$

18. $2 - x^{-1} - 15x^{-2} = 0$

For Exercises 19–30, solve the equations with radical expressions.

19. $x - 8\sqrt{x} + 15 = 0$

20. $x - 3\sqrt{x} + 2 = 0$

21. $2x - 5\sqrt{x} - 7 = 0$

22. $3x + 4\sqrt{x} - 4 = 0$

23. $\sqrt{2a + 5} = 3a - 3$

24. $\sqrt{3b + 1} = 5b - 3$

25. $\sqrt{2m - 8} - m - 1 = 0$

26. $\sqrt{8x - 12x} - 2x + 3 = 0$

27. $\sqrt{4x + 1} = \sqrt{x + 2} + 1$

28. $\sqrt{3x + 7} = \sqrt{2x + 3} + 1$

29. $\sqrt{2x + 1} - \sqrt{3x + 4} = -1$

30. $\sqrt{2x - 1} - \sqrt{4x + 5} = -2$

For Exercises 31–52, solve using substitution.

31. $x^4 - 10x^2 + 9 = 0$

32. $x^4 - 13x^2 + 36 = 0$

33. $4x^4 - 13x^2 + 9 = 0$

34. $9x^4 - 13x^2 + 4 = 0$

35. $x^4 + 5x^2 - 36 = 0$

36. $x^4 - 3x^2 - 4 = 0$

37. $(x + 2)^2 + 6(x + 2) + 8 = 0$

38. $(x - 3)^2 + 2(x - 3) - 15 = 0$

39. $2(x + 3)^2 - 9(x + 3) - 5 = 0$

40. $3(x - 1)^2 + 4(x - 1) - 4 = 0$

41. $\left(\dfrac{x - 1}{2}\right)^2 + 8\left(\dfrac{x - 1}{2}\right) + 15 = 0$

42. $\left(\dfrac{x + 2}{3}\right)^2 - 5\left(\dfrac{x + 2}{3}\right) - 6 = 0$

43. $2\left(\dfrac{x + 2}{2}\right)^2 + \left(\dfrac{x + 2}{2}\right) - 3 = 0$

44. $3\left(\dfrac{x - 1}{3}\right)^2 + 10\left(\dfrac{x - 1}{3}\right) - 8 = 0$

45. $x^{2/3} - 5x^{1/3} + 6 = 0$

46. $x^{2/3} + 3x^{1/3} - 10 = 0$

47. $2x^{2/3} - 3x^{1/3} - 2 = 0$

48. $3x^{2/3} - 4x^{1/3} - 4 = 0$

49. $x^{1/2} - 7x^{1/4} + 12 = 0$

50. $x^{1/2} - 4x^{1/4} + 3 = 0$

51. $5x^{1/2} + 8x^{1/4} - 4 = 0$

52. $2x^{1/2} - x^{1/4} - 3 = 0$

For Exercises 53–64, solve.

53. The average rate of a bus is 15 miles per hour more than the average rate of a truck. The truck takes 1 hour longer than the bus to travel 180 miles. How long does it take the bus to travel 180 miles?

54. A charter business has two types of small plane, a jet and a twin propeller plane. The prop-plane's average air speed is 120 miles per hour less than the jet's. The prop-plane takes 1 hour longer to travel 720 miles than the jet. How long does it take the prop-plane to travel the 720 miles?

55. The average speed of the winner of the Boston Marathon in 2003 was 3.8 miles per hour faster than the person who finished 1 hour behind. Find the time it took the winner to run the 26-mile course.

56. In the 2003 Tour de France, Lance Armstrong's average speed was 0.0081 kilometers per hour faster than the second-place finisher, who finished 1 minute behind Armstrong. Find Armstrong's time to complete the 3415-kilometer race.

57. Suppose the time required for a bus to travel 360 miles is 3 hours more than the time required for a motorcycle to travel 300 miles. If the average rate of the motorcycle is 15 miles per hour more than the average rate of the bus, find the average rate of the bus.

58. Suppose the time required for a truck to travel 400 miles is 3 hours more than the time required for a car to travel 300 miles. If the average rate of the car is 10 miles per hour more than the average rate of the truck, find the average rate of the truck.

59. After training hard for a year, a novice cyclist discovers that she has increased her average rate by 6 miles per hour and can travel 36 miles in 1 less hour than a year ago.

    a. What was her old average rate? What is her new average rate?

    b. What was her old time to travel 36 miles? What is her new time for 36 miles?

60. A high school track coach determines that his fastest long-distance runner runs 2 miles per hour faster than his slowest runner. The slower runner takes a half-hour longer to run 12 miles than the faster runner.

    a. Find the average rates of both runners.

    b. Find the time for both runners to run 12 miles.

61. Billy and Jody are commercial fishermen. Working alone it takes Billy 2 hours longer to run the hoop nets than it takes Jody working alone. Together they can run the hoop nets in $2\frac{2}{5}$ hours. How long does it take each working alone?

62. Using a riding mower, Fran can mow the grass at the campground in 4 hours less time than it takes Donnie using a push mower. Together they can mow the grass in $2\frac{2}{3}$ hours. How long does it take each working alone?

63. A newspaper has two presses, one of which is older than the other. Working alone, the newer press can print all the copies for a typical day in a half-hour less time than the older press. When running at the same time, they print all of a typical day's copies in 2 hours. How long would it take to print a day's worth if they worked alone?

Of Interest

Lance Armstrong has won the Tour de France a record 5 times each year from 1999 to 2003. This seemingly superhuman feat is all the more incredible given that in 1996, he was diagnosed with advanced testicular cancer, which had already spread to his lungs and brain. After surgery, doctors gave him less than a 50/50 chance of recovery even with intensive chemotherapy. Only five months after being diagnosed, Armstrong began training and he returned to professional racing in 1998, winning several prestigious events that year. (*Source:* www.lancearmstrong.com/lance/online2.nsf/html/bio)

64. A school copy center prints the school newsletter in a half-hour using two copiers running at the same time. If the center uses only one copier, the faster of the two copiers takes 6 minutes less time than the other copier to print all of the newsletters. How long does each copier take to print the newsletters working alone?

 On a guitar, the frequency of the vibrating string is related to the tension on the string. Suppose two strings of the same diameter and length are placed on an instrument and wound to different tensions. The formula $\dfrac{F_1^2}{F_2^2} = \dfrac{T_1}{T_2}$ describes the relationship between their frequency and tension.

65. One string is wound on a guitar to a tension of 50 pounds. A second string of the same diameter is wound to 60 pounds. If the string with the greater tension has a frequency that is 40 vibrations per second greater than the other string, what are the frequencies of the two strings? Round to the nearest ten.

66. Two strings of the same diameter are wound on a banjo to different tensions: one at 80 pounds and the other at 90 pounds. The frequency of the string under less tension is 20 vibrations per second less than the other string. What are the frequencies of the two strings? Round to the nearest ten.

> **Of Interest**
>
> In addition to tension, the diameter and length of a string also affect its frequency. Increasing the diameter or length of a string under the same tension decreases its frequency.

REVIEW EXERCISES

1. Evaluate $-\dfrac{b}{2a}$ when a is 2 and b is 8.

2. Find the x- and y-intercepts for $2x + 3y = 6$.

3. If $f(x) = 2x^2 - 3x + 1$, find $f(-1)$.

For Exercises 4–6, graph.

4. $f(x) = |x - 3| + 1$ **5.** $f(x) = x^2 - 3$ **6.** $f(x) = 3x^2$

11.4 Graphing Quadratic Functions

OBJECTIVE

1. Graph quadratic functions of the form $f(x) = ax^2$.
2. Graph quadratic functions of the form $f(x) = ax^2 + k$.
3. Graph quadratic functions of the form $f(x) = a(x - h)^2$.
4. Graph quadratic functions of the form $f(x) = a(x - h)^2 + k$.
5. Graph quadratic functions in the form $f(x) = ax^2 + bx + c$.
6. Solve applications involving parabolas.

In Sections 6.7 and 8.4 we learned that quadratic functions have the form $f(x) = ax^2 + bx + c$ and that the graphs of these functions are parabolas. We graphed these functions by finding many ordered-pair solutions, as in the following graph of $f(x) = x^2$.

Note: Remember, $f(x)$ and y are interchangeable.

x	$f(x)$
-3	9
-2	4
-1	1
0	0
1	1
2	4
3	9

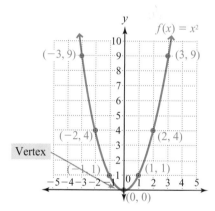

In this section, we will learn more about parabolas in order to graph quadratic functions more efficiently. We will start with simple forms of quadratic functions and build to the $f(x) = ax^2 + bx + c$ form.

OBJECTIVE 1 Graph quadratic functions of the form $f(x) = ax^2$.

First we consider $f(x) = ax^2$ to discover how a affects the parabola. Recall from Section 6.7 that the sign of a determines whether the parabola opens up or down. If $a > 0$, the parabola opens up and if $a < 0$, then the parabola opens down. For example, the graph of $f(x) = x^2$ opens up because a is positive 1, whereas, the graph of $f(x) = -x^2$ opens down because a is -1.

x	f(x)
-3	-9
-2	-4
-1	-1
0	0
1	-1
2	-4
3	-9

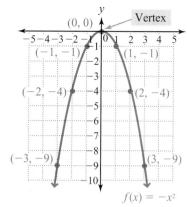

$f(x) = -x^2$

The value of a also affects the width of a parabola.

EXAMPLE 1 Compare the graphs of each equation.

a. $f(x) = \dfrac{1}{4}x^2$, $g(x) = \dfrac{1}{3}x^2$, $h(x) = \dfrac{1}{2}x^2$, $k(x) = x^2$, and $m(x) = 2x^2$

Solution We will graph all five functions on the same grid.

Note: All these graphs open up because a is positive. Also notice that as a increases, the parabolas appear narrower.

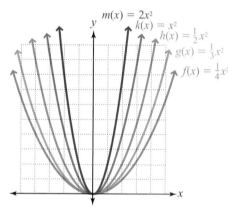

$m(x) = 2x^2$
$k(x) = x^2$
$h(x) = \frac{1}{2}x^2$
$g(x) = \frac{1}{3}x^2$
$f(x) = \frac{1}{4}x^2$

b. $f(x) = -\dfrac{1}{4}x^2$, $g(x) = -\dfrac{1}{3}x^2$, $h(x) = -\dfrac{1}{2}x^2$, $k(x) = -x^2$, and $m(x) = -2x^2$

Note: All these graphs open down because a is negative. Also notice that as $|a|$ increases, the parabolas appear narrower.

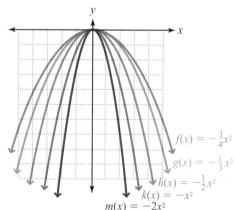

$f(x) = -\frac{1}{4}x^2$
$g(x) = -\frac{1}{3}x^2$
$h(x) = -\frac{1}{2}x^2$
$k(x) = -x^2$
$m(x) = -2x^2$

Example 1 suggests the following conclusion about a.

Conclusion If $a > 0$, then the greater the value of a, the narrower the parabola. If $a < 0$, then the greater the $|a|$, then narrower the parabola.

Graph the following equations.

a. $f(x) = -4x^2$

b. $f(x) = \dfrac{3}{4}x^2$

OBJECTIVE **2** **Graph functions of the form $f(x) = ax^2 + k$.** Recall from Section 6.7 that the vertex of a parabola is the *lowest* point in a parabola that opens up and the *highest* point in a parabola that opens down. Notice the vertex of every parabola in Example 1 is at the origin. We'll now consider parabolas with vertices shifted up or down from the origin. The equations of these parabolas have the form $f(x) = ax^2 + k$, where k is a constant.

EXAMPLE **2** Graph the following equations.

a. $g(x) = 2x^2 + 3$ and $h(x) = 2x^2 - 4$

Solution We will graph both functions on the same grid.

Note: The vertex of the "basic" function $f(x) = 2x^2$ is the origin $(0, 0)$. The vertex of $g(x) = 2x^2 + 3$ is shifted up 3 and the vertex of $h(x) = 2x^2 - 4$ is shifted down 4 from the origin.

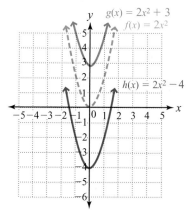

b. $g(x) = -x^2 + 4$ and $h(x) = -x^2 - 2$

Solution

Note: The vertex of the "basic" function $f(x) = -x^2$ is the origin $(0, 0)$. The vertex of $g(x) = -x^2 + 4$ is shifted up 4 and the vertex of $h(x) = -x^2 - 2$ is shifted down 2 from the origin.

ANSWERS

a.

b.

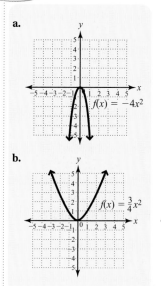

Example 2 suggests the following conclusion about the constant k.

Conclusion Given a function in the form $f(x) = ax^2 + k$, if $k > 0$, then the graph of $f(x) = ax^2$ is shifted k units *up* from the origin. If $k < 0$, then the graph of $f(x) = ax^2$ is shifted k units *down* from the origin. The new position of the vertex is $(0, k)$.

YOUR TURN Graph the following equations.

    a. $f(x) = 3x^2 - 2$ **b.** $f(x) = -\dfrac{1}{2}x^2 + 4$

OBJECTIVE **3** **Graph quadratic functions of the form $f(x) = a(x - h)^2$.** We now consider the form $f(x) = a(x - h)^2$ and we will see that the constant h causes the parabola to shift right or left.

EXAMPLE **3** Graph $m(x) = 2(x - 3)^2$ and $n(x) = 2(x + 4)^2$.

Solution

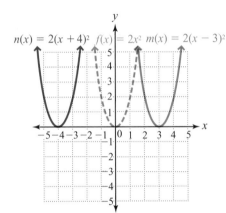

Note: The vertex of the "basic" function $f(x) = 2x^2$ is the origin $(0, 0)$. The vertex of $m(x) = 2(x - 3)^2$ is shifted right 3 units, and the vertex of $n(x) = 2(x + 4)^2$ is shifted left 4 units from the origin.

Example 3 suggests the following conclusion about h in $f(x) = a(x - h)^2$.

Conclusion   Given a function in the form $f(x) = a(x - h)^2$, if $h > 0$, then the graph of $f(x) = ax^2$ is shifted h units *right* from the origin. If $h < 0$, then the graph of $f(x) = ax^2$ is shifted h units *left* from the origin. The new position of the vertex is $(h, 0)$.

YOUR TURN Graph the following equations.

    a. $f(x) = -2(x + 1)^2$ **b.** $g(x) = \dfrac{1}{3}(x - 4)^2$

OBJECTIVE **4** **Graph quadratic functions of the form $f(x) = a(x - h)^2 + k$.** From Examples 2 and 3, we can conclude that in a function of the form $f(x) = a(x - h)^2 + k$, the constants h and k shift the vertex of the parabola from the origin to the position (h, k). Another feature of parabolas we need to consider is the **axis of symmetry**.

DEFINITION   *Axis of symmetry:* A line that divides a graph into two symmetrical halves.

To divide a parabola into two symmetrical halves, the axis of symmetry must pass through the vertex, so the equation of the axis of symmetry will have x equal to the x-coordinate of the vertex. For example, the vertex of $f(x) = 2(x - 3)^2$ is $(3, 0)$, so the axis of symmetry is $x = 3$. We can summarize all that we have learned in the following rule.

RULE **PARABOLA WITH VERTEX (h, k).**

The graph of a function in the form $f(x) = a(x - h)^2 + k$ is a parabola with vertex at (h, k). The equation of the axis of symmetry is $x = h$. The parabola opens up if $a > 0$ and down if $a < 0$. The larger the $|a|$, the narrower the graph.

Now let's graph quadratic equations of the form $f(x) = a(x - h)^2 + k$.

EXAMPLE 4 Given $f(x) = 2(x - 3)^2 + 1$, determine whether the graph opens up or down, find the vertex and axis of symmetry, and draw the graph.

Solution We see that $f(x) = 2(x - 3)^2 + 1$ is in the form $f(x) = a(x - h)^2 + k$, where $a = 2$, $h = 3$, and $k = 1$. Since a is positive 2, the parabola opens up. The vertex is at $(3, 1)$ and the axis of symmetry is $x = 3$. To complete the graph, we find a few points on either side of the axis of symmetry.

Note: We find these additional points by choosing x-values on either side of the axis of symmetry and by using the equation to find the corresponding y-values.

x	y
2	3
1	9
4	3
5	9

YOUR TURN

Given the equation $f(x) = -\dfrac{1}{2}(x - 1)^2 + 3$, determine whether the graph opens up or down, find the vertex and axis of symmetry, and draw the graph.

OBJECTIVE **5** **Graph quadratic functions in the form $f(x) = ax^2 + bx + c$.** The advantage of the form $f(x) = a(x - h)^2 + k$ is that we can "see" the vertex, axis of symmetry, and whether the parabola opens up or down by looking at the equation. If an equation is in the form $f(x) = ax^2 + bx + c$, it can be transformed into $f(x) = a(x - h)^2 + k$ by completing the square.

EXAMPLE **5** Rewrite $f(x) = x^2 - 2x - 8$ in the form $f(x) = a(x - h)^2 + k$. Determine whether the graph opens up or down, find the vertex and axis of symmetry, and draw the graph.

Solution

$y = x^2 - 2x - 8$	Replace $f(x)$ with y to make manipulations more workable.
$y + 8 = x^2 - 2x$	Add 8 to both sides of the equation to isolate the x terms.
$y + 8 + 1 = x^2 - 2x + 1$	Add 1 to both sides of the equation to complete the square.
$y + 9 = (x - 1)^2$	Write the right side as a square.
$y = (x - 1)^2 - 9$	Subtract 9 from both sides to solve for y.

Since $a = 1$ and 1 is positive, the graph opens up. The vertex is at $(1, -9)$ and the axis of symmetry is $x = 1$. Plot a few points on either side of the axis of symmetry and graph.

x	$f(x)$
0	-8
-1	-5
2	-8
3	-5

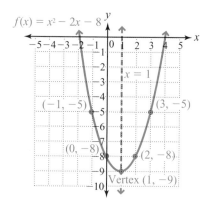

ANSWERS

$f(x) = (x - 1)^2 - 4$,
Opens up, Vertex: $(1, -4)$,
Axis: $x = 1$

YOUR TURN

Rewrite $f(x) = x^2 - 2x - 3$ in the form $f(x) = a(x - h)^2 + k$. Determine whether the graph opens up or down, find the vertex and axis of symmetry, and draw the graph.

We can also find the vertex using a formula, which we can derive by completing the square on the general form $f(x) = ax^2 + bx + c$.

Note: Adding $\left[\dfrac{1}{2}\left(\dfrac{b}{a}\right)\right]^2$ or $\dfrac{b^2}{4a^2}$ completes the square inside the parentheses. Since those parentheses are multiplied by a, we add $a \cdot \dfrac{b^2}{4a^2}$ to the left side to keep the equation balanced.

$y = ax^2 + bx + c$	Replace $f(x)$ with y.
$y - c = ax^2 + bx$	Isolate the x terms.
$y - c = a\left(x^2 + \dfrac{b}{a}x\right)$	Factor out a.
$y - c + \dfrac{ab^2}{4a^2} = a\left(x^2 + \dfrac{b}{a}x + \dfrac{b^2}{4a^2}\right)$	Complete the square.
$y - \dfrac{4a^2c - ab^2}{4a^2} = a\left(x + \dfrac{b}{2a}\right)^2$	Find the LCD of the two terms on the left. Factor on the right.

$$y - \frac{a(4ac - b^2)}{4a^2} = a\left(x + \frac{b}{2a}\right)^2 \qquad \text{Remove a common factor of } a \text{ on the left side.}$$

$$y - \frac{4ac - b^2}{4a} = a\left(x + \frac{b}{2a}\right)^2 \qquad \text{Reduce to lowest terms.}$$

$$y = a\left(x + \frac{b}{2a}\right)^2 + \frac{4ac - b^2}{4a} \qquad \text{Add } \frac{4ac - b^2}{4a} \text{ to both sides.}$$

Notice that if we rewrite the last line above as $y = a\left(x - \left(-\frac{b}{2a}\right)\right)^2 + \frac{4ac - b^2}{4a}$, we see that $h = -\frac{b}{2a}$ and $k = \frac{4ac - b^2}{4a}$ so that the vertex is $\left(-\frac{b}{2a}, \frac{4ac - b^2}{4a}\right)$.

Notice this also means that the axis of symmetry is $x = -\frac{b}{2a}$. Also, though the y-coordinate of the vertex is $\frac{4ac - b^2}{4a}$, it is usually easier to find it by substituting the x-coordinate into the original function.

PROCEDURE **FINDING THE VERTEX OF A QUADRATIC FUNCTION IN THE FORM $f(x) = ax^2 + bx + c$**

Given an equation in the form $f(x) = ax^2 + bx + c$, to determine the vertex,

1. Find the x-coordinate using the formula $-\frac{b}{2a}$.

2. Find the y-coordinate by evaluating $f\left(-\frac{b}{2a}\right)$.

EXAMPLE **6** For the function $f(x) = 2x^2 - 8x + 5$, find the coordinates of the vertex.

Solution First, find the x-coordinate of the vertex using $-\frac{b}{2a}$.

$$x\text{-coordinate of the vertex: } -\frac{b}{2a} = -\frac{(-8)}{2(2)} = \frac{8}{4} = 2$$

Now find the y-coordinate by evaluating $f(2)$.

$$y\text{-coordinate of the vertex: } f(2) = 2(2)^2 - 8(2) + 5 = 8 - 16 + 5 = -3$$

The vertex is $(2, -3)$.

Note: Using $\frac{4ac - b^2}{4a}$ also gives the y-coordinate:
$$\frac{4(2)(5) - (-8)^2}{4(2)} = \frac{-24}{8} = -3$$
However, substituting the x-value, 2, into the original function is easier.

ANSWER

$(-3, -46)$

YOUR TURN

For the equation $f(x) = 3x^2 + 18x - 19$, find the vertex.

OBJECTIVE **6** **Solve applications involving parabolas.** Because the vertex of a parabola is the highest point in parabolas that open down or the lowest point in parabolas that open up, we can find a maximum or minimum value in applications involving quadratic functions.

EXAMPLE 7

A toy rocket is launched straight up with an initial velocity of 40 feet per second. The equations $h = -16t^2 + 40t$ describes the height, h, of the rocket t seconds after being launched.

a. Find the maximum height that the rocket reaches.
b. Find the amount of time that the rocket is in the air.

Solution

a. Since the graph of $h = -16t^2 + 40t$ is a parabola that opens down ($a = -16$), the maximum height occurs at its vertex.

$$t\text{-coordinate of the vertex: } -\frac{b}{2a} = -\frac{40}{2(-16)} = -\frac{40}{-32} = \frac{5}{4} = 1.25 \text{ seconds}$$

To find the h-coordinate of the vertex we replace t with 1.25 in $h = -16t^2 + 40t$.

$$h = -16(1.25)^2 + 40(1.25) = 25 \text{ feet}$$

The vertex is (1.25, 25), so the maximum height is 25 feet, which occurs 1.25 seconds after the rocket is launched.

> **Note:** We could have found the vertex by writing the equation in $y = a(x - h)^2 + k$ form, which would be $h = -16(t - 1.25)^2 + 25$.

b. The time the rocket is in the air is from launch until it returns to the ground. At launch and upon returning to the ground, the rocket's height is 0, so we need to find t when $h = 0$.

$0 = -16t^2 + 40t$		Replace h with 0.
$0 = -8t(2t - 5)$		Factor out a common factor of $-8t$.
$-8t = 0$ or	$2t - 5 = 0$	Use the zero-factor theorem.
$t = 0$	$2t = 5$	
	$t = 2.5$	

This means the height is 0 when $t = 0$ and when $t = 2.5$ seconds, so the rocket is in the air for 2.5 seconds.

> **Note:** The graph of $h = -16t^2 + 40t$ is *not* the flight path of the rocket.

The graph of $h = -16t^2 + 40t$ shows that at 0 seconds the rocket is at 0 feet, which is when it is launched. Its height increases to a maximum of 25 feet 1.25 seconds after launch. Then the height decreases back to 0 feet 2.5 seconds after launch.

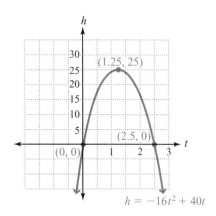

$$h = -16t^2 + 40t$$

YOUR TURN

A soccer player kicks the ball straight up with an initial velocity of 56 feet per second. The equation $h = -16t^2 + 56t$ describes the height, h, of the ball t seconds after being kicked.

a. After how many seconds is the ball at its maximum height?
b. What is the maximum height that the ball reaches?
c. How long is the ball in the air?

If an object is thrown or shot upwards and outwards, gravity causes its path, or trajectory, to be in the shape of a parabola.

Trajectory is the path the object travels

Initial height

Note: x represents the horizontal distance the object travels, and y represents the vertical distance the object travels.

EXAMPLE 8 The equation $y = -0.8x^2 + 2.4x + 6$ when $x \geq 0$ and $y \geq 0$ describes the trajectory of a ball thrown upwards and outwards from an initial height of 6 feet.

a. What is the maximum height that the ball reaches?
b. How far does the ball travel horizontally?
c. Graph the trajectory of the ball.

Solution

a. The maximum height will be the y-coordinate of the vertex.

x-coordinate of the vertex: $-\dfrac{b}{2a} = -\dfrac{(2.4)}{2(-0.8)} = 1.5$

y-coordinate of the vertex: $y = -0.8(1.5)^2 + 2.4(1.5) + 6 = 7.8$

The vertex is (1.5, 7.8), so the maximum height is 7.8 feet.

b. To find how far the ball travels, we need to find the x-value when it hits the ground, which is where $y = 0$.

$0 = -0.8x^2 + 2.4x + 6$ Substitute 0 for y in the equation.

$x = \dfrac{-(2.4) \pm \sqrt{(2.4)^2 - 4(-0.8)(6)}}{2(-0.8)}$ Substitute $a = -0.8$, $b = 2.4$, and $c = 6$ into the quadratic formula.

$= \dfrac{-2.4 \pm \sqrt{24.96}}{-1.6} \approx -1.62 \text{ or } 4.62$

The negative value does not make sense in the context of this problem, so the distance the ball travels must be approximately 4.62 feet.

Note: Because the equation describes the trajectory when $x \geq 0$ and $y \geq 0$, we only consider the part of the graph in the first quadrant.

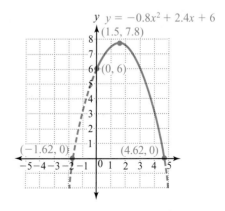

Besides being able to graph parabolas on the graphing calculator, you can also find the x-intercepts and the vertex.

Let's work through Example 8 using a graphing calculator. First, press the [Y=] key and enter the equation $y = -0.8x^2 + 2.4x + 6$ as Y1, then press GRAPH to see the graph.

Now, to find the x-intercepts, press [2nd] [TRACE], which will bring up the CALC menu. Use the arrow keys to highlight ZERO in the CALC menu, then press [ENTER] to select it. This will return you to the graph screen where you will be prompted for the Left Bound. Use the left-right arrow keys to move the cursor along the parabola until it is to the left of one of the x-intercepts and press [ENTER]. Next you will be prompted for the Right Bound. Again, use the arrow keys to move the cursor to the right of the x-intercept, then press [ENTER]. You will then be prompted for a Guess. Press [ENTER] and you will see the coordinates of this x-intercept. Follow the same procedure for the other x-intercept.

To find the vertex, we use the MAXIMUM or MINIMUM functions under the CALC menu. Since our graph opens down, the vertex will be a maximum, so we will use the MAXIMUM function. After selecting the MAXIMUM function, you will be prompted for the left bound, then the right bound, then a guess. After entering the left bound, right bound, and pressing [ENTER] for the guess, you will see the coordinates of the vertex.

11.4 EXERCISES

For Extra Help

▲ **InterAct Math**

Tutor Center · **Addison-Wesley Math Tutor Center**

▭ **Video 17/DVT 6**

▯ **MyMathLab**

Math XL · **MathXL**

1. In an equation in the form $f(x) = a(x - h)^2 + k$, what determines whether the parabola opens upwards or downwards?

2. In an equation in the form $f(x) = a(x - h)^2 + k$, what are the coordinates of the vertex?

3. Describe the axis of symmetry in a parabola that opens up or down. What is its equation?

4. Suppose the y-intercept for a given quadratic equation is $(0, -5)$. If the vertex is at $(3, -2)$, what can you conclude about the x-intercepts?

5. If the solutions for a quadratic equation are imaginary numbers, then what does that indicate about its graph?

6. Given that the x-coordinate of the vertex is $-\dfrac{b}{2a}$, what must be true about a and b if the x-coordinate of the vertex is -1?

For Exercises 7–16, state whether the parabola opens up or down.

7. $f(x) = 2x^2$

8. $g(x) = -3x^2$

9. $h(x) = -x^2 + 3$

10. $k(x) = 4x^2 - 5$

11. $f(x) = 2(x - 1)^2 - 4$

12. $h(x) = -0.8(x + 3)^2 - 1$

13. $g(x) = -\dfrac{2}{3}x^2 + 4x - 1$

14. $f(x) = -x^2 - 8x - 9$

15. $k(x) = 3x - 0.5x^2$

16. $g(x) = -2x + \dfrac{3}{4}x^2$

For Exercises 17–30, find the coordinates of the vertex and write the equation of the axis of symmetry.

17. $f(x) = 5(x - 2)^2 - 3$

18. $h(x) = 2(x - 1)^2 + 4$

19. $g(x) = -2(x + 1)^2 - 5$

20. $k(x) = -(x + 4)^2 - 3$

21. $k(x) = x^2 + 2$

22. $g(x) = x^2 - 6$

23. $h(x) = 3(x - 7)^2$

24. $f(x) = 5(x + 1)^2$

25. $f(x) = -0.5x^2$

26. $k(x) = 2.5x^2$

27. $g(x) - 3x^2 + 2x + 1$

28. $h(x) = 5x^2 - 3x + 2$

29. $k(x) - 0.5x^2 - 0.4x + 1$

30. $f(x) = 0.2x^2 - 0.3x - 5$

For Exercises 31–50: a. State whether the parabola opens upwards or downwards.
b. Find the coordinates of the vertex.
c. Write the equation of the axis of symmetry.
d. Graph.

31. $h(x) = -3x^2$

32. $f(x) = 2x^2$

33. $k(x) = \dfrac{1}{4}x^2$

34. $g(x) = -\dfrac{1}{3}x^2$

35. $f(x) = 4x^2 - 3$

36. $h(x) = -x^2 + 4$

37. $g(x) = -0.5x^2 + 2$

38. $k(x) = 0.5x^2 - 3$

39. $f(x) = (x - 3)^2 + 2$

40. $h(x) = (x - 2)^2 - 1$

41. $k(x) = -2(x + 1)^2 - 3$

42. $g(x) = -(x + 3)^2 + 1$

43. $h(x) = \dfrac{1}{3}(x - 2)^2 - 1$

44. $f(x) = -\dfrac{1}{4}(x - 3)^2 - 2$

45. $k(x) = -x^2 - 2x + 3$

46. $h(x) = x^2 + 6x + 9$

47. $f(x) = 2x^2 + 6x + 3$

48. $g(x) = -3x^2 + 6x - 5$ **49.** $h(x) = -3x^2 - 6x + 4$ **50.** $k(x) = 3x^2 - 6x + 1$

For Exercises 51–62, solve.

51. A toy rocket is launched with an initial velocity of 45 meters per second. The equation $h = -4.9t^2 + 45t$ describes the height, h, of the rocket t seconds after being launched.

    a. After how many seconds does the rocket reach its maximum height?

    b. What is the maximum height the rocket reaches?

    c. How long is the rocket in the air?

52. A ball is drop-kicked straight up with an initial velocity of 36 feet per second. The equation $h = -16t^2 + 36t$ describes the height, h, of the ball t seconds after being kicked.

    a. After how many seconds does the ball reach its maximum height?

    b. What is the maximum height the ball reaches?

    c. How long is the ball in the air?

53. The equation $y = -0.8x^2 + 2.4x + 6$ models the trajectory of a ball thrown upwards and outwards from a height of 6 feet (assume that $x \geq 0$ and $y \geq 0$).

    a. What is the maximum height that the ball reaches?

    b. How far does the ball travel horizontally?

    c. Graph the trajectory of the ball.

54. The javelin toss is an event in track and field in which participants try to throw a javelin the farthest. Suppose the equation $y = -0.02x^2 + 1.3x + 8$ models the trajectory of one particular throw (assume that x and y represent distances in meters and that $x \geq 0$ and $y \geq 0$).

    a. What is the maximum height that the javelin reaches?

    b. How far does the javelin travel horizontally?

    c. Graph the trajectory of the javelin.

55. A record company discovers that the number of CDs sold each week after release follows a parabolic pattern. The function $n(t) = -200t^2 + 4000t$ describes the number, n, of CDs an artist sold each of t weeks after the release of the album.

    a. Which week had the greatest number of CDs sold?

    b. How many CDs sold that week?

56. The function $n(t) = -3t^2 + 42t$ describes the number, n, of tickets sold for a play each of t days after tickets went on sale.

    a. What day had the greatest number of tickets sold?

    b. How many tickets sold that day?

57. A farmer has enough materials to build a fence with a total length of 400 feet. He wants the enclosed space to be rectangular and also wants to maximize the area enclosed. Find the length and width so that the area is maximized.

58. An architect wants the length and width of a rectangular building to be a total of 150 feet long. She also wants to maximize the rectangular area that the building occupies. Find the length and width so that the area is maximized.

59. The function $C(n) = n^2 - 110n + 1025$ describes a company's cost, C, of producing n units of its product. Find the number of units the company should produce to minimize its cost.

60. The function $P(t) = 0.001t^2 - 0.24t + 59.90$ roughly models a particular stock's closing price, P, each of t days of trading during one year.

    a. After how many days is the price at its lowest?

    b. What was the lowest price during that year?

61. One integer is 12 more than another. If their product is minimized, find the integers and their product.

62. The greater of two integers minus the smaller gives a result of 20. Find the two integers so that their product is minimized. Also find their product.

Given the vertex and the x- and y-intercepts of a parabola, reconstruct the equation.

a. *Vertex $(-3, -1)$*
   Intercepts $(-2,0), (-4,0), (0,8)$

b. *Vertex $(-1,0)$*
   Intercepts $(-1, 0), (0,1)$

c. *Vertex $(-2,1)$*
   Intercept $(0,5)$

d. *Vertex $(1,0)$*
   Intercepts $(1,0), (0,1)$

Collaborative Exercises

Arch Span

The Gateway Arch, located in St. Louis, Missouri, was built from 1963 to 1965 and is the nation's tallest memorial. The equation $h(x) = -.0063492063x^2 + 630$ can be used to approximate the height of the arch, where x represents the distance from its axis of symmetry and h(x) represents its height above the ground.

1. Using the equation, find the maximum height of the structure.

2. The span of the arch at a given height is the horizontal distance between the two opposing points on the parabola at that height. Find the span of the arch at ground level. (*Hint:* Think of ground level as the x-axis. The height is 0 along the x-axis.)

3. How does the span at ground level relate to the maximum height of the arch?

4. The arch has foundations 60 feet below ground level. What is the span of the arch between its foundations?

REVIEW EXERCISES

1. Evaluate: $-|-2 + 3 \cdot 4| - 4^0$

2. Solve $x^2 + 2x = 15$ by factoring.

3. Solve $x^2 = -20$ using the square root principle.

For Exercises 4–6, solve, then graph the solution set.

4. $4x - 8 \le 2x + 1$

5. $|x - 3| \ge 8$

6. $\begin{cases} x + y > 2 \\ 2x - 3y \le 6 \end{cases}$

11.5 Solving Nonlinear Inequalities

OBJECTIVES

1. Solve quadratic and other inequalities.
2. Solve rational inequalities.

OBJECTIVE 1 **Solve quadratic and other inequalities.** Now that we have solved quadratic equations, we can learn how to solve **quadratic inequalities**.

> **DEFINITION**
>
> **Quadratic inequality:** An inequality that can be written in the form $ax^2 + bx + c > 0$ or $ax^2 + bx + c < 0$, where $a \neq 0$.

Note: In this definition, the symbols $<$ and $>$ can be replaced with $\leq$ or $\geq$.

For example, $2x^2 - x - 3 < 0$ and $2x^2 - x - 3 \geq 0$ are quadratic inequalities. Let's see what we can learn about these inequalities by looking at the graph of the corresponding quadratic function: $y = 2x^2 - x - 3$. Note that by letting $y = 0$, we have the corresponding equation $2x^2 - x - 3 = 0$ and solving this equation gives the x-intercepts $x = -1$ and $x = \dfrac{3}{2}$. Also notice that those x-intercepts divide the x-axis into three intervals: $(-\infty, -1)$, $\left(-1, \dfrac{3}{2}\right)$, and $\left(\dfrac{3}{2}, \infty\right)$.

Notice the parabola is above the x-axis in intervals $(-\infty, -1)$ and $\left(\dfrac{3}{2}, \infty\right)$, shown in blue. Evaluating $y = 2x^2 - x - 3$ using any x-value in these intervals produces a y-value that is greater than 0, so these intervals are the solution sets for $2x^2 - x - 3 > 0$.

The parabola is below the x-axis in the interval $\left(-1, \dfrac{3}{2}\right)$, shown in red. Evaluating $y = 2x^2 - x - 3$ using any x-value in this interval produces a y-value that is less than 0, so this interval is the solution set for $2x^2 - x - 3 < 0$.

The following number lines indicate various solution sets based on the inequality.

$2x^2 - x - 3 > 0$

Solution set: $(-\infty, -1) \cup \left(\dfrac{3}{2}, \infty\right)$

$2x^2 - x - 3 < 0$

Solution set: $\left(-1, \dfrac{3}{2}\right)$

$2x^2 - x - 3 \geq 0$

Solution set: $(-\infty, -1] \cup \left[\dfrac{3}{2}, \infty\right)$

$2x^2 - x - 3 \leq 0$

Solution set: $\left[-1, \dfrac{3}{2}\right]$

Based on these observations, we use the following procedure to solve quadratic inequalities.

> **PROCEDURE** SOLVING QUADRATIC INEQUALITIES
>
> 1. Solve the related equation $ax^2 + bx + c = 0$.
> 2. Plot the solutions of $ax^2 + bx + c = 0$ on a number line. These solutions will divide the number line into intervals.
> 3. Choose a test number from each interval and substitute the number into the inequality. If the test number makes the inequality *true*, then all numbers in that interval will solve the inequality. If the test number makes the inequality *false*, then no numbers in that interval will solve the inequality.
> 4. State the solution set of the inequality: It is the union of all the intervals that solve the inequality. If the inequality symbols are $\leq$ or $\geq$, then the values from step 2 are included. If the symbols are $<$ or $>$, they are not solutions.

EXAMPLE 1 Solve $x^2 - x < 6$. Write the solution set using interval notation, then graph the solution set on a number line.

Solution

$x^2 - x = 6$	Write the related equation.
$x^2 - x - 6 = 0$	Subtract 6 from both sides to get quadratic form.
$(x - 3)(x + 2) = 0$	Factor.
$x - 3 = 0$　or　$x + 2 = 0$	Use the zero-factor theorem.
$x = 3$　　　　$x = -2$	These are the x-intercepts of the graph of $y = x^2 - x - 6$.

Plot -2 and 3 on a number line (x-axis), which divides it into three intervals.

Note: -2 and 3 are not part of the solution set because the inequality symbol is $<$.

CONNECTION

Notice the solution set for $x^2 - x - 6 < 0$ corresponds to the interval where the graph of $y = x^2 - x - 6$ is below the x-axis and the x-intercepts are the endpoints of the interval.

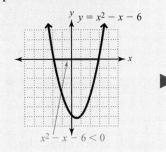

Choose a test number from each interval and substitute that value into $x^2 - x < 6$.

For $(-\infty, -2)$, we choose $x = -3$.

$$(-3)^2 - (-3) < 6$$
$$9 + 3 < 6$$
$$12 < 6$$

This is false, so $(-\infty, -2)$ is not in the solution set.

For $(-2, 3)$, we choose $x = 0$.

$$0^2 - 0 < 6$$
$$0 < 6$$

This is true, so $(-2, 3)$ is in the solution set.

For $(3, \infty)$, we choose $x = 4$.

$$4^2 - 4 < 6$$
$$16 - 4 < 6$$
$$12 < 6$$

This is false, so $(3, \infty)$ is not in the solution set.

Since $(-2, 3)$ is the only interval that has solutions to the inequality, it is the solution set. Following is the graph of the solution set.

11.5 Solving Nonlinear Inequalities 959

EXAMPLE 2 Solve $x^2 + 3x \geq 0$. Write the solution set using interval notation, then graph the solution set.

Solution

$$x^2 + 3x = 0 \qquad \text{Write the related equation.}$$
$$x(x + 3) = 0 \qquad \text{Factor.}$$
$$x = 0 \quad \text{or} \quad x + 3 = 0 \qquad \text{Use the zero-factor theorem.}$$
$$x = -3$$

Plot -3 and 0 on a number line and note the intervals.

$(-\infty, -3)$ $(-3, 0)$ $\qquad$ $(0, \infty)$

Note: Since we have already determined that -3 and 0 are in the solution set, we will not include them in any of our test intervals, which is why we use parentheses for each interval.

Note: -3 and 0 are included in the solution set because the inequality symbol is $\geq$.

Choose a test number from each interval and test in $x^2 + 3x \geq 0$.

For $(-\infty, -3)$, we choose $x = -4$.

$$(-4)^2 + 3(-4) \geq 0$$
$$16 - 12 \geq 0$$
$$4 \geq 0$$

This is true, so $(-\infty, -3)$ is in the solution set.

For $(-3, 0)$, we choose $x = -1$.

$$(-1)^2 + 3(-1) \geq 0$$
$$1 - 3 \geq 0$$
$$-2 \geq 0$$

This is false, so $(-3,0)$ is not in the solution set.

For $(0, \infty)$, we choose $x = 1$.

$$1^2 + 3(1) \geq 0$$
$$1 + 3 \geq 0$$
$$4 \geq 0$$

This is true, so $(0, \infty)$ is in the solution set.

The solution set is $(-\infty, -3] \cup [0, \infty)$, which is graphed next.

EXAMPLE 3 Solve. Write the solution set using interval notation, then graph the solution set.

a. $(x - 1)^2 > -2$

Solution Since $(x - 1)^2$ is always 0 or positive, every real number is a solution.

Solution set: $\mathbb{R}$ or $(-\infty, \infty)$

b. $(x - 1)^2 < -2$

Solution Since $(x - 1)^2$ is never negative, its value can never be less than -2, so there are no real solutions for $(x - 1)^2 < -2$.

Solution set: $\varnothing$

CONNECTION

Look at the graph of $y = (x - 1)^2$.

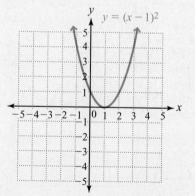

Notice that the parabola touches the x-axis at 1 but does not go below the x-axis.

ANSWERS

a. $[-2, 4]$

b. $(-\infty, 0) \cup (4, \infty)$

c. $\varnothing$

YOUR TURN Solve. Write the solution set using interval notation, then graph the solution set on a number line.

a. $x^2 - 2x - 8 \leq 0$ $\qquad$ **b.** $x^2 - 4x > 0$ $\qquad$ **c.** $(3x - 8)^2 < -5$

Other Polynomial Inequalities

We use a similar procedure for expressions with more than two factors. In the following example, we condense the procedure to a table.

EXAMPLE 4 Solve $(x + 4)(x + 1)(x - 3) \leq 0$. Write the solution set in interval notation, then graph the solution set on a number line.

Solution $(x + 4)(x + 1)(x - 3) = 0$ Write the related equation.

$x + 4 = 0$ or $x + 1 = 0$ or $x - 3 = 0$ Set each factor equal to 0.

$x = -4$ $x = -1$ $x = 3$ Solve the equations.

Plot -4, -1, and 3 on a number line and note the intervals.

$$-6\ -5\ -4\ -3\ -2\ -1\ 0\ 1\ 2\ 3\ 4\ 5\ 6$$
$$(-\infty, -4)\ (-4, -1)\ \ \ (-1, 3)\ \ \ \ \ (3, \infty)$$

Note: -4, -1, and 3 are included in the solution set because the inequality symbol is $\leq$. Remember that we test intervals around these values, so they are not included in those intervals.

Interval	$(-\infty, -4)$	$(-4, -1)$	$(-1, 3)$	$(3, \infty)$
Test Number	-5	-2	0	4
Test Results	$-32 \leq 0$	$10 \leq 0$	$-12 \leq 0$	$40 \leq 0$
True or False	True	False	True	False

Note: Because the inequality is $\leq$, -4, -1, and 3 are included in the solution set.

Therefore, the solution set is $(-\infty, -4] \cup [-1, 3]$, which is graphed next.

$$-6\ -5\ -4\ -3\ -2\ -1\ 0\ 1\ 2\ 3\ 4\ 5\ 6$$

ANSWER

$(-\infty, -5] \cup [-3, 2]$

$$-7\ -6\ -5\ -4\ -3\ -2\ -1\ 0\ 1\ 2\ 3$$

YOUR TURN Solve. Write the solution set using interval notation, then graph the solution set on a number line.

$$(x + 3)(x - 2)(x + 5) \leq 0$$

OBJECTIVE **2** Solve rational inequalities. Now we consider solving **rational inequalities**.

DEFINITION   *Rational inequality:* An inequality containing a rational expression.

For example, $\dfrac{x + 2}{x - 3} > 4$ is a rational inequality. Recall that in solving a polynomial inequality, we divided the number line into intervals using x-values we found from solving its related equation. With rational inequalities, we not only find values that solve the related equation but we must also consider values that make the rational expression undefined (denominator $= 0$).

PROCEDURE	**SOLVING RATIONAL INEQUALITIES**

1. Solve the related equation.
2. Find all values that make any denominator equal to 0.
3. Plot the numbers found in steps 1 and 2 on a number line.
4. Choose a test number from each interval and determine whether it solves the inequality.
5. The solution set is the union of all the regions whose test number solves the inequality. If the inequality symbol is ≤ or ≥, include the values found in step 1. The solution set never includes the values found in step 2 because they make a denominator equal to 0.

EXAMPLE 5 Solve. Write the solution set using interval notation, then graph the solution set on a number line.

a. $\dfrac{x - 4}{x + 2} \geq 0$

Solution First, solve the related equation.

$$\dfrac{x - 4}{x + 2} = 0 \qquad \text{Write the related equation.}$$

$$(x + 2)\dfrac{x - 4}{x + 2} = (x + 2)(0) \qquad \text{Multiply both sides by } x + 2.$$

$$x - 4 = 0 \qquad \text{Simplify.}$$

$$x = 4$$

Now we find the values that make the denominator equal to 0.

$$x + 2 = 0$$
$$x = -2$$

Plot -2 and 4 on a number line and note the intervals.

Note: Since -2 makes the expression undefined, it is not included in the solution set.

Again, we use a table.

Interval	$(-\infty, -2)$	$(-2, 4)$	$(4, \infty)$
Test Number	-3	0	5
Test Results	$7 \geq 0$	$-2 \geq 0$	$\dfrac{1}{7} \geq 0$
True or False	True	False	True

The solution set is $(-\infty, -2) \cup [4, \infty)$ and is graphed next.

b. $\dfrac{x+5}{x-1} < 4$

Solution Solve the related equation.

$$\dfrac{x+5}{x-1} = 4 \qquad \text{Write the related equation.}$$

$$(x-1)\dfrac{x+5}{x-1} = (x-1)(4) \qquad \text{Multiply both sides by the LCD.}$$

$$x + 5 = 4x - 4 \qquad \text{Simplify.}$$

$$-3x = -9 \qquad \text{Subtract } 4x \text{ and 5 from both sides.}$$

$$x = 3$$

Find the values that make
the denominator equal to 0. $x - 1 = 0$

$$x = 1$$

Plot 1 and 3 on a number line.

Interval	$(-\infty, 1)$	$(1, 3)$	$(3, \infty)$
Test Number	0	2	4
Test Results	$-5 < 4$	$7 < 4$	$3 < 4$
True or False	True	False	True

The solution set is $(-\infty, 1) \cup [3, \infty)$ and is graphed next.

YOUR TURN   Solve. Write the solution set using interval notation, then graph the solution set on a number line.

a. $\dfrac{x+3}{x-1} < 0$

b. $\dfrac{x-2}{x+4} \le 3$

1. Explain how you would find the solution set of $x^2 + 2x - 15 \geq 0$.

2. If the graph of $y = ax^2 + bx + c$ intersects the x-axis at -2 and 2, what regions would you check to solve $ax^2 + bx + c > 0$?

3. Is it possible to have a quadratic inequality whose solution set is the empty set? Explain.

4. Is it possible to have a quadratic inequality whose solution set is one number? Explain.

5. The quadratic inequality $(x - 2)(x + 5) < 0$ and the rational inequality $\dfrac{x - 2}{x + 5} < 0$ have the same solution sets. Explain how this is possible.

6. If you were to solve $\dfrac{x + 2}{(x + 5)(x - 3)} \geq 0$, what are the regions that you would need to test?

For Exercises 7–14, the graph of a quadratic function is given. Use the graph to solve each equation and inequality. For solution sets that involve intervals, use interval notation.

7. **a.** $x^2 + 6x + 5 = 0$

   b. $x^2 + 6x + 5 < 0$

   c. $x^2 + 6x + 5 > 0$

8. **a.** $x^2 - 4 = 0$

   b. $x^2 - 4 > 0$

   c. $x^2 - 4 < 0$

$y = x^2 + 6x + 5$

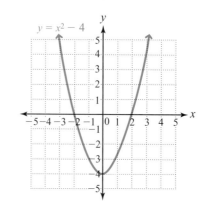

$y = x^2 - 4$

9. **a.** $-x^2 + 2x + 3 = 0$

b. $-x^2 + 2x + 3 \leq 0$

c. $-x^2 + 2x + 3 \geq 0$

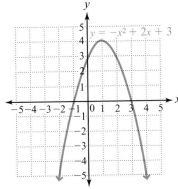

10. **a.** $-x^2 + 5x = 0$

b. $-x^2 + 5x \leq 0$

c. $-x^2 + 5x \geq 0$

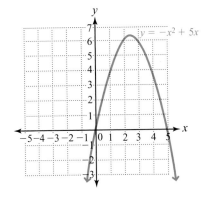

11. **a.** $x^2 + 4x + 4 = 0$

b. $x^2 + 4x + 4 > 0$

c. $x^2 + 4x + 4 < 0$

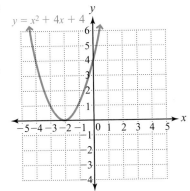

12. **a.** $x^2 - 6x + 9 = 0$

b. $x^2 - 6x + 9 < 0$

c. $x^2 - 6x + 9 > 0$

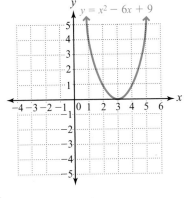

13. **a.** $x^2 + x + 4 = 0$

b. $x^2 + x + 4 \leq 0$

c. $x^2 + x + 4 \geq 0$

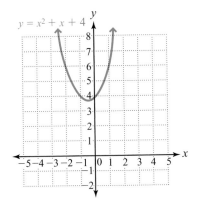

14. **a.** $-x^2 - x - 2 = 0$

b. $-x^2 - x - 2 \leq 0$

c. $-x^2 - x - 2 \geq 0$

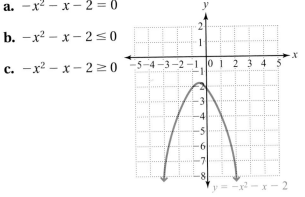

For Exercises 15–44, solve. Write the solution set using interval notation, then graph the solution set on a number line.

15. $(x + 4)(x + 2) < 0$

16. $(x - 3)(x + 1) < 0$

17. $(x - 2)(x - 5) > 0$

18. $(x + 4)(x - 3) > 0$

19. $x^2 + 5x + 4 < 0$

20. $x^2 + 6x + 5 < 0$

21. $x^2 - 4x + 3 > 0$

22. $x^2 - 8x + 7 > 0$

23. $b^2 - 6b + 8 \leq 0$

24. $c^2 - 9c + 14 \leq 0$

25. $y^2 - 5 \geq 4y$

26. $z^2 - 3 \leq 2z$

27. $a^2 - 3a < 10$

28. $b^2 + 4b \geq 21$

29. $y^2 + 6y + 9 \geq 0$

30. $x^2 + 10x + 25 \geq 0$

31. $c^2 - 4c + 5 < 0$

32. $2a^2 - 3a + 5 \geq 0$

33. $4r^2 + 21r + 5 > 0$

34. $4s^2 + 5s - 6 < 0$

35. $x^2 - 5x > 0$

36. $x^2 + 3x > 0$

37. $x^2 \leq 6x$

38. $x^2 \geq 3x$

39. $(x - 3)^2 \geq -1$

40. $(2x - 1)^2 < -4$

41. $(x - 4)(x + 2)(x + 4) \geq 0$

42. $(x + 5)(x - 3)(x + 1) \geq 0$

43. $(x + 2)(x + 6)(x - 1) < 0$

44. $(x - 3)(x + 3)(x + 1) < 0$

For Exercises 45–50, solve using the quadratic formula.

45. $3c^2 + 4c - 1 < 0$

46. $5a^2 - 10a + 2 < 0$

47. $4r^2 + 8r - 3 \geq 0$

48. $-2y^2 + 6y + 5 < 0$

49. $-0.2a^2 - 1.6a - 2 \leq 0$

50. $-0.1x^2 - 1.2x + 4 > 0$

For Exercises 51–68, solve the rational inequalities. Write the solution set using interval notation, then graph the solution set on a number line.

51. $\dfrac{a + 4}{a - 1} > 0$

52. $\dfrac{b - 2}{b + 3} < 0$

53. $\dfrac{n + 1}{n + 5} \leq 0$

54. $\dfrac{m - 6}{m + 2} \geq 0$

55. $\dfrac{6}{x + 4} > 0$

56. $\dfrac{3}{x - 3} < 0$

57. $\dfrac{c}{c+3} < 3$

58. $\dfrac{m}{m-2} < 2$

59. $\dfrac{a+5}{a-4} > 4$

60. $\dfrac{j+4}{j-4} > 5$

61. $\dfrac{p+2}{p-3} \geq 4$

62. $\dfrac{c+1}{c-4} \leq 6$

63. $\dfrac{(k+3)(k-2)}{k-5} \leq 0$

64. $\dfrac{(m+1)(m-3)}{m+4} \geq 0$

65. $\dfrac{(2x-1)^2}{x} \geq 0$

66. $\dfrac{(4x+3)^2}{x} < 0$

67. $\dfrac{x^2-7x+10}{x+1} < 0$

68. $\dfrac{x^2-2x-8}{x-1} \geq 0$

For Exercises 69–72, solve.

69. If a ball is thrown upward with an initial velocity of 80 feet per second from the top of a building 96 feet high, then the height, h, above the ground after t seconds is given by $h = -16t^2 + 80t + 96$, where h is in feet.

    a. After how many seconds will the ball hit the ground? (*Hint:* Think about the value of h when the ball is on the ground.)

    b. After how many seconds is the ball 192 feet above the ground?

c. Find the interval of time when the ball is more than 192 feet above the ground.

d. Use the answers to parts a, b, and c to find the intervals of time when the ball is less than 192 feet above the ground.

70. If an object is dropped from the top of a cliff that is 144 feet high, the equation giving the height, h, above the ground is $h = 144 - 16t^2$, where h is in feet and t is in seconds.

a. After how many seconds will the object hit the ground?

b. After how many seconds is the object 80 feet above the ground?

c. Find the interval of time when the object is more than 80 feet above the ground.

d. Find the interval of time when the ball is less than 80 feet above the ground.

71. In the parallelogram shown, the height is to be 2 inches less than x. The base is to be 6 inches more than x.

a. Find the range of values for x so that the area of the parallelogram is at least 20 square inches.

b. Find the range of values for the base and the height.

72. An auditorium is to be designed roughly in the shape of a box. The height is set to be 40 feet. It is desired that the length of the space be 20 feet more than the width.

a. Find the range of values for the width so that the volume of the space is at least 140,000 cubic feet.

b. Find the range of values for the length.

For Exercises 73 and 74, use the formula for the slope of a line: $m = \dfrac{y_2 - y_1}{x_2 - x_1}$.

73. Suppose a line is to be drawn in the coordinate plane so that it passes through the point at (2, 5). Find the range of values for the second point (x_2, y_2) so that $x_2 = y_2$ and the slope of the line is at most $\dfrac{1}{2}$.

74. Suppose a line is to be drawn in the coordinate plane so that it passes through the point at (−3, 2). Find the range of values for the second point (x_2, y_2) so that $x_2 = y_2$ and the slope of the line is at least $\dfrac{1}{4}$.

REVIEW EXERCISES

1. Solve: $\sqrt{x-2} = 4$

For Exercises 2 and 3, determine the constant that would need to be added to the given binomial in order for the resulting trinomial to be a perfect square trinomial. Write the resulting expression as the square of a binomial.

2. $x^2 - 6x$

3. $x^2 + 5x$

4. What are the x-intercepts of the graph of $y = x^2 + 4x - 12$?

5. What are the coordinates of the vertex of the graph of $f(x) = x^2 - 6x + 5$?

6. Does the graph of $f(x) = -x^2 + 2x - 3$ open up or down? Why?

Defined Terms

Section 11.2
  Discriminant (p. 922)

Section 11.3
  Equations quadratic in
  form (p. 933)

Section 11.4
  Axis of symmetry
  (p. 946)

Section 11.5
  Quadratic Inequality
  (p. 958)
  Rational Inequality
  (p. 961)

Procedures, Rules, and Key Examples

Procedures/Rule	Key Examples

Section 11.1 Completing the Square
The square root principle:
If $x^2 = a$ and $a \geq 0$, then $x = \sqrt{a}$ or $x = -\sqrt{a}$.

It is common to indicate the positive and negative solutions by writing $\pm\sqrt{a}$.

To solve a quadratic equation by completing the square,
1. Write the equation in the form $x^2 + bx = c$.
2. Complete the square by adding $\left(\dfrac{b}{2}\right)^2$ to both sides.
3. Write the completed square in factored form.
4. Use the square root principle to eliminate the square.
5. Isolate the variable.
6. Simplify as needed.

Example 1: Solve.
a. $x^2 = 36$
Solution: $x = \pm\sqrt{36}$
$\qquad\quad x = \pm 6$

b. $(x - 3)^2 = 12$
Solution: $x - 3 = \pm\sqrt{12}$
$\qquad\qquad x = 3 \pm 2\sqrt{3}$

Example 2: Solve by completing the square.
a. $x^2 + 6x - 7 = 8$

$x^2 + 6x = 15$	Add 7 to both sides.
$x^2 + 6x + 9 = 15 + 9$	Complete the square.
$(x + 3)^2 = 24$	Factor.
$x + 3 = \pm\sqrt{24}$	Use the square root principle.
$x = -3 \pm 2\sqrt{6}$	Subtract 3 from both sides and simplify the square root.

b. $3x^2 - 9x - 2 = 10$

$3x^2 - 9x = 12$	Add 2 to both sides.
$x^2 - 3x = 4$	Divide both sides by 3.
$x^2 - 3x + \dfrac{9}{4} = 4 + \dfrac{9}{4}$	Complete the square.
$\left(x - \dfrac{3}{2}\right)^2 = \dfrac{25}{4}$	Factor.
$x - \dfrac{3}{2} = \pm\sqrt{\dfrac{25}{4}}$	Use the square root principle.
$x = \dfrac{3}{2} \pm \dfrac{5}{2}$	Add $\dfrac{3}{2}$ to both sides and simplify the square root.

$x = \dfrac{3}{2} + \dfrac{5}{2} = 4$ $\quad$ or $\quad$ $x = \dfrac{3}{2} - \dfrac{5}{2} = -1$

Procedures/Rule	Key Examples

Section 11.2 Solving Quadratic Equations Using the Quadratic Formula

To solve a quadratic equation in the form $ax^2 + bx + c = 0$, where $a \neq 0$, use the quadratic formula:

$$x = \frac{-b \pm \sqrt{b^2 - 4ac}}{2a}$$

Example 1: Solve: $3x^2 - 4x + 2 = 0$

$$x = \frac{-(-4) \pm \sqrt{(-4)^2 - 4(3)(2)}}{2(3)}$$

In the quadratic formula, replace a with 3, b with -4, and c with 2.

$$= \frac{4 \pm \sqrt{16 - 24}}{6} = \frac{4 \pm \sqrt{-8}}{6}$$

$$= \frac{4 \pm 2\sqrt{-2}}{6} = \frac{2}{3} \pm \frac{\sqrt{2}}{3}i$$

Given a quadratic equation in the form $ax^2 + bx + c = 0$, where $a \neq 0$, to determine the number and type of solutions a quadratic equation will have, evaluate the discriminant $b^2 - 4ac$.

If the **discriminant is positive**, then the equation has two real-number solutions. They will be rational if the discriminant is a perfect square and irrational otherwise.

If the **discriminant is 0**, then the equation has one real solution. The solution is rational if $-\dfrac{b}{2a}$ is a rational number.

If the **discriminant is negative**, then the equation has two nonreal complex solutions.

Example 2: Use the discriminant to determine the number and type of solutions for $5x^2 - 7x + 8 = 0$.

$(-7)^2 - 4(5)(8) = 49 - 160$

In the discriminant, replace a with 5, b with -7, and c with 8.

$= -111$

Since the discriminant is negative, the equation has two nonreal complex solutions.

Section 11.3 Solving Equations That Are Quadratic in Form

If the equation involves rational expressions, multiply both sides by the LCD. Be sure to check for extraneous solutions.

Example 1: Solve $\dfrac{6}{x - 2} + \dfrac{6}{x - 1} = 5$.

$6(x - 1) + 6(x - 2) = 5(x - 2)(x - 1)$

Multiply both sides by the LCD.

$12x - 18 = 5x^2 - 15x + 10$

Simplify.

$0 = 5x^2 - 27x + 28$

Write in $ax^2 + bx + c = 0$ form.

$0 = (5x - 7)(x - 4)$

Factor.

$5x - 7 = 0$ or $x - 4 = 0$

Use the zero-factor theorem.

$x = \dfrac{7}{5}$ $x = 4$

Solve each equation.

Check: Verify that $\dfrac{7}{5}$ and 4 solve the original equation.

Procedures/Rule

If an equation has a radical, isolate the radical and then square both sides. Continue the process as needed until all radicals have been eliminated. Solve the resulting equation and check for extraneous solutions.

To solve equations that are quadratic in form using substitution,
1. Rewrite the equation so that it is in the form $au^2 + bu + c = 0$.

2. Solve the quadratic equation for u.

3. Substitute for u and solve.

4. Check the solutions.

Section 11.4 Graphing Quadratic Functions
The graph of a function in the form $f(x) = a(x - h)^2 + k$ is a parabola with vertex at (h, k). The equation of the axis of symmetry is $x = h$. The parabola opens up if $a > 0$ and down if $a < 0$.

Key Examples

Example 2: Solve $\sqrt{4x + 4} = 2x - 2$.

$4x + 4 = 4x^2 - 8x + 4$ Square both sides

$0 = 4x^2 - 12x$ Write in $ax^2 + bx + c = 0$ form.

$0 = 4x(x - 3)$ Factor.

$4x = 0$ or $x - 3 = 0$ Set each factor equal to 0.

$x = 0$ $x = 3$ Solve each equation.

Check:

$x = 0$ $x = 3$

$\sqrt{4(0) + 4} = 2(0) - 2$ $\sqrt{4(3) + 4} = 2(3) - 2$

$\sqrt{0 + 4} = 0 - 2$ $\sqrt{12 + 4} = 6 - 2$

$2 = -2$ False $4 = 4$ True

$x = 3$ is the only solution.

Example 3: Solve
$$(a + 2)^2 + 7(a + 2) + 12 = 0.$$

$u^2 + 7u + 12 = 0$ Substitute u for $a + 2$.

$(u + 3)(u + 4) = 0$ Factor.

$u + 3 = 0$ or $u + 4 = 0$

                      Set each factor equal to 0.

$u = -3$ $u = -4$

                      Solve each equation.

$a + 2 = -3$ $a + 2 = -4$

                      Substitute $a + 2$ for u.

$a = -5,$ $a = -6$

                      Solve for a.

Check: Verify that -5 and -6 satisfy the original equation.

Example 1: For $f(x) = -3(x + 1)^2 - 2$,
a. Determine whether the graph opens up or down.
b. Find the vertex.
c. Write the equation of the axis of symmetry.
d. Graph.
Answers:
a. Down   **b.** $(-1, -2)$ **c.** $x = -1$
d.

continued

Procedures/Rule	Key Examples

Procedures/Rule

Section 11.4 (continued)

Given an equation in the form $f(x) = ax^2 + bx + c$, to determine the vertex of the corresponding parabola,

1. Find the x-coordinate using the formula $-\dfrac{b}{2a}$.

2. Find the y-coordinate by evaluating $f\left(-\dfrac{b}{2a}\right)$.

Note: As an alternative approach to Example 3, we could have transformed $f(x) = 2x^2 - 12x + 19$ to $f(x) = 2(x - 3)^2 + 1$ by completing the square.

Key Examples

Example 2: For $f(x) = 2x^2 - 12x + 19$,
a. Determine whether the graph opens up or down.
b. Find the vertex.
c. Write the equation of the axis of symmetry.
d. Graph.

Answers:
a. Up
b. For the x-coordinate, use $-\dfrac{b}{2a}$, replacing a with 2 and b with -12.

$$x = -\frac{(-12)}{2(2)} = 3$$

For the y-coordinate, evaluate $f(3)$.
$$f(3) = 2(3)^2 - 12(3) + 19$$
$$= 18 - 36 + 19 = 1$$
Vertex: $(3, 1)$

c. $x = 3$

d.

Section 11.5 Solving Nonlinear Inequalities

Solving quadratic inequalities.

1. Solve the related equation $ax^2 + bx + c = 0$.
2. Plot the solutions of $ax^2 + bx + c = 0$ on a number line. These solutions will divide the number line into intervals.
3. Choose a test number from each interval and substitute the number into the inequality. If the test number makes the inequality true, then all numbers in that interval will solve the inequality. If the test number makes the inequality false, then no numbers in that interval will solve the inequality.
4. State the solution set of the inequality: It is the union of all the intervals that solve the inequality. If the inequality symbols are $\leq$ or $\geq$, then the values from step 2 are included. If the symbols are $<$ or $>$, they are not solutions.

Example 1: Solve $x^2 + 2x - 15 \geq 0$.

$x^2 + 2x - 15 = 0$ **Write the related equation.**

$(x + 5)(x - 3) = 0$ **Factor.**

$x + 5 = 0$ or $x - 3 = 0$

     Use the zero-factor theorem.

$x = -5$ $x = 3$

     Solve each equation.

Plot -5 and 3 on a number line and label the intervals.

Choose a number from each interval and test it in the original inequality.

Interval	Test value	Result	True–False
$(-\infty, -5)$	$x = -6$	$9 \geq 0$	True
$(-5, 3)$	$x = 0$	$-15 \geq 0$	False
$(3, \infty)$	$x = 4$	$9 \geq 0$	True

Procedures/Rule	Key Examples

Section 11.5 (continued)

Solving rational inequalities.
1. Solve the related equation.
2. Find all values that make any denominator equal to 0.
3. Plot the numbers found in steps 1 and 2 on a number line and label the regions.
4. Choose a test number from each interval and determine whether it solves the inequality.
5. The solution set is the union of all the regions whose test number solves the inequality. If the inequality symbol is $\leq$ or $\geq$, include the values found in step 1. The solution set never includes the values found in step 2 because they make a denominator equal to 0.

Key Examples

The solution set is $(-\infty, -5] \cup [3, \infty)$.

Example 2: Solve $\dfrac{x + 5}{x - 1} > 4$

$$\dfrac{x + 5}{x - 1} = 4$$

Write the related equation.

$$(x - 1)\dfrac{x + 5}{x - 1} = (x - 1)(4)$$

Multiply both sides by the LCD.

$$x + 5 = 4x - 4 \quad \text{Simplify.}$$
$$9 = 3x \quad \text{Subtract } x \text{ and add 5 on both sides.}$$
$$3 = x \quad \text{Divide by 3.}$$

Find the value(s) that make any denominator equal to 0.

$$x - 1 = 0$$
$$x = 1$$

Plot 1 and 3 on the number line and label the regions.

Choose a number from each interval and test it in the original inequality.

Interval	Test value	Result	True/False
$(-\infty, 1)$	$x = 0$	$-5 > 4$	False
$(1, 3)$	$x = 2$	$7 > 4$	True
$(3, \infty)$	$x = 4$	$3 > 4$	False

The solution set is $(1, 3)$.

Formulas

The quadratic formula: $x = \dfrac{-b \pm \sqrt{b^2 - 4ac}}{2a}$

The x-coordinate of the vertex of a parabola is $-\dfrac{b}{2a}$

and the equation of the axis of symmetry is $x = -\dfrac{b}{2a}$.

For Exercises 1–5, answer true or false.

1. The notation $\sqrt{a}$ has the same meaning as the notation $\pm\sqrt{a}$.

2. Equations that are quadratic in form never have extraneous solutions.

3. Every quadratic equation has a real-number solution.

4. If the solutions for a quadratic equation in the form $ax^2 + bx + c = 0$ are complex, the discriminant can be used to find those solutions.

5. Given an equation in the form $y = ax^2 + bx + c$, if $a > 0$, then the parabola opens up.

For Exercises 6–10, complete the rule.

6. If $x^2 = a$ and $a \geq 0$, then $x = $ _____ or _____ .

7. Given an equation in the form $x^2 + bx = c$, to complete the square we add the constant term _____ to both sides of the equation.

8. The quadratic formula is $x = $ _____ .

9. The discriminant is _____ .

10. Given a function in the form $f(x) = a(x - h)^2 + k$, the coordinates of the vertex are _____ and the axis of symmetry is _____ .

For Exercises 11–18, solve and check.

11. $x^2 = 16$

12. $y^2 = \dfrac{1}{36}$

13. $k^2 + 2 = 30$

14. $3x^2 = 42$

15. $5h^2 + 24 = 9$

16. $(x + 7)^2 = 25$

17. $(x - 9)^2 = -16$

18. $\left(m + \dfrac{3}{5}\right)^2 = \dfrac{16}{25}$

For Exercises 19–22, solve by completing the square.

19. $m^2 + 8m = -7$ **20.** $u^2 - 6u - 12 = 100$ **21.** $2b^2 - 6b + 7 = 0$ **22.** $u^2 + \dfrac{1}{4}u = \dfrac{3}{4}$

For Exercises 23–26, solve using the quadratic formula.

23. $2t^2 + t - 5 = 0$ **24.** $3x^2 - 2x + 1 = 0$

25. $p^2 - 5 = -2p$ **26.** $2x^2 + 0.1x - 0.03 = 0$

For Exercises 27–30, use the discriminant to determine the number and type of solutions for the equation.

27. $b^2 - 4b - 12 = 0$ **28.** $6z^2 - 7z + 5 = 0$

29. $0.8x^2 + 1.2x + 0.3 = 0$ **30.** $k^2 + 6k + 9 = 0$

For Exercises 31–34, solve.

31. $\dfrac{1}{y} + \dfrac{1}{y+3} = \dfrac{2}{3}$ **32.** $\dfrac{1}{u} + \dfrac{1}{u-5} = \dfrac{10}{u^2 - 25}$

33. $6 - 5x^{-1} + x^{-2} = 0$ **34.** $2 - 3x^{-1} - x^{-2} = 0$

For Exercises 35–38, solve.

35. $14\sqrt{x} + 45 = 0$ **36.** $\sqrt{4m} = 3m - 1$

37. $\sqrt{6r + 13} = 2r + 1$ **38.** $\sqrt{21t + 2} + t = 2 + 4t$

For Exercises 39–44, solve the equations using substitution.

39. $x^4 - 5x^2 + 6 = 0$ **40.** $2m^4 - 3m^2 + 1 = 0$

41. $6(x + 5)^2 - 5(x + 5) + 1 = 0$ **42.** $\left(\dfrac{x-1}{3}\right)^2 + 10\left(\dfrac{x-1}{3}\right) + 9 = 0$

43. $p^{2/3} - 11p^{1/3} + 24 = 0$ **44.** $5a^{1/2} + 13a^{1/4} - 6 = 0$

For Exercises 45–50: a. Find the x- and y-intercepts.
b. State whether the parabola opens up or down.
c. Find the coordinates of the vertex.
d. Write the equation of the axis of symmetry.
e. Graph.

45. $f(x) = -2x^2$

46. $g(x) = \dfrac{1}{2}x^2 + 1$

47. $h(x) = -\dfrac{1}{3}(x - 2)^2$

48. $k(x) = 4(x + 3)^2 - 2$

49. $m(x) = x^2 + 2x - 1$

50. $p(x) = -0.5x^2 + 4x - 6$

For Exercises 51–54, solve the following inequalities.

51. $(x + 5)(x - 3) > 0$

52. $(x + 3)(x - 1)(x - 2) \geq 0$

53. $x^2 + 9x + 14 < 0$

54. $n^2 - 6n \leq -8$

For Exercises 55–58, solve the rational inequalities.

55. $\dfrac{a + 3}{a - 1} \geq 0$

56. $\dfrac{r}{r + 2} < 2$

57. $\dfrac{n - 3}{n - 4} \leq 5$

58. $\dfrac{(k + 2)(k - 3)}{k - 5} < 0$

59. A crop circle appeared July 7, 2003 in Windham Hill, England, and covered $22{,}500\pi$ square feet. Find the radius of the circle.

60. Using the formula $E = \dfrac{1}{2}mv^2$, where E represents the kinetic energy in joules of an object with a mass of m kilograms and a velocity of v meters per second, find the velocity of an object with a mass of 50 kilograms and 400 joules of kinetic energy.

61. The length of a small rectangular shed is 4 feet more than its width. If the area is 285 square feet, what are the dimensions of the shed?

62. A cylinder is to be constructed so that its volume is equal to that of a sphere with a radius of 9 inches. If the cylinder is to have a height of 4 inches, find the radius of the cylinder.

63. A ramp is constructed so that it is a right triangle with a base that is 9 feet longer than its height. If the hypotenuse is 17 feet, find the dimensions of the base and height.

64. An acrobat is launched upwards from one end of a lever with an initial velocity of 24 feet per second. The function $h = -16t^2 + 24t$ describes the height, h, of the acrobat t seconds after being launched.

   a. After how many seconds does the acrobat reach maximum height?

   b. What is the maximum height that the acrobat reaches?

   c. How long is the acrobat in the air?

   d. Graph the function.

65. The longest punt on record in the NFL was by Steve O'Neal in a game between the New York Jets and Denver Broncos on September 21, 1969. The function $y = -0.03x^2 + 2.16x$ models the trajectory of the punt. (Note that x and y are distances in yards.)

   a. Find the maximum height the punt reached.

   b. Find the distance of the punt.

66. In the triangle shown, the height is to be 2 inches less than x. The base is to be 4 inches more than x.

   a. Find the range of values for x so that the area of the triangle is at least 56 square inches.

   b. Find the range of values for the base and the height.

Of Interest

The yardage you found is how far the punt carried in the air. It was actually recorded as a 98 yard punt, which includes the amount of roll after the punt landed and was downed.

For 1 and 2, use the square root principle to solve and check.

1. $x^2 = 81$

2. $(x - 3)^2 = 20$

For 3 and 4, solve by completing the square.

3. $x^2 - 8x = -4$

4. $3m^2 - 6m = 5$

For 5 and 6, solve using the quadratic formula.

5. $2x^2 + x - 6 = 0$

6. $x^2 - 8x + 15 = 0$

For 7–12, solve using any method.

7. $u^2 - 16 = -6u$

8. $x^2 = 81$

9. $4w^2 + 6w + 3 = 0$

10. $2x^2 + 4x = 0$

11. $x^2 + 16 = 0$

12. $3k^2 = -5k$

For 13–16, solve.

13. $\dfrac{1}{x + 2} + \dfrac{1}{x} = \dfrac{5}{12}$

14. $3 - x^{-1} - 2x^{-2} = 0$

15. $9\sqrt{x} + 8 = 0$

16. $\sqrt{x + 8} - x = 2$

For 17 and 18, solve the equations using substitution.

17. $9a^4 + 26a^2 - 3 = 0$

18. $(x + 1)^2 + 3(x + 1) - 4 = 0$

19. For $f(x) = -\dfrac{1}{2}(x - 3)^2 + 1,$

    a. State whether the parabola opens up or down.

    b. Find the x- and y-intercepts.

    c. Find the coordinates of the vertex.

    d. Write the equation of the axis of symmetry.

    e. Graph.

For 20 and 21: a. Solve the inequality.
                 b. Graph the solution set on a number line.

20. $(x + 1)(x - 4) \le 0$ **21.** $\dfrac{x + 2}{x - 1} > 0$

For 22–25, solve.

22. A ball is thrown downward from a window in a tall building. The distance, d, traveled by the ball is given by the equation $d = 16t^2 + 32t$, where t is the time traveled in seconds. How long will it take the ball to fall 180 feet?

23. A rectangular parking lot needs to have an area of 400 square feet. The length is to be 20 feet more than the width. Find the dimensions of the parking lot.

24. An archer shoots an arrow in a field. Suppose the equation $y = -0.02x^2 + 1.3x + 8$ models the trajectory of the arrow (assume that x and y represent distances in meters and that $x \geq 0$ and $y \geq 0$).

   a. What is the maximum height that the arrow reaches?

   b. How far does the arrow travel horizontally?

   c. Graph the trajectory of the arrow.

25. A zoo is planning to install a new aquarium tank. The tank is to be in the shape of a box with a height of 12 feet. It is desired that the length be 15 feet more than the width.

   a. Find the range of values for the width so that the volume of the space is at most 12,000 cubic feet.

   b. Find the range of values for the length.

CHAPTER 12

Exponential and Logarithmic Functions

We begin this chapter with composite and inverse functions. Exponential and logarithmic functions, the subject of the rest of the chapter, are inverses of each other. These functions are used in modeling population growth, electrical circuits, sound decibels, pH, and the Richter scale.

"The mathematics of uncontrolled growth are frightening. A single cell of the bacterium *E. coli* would, under ideal circumstances, divide every twenty minutes. . . . one cell becomes two, two become four, four become eight and so on. In this way it can be shown that in a single day, one cell of *E. coli* could produce a super-colony equal in size and weight to the entire planet Earth."

—*MICHAEL CRICHTON*
The Andromeda Strain (1971)

12.1 Composite and Inverse Functions

OBJECTIVES

1 Find the composition of two functions.

2 Show that two functions are inverses.

3 Show that a function is one to one.

4 Find the inverse of a function.

5 Graph a given function's inverse function.

OBJECTIVE 1 Find the composition of two functions. In Section 8.5, we learned to perform the basic operations with functions. We now explore a new operation. Recall that a function pairs elements from an "input" set called the *domain* with elements in an "output" set called the *range*. If the output elements of one function are used as input elements in a second function, the resulting function is the **composition** of the two functions. Composition occurs when one quantity depends on a second quantity, which, in turn, depends on a third quantity.

We can visualize the composition of two functions as a machine with two stages, which are the two functions. For example, some coffee machines have two stages. We put in roasted coffee beans, which are ground in the first stage (first function). The grounds are then fed into the second stage (second function), which runs hot water over the grounds to produce coffee.

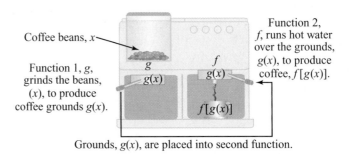

Coffee beans, x

Function 1, g, grinds the beans, (x), to produce coffee grounds $g(x)$.

Function 2, f, runs hot water over the grounds, $g(x)$, to produce coffee, $f[g(x)]$.

Grounds, $g(x)$, are placed into second function.

Note: The notation $(g \circ f)(x)$ is read "g composed with f of x," or simply "g of f of x."

DEFINITION

Warning: Do not confuse $(f \circ g)(x)$ with $(f \cdot g)(x)$

Composition of functions: If f and g are functions, then the composition of f and g is defined as $(f \circ g)(x) = f[g(x)]$ for all x in the domain g for which $g(x)$ is in the domain of f. The composition of g and f is defined as $(g \circ f)(x) = g[f(x)]$ for all x in the domain of f for which $f(x)$ is in the domain of g.

For example, suppose a pebble is dropped into a calm lake causing ripples to form in concentric circles. If the radius of the outer ripple increases at a rate of 2 feet per second, then we can describe the length of the radius as a function of t by using $r(t) = 2t$, where t is in seconds. The area of the circle formed by the outer ripple is also a function described by $A(r) = \pi r^2$. If we substitute the output of $r(t)$, which is $2t$ for r into $A(r)$, then we form the composite function $(A \circ r)(t)$.

$$(A \circ r)(t) = A[r(t)]$$
$$= \pi (2t)^2 \quad \text{Substitute } 2t \text{ for } r \text{ in } A(r) = \pi r^2.$$
$$= 4\pi t^2 \quad \text{Simplify.}$$

The composite function $(A \circ r)(t)$ gives the area of the circle as a function of time.

EXAMPLE 1 If $f(x) = 2x - 3$ and $g(x) = x^2 + 1$, find the following.

a. $(f \circ g)(2)$

Solution
$$\begin{aligned}(f \circ g)(2) &= f[g(2)] & \text{Definition of composition.}\\ &= f(5) & g(2) = 2^2 + 1 = 4 + 1 = 5.\\ &= 2(5) - 3 & \text{In } f(x), \text{ replace } x \text{ with } 5.\\ &= 7 & \text{Simplify.}\end{aligned}$$

Note: We first find $g(2)$, which is 5, then substitute that result into f.

b. $(g \circ f)(2)$

Solution
$$\begin{aligned}(g \circ f)(2) &= g[f(2)] & \text{Definition of composition.}\\ &= g(1) & f(2) = 2(2) - 3 = 4 - 3 = 1.\\ &= 1^2 + 1 & \text{In } g(x), \text{ replace } x \text{ with } 1.\\ &= 2 & \text{Simplify.}\end{aligned}$$

c. $(f \circ g)(x)$

Solution
$$\begin{aligned}(f \circ g)(x) &= f[g(x)] & \text{Definition of composition.}\\ &= f(x^2 + 1) & \text{Replace } g(x) \text{ with } x^2 + 1.\\ &= 2(x^2 + 1) - 3 & \text{In } f(x), \text{ replace } x \text{ with } x^2 + 1.\\ &= 2x^2 + 2 - 3 & \text{Simplify.}\\ &= 2x^2 - 1\end{aligned}$$

Note: We could have found $(f \circ g)(2)$ from part a, by first finding $(f \circ g)(x) = 2x^2 - 1$, then substituting 2 for x:
$$(f \circ g)(2) = 2(2)^2 - 1$$
$$= 8 - 1$$
$$= 7$$

d. $(g \circ f)(x)$

Solution
$$\begin{aligned}(g \circ f)(x) &= g[f(x)] & \text{Definition of composition.}\\ &= g(2x - 3) & \text{Replace } f(x) \text{ with } 2x - 3.\\ &= (2x - 3)^2 + 1 & \text{In } g(x), \text{ replace } x \text{ with } 2x - 3.\\ &= 4x^2 - 12x + 9 + 1 & \text{Simplify.}\\ &= 4x^2 - 12x + 10\end{aligned}$$

Note: Generally, $(f \circ g)(x) \neq (g \circ f)(x)$ as in Examples 1c and d, but there are special functions for which $(f \circ g)(x) = (g \circ f)(x)$.

ANSWERS

a. $f[g(-3)] = 18$
b. $g[f(2)] = 23$
c. $f[g(x)] = 9x^2 + 30x + 27$

YOUR TURN If $f(x) = x^2 + 2$ and $g(x) = 3x + 5$, find the following:

a. $f[g(-3)]$ **b.** $g[f(2)]$ **c.** $f[g(x)]$

OBJECTIVE 2 Show that two functions are inverses. Operations, such as addition and subtraction, that undo each other are called *inverse operations*. For example, if we begin with a number x and add a second number y, we have $x + y$. Now subtract y (the number we just added) from that result and we have $x + y - y = x$, which is the beginning number.

Likewise, two functions that undo each other under composition are called **inverse functions**.

> ***Inverse Functions:*** Two functions f and g are inverses if and only if $(f \circ g)(x) = x$ for all x in the domain of g and $(g \circ f)(x) = x$ for all x in the domain of f.

Loosely speaking, f and g are inverse functions if you evaluate f for a value x in its domain, substitute that result into g and evaluate and you get x again and vice versa.

PROCEDURE **INVERSE FUNCTIONS**

To determine whether two functions f and g are inverses of each other,

1. Show that $f[g(x)] = x$ for all x in the domain of g.
2. Show that $g[f(x)] = x$ for all x in the domain of f.

EXAMPLE 2 Verify that f and g are inverses.

a. $f(x) = 3x + 2$, $g(x) = \dfrac{x - 2}{3}$

Solution We need to show that $f[g(x)] = x$ and $g[f(x)] = x$.

$$f[g(x)] = f\left(\frac{x - 2}{3}\right) \qquad \text{Substitute } \frac{x-2}{3} \text{ for } g(x).$$

$$= 3\left(\frac{x - 2}{3}\right) + 2 \qquad \text{Substitute } \frac{x-2}{3} \text{ for } x \text{ in } f(x).$$

$$= x - 2 + 2 \qquad \text{Simplify.}$$

$$= x \qquad \text{So } f[g(x)] = x.$$

$$g[f(x)] = g(3x + 2) \qquad \text{Substitute } 3x + 2 \text{ for } f(x).$$

$$= \frac{3x + 2 - 2}{3} \qquad \text{Replace } x \text{ with } 3x + 2 \text{ in } g(x).$$

$$= \frac{3x}{3} \qquad \text{Simplify.}$$

$$= x \qquad \text{So } g[f(x)] = x.$$

Since $f[g(x)] = x$ and $g[f(x)] = x$, f and g are inverses.

b. $f(x) = x^3 + 5$ and $g(x) = \sqrt[3]{x - 5}$.

Solution We need to show that $f[g(x)] = x$ and $g[f(x)] = x$.

$$f[g(x)] = f(\sqrt[3]{x - 5}) \qquad \text{Replace } g(x) \text{ with } \sqrt[3]{x - 5}.$$

$$= (\sqrt[3]{x - 5})^3 + 5 \qquad \text{Replace } x \text{ with } \sqrt[3]{x - 5} \text{ in } f(x).$$

$$= x - 5 + 5 \qquad \text{Simplify.}$$

$$= x \qquad \text{So } f[g(x)] = x.$$

$$g[f(x)] = g(x^3 + 5) \qquad \text{Replace } f(x) \text{ with } x^3 + 5.$$

$$= \sqrt[3]{x^3 + 5 - 5} \qquad \text{Replace } x \text{ with } x^3 + 5 \text{ in } g(x).$$

$$= \sqrt[3]{x^3} \qquad \text{Simplify.}$$

$$= x \qquad \text{So } g[f(x)] = x.$$

Since $f[g(x)] = x$ and $g[f(x)] = x$, f and g are inverses.

$$f[g(x)] = f\left[\frac{x-6}{5}\right]$$
$$= 5\left(\frac{x-6}{5}\right) + 6$$
$$= x - 6 + 6$$
$$= x$$
$$g[f(x)] = g[5x + 6]$$
$$g[f(x)] = \frac{5x + 6 - 6}{5}$$
$$= \frac{5x}{5}$$
$$= x$$

Since $f[g(x)] = x$ and $g[f(x)] = x$, f and g are inverses.

YOUR TURN

Verify that $f(x) = 5x + 6$ and $g(x) = \dfrac{x-6}{5}$ are inverse functions.

Calculator TIPS

We can verify that $f[g(x)] = g[f(x)] = x$ without actually finding the composition functions by using a graphing utility. We illustrate using $f(x) = \dfrac{x+3}{x}$ and $g(x) = \dfrac{3}{x-1}$, which are inverse functions for $x \neq 0$ and $x \neq 1$. Enter $f(x) = Y_1 = \dfrac{x+3}{x}$, $g(x) = Y_2 = \dfrac{3}{x-1}$.

Since $f[g(x)] = Y_1(Y_2)$, enter $Y_3 = Y_1(Y_2)$. To enter Y_1 and Y_2 in the $\boxed{Y=}$ menu on the TI-83 plus, put the cursor at Y_3, press $\boxed{\text{VARS}}$, select Y-VARS, then select Function and press $\boxed{\text{ENTER}}$. Select your choice and press $\boxed{\text{ENTER}}$. Graph Y_3 without graphing Y_1 or Y_2. To graph Y_3 only, press $\boxed{Y=}$, put the cursor on the $=$ sign of Y_1 and press $\boxed{\text{ENTER}}$. Repeat for Y_2. Press $\boxed{\text{GRAPH}}$. Trace to verify that the graph is $y = x$.

```
Y₁ ▨ (X + 3)/X
Y₂ ▨ 3/(X −1)
Y₃ ▨ Y₁ (Y₂)▨
Y₄ =
Y₅ =
Y₆ =
Y₇ =
Y₈ =
```

X= 2.6315789   Y= 2.6315789

Repeat for $g[f(x)]$.

Before finding inverse functions, we need to determine what types of functions have inverses. Consider the following:

$f = \{(1, 2), (3, 4), (5, 6), (x, y)\}$ Domain $= \{1, 3, 5, x\}$ and range $= \{2, 4, 6, y\}$
$g = \{(2, 1), (4, 3), (6, 5), (y, x)\}$ Domain $= \{2, 4, 6, y\}$ and range $= \{1, 3, 5, x\}$

Note the ordered pairs in g are the ordered pairs of f with x and y interchanged. Since $(1, 2)$ is in f, by definition $f(1) = 2$. Since $(2, 1)$ is in g, $g(2) = 1$. Therefore, $f[g(2)] = f(1) = 2$ and $g[f(1)] = g(2) = 1$. Similarly, $f[g(4)] = f(3) = 4$ and $g[f(3)] = g(4) = 3$. In particular, $f[g(y)] = f(x) = y$ and $g[f(x)] = g(y) = x$ for all real numbers x and y. Consequently, f and g are inverses.

In general, to find the inverse of a function we reverse the ordered pairs by interchanging x and y. To indicate the inverse of the function f, we use a special notation f^{-1} instead of g. The following figure illustrates inverse functions.

Warning: Do not confuse f^{-1} with raising an expression to the negative 1 power. It is usually clear from the context that f^{-1} means an inverse function.

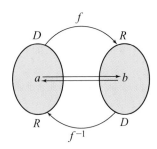

Note: Notice that f sends a to b and f^{-1} sends b back to a. Hence, $f^{-1}[f(a)] = a$. Also, f^{-1} sends b to a and f sends a back to b. Hence, $f[f^{-1}(b)] = b$. We also see that the domain of f is the range of f^{-1} and the range of f is the domain of f^{-1}.

Before we can find inverse functions, we need to explore one more concept.

OBJECTIVE **3** **Show that a function is one-to-one.** If we merely reverse the ordered pairs of a function, we do not always get an inverse function. Consider the following sets in which the ordered pairs of B are the ordered pairs of A with x and y interchanged:

$$A = \{(1, 2), (2, 4), (3, 2), (-2, 5)\}$$
$$B = \{(2, 1), (4, 2), (2, 3), (5, -2)\}$$

Note that A represents a function, but B does not because 2 in the domain of B is paired with 1 and 3. How did this happen? Since the ordered pairs of B are those of A with x and y interchanged, two ordered pairs of A having the same y-value of 2 [(1, 2) and (3, 2)] became two ordered pairs with the same x-value in B. Therefore, if the inverse of a function is to be a function, no two ordered pairs of the function may have the same y-value. Such a function is called a **one-to-one function**.

DEFINITION

Note: The definition says two different x-values must result in two different y-values. We can also say if $f(a) = f(b)$, then $a = b$, and if $a \neq b$, then $f(a) \neq f(b)$.

One-to-One Function: A function is one-to-one if each input (value from the domain) corresponds to only one output (value from the range) and every output corresponds to only one input.

There is a way of determining if a function is one-to-one by looking at the graph. If two ordered pairs have the same y-value, their graphs lie on the same horizontal line. Consequently, if a function is one-to-one, then the graph cannot be intersected by any horizontal line in more than one point.

RULE **HORIZONTAL LINE TEST FOR ONE-TO-ONE FUNCTIONS**

A function is one-to-one if any horizontal line can intersect the graph in at most one point.

EXAMPLE **3** Determine whether the following are graphs of one-to-one functions.

a. $f(x) = x^3$

b. $f(x) = x^2 + 2$

c. $f(x) = \sqrt{36 - x^2}$

Solution

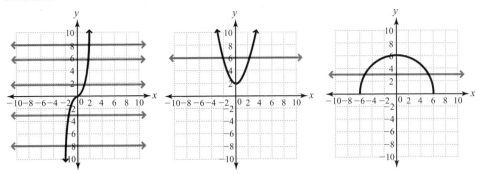

a. Horizontal lines can intersect this graph in at most one point, so the function is one-to-one.

b. A horizontal line can intersect this graph in more than one point, so the function is not one-to-one.

c. A horizontal line can intersect this graph in more than one point, so the function is not one-to-one.

> **YOUR TURN** Determine whether the following are graphs of one-to-one functions.

a.

b.

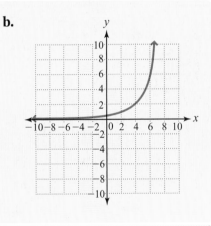

ANSWERS

a. no   **b.** yes

OBJECTIVE ④ **Find the inverse of a function.** Interchanging the ordered pairs of a function that is not one-to-one results in a relation that is not a function. Consequently, we have the following.

> **RULE** **EXISTENCE OF INVERSE FUNCTIONS**
>
> A function has an inverse function if and only if the function is one to one.

We have already seen that in order to find the inverse of a function, we need to interchange the *x*- and *y*-values of all the ordered pairs of *f*. That is, we replace *x* with *y* and *y* with *x* in the equation defining the function. However, we are not finished. We now have the inverse in the form $f(y) = x$ so we must solve this expression for *y*. The steps are summarized as follows.

PROCEDURE FINDING THE INVERSE FUNCTION OF A ONE-TO-ONE FUNCTION

1. If necessary, replace $f(x)$ with y.
2. Replace all x's with y and y's with x.
3. Solve the expression from step 2 for y.
4. Replace y with $f^{-1}(x)$.

EXAMPLE 4 Find the inverse of each of the following:

a. $f(x) = 2x + 4$

The domain and range of f is the set of all real numbers, so the domain and range of f^{-1} will also be the set of all real numbers.

Solution

$$y = 2x + 4 \qquad \text{Replace } f(x) \text{ with } y.$$
$$x = 2y + 4 \qquad \text{Replace } x \text{ with } y \text{ and } y \text{ with } x.$$
$$\frac{x - 4}{2} = y \qquad \text{Solve for } y.$$
$$f^{-1}(x) = \frac{x - 4}{2} \qquad \text{Replace } y \text{ with } f^{-1}(x).$$

In order to verify that we have found the inverse, we need to show that $f[f^{-1}(x)] = x$ and $f^{-1}[f(x)] = x$.

$$f[f^{-1}(x)] = f\left(\frac{x-4}{2}\right) = 2\left(\frac{x-4}{2}\right) + 4 = x - 4 + 4 = x$$

$$f^{-1}[f(x)] = f^{-1}(2x + 4) = \frac{(2x+4) - 4}{2} = \frac{2x + 4 - 4}{2} = \frac{2x}{2} = x$$

Since $f[f^{-1}(x)] = x$ and $f^{-1}[f(x)] = x$, they are inverses.

b. $f(x) = x^3 + 2$

The domain and range of f is the set of all real numbers, so the domain and range of f^{-1} will also be the set of all real numbers.

Solution

$$y = x^3 + 2 \qquad \text{Replace } f(x) \text{ with } y.$$
$$x = y^3 + 2 \qquad \text{Replace } x \text{ with } y \text{ and } y \text{ with } x.$$
$$x - 2 = y^3 \qquad \text{Solve for } y.$$
$$\sqrt[3]{x - 2} = y \qquad \text{Take the cube root of each side.}$$
$$f^{-1}(x) = \sqrt[3]{x - 2} \qquad \text{Replace } y \text{ with } f^{-1}(x).$$

We can verify that f and f^{-1} are inverses as in part a above.

c. $f(x) = \sqrt{x - 3}$

The domain of f is $[3, \infty)$ and the range is $[0, \infty)$, Therefore, the domain of f^{-1} is $[0, \infty)$ and the range of f^{-1} is $[3, \infty)$.

Solution $y = \sqrt{x - 3}$ Replace $f(x)$ with y.

$\qquad\qquad x = \sqrt{y - 3}$ Replace x with y and y with x.

$\qquad\qquad x^2 = y - 3$ Solve for y.

$\qquad\quad x^2 + 3 = y$

The domain of $y = x^2 + 3$ is all real numbers, but the domain of f^{-1} is $[0, \infty)$. Therefore, we write $f^{-1}(x) = x^2 + 3$, $x \geq 0$.

In order to verify that these are inverses, we need to show that $f[f^{-1}(x)] = x$ and $f^{-1}[f(x)] = x$.

$$f[f^{-1}(x)] = f(x^2 + 3) = \sqrt{(x^2 + 3) - 3} = \sqrt{x^2 + 3 - 3} = \sqrt{x^2}.$$

Since the domain of f^{-1} is $[0, \infty)$, $\sqrt{x^2} = x$. So, $f[f^{-1}(x)] = x$.

$$f^{-1}[f(x)] = f^{-1}(\sqrt{x - 3}) = (\sqrt{x - 3})^2 + 3 = x - 3 + 3 = x$$

Since $f[f^{-1}(x)] = x$ and $f^{-1}[f(x)] = x$, they are inverses.

YOUR TURN Find $f^{-1}(x)$ for each of the following:

a. $f(x) = 5x + 2$

b. $f(x) = \dfrac{x + 3}{x}$

OBJECTIVE 5 **Graph a given function's inverse function.** In the following figure, we have plotted pairs of points whose coordinates are interchanged and graphed the line $y = x$.

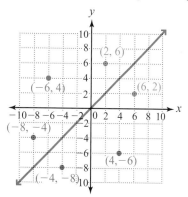

If the graph were folded along the line $y = x$, the points whose coordinates are interchanged would fall on top of each other. These points, therefore, are symmetric with respect to the line $y = x$. It can be shown that the graphs of any two points of the form (a, b) and (b, a) are symmetric with respect to the graph of $y = x$. Similarly, since the ordered pairs of f and f^{-1} have interchanged coordinates, the graphs of f and f^{-1} are symmetric with respect to the line $y = x$.

RULE **GRAPHS OF INVERSE FUNCTIONS**

The graphs of f and f^{-1} are symmetric with respect to the graph of $y = x$.

Following are the graphs of f and f^{-1} for Example 4 along with the graph of $y = x$.

a. $f(x) = 2x + 4$
$$f^{-1}(x) = \frac{x - 4}{2}$$

b. $f(x) = x^3 + 2$
$$f^{-1}(x) = \sqrt[3]{x - 2}$$

c. $f(x) = \sqrt{x - 3}$
$$f^{-1}(x) = x^2 + 3, \ x \geq 0$$

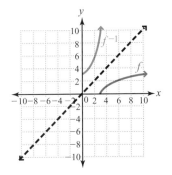

EXAMPLE 5 Sketch the inverse of the functions whose graphs are shown in parts a and b.

a.

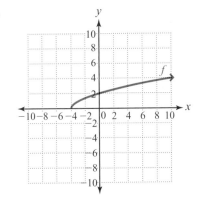

Solution Draw the line $y = x$ and reflect the graph in the line.

b.

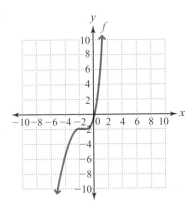

Solution Draw the line $y = x$ and reflect the graph in the line.

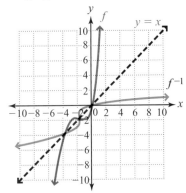

a.

b.

YOUR TURN Sketch the graph of the inverse of the functions whose graphs are given.

a.

b.

12.1 EXERCISES

1. Does $f \circ g = g \circ f$ for any two functions f and g? Why or why not?

2. How are the domains and ranges of inverse functions related? Why?

3. How is the graph of $(-2, 3)$ related to the graph of $(3, -2)$?

4. If the coordinates of the ordered pairs of a function are interchanged, is the result always the inverse function? Why or why not?

5. What needs to be shown to prove that two functions, f and g, are inverses?

6. How can you determine whether or not a function is one-to-one?

For Exercises 7–18, if $f(x) = 3x + 5$, $g(x) = x^2 + 3$, and $h(x) = \sqrt{x + 1}$, find each composition.

7. $(f \circ g)(0)$

8. $(g \circ f)(0)$

9. $(h \circ f)(1)$

10. $(h \circ g)(3)$

11. $(f \circ g)(-2)$

12. $(g \circ f)(-1)$

13. $(f \circ g)(x)$

14. $(g \circ f)(x)$

15. $(f \circ h)(x)$

16. $(h \circ g)(x)$

17. $(h \circ f)(0)$

18. $(g \circ h)(0)$

For Exercises 19–28, find $(f \circ g)(x)$ and $(g \circ f)(x)$.

19. $f(x) = 2x - 2$, $g(x) = 3x + 4$

20. $f(x) = 4x + 7$, $g(x) = 3x - 2$

21. $f(x) = x + 2$, $g(x) = x^2 + 1$

22. $f(x) = x^2 - 3$, $g(x) = x + 5$

23. $f(x) = x^2 + 3x - 4$, $g(x) = 3x$

24. $f(x) = -2x$, $g(x) = x^2 - 2x + 4$

25. $f(x) = \sqrt{x + 2}$, $g(x) = 2x - 5$

26. $f(x) = 5x + 2$, $g(x) = \sqrt{x - 5}$

27. $f(x) = \dfrac{x + 1}{x}$, $g(x) = \dfrac{x - 3}{x}$

28. $f(x) = \dfrac{x + 4}{x}$, $g(x) = \dfrac{2 - x}{x}$

For Exercises 29–30, answer each question.

29. If the domain of f is $[3, \infty)$ and the range is $[0, \infty)$, what are the domain and range of f^{-1}?

30. If f and g are inverse functions and $f(2) = 5$, then $g(5) = $ _____.

For Exercises 31–34, determine whether the following functions f and g are inverses.

31. $f = \{(1, 2), (-1, -3), (3, 4), (2, -5)\}$, $g = \{(2, 1), (-3, -1), (4, 3), (-5, 2)\}$

32. $f = \{(4, -3), (1, 4), (5, 2), (-3, 1), (-1, 3)\}$, $g = \{(-3, 4), (4, 1), (2, 5), (1, -3), (3, -1)\}$

33. $f = \{(-2, -2), (3, -3), (-4, 4), (-6, -6)\}$, $g = \{(-2, -2), (-3, 3), (-4, 4), (-6, -6)\}$

34. $f = \{(5, 5), (-4, 4), (2, -2), (-7, -7)\}$, $g = \{(-5, -5), (4, -4), (-2, 2), (7, 7)\}$

For Exercises 35–48, determine if f and g are inverses by determining whether
$(f \circ g)(x) = x$ **and** $(g \circ f)(x) = x$.

35. $f(x) = x + 5$, $g(x) = x - 5$

36. $f(x) = x - 1$, $g(x) = x + 1$

37. $f(x) = 6x$, $g(x) = \dfrac{x}{6}$

38. $f(x) = -\dfrac{x}{3}$, $g(x) = -3x$

39. $f(x) = 2x - 3$, $g(x) = \dfrac{x - 3}{2}$

40. $f(x) = \dfrac{x - 4}{5}$, $g(x) = 5x + 4$

41. $f(x) = x^3 - 4$, $g(x) = \sqrt[3]{x + 4}$

42. $f(x) = x^5 + 3$, $g(x) = \sqrt[5]{x - 3}$

43. $f(x) = x^2$, $g(x) = \sqrt{x}$

44. $f(x) = x^4$, $g(x) = \sqrt[4]{x}$

45. $f(x) = x^2$, $x \geq 0$; $g(x) = \sqrt{x}$

46. $f(x) = x^2 + 2$, $x \geq 0$; $g(x) = \sqrt{x - 2}$

47. $f(x) = \dfrac{3}{x + 5}$, $g(x) = \dfrac{3 - 5x}{x}$

48. $f(x) = \dfrac{x}{x - 3}$, $g(x) = \dfrac{3x}{x - 1}$

For Exercises 49–50, answer each question.

49. Is $f(x) = \dfrac{1}{x}$ its own inverse? Explain.

50. Is $f(x) = x$ its own inverse? Explain.

For Exercises 51–62, sketch the graph of the inverse of each of the following functions.

51.

52.

53.

54.

55.

56.

57.

58.

59.

60.

61.

62.

For Exercises 63–66, determine whether the function is one-to-one.

63.

64.

65.

66.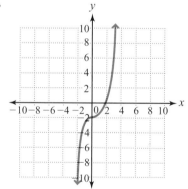

For Exercises 67–88, find $f^{-1}(x)$ for each of the following one-to-one functions f:

67. $f = \{(-3, 2), (-1, -3), (0, 4), (4, 6)\}$

68. $f = \{(-4, 2), (-1, -3), (2, 3), (5, 7)\}$

69. $f = \{(7, -2), (9, 2), (-4, 1), (3, 3)\}$

70. $f = \{(1, 2), (4, 5), (-3, 2), (-9, 0)\}$

71. $f(x) = x + 6$

72. $f(x) = x - 4$

73. $f(x) = 2x + 3$

74. $f(x) = -3x + 2$

75. $f(x) = x^3 + 2$

76. $f(x) = x^3 - 3$

77. $f(x) = \dfrac{2}{x + 2}$

78. $f(x) = \dfrac{-3}{x - 3}$

79. $f(x) = \dfrac{x + 2}{x - 3}$

80. $f(x) = \dfrac{x - 4}{x + 2}$

81. $f(x) = \sqrt{x - 2}$

82. $f(x) = \sqrt{2x - 4}$

83. $f(x) = 2x^3 + 4$

84. $f(x) = 4x^3 - 5$

85. $f(x) = \sqrt[3]{x + 2}$

86. $f(x) = \sqrt[3]{x - 5}$

87. $f(x) = 2\sqrt[3]{2x + 4}$

88. $f(x) = 3\sqrt[3]{4x - 3}$

For Exercises 89–92, solve each problem.

89. If a salesperson worked for $100 per week plus 5% commission on their sales, their weekly salary is $y = 0.05x + 100$, where y is the salary and x is the sales.

   a. Find the inverse function.

   b. What does each variable of the inverse function represent?

   c. Use the inverse function to find the sales for a week in which the salary was $350.

90. A painting contractor purchased 24 gallons of paint for the interior and exterior of a house. If he paid $12 per gallon for the interior paint and $18 per gallon for the exterior paint, then the amount that he paid is $y = 12x + 18(24 - x)$, where y is the amount paid and x is the number of gallons of interior paint.

   a. Find the inverse function.

   b. What does each variable of the inverse function represent?

   c. Use the inverse function to find the number of gallons of interior paint if he paid $372 total for the 24 gallons of paint.

91. An office building installed 105 fans, some of which measure 36 inches and the remainder measure 54 inches. If the 36 inch fans cost $45 each and the 54 inch fans cost $65 each, then the cost of the fans is $y = 45x + 65(105 - x)$, where y is the cost and x is the number of 36 inch fans.

  a. Find the inverse function.

  b. What does each variable of the inverse function represent?

  c. Use the inverse function to find the number of 36 inch fans if the total cost was $5225.

92. If a Toyota Prius averages 50 miles per hour on a trip, then $y = 50x$, where y is the number of miles traveled and x is the number of hours.

  a. Find the inverse function.

  b. What does each variable of the inverse function represent?

  c. Use the inverse function to find the number of hours to travel 210 miles.

For Exercises 93–98, answer the question.

93. If $(-4, 5)$ is an ordered pair on the graph of g, what are the coordinates of an ordered pair on the graph of g^{-1}?

94. If $f(2) = 4$ and $f^{-1}(a) = 2$, find a.

95. A linear function is of the form $f(x) = ax + b$, $a \neq 0$. Find $f^{-1}(x)$.

96. The square root function is defined by $f(x) = \sqrt{x}$. Find $f^{-1}(x)$. Be careful!

97. The graph of an even function is always symmetric with respect to the y-axis. Is the inverse of an even function also a function? Explain.

98. The graph of an odd function is always symmetric with respect to the origin. Is the inverse of an odd function also a function? Explain.

REVIEW EXERCISES

For Exercises 1–4, evaluate each expression.

1. 2^3

2. $4^{3/2}$

3. $\left(\dfrac{1}{3}\right)^3$

4. $\left(\dfrac{1}{2}\right)^4$

5. What would you do to the graph of $f(x) = x^2$ to get the graph of $g(x) = 2(x - 3)^2 + 1$?

6. Write 32 as 2 to a power.

12.2 Exponential Functions

OBJECTIVES

1. Define and graph exponential functions.
2. Solve equations of the form $b^x = b^a$ for x.
3. Use exponential functions to solve application problems.

OBJECTIVE 1 **Define and graph exponential functions.** Previously we defined rational-number exponents. For example, we know that $b^3 = b \cdot b \cdot b$, $b^{2/3} = \sqrt[3]{b^2}$, $b^0 = 1$, and $b^{-3} = \dfrac{1}{b^3}$. It can be shown that irrational-number exponents have meaning as well, and we can approximate expressions like $2^{\sqrt{3}}$ and 5^π by using rational approximations for the exponents. Therefore, the exponential expression b^x has meaning if x is any real number (rational or irrational), so we can define the **exponential function** as follows.

DEFINITION

> **Exponential Function:** If $b > 0$, $b \neq 1$, and x is any real number, then the exponential function is $f(x) = b^x$.

Note: The definition of the exponential function has two restrictions on b. If $b = 1$, then $f(x) = b^x = 1^x = 1$, which is a linear function. If $b < 0$, then we could get values for which the function is not defined as a real number. For example, if $f(x) = (-4)^x$, then
$$f\left(\frac{1}{2}\right) = (-4)^{1/2} = \sqrt{-4},$$
which is not a real number.

We graph exponential functions by plotting enough points to determine the graph's shape. We will find that the graph has one typical shape if $b > 1$ and a different typical shape if $0 < b < 1$.

EXAMPLE 1 Graph $f(x) = 2^x$.

Solution Choose some values of x and find the corresponding values of $f(x)$ (which is y).

x	-3	-2	-1	0	1	2	3
y = f(x)	$\frac{1}{8}$	$\frac{1}{4}$	$\frac{1}{2}$	1	2	4	8

Plotting the points gives a smooth curve. The shape of this graph is typical of the graphs of $f(x) = b^x$ with $b > 1$. Comparing the graphs of $f(x) = 2^x$ and $f(x) = 4^x$, we can see that the greater the value of b, the steeper the graph.

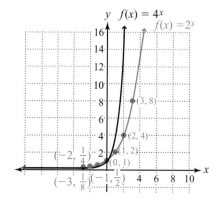

12.2 Exponential Functions **999**

EXAMPLE 2 Graph $f(x) = \left(\dfrac{1}{2}\right)^x$.

Solution Choose some values of x and find the corresponding values of $f(x)$. (which is y).

x	-3	-2	-1	0	1	2	3
$y = f(x)$	8	4	2	1	$\frac{1}{2}$	$\frac{1}{4}$	$\frac{1}{8}$

Note: From the graphs, we see that b^x is never negative.

After plotting the points, we determine a graph that is typical of exponential functions in the form $f(x) = b^x$ with $0 < b < 1$.

Comparing the graphs of $f(x) = \left(\dfrac{1}{2}\right)^x$ and $f(x) = \left(\dfrac{1}{4}\right)^x$, we see that the smaller the value of b, the steeper the graph.

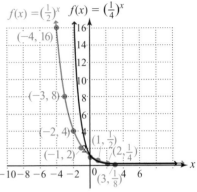

We summarize the graphs of $f(x) = b^x$ as follows.

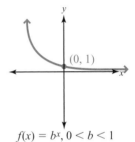

$f(x) = b^x, b > 1$

$f(x) = b^x, 0 < b < 1$

The function is increasing and one-to-one. The graph always passes through $(0, 1)$. For negative values of x, the graph approaches the x-axis but never touches it. The larger the values of b, the steeper the graph. The domain is $(-\infty, \infty)$ and the range is $(0, \infty)$.

The function is decreasing and one-to-one. The graph always passes through $(0, 1)$. For positive values of x, the graph approaches the x-axis but never touches it. The smaller the values of x, the steeper the graph. The domain is $(-\infty, \infty)$ and the range is $(0, \infty)$.

More complicated exponential functions are graphed in the same manner.

EXAMPLE 3 Graph $f(x) = 3^{2x-1}$.

Solution Find some ordered pairs, plot them, and draw the graph.

x	y = f(x)
−1	$3^{-3} = \dfrac{1}{27}$
0	$3^{-1} = \dfrac{1}{3}$
1	$3^1 = 3$
2	$3^3 = 27$

ANSWER

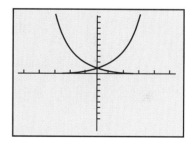

YOUR TURN

Graph $f(x) = 2^{x+1}$.

Calculator TIPS

To graph $y = 2^x$ on a graphing calculator, we enter $Y_1 = 2^x$. Set the window to $[-5, 5]$ for x and $[-1, 10]$ for y and press GRAPH *. We graph $y = \left(\dfrac{1}{2}\right)^x$ in the same window by entering $Y_2 = \left(\dfrac{1}{2}\right)^x$. Notice that if we folded the graph along the y-axis, the graphs would lie on top of each other, indicating they are symmetric with respect to the y-axis.*

```
Plot1 Plot2 Plot3
Y₁ ▧ 2^X
Y₂ ▧ (1/2)^X ▮
Y₃ =
Y₄ =
Y₅ =
Y₆ =
Y₇ =
```

OBJECTIVE 2 **Solve equations of the form $b^x = b^a$ for x.** Earlier we solved equations containing expressions having a variable base and a constant exponent like $x^2 = 4$. We will now solve equations that contain expressions having a constant base and a variable exponent like $2^x = 16$. We need a new method to solve these equations. Previously we noted that the exponential function is one-to-one, so for each x-value there is a unique y-value and vice versa. Consequently, we have the following rule.

RULE **THE ONE-TO-ONE PROPERTY OF EXPONENTIALS**

Given $b > 0$ and $b \neq 1$, if $b^x = b^y$, then $x = y$.

To solve some types of exponential equations, we use the following procedure.

PROCEDURE SOLVING EXPONENTIAL EQUATIONS

1. If necessary, write both sides of the equation as a power of the same base.
2. If necessary, simplify the exponents.
3. Set the exponents equal to each other.
4. Solve the resulting equation.

EXAMPLE 4 Solve the following equations for x.

a. $3^x = 81$

Solution $3^x = 3^4$ Write 81 as 3^4 so both sides have the same base.

$\qquad\quad x = 4$ Set the exponents equal to each other.

Check: $\quad 3^x = 81$

$\qquad\qquad 3^4 = 81$ Replace x with 4.

$\qquad\qquad 81 = 81$ True, so $x = 4$ is the solution.

b. $4^x = 32$

Solution $(2^2)^x = 2^5$ Write 4 as 2^2 and 32 as 2^5 so both sides have the same base.

$\qquad\quad\ 2^{2x} = 2^5$ Simplify $(2^2)^x$ by applying $(a^m)^n = a^{mn}$.

$\qquad\qquad 2x = 5$ Set the exponents equal to each other.

$\qquad\qquad\ x = \dfrac{5}{2}$ Solve for x.

c. $8^{x+4} = 4^{2x+3}$

Solution $(2^3)^{x+4} = (2^2)^{2x+3}$ Write 8 as 2^3 and 4 as 2^2 so both sides have the same base.

$\qquad\quad\ 2^{3x+12} = 2^{4x+6}$ Simplify the exponents by applying $(a^m)^n = a^{mn}$.

$\qquad 3x + 12 = 4x + 6$ Set the exponents equal to each other.

$\qquad\qquad\quad 6 = x$ Solve for x.

d. $\left(\dfrac{1}{5}\right)^x = 25$

Solution $(5^{-1})^x = 5^2$ Write $\dfrac{1}{5}$ as 5^{-1} and 25 as 5^2 so both sides have the same base.

$\qquad\quad\ 5^{-x} = 5^2$ Simplify $(5^{-1})^x$ by applying $(a^m)^n = a^{mn}$.

$\qquad\qquad -x = 2$ Set the exponents equal to each other.

$\qquad\qquad\ x = -2$ Solve for x.

ANSWERS

a. 2
b. 6
c. -4

YOUR TURN Solve the following equations for x.

a. $2^{x+1} = 8$ **b.** $27^{x-2} = 9^x$ **c.** $\left(\dfrac{1}{2}\right)^x = 16$

Compound Interest Formula

Exponential functions occur in many areas, especially the sciences and business. If P dollars is invested at an annual interest rate of r (written as a decimal) compounded n times per year for t years, the accumulated amount A in the account is given by the formula $A = P\left(1 + \dfrac{r}{n}\right)^{nt}$.

EXAMPLE 5 Find the accumulated amount in an account if $5000 is deposited at 6% compounded quarterly for 10 years. Round the answer to the nearest cent.

Solution We are asked to find A given that $P = \$5000$, $r = 0.06$, $n = 4$, and $t = 10$.

$$A = P\left(1 + \frac{r}{n}\right)^{nt}$$

$$A = 5000\left(1 + \frac{.06}{4}\right)^{4(10)} \qquad \text{Substitute for } P, r, n, \text{ and } t.$$

$$A = 5000(1 + 0.015)^{40} \qquad \text{Simplify.}$$

$$A = 5000(1.015)^{40}$$

$$A = \$9070.09 \qquad \text{Evaluate using a calculator.}$$

Half-Life

The *half-life* of a radioactive substance is the amount of time it takes until only half the original amount of the substance remains. Suppose we begin with 100 grams of a substance that has a half-life of 10 days. After 10 days (one half-life) 50 grams will remain, after 20 days (two half-lives) 25 grams will remain, after 30 days (three half-lives) 12.5 grams will remain, and so forth. The formula $A = A_0\left(\dfrac{1}{2}\right)^{t/h}$ gives the amount remaining, where A_0 is the initial amount, t is the time, and h is the half-life.

EXAMPLE 6 The isotope ^{45}Ca has a half-life of 165 days. How many grams of a 50-gram sample of ^{45}Ca will remain after 825 days? (*Source: CRC Handbook of Chemistry and Physics*, 62nd edition, CRC Press, 1981)

Solution $A = A_0\left(\dfrac{1}{2}\right)^{t/h}$

$$A = 50\left(\frac{1}{2}\right)^{825/165} \qquad \text{Substitute 50 for } A_0, 165 \text{ for } h, \text{ and } 825 \text{ for } t.$$

$$A = 50\left(\frac{1}{2}\right)^{5} \qquad \text{Simplify.}$$

$$A = 1.5625 \text{ grams.} \qquad \text{Evaluate using a calculator.}$$

Aging

EXAMPLE 7 The number of people in the United States age 65 or over (in millions) is given in the following table:

Year	Number 65 or over (in millions)
1900	3.1
1910	4.0
1920	4.9
1930	6.7
1940	9.0
1950	12.4
1960	16.7
1970	20.1
1980	25.5
1990	31.4
2000	35.6

(*Source:* U.S. Bureau of the Census)

This data can be modeled by the function $y = 3.17(1.026)^x$, where x is the number of years after 1900 and y is the number of people in millions. Use the model to estimate the number of people in the United States age 65 or over in the year 2020. Round the answer to the nearest tenth of a million.

Solution $x = 2020 - 1900 = 120$

$y = 3.17(1.026)^x$

$y = 3.17(1.026)^{120}$ **Substitute 120 for x.**

$y = 69.0$ million   **Evaluate using a calculator.**

Following is a graph of the function with the point corresponding to the solution of Example 7 indicated.

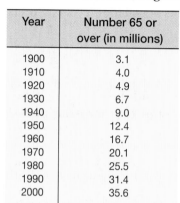

Years After 1900

YOUR TURN

a. If $3000 is invested at 4% compounded semiannually (twice per year), how much money is in the account at the end of 8 years? Round to the nearest cent.

b. The radioactive isotope ^{61}Cr has a half-life of 26 days. How much of a 10-gram sample would remain after 208 days? Give the answer to the nearest thousandth of a gram. (*Source: CRC Handbook of Chemistry and Physics*, 62nd edition, CRC Press, 1981)

contiuned

c. The following table shows the number of computers (in millions) in use in the United States for selected years since 1984:

Year	Number of computers (in millions)
1985	21.5
1988	40.8
1989	47.6
1991	62.0
1992	68.2
1993	76.5
1994	85.8
1995	96.2
2000	160.5

This data can be modeled by $y = 23.14(1.138)^x$, where x is the number of years after 1984 and y is the number of computers (in millions). Use the model to approximate the number of computers in use in the United States in 2010.

ANSWERS

a. $4118.36

b. 0.039 gm

c. 666.9 million

12.2 EXERCISES

For Extra Help

InterAct Math

Tutor Center — Addison-Wesley Math Tutor Center

Video 18/DVT 6

MyMathLab

MathXL

1. Are the graphs of $f(x) = 2^x$ and $g(x) = \left(\frac{1}{2}\right)^x$ symmetric with respect to the y-axis? Explain.

2. As x gets larger, which graph is steeper: $f(x) = 3^x$ or $g(x) = 1.5^x$? Why?

3. Find the domain and range of $f(x) = 2^{x-4}$.

4. Is $f(x) = (-2)^x$ an exponential function? Why or why not?

5. For $b > 0$ and $b \neq 1$ if $b^x = b^y$, why does $x = y$?

6. If we have 10 grams of a substance that has a half-life of 50 days, how many grams will be present after 200 days?

For Exercises 7–24, sketch the graph of the function.

7. $f(x) = 3^x$

8. $f(x) = 4^x$

9. $f(x) = 4^x - 3$

10. $f(x) = 3^x + 1$

11. $f(x) = \left(\dfrac{1}{3}\right)^x$

12. $f(x) = \left(\dfrac{1}{4}\right)^x$

13. $f(x) = \left(\dfrac{2}{3}\right)^x + 2$

14. $f(x) = \left(\dfrac{3}{2}\right)^x - 1$

15. $f(x) = -2^x$

16. $f(x) = -3^x$

17. $f(x) = 2^{x-2}$

18. $f(x) = 3^{x+1}$

19. $f(x) = 3^{-x}$

20. $f(x) = 2^{-x}$

21. $f(x) = 2^{2x-3}$

22. $f(x) = 3^{2x+1}$

23. $f(x) = 3^{-x+2}$

24. $f(x) = 2^{-x-1}$

For Exercises 25–42, solve each equation.

25. $2^x = 8$

26. $3^x = 81$

27. $8^x = 32$

28. $27^x = 81$

29. $16^x = 4$

30. $36^x = 216$

31. $5^x = \dfrac{1}{25}$

32. $3^x = \dfrac{1}{27}$

33. $\left(\dfrac{1}{3}\right)^x = 9$

34. $\left(\dfrac{1}{5}\right)^x = 125$

35. $\left(\dfrac{2}{3}\right)^x = \dfrac{8}{27}$

36. $\left(\dfrac{3}{2}\right)^x = \dfrac{9}{4}$

37. $\left(\dfrac{1}{2}\right)^x = 16$

38. $\left(\dfrac{1}{3}\right)^x = 27$

39. $25^{x+1} = 125$

40. $9^{x+2} = 81$

41. $8^{2x-1} = 32^{x-3}$

42. $5^{2x+1} = 125^{2x}$

For Exercises 43–46, graph each and answer the questions.

43. Graph $f(x) = 2^x$ and $g(x) = 2^{x+2}$ in the window $[-5, 5]$ for x and $[-1, 10]$ for y. How does the graph of g compare with the graph of f?

44. Graph $f(x) = 3^x$ and $g(x) = 3^x - 3$ in the window $[-3, 3]$ for x and $[-4, 10]$ for y. How does the graph of g compare with the graph of f?

45. Graph $f(x) = 2^x$ and $g(x) = 2^{-x}$ in the window $[-5, 5]$ for x and $[-1, 10]$ for y. How does the graph of g compare with the graph of f?

46. Graph $f(x) = 2^x$ and $g(x) = -2^x$ in the window $[-10, 10]$ for x and $[-10, 10]$ for y. How does the graph of g compare with the graph of f?

For Exercises 47–50, use the following.
Under ideal conditions a culture of *E. coli* bacteria doubles in size every 20 minutes. If A_0 is the initial amount and t is the number of minutes passed, the amount present is A, where $A = A_0(2)^{t/20}$.

47. If a culture of *E. coli* began with 100 cells, how many cells would be present after 120 minutes?

48. If a culture of *E. coli* currently has 500 cells, how many cells were present 90 minutes earlier? (*Hint:* Let $A_0 = 500$ and the time will be negative.)

49. Under ideal conditions, human beings could double their population every 50 years. If A_0 is the initial population and t is the number of years passed, the current population is A, where $A = A_0(2)^{t/50}$. If there were 6 billion humans on Earth in the year 2000, how many would there be in the year 2500 if their growth is uncontrolled?

50. Using the information from Exercise 49, how many human beings were on the Earth in 1900?

For Exercises 51–52, use the formula $A = P\left(1 + \dfrac{r}{n}\right)^{nt}$.

51. If $10,000 is deposited into an account paying 8% interest compounded quarterly, how much would be in the account after 12 years?

52. If $15,000 is deposited into an account paying 6% interest compounded semiannually, how much would be in the account after 9 years?

For Exercises 53–54, use the formula $A = A_0\left(\dfrac{1}{2}\right)^{t/h}$ from Example 6.

53. Einsteinium (^{254}Es) has a half-life of 270 days. How much of a 5-gram sample would remain after 2160 days? Give the answer to the nearest thousandth of a gram. (*Source: CRC Handbook of Chemistry and Physics*, 62nd Edition, CRC Press, 1981)

54. Nobelium (^{257}No) has a half-life of 23 seconds. How much of a 100-gram sample would remain after 275 seconds? Give the answer to the nearest thousandth of a gram. (*Source: CRC Handbook of Chemistry and Physics*, 62nd Edition, CRC Press, 1981)

55. Since the first Super Bowl in 1967, ticket prices for the game have risen exponentially. Ticket prices can be modeled by the function $P(t) = 9.046(1.114)^t$, where $t = 1$ is the year 1967 and ticket prices are in dollars. Estimate the price of a Super Bowl ticket in the year 2010. (*Source: Orlando Sentinel*, January 21, 2003, p. A8)

56. The average cost of a 30-second commercial during the Super Bowl has increased exponentially over the years and can be modeled by $C(t) = 55.66(1.114)^t$, where t is the number of years after 1967 and C is the cost in thousands of dollars. Estimate the cost of a 30-second commercial in 1975; in 1985; 2005. (*Source: Orlando Sentinel*, January 21, 2003, p. A8)

57. Chlorine is frequently used to disinfect swimming pools. The concentration should remain between 1.5 and 2.5 parts per million. On a warm, sunny day, 30% of the chlorine can dissipate into the air or combine with other chemicals. If the initial amount of chlorine is 2.5 million parts per million, the function $f(x) = 2.5(0.7)^x$ models the amount of chlorine after x days. How much chlorine is in the pool after 2 days? (*Source:* D. Thomas, *Swimming Pool Operations Handbook*, 1972)

58. It is estimated that the value of a car depreciates 20% per year for the first five years. If the original price of a car is P, the value, A, of a car after t years is given by $A = P(.8)^t$. If a car originally cost \$25,960, find the value of the car after 3 years to the nearest dollar.

59. Between 1971 and 1995 the number of transistors that can be placed on a single chip has grown significantly, as indicated in the following table:

Year	Chip	Transistors (millions)
1971	4004	0.0023
1986	386DX	0.275
1989	486DX	1.2
1993	Pentium	3.3
1995	P6	5.5

This data can be modeled by the function $T(x) = 0.001668(1.39)^x$, where x is the number of years after 1970 and $T(x)$ is the number of transistors in millions. If the current trend continues, estimate the number of transistors that could be put on a single chip in 2005. (*Source:* Intel)

60. Suppose that Dave fixes a cup of coffee with cream and places it on the counter to cool. The temperature of the coffee at various times is given in the following table:

t (seconds)	T (°F)
0.2	155.8
8.4	133.2
16.6	117.9
24.8	107.9
33	100.7
41.1	94.9
49.3	90.5

This data can be modeled by the function $T(t) = 146.9(.989)^t$, where t is time in seconds and T is the temperature in °F. Estimate the temperature of the coffee after 1 minute.

PUZZLE PROBLEM

What is the greatest number that can be written using three numerals?

REVIEW EXERCISES

1. If $f(x) = 3x + 4$, find $f^{-1}(x)$.

2. Graph the inverse of the function whose graph is shown.

For Exercises 3–5, evaluate.

3. 5^3

4. $\left(\dfrac{1}{3}\right)^{-2}$

5. 4^{-3}

6. Write $5^{2/3}$ in radical notation.

12.3 — Logarithmic Functions

OBJECTIVES

1. Convert between exponential and logarithmic forms.
2. Solve logarithmic equations by changing to exponential form.
3. Graph logarithmic functions.
4. Solve applications involving logarithms.

OBJECTIVE 1 Convert between exponential and logarithmic forms. In Section 12.2, we defined the exponential function as $f(x) = b^x$ with $b > 0$ and $b \neq 1$. Since the exponential function is a one-to-one function, it has an inverse. Following is the graph of $f(x) = 2^x$ and its inverse, which we find by reflecting the graph in the line $y = x$.

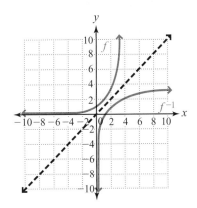

To find the inverse of $f(x) = 2^x$, we use the procedure from Section 12.1.

$$f(x) = 2^x$$
$$y = 2^x \qquad \text{Replace } f(x) \text{ with } y.$$
$$x = 2^y \qquad \text{Replace } y \text{ with } x \text{ and } x \text{ with } y.$$

The next step is to solve for y, but we haven't learned how to isolate a variable that is an exponent. To solve for y we must define a **logarithm**.

DEFINITION   **Logarithm:** If $b > 0$ and $b \neq 1$, then $y = \log_b x$ is equivalent to $x = b^y$.

If we apply this definition to $x = 2^y$ in the preceding equation, we get $y = \log_2 x$. Replacing y with $f^{-1}(x)$, we get $f^{-1}(x) = \log_2 x$, which means that exponential and **logarithmic functions** are inverses. Consequently, if $y = \log_b x$, the domain is $(0, \infty)$ and the range is $(-\infty, \infty)$ with $b > 0$ and $b \neq 1$.

The expression $\log_b x$ is read "the logarithm base b of x" and *is the exponent to which b must be raised to get x*. Compare the two forms.

The exponent is the logarithm

$$x = b^y \qquad y = \log_b x$$

The base is the base of the logarithm.

Note: $y = \log_b x$ means y is the power to which we raise b to get x.

The definition of a logarithm allows us to convert from one form to another. The following table contains pairs of equivalent forms.

Logarithmic Form	Exponential Form
$\log_2 8 = 3$	$2^3 = 8$
$\log_{10} \dfrac{1}{10} = -1$	$10^{-1} = \dfrac{1}{10}$
$\log_5 1 = 0$	$5^0 = 1$
$\log_{16} 4 = \dfrac{1}{2}$	$16^{1/2} = 4$
$\log_{1/2} 8 = -3$	$\left(\dfrac{1}{2}\right)^{-3} = 8$

Note: The logarithmic equations are in the form $\log_b x = y$. The values of x are positive numbers only and the values of y are both positive and negative.

EXAMPLE 1 Write in logarithmic form.

a. $3^4 = 81$

Solution $\log_3 81 = 4$ 　The base of the exponent is the base of the logarithm, and the exponent is the logarithm.

b. $\left(\dfrac{1}{2}\right)^{-3} = 8$

Solution $\log_{1/2} 8 = -3$ 　The base of the exponent is the base of the logarithm, and the exponent is the logarithm.

c. $9^{1/2} = 3$

Solution $\log_9 3 = \dfrac{1}{2}$ 　The base of the exponent is the base of the logarithm, and the exponent is the logarithm.

ANSWERS

a. $\log_4 64 = 3$

b. $\log_2 \dfrac{1}{8} = -3$

c. $\log_{27} 3 = \dfrac{1}{3}$

YOUR TURN Write in logarithmic form.

a. $4^3 = 64$ 　　　**b.** $2^{-3} = \dfrac{1}{8}$ 　　　**c.** $27^{1/3} = 3$

Learning Strategy

If you are a visual learner, try visualizing the following "loop" for rewriting logarithms in exponential form.

$\log_6 36 = 2$ 　equals
　　raised to

EXAMPLE 2 Write in exponential form.

a. $\log_6 36 = 2$

Solution $6^2 = 36$ 　The base of the logarithm is the base of the exponent and the logarithm is the exponent.

b. $\log_{16} 2 = \dfrac{1}{4}$

Solution $16^{1/4} = 2$ The base of the logarithm is the base of the exponent, and the logarithm is the exponent.

c. $\log_{1/2} 16 = -4$

Solution $\left(\dfrac{1}{2}\right)^{-4} = 16$ The base of the logarithm is the base of the exponent, and the logarithm is the exponent.

YOUR TURN Write in exponential form.

a. $\log_4 64 = 3$ **b.** $\log_{16} 2 = \dfrac{1}{4}$ **c.** $\log_5 \dfrac{1}{125} = -3$

OBJECTIVE ② Solve logarithmic equations by changing to exponential form. A logarithmic equation in the form $\log_b a = c$ has three unknowns: a, b, and c. If we know any two, we can often change the equation into exponential form and find the third.

EXAMPLE 3 Solve.

a. $\log_b 16 = 2$

Solution $b^2 = 16$ Change into exponential form
$\qquad\quad b = \pm 4$ Find the positive and negative square roots of 16.
$\qquad\quad b = 4$ The base must be positive and not 1, so $b = 4$.

b. $\log_3 \dfrac{1}{27} = c$

Solution $3^c = \dfrac{1}{27}$ Change into exponential form.

$\qquad\quad 3^c = \dfrac{1}{3^3}$ Write 27 as 3^3.

$\qquad\quad 3^c = 3^{-3}$ Write $\dfrac{1}{3^3}$ as 3^{-3}.

$\qquad\quad c = -3$ Set the exponents equal to each other.

c. $\log_{25} a = \dfrac{1}{2}$

Solution $25^{1/2} = a$ Change into exponential form

$\qquad\quad 5 = a$ $25^{1/2} = \sqrt{25} = 5$.

d. $\log_{36} \sqrt[4]{6} = c$

Solution $36^c = \sqrt[4]{6}$

$(6^2)^c = 6^{1/4}$ Write 36 as 6^2 and $\sqrt[4]{6}$ as $6^{1/4}$.

$6^{2c} = 6^{1/4}$ $(6^2)^c = 6^{2c}$.

$2c = \dfrac{1}{4}$ Set the exponents equal to each other.

$c = \dfrac{1}{8}$ Solve for c.

ANSWERS

a. $b = 3$

b. $c = -2$

c. $a = 3$

YOUR TURN   Solve.

a. $\log_b \dfrac{1}{9} = -2$ **b.** $\log_5 \dfrac{1}{25} = c$ **c.** $\log_{27} a = \dfrac{1}{3}$

The definition of logarithms leads to the following two properties.

RULE   For any real number $b > 0$ and $b \neq 1$, $\log_b b = 1$ and $\log_b 1 = 0$.

Based on the definition of a logarithm, $\log_b b = 1$ because $b^1 = b$, and $\log_b 1 = 0$ because $b^0 = 1$.

EXAMPLE 4   Use the preceding property to find the value of the following.

a. $\log_5 5$

Solution $\log_5 5 = 1$

b. $\log_e e$

Solution $\log_e e = 1$

c. $\log_{10} 1$

Solution $\log_{10} 1 = 0$

ANSWERS

a. 1   **b.** 0

YOUR TURN   Find the value of the following.

a. $\log_{10} 10$ **b.** $\log_{\sqrt{3}} 1$

Graph logarithmic functions. To graph logarithmic functions, which have the form $f(x) = \log_b x$, where $b > 0$ and $b \neq 1$, we will first change to exponential form so it will be easier to find ordered pairs.

PROCEDURE **GRAPHING LOGARITHMIC FUNCTIONS**

1. Change the logarithmic function $y = \log_b x$ to the exponential form $x = b^y$.
2. Find ordered pairs that satisfy the equation by assigning values to y and finding x.
3. Plot the ordered pairs and draw a smooth curve through the points.

EXAMPLE 5 Graph.

a. $f(x) = \log_2 x$

Solution $y = \log_2 x$ Replace $f(x)$ with y.
$\qquad 2^y = x$ Write in exponential form.

Choose values for y and find x.

x	y
$2^0 = 1$	0
$2^1 = 2$	1
$2^2 = 4$	2
$2^3 = 8$	3
$2^{-1} = \dfrac{1}{2}$	-1
$2^{-2} = \dfrac{1}{4}$	-2

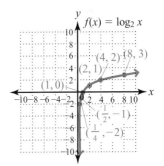

b. $f(x) = \log_{1/2} x$

Solution $y = \log_{1/2} x$ Replace $f(x)$ with y.
$\qquad x = \left(\dfrac{1}{2}\right)^y$ Write in exponential form.

Choose values for y and find x.

x	y
$\left(\dfrac{1}{2}\right)^0 = 1$	0
$\left(\dfrac{1}{2}\right)^1 = \dfrac{1}{2}$	1
$\left(\dfrac{1}{2}\right)^2 = \dfrac{1}{4}$	2
$\left(\dfrac{1}{2}\right)^3 = \dfrac{1}{8}$	3
$\left(\dfrac{1}{2}\right)^{-1} = 2$	-1
$\left(\dfrac{1}{2}\right)^{-2} = 4$	-2
$\left(\dfrac{1}{2}\right)^{-3} = 8$	-3

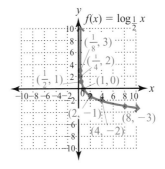

Following is a summary of the key features of the graphs of logarithmic functions.

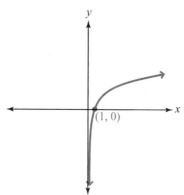

$f(x) = \log_b x$, $b > 1$

Graph passes through $(1, 0)$; domain is $x > 0$; range is all real numbers. Graph approaches the y-axis and increases.

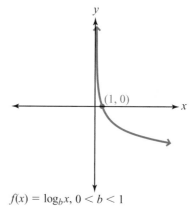

$f(x) = \log_b x$, $0 < b < 1$

Graph passes through $(1, 0)$; domain is $x > 0$; range is all real numbers. Graph approaches the y-axis and decreases.

YOUR TURN **Graph.**

$f(x) = \log_3 x$.

OBJECTIVE **4** Solve applications involving logarithms.

EXAMPLE **6** The function $P = 95 - 30 \log_2 x$ models the percent, P, of students who recall the important features of a classroom lecture over time, where x is the number of days that have elapsed since the lecture was given. What percent of the students recall the important features of a lecture 8 days after it was given? (*Source: Psychology for the New Millennium*, 8th Edition, Spencer A. Rathos, Thomson Publishing Company)

Solution $P = 95 - 30 \log_2 x$

$P = 95 - 30 \log_2 8$ Substitute 8 for x.

$P = 95 - 30(3)$ $\log_2 8 = 3$. Since $2^3 = 8$.

$P = 95 - 90$

$P = 5$ Simplify.

Therefore, 5% of the students remember the important features of a lecture 8 days after it is given.

YOUR TURN **Refer to Example 6.**

a. Find the percent of students who remember the important features of a lecture 2 days after it was given.

b. When the number of days was decreased by a third (from 8 to 2), was the amount retained also decreased by one-third?

1. If $f(x) = 3^x$, what is $f^{-1}(x)$? Why?

2. If $f(x) = \log_b x$, what are the restrictions on b?

3. If $f(x) = \log_b x$, what are the restrictions on x?

4. What is the relationship between the graphs of $f(x) = 5^x$ and $g(x) = \log_5 x$? Why?

5. Why is a logarithm an exponent?

6. The $\log_b x$ is the _____ to which _____ must be raised to get _____.

For Exercises 7–22, write in logarithmic form.

7. $2^5 = 32$
8. $4^3 = 64$
9. $10^3 = 1000$
10. $10^4 = 10{,}000$

11. $e^4 = x$
12. $e^{-2} = z$
13. $5^{-3} = \dfrac{1}{125}$
14. $6^{-3} = \dfrac{1}{216}$

15. $10^{-2} = \dfrac{1}{100}$
16. $10^{-4} = \dfrac{1}{10000}$
17. $625^{1/4} = 5$
18. $343^{1/3} = 7$

19. $\left(\dfrac{1}{4}\right)^2 = \dfrac{1}{16}$
20. $\left(\dfrac{3}{4}\right)^3 = \dfrac{27}{64}$
21. $7^{1/2} = \sqrt{7}$
22. $9^{1/3} = \sqrt[3]{9}$

For Exercises 23–38, write in exponential form.

23. $\log_3 81 = 4$
24. $\log_4 64 = 3$
25. $\log_4 \dfrac{1}{16} = -2$
26. $\log_3 \dfrac{1}{243} = -5$

27. $\log_{10} 100 = 2$
28. $\log_{10} \dfrac{1}{100} = -2$
29. $\log_e a = 5$
30. $\log_e y = -4$

31. $\log_e \dfrac{1}{e^4} = -4$
32. $\log_e \dfrac{1}{e} = -1$
33. $\log_{1/8} \dfrac{1}{64} = 2$
34. $\log_{1/4} \dfrac{1}{256} = 4$

35. $\log_{1/5} 25 = -2$
36. $\log_{1/3} 81 = -4$
37. $\log_7 \sqrt{7} = \dfrac{1}{2}$
38. $\log_6 \sqrt[4]{6} = \dfrac{1}{4}$

For Exercises 39–58, solve for x.

39. $\log_2 x = 3$

40. $\log_3 x = 4$

41. $\log_5 x = -2$

42. $\log_6 x = -1$

43. $\log_3 81 = x$

44. $\log_2 32 = x$

45. $\log_5 \dfrac{1}{5} = x$

46. $\log_4 \dfrac{1}{16} = x$

47. $\log_x 1000 = 3$

48. $\log_x 81 = 4$

49. $\log_x \dfrac{1}{16} = -4$

50. $\log_x \dfrac{1}{36} = -2$

51. $\log_{1/2} x = 2$

52. $\log_{1/5} x = 3$

53. $\log_{1/3} x = -5$

54. $\log_{1/6} x = -2$

55. $\log_{1/3} \dfrac{1}{9} = x$

56. $\log_{1/4} \dfrac{1}{64} = x$

57. $\log_{1/2} 64 = x$

58. $\log_{1/5} 125 = x$

For Exercises 59–62, graph.

59. $f(x) = \log_4 x$

60. $f(x) = \log_5 x$

61. $f(x) = \log_{1/3} x$

62. $f(x) = \log_{1/4} x$

63. Why does $\log_b b = 1$ for any value of $b > 0$ and $b \neq 1$?

64. In the definition of a logarithm, $y = \log_b x$, why must $b \neq 1$?

65. Why does $\log_b 1 = 0$?

66. If $f(x) = b^x$, the domain is $(-\infty, \infty)$ and the range is $(0, \infty)$. What are the domain and range of $f(x) = \log_b x$? Why?

67. The percent of adult height attained by a 5- to 15-year-old girl can be modeled by $f(x) = 62 + 35 \log_{10}(x - 4)$, where x is the age in years and $f(x)$ is the percent. At age 14, what percent of her adult height has a girl reached?

68. The percent of adult height attained by a 5- to 15-year-old boy can be modeled by $f(x) = 29 + 48.8 \log_{10}(x + 1)$, where x is the age in years and $f(x)$ is the percent. At age 9, what percent of his adult height has a boy reached?

69. In Example 6 we learned that the percent of students who recall the important features of a lecture is given by $P = 95 - 30 \log_2 x$, where P is the percent and x is the number of days that have elapsed since the lecture was given. After how many days will 35% of the students recall the important features?

70. Using the formula from Exercise 69, find the percent of the students who recall the important features of a lecture 4 days after it was given.

Collaborative Exercises

Exploring Graphs of Logarithms

Using a graphing calculator, draw the graph of $f(x) = \log x$ and $g(x) = \log(10x)$ in the window $[0, 100]$ for x and $[-2, 4]$ for y.

1. Press the $\boxed{\text{TRACE}}$ key and using up or down arrow keys move from one graph to the other and observe the y-values of several points with the same x-values. What do you observe?

2. How is the graph of $g(x) = \log(10x)$ related to the graph of $f(x) = \log x$? Why?

3. In general, how is the graph of $g(x) = \log(kx)$ for $k > 0$ related to the graph of $f(x) = \log x$?

4. Repeat step 1 using $f(x) = \log x$ and $g(x) = \log \dfrac{x}{10}$.

5. How is the graph of $g(x) = \log \dfrac{x}{10}$ related to the graph of $f(x) = \log x$? Why?

6. In general, how is the graph of $g(x) = \log \dfrac{x}{k}$ for $k > 0$ related to the graph of $f(x) = \log x$?

REVIEW EXERCISES

For Exercises 1–4, simplify using the rules of exponents.

1. $x^4 \cdot x^2$

2. $\dfrac{x^6}{x^3}$

3. $(x^3)^5$

4. $\dfrac{(x^3)^2 \cdot x^4}{x^5}$

5. Write $\sqrt[4]{x^3}$ in exponential form.

6. Write $\dfrac{1}{x^6}$ as x to a negative power.

12.4 Properties of Logarithms

OBJECTIVES

1. Apply the inverse properties of logarithms.
2. Apply the product, quotient, and power properties of logarithms.

OBJECTIVE 1 **Apply the inverse properties of logarithms.** Earlier we developed two properties of logarithms, $\log_b b = 1$ and $\log_b 1 = 0$, which were based on the definition of a logarithm. The fact that $f[f^{-1}(x)] = x$ and $f^{-1}[f(x)] = x$ gives us the following properties of logarithms.

RULE **INVERSE PROPERTIES OF LOGARITHMS**

For $b > 0$, $b \neq 1$ and $x > 0$,

1. $b^{\log_b x} = x$ 2. $\log_b b^x = x$

To prove $b^{\log_b x} = x$, let $f(x) = b^x$ so $f^{-1}(x) = \log_b x$.

$f[f^{-1}(x)] = x$ Composition of a function with its inverse is x.

$f[\log_b x] = x$ Replace $f^{-1}(x)$ with $\log_b x$.

$b^{\log_b x} = x$ Replace x with $\log_b x$ in $f(x)$.

To prove $\log_b b^x = x$, use the definition of a logarithm. $\log_a b = c$ means $a^c = b$, so $\log_b b^x = x$, because $b^x = b^x$.

For example, using $b^{\log_b x} = x$ we have $3^{\log_3 8} = 8$, $6^{\log_6 x} = x$, and $e^{\log_e 4} = 4$. Using $\log_b b^x = x$, we have $\log_3 3^6 = 6$, $\log_5 5^a = a$, and $\log_e e^{x^2} = x^2$.

YOUR TURN   **Find the value of each of the following:**

a. $8^{\log_8 4}$ **b.** $\log_3 3^{-2}$

ANSWERS

a. 4   **b.** -2

OBJECTIVE 2 **Apply the product, quotient, and power properties of logarithms.** Logarithms were invented to perform operations on very large or very small numbers. With the invention of hand-held calculators, they are no longer used for this purpose. However, we still use the properties of logarithms, which are based on the fact that logarithms are exponents.

FURTHER PROPERTIES OF LOGARITHMS

For $x > 0$, $y > 0$, $b > 0$, and $b \neq 1$:

Product Rule of Logarithms: $\quad \log_b xy = \log_b x + \log_b y$

(The logarithm of the product of two numbers is equal to the sum of the logarithms of the numbers.)

Quotient Rule of Logarithms: $\quad \log_b \dfrac{x}{y} = \log_b x - \log_b y$

(The logarithm of the quotient of two numbers is equal to the difference of the logarithms of the numbers.)

Power Rule of Logarithms: $\quad \log_b x^r = r \log_b x$

(The logarithm of a number raised to a power is equal to the exponent times the logarithm of the number.)

To prove $\log_b xy = \log_b x + \log_b y$, let

$M = \log_b x$ and $N = \log_b y$

$x = b^M$ and $y = b^N$ Write each in exponential form.

$xy = b^M \cdot b^N$ Multiply the left and right sides of the exponential forms.

$xy = b^{M+N}$ Add exponents.

$\log_b xy = M + N$ Write in logarithmic form.

$\log_b xy = \log_b x + \log_b y$ Substitute $\log_b x$ for M and $\log_b y$ for N and the proof is complete.

The proof of $\log_b \dfrac{x}{y} = \log_b x - \log_b y$ is left as an exercise. The proof of $\log_b x^r = r \log_b x$ will be a collaborative exercise.

EXAMPLE 1   Use the product rule of logarithms to write the following as the sum of logarithms:

a. $\log_{10} xyz$

Solution $\log_{10} xyz = \log_{10} x + \log_{10} y + \log_{10} z$

b. $\log_b x(x + 3)$

Solution $\log_b x(x + 3) = \log_b x + \log_b(x + 3)$ **Note:** $\log_b(x + 3) \neq \log_b x + \log_b 3$.

EXAMPLE 2   Use the product rule of logarithms in the form $\log_b x + \log_b y = \log_b xy$ to write the following as a single logarithm:

a. $\log_3 8 + \log_3 2$

Solution $\log_3 8 + \log_3 2 = \log_3 8 \cdot 2 = \log_3 16$

b. $\log_8 5 + \log_8 x + \log_8(2x - 3)$

Solution $\log_8 5 + \log_8 x + \log_8(2x - 3) = \log_8 5x(2x - 3)$
$$= \log_8(10x^2 - 15x)$$

Use the product rule of logarithms to write each of the following as a sum of logarithms:

a. $\log_a 6x$

b. $\log_2 x(3x + 2)$

Use the product rule of logarithms to write each of the following as a single logarithm:

c. $\log_6 7 + \log_6 9$

d. $\log_7 2 + \log_7 x + \log_7(2x + 4)$

EXAMPLE 3 Use the quotient rule of logarithms to write each of the following as the difference of logarithms.

a. $\log_5 \dfrac{5}{11}$

Solution $\log_5 \dfrac{5}{11} = \log_5 5 - \log_5 11$

$\qquad\qquad\quad = 1 - \log_5 11 \qquad$ Remember, $\log_5 5 = 1$.

b. $\log_4 \dfrac{x}{x - 5}$

Note: $\log_4(x - 5) \neq \log_4 x - \log_4 5$.

Solution $\log_4 \dfrac{x}{x - 5} = \log_4 x - \log_4 (x - 5)$

EXAMPLE 4 Use the quotient rule of logarithms in the form $\log_b x - \log_b y = \log_b \dfrac{x}{y}$ to write each of the following as a single logarithm:

a. $\log_9 3 - \log_9 x$

Solution $\log_9 3 - \log_9 x = \log_9 \dfrac{3}{x}$

b. $\log_{10} x - \log_{10}(x^2 + 4)$

Solution $\log_{10} x - \log_{10}(x^2 + 4) = \log_{10} \dfrac{x}{x^2 + 4}$

Use the quotient rule of logarithms to write each of the following as the difference of logarithms:

a. $\log_9 \dfrac{9}{10}$

b. $\log_5 \dfrac{x}{x + 2}$

Use the quotient rule of logarithms to write each of the following as a single logarithm:

c. $\log_3 15 - \log_3 5$

d. $\log_4(x^2 + 2) - \log_4(x + 1)$

EXAMPLE 5 Use the power rule of logarithms to write each of the following as a multiple of a logarithm:

a. $\log_4 a^6$

Solution $\log_4 a^6 = 6 \log_4 a$

b. $\log_b \sqrt[4]{x^3}$

Solution $\log_b \sqrt[4]{x^3} = \log_b x^{3/4}$ Write $\sqrt[4]{x^3}$ as $x^{3/4}$.

$$= \frac{3}{4} \log_b x$$

c. $\log_b \dfrac{1}{x^3}$

Solution $\log_b \dfrac{1}{x^3} = \log_b x^{-3}$ Write $\dfrac{1}{x^3}$ as x^{-3}.

$$= -3 \log_b x$$

EXAMPLE 6 Use the power rule of logarithms in the form $r \log_b x = \log_b x^r$ to write the following as a logarithm of a quantity to a power. Leave answers in simplest form without negative or fractional exponents.

a. $5 \log_4 x$

Solution $5 \log_4 x = \log_4 x^5$

b. $-2 \log_b 5$

Solution $-2 \log_b 5 = \log_b 5^{-2}$

$$= \log_b \frac{1}{5^2}$$ Write 5^{-2} as $\dfrac{1}{5^2}$.

$$= \log_b \frac{1}{25}$$ $5^2 = 25$.

c. $\dfrac{2}{3} \log_7 y$

Solution $\dfrac{2}{3} \log_7 y = \log_7 y^{2/3}$

$$= \log_7 \sqrt[3]{y^2}$$ Write $y^{2/3}$ as $\sqrt[3]{y^2}$.

YOUR TURN

Use the power rule of logarithms to write each of the following as a multiple of a logarithm:

a. $\log_a z^4$

b. $\log_a \sqrt[3]{x^2}$

Use the power rule of logarithms to write each of the following as a logarithm of a quantity to a power:

c. $5 \log_7 x$

d. $-3 \log_b z$

Often it is necessary to use more than one rule to simplify a logarithmic expression.

EXAMPLE 7 Write each of the following as a sum or difference of multiples of logarithms:

a. $\log_b \dfrac{z^3}{yz}$

Solution

$$\log_b \dfrac{z^3}{yz} = \log_b z^3 - \log_b yz \qquad \text{Use quotient rule.}$$

$$= 3\log_b z - (\log_b y + \log_b z) \qquad \begin{array}{l}\text{Use power rule and product rule.}\\ \text{Note the use of parentheses.}\end{array}$$

$$= 3\log_b z - \log_b y - \log_b z \qquad \text{Remove the parentheses.}$$

b. $\log_3 \sqrt{\dfrac{a^3}{b}}$

Solution

$$\log_3 \sqrt{\dfrac{a^3}{b}} = \log_3 \left(\dfrac{a^3}{b}\right)^{1/2} \qquad \text{Write } \sqrt{\dfrac{a^3}{b}} \text{ as } \left(\dfrac{a^3}{b}\right)^{1/2}.$$

$$= \dfrac{1}{2}\log_3\left(\dfrac{a^3}{b}\right) \qquad \text{Use power rule.}$$

$$= \dfrac{1}{2}(\log_3 a^3 - \log_3 b) \qquad \text{Use quotient rule.}$$

$$= \dfrac{1}{2}(3\log_3 a - \log_3 b) \qquad \text{Use power rule.}$$

$$= \dfrac{3}{2}\log_3 a - \dfrac{1}{2}\log_3 b \qquad \text{Distribute } \dfrac{1}{2}.$$

c. $\log_5 5^2 b^3$

Solution

$$\log_5 5^2 b^3 = \log_5 5^2 + \log_5 b^3 \qquad \text{Use product rule.}$$

$$= 2\log_5 5 + 3\log_5 b \qquad \text{Use power rule.}$$

$$= 2 + 3\log_5 b \qquad \log_5 5 = 1.$$

EXAMPLE 8 Write each of the following as a single logarithm. Leave answers in simplest form without negative or fractional exponents.

a. $4\log_b 2 - 2\log_b 3$

Solution

$$4\log_b 2 - 2\log_b 3 = \log_b 2^4 - \log_b 3^2 \qquad \text{Use power rule.}$$

$$= \log_b \dfrac{2^4}{3^2} \qquad \text{Use quotient rule.}$$

$$= \log_b \dfrac{16}{9} \qquad \text{Simplify.}$$

b. $\dfrac{1}{2}(\log_2 5 - \log_2 b)$

Solution

$$\dfrac{1}{2}(\log_2 5 - \log_2 b) = \dfrac{1}{2}\log_2 \dfrac{5}{b} \qquad \text{Use quotient rule.}$$

$$= \log_2 \left(\dfrac{5}{b}\right)^{1/2} \qquad \text{Use power rule.}$$

$$= \log_2 \sqrt{\dfrac{5}{b}} \qquad \text{Write } \left(\dfrac{5}{b}\right)^{1/2} \text{ as } \sqrt{\dfrac{5}{b}}.$$

c. $\log_a(x + 2) + \log_a(x - 3)$

Solution $\log_a(x + 2) + \log_a(x - 3) = \log_a(x + 2)(x - 3)$ Apply $\log_b xy = \log_b x + \log_b y$.

$$= \log_a(x^2 - x - 6)$$ Multiply.

ANSWERS

a. $\log_a \dfrac{16}{x^3}$

b. $\log_a \sqrt[3]{xy^2}$

c. $\log_6(x^2 - 2x)$

YOUR TURN   Write the following as a single logarithm:

a. $2 \log_a 4 - 3 \log_a x$ **b.** $\dfrac{1}{3}(\log_a x + 2 \log_a y)$ **c.** $\log_6 x + \log_6(x - 2)$

12.4 EXERCISES

For Extra Help

 InterAct Math

 Addison-Wesley Math Tutor Center

Video 18/DVT 6

 MyMathLab

MathXL

1. Why does $b^{\log_b x} = x$?

2. How can you use the facts that $\log_b x^r = r \log_b x$ and $\log_b b = 1$ to show that $\log_3 81 = 4$?

3. $\log_b(x + y) = $ _____.

4. Why does $\log_a \sqrt[3]{x^4} = \dfrac{4}{3} \log_a x$?

For Exercises 5–16, find the value of each.

5. $8^{\log_8 2}$ **6.** $5^{\log_5 6}$ **7.** $a^{\log_a r}$ **8.** $b^{\log_b a}$

9. $a^{\log_a 4x}$ **10.** $b^{\log_b 5a}$ **11.** $\log_3 3^5$ **12.** $\log_7 7^3$

13. $\log_b b^a$ **14.** $\log_c c^x$ **15.** $\log_a a^{7x}$ **16.** $\log_b b^{6y}$

For Exercises 17–24, use the product rule to write as a sum of logarithms.

17. $\log_2 5y$ **18.** $\log_3 4z$ **19.** $\log_a pq$ **20.** $\log_b rs$

21. $\log_4 mnp$ **22.** $\log_4 pqr$ **23.** $\log_a x(x - 5)$ **24.** $\log_b x(2x + 6)$

For Exercises 25–36, use the product rule to write each as a single logarithm.

25. $\log_3 5 + \log_3 8$ **26.** $\log_6 4 + \log_6 7$ **27.** $\log_4 2 + \log_4 8$ **28.** $\log_9 3 + \log_9 27$

29. $\log_a 7 + \log_a m$

30. $\log_b 2 + \log_b n$

31. $\log_4 a + \log_4 b$

32. $\log_6 r + \log_6 m$

33. $\log_a 2 + \log_a x + \log_a (x + 5)$

34. $\log_a 4 + \log_a y + \log_a (y - 5)$

35. $\log_4 (x + 1) + \log_4 (x + 3)$

36. $\log_6 (x - 3) + \log_6 (x + 3)$

For Exercises 37–46, use the qoutient rule to write each as a difference of logarithms.

37. $\log_2 \dfrac{7}{9}$

38. $\log_4 \dfrac{5}{12}$

39. $\log_a \dfrac{x}{5}$

40. $\log_b \dfrac{y}{3}$

41. $\log_a \dfrac{a}{b}$

42. $\log_b \dfrac{a}{b}$

43. $\log_a \dfrac{x}{x - 3}$

44. $\log_b \dfrac{y}{2y + 5}$

45. $\log_4 \dfrac{2x - 3}{4x + 5}$

46. $\log_6 \dfrac{4x - 1}{x^2 + 3}$

For Exercises 47–58, use the quotient rule to write each as a single logarithm.

47. $\log_6 24 - \log_6 3$

48. $\log_8 12 - \log_8 3$

49. $\log_2 24 - \log_2 12$

50. $\log_3 48 - \log_3 16$

51. $\log_a x - \log_a 3$

52. $\log_a r - \log_a 5$

53. $\log_4 p - \log_4 q$

54. $\log_a m - \log_a n$

55. $\log_b x - \log_b (x - 4)$

56. $\log_b y - \log_b (y - 5)$

57. $\log_x (x^2 - x) - \log_x (x - 1)$

58. $\log_a (a^2 + 2a) - \log_a (a + 2)$

For Exercises 59–70, use the power rule to write each as a multiple of a logarithm.

59. $\log_7 4^3$

60. $\log_3 5^4$

61. $\log_a x^7$

62. $\log_b y^8$

63. $\log_a \sqrt{3}$

64. $\log_a \sqrt{6}$

65. $\log_3 \sqrt[3]{x^2}$

66. $\log_4 \sqrt[5]{a^3}$

67. $\log_a \dfrac{1}{y^2}$

68. $\log_a \dfrac{1}{r^4}$

69. $\log_a \dfrac{3}{y^2}$

70. $\log_a \dfrac{6}{x^5}$

For Exercises 71–82, use the power rule to write each as a logarithm of a quantity to a power. Leave answers in simplest form without negative or fractional exponents.

71. $4 \log_3 5$

72. $5 \log_3 4$

73. $-3 \log_2 x$

74. $-4 \log_3 y$

75. $\dfrac{1}{2} \log_5 4$

76. $\dfrac{1}{3} \log_3 8$

77. $\dfrac{3}{4} \log_a x$

78. $\dfrac{5}{6} \log_b y$

79. $\dfrac{2}{3} \log_a 8$

80. $\dfrac{3}{4} \log_a 16$

81. $-\dfrac{1}{2} \log_3 x$

82. $-\dfrac{1}{3} \log_2 y$

For Exercises 83–94, write each as the sum or difference of multiples of logarithms.

83. $\log_a \dfrac{x^3}{y^4}$

84. $\log_a \dfrac{x^6}{y^5}$

85. $\log_3 a^4 b^2$

86. $\log_7 m^3 n^5$

87. $\log_a \dfrac{xy}{z}$

88. $\log_a \dfrac{pq}{r}$

89. $\log_x \dfrac{a^2}{bc^3}$

90. $\log_x \dfrac{c^2}{m^2 n}$

91. $\log_4 \sqrt{\dfrac{x^3}{y}}$

92. $\log_5 \sqrt[3]{\dfrac{a^5}{b^2}}$

93. $\log_a \sqrt{\dfrac{x^2 y}{z^3}}$

94. $\log_3 \sqrt[3]{\dfrac{c^2 d^3}{m^4}}$

For Exercises 95–108, write each as a single logarithm.

95. $3 \log_3 2 - 2 \log_3 4$

96. $2 \log_4 5 - 3 \log_4 2$

97. $4 \log_b x + 3 \log_b y$

98. $5 \log_b a + 4 \log_b c$

99. $\dfrac{1}{2}(\log_a 5 - \log_a 7)$

100. $\dfrac{1}{2}(\log_4 6 - \log_4 5)$

101. $\dfrac{2}{3}(\log_a x^2 + \log_a y)$

102. $\dfrac{3}{4}(\log_a m^3 + \log_a n^2)$

103. $\log_b x + \log_b(3x - 2)$

104. $\log_a y + \log_b(2y - 4)$

105. $3 \log_a(x - 2) - 4 \log_a(x + 1)$

106. $2 \log_b(x + 4) - 3 \log_b(2x - 1)$

107. $2 \log_a x + 4 \log_a z - 3 \log_a w - 6 \log_a u$

108. $5 \log_a b + 2 \log_a c - 4 \log_a d - \log_a e$

Proving the Quotient and Power Rules

1. Using the proof of $\log_b xy = \log_b x + \log_b y$ as a guide, prove
$$\log_b \frac{x}{y} = \log_b x - \log_b y.$$

2. The proof that $\log_b x^r = r \log_b x$ was not given. For r a positive integer, the following use of the product rule suggests (but does not prove) this rule:
$$\log_b x^3 = \log_b x \cdot x \cdot x = \log_b x + \log_b x + \log_b x = 3 \log_b x.$$
Using the preceding as a guide, prove that $\log_b x^5 = 5 \log_b x$.

3. Complete the following steps to prove $\log_b x^r = r \log_b x$ for all values of r:

   Let $N = \log_b x$.

   Rewrite in exponential form.

   Raise both sides to the r power.

   Simplify the exponents, if necessary.

   Rewrite in logarithmic form.

   Substitute for N.

REVIEW EXERCISES

Simplify each of the following:

1. $\dfrac{10^{-3}}{10^{-12}}$

2. $(10^{9.5})(10^{-12})$

Convert the following from logarithms to exponential form:

3. $\log_{10} 45 = 1.6532$

4. $\log_e 0.25 = -1.3863$

5. Change $10^{1.6990} = 50$ into logarithmic form.

6. If $10,000 is deposited at 6% compounded quarterly for 5 years, how much will be in the account?

12.5 Common and Natural Logarithms

OBJECTIVES

1. Define common logarithms and evaluate them using a calculator.
2. Solve applications using common logarithms.
3. Define natural logarithms and evaluate them using a calculator.
4. Solve applications using natural logarithms.

OBJECTIVE 1 **Define common logarithms and evaluate them using a calculator.** Of all the possible bases of logarithms, two are the most useful. As previously mentioned, logarithms were invented to do computations on very large and very small numbers. Since ours is a base-10 number system, base-10 logarithms were commonly used for this purpose. Consequently, base-10 logarithms are called **common logarithms** and $\log_{10} x$ is written as $\log x$, where the base 10 is understood. Base-10 logarithms are found in engineering, economics, social sciences, and in the natural sciences.

> **DEFINITION**
> Common logarithms are logarithms with a base of 10. $\text{Log}_{10}\, x$ is written as $\log x$.

> **CONNECTION**
> Since logarithmic and exponential functions are inverse functions, if $f(x) = 10^x$, then $f^{-1}(x) = \log x$ and vice versa.

To evaluate common logarithms, we use a calculator. We will round all results to four places.

Calculator TIPS

To evaluate common logarithms, use the [LOG] *key. For most scientific calculators, enter the number and then press the* [LOG] *key. For most graphing calculators, press the* [LOG] *key, enter the number, then press the* [ENTER] *key.*

EXAMPLE 1 Evaluate each common logarithm to four decimal places using a calculator.

a. $\log 23$

Solution $\log 23 \approx 1.3617$

> **CONNECTION**
> Remember that $\log 23 \approx 1.3617$ means $10^{1.3617} \approx 23$, which provides a way to check. Because we rounded the decimal value, evaluating $10^{1.3617}$ may not give exactly 23.

b. $\log 0.00236$

Solution $\log 0.00236 \approx -2.6271$

Evaluate each common logarithm to four decimal places using a calculator.

a. log 436

b. log 0.0724

OBJECTIVE 2 Solve applications using common logarithms. The intensity of sound is defined as the sound intensity per unit area. Sound intensity is measured in decibels, where $d = 10 \log \dfrac{I}{I_0}$, where I_0 is the intensity of a sound at the threshold of hearing. The threshold of hearing represents the faintest sound audible to the average human ear and $I_0 = 10^{-12}$ watts/m².

EXAMPLE 2

a. The sound intensity of a motorcycle is about $10^{-2.5}$ watts per square meter. Find the decibel reading for the motorcycle.

Solution We substitute $10^{-2.5}$ for I and 10^{-12} for I_0 and solve for d.

$$d = 10 \log \dfrac{I}{I_0}$$

$$d = 10 \log \dfrac{10^{-2.5}}{10^{-12}} \qquad \text{Substitute for } I \text{ and } I_0.$$

$$d = 10 \log 10^{\,9.5} \qquad \text{Subtract exponents } [-2.5 - (-12) = 9.5].$$

$$d = 10(9.5) \qquad \text{Use } \log_b b^x = x.$$

$$d = 95 \qquad \text{The motorcycle has a decibel reading of 95.}$$

b. The minimum length of airport runway needed for a plane to take off is related to the weight of the plane. For some planes, the minimum runway length may be modeled by the function $y = 3 \log x$, where x is the plane's weight in thousands of pounds and y is the length of the runway in thousands of feet. Find the minimum length of a runway needed by a Boeing 737 whose maximum takeoff weight is 174,200 pounds.

Solution Since x is the plane's weight in thousands of pounds,

$$x = \dfrac{174{,}200}{1{,}000} = 174.2.$$

$$y = 3 \log x$$

$$y = 3 \log 174.2 \qquad \text{Substitute 174.2 for } x.$$

$$y = 6.723 \qquad \text{Evaluate using a calculator.}$$

Since y is in thousands of feet, the minimum runway length is $(6.723)(1000) = 6723$ ft.

a. The sound intensity of a rock band often exceeds $10^{-0.5}$ watts per square meter. Find the decibel rating of this band.

b. Find the minimum runway length needed for a B-52 Stratofortress whose maximum takeoff weight is 488,000 pounds.

Define natural logarithms and evaluate them using a calculator. The number *e* is an irrational number whose approximate value is 2.7182818285. It is a universal constant like π. In the natural sciences, base-*e* logarithms are much more prevalent than base 10. Since base-*e* logarithms occur in so many "natural" situations, they are called **natural logarithms**. The notation for $\log_e x$ is ln *x*, which is read "el en *x*."

DEFINITION | *Natural Logarithms:* Base-e logarithms are called natural logarithms and $\log_e x$ is written as ln *x*. Note that ln *e* = 1.

To find natural logarithms, we use a calculator.

> *Calculator*
> **TIPS**
>
> *When using a calculator to evaluate natural logarithms, we use the* [ln] *key. For most scientific calculators, enter the number and then press the* [ln] *key. For most graphing calculators, press the* [ln] *key, enter the number, then press the* [ENTER] *key.*

EXAMPLE **3** Evaluate each natural logarithm using a calculator and round to four decimal places.

a. ln 83

Solution ln 83 ≈ 4.4188

b. ln 0.0055

Solution ln 0.0055 ≈ −5.2030

> **CONNECTION**
>
> In Example 3, ln 83 ≈ 4.4188 means that $e^{4.4188} \approx 83$. Similarly, ln 0.0055 = −5.2030 means that $e^{-5.2030} \approx 0.0055$.

YOUR TURN Evaluate each natural logarithm using a calculator and round to four decimal places.

a. ln 102   **b.** ln 0.0573

ANSWERS

a. 4.6250   **b.** −2.8595

Solve applications using natural logarithms. If money is deposited into an account and the interest is compounded continuously, then the time *t* (in years) that it would take an investment of *P* dollars to grow into *A* dollars at an interest rate *r* (written as a decimal) is given by $t = \dfrac{1}{r} \ln \dfrac{A}{P}$.

EXAMPLE 4

An amount of $5000 is deposited into an account earning 5% annual interest compounded continuously. How many years would it take until the account has reached $10,000?

Solution We are given $P = 5000$, $r = 0.05$, and $A = 10,000$.

$$t = \frac{1}{r} \ln \frac{A}{P}$$

$$t = \frac{1}{0.05} \ln \frac{10,000}{5000} \qquad \text{Substitute for } P, r, \text{ and } A.$$

$$t = 20 \ln 2 \qquad \qquad \text{Divide.}$$

$$t = 13.86 \text{ years} \qquad \text{Multiply using a calculator.}$$

YOUR TURN

ANSWER

11.5 yr.

How long would it take $2000 to grow into $5000 at 8% interest if the interest is compounded continuously?

12.5 EXERCISES

For Extra Help

InterAct Math

Tutor Center
Addison-Wesley
Math Tutor Center

Video 18/DVT 6

MyMathLab

MathXL

1. If $y = \log_b x$, what values are allowed for b? Why?

2. Is $\log x$ positive or negative for $0 < x < 1$? Why?

3. Is $\ln x$ positive or negative for $x > 1$? Why?

4. Give the symbol and meaning for **a.** common logarithms, **b.** natural logarithms.

5. If $e^{0.91629} = 2.5$, find $\ln 2.5$.

6. Without using a calculator, find the exact value of $\log_{10} 10^{\sqrt{5}}$.

For Exercises 7–26, find each logarithm. Approximate your answers to four decimal places.

7. $\log 64$

8. $\log 82$

9. $\log 0.0067$

10. $\log 0.00087$

11. $\log 247.8$

12. $\log 785.4$

13. $\log(1.5 \times 10^4)$

14. $\log(5.7 \times 10^7)$

15. $\log(1.6 \times 10^{-6})$

16. $\log(7.5 \times 10^{-5})$

17. $\ln 9.34$

18. $\ln 5.33$

19. ln 79.2

20. ln 765.4

21. ln 0.034

22. ln 0.923

23. ln(5.4 × e^4)

24. ln(7.3 × e^6)

25. log e

26. ln 10

27. Use your calculator to find log 0. What happened? Why?

28. Use your calculator to find ln −1. What happened? Why?

For Exercises 29–40, find the exact value of each logarithm using $\log_b b^x = x$.

29. log 100

30. log 1000

31. $\log \dfrac{1}{10}$

32. $\log \dfrac{1}{1000}$

33. $\log \sqrt[3]{10}$

34. $\log \sqrt{10}$

35. log 0.001

36. log 0.00001

37. ln e^3

38. ln e^5

39. ln $\sqrt{e}$

40. ln $\sqrt[5]{e}$

For Exercises 41–44, use the formula $d = 10 \log \dfrac{I}{I_0}$, *where* $I_0 = 10^{-12}$ *watts/m^2.*

41. The sound intensity of a firecracker is 10^{-3} watts per square meter. What is the decibel reading for the firecracker?

42. The sound intensity of a race car is 10^{-1} watts per square meter. What is the decibel reading for the race car?

43. If a noisy office has a decibel reading of 60, what is the sound intensity?

44. If loud thunder has a decibel reading of 90, what is the sound intensity?

For Exercises 45–48, use the following:

In chemistry, the pH of a substance determines whether it is a base (pH > 7) or an acid (pH < 7). To find the pH of a solution, we use the formula pH = $-\log [H_3O^+]$, where $[H_3O^+]$ is the hydronium ion concentration in moles per liter. Note that the pH is unitless.

45. Find the pH of vinegar if $[H_3O^+] = 1.6 \times 10^{-3}$ moles per liter. (*Source: CRC Handbook of Chemistry and Physics*, 62nd Edition, CRC Press, 1981)

46. Find the pH of maple syrup if $[H_3O^+] = 2.3 \times 10^{-7}$. (*Source: CRC Handbook of Chemistry and Physics*, 62nd Edition, CRC Press, 1981)

47. Find the hydronium ion concentration of sauerkraut, which has a pH of 3.5.

48. Find the hydronium ion concentration of blood, which has a pH of 7.4.

For Exercises 49 and 50, use the formula $t = \dfrac{1}{r}\ln\dfrac{A}{P}$.

49. Find how long it will take $2000 to grow to $5000 at 4% interest if the interest is compounded continuously.

50. Find how long it will take $5000 to grow to $8000 at 5% interest if the interest is compounded continuously.

51. The magnitude of an earthquake is given by the Richter scale, whose formula is $R = \log\dfrac{I}{I_0}$, where I is the intensity of the earthquake and I_0 is the intensity of a minimal earthquake and is used for comparison purposes. The 1906 San Francisco earthquake had a magnitude of 7.8, and the 1964 Alaska earthquake had a magnitude of 8.4. Compare the intensity of the two earthquakes. (*Hint:* Express the intensity of each in terms of I_0.)

52. Using the formula from Exercise 51, compare the intensity of the 1949 Queen Charlotte Islands earthquake, whose magnitude was 8.1, with the 1700 earthquake at Cascadia whose magnitude was estimated at 9.0.

53. During an earthquake, energy is released in various forms. The amount of energy radiated from the earthquake as seismic waves is given by $\log E_s = 11.8 + 1.5\,M$, where E_s is measured in ergs and M is the magnitude of the earthquake as given by the Richter scale. Vancouver Island had an earthquake whose magnitude was 7.3. How much energy was released in the form of seismic waves?

54. Using the formula from Exercise 53, find the energy released in the form of seismic waves from the Double Springs Flat earthquake of 1994, whose magnitude was 6.1 on the Richter scale.

55. Since 1995, the number of adult deaths from AIDS in the United States has been declining and can be modeled by $y = 49{,}971.5 - 21{,}298.87\ln x$, where y is the number of deaths and $x = 1$ corresponds to 1995. (*Source:* CDC Division of HIV/AIDS Prevention, *Surveillance Report*, Vol. 13, No. 2)

  a. Estimate the number of adult AIDS deaths in 2000.

  b. Find the year when the number of adult deaths by AIDS was 26,572.

56. The number of adult deaths by AIDS in the United States as a percent of the number of diagnosed cases peaked in 1995 at 73.4%. Since then, the percent has declined and can be modeled by $y = 0.7216 - 0.1911\ln x$, where y is the percent of deaths and $x = 1$ corresponds to 1995. (*Data Source:* CDC Division of HIV/AIDS Prevention, *Surveillance Report*, Vol. 13, No. 2)

  a. Find the percent of deaths among diagnosed cases in 1999.

  b. In what year did 35% of the diagnosed cases of AIDS result in death?

57. Using the formula from Example 2b, find the minimum runway length needed for a Boeing 717 whose maximum weight is 110,000 pounds at takeoff.

58. Walking speeds in various cities is a function of the population, since as populations increase, so does the pace of life. Average walking speeds can be modeled by the function $W = 0.35 \ln P + 2.74$, where P is the population, in thousands, and W is the walking speed in feet per second. Find the average walking speed in Chicago whose population was approximately 8,500,000 in 2002.

REVIEW EXERCISES

For Exercises 1 and 2, find the exact value of the logarithm.

1. $\ln e^{4x}$

2. $\ln e^{0.05t}$

For Exercises 3 and 4, write each as a single logarithm.

3. $\log_5 x + \log_5(x + 3)$

4. $\log_3(2x + 1) - \log_3(x - 1)$

5. Write $\log_3(2x + 5) = 2$ in exponential form.

6. Given $\log_4(x - 3)$, can $x = 0$? Why or why not?

12.6 Exponential and Logarithmic Equations with Applications

OBJECTIVES

1. Solve equations that have variables as exponents.
2. Solve equations containing logarithms.
3. Solve applications involving exponential and logarithmic functions.
4. Use the change-of-base formula.

In Section 12.2, we solved equations that could be put in the form $b^x = b^y$. For example, if $3^x = 81$, then $3^x = 3^4$, so $x = 4$. In order to use this form, we had to write both sides of the equation as the same base raised to a power. In this section, we will solve equations like $3^x = 16$, which is the same as $3^x = 2^4$, where the bases are not the same. To solve this and other exponential and logarithmic equations, we need the following properties.

RULE

PROPERTIES FOR SOLVING EXPONENTIAL AND LOGARITHMIC EQUATIONS

For $b > 0$ and $b \neq 1$, and x and y real numbers,

1. If $b^x = b^y$, then $x = y$.
2. If $x = y$, then $b^x = b^y$.
3. For $x > 0$ and $y > 0$, if $\log_b x = \log_b y$, then $x = y$.
4. For $x > 0$ and $y > 0$, if $x = y$, then $\log_b x = \log_b y$.
5. For $x > 0$, if $\log_b x = y$, then $b^y = x$.

These properties are true because the exponential and logarithmic functions are one-to-one.

OBJECTIVE 1 Solve equations that have variables as exponents. To solve equations that have variables as exponents, we use property 4, which says that if two positive numbers are equal, then so are their logarithms.

EXAMPLE 1 Solve $5^x = 16$

Solution
$\log 5^x = \log 16$ Use if $x = y$, then $\log_b x = \log_b y$ (property 4).
$x \log 5 = \log 16$ Use $\log_b x^r = r \log_b x$.
$x = \dfrac{\log 16}{\log 5}$ Divide both sides of the equation by log 5.

The exact solution is $x = \dfrac{\log 16}{\log 5}$. Using a calculator, we find $x \approx 1.7227$ correct to four decimal places.

Check: $5^x = 16$

$5^{1.7227} = 16$ Substitute 1.7727 for x.
$15.9998 \approx 16$ The answer is correct.

In Example 1, we took the common logarithm of both sides, but we could have used natural logarithms (or any other base) instead. If one side of the equation contains a power of e, natural logs are preferred so that we can use the fact that $\log_e e^x = x$ or, more simply, $\ln e^x = x$.

EXAMPLE 2 Solve $e^{4x} = 23$.

Solution $\ln e^{4x} = \ln 23$ Use if $x = y$, then $\log_b x = \log_b y$.

$\qquad\qquad 4x = \ln 23$ Use $\log_b b^x = x$.

$$x = \frac{\ln 23}{4}$$ Divide both sides by 4.

$\qquad\qquad x \approx 0.7839$

We will leave the check to the reader.

YOUR TURN   **Solve the following for x. Round answers to four decimal places.**

   a. $6^{2x} = 42$ **b.** $e^{3x} = 5$

OBJECTIVE 2 **Solve equations containing logarithms.** To solve equations containing logarithms, we first use the properties of logarithms to simplify each side of the equation, if necessary. If the simplification results in an equation of the form $\log_b x = \log_b y$, then use property 3 (if $\log_b x = \log_b y$, then $x = y$). If the simplification results in an equation of the form $\log_b x = y$, then we use the definition of a logarithm and write $\log_b x = y$ as $b^y = x$.

EXAMPLE 3 Solve.

   a. $\log_5 x + \log_5(x + 3) = \log_5 4$

Solution $\log_5 x(x + 3) = \log_5 4$ Use $\log_b xy = \log_b x + \log_b y$.

$\qquad\qquad\quad x(x + 3) = 4$ Use if $\log_b x = \log_b y$, then $x = y$.

$\qquad\qquad\quad x^2 + 3x = 4$ Simplify.

$\qquad\qquad x^2 + 3x - 4 = 0$ Write in $ax^2 + bx + c = 0$ form.

$\qquad\quad (x + 4)(x - 1) = 0$ Factor.

$\quad x + 4 = 0 \quad$ or $\quad x - 1 = 0$ Set each factor equal to 0.

$\qquad x = -4 \quad$ or $\quad\; x = 1$ Possible solutions.

If -4 is substituted into the original equation, we have $\log_5(-4) + \log_5(-1) = \log_5 4$, but logarithms are defined for positive numbers only. So $x = -4$ is not a solution. A check will show that $x = 1$ is a solution.

   b. $\log_3(2x + 5) = 2$

Solution $3^2 = 2x + 5$ Use if $\log_b x = y$, then $b^y = x$.

$\qquad\qquad 9 = 2x + 5$

$\qquad\qquad 4 = 2x$

$\qquad\qquad 2 = x$ Solve for x.

Check that $x = 2$ is a solution.

c. $\log_2(5x + 1) - \log_2(x - 1) = 3$

Solution $\log_2 \dfrac{5x + 1}{x - 1} = 3$ Use $\log_b \dfrac{x}{y} = \log_b x - \log_b y$.

$\dfrac{5x + 1}{x - 1} = 2^3$ Use if $\log_b x = y$, then $b^y = x$.

$5x + 1 = 8x - 8$ Multiply both sides by $x - 1$.

$9 = 3x$ Isolate the x term.

$x = 3$ Solve for x.

Check that $x = 3$ is a solution.

YOUR TURN   **Solve.**

a. $\ln x + \ln(x + 2) = \ln 8$ **b.** $\log_3(4x + 2) - \log_3(x - 2) = 2$

OBJECTIVE **3** **Solve applications involving exponential and logarithmic functions.** A wide variety of problems from business and the sciences can be solved with exponential or logarithmic functions. Earlier, we used the formula for compound interest,

$A = P\left(1 + \dfrac{r}{n}\right)^{nt}$, to find A when given P, r, n, and t. Using logarithms, it is also possible to find t when given A, P, r, and n.

EXAMPLE 4 How long will it take $6000 invested at 6% interest compounded quarterly to grow to $10,000?

Solution $A = P\left(1 + \dfrac{r}{n}\right)^{nt}$ We are given $A = 10{,}000$, $P = 6000$, $r = 0.06$, and $n = 4$.

$10{,}000 = 6000\left(1 + \dfrac{.06}{4}\right)^{4t}$ Substitute for A, P, r, and n.

$\dfrac{5}{3} = (1.015)^{4t}$ Divide both sides by 6000 and simplify inside the parentheses.

$\log \dfrac{5}{3} = \log 1.015^{4t}$ Use if $x = y$, then $\log_b x = \log_b y$.

$\log \dfrac{5}{3} = 4t \log 1.015$ Use $\log_b x^r = r \log_b x$.

$\dfrac{\log \dfrac{5}{3}}{4 \log 1.015} = t$ Divide both sides by $4 \log 1.015$

$8.58 \approx t$ Evaluate using a calculator.

It would take about $8\dfrac{1}{2}$ years.

Many banks compound interest continuously. The formula for interest compounded continuously is $A = Pe^{rt}$, where A is the amount in the account, P is the amount deposited, r is the interest rate as a decimal, and t is the time in years.

EXAMPLE 5 If $5000 is deposited into an account at 5% interest compounded continuously:

a. How much will be in the account after 9 years?

Solution $A = Pe^{rt}$ We are given $P = \$5000$, $r = 0.05$, and $t = 9$ and are asked to find A.

$A = 5000e^{0.05(9)}$ Substitute for P, r, and t.

$A = 5000e^{0.45}$

$A = \$7841.56$ Evaluate using a calculator.

There will be $7841.56 in the account.

b. How long will it take the $5000 to grow into $12,000?

Solution $A = Pe^{rt}$ We are given $A = \$12,000$, $P = \$5000$, $r = 0.05$ and asked to find t.

$12,000 = 5000e^{0.05t}$ Substitute for A, P, and r.

$\dfrac{12}{5} = e^{0.05t}$ Simplify.

$\ln \dfrac{12}{5} = \ln e^{0.05t}$ Use if $x = y$, then $\log_b x = \log_b y$.

$\ln \dfrac{12}{5} = 0.05t$ Use $\ln e^x = x$.

$\dfrac{\ln \dfrac{12}{5}}{0.05} = t$ Divide both sides by 0.05.

$17.51 \approx t$ Evaluate using a calculator.

It would take about $17\dfrac{1}{2}$ years for $5000 to grow to $12,000.

EXAMPLE 6 Since the 1970s, Orlando, Florida, has been one of the fastest-growing cities in the United States. Following is a chart showing the population for selected years:

Year	Population in millions
1975	0.6896
1980	0.805
1985	0.996
1990	1.225
1995	1.428
2000	1.645

This data can be modeled by $y = 0.6896e^{0.0383t}$, where y is the population in millions and t is the number of years after 1975. (*Source*: U.S. Bureau of the Census)

a. Assuming the population continues to grow in the same manner, use the model to estimate the population of Orlando in 2005.

Solution Since t is the number of years after 1975, $t = 2005 - 1975 = 30$.

$$y = 0.6896e^{0.0383t} \qquad \text{Formula.}$$
$$y = 0.6896e^{0.0383(30)} \qquad \text{Substitute 30 for } t.$$
$$y = 2.176 \qquad \text{Evaluate using a calculator.}$$

Since y is in millions, the estimated population is 2,176,000 in 2005.

b. Find the year in which the population will be 2,500,000.

Solution Since y is in millions, $y = \dfrac{2,500,000}{1,000,000} = 2.5$.

$$2.5 = 0.6896e^{0.0383t} \qquad \text{Substitute for } y.$$

$$\frac{2.5}{0.6896} = e^{0.0383t} \qquad \text{Divide both sides by 0.6896.}$$

$$\ln \frac{2.5}{0.6896} = \ln e^{0.0383t} \qquad \text{Use if } x = y, \text{ then } \log_b x = \log_b y.$$

$$\ln \frac{2.5}{0.6896} = 0.0383t \qquad \text{Use } \log_b b^x = x.$$

$$\frac{\ln \dfrac{2.5}{0.6896}}{0.0383} = t \qquad \text{Divide both sides by 0.0383.}$$

$$33.6 = t \qquad \text{Evaluate with a calculator.}$$

Since $1975 + 33.6 = 2008.6$, the population will reach 2,500,000 in 2008.

YOUR TURN

a. How long will it take $8000 invested at 4% annual interest compounded semiannually to grow to $12,000?

b. Between April 2, 2003, and April 8, 2003, the reported number of SARS (severe acute respiratory syndrome) cases was increasing at a rate of 3.1% per day. As of April 2, there were 2671 reported cases of SARS. If the reported number of SARS cases continues to increase at this rate, the function $A = 2671e^{0.031t}$ models the spread of SARS, where A represents the number of cases and t is the number of days after April 2. How many cases would there be on April 10?

c. How many days after April 2 would it take for there to be 8000 cases?

Exponential growth and decay can be represented by the equation $A = A_0 e^{kt}$, where A is the amount present, A_0 is the initial amount, t is the time, and k is a constant that is determined by the substance. There is exponential growth indicated if $k > 0$ and exponential decay if $k < 0$. Recall that the half-life of a substance is the amount of time until only one-half the original amount is present.

Plutonium-239 is frequently used as fuel in nuclear reactors to generate electricity. One of the problems with using plutonium is disposing of the radioactive waste, which is extremely dangerous for a very long period of time.

ANSWERS

a. 10.14 years

b. 3423 cases

c. Approximately 35.39 days

EXAMPLE 7 A nuclear reactor contains 10 kg of radioactive plutonium ^{239}P. Plutonium disintegrates according to the formula $A = A_0 e^{-0.0000284t}$.

a. How much will remain after 10,000 years?

Solution $A = A_0 e^{-0.0000284t}$ We are given $A_0 = 10$ kg and $t = 10{,}000$ years and asked to find A.

$A = 10 e^{-0.0000284(10{,}000)}$ Substitute for A_0 and t.

$A = 10 e^{-0.284}$ Simplify.

$A \approx 7.53$ Evaluate using a calculator.

About 7.53 kg (or about $\frac{3}{4}$ of the original amount) will remain after 10,000 years.

b. Find the half-life of ^{239}P.

Solution $A = A_0 e^{-0.0000284t}$ After one half-life, 5 kg will remain so that $A = 5$. Find t.

$5 = 10 e^{-0.0000284t}$ Substitute for A and A_0.

$\dfrac{1}{2} = e^{-0.0000284t}$ Divide both sides by 10.

$\ln \dfrac{1}{2} = \ln e^{-0.0000284t}$ Use if $x = y$, then $\log_b x = \log_b y$.

$\ln \dfrac{1}{2} = -0.0000284t$ Use $\ln e^x = x$.

$\dfrac{\ln \dfrac{1}{2}}{-0.0000284} = t$ Divide both sides by -0.0000284.

$24{,}406.59 \approx t$ Evaluate using a calculator.

The half-life is approximately 24,400 years.

YOUR TURN Carbon-14 is a radioactive form of carbon that is present in all living things. Archaeologists and paleontologists frequently use carbon-14 dating in estimating the age of organic fossils. Carbon-14 disintegrates according to the formula $A = A_0 e^{-0.000121t}$.

a. If a sample contains 5 grams of carbon-14, how much will be present after 1500 years?

b. What is the half-life of carbon-14?

ANSWERS

a. 4.17 grams
b. About 5700 years

OBJECTIVE 4 **Use the change-of-base formula.** Sometimes we need logarithms other than common or natural logarithms. For example, earlier we were given the formula $P = 95 - 30 \log_2 x$, where P is the percent of students who recall the important features of a lecture after x days. For example, to find the percent after 5 days, we need to find $\log_2 5$. The following rule allows conversion from any base into any other base.

We usually use common or natural logarithms since most calculators have only base-10 and base-e logarithms.

To prove the formula, let

$$
\begin{aligned}
y &= \log_a x \\
a^y &= x && \text{Use if } \log_b x = y, \text{ then } b^y = x. \\
\log_b a^y &= \log_b x && \text{Take } \log_b \text{ of both sides.} \\
y \log_b a &= \log_b x && \text{Use } \log_b x^r = r \log_b x. \\
y &= \frac{\log_b x}{\log_b a} && \text{Divide both sides by } \log_b a. \\
\log_a x &= \frac{\log_b x}{\log_b a} && \text{Substitute } \log_a x \text{ for } y.
\end{aligned}
$$

EXAMPLE 8 Find $\log_5 19$.

Solution $\log_5 19 = \dfrac{\log 19}{\log 5}$ Use $\log_a x = \dfrac{\log_b x}{\log_b a}$.

Note: We could have used ln rather than log.

$\log_5 19 \approx 1.8295$ Evaluate using a calculator.

Check: $5^{1.8295} = 19.0005 \approx 19$, so the answer is correct.

EXAMPLE 9 Use the change-of-base formula to find the percent of students who retained the main points of a lecture after 5 days.

Solution $P = 95 - 30 \log_2 x$

$P = 95 - 30 \log_2 5$ Substitute 5 for x.

$P = 95 - 30 \dfrac{\ln 5}{\ln 2}$ Use $\log_a x = \dfrac{\log_b x}{\log_b a}$.

$P = 95 - 30(2.3219)$ Evaluate $\dfrac{\ln 5}{\ln 2}$.

$P = 25.34$ Evaluate using a calculator.

About 25% of the students remember the main points of a lecture 5 days later.

YOUR TURN

a. Find $\log_6 25$.

b. Use the formula from Example 9 to find the percent of students who retain the main points of a lecture 8 days later.

ANSWERS

a. 1.7965   b. 5%

12.6 EXERCISES

For Extra Help

 InterAct Math

 Addison-Wesley Math Tutor Center

Video 18/DVT 6

MyMathLab

MathXL

1. If $\log_a m = \log_a n$, then _____.

2. When solving $10^{x+2} = 45$, would natural or common logarithms be the better choice? Why?

3. What principle would you use in solving $\log(x - 3) = \log(3x - 13)$?

4. What procedure should be used to solve an equation with a variable exponent?

5. In solving $\log x + \log(x + 2) = \log 15$, we get possible solutions of $x = 3$ and $x = -5$. Why must $x = -5$ be rejected?

6. Suppose you were solving an equation that contained $\log(5 - x)$. You did all the algebraic work correctly and you obtained $x = -3$ as an apparent solution. Would you have to reject $x = -3$ as a solution? Why or why not?

For Exercises 7–18, solve. Round answers to four decimal places.

7. $2^x = 9$

8. $3^x = 20$

9. $5^{2x} = 32$

10. $4^{3x} = 13$

11. $5^{x+3} = 10$

12. $7^{x+1} = 41$

13. $8^{x-2} = 6$

14. $5^{x-3} = 12$

15. $4^{x+2} = 5^x$

16. $6^{x+4} = 10^x$

17. $2^{x+1} = 3^{x-2}$

18. $5^{x-3} = 3^{x+1}$

For Exercises 19–26, solve. Round answers to four decimal places.

19. $e^{3x} = 5$

20. $e^{2x} = 7$

21. $e^{0.03x} = 25$

22. $e^{0.002x} = 12$

23. $e^{-0.022x} = 5$

24. $e^{-0.0032x} = 8$

25. $\ln e^{4x} = 24$

26. $\ln e^{5x} = 35$

For Exercises 27–60, solve. Give exact answers.

27. $\log_2(x - 3) = 3$

28. $\log_3(x + 4) = 2$

29. $\log_4(4x - 8) = 2$

30. $\log_3(3x + 9) = 3$

31. $\log_4 x^2 = 2$

32. $\log_2 x^2 = 6$

33. $\log_6(x^2 + 5x) = 2$

34. $\log_4(x^2 + 6x) = 2$

35. $\log (3x - 2) = \log (2x + 5)$

36. $\log (4x + 1) = \log (2x + 7)$

37. $\ln (4x + 6) = \ln (2x - 8)$

38. $\ln (5x + 6) = \ln (3x - 8)$

39. $\log_9 (x^2 + 4x) = \log_9 12$

40. $\log_8 (x^2 + x) = \log_8 30$

41. $\log_4 x + \log_4 8 = 2$

42. $\log_6 x + \log_6 4 = 2$

43. $\log_5 x - \log_5 2 = 1$

44. $\log_5 x - \log_5 3 = 2$

45. $\log_3 x + \log_3 (x + 6) = 3$

46. $\log_2 x + \log_2 (x - 3) = 2$

47. $\log_3 (2x + 15) + \log_3 x = 3$

48. $\log_2 (3x - 2) + \log_2 x = 4$

49. $\log_2 (7x + 3) - \log_2 (2x - 3) = 3$

50. $\log_3 (3x + 3) - \log_3 (x - 3) = 2$

51. $\log (x + 2) - \log (2x - 2) = 0$

52. $\log_3 (2x + 1) - \log_3 (x - 1) = 1$

53. $\log_8 2x + \log_8 6 = \log_8 10$

54. $\log_5 3x + \log_5 2 = \log_5 17$

55. $\ln x + \ln (2x - 1) = \ln 10$

56. $\ln x + \ln (3x - 5) = \ln 12$

57. $\log x - \log (x - 5) = \log 6$

58. $\log x - \log (x - 2) = \log 3$

59. $\log_6 (3x + 4) - \log_6 (x - 2) = \log_6 8$

60. $\log_7 (5x + 2) - \log_7 (x - 2) = \log_7 9$

61. How long will it take $5000 invested at 5% compounded quarterly to grow to $8000? Round your answer to the nearest tenth of a year.

62. How long will it take $7000 invested at 6% compounded monthly to grow to $12,000? Round your answer to the nearest tenth of a year.

63. If $8000 is deposited into an account at 6% annual interest compounded continuously:

  a. How much money will be in the account after 15 years?

  b. How long will it take the $8000 to grow into $14,000?

64. If $4000 is deposited into an account at 5% annual interest compounded continuously:

   a. How much money will be in the account after 10 years?

   b. How long will it take the $4000 to grow into $10,000?

65. The world population in 2002 was about 6 billion and is increasing at a rate of 1% per year. The world population after 2002 can be modeled by the equation $A = A_0 e^{0.01t}$, where A_0 is the world population in 2002 and t is the number of years after 2002.

   a. Assuming the growth rate follows the same trend, find the world population in 2020.

   b. In what year will the world population reach 6.5 billion?

66. In 1996, Africa had a population of 731.5 million and a natural growth rate of 2.5% (two and one-half times the world's growth rate). The population growth can be modeled by $A = A_0 e^{0.025t}$, where A_0 is the population in millions in 1996 and t is the number of years after 1996.

   a. Excluding immigration, what would the population of Africa be in 2025?

   b. Excluding immigration, in what year would Africa's population reach 1 billion?

67. The barometric pressure x miles from the eye of a hurricane is approximated by the function

$$P = 0.48 \ln(x + 1),$$

where P is inches of mercury. (*Source:* A. Miller and R. Anthes, *Meteorology*, 4th edition, Merrill Publishing)

   a. Find the barometric pressure 50 miles from the center of the hurricane.

   b. Find the distance from the center where the pressure is 1.5 inches of mercury.

68. The percent, $f(x)$, of adult height attained by a boy who is x years old is modeled by

$$f(x) = 29 + 48.8 \log(x + 1),$$

where $5 \le x \le 15$.

   a. At age 10, about what percent of his adult height has a boy reached?

   b. At what age will a boy attain 75% of his adult height?

69. The first two-year college was founded in 1901 with Joliet Junior College in Chicago. The number of two-year colleges in the United States grew rapidly, especially during the 1960s, but growth has tapered off. The total number of two-year colleges in the United States since 1960 can be modeled by the function

$$y = 175.6 \ln x + 513,$$

where x is the number of years after 1960 and y is the number of two-year colleges. Use the model to estimate the number of two-year colleges in the United States in 2003. (*Source:* American Association of Two-Year Colleges)

70. The annual depreciation rate r of a car purchased for P dollars and worth A dollars after t years can be found by the formula

$$\log (1 - r) = \frac{1}{t} \log \frac{A}{P}.$$

Find the depreciation rate of a car that is purchased for \$22,500 and is sold 3 years later for \$10,000.

71. The probability that a person will have an accident while driving at a given blood alcohol level is modeled by

$$P(b) = e^{21.5b},$$

where b is the blood alcohol level ($0 \le b \le 0.4$) and P is the percent probability of having an accident.

a. What is the probability of an accident if the blood alcohol level is 0.08, which is legally drunk in many states?

b. Estimate the blood alcohol level when the probability of an accident is 50%.

72. Atmospheric pressure (in pounds per square inch) is a function of the altitude above sea level and can be modeled by

$$P(a) = 14.7e^{-0.21a},$$

where P is the pressure and a is the altitude above sea level in miles.

a. Find the atmospheric pressure at the peak of Mt. McKinley, Alaska, which is 3.85 miles above sea level.

b. If the atmospheric pressure at the peak of Mt. Everest in Nepal, is 4.68 pounds per square inch, find the height of Mt. Everest.

73. Housing prices in Volusia County, Florida, have risen exponentially. The average sales price of existing houses sold in Volusia County can be approximated by $S(t) = 51.6(1.06)^t$, where t is the number of years after 1990 and S is the price in thousands of dollars.

a. Estimate the average price of existing homes sold in 1998.

b. Find the number of years after 1990 when the average price of an existing home will be \$100,000.

74. Since the first Super Bowl in 1967, ticket prices have risen exponentially according to the formula $P(t) = 9.046(1.114)^t$, where $t = 1$ is the year 1967 and ticket prices are in dollars. (*Source: Orlando Sentinel*, January 21, 2003, p. A8)

a. Estimate the price of a Super Bowl ticket in 1990.

b. Find the year when Super Bowl prices were $230.

75. Following is a chart giving the number of registered cocaine users in the United Kingdom from 1987 to 1995.

Year	Number of Users
1987	431
1988	462
1989	527
1990	633
1991	882
1992	1131
1993	1375
1994	1636
1995	1809

These data can be modeled by the exponential function $y = 387.7(1.222)^t$, where y is the number of registered users and t is the number of years after 1987. If the current trend continues, use the model to estimate the number of users in 2004.

76. The thickness of a runway's pavement is determined by the weight of the planes that will be using it. The relationship between runway thickness, x, in inches and gross airplane weight, y, in thousands of pounds can be modeled by $y = 18.29(1.279)^x$. If a partially loaded B-52 Stratofortress weighs 350,000 pounds, will a runway that is 12 inches thick support it? (*Source:* Federal Aviation Administration)

77. The number of females working in automotive repair has been increasing, as shown in the table.

Year	Number of Female Technicians
1988	556
1989	614
1990	654
1991	737
1992	849
1993	1086
1994	1329
1995	1592

This data can be modeled by $y = 507(1.116)^t$, where y is the number of technicians and t is the number of years after 1988. If the trend continues in the same manner, use the model to estimate the number of female technicians in 2005. (*Source:* National Institute for Automotive Service Excellence)

78. The gap between available organs for liver transplants and people who need them has widened in recent years. The equation $y = 2329(1.2406)^x$, where y is the number of persons waiting and x is the number of years after 1988, models the gap. Assuming the same trend continues, use the model to estimate the number of people waiting for a transplant in 2006.

79. The population of a mosquito colony increases at a rate of 4% per day. If the initial number of mosquitoes is 500, the number present, A, at the end of t days is given by $A = 500e^{0.04t}$.

   a. How many mosquitoes are present after 2 weeks?

   b. Find the number of days until 10,000 mosquitoes are present.

80. A cake is removed from the oven at a temperature of 210°F and is left to cool on a counter where the room temperature is 70°F. The cake cools according to the function $T = 70 + 140e^{-0.0231t}$, where T is the temperature of the cake and t is in minutes.

   a. What is the temperature of the cake after 40 minutes?

   b. After how many minutes will the cake's temperature be 75°F?

For Exercises 81–88, use the change-of-base formula to find the logarithms. Round answers to four decimal places.

81. $\log_4 12$ **82.** $\log_5 23$ **83.** $\log_8 3$ **84.** $\log_9 5$

85. $\log_{1/2} 5$ **86.** $\log_{1/3} 4$ **87.** $\log_{1/4} \dfrac{3}{5}$ **88.** $\log_{1/3} \dfrac{4}{7}$

REVIEW EXERCISES

Evaluate the following for $n = 1, 2, 3,$ and 4:

1. $2n^2 - 3$ **2.** $\dfrac{(-1)^n}{3n - 2}$

3. Find the value of $(3 \cdot 1 + 2) + (3 \cdot 2 + 2) + (3 \cdot 3 + 2) + (3 \cdot 4 + 2)$.

4. Find the value of $\dfrac{n}{2}(a_1 + a_n)$ if $n = 12$, $a_1 = -8$, and $a_n = 60$.

5. Find 5% of 550.

6. Find the value of $a_1 + (n - 1)d$ if $a_1 = -12$, $n = 25$, and $d = -3$.

Defined Terms

Section 12.1
Composition of
functions (p. 984)
Inverse functions
(p. 986)
One-to-one functions
(p. 988)

Section 12.2
Exponential function
(p. 999)

Section 12.3
Logarithm (p. 1011)
Logarithmic function
(p. 1011)

Section 12.5
Common logarithms
(p. 1030)
Natural logarithms
(p. 1032)

Procedures, Rules, and Key Examples

Procedures/Rules	Key Examples

Section 12.1 Composite and Inverse Functions

Composition of Functions

$(f \circ g)(x) = f[g(x)]$ for all x in the domain of g such that $g(x)$ is in the domain of f.

$(g \circ f)(x) = g[f(x)]$ for all x in the domain of f such that $f(x)$ is in the domain of g.

Example 1: If $f(x) = x^2 - 3$ and $g(x) = 3x + 4$ find (a) $(f \circ g)(x)$ and (b) $(g \circ f)(x)$.

a. $(f \circ g)(x) = f[g(x)]$
$= f(3x + 4)$
    Substitute $3x + 4$ for $g(x)$.
$= (3x + 4)^2 - 3$
    Replace x with $3x + 4$ in $f(x)$.
$= 9x^2 + 24x + 13$
    Square $3x + 4$ and add like terms.

b. $(g \circ f)(x) = g[f(x)]$
$= g(x^2 - 3)$
    Substitute $x^2 - 3$ for $f(x)$.
$= 3(x^2 - 3) + 4$
    Replace x with $x^2 - 3$ in $g(x)$.
$= 3x^2 - 5$
    Distribute 3 and add.

Inverse Functions

Two functions f and g are inverses if and only if $f[g(x)] = x$ for all x in the domain of g and $g[f(x)] = x$ for all x in the domain of f.

Example 2: Show that $f(x) = 5x - 6$ and $g(x) = \dfrac{x + 6}{5}$ are inverse functions.

$$f[g(x)] = f\left(\frac{x + 6}{5}\right) = 5\left(\frac{x + 6}{5}\right) - 6$$
$$= x + 6 - 6 = x.$$
$$g[f(x)] = g(5x - 6) = \frac{5x - 6 + 6}{5}$$
$$= \frac{5x}{5} = x.$$

Since $f[g(x)] = g[f(x)] = x$, f and g are inverse functions.

continued

Procedures/Rules

One-to-One Functions
A function is one-to-one if for any numbers a and b in its domain, $f(a) = f(b)$ implies that $a = b$. This means that two different x-values result in two different y-values.

Horizontal Line Test for One-to-One Functions
If the graph of a function can be intersected in at most one point by any horizontal line, then the function is one-to-one.

Note: Only one-to-one functions have inverses that are functions.

Finding the Inverse of a One-to-One Function
1. If necessary, replace $f(x)$ with y or solve for y.
2. Replace all x's with y and all y's with x.
3. Solve the expression from step 2 for y.
4. Replace y with $f^{-1}(x)$.

Graphs of Inverse Functions
The graphs of inverse functions are symmetric with respect to the line $y = x$.

Key Examples

Example 3: Determine whether the functions whose graphs follow are one to one.

a.

b.

Solution: No. This graph can be crossed by a horizontal line in more than one point.

Solution: Yes. No horizontal line intersects the graph in more than one point.

Example 4: If $f(x) = 3x - 7$, find $f^{-1}(x)$.
Solution:

$$f(x) = 3x - 7$$
$$y = 3x - 7 \qquad \text{Replace } f(x) \text{ with } y.$$
$$x = 3y - 7 \qquad \text{Interchange } x \text{ and } y.$$
$$\frac{x + 7}{3} = y \qquad \text{Solve for } y.$$
$$f^{-1}(x) = \frac{x + 7}{3} \qquad \text{Replace } y \text{ with } f^{-1}(x).$$

Verify by showing that $f[f^{-1}(x)] = x$ and $f^{-1}[f(x)] = x$.

Example 5: Sketch the inverse of the function whose graph follows.
Solution: Draw the line $y = x$ and reflect the graph in the line.

continued

Procedures/Rules

Section 12.2 Exponential Functions

Definition of Exponential Function
If $b > 0$ and $b \neq 1$ and x is a real number, then $f(x) = b^x$ is the exponential function.

Graphs of Exponential Functions

$f(x) = b^x, b > 1$

$f(x) = b^x, 0 < b < 1$

Solving Exponential Equations
1. If necessary, write both sides as a power of the same base.
2. If necessary, simplify the exponents.
3. Set the exponents equal to each other.
4. Solve the resulting equation.

Section 12.3 Logarithmic Functions

Definition of Logarithm
If $b > 0$ and $b \neq 1$, then $y = \log_b x$ is equivalent to $x = b^y$. The expression $\log_b x$ is the exponent to which b is raised to get x.

Solve Logarithmic Equations by Changing into Exponential Form
To solve an equation of the form $\log_b x = y$ for x, change the equation into exponential form. Then solve for x.

Key Examples

Example 1: Draw the graph of $f(x) = 2^x$. Choose some values for x and find y.

x	y
-2	$\frac{1}{4}$
-1	$\frac{1}{2}$
0	1
1	2
2	4

Plot the points and draw the graph.

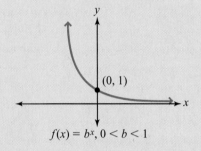

Example 2: Solve $8^{x-3} = 16^{2x}$
Solution: $8^{x-3} = 16^{2x}$

$(2^3)^{x-3} = (2^4)^{2x}$ Rewrite 8 and 16 as powers of 2.

$2^{3x-9} = 2^{8x}$ Multiply exponents.

$3x - 9 = 8x$ Set exponents equal.

$-\dfrac{9}{5} = x$ Solve for x.

Example 1: Change each into logarithmic form.
a. $3^4 = 81$ **b.** $\left(\dfrac{1}{4}\right)^{-3} = 64$

Solution:
$\log_3 81 = 4$

Solution:
$\log_{1/4} 64 = -3$

Example 2: Change each into exponential form.
a. $\log_5 125 = 3$ **b.** $\log_3 \dfrac{1}{27} = -3$

Solution: $5^3 = 125$ Solution: $3^{-3} = \dfrac{1}{27}$

Example 3: Solve the following for x:
a. $\log_2 \dfrac{1}{16} = x$

continued

Procedures/Rules	Key Examples

Procedures/Rules

Section 12.3 (Continued)

Key Examples

Solution: $2^x = \dfrac{1}{16}$ Change into exponential form.

$\qquad 2^x = \dfrac{1}{2^4}$ $16 = 2^4$.

$\qquad 2^x = 2^{-4}$ $\dfrac{1}{2^4} = 2^{-4}$

$\qquad x = -4$

b. $\log_x 49 = 2$

Solution: $x^2 = 49$ Change into exponential form.

$\qquad x = \pm 7$ The square roots of 49 are ±7.

$\qquad x = 7$ The base must be positive.

Two Properties of Logarithms
For $b \neq 1$ and $b > 0$, $\log_b b = 1$ and $\log_b 1 = 0$

Example 4: Find the following logarithms:
a. $\log_8 8$ **b.** $\log_5 1$
Solution: $\log_8 8 = 1$ Solution: $\log_5 1 = 0$

Procedure for Graphing Logarithmic Functions
a. Convert to exponential form.

b. Find ordered pairs that satisfy the equation by assigning values to y and finding x.

c. Plot the ordered pairs and draw a smooth curve through the points.

Example 5: Graph $y = \log_3 x$.
Solution: Write as $3^y = x$ and let y have values.

x	y
$\dfrac{1}{9}$	-2
$\dfrac{1}{3}$	-1
1	0
3	1
9	2

Section 12.4 Properties of Logarithms
Inverse Properties of Logarithms
For $b > 0$, $x > 0$, and $b \neq 1$, $b^{\log_b x} = x$ and $\log_b b^x = x$.

Further Properties of Logarithms
For $x > 0$, $y > 0$, $b > 0$, and $b \neq 1$:

Product Rule: $\log_b xy = \log_b x + \log_b y$

Quotient Rule: $\log_b \dfrac{x}{y} = \log_b x - \log_b y$

Power Rule: $\log_b x^r = r \log_b x$

Example 1: Find the value of each.
a. $6^{\log_6 5}$ **b.** $3^{\log_3 8}$
Solution $6^{\log_6 5} = 5$ Solution: $3^{\log_3 8} = 8$

Example 2: Use the product rule for the following:
a. Write $\log_4 4x$ as the sum of logarithms.
Solution:

$\log_4 4x = \log_4 4 + \log_4 x$ Use $\log_b xy = \log_b x + \log_b y$.

$\qquad\qquad = 1 + \log_4 x$ $\log_4 4 = 1$

b. Write $\log_7 4 + \log_7 2$ as a single logarithm.

Solution:

$\log_7 4 + \log_7 2 = \log_7 4 \cdot 2$ Use $\log_b xy = \log_b x + \log_b y$.

$\qquad\qquad = \log_7 8$

continued

Procedures/Rules	Key Examples

Procedures/Rules

Section 12.4 (Continued)

Key Examples

Example 3: Use the quotient rule for the following:

a. Write $\log_a \dfrac{w}{7}$ as the difference of logarithms.

Solution:

$$\log_a \frac{w}{7} = \log_a w - \log_a 7 \qquad \text{Use } \log_b \frac{x}{y} = \log_b x - \log_b y.$$

b. Write $\log_7 (x + 5) - \log_7 (x - 3)$ as a single logarithm.

Solution:

$$\log_7 (x + 5) - \log_7 (x - 3) = \log_7 \frac{x + 5}{x - 3}$$

$$\text{Use } \log_b \frac{x}{y} = \log_b x - \log_b y$$

Example 4: Use the Power Rule for the following:

c. Write $\log_b \sqrt[5]{x^3}$ as a multiple of a logarithm.

Solution:

$$\log_b \sqrt[5]{x^3} = \log_b x^{3/5} = \frac{3}{5} \log_b x \qquad \text{Use } \log_b x^r = r \log_b x.$$

d. Write $-3 \log_2 y$ as a logarithm of a quantity to a power.

Solution:

$$-3 \log_2 y = \log_2 y^{-3} \qquad \text{Use } \log_b x^r = r \log_b x.$$

$$= \log_2 \frac{1}{y^3}$$

Example 5: Write $\log_a \dfrac{x^2 y}{z^4}$ as the sum or difference of multiples of logarithms.

Solution:

$$\log_a \frac{x^2 y}{z^4} = \log_a x^2 y - \log_a z^4 \qquad \text{Quotient Rule.}$$

$$= \log_a x^2 + \log_a y - \log_a z^4$$

$$\text{Product Rule.}$$

$$= 2 \log_a x + \log_a y - 4 \log_a z$$

$$\text{Power Rule.}$$

Example 6: Write $\dfrac{1}{4}(2 \log_5 x - 3 \log_5 y)$ as a single logarithm.

Solution: $\dfrac{1}{4}(2 \log_5 x - 3 \log_5 y)$

$$= \frac{1}{4}(\log_5 x^2 - \log_5 y^3) \qquad \text{Power Rule.}$$

$$= \frac{1}{4} \log_5 \frac{x^2}{y^3} \qquad \text{Quotient Rule.}$$

$$= \log_5 \left(\frac{x^2}{y^3}\right)^{1/4} \qquad \text{Power Rule.}$$

$$= \log_5 \sqrt[4]{\frac{x^2}{y^3}} \qquad \qquad \textit{continued}$$

Procedures/Rules	Key Examples

Procedures/Rules

Section 12.5 Common and Natural Logarithms

Base-10 logarithms are common logarithms and $\log_{10} x$ is written as $\log x$. Common logarithms can be evaluated using the $\boxed{\text{LOG}}$ key on a calculator.

Base-e logarithms are called natural logarithms and $\log_e x$ is written as $\ln x$. Natural logarithms can be evaluated using the $\boxed{\text{ln}}$ key on a calculator.

Section 12.6 Exponential and Logarithmic Equations with Applications

Properties for Solving Exponential and Logarithmic Equations.
For $b > 0$ and $b \neq 1$, and x and y real numbers:
1. If $b^x = b^y$, then $x = y$.
2. If $x = y$, then $b^x = b^y$.
3. For $x > 0$ and $y > 0$, if $\log_b x = \log_b y$, then $x = y$.
4. For $x > 0$ and $y > 0$, if $x = y$, then $\log_b x = \log_b y$.
5. For $x > 0$, if $\log_b x = y$, then $b^y = x$.

Solve equations containing logarithms.
Use the properties of logarithms to simplify both sides of the equation. If the simplification results in an equation of the form $\log_b x = \log_b y$, then $x = y$. If the simplification results in an equation of the form $\log_b x = y$, then write it as $b^y = x$.

Key Examples

Example 1: Evaluate using a calculator and round to 4 decimal places.
a. $\log 356$
Solution: $\log 356 = 2.5514$

b. $\log 0.0059$
Solution: $\log 0.0059 = -2.2291$

Example 2: Evaluate using a calculator and round to four decimal places.
a. $\ln 72$
Solution: $\ln 72 = 4.2767$

b. $\ln 0.097$
Solution: $\ln 0.097 = -2.3330$

Example 1: Solve $3^x = 4$.
Solution:

$$\log 3^x = \log 4 \qquad \text{Use if } x = y, \text{ then } \log_b x = \log_b y.$$
$$x \log 3 = \log 4 \qquad \text{Power Rule.}$$
$$x = \frac{\log 4}{\log 3} \qquad \text{Divide by } \log 3.$$
$$x \approx 1.2619 \qquad \text{Evaluate using a calculator.}$$

Example 2: Solve $e^{3x} = 12$.
Solution: Since the base of the exponent is e, take the natural logarithm of both sides.

$$\ln e^{3x} = \ln 12 \qquad \text{Use if } x = y, \text{ then } \log_b x = \log_b y.$$
$$3x = \ln 12 \qquad \text{Use } \log_b b^x = x.$$
$$x = \frac{\ln 12}{3} \approx 0.8283 \qquad \text{Divide by 3 and approximate.}$$

Example 3: Solve
$\log_3 x + \log_3 (x - 3) = \log_3 10$.
Solution:

$$\log_3 x(x - 3) = \log_3 10$$
$$\text{Use the product rule.}$$
$$x(x - 3) = 10$$
$$\text{Use if } \log_b x = \log_b y, \text{ then } x = y.$$
$$x^2 - 3x - 10 = 0$$
$$\text{Multiply then subtract 10.}$$
$$(x - 5)(x + 2) = 0 \qquad \text{Factor.}$$
$$x - 5 = 0 \quad \text{or} \quad x + 2 = 0$$
$$\text{Set each factor equal to 0.}$$
$$x = 5 \quad \text{or} \quad x = -2$$
$$\text{Solve each equation.}$$

We must reject $x = -2$ because it results in $\log_3 (-2)$ and $\log_3 (-7)$ in the original equation, which do not exist. ***continued***

Procedures/Rules	Key Examples
Section 12.6 (Continued)	**Example 4:** Solve $\log_2(2x + 2) - \log_2(x - 4) = 2$. Solution: $\log_2 \dfrac{2x + 2}{x - 4} = 2$ Use the quotient rule. $\dfrac{2x + 2}{x - 4} = 2^2$ Use if $\log_b x = y$, then $b^y = x$. $\dfrac{2x + 2}{x - 4} = 4$ $2^2 = 4$. $2x + 2 = 4x - 16$ Multiply by $x - 4$. $18 = 2x$ $9 = x$ Solve for x.
Solve applications problems involving exponentials and logarithms	**Example 5:** How long will it take \$12,000 invested at 4% compounded quarterly to grow to \$15,000? Solution: Use $A = P\left(1 + \dfrac{r}{n}\right)^{nt}$ with $A = 15{,}000$, $P = 12{,}000$, $r = 4\% = 0.04$, and $n = 4$. Find t. $A = P\left(1 + \dfrac{r}{n}\right)^{nt}$ $15{,}000 = 12{,}000\left(1 + \dfrac{0.04}{4}\right)^{4t}$ Substitute. $\dfrac{5}{4} = (1.01)^{4t}$ Divide by 12,000. $\log \dfrac{5}{4} = \log 1.01^{4t}$ Apply if $x = y$, then $\log_b x = \log_b y$. $\log \dfrac{5}{4} = 4t \log 1.01$ Power rule. $\dfrac{\log \dfrac{5}{4}}{4 \log 1.01} = t$ Divide by 4 log 1.01. $5.61 \approx t$ Evaluate. It would take about 5.6 years. *continued*

Procedures/Rules	Key Examples
Section 12.6 (Continued)	**Example 6:** If \$3000 is invested in an account paying 4.5% compounded continuously:

Section 12.6 (Continued)

Example 6: If \$3000 is invested in an account paying 4.5% compounded continuously:

a. How much will be in the account after 10 years?

Solution: Use $A = Pe^{rt}$ with $P = 3000$, $r = 0.045$, and $t = 10$.

$$A = Pe^{rt}$$
$$A = 3000e^{(0.045)(10)} \quad \text{Substitute.}$$
$$A = 3000e^{0.45} \quad 10(0.045) = 0.45.$$
$$A = 4704.94 \quad \text{Evaluate using a calculator.}$$

b. How long would it take for the \$3000 to grow to \$5000?

Solution: Use $A = Pe^{rt}$ with $A = 5000$, $P = 3000$, and $r = 0.045$. Find t.

$$A = Pe^{rt}$$
$$5000 = 3000e^{0.045t} \quad \text{Substitute}$$
$$\frac{5}{3} = e^{0.045t} \quad \text{Divide by 3000.}$$
$$\ln\frac{5}{3} = \ln e^{0.045t} \quad \text{Use if } x = y, \text{ then } \log_b x = \log_b y.$$
$$\ln\frac{5}{3} = 0.045t \quad \text{Use if } x = y, \text{ then } \log_b x = \log_b y$$
$$\frac{\ln\frac{5}{3}}{0.045} = t \quad \text{Divide by 0.045.}$$
$$11.35 \approx t \quad \text{Evaluate using a calculator.}$$

It would take about $11\frac{1}{3}$ years.

Exponential Growth and Decay

Exponential growth and decay can be represented by the equation $A = A_0 e^{kt}$, where A_0 is the initial amount, t is the time, and k is a constant that is determined by the substance. If $k > 0$, there is exponential growth, and if $k < 0$, there is exponential decay. The half-life of a substance is the amount of time that must pass until one-half of the original substance remains.

Example 7: The element bismuth has an isotope, ^{200}Bi, that disintegrates according to the formula $A = A_0 e^{-0.0198t}$, where t is in minutes.

a. How much of a 100-gram sample remains after 120 minutes?

Solution: We are given $A_0 = 100$ and $t = 120$.

$$A = A_0 e^{-0.0198t}$$
$$A = 100e^{-0.0198(120)} \quad \text{Substitute for } A_0 \text{ and } t.$$
$$A = 9.29 \quad \text{Evaluate using a calculator.}$$

There would be about 9.29 grams left after 120 minutes.

continued

Procedures/Rules	Key Examples

Procedures/Rules

Section 12.6 (Continued)

Key Examples

b. Find the half-life of ^{200}Bi.

Solution: After one half-life, 50 of the original 100 grams remain, so $A = 50$.

$$A = A_0 e^{-0.0198t}$$
$$50 = 100 e^{-0.0198t} \quad \text{Substitute for } A \text{ and } A_0.$$
$$\frac{1}{2} = e^{-0.0198t} \quad \text{Divide by 100.}$$
$$\ln \frac{1}{2} = \ln e^{-0.0198t} \quad \text{Use if } x = y, \text{ then } \log_b x = \log_b y.$$
$$\ln \frac{1}{2} = -0.0198t \quad \text{Use } \log_b b^x = x.$$
$$\frac{\ln \frac{1}{2}}{-0.0198} = t \quad \text{Divide by } -0.0198$$
$$35 = t \quad \text{Evaluate using a calculator.}$$

The half-life is 35 minutes.

Change-of-Base Formula

If $a > 0$, $a \neq 1$, $b > 0$, $b \neq 1$, and $x > 0$, then $\log_a x = \dfrac{\log_b x}{\log_b a}$.

Example 8: Find $\log_5 16$.
Solution:

$$\log_5 16 = \frac{\ln 16}{\ln 5} \quad \text{Use } \log_a x = \frac{\log_b x}{\log_b a}$$
$$= 1.7227 \quad \text{Evaluate.}$$

For Exercises 1–6, answer true or false.

1. If f and g are inverse functions and $f(2) = 5$, then $g(5) = 2$.

2. If $a^x = b^x$, then $a = b$.

3. If $y = \log_a (x - 4)$, it is possible for x to equal 3.

4. If $\log_a b = c$, then $c^a = b$.

5. Logarithms are exponents.

6. Base-e logarithms are called common logarithms.

For Exercises 7–10, find each composition if $f(x) = 3x + 4$ and $g(x) = x^2 - 2$.

7. $(f \circ g)(3)$ **8.** $(g \circ f)(3)$ **9.** $f[g(0)]$ **10.** $g[f(0)]$

For Exercises 11–14, find $(f \circ g)(x)$ and $(g \circ f)(x)$.

11. $f(x) = 3x - 6$, $g(x) = 2x + 3$ **12.** $f(x) = x^2 + 4$, $g(x) = 3x - 7$

13. $f(x) = \sqrt{x - 3}$, $g(x) = 2x - 1$ **14.** $f(x) = \dfrac{x + 3}{x}$, $g(x) = \dfrac{x - 4}{x}$

For Exercises 15 and 16, determine if each is the graph of a one-to-one function. If the function is one-to-one, sketch the graph of the inverse function.

15.

16.

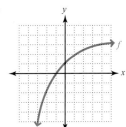

For Exercises 17–20, determine whether f and g are inverse functions.

17. $f(x) = 2x - 9$, $g(x) = \dfrac{x + 9}{2}$

18. $f(x) = x^3 + 6$, $g(x) = \sqrt[3]{x - 6}$

19. $f(x) = x^2 - 3$, $x \geq 0$; $g(x) = \sqrt{x + 2}$

20. $f(x) = \dfrac{x}{x + 4}$, $g(x) = \dfrac{-4x}{x + 1}$

For Exercises 21–24, find $f^{-1}(x)$ for each of the one-to-one functions.

21. $f(x) = 5x + 4$

22. $f(x) = x^3 + 6$

23. $f(x) = \dfrac{4}{x + 5}$

24. $f(x) = \sqrt[3]{3x + 2}$

25. If $(4, -6)$ is an ordered pair on the graph of f, what ordered pair is on the graph of f^{-1}?

26. If $f(a) = b$, then $f^{-1}(\ \ ) =$ _____.

For Exercises 27–30, draw the graph of each.

27. $f(x) = 3^x$

28. $f(x) = \left(\dfrac{1}{2}\right)^x$

29. $f(x) = 2^{x-3}$

30. $f(x) = 3^{-x+2}$

For Exercises 31–38, solve each equation.

31. $5^x = 625$

32. $9^x = 27$

33. $6^x = \dfrac{1}{36}$

34. $\left(\dfrac{3}{4}\right)^x = \dfrac{16}{9}$

35. $3^{x-2} = 9$

36. $5^{x+2} = 25^x$

37. $\left(\dfrac{1}{3}\right)^{-x} = 27$

38. $8^{3x-2} = 16^{4x}$

39. The median doubling time for a malignant tumor is about 100 days. If there are 500 cells initially, then the number of cells, A, after t days is given by $A = A_0 2^{t/100}$. How many cells are present after 1 year?

40. If \$25,000 is deposited into an account paying 6% interest compounded monthly, how much will be in the account after 8 years?

41. The radioactive isotope ^{82}R has a half-life of 107 days. How much of a 50 gram sample remains after 300 days? (Use $A = A_0\left(\dfrac{1}{2}\right)^{t/h}$.)

42. Cocaine use in the United Kingdom skyrocketed between 1987 and 1995 and can be modeled by the exponential function $N(t) = 387.7(1.222)^t$, where $N(t)$ is the number of addicts and t is the number of years after 1987. Estimate the number of addicts in 1991.

For Exercises 43–46, convert from exponential form into logarithmic form.

43. $7^3 = 343$

44. $4^{-3} = \dfrac{1}{64}$

45. $\left(\dfrac{3}{2}\right)^4 = \dfrac{81}{16}$

46. $11^{1/3} = \sqrt[3]{11}$

For Exercises 47–50, convert from logarithmic form to exponential form.

47. $\log_9 81 = 2$

48. $\log_{1/5} 125 = -3$

49. $\log_a 16 = 4$

50. $\log_e c = b$

For Exercises 51–58, solve each equation. Find exact solutions.

51. $\log_3 x = -4$

52. $\log_{1/2} x = -2$

53. $\log_2 32 = x$

54. $\log_{1/4} 16 = x$

55. $\log_x 81 = 4$

56. $\log_x \dfrac{1}{1000} = 3$

57. $\log_{3/4} \dfrac{9}{16} = x$

58. $\log_{121} x = \dfrac{1}{2}$

For Exercises 59–60, graph each function.

59. $f(x) = \log_4 x$

60. $f(x) = \log_{1/3} x$

61. If $f(x) = \log_b x$, the domain is $(0, \infty)$, and the range is $(-\infty, \infty)$, what are the domain and range of $g(x) = b^x$? Why?

62. The formula for the number of decibels in a sound is $d = 10 \log \dfrac{I}{I_0}$. Find the decibel reading of a sound whose intensity is $I = 1000 \, I_0$.

For Exercises 63 and 64, find the value of each.

63. $3^{\log_3 8}$

64. $\log_9 9^6$

For Exercises 65 and 66, use the product rule to write each as a sum of logarithms.

65. $\log_6 6x$

66. $\log_4 x(2x - 5)$

For Exercises 67 and 68, use the product rule to write each as a single logarithm.

67. $\log_3 4 + \log_3 8$

68. $\log_5 3 + \log_5 x + \log_5(x - 2)$

For Exercises 69 and 70, use the quotient rule to write each as a difference of logarithms.

69. $\log_b \dfrac{x}{5}$

70. $\log_a \dfrac{3x - 2}{4x + 3}$

For Exercises 71 and 72, use the quotient rule to write each as a single logarithm.

71. $\log_5 24 - \log_5 4$

72. $\log_2(x + 5) - \log_2(2x - 3)$

For Exercises 73–76, use the power rule to write each as a multiple of a logarithm.

73. $\log_3 7^4$

74. $\log_a \sqrt[3]{x}$

75. $\log_4 \dfrac{1}{a^4}$

76. $\log_a \sqrt[5]{a^4}$

For Exercises 77 and 78, use the power rule to write each as the logarithm of a quantity to a power. Simplify the answer, if possible.

77. $4 \log_a x$

78. $\dfrac{3}{5} \log_a y$

For Exercises 79–82, write each as the sum or differences of multiples of logarithms.

79. $\log_a x^2 y^3$

80. $\log_a \dfrac{c^4}{d^3}$

81. $\log_a \dfrac{x^2 y^3}{z^4}$

82. $\log_a \sqrt{\dfrac{a^3}{b^4}}$

For Exercises 83–86, write each as a single logarithm.

83. $2 \log_a b + 4 \log_a c$

84. $3 \log_a 4 - 2 \log_a 3$

85. $\dfrac{1}{4}(2 \log_a x + 3 \log_a y)$

86. $4 \log_a (x + 5) + 2 \log_a (x - 3)$

For Exercises 87–90, use a calculator to approximate each logarithm to four decimal places.

87. $\log 326$

88. $\log 0.0035$

89. $\ln 0.043$

90. $\ln 92$

For Exercises 91 and 92, find the exact value without using a calculator.

91. $\log 0.00001$

92. $\ln \sqrt[4]{e}$

93. The sound intensity of a clap of thunder was $10^{-3.5}$ watts per square meter. What is the decibel reading?

94. Using $pH = -\log [H_3O^+]$ (the hydronium ion concentration), find the pH of an apple whose $[H_3O^+]$ is 0.001259.

95. How long will it take \$6000 to grow to \$10,000 if it is invested at 3% annual interest compounded continuously?

96. Using $R = \log \dfrac{I}{I_0}$, for Richter scale readings, compare the intensity of an earthquake whose Richter scale reading was 7.8 with one whose reading was 6.8. What do you notice? (*Hint:* Solve for I in terms if I_0.)

For Exercises 97–102, solve each equation. Give answers correct to four decimal places.

97. $9^x = 32$

98. $3^{5x} = 19$

99. $6^{2x-1} = 22$

100. $4^{2x-3} = 5^{x+1}$

101. $e^{4x} = 11$

102. $e^{-0.003x} = 5$

For Exercises 103–110, solve each equation. Give exact answers.

103. $\log_3(x - 4) = 2$

104. $\log_2 (x^2 + 2x) = 3$

105. $\log (3x - 8) = \log (x - 2)$

106. $\log 5 + \log x = 2$

107. $\log_3 x - \log_3 4 = 2$

108. $\log_2 x + \log_2 (x - 6) = 4$

109. $\log_4 x + \log_4 (6x - 9) = \log_4 15$

110. $\log_3 (5x + 2) - \log_3 (x - 2) = 2$

111. How long will it take \$15,000 invested at 3% compounded monthly to grow to \$18,000? Round the answer to the nearest tenth of a year.

112. If \$7000 is deposited at 7% annual interest compounded continuously:

    a. How much will be in the account after 8 years?

    b. How long will it take until there is \$12,000 in the account?

113. The population of an ant colony is 400 and is increasing at the rate of 3% per month. Use $A = A_0 e^{0.03t}$ to answer the following:

   a. How many ants will be in the colony after one year?

   b. After how many months will there be 800 ants in the colony?

114. The number of diagnosed AIDS cases among adults in the United States since 1992 has declined and can be modeled by the function $y = 87{,}419 - 24{,}647.9 \ln x$, where y is the number of diagnosed cases and x is the number of years after 1992. (*Data Source:* CDC Division of HIV/AIDS Prevention, *Surveillance Report*, Vol. 13, No. 2)

   a. Estimate the number of cases of diagnosed AIDS in 2000.

   b. In what year were 43,256 new cases of AIDS diagnosed?

For Exercises 115 and 116, use the change-of-base formula to approximate each logarithm to four decimal places.

115. $\log_4 15$

116. $\log_{1/2} 6$

1. If $f(x) = x^2 - 6$ and $g(x) = 3x - 5$, find $f[g(x)]$.

2. If $f(x) = 4x - 3$, find the following:
   a. $f^{-1}(x)$

   b. Verify that $f[f^{-1}(x)] = x$ and $f^{-1}[f(x)] = x$.

   c. What is the relationship between the graphs of $f(x)$ and $f^{-1}(x)$?

3. Following is the graph of a function f. On the same set of axes, sketch the graph of f^{-1}.

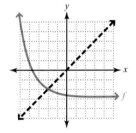

4. Sketch the graph of $f(x) = 2^{x-1}$.

5. Solve $32^{x-2} = 8^{2x}$ for x.

6. A sum of $20,000 is invested at 5% compounded quarterly.
   a. Find the amount in the account after 15 years.

   b. Find the number of years until there is $30,000 in the account.

7. The isotope ^{98}Nb has a half-life of 30 minutes.
   a. How much of a 100 gram sample remains after 3.6 hours?

   b. How long will it take until there are only 20 grams remaining?

8. The number of people in the United States aged 65 or over has increased rapidly since 1900 and can be modeled by $y = 3.17(1.026)^x$, where y is the number in millions aged 65 or older and x is the number of years after 1900. (*Source:* U.S. Bureau of the Census)

  a. Find the number aged 65 or older in 1960. In 2005.

  b. In what year were there 31.4 million people aged 65 or older?

9. Convert $\log_{1/3} 81 = -4$ into exponential form.

For Exercises 10 and 11, find the exact solution of each equation.

10. $\log_6 \dfrac{1}{216} = x$

11. $\log_x 625 = 4$

12. Graph $y = \log_2 x$.

For Exercises 13 and 14, write as the sum or difference of multiples of logarithms.

13. $\log_b \dfrac{x^4 y^2}{z}$

14. $\log_b \sqrt[4]{\dfrac{x^5}{y^7}}$

15. Write $\dfrac{3}{4}(2\log_b x + 3\log_b y)$ as a single logarithm.

16. The isotope ^{119}Sn disintegrates according to the function $A = A_0 e^{-0.0028t}$, where t is the time in days.

  a. How much of a 300 gram sample remains after 500 days?

  b. What is the half-life of ^{119}Sn?

For Exercises 17–19, solve each equation for x. Approximate to four decimal places.

17. $6^{x-3} = 19$

18. $\log_3 x + \log_3(x + 6) = 3$

19. $\log(4x + 2) - \log(3x - 2) = \log 2$

20. The number of children younger than 13 diagnosed with AIDS has declined since 1992 and can be modeled by $y = 1141.7 - 428.64 \ln x$, where y is the number of diagnosed cases and $x = 1$ is 1992. (*Data Source:* CDC Divisions of HIV/AIDS Prevention, *Surveillance Report*, Vol. 13, no. 2)

  a. Find the number of diagnosed cases in 1998.

  b. In what year were there 155 cases diagnosed?

CHAPTER 13

Conic Sections

In this chapter, we expand our understanding of equations, inequalities, and graphs and learn about conic sections. We will see that the conic sections are parabolas, circles, ellipses, and hyperbolas. These shapes are called conic sections because they are formed by the intersection of a cone with a plane. Later, we will study the equations of conic sections, nonlinear systems, and nonlinear inequalities.

"The universe . . . is written in the language of mathematics, and its characters are triangles, circles, and other geometrical figures."

—*GALILEO GALILEI (1564–1642)*
Italian mathematician, astronomer, physicist, and philosopher. *Il Saggiatore* (1623).

"Mankind is not a circle with a single center but an ellipse with two focal points of which facts are one and ideas the other."

—*VICTOR HUGO (1802–1883)*
French poet, dramatist, novelist. *Les Miserables* (1862)

13.1 The Parabola and Circle

OBJECTIVES

1. Graph parabolas of the form $x = a(y - k)^2 + h$.
2. Find the distance between two points.
3. Graph circles of the form $(x - h)^2 + (y - k)^2 = r^2$.
4. Find the equation of a circle with a given center and radius.
5. Graph circles of the form $x^2 + y^2 + dx + ey + f = 0$.

The intersection of a plane with a cone will be a circle, ellipse, parabola, or hyperbola. For that reason, these curves are called **conics** or **conic sections**.

Circle   Ellipse   Parabola   Hyperbola

Recall from Section 11.4 that we graphed parabolas in the form $y = a(x - h)^2 + k$. The graph opened up if $a > 0$, down if $a < 0$, had a vertex at (h, k), and $x = h$ as the axis of symmetry.

EXAMPLE 1   Given $y = 2(x - 3)^2 + 1$, determine whether the graph opens up or down, find the vertex and axis of symmetry, and draw the graph.

Solution The graph opens up since $a = 2$ and 2 is positive. We compare the equation with the form $y = a(x - h)^2 + k$ and observe that the vertex is at the point with coordinates $(3, 1)$ and the axis of symmetry is $x = 3$. Plot a few points on either side of the axis of symmetry by letting x have values on either side of 3 and finding y.

x	y
2	3
1	9
4	3
5	9

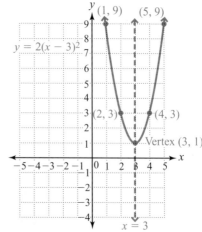

Given the equation $y = -2(x - 1)^2 + 3$, determine whether the graph opens up or down, find the vertex and axis of symmetry, and draw the graph.

OBJECTIVE **1** **Graph parabolas of the form $x = a(y - k)^2 + h$.** If we interchange x and y in the equations of parabolas that open up and down, we get the equations of parabolas that open to the left or right. To keep the vertex at (h, k), we also interchange h and k.

RULE **EQUATIONS OF PARABOLAS OPENING LEFT OR RIGHT**

The graph of an equation in the form $x = a(y - k)^2 + h$ is a parabola with vertex at (h, k). The parabola opens to the right if $a > 0$ and to the left if $a < 0$. The equation of the axis of symmetry is $y = k$.

$x = a(y - k)^2 + h$ where $a > 0$ $x = a(y - k)^2 + h$ where $a < 0$

EXAMPLE **2** For each equation, determine whether the graph opens left or right, find the vertex and axis of symmetry, and draw the graph.

a. $x = -2(y + 3)^2 - 2$

Solution This parabola opens to the left because $a = -2$, which is negative. Rewrite the equation as $x = -2(y - (-3))^2 - 2$. Comparing this equation with $x = a(y - k)^2 + h$, we see that $h = -2$ and $k = -3$. The vertex is at the point with coordinates $(-2, -3)$ and the axis of symmetry is $y = -3$. To graph, plot a few points on either side of the axis of symmetry by letting y equal values on either side of -3 and finding x.

ANSWER

Opens down. Vertex: $(1, 3)$; axis: $x = 1$.

$y = -2(x - 1)^2 + 3$

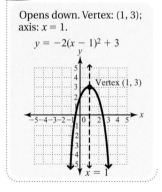

x	y
−4	−2
−10	−1
−4	−4
−10	−5

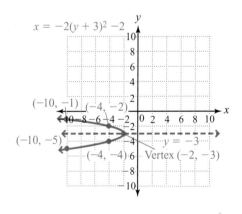

13.1 The Parabola and Circle   **1069**

b. $x = 3y^2 + 12y + 8$

Solution This parabola opens to the right because $a = 3$, which is positive. To find the vertex and axis of symmetry we need to write the equation in the form $x = a(y - k)^2 + h$.

$x = 3y^2 + 12y + 8$	Original equation.
$x - 8 = 3y^2 + 12y$	Subtract 8 from both sides.
$x - 8 = 3(y^2 + 4y)$	Factor out the common factor, 3.
$x - 8 + \mathbf{12} = 3(y^2 + 4y + \mathbf{4})$	Complete the square. Note that we added $3 \cdot 4 = 12$ to both sides of the equation.
$x + 4 = 3(y + 2)^2$	Simplify the left side and factor the right.
$x = 3(y + 2)^2 - 4$	Subtract 4 from both sides of the equation.

The vertex is at the point with coordinates $(-4, -2)$ and the axis of symmetry is $y = -2$.

To draw the graph, let y equal values on either side of -2 and find x.

x	y
-1	-1
8	0
-1	-3
8	-4

ANSWERS

a. Left, vertex at $(1, 2)$, axis: $y = 2$.

$x = -(y - 2)^2 + 1$

b. Right, vertex at $(-1, -2)$, axis is $y = -2$.

$x = y^2 + 4y + 3$

YOUR TURN For each equation, determine whether the graph opens left or right, find the vertex and axis of symmetry, and draw the graph.

a. $x = -(y - 2)^2 + 1$

b. $x = y^2 + 4y + 3$

OBJECTIVE ② Find the distance between two points. To derive the other conic's general equations, we need to be able to find the distance between any two points in the coordinate plane. Consider the two points (x_1, y_1) and (x_2, y_2) shown in the following graph.

Note: If y_2 were 6 and y_1 were 2, the distance between would be $6 - 2 = 4$. Therefore, to calculate the length of the vertical leg we calculate $y_2 - y_1$.

Since the distance between the two points is measured along a line segment that is the hypotenuse of a right triangle, we can use the Pythagorean theorem (see below) to find the distance.

$$c^2 = a^2 + b^2$$

The lengths of the legs of the triangle are $x_2 - x_1$ and $y_2 - y_1$, as illustrated.

Note: If x_2 were 4 and x_1 were 1, the distance between would be $4 - 1 = 3$. Therefore, to calculate the length of the horizontal leg we calculate $x_2 - x_1$.

Now we can use the Pythagorean theorem, replacing a with $x_2 - x_1$, b with $y_2 - y_1$, and c with d.

$$d^2 = (x_2 - x_1)^2 + (y_2 - y_1)^2$$
$$d = \pm\sqrt{(x_2 - x_1)^2 + (y_2 - y_1)^2}$$

Use the square root principle to isolate d.

Because d is a distance, it must be positive, so we use only the positive value.

RULE | **THE DISTANCE FORMULA**

The distance, d, between two points with coordinates (x_1, y_1) and (x_2, y_2) is given by the formula
$$d = \sqrt{(x_2 - x_1)^2 + (y_2 - y_1)^2}$$

EXAMPLE 3 Find the distance between $(4, 2)$ and $(-3, -1)$. If the distance is an irrational number, also give a decimal approximation rounded to three places.

Solution
$$d = \sqrt{(x_2 - x_1)^2 + (y_2 - y_1)^2}$$
Use the distance formula.
$$d = \sqrt{(-3 - 4)^2 + (-1 - 2)^2}$$
Let $(4, 2) = (x_1, y_1)$ and $(-3, -1) = (x_2, y_2)$.
$$d = \sqrt{(-7)^2 + (-3)^2}$$
$$d = \sqrt{49 + 9}$$
$$d = \sqrt{58}$$
$$d \approx 7.616$$

Note: It does not matter which ordered pair is (x_1, y_1) and which is (x_2, y_2). Consider Example 3 again with $(-3, -1)$ as (x_1, y_1) and $(4, 2)$ as (x_2, y_2):
$$d = \sqrt{(4 - (-3))^2 + (2 - (-1))^2}$$
$$= \sqrt{(7)^2 + (3)^2} = \sqrt{49 + 9} = \sqrt{58} \approx 7.616$$

Note: The distance formula holds no matter what quadrants the points are in.

ANSWERS

a. $\sqrt{61} \approx 7.810$
b. $2\sqrt{13} \approx 7.211$

YOUR TURN Determine the distance between the given points. If the distance is an irrational number, also give a decimal approximation rounded to three places.

a. $(8, 2)$ and $(3, -4)$ b. $(6, -5)$ and $(0, -1)$

OBJECTIVE 3 Graph circles of the form $(x - h)^2 + (y - k)^2 = r^2$.

DEFINITIONS

Circle: The set of points that is equally distant from a central point. The central point is the center.
Radius: The distance from the center of a circle to any point on the circle.

If the **center of a circle** is (h, k) and the **radius** is r, we can use the distance formula to derive the equation of the **circle**. If (x, y) is any point on the circle, the distance between (x, y) and (h, k) must be the radius, r.

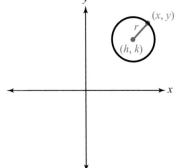

$$\sqrt{(x_2 - x_1)^2 + (y_2 - y_1)^2} = r$$

$$\sqrt{(x - h)^2 + (y - k)^2} = r \qquad \text{Substitute } (x, y) \text{ for } (x_2, y_2) \text{ and } (h, k) \text{ for } (x_1, y_1).$$

$$(x - h)^2 + (y - k)^2 = r^2 \qquad \text{Square both sides.}$$

RULE   **STANDARD FORM OF THE EQUATION OF A CIRCLE**

The equation of a circle with center at (h, k) and radius of r is $(x - h)^2 + (y - k)^2 = r^2$.

Note: If the center of a circle is at the origin, then $(h, k) = (0, 0)$ and the equation of the circle becomes $(x - 0)^2 + (y - 0)^2 = r^2$, which simplifies to $x^2 + y^2 = r^2$.

EXAMPLE 4   Find the center and radius of each circle and draw their graphs.

a. $(x - 3)^2 + (y + 2)^2 = 36$

Solution $(x - 3)^2 + (y - (-2))^2 = 6^2$ Write in the form $(x - h)^2 + (y - k)^2 = r^2$.

Since $h = 3$ and $k = -2$, the center is $(3, -2)$ and $r = 6$.

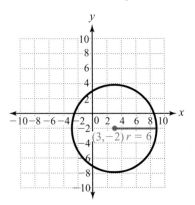

b. $(x + 4)^2 + (y + 1)^2 = 28$

Solution $(x - (-4))^2 + (y - (-1))^2 = (\sqrt{28})^2$ Write in the form $(x - h)^2 + (y - k)^2 = r^2$.

Since $h = -4$ and $k = -1$, the center is $(-4, -1)$ and $r = \sqrt{28} = 2\sqrt{7} \approx 5.292$.

YOUR TURN   Find the center and radius of the following circles and draw their graphs.

a. $(x - 3)^2 + (y + 5)^2 = 4$ **b.** $(x + 1)^2 + (y - 1)^2 = 18$

OBJECTIVE ④ **Find the equation of a circle with a given center and radius.** We can also use the standard form of a circle to write equations of circles.

EXAMPLE 5   Write the equations of the following circles in standard form.

a.   Center at (−4, 2) and radius of 8.

Solution Since the center is at (−4, 2), *h* = −4 and *k* = 2.

$$(x - h)^2 + (y - k)^2 = r^2$$
$$(x - (-4))^2 + (y - 2)^2 = 8^2 \quad \text{Substitute for } h, k, \text{ and } r.$$
$$(x + 4)^2 + (y - 2)^2 = 64 \quad \text{Simplify}$$

b.   Center at (0, 0) and radius of 2.

Solution Since the center is at (0, 0), the standard form of the equation is $x^2 + y^2 = r^2$.

$$x^2 + y^2 = 2^2 \quad \text{Substitute for } r.$$
$$x^2 + y^2 = 4$$

YOUR TURN   Write the equations of the following circles.

a. Center at (4, −2) and radius of 5.   **b.** Center at (0, 0) and radius of 9.

Learning Strategy

Note how the signs of the numbers in the parentheses of the general equation are the opposite of the signs of the center coordinates.

$$(x + 4)^2 + (y - 2)^2 = 8^2$$

Center: (−4, 2).

OBJECTIVE ⑤ **Graph circles of the form** $x^2 + y^2 + dx + ey + f = 0$. If the equation of a circle is not given in standard form, we complete the square to write the equation in the form $(x - h)^2 + (y - k)^2 = r^2$.

EXAMPLE 6   Find the center and radius of the circle whose equation is $x^2 + y^2 - 6x + 8y + 9 = 0$ and draw the graph.

Solution $x^2 + y^2 - 6x + 8y = -9$ Subtract 9 from both sides to isolate the variable terms.

$$(x^2 - 6x) + (y^2 + 8y) = -9 \quad \text{Group the terms in } x \text{ and } y.$$

$$(x^2 - 6x + 9) + (y^2 + 8y + 16) = -9 + 9 + 16 \quad \begin{array}{l}\text{Complete the square for } x \text{ and } y \text{ by} \\ \text{adding 9 and 16 to both sides of the} \\ \text{equation.}\end{array}$$

$$(x - 3)^2 + (y + 4)^2 = 16 \quad \text{Factor and simplify.}$$

The center is at (3, −4) and the radius is 4.

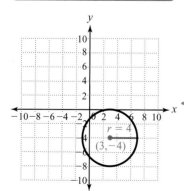

The graph of a circle fails the vertical line test, so the equation of a circle is not a function. To graph equations that are not functions, we solve for y and graph the two resulting functions on the same screen.

EXAMPLE Graph $(x - 2)^2 + (y + 1)^2 = 16$.

Solution We solve $(x - 2)^2 + (y + 1)^2 = 16$ for y.

$(x - 2)^2 + (y + 1)^2 = 16$

$(y + 1)^2 = 16 - (x - 2)^2$ Subtract $(x - 2)^2$ from both sides.

$y + 1 = \pm\sqrt{16 - (x - 2)^2}$ Apply the square root principle.

$y = -1 \pm \sqrt{16 - (x - 2)^2}$ Subtract 1 from both sides.

This equation defines two equations. On your graphing calculator define $Y_1 = -1 + \sqrt{16 - (x - 2)^2}$ and $Y_2 = -1 - \sqrt{16 - (x - 2)^2}$ and graph both in the window $[-8, 8]$ for x and $[-8, 8]$ for y. The resulting graph is labeled "**a**" below; note that the graph does not look like the graph of a circle. To make the graph look like a circle, select zoom from the GRAPH menu and then choose ZSQR. This results in the graph labeled "**b**" below.

a.

b.

ANSWER

Center: $(-1, 4)$, $r = 3$.

YOUR TURN

Find the center and radius of the circle whose equation is $x^2 + y^2 + 2x - 8y + 8 = 0$ and draw the graph.

13.1 EXERCISES

1. Is $x^2 + 2x - 3 + y = 0$ the equation of a circle or a parabola? Explain.

2. Find the coordinates of the vertex of the parabola $x = 3y^2 - 12y + 2$.

3. In which direction does the graph of the parabola in Exercise 2 open? Why?

4. Explain how the Pythagorean theorem is used to derive the distance formula.

5. Complete the following: The set of all points that are four units from the point whose coordinates are $(-3, 2)$ is a _____ with _____ and _____.

6. How is the distance formula related to the equation of a circle?

For Exercises 7–28, find the direction the parabola opens, the coordinates of the vertex, the equation of the axis of symmetry and draw the graph.

7. $y = (x - 1)^2 + 2$

8. $y = (x + 2)^2 - 3$

9. $y = x^2 - 4x + 1$

10. $y = x^2 + 2x - 3$

11. $x = (y + 2)^2 - 2$

12. $x = (y + 3)^2 + 2$

13. $x = -(y - 1)^2 + 3$

14. $x = -(y - 3)^2 + 3$

15. $x = 2(y + 2)^2 - 4$

16. $x = 3(y + 3)^2 - 4$

17. $x = -3(y + 2)^2 - 5$

18. $x = -2(y - 4)^2 + 1$

19. $x = y^2 + 4y + 3$

20. $x = y^2 - 2y - 8$

21. $x = -y^2 + 6y - 5$

22. $x = -y^2 - 4y - 3$

23. $x = 2y^2 + 8y + 3$

24. $x = 2y^2 - 4y + 1$

25. $x = 3y^2 - 6y + 3$

26. $x = 3y^2 - 12y + 9$

27. $x = -2y^2 + 4y + 5$

28. $x = -3y^2 - 6y - 2$

For Exercises 29–32, match the equation with the correct graph.

29. $x = (y + 3)^2 - 2$

a.

b.

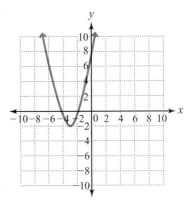

30. $y = (x + 3)^2 - 2$

c.

d.

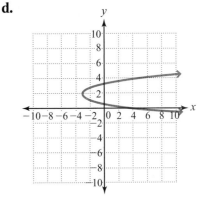

31. $y = 2x^2 - 8x + 5$

32. $x = 2y^2 - 8y + 5$

For Exercises 33–44, find the distance between the two points.

33. $(-4, 2)$ and $(-1, 6)$ **34.** $(5, -1)$ and $(1, 2)$ **35.** $(-8, -4)$ and $(-3, 8)$ **36.** $(-3, 2)$ and $(3, -6)$

37. $(-8, -10)$ and $(4, -5)$ **38.** $(4, -6)$ and $(10, 2)$ **39.** $(2, 4)$ and $(4, 8)$ **40.** $(-3, 2)$ and $(1, -4)$

41. $(-5, 2)$ and $(3, -2)$ **42.** $(3, -4)$ and $(7, 2)$ **43.** $(6, -2)$ and $(1, -5)$ **44.** $(3, -6)$ and $(-2, 1)$

For Exercises 45–48, the coordinates of the center of a circle and a point on the circle are given. Find the radius of the circle.

45. Center: $(4, 2)$, point on the circle: $(8, -1)$.

46. Center: $(-4, 6)$, point on the circle: $(2, -2)$.

47. Center: $(2, -6)$, point on the circle: $(10, -1)$.

48. Center: $(3, -4)$, point on the circle: $(-6, 8)$.

For Exercises 49–64, find the center and radius and draw the graph.

49. $(x - 2)^2 + (y - 1)^2 = 4$

50. $(x - 1)^2 + (y - 3)^2 = 25$

51. $(x + 3)^2 + (y + 2)^2 = 81$

52. $(x + 5)^2 + (y + 4)^2 = 36$

53. $(x - 5)^2 + (y + 3)^2 = 49$

54. $(x + 6)^2 + (y - 2)^2 = 1$

55. $(x + 1)^2 + (y - 3)^2 = 18$

56. $(x - 6)^2 + (y + 1)^2 = 12$

57. $(x + 4)^2 + (y + 2)^2 = 32$

58. $(x - 6)^2 + (y + 5)^2 = 8$

59. $x^2 + y^2 - 2x - 6y - 39 = 0$

60. $x^2 + y^2 + 8x - 6y + 16 = 0$

61. $x^2 + y^2 + 10x - 4y - 35 = 0$

62. $x^2 + y^2 + 12x + 10y + 60 = 0$

63. $x^2 + y^2 + 14x - 4y + 49 = 0$

64. $x^2 + y^2 + 8x - 10y + 16 = 0$

For Exercises 65–68, match the equation with the correct graph.

65. $(x - 2)^2 + (y + 3)^2 = 25$

a.

b.

66. $(x + 2)^2 + (y - 3)^2 = 25$

c.

67. $x^2 + y^2 + 8x + 2y - 8 = 0$

d.
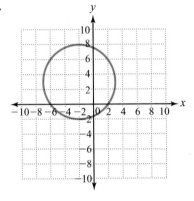

68. $x^2 + y^2 + 2x - 8y - 8 = 0$

 For Exercises 69–72, graph using a graphing calculator.

69. $x^2 + y^2 = 49$

70. $x^2 + y^2 = 36$

71. $(x - 2)^2 + (y + 3)^2 = 25$

72. $(x + 4)^2 + (y - 1)^2 = 9$

For Exercises 73–80, the center and radius of a circle are given. Write the equation of each circle in standard form.

73. Center: $(8, -1)$, $r = 6$

74. Center: $(-3, 2)$, $r = 6$

75. Center: $(-4, -3)$, $r = 5$

76. Center: $(-6, -5)$, $r = 2$

77. Center: $(6, -2)$, $r = \sqrt{14}$

78. Center: $(-3, -3)$, $r = \sqrt{26}$

79. Center: $(-5, 2)$, $r = 3\sqrt{5}$

80. Center: $(6, 2)$, $r = 2\sqrt{6}$

For Exercises 81–84, the coordinates of the center of a circle and a point on the circle are given. Write the equation of the circle in standard form.

81. Center: $(2, 4)$, point on the circle: $(5, 8)$.

82. Center: $(-4, 3)$, point on the circle: $(4, 9)$.

83. Center: $(2, 4)$, point on the circle: $(7, 16)$.

84. Center: $(-6, 8)$, point on the circle: $(-12, 0)$.

85. Write the equation of the set of all points that are a distance of 8 units from $(2, -5)$.

86. Write the equation of the set of all points that are a distance of 10 units from $(-3, -7)$.

87. If a rock is thrown vertically upward from the top of a building 112 feet high with an initial velocity of 96 feet per second, the height, h, above ground level after t seconds is given by $h = -16t^2 + 96t + 112$, where h is in feet and t is in seconds.

a. What is the maximum height the rock will reach?

b. How many seconds will it take the rock to reach its maximum height?

c. How many seconds will it take the rock to hit the ground?

88. The path of a shell fired from ground level is in the shape of the parabola $y = 4x - x^2$, where x and y are given in kilometers.

a. How high does the shell go?

b. How many seconds does it take the shell to reach its maximum height?

c. How far from its firing point does the shell land?

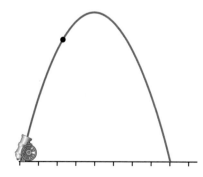

89. Arches in the shapes of parabolas are often used in construction. Find the equation of a parabolic arc that is 18 feet high at its highest point and 30 feet wide at the base, as illustrated in the following figure. Place the origin at the midpoint of the base.

18 ft.

30 ft.

90. The cross sections of satellite dishes are in the shape of parabolas. Find the equation of a dish that is 6 feet across and 1 foot deep if the vertex is at the origin and the parabola is opening upward.

6 ft.

1 ft.

91. The percent of deaths by age per million miles driven can be approximated by the equation $y = .0038x^2 - .3475x + 8.316$, where x is the age and y is the percent. Find the percent of deaths per million miles for drivers 17 years old. (*Source:* National Highway Traffic Safety Administration)

92. The number of drivers involved in fatal accidents for a given blood alcohol content (BAC) can be approximated by $y = -8862.5x^2 + 26622.6x + 332$, where x is the BAC and y is the number of drivers involved in fatal accidents. Find the number of drivers involved in fatal accidents who had a BAC of .20. (*Source:* National Highway Traffic Safety Administration)

93. Bill and Don are fishing in the Gulf of Mexico in separate boats that are equipped with radios with a range of 20 miles. If we put Bill's radio at the origin of a coordinate system, what is the equation of all possible locations of Don's boat where the radios would be at their maximum range.

94. A toy plane is attached to a string pinned to the ceiling so that the plane flies in a circle. If the string is 4 feet long, write an equation that describes the path of the plane if the pin is at the origin.

95. A Ferris wheel has a diameter of 200 feet and the bottom of the Ferris wheel is 10 feet above the ground. Find the equation of the wheel if the origin is placed on the ground directly below the center of the wheel, as illustrated.

96. The Fermilab Tunnel houses the world's largest superconducting synchrotron. A cross section of the tunnel is a circle with radius of 1000 meters. Find the equation of a cross section of the tunnel if the center of the circle is at the center of the tunnel.

y

200 ft.

10 ft.

x

97. If a satellite is placed in an orbit of 230 kilometers above the Earth, what is the equation of the path of the satellite if the origin is placed at the center of the Earth (the radius of the Earth is approximately 6370 kilometers)?

98. The minute hand of Big Ben is 14 feet long. What is the equation of the circle swept out by the tip of the hand as it makes one complete revolution?

PUZZLE PROBLEM

Using only a pencil, a rough circle and its center can both be drawn without the point of the pencil losing contact with the paper resulting in a picture like the one shown. Explain how.

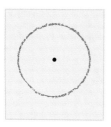

REVIEW EXERCISES

1. Solve $\dfrac{x^2}{16} = 1$.

2. Find the x- and y-intercepts of $9x^2 + 16y^2 = 144$.

3. Find the x- and y-intercepts of $25x^2 - 9y^2 = 225$.

4. Graph $y = \dfrac{2}{3}x$ and $y = -\dfrac{2}{3}x$ on the same set of axes.

5. Following is a coordinate system with the point $(2, 3)$ plotted. Plot the points that are 4 units to the left and right of $(2, 3)$ and give their coordinates.

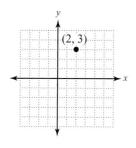

6. Factor $25x^2 - 9y^2$.

13.2 Ellipses and Hyperbolas

OBJECTIVES

1. Graph ellipses.
2. Graph hyperbolas.

OBJECTIVE 1 **Graph ellipses.** Suppose you drive two nails in a board and tie a string to the two nails. Now take a pencil, pull the string taut, and draw a figure around the two nails.

The figure you've drawn is called an **ellipse**. The locations of the two nails are the *focal points*.

> **DEFINITION**
>
> ***Ellipse:*** The set of all points the sum of whose distance from two fixed points is constant.

Ellipses occur in many situations. The orbits of the planets about the Sun are elliptical with the Sun at one focal point. The orbits of satellites about the Earth are also elliptical. The cams of compound bows are elliptical, which allows a decrease in the amount of effort required to hold the bow at full draw.

In the definition of an ellipse, the two fixed points are the *foci* (plural of *focus*) and the point halfway between the foci is the *center*. The following figure shows the graph of an ellipse with foci at $(c, 0)$ and $(-c, 0)$, x-intercepts at $(a, 0)$ and $(-a, 0)$, and y-intercepts at $(0, b)$ and $(0, -b)$. Consequently, the center is at the origin, $(0, 0)$. It can be shown that $c^2 = a^2 - b^2$ for ellipses in which $a > b$.

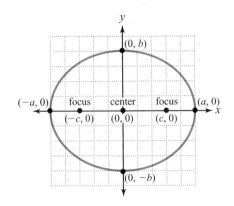

Using the distance formula, it can be shown that an ellipse with these characteristics has the following equation.

RULE   STANDARD FORM OF AN EQUATION OF AN ELLIPSE

The equation of an ellipse with x-intercepts (a, 0) and (−a, 0) and y-intercepts (0, b) and (0, −b) is

$$\frac{x^2}{a^2} + \frac{y^2}{b^2} = 1$$

EXAMPLE 1   Graph the following ellipses.

a. $\dfrac{x^2}{16} + \dfrac{y^2}{9} = 1$

Solution The equation can be rewritten as $\dfrac{x^2}{4^2} + \dfrac{y^2}{3^2} = 1$, so $a = 4$ and $b = 3$.

Consequently, the x-intercepts are (4, 0) and (−4, 0) and the y-intercepts are (0, 3) and (0, −3).

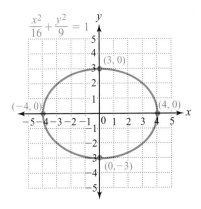

b. $25x^2 + 9y^2 = 225$

Solution We first need to write the equation in standard form: $\dfrac{x^2}{a^2} + \dfrac{y^2}{b^2} = 1.$

$$\frac{25x^2}{225} + \frac{9y^2}{225} = \frac{225}{225} \qquad \text{Divide both sides by 225.}$$

$$\frac{x^2}{9} + \frac{y^2}{25} = 1 \qquad \text{Simplify both sides.}$$

We now see that this is an equation of an ellipse with $a = 3$ and $b = 5$, so the x-intercepts are (3, 0) and (−3, 0) and the y-intercepts are (0, 5) and (0, −5).

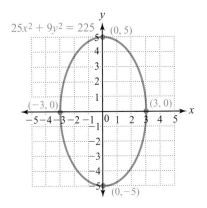

If the center of an ellipse is not at the origin, it can be shown that the equation has the following form.

GENERAL EQUATION FOR AN ELLIPSE

$$\frac{(x-h)^2}{a^2} + \frac{(y-k)^2}{b^2} = 1,$$ where (h, k) is the center. The ellipse passes through two points that are a units to the right and left of the center, and two points that are b units above and below the center.

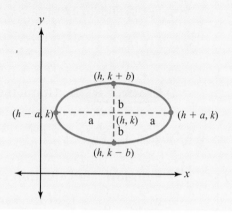

EXAMPLE 2 Graph $\dfrac{(x-2)^2}{36} + \dfrac{(y+3)^2}{16} = 1.$

Solution Since $h = 2$, and $k = -3$, the center of the ellipse is $(2, -3)$. Also, we see that $a = 6$, which means the ellipse passes through two points 6 units to the right and left of $(2, -3)$. These points are $(8, -3)$ and $(-4, -3)$. Since $b = 4$, the ellipse passes through two points 4 units above and below the center. These points are $(2, 1)$ and $(2, -7)$.

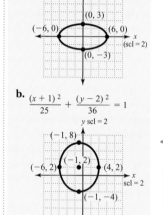

a. $\dfrac{x^2}{36} + \dfrac{y^2}{9} = 1$

b. $\dfrac{(x+1)^2}{25} + \dfrac{(y-2)^2}{36} = 1$

YOUR TURN **Graph the following ellipses.**

a. $\dfrac{x^2}{36} + \dfrac{y^2}{9} = 1$

b. $\dfrac{(x+1)^2}{25} + \dfrac{(y-2)^2}{36} = 1$

OBJECTIVE 2 Graph hyperbolas. The last conic is the **hyperbola**. Applications of hyperbolas include the LORAN tracking system and the orbits of some comets.

DEFINITION

Hyperbola: The set of all points the difference of whose distances from two fixed points remains constant.

Just as with the ellipse, the two fixed points are the *foci* and the point halfway between the foci is the *center*.

EQUATIONS OF HYPERBOLAS IN STANDARD FORM

The hyperbola with center (0, 0) and x-intercepts at $(a, 0)$ and $(-a, 0)$ and foci at $(c, 0)$ and $(-c, 0)$ has the equation $\dfrac{x^2}{a^2} - \dfrac{y^2}{b^2} = 1$.

The hyperbola with center (0, 0) and y-intercepts at $(0, b)$ and $(0, -b)$ and foci $(0, c)$ and $(0, -c)$ has the equation $\dfrac{y^2}{b^2} - \dfrac{x^2}{a^2} = 1$.

The dashed lines that intersect at the center of a hyperbola are *asymptotes* and are not a part of the graph, but are used as an aid in graphing. An asymptote is line that the graph approaches, but does not cross, as the graph goes away from the origin. The rectangle whose vertices are (a, b), $(-a, b)$, $(a, -b)$, and $(-a, -b)$ is called the *fundamental rectangle* and the asymptotes are the extended diagonals of the fundamental rectangle, as in the following illustration.

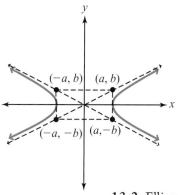

PROCEDURE GRAPHING A HYPERBOLA

1. Find the intercepts. If the x^2 term is positive, the x-intercepts are $(a, 0)$ and $(-a, 0)$ and there are no y-intercepts. If the y^2 term is positive, the y-intercepts are $(0, b)$ and $(0, -b)$ and there are no x-intercepts.
2. Draw the fundamental rectangle. The vertices are (a, b), $(-a, b)$, $(a, -b)$, and $(-a, -b)$.
3. Draw the asymptotes, which are the extended diagonals of the fundamental rectangle.
4. Sketch the graph. Each branch of the graph passes through an intercept and approaches the asymptotes the farther the graph goes from the origin.

EXAMPLE 3   Draw the graph of the following hyperbolas.

a. $\dfrac{x^2}{16} - \dfrac{y^2}{9} = 1$

Solution

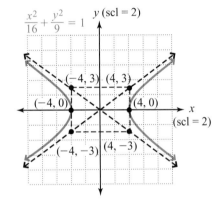

1. This equation can be written as $\dfrac{x^2}{4^2} - \dfrac{y^2}{3^2} = 1$, so $a = 4$ and $b = 3$. Since the x term is positive, the graph has x-intercepts at $(4, 0)$ and $(-4, 0)$.
2. The fundamental rectangle has vertices at $(4, 3)$, $(4, -3)$, $(-4, 3)$ and $(-4, -3)$.
3. Draw the asymptotes.
4. Sketch the graph so it passes through the x-intercepts and then approaches the asymptotes.

b. $\dfrac{y^2}{4} - \dfrac{x^2}{9} = 1$

Solution

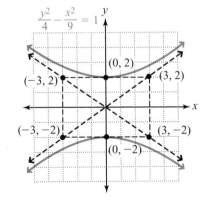

1. This equation can be written as $\dfrac{y^2}{2^2} - \dfrac{x^2}{3^2} = 1$, so $b = 2$ and $a = 3$. Since the y term is positive, the graph has y-intercepts at $(0, 2)$ and $(0, -2)$.
2. The fundamental rectangle has vertices at $(3, 2)$, $(3, -2)$, $(-3, 2)$ and $(-3, -2)$.
3. Draw the asymptotes.
4. Sketch the graph so it passes through the y-intercepts and then approaches the asymptotes.

YOUR TURN   Graph the following hyperbola.

$$\frac{x^2}{25} - \frac{y^2}{16} = 1$$

Graph the ellipse $\dfrac{x^2}{9} + \dfrac{y^2}{16} = 1$. We solve for y just as when graphing a circle.

$$\frac{x^2}{9} + \frac{y^2}{16} = 1$$

$$\frac{x^2}{16} = 1 - \frac{x^2}{9} \qquad \text{Subtract } \frac{x^2}{9} \text{ from both sides.}$$

$$y^2 = 16\left(1 - \frac{x^2}{9}\right) \qquad \text{Multiply both sides by 16.}$$

$$y = \pm 4\sqrt{1 - \frac{x^2}{9}} \qquad \text{Apply the square root principle.}$$

Define $\mathrm{Y}_1 = 4\sqrt{1 - \dfrac{x^2}{9}}$ and $\mathrm{Y}_2 = -4\sqrt{1 - \dfrac{x^2}{9}}$ and graph in a window $[-5, 5]$ for x and $[-5, 5]$ for y and square the window.

Following is a summary of the conic sections.

Standard Forms

Parabola $y = a(x - h)^2 + k$

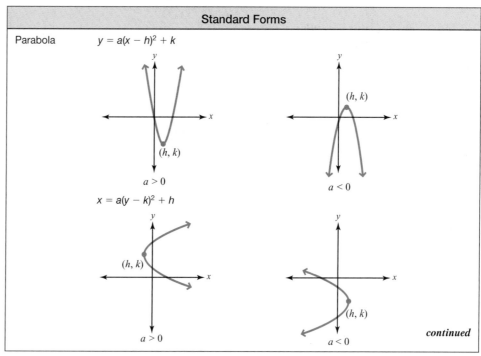

$x = a(y - k)^2 + h$

continued

Standard Forms continued		
Circle	$(x - h)^2 + (y - k)^2 = r^2$	

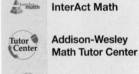

| Ellipse | $\dfrac{x^2}{a^2} + \dfrac{y^2}{b^2} = 1$ | |

| Hyperbola | $\dfrac{x^2}{a^2} - \dfrac{y^2}{b^2} = 1$ | $\dfrac{y^2}{b^2} - \dfrac{x^2}{a^2} = 1$ |

13.2 EXERCISES

For Extra Help

 InterAct Math

Tutor Center Addison-Wesley Math Tutor Center

Video 19/DVT 7

MyMathLab

 MathXL

1. What is the definition of an ellipse?

2. In the definitions of the ellipse and hyperbola, two fixed points are mentioned. What are these two points called?

3. How can you tell the equation of an ellipse from the equation of a hyperbola?

4. How do you determine whether a hyperbola opens up and down or left and right?

5. Which one conic has an equation that has either x^2 or y^2, but not both?

6. What do the graphs of $\dfrac{x^2}{9} - \dfrac{y^2}{16} = 1$ and $\dfrac{x^2}{9} + \dfrac{y^2}{16} = 1$ have in common?

For Exercises 7–18, sketch the graph of each equation and label the relevant points.

7. $\dfrac{x^2}{25} + \dfrac{y^2}{4} = 1$

8. $\dfrac{x^2}{81} + \dfrac{y^2}{64} = 1$

9. $\dfrac{x^2}{36} + y^2 = 1$

10. $\dfrac{x^2}{9} + \dfrac{y^2}{16} = 1$

11. $4x^2 + 9y^2 = 36$

12. $25x^2 + 4y^2 = 100$

13. $36x^2 + 4y^2 = 144$

14. $x^2 + 9y^2 = 36$

15. $\dfrac{(x-1)^2}{49} + \dfrac{(y+3)^2}{25} = 1$

16. $\dfrac{(x+3)^2}{25} + \dfrac{(y-2)^2}{64} = 1$

17. $\dfrac{(x-4)^2}{4} + \dfrac{(y+3)^2}{36} = 1$

18. $\dfrac{(x+5)^2}{9} + \dfrac{(y+3)^2}{25} = 1$

19. $\dfrac{x^2}{25} + \dfrac{y^2}{4} = 1$

20. $\dfrac{y^2}{12} + \dfrac{x^2}{8} = 1$

21. $\dfrac{(y-2)^2}{36} + \dfrac{(x+4)^2}{25} = 1$

22. $\dfrac{(x+1)^2}{16} + \dfrac{(y+2)^2}{4} = 1$

23. If the x-intercepts of an ellipse are $(-3, 0)$ and $(3, 0)$ and the y-intercepts are $(0, 4)$ and $(0, -4)$, what is the equation of the ellipse?

24. If the x-intercepts of an ellipse are $(-6, 0)$ and $(6, 0)$ and the y-intercepts are $(0, 3)$ and $(0, -3)$, what is the equation of the ellipse?

For Exercises 25–32, sketch the graph of each equation and label relevant points.

25. $\dfrac{x^2}{9} - \dfrac{y^2}{4} = 1$

26. $\dfrac{x^2}{16} - \dfrac{y^2}{25} = 1$

27. $\dfrac{y^2}{36} - \dfrac{x^2}{9} = 1$

28. $\dfrac{y^2}{4} - \dfrac{x^2}{25} = 1$

29. $9x^2 - y^2 = 36$

30. $x^2 - 4y^2 = 16$

31. $16y^2 - 4x^2 = 64$

32. $9y^2 - 25x^2 = 225$

For Exercises 33–34, graph using a graphing calculator.

33. $\dfrac{x^2}{36} - \dfrac{y^2}{4} = 1$

34. $\dfrac{y^2}{9} - \dfrac{x^2}{25} = 1$

35. If a hyperbola opens left and right and the vertices of the fundamental rectangle are $(3, 2)$, $(3, -2)$, $(-3, 2)$ and $(-3, -2)$, what is the equation of the hyperbola?

36. If a hyperbola opens up and down and the vertices of the fundamental rectangle are $(5, 3)$, $(5, -3)$, $(-5, 3)$ and $(-5, -3)$, what is the equation of the hyperbola?

For Exercises 37–40, match the equation with the graph.

37. $\dfrac{x^2}{9} + \dfrac{y^2}{25} = 1$

a.

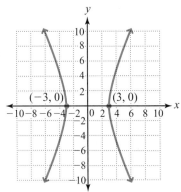

38. $\dfrac{x^2}{9} - \dfrac{y^2}{25} = 1$

b.

c.

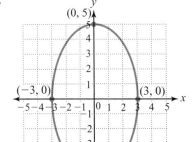

39. $\dfrac{x^2}{25} + \dfrac{y^2}{9} = 1$

d.

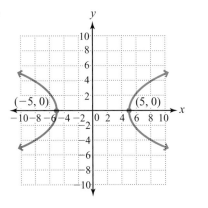

40. $\dfrac{x^2}{25} - \dfrac{y^2}{9} = 1$

For Exercises 41–44, determine whether the graph of the equation is a circle, parabola, ellipse, or hyperbola. Do not draw the graph.

41. $9x^2 + 16y^2 = 144$

42. $16y^2 - 9x^2 = 144$

43. $x^2 + y^2 - 6x + 8y - 75 = 0$

44. $2x^2 - 12x + 23 - y = 0$

For Exercises 45–58, determine whether the graph of the equation is a circle, parabola, ellipse, or hyperbola. Sketch the graph of each.

45. $x^2 + y^2 = 81$

46. $y = 2(x + 1)^2 + 3$

47. $\dfrac{x^2}{36} + \dfrac{y^2}{16} = 1$

48. $(x - 2)^2 + (y + 2)^2 = 49$

49. $x = (y + 3)^2 - 4$

50. $\dfrac{y^2}{49} - \dfrac{x^2}{4} = 1$

51. $\dfrac{x^2}{4} - \dfrac{y^2}{25} = 1$

52. $\dfrac{x^2}{36} + \dfrac{y^2}{81} = 1$

53. $y = 2x^2 + 8x + 6$

54. $x = y^2 + 6$

55. $\dfrac{x^2}{16} - \dfrac{y^2}{16} = 1$

56. $x = -2y^2 + 4y - 3$

57. $\dfrac{(x+3)^2}{9} + \dfrac{(y+2)^2}{36} = 1$

58. $\dfrac{(x-1)^2}{4} + (y+3)^2 = 1$

For Exercises 59–66, solve.

59. A bridge over a waterway has an arch in the form of half an ellipse. The equation of the ellipse is $400x^2 + 256y^2 = 102{,}400$.

   a. A sailboat, the top of whose mast is 18 feet above the water, is approaching the arch. Will the mast clear the bridge? Why or why not?

   b. How wide is the base of the arch?

60. A highway passes beneath an overpass that is in the shape of half an ellipse. The overpass is 15 feet high at the center and 40 feet wide at the base.

   a. What is the equation of the ellipse?

   b. A truck that is 10 feet wide and carrying a load that is 14 feet high is approaching the bridge. If the truck goes down the middle of the road, will the load clear the bridge? Why or why not?

61. The comet Epoch has an orbit that is in the shape of an ellipse with the Sun at one of the foci. The equation is approximately $\dfrac{x^2}{3.6^2} + \dfrac{y^2}{2.88^2} = 1$, where x and y are in astronomical units (an astronomical unit is 93,000,000 miles). Sketch the graph of the comet Epoch. (*Source: Orbital Motion,* A. E. Roy, Institute of Physics Publishing, London.)

62. The planet Pluto has an orbit that is that is in the shape of an ellipse with the Sun at one of the foci. The equation is approximately $\dfrac{x^2}{39.4^2} + \dfrac{y^2}{38.2^2} = 1$, where x and y are measured in astronomical units. Sketch the graph of the orbit of Pluto. (*Hint:* Make each unit on the x- and y-axes equal to 10.) (*Source: Orbital Motion,* A. E. Roy, Institute of Physics Publishing, London.)

63. Compound bows have elliptical cams that decrease the amount of effort required to hold the bow at full draw. If a cam on a bow is 4 inches from top to bottom and 3 inches across, what is the equation of the ellipse if the origin is at the center of the cam?

64. One of the most popular exercise machines is an elliptical trainer in which your foot moves in an elliptical path. On a typical machine, the length of the stride is about 19 inches and the height varies from 3 to 5 inches depending on the settings. Write the equation for the path that your foot takes if the total length of the stride is 19 inches and the total height is 4 inches. (*Source:* Precor National Headquarters.)

65. If a source of light or sound is placed at one focal point of an elliptic reflector, the light or sound is reflected through the other focal point. This principle is used in a lithotripter that uses sound waves to crush kidney stones by placing a source of sound at one focal point and the kidney stone at the other. If the elliptic reflector is based on the ellipse $\dfrac{x^2}{25} + \dfrac{y^2}{9} = 1$, how many units from the center should the kidney stone be placed? (*Hint:* $c^2 = a^2 - b^2$.)

66. The same principle used in Exercise 65 is also used in whispering rooms. The room is in the shape of an elliptic reflector where one person speaks at one focal point and the other places his/her ear at the other and can hear sounds as faint as a whisper. If a whispering room is based on the ellipse $\dfrac{x^2}{169} + \dfrac{y^2}{144} = 1$, how many units from the center of the ellipse would the two people stand? (*Hint:* $c^2 = a^2 - b^2$.)

Collaborative Exercises

The Elliptical Table Cloth

Recall that an ellipse can be drawn by fixing the ends of a string to the foci and then tracing out the ellipse. By considering the following figures we can determine an expression for the length of the string and also a relationship between a, b, and c.

1. Use the figure to write a formula for the length of the string.

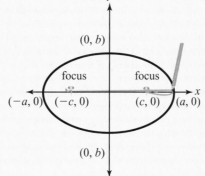

2. In the following figure, notice that the string forms an isosceles triangle and the y-axis splits that triangle into two identical right triangles.

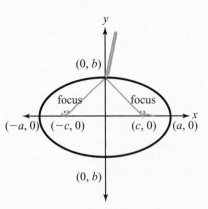

a. Find the length of the hypotenuse of each of those right triangles.

b. What expression describes the length of the string?

c. Use the Pythagorean theorem to write a formula relating a, b, and c.

d. Solve the formula for c.

Suppose we are to make an elliptical tablecloth for an elliptical table that is 76 inches long and 58 inches wide. The tablecloth is to drape over the edge of the table all the way around plus we need an additional inch for the hem. We have a large rectangular piece of cloth to make the tablecloth. To trace the ellipse on the cloth, we need to know the string length and the location of the foci.

3. Find the dimensions of the elliptical tablecloth taking into account the amount it needs to drape and also the hem.

4. How long must the string be in order to trace the ellipse?

5. How far from the center are the foci located?

PUZZLE PROBLEM

Suppose an ellipse is drawn using the method described on p. xxx. Assuming the ellipse is centered at the origin, if the nails are 8 inches apart and the string is 10 inches in length, what are the x- and y-intercepts?

REVIEW EXERCISES

1. Solve the system $\begin{cases} 2x + y = 1 \\ 3x + 4y = -6 \end{cases}$ using the substitution method.

2. Solve the system $\begin{cases} 2x + 3y = 6 \\ 3x - 4y = -25 \end{cases}$ using the elimination method.

3. Solve the system $\begin{cases} x + y = 3 \\ 2x + y = 4 \end{cases}$ using the graphical method.

4. Solve $3x^2 + y = 6$ for y.

For Exercises 5 and 6, solve.

5. $4x^2 = 36$

6. $3x^2 + 10x - 8 = 0$

13.3 Nonlinear Systems of Equations

OBJECTIVES

1 Solve nonlinear systems of equations using substitution.

2 Solve nonlinear systems of equations using elimination.

If an equation has a variable with an exponent of degree 2 or more such as equations of conic sections, then the equation is **nonlinear**. A **nonlinear system** of equations contains at least one equation that is nonlinear.

    When solving a system of nonlinear equations, it is helpful to consider the graphs of the equations of the system in order to determine the possible number of solutions. For example, a system whose graph is a parabola and a line can have 0, 1, or 2 solutions, as shown by the following figures.

CONNECTION

In Chapter 9, we learned that the number of solutions for a system of linear equations depended on the relative positions of the graphs of each of the equations. We see a similar relationship between the graphs and the number of solutions here.

No points of intersection:
No solutions.

One point of intersection:
One solution.

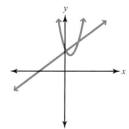

Two points of intersection:
Two solutions.

OBJECTIVE **1** **Solve nonlinear systems of equations using substitution.** Nonlinear systems can be solved by the substitution or elimination method. When one equation is linear, the substitution method is usually preferred.

EXAMPLE 1 Solve the following systems using the substitution method.

a. $\begin{cases} y = 2(x + 2)^2 - 2 \\ 2x + y = -2 \end{cases}$ **Note:** The graphs are a parabola and a line, so there will be 0, 1, or 2 solutions.

Solution It is easier to solve the linear equation for one of its variables and substitute into the nonlinear equation. Since the coefficient of y is 1, we will solve $2x + y = -2$ for y.

$y = -2x - 2$	Subtract $-2x$ from both sides.
$-2x - 2 = 2(x + 2)^2 - 2$	Substitute $-2x - 2$ for y in $y = 2(x + 2)^2 - 2$.
$-2x - 2 = 2(x^2 + 4x + 4) - 2$	Since the equation is quadratic, we need to write it in the form $ax^2 + bx + c = 0$. First, we simplify the right side.
$-2x - 2 = 2x^2 + 8x + 6$	
$0 = 2x^2 + 10x + 8$	Add $2x$ and 2 to both sides.
$0 = x^2 + 5x + 4$	Divide both sides by 2.
$0 = (x + 1)(x + 4)$	Factor.
$0 = x + 1$ or $0 = x + 4$	Use the zero-factor theorem.
$-1 = x$ or $-4 = x$	Solve each equation.

To find y, substitute -1 and -4 for x in $y = -2x - 2$.

$$y = -2\,(-\mathbf{1}) - 2 \qquad\qquad y = -2\,(-\mathbf{4}) - 2$$
$$y = 2 - 2 \qquad\qquad\qquad y = 8 - 2$$
$$y = 0 \qquad\qquad\qquad\qquad y = 6$$

The solutions are $(-1, 0)$ and $(-4, 6)$, as verified by the following graph.

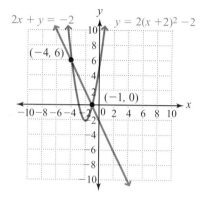

Note: A system containing a root function and a circle, as in Example 1b, can have 0, 1, or 2 solutions, as shown below.

b. $\begin{cases} y = \sqrt{x + 2} \\ x^2 + y^2 = 8 \end{cases}$

Solution Since $y = \sqrt{x + 2}$ is already solved for y, substitute $\sqrt{x + 2}$ for y in $x^2 + y^2 = 8$.

$x^2 + (\sqrt{x + 2})^2 = 8$	Substitute $\sqrt{x + 2}$ for y in $x^2 + y^2 = 8$.
$x^2 + x + 2 = 8$	Square $\sqrt{x + 2}$.
$x^2 + x - 6 = 0$	Since the equation is now quadratic, we subtract 8 from both sides to get the form $ax^2 + bx + c = 0$.
$(x + 3)(x - 2) = 0$	Factor.
$x + 3 = 0 \quad$ or $\quad x - 2 = 0$	Use the zero-factor theorem.
$x = -3 \quad$ or $\qquad x = 2$	Solve each equation.

To find y, substitute -3 and 2 for x in $y = \sqrt{x + 2}$.

$$y = \sqrt{-\mathbf{3} + 2} \qquad\qquad y = \sqrt{\mathbf{2} + 2}$$
$$y = \sqrt{-1} \qquad\qquad\qquad y = \sqrt{4}$$
$$\qquad\qquad\qquad\qquad\qquad y = 2$$

Since $\sqrt{-1}$ is imaginary, the only real solution is $(2, 2)$, as verified by the following graph.

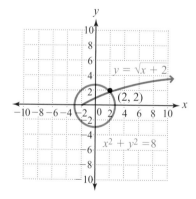

YOUR TURN Solve the following systems using the substitution method.

a. $\begin{cases} 2x - y = -1 \\ x^2 + y^2 = 1 \end{cases}$

b. $\begin{cases} xy = 3 \\ 4x - y = 1 \end{cases}$

OBJECTIVE 2 Solve nonlinear systems of equations using elimination. If neither equation contains a radical expression or both equations contain the same powers of the variables, the elimination method can be used.

EXAMPLE 2 Solve $\begin{cases} 9x^2 - 4y^2 = 20 \\ x^2 + y^2 = 8 \end{cases}$ using the elimination method.

Note: If the graphs are a hyperbola and a circle both centered at the origin as in Example 2, there will be 0, 2, or 4 solutions, as shown.

Solution $\begin{cases} 9x^2 - 4y^2 = 20 & \text{(Equation 1)} \\ x^2 + y^2 = 8 & \text{(Equation 2)} \end{cases}$

To eliminate y^2, multiply equation 2 by 4 and add the equations.

$$\begin{array}{ll} 9x^2 - 4y^2 = 20 \\ x^2 + \ y^2 = 8 \end{array} \xrightarrow[\text{Multiply by 4}]{}$$

$$\begin{array}{ll} 9x^2 - 4y^2 = 20 \\ \underline{4x^2 + 4y^2 = 32} \\ 13x^2 \qquad = 52 & \text{Add the equations.} \\ \quad x^2 = 4 & \text{Divide both sides by 13.} \\ \quad x = \pm 2 & \text{Find the square roots of 4.} \end{array}$$

Four Solutions

To find y, substitute 2 and -2 for x in one of the original equations. We will use $x^2 + y^2 = 8$.

$$\begin{array}{ll} 2^2 + y^2 = 8 \\ 4 + y^2 = 8 \\ y^2 = 4 \\ y = \pm 2 \end{array} \qquad \begin{array}{ll} (-2)^2 + y^2 = 8 \\ 4 + y^2 = 8 \\ y^2 = 4 \\ y = \pm 2 \end{array}$$

Two Solutions

Therefore, $(2, 2)$ and $(2, -2)$ are solutions.

Therefore, $(-2, 2)$ and $(-2, -2)$ are solutions.

This system has four solutions: $(2, 2)$, $(2, -2)$, $(-2, 2)$, and $(-2, -2)$, as verified by the graph to the right.

No Solutions

A system of equations can be solved using a graphing calculator and the "isect" feature.

Solve $\begin{cases} x^2 + y^2 = 10 \\ 3x + y = 6 \end{cases}$ *using a graphing calculator.*

Solve each equation for y

$x^2 + y^2 = 10$ $3x + y = 6$

$y^2 = 10 - x^2$ $y = 6 - 3x$

$y = \pm\sqrt{10 - x^2}$

Enter $Y_1 = \sqrt{10 - x^2}$, $Y_2 = -\sqrt{10 - x^2}$, *and* $Y_3 = 6 - 3x$

Graph in a window $[-10, 10]$ *for x and* $[-10, 10]$ *for y and find the intersection of* y_1 *and* y_3 *and also* y_2 *and* y_3.

Intersection
X=1   Y=3

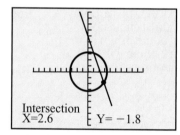

Intersection
X=2.6   Y= -1.8

YOUR TURN

ANSWERS

(3, 2), (3, −2), (−3, 2),
(−3, −2)

Solve $\begin{cases} 4x^2 + 9y^2 = 72 \\ x^2 + y^2 = 13 \end{cases}$ using the elimination method.

13.3 EXERCISES

1. How many real solutions are possible for a system of equations whose graphs are a circle and a parabola? Draw a figure illustrating each case.

2. How many real solutions are possible for a system of equations whose graphs are a line and a hyperbola? Draw a figure illustrating each case.

3. Which method (substitution or elimination) would you use to solve the system $\begin{cases} 2x + y = 8 \\ 4x^2 + 3y^2 = 24 \end{cases}$? Why? (Do not attempt to solve the system.)

4. Which method (substitution or elimination) would you use to solve the system $\begin{cases} 3x^2 - 4y^2 = 12 \\ 2x^2 + 3y^2 = 24 \end{cases}$? Why? (Do not attempt to solve the system.)

5. Without solving the system, what is the number of possible solutions of the system $\begin{cases} 3x - y = 6 \\ 4x^2 + 9y^2 = 36 \end{cases}$?

6. Is it possible for a system that consists of an ellipse and a hyperbola (both centered at the origin) to have three solutions? Why or why not?

For Exercises 7–42, solve each system of equations.

7. $\begin{cases} y = 2x^2 \\ 2x + y = -4 \end{cases}$

8. $\begin{cases} x^2 + 2y = 1 \\ 2x + y = 2 \end{cases}$

9. $\begin{cases} y = x^2 + 4x + 4 \\ 3x - y = -6 \end{cases}$

10. $\begin{cases} y = 6x - x^2 \\ 2x - y = -3 \end{cases}$

11. $\begin{cases} x^2 + y^2 = 25 \\ x - y = -1 \end{cases}$

12. $\begin{cases} x^2 + y^2 = 25 \\ x - 7y = -25 \end{cases}$

13. $\begin{cases} x^2 + y^2 = 13 \\ 2x + y = 7 \end{cases}$

14. $\begin{cases} x^2 + y^2 = 10 \\ 3x + y = 6 \end{cases}$

15. $\begin{cases} x^2 + 2y^2 = 4 \\ x + y = 5 \end{cases}$

16. $\begin{cases} 3x^2 + y^2 = 9 \\ 2x + y = 11 \end{cases}$

17. $\begin{cases} y = 2x^2 + 3 \\ 2x - y = -3 \end{cases}$

18. $\begin{cases} y = -3x^2 - 2 \\ 3x + y = -2 \end{cases}$

19. $\begin{cases} y = (x - 3)^2 + 2 \\ y = -(x - 2)^2 + 3 \end{cases}$

20. $\begin{cases} y = 2(x + 4)^2 + 2 \\ y = -2(x + 3)^2 + 4 \end{cases}$

21. $\begin{cases} x^2 + y^2 = 20 \\ x^2 - y^2 = 12 \end{cases}$

22. $\begin{cases} x^2 + y^2 = 48 \\ x^2 - y^2 = 24 \end{cases}$

23. $\begin{cases} 9x^2 + 4y^2 = 145 \\ x^2 + y^2 = 25 \end{cases}$

24. $\begin{cases} 4x^2 + 9y^2 = 72 \\ x^2 + y^2 = 13 \end{cases}$

25. $\begin{cases} 4x^2 - y^2 = 15 \\ x^2 + y^2 = 5 \end{cases}$

26. $\begin{cases} 9x^2 - 4y^2 = 32 \\ x^2 + y^2 = 5 \end{cases}$

27. $\begin{cases} x^2 + 3y^2 = 36 \\ x = y^2 - 6 \end{cases}$

28. $\begin{cases} 4x^2 + y^2 = 16 \\ y = x^2 - 4 \end{cases}$

29. $\begin{cases} 9x^2 - 16y^2 = 144 \\ 4x^2 + 9y^2 = 36 \end{cases}$

30. $\begin{cases} 9x^2 + 4y^2 = 36 \\ x - y = -4 \end{cases}$

31. $\begin{cases} y = x^2 \\ x^2 + y^2 = 20 \end{cases}$

32. $\begin{cases} 16x^2 + y^2 = 128 \\ y = 2x^2 \end{cases}$

33. $\begin{cases} 25x^2 - 16y^2 = 400 \\ x^2 + 4y^2 = 16 \end{cases}$

34. $\begin{cases} 9y^2 - 25x^2 = 225 \\ 4y^2 + 25x^2 = 100 \end{cases}$

35. $\begin{cases} xy = 2 \\ 4x^2 + y^2 = 8 \end{cases}$

36. $\begin{cases} y = x^2 + 2 \\ 2x^2 - y^2 = -7 \end{cases}$

37. $\begin{cases} y = x^2 - 2x + 3 \\ y = -x^2 + 6x - 7 \end{cases}$

38. $\begin{cases} y = \dfrac{1}{3}x^2 - 2 \\ 3x^2 + 9y^2 = 36 \end{cases}$

39. $\begin{cases} 4x^2 + 5y^2 = 36 \\ 4x^2 - 3y^2 = 4 \end{cases}$

40. $\begin{cases} 4x^2 + 7y^2 = 64 \\ 4x^2 - 3y^2 = 24 \end{cases}$

41. $\begin{cases} x = -y^2 + 2 \\ x^2 - 3y^2 = 4 \end{cases}$

42. $\begin{cases} x = -y^2 + 2 \\ 9x^2 - 45y^2 = 36 \end{cases}$

43. Give the equations of a system of equations that consists of a circle and a line for which there is no solution. Include the graphs.

44. Give the equations of a system of equations that consists of a circle and a hyperbola for which there are two solutions. Include the graphs.

45. The sum of the squares of two integers is 34 and the difference of their squares is 16. Find the integers.

46. The difference of the squares of two integers is 32 and their product is 12. Find the integers.

47. A computer keyboard has an area of 144 square inches and a perimeter of 52 inches. Find the length and width.

48. A rectangular living room has an area of 48 square meters and a perimeter of 28 meters. Find the length and width.

49. If p is in dollars and x is in hundreds of units, the demand function for a certain style of chair is given by

$$p = -3x^2 + 120$$

and the supply function is given by

$$p = 11x + 28$$

The *market equilibrium* occurs when the number produced is equal to the number demanded. Find the number of chairs and the price per chair when market equilibrium is reached.

50. If y is in dollars and x is the number of cell phones manufactured (in thousands), a cell phone manufacturer has determined that the cost y to manufacture x cell phones is given by $y = 5x^2 + 30x + 50$ and the revenue from the sales is given by $y = 13x^2$. The *break-even point* is the point (x, y) for which the cost equals the revenue. Find the number of units necessary to break even by solving the system.

51. Exercise 7

52. Exercise 17

53. Exercise 19

54. Exercise 37

REVIEW EXERCISES

For Exercises 1–4, graph.

1. $y \geq 3$

2. $y < 2x + 3$

3. $2x + 3y < -6$

4. $x \geq -2$

For Exercises 5 and 6, determine if the ordered pair is a solution for $x + 3y \leq 6$.

5. $(0, 0)$

6. $(3, 4)$

13.4 Nonlinear Inequalities and Systems of Inequalities

OBJECTIVES

1 Graph nonlinear inequalities.

2 Solve systems of nonlinear inequalities.

OBJECTIVE 1 Graph nonlinear inequalities. In Section 4.6, we graphed linear inequalities like $x + 2y > 6$ by first graphing the line corresponding to $x + 2y = 6$ and then shading the solution which is the region on one side of that boundary line. Recall that we used a dashed line for $<$ and $>$ and a solid line for $\le$ and $\ge$. We determined which side to shade by using a test point on one side of the boundary.

We use a similar procedure to graph nonlinear inequalities such as $x^2 + y^2 \le 25$. The boundary is the graph of $x^2 + y^2 = 25$, which is a circle with center at $(0,0)$ and radius of 5, as shown in the following figure. The solution set of $x^2 + y^2 \le 25$ must be either the circle (boundary) and all points outside it or the circle and all points inside it. To determine which, choose a test point like $(0, 0)$ and substitute into the inequality. This gives $0^2 + 0^2 \le 25$. Since $0 \le 25$ is true, the solution set is the circle and all points inside it.

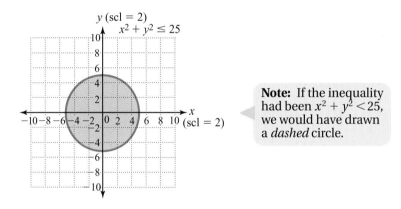

Note: If the inequality had been $x^2 + y^2 < 25$, we would have drawn a *dashed* circle.

This suggests the following procedure for graphing nonlinear inequalities.

PROCEDURE GRAPHING NONLINEAR INEQUALITIES

1. Graph the related equation. If the inequality symbol is $\le$ or $\ge$, draw the graph as a solid curve. If the inequality symbol is $>$ or $<$, use a dashed curve.
2. The graph will divide the coordinate plane into at least two regions. Test an ordered pair from each region by substituting it into the inequality. If the ordered pair satisfies the inequality, then shade the region containing the test point.
3. The solution set is the union of all the regions that solve the inequality.

EXAMPLE 1 Graph each of the following nonlinear inequalities.

a. $y > 2(x - 1)^2 + 3$

Solution We first draw the graph of the related equation, $y = 2(x - 1)^2 + 3$.

Note: The parabola is dashed because the inequality is $>$. Also, notice that the graph divides the coordinate plane into two regions: region 1 above the boundary and region 2 below.

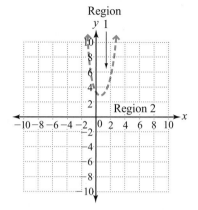

We now test a point from each region.

Note: To make computations easier, choose points with 0 for at least one of the coordinates.

Region 1: We choose (0, 6).

$6 > 2(0 - 1)^2 + 3$ Substitute.
$6 > 2(-1)^2 + 3$ Simplify
$6 > 5$
True: Region 1 is in solution set.

Region 2: We choose (0, 0).

$0 > 2(0 - 1)^2 + 3$ Substitute.
$0 > 2(-1)^2 + 3$ Simplify.
$0 > 5$
False: Region 2 is not in the the solution set.

Since points in only region 1 solve the inequality, we shade it.

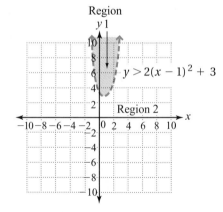

b. $\dfrac{x^2}{9} - \dfrac{y^2}{25} \leq 1$

Solution First graph the related equation, which is $\dfrac{x^2}{9} - \dfrac{y^2}{25} = 1$.

Note: The hyperbola is solid because the inequality is $\leq$. Also, notice that the hyperbola divides the coordinate plane into three regions.

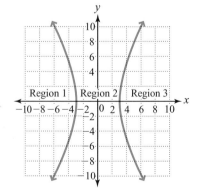

We test a point from each region in the inequality.

Region 1: We choose $(-5, 0)$.

$$\frac{(-5)^2}{9} - \frac{0^2}{25} \leq 1 \quad \text{Substitute.}$$

$$\frac{25}{9} \leq 1 \quad \text{Simplify.}$$

False, so region 1 is not in the solution set.

Region 2: We choose $(0, 0)$.

$$\frac{0^2}{9} - \frac{0^2}{25} \leq 1 \quad \text{Substitute.}$$

$$0 \leq 1 \quad \text{Simplify.}$$

True, so region 2 is in the solution set.

Region 3: We choose $(5, 0)$.

$$\frac{(5)^2}{9} - \frac{0^2}{25} \leq 1 \quad \text{Substitute}$$

$$\frac{25}{9} \leq 1 \quad \text{Simplify.}$$

False, so region 3 is not in the solution set.

Since region 2 is the only region that solves the inequality, we shade it.

ANSWERS

a.

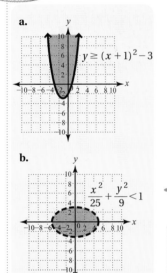
$y \geq (x + 1)^2 - 3$

b.

$\frac{x^2}{25} + \frac{y^2}{9} < 1$

YOUR TURN   Graph each of the following nonlinear inequalities.

a. $y \geq (x + 1)^2 - 3$ **b.** $\frac{x^2}{25} + \frac{y^2}{9} < 1$

OBJECTIVE 2 Solve systems of nonlinear inequalities.   In Section 9.7, we solved systems of linear inequalities by graphing each inequality. The solution set was the region where the graphs overlapped. We use a similar procedure for systems of nonlinear inequalities.

EXAMPLE 2   Graph the solution set of the following systems of inequalities.

a. $\begin{cases} y \geq x^2 - 4 \\ 2x - y < 2 \end{cases}$

Solution   We begin by graphing $y \geq x^2 - 4$ and $2x - y < 2$.

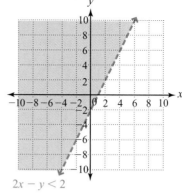

$2x - y < 2$

If we place both graphs on the same coordinate system, their intersection (purple shading) is the solution set for the system.

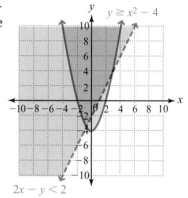

$y \geq x^2 - 4$

$2x - y < 2$

b. $\begin{cases} x^2 + y^2 \leq 49 \\ \dfrac{x^2}{16} - \dfrac{y^2}{9} < 1 \\ y \geq 2x + 2 \end{cases}$

Solution Graph each inequality on the same coordinate system.

The graph of $x^2 + y^2 \leq 49$ is a circle and its interior. The graph of $\dfrac{x^2}{16} - \dfrac{y^2}{9} < 1$ is the region between the branches of the hyperbola with the curve dashed. The graph of $y \geq 2x + 2$ is a solid line and the region above the line. The solution set for the system is the region where the three graphs overlap.

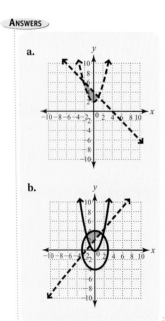
YOUR TURN Graph the solution set of the following systems of inequalities.

a. $\begin{cases} y > x^2 + 2 \\ x + y < 4 \end{cases}$

b. $\begin{cases} \dfrac{x^2}{9} + \dfrac{y^2}{16} \leq 1 \\ y \geq x^2 \\ -x + y > 2 \end{cases}$

1. Do the points that lie on the boundary curve solve the inequality $9x^2 + 4y^2 > 36$? Why or why not? Would the graph be drawn solid or dashed? Why?

2. Discuss how graphing linear inequalities like $x + 2y \geq 4$ is similar to graphing a nonlinear inequality like $y \geq x^2 + 3$.

3. Draw the graph of a system of nonlinear inequalities for which there is no solution and explain why there is no solution.

4. Describe the procedure you would use to graph $\dfrac{x^2}{9} - \dfrac{y^2}{16} > 1$.

For Exercises 5–24, graph the inequality.

5. $x^2 + y^2 \geq 4$

6. $x^2 + y^2 \leq 9$

7. $y > x^2$

8. $y < -x^2$

9. $y < (x - 1)^2 + 2$

10. $y > 2(x + 2)^2 - 3$

11. $\dfrac{x^2}{16} + \dfrac{y^2}{9} \leq 1$

12. $\dfrac{x^2}{4} + \dfrac{y^2}{9} \geq 1$

13. $\dfrac{x^2}{25} - \dfrac{y^2}{4} > 1$

14. $\dfrac{x^2}{36} - \dfrac{y^2}{16} < 1$

15. $x^2 + y^2 < 16$

16. $x^2 + y^2 > 25$

17. $y \geq -2(x-3)^2 + 3$

18. $y < -(x+3)^2 - 2$

19. $\dfrac{x^2}{49} + \dfrac{y^2}{25} > 1$

20. $\dfrac{x^2}{16} + \dfrac{y^2}{9} \leq 1$

21. $\dfrac{y^2}{25} - \dfrac{x^2}{4} \leq 1$

22. $\dfrac{y^2}{36} - \dfrac{x^2}{25} > 1$

23. $y < x^2 + 4x - 5$

24. $y > x^2 - 6x - 6$

For Exercises 25–50, graph the solution set of each system of inequalities.

25. $\begin{cases} y \geq x^2 \\ x + y \leq 3 \end{cases}$

26. $\begin{cases} y < -x^2 \\ 2x - y < 4 \end{cases}$

27. $\begin{cases} x - 2y < -4 \\ x^2 + y^2 < 16 \end{cases}$

28. $\begin{cases} 3x + 2y \geq -6 \\ x^2 + y^2 \leq 25 \end{cases}$

29. $\begin{cases} x^2 + y^2 \geq 9 \\ x^2 + y^2 \leq 25 \end{cases}$

30. $\begin{cases} x^2 + y^2 > 4 \\ x^2 + y^2 > 9 \end{cases}$

31. $\begin{cases} y > x^2 + 1 \\ 2x + y < 3 \end{cases}$

32. $\begin{cases} 3x - y \geq 2 \\ y \leq -x^2 + 2 \end{cases}$

33. $\begin{cases} y < -x^2 + 3 \\ y > x^2 - 2 \end{cases}$

34. $\begin{cases} y > -x^2 + 2 \\ y < x^2 + 5 \end{cases}$

35. $\begin{cases} \dfrac{x^2}{25} + \dfrac{y^2}{9} \leq 1 \\ x^2 + y^2 \geq 4 \end{cases}$

36. $\begin{cases} \dfrac{x^2}{9} + \dfrac{y^2}{4} \leq 1 \\ x^2 + y^2 \leq 4 \end{cases}$

37. $\begin{cases} \dfrac{x^2}{25} + \dfrac{y^2}{9} < 1 \\ y > x^2 + 1 \end{cases}$

38. $\begin{cases} \dfrac{x^2}{9} + \dfrac{y^2}{4} < 1 \\ y < -x^2 + 2 \end{cases}$

39. $\begin{cases} \dfrac{x^2}{9} - \dfrac{y^2}{4} \le 1 \\ \dfrac{x^2}{25} + \dfrac{y^2}{9} \le 1 \end{cases}$

40. $\begin{cases} \dfrac{x^2}{16} - \dfrac{y^2}{9} \ge 1 \\ \dfrac{x^2}{36} + \dfrac{y^2}{16} \le 1 \end{cases}$

41. $\begin{cases} \dfrac{x^2}{4} - \dfrac{y^2}{4} > 1 \\ y > 2 \end{cases}$

42. $\begin{cases} \dfrac{x^2}{9} - \dfrac{y^2}{9} > 1 \\ x > 3 \end{cases}$

43. $\begin{cases} 3x + 2y \le 6 \\ x - y > -3 \\ x + 6y \ge 2 \end{cases}$

44. $\begin{cases} 2x + 3y < 6 \\ x - 2y > -4 \\ x + 5y \ge -4 \end{cases}$

45. $\begin{cases} \dfrac{x^2}{16} + \dfrac{y^2}{4} \le 1 \\ x^2 + y^2 \le 9 \\ y \le x \end{cases}$

46. $\begin{cases} \dfrac{x^2}{9} + \dfrac{y^2}{16} < 1 \\ x^2 + y^2 < 9 \\ y > x + 1 \end{cases}$

47. $\begin{cases} \dfrac{x^2}{49} - \dfrac{y^2}{16} \le 1 \\ \dfrac{x^2}{64} + \dfrac{y^2}{36} \le 1 \\ 2x - y \le -3 \end{cases}$

48. $\begin{cases} \dfrac{x^2}{9} + \dfrac{y^2}{49} \le 1 \\ y \ge x^2 + 3 \\ x + y \le 6 \end{cases}$

49. $\begin{cases} y < 2x^2 + 8 \\ 2x + y > 3 \\ x - y < 4 \end{cases}$

50. $\begin{cases} y < x^2 + 2 \\ 2x + y < 4 \\ 2x - y > -5 \end{cases}$

REVIEW EXERCISES

For Exercises 1 and 2, use $f(x) = x^2 + 3x + 2$.

1. Find $f(3)$.

2. Find $f(a)$.

For Exercises 3–5, simplify.

3. $(3x + 2) - (2x - 1)$

4. $(3x + 2)(2x - 1)$

5. $(x - 2)^2 + 3(x - 2) + 1$

6. Given $a^2 + 4$, substitute $2x + 1$ for a and simplify.

Defined Terms

Section 13.1
  Conic (p. 1068)
  Conic Section (p. 1068)
  Center of circle (p. 1072)
  Circle (p. 1072)
  Radius (p. 1072)

Section 13.2
  Ellipse (p. 1084)
  Hyperbola (p. 1086)

Section 13.3
  Nonlinear equation (p. 1099)
  Nonlinear system (p. 1099)

Procedures, Rules, and Key Examples

Procedures/Rules	Key Example(s)

Section 13.1 The Parabola and Circle

Graph parabolas in the form $x = a(y - k)^2 + h$.

If $x = a(y - k)^2 + h$, the parabola opens to the right if $a > 0$ and to the left if $a < 0$. The vertex is at (h, k) and the equation of the axis of symmetry is $y = k$.

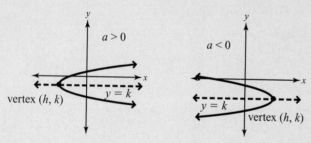

Example 1: For the equation $x = -2(y - 3)^2 - 2$, determine whether the graph opens left or right, find the vertex and the axis of symmetry, and draw the graph.

Solution: Since $a = -2$ and $-2 < 0$, the graph opens to the left. The vertex is $(-2, 3)$ and the equation of the axis of symmetry is $y = 3$.

Choose values for y near the axis of symmetry and plot the graph.

x	y
−10	1
−4	2
−4	4
−10	5

continued

Procedures/Rules	Key Example(s)

Example 2: For $x = y^2 - 4y - 5$, determine whether the graph opens left or right, find the vertex and axis of symmetry, and draw the graph.

Solution: Opens to the right since $a = 1$ and $1 > 0$. To find the vertex, axis of symmetry, and to graph, complete the square.

$$x = y^2 - 4y - 5$$

$$x + 5 = y^2 - 4y \qquad \text{Add 5 to both sides.}$$

$$x + 5 + 4 = y^2 - 4y + 4 \qquad \text{Add 4 to both sides.}$$

$$x + 9 = (y - 2)^2 \qquad \text{Simplify both sides.}$$

$$x = (y - 2)^2 - 9 \qquad \text{Subtract 9 from both sides.}$$

The vertex is $(-9, 2)$ and the equation of the axis of symmetry is $y = 2$.

To graph, choose values of y near the axis of symmetry and plot the graph.

x	y
−5	0
−8	1
−8	3
−5	4

To find the distance between two points whose coordinates are (x_1, y_1) and (x_2, y_2), use the formula
$$d = \sqrt{(x_2 - x_1)^2 + (y_2 - y_1)^2}.$$

Example 3: Find the distance between the points whose coordinates are $(-4, 3)$ and $(2, -1)$.

Solution: Let $(x_1, y_1) = (-4, 3)$ and $(x_2, y_2) = (2, -1)$.

$$d = \sqrt{(x_2 - x_1)^2 + (y_2 - y_1)^2}$$

$$d = \sqrt{(2 - (-4))^2 + (-1 - 3)^2}$$

Substitute.

$$d = \sqrt{52} \qquad \text{Simplify.}$$

$$d = \sqrt{4 \cdot 13} \qquad \text{Rewrite 52 as } 4 \cdot 13.$$

$$d = 2\sqrt{13}$$

continued

Procedures/Rules	Key Example(s)

Section 13.1 (continued)

Graph circles in the form $(x - h)^2 + (y - k)^2 = r^2$.

The graph of an equation of the form $(x - h)^2 + (y - k)^2 = r^2$ is a circle with center (h, k) and radius r.

Example 4: Find the center and radius of the circle whose equation is $(x - 2)^2 + (y + 4)^2 = 25$ and draw the graph.

Solution: Rewrite the equation as
$$(x - 2)^2 + (y - (-4))^2 = 5^2.$$
Since $h = 2$ and $k = -4$, the center is $(2, -4)$ and $r = 5$.

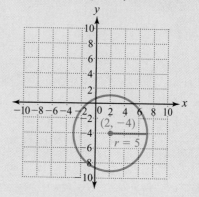

To write the equation of a circle, substitute the appropriate values into $(x - h)^2 + (y - k)^2 = r^2$.

Example 5: Write the equation of the circle whose center is $(5, -6)$ and whose radius is 8.

Solution: $(x - h)^2 + (y - k)^2 = r^2$.
Substitute for h, k, and r.
$$(x - 5)^2 + (y - (-6))^2 = 8^2$$
$$(x - 5)^2 + (y + 6)^2 = 64 \quad \text{Simplify.}$$

continued

Procedures/Rules	Key Example(s)

Procedures/Rules

Graph circles in the form
$x^2 + y^2 + dx + ey + f = 0$.

To graph circles in the form
$x^2 + y^2 + dx + ey + f = 0$, complete the square in x and y to put the equation in the form $(x - h)^2 + (y - k)^2 = r^2$.

Key Example(s)

Example 6: Find the center and radius of the circle whose equation is
$x^2 + y^2 + 8x - 2y + 8 = 0$.
Solution: Complete the square in x and y.

$x^2 + y^2 + 8x - 2y + 8 = 0$

$x^2 + 8x + y^2 - 2y = -8$ Group the terms in x and y and subtract 8.

$x^2 + 8x + 16 + y^2 - 2y + 1 = -8 + 16 + 1$
     Complete the square.

$(x + 4)^2 + (y - 1)^2 = 9$
     Factor on the left and simplify on the right.

The center is $(-4, 1)$ and the radius is 3.

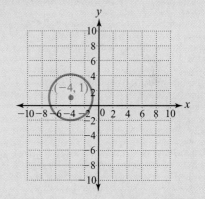

continued

Procedures/Rules	Key Example(s)

Section 13.2 Ellipses and Hyperbolas

Graph of an ellipse: The graph of an equation of the form $\dfrac{x^2}{a^2} + \dfrac{y^2}{b^2} = 1$ is an ellipse with x-intercepts at $(a, 0)$ and $(-a, 0)$ and y-intercepts at $(0, b)$ and $(0, -b)$.

Example 1: Graph $\dfrac{x^2}{49} + \dfrac{y^2}{36} = 1$.

Solution: The equation can be rewritten as $\dfrac{x^2}{7^2} + \dfrac{y^2}{6^2} = 1$, so $a = 7$ and $b = 6$. Consequently the x-intercepts are $(7, 0)$ and $(-7, 0)$ and the y-intercepts are $(0, 6)$ and $(0, -6)$.

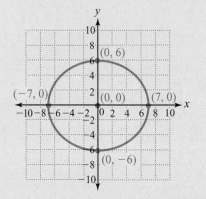

If the center of an ellipse is (h, k), the equation is of the form
$$\dfrac{(x - h)^2}{a^2} + \dfrac{(y - k)^2}{b^2} = 1.$$

To graph, plot the center and go a units left and right of the center and b units above and below the center to find additional points.

Example 2: Graph $\dfrac{(x + 2)^2}{9} + \dfrac{(y - 1)^2}{25} = 1$.

Solution: The equation can be rewritten as
$$\dfrac{(x - (-2))^2}{3^2} + \dfrac{(y - 1)^2}{5^2} = 1.$$
So $h = -2$, $k = 1$, $a = 3$, and $b = 5$. The center is $(-2, 1)$. To find other points on the ellipse, go 3 units left and right of the center and 5 units up and down from the center.

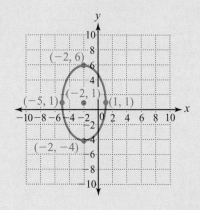

continued

1120   **CHAPTER 13** Conic Sections

Procedures/Rules	Key Example(s)

Procedures/Rules

Graphs of hyperbolas: The graph of $\dfrac{x^2}{a^2} - \dfrac{y^2}{b^2} = 1$ is a hyperbola with x-intercepts at $(a, 0)$ and $(-a, 0)$ and no y-intercepts.

The graph of $\dfrac{y^2}{b^2} - \dfrac{x^2}{a^2} = 1$ is a hyperbola with y-intercepts at $(0, b)$ and $(0, -b)$ and no x-intercepts.

To graph a hyperbola:
1. Find the intercepts which are the vertices.
2. Draw the fundamental rectangle whose vertices are (a, b), $(-a, b)$, $(a, -b)$, and $(-a, -b)$.
3. Draw the asymptotes, which are the extended diagonals of the fundamental rectangle.
4. Sketch the graph. Each branch will pass through an intercept and approach the asymptotes as the graph goes away from the origin.

Key Example(s)

Example 3: Graph $\dfrac{x^2}{36} - \dfrac{y^2}{16} = 1$.

Solution: The equation can be rewritten as $\dfrac{x^2}{6^2} - \dfrac{y^2}{4^2} = 1$, so $a = 6$ and $b = 4$. Since the x term is positive, the x-intercepts are $(6, 0)$ and $(-6, 0)$ and there are no y-intercepts. The fundamental rectangle has vertices $(6, 4)$, $(6, -4)$, $(-6, 4)$, and $(-6, -4)$.

Section 13.3 Nonlinear Systems of Equations

If a nonlinear system has a linear equation, the substitution method is usually preferred. Solve the linear equation for one of its variables and substitute for that variable in the other equation.

Example 1: Solve the system
$$\begin{cases} y = (x - 3)^2 + 2 \\ 2x - y = -4 \end{cases}$$
by substitution.
(**Note:** This is a parabola and a line.)

Solution:

$2x - y = -4$	Solve for this equation for y.
$y = 2x + 4$	
$2x + 4 = (x - 3)^2 + 2$	Substitute $2x + 4$ for y in the other equation $y = (x - 3)^2 + 2$.
$2x + 4 = x^2 - 6x + 11$	Simplify the right side.
$0 = x^2 - 8x + 7$	Write in $ax^2 + bx + c$ form.
$0 = (x - 7)(x - 1)$	Factor.
$x - 7 = 0$ or $x - 1 = 0$	Set each factor equal to 0.
$x = 7$ or $x = 1$	Solve each equation.

To find y, substitute 7 and 1 into $y = 2x + 4$.

$$\begin{array}{ll} y = 2(7) + 4 & y = 2(1) + 4 \\ y = 18 & y = 6 \end{array}$$

Solutions: $(7, 18)$, $(1, 6)$. *continued*

Procedures/Rules	Key Example(s)

Section 13.3 (continued)

If neither equation is linear, the elimination method is often preferred.

Example 2: Solve the system
$$\begin{cases} 4x^2 + 5y^2 = 36 & \text{(Equation 1)} \\ x^2 + y^2 = 8 & \text{(Equation 2)} \end{cases}$$
by elimination.

(**Note:** This is an ellipse and a circle.)

Solution: To eliminate x^2, multiply Equation 2 by -4 and add the equations.

$$4x^2 + 5y^2 = 36 \qquad\qquad 4x^2 + 5y^2 = 36$$
$$x^2 + y^2 = 8 \xrightarrow[\text{Multiply by } -4]{} \quad \underline{-4x^2 - 4y^2 = -32}$$
$$y^2 = 4$$
$$y = \pm 2$$

Substitute 2 and -2 for y into either equation to find x.
Let's use $x^2 + y^2 = 8$.

$$x^2 + 2^2 = 8 \qquad x^2 + (-2)^2 = 8$$
$$x^2 + 4 = 8 \qquad x^2 + 4 = 8$$
$$x^2 = 4 \qquad\qquad x^2 = 4$$
$$x = \pm 2 \qquad\qquad x = \pm 2$$

Solutions: $(2, 2)$, $(2, -2)$, $(-2, 2)$, and $(-2, -2)$.

Section 13.4 Nonlinear Inequalities and Systems of Inequalities

Graphing nonlinear inequalities.

1. Graph the related equation. If the inequality symbol is $\leq$ or $\geq$, draw the graph as a solid curve. If the inequality symbol is $<$ or $>$, draw the graph as a dashed curve.

2. The graph divides the coordinate plane into at least two regions. Test an ordered pair from each region by substituting it into the inequality. If the ordered pair satisfies the inequality, then shade the region that contains it.

3. The solution set is the union of all the regions that solve the inequality.

Example 1: Graph $y \geq (x - 2)^2 - 1$.

Solution: First graph the related curve.

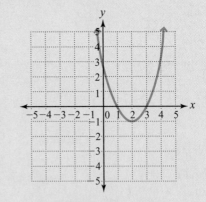

Note the graph is solid since the curve is part of the solution set. Choose $(0, 0)$ as a test point.

continued

Procedures/Rules	Key Example(s)

$0 \geq (0 - 2)^2 - 1$

$0 \geq 3$ which is false, so $(0, 0)$ is not in the solution set.

Shade the region that does not contain $(0, 0)$.

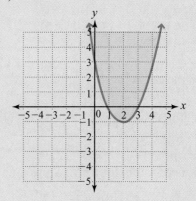

Graph systems of inequalities: To graph a system of inequalities, graph the solution sets of the individual inequalities. The solution set of the system is the intersection of the solution sets of the individual inequalities.

Example 2: Graph the solution set of

$$\begin{cases} \dfrac{x^2}{4} + \dfrac{y^2}{16} \leq 1 \\ x^2 + y^2 \geq 9 \end{cases}.$$

Solution: The graph of $\dfrac{x^2}{4} + \dfrac{y^2}{16} \leq 1$ is an ellipse and all points inside the ellipse. The graph of $x^2 + y^2 \geq 9$ is a circle and all points outside the circle. So, the solution set of the system is the set of all points inside the ellipse and outside the circle.

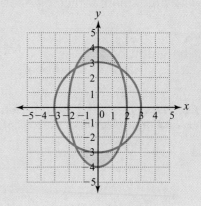

For Exercises 1–6, answer true or false.

1. The graph of $x = -3(y - 2)^2 + 5$ opens down.

2. The graph of $(x + 2)^2 + (y - 3)^2 = 36$ is a circle with center at $(-2, 3)$ and radius 6.

3. A parabola has asymptotes.

4. Choose any two points on an ellipse. The sum of their distances from the foci is the same.

5. It is possible for a circle and a parabola to intersect in three points.

6. The graph of the system $\begin{cases} x^2 + y^2 > 36 \\ 2x + y > 3 \end{cases}$ lies inside the circle and above the line.

For Exercises 7–10, fill in the blanks.

7. The graph of $\dfrac{x^2}{16} + \dfrac{y^2}{36} = 1$ is a(n) _____ with x-intercepts at _____.

8. The distance between the points whose coordinates are $(-3, 4)$ and $(3, -4)$ is _____.

9. The fundamental rectangle for the graph of $\dfrac{x^2}{25} - \dfrac{y^2}{16} = 1$ has vertices at _____.

10. The graph of $x^2 + y^2 + 8x - 2y + 13 = 0$ is a circle with center at _____ and radius of _____.

For Exercises 11–14, find the direction the parabola opens, the coordinates of the vertex, and the equation of the axis of symmetry. Draw the graph.

11. $y = 2(x - 3)^2 - 5$

12. $y = -2(x + 2)^2 + 3$

13. $x = -(y - 2)^2 + 4$ **14.** $x = 2(y + 3)^2 - 2$

For Exercises 15–18, find the direction the parabola opens, the coordinates of the vertex, and the equation of the axis of symmetry. Draw the graph.

15. $x = y^2 + 6y + 8$ **16.** $x = 2y^2 - 8y - 6$

17. $x = -y^2 - 2y + 3$ **18.** $x = -3y^2 - 12y - 9$

For Exercises 19 and 20, find the distance between the pairs of points.

19. $(-1, -2)$ and $(-5, 1)$ **20.** $(2, -5)$ and $(-2, 3)$

Answer the following.

21. The center of a circle is at $(6, -4)$ and the circle passes through $(-2, 2)$. What is the radius of the circle?

22. The center of a circle is at $(-6, 8)$ and the circle passes through $(3, -4)$. What is the equation of the circle?

For Exercises 23–26, find the center and radius. Draw the graph.

23. $(x - 3)^2 + (y + 2)^2 = 25$

24. $(x + 5)^2 + (y - 1)^2 = 4$

25. $x^2 + y^2 - 4x + 8y + 11 = 0$

26. $x^2 + y^2 + 10x + 2y + 22 = 0$

For Exercises 27–28, the center and radius of a circle are given. Write the equation of each circle in standard form.

27. Center: $(6, -8)$, $r = 9$

28. Center: $(-3, -5)$, $r = 10$

29. If a heavy object is thrown vertically upward with an initial velocity of 32 feet per second from the top of a building 128 feet high, its height above the ground after t seconds is given by $h = -16t^2 + 32t + 128$, where h is the height in feet and t is the time in seconds.

  a. What is the maximum height the object will reach?

  b. How many seconds will it take the object to reach its maximum height?

  c. How many seconds will it take for the object to strike the ground?

30. To lay out the border of a circular flower bed, sticks are tied to each end of a rope that is 20 feet long. One person holds one of the sticks stationary while another person uses the other stick to trace out a circular path while keeping the rope taut. If the center of the circle is at the stationary stick, what is the equation of the circle?

For Exercises 31–36, sketch the graph of each ellipse, and label relevant points.

31. $\dfrac{x^2}{49} + \dfrac{y^2}{25} = 1$

32. $\dfrac{x^2}{9} + \dfrac{y^2}{25} = 1$

33. $4x^2 + 9y^2 = 36$

34. $25x^2 + 9y^2 = 225$

35. $\dfrac{(x+2)^2}{16} + \dfrac{(y-3)^2}{4} = 1$

36. $\dfrac{(x-1)^2}{9} + \dfrac{(y+4)^2}{25} = 1$

For Exercises 37–40, sketch the graph of each hyperbola and label relevant points.

37. $\dfrac{x^2}{25} - \dfrac{y^2}{16} = 1$

38. $\dfrac{x^2}{36} - \dfrac{y^2}{9} = 1$

39. $y^2 - 9x^2 = 36$

40. $25y^2 - 9x^2 = 225$

41. A bridge over a canal is in the shape of half an ellipse. If the highest point of the bridge is 15 feet above the water and the base of the bridge is 50 feet across, what is the equation of the ellipse, half of which forms the bridge?

42. The cam of a compound bow is elliptical and is 4 inches long and 3.25 inches wide, as shown in the figure. What is the equation of the ellipse forming the shape of the cam?

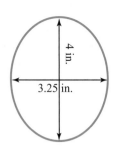

For Exercises 43–50, find the solution set of each system of equations.

43. $\begin{cases} y = 2x^2 - 3 \\ 2x - y = -1 \end{cases}$

44. $\begin{cases} x^2 + y^2 = 17 \\ x - y = 3 \end{cases}$

45. $\begin{cases} 25x^2 + 3y^2 = 100 \\ 2x - y = -3 \end{cases}$

46. $\begin{cases} x^2 + y^2 = 64 \\ x^2 - y^2 = 64 \end{cases}$

47. $\begin{cases} x^2 + y^2 = 25 \\ 25y^2 - 16x^2 = 256 \end{cases}$

48. $\begin{cases} x^2 + 5y^2 = 36 \\ 4x^2 - 7y^2 = 36 \end{cases}$

49. $\begin{cases} y = x^2 - 2 \\ 4x^2 + 5y^2 = 36 \end{cases}$

50. $\begin{cases} y = x^2 - 1 \\ 4y^2 - 5x^2 = 16 \end{cases}$

51. The sum of the squares of two integers is 89 and the difference of their squares is 39. Find the integers.

52. A rectangular rug has a perimeter of 36 feet and an area of 80 square feet. Find the length and width of the rug.

For Exercises 53 and 54, use a graphing calculator to verify the results of the exercise given.

53. Exercise 45.

54. Exercise 49.

For Exercises 55–58, draw the graph.

55. $x^2 + y^2 \leq 64$

56. $y < 2(x - 3)^2 + 4$

57. $\dfrac{y^2}{25} + \dfrac{x^2}{49} > 1$

58. $\dfrac{x^2}{25} - \dfrac{y^2}{36} < 1$

For Exercises 59– 64, graph the solution set of the system of inequalities.

59. $\begin{cases} y \geq x^2 - 3 \\ 2x + y < 2 \end{cases}$

60. $\begin{cases} x + 2y < 4 \\ x^2 + y^2 \leq 25 \end{cases}$

61. $\begin{cases} y > x^2 - 2 \\ y < -x^2 + 1 \end{cases}$

62. $\begin{cases} \dfrac{x^2}{9} + \dfrac{y^2}{25} \leq 1 \\ x^2 + y^2 \geq 9 \end{cases}$

63. $\begin{cases} \dfrac{y^2}{4} - \dfrac{x^2}{9} \leq 1 \\ \dfrac{x^2}{9} + \dfrac{y^2}{25} \leq 1 \end{cases}$

64. $\begin{cases} \dfrac{y^2}{4} + \dfrac{x^2}{16} \leq 1 \\ x^2 + y^2 \leq 9 \\ y \leq x + 1 \end{cases}$

For Exercises 1 and 2, find the direction the parabola opens, the coordinates of the vertex, the equation of the axis of symmetry, and then draw the graph.

1. $y = 2(x + 1)^2 - 4$

2. $x = -2(y - 3)^2 + 1$

3. Given $x = y^2 + 4y - 3$, find the direction the parabola opens, the coordinates of the vertex, and the equation of the axis of symmetry. Draw the graph.

4. Find the distance between the points whose coordinates are $(2, -3)$ and $(6, -5)$.

For Exercises 5 and 6, find the center, the radius, and then draw the graph.

5. $(x + 4)^2 + (y - 3)^2 = 36$

6. $x^2 + y^2 + 4x - 10y + 20 = 0$

7. Write the equation of the circle with center $(2, -4)$ and that passes through the point $(-4, 4)$.

For Exercises 8–11, sketch the graph of each equation and label relevant points.

8. $\dfrac{x^2}{36} + \dfrac{y^2}{16} = 1$

9. $4y^2 + 25x^2 = 100$

10. $\dfrac{x^2}{49} - \dfrac{y^2}{25} = 1$

11. $16x^2 - 64y^2 = 1024$

For Exercises 12–15, find the solution set of each system of equations.

12. $\begin{cases} y = (x + 1)^2 + 2 \\ 2x + y = 8 \end{cases}$

13. $\begin{cases} 3x - y = 4 \\ x^2 + y^2 = 34 \end{cases}$

14. $\begin{cases} x^2 + y^2 = 13 \\ 3x^2 + 4y^2 = 48 \end{cases}$

15. $\begin{cases} x^2 - 2y^2 = 1 \\ 4x^2 + 7y^2 = 64 \end{cases}$

For Exercises 16 and 17, draw the graph.

16. $y \le -2(x + 3)^2 + 2$

17. $\dfrac{x^2}{4} + \dfrac{y^2}{9} > 1$

18. Graph the solution set of the following system of inequalities.

$$\begin{cases} y \geq x^2 - 4 \\ \dfrac{x^2}{9} + \dfrac{y^2}{16} \leq 1 \end{cases}$$

19. An arch is in the shape of a parabola as shown in the figure. If we place the origin, O, as indicated, find the height of the arch and the distance across the base if the equation is $y = -\dfrac{1}{2}x^2 + 18$.

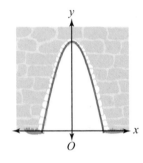

20. The path of a person's foot using an elliptical exercise machine on a particular setting is an ellipse. If the length of the stride is 19 inches and the height of the stride is 2 inches, what is the equation of the path of the foot?

2 in.

19 in.

For Exercises 1–6, answer true or false.

1. If $\dfrac{a}{b} = \dfrac{c}{d}$, then $ad = bc$ where $b \neq 0$ and $d \neq 0$.

2. A function must be one-to-one to have an inverse that is a function.

3. Logarithms are exponents.

4. The vertical line test can help determine if a graph is a function.

5. A positive base raised to a negative exponent simplifies to a negative number.

6. A circle has asymptotes.

For Exercises 7–10, fill in the blank.

7. The equation of a circle with center at (h, k) and radius of r is
_____.

8. $\sqrt{-1} = $ _____

9. The formula for finding the distance between two points (x_1, y_1) and (x_2, y_2)
is _____.

10. If $A = \begin{bmatrix} a_1 & b_1 \\ a_2 & b_2 \end{bmatrix}$, then $\det(A) = \begin{vmatrix} a_1 & b_1 \\ a_2 & b_2 \end{vmatrix} = $ _____.

For Exercises 11–16, simplify.

11. $6 - (-3) + 2^{-4}$

12. $\sqrt{-20}$

13. $7x \cdot \dfrac{8}{21x^2}$

14. $\dfrac{\dfrac{1}{x}}{3 + \dfrac{1}{x^2}}$

15. $\dfrac{x - y}{4y - 4x}$

16. $(5m^3n^{-2})^{-3}$

For Exercises 17 and 18, factor completely.

17. $k^4 - 81$

18. $ax + bx + ay + by$

For Exercises 19–28, solve.

19. $\begin{cases} 2x + y = 3 \\ 2x - y = 5 \end{cases}$

20. $4x^2 - 2x + 1 = 0$

21. $|2x + 1| = |x - 3|$

22. $\sqrt{x - 3} = 7$

23. $u^4 - 3u^2 - 4 = 0$

24. $9 + 24x^{-1} + 16x^{-2} = 0$

25. $9^x = 27$

26. $\begin{cases} x + y = 5 \\ x^2 + y^2 = 4 \end{cases}$

27. $\log_3 x = -2$

28. $\log_3(x + 1) - \log_3 x = 2$

For Exercises 29 and 30, solve the inequality, then: a. Graph the solution set.
b. Write the solution set in set-builder notation.
c. Write the solution set in interval notation.

29. $|3x + 1| > 4$

30. $5m^2 - 3m < 0$

31. Given $f(x) = 2x + 1$ and $g(x) = x^2 - 1$, find

   a. $(f + g)(x)$ **b.** $(f - g)(x)$ **c.** $(f \cdot g)(x)$

   d. $(f/g)(x)$ **e.** $(f \circ g)(x)$ **f.** $(g \circ f)(x)$

32. Determine whether $f(x) = \dfrac{x - 4}{5}$ and $g(x) = 5x + 4$ are inverses.

33. Write the equation of a line perpendicular to $3x + 4y = 8$ and passing through the point $(3, -8)$. Write the equation in slope-intercept form and in standard form.

34. Convert $5^3 = 125$ into logarithmic form.

35. Write as a sum or difference of multiples of logarithms: $\log_5 x^5 y$

36. Write the equation of a circle in standard form with a center at $(2, 4)$ passing through the point $(7, 16)$.

For exercises 37–41, graph.

37. $\begin{cases} y < -x + 2 \\ y \ge x - 4 \end{cases}$

38. $f(x) = \log_3 x$

39. $\dfrac{x^2}{4} + \dfrac{y^2}{9} = 1$

40. $\dfrac{x^2}{16} - \dfrac{y^2}{36} = 1$

41. $\begin{cases} y < x + 2 \\ x^2 + y^2 < 9 \end{cases}$

For Exercises 42–50, solve.

42. The percentage of infants regularly sleeping in a bed with parents or caregivers increased from 5.5% in 1993 to 12.8% in 2000. By how much did the percent increase? (*Source:* Time, January 27, 2003)

43. The volume of a cylinder can be found using the formula $V = \pi r^2 h$.

    a. Solve the formula for h.

    b. Suppose the volume of a cylinder is 15π cubic inches. Find its height.

44. During 2001, shrimp overtook tuna as the top-selling seafood in the United States. It is estimated that each person consumed a combined total of 6.3 pounds of these two seafoods and that each person ate $\frac{1}{2}$ pound more of shrimp than tuna. How many pounds of each were consumed? *(Source: Good Housekeeping, March 2003)*

45. Karen and Eric operate a landscaping company. Karen can mow, edge, and trim hedges on a quarter acre lot in 2 hours. Eric takes 1.5 hours to do the same work. How much time would it take them working together?

46. A total of $5000 is invested in three funds. The money market fund pays 5%, the income fund pays 6%, and the growth fund pays 3%. The total annual interest from the three accounts is $255 and the amount in the growth fund is $500 less than the amount invested in the money market account. Find the amount invested in each of the funds.

47. A bridge over a canal is in the shape of half an ellipse. If the highest point of the bridge is 20 feet above the water and the base of the bridge is 150 feet across, what is the equation of the ellipse, half of which forms the bridge?

48. What is the pH of a solution whose concentration of hydrogen ions is 4.23×10^{-8} moles per liter?

49. If $15,000 is deposited into an account paying 4% interest compounded monthly, how much will be in the account after 3 years?

50. The magnitude of an earthquake was defined in 1935 by Charles Richter as $M = \log\dfrac{I}{I_0}$. We let I be the intensity of the earthquake and measured by the amplitude of a seismograph reading taken 100 km from the epicenter of the earthquake and we let I_0 be the intensity of a "standard earthquake" whose amplitude is 10^{-4} cm. On January 26, 2001, an earthquake in Rann of Kutch, Gujarat was measured with an intensity of $10^{2.9}$. What was the magnitude of this earthquake?

Arithmetic Sequences and Series

OBJECTIVES

1. Find the terms of a sequence given the general term.
2. Define and write arithmetic sequences, find their common difference, and find a particular term.
3. Define and write series, find partial sums, and use summation notation
4. Write arithmetic series and find their sums.

The word **sequence** is used in mathematics much as it is in everyday life. For example, the classes that you attend on a given day occur in a particular order or sequence. As such, a sequence is an ordered list. Suppose we have a bacteria colony that has an initial population of 10,000 and increases at a rate of 10% each day.

On the second day, we would have
$10,000 + 0.10(10,000) = 10,000 + 1000 = 11,000$.
On the third day we have $11,000 + 0.10(11,000) = 11,000 + 1100 = 12,100$.
On the fourth day we have $12,100 + 0.10(12,100) = 12,100 + 1210 = 13,310$.
On the fifth day we have $13,310 + 0.10(13,310) = 13,310 + 1331 = 14,641$.

The number of bacteria present after each day forms a sequence which we can summarize as follows.

Days	1	2	3	4	5
Number of Bacteria	10,000	11,000	12,100	13,310	14,641

Based on this, we make the following definition.

DEFINITION

Sequence: An ordered list whose domain is a subset of the natural numbers.

The input values or *domain* of our bacteria example is the numbers of the days (1, 2, ... 5). Each number in a sequence (output values) is called a *term*, so the terms of the sequence are the numbers of bacteria (10,000, 11,000, . . ., 14,641). Sequences can be finite or infinite depending on the number of terms. Our bacteria example is a **finite sequence** because it has a finite number of terms. The sequence 2, 4, 6, 8, 10, . . . is an **infinite sequence** because it has an infinite number of terms.

Finite sequence: A function with a domain that is the set of natural numbers from 1 to n.

Infinite sequence: A function with a domain that is the set of natural numbers.

OBJECTIVE **1** **Find the terms of a sequence when given the general term.** Since a sequence is a function, we could describe sequences with functional notation. Instead, we use a different notation that emphasizes the fact that the domain is a subset of the natural numbers. We think of the terms of a sequence as $a_1, a_2, a_3, \ldots, a_n$, where the subscript gives the number of the term. Thus, a_n is the nth or **general term** of the sequence and we represent a sequence by giving a formula for a_n. Consider the following sequence:

Term	a_1	a_2	a_3	a_4	a_5	a_n
Term of Sequence	1	4	9	16	25	n^2

We represent this sequence by writing $a_n = n^2$, which means $a_1 = 1^2$, $a_2 = 2^2$, etc.

EXAMPLE **1** Find the first three terms of the following sequences and the 25th term.

a. $a_n = 3n - 1$

Solution We let $n = 1, 2, 3$, and 25 and evaluate.

$$\begin{aligned} a_1 &= 3(1) - 1 = 3 - 1 = 2 & \text{Let } n = 1 \text{ and evaluate.} \\ a_2 &= 3(2) - 1 = 6 - 1 = 5 & \text{Let } n = 2 \text{ and evaluate.} \\ a_3 &= 3(3) - 1 = 9 - 1 = 8 & \text{Let } n = 3 \text{ and evaluate.} \\ a_{25} &= 3(25) - 1 = 75 - 1 = 74 & \text{Let } n = 25 \text{ and evaluate.} \end{aligned}$$

The first three terms of the sequence are 2, 5, and 8. The 25th term is 74.

b. $a_n = \dfrac{(-1)^n}{n^2 + 1}$

Solution We let $n = 1, 2, 3$, and 25 and evaluate.

$$\begin{aligned} a_1 &= \frac{(-1)^1}{1^2 + 1} = \frac{-1}{1 + 1} = -\frac{1}{2} & \text{Let } n = 1 \text{ and evaluate.} \\ a_2 &= \frac{(-1)^2}{2^2 + 1} = \frac{1}{4 + 1} = \frac{1}{5} & \text{Let } n = 2 \text{ and evaluate.} \\ a_3 &= \frac{(-1)^3}{3^2 + 1} = \frac{-1}{9 + 1} = -\frac{1}{10} & \text{Let } n = 3 \text{ and evaluate.} \\ a_{25} &= \frac{(-1)^{25}}{25^2 + 1} = \frac{-1}{625 + 1} = -\frac{1}{626} & \text{Let } n = 25 \text{ and evaluate.} \end{aligned}$$

The first three terms of the sequence are $-\dfrac{1}{2}, \dfrac{1}{5}$, and $-\dfrac{1}{10}$. The 25th term is $-\dfrac{1}{626}$.

OBJECTIVE **2** Define and write arithmetic sequences, find their common difference, and find a particular term. In the sequence $-3, 1, 5, 9, 13, \ldots$, notice that each term after the first is found by adding 4 to the previous term. This is an example of an **arithmetic sequence** or *arithmetic progression*. Any two successive terms of an arithmetic sequence differ by the same amount, which is called the **common difference** and is denoted by d. To find d, choose any term (except the first) and subtract the previous term.

DEFINITION
Arithmetic sequence: A sequence in which each term after the first is found by adding the same number to the previous term. The common difference is $d = a_n - a_{n-1}$.

EXAMPLE **2** Write the first four terms of the following arithmetic sequences.

a. The first term is 2 and the common difference is 5.

Solution Begin with 2 and find each successive term by adding 5 to the previous term.

$$a_1 = 2, a_2 = 2 + 5 = 7, a_3 = 7 + 5 = 12, a_4 = 12 + 5 = 17$$

The first four terms of the sequence are 2, 7, 12, 17.

b. $a_1 = -1, d = -2$

Solution Begin with -1 and find each successive term by adding -2 to the previous term.

$$a_1 = -1, a_2 = -1 - 2 = -3, a_3 = -3 - 2 = -5, a_4 = -5 - 2 = -7$$

The first four terms of the sequence are $-1, -3, -5, -7$.

EXAMPLE **3** Find the common difference, d, for the following arithmetic sequence.

a. $-4, -1, 2, 5, 8, \ldots$

Solution Pick any term (except the first) and subtract the term before it.

$$d = -1 - (-4) = -1 + 4 = 3$$

or we could use $d = 5 - 2 = 3$, etc.

b. $8, 6, 4, 2, 0, \ldots$

Solution Pick any term (except the first) and subtract the term before it.

$$d = 6 - 8 = -2$$

or we could use $d = 2 - 4 = -2$, etc.

If the first term of an arithmetic sequence is a_1 and the common difference is d, then the arithmetic sequence can be written as a_1, $a_1 + d$, $a_1 + 2d$, $a_1 + 3d$, $a_1 + 4d$, ... Note that the coefficient of d is one less than the number of the term, so we have the following rule.

RULE   **THE n^{TH} TERM OF AN ARITHMETIC SEQUENCE**

The formula for finding the nth term of an arithmetic sequence is $a_n = a_1 + (n - 1)d$, where a_1 is the first term and d is the common difference.

EXAMPLE 4   Find the 23rd term and an expression for the nth term of an arithmetic sequence in which $a_1 = -8$ and $d = 3$.

Solution

$a_n = a_1 + (n - 1)d$

$a_{23} = -8 + (23 - 1)(3)$ Substitute 23 for n, -8 for a_1, 3 for d.

$a_{23} = 58$ Evaluate.

n^{th} term: $a_n = -8 + (n - 1)3$ Substitute for a_1 and d.

$a_n = -11 + 3n$ Simplify.

Calculator TIPS

Some graphing calculators have a sequence mode that allows you to display the terms of a sequence in a table. To generate the sequence $a_n = n^2$ on a TI-83, press MODE *and select seq then follow the steps below.*

Press Y = *and enter as below.*

Press 2nd WINDOW *to set the table as below.*

Press 2nd GRAPH *to display the table.*

```
Plot1   Plot2   Plot3
nMin =1
\u(n)◼ n²
u(nMin)◼ (1)
\v(n) =
v(nMin) =
\w(n) =
w(nMin) =
```

```
TABLE SETUP
 TblStart=1
 ▵Tbl=1
Indpnt: Auto Ask
Depend: Auto Ask
```

n	$u(n)$
1	1
2	4
3	9
4	16
5	25
6	36
7	49

$n=1$

The numbers of the terms are in the n column and the terms of the sequence are in the $u(n)$ column.

If we know the first term and one other term, we can use the formula for the nth term of an arithmetic sequence to find the common difference. Consequently, we can find the sequence.

EXAMPLE 5   The first term of an arithmetic sequence is 1 and the 20th term (a_{20}) is 58. Find the common difference and the first four terms of the sequence.

Solution We will use the fact that we know the a_{20} and the formula for the nth term to find d.

$a_n = a_1 + (n - 1)d$

$a_{20} = 1 + (20 - 1)d$ Substitute for a_1 and n.

$58 = 1 + 19d$ Substitute for a_{20} and solve.

$3 = d$

The first four terms are 1, $1 + 3 = 4$, $4 + 3 = 7$, $7 + 3 = 10$.

OBJECTIVE 3 Define and write series, find partial sums, and use summation notation. When the terms of a sequence are added, the sum is called a **series**.

> **DEFINITION**
>
> **Series:** The sum of the terms of a sequence.

Given a sequence $a_1, a_2, a_3, \ldots, a_n$, then $a_1 + a_2 + a_3 + \cdots + a_n$ is the corresponding series. Finite series correspond to finite sequences and infinite series correspond to infinite sequences. If an expression for the nth term is known, we can write the series in *summation notation* using the capital Greek letter *sigma* (Σ) as follows:

$$a_1 + a_2 + a_3 + \cdots + a_n = \sum_{i=1}^{n} a_i$$

The i is called the *index of summation*; 1 is the *lower limit* of i and n is the *upper limit* of i. If the upper limit is a natural number, the series is finite, and if the upper limit is ∞, the series is infinite. To find a finite series from summation notation, first replace the index of summation (usually i) with its lower limit and evaluate that term, replace i with the lower limit plus 1 and evaluate that term, and so on until you reach the upper limit. This gives the series. To find the sum, add the terms of the series.

EXAMPLE 6

Warning: Do not confuse the use of i in sigma notation, as in $\sum_{i=1}^{5}(3i+2)$, with its use as the imaginary unit, as in $7 \pm 5i$. In sigma notation, i is a variable representing natural numbers whereas as the imaginary unit it represents $\sqrt{-1}$.

a. Write the terms of $\sum_{i=1}^{5}(3i + 2)$ and find the sum of the series.

Solution Replace i with 1, 2, 3, 4, and 5. Then evaluate and find the sum of the results.

$$\sum_{i=1}^{5}(3i + 2) = (3 \cdot 1 + 2) + (3 \cdot 2 + 2) + (3 \cdot 3 + 2) + (3 \cdot 4 + 2) + (3 \cdot 5 + 2)$$
$$= 5 + 8 + 11 + 14 + 17$$
$$= 55$$

b. Write the first five terms of the infinite series $\sum_{i=1}^{\infty} 2i^2$.

Solution Replace i with 1, 2, 3, 4, and 5 evaluate.

$$\sum_{i=1}^{\infty} 2i^2 = 2 \cdot 1^2 + 2 \cdot 2^2 + 2 \cdot 3^2 + 2 \cdot 4^2 + 2 \cdot 5^2 + \cdots$$
$$= 2 + 8 + 18 + 32 + 50 + \cdots$$

OBJECTIVE 4 Write arithmetic series and find their sums. If the terms of an arithmetic sequence are added, it is called an **arithmetic series**.

> **DEFINITION**
>
> **Arithmetic series:** The sum of an arithmetic sequence.

Consequently, an arithmetic series has the form $a_1 + (a_1 + d) + (a_1 + 2d) + (a_1 + 3d) + \cdots [a_1 + (n-1)d]$.

When a finite number of terms of an infinite series are added, it is called a *partial sum* and the symbol S_n is used to indicate the sum of the first n terms. For example, S_5 means add the first five terms. We develop a formula for S_n for an arithmetic sequence as follows:

$$S_n = a_1 + (a_1 + d) + (a_1 + 2d) + (a_1 + 3d) + \cdots + a_n$$

Now write S_n by beginning with the last term, a_n, and subtracting the common difference, d, from the previous term.

$$S_n = a_n + (a_n - d) + (a_n - 2d) + (a_n - 3d) + \cdots + a_1$$

Add the two series together and we get

$$
\begin{aligned}
S_n &= a_1 &&+ (a_1 + d) &&+ (a_1 + 2d) &&+ (a_1 + 3d) &&+ \cdots &&+ a_n \\
+\ S_n &= a_n &&+ (a_n - d) &&+ (a_n - 2d) &&+ (a_n - 3d) &&+ \cdots &&+ a_1 \\
\hline
2S_n &= (a_1 + a_n) &&+ (a_1 + a_n) &&+ (a_1 + a_n) &&+ (a_1 + a_n) &&+ \cdots &&+ (a_1 + a_n)
\end{aligned}
$$

Since S_n has n terms, there are n terms of $(a_1 + a_n)$, so

$$2S_n = n(a_1 + a_n), \text{ or}$$
$$S_n = \frac{n}{2}(a_1 + a_n)$$

RULE   **PARTIAL SUM S_n OF AN ARITHMETIC SERIES**

The sum of the first n terms of an arithmetic series, S_n, called the nth partial sum, is given by

$$S_n = \frac{n}{2}(a_1 + a_n),$$

where n is the number of terms, a_1 is the first term, and a_n is the nth term.

EXAMPLE 7   Find the sum of the first 20 terms (S_{20}) of the arithmetic series $-6 - 2 + 2 + 6 + \cdots$.

Solution   To find the S_{20}, we first need to find the 20th term. Since $d = 4$,

$$a_{20} = -6 + (20 - 1)(4) = 70$$
$$S_n = \frac{n}{2}(a_1 + a_n) \qquad \text{Rule for } n^{\text{th}} \text{ partial sum.}$$
$$S_{20} = \frac{20}{2}(-6 + 70) \qquad \text{Substitute for } n, a_1, \text{ and } a_n.$$
$$S_{20} = 10(64) = 640 \qquad \text{Evaluate.}$$

1. A finite sequence is a function whose domain is _____.

2. What name do we give a sequence with an unlimited number of terms?

3. What is a series?

4. What is an arithmetic sequence? How do you find the common difference?

5. The series $-10 - 5 + 0 + 5 + 10 + \cdots$ is an example of a(n) _____ series with a common difference of _____.

6. What does S_n represent?

For Exercises 7–14, write the first four terms of the sequence and the indicated term.

7. $a_n = 2n + 1$, 20th term

8. $a_n = 3n - 4$, 18th term

9. $a_n = n^2 + 2$, 15th term

10. $a_n = n^2 - 3$, 12th term

11. $a_n = \dfrac{n}{n + 2}$, 22nd term

12. $a_n = \dfrac{2n}{n + 3}$, 10th term

13. $a_n = \dfrac{(-1)^n}{n^2 + 1}$, 15th term

14. $a_n = \dfrac{(-1)^n}{n^2 - 3}$, 26th term

For Exercises 15–20, find the common difference, d, for each arithmetic sequence.

15. $2, 7, 12, 17, \ldots$

16. $3, 11, 19, 27, \ldots$

17. $25, 22, 19, 16, \ldots$

18. $42, 36, 30, 24, \ldots$

19. $-12, -5, 2, 9, \ldots$

20. $-24, -27, -30, -33, \ldots$

For Exercises 21–26, find the indicated term and an expression for the nth term of the given arithmetic sequence.

21. a_{14} if $a_1 = 14$ and $d = 4$

22. a_{24} if $a_1 = 16$ and $d = 6$

23. a_{28} if $a_1 = -8$ and $d = -3$

24. a_{30} if $a_1 = -7$ and $d = -6$

25. a_{34} of $-5, -1, 3, 7, \ldots$

26. a_{21} of $8, 15, 22, 29, \ldots$

For Exercises 27–32, write the first four terms of the arithmetic sequence with the given characteristics.

27. $a_1 = -6, d = 7$

28. $a_1 = -2, d = 6$

29. $a_1 = -5, d = -3$

30. $a_1 = -13, d = -5$

31. $a_1 = 7, a_{18} = 75$

32. $a_1 = 6, a_{22} = 153$

33. Find the first term of an arithmetic sequence if $a_{45} = 143$ and $d = 3$.

34. Find the first term of an arithmetic sequence if $a_{39} = 181$ and $d = 6$.

35. Find the common difference of an arithmetic sequence if the first term is -110 and the 29th term is 2.

36. Find the common difference of an arithmetic sequence if the first term is 78 and the 15th term is -76.

For Exercises 37–40, write the first four terms of the arithmetic sequence with the given d and a_n.

37. $d = 7, a_8 = 41$

38. $d = 4, a_{11} = 27$

39. $d = -3, a_7 = 9$

40. $d = -6, a_{10} = -36$

For Exercises 41–48, write the series and find the sum.

41. $\displaystyle\sum_{i=1}^{6} i^2$

42. $\displaystyle\sum_{i=1}^{3} 3i^2$

43. $\displaystyle\sum_{i=1}^{4} (2i - 5)$

44. $\displaystyle\sum_{i=1}^{5} (4i + 1)$

45. $\displaystyle\sum_{i=1}^{3} (3i^2 - 4)$

46. $\displaystyle\sum_{i=1}^{4} (-2i^2 + 5)$

47. $\displaystyle\sum_{i=3}^{6} (4i - 3)$

48. $\displaystyle\sum_{i=2}^{5} (4i - 2)$

For Exercises 49–56, find the given S_n for the arithmetic series.

49. If $a_1 = 10$ and $d = 4$, find S_{25}.

50. If $a_1 = -8$ and $d = 2$, find S_{30}.

51. If $a_1 = 12$ and $d = -3$, find S_{22}.

52. If $a_1 = 18$ and $d = -2$, find S_{30}.

53. $3 + 9 + 15 + 21 + \cdots$. Find S_{15}.

54. $5 + 9 + 13 + 17 + \cdots$. Find S_{25}.

55. $54 + 46 + 38 + 30 + \cdots$. Find S_{12}.

56. $44 + 39 + 34 + 29 + \cdots$. Find S_{18}.

For Exercises 57–64, answer each question.

57. Find the sum of the first 100 natural numbers.

58. Find the sum of the even integers 2 through 200.

59. A concert hall has 60 seats in the first row, 64 in the second, 68 in the third, and so on.

   a. How many seats are in the 22nd row?

   b. How many seats are in the concert hall if there are 35 rows?

60. During the year 2003, many car dealers were offering loans at 0% interest. Johanna bought a car for $24,000 and made monthly payments of $400. Using 24,000 as the first term, write the first five terms of an arithmetic sequence that gives the amount that she still owes at the end of each month. How much will she owe after 25 months?

61. Fence posts are arranged in a triangular stack with 25 on the bottom row, 24 on the next, 23 on the next, and so forth until there is a single post on the top.

   a. How many posts are on the 10th row from the bottom?

   b. How many posts are in the stack?

62. Carlos is doing sit-ups to flatten his stomach. He did 25 the first night and plans to add 1 each night.

   a. Write the first five terms of an arithmetic sequence that gives the number of sit-ups that he does each night.

   b. Find the number he does on the 30th night.

   c. Find the total number of sit-ups that he has done after 30 nights.

63. Tanisha takes a job that pays $28,000 the first year and a raise of $1500 per year.

   a. Write the first five terms of an arithmetic sequence that gives her salary at the end of each year.

   b. What will her salary be for the 10th year?

   c. What are her total earnings for her first 10 years?

64. You have a choice of two jobs. Job A has a starting salary of $25,000 with raises of $900 per year, and job B has a starting salary of $28,000 with raises of $600 per year.

a. Which job will pay the most during the 10th year?

b. Which job will pay the most total amount during the first 15 years?

Geometric Sequences and Series

OBJECTIVES

1. Write a geometric sequence and find its common ratio and a specified term.
2. Find partial sums of a geometric series.
3. Find the sums of infinite geometric series.
4. Solve applications using geometric series.

OBJECTIVE **1** **Write a geometric sequence and find its common ratio and a specified term.** In Appendix A, we generated an arithmetic sequence by adding the same number to each term to get the next term. In this section, we will multiply each term by the same number to generate a **geometric sequence**. The number we multiply by is called the **common ratio** and is denoted as r. We can find r by dividing any term (except the first) by the term before it.

DEFINITION

> *Geometric sequence:* A sequence in which every term after the first is found by multiplying the previous term by the same number, called the common ratio, r, where $r = \dfrac{a_n}{a_{n-1}}$.

Note: In geometric series, the first term is usually denoted as a rather than a_1.

If the first term is a and the common ratio is r, a geometric sequence has the form $a, ar, ar^2, ar^3, ar^4, \ldots, ar^{n-1}$. Thus, the general term of a geometric series is $a_n = ar^{n-1}$.

We notice that the exponent of r is one less than the number of the term, so we have the following rule.

RULE **THE nᵀᴴ TERM OF A GEOMETRIC SEQUENCE**

The formula for finding the nth term of a geometric sequence is $a_n = ar^{n-1}$, where a is the first term and r is the common ratio.

EXAMPLE **1** A geometric sequence has a first term of 3 and a common ratio of 4.

a. Write the first five terms

Solution To find each term, multiply the term before it by 4. So,

$$a_1 = 3, a_2 = 3(4) = 12, a_3 = 12(4) = 48, a_4 = 48(4) = 192, a_5 = 192(4) = 768$$

The first five terms of the sequence are 3, 12, 48, 192, 768.

b. Find the 10th term.

Solution $a_n = ar^{n-1}$ Formula for a_n.
           $a_{10} = 3(4)^{10-1}$ Substitute $a = 3$, $r = 4$, and $n = 10$.
           $a_{10} = 786{,}432$ Evaluate.

c. Find the general term.

Solution $a_n = ar^{n-1}$ Formula for a_n.
           $a_n = 3(4)^{n-1}$ Substitute for a and r.

EXAMPLE 2   Given the geometric sequence $32, -16, 8, -4, \ldots$, find the common ratio r.

Solution To find r, divide any term except the first by the term before it.

$$r = \frac{-16}{32} = -\frac{1}{2} \quad \text{or} \quad r = \frac{8}{-16} = -\frac{1}{2}, \text{ etc.}$$

Geometric sequences often occur in populations and other applications.

EXAMPLE 3   The population of rabbits in a large pen increases at a rate of 12% per month. If there are currently 50 rabbits, find the population after 15 months.

Solution Let P_0 be the initial number of rabbits, P_1 the number after one month, P_2 the number after two months, etc. The number of rabbits at the end of each month is the number at the beginning of the month plus an increase of 12% of that number. So the number at the end of each month can be found as follows:

$$P_1 = P_0 + 0.12P_0 = 1.12P_0$$
$$P_2 = P_1 + 0.12P_1 = 1.12P_1 = 1.12(1.12P_0) = (1.12)^2 P_0$$
$$P_3 = P_2 + 0.12P_2 = 1.12P_2 = 1.12[(1.12)^2 P_0] = (1.12)^3 P_0$$

The number of rabbits at the end of each month can be represented by the sequence $P_0, P_1, P_2, P_3, \ldots P_n = P_0, 1.12P_0, (1.12)^2 P_0, (1.12)^3 P_0, \ldots, (1.12)^{n-1} P_0$, which is a geometric sequence whose first term is P_0 and $r = 1.12$. Consequently, the nth term is $P_0(1.12)^{n-1}$. The number of rabbits at the end of 15 months is the 16th term of the sequence, so

$$P_{16} = P_0(1.12)^{16-1}$$
$$P_{16} = 50(1.12)^{15}$$
$$P_{16} = 273.68$$

There are about 274 rabbits at the end of 15 months.

Note: It can be shown using this procedure that if a population is growing at $p\%$ per unit time and the initial population is P_0, then the population at the end of each unit of time forms a geometric sequence whose first term is P_0 and whose common ratio is $(1 + p)$ where p is $p\%$ written as a decimal. So the geometric sequence is $P_0, P_0(1 + p), P_0(1 + p)^2, P_0(1 + p)^3, \ldots, P_0(1 + p)^{n-1}$. The population after n time periods is the $(n + 1)$st term of the sequence, which is $P_0(1 + p)^n$.

OBJECTIVE 2 **Find partial sums of geometric series.** If we add the terms of an arithmetic sequence, we get an arithmetic series. Likewise, if we add the terms of a geometric sequence, we get a **geometric series**.

> **DEFINITION**
>
> **Geometric series:** The sum of the terms of a geometric sequence. A geometric series is of the form $a + ar + ar^2 + \cdots + ar^{n-1}$ if the series is finite, and $a + ar + ar^2 + \cdots$ if the series is infinite.

In the geometric series $3 + 6 + 12 + 24 + \cdots$, we have $a = 3$ and $r = 2$. In $243 - 81 + 27 - 9 + \cdots$, we have $a = 243$ and $r = -\dfrac{1}{3}$. Just as with arithmetic series, we can find a formula for partial sum, S_n. Begin with

$$
\begin{aligned}
S_n &= a + ar + ar^2 + \cdots + ar^{n-1} \\
-rS_n &= \quad - ar - ar^2 - \cdots - ar^{n-1} - ar^n \qquad \text{Multiply both sides of the equation by } -r. \\
\hline
S_n - rS_n &= a - ar^n \qquad \text{Add the equations.} \\
S_n(1 - r) &= a(1 - r^n) \qquad \text{Factor both sides.} \\
S_n &= \frac{a(1 - r^n)}{1 - r} \qquad \text{Divide both sides by } 1 - r.
\end{aligned}
$$

RULE **THE PARTIAL SUM, S_n, OF A GEOMETRIC SERIES**

The sum of the first n terms of a geometric series, S_n, called the nth partial sum, is given by

$$S_n = \frac{a(1 - r^n)}{1 - r},$$

where n is the number of terms, a is the first term, and r is the common ratio ($r \neq 1$).

EXAMPLE 4 Find the sum of the first 10 terms of the geometric series $1 - 3 + 9 - 27 + \cdots$.

Solution We first find r: $r = \dfrac{-3}{1} = -3$.

$$
\begin{aligned}
S_n &= \frac{a(1 - r^n)}{1 - r} &&\text{Formula for } S_n. \\
S_{10} &= \frac{1(1 - (-3)^{10})}{1 - (-3)} &&\text{Substitute } a = 1, r = -3, \text{ and } n = 10 \\
S_{10} &= -14{,}762 &&\text{Evaluate.}
\end{aligned}
$$

OBJECTIVE 3 **Find the sums of infinite geometric series.** If the common ratio satisfies $|r| > 1$, the partial sums become infinitely large as n becomes infinitely large. However, if $|r| < 1$, the partial sums approach a value as n becomes infinitely large. This value is called the *limit* of the partial sums and is the sum of the infinite series.

For example, for the series $2 + 1 + \dfrac{1}{2} + \dfrac{1}{4} + \cdots$, we have

$$S_5 = \frac{2\left(1 - \left(\frac{1}{2}\right)^5\right)}{1 - \frac{1}{2}} = 3.875$$

$$S_{10} = \frac{2\left(1 - \left(\frac{1}{2}\right)^{10}\right)}{1 - \frac{1}{2}} \approx 3.996$$

$$S_{15} = \frac{2\left(1 - \left(\frac{1}{2}\right)^{15}\right)}{1 - \frac{1}{2}} \approx 3.9998779$$

The partial sums are approaching 4. The reason is that as n gets larger, $\left(\dfrac{1}{2}\right)^n$ approaches 0. In the preceding example, $\left(\dfrac{1}{2}\right)^5 = 0.03125$, $\left(\dfrac{1}{2}\right)^{10} \approx 0.000977$, and $\left(\dfrac{1}{2}\right)^{15} \approx 0.0000305$. In general, if $|r| < 1$, then r^n approaches 0 as n becomes large. Consequently, if $|r| < 1$ the formula $S_n = \dfrac{a(1 - r^n)}{1 - r}$ becomes $S_\infty = \dfrac{a(1 - 0)}{1 - r} = \dfrac{a}{1 - r}$ as n becomes infinitely large. We denote the sum as S_∞ rather than S_n.

RULE   **THE SUM OF AN INFINITE GEOMETRIC SERIES**

If $|r| < 1$, the sum of an infinite geometric series, S_∞, is given by the formula

$$S_\infty = \frac{a}{1 - r},$$

where a is the first term and r is the common ratio. If $|r| > 1$, S_∞ does not exist.

EXAMPLE 5   Find the sum of the infinite geometric series

$$2 + 1 + \frac{1}{2} + \frac{1}{4} + \cdots.$$

Solution We know that $a = 2$ and $r = \dfrac{1}{2}$. Since $|r| < 1$, the sum exists.

$$S_\infty = \frac{a}{1 - r}$$

$$S_\infty = \frac{2}{1 - \dfrac{1}{2}} \qquad \text{Substitute for } a \text{ and } r.$$

$$S_\infty = 4 \qquad \text{Evaluate.}$$

Infinite geometric series also provides us with a method of changing a repeating decimal into a fraction.

EXAMPLE 6 Write $0.\overline{37}$ as a fraction.

Solution First write $0.\overline{37}$ as an infinite geometric series as follows:

$$0.\overline{37} = 0.37 + 0.0037 + 0.000037 + \cdots$$
$$0.\overline{37} = \frac{37}{100} + \frac{37}{10000} + \frac{37}{1000000} + \cdots$$

The last line is an infinite geometric series with $a = \frac{37}{100}$ and $r = \frac{1}{100}$. Since $|r| < 1$, the sum of this series exists.

$$S_\infty = \frac{a}{1 - r}$$

$$S_\infty = \frac{\dfrac{37}{100}}{1 - \dfrac{1}{100}} \qquad \text{Substitute } a = \frac{37}{100} \text{ and } r = \frac{1}{100}.$$

$$S_\infty = \frac{\dfrac{37}{100}}{\dfrac{99}{100}} \qquad \text{Simplify.}$$

$$S_\infty = \frac{37}{99}$$

So $0.\overline{37} = \frac{37}{99}$.

OBJECTIVE 4 Solve applications using geometric series.

EXAMPLE 7 To save for their child's college education, the McBride family put $1000 into a savings account the first year, and each year thereafter deposited 10% more than the previous year.

a. Write the first five terms of the geometric sequence that gives the amount of money deposited into the account each year.

Solution From the note following Example 3, this is a geometric series in which $a = 1000$ and $r = (1 + .10) = 1.1$

$$a_1 = 1000, \; a_2 = 1000(1.1) = 1100, \; a_3 = 1000(1.1)^2 = 1210,$$
$$a_4 = 1000(1.10)^3 = 1331, \; a_5 = 1000(1.10)^4 = 1464.10$$

The first five terms of the sequence are 1000, 1100, 1210, 1331, 1464.10.

b. How much money will have been deposited into the account at the end of 15 years?

Solution The series $1000 + 1100 + 1210 + 1331 + 1464.10 + \cdots$ is geometric with $a = 1000$ and $r = 1.1$.

$$S_n = \frac{a(1 - r^n)}{1 - r}$$

$$S_{15} = \frac{1000(1 - 1.1^{15})}{1 - 1.1} \qquad \text{Substitute for } a, r, \text{ and } n.$$

$$S_{15} = 31{,}772.48 \qquad \text{Evaluate.}$$

They will have deposited $31,772.48 in the account after 15 years.

APPENDIX B
Exercises

1. If you divide a term of a geometric sequence by the term before it, what do you get?

2. What is the form of a geometric sequence?

3. Can you find the sum of the first n terms of the geometric series $2 + 6 + 18 + \cdots$ using the formula $S_n = \dfrac{a(1 - r^n)}{1 - r}$? Why or why not?

4. Can you find the sum of the infinite geometric series $2 + 6 + 18 + \cdots$ using the formula $S_\infty = \dfrac{a}{1 - r}$? Why or why not?

5. The fourth partial sum, S_4, of $2 + 6 + 18 + \cdots$ is _____ + _____ + _____ + _____ = _____.

6. Write $0.\overline{23}$ as an infinite geometric series. What are a and r?

For Exercises 7–14: a. Find the common ratio, r, for the given geometric sequence.
b. Find the indicated term.
c. Find an expression for the general term, a_n.

7. $1, 3, 9, 27, \ldots$; 9th term

8. $7, 14, 28, 56, \ldots$; 11th term

9. $-2, 4, -8, 16, \ldots$; 15th term

10. $-8, 24, -72, 216, \ldots$; 9th term

11. $243, 81, 27, 9, \ldots$; 10th term

12. $8, 4, 2, 1, \ldots$; 8th term

13. $128, -32, 8, -2, \ldots$; 7th term

14. $125, -25, 5, -1, \ldots$; 9th term.

For Exercises 15–18: a. Write the first five terms of the geometric sequence
satisfying the given conditions.
b. Find the indicated term.

15. $a = -4$, $r = -3$; 8th term

16. $a = -6$, $r = -2$; 10th term

17. $a = 243$, $r = \dfrac{1}{3}$; 7th term

18. $a = 256$, $r = -\dfrac{1}{2}$; 10th term

For Exercises 19–24: a. Use the formula for the nth term to find r.
b. Write the first four terms of the geometric sequence.

19. $a = 1$, $a_5 = 16$

20. $a = 3$, $a_4 = -24$

21. $a = -6$, $r > 0$, $a_5 = -96$

22. $a = -4$, $r > 0$, $a_4 = -32$

23. $a = 128$, $r > 0$, $a_4 = 2$

24. $a = -243$, $r < 0$, $a_6 = 1$

25. Find the first term of a geometric sequence in which $r = 3$ and the fifth term is -486.

26. Find the first term of a geometric sequence in which $r = -\dfrac{1}{2}$ and the fifth term is $-\dfrac{1}{8}$.

For Exercises 27–32, find the sum of the first n terms of each geometric series for the given value of n.

27. $3 + 9 + 27 + 81 + \cdots$, $n = 11$

28. $-1 - 2 - 4 - 8 - \cdots$, $n = 9$

29. $32 + 16 + 8 + 4 + \cdots$, $n = 9$

30. $81 + 27 + 9 + 3 + \cdots$, $n = 7$

31. $128 - 32 + 8 - 2 + \cdots$, $n = 9$

32. $625 - 125 + 25 - 5 + \cdots$, $n = 8$

For Exercises 33–38, find the sum of the infinite geometric series, if possible. If it is not possible, explain why.

33. $27 + 9 + 3 + \cdots$

34. $8 + 4 + 2 + \cdots$

35. $15 - 9 + \dfrac{27}{5} - \cdots$

36. $15 - 10 + \dfrac{20}{3} - \cdots$

37. $9 + 12 + 16 + \cdots$

38. $16 + 20 + 25 + \cdots$

For Exercises 39 and 42, write each repeating decimal as a fraction.

39. $0.\overline{4}$

40. $0.\overline{7}$

41. $0.\overline{17}$

42. $0.2\overline{5}$

For Exercises 43 and 44, answer each question.

43. If $a_n = 500(1.04)^n$:

    a. Find the first five terms of the sequence.

    b. Find the 10th term of the sequence.

44. If $a_n = 350(1.06)^n$:

    a. Find the first five terms of the sequence.

    b. Find the 8th term of the sequence.

For Exercises 45–50, solve.

45. A population of mink is increasing at a rate of 8% per month. The current mink population is 100.

   a. Using 100 as the first term, find the first four terms of the geometric sequence that gives the number of mink at the end of each month.

   b. Find the number of mink present after eight months.

   c. Find the expression for the general term, a_n.

46. The generation time (the time required for the number present to double) for a particular bacteria is 1 hour. Suppose initially one bacteria was present.

   a. Using 1 as the first term, write the first five terms of the geometric sequence giving the number of bacteria present after each hour.

   b. Find an expression for the number present after the nth hour.

   c. How many bacteria are present after 1 day?

47. Suppose you took a job for a month (20 working days) that paid $0.01 the first day and your salary doubled each day.

   a. Write the first five terms of the geometric sequence that gives your salary each day.

   b. Find an expression for the amount earned on the nth day.

   c. How much would you earn on the 20th day?

   d. What are your total earnings for the month?

48. Damarys deposits $200 in the bank and each month thereafter deposits 5% more than the month before. She does this for 1 year.

   a. Find an expression for the amount she deposits on the nth month.

   b. Write the first four terms of the geometric sequence that gives the amount of her deposit each month.

   c. How much did she deposit on the 10th month?

   d. How much does she deposit for the year?

49. A new boat costs $20,000 and depreciates by 7% each year. What will the boat be worth in eight years?

50. The isotope $_{15}P^{33}$ has a half-life of 25 days. A sample has 400 grams.

    a. Find the first five terms of the geometric sequence that gives the amount present at the end of each half-life.

    b. Find an expression for the amount present after the nth half-life.

    c. Find the amount present after the 10th half-life.

APPENDIX C
The Binomial Theorem

OBJECTIVES

1. Expand a binomial using Pascal's triangle.
2. Evaluate factorial notation and binomial coefficients.
3. Expand a binomial using the binomial theorem.
4. Find a particular term of a binomial expansion.

In Section 5.5 we learned to square binomials. In this section, we will learn to raise binomials to natural-number powers.

OBJECTIVE 1 **Expand a binomial using Pascal's triangle.** We begin by writing out $(a + b)^n$, where n is a natural number, and look for patterns. These products are called **binomial expansions**.

$$(a + b)^0 = 1$$
$$(a + b)^1 = a + b$$
$$(a + b)^2 = a^2 + 2ab + b^2$$
$$(a + b)^3 = a^3 + 3a^2b + 3ab^2 + b^3$$
$$(a + b)^4 = a^4 + 4a^3b + 6a^2b^2 + 4ab^3 + b^4$$
$$(a + b)^5 = a^5 + 5a^4b + 10a^3b^2 + 10a^2b^3 + 5ab^4 + b^5$$

Conclusions: Several patterns can be observed from the preceding expansions.

1. The first term in the expansion, a, is raised to the same power as the binomial and the power of a decreases by 1 in each successive term. Note that the last term does not contain an a.
2. The exponent of b is 0 in the first term and increases by 1 on each successive term.
3. The sum of the exponents of the variables of each term equals the exponent of the binomial.
4. The number of terms in the expansion is one more than the exponent of the binomial.

Now consider the coefficients of the terms in these expansions.

	Coefficients of Expansions
$(a + b)^0$	1
$(a + b)^1$	1 1
$(a + b)^2$	1 2 1
$(a + b)^3$	1 3 3 1
$(a + b)^4$	1 4 6 4 1
$(a + b)^5$	1 5 10 10 5 1

If we arrange the coefficients of each expansion in a triangular array, we see an interesting pattern. Each row begins and ends with 1. Each number inside a row is the sum of the two numbers in the row above it. For example, each 10 in the bottom row comes from adding the 4 and 6 directly above.

This triangular array of numbers is called **Pascal's triangle** in honor of the French mathematician Blaise Pascal. Using these observations, the expansion of $(a + b)^6$ would have seven terms and the variable portions of the terms would be a^6, a^5b, a^4b^2, a^3b^3, a^2b^4, ab^5, b^6. By continuing the pattern in Pascal's triangle, we can find the coefficients for each of those terms.

$(a + b)^5$ 1   5   10   10   5   1

$(a + b)^6$ 1   6   15   20   15   6   1

Using the coefficients from the last line of Pascal's triangle and the variables previously listed gives us

$$(a + b)^6 = a^6 + 6a^5b + 15a^4b^2 + 20a^3b^3 + 15a^2b^4 + 6ab^5 + b^6$$

OBJECTIVE 2   **Evaluate factorial notation and binomial coefficients.**   Although Pascal's triangle is easy to use, it isn't practical, especially for binomials raised to large powers. Consequently, another method called the *binomial theorem* is often used. Before introducing the binomial theorem, we need **factorial notation**.

DEFINITION

Factorial notation: For any natural number n, the symbol $n!$ (read "n factorial") means $n(n - 1)(n - 2) \ldots 3 \cdot 2 \cdot 1$. 0! is defined to be 1, so $0! = 1$.

EXAMPLE 1   Evaluate the following factorials.

a. 5!

Solution $5! = 5 \cdot 4 \cdot 3 \cdot 2 \cdot 1 = 120$

b. 7!

Solution $7! = 7 \cdot 6 \cdot 5 \cdot 4 \cdot 3 \cdot 2 \cdot 1 = 5040$

Notice the following characteristic of factorials.

$7! = 7 \cdot 6 \cdot 5 \cdot 4 \cdot 3 \cdot 2 \cdot 1 = 7 \cdot 6!$ and $7! = 7 \cdot 6 \cdot 5!$ and $7! = 7 \cdot 6 \cdot 5 \cdot 4!$

This characteristic means that when writing a factorial, we can stop at any point and write the last factor as a factorial. So $10! = 10 \cdot 9!$ and $10! = 10 \cdot 9 \cdot 8 \cdot 7 \cdot 6!$, etc.

The coefficients of a binomial expansion can be expressed in terms of factorials using a special notation called the **binomial coefficient**. We will see exactly how the binomial coefficient is used later.

DEFINITION

Binomial coefficient: A number written as $\binom{n}{r}$ and defined as $\dfrac{n!}{r!(n - r)!}$.

EXAMPLE 2 Evaluate the following binomial coefficients.

a. $\dbinom{6}{2}$

Solution $\dbinom{n}{r} = \dfrac{n!}{r!(n-r)!}$ The binomial coefficient.

$\dbinom{6}{2} = \dfrac{6!}{2!(6-2)!}$ Substitute 6 for n and 2 for r.

$= \dfrac{6 \cdot 5 \cdot 4 \cdot 3 \cdot 2 \cdot 1}{(2 \cdot 1)(4 \cdot 3 \cdot 2 \cdot 1)}$ Expand the factorials.

$= 15$ Simplify.

Note: These factorials could have been evaluated as follows:

$\dfrac{6!}{2! \cdot 4!} = \dfrac{6 \cdot 5 \cdot \cancel{4!}}{2 \cdot 1 \cdot \cancel{4!}} = \dfrac{6 \cdot 5}{2} = \dfrac{30}{2} = 15$

b. $\dbinom{8}{5}$

Solution $\dbinom{8}{5} = \dfrac{8!}{5!(8-5)!}$ Substitute 8 for n and 5 for r in $\dfrac{n!}{r!(n-r)!}$.

$= \dfrac{8!}{5! \cdot 3!}$ Simplify.

$= \dfrac{8 \cdot 7 \cdot 6 \cdot 5 \cdot 4 \cdot \cancel{3!}}{5 \cdot 4 \cdot 3 \cdot 2 \cdot 1 \cdot \cancel{3!}}$ Rewrite 8! as $8 \cdot 7 \cdot 6 \cdot 5 \cdot 4 \cdot 3!$

$= 56$ Evaluate.

Using the formula for evaluating binomial coefficients, we can prove two special cases: $\dbinom{n}{0} = 1$ and $\dbinom{n}{n} = 1$.

Calculator TIPS

Factorials can be found on most graphing calculators. To find 10! on the TI-83 Plus, enter [1] [0] *, go to the Math menu, select PRB, and press* [4] *.*

```
10!
          3628800
```

Let's look at the expansion of $(a + b)^6$ and make another observation.

$$(a + b)^6 = a^6 + 6a^5b + 15a^4b^2 + 20a^3b^3 + 15a^2b^4 + 6ab^5 + b^6$$

Look at the third term of the expansion, $15a^4b^2$. The coefficient is 15 and from Example 2a, we see that $\binom{6}{2} = 15$. The coefficient of the fourth term is 20 and $\binom{6}{3} = 20$. In both cases, the n value of the binomial coefficient, $\binom{n}{r}$, is the exponent of the binomial and the r value is the exponent of b. This observation leads to the **binomial theorem**.

DEFINITION

The Binomial Theorem: For any positive integer n,

$$(a + b)^n = \binom{n}{0}a^n + \binom{n}{1}a^{n-1}b + \binom{n}{2}a^{n-2}b^2 + \binom{n}{3}a^{n-3}b^3 + \cdots + \binom{n}{n}b^n$$

EXAMPLE 3 Expand each of the following binomials using the binomial theorem.

a. $(a + b)^6$

Solution

$$(a + b)^6 = \binom{6}{0}a^6 + \binom{6}{1}a^5b + \binom{6}{2}a^4b^2 + \binom{6}{3}a^3b^3 + \binom{6}{4}a^2b^4 + \binom{6}{5}ab^5 + \binom{6}{6}b^6$$

$$= \frac{6!}{0!6!}a^6 + \frac{6!}{1!5!}a^5b + \frac{6!}{2!4!}a^4b^2 + \frac{6!}{3!3!}a^3b^3 + \frac{6!}{4!2!}a^2b^4 + \frac{6!}{5!1!}ab^5 + \frac{6!}{6!0!}b^6$$

$$= a^6 + 6a^5b + 15a^4b^2 + 20a^3b^3 + 15a^2b^4 + 6ab^5 + b^6$$

Note: This is the same result we got using Pascal's triangle earlier. Also note that

$$\binom{6}{3} = \frac{6!}{3!3!} = \frac{6 \cdot 5 \cdot 4 \cdot 3!}{3!3!} = \frac{6 \cdot 5 \cdot 4 \cdot \cancel{3!}}{3 \cdot 2 \cdot 1 \cdot \cancel{3!}} = \frac{\cancel{6} \cdot 5 \cdot 4}{\cancel{3} \cdot \cancel{2} \cdot 1} = 20, \text{ etc.}$$

b. $(a - 3b)^5$

Solution Write $(a - 3b)^5$ as $[a + (-3b)]^5$

$$[a + (-3b)]^5 = \binom{5}{0}a^5 + \binom{5}{1}a^4(-3b) + \binom{5}{2}a^3(-3b)^2 +$$

$$\binom{5}{3}a^2(-3b)^3 + \binom{5}{4}a(-3b)^4 + \binom{5}{5}(-3b)^5$$

$$= a^5 + 5a^4(-3b) + 10a^3(9b^2) + 10a^2(-27b^3) + 5a(81b^4) + (-243b^5)$$

$$= a^5 - 15a^4b + 90a^3b^2 - 270a^2b^3 + 405ab^4 - 243b^5$$

Note: $\binom{5}{2} = \frac{5!}{2!3!} = \frac{5 \cdot 4 \cdot 3!}{2 \cdot 1 \cdot 3!} = \frac{5 \cdot 4 \cdot \cancel{3!}}{2 \cdot 1 \cdot \cancel{3!}} = \frac{5 \cdot 4}{2} = \frac{20}{2} = 10, \text{ etc.}$

OBJECTIVE 4 **Find a particular term of a binomial expansion.** Sometimes, it is necessary to find only a specific term of a binomial expansion without writing out the entire expansion. Look again at the binomial expansion. Note that the third term (which we will call the $(2 + 1)^{\text{st}}$ term) is $\binom{n}{2}a^{n-2}b^2$ and the fourth term (which we will call the $(3 + 1)^{\text{st}}$ term) is $\binom{n}{3}a^{n-3}b^3$. Similarly, the $(m + 1)^{\text{st}}$ term is $\binom{n}{m}a^{n-m}b^m$. These observations lead to the following.

RULE **FINDING THE $(m + 1)^{\text{ST}}$ TERM OF A BINOMIAL EXPANSION**

The $(m + 1)^{\text{st}}$ term of the expansion $(a + b)^n$ is $\binom{n}{m}a^{n-m}b^m$.

EXAMPLE 4 Find the indicated term of each of the following binomial expansions.

a. $(a + b)^{11}$, seventh term

Solution Use the formula for the $(m + 1)^{\text{st}}$ term with $n = 11$, and $m = 6$ (to find the seventh term).

$$\binom{n}{m}a^{n-m}b^m = \binom{11}{6}a^{11-6}b^6 = 462a^5b^6$$

b. $(2x - 5y)^8$, fourth term

Solution Write $(2x - 5y)^8$ as $[2x + (-5y)]^8$. Use the formula for the $(m + 1)^{\text{st}}$ term with $n = 8$, $m = 3$ (to find the fourth term), $a = 2x$, and $b = -5y$.

$$\binom{n}{m}a^{n-m}b^m = \binom{8}{3}(2x)^{8-3}(-5y)^3 = 56(32x^5)(-125y^3) = -224{,}000x^5y^3$$

For Extra Help

 InterAct Math

 Addison-Wesley
Math Tutor Center

 Video 20/DVT 7

MyMathLab

MathXP MathXL

1. How many terms are in the expansion of $(5a + b)^{12}$?

2. What is the sum of the exponents on x and y for any term in the expansion of $(x + y)^9$?

3. How is the symbol 8! read?

4. What is the meaning of 8!? What is its value?

5. What is the exponent of b in the sixth term of the expansion of $(a + b)^{13}$? What is the exponent of a in that term?

6. On a term in the expansion of $(x + y)^n$ the exponent of x is 4. Find the exponent of y.

For Exercises 7–18, evaluate each expression.

7. 4!

8. 7!

9. 4!(3!)

10. 3!2!

11. 6(5!)

12. 4!(7)

13. $\dfrac{8!}{10!}$

14. $\dfrac{7!}{9!}$

15. $\dfrac{10!}{9!}$

16. $\dfrac{12!}{11!}$

17. $\dfrac{8!}{6!(8 - 6)!}$

18. $\dfrac{10!}{6!(10 - 6)!}$

For Exercises 19–26, evaluate each binomial coefficient.

19. $\dbinom{7}{3}$

20. $\dbinom{5}{2}$

21. $\dbinom{10}{4}$

22. $\dbinom{6}{5}$

23. $\dbinom{7}{7}$

24. $\dbinom{4}{4}$

25. $\dbinom{8}{0}$

26. $\dbinom{9}{0}$

For Exercises 27–38, use the binomial theorem to expand each of the following.

27. $(a + b)^5$

28. $(a + b)^7$

29. $(x - y)^4$

30. $(x - y)^3$

31. $(2a + b)^3$

32. $(a + 2b)^3$

33. $(x - 2y)^5$

34. $(x - 3y)^4$

35. $(2m + 3n)^6$

36. $(3c + 2d)^5$

37. $(3x - 4y)^4$

38. $(4a - b)^7$

For Exercises 39–46, find the indicated term of each binomial expansion.

39. $(x + y)^8$, fifth term

40. $(a + b)^9$, fourth term

41. $(a - b)^{10}$, third term

42. $(m - n)^7$, second term

43. $(4x + y)^9$, sixth term

44. $(3a + b)^{11}$, seventh term

45. $(3m - 2n)^7$, fourth term.

46. $(5x - 3y)^{12}$, fifth term

APPENDIX D
Permutations and Combinations

OBJECTIVES

1. Use the multiplication principle.
2. Calculate permutations.
3. Calculate combinations.

In this section, we will be counting the number of different ways in which tasks can be performed. For example, in a race with 8 people, we might count the possible number of first, second, and third place finishers. Or in a club with 25 members, how many different ways can we choose a committee of 5 to plan a party? Before we answer these questions, we need to establish the *multiplication principle*.

OBJECTIVE 1 **Use the multiplication principle.** Murlene goes into an ice cream shop to buy some fat-free frozen yogurt. For a container, she has a choice of a cup, sugar cone, or waffle cone. Her choices in flavors are vanilla, chocolate, strawberry, or peach. In how many ways can she choose a container and flavor? We use a tree diagram to answer this question.

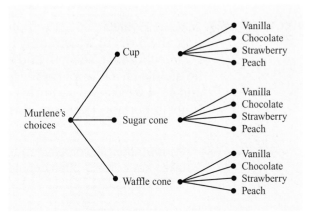

From the diagram, we see that for each of the 3 choices of containers there are 4 choices of flavors, so there are $3 \cdot 4 = 12$ different choices of containers and flavors. Our example suggests the multiplication principle.

RULE **THE MULTIPLICATION PRINCIPLE**

To find the number of ways in which successive tasks can be performed, multiply the number of ways in which each task can be performed. Symbolically, if task A can be performed in m ways and task B can be performed in n ways, then the two tasks can be performed in mn ways.

EXAMPLE 1

a. How many different three-digit numbers can be formed using the digits 2, 3, 5, and 7 if no digit may be repeated?

Solution We have four choices of the first digit (2, 3, 5 or 7), three for the second (can't use the one chosen as the first digit), and two for the third, so the total number of possible different three digit numbers is $4 \cdot 3 \cdot 2 = 24$.

b. In how many different ways (orders) can 6 people sit on a bench?

Solution Six choices are possible for the first person. After that person is seated, 5 choices are possible for the second person. Similarly, there are 4 choices for the third, 3 for the fourth, 2 for the fifth and 1 for the sixth. Therefore, 6 people can sit on a bench in $6 \cdot 5 \cdot 4 \cdot 3 \cdot 2 \cdot 1 = 720$ different ways.

c. A race has 8 participants. How many different first-, second-, and third-place finishers are possible?

Solution Any one of the 8 could finish first. Any one of the remaining 7 could finish second. Similarly, any one of the remaining 6 could finish third, so there are $8 \cdot 7 \cdot 6 = 336$ different first-, second-, and third-place finishers possible.

d. How many different phone numbers are possible for a given area code and prefix such as (920) 423-__ __ __ __?

Solution The telephone numbers have four more digits. Each digit has 10 possible choices, so there are $10 \cdot 10 \cdot 10 \cdot 10 = 10,000$ possible phone numbers in area code 920 and prefix 423.

OBJECTIVE 2 **Calculate permutations.** Examples 1a–c illustrate **permutations**. In a permutation, the *order* in which things occur is important. For example, if points are awarded for first, second, and third place, the order in which the participants finish in a race is important. However, if a troupe of dancers has 15 members and 4 are selected to perform a routine, the order in which dancers are selected is not important. Therefore, this is not a permutation.

DEFINITION

Permutation: An ordered arrangement of objects. The permutation of n things taken r at a time is often denoted as $_nP_r$.

In Example 1b, we found $_6P_6 = 720$. In Example 1c, we found $_8P_3 = 336$. Notice from Example 1b that $_6P_6 = 6 \cdot 5 \cdot 4 \cdot 3 \cdot 2 \cdot 1 = 6!$ Instead of finding $_8P_3 = 336$ as $8 \cdot 7 \cdot 6$, we could have used $\dfrac{8 \cdot 7 \cdot 6 \cdot 5 \cdot 4 \cdot 3 \cdot 2 \cdot 1}{5 \cdot 4 \cdot 3 \cdot 2 \cdot 1} = \dfrac{8!}{(8-3)!} = \dfrac{8!}{5!}$. This leads to the following formula.

RULE **EVALUATING $_nP_r$.**

The formula for the number of permutations of n things taken r at a time is
$$_nP_r = \frac{n!}{(n-r)!}$$

EXAMPLE 2

a. A horse race has 10 horses entered. How many different first-, second-, and third-place finishers are possible?

Solution We need to find the number of permutations of 10 horses taken 3 at a time, $_{10}P_3$.

$$_nP_r = \frac{n!}{(n-r)!} \qquad \text{Formula for } _nP_r.$$

$$_{10}P_3 = \frac{10!}{(10-3)!} \qquad \text{Substitute for } n \text{ and } r.$$

$$= \frac{10 \cdot 9 \cdot 8 \cdot 7!}{7!} \qquad \text{Replace } 10! \text{ with } 10 \cdot 9 \cdot 8 \cdot 7! \text{ and } (10-3)! = 7!.$$

$$= 10 \cdot 9 \cdot 8 \qquad \text{Divide the } 7!s.$$

$$= 720 \qquad \text{Evaluate.}$$

There are 720 different possibilities for first-, second-, and third-place finishers.

b. Johann is going on a short trip and selects 4 CDs to play while driving. In how many different orders can he play the 4 CDs?

Solution We need to find the number of permutations of 4 things taken 4 at a time, $_4P_4$?

$$_nP_r = \frac{n!}{(n-r)!}$$

$$_4P_4 = \frac{4!}{(4-4)!} = \frac{4!}{0!} = \frac{4 \cdot 3 \cdot 2 \cdot 1}{1} = 24$$

From Example 2, we can derive the following.

RULE **THREE SPECIAL PERMUTATION FORMULAS**

$$_nP_n = n! \qquad _nP_0 = 1 \qquad _nP_1 = n$$

Also note that Example 2 could have been done using the multiplication principle.

Graphing Calculator
TIPS

Most graphing calculators can evaluate $_nP_r$. To evaluate $_{15}P_5$ on the TI-83 Plus, press 1 5 *, MATH, select PRB, select 2, press 5, and press* ENTER *. The screen appears as follows.*

```
15 nPr 5
              360360
```

Calculate combinations. Suppose a student governance association has 12 members and selects a committee of 4 to organize the upcoming election. The order in which the members of the committee are selected is not important, since an individual is either on the committee or not. So this situation is not a permutation. It is an example of a **combination**.

> **DEFINITION**
>
> **Combination:** An unordered arrangement of objects. The combination of n things taken r at a time is often denoted as nCr.

To determine the number of different ways the committee of 4 can be formed from the 12 members, let's first calculate the number if order did matter (that is, a permutation).

$$_{12}P_4 = \frac{12!}{(12-4)!} = \frac{12!}{8!} = \frac{12 \cdot 11 \cdot 10 \cdot 9 \cdot 8!}{8!} = 12 \cdot 11 \cdot 10 \cdot 9 = 11{,}880$$

Since order is not important, many of these committees have the same members. For example, a committee made up of Aricellis, Ebony, Hector, and Carol is the same as the committee made up of Ebony, Aricellis, Carol, and Hector. Since each committee has 4 members, there are $4! = 24$ different orders in which the same committee can be written. To eliminate these duplicate committees, divide 11,880 by 24, which gives 495 different committees.

Summarizing, to find the number of combinations of 12 things taken 4 at a time, we found $\dfrac{_{12}P_4}{4!} = \dfrac{\frac{12!}{(12-4)!}}{4!} = \dfrac{12!}{4!(12-4)!}$, which suggests the following rule.

PROCEDURE **PROCEDURE EVALUATING nCr**

CONNECTION

$nCr = \dbinom{n}{r}$, which is the binomial coefficient.

▶

The formula for the number of combinations of n things taken r at a time, is

$$nCr = \frac{n!}{r!(n-r)!}$$

Note that $nCn = 1$, $nC0 = 1$, and $nC1 = n$.

EXAMPLE 3 In a BMX dirt bike race, 4 of the 8 bikes in a preliminary heat can advance to the next heat. How many different groups of 4 can advance?

Solution Since the order in which the top 4 finishers is not important (all 4 go on to the next heat), this is a combination of 8 things taken 4 at a time or $8C4$.

$$nCr = \frac{n!}{r!(n-r)!} \qquad \text{Combination formula.}$$

$$8C4 = \frac{8!}{4!(8-4)!} = \frac{8!}{4!(4!)} = \frac{8 \cdot 7 \cdot 6 \cdot 5 \cdot 4!}{4 \cdot 3 \cdot 2 \cdot 1(4!)} = \frac{8 \cdot 7 \cdot 6 \cdot 5}{4 \cdot 3 \cdot 2 \cdot 1} = 70$$

There are 70 possible combinations of four riders who can advance to the next heat.

Evaluating nCr is done exactly the same way as evaluating nPr except choose 3 under the PRB menu instead of 2.

APPENDIX D
Exercises

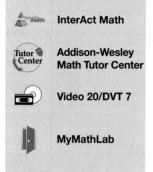

For Extra Help

InterAct Math

Addison-Wesley
Math Tutor Center

Video 20/DVT 7

MyMathLab

MathXL

1. If one task can be performed in x ways and a second task can be performed in y ways, what does $x \cdot y$ represent?

2. How do permutations differ from combinations?

3. A family is going on vacation and can visit any 5 of 12 amusement parks. To determine the number of different groups of 5 parks they could visit, would you use the formula for nPr or nCr? Why?

4. Fredrica has a choice of 5 pies and a choice of 3 different flavors of ice cream. What formula or principle would you use to determine the total number of different choices of pie and ice cream?

5. The combination to a combination lock consists of 3 different numbers chosen from a possible 50. To determine the number of possible combinations would you use the formula for nPr or nCr? Why?

6. In deriving the formula nCr, why did we calculate $\dfrac{nPr}{r!}$?

For Exercises 7–18, evaluate each.

7. $7P2$

8. $9P5$

9. $6P6$

10. $5P5$

11. $3P0$

12. $8P0$

13. $6C4$

14. $10C4$

15. $5C5$

16. $9C9$

17. $7C0$

18. $8C0$

19. John works as a salesperson and has to wear a sports coat and tie to work each day. Assuming all his clothing matches, find the number of possible outfits that he can wear if he has 4 coats, 6 pairs of pants, 5 shirts, and 3 ties.

20. The executive board of a corporation has 8 members. If one member is to be in charge of public relations, a different member in charge of marketing, and a third member in charge of research, how many different ways can these 3 positions be filled?

21. Jorge plans to take one course each in the subject areas of literature, computers, history, and humanities. If he has a choice of 4 literature courses, 3 computer courses, 5 history courses, and 2 humanities courses that fit his schedule, in how many different ways can he build his schedule?

22. At a sandwich shop, you have a choice of 3 breads, one of 3 dressings, with or without lettuce, with or without tomato, with or without pickles, and 5 meats. How many different sandwiches are possible?

23. How many different three-digit numbers can be formed from the digits 4, 6, and 7 (a) if no digit may be repeated? (b) If the digits may be repeated?

24. A club consists of 8 women and 5 men. In how many different ways can the club elect a president and a secretary if the president must be a woman and the secretary must be a man?

25. Some states' license plates have 3 numbers followed by 3 letters. How many different license plates are possible in one of these states?

26. Postal Zip codes have 5 digits. How many Zip codes are possible?

27. A telephone calling card requires a four-digit PIN (personal identification number). How many different PINs are possible? How many different PINs have no repeated digits?

28. A singer has 12 songs to record on a CD. In how many different orders can the songs be recorded?

29. In how many different orders can the 11 members of the starting offensive lineup of a football team be introduced?

30. A television executive has decided on 4 shows to air on Thursday evenings. In how many different orders can the shows be shown?

31. To win a state lottery, you must correctly select 5 numbers, from 1 to 49 inclusive. Since the numbers are drawn using ping-pong balls, no number can be repeated. What is the total number of possible winning tickets?

32. Basketball teams in the National Basketball Association have 12 members. How many different starting lineups of 5 players are possible?

33. A singer has been asked to perform 3 songs at a benefit. If she has 18 songs to choose from, how many different groups of 3 songs are possible?

34. A poet is putting together a collection for publication. He has 30 poems to choose from, but can include only 20 in the collection. How many different collections are possible?

APPENDIX E
Probability

1 Find sample spaces and events.

2 Calculate probabilities.

We often hear the word **probability** in everyday language. For example, a weather forecaster might say that the probability of rain is 40% (0.40). In this usage, probability measures the likelihood that the event will occur. In this section, we will develop a more precise way of defining probability.

OBJECTIVE 1 **Find sample spaces and events.** To talk about probability, we need to define an **experiment** and **outcomes**.

DEFINITIONS

Experiment: Any act or process whose result is not known in advance.
Outcomes: The possible results of an experiment.

If an experiment is tossing a coin, then an outcome would be getting a head. If an experiment is rolling two dice (plural of *die*), then an outcome would be getting a total of 7. If an experiment is drawing a card from a deck of cards, then getting the King of Hearts is an outcome. Associated with every experiment is the **sample space**.

DEFINITION

Sample space: The set of all possible outcomes of an experiment.
The sample space is denoted by S.

EXAMPLE 1 Find the sample spaces for each of the following experiments.

a. Toss a single coin.

Solution $S = \{H, T\}$, where H means heads; T tails

b. Toss two coins (or one coin twice).

Solution $S = \{HH, HT, TH, TT\}$, where HT means heads on the first toss and tails on the second

c. Roll a single die.

Solution $S = \{1, 2, 3, 4, 5, 6\}$

d. Roll two dice.

Solution It is convenient to think of the dice as having different colors. Suppose one is black and the other white. The following ordered pairs are written in the form of (B, W) meaning the first number is the number on the black die and the second is the number on the white die.

$$\{(1, 1)\ (1, 2)\ (1, 3)\ (1, 4)\ (1, 5)\ (1, 6)$$
$$(2, 1)\ (2, 2)\ (2, 3)\ (2, 4)\ (2, 5)\ (2, 6)$$
$$(3, 1)\ (3, 2)\ (3, 3)\ (3, 4)\ (3, 5)\ (3, 6)$$
$$(4, 1)\ (4, 2)\ (4, 3)\ (4, 4)\ (4, 5)\ (4, 6)$$
$$(5, 1)\ (5, 2)\ (5, 3)\ (5, 4)\ (5, 5)\ (5, 6)$$
$$(6, 1)\ (6, 2)\ (6, 3)\ (6, 4)\ (6, 5)\ (6, 6)\}$$

Note that there are 36 outcomes in the sample space.

To calculate probabilities, we need to discuss **events**.

DEFINITION *Event:* Any subset of a sample space.

EXAMPLE 2 Find the events that correspond with the following outcomes.

a. Having exactly one head on the toss of two coins.

Solution $E = \{HT, TH\}$

b. Rolling a number greater than 2 with a single roll of a die.

Solution $E = \{3, 4, 5, 6\}$

c. Rolling two dice for a total of 7.

Solution $\{(1, 6)\ (2, 5)\ (3, 4)\ (4, 3)\ (5, 2)\ (6, 1)\}$

OBJECTIVE 2 **Calculate probabilities.** We are now ready to give a more formal definition of probability.

DEFINITION *Probability of an Event* **E:** Given an experiment, the probability of an event E, written as $P(E)$, is the number of outcomes in event E divided by the number of outcomes in the sample space S of the experiment. That is,

$$P(E) = \frac{\text{the number of outcomes in } E}{\text{the number of outcomes in } S}$$

From the definition, we can conclude that the probability of an event must be between 0 and 1 inclusive, that is, $0 \leq P(E) \leq 1$. If an event cannot occur, then the number of outcomes in the event is 0, so the probability of that event is 0. For example, the probability of rolling a die and getting a 7 is 0 since this cannot occur. The event is $\{7\}$, which can occur in 0 ways, and the sample space is $S = \{1, 2, 3, 4, 5, 6\}$, so $P(E) = \dfrac{0}{6} = 0$. If an event is certain to occur, then the number of outcomes in the event is the same as the number of outcomes in the sample space, so the probability is 1. For example, the probability of rolling a die and getting a number less than 7 is 1 since this event is certain to occur. The event is $E = \{1, 2, 3, 4, 5, 6\}$ and the sample space is $S = \{1, 2, 3, 4, 5, 6\}$, so $P(E) = \dfrac{6}{6} = 1$.

Many probability problems involve a standard deck of cards, which consist of 52 cards with 4 suits of 13 cards each. The suits are diamonds and hearts, which are red, and spades and clubs, which are black. Each suit has the cards 2, 3, 4, 5, 6, 7, 8, 9, 10, Jack, Queen, King, and Ace. The cards Jack, Queen, and King are called face cards.

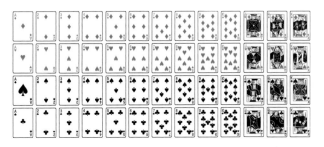

EXAMPLE 3 Find the probabilities of each of the following events.

a. Drawing a King from a standard deck on a single draw.

Solution Since there are 52 cards in the deck, there are 52 possible outcomes in the sample space. Since there are 4 Kings in the deck, the event has 4 outcomes. Therefore, $P(E) = \dfrac{4}{52} = \dfrac{1}{13}$.

b. Rolling a total of 7 on a single roll of two dice

Solution We found in Example 1d that the sample space for rolling two dice has 36 outcomes, and in Example 2c we found 6 different ways of getting a total of 7. Thus, $P(E) = \dfrac{6}{36} = \dfrac{1}{6}$.

c. A club has 20 members, of which 8 are men and 12 are women. If a committee of 5 is chosen to revise the bylaws, what is the probability that all 5 are women?

Solution $P(E) = \dfrac{\text{number of ways of selecting 5 women from 12}}{\text{number of ways of selecting 5 members from 20}}$

The number of ways of selecting 5 women from 12 is $_{12}C_5 = \dfrac{12!}{5!(12 - 5)!} = \dfrac{12!}{5!7!} = 792$. The number of ways of selecting 5 members from 20 is $_{20}C_5 = \dfrac{20!}{5!(20 - 5)!} = \dfrac{20!}{5!15!} = 15{,}504$. Therefore, $P(E) = \dfrac{792}{15{,}504} = \dfrac{33}{646}$.

1. What is the sample space for an experiment?

2. What is an event?

3. The probability of any event lies between what two numbers inclusively?

4. What is the probability of an event that cannot happen?

5. Define the probability of an event.

6. How many face cards are in a standard deck of cards?

For Exercises 7–10, find the sample space for each experiment.

7. Selecting a vowel from the English alphabet

8. A jar contains 1 red, 1 blue, and 1 white ball; draw a ball, replace it, and draw again

9. Tossing 3 coins (or 1 coin 3 times)

10. Jar 1 contains a blue marble and a white marble; jar 2 contains a red marble and a black marble; select a marble from jar 1 and then select a marble from jar 2.

For Exercises 11–14, find the number of elements in the sample space of the experiment.

11. Selecting a two-digit number

12. Rolling a die and tossing a coin

13. Choosing a committee of 4 from a club with 16 members

14. Drawing 5 cards from a standard deck of cards

For Exercises 15–18, find the event that corresponds with the following outcomes.

15. Having 1 head if 3 coins are tossed

16. Having a number greater than 6 on a single roll of a die

17. Rolling a total of 4 on a single roll of 2 dice

18. Rolling a total of 13 on a single roll of 2 dice

For Exercises 19 and 20, a single die is rolled. Find the probability of each of the following events.

19. Having a 5

20. Having a number greater than 3

For Exercises 21–24, two coins are tossed. Find the probability of each of the following events.

21. Having exactly 1 head

22. Having exactly 2 tails

23. Having at least 1 tail

24. Having more than 2 heads

For Exercises 25–28, two dice are rolled. Find the probability of each of the following events.

25. Having a total of 4

26. Having a total of 8

27. Having a total less than 4

28. Having a total greater than 7

For Exercises 29–32, a single card is drawn from a standard deck. Find the probability of each of the following events.

29. Drawing a 10

30. Drawing the Jack of spades

31. Drawing a face card

32. Drawing a number less than 5

For Exercises 33–34, use the spinner shown to find the probability of each of the following events.

33. The spinner stopping on red

34. The spinner stopping on blue

For Exercises 35–40, we have a jar that contains 1 blue, 1 red, and 1 pink ball. Find the probability of each of the following events.

35. Drawing a pink ball on a single draw

36. Drawing a blue ball on a single draw

37. Draw a ball, do not replace it, and then draw a second ball. What is the probability that both balls are blue?

38. Draw a ball, do not replace it, and draw a second ball. What is the probability that both balls are red?

39. Draw a ball, replace it, and draw again. What is the probability that both balls are red?

40. Draw a ball, replace it, and draw again. What is the probability that the first ball is blue and the second ball is pink?

41. A farm pond is stocked with 80 warmouth perch, 140 bluegills, and 200 shellcrackers. If we assume that each fish is equally likely to be caught, what is the probability that the first fish caught is a shellcracker?

42. The quality-control department of a company that manufactures car batteries has determined that 3% of the batteries produced are defective. If one battery that is manufactured by this company is selected at random, what is the probability that it is defective?

43. A committee has 12 members of which 8 are Democrats and 4 are Republicans. A subcommittee of 3 is appointed. What is the probability that all 3 are Republicans?

44. A club has 20 members of which 13 are women and 7 are men. If a committee of 6 is appointed, what is the probability that all 6 are women?

45. A jar contains 14 balls of which 8 are pink and 6 are purple. If 4 balls are drawn at random, what is the probability that all 4 are pink?

46. A box contains 18 pieces of paper of which 9 are blue and 9 are white. If 4 pieces are selected at random, what is the probability that all 4 are blue?

GLOSSARY

Absolute value: A given number's distance from 0 on a number line.

Additive inverses: Two numbers whose sum is 0.

Area: The total number of square units that fill a figure.

Arithmetic sequence: A sequence in which each term after the first is found by adding the same number to the previous term. The common difference is $d = a_n - a_{n-1}$.

Arithmetic series: The sum of an arithmetic sequence.

Axis: A number line used to locate a point in a plane.

Axis of symmetry: A line that divides a graph into two symmetrical halves.

Base: The number that is repeatedly multiplied.

Binomial: A polynomial containing two terms.

Binomial coefficient: A number written as $\binom{n}{r}$ and defined as $\dfrac{n!}{r!(n-r)!}$.

Circle: The set of points that are equally distant from a central point. The central point is the center.

Circumference: The distance around a circle.

Coefficient: The numerical factor in a term.

Combination: An unordered arrangement of objects. The combination of n things taken r at a time is often denoted as nCr.

Common logarithms: Logarithms with a base of 10. $\log_{10} x$ is written $\log x$.

Complementary angles: Two angles are complementary if the sum of their measures is 90°.

Complex conjugates: The complex conjugate of a complex number $a + bi$ is $a - bi$.

Complex number: A number that can be expressed in the form $a + bi$, where a and b are real numbers and i is the imaginary unit.

Complex rational expression: A rational expression that contains rational expressions in the numerator or denominator.

Compound inequality: An inequality containing two inequality symbols.

Composition of functions: If f and g are functions, then the composition of f and g is defined as $(f \circ g)(x) = f[g(x)]$ for all x in the domain g for which $g(x)$ is in the domain of f. The composition of g and f is defined as $(g \circ f)(x) = g[f(x)]$ for all x in the domain of f for which $f(x)$ is in the domain of g.

Congruent angles: Angles that have the same measure. The symbol for congruent is $\cong$.

Conjugates: Binomials that differ only in the sign separating the terms.

Consecutive integers: $\ldots -3, -2, -1, 0, 1, 2, 3, \ldots$

Consistent system of equations: A system of equations that has at least one solution.

Constant: A symbol that does not vary in value.

Cramer's Rule:

The solution to the system of linear equations

$$\begin{cases} a_1 x + b_1 y = c_1 \\ a_2 x + b_2 y = c_2 \end{cases} \text{ is }$$

$$x = \frac{\begin{vmatrix} c_1 & b_1 \\ c_2 & b_2 \end{vmatrix}}{\begin{vmatrix} a_1 & b_1 \\ a_2 & b_2 \end{vmatrix}} = \frac{D_x}{D} \quad \text{and} \quad x = \frac{\begin{vmatrix} a_1 & c_1 \\ a_2 & c_2 \end{vmatrix}}{\begin{vmatrix} a_1 & b_1 \\ a_2 & b_2 \end{vmatrix}} = \frac{D_y}{D}$$

The solution to the system of linear equations

$$\begin{cases} a_1 x + b_1 y = c_1 z = d_1 \\ a_2 x + b_2 = c_2 z = d_2 \text{ is} \\ a_3 x + b_3 y = c_3 z = d_3 \end{cases}$$

Note: Each denominator, D, is the determinant of a matrix containing only the coefficients in the system. To find D_x, we replace the column of x-coefficients in the coefficient matrix with the constants from the system. To find D_y, we replace the column of y-coefficients in the coefficient matrix with the constant terms, and do likewise to find D_z.

$$x = \frac{\begin{vmatrix} d_1 & b_1 & c_1 \\ d_2 & b_2 & c_2 \\ d_3 & b_3 & c_3 \end{vmatrix}}{\begin{vmatrix} a_1 & b_1 & c_1 \\ a_2 & b_2 & c_2 \\ a_3 & b_3 & c_3 \end{vmatrix}} = \frac{D_x}{D}, \quad y = \frac{\begin{vmatrix} a_1 & d_1 & c_1 \\ a_2 & d_2 & c_2 \\ a_3 & d_3 & c_3 \end{vmatrix}}{\begin{vmatrix} a_1 & b_1 & c_1 \\ a_2 & b_2 & c_2 \\ a_3 & b_3 & c_3 \end{vmatrix}} = \frac{D_y}{D}, \text{ and}$$

$$z = \frac{\begin{vmatrix} a_1 & b_1 & d_1 \\ a_2 & b_2 & d_2 \\ a_3 & b_3 & d_3 \end{vmatrix}}{\begin{vmatrix} a_1 & b_1 & c_1 \\ a_2 & b_2 & c_2 \\ a_3 & b_3 & c_3 \end{vmatrix}} = \frac{D_z}{D},$$

Degree: The sum of the exponents of all variables in a monomial.

Degree of a polynomial: The greatest degree of any of the terms in the polynomial.

Dependent equations: Equations with identical graphs.

Diameter: The distance across a circle through its center.

Direct variation: Two variables, y and x, are in direct variation if $y = kx$, where k is a constant.

Discriminant: The radicand $b^2 - 4ac$ in the quadratic formula.

Domain: The set of all input values for a relation.

Echelon form: An augmented matrix whose coefficient portion has 1's on the diagonal from upper left to lower right and 0's below the 1's.

Ellipse: The set of all points the sum of whose distance from two fixed points is constant.

Equation: Two expressions set equal to each other.

Event: Any subset of a sample space.

Experiment: Any act or process whose result is not known in advance. The possible results of an experiment are called outcomes.

Exponent: A symbol written to the upper right of a base number that indicates how many times to use the base as a factor.

Exponential function: If $b > 0$, $b \neq 1$, and x is any real number, then the exponential function is $f(x) = b^x$.

Expression: A collection of constants, variables, and arithmetic symbols.

Extraneous solution: An apparent solution that does not solve its equation.

Factored form: A number or expression written as a product of factors.

Factorial notation: For any natural number n, the symbol $n!$ (read n factorial) means $n(n - 1)(n - 2)\ldots 3 \cdot 2 \cdot 1$. 0! is defined to be 1, so $0! = 1$.

Factors: If $a \cdot b = c$, then a and b are factors of c.

Finite sequence: A function with a domain that is the set of natural numbers from 1 to n.

Formula: An equation that describes a mathematical relationship.

Fraction: A quotient of two numbers.

Function: A relation in which every value in the domain is paired with exactly one value in the range.

Geometric sequence: A sequence in which every term after the first is found by multiplying the previous term by the same number. The common ratio is $r = \dfrac{a_n}{a_{n-1}}$.

Geometric series: The sum of the terms of a geometric sequence.

Greatest common factor (GCF): The largest natural number that divides all given numbers evenly.

Hyperbola: The set of all points the difference of whose distances from two fixed points remains constant.

Identity: An equation that has every real number as a solution (excluding any numbers that cause an expression in the equation to be undefined).

Imaginary number: A number that can be expressed in the form bi, where b is a real number and i is the imaginary unit.

Imaginary unit: The number represented by i, where $i = \sqrt{-1}$, and $i^2 = -1$.

Inconsistent system of equations: A system of equations that does not have a solution.

Independent equations: Equations that have different graphs.

Inequality: Two expressions separated by $\neq$, $<$, $>$, $\leq$, or $\geq$.

Infinite sequence: A function with a domain that is the set of natural numbers.

Intersection: For two sets A and B, the intersection of A and B, symbolized by $A \cap B$, is a set containing only elements that are in both A and B.

Inverse functions: Two functions f and g are inverses if and only if $(f \circ g)(x) = x$ for all x in the domain of g and $(g \circ f)(x) = x$ for all x in the domain of f.

Inverse variation: Two variables, y and x, are in inverse variation if $y = \dfrac{k}{x}$, where k is a constant.

Irrational number: Any real number that is not rational.

Least common denominator (LCD): The least common multiple of the denominators of a given set of fractions.

Least common multiple (LCM): The smallest number that is evenly divisible by each number in a given set of numbers.

Like radicals: Two radical expressions with identical radicands (and indexes).

Like terms: Variable terms that have the same variable(s) raised to the same exponents, or constant terms.

Linear equation: An equation in which each variable term contains a single variable raised to an exponent of 1.

Linear equation in one variable: An equation that can be written in the form $ax + b = c$, where a, b, and c are real numbers and $a \neq 0$.

Linear inequality: An inequality containing expressions in which each variable term contains a single variable with an exponent of 1.

Lowest terms: Given a fraction $\dfrac{a}{b}$ and $b \neq 0$, if the only factor common to both a and b is 1, then the fraction is in lowest terms.

Matrix: A rectangular array of numbers.

Minor: The determinant of the remaining matrix when the row and column in which the element is located are ignored.

Monomial: An expression that is a constant or a product of a constant and variables that are raised to whole number powers.

Multiple: A number that is evenly divisible by a given number.

Multiplicative inverses: Two numbers whose product is 1.

Natural logarithm: Base-e logarithms are called natural logarithms and $\log_x$ is written as $\ln x$. Note that $\ln e = 1$.

***n*th root:** The number b is the nth root of a number a if $b^n = a$.

One-to-one function: A function is one to one if each input (value from the domain) corresponds to only one output (value from the range) and every output corresponds to only one input.

Percent: A ratio representing some part out of 100.

Perimeter: The distance around a figure.

Permutation: An ordered arrangement of objects. The permutation of n things taken r at a time is often denoted as nPr.

Polynomial: A monomial or an expression that can be written as a sum of monomials.

Polynomial in one variable: A polynomial in which every variable term has the same variable.

Prime factorization: A factorization that contains only prime factors.

Prime number: A natural number that has exactly two different factors, 1 and the number itself.

Proportion: An equation in the form $\frac{a}{b} = \frac{c}{d}$, where $b \neq 0$ and $d \neq 0$.

Quadrant: One of four regions created by the intersection of the axes in the coordinate plane.

Quadratic equation: An equation that can be written in the form $ax^2 + bx + c = 0$, where a, b, c are all real numbers and $a = 0$.

Quadratic inequality: An inequality that can be written in the form $ax^2 + bx + c = 0$ or $ax^2 + bx + c \leq 0$, where a, b, and c are all real numbers and $a \neq 0$.

Radical equation: An equation containing at least one radical expression whose radicand has a variable.

Radical function: A function of the form $f(x) = \sqrt[n]{P}$, where P is a polynomial.

Radius: The distance from the center of a circle to any point on the circle.

Range: The set of all output values for a relation.

Ratio: A comparison of two quantities using a quotient.

Rational expression: An expression that can be written in the form $\frac{P}{Q}$, where P and Q are polynomials and $Q \neq 0$.

Rational inequality: An inequality containing a rational expression.

Rational number: Any real number that can be expressed in the form $\frac{a}{b}$, where a and b are integers and $b \neq 0$.

Relation: A set of ordered pairs.

Sample space: The set of all possible outcomes of an experiment. The sample space is denoted by S.

Scientific notation: A number expressed in the form $a \times 10^n$, where a is a decimal number with $1 \leq |a| < 10$ and n is an integer.

Sequence: An ordered list whose domain is a subset of the natural numbers.

Set: A collection of objects.

Similar figures: Figures with congruent angles and proportional side lengths.

Slope: The ratio of a line's vertical change to its horizontal change.

Solution: A number that makes an equation true when it replaces the variable in the equation.

Solution for a system of equations: An ordered set of numbers that makes all equations in the system true.

Supplementary angles: Two angles are supplementary if the sum of their measures is 180°.

System of equations: A group of two or more equations.

Term: An expression that is separated by addition or subtraction.

Trinomial: A polynomial containing three terms.

Unit ratio: A ratio with a denominator of 1.

Union: For two sets A and B, the union of A and B, symbolized by $A \cup B$, is a set containing every element in A or in B.

Variable: A symbol that can vary in value.

Vertex: The lowest point on a parabola that opens up or highest point on a parabola that opens down.

Volume: The total number of cubic units that fill a space.

***x*-intercept:** A point where a graph intersects the x-axis.

***y*-intercept:** A point where a graph intersects the y-axis.

INDEX

B

C

Z

Zero
  absolute, 40
  as an exponent, 379, 835
  division involving, 51–52
  multiplicative property of, 43
  as subscript, 149
Zero factor theorem, 82, 508, 904
Zero function on a graphing calculator, 924
Zero slope, 307

PHOTOGRAPH CREDITS

ANSWERS

Chapter 1

1.1 Exercises
1. A collection of objects **3.** Rational numbers can be expressed as a ratio of integers and irrational numbers cannot. **5.** true **7.** true **9.** False, all real numbers are either rational or irrational. **11.** {Sunday, Monday, Tuesday, Wednesday, Thursday, Friday, Saturday} **13.** {a, e, i, o, u} **15.** {5, 10, 15, 20, ...} **17.** {9, 11, 13, 15, ...} **19.** {−6, −5, −4, −3} **21.** rational **23.** irrational **25.** irrational **27.** rational

29. **31.**

33. **35.**

37. 5.4;

39. −3.7;

41. 4.5 **43.** $3\frac{1}{8}$ **45.** 2 **47.** 67.8 **49.** > **51.** = **53.** <

55. = **57.** < **59.** > **61.** $-4.7, -2.56, 5.4, \left|-7\frac{1}{2}\right|, |8.3|$

63. $-0.6, -0.44, 0, |-0.02|, 0.4, \left|1\frac{2}{3}\right|, 3\frac{1}{4}$

1.2 Exercises
1. Multiply or divide the numerator and denominator by the same number. **3.** No, all numbers have only one prime factorization. **5.** Find all of the prime factors of a given number. **7.** True, prime number are natural numbers and every natural number is a whole number. **9.** True, composite numbers can be factored into a combination of prime numbers. **11.** $\frac{3}{10}$ **13.** $\frac{1}{3}$ **15.** $\frac{3}{4}$ **17.** $\frac{15}{16}$ **19.** 12 **21.** −35 **23.** 15 **25.** 16 **27.** $\frac{12}{15}$ and $\frac{10}{15}$ **29.** $\frac{8}{18}$ and $\frac{21}{18}$ **31.** $-\frac{44}{72}$ and $-\frac{51}{72}$ **33.** $\frac{9}{144}$ and $\frac{60}{144}$ **35.** $2 \cdot 2 \cdot 3 \cdot 3$

37. $2 \cdot 5 \cdot 5 \cdot 5$ **39.** $2 \cdot 2 \cdot 11$ **41.** $2 \cdot 2 \cdot 2 \cdot 2 \cdot 2 \cdot 2$ **43.** $\frac{4}{5}$ **45.** $-\frac{3}{4}$ **47.** $-\frac{3}{4}$ **49.** $\frac{5}{7}$ **51.** Incorrect, you may only divide out factors, not addends. **53.** $\frac{7}{8}$ **55.** $\frac{12}{23}$ **57.** $\frac{5}{21}$ **59.** $\frac{2}{3}$ **61.** $\frac{13}{25}$ **63.** $\frac{31}{50}$ **65. a.** 1975 **b.** $\frac{277}{500}$ **67.** $\frac{14}{69}$ **69.** $\frac{6}{73}$ **71.** $\frac{2}{3}$ **73.** $\frac{71}{145}$ **75.** $\frac{4}{15}$

Review Exercises
1. Expression because it does not contain an equals symbol. **2.** {Mercury, Venus, Earth, Mars} **3.** Rational because it can be written as a ratio of the integers 8 and 10. **4.** **5.** 27 **6.** =

1.3 Exercises
1. The commutative property of addition changes the order while the associative property of addition changes the grouping. **3.** Adding 0 to a number or expression gives a number or expression identical to the original. **5.** When adding two numbers that have the same sign, add their absolute values and keep the same sign. **7.** To write a subtraction statement as an equivalent addition statement, change the operation symbol from a minus sign to a plus sign and change the subtrahend to its additive inverse. **9.** commutative property of addition **11.** additive inverse **13.** commutative property of addition **15.** associative property of addition **17.** additive inverse **19.** commutative property of addition **21.** −6 **23.** −4 **25.** 14 **27.** $-\frac{2}{3}$ **29.** $-\frac{2}{3}$ **31.** $\frac{11}{12}$ **33.** $-\frac{3}{4}$ **35.** $-\frac{9}{14}$ **37.** 0.26 **39.** 6.52 **41.** −4.38 **43.** −5 **45.** 0 **47.** $-\frac{5}{6}$ **49.** 0.29 **51.** x **53.** $-\frac{m}{n}$ **55.** 7 **57.** −6 **59.** −9 **61.** $-\frac{18}{35}$ **63.** $\frac{13}{10}$ **65.** 0 **67.** 0.36 **69.** −1.6 **71.** −5.75 **73.** −6 **75.** −56 **77.** 1.62 **79.** $-\frac{11}{24}$ **81.** $21,687 million profit **83.** −268.2 N; The negative indicates the beam is moving downward. **85.** 96.25 **87.** $69\frac{3}{4}$ **89.** $-208 + x = -196; x = 12$ **91.** $x = 20.9 - 19.1$ **93.** 1989, ≈ $16,000

Review Exercises   1. {Washington, Adams, Jefferson, Madison}   2. 7   3. 5   4. <   5. $2 \cdot 2 \cdot 5 \cdot 5$ 6. $\dfrac{6}{7}$

1.4 Exercises   1. positive   3. negative   5. There is no quotient that we can multiply by 0 to get 5.   7. Their product is 1.   9. An odd number of negatives in division will result in a negative answer.   11. distributive property   13. associative property of multiplication   15. multiplicative identity   17. commutative property of multiplication   19. multiplicative property of 0   21. multiplicative identity   23. commutative property of multiplication   25. -18 27. -20 29. 54   31. 720   33. $-\dfrac{1}{4}$ 35. $\dfrac{1}{2}$ 37. $-\dfrac{4}{9}$ 39. -38 41. 6.72   43. -6.536 45. $-\dfrac{1}{5}$ 47. $\dfrac{3}{2}$ 49. $-\dfrac{2}{5}$ 51. There is no multiplicative inverse of 0.   53. -3 55. -6 57. 2   59. undefined   61. indeterminate   63. 9   65. $-\dfrac{2}{3}$ 67. $\dfrac{3}{4}$ 69. $-\dfrac{14}{15}$ 71. 2.1   73. 91.8   75. $-46.\overline{6}$ 77. $\dfrac{1}{2}$ 79. $-\$8680$ 81. 612.8   83. -402.5 lbs.; the force is downward   85. 169,202.2245 kg   87. -51.2 V   89. 3600 W

Review Exercises   1. irrational

2.   3. -22 4. $-\dfrac{2}{3}$

5. $-\dfrac{11}{24}$ 6. 6.8

1.5 Exercises   1. two cubed   3. A square root is the inverse of a square.   5. either   7. Base: -5 Exponent: 3 "negative 5 cubed"   9. Base: 2 Exponent: 7 "additive inverse of two to the seventh power"   11. 81   13. -64 15. -8 17. 1   19. $\dfrac{1}{25}$ 21. $-\dfrac{27}{64}$ 23. 0.008   25. -68.921 27. ± 11 29. ± 9 31. ± 14 33. ± 16 35. 4   37. 12   39. 0.7   41. Not a real number.   43. $\dfrac{8}{9}$ 45. 5   47. 19   49. 15   51. 7   53. -7 55. 26.8   57. 10   59. $-\dfrac{253}{300}$ 61. 22   63. 1   65. 12   67. -100.8 69. 0   71. 4   73. undefined   75. associative property of multiplication The multiplication was not performed from left to right.   77. distributive property The parentheses were not simplified first.   79. Mistake: Multiplied before dividing. Correct: 1   81. Mistake: Did not subtract under radical. Correct: 24   83. 77.2   85. 96   87. $1.75625   89. $\approx \$51.3$

Review Exercises   1. Expression because there is no equals symbol.   2. {1, 2, 3, 4, 5, 6, 7, 8, 9, 10}

3.   4. 24   5. 3   6. $-\dfrac{1}{9}$

1.6 Exercises   1. sum, plus, added   3. product, times, twice   5. less than, subtracted from   7. $4x$ 9. $4x + 16$ 11. $5 + y$ 13. n^5 15. $\dfrac{m}{14}$ 17. $7w - \dfrac{1}{5}$ 19. $(2 - l)^3$

21. $mn - 5$ 23. $-27 - (a + b)$ 25. $\dfrac{4}{5}n - 7$ 27. $\sqrt{y} - mn$ 29. $n - 3(n - 6)$ 31. Mistake: Multiplied x by 9 instead of the sum of x and y. Correct: $9(x + y)$ 33. Mistake: Order is incorrect. Correct: $m^2 - 4$ 35. $5 + w$ 37. $3w$ 39. $\dfrac{1}{2}d$ 41. $42 - n$ 43. $t + \dfrac{1}{4}$ 45. $2w + 2l$ 47. $\dfrac{1}{3}\pi r^2 h$ 49. $\sqrt{(x_2 - x_1)^2 + (y_2 - y_1)^2}$ 51. mc^2 53. One half the product of the base and height.   55. The product of the length, width, and height.   57. Product of the length and width added to the product of the length and height added to the product of the width and height, all doubled.   59. The ratio of the difference of y_2 and y_1 to the difference of x_2 and x_1.

Review Exercises   1. Any real number that can be expressed in the form $\dfrac{a}{b}$, where a and b are integers and $b \neq 0$.   2. additive identity   3. distributive property   4. -28 5. 12   6. -9

1.7 Exercises   1. To evaluate an expression: 1. Replace the variables with their corresponding given numbers. 2. Calculate the numerical expression using the order of operations.   3. An expression is undefined when the denominator is equal to 0.   5. The numerical factor in a term.   7. 9   9. 3.2   11. $\dfrac{11}{2}$ 13. -14 15. -12 17. -48 19. a. 25   b. -12 21. a. 2   b. $\dfrac{2}{3}$ 23. -5 25. $-1, 3$ 27. 0   29. $\dfrac{2}{3}$ 31. $6a + 12$ 33. $-32 + 24y$ 35. $\dfrac{7}{16}c - 14$ 37. $0.6n - 1.6$ 39. -6 41. 1   43. -1 45. $\dfrac{1}{5}$ 47. $8y$ 49. $-\dfrac{1}{3}b^2$ 51. $-1.9n$ 53. $-26c$ 55. $5x - 1$ 57. $-10x + 4y - 1$ 59. $-9m + 5n + y - 11$ 61. $\dfrac{5}{6}c + \dfrac{5}{2}d + \dfrac{1}{7}$ 63. $\dfrac{25}{4}a + \dfrac{18}{35}b^2 + 4$ 65. $-1.3x - 0.4$ 67. a. $14 + (6n - 8n)$ b. $14 - 2n$ c. 20

Review Exercises   1. -22 2. 6.1   3. $\dfrac{3}{2}$ 4. 0   5. 4   6. $16 + 3n$

Chapter 1 Review Exercises   1. false   2. false   3. false   4. true   5. true   6. true   7. positive   8. prime   9. keep   10. even   11. {March, May}   12. {a, l, g, e, b, r}   13. {2, 4, 6, ...}   14. {..., $-6, -4, -2, 0, 2, 4, ...$}

15.

16.

17.

18.

19. 6.3 **20.** 8.46 **21.** $2\frac{1}{6}$ **22.** $4\frac{3}{8}$ **23.** $=$ **24.** $<$ **25.** $<$
26. $=$ **27.** $\frac{11}{15}$ **28.** $\frac{3}{8}$ **29.** 12 **30.** 25 **31.** 5 **32.** -10
33. $3 \cdot 17$ **34.** $2 \cdot 2 \cdot 5 \cdot 5$ **35.** $2 \cdot 2 \cdot 3 \cdot 3 \cdot 3$ **36.** $2 \cdot 2 \cdot 3 \cdot 7$
37. $\frac{19}{25}$ **38.** $\frac{1}{2}$ **39.** $\frac{25}{36}$ **40.** $\frac{2}{3}$ **41.** associative property of
addition **42.** distributive property **43.** commutative property
of multiplication **44.** commutative property of addition
45. commutative property of multiplication **46.** associative
property of multiplication **47.** -3.8 **48.** 15.1 **49.** $\frac{11}{15}$
50. $\frac{5}{12}$ **51.** 91 **52.** -48 **53.** -5 **54.** 10 **55.** -36 **56.** 9
57. 52 **58.** 9 **59.** 160 **60.** 1 **61.** $-\frac{28}{3}$ **62.** -8.6 **63.** -1
64. 27 **65.** 7 **66.** -103 **67.** 3 **68.** 24 **69.** $14 - 2n$ **70.** $\frac{y}{7}$
71. $y - 2(y + 4)$ **72.** $7(m - n)$ **73.** 61 **74.** 52 **75.** -8
76. 7 **77.** 5 **78.** -4 **79.** -6 **80.** $-3, 4$ **81.** $5x + 30$
82. $-15n + 24$ **83.** $2y + \frac{3}{4}$ **84.** $2.7m - 1.26$
85. $-3x - 3y - 15$ **86.** $4y^2 + 7y$ **87.** $2xy$ **88.** $-2x^3 - 6x^2$
89. $-8m$ **90.** $-14x - 6y + 6$ **91.** $-\$503.59$ **92. a.** \$1.50
b. \$6.50 profit **93. a.** \$9700 **b.** \$1300 **94.** -152 V
95. ≈ 6.2 **96.** $\$54{,}158\frac{1}{3}$ or \$54,158.33

Chapter 1 Practice Test

1. **2.** 3.67 **3.** $2 \cdot 2 \cdot 5 \cdot 5$ **4.** $\frac{1}{3}$

5. 12 **6.** distributive property **7.** commutative property of
addition **8.** 4 **9.** $1\frac{17}{24}$ or $\frac{41}{24}$ **10.** 0.6 **11.** $-\frac{5}{4}$ **12.** -64
13. ± 9 **14.** -8 **15.** 22 **16.** 2 **17.** 5 **18.** 2 **19.** 10
20. \$250 **21.** $5.58\overline{3}$ **22.** -11 **23.** 10 **24.** $-20y - 45$
25. $5.6x + 1.3$

Chapter 2

2.1 Exercises
1. An expression does not have an equal
symbol. **3.** To check a value to see if it is a solution to a given
equation: 1. Replace the variable(s) in the equation with the
value. 2. If the resulting equation is true, then the value is a
solution. **5.** no **7.** yes **9.** no **11.** yes **13.** no **15.** no
17. yes **19.** yes **21.** yes **23.** yes **25.** no **27.** no
29. no **31. a.** 66 ft. **b.** 6 **c.** \$53.94 **33.** ≈ 12.57 km
35. a. 78 ft.^{2} **b.** 8 bails **c.** \$36 **37.** 7000 ft.^{2} **39.** \$6926.40
41. 82.25 ft.^{2} **43.** ≈ 6.9 ft.^{2} **45.** 47.5 m^3 **47.** ≈ 1005.3 cm^3
49. 83,068,741.$\overline{6}$ ft.^{3} **51.** 1747.5 miles **53.** 64.5 mph
55. ≈ 54.3 mph **57.** -52 V **59.** 93.2°F **61.** 2795°F
63. -60°C

Review Exercises **1.** $8x - 5$ **2.** $5.4y + 5.5$ **3.** $\frac{1}{3}x + 2$
4. $-4w + 6$ **5.** 59 **6.** -34

2.2 Exercises **1.** The addition principle of equality says
that we can add (or subtract) the same amount on both sides of
an equation without affecting its solution(s). **3.** Add the same
number to both sides of the equation. **5.** Get the variable
terms together on the same side of the equal sign. Use the
addition principle. **7.** yes **9.** no **11.** no **13.** yes **15.** yes
17. no **19.** yes **21.** 9 **23.** -25 **25.** $-\frac{3}{40}$ **27.** $-\frac{11}{24}$
29. -8.2 **31.** 6.2 **33.** 21 **35.** 7 **37.** 13 **39.** 6 **41.** 13.2
43. -1 **45.** -42 **47.** -10 **49.** 16 **51.** -8.8
53. No solution. **55.** All real numbers.
57. $1947 + x = 2373$; \$426 **59.** $16 + x = 42$; 26 mi.
61. $23 + x = 28$; 5 **63.** $x + 12.4 + 16.3 + 27.2 = 67.2$; 11.3 cm
65. $x + 19 + 10 = 54$; 25 ft. **67.** $x + \frac{1}{3} + \frac{2}{5} = 1$; $\frac{4}{15}$

Review Exercises **1.** -33.2 **2.** -9 **3.** 37 **4.** -19
5. $8x^2 - 5x - 9$ **6.** $12x - 20$

2.3 Exercises **1.** We can multiply (or divide) both sides of
an equation by the same amount without affecting its solution(s).
3. Multiply both sides of the equation by the multiplicative inverse of that coefficient. **5.** Multiply both sides of the equation
by a number that will clear the denominators. **7.** -3 **9.** 7
11. -5 **13.** $\frac{20}{3}$ **15.** 16 **17.** 0 **19.** 28 **21.** 16 **23.** $\frac{48}{13}$
25. -9 **27.** $-\frac{2}{3}$ **29.** All real numbers. **31.** -2
33. Mistake: The minus sign was dropped from in front of 11.
Correct: 2 **35.** Mistake: Did not distribute the minus sign to
the 2. Correct: $\frac{13}{3}$ **37.** 14 ft. **39.** 19 cm **41.** 52 ft. **43.** 110 ft.
45. ≈ 2 in. **47.** ≈ 6374.8 km **49.** ≈ 4.8 m **51.** -9.5 A
53. 4 m/sec.^{2} **55.** 945.5 kg **57.** 413.6 m/sec. **59.** \$286.67

Review Exercises **1.** 70 **2.** $3.8x^2 - 7.63x + 1.5$
3. $-\frac{9}{2}x - 6$ **4.** 20 **5.** -25 **6.** -10

2.4 Exercises **1.** Add 2 on both sides. Since 2 is sub-
tracted from w, adding 2 eliminates the 2's leaving w isolated.
3. Subtract $2x$ from both sides. Subtracting the $2x$ isolates the
$3y$ term, which contains the variable that we wish to isolate.
5. $t = 4u + v$ **7.** $y = -\frac{x}{5}$ **9.** $x = \frac{b + 3}{2}$ **11.** $b = -\frac{c}{a}$
13. $y = \frac{8 - 3x}{4}$ **15.** $Y = \frac{mn}{4} - f$ **17.** $c = \frac{7 - 6d}{6}$
19. $n = c\left(\frac{3}{4} + 5a\right)$ **21.** $x = \frac{2c - 60}{15}$ **23.** $a = P - b - c$
25. $l = \frac{A}{w}$ **27.** $r^2 = \frac{V}{\pi h}$ **29.** $d = \frac{C}{\pi}$ **31.** $b = \frac{2A}{h}$
33. $w = \frac{P - 2l}{2}$ **35.** $l = \frac{2S - na}{n}$ **37.** $C = R - P$
39. $r = \frac{I}{Pt}$ **41.** $r = \frac{A - P}{Pt}$ **43.** $t = \frac{d}{r}$ **45.** $d = \frac{W}{F}$

47. $t = \dfrac{v - v_0}{-32}$ or $t = \dfrac{v_0 - v}{32}$ **49.** $C = \dfrac{5}{9}(F - 32)$

51. $t = \dfrac{W}{P}$ **53.** $m = \dfrac{FR^2}{GM}$ **55.** Mistake: Subtracted the

coefficient 7 instead of dividing. Correct: $t = \dfrac{54 - 3n}{7}$

57. Mistake: Multiplied 5 times -2. Correct: $m = \dfrac{5nk + 2}{3}$

Review Exercises
1. $>$ **2.** 0.36 **3.** -16 **4.** -6
5. $4 + 7n$ **6.** $3(x + 2) - 9$

2.5 Exercises
1. sum, plus, added **3.** of, times, twice
5. The order of the translation is different. **7.** $6 - x = -3; 9$
9. $c \div 4 = 7; 28$ **11.** $y - 5 = -1; 4$ **13.** $7 + 3w = 34; 9$

15. $5a - 9 = 76; 17$ **17.** $8(8 + t) = 160; 12$

19. $3m - 11 = m + 5; 8$ **21.** $3(a + 4) + 2a = -3; -3$

23. $60 = \dfrac{4y}{3}; 45$ **25.** $m \div 4 = 1.6; 6.4$

27. $k \div (-6.5) = 4.2; -27.3$ **29.** $10 = \dfrac{r}{4} - 1; 44$

31. $(2d - 8) - (d - 12) = 11; 7$ **33.** $\dfrac{1}{3}z = \dfrac{5}{6}; \dfrac{5}{2}$

35. $\dfrac{1}{3}(g + 2) = 1; 1$ **37.** $5\left(x + \dfrac{1}{3}\right) = 3x + \dfrac{2}{3}; -\dfrac{1}{2}$

39. $j + 4 = 3j + 16; -6$ **41.** $\dfrac{n - 3}{3} = n \div 6; 6$

43. Four times a number added to three is seven. **45.** Six times the sum of a number and four is equal to the product of negative 10 and the number. **47.** Half a number less three will result in two-thirds the difference of the number and eight.
49. Five hundredths of a number added to six hundredths of the difference of the number and eleven is twenty-two.
51. Four-thirds of the sum of a number and one will result in the number added to four. **53.** Mistake: Subtraction translated in reverse order. Correct: $n - 10 = 40$ **55.** Mistake: Multiplied 5 times the unknown number instead of the difference, which requires parentheses. Correct: $5(x - 6) = -2$
57. Mistake: Subtracted the unknown number instead of the sum, which requires parentheses. Correct: $2t - (t + 3) = -6$

59. Translation: $P = 2(l + w)$ **a.** 84 ft. **b.** 61 cm **c.** $37\dfrac{1}{4}$ in.

61. Translation: $P = b + 2s$ **a.** 27 in. **b.** 2.7 m **c.** $34\dfrac{1}{4}$ in.

63. Translation: $I = Prt$ **a.** \$400 **b.** \$7.50 **c.** \$30

65. Translation: $t = \dfrac{e - d}{153.8}$ **a.** ≈ 49 sec. **b.** ≈ 52 sec.

c. ≈ 62 sec.

Review Exercises
1. True, 19 is equal to 19. **2.** $=$
3. $>$ **4.** -12 **5.** -25 **6.** -30

2.6 Exercises
1. Any number that can replace the variable(s) in the inequality and make it true. **3.** Any number greater than -5 and -5 itself.

5. $\{x \mid x \geq -3\}$

7. $\{h \mid h < 6\}$

9. $\left\{n \mid n < -\dfrac{2}{3}\right\}$

11. $\{t \mid t \leq 2.4\}$

13. $\{x \mid -3 < x < 6\}$

15. $\{n \mid 0 \leq n \leq 5\}$

17. a. $m < 25$ **b.** $\{m \mid m < 25\}$ **c.**

19. a. $y \geq 2$ **b.** $\{y \mid y \geq 2\}$ **c.**

21. a. $y > 3$ **b.** $\{y \mid y > 3\}$ **c.**

23. a. $x > -4$ **b.** $\{x \mid x > -4\}$ **c.**

25. a. $a < 1$ **b.** $\{a \mid a < 1\}$ **c.**

27. a. $f < -\dfrac{1}{2}$ **b.** $\left\{f \mid f < -\dfrac{1}{2}\right\}$

c.

29. a. $u \leq 2$ **b.** $\{u \mid u \leq 2\}$ **c.**

31. a. $c > -12$ **b.** $\{c \mid c > -12\}$

c. **33. a.** $w \geq 3$ **b.** $\{w \mid w \geq 3\}$

c. **35. a.** $x \leq -13$

b. $\{x \mid x \leq -13\}$ **c.**

37. a. $n < 7$ **b.** $\{n \mid n < 7\}$ **c.**

39. a. $l < 10$ **b.** $\{l \mid l < 10\}$ **c.**

41. a. $s \geq 4$ **b.** $\{s \mid s \geq 4\}$ **c.**

43. $4 + n > 24; n > 20$ **45.** $8y - 36 > 60; y > 12$

47. $5 + \dfrac{1}{2}a \leq 2; a \leq -6$ **49.** $\dfrac{2}{5}x + 2 \leq 8; x \leq 15$

51. $x \le 10$ ft. **53.** $h \ge 15$ in. **55.** $t \ge 6\frac{4}{13}$ **57.** $i \le 1.5$ A

59. $R \ge \$1,475,000$ **61.** $x \ge 98$

Review Exercises **1.** undefined **2.** 16 **3.** -14
4. $\{1, 2, 3, 4, 5, 6, 7, 8, 9, 10\}$ **5.** commutative **6.** $6 + 2(x - 5)$

Chapter 2 Review Exercises **1.** true **2.** true
3. false **4.** false **5.** true **6.** true **7.** identity **8.** linear
9. Simplify; Distribute; multiplying; LCD; Combine; addition;
multiplication **10.** Replace the variable(s) in the equation with
the value. If the resulting equation is true, then the value is a
solution. **11.** yes **12.** no **13.** no **14.** yes **15.** no
16. yes **17.** $\frac{9}{5}$ **18.** 5 **19.** $-\frac{1}{2}$ **20.** $\frac{3}{4}$ **21.** 7 **22.** -18
23. No solution. **24.** $\frac{9}{4}$ **25.** All real numbers. **26.** No
solution. **27.** No solution. **28.** -2.7 **29.** $b = \frac{A}{h}$ **30.** $d = \frac{C}{\pi}$
31. $a^2 = c^2 - b^2$ **32.** $x = 1 - y$ **33.** $h = \frac{2A}{b}$ **34.** $m = \frac{Fr}{v^2}$
35. $m = \frac{E}{c^2}$ **36.** $h = \frac{3V}{\pi r^2}$ **37.** $w = \frac{P - 2l}{2}$ **38.** $h = \frac{2A}{a + b}$
39. a. $x < -3$ **b.** $\{x \mid x < -3\}$ **c.**
40. a. $x \ge -5$ **b.** $\{x \mid x \ge -5\}$ **c.**
41. a. $z \le -1$ **b.** $\{z \mid z \le -1\}$ **c.**
42. a. $v > -4$ **b.** $\{v \mid v > -4\}$ **c.**
43. a. $m > 4$ **b.** $\{m \mid m > 4\}$ **c.**
44. a. $c \ge -9$ **b.** $\{c \mid c \ge -9\}$ **c.**
45. $6x = -18; x = -3$ **46.** $\frac{1}{2}m - 3 = \frac{1}{4}m - 9; m = -24$
47. $2(v + 4) - 1 = 1; v = -3$
48. $2(y - 1) + 3y = 6y - 20; y = 18$
49. $13 > 3 - 10p; p > -1$ **50.** $3 + 2z < 3(z - 5); z > 18$
51. $-\frac{1}{2}x \ge 4; x \le -8$ **52.** $-2 \le 1 - \frac{1}{4}k; k \le 12$ **53.** 1.2 mi.
54. 80 in. **55.** 19.75 ft.^{2} **56.** 102.6 in.^{2} **57.** $23.75
58. 15 in. **59.** 7.1 in. **60.** $d \le 3.18$ ft. **61.** $t \ge 5\frac{5}{6}$ hr.
62. $R \ge \$825,000$

Chapter 2 Practice Test **1.** Equation because there
is an equals symbol **2.** Nonlinear because there is a variable
raised to an exponent other than 1 **3.** yes **4.** no **5.** -3
6. 13.2 **7.** -26 **8.** No solution. **9.** $-\frac{2}{3}$ **10.** 4

11. $x = \frac{7 - y}{2}$ **12.** $r = \frac{I}{pt}$ **13.** $h = \frac{2A}{b}$ **14.** $F = \frac{9C + 160}{5}$
15. a. $m < 3$ **b.** $\{m \mid m < 3\}$ **c.**
16. a. $x \le -8$ **b.** $\{x \mid x \le -8\}$ **c.**
17. a. $y \le -12$ **b.** $\{y \mid y \le -12\}$
c. **18. a.** $p > -\frac{2}{5}$
b. $\left\{p \mid p > -\frac{2}{5}\right\}$ **c.**
19. a. $x > -10$ **b.** $\{x \mid x > -10\}$
c. **20. a.** $l \le -\frac{5}{3}$
b. $\left\{l \mid l \le -\frac{5}{3}\right\}$ **c.**
21. $\frac{2}{3}n - \frac{1}{6} = 2n; n = -\frac{1}{8}$ **22.** $-7x + 12 = 5; x = 1$
23. $5(n - 2) - 3 = 10 - 4(n - 1); n = 3$
24. $1 - n > 2n; n < \frac{1}{3}$ **25.** 20 ft. **26.** 68.25 in.^{2} **27.** 7 cm
28. $530 **29.** $l \le 3$ ft. **30.** $x \ge 101$

Chapter 3
3.1 Exercises **1.** The quantity preceding *to* is written in
the numerator, and the quantity following the word *to* is written
in the denominator. **3.** When fractions are equivalent, the
cross products are equal. **5.** To solve proportion problems:
a. Set up the given ratio any way you wish. b. Set up the ratio
with the unknown so that it logically corresponds to the way
you set up the given ratio. c. Solve using cross products.
7. a. $\frac{4}{15}$ **b.** $\frac{6}{11}$ **c.** $\frac{2}{1}$ **9.** $\frac{2}{3}$ **11.** $\approx \frac{12.26}{1}$; the price of the stock
is $12.26 for every $1 earned in 2000. **13.** The 3-liter bottle is
better because it costs less per liter (the unit ratio of price to
quantity is less). **15. a.** $\frac{0.73}{1}$ (answers may vary)
b. $\frac{0.86}{1}$ (answers may vary) **c.** Varies **17. a.** $\frac{0.70}{1}$; women
earn $0.70 for every $1 men earn. **b.** $\frac{0.74}{1}$ **c.** Answers will vary.
19. yes **21.** yes **23.** no **25.** no **27.** 10.5 **29.** -14
31. 10.5 **33.** -8 **35.** $3\frac{3}{5}$ **37.** 115 **39.** $183\frac{1}{3}$ mi.
41. 548 pounds **43.** $1350.65 **45.** 4.875 in.
47. 420 mi. **49.** 617,647 **51.** 11.25 cm
53. $a = 4\frac{8}{19}$ ft. $b = 5\frac{1}{4}$ ft. $c = 5\frac{17}{19}$ ft. **55.** 300 m **57.** 26.4 m

Review Exercises **1.** The set of integers contains all the whole numbers and the opposite (or negative) of every natural number. **2.** 10 **3.** 75 **4.** 7 **5.** $\frac{1}{2}$

3.2 Exercises **1.** To write a percent as a fraction or decimal: a. Write the percent as a ratio with 100 in the denominator. b. Simplify to the desired form. **3.** multiplication **5.** In a word-for-word translation, the division yields a decimal number that must be written as a percent. **7.** $0.2, \frac{1}{5}$ **9.** $0.15, \frac{3}{20}$
11. $0.148, \frac{37}{250}$ **13.** $0.0375, \frac{3}{80}$ **15.** $0.455, \frac{91}{200}$ **17.** $0.\overline{3}, \frac{1}{3}$
19. 60% **21.** 37.5% **23.** 83.3% **25.** 44.4% **27.** 96%
29. 80% **31.** 9% **33.** 120% **35.** 2.8% **37.** 405.1%
39. 311.25 **41.** 103.2 **43.** 50 **45.** 62.4 **47.** $33\frac{1}{3}$%
49. $\approx 70.4\%$ **51.** \$4800, \$400 **53.** \$462.80
55. \$1155.60, \$256.80 **57.** \$854.10 **59.** ≈ 18.6 million
61. 23.7% **63.** 42% **65.** 20% **67.** \$1950, 75%
69. $34\frac{1}{6}$% **71.** 0.3% **73.** \$70.34

Review Exercises **1.** $-3x - 21$ **2.** $-xy + 4x + 3y$
3. -1 **4.** $3(n + 9) = n - 11; -19$ **5.** 2.1 **6.** 10.5 cm

3.3 Exercises **1.** "The second number;" $5n$ **3.** There is always a difference of 2 between consecutive even or odd numbers. **5.** 11, 17 **7.** 6, 8 **9.** 13, 39 **11.** Marie is 16, Susan is 4 **13.** Arianna is 50, Emma is 25. **15.** Length is 70 ft., width is 50 ft. **17.** Length is 11 cm, width is 7 cm **19.** Length is 7.5 ft., width is 5 ft. **21.** 26 m, 26 m, 17 m **23.** $53\frac{1}{3}°, 126\frac{2}{3}°$
25. 35°, 55° **27.** 73, 74 **29.** 76, 78, 80 **31.** 9, 11, 13, 15
33. 69, 70, 71 **35.** 15, 17, 19 **37.** 70 public tickets, 180 student tickets **39.** 10 \$5 bills, 6 \$10 bills **41.** \$5, \$50 **43.** \$3, \$8

Review Exercises **1.** yes **2.** yes **3.** -16 **4.** -22
5. 3 **6.** 6.75 mi.

3.4 Exercises **1.** They will travel the same amount of time.
3. $45t + 24t = 10$ **5.**

1st car
2nd car
Bridge   2nd car catches up
7. 4 hr.

9. 3.5 hr. **11.** 2 hr. **13.** 2 hr. **15.** 9:36 A.M. **17.** Car 1: 22 mph, car 2: 33 mph **19.** 4.5 hr.; 247.5 mi. **21.** 5:30 P.M.
23. 5:00 **25. a.** 75 mph **b.** 300 miles **27.** 6 hr. **29.** 4 hr.
31. 79 mph **33.** 3 mph, 67 mph

Review Exercises **1.** 38 **2.** -43 **3.** $8x + 7y + 6$
4. commutative property of multiplication **5.** $90.\overline{90}$ **6.** 700

3.5 Exercises **1. a.** \$1500, \$3600, \$6800 **b.** 10,000 − known amount **c.** $10,000 - P$ **3.** Each interest rate times the principal amounts should be added to get the total interest.
5. a. 120, 200, 300 **b.** twice solution 1 **c.** $2n$ **d.** $n + 2n$

7. \$4600 at 7%, \$13,800 at 10% **9.** \$1250 at 8%, \$1900 at 12%
11. \$1800 in plan 1, \$2700 in plan 2 **13.** \$1000 at 5.5%, \$4000 at 7% **15.** \$8500 at 4%, \$3500 at 5.5% **17.** \$4200 at 6%, \$8400 at −2% **19.** \$4000 in plan A, \$6000 in plan B
21. 30 ml **23.** 1.8 gal. **25.** 1.2 qt.
27. $166\frac{2}{3}$ lb. of 26%, $133\frac{1}{3}$ lb. of 35% **29.** 52.5 ml **31.** 50 g of 6%, 25 g of 18% **33.** 56 lb. at \$1.20, 24 lb. at \$0.90 **35.** 3 lb. at \$1.65, 2 lb. at \$1.25 **37.** Approximately 6 lb. of peanuts and 4 lb. of cashews.

Review Exercises **1.** $<$ **2.** $\frac{41}{108}$ **3.** -64 **4.** $12x$
5. -2 **6.** 72

Chapter 3 Review Exercises **1.** true **2.** true
3. false **4.** false **5.** false **6.** false **7.** 100 **8.** congruent
9. complementary **10.** proportion **11.** $\frac{3}{8}$ **12.** $\approx \frac{\$0.085}{1 \text{ oz.}}$
13. no **14.** yes **15.** yes **16.** no **17.** 25 **18.** 22 **19.** $78\frac{3}{4}$
20. 9.5625 **21.** -120 **22.** $21.1\overline{6}$ **23.** 216,606 **24.** 300 mi.
25. 5 ft. **26.** \$1334 **27.** \$5.33 **28.** ≈ 200.2 mi. **29.** 12 in.
30. 43.2 m **31.** $x = 12$ ft., $y = 6$ ft.
32. $a = 4\frac{5}{7}$ in., $b = 4\frac{2}{3}$ in., $c = 13\frac{1}{2}$ in. **33.** $0.15, \frac{3}{20}$
34. $0.825, \frac{33}{40}$ **35.** $0.125, \frac{1}{8}$ **36.** $0.0245, \frac{49}{2000}$ **37.** $0.1, \frac{1}{10}$
38. $0.\overline{3}, \frac{1}{3}$ **39.** 40% **40.** $33\frac{1}{3}$% or $33.\overline{3}$% **41.** $36\frac{4}{11}$% or $36.\overline{36}$%
42. 35% **43.** 120% **44.** 201.6% **45.** 30% **46.** 289.7
47. 14.56 **48.** \$13,000 **49.** \$44,176 **50.** \$673.20 **51.** 478
52. 2.5, 12.5 **53.** $-2, -1, 0$ **54.** Length is 45 ft., width is 15 ft.
55. 17, 19, 19 **56.** 106°, 74° **57.** 31°, 59° **58.** 3 regular, 2 control top **59.** Pepsi is \$46, Coca-Cola is \$58. **60.** 3.5 hr.
61. 2.6 hr. **62.** 4:00 **63.** 75 mph **64.** \$20,000 at 8%, \$40,000 at 10% **65.** \$10,200 at 7%. \$30,600 at −2% **66.** 40 L of 90%, 60 L of 40% **67.** $16\frac{2}{3}$ lb. **68.** 60 oz.

Chapter 3 Practice Test **1.** no **2.** $3\frac{1}{3}$ **3.** 2.25
4. $0.22, \frac{11}{50}$ **5.** $0.0\overline{3}, \frac{1}{30}$ **6.** 320% **7.** 40% **8.** 20%
9. 14.07 **10.** 175 **11.** 185.5 or about 186 **12.** \$717.98
13. 54, 56 **14.** 8 **15.** 5, 11 **16.** \$41 Kmart, \$52 Wal-Mart
17. 0.5 hr. **18.** 0.625 hr. **19.** \$3000 in plan A, \$1500 in plan B
20. 30 oz.

Chapters 1–3 Cumulative Review **1.** false **2.** true
3. true **4.** true **5.** false **6.** false **7.** itself **8.** LCD
9. 100 **10.** cross **11.** Set the cross products equal to each other and solve. **12.** Answers will vary. (Most commonly: {1, 2, 3, ...})
13. $-3\frac{1}{4}$
←+++�→+++|+++|+++|→   **14.** 3 **15.** Use the
  -4 -3 -2 -1 0
distributive property. **16.** $2^3 \cdot 3 \cdot 5$ **17.** $\frac{4}{9}$ **18.** $\frac{2}{9}$
19. Not a real number. **20.** -6.5 **21.** 7 **22.** $\frac{1}{10}$ **23.** -14

24. $\dfrac{11}{10}$ **25.** -26.48 **26.** undefined **27.** $-\dfrac{1}{2}x + 27$

28. additive inverse **29.** commutative property of addition
30. associative property of addition **31.** additive inverse

32. -5 **33.** 9 **34.** $\dfrac{2}{5}$ **35.** $\dfrac{32}{19}$ **36.** All real numbers. **37.** 5

38. $b = \dfrac{2A}{h}$ **39.** $r = \dfrac{A - P}{Pt}$ **40.** $c = P - a - b$ **41.** $t = \dfrac{d}{r}$

42. 12 **43.** 6.345 **44.** $-3, -1, 1$ **45.** $t \geq 88$ **46.** 20 L
47. 15 dimes, 20 quarters **48.** 4:00 **49.** 3.75 lb. of cashews,
6.25 lb. of peanuts **50.** 300 miles

Chapter 4

4.1 Exercises **1.** horizontal axis coordinate

3.

5. $A(4,3), B(-3,1), C(0, -2), D(2, -5)$ **7.** $A(-4,0), B(2,4),$
$C(-2, -3), D(4, -4)$

9.

11.

13. II **15.** I **17.** IV **19.** III **21.** y-axis **23.** IV
25. linear **27.** nonlinear **29.** linear **31.** $(-1, -5),$
$(0, -3), (2,1)$ **33. a.** $(-3, 1), (2, 1), (1, -3), (-4, -3)$
b. $(0, 2), (5, 2), (4, -2), (-1, -2)$ **c.** $(x + 3, y + 1)$
35. a.

b. 20 **c.** 21

2. 2 **3.** 12 **4.** $\dfrac{15}{2}$ **5.** $-\dfrac{9}{10}$ **6.** 2

4.2 Exercises **1. a.** Replace the variables in the equation
with the corresponding coordinates. **b.** Verify that the equation is true. **3. a.** Choose a value for one of the variables.
b. Replace the corresponding variable with your chosen value.
c. Solve the equation for the value of the other variable.
5. A minimum of two ordered pairs are needed. **7.** yes
9. no **11.** yes **13.** no **15.** no **17.** no

19. $(3, -5), (5, -3), (6, -2)$

21. $(1, 4), (3, 0), (5, -4)$

23. $(-1, -1), (0, 0), (1, 1)$

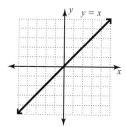

25. $(-1, -2), (0, 0), (1, 2)$

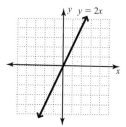

27. $(-1, 5), (0, 0), (1, -5)$

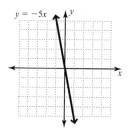

29. $(0, -3), (2, -1), (4, 1)$

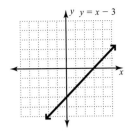

31. $(-1, 6), (0, 4), (1, 2)$

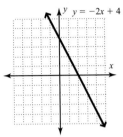

33. $(-1, -1), (0, 2), (1, 5)$

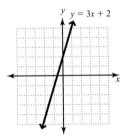

35. $(-2, -1), (0, 0), (2, 1)$

37. $(-3, 6), (0, 4), (3, 2)$

39. $(1, -4), (2, 0), (3, 4)$

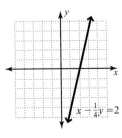

41. $(-1, -2.9), (0, -2.5), (1, -2.1)$

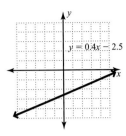

43. $(-1, -5), (0, -5), (1, -5)$

45. $(7, -1), (7, 0), (7, 1)$

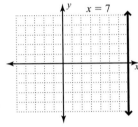

47. As a gets larger, the graph gets steeper.

49. The graph will shift up b units on the y-axis.

51. a. $125   **b.** 4 hr.   **c.**

53. a. $15   **b.** 2 days   **c.**

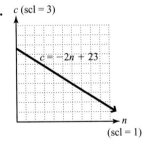

55. a. $212.50   **b.** 2 hr.   **c.**

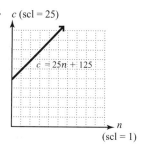

Review Exercises   **1.** -64.4 **2.** 40 m   **3.** 5.67 in.^{2}
4. $x = \dfrac{v - k}{m}$ **5.** 3

4.3 Exercises   **1.** The point where a graph intersects the
x-axis.   **3.** 1. Replace y with 0 in the given equation.   2. Solve
for x.   **5.** The graph of any equation in the form $y = mx$ is
always a line that passes through the origin, which means the
x- and y-intercepts are both at $(0,0)$.   **7.** $(3, 0), (0, 2)$ **9.** $(2, 0)$,
$(0, -3)$ **11.** $\left(-\dfrac{10}{3}, 0\right), \left(0, -\dfrac{5}{2}\right)$ **13.** $(4, 0), (0, -4)$
15. $(0, 0)$ for both   **17.** $(5, 0)$, no y-intercept   **19.** $(0, 0)$ for both

21.   **23.**

25.   **27.**

29.   **31.**

33.   **35.**

37.   **39.**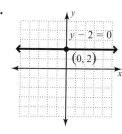

41. d, the x-intercept is $+$ and the y-intercept is $-$.
43. $(-2.7, 0), (2.7, 0), (0, -2.7), (0, 2.7)$ **45. a.** $15x + 20y$
b. $15x + 20y = 2000$ **c.** $\left(133\dfrac{1}{3}, 0\right), (0, 100)$

d. The number of units required to meet the goal if only one
size is sold.   **e.** varies

f.

Review Exercises   **1.** $w = \dfrac{P - 2l}{2}$

2. $4 + 3x = 5(x + 6); -13$
3. $x + x + 2 + x + 4 = 141; 45, 47, 49$
4. $37.5°, 52.5°$ **5.** I   **6.** no

4.4 Exercises   **1.** How steep a line is.   **3.** Uphill, because
the slope is positive.   **5.** horizontal
7.

9.

11.

13.

15. $m = 2 \ (0, 3)$

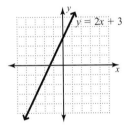

17. $m = 1 \ (0, -6)$

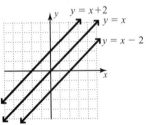

19. $m = -\dfrac{2}{3} \left(0, \dfrac{7}{3}\right)$

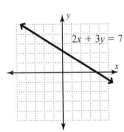

21. $m = -2 \ (0, 4)$

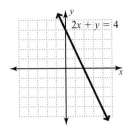

23. $m = -\dfrac{2}{3} \ (0, 8)$

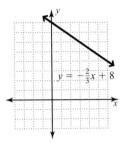

25. $m = -3 \ (0, -1)$

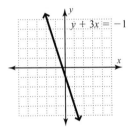

27. $m = -\dfrac{5}{12} \left(0, \dfrac{1}{6}\right)$

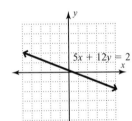

29. $m = \dfrac{5}{2} \ (0, -1)$

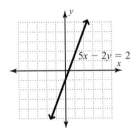

31. $m = \dfrac{2}{3} \left(0, \dfrac{8}{3}\right)$

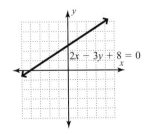

33. $\dfrac{2}{7}$ **35.** $-\dfrac{5}{4}$ **37.** $\dfrac{5}{2}$ **39.** undefined **41.** 0 **43.** -3

45. d **47.** a **49.** f **51.** $m = \frac{4}{3}$; $(0,0)$; $y = \frac{4}{3}x$

53. $m = -2$; $(0, -1)$; $y = -2x - 1$ **55.** $y = -2$ **57.** $y = 0$

59. a.   **b.** 3 and 0

c. They are the same.

61. $0.58\overline{3}$, $\frac{7}{12}$ **63.** 1.27

65. a. **T-bond**
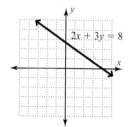

b. ≈ -0.005 **c.** ≈ 5.53

Review Exercises **1.** $11x + 7$ **2.** $x = \dfrac{C - By}{A}$

3. $-2x - 14$ **4.** $\dfrac{21}{2}$

5.

6. $\left(\dfrac{7}{3}, 0\right), \left(0, -\dfrac{7}{5}\right)$

4.5 Exercises **1.** slope-intercept form **3.** Find the slope.
5. Solve for y. **7.** $y = 2x + 3$ **9.** $y = -3x - 9$

11. $y = -\dfrac{2}{5}x + \dfrac{7}{8}$ **13.** $y = -9x - \dfrac{3}{4}$ **15.** $y = 0.8x - 5.1$

17. $y = 4.1x + 5.85$ **19.** $y = \dfrac{5}{2}x - \dfrac{13}{2}$ **21.** $y = \dfrac{7}{12}x + \dfrac{11}{12}$

23. $y = 2.42x - 5.16$ **25.** $y = 7x$ **27.** $x = 8$ **29.** $y = -1$

31. $y = 3x - 13$ **33.** $y = \dfrac{2}{3}x + 3$ **35.** $y = \dfrac{3}{4}x - \dfrac{17}{4}$

37. $y = -\dfrac{4}{3}x + \dfrac{4}{3}$ **39.** $y = -x - 2$ **41.** $y = \dfrac{2}{9}x - \dfrac{61}{9}$

43. $5x - 2y = 4$ **45.** $3x - 2y = 6$ **47.** $x - y = -6$

49. $x = -1$ **51.** $3x - 8y = 37$ **53.** $y = -1$ **55.** parallel

57. perpendicular **59.** parallel **61.** perpendicular

63. a. $y = 4x - 14$ **b.** $4x - y = 14$ **65. a.** $y = -3x - 5$

b. $3x + y = -5$ **67. a.** $y = \dfrac{1}{3}x + \dfrac{1}{3}$ **b.** $x - 3y = -1$

69. a. $y = -2x + 1$ **b.** $2x + y = 1$ **71. a.** $y = -2x + 11$
b. $2x + y = 11$ **73. a.** $y = x + 3$ **b.** $x - y = -3$
75. a. $y = \dfrac{1}{2}x - 5$ **b.** $x - 2y = 10$ **77. a.** $y = \dfrac{3}{2}x - \dfrac{17}{2}$
b. $3x - 2y = 17$ **79. a.** $y = -x + 1$ **b.** $x + y = 1$
81. a. $y = -\dfrac{1}{3}x + \dfrac{25}{3}$ **b.** $x + 3y = 25$

83. a.
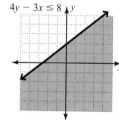

b. 107 **c.** $y = 107x + 855$ **d.** $2995

Review Exercises **1.** $\dfrac{36}{5}$ **2.** 10.32 **3.** $<$

4.

5.

6. $(7,0)$ $(0, -2)$

4.6 Exercises **1.** To determine whether an ordered pair is a solution for an inequality, replace the variables with the corresponding coordinates and see if the resulting inequality is true. If so, the ordered pair is a solution. **3.** A solid line indicates the equality is included. A dashed line indicates the equality is excluded. **5.** yes **7.** no **9.** no **11.** yes

13. $y \leq -x + 1$

15. $4y - 3x \leq 8$
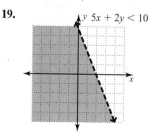

17. $y > \dfrac{1}{3}x$

19. $5x + 2y < 10$

21.
$y > 2x$

23.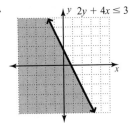
$2y + 4x \leq 3$

25.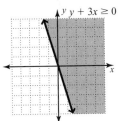
$y + 3x \geq 0$

27.
$\frac{1}{4}x + 2y > -8$

29.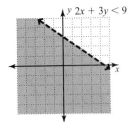
$2x + 3y < 9$

31.
$x + y > -2$

33.
$x > -4$

35.
$y \leq 2$

37.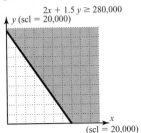
$y > -x$

39. a. x represents large bottles sold, y represents small bottles sold. **b.**

$2x + 1.5y \geq 280,000$
y (scl = 20,000)
(scl = 20,000)

c. The exact combination of bottle sizes sold to break even.
d. varies **e.** varies **f.** No, only combinations of natural numbers are possible because we cannot sell negative numbers of bottles nor fractions of a bottle. **41. a.** $5.5x + 12.5y \leq 100$

b.

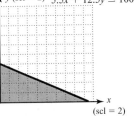
y (scl = 2) $5.5x + 12.5y \leq 100$
(scl = 2)

c. The exact combination of plant purchases that would be \$100. **d.** The combinations of plants sold not to exceed \$100.
e. varies **f.** Yes, $(0, 8)$. Since she cannot purchase negative or fraction numbers of plants, the combination must be natural numbers and $(0, 8)$ is the only combination of natural numbers that costs exactly \$100. **43. a.** $2l + 2w \leq 180$

b.

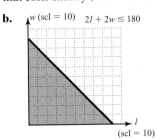
w (scl = 10) $2l + 2w \leq 180$
(scl = 10)

c. The exact combination of lengths and widths that would yield a perimeter of 180 ft. **d.** The combinations of length and width that do not exceed a perimeter of 180 ft. **e.** varies
f. varies

Review Exercises 1. $\{a, e, i, o, u\}$ **2.** 1 **3.** -13
4. 20 ft. by 24. ft. **5.** $\left(\frac{8}{3}, 0\right)$, $(0, -4)$

6.

$-x + y = 7$

4.7 Exercises 1. The domain is a set containing all x-values. **3.** Trail along the x axis and determine what values of x have a corresponding y-value. **5.** Draw or imagine a vertical line through every point. If each vertical line intersects the graph at only one point, the relation is a function.
7. Domain: $\{2, 3, 1, -1\}$ Range: $\{1, -2, 4, -1\}$ **9.** Domain: $\{4, 2, 3, -4\}$ Range: $\{1, -5,0\}$ **11.** Domain: $\{0, -3, 2, 8\}$
Range: $\{0, -3, 9, -16\}$ **13.** Yes. Every element in the domain

is assigned to one element in the range.   **15.** No. An element in the domain is assigned to more than one element in the range (8 is assigned to both 2 and −1).   **17.** Yes. Every element in the domain is assigned to one element in the range.   **19.** No. An element in the domain is assigned to more than one element in the range.   **21.** No. It fails the vertical line test.   **23.** Yes. It passes the vertical line test.   **25.** no   **27.** no   **29.** yes   **31.** no **33.** yes   **35.** Domain: $\{x\,|\,1900 \leq x \leq 2001\}$ Range: $\{y\,|\,13 \text{ million} \leq y \leq 63 \text{ million}\}$; Function   **37.** Domain: {*Charleston, Beaufort, Georgetown, Horry*} Range: {756, 974, 1567, 3860}; Function   **39.** Domain: $\{x\,|\,1991 \leq x \leq 2002\}$ Range: {29, 32, 33, 34, 37}; Function   **41.** Domain: $\{x\,|\,x \leq 0\}$ Range: All real numbers; Not a function.   **43.** Domain: all real numbers   Range: $\{y\,|\,y \geq -1\}$; Function   **45. a.** −5   **b.** −3 **c.** −9   **d.** −4   **47. a.** 7   **b.** 5   **c.** 11   **d.** 17   **49. a.** 4   **b.** 5 **c.** $4\frac{1}{16}$ **d.** 5   **51. a.** −1

b. undefined   **c.** $-\frac{1}{2}$ **d.** $-\frac{1}{3}$ **53. a.** −3   **b.** −3.19 **c.** $a^2 - 2a - 3$ **d.** $a^2 - 8a + 12$ **55. a.** $\sqrt{3}$ **b.** $\sqrt{5}$ **c.** $\sqrt{2}$ **d.** 1   **57. a.** −2   **b.** $-\frac{5}{3}$ **c.** $-\frac{7}{3}$ **d.** −3   **59. a.** undefined

b. $\frac{1}{\sqrt{3}}$ **c.** 1   **d.** $\frac{1}{\sqrt{2}}$ **61. a.** $-\frac{3}{4}$ **b.** $-\frac{4}{3}$ **c.** $-\frac{1}{6}$ **d.** −6   **63. a.** 2   **b.** 0.5   **c.** 3   **d.** 4   **65. a.** 3   **b.** 4   **c.** 2 **67. a.** −5   **b.** −2   **c.** 1

69.

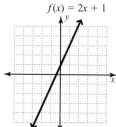

$f(x) = 2x + 1$

71.

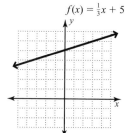

$f(x) = \frac{1}{3}x + 5$

73.

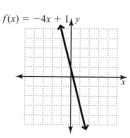

$f(x) = -4x + 1$

75.

$f(x) = -\frac{2}{3}x$

77.

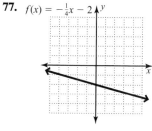

$f(x) = -\frac{1}{4}x - 2$

Answers   A-13

28. $(0, 2), (3, 0), (6, -2)$

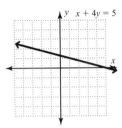

29. $(5, 0), (1, 1), (-3, 2)$

30. $(0, 7), (1, 6), (3, 4)$

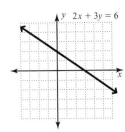

31. $(12, 0), (0, 6)$ **32.** No x-intercept, $(0, 7)$ **33.** $(-2, 0)$,
no y-intercept   **34.** $(3.1, 0), (0, -2)$ **35.** $\left(-\dfrac{15}{2}, 0\right), (0, 5)$
36. $(0, 0)$ for both

37.

38.

39.

40.

41.

42.

43.

44.

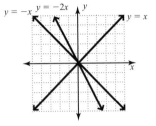

45. $m = -2; (0, 3)$

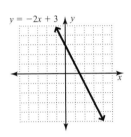

46. $m = \dfrac{2}{3}; \left(0, -\dfrac{8}{3}\right)$

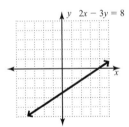

47. $m = -\dfrac{1}{4}; (0, 2)$

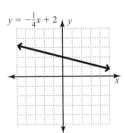

48. $m = -1; (0, 5)$

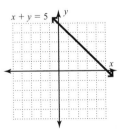

49. $m = \dfrac{1}{3}; \left(0, -\dfrac{7}{3}\right)$

50. $m = 3; (0, 0)$

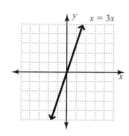

51. 3 **52.** 1 **53.** $-\dfrac{1}{3}$ **54.** $-\dfrac{3}{5}$ **55.** 0 **56.** undefined

57. $y = -x + 7$ **58.** $y = -\dfrac{1}{5}x - 8$ **59.** $y = 0.2x + 6$

60. $y = x$ **61.** $y = -2x + 9$ **62.** $y = 0.4x - 1.6$

63. $y = \dfrac{1}{3}x + \dfrac{5}{3}$ **64.** $y = x - 9$ **65.** $y = 6.2x + 9.4$

66. $y = -\dfrac{2}{5}x + \dfrac{21}{5}$

67. $y = -\dfrac{1}{2}x + \dfrac{7}{2}$
$x + 2y = 7$

68. $y = \dfrac{5}{4}x - \dfrac{37}{4}$
$5x - 4y = 37$

69. $y = -x + 4$
$x + y = 4$

70. $y = -\dfrac{1}{3}x - \dfrac{4}{3}$
$x + 3y = -4$

71. $y = -\dfrac{1}{5}x - \dfrac{13}{5}$
$x + 5y = -13$

72. $y = -\dfrac{6}{5}x + 6$
$6x + 5y = 30$

73. $y = 2x - 2$ **74.** $y = -\dfrac{2}{3}x + \dfrac{17}{3}$ **75.** $y = -\dfrac{5}{3}x - \dfrac{11}{3}$

76. $y = \dfrac{5}{2}x + 9$ **77.** yes **78.** yes **79.** yes **80.** no

81.

82.

83.

84.

85.

86.

87. Domain: $\{2, -1, 3, -3\}$ Range: $\{3, 5, 6, -4\}$ **88.** Domain: {Volkswagen, Honda, Toyota, Subaru, Nissan} Range: {52.2, 49.7, 49.0, 47.8, 45.8} **89.** no **90.** yes **91.** Domain: all real numbers Range: all real numbers; It is a function. **92.** Domain: $\{x \mid x \le 0\}$ Range: all real numbers; It is not a function. **93.** Domain: $\{x \mid -4 \le x \le 5\}$ Range: $\{1, 2, 3\}$; It is a function. **94. a.** 1 **b.** 3 **c.** 3 **d.** 2 **95. a.** 0 **b.** 2 **c.** −10 **96.** −3, 3

Chapter 4 Practice Test
1. $A(2,4)$ $B(-3,2)$ $C(0, -5)$ $D(4, -2)$

2.

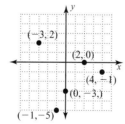

3. I **4.** no

5. $(4, 0), (0, -2)$

6. $\left(\dfrac{9}{2}, 0\right), (0, 3)$

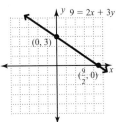

7. $(-8, 0), (0, 4)$

8. $(0, 0)$

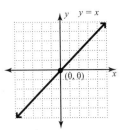

9. $(4, 0), (0, -8)$

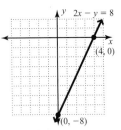

10. $\left(\dfrac{4}{3}, 0\right), (0, -1)$

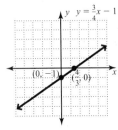

11. $m = \dfrac{3}{4}; (0, 11)$ **12.** $m = \dfrac{5}{3}; \left(0, -\dfrac{8}{3}\right)$ **13.** $-\dfrac{10}{3}$ **14.** 0

15. $y = \dfrac{3}{5}x + 4$ **16.** $y = -x - 5$ **17.** $y = \dfrac{9}{2}x + \dfrac{5}{2}$

18. $y = -\dfrac{2}{5}x + \dfrac{11}{5}$ **19.** $2x - 3y = 1$ **20.** $2x - 3y = 9$

21. $x - 3y = 10$ **22.** yes

23.

24.

25.

26. yes **27.** no **28.** 4 **29. a.** $-\dfrac{2}{3}$ **b.** -9 **c.** 0

30. a. Domain: $\{x \mid -3 \le x \le 3\}$ Range: $\{-2, 1, 2\}$ **b.** 2

5.1 Exercise

1. If a is a real number, where $a \ne 0$ and n is a natural number, then $a^{-n} = \dfrac{1}{a^n}$. **3.** The result is positive. We simplify an expression with a negative exponent by inverting the expression and changing the sign of the exponent. $\left(2^{-3} = \dfrac{1}{2^3} = \dfrac{1}{8}\right)$

5. To write a number greater than 1 in scientific notation: **a.** Move the decimal point so that the number is greater than or equal to 1 but less than 10. **b.** Write the decimal number multiplied by 10^n, where n is the number of places between the new decimal position and the original decimal position. **c.** Delete 0's to the right of the last non-zero digit. **7.** 1 **9.** -8

11. -125 **13.** -16 **15.** $-\dfrac{27}{64}$ **17.** 0.008 **19.** 5.29 **21.** $\dfrac{1}{8}$

23. $\dfrac{49}{25}$ **25.** $-\dfrac{1}{64}$ **27.** $-\dfrac{1}{16}$ **29.** 125 **31.** $\dfrac{-1}{7,962,624}$

33. $\approx -77{,}160$ **35.** $-\dfrac{512}{125}$

37. Mistake: The expression was interpreted to mean $(-3)^4$, but it means the additive inverse of 3^4. Correct: -81 **39.** Mistake: The negative sign in the exponent was passed to the product instead of inverting the original expression like it should have. Correct: $\dfrac{1}{25}$ **41.** $16{,}500{,}000{,}000$ **43.** $299{,}800{,}000$

45. 6,378,000   **47.** 92,920,000   **49.** 0.0000025
51. 0.00000000000000000000000000000167   **53.** 0.00000000295
55. 0.00000000000000000000000000006645   **57.** 2.5×10^{13}
59. 5.3×10^9 **61.** 2.342×10^7 **63.** 1.008×10^9
65. 5.86×10^{-7} **67.** 1.67×10^{-27} **69.** 6.25×10^{-5}
71. 2.5×10^{-5} **73.** $8.95 \times 10^5, 7.2 \times 10^6, 7.5 \times 10^6, 9 \times 10^7,$
1.3×10^8 **75.** The wavelength of ultraviolet light is about 40 times the wavelength of x-rays.

Review Exercise
1. $-4y + 12$ **2.** slope $\frac{2}{3}$, y-intercept:
$(0, -2)$ **3.** $266\frac{2}{3}$ **4.** $3\frac{1}{3}$ **5.** $\frac{1}{5}$ **6.** 43

5.2 Exercise
1. The sum of the exponents on all variables in a monomial.   **3.** The greatest degree of any of the terms in the polynomial.   **5.** Place the highest degree term first, then the next highest degree and so on.   **7.** monomial   **9.** monomial   **11.** not   **13.** not   **15.** monomial   **17.** monomial
19. c: -5 d:3   **21.** c: -1 d:5   **23.** c: -9 d:0   **25.** c: 4.2 d:4
27. c: 16 d:3   **29.** c: 1 d: 1   **31.** binomial   **33.** monomial
35. none   **37.** binomial   **39.** monomial   **41.** binomial
43. 4   **45.** 5   **47.** 7   **49.** 60   **51.** -8 **53.** -4.8 **55. a.** 46 ft.
b. 26.96 ft.   **57. a.** 30 units   **b.** 43 units   **59. a.** ≈ 1273.4 ft.^{3}
b. ≈ 537.2 ft.^{3} **61.** $5x^7 - 3x^5 + 7x^4 - 8x + 14$ **63.** $-8r^6 + 18r^5 + 7r^3 - 3r^2 + 4r$ **65.** $-12w^4 - w^3 + 11w^2 + 5w + 20$
67. $-5x^2 + 8$ **69.** $5y^2 - 9y + 22$ **71.** $k^4 + 6k^3 - 3k^2 + 6k$
73. $-10a^4 + 8a^2 + 12$ **75.** $-34v^5 - 5v^3 + 35v^2 - 20$
77. $-9x^2 + 9xy$ **79.** $2a^2b^3 + 5a^2b^2$ **81.** $-4w^2z + 6wz^2 + 12w^2 - 1$ **83.** $18k^4 - 8l^4 + 4k^2l^2 - 3k^3l + 3kl + 5$
85. $x^2yz^3 - 4xy^2z + 18xyz - 21$

Review Exercise
1. -1 **2.** 2×10^{11} **3.** 12   **4.** 600
liters   **5.** $\frac{71}{500}$ **6.** -179

5.3 Exercise
1. To add polynomials, combine like terms.
3. Change the signs of each term in the polynomial.   **5.** $8x + 1$
7. $10y + 12$ **9.** $7x$ **11.** $3x^2 - 2x + 12$ **13.** $6z^2 - 11z - 2$
15. $5r^2 - 7r - 1$ **17.** $9y^2 - 5y + 3$ **19.** $5a^3 + 3a^2 + 3a - 3$
21. $3p^3 - p^2 + 7p + 9$ **23.** $-8w^4 + 3w^3 - 7w^2 + 13w - 9$
25. $8x^5 - 2x^4 - 3x^2 - 12x + 6$ **27.** $8m^3n^4 - 4n^4m + 4m^2n^2 - 12mn^2 + 14m^2n - 31n + 9$ **29.** $-7u^4 + u^2v^3 + 7u^3v + 7u - 13v^2 - 3$ **31.** $5x - 2$ **33.** $14a - 12$ **35.** $4x + 2$ **37.** $13a^2 - 1$
39. $2x^2 - 9x + 2$ **41.** $2z^2 - 7z + 9$ **43.** $12a^2 + 10a - 15$
45. $-u^3 - 2u^2 - 6u + 8$ **47.** $-11w^5 + 3w^4 + 5w^2 + 2$
49. $-11p^4 + 3p^3 - 13p^2 + 5p - 6$ **51.** $11v^4 - 26v^3 + 23v^2 - 5v - 9$ **53.** $2xy$ **55.** $10a^2 - 5ab + 2b^2$ **57.** $-5r^5 + 2r^3s^2 + 3r^2s^3 + 3r^3 - s^2 + 12rs + 15$ **59.** $32.1x + 35.45y + 6.01z$
61. $4.22a + 1.63b + 12.13c$ **63.** Mistake: Only the sign of the first term in the subtrahend was changed instead of all three terms. Correct: $-4x^2 - 2x + 5$

Review Exercises
1. -682 **2.** -32 **3.** 0.0001   **4.** 128
5. 58,900,000   **6.** 300 m^2

5.4 Exercise
1. Product rule for exponents.   **3.** To multiply monomials: a. Multiply coefficients.   b. Add the exponents of the like bases.   c. Write any unlike variable bases unchanged in the product.   **5.** Multiply the exponents.   **7.** a^5 **9.** 2^{13}

11. a^5b **13.** $6x^3$ **15.** $-8m^3n^2$ **17.** $-\frac{5}{28}s^3t^8$ **19.** $a^3b^5c^6$
21. $-6r^5s^{10}$ **23.** $-x^4t^2z^9$ **25.** $90x^6y^5$ **27.** $-\frac{1}{4}a^6b^3c^2$
29. $w^3x^3y^5$ **31.** $5q^5r^{10}s^{21}$ **33.** Mistake: The exponents of x were multiplied instead of added. Correct: $35x^7y$ **35.** Mistake: The exponents of x were not added. Correct: $54x^4y^6z$ **37.** $5w^2$
39. $4.5w^3$ **41. a.** height $= \frac{1}{2}b$, top side $= \frac{1}{4}b$ **b.** $A = \frac{5}{16}b^2$
43. 1.2×10^9 **45.** 2.4×10^{14} **47.** 2.4985×10^{19}
49. 4.2×10^4 mi.^{2} **51.** 5.8404×10^8 mi.   **53.** x^6 **55.** 3^8
57. h^8 **59.** $-y^6$ **61.** x^3y^3 **63.** $\frac{1}{64}m^6n^3$ **65.** $-0.008x^6y^{15}z^3$
67. $64p^{24}q^{24}$ **69.** $18x^7$ **71.** r^5s^4 **73.** $100a^6b^4$ **75.** $\frac{1}{1024}a^{10}b^{15}c^5$
77. $3.24u^6v^3$ **79.** $324u^6v^3$ **81.** Mistake: The exponents were added instead of multiplied. Correct: $25x^6$ **83.** Mistake: The coefficient was multiplied by 4 instead of being raised to the 4th power. Also, the exponents were not multiplied properly. Correct: $16x^{16}y^4$ **85.** $\approx 1.412 \times 10^{27}$ m^3

Review Exercises
1. $-4x - 4z$ **2.** -16 **3.** -61
4. 4, 5, 6   **5.** 6   **6.** $615,000

5.5 Exercise
1. distributive property   **3.** For binomials, the first terms are the same and the second terms are opposite signs.   **5.** The middle term is missing.   **7.** $5x + 15$
9. $-12x - 24$ **11.** $28n^2 - 8n$ **13.** $9a^2 - 9ab$ **15.** $\frac{1}{4}m^2 - \frac{5}{8}m$
17. $-0.3p^5 + 0.6p^2$ **19.** $-20x^3 - 10x^2 + 25x$
21. $-7x^5 + 5x^3 - x^2$ **23.** $-r^4s^4 - 3r^3s^3 + r^2s^2$
25. $-4a^3b^2 + 12a^2b^3 - 6a^3b^3 + 2a^4b^4$ **27.** $a^3b^3c^2 - a^2b^2c^2 + abc$ **29.** $4b^6 - b^4 + \frac{1}{5}b^3 - \frac{1}{2}b^2 + 16b$ **31.** $-r^4t^3 + 3.2r^3t^4 - 1.64r^2t^5 + 0.8rt^2$ **33.** $-25x^6y + 15x^5y^2 - 50x^3y^9$
35. a. $x + 5$ **b.** $2x$ **c.** $2x(x + 5)$ **d.** $2x^2 + 10x$ **e.** They describe the same area.   **37.** $x^2 - x - 12$ **39.** $x^2 - 5x - 14$
41. $6y^2 - 11y - 10$ **43.** $15m^2 + 11m - 12$ **45.** $12t^2 - 26t + 10$
47. $15q^2 + 11qt - 12t^2$ **49.** $14y^2 + 34xy + 12x^2$
51. $a^4 - a^2b - 2b^2$ **53. a.** $x + 3$ **b.** $x + 2$ **c.** $(x + 2)(x + 3)$
d. $x^2 + 5x + 6$ **e.** They describe the same area.
55. $x^3 - 3x^2 + 3x - 1$ **57.** $3a^3 - 4a^2 - a + 2$
59. $2c^3 + 3c^2 - 5c - 6$ **61.** $15p^3 + 17p^2 + 46p - 10$
63. $12y^3 - 24y^2z + 33yz^2 - 12z^3$
65. $x^4 + 3x^3y + 2x^2y^2 + xy^3 - y^4$ **67.** $u^4 + u^3 - 2u^2 - 3u - 1$
69. $x + 3$ **71.** $4x + 2y$ **73.** $4d + 3c$ **75.** $-3j + k$
77. $x^2 - 25$ **79.** $4m^2 - 25$ **81.** $x^2 - y^2$ **83.** $64r^2 - 100s^2$
85. $4x^2 - 9$ **87.** $x^2 - 6x + 9$ **89.** $16t^2 + 8t + 1$
91. $m^2 + 2mn + n^2$ **93.** $81 - 90y + 25y^2$
95. $4u^2 + 12uv + 9v^2$ **97.** $81w^2 - 72wz + 16z^2$
99. $10xyz + 2y^2z$ **101.** $h^2 + 4h$ **103.** $x^2 + 3x$
105. $4x^3 + 24x^2 + 20x$ **107.** $4w^3 + 6w^2$
109. $V = 3.14r^3 + 6.28r^2$ or $\pi r^3 + 2\pi r^2$

Review Exercise
1. -5.25 **2.** x^7 **3.** 9, 36
4. 6,304,000   **5.** -1 **6.** $-\dfrac{5 - d}{m}$

5.6 Exercise

1. When dividing exponential forms that have the same base we can subtract the divisor's exponent from the dividend's exponent and keep the same base. **3.** The decimal factors and powers of 10 can be separated into a product of two fractions. This allows us to calculate the decimal division and divide the powers of 10 separately. **5.** To divide a polynomial by a monomial, divide each term in the polynomial by the monomial. **7.** $\frac{1}{a^3}$ **9.** $\frac{1}{2^5}$ **11.** 3

13. $\frac{1}{4^6}$ **15.** $\frac{1}{x^2}$ **17.** y^5 **19.** a^2 **21.** $\frac{1}{w^3}$ **23.** $\frac{1}{a^7}$ **25.** y^8

27. 1 **29.** $\frac{1}{s^9}$ **31.** 3.75×10^{-2} **33.** $\approx 6.67 \times 10^{-8}$ **35.** 7.1×10^{10}

37. $\approx 2.734 \times 10^4$ **39.** 500 seconds or $8\frac{1}{3}$ minutes

41. \$21,237.83 **43.** x^3 **45.** $-3m^2n^4$ **47.** $\frac{-8q^3}{5}$ **49.** $\frac{4c}{3a^2}$

51. $a + 2b$ **53.** $xy - 3y^2$ **55.** $-c + 4ab$ **57.** $x^2 + x - 1$

59. $y^2 + y^4 - 3y^5 + \frac{1}{y^2}$ **61.** $12u^2v^3 + 4v^4 - 5uv$ **63.** $\frac{5mn}{2}$

65. a. $3x - 5 + \frac{4}{x}$ **b.** area = 18, length = 6, width = 3

67. $x + 4$ **69.** $x - 12 + \frac{4}{x - 5}$ **71.** $m + 5$

73. $2x^2 + 3x + 6 + \frac{17}{x - 2}$ **75.** $4p^2 - 10p + 25$

77. $x^2 - 2x + 4$ **79.** $2y - 2 + \frac{23}{7y + 3}$

81. $b^2 - 3b + 5 + \frac{3}{b - 3}$ **83.** $4k^2 - 1 + \frac{2}{k + 2}$

85. $2x + 3 + \frac{8}{3x - 1}$ **87.** $w^2 + 8 + \frac{1}{2w - 1}$

89. $y^3 + 2 + \frac{2}{y + 2}$ **91. a.** $x + 4$ **b.** length = 7 ft

height = 8 ft   volume = 336 ft³ **93.** $\frac{-3}{x^2}$ **95.** $\frac{6}{x^4}$ **97.** $-\frac{27}{8}$

99. $\frac{b^4}{a^{12}}$ **101.** $\frac{27y^6}{x^3z^6}$ **103.** $\frac{4x^4z^4}{y^5}$ **105.** $\frac{2}{3ab^2}$ **107.** $\frac{36m^4}{n^4}$

109. $16r^9s^7$ **111.** $-x^{10}y^4$ **113.** 2.25×10^{11} joules

115. $\approx 9.3 \times 10^{-4}$ kg

Review Exercise

1. 1.67×10^{-24} **2.** 29,980,000,000

3. $x + \frac{1}{2}y$ **4.** 340 ft.² **5.** -24 **6.** No, it does not pass the vertical line test.

Chapter 5 Review Exercises

1. false **2.** false **3.** true **4.** false **5.** false **6.** false **7.** combine

8. a. coefficients **b.** exponents **9.** $(a^m)^n = a^{mn}$ **10.** $2ab$

11. $\frac{8}{125}$ **12.** -16 **13.** $\frac{1}{25}$ **14.** $\frac{1}{13}$

15. 4,500,000,000 **16.** 13,800,000

17. 0.00000000000000000000001663

18. 0.0000000000000000000000000002006

19. 1.661×10^{-24} **20.** 9.63×10^{-23} **21.** 3×10^8

22. 6.37×10^6 **23.** yes **24.** yes **25.** no **26.** no **27.** c: 6 d: 4 **28.** c: 27 d: 0 **29.** c: -2.6 d: 4 **30.** c: -1 d: 1

31. binomial **32.** monomial **33.** trinomial **34.** none of these **35.** 9: $-2x^9 + 21x^5 - 19x^3 - 3x + 15$

36. 4: $-j^4 + 22j^2 + 5j - 19$ **37.** 5: $4u^5 + 13u^4 - 18u^3 - u - 21$

38. 8: $6v^8 + 21v^4 + 16v^3 - v^2 - 2v - 19$ **39.** $4y^5 + 4y^4 + 8$

40. $18l^5 + 5l^4 - 2l^3 + 6l^2 + 7l + 20$ **41.** $-2m^3 + 7m^2 + m + 3$

42. $-3y^2 + 7y - 15$ **43.** $a^2bc - ab^2c - 6$

44. $-11x^2yz - 2xyz^2$ **45.** $-4c^3 - cd^2 + 11cd + 2d + 8$

46. $-8a^5 + 3abc^3 - 8abc + 8$ **47.** $-3j^4 + 3jk^3 + k^2 + 10jk + 12$

48. $3mn^2 - 8n^2 - 9mn + 7$ **49.** $-4ab - 18a - 19c + 8$

50. $11y - 10$ **51.** $10x^2 + 3x - 4$ **52.** $3m - 8$

53. $-2n^3 + 3n^2 + 4n + 3$ **54.** 9 **55.** $-x^3 + 4x^2 + 17$

56. $-p^3 + 10p^2 - 10p + 9$ **57.** $-16x^3 - 4x - 2$

58. $8x^2 - xy + 6y^2$ **59.** $-3m^2 + 8mn - 2$ **60.** m^5 **61.** $10a^9$

62. $-6x^6y^6$ **63.** $25x^4y^2$ **64.** $72u^{10}$ **65.** $32x^{10}$ **66.** x^8

67. u^7 **68.** $\frac{1}{s^6}$ **69.** x^6 **70.** $\frac{b^6}{8a^9}$ **71.** $\frac{4j^8}{9hk^2}$ **72.** $\frac{-5}{x^2}$ **73.** 7

74. 1 **75.** $2x - 42$ **76.** $-16x + 8$ **77.** $4a^2 - 12a$

78. $-12b^3 + 24b^2 + 6b$ **79.** $-8m^7 - 12m^5 + 4m^4 + 20m^3$

80. $-12a^2bc^2 + 16a^2b^4c^2 - 8a^2b^2c^4 - 32abc^2$ **81.** $x^2 + 4x - 5$

82. $y^2 - 11y + 24$ **83.** $12m^2 + 28m - 5$ **84.** $15a^2 - 4ab - 4b^2$

85. $4x^2 - 1$ **86.** $x^3 - x^2 - x - 15$

87. $3y^3 - 7y^2 + 14y - 4$ **88.** $a^4 - a^3b - 2a^2b^2 - 3ab^3 - b^4$

89. $16 - x^2$ **90.** $x^2 - 12x + 36$ **91.** $9r^2 - 30r + 25$

92. $36s^2 - 4r^2$ **93. a.** $20x^2 - 28x$ **b.** 96 ft.² **94.** 2.25×10^8 m²

95. 2.94×10^{-4} volts **96.** $x - 5$ **97.** $-y^2 + 2y - 3$

98. $1 + 10st^3 - 2t^4$ **99.** $a^2 - \frac{1}{c} + 2ac$ **100.** $2 + b$

101. $\frac{2z}{5y^2} - \frac{xz}{y} + \frac{2}{y}$ **102. a.** $9x - 10$ **b.** 24 in., 26 in., and 624 in.² **103.** ≈ 4.2 years **104.** $z + 5$

105. $3m - 2 + \frac{1}{m - 5}$ **106.** $y - 4 - \frac{4}{2y + 1}$

107. $2m - 2 + \frac{1}{3m - 2}$ **108.** $x^2 - x + 1 - \frac{2}{x + 1}$

109. $4x^2 + 6x + 9$ **110.** $2s^2 + 2s - \frac{3}{s - 1}$ **111.** $5y^3 - 3y + 2$

112. $4x - 5$

Chapter 5 Practice Test

1. $\frac{1}{8}$ **2.** $\frac{9}{4}$ **3.** 0.006201

4. 2.75×10^8 **5.** 3 **6.** 4 **7.** 56 **8.** $-10x^4 - 6x^3 + 14x^2 + 12$

9. $8x^2 + x - 4$ **10.** $5x^4 - 3x^2 - x + 8$ **11.** $12x^7$

12. $-2a^6b^4c^7$ **13.** $16x^2y^6$ **14.** $3x^3 - 12x^2 + 15x$

15. $-24t^5u + 48t^3u^3$ **16.** $n^2 + 3n - 4$ **17.** $4x^2 - 12x + 9$

18. $x^3 - 2x^2 - 5x + 6$ **19.** $6n^2 - 7n - 20$ **20.** x^5 **21.** $\frac{1}{x^5}$

22. $\frac{x^9y^4}{9}$ **23.** $4x^3 + 3x$ **24.** $x - 4$ **25.** $5x - 4 + \frac{6}{3x - 2}$

Chapter 6

6.1 Exercises

1. To list all of the possible factors of a number, we can divide by 1, 2, 3 and so on, writing each divisor and quotient pair as a product until we have all possible combinations. **3.** To find the GCF of a given set of numbers:

a. Write the prime factorization of each given number in exponential form.   **b.** For each prime base common to all factorizations, compare exponents and use the smallest exponent in the GCF factorization.   **c.** Multiply.   **5.** A four-term polynomial
7. 1, 2, 3, 4, 5, 6, 10, 12, 15, 20, 30, 60   **9.** 1, 2, 3, 6, 9, 18
11. 1, 3, 5, 9, 15, 45   **13.** 1, 3, 9   **15.** 1, 2, 4, 11, 22, 44,
17. 1, 2, 3, 5, 6, 9, 10, 15, 18, 30, 45, 90   **19.** 1, 2, 4, 7, 8, 14, 28,
56   **21.** 3   **23.** 17   **25.** 6   **27.** $2x$ **29.** xy **31.** $3a^4b$
33. $5(c - 4)$ **35.** $x(x - 1)$ **37.** $3ab(2a - b)$ **39.** $7uv(v - 1)$
41. $25x(y - 2z + 4)$ **43.** $w(w^2 + 3wv + 3v^2)$
45. $18ab^2c(b - 2a)$ **47.** $xy(4 - 8x + 15y)$
49. $8(3m - 2n + 3b)$ **51.** $3lmn(ln^4 + 6n^2 - 2)$
53. $21a^2b^2(5a - 3b + 4a^4b^2)$
55. $3x(6x^3 - 3x^2 + 10x - 4)$ **57.** $(a - 3)(2 + y)$
59. $(4x - 3)(ay - 2)$ **61.** $(b + c)(x + 2)$
63. $(y + 4)(y^3 - b)$ **65.** $(3x + 5)(y + 2)$
67. $(x^2 + y^2)(a - 5)$ **69.** $(u - 6)(u - 4)$
71. $(2 - a)(3 + b)$ **73.** $(m - n)(a - b)$
75. $(y - s)(x^2 - r)$ **77.** $(w + 3z)(w + 5)$
79. $(s + y)(3t - 2)$ **81.** $(ab + x)(4 + c)$
83. $3m(m + 1)(n - 4m)$ **85.** $2x(x + 3y)(x + 5)$
87. $h^2(g + h)(7 + h)$ **89.** $3a(y - 4)(a + 3)$ **91.** $6x(x + 6)$
93. $3x(7x + 9)$ **95.** $12\pi r^2(3r + 2)$

Review Exercises
1. 3,740,000,000   **2.** 4.56×10^7
3. $x^2 + 7x + 10$ **4.** $x^2 - x - 12$ **5.** $x^2 - 8x + 15$
6. $x^2 - 7x$

6.2 Exercises
1. Both are positive.   **3.** The number with the greater absolute value is positive and the other number is negative.   **5.** 5   **7.** 2   **9.** 6   **11.** 2   **13.** $(a + 3)(a + 4)$
15. $(x - 5)(x - 6)$ **17.** $(y - 6)^2$ **19.** $(n - 6)(n - 5)$
21. $(x - 6)(x + 5)$ **23.** $(r + 3)(r + 1)$
25. $(u - 2)(u - 4)$ **27.** $(k - 5)(k + 2)$
29. $(u + 3)(u - 2)$ **31.** $(b - 3)(b - 7)$
33. $(r - 3)(r - 6)$ **35.** $(a - 12)(a + 2)$ **37.** prime
39. prime   **41.** $(p + 3q)^2$ **43.** $(a - 9b)(a + 3b)$
45. $(x - 12y)(x - 2y)$ **47.** $(r - 6s)(r + 5s)$
49. $2m(m - 1)(m - 6)$ **51.** $3b^2(a - 8)(a + 3)$
53. $4(x - 3)^2$ **55.** $n^2(n + 2)(n + 3)$ **57.** $6r^2(a + 1)^2$
59. $7u^2(u + 5)(u - 1)$ **61.** $2x(x + 3)(x + 5)$
63. $3x(x + y)^2$ **65.** $2b(a - 6b)(a + 3b)$
67. $(x - 6)(x + 1)$ **69.** $(x - 4)(x + 1)$ **71.** 7, 8, 13
73. 4, 20   **75.** 9, 16, 21, 24, 25   **77.** 6, 10, 12

Review Exercises
1. commutative property of multiplication   **2.** $10x^2 + 17x + 3$ **3.** $15y^2 - 2y - 24$
4. $72n^3 - 60n^2 + 8n$ **5.** 1, 2, 3, 4, 6, 8, 12, 16, 24, 48   **6.** 1

6.3 Exercises
1. monomial GCF in all the terms
3. Write a pair of last terms whose product is c.   **5.** 2, 4
7. $t, 4t$ **9.** $7, 3x$ **11.** $(2y - 5)(y + 1)$
13. $(3m - 4)(m - 2)$ **15.** $(4a - 3)(a - 4)$
17. $(6a + 7)(a + 1)$ **19.** prime   **21.** $(2k - 5)(k + 6)$
23. $(2j + 1)(j + 2)$ **25.** $(3p - q)(p - 4q)$
27. $(k + h)(5k - 12h)$ **29.** $(2a - 5b)(6a - 5b)$

31. $(4w + 3v)(w + v)$ **33.** $y(8x^2 - 4x - 1)$
35. $2a(3b - 7)(b - 1)$ **37.** $2v^2(3w + 2)(w + 1)$
39. $(a + 1)(3a + 1)$ **41.** $(y - 2)(2y + 3)$
43. $(4r + 3)(2r - 3)$ **45.** $(2t - 1)(t - 1)$
47. $2(k - 1)(2k - 3)$ **49.** $(5x - 2)(4x - 3)$
51. $(3h - 2)(h - 1)$ **53.** $(3u - 5)(2u + 1)$
55. $(x - 5y)(5x - y)$ **57.** $(2k + j)(3k + 2j)$
59. $(5s - 2t)(2s + t)$ **61.** $2(u - v)(7u + 4v)$
63. $6y(3x - 2)(2x - 3)$ **65.** $2n(7m + 4)(2m - 1)$
67. Mistake: Using FOIL to check, we see that the first and last terms check but the inner and outer terms combine to give $12x$ instead of $13x$. Correct: $(2x + 1)(3x + 5)$
69. Mistake: Did not factor completely. Correct:
$4(n + 2)(n + 1)$ **71.** $(5x - 2)(3x - 1)$ **73.** 7, 8, or 13   **75.** 4, 12, or 44   **77.** 1, 11, 19, or 41

Review Exercises
1. $9x^2 - 25$ **2.** $x^3 - x^2 + 2x - 8$
3. $4x^2 - 20x + 25$ **4.** $36y^2 + 12y + 1$ **5.** $16x^2$ **6.** $8y^3$

6.4 Exercises
1. The first and last terms are perfect squares and that twice the product of their roots equals the middle term.   **3.** conjugates   **5.** The trinomial factor.
7. $(x + 7)^2$ **9.** $(b - 4)^2$ **11.** $(5u - 3)^2$ **13.** $(n + 12)^2$
15. $(y + 6)^2$ **17.** $(2p - 7q)^2$ **19.** $(4g + 3h)^2$
21. $(3r - 2s)^2$ **23.** $(10w + 1)^2$ **25.** $(x + 2)(x - 2)$
27. $(4 + y)(4 - y)$ **29.** $(5u + 4)(5u - 4)$
31. $(3x + b)(3x - b)$ **33.** $(5x^3 + 3y)(5x^3 - 3y)$
35. $(u^2 + v)(u^2 - v)$ **37.** $(10w + 1)(10w - 1)$
39. $(7j + k)(7j - k)$ **41.** $(n - 3)(n^2 + 3n + 9)$
43. $(y - 4)(y^2 + 4y + 16)$ **45.** $(3k - 2)(9k^2 + 6k + 4)$
47. $(3x - 4y)(9x^2 + 12xy + 16y^2)$
49. $(10a - 3b)(100a^2 + 30ab + 9b^2)$
51. $(x + 3)(x^2 - 3x + 9)$ **53.** $(u + 5)(u^2 - 5u + 25)$
55. $(5x + 4y)(25x^2 - 20xy + 16y^2)$
57. $(m + n)(m^2 - mn + n^2)$
59. $(2p + qz)(4p^2 - 2pqz + q^2z^2)$
61. $\left(4x + \dfrac{5}{7}\right)\left(4x - \dfrac{5}{7}\right)$ **63.** $2x(5x + 1)(5x - 1)$
65. $y^3(y + 4b)(y - 4b)$ **67.** $2u(u + 1)(u - 1)$
69. $2(x + 5)(x - 5)$ **71.** $2c(2c + d)(4c^2 - 2cd + d^2)$
73. $3(y - 2z)(y^2 + 2yz + 4z^2)$
75. $\left(c - \dfrac{2}{3}\right)\left(c^2 + \dfrac{2}{3}c + \dfrac{4}{9}\right)$ **77.** $(2a - b + c)(2a - b - c)$
79. $(1 - 3x + 3y)(1 + 3x - 3y)$
81. $(x + 3y + 3z)(x^2 - 3xy - 3xz + 9y^2 + 18yz + 9z^2)$
83. $(4d - x - y)(16d^2 + 4dx + 4dy + x^2 + 2xy + y^2)$
85. 24   **87.** 36   **89.** 25   **91.** 16   **93.** $10(x + 2)(x - 2)$
95. $(2x - 3)(4x^2 + 6x + 9)$

Review Exercises
1. 1, 2, 3, 4, 6, 9, 12, 18, 36
2. $2 \cdot 2 \cdot 5 \cdot 5$ **3.** $4x^3 + 2x^2 - 3x - 26$
4. $10n^4 - 18n^3 + 12n^2 - 16n - 6$ **5.** $16x^2 - 40x + 25$
6. $5x^2 - 6x + 7 + \dfrac{3}{x + 2}$

6.5 Exercises

1. A monomial GCF. **3.** Sum of squares.
5. Try to factor by grouping if there are four terms. Grouping may also be used as a method to factor trinomials.
7. $3xy(y + 2x)$ **9.** $2(x + 4)(x - 4)$ **11.** $(a + b)(x + y)$
13. $(x + 3)(x + 5)$ **15.** $(3b + 2)(2b - 1)$ **17.** prime
19. $3(y - 2)(y^2 + 2y + 4)$ **21.** $6ab(1 - 6b)$
23. $(x + 7)(x - 7)$ **25.** $(2x - 3y)(4x^2 + 6xy + 9y^2)$
27. $(5x - 1)(3x + 2)$ **29.** $2(b + 3)(b + 4)$
31. $x(y - 5)(y + 4)$ **33.** $7(u - 5)(u + 3)$
35. $(t^2 + 9)(t + 3)(t - 3)$ **37.** $(y - 10)(y + 4)$
39. $5(p + 4)(p - 4)$ **41.** $(q^2 - 8)(q^2 + 2)$
43. $(3v + 2)(4v + 5)$ **45.** $2(1 + 5x)(1 - 5x)$
47. $a(1 + 9x^2)(1 + 3x)(1 - 3x)$ **49.** $(2j + 3)(j - 1)$
51. $2(5 - t)^2$ **53.** $5ax(1 - a)$ **55.** $(x - 2y)(3a + 4b)$
57. $(3m - 5)(m - 1)$ **59.** $2(x^2 + 9)(x + 3)(x - 3)$
61. $5x(x - 2)$ **63.** $(2x - 3)(3x + 2)$
65. $x(x^2 + 2y)(x^2 - 2y)$ **67.** $3x(x + 2)(x - 2)$
69. $(a + 3b)(2x - y)$ **71.** $2x(3 - y)(9 + 3y + y^2)$
73. $3(m^2 + 3m + 9)$ **75.** $y^3(y - 6)$
77. $l = 2x + 1$
    $w = 3x - 7$
79. $l = 2x$
    $w = 3x + 5$
    $h = 3x + 5$

Review Exercises
1. $17, 18, 19$ **2.** 224 ft.^{2} **3.** -42
4. $x - 4 - \dfrac{4}{2x + 1}$ **5.** $6x^2$

6.6 Exercises

1. If a and b are real numbers and $ab = 0$, then $a = 0$ or $b = 0$. **3.** Make it equal to 0. **5.** $-5, -2$
7. $-3, 4$ **9.** $-\dfrac{3}{2}, \dfrac{4}{3}$ **11.** $0, 7$ **13.** $-2, 1, 0$ **15.** $0, 2$
17. $-2, 2$ **19.** $-3, 2$ **21.** $-9, 3$ **23.** $-3, 7$ **25.** $1, 2$
27. $-3, 3$ **29.** $-4, \dfrac{15}{4}$ **31.** $-6, 3$ **33.** $-2, 7$ **35.** $-\dfrac{7}{2}$
37. $5, 11$ **39.** $13, 52$ **41.** $13, 14$ **43.** 12m by 21m
45. 14 ft. **47.** 9 in. **49.** 1.25 sec. **51.** 10% **53.** 15 **55.** 25
57. 37 in. **59.** 30 blocks

Review Exercises
1. $(a + b)(x + y)$
2. $(x + a)(x - a)$ **3.** -3 **4.** -4 **5.** 4.203×10^{-4}
5. $(8, 0)(0, 6)$

6.7 Exercises

1. If a vertical line is drawn through the vertex, the sides of the parabola are mirror images. **3.** The y-coordinate is c. **5.** $(2, -1)$ **7.** $(3, 0)$ **9.** $(0, -4)$
11.

13.

15.

17.

19.

21.

23.

25.

27.

29.

31.

33.

$f(x) = -x^2 - 4x - 3$

35.

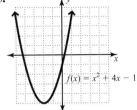

$f(x) = x^2 + 4x - 1$

37.

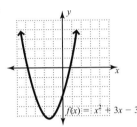

$f(x) = x^2 + 3x - 3$

39.

$f(x) = 2x^2 - 3x + 1$

41.

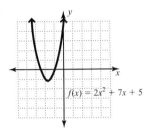

$f(x) = 2x^2 + 7x + 5$

43. A function. **45.** Not a function. **47. a.** $(0, 0)$ **b.** $(0, 2)$
c. c is the y-intercept. **d.** The original graph moved downward
3 units. **49. a.** $y = x^2$ opens up and $y = -x^2$ opens down. The
sign of a indicates whether the parabola will open up or down.
b. yes. **c.** These graphs are "wider." **d.** The smaller the
absolute value of a, the wider the parabola.

Review Exercises
1. $3m^3n - 12m^2n^2 + 15mn$
2. $14x^3 - 35x^2 - 21x$ **3.** $(x + 5)(x - 5)$
4. $(3n - 5)(n - 4)$ **5.** $0, 3$ **6.** $-6, 3$

Chapter 6 Review Exercises
1. true **2.** false
3. false **4.** true **5.** true **6.** GCF, $\dfrac{\text{given polynomial}}{\text{GCF}}$
7. $(a + b)(a - b)$ **8.** $(a - b)(a^2 + ab + b^2)$ **9.** $(a - b)^2$
10. c, b **11.** $1, 2, 3, 4, 5, 6, 10, 12, 15, 20, 30, 60$ **12.** $1, 3, 9$
13. $1, 3, 9, 27, 81$ **14.** $1, 2, 4, 11, 22, 44$ **15.** $1, 3, 5, 9, 15, 45$
16. $1, 7, 49$ **17.** $1, 2, 3, 5, 6, 10, 15, 30$ **18.** $1, 3, 11, 33$ **19.** 3
20. 17 **21.** 6 **22.** 5 **23.** $5y^2$ **24.** $4x$ **25.** $3mn$ **26.** $2k$
27. $2(2x - 1)$ **28.** $5m(1 - 7m^2)$ **29.** $y(y^2 - 1)$
30. $abc(1 + abc - a^2b^2c^2)$ **31.** $xy(x + y + x^2y^2)$
32. $21a^2b^2(5a - 3b + 4a^4b^2)$
33. $10k(10k^3 + 12k^4 - 1 + 4k^2)$ **34.** $18ab^2c(b - 2a)$
35. $(x + y)(a + b)$ **36.** $(x + 2)(a + b)$
37. $(y + 2)(y^2 + 3)$ **38.** $(y + 4)(y^3 - b)$
39. $(x + 1)(y + 1)$ **40.** $(x^2 + y^2)(a - 5)$
41. $(2b^2 + 1)(b - 1)$ **42.** $(u - 3)(u + 4v)$
43. $(x - 4)(x + 3)$ **44.** $(x + 5)(x + 9)$

45. $(n - 2)(n - 4)$ **46.** prime **47.** $(h + 3)(h + 48)$
48. $(y - 12)(y + 2)$ **49.** $4(x - 3)^2$ **50.** $3(m - 9)(m - 2)$
51. prime **52.** $(2u + 1)(u + 2)$ **53.** $(3m - 1)(m - 3)$
54. $(5k - 12h)(k + h)$ **55.** $2(3a - 4)(a - 2)$
56. $y(8x^2 - 4x - 1)$ **57.** $(3p - q)(p - 4q)$
58. $8(3x + y)(x - 3y)$ **59.** $(v - 4)^2$ **60.** $(u + 3)^2$
61. $(2x + 5)^2$ **62.** $(3y - 2)^2$ **63.** $(x + 2)(x - 2)$
64. $(5 + y)(5 - y)$ **65.** $(x - 1)(x^2 + x + 1)$
66. $(x + 3)(x^2 - 3x + 9)$ **67.** $(3b + 2)(2b - 1)$
68. $4ab(1 - 6b)$ **69.** prime **70.** $3(x + y)(x - y)$
71. $(x^2 + 9)(x + 3)(x - 3)$ **72.** $7(u - 5)(u + 3)$
73. $(2x - 3y)(4x^2 + 6xy + 9y^2)$ **74.** $2(1 + 5y)(1 - 5y)$
75. $(m - 2n)(3a + 4b)$ **76.** $3(m^2 + 3m + 9)$ **77.** $-3, 4$
78. $-2, 2$ **79.** $2, 3$ **80.** $-3, 7$ **81.** $-3, 3$ **82.** $-6, 3$ **83.** -3
84. $-2, 7$ **85.** $5x^2 = 2x; 0, \dfrac{2}{5}$ **86.** $x^2 + (x + 1)^2 = 61; 5, 6$
87. $x \cdot 4x = 100; -5, 5$ **88.** $(w + 5)w = 91; 13 \text{ in. by } 7 \text{ in.}$
89. $x(x + 1) = 110; 10, 11$ **90.** $3^2 + b^2 = 5^2; 4$

91.

$y = -5x^2$

92.

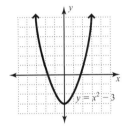

$y = x^2 - 3$

93.

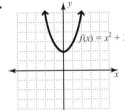

$f(x) = x^2 + 2$

94.

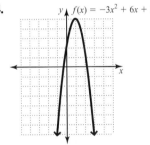

$y\ f(x) = -3x^2 + 6x + 2$

Chapter 6 Practice Test
1. $1, 2, 4, 5, 8, 10, 16, 20, 40, 80$
2. $5m$ **3.** $5(y - 6)$ **4.** $2y(3x^2 - y)$ **5.** $(x - y)(a - b)$
6. $(m + 3)(m + 4)$ **7.** $(c + 5)(c - 5)$ **8.** $a(x + 3)(x - 8)$
9. $(m + 5)(m^2 - 5m + 25)$ **10.** $(r - 4)^2$
11. $2(1 + 5u)(1 - 5u)$ **12.** $(q - 2)(3q - 4)$ **13.** prime
14. $-4(x + 2)(x - 2)$ **15.** $(x - 2)(x^2 + 2x + 4)$ **16.** $-3, 0$
17. $-9, 3$ **18.** 9 ft., 4 ft.
19.

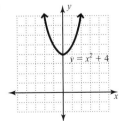

$y = x^2 + 4$

20.

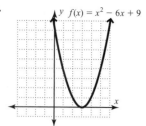
$y \quad f(x) = x^2 - 6x + 9$

Chapters 1–6 Cumulative Review Exercises

1. false   **2.** true   **3.** false   **4.** false   **5.** false   **6.** false
7. GCF, grouping   **8.** highest   **9.** It cannot be factored any further.   **10.** 0, factored   **11.** The x-intercept is the point where the graph crosses the x-axis and the y-intercept is where the graph crosses the y-axis.   **12.** A polynomial containing 2 terms.   **13.** 42   **14.** 10   **15.** 16   **16.** -89 **17.** 10
18. $\dfrac{17}{12}$ **19.** 13   **20.** 13.2   **21.** $x = \dfrac{7 + 3y}{2}$ **22.** no   **23.** 0
24. $(3, 0), (0, -6)$

25.   **26.**

$y = 3x + 2$ $2x - y = 6$

27. $5x^2 + 4x + 2$ **28.** $12m^2 - 6m - 7$ **29.** $18x^4y$ **30.** $48x^7$
31. $2y^2 - 15y - 8$ **32.** $4x^2 + 20x + 25$ **33.** u
34. $4h + 1 - \dfrac{2}{h}$ **35.** $x + 4$ **36.** $x^2 - 2x + 4 - \dfrac{56}{x + 2}$
37. $(x + 11)(x - 11)$ **38.** $10(x^2 + 4)$ **39.** $(x - 4)^2$
40. $(2x - 3)(x + 4)$ **41.** $(x - 5)(x^2 + 5)$
42. $(n - 2)(7m + 6)$ **43.** $(2x - 3)(4x^2 + 6x + 9)$
44. $(3x + 2)(x + 2)$ **45.** $-10, 10$ **46.** $-4, 7$ **47.** 6, 8, 10
48. \$2500 in 5%, \$5500 in 8%   **49.** $33\dfrac{1}{3}$ ml   **50.** ≈ 686

7.1 Exercises

1. We replace the variables with given values, and then simplify the numerical expression.   **3.** Write an equation that has the denominator set equal to zero, and solve the equation.   **5.** y is not a factor in $x + y$.   **7. a.** $\dfrac{16}{45}$ **b.** -1 **c.** 0
9. a. $\dfrac{11}{8}$ **b.** ≈ 0.49 **c.** $-\dfrac{1}{7}$ **11. a.** $-\dfrac{3}{2}$ **b.** $-11.3\overline{6}$ **c.** $-\dfrac{4}{3}$
13. a. $\dfrac{1}{6}$ **b.** $-\dfrac{2}{21}$ **c.** ≈ 0.063 **15.** 3   **17.** $-3, 3$ **19.** 5, 9
21. $-\dfrac{3}{4}, 5$ **23.** $-1, \dfrac{5}{2}$ **25.** $\dfrac{3x}{4y}$ **27.** $-\dfrac{m^2}{n}$ **29.** $\dfrac{3}{2x^2z^2}$
31. $-\dfrac{3m^2n}{m + 2n}$ **33.** $\dfrac{x + 5}{2}$ **35.** $\dfrac{3}{5}$ **37.** $\dfrac{2}{3}$ **39.** $\dfrac{5}{3}$ **41.** $\dfrac{x}{4}$ **43.** $\dfrac{5}{x}$
45. $\dfrac{x - y}{x + y}$ **47.** $\dfrac{x}{x - 2}$ **49.** $\dfrac{3}{(x^2 + 4)(x - 2)}$ **51.** $\dfrac{3(x + y)}{2x}$

---

53. x **55.** $\dfrac{a - b}{a + b}$ **57.** $\dfrac{t - 3}{t + 2}$ **59.** $\dfrac{k}{k + 2}$ **61.** $\dfrac{1 - 2m}{1 - 2m + 4m^2}$
63. $\dfrac{2u^2 - 1}{u + 1}$ **65.** -4 **67.** $\dfrac{-1}{m^2 + 3m + 9}$ **69.** $\dfrac{-1}{w + 1}$ **71.** $-\dfrac{5}{2}$
73. Mistake: Placed remaining x in numerator. Correct: $-\dfrac{1}{2x}$
75. Mistake: Divided out a part of a multiform polynomial. Correct: Can't reduce   **77.** Mistake: $y - x$ and $x - y$ are not identical factors until -1 is factored out of one of them. Correct: $-x$
79. a. 23.7, 22.9, 28.3   **b.** The first two are normal weight and the third is overweight.   **81.** $533.\overline{3}\pi, 666.\overline{6}\pi, 666.\overline{6}\pi, 833.\overline{3}\pi$
83. a. $h = \dfrac{3V}{\pi r^2}$ **b.** 9 cm   **85.** $\dfrac{s - 3}{s + 4}$

Review Exercises   **1.** 1   **2.** $-\dfrac{3}{4}$ **3.** $\dfrac{9}{20}$ **4.** $\dfrac{28}{15}$
5. $(x - y)(a + b)$ **6.** -1

7.2 Exercises   **1.** To eliminate any common factors.
3. Their product is 1.   **5.** A fraction with a ratio equivalent to 1. Examples will vary.
7. $\dfrac{m}{n^2}$ **9.** $\dfrac{m}{n}$ **11.** y^2 **13.** $\dfrac{bp}{2c}$ **15.** $\dfrac{12}{5}$ **17.** r
19. $\dfrac{w - 1}{(w + 1)(w - 5)}$ **21.** $\dfrac{n + 1}{n}$ **23.** $\dfrac{q + 1}{2q(q - 1)}$
25. $\dfrac{(r + s)(r + 5)}{(r - s)(r - 2)}$ **27.** $\dfrac{1}{y}$ **29.** $\dfrac{ad}{bc}$ **31.** $\dfrac{y^2}{bc}$ **33.** $\dfrac{5}{4m^2n}$
35. $\dfrac{y^2}{5z^2}$ **37.** 1   **39.** $\dfrac{a(a - 4)(a + 4)}{b}$ **41.** $\dfrac{3}{2}$ **43.** $\dfrac{2(k + 1)}{k + 2}$
45. $\dfrac{a + 2b}{a - 2b}$ **47.** $\dfrac{2k + 1}{k(2k - 1)}$ **49.** $\dfrac{x + 2}{(x + 1)(3x - 1)}$
51. $\dfrac{-3(y - 2)}{y + 2}$ **53.** $\dfrac{7 + y}{y(y + 4)}$ **55.** $\dfrac{x}{y}$ **57.** Mistake: Did not invert the divisor. Correct: $\dfrac{50}{27}$ **59.** Mistake: Divided out y, which is illegal because it is not a factor in $y - 2$. Correct: $\dfrac{(x - 3)(y - 2)}{15y}$ **61.** 222 in.   **63.** 11,000 yd.
65. 88 oz.   **67.** 12.7 T   **69.** $1087.97\overline{3}$ ft./sec.   **71.** ≈ 19.03 mi./hr.
73. $\approx 5.87 \times 10^{12}$ mi./yr.   **75.** 333.74 £   **77.** \$434.07
79. 1009.57 €

Review Exercises   **1.** 120   **2.** $4x^3y^2$ **3.** 1   **4.** $-\dfrac{1}{2}$ **5.** $2x$
6. $\dfrac{1}{12}y - 4$

7.3 Exercises   **1.** The numerator may not contain like terms.   **3.** Write an equivalent addition with the additive inverse of the subtrahend.   **5.** $\dfrac{x}{2}$ **7.** $\dfrac{a^2}{3}$ **9.** $\dfrac{5}{a^2 + a}$ **11.** 3
13. $\dfrac{3r}{r + 2}$ **15.** $\dfrac{1}{m + 1}$ **17.** 1   **19.** $\dfrac{2(x^2 - 2)}{x + 1}$ **21.** $\dfrac{-11t + 3}{2t - 3}$
23. $\dfrac{11}{w + 1}$ **25.** $\dfrac{11x^2 - 21x + 7}{2x + 1}$ **27.** $\dfrac{t + 11}{u}$ **29.** $\dfrac{1}{g + 3}$

31. $\frac{y-3}{y}$ 33. $\frac{6w-5}{5}$ 35. h^2-1 37. $\frac{1}{x^2+1}$ 39. $\frac{3x+9}{x^2-25}$

41. Mistake: Combined terms that are not alike. Correct: $\frac{4y+1}{5}$
43. Mistake: Did not change sign of −5. Correct: 2
45. $\frac{5x+9}{2}$ 47. $\frac{3x+2}{2}$

Review Exercises 1. 72 2. $\frac{11}{10}$ 3. $\frac{11}{24}$ 4. $1\frac{6}{35}$
5. $\frac{11}{15}x+\frac{7}{12}$ 6. $\frac{15}{2}$

7.4 Exercises 1. 1. Factor each denominator. 2. For each unique factor, compare the number of times it appears in each factorization. Write a factored form that includes each factor the greatest number of times it appears in the denominator factorizations. 3. Factor the denominators. 5. Rewrite the numerator as an equivalent addition sentence.

7. $15a^2;\ \frac{10a}{15a^2};\ \frac{9}{15a^2}$

9. $(y+2)(y-2);\ \frac{3y-6}{(y+2)(y-2)};\ \frac{4y^2+8y}{(y+2)(y-2)}$

11. $p^2-4;\ \frac{2}{p^2-4};\ \frac{p^2+p-6}{p^2-4}$

13. $3(u+2)^2;\ \frac{12u}{3(u+2)^2};\ \frac{2u+4}{3(u+2)^2}$

15. $c^3+8;\ \frac{c}{c^3+8};\ \frac{2c^2-4c+8}{c^3+8}$

17. $(x-1)(x-2);\ \frac{2x-4}{(x-1)(x-2)};\ \frac{3}{(x-1)(x-2)};$ $\frac{5x^2-5x}{(x-1)(x-2)}$

19. $\frac{-31}{18}$ 21. $\frac{8}{3c}$ 23. $\frac{5x^2-2}{x^3}$ 25. $\frac{2a-b-5a^2-2ab}{a^2}$

27. $\frac{8y-2}{(y+2)(y-4)}$ 29. $\frac{5t+16}{(t+4)(t+6)}$ 31. $\frac{18m-3m^2}{2(m-2)^2}$

33. $\frac{x+y}{x-y}$ 35. $\frac{-3}{t-3}$ 37. $\frac{y+2}{y}$ 39. $\frac{w-6}{6(w+1)}$ 41. $\frac{3}{2(c-5)}$

43. $\frac{z}{2(z+3)}$ 45. $\frac{3m^2-3mn+4n^2}{(m+n)(m-n)}$ 47. $\frac{a^2+2ab}{a^2-b^2}$

49. $\frac{1}{(w-1)(w-2)}$ 51. $\frac{4-x}{2x-1}$ 53. $\frac{y+6}{3(y-2)}$

55. $\frac{r^2-6r-1}{(r+1)(r-1)^2}$ 57. $\frac{q+1}{q(q-3)}$ 59. $\frac{1+2d}{d^2(d+1)}$

61. Mistake: Did not find a common denominator.
Correct: $\frac{6-x^2}{2x}$ 63. Mistake: Added denominators. Correct: $\frac{7w}{x}$ 65. Mistake: Did not change the sign of 7. Correct: $\frac{-x-7}{x+2}$

67. $\frac{t}{2}+\frac{t}{3}$ 69. $\frac{2}{x+3}+\frac{2}{x}$ 71. $\frac{6x^2+8x+2}{(2x-1)(2x+1)}$

73. $\frac{2a^2+2b^2}{(a+b)^2}$ 75. $\frac{x(x^2+12)}{(x+3)(3x-4)}$ 77. $\frac{m^2+1}{m^4}$ 79. $\frac{32}{ax^2}$

Review Exercises 1. $\frac{3}{8}$ 2. $\frac{m^2}{108}$ 3. $4(x+2)(x-2)$
4. $3y(2y+1)(y-5)$ 5. $\frac{x-1}{x+2}$ 6. $\frac{1}{x^2-3x+9}$

7.5 Exercises 1. $\frac{4}{x+3}$ is the numerator, $\frac{3}{x}$ is the denominator 3. $\dfrac{\frac{2x}{x-3}}{\frac{4-x}{2x-6}}$ 5. Answers will vary.

7. $\frac{1}{2}$ 9. $\frac{mp}{nq}$ 11. 5 13. $1\frac{7}{8}$ 15. $\frac{a}{b(a-b)}$ 17. $\frac{x}{5(2x+1)}$

19. $\frac{x}{x-2}$ 21. xy 23. $\frac{1-y}{1+y}$ 25. u 27. $\frac{4r-3}{2-4r}$

29. $\frac{1}{2a-1}$ 31. $-w$ 33. 2 35. $\frac{4y^2-11y-3}{-2y}$

37. $\frac{1}{n-1}$ 39. $\frac{-2}{x(x+h)}$ 41. $\frac{m+n}{m-n}$ 43. $y-1$

45. Mistake: Divided out $\frac{1}{2}$, which is not a common factor.
Correct: $\frac{2x-1}{2y-1}$ 47. Mistake: Did not write as an equivalent multiplication. Correct: $\frac{m^2}{18}$ 49. Mistake: Did not distribute c in the numerator. Correct: c^2+1 51. Mistake: Added the x's instead of multiplying them. Correct: $\frac{18}{x^2}$ 53. $34\frac{2}{7}$ mph

55. $\frac{3(5n-2)}{4(n-1)}$ 57. a. $\frac{R_1R_2}{R_2+R_1}$ b. ≈ 9.1 c. 8

Review Exercises 1. −7 2. 3 3. 2.5 4. 4
5. 4.8 mph 6. 0.09 hr. or 5.4 min.

7.6 Exercises 1. 1. Factor each denominator. 2. Write a factored form that includes each factor the greatest number of times it appears in the denominator factorizations. 3. Eliminates the denominators. 5. 4 and −2 when substituted for x would cause the denominators to become zero, causing the rational expression to become undefined. 7. no

9. yes 11. yes 13. 2 15. 10 17. 15 19. $-\frac{1}{9}$ 21. 3

23. 1 25. 6 27. 10 29. 8 31. −1 33. $\frac{1}{2}$ 35. $-\frac{3}{4}$ 37. −5

39. $-\frac{1}{4}, 5$ 41. 9 43. $-\frac{1}{3}, 3$ 45. −2 47. $-\frac{1}{4}$ 49. $-\frac{20}{3}$

51. $\frac{1}{3}$ 53. 9 55. 5 57. −2 59. No solution.

61. −3 is the only answer, 3 is extraneous. 63. In an expression, the LCD is used to combine the numerator over a single denominator. In an equation, the LCD is used to eliminate the denominator. 65. 400 67. 20 and 30 69. ≈ 2.1 ft.
71. $f = 15$ mm, $i = 20$ mm

Review Exercises 1. 13 quarters and 17 nickels
2. Katlin is 16 and her brother is 23. **3.** The length is 11 mm
and the width is 7 mm. **4.** The numbers are 15 and 60.
5. The numbers are 7 and 16. **6.** The numbers are 40 and 76.

7.7 Exercises 1. $\dfrac{1}{x}$ 3. $\dfrac{100}{r}$ 5. increases 7. $4\dfrac{4}{5}$ days

9. 12 min. **11.** $2\dfrac{2}{5}$ hrs. **13.** $2\dfrac{8}{11}$ hr. **15.** 6 mph, 4 mph

17. 340 mph **19.** 60 mph, 20 mph **21.** 16 mph, 20 mph
23. 30 **25.** 4.5 **27.** \$17.16 **29.** 15 gal. **31.** 144.9 lb.
33. 660 vps **35.** 420 cu. in. **37.** 5 **39.** 14 **41.** 30 psi **43.** 4 A
45. $\approx$ 4.39 million pounds **47.** $2.\overline{3}$ cm/sec.

49. a.

b. directly proportional. **c.** Yes; each value in the domain
(supply) is paired with one value in the range (price per unit).

Review Exercises 1. no 2. $\left(-\dfrac{3}{2}, 0\right)$, $(0, 6)$

3.

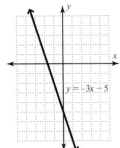

4. $4x^2 - 13x + 12$ **5.** $\dfrac{49}{16}$ **6.** 5, 6

Chapter 7 Review Exercises 1. false 2. true
3. true **4.** true **5.** true **6.** multiplication, Divide **7.** LCD
8. numerators, denominator **9.** complex **10.** numerator,
denominator **11. a.** $\dfrac{17}{6}$ **b.** ≈ 0.077 **c.** $\dfrac{3}{4}$ **12. a.** 3 **b.** 15
c. 1.875 **13.** 5 **14.** $-2, 2$ **15.** $-6, 1$ **16.** $0, -6$ **17.** $\dfrac{x + 5}{3}$

18. $\dfrac{2}{3}$ **19.** $\dfrac{x}{4}$ **20.** $\dfrac{y}{3}$ **21.** $\dfrac{a + b}{a - b}$ **22.** $\dfrac{3}{(x^2 + 9)(x - 3)}$

23. $-(x + y)$ **24.** $\dfrac{-1}{w + 3}$ **25.** $\dfrac{7x^2}{4y^2}$ **26.** $\dfrac{3}{5mnp^4}$ **27.** $\dfrac{3}{5}$

28. $\dfrac{x^2}{(x + 4)(x - 3)}$ **29.** $\dfrac{n + 1}{n}$ **30.** $\dfrac{2}{r - 1}$ **31.** $\dfrac{m + 1}{4m(m - 1)}$

32. $\dfrac{w + 4}{w - 4}$ **33.** $\dfrac{y^3}{abx}$ **34.** $\dfrac{5}{6}$ **35.** $\dfrac{x + y}{c - d}$

36. $\dfrac{5(b + 4)(b - 4)}{6}$ **37.** 2 **38.** $\dfrac{2j + 1}{j(2j - 1)}$ **39.** $\dfrac{1}{(u + 1)^2}$

40. $\dfrac{5a}{2}$ **41.** $\dfrac{x}{3}$ **42.** $\dfrac{3r}{r + 3}$ **43.** 1 **44.** 1 **45.** $\dfrac{11x^2 - 21x + 7}{2x + 1}$

46. $q - 3$ **47.** $\dfrac{s}{r^2 - s^2}$ **48.** 1 **49.** $\dfrac{6c^2}{12c^3}; \dfrac{12}{12c^3}$ **50.** $\dfrac{5x}{x^2y^3}; \dfrac{3y}{x^2y^3}$

51. $\dfrac{4m + 4}{(m - 1)(m + 1)}; \dfrac{4my - 4y}{(m - 1)(m + 1)}$

52. $\dfrac{x}{(x + 1)(x - 1)}; \dfrac{x^2 + x - 2}{(x + 1)(x - 1)}$

53. $\dfrac{2p + 4}{(p - 2)(p + 2)^2}; \dfrac{p^2 - 2p}{(p - 2)(p + 2)^2}$

54. $\dfrac{3h}{4h(h - 2)}; \dfrac{20}{4h(h - 2)}$ **55.** $-\dfrac{23}{36}$ **56.** $\dfrac{-x - 27}{(x + 3)(x - 3)}$

57. $\dfrac{8y - 22}{(y - 2)(y - 4)}$ **58.** $\dfrac{4x - 3}{(x + 2)(x - 2)}$ **59.** $\dfrac{-3t^2 + 21t}{2(t - 3)^2}$

60. $\dfrac{-a^2 + 18}{(a + 3)(a + 4)}$ **61.** $\dfrac{15}{2y - 1}$ **62.** $\dfrac{3x + 57}{3(x + 4)(x - 4)}$

63. $\dfrac{ay}{bx}$ **64.** -1 **65.** $\dfrac{5(x - 2)}{x(3x + 5)}$ **66.** $\dfrac{x}{x - 3}$ **67.** $\dfrac{x}{5(2x - 1)}$

68. 1 **69.** $\dfrac{y^2 + x^2}{7xy}$ **70.** $-h$ **71.** $\dfrac{38}{13}$ **72.** 6 **73.** 1 **74.** 4

75. 14 **76.** -1 **77.** 25 **78.** 23 **79.** 31.875 min. **80.** 8 yd./sec.
and 10 yd./sec. **81.** 30 **82.** 2 **83.** 10 gallons **84.** 4.8 A

Chapter 7 Practice Test 1. 7 2. $-4, 4$ 3. $\dfrac{-3}{x - 4}$

4. $\dfrac{1}{x + y}$ **5.** $\dfrac{6f}{f^2g^3}; \dfrac{2g^3}{f^2g^3}$ **6.** $\dfrac{5x(x + 3)}{(x + 3)^2(x - 3)}; \dfrac{2(x - 3)}{(x + 3)^2(x - 3)}$

7. $\dfrac{5ab^2x^2}{12}$ **8.** $\dfrac{7a^6b^7y^2}{40x^2}$ **9.** $-4x$ **10.** $\dfrac{12x^2}{(x + 1)(2x + 1)}$

11. $\dfrac{7x}{2x + 3}$ **12.** $\dfrac{1}{x + 3}$ **13.** $\dfrac{x^2 + 5x - 6}{x(x + 2)(x - 2)}$

14. $\dfrac{9x + 3}{(x - 1)(x + 2)}$ **15.** $\dfrac{3x + 13}{(x - 4)(x + 1)}$ **16.** $\dfrac{x - 2}{2x}$

17. $\dfrac{2y + 1}{3y - 1}$ **18.** $\dfrac{2}{x + y}$ **19.** $\dfrac{25}{9}$ **20.** -1 **21.** -1
22. 2, 4 **23.** 1.2 hr. **24.** 6 mph, 4 mph **25.** 529.2 N

Chapter 8

8.1 Exercises

1. An inequality containing two inequality symbols. **3.** For two sets A and B, the intersection of A and B, symbolized by $A \cap B$, is a set containing only elements that are in both A and B. **5.** We graph the area between the two values.

7. a. $\{1, 3, 5\}$ **b.** $\{1, 3, 5, 7, 9\}$

9. a. $\{7\}$ **b.** $\{5, 6, 7, 8, 9\}$

11. a. $\varnothing$ **b.** $\{a, c, d, g, o, t\}$

13. a. $\{x, y, z\}$ **b.** $\{w, x, y, z\}$

15. $-4 < x < 5$ **17.** $-2 < y \leq 0$

19. $-7 < w < 3$ **21.** $0 \leq u \leq 2$

23.

25.

27.

29.

31. a. **b.** $\{x \mid -3 < x < -1\}$ **c.** $(-3, -1)$

33. a. **b.** $\{x \mid 3 < x \leq 6\}$ **c.** $(3, 6]$

35. a. **b.** $\varnothing$ **c.** No solution.

37. a. **b.** $\{x \mid -5 \leq x < 3\}$ **c.** $[-5, 3)$

39. a. **b.** $\{x \mid x \geq 2\}$ **c.** $[2, \infty)$

41. a. **b.** $\{x \mid -7 < x < -3\}$ **c.** $(-7, -3)$

43. a. **b.** $\{x \mid -1 \leq x \leq 2\}$ **c.** $[-1, 2]$

45. a. **b.** $\left\{ x \mid -\dfrac{2}{3} \leq x < 2 \right\}$ **c.** $\left[-\dfrac{2}{3}, 2 \right)$

47. a. **b.** $\{x \mid 0 \leq x < 3\}$ **c.** $[0, 3)$

49. a. **b.** $\{x \mid 0 \leq x \leq 3\}$ **c.** $[0, 3]$

51.

53.

55.

57.

59. a. **b.** $\{y \mid y < -9 \text{ or } y > 5\}$ **c.** $(-\infty, -9) \cup (5, \infty)$

61. a. **b.** $\left\{ x \mid x \leq -\dfrac{3}{5} \text{ or } x \geq \dfrac{1}{5} \right\}$ **c.** $\left(-\infty, -\dfrac{3}{5} \right] \cup \left[\dfrac{1}{5}, \infty \right)$

63. a. **b.** $\left\{ r \mid r < -\dfrac{5}{3} \text{ or } r > -\dfrac{1}{3} \right\}$ **c.** $\left(-\infty, -\dfrac{5}{3} \right) \cup \left(-\dfrac{1}{3}, \infty \right)$

65. a. **b.** $\{w \mid w \leq -1 \text{ or } w \geq 7\}$ **c.** $(-\infty, -1] \cup [7, \infty)$

67. a. **b.** $\left\{ c \mid c > \dfrac{7}{2} \right\}$ **c.** $\left(\dfrac{7}{2}, \infty \right)$

69. a. **b.** $\{x \mid x \text{ is a real number}\}$ or $\mathbb{R}$ **c.** $(-\infty, \infty)$

71. a. **b.** $\{x \mid x \leq -4\}$ **c.** $(-\infty, 4]$

73. a. **b.** $\left\{ m \mid m \leq 1 \text{ or } m \geq \dfrac{8}{3} \right\}$ **c.** $(-\infty, 1] \cup \left[\dfrac{8}{3}, \infty \right)$

75. a. **b.** $\{k \mid k \leq 2 \text{ or } k \geq 3\}$ **c.** $(-\infty, 2] \cup [3, \infty)$

77. a.

$-6\,-5\,-4\,-3\,-2\,-1\ 0\ 1\ 2\ 3\ 4\ 5\ 6$

b. $\{x\,|-3 < x < -2\}$ **c.** $(-3, -2)$

79. a.

$-7\,-6\,-5\,-4\,-3\,-2\,-1\ 0\ 1\ 2\ 3\ 4\ 5\ 6$

b. $\{x\,|-7 < x < 3\}$ **c.** $(-7, 3)$

81. a.

$-6\,-5\,-4\,-3\,-2\,-1\ 0\ 1\ 2\ 3\ 4\ 5\ 6$

b. $\{x\,|\,x < -2 \text{ or } x > 2\}$ **c.** $(-\infty, -2)\cup(2, \infty)$

83. a.

$-6\,-5\,-4\,-3\,-2\,-1\ 0\ 1\ 2\ 3\ 4\ 5\ 6$

b. $\{x\,|\,3 \le x \le 6\}$ **c.** $[3, 6]$

85. a.

$-6\,-5\,-4\,-3\,-2\,-1\ 0\ 1\ 2\ 3\ 4\ 5\ 6$

b. $\{x\,|\,1 < x < 3\}$ **c.** $(1, 3)$

87. a.

$-6\,-5\,-4\,-3\,-2\,-1\ 0\ 1\ 2\ 3\ 4\ 5\ 6$

b. $\{x\,|\,x > 1\}$ **c.** $(1, \infty)$

89. a.

$333.\overline{3}$

$250\quad 300\quad 350\quad 400\quad 450\quad 500\quad 550$

b. $\{x\,|\,333.\overline{3} \le x \le 500\}$

91. a.

$55\qquad\qquad\qquad 105$

$40\quad 50\quad 60\quad 70\quad 80\quad 90\quad 100\quad 110$

b. $\{x\,|\,55 \le x \le 105\}$

93. a.

$66\quad 68\quad 70\quad 72\quad 74\quad 76\quad 78$

b. $\{x\,|\,68° \le x \le 78°\}$

95. a.

$68\quad 70\quad 72\quad 74\quad 76\quad 78\quad 80\quad 82$

b. $\{x\,|\,72° \le x \le 80°\}$

97. a.

$170\ 180\ 190\ 200\ 210\ 220\ 230\ 240\ 250\ 260\ 270$

b. $\{x\,|\,180 \text{ ft.} \le x \le 260 \text{ ft.}\}$

99. $\{x\,|\,1996 \le x \le 2000\}; [1996, 2000]$

101. $\{x\,|\,1980 \le x < 1988\}; [1980, 1988)$

103. $\{x\,|\,1989 \le x < 2000\}; [1989, 2000)$

Review Exercises
1. -11 **2.** -16 **3.** 6 **4.** $\dfrac{1}{5}$

5. No, the absolute value of zero is zero and zero has no sign.

6. -2

8.2 Exercises
1. The units a number is from zero.

3. If $|n| = a$ where n is a variable or an expression and $a \ge 0$, then

$n = a$ or $n = -a$. **5.** Separate the absolute value equation into two equations: $ax + b = cx + d$ and $ax + b = -(cx + d)$

7. $-2, 2$ **9.** No solution. **11.** $-11, 5$ **13.** $2, 3$ **15.** $1, \dfrac{7}{5}$

17. $-1, \dfrac{7}{3}$ **19.** No solution. **21.** $\dfrac{3}{4}$ **23.** $-1, 3$ **25.** $-12, 4$

27. $-4, 4$ **29.** $-\dfrac{3}{5}, 1$ **31.** $-4, 1$ **33.** $-2, 6$ **35.** $-\dfrac{9}{2}, \dfrac{15}{2}$

37. $0, 20$ **39.** $\dfrac{1}{3}, 7$ **41.** $-5, -\dfrac{3}{5}$ **43.** $-2, 4$

45. All real-numbers. **47.** 1. **49.** $\dfrac{5}{6}, \dfrac{11}{6}$ **51.** $-6, 10$ **53.** $-\dfrac{7}{2}, 2$

55. $-3, 13$

Review Exercises
1. $x > -7$ **2.** $x \ge -3$

3.

$-6\,-5\,-4\,-3\,-2\,-1\ 0\ 1\ 2\ 3\ 4\ 5\ 6$

4. $-2 < x < \dfrac{14}{3}$ **5.** $(-2, 0]$ **6.** $x < 7$

8.3 Exercises
1. $x \ge -a$ and $x \le a$ **3.** Shade between the values of $-a$ and a. **5.** When $a < 0$.

7. a.

$-6\,-5\,-4\,-3\,-2\,-1\ 0\ 1\ 2\ 3\ 4\ 5\ 6$

b. $\{x\,|-5 < x < 5\}$ **c.** $(-5, 5)$

9. a.

$-12\,-10\,-8\,-6\,-4\,-2\ 0\ 2\ 4$

b. $\{x\,|-10 \le x \le 4\}$ **c.** $[-10, 4]$

11. a.

$-6\,-5\,-4\,-3\,-2\,-1\ 0\ 1\ 2\ 3\ 4\ 5\ 6$

b. $\{s\,|-6 < s < 0\}$ **c.** $(-6, 0)$

13. a.

$-3\,-2\,-1\ 0\ 1\ 2\ 3\ 4\ 5\ 6\ 7\ 8$

b. $\{m\,|-2 < m < 7\}$ **c.** $(-2, 7)$

15. a.

$\dfrac{4}{3}$

$0\qquad 1\qquad 2\qquad 3$

b. $\left\{k\,\Big|\,\dfrac{4}{3} \le k \le 2\right\}$ **c.** $\left[\dfrac{4}{3}, 2\right]$

17. a.

$-6\,-5\,-4\,-3\,-2\,-1\ 0\ 1\ 2\ 3\ 4\ 5\ 6$

b. $\{x\,|-5 \le x \le 5\}$ **c.** $[-5, 5]$

19. a.

$-6\,-5\,-4\,-3\,-2\,-1\ 0\ 1\ 2\ 3\ 4\ 5\ 6$

b. $\{w\,|\,0 < w < 6\}$ **c.** $(0, 6)$

21. a.

$-12\,-10\,-8\,-6\,-4\,-2\ 0\ 2\ 4\ 6\ 8\ 10\ 12$

b. $\{c\,|\,c < -12 \text{ or } c > 12\}$ **c.** $(-\infty, -12)\cup(12, \infty)$

23. a.

$-9\,-8\,-7\,-6\,-5\,-4\,-3\,-2\,-1\ 0\ 1\ 2\ 3\ 4\ 5\ 6$

b. $\{y\,|\,y \le -9 \text{ or } y \ge 5\}$ **c.** $(-\infty, -9]\cup[5, \infty)$

25. a.

–4 –2 0 2 4 6 8 10 12 14 16 18 20

b. $\{p\,|\,p < -2 \text{ or } p > 14\}$ **c.** $(-\infty, -2) \cup (14, \infty)$

27. a.

–6 –5 –4 –3 –2 –1 0 1 2 3 4   ($\frac{11}{3}$)

b. $\left\{x\,\middle|\,x \le -5 \text{ or } x \ge \dfrac{11}{3}\right\}$ **c.** $(-\infty, -5] \cup \left[\dfrac{11}{3}, \infty\right)$

29. a.

–4 –3 –2 –1 0 1 2   ($-\frac{5}{2}$)

b. $\left\{n\,\middle|\,n < -\dfrac{5}{2} \text{ or } n > 0\right\}$ **c.** $\left(-\infty, -\dfrac{5}{2}\right) \cup (0, \infty)$

31. a.

–6 –5 –4 –3 –2 –1 0 1 2 3 4 5 6

b. $\{v\,|\,v \le -1 \text{ or } v \ge 1\}$ **c.** $(-\infty, -1] \cup [1, \infty)$

33. a.

–6 –5 –4 –3 –2 –1 0 1 2 3 4 5 6

b. $\{y\,|\,y < -3 \text{ or } y > -1\}$ **c.** $(-\infty, -3) \cup (-1, \infty)$

35. a.

–6 –5 –4 –3 –2 –1 0 1 2 3 4 5 6   ($\frac{3}{2}$)

b. $\left\{m\,\middle|\,m < -3 \text{ or } m > \dfrac{3}{2}\right\}$ **c.** $(-\infty, -3) \cup \left(\dfrac{3}{2}, \infty\right)$

37. a.

–6 –5 –4 –3 –2 –1 0 1 2 3 4 5 6   ($\frac{4}{3}$)

b. $\left\{x\,\middle|\,-2 < x < \dfrac{4}{3}\right\}$ **c.** $\left(-2, \dfrac{4}{3}\right)$

39. a.

–6 –5 –4 –3 –2 –1 0 1 2 3 4 5 6

b. $\{x\,|\,x \text{ is a real number}\}$ **c.** $(-\infty, \infty)$

41. a.

–6 –5 –4 –3 –2 –1 0 1 2 3 4 5 6

b. $\{x\,|\,-4 < x < -2\}$ **c.** $(-4, -2)$

43. a.

–6 –5 –4 –3 –2 –1 0 1 2 3 4 5 6

b. $\{x\,|\,0 < x < 1\}$ **c.** $(0, 1)$

45. a.

–14 –12 –10 –8 –6 –4 –2 0 2 4 6 8 10

b. $\{w\,|\,-12 < w < 4\}$ **c.** $(-12, 4)$

47. a.

–6 –5 –4 –3 –2 –1 0 1 2 3 4 5 6   ($\frac{14}{3}$)

b. $\left\{k\,\middle|\,-2 \le k \le \dfrac{14}{3}\right\}$ **c.** $\left[-2, \dfrac{14}{3}\right]$

49. a.

0 4 8 12 16 20 24 28

b. $\{x\,|\,x < 4 \text{ or } x > 20\}$ **c.** $(-\infty, 4) \cup (20, \infty)$

51. a.

–6.4   12.8
–8 –6 –4 –2 0 2 4 6 8 10 12 14

b. $\{y\,|\,-6.4 \le y \le 12.8\}$ **c.** $[-6.4, 12.8]$

53. a.

–5 –4 –3 –2 –1 0 1 2 3 4 5

b. $\varnothing$ **c.** No solution.

55. $|x + 1| > 1$ **57.** $|x - 3| \le 2$ **59.** $|x| < 3$

61. $|x| >$ any negative number

Review Exercises

1.

$2x + y = 7$

2.

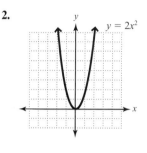

$y = 2x^2$

3. $(6, 0)$ and $(0, 6)$ **4.** $(-9, 0), (-5, 0),$ and $(0, 45)$

5. -26 **6.** -44

8.4 Exercises

1. A relation in which every value in the domain is paired with exactly one value in the range.

3. *Domain*: The set of all input values for a relation. *Range*: The set of all output values for a relation. **5.** A parabola. You need to find many ordered pairs and look for the symmetry.

7. Domain: $\{1, 2, 3, 4, 5\}$
Range: {Hawaii, Florida, Maine, Alaska, California}
It is a function.

9. Domain: $\{2\%, 3\%, 5\%, 14\%, 15\%, 26\%, 33\%\}$
Range:
Work at home
Midnight to 4:59 A.M.
10 A.M. to 11:59 A.M.
9:A.M. to 9:59 A.M.
Noon to 11:59 A.M.
8 A.M. to 8:59 A.M.
5 A.M. to 6:59 A.M.
7 A.M. to 7:59 A.M.

It is not a function.

11. Domain: $\{-3, 0, 1, 2\}$
Range: $\{1, 2, 5, 8\}$
It is a function.

13. Domain: $\{-3, 2, 4\}$
Range: $\{1, 2, 5, 8\}$
It is not a function.

15. Domain: $\{1970, 1975, 1980, 1985, 1990, 1995, 2000\}$
Range: $\{21.4, 21.8, 22.7, 23.7, 24.2, 24.5, 24.9\}$
It is a function.

17. Domain: $\{x\,|\,-3 \le x \le 3\}$
Range: $\{y\,|\,-3 \le y \le 3\}$
It is not a function.

19. Domain: {1988, 1989, 1990, 1991, 1992, 1993, 1994, 1995, 1996, 1997, 1998, 1999, 2000, 2001, 2002, 2003}

Range: {$1.63, $1.71, $1.59, $1.41, $1.31, $1.23, $1.28, $1.24, $1.36, $1.27, $1.03, $1.11, $1.08, $1.25}

It is a function.

21. Domain: $\{x \mid x \geq -4\}$ Range: All real numbers. It is not a function. **23.** 15 **25.** $\dfrac{2}{9}$ **27.** 5 **29.** $\dfrac{28}{5}$ **31.** 3 **33.** $\dfrac{185}{64}$

35. undefined **37.** –4 **39. a.** 2 **b.** 1 **c.** 0 **41. a.** 1 **b.** 9 **c.** 4 **43. a.** 8 **b.** 5 **c.** 0

45.

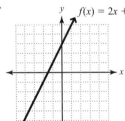

$f(x) = 2x + 3$

47.

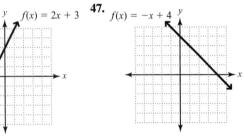

$f(x) = -x + 4$

49.

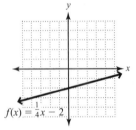

$f(x) = \frac{1}{4}x - 2$

51.

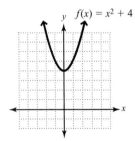

$f(x) = x^2 + 4$

53.

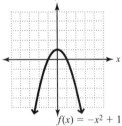

$f(x) = -x^2 + 1$

55.

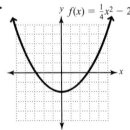

$f(x) = \frac{1}{4}x^2 - 2$

57.

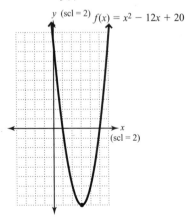

y (scl = 2) $f(x) = x^2 - 12x + 20$ (scl = 2)

59.

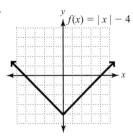

$f(x) = |x| - 4$

61.

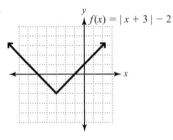

$f(x) = |x + 3| - 2$

63.

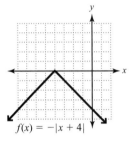

$f(x) = -|x + 4|$

65.

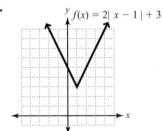

$f(x) = 2|x - 1| + 3$

67.

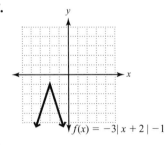

$f(x) = -3|x + 2| - 1$

69.

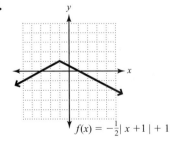

$f(x) = -\frac{1}{2}|x + 1| + 1$

Review Exercises   **1.** $5x^2 + 6$ **2.** $7x^2 - 3x - 7$
3. $4x^2 - 16x + 16$ **4.** $6x^2 - 5x - 4$ **5.** $2x^2 + 3x - 5$
6. $x^3 + 3x^2 + 9x + 27$

8.5 Exercises   **1.** $(p + q)(x) = p(x) + q(x)$

3. Use FOIL to simplify the binomials.   **5.** $5x - 1; -x + 1$

7. $3x - 8; -x - 2$ **9.** $-3x + 8; -5x + 6$

11. $2x^2 - 4x + 2; 2x^2 - 6x - 8$

13. $-2x^2 - 2x + 7; -8x^2 + 2x + 9$ **15.** $x^2 - 5x$

17. $x^2 - x - 2$ **19.** $3x^2 + 8x + 4$

21. $x^4 - x^3 - 11x^2 + 25x - 14$

23. $6x^4 - 11x^3 + 4x^2 + 2x - 1$ **25.** $x - 3$ **27.** $2x - 1 + \dfrac{2}{x}$

29. $x - 5$ **31.** $x + 1$ **33.** $2x - 15 + \dfrac{68}{x + 5}$

35. a. $7x + 2$

   b. $3x - 4$

   c. $10x^2 + 13x - 3$

   d. $\dfrac{5x - 1}{2x + 3}, x \neq \dfrac{-3}{2}$

37. a. $2x^2 - 2x + 4$

   b. $2x^2 - 4x + 6$

   c. $2x^3 - 5x^2 + 8x - 5$

   d. $2x - 1 + \dfrac{4}{x - 1}, x \neq 1$

39. a. -12 **b.** 11 **c.** -6 **d.** 2 **41. a.** 16 **b.** 54 **c.** 7 **d.** -8

43. a. $4x + 6$

   b.

x	$f(x)$	$g(x)$	$(f + g)(x)$
-4	-11	1	-10
-2	-5	3	-2
0	1	5	6
2	7	7	14
4	13	9	22

   c.

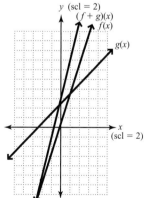

   d. The output values for $f(x)$ added to the output values of $g(x)$ are equal to the output values of $(f + g)(x)$.

45. a. $-x - 14$

   b.

x	$h(x)$	$k(x)$	$(h - k)(x)$
-4	-17	-7	-10
-2	-13	-1	-12
0	-9	5	-14
2	-5	11	-16
4	-1	17	-18

   c.

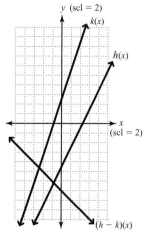

   d. The output values for $k(x)$ subtracted from the output values of $h(x)$ are equal to the output values of $(h - k)(x)$.

47. a. $t(x) = 4x + 4$ **b.** $804

   c.

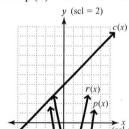

49. a. $p(x) = x^2 - 4x - 12$ **b.** $9588 profit

   c.

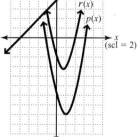

   d. 4 units

51. a. $A(x) = 6x^2 + 4x$ **b.** 66 ft.^{2} **53. a.** $h(x) = 4x + 1$

   b. 17 cm

Review Exercises **1.** $-4, 4$ **2.** No solution.

3.

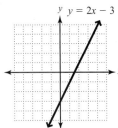

$y = 2x - 3$

4.

$y = x^2 - 9$

5. $(3 - y)(x - y)$

6. $-3(x + 3)(x - 3)$

Chapter 8 Review Exercises **1.** true **2.** true
3. false **4.** true **5.** true **6.** true **7.** $\cap$
8. $ax + b = -c, ax + b = c$ **9.** The range. **10.** $f(x) + g(x)$

11. a. $\{1, 5, 9\}$ **12. a.** $\{\}$
   b. $\{1, 5, 7, 9\}$ **b.** $\{1, 2, 3, 4, 5, 6, 7\}$

13. a.

number line: -4 to 1, open circles at -3 and -2

b. $\{x \mid -3 < x < -2\}$ **c.** $(-3, -2)$

14. a.

number line: -7 to 1, closed circles at -5 and -3

b. $\{x \mid -5 \le x \le -3\}$ **c.** $[-5, -3]$

15. a.

number line: -8 to 4, open circles at -7 and 3

b. $\{x \mid -7 < x < 3\}$ **c.** $(-7, 3)$

16. a.

number line: -6 to 6, open circle at 1, closed circle at 4

b. $\{x \mid 1 < x \le 4\}$ **c.** $(1, 4]$

17. a.

number line: -6 to 6, closed circle at -2, open circle at 0

b. $\{x \mid -2 \le x < 0\}$ **c.** $[-2, 0)$

18. a.

number line: -6 to 6, open circles at -1 and 3

b. $\{x \mid -1 < x < 3\}$ **c.** $(-1, 3)$

19. a.

number line: -6 to 6, closed circles at -6 and -2

b. $\{w \mid w \le -6 \text{ or } w \ge -2\}$ **c.** $(-\infty, -6] \cup [-2, \infty)$

20. a.

number line: -6 to 6, closed circle at $-\frac{4}{3}$, closed circle at 2

b. $\left\{x \mid x \le -\dfrac{4}{3} \text{ or } x \ge 2\right\}$ **c.** $\left(-\infty, -\dfrac{4}{3}\right] \cup [2, \infty)$

21. a.

number line: -6 to 6, open circle at $\frac{5}{2}$, open circle at 5

b. $\left\{m \mid m < \dfrac{5}{2} \text{ or } m > 5\right\}$ **c.** $\left(-\infty, \dfrac{5}{2}\right) \cup (5, \infty)$

22. a.

number line: -6 to 6

b. $\{x \mid x \text{ is a real number}\}$ **c.** $(-\infty, \infty)$

23. a.

number line: -9 to 6, closed circles at -9 and -4

b. $\{x \mid x \le -9 \text{ or } x \ge -4\}$ **c.** $(-\infty, -9] \cup [-4, \infty)$

24. a.

number line: -6 to 6, closed circles at -1 and 1

b. $\{w \mid w \le -1 \text{ or } w \ge 1\}$ **c.** $(-\infty, -1] \cup [1, \infty)$

25. $-4, 4$ **26.** $-3, 11$ **27.** $-1, 2$ **28.** $\dfrac{2}{3}, 2$ **29.** No solution.

30. $-7, 15$ **31.** $-3, 3$ **32.** $0, \dfrac{8}{3}$ **33.** $-\dfrac{7}{2}, \dfrac{1}{2}$ **34.** $-1, 11$

35. a.

number line: -6 to 6, open circles at -5 and 5

b. $\{x \mid -5 < x < 5\}$ **c.** $(-5, 5)$

36. a.

number line: -6 to 6, closed circles at -4 and 4

b. $\{p \mid -4 \le p \le 4\}$ **c.** $[-4, 4]$

37. a.

number line: -6 to 11, open circles at -4 and 10

b. $\{x \mid x < -4 \text{ or } x > 10\}$ **c.** $(-\infty, -4) \cup (10, \infty)$

38. a.

number line: -6 to 6, open circles at -5 and -1

b. $\{m \mid -5 < m < -1\}$ **c.** $(-5, -1)$

39. a.

number line: -6 to 6, no shading

b. $\varnothing$ **c.** No solution.

40. a.

number line: -6 to 6, open circles at -1 and 1

b. $\{b \mid b < -1 \text{ or } b > 1\}$ **c.** $(-\infty, -1) \cup (1, \infty)$

41. a.

number line: -6 to 6, closed circles at -6 and 0

b. $\{m \mid -6 \le m \le 0\}$ **c.** $[-6, 0]$

42. a.

number line: -6 to 11, open circles at 0 and 10

b. $\{t \mid t < 0 \text{ or } t > 10\}$ **c.** $(-\infty, 0) \cup (10, \infty)$

43. a.

number line: -6 to 8, closed circle at $-\frac{7}{2}$, closed circle at $\frac{13}{2}$

b. $\left\{k \mid k \le -\dfrac{7}{2} \text{ or } k \ge \dfrac{13}{2}\right\}$ **c.** $\left(-\infty, -\dfrac{7}{2}\right] \cup \left[\dfrac{13}{2}, \infty\right)$

44. a.

number line: -6 to 6

b. $\{x \mid x \text{ is a real number}\}$ **c.** $(-\infty, \infty)$

45. Domain: $\{1, 2, 3, 4, 5, 6\}$

Range: $\left\{\begin{array}{l}\text{Batman No. 613}\\\text{Ultimate Spider-man No. 38}\\\text{Ultimate Spider-man No. 37}\\\text{Amazing Spider-man No. 51}\\\text{New X-Men No. 138}\\\text{Ultimate X-Men No. 31}\end{array}\right\}$

It is a function.

46. Domain: $\{1, 2, 3, 4\}$

Range: $\left\{\begin{array}{l}\text{Harvard University (MA)}\\\text{Stanford University (CA)}\\\text{University of Pennsylvania (Wharton)}\\\text{Massachusetts Institute of Technology (Sloan)}\\\text{Northwestern University (Kellogg) (IL)}\\\text{Columbia University (NY)}\end{array}\right\}$

It is not a function.

47. Domain: $\{2, 3, 4\}$ Range: $\{-3, 1, 4\}$ It is not a function.
48. Domain: $\{-2, 0, 1\}$ Range: $\{3, 4\}$ It is a function.
49. Domain: All real numbers. Range: $\{y \mid y \geq -4\}$
It is a function. **50.** Domain: All real numbers.
Range: $\{y \mid y \leq 4\}$ It is a function. **51.** 17 **52.** -9

53.

$f(x) = -3x + 2$

54.
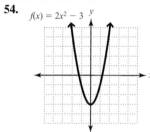
$f(x) = 2x^2 - 3$

55.
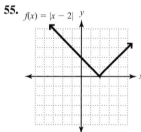
$f(x) = |x - 2|$

56.
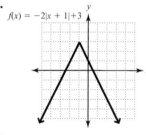
$f(x) = -2|x + 1| + 3$

57. a. $x + 10$
b. $-7x + 8$
c. $-12x^2 + 33x + 9$

58. a. $7x$
b. $-9x - 4$
c. $-8x^2 - 18x - 4$

59. a. $3x^2 + 7$
b. $3x^2 - 2x + 3$
c. $3x^3 + 5x^2 + 3x + 10$

60. a. $x^2 + x$
b. $x^2 - 5x - 2$
c. $3x^3 - 5x^2 - 5x - 1$

61. $(f/g)(x) = 2x^2 - 5x + 4, x \neq 0$

62. $(f/g)(x) = x^2 - 6x + 3 + \dfrac{-2}{2x - 5}, x \neq \dfrac{5}{2}$

63. $68 \leq x \leq 108$ **64.** $150 \leq l \leq 200$ **65. a.** $c(x) = 4x + 4$

b. \$904

c.
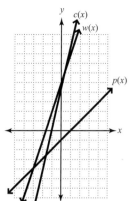

66. a. $2x - 1$ **b.** 17 in.

Chapter 8 Practice Test
1. $\{e, h, o\}$ **2.** $-8, 2$
$\{e, h, m, o, s, u\}$

3. $-3, 6$ **4.** $-8, \dfrac{2}{3}$

5. No solution.

6. a.

b. $\{x \mid -7 < x < 3\}$ **c.** $(-7, 3)$

7. a.

b. $\{x \mid -3 \leq x < -2\}$ **c.** $[-3, -2)$

8. a.

b. $\{x \mid x \leq -1 \text{ or } x \geq 2\}$ **c.** $(-\infty, -1] \cup [2, \infty)$

9. a.

b. $\{n \mid n > 1\}$ **c.** $(1, \infty)$

10. a.

b. $\{x \mid -13 < x < 5\}$ **c.** $(-13, 5)$

11. a.

b. $\{x \mid x < -1 \text{ or } x > 3\}$ **c.** $(-\infty, -1) \cup (3, \infty)$

12. a.

b. $\{x \mid -7 < x < -1\}$ **c.** $(-7, -1)$

13. a.

b. $\varnothing$ **c.** No solution.

14. a.

b. $\{x \mid x \text{ is a real number}\}$ **c.** $(-\infty, \infty)$

15. a.

b. $\left\{x \mid -\dfrac{1}{3} \leq x \leq 3\right\}$ **c.** $\left[-\dfrac{1}{3}, 3\right]$

16. Domain: All real numbers. Range: $\{y \mid y \le 0\}$ It is a function.
17. 18
18.

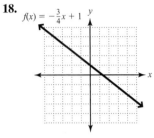

$f(x) = -\frac{3}{4}x + 1$

19.

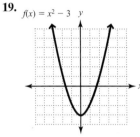

$f(x) = x^2 - 3$

20.

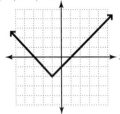

$f(x) = |x + 1| - 2$

21. a. $7x - 7$ **22. a.** $4x^2 - 2$
   b. $5x - 11$ **b.** $-2x^2 + 2$
   c. $6x^2 + 3x - 18$ **c.** $3x^4 - 2x^2$

23. $5x^2 + 9x - 7$ **24.** $4 \le x \le 10$ **25. a.** $h(x) = 9x + 2$
b. 137 cm

Chapter 9

9.1 Exercises
1. 1. Replace each variable in each equation with its corresponding value. 2. Verify that each equation is true. **3.** The graphs are parallel. **5.** Yes, the graphs are identical. **7.** yes **9.** yes **11.** no **13.** no **15.** yes **17. a.** Inconsistent with independent equations. **b.** No solution. **19. a.** Consistent with independent equations. **b.** One solution. **21. a.** Consistent with dependent equations. **b.** Infinite number of solutions. **23. a.** Consistent with independent equations. **b.** One solution. **25. a.** Inconsistent with independent equations. **b.** No solution. **27. a.** Consistent with dependent equations. **b.** Infinite number of solutions. **29. a.** Consistent with independent equations **b.** One solution. **31.** $(4, 1)$ **33.** $(3, 2)$ **35.** $(4, 7)$ **37.** $(2, -3)$ **39.** $(-3, 4)$ **41.** $(-1, 3)$ **43.** $(3, -4)$ **45.** $(-2, 3)$ **47.** $(5, -2)$ **49.** No solution (inconsistent with independent equations). **51.** Infinite number of solutions along $y = 3x$ (consistent with dependent equations). **53. a.** 5000 units   **b.** \$4000   **c.** $n \ge 5000$ **55. a.** 160 minutes   **b.** yes   **c.** Calling primarily AT&T customers.

Review Exercises   **1.** $x^2 + 9x + 20$ **2.** 11, 73

3. $(w - 2)(3w + 1)$ **4.** $4(x^3 - 2x + 3)$ **5.** $y = \dfrac{10 - x}{3}$

6. $x = 30 + 6y$

9.2 Exercises
1. If the point of intersection is in between grid lines, the method of substitution does not require guessing. **3.** y in the first equation, since it can be isolated without using division. **5.** The resulting equation will no longer contain variables and become a false equation. **7.** $(9, 3)$ **9.** $(-10, -24)$

11. $(-3, 2)$ **13.** $(2, 9)$ **15.** $(-1, -3)$ **17.** $(12, 4)$ **19.** $(3, 4)$

21. $(1, 1)$ **23.** Infinite number of solutions along $y = -2x$ (consistent with dependent equations). **25.** $\left(-\dfrac{1}{4}, \dfrac{2}{3}\right)$ **27.** $(4, -2)$

29. $\left(\dfrac{3}{28}, -\dfrac{29}{84}\right)$ **31.** $(2, 5)$ **33.** $(1, 2)$ **35.** Mistake: 3 was not distributed to $3y$. Correct: $\left(\dfrac{27}{11}, -\dfrac{13}{11}\right)$ **37.** 155, 93 **39.** $-6, -2$

41. J = 20, T = 7 **43.** P = 20, R = 17 **45.** $l = 11$ m, $w = 5$ m
47. 22 in., 8 in. **49.** $l = 36$ ft., $w = 30$ ft. **51.** 78 ft. by 36 ft.
53. 75 ft. by 75 ft. **55.** $135°, 45°$ **57.** $60°, 30°$ **59.** $37.5°, 52.5°$
61. \$60,567 **63.** 250 tons paper, 125 tons cardboard
65. \$2025, \$675 **67.** 4:45 P.M. **69.** 1:33 P.M.

Review Exercises   **1.** $9x$ **2.** $13y$ **3.** $5x - 4y$
4. $-3x + 21y$ **5.** 80   **6.** $-\dfrac{3}{2}$

9.3 Exercises
1. The substitution method is advantageous over graphing when the solution involves fractions. The elimination method is advantageous over substitution when no coefficients are 1. **3. a.** x, because multiplying the second equations by -2 will allow the x to be eliminated. **b.** Add the two new equations. **5.** Both variables have been eliminated and the resulting equation is false. **7.** $(6, -8)$ **9.** $(2, 3)$
11. $(-1, -2)$ **13.** $(2, 3)$ **15.** $(3, -1)$ **17.** $(-2, 4)$ **19.** $(1, 1)$
21. $(0.2, 0.3)$ **23.** $(6, 4)$ **25.** $(7, -9)$ **27.** Infinite number of solutions (dependent). **29.** $(-4, -13)$ **31.** $\left(2, \dfrac{5}{2}\right)$ **33.** $(4, 4)$
35. Mistake: Did not multiply the right side of $x - y = 1$ by -9. Correct: $(5, 4)$ **37.** 11, 7 **39.** 5.5 yr., 9.5 yr. **41.** 5, 12
43. $56°$ and $34°$ **45.** $126°$ and $54°$ **47.** G = 35, U = 34
49. 13 \$5 bills, 6 \$10 bills **51.** 12 algebra, 8 calculus **53.** 8 \$350 tickets, 2 \$250 tickets **55.** 36 adult, 78 child **57.** 27 3-hr. classes, 15 5-hr. classes **59.** 10 mph in still water, current is 2 mph
61. 470 mph in still air, 30 mph wind **63.** \$12,000 at 4%, \$3500 at 3% **65.** \$11,100 at 5%, \$2150 at 8% **67.** \$10,253 at 2.9%, \$5997 at 6% **69.** \$12,500 at 7%, \$7500 at 9% **71.** 200 ml of 15%, 100 ml of 45% **73.** 120 ml **75.** 11.5 lb. of Italian Roast, 8.5 lb. of GCB **77.** 5 lb. of peanuts, 4 lb. of cashews

Review Exercises   **1.** \$500 in checking, \$600 in CD
2. 13 dimes, 11 quarters   **3.** 24 dimes, 17 nickels   **4.** 3:12 P.M.
5. $(1, -6)$ **6.** $\left(-\dfrac{5}{4}, \dfrac{5}{2}\right)$

9.4 Exercises
1. No. From two pairs of equations we can eliminate the same variables from each pair. **3.** I would eliminate the variable z. Multiplying equation 1 by -3 and then

adding this result to both equations 2 and 3 takes very few steps. The pairs of equations are 1 and 2 and 1 and 3.
5. Two parallel planes intersecting a third plane

7. no   **9.** no   **11.** yes   **13.** $(2, -1, 4)$ **15.** $(-1, 0, 3)$

17. $(-1, 1, 3)$ **19.** No solution.   **21.** $(2, -3, 1)$ **23.** $\left(-2, -\dfrac{1}{2}, 1\right)$

25. $(2, -1, 1)$ **27.** $(2, -4, 1)$ **29.** $(4, -4, 2)$ **31.** $6, 2, 8$
33. $105°, 35°, 40°$ **35.** 8 in., 10 in., and 15 in.   **37.** Burger = $2.50, Fry = $1.50, Drink = $1.00   **39.** 150 children, 250 students, 100 adults   **41.** 7 free throws, 6 field goals, 3 three pointers   **43.** 8 nickels, 7 dimes, 10 quarters   **45.** 6 pennies, 12 nickels, 7 dimes   **47.** Zinc = 20 lbs., tin = 60 lbs., copper = 920 lbs.   **49.** Ham = 3 lbs., turkey = 5 lbs., beef = 2 lbs.
51. $2000 at 4%, $2500 at 6%, $3500 at 7%   **53.** $30,000 in stocks, $50,000 in bonds, $20,000 in certificates   **55.** Jumbo: 30 oz., extra large: 27 oz., large: 24 oz.   **57.** Bicycling = 420, walking = 300, stair = 600   **59.** Angola = 19, Burundi = 18, Afghanistan = 21   **61.** $a = -16, v_0 = 100, h = 150; h = -16t^2 + 100t + 150$

Review Exercises   **1.** $-\dfrac{5}{2}y$ **2.** $-y - 6z$ **3.** 9 **4.** 2

5. $\left(-\dfrac{53}{19}, \dfrac{25}{19}\right)$ **6.** $(-2, 3)$

9.5 Exercises   **1.** 4 rows, 2 columns   **3.** The dashed line separates the coefficients from the constants.   **5.** A matrix is in echelon form when the coefficient portion of the augmented matrix has 1's on the diagonal from upper left to lower right and 0's below the 1's.   **7.** $\begin{bmatrix} 14 & 7 & 6 \\ 7 & 6 & 8 \end{bmatrix}$ **9.** $\begin{bmatrix} 7 & -6 & 1 \\ 8 & -12 & 6 \end{bmatrix}$

11. $\begin{bmatrix} 1 & -3 & 1 & 4 \\ 2 & -4 & 2 & -4 \\ 6 & -2 & 5 & -4 \end{bmatrix}$ **13.** $\begin{bmatrix} 4 & 6 & -2 & -1 \\ 8 & 3 & 0 & -12 \\ 0 & -1 & 2 & 4 \end{bmatrix}$

15. $(4, -2)$ **17.** $(-6, -5, 1)$ **19.** $\begin{bmatrix} 1 & 3 & -1 \\ 0 & 11 & 4 \end{bmatrix}$

21. $\begin{bmatrix} 1 & -2 & 4 & 6 \\ 0 & 2 & -1 & -5 \\ 0 & 0 & -2 & 17 \end{bmatrix}$ **23.** $\begin{bmatrix} 1 & 2 & -2.5 \\ -1 & 3 & 2 \end{bmatrix}$

25. $3R_1 + R_2$ **27.** $-2R_2 + R_3$ **29.** $(2, 0)$ **31.** $(1, 2)$ **33.** $(-3, -2)$
35. $(-2, 1)$ **37.** $(4, 6)$ **39.** $(4, -4)$ **41.** $(15, -15, 6)$
43. $(2, 1, -2)$ **45.** $(2, -3, 1)$ **47.** $(-1, -1, -2)$ **49.** $(-3, 1, 5)$
51. $(8, -2, -4)$ **53.** $(3, -4)$ **55.** $(-6, 8)$ **57.** $(4, -3, 5)$
59. $(7, -6, 5)$ **61.** Mistake: In the second step, $4R_1 + R_2$ is calcu-

lated incorrectly. Correct: The correct calculation is $\begin{bmatrix} 1 & 3 & 13 \\ 0 & 11 & 26 \end{bmatrix}$.
The solution is $(65/11, 26/11)$.   **63.** $\begin{cases} 3x + 2y = 9.90 \\ 7x + 4y = 22.30; \end{cases}$ Chicken
sandwich = $2.50; drink = $1.20. Cost for (chicken, drinks) is ($2.50, $1.20).   **65.** $58°, 32°$ **67.** Nile = 4150 mi., Amazon = 3900 mi.   **69.** $6400 in CD, $3600 in MM
71. CD = $16, tape = $12, DVD = $28   **73.** 6 TD, 6 EP, 2 FG
75. $v_1 = 9, v_2 = 6, v_3 = 2$

Review Exercises   **1.** 8   **2.** 11   **3.** 2   **4.** 2   **5.** 3   **6.** -1

9.6 Exercises   **1.** No, it is not a square matrix.
3. Eliminate the elements in the same row and column as the element.   **5.** 13   **7.** -22 **9.** 24   **11.** -7 **13.** -22 **15.** 15

17. 1   **19.** -12 **21.** -14 **23.** -58 **25.** $\dfrac{13}{30}$ **27.** 0.47

29. $-\dfrac{317}{2400}$ **31.** 14.451   **33.** $-x - 8y - 2$ **35.** $(-4, -1)$

37. $(3, 4)$ **39.** $(4, -4)$ **41.** $(-3, -2)$ **43.** $(2, 2)$ **45.** $(2, 4)$
47. $(2, 6)$ **49.** $(2, 1, 3)$ **51.** $(-2, 1, -1)$ **53.** $(2, -1, 1)$

55. $\left(\dfrac{2}{3}, -7, \dfrac{1}{2}\right)$ **57.** $(1, -2, 3)$ **59.** $(2, 2, -2)$ **61.** $x = -4$

63. $x = -5$ **65.** 9   **67.** 21.5   **69.** 19   **71.** Hoba is 60 tons, Ahnighito is 30 tons   **73.** 6 lbs. of garbanzo, 4 lbs. of black turtle
75. 1% zinc, 4% tin, 95% copper

Review Exercises   **1.** 19, 20
2. $x < 3$;

3. $x \geq 1$;

4.

5.

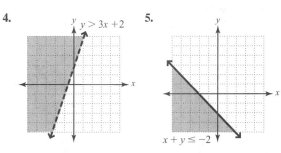

6. $(-2, 5)$

9.7 Exercises   **1.** Select a point in the solution region and verify that it makes both inequalities true.   **3.** Parallel lines and shaded regions that do not overlap. **5.** $\begin{cases} x < 0 \\ y < 0 \end{cases}$

7.
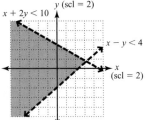
$x + 2y < 10$
y (scl = 2)
$x - y < 4$
x (scl = 2)

9.
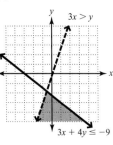
$3x > y$
$3x + 4y \leq -9$

23.
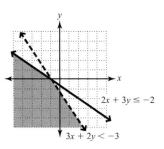
$2x + 3y \leq -2$
$3x + 2y < -3$

25.
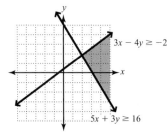
$3x - 4y \geq -2$
$5x + 3y \geq 16$

11.
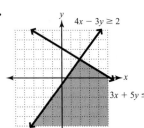
$4x - 3y \geq 2$
$3x + 5y \leq 16$

13.
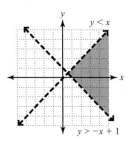
$y < x$
$y > -x + 1$

27.
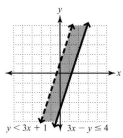
$y < 3x + 1$ $3x - y \leq 4$

29.
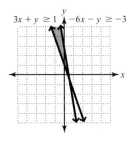
$3x + y \geq 1$ $-6x - y \geq -3$

15.
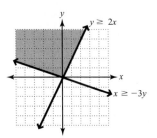
$y \geq 2x$
$x \geq -3y$

17.
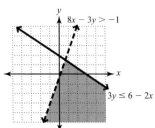
$8x - 3y > -1$
$3y \leq 6 - 2x$

31.
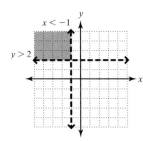
$x < -1$
$y > 2$

33. $x + y > 3$ should be a dashed line. **35.** The wrong area is shaded.

37. a. $\begin{array}{l} V + M \geq 1100 \\ M \geq 500 \end{array}$ **b.** $\begin{array}{l} V \leq 800 \\ M \leq 800 \end{array}$

19.
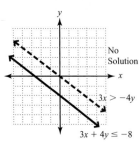
No Solution
$3x > -4y$
$3x + 4y \leq -8$

21.
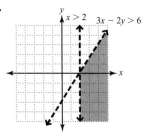
$x > 2$ $3x - 2y > 6$

c.

M
$V \leq 800$
$M \leq 800$
$M \geq 500$
$V + M \geq 1100$
V

d. Answers will vary.

39. a. $R + D \geq 100$
$15.95R + 20.95D \geq 1700$

b.

c. Answers will vary.

Review Exercises
1. 9 **2.** -100 **3.** $-x - 14$
4. $-8a^3b^5$ **5.** $\dfrac{-5b^5}{a^6}$ **6.** 2, 3

Chapter 9 Review Exercises
1. false **2.** true
3. true **4.** true **5.** false **6.** Replace **7.** slope, y-intercepts **8.** the intersection (overlap) of the two solutions
9. the coefficient portion of the augmented matrix has 1's on the diagonal from upper left to lower right and 0's below the 1's
10. determinant **11.** yes **12.** no **13. a.** Consistent with independent equations. **b.** One **14. a.** Inconsistent with independent equations. **b.** Zero **15. a.** Consistent with independent equations. **b.** One **16. a.** Consistent with independent equations. **b.** Infinite solutions along $x - 5y = 10$
17. $(-1, -4)$ **18.** $(3, -1)$ **19.** No solution **20.** Infinite solutions along $y = \dfrac{3}{2}x - 3$ **21.** $(4, -1)$ **22.** $-\dfrac{2}{5}$
23. Infinite solutions along $y = \dfrac{4}{3}x + 2$ **24.** No solution.
25. $(1, 3)$ **26.** $(2, -1)$ **27.** $(1, 5)$ **28.** $(5, -3)$ **29.** $(1, 0, -3)$
30. No solution. **31.** $(3, -3, 2)$ **32.** $(4.2, 0, 1.8)$ **33.** $(4, -1)$
34. $(-3, -2)$ **35.** $(-2, 3, 1)$ **36.** $(-1, 2, -4)$ **37.** $(-2, 9)$
38. No solution, the lines are parallel.
39. $(2, 4, -1)$ **40.** $(-2, 4, -2)$

41.

42.

43.

44.

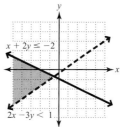

45. 47 adult, 92 children **46.** 5805 **47.** 222 black cars, 187 silver cars **48.** 10, 6 **49.** 75°, 15° **50.** \$3500 at 7% and \$1500 at 9%. **51.** 5:45 P.M. **52.** 120 miles per hour
53. 30 ml **54.** 1 vitamin supplement, 3 rolls of film, 4 bags of candy **55.** 4 first place, 3 second place, 9 third place
56. a. $x + y \leq 16, 250x + 400y \geq 2750, x \geq 0, y \geq 0$
b.

Chapter 9 Practice Test
1. yes **2.** no **3.** $(-2, 1)$
4. $(8, -1)$ **5.** $(3, 1)$ **6.** $(-4, -2)$ **7.** Infinite solutions along $2x + 3y = 1$. **8.** $(-2, 3, -2)$ **9.** $(-4, 2, 3)$ **10.** $(2, -4)$ **11.** $(-1, 5, 2)$ **12.** $(-2, 5)$ **13.** $(-4, 0, 3)$
14.

15. Accountants 86; waiters/waitresses 77 **16.** 180
17. Rate is 7 mph (current is 3 mph) **18.** \$4000 at 6%, \$8000 at 8% **19.** Children 150, students 300, adult 350
20. a. $x \geq 0, y \geq 0, x + y \leq 100, x \geq y + 10$

b.

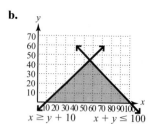

$x \geq y + 10$ $x + y \leq 100$

Chapters 1–9 Cumulative Review   **1.** false

2. true   **3.** true (though the set might be empty)   **4.** true
5. true   **6.** false   **7.** variable   **8.** $8x + 12$ **9.** ratios   **10.** sum
11. 11   **12.** undefined   **13.** 0   **14.** $-6x - 6$
15. $4x^3 - 14x^2 + 7x - 3$ **16.** $(x^3 + 5x^2 + 10x) \div 5x$

17. $5m - 4 + \dfrac{6}{3m - 2}$ **18.** $(m + 2)(m^2 - 2m + 4)$

19. $(x + 5)(x + 5)$ **20.** $(x^2 + 9)(x + 3)(x - 3)$ **21.** $\dfrac{1}{u - 1}$

22. $\dfrac{a - b}{x + y}$ **23.** $\dfrac{x}{10x + 5}$ **24.** $\dfrac{y^2 + y + 9}{(y - 2)(y + 3)}$ **25.** $\dfrac{1}{g + 3}$

26. $\dfrac{x + 10}{(x - 2)(x + 2)}$ **27.** All real numbers.   **28.** $-6 < x < 1$

29. ± 6 **30.** 8   **31.** $-\dfrac{8}{3}, 0$ **32.** $m = \dfrac{2E}{v^2}$

33. The data does represent a function. Domain = {CPUs, Monitors, Keyboards, Printers, Peripherals, Macintoshes, Typewriters, Laptops, Televisions, Other} Range = {1163, 775, 706, 280, 81, 67, 62, 25, 21, 228}

34.

$y = -2x$

35.

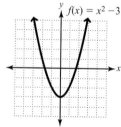

$f(x) = x^2 - 3$

36.

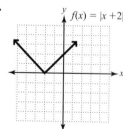

$f(x) = |x + 2|$

37. $2x + 3y = 8$ **38.** $(-3.5, -4)$

39. $(2, -1, 4)$ **40.** $(2, -3)$ **41.**

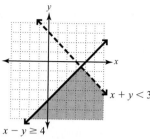

$x + y < 3$
$x - y \geq 4$

42. $\approx 51\%$ **43.** Women \$531, men \$685   **44.** \$75 ml
45. a. x: number of small events y: number of large events
$250x + 500y \geq 50{,}000; x \geq 0; y \geq 0$
b.

$y \geq 0$
$x \geq 0$
$250x + 500y \geq 50{,}000$

c. 160 small events, 20 large events
d. 40 small events, 85 large events

46. ≈ 5.5 hrs.   **47.** 3.2 cm/sec.   **48.** 1.2 hours, or 1 hour and 12 minutes   **49.** 100 miles per hour   **50.** Pens at \$2.00, erasers at \$0.50, reams of paper at \$1.00

Chapter 10

10.1 Exercises   **1.** Answers will vary. Rational roots can be expressed as whole numbers or fractions with integers in the numerator and denominator, whereas irrational roots cannot.
3. Squaring a number or its additive inverse results in the same positive number.   **5.** You cannot raise a number to an even power and get a negative value.   **7.** ± 6 **9.** ± 11 **11.** ± 14
13. ± 15 **15.** 5   **17.** -5 **19.** ± 5 **21.** 1.2   **23.** Not real.
25. -0.11 **27.** $\dfrac{7}{9}$ **29.** $-\dfrac{12}{13}$ **31.** 3   **33.** -3 **35.** 6   **37.** 5
39. -3 **41.** 2   **43.** 2   **45.** 2   **47.** $\dfrac{2}{3}$ **49.** $\dfrac{2}{3}$ **51.** 2.236
53. -3.317 **55.** 3.684   **57.** -3.756 **59.** 2.839   **61.** -3.036
63. 2.454   **65.** 2.295   **67.** b^2 **69.** $4x$ **71.** $10rs^3$ **73.** $0.5a^3b^6$
75. m **77.** $3a^3b^2$ **79.** $-4ab^4$ **81.** $0.2x^6$ **83.** a **85.** $2x^4$
87. $2x^2$ **89.** x^2y **91.** $6|m|$ **93.** $|r - 1|$ **95.** $4|y^3|$ **97.** $3y$
99. $3x^2$ **101.** $y - 4$ **103.** 2   **105.** $\sqrt{15}$
107. $\{x \mid x \geq 4\}$ or $[4, \infty)$ **109.** $\{x \mid x \leq 4\}$ or $(-\infty, 4]$

111. a.

b. $\{x \mid x \geq 2\}$ or $[2, \infty)$

113. a. 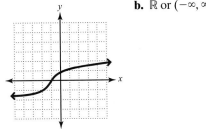 b. $\mathbb{R}$ or $(-\infty, \infty)$

115. ≈ -17.709 m/sec. 117. ≈ 3.476 sec. 119. ≈ 19.170 mph
121. 13 ft. 123. 15 N

Review Exercises 1. x^8 2. $-45m^4n^3$ 3. $12x^4 - 20x^3 + 4x^2$ 4. $21y^2 + 23y - 20$ 5. $2x^2 - 17x - 9$ 6. a. 12 b. 12

10.2 Exercises
1. 4 3. a must be positive, because with an even index the radicand must be positive. 5. Yes, because $100^{1/4} = (10^2)^{1/4} = 10^{1/2}$. 7. $\sqrt{25} = 5$ 9. $-\sqrt{100} = -10$
11. $\sqrt[3]{27} = 3$ 13. $\sqrt[3]{-64} = -4$ 15. $\sqrt[4]{y}$ 17. $\sqrt{144x^8} = 12x^4$
19. $18\sqrt{r}$ 21. $\sqrt{\frac{x^4}{121}} = \frac{x^2}{11}$ 23. $\sqrt[3]{8^2} = 4$ 25. $-\sqrt[4]{81^3} = -27$
27. $\sqrt[3]{(-8)^4} = 16$ 29. $\frac{1}{\sqrt{16^3}} = \frac{1}{64}$ 31. $\sqrt[5]{x^4}$ 33. $8\sqrt[3]{n^2}$
35. $\frac{1}{\sqrt[5]{(-32)^2}} = \frac{1}{4}$ 37. $\sqrt{\frac{1}{25}^3} = \frac{1}{125}$ 39. $\sqrt[6]{(2a+4)^5}$
41. $25^{1/4}$ 43. $z^{5/6}$ 45. $5^{-5/6}$ 47. $5x^{-4/5}$ 49. $5^{7/3}$ 51. $x^{2/7}$
53. $(4a-7)^{7/4}$ 55. $(2r-5)^{8/5}$ 57. $x^{4/5}$ 59. $x^{7/6}$ 61. $a^{17/12}$
63. $12w^{3/5}$ 65. $-12a^{17/12}$ 67. $7^{3/5}$ 69. $x^{-2/5}$ 71. $x^{1/4}$ 73. $r^{1/12}$
75. $x^{-5/7}$ 77. $a^{9/4}$ 79. $20s^{3/7}$ 81. $-24b^{1/4}$ 83. x^2 85. $a^{5/3}$
87. $b^{2/5}$ 89. $64x^4y^3$ 91. $2q^{1/2}t^{1/4}$ 93. $81a$ 95. $3z^{1/3}$ 97. $\sqrt{2}$
99. $\sqrt[3]{5}$ 101. $\sqrt{x}$ 103. $\sqrt[4]{r^3}$ 105. $\sqrt[4]{x^3y}$ 107. $\sqrt[5]{m^3n^3}$
109. $\sqrt[6]{x^5}$ 111. $\sqrt[6]{y^7}$ 113. $\sqrt[6]{x^5}$ 115. $\sqrt[15]{n^2}$ 117. $\sqrt[6]{1125}$
119. $\sqrt[12]{3456}$ 121. $\sqrt[9]{x}$ 123. $\sqrt[6]{n}$

Review Exercises 1. 12 2. 15 3. $2^4 \cdot 3 \cdot 5$
4. 8×10^{11} 5. $-\frac{5}{8}x^4y^2z^2$ 6. $2x^2 - 8x + 5 + \frac{3}{x+3}$

10.3 Exercises
1. If we approximate both $\sqrt{8}$ and $\sqrt{18}$ to three places as 2.828 and 4.243 and multiply these, we get 11.999. On the other hand, if we multiply 8 and 18 first, then we get $\sqrt{(8 \cdot 18)} = \sqrt{(144)}$, which we recognize as exactly 12. The important thing to remember is that it is best to estimate only once rather than twice. Each time we estimate, we increase the possibility of an error in the answer. 3. Rewrite the expression as a product of two radicals, the first containing the perfect cube and the second containing no perfect cubes. Then simplify the first radical. For example, $\sqrt[3]{54x^8}$ can be rewritten as $\sqrt[3]{27x^6} \cdot \sqrt[3]{2x^2}$, then simplified to $3x^2 \cdot \sqrt[3]{2x^2}$ 5. 8 7. $9x^3$
9. $12xy^2$ 11. $\sqrt{14}$ 13. $\sqrt{15x}$ 15. 3 17. $\sqrt[3]{10y^2}$ 19. $\sqrt[4]{21}$
21. $w\sqrt[4]{72}$ 23. $\sqrt[4]{15x^3y^3}$ 25. $\sqrt[5]{30x^4}$ 27. $\sqrt[6]{8x^5y^4}$
29. $\frac{\sqrt{210}}{10}$ 31. $\frac{\sqrt{30xy}}{5x}$ 33. $\frac{5}{6}$ 35. $\frac{\sqrt{10}}{3}$ 37. 7 39. $\sqrt{3}$
41. $\frac{\sqrt[3]{4}}{w^2}$ 43. $\frac{\sqrt[3]{5y^2}}{3x^3}$ 45. 4 47. $\frac{\sqrt[4]{3u^3}}{2x^2}$ 49. $7\sqrt{2}$ 51. $8\sqrt{2}$
53. $24\sqrt{5}$ 55. $20\sqrt{7}$ 57. $a\sqrt{a}$ 59. xy^2 61. $x^3y^4z^5$

63. $r^5s^4\sqrt{rs}$ 65. $18x^2\sqrt{2x}$ 67. $2\sqrt[3]{4}$ 69. $x^2\sqrt[3]{x}$ 71. $x^2y\sqrt[3]{y^2}$
73. $4z^2\sqrt[3]{2z^2}$ 75. $4\sqrt[3]{3}$ 77. $2\sqrt[4]{5}$ 79. $3x^4\sqrt[4]{135x}$ 81. $3x^3\sqrt[5]{2x}$
83. $xy^2z\sqrt[6]{x^2y^2z^5}$ 85. $3\sqrt{7}$ 87. $30\sqrt{35}$ 89. $y^2\sqrt[4]{y}$ 91. x^3y^3
93. $24c^4\sqrt{15}$ 95. $60\sqrt{2}$ 97. $2\sqrt{2}$ 99. $6\sqrt{5}$ 101. $c^2\sqrt{d}$
103. $12a^2$ 105. $6c\sqrt{3c}$ 107. $36x^2y^3\sqrt{2y}$ 109. $\frac{2\sqrt{6}}{7}$
111. $\frac{a^4}{2}$ 113. $\frac{3x^5\sqrt{5}}{4}$ 115. $\frac{3}{8}\sqrt{5}$

Review Exercises 1. no 2. yes 3. $8x^2 - 7x$
4. $8a^2 - 6ab - 9b^2$ 5. $9m^2 - 25n^2$ 6. $4x^2 - 12xy + 9y^2$

10.4 Exercises
1. Like radicals have the same index and the same radicand, but their coefficients may be different. 3. $(x + 3)(x + 2) = x^2 + 2x + 3x + 6 = x^2 + 5x + 6$, $(\sqrt{5} + 3)(\sqrt{5} + 2) = \sqrt{5}^2 + 2\sqrt{5} + 3\sqrt{5} + 6 = 5 + 5\sqrt{5} + 6 = 11 + 5\sqrt{5}$ In both cases we use the FOIL method and then simplify the expansion. 5. $-6\sqrt{6}$ 7. $9\sqrt{a}$
9. $12\sqrt{5} - 8\sqrt{6}$ 11. $11a\sqrt{5a} - 2b\sqrt{7b}$ 13. $3x\sqrt[3]{9}$
15. $18x^2\sqrt[4]{5x}$ 17. Cannot combine because the radicals are not like. 19. $4\sqrt{3} - 5\sqrt{3} = -\sqrt{3}$ 21. $4\sqrt{5y} - 5\sqrt{5y} = -\sqrt{5y}$
23. $4\sqrt{5} - 12\sqrt{5} = -8\sqrt{5}$ 25. $12\sqrt{6} - 6\sqrt{6} = 6\sqrt{6}$
27. $14a\sqrt{3a}$ 29. $4\sqrt{6}$ 31. $18\sqrt{2} - 17\sqrt{3}$ 33. $5\sqrt[3]{2}$
35. $-12x\sqrt[3]{5x^2}$ 37. $-4x^2\sqrt[4]{x} + 4x\sqrt[4]{x^3}$ 39. $3\sqrt{2} + 2$
41. $3 - 3\sqrt{5}$ 43. $\sqrt{15} + 10\sqrt{3}$ 45. $24x - 48x\sqrt{2}$
47. $12 - 3\sqrt{2} + 4\sqrt{5} - \sqrt{10}$ 49. $6 + 5\sqrt{x} + x$
51. $6 + 10\sqrt{2} + 9\sqrt{3} + 15\sqrt{6}$ 53. $\sqrt{6} + \sqrt{10} + 3 + \sqrt{15}$
55. $x - \sqrt{xy} - 2y$ 57. $12\sqrt{14} - 12\sqrt{6} + 6\sqrt{35} - 6\sqrt{15}$
59. $8a + 10\sqrt{ab} - 3b$ 61. $2\sqrt[3]{2} - 3\sqrt[3]{4} - 40$
63. $1 - \sqrt[3]{18} + \sqrt[3]{12}$ 65. $x + 8$ 67. $22 + 8\sqrt{6}$
69. $39 + 12\sqrt{3}$ 71. $16 + 8\sqrt{3}$ 73. $30 + 12\sqrt{6}$ 75. 1
77. -14 79. $36 - x$ 81. 1 83. $x - y$ 85. 4 87. 50
89. $7\sqrt{5}$ 91. $-15\sqrt{6}$ 93. $7\sqrt{2}$ 95. $-14\sqrt{5}$ 97. $25\sqrt{3}$
99. a. $42 + 2\sqrt{5}$ ft. b. 46.5 ft. c. \$87.89

Review Exercises 1. slope is $\frac{2}{5}$, y-intercept is (0, 2)
2. 3. $2x + 5$ 4. $16x^2 - 9$ 5. $\sqrt{2}$ 6. $\sqrt[3]{4}$

10.5 Exercises
1. a. $\sqrt{16}$ is rational; rule 1. b. There is a radical, $\sqrt{3}$, in the denominator; rule 4. 3. Multiply both the numerator and denominator by the same square root as appears in the denominator. 5. $\frac{\sqrt{3}}{3}$ 7. $\frac{2\sqrt{3}}{3}$ 9. $\frac{6\sqrt{7}}{7}$ 11. $\frac{x\sqrt{14}}{10}$
13. $\frac{\sqrt{7}}{7}$ 15. $\frac{\sqrt{2}}{2}$ 17. $\frac{5\sqrt{3a}}{3a}$ 19. $\frac{\sqrt{33mn}}{11n}$ 21. $\frac{2\sqrt{5x}}{x}$
23. $\frac{\sqrt{3}}{4}$ 25. $\frac{3\sqrt{x}}{x^2}$ 27. $4x\sqrt{2x}$ 29. Multiply both the numerator

and the denominator by $\sqrt{2}$, not 2. Answer is $\dfrac{\sqrt{6}}{2}$. **31.** $\dfrac{5\sqrt[3]{9}}{3}$

33. $\dfrac{\sqrt[3]{20}}{2}$ **35.** $3\sqrt[3]{2}$ **37.** $\dfrac{m\sqrt[3]{n^2}}{n}$ **39.** $\dfrac{\sqrt[3]{ab}}{b}$ **41.** $\dfrac{2\sqrt[3]{4x^2}}{x}$

43. $\dfrac{\sqrt[3]{30a}}{5a}$ **45.** $\dfrac{5\sqrt[4]{8}}{2}$ **47.** $\dfrac{\sqrt[4]{3x^2}}{x}$ **49.** $\dfrac{3\sqrt[4]{27x}}{x}$ **51.** $\sqrt{6}-\sqrt{3}$

53. $8+4\sqrt{3}$ **55.** $5\sqrt{3}-5\sqrt{2}$ **57.** $-1-\sqrt{5}$ **59.** $\dfrac{3+\sqrt{3}}{2}$

61. $\dfrac{-6-8\sqrt{3}}{13}$ **63.** $\dfrac{4\sqrt{21}-4\sqrt{6}}{5}$ **65.** $4\sqrt{2}-2\sqrt{6}$

67. $\dfrac{6y-6\sqrt{y}}{y-1}$ **69.** $\dfrac{3t-6\sqrt{tu}}{t-4u}$ **71.** $\dfrac{\sqrt{2xy}+2y\sqrt{3}}{x-6y}$ **73.** $\dfrac{3}{2\sqrt{3}}$

75. $\dfrac{2x}{5\sqrt{2x}}$ **77.** $\dfrac{2n}{3\sqrt{2n}}$ **79.** $\dfrac{1}{10-5\sqrt{3}}$ **81.** $\dfrac{5x-36}{9\sqrt{5x}+54}$

83. $\dfrac{19n}{10n\sqrt{n}-2n\sqrt{6n}}$ **85. a.** $\dfrac{5\sqrt{3}}{3}$ **b.** $\sqrt{5}$ **c.** $\dfrac{5\sqrt{11}}{11}$

87. a. They are the same. The functions are the same.
b. $f(x)=g(x)$

89. a. $T=\dfrac{2\pi\sqrt{9.8L}}{9.8}$ **b.** $T=\dfrac{2\pi L}{\sqrt{9.8L}}$ **91. a.** $s=\dfrac{\sqrt{3Vh}}{h}$

b. 745 ft. **93. a.** $V_{\text{rms}}=\dfrac{\sqrt{2}}{2}V_{\text{m}}$ or $\dfrac{\sqrt{2}V_{\text{m}}}{2}$ **b.** $V_{\text{rms}}=\dfrac{163\sqrt{2}}{2}$

c. ≈ 115.3 **95.** $\dfrac{15\sqrt{2}-10\sqrt{3}}{3}$

Review Exercises
1. $-6, 6$ **2.** $2, 3$ **3.** $(x-3)(x-3)$
4. $\pm 2\sqrt{7}$ **5.** 4 **6.** -1

10.6 Exercises
1. Some of the answers may be extraneous.
3. The principal square root of a number cannot equal a negative. **5.** Subtract $3x$ from both sides to isolate the radical. This allows us to use the squaring principle of equality to eliminate the radical. **7.** 4 **9.** No real-number solution.

11. 27 **13.** -8 **15.** 17 **17.** 11 **19.** 6 **21.** $-\dfrac{1}{2}$ **23.** No real-number solution. **25.** 11 **27.** -2 **29.** 124 **31.** 55 **33.** 30 **35.** -11 **37.** No real-number solution. **39.** -1 **41.** No real-number solution. **43.** 3 **45.** 3 **47.** 12 (5 is an extraneous solution) **49.** 1, 3 **51.** $-3, -2$ **53.** $\dfrac{1}{2}$ **55.** 5 **57.** $-1, 3$ **59.** 0, 4 **61.** $\dfrac{3}{16}$ **63.** 1 **65.** Mistake: You cannot take the principal square root of a number and get a negative. Correct: No real-number solution. **67.** Mistake: The binomial $x-3$ was not squared correctly. Correct: $x=6$ with $x=1$ an extraneous solution. **69.** 9.8 m **71.** 2.45 m **73.** 1.44 ft. **75.** 144 ft. **77.** 36.73 ft. **79.** 82.65 ft. **81.** 4 N **83.** 6 N **85. a.** 3, 4, 25 **b.**

c. No. The x-values must be 0 or positive because real square roots exist only when $x \geq 0$. The y-values must be 0 or positive because by definition the principal square root is either 0 or positive. **d.** Yes, because it passes the vertical line test. **87.** The graph becomes steeper. **89.** The graph rises or lowers according to the value of the constant.

Review Exercises
1. 81 **2.** -0.008 **3.** 39.0625 or $\dfrac{625}{16}$ **4.** x^8 **5.** y^4 **6.** n^{24}

10.7 Exercises
1. $\sqrt{-1}$ **3.** No. The set of complex numbers contains both the real and the imaginary numbers. **5.** We subtract complex numbers just like we subtract polynomials by writing an equivalent addition and changing the signs in the second complex number. **7.** $4i$ **9.** $2i\sqrt{5}$ **11.** $3i\sqrt{3}$ **13.** $5i\sqrt{5}$ **15.** $3i\sqrt{7}$ **17.** $3i\sqrt{5}$ **19.** $7i\sqrt{3}$ **21.** $i\sqrt{2}$ **23.** $1+5i$ **25.** $8-4i$ **27.** $-6+5i$ **29.** $10+2i$ **31.** $-13-2i$ **33.** 0 **35.** $8-4i$ **37.** $2+8i$ **39.** -6 **41.** 16 **43.** $12+10i$ **45.** $12-6i$ **47.** $7+i$ **49.** $45+7i$ **51.** 10 **53.** $8+6i$ **55.** $-2i$ **57.** $-\dfrac{4i}{5}$ **59.** $-3i$ **61.** $-\dfrac{1}{2}-i$ **63.** $\dfrac{1}{2}-i$ **65.** $5-2i$ **67.** $\dfrac{2}{5}+\dfrac{1}{5}i$ **69.** $\dfrac{6}{5}-\dfrac{3}{5}i$ **71.** $-\dfrac{8}{25}+\dfrac{6}{25}i$ **73.** $\dfrac{10}{17}+\dfrac{11}{17}i$ **75.** $-i$ **77.** -1 **79.** -1 **81.** 1 **83.** 1 **85.** -1 **87.** $-i$ **89.** i

Review Exercises
1. $(2x-3)(2x-3)$ **2.** $\pm 4\sqrt{3}$
3. $x=2$ **4.** $x=-\dfrac{3}{2}, 2$ **5.** $(0,2),(3,0)$

6.

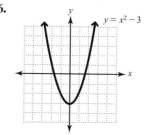

Chapter 10 Review Exercises
1. true **2.** false **3.** false **4.** false **5.** true **6.** radicands **7.** coefficients **8.** multiply **9.** conjugate **10.** extraneous **11.** ± 11 **12.** ± 7 **13.** 13 **14.** -7 **15.** $6i$ **16.** $\dfrac{1}{5}$ **17.** 2.646 **18.** 9.487 **19.** $9|x|$ **20.** $(x-1)^2$ **21.** $\dfrac{5a^2b^5}{4}$ **22.** $\dfrac{9x^2y}{8a^4}$ **23.** 3 **24.** $2y\sqrt[4]{x^2}$ **25.** $6m^2$ **26.** $10xyz^2$ **27.** $3\sqrt{2}$ **28.** $2\sqrt{30}$ **29.** $x^2b\sqrt{6b}$ **30.** $-4ab^4$ **31.** $-0.2x^6$ **32.** $\dfrac{1}{4m^2}$ **33.** $\dfrac{h^2}{2}$ **34.** $-\dfrac{4}{5}$ **35.** $\dfrac{2}{3}$ **36.** $\dfrac{a\sqrt{17}}{c}$ **37.** -4 **38.** $2a^2\sqrt{6}$ **39.** 1/8 **40.** $\sqrt[7]{(5r-2)^5}$ **41.** $33^{1/8}$ **42.** $8n^{-3/7}$ **43.** $8^{3/5}$ **44.** $m^{8/3}$ **45.** $(3xw)^{3/4}$ **46.** $(a+b)^{4/3}$ **47.** x^2 **48.** $32m^{3/2}$ **49.** $y^{-1/5}$ **50.** b **51.** $k^{1/2}$ **52.** $2^{3/4}x^{3/4}y^{3/20}$ **53.** $\sqrt{15}+\sqrt{10}$ **54.** $\sqrt[3]{21}+2\sqrt[3]{49}$

55. $6\sqrt{6} - 54$ **56.** $\sqrt{10} + \sqrt{14} - \sqrt{15} - \sqrt{21}$
57. $-6 + 2\sqrt[3]{2} - 4\sqrt[3]{4}$ **58.** $\sqrt[4]{12x^2} - \sqrt[4]{6x} + 2\sqrt[4]{2x} - 2$
59. $5a - 3b$ **60.** 7 **61.** $3 + 2\sqrt{2}$ **62.** $7 - 2\sqrt{10}$ **63.** $\dfrac{\sqrt{2}}{2}$
64. $\sqrt[3]{9}$ **65.** $\dfrac{\sqrt[2]{7}}{7}$ **66.** $\dfrac{\sqrt[4]{40x^2}}{2}$ **67.** $\dfrac{\sqrt[3]{153y}}{3y}$ **68.** $\dfrac{\sqrt[4]{27}}{3}$
69. $-4\sqrt{2} - 4\sqrt{3}$ **70.** $\dfrac{4 - \sqrt{3}}{13}$ **71.** $\dfrac{1}{2 - \sqrt{n}}$ **72.** $\sqrt{6} + 2$
73. $\dfrac{5}{3\sqrt{10}}$ **74.** $\dfrac{3x}{5\sqrt{3x}}$ **75.** $\dfrac{1}{16 + 8\sqrt{3}}$ **76.** $\dfrac{1}{10\sqrt{t} - 5\sqrt{3t}}$
77. $x = 81$ **78.** No solution.   **79.** $w = 10$ **80.** $x = -2$
81. $x = 18$ **82.** $y = 5$ **83.** No real-number solutions.
84. $\dfrac{719}{3}$ **85.** 7 **86.** $1 \left(\dfrac{17}{4} \text{ is extraneous} \right)$ **87.** $0, 4$
88. $\dfrac{3}{16}$ **89.** $3i$ **90.** $2i\sqrt{5}$ **91.** $8 - 6i$ **92.** $9 - 7i$ **93.** -12
94. $2 + 8i$ **95.** $26 + 2i$ **96.** $24 - 10i$ **97.** $-5i$ **98.** $3i$
99. $\dfrac{-4i}{3}$ **100.** $\dfrac{1}{5} - \dfrac{7}{5}i$ **101.** $\dfrac{6}{5} - \dfrac{3}{5}i$ **102.** $\dfrac{7}{13} + \dfrac{17}{13}i$ **103.** 1
104. $-i$ **105.** $\dfrac{\sqrt{6}}{4}$ **106. a.** $8\sqrt{5}$ mph   **b.** 17.9 mph
107. 312.5 ft.   **108.** π sec.   **109.** 0.272 m   **110. a.** $\dfrac{\sqrt{10}}{2}$ sec.
b. 1.58 sec.   **111.** 1.44 ft.   **112.** 5 ft.

Chapter 10 Practice Test   **1.** 6 **2.** $7i$ **3.** $9xy^2\sqrt{y}$

4. $3\sqrt[3]{2}$ **5.** $2x\sqrt[4]{x^2}$, or $2x\sqrt{x}$ **6.** $3r^5$ **7.** $\dfrac{1}{3}$ **8.** $\dfrac{1}{3}$ **9.** $5\sqrt{7}$
10. $4 - 2\sqrt{3}$ **11.** $x^{-2/3}$ **12.** $-6 + 2\sqrt[3]{2} - 4\sqrt[3]{4}$
13. $8^{1/5}x^{3/5}$ **14.** $(2xy)^{2/3}$ **15.** $\dfrac{\sqrt[3]{2}}{2}$ **16.** $\dfrac{x - \sqrt{xy}}{x - y}$ **17.** 22
18. 7 **19.** $-2 - 4i$ **20.** $2 + 0i$ or 2 **21.** $17 + 0i$ or 17
22. $\dfrac{8}{25} + \dfrac{6}{25}i$ **23.** 60 **24. a.** $\dfrac{\sqrt{3}}{2}$ **b.** 64 ft.   **25. a.** $3.5\sqrt{10}$
b. 293.88 ft. or $\dfrac{14400}{49}$ ft.

Chapter 11

11.1 Exercises   **1.** Because you can square both a
negative and a positive number to get a.   **3.** $x = \dfrac{b \pm \sqrt{c}}{a}$
5. Factoring, because it's faster. $x^2 - 6x + 8 = (x - 4)(x - 2)$
7. ± 4 **9.** $\pm \dfrac{2}{3}$ **11.** ± 0.9 **13.** $\pm 5\sqrt{2}$ **15.** $\pm 5i$ **17.** ± 7
19. ± 6 **21.** ± 3 **23.** $\pm \dfrac{3}{5}$ **25.** $\pm 2i$ **27.** $\pm \dfrac{3\sqrt{5}}{10}$ **29.** ± 1
31. $\pm i\sqrt{2}$ **33.** $\pm \dfrac{4\sqrt{6}}{3}$ **35.** ± 1.2 **37.** $-7, 1$
39. $-\dfrac{9}{2}, \dfrac{3}{2}$ **41.** $7 \pm 2i\sqrt{3}$ **43.** $\dfrac{1 \pm 2\sqrt{10}}{4}$ **45.** $8 \pm i$

47. $0, \dfrac{4}{3}$ **49.** $-\dfrac{1}{8}, \dfrac{7}{8}$ **51.** $-5, -9.\overline{6}$ **53.** Mistake: Gave only
the positive square root. Correct: ± 3 **55.** Mistake: The second
line dropped the negative sign. Correct: $3 \pm i\sqrt{2}$
57. $\pm 3\sqrt{10} = \pm 9.487$ **59.** $\pm\sqrt{15} = \pm 3.873$
61. $2 \pm \sqrt{3} = 0.268, 3.732$ **63. a.** $x^2 + 14x + 49$ **b.** $(x + 7)^2$
65. a. $n^2 - 10n + 25$ **b.** $(n - 5)^2$ **67. a.** $y^2 - 7y + \dfrac{49}{4}$
b. $\left(y - \dfrac{7}{2} \right)^2$ **69. a.** $s^2 - \dfrac{2}{3}s + \dfrac{1}{9}$ **b.** $\left(s - \dfrac{1}{3} \right)^2$
71. a. $b^2 + \dfrac{3}{2}b + \dfrac{9}{16}$ **b.** $\left(b + \dfrac{3}{4} \right)^2$ **73. a.** $p^2 + 9p + \dfrac{81}{4}$
b. $\left(p + \dfrac{9}{2} \right)^2$ **75.** $-5, 3$ **77.** $-8, -2$ **79.** $1 \pm 7i$ **81.** $3, 6$
83. $-12, 2$ **85.** $-3, 5$ **87.** $3 \pm 2i\sqrt{5}$ **89.** $-\dfrac{3}{2}$ **91.** $-\dfrac{7}{2}, \dfrac{1}{2}$
93. $0, 3$ **95.** $\dfrac{-1 \pm \sqrt{41}}{10}$ **97.** $-2, \dfrac{3}{2}$ **99.** $-\dfrac{3}{2}, \dfrac{2}{3}$ **101.** $-1, \dfrac{3}{2}$
103. $-\dfrac{4}{3}, \dfrac{7}{2}$ **105.** Did not divide by 3 so that x^2 has a coefficient
of 1. Then wrote an incorrect factored form.   **107.** 14 in.
109. 3 ft., 6 ft.   **111.** 305 m   **113.** 18 in.   **115.** 12 ft. by 14 ft.
117. 12 ft. by 5 ft.   **119.** 15 in.   **121. a.** 5 in.
b. 30 in. by 20 in.   **123.** $\dfrac{3}{4}$ sec.   **125.** 4.9 sec.   **127.** 4 m/sec.

Review Exercises   **1.** $(x - 5)^2$ **2.** $(x + 3)^2$ **3.** $6x$

4. $2x + 3$ **5.** $\dfrac{-2 + \sqrt{6}}{2}$ or $-1 + \dfrac{\sqrt{6}}{2}$ **6.** $\dfrac{-3 - i}{5}$ or $-\dfrac{3}{5} - \dfrac{1}{5}i$

11.2 Exercises   **1.** Follow the procedure for solving a qua-
dratic equation by completing the square on $ax^2 + bx + c = 0$.
3. Yes, if they have irrational or nonreal complex solutions.
5. When the discriminant is zero. The quadratic formula
becomes $x = -\dfrac{b}{2a}$.   **7.** $a = 1, b = -4, c = 9$ **9.** $a = 2,$
$b = -7, c = -5$ **11.** $a = 0.5, b = -1, c = 2.4$ or
$a = -0.5, b = 1, c = -2.4$ **13.** $a = -\dfrac{1}{2}, b = -\dfrac{2}{3}, c = 8$ or
$a = \dfrac{1}{2}, b = \dfrac{2}{3}, c = -8$ **15.** $-6, -3$ **17.** $\dfrac{1 \pm \sqrt{5}}{2}$ **19.** $-8, 2$
21. $0, 3$ **23.** 4 **25.** $1 \pm i$ **27.** $-\dfrac{1}{3}, 1$ **29.** $\dfrac{-5 \pm \sqrt{10}}{3}$
31. $\dfrac{3 \pm i\sqrt{11}}{4}$ **33.** $-2, -\dfrac{2}{3}$ **35.** $\dfrac{3 \pm 6i}{5}$ **37.** $\dfrac{3 \pm \sqrt{105}}{12}$
39. $-0.15, 0.1$ **41.** $-2, \dfrac{3}{2}$ **43.** $\pm \dfrac{3}{2}$ **45.** $1 \pm i\sqrt{2}$
47. $\dfrac{-3 \pm \sqrt{5009}}{100} \approx -0.738, 0.678$ **49.** $\dfrac{3 \pm i\sqrt{51}}{12}$
51. Mistake: The b was not properly used. Correct: $\dfrac{7 \pm \sqrt{37}}{6}$
53. Mistake: Not completely simplified. Correct: $1 \pm i\sqrt{2}$
55. 2 nonreal complex.   **57.** 1 rational.   **59.** 2 nonreal complex.
61. 1 rational.   **63.** 2 irrational.   **65.** square root principle
or factoring   **67.** Square root principle or factoring
69. quadratic formula   **71.** factoring   **73.** quadratic formula

75. $(2, 0), (-1, 0), (0, -2)$ **77.** $(-4, 0), (2, 0), (0, 8)$ **79.** $\left(\frac{1}{2}, 0\right),$
$(-8, 0), (0, -8)$ **81.** No x-intercepts, $(0, -6)$. **83.** $6, 7$
85. $3, 4, 5$ **87.** 14 in. by 25 in. **89. a.** 12 ft. **b.** 18 ft., 12 ft., 6 ft.,
8 ft. (from bottom clockwise) **91.** ≈ 3.43 sec. **93. a.** $h = -4.9t^2$
$+ 10$ **b.** ≈ 1.01 sec. **c.** ≈ 1.43 sec. **95.** 8000 units
97. a. $\frac{25}{8}$ **b.** $c < \frac{25}{8}$ **c.** $c > \frac{25}{8}$ **99. a.** 9 **b.** $a < 9$ **c.** $a > 9$
Review Exercises **1.** $2\sqrt{5}$ **2.** $-x + 7$ **3.** $-x - 6$
4. $-6x^2 + 3x$ **5.** $y^2 - 49$ **6.** $3 - \sqrt{3}$

11.3 Exercises **1.** It can be rewritten as a quadratic

equation. **3.** $u = \dfrac{x + 2}{3}$ **5.** $\dfrac{1}{x}$ **7.** $-\dfrac{4}{3}, 2$ **9.** $-6, 4$
11. -8 **13.** $-7.5, 5$ **15.** $-4, 2$ **17.** $-5, \dfrac{2}{3}$ **19.** $9, 25$
21. $\dfrac{49}{4}$ **23.** 2 **25.** $3i, -3i$ **27.** 2 **29.** $0, 4$ **31.** $-3, -1,$
$1, 3$ **33.** $-1.5, -1, 1, 1.5$ **35.** $2, -2, 3i, -3i$ **37.** $-6, -4$
39. $-3.5, 2$ **41.** $-9, -5$ **43.** $-5, 0$ **45.** $8, 27$ **47.** $-\dfrac{1}{8}, 8$
49. $81,256$ **51.** $\dfrac{16}{625}$ **53.** 3 hr. **55.** ≈ 2.16 hr.

57. 45 mph **59. a.** old rate $= 12$ mph; new rate $= 18$ mph
b. old time $= 3$ hr.; new time $= 2$ hr. **61.** Jody: 4 hr.;
Billy: 6 hr. **63.** ≈ 4.27 hr., 3.77 hr. **65.** 420 vps; 460 vps

Review Exercises **1.** -2 **2.** $(3, 0), (0, 2)$ **3.** 6
4.

$f(x) = |x - 3| + 1$

5.

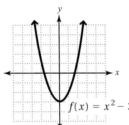

$f(x) = x^2 - 3$

6.

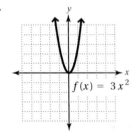

$f(x) = 3x^2$

11.4 Exercises **1.** The sign of a. **3.** The axis of symmetry
is the vertical line through the vertex. If $y = ax^2 + bx + c$, then
the equation of symmetry is $x = -\dfrac{b}{2a}$. **5.** There are no
x-intercepts. **7.** up **9.** down **11.** up **13.** down **15.** down
17. V: $(2, -3)$; Axis: $x = 2$ **19.** V: $(-1, -5)$; Axis: $x = -1$
21. V: $(0, 2)$; Axis: $x = 0$ **23.** V: $(7, 0)$; Axis: $x = 7$
25. V: $(0, 0)$; Axis: $x = 0$ **27.** V: $\left(\dfrac{1}{3}, \dfrac{4}{3}\right)$; Axis: $x = \dfrac{1}{3}$
29. V: $(-0.4, 1.08)$; Axis: $x = -0.4$ **31. a.** down **b.** V: $(0, 0)$
c. $x = 0$

d.

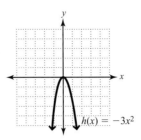

$h(x) = -3x^2$

33. a. up **b.** V: $(0, 0)$ **c.** $x = 0$ **d.**

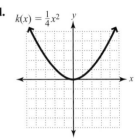

$k(x) = \frac{1}{4}x^2$

35. a. up **b.** V: $(0, -3)$ **c.** $x = 0$ **d.**

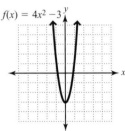

$f(x) = 4x^2 - 3$

37. a. down **b.** V: $(0, 2)$ **c.** $x = 0$
d.

$g(x) = -0.5x^2 + 2$

39. a. up **b.** V: $(3, 2)$ **c.** $x = 3$
d.

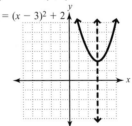

$f(x) = (x - 3)^2 + 2$

41. a. down **b.** V: $(-1, -3)$ **c.** $x = -1$
d.

$k(x) = -2(x + 1)^2 - 3$

43. a. up **b.** V: $(2, -1)$ **c.** $x = 2$
d. $h(x) = \frac{1}{3}(x - 2)^2 - 1$

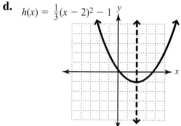

45. a. down **b.** V: $(-1, 4)$ **c.** $x = -1$
d.

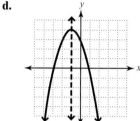

$k(x) = -x^2 - 2x + 3$

47. a. up **b.** V: $\left(-\frac{3}{2}, -\frac{3}{2}\right)$ **c.** $x = -\frac{3}{2}$

d. $f(x) = 2x^2 + 6x + 3$

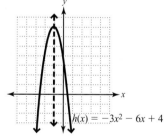

49. a. down **b.** V: $(-1, 7)$ **c.** $x = -1$
d.

$h(x) = -3x^2 - 6x + 4$

51. a. 4.59 sec. **b.** 103.32 m **c.** 9.18 sec. **53. a.** 7.8 ft.
b. ≈ 4.62 ft. **c.**

$y = -0.8x^2 + 2.4x + 6$

(scl = 4)

55. a. The tenth week. **b.** 20,000 **57.** Length and width of
100 ft. **59.** 55 units **61.** $-6, 6$; product is -36.

Review Exercises **1.** -11 **2.** $x = -5, 3$ **3.** $x = \pm 2i\sqrt{5}$
4. $x \le 4.5$;

$-6\ -5\ -4\ -3\ -2\ -1\ \ 0\ \ 1\ \ 2\ \ 3\ \ 4\ \ 5\ \ 6$

5. $x \le -5$ or $x \ge 11$;

$-12\ -10\ -8\ -6\ -4\ -2\ \ 0\ \ 2\ \ 4\ \ 6\ \ 8\ \ 10\ \ 12$

6.

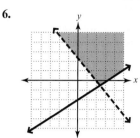

Exercise Set 11.5 **1.** Solve the related equation, then
determine which of the three intervals satisfy the inequality by
examining one value in the interval. If this one value satisfies
the inequality, then all points in the interval satisfy the inequality. **3.** Yes, for example $x^2 + 2 \le 0$ has an empty solution set
since $x^2 + 2$ is always positive. **5.** In both cases the intervals
to check are $(-\infty, -5)$, $(-5, 2)$, and $(2, \infty)$ and in both cases only
x-values in $(-5, 2)$ satisfy the original inequalities.
7. a. $x = -5, -1$; **b.** $(-5, -1)$; **c.** $(-\infty, -5) \cup (-1, \infty)$
9. a. $x = -1, 3$; **b.** $(-\infty, -1] \cup [3, \infty)$; **c.** $[-1, 3]$
11. a. $x = -2$; **b.** $(-\infty, -2) \cup (-2, \infty)$; **c.** $\varnothing$ **13. a.** $\varnothing$;
b. $\varnothing$; **c.** $\mathbb{R}$ or $(-\infty, \infty)$
15. $(-4, -2)$

$-6\ -5\ -4\ -3\ -2\ -1\ \ 0\ \ 1\ \ 2\ \ 3\ \ 4\ \ 5\ \ 6$

17. $(-\infty, 2) \cup (5, \infty)$

$-6\ -5\ -4\ -3\ -2\ -1\ \ 0\ \ 1\ \ 2\ \ 3\ \ 4\ \ 5\ \ 6$

19. $(-4, -1)$

$-6\ -5\ -4\ -3\ -2\ -1\ \ 0\ \ 1\ \ 2\ \ 3\ \ 4\ \ 5\ \ 6$

21. $(-\infty, 1) \cup (3, \infty)$

$-6\ -5\ -4\ -3\ -2\ -1\ \ 0\ \ 1\ \ 2\ \ 3\ \ 4\ \ 5\ \ 6$

23. $[2, 4]$

$-6\ -5\ -4\ -3\ -2\ -1\ \ 0\ \ 1\ \ 2\ \ 3\ \ 4\ \ 5\ \ 6$

25. $(-\infty, -1] \cup [5, \infty)$

$-6\ -5\ -4\ -3\ -2\ -1\ \ 0\ \ 1\ \ 2\ \ 3\ \ 4\ \ 5\ \ 6$

27. $(-2, 5)$

$-6\ -5\ -4\ -3\ -2\ -1\ \ 0\ \ 1\ \ 2\ \ 3\ \ 4\ \ 5\ \ 6$

29. $(-\infty, \infty)$ or $\mathbb{R}$

$-6\ -5\ -4\ -3\ -2\ -1\ \ 0\ \ 1\ \ 2\ \ 3\ \ 4\ \ 5\ \ 6$

31. $\varnothing$, no solution

$-6\ -5\ -4\ -3\ -2\ -1\ \ 0\ \ 1\ \ 2\ \ 3\ \ 4\ \ 5\ \ 6$

33. $(-\infty, -5) \cup (-0.25, \infty)$

$-6\ -5\ -4\ -3\ -2\ -1\ \ 0\ \ 1\ \ 2\ \ 3\ \ 4\ \ 5\ \ 6$

35. $(-\infty, 0) \cup (5, \infty)$

$-6\ -5\ -4\ -3\ -2\ -1\ \ 0\ \ 1\ \ 2\ \ 3\ \ 4\ \ 5\ \ 6$

37. $[0, 6]$

39. $(-\infty, \infty)$ or $\mathbb{R}$

41. $[-4, -2] \cup [4, \infty)$

43. $(-\infty, -6) \cup (-2, 1)$

45. $\left(\dfrac{-2 - \sqrt{7}}{3}, \dfrac{-2 + \sqrt{7}}{3}\right)$

47. $\left(-\infty, \dfrac{-2 - \sqrt{7}}{2}\right] \cup \left[\dfrac{-2 + \sqrt{7}}{2}, \infty\right)$

49. $[-4 + \sqrt{6}, \infty) \cup (-\infty, -4 - \sqrt{6}]$

51. $(-\infty, -4) \cup (1, \infty)$

53. $(-5, -1]$

55. $(-4, \infty)$

57. $\left(-\infty, -\dfrac{9}{2}\right) \cup (-3, \infty)$

59. $(4, 7)$

61. $\left(3, \dfrac{14}{3}\right]$

63. $(-\infty, -3] \cup [2, 5)$

65. $(0, \infty)$

67. $(-\infty, -1) \cup (2, 5)$

69. a. 6 sec. **b.** at 2 sec. and again at 3 sec. **c.** $(2, 3)$ sec. **d.** $[0, 2) \cup (3, 6]$ **71. a.** $x \geq 4$
b. base ≥ 10, height ≥ 2 **73.** $2 < x \leq 8$

Review Exercises
1. $x = 18$ **2.** $9, (x - 3)^2$
3. $\dfrac{25}{4}, \left(x + \dfrac{5}{2}\right)^2$ **4.** $(-6, 0), (2, 0)$ **5.** V: $(3, -4)$ **6.** Down, because the coefficient of x^2 is negative.

Chapter 11 Review Exercises
1. false **2.** false
3. false **4.** false **5.** true **6.** $+\sqrt{a}, -\sqrt{a}$ **7.** $\left(\dfrac{b}{2}\right)^2$
8. $\dfrac{-b \pm \sqrt{b^2 - 4ac}}{2a}$ **9.** $b^2 - 4ac$ **10.** V: $(h, k); x = h$
11. ± 4 **12.** $\pm\dfrac{1}{6}$ **13.** $\pm 2\sqrt{7}$ **14.** $\pm\sqrt{14}$ **15.** $\pm i\sqrt{3}$
16. $-12, -2$ **17.** $9 \pm 4i$ **18.** $-\dfrac{7}{5}, \dfrac{1}{5}$ **19.** $-7, -1$ **20.** $-8, 14$
21. $\dfrac{3 \pm i\sqrt{5}}{2}$ **22.** $-1, \dfrac{3}{4}$ **23.** $\dfrac{-1 \pm \sqrt{41}}{4}$ **24.** $\dfrac{1 \pm i\sqrt{2}}{3}$

25. $-1 \pm \sqrt{6}$ **26.** $-\dfrac{3}{20}, \dfrac{1}{10}$ **27.** $D = 64$; 2 rational.
28. $D = -71$; 2 nonreal complex. **29.** $D = 0.48$; 2 irrational.
30. $D = 0$; 1 rational. **31.** $\pm\dfrac{3\sqrt{2}}{2}$ **32.** $-\dfrac{5}{2}$ **33.** $\dfrac{1}{3}, \dfrac{1}{2}$
34. $\dfrac{3 \pm \sqrt{17}}{4}$ **35.** No solution. **36.** 1 **37.** 2 **38.** $\dfrac{1}{3}, \dfrac{2}{3}$
39. $\pm\sqrt{2}, \pm\sqrt{3}$ **40.** $\pm 1, \pm\dfrac{\sqrt{2}}{2}$ **41.** $-\dfrac{9}{2}, -\dfrac{14}{3}$ **42.** $-26, -2$
43. 27,512 **44.** $\dfrac{16}{625}$ **45. a.** $(0, 0)$ **b.** down **c.** $(0, 0)$
d. $x = 0$ **e.**

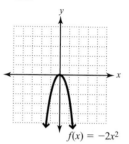

$f(x) = -2x^2$

46. a. $(0, 1)$ **b.** up **c.** $(0, 1)$ **d.** $x = 0$
e.

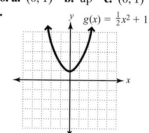

$g(x) = \frac{1}{2}x^2 + 1$

47. a. $(2, 0), \left(0, -\dfrac{4}{3}\right)$ **b.** down **c.** $(2, 0)$ **d.** $x = 2$
e.

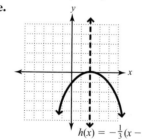

$h(x) = -\frac{1}{3}(x - 2)^2$

48. a. $\left(-3 + \dfrac{\sqrt{2}}{2}, 0\right), \left(-3 - \dfrac{\sqrt{2}}{2}, 0\right), (0, 34)$ **b.** up
c. $(-3, -2)$ **d.** $x = -3$ **e.**

$k(x) = 4(x + 3)^2 - 2$

49. a. $(-1 + \sqrt{2}, 0), (-1 - \sqrt{2}, 0), (0, -1)$ **b.** up

c. $(-1, -2)$ **d.** $x = -1$ **e.**

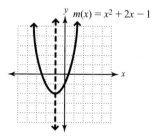

y $m(x) = x^2 + 2x - 1$

50. a. $(2, 0), (6, 0), (0, -6)$ **b.** down   **c.** $(4, 2)$ **d.** $x = 4$
e.

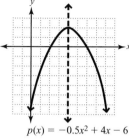

$p(x) = -0.5x^2 + 4x - 6$

51. $(-\infty, -5) \cup (3, \infty)$ **52.** $[-3, 1] \cup [2, \infty)$ **53.** $(-7, -2)$
54. $[2, 4]$ **55.** $(-\infty, -3] \cup (1, \infty)$ **56.** $(-\infty, -4) \cup (-2, \infty)$
57. $(-\infty, 4) \cup \left[\dfrac{17}{4}, \infty\right)$ **58.** $(-\infty, -2] \cup (3, 5)$ **59.** 150 ft.

60. 4 m/sec.   **61.** 15 ft. by 19 ft.   **62.** $9\sqrt{3}$
63. 6.65 ft. and 15.65 ft.   **64. a.** 0.75 sec.   **b.** 9 ft.   **c.** 1.5 sec.
d.

$h = -16t^2 + 24t$

65. a. 38.88 yd.   **b.** 72 yd.
66. a. $x \geq 10$ in.   **b.** height ≥ 8 in., base ≥ 14 in.

Chapter 11 Practice Test   **1.** ± 9 **2.** $3 \pm 2\sqrt{5}$

3. $4 \pm 2\sqrt{3}$ **4.** $\dfrac{3 \pm 2\sqrt{6}}{3}$ **5.** $-2, \dfrac{3}{2}$ **6.** 3, 5   **7.** $-8, 2$

8. ± 9 **9.** $\dfrac{-3 \pm i\sqrt{3}}{4}$ **10.** $0, -2$ **11.** $\pm 4i$ **12.** $0, -\dfrac{5}{3}$

13. $-\dfrac{6}{5}, 4$ **14.** $-\dfrac{2}{3}, 1$ **15.** No solution.   **16.** 1   **17.** $\pm\dfrac{1}{3}, \pm i\sqrt{3}$

18. $-5, 0$ **19. a.** down   **b.** $(3 \pm \sqrt{2}, 0), (0, -3.5)$
c. $(3, 1)$ **d.** $x = 3$ **e.**

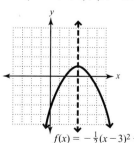

$f(x) = -\frac{1}{2}(x - 3)^2 + 1$

20. $[-1, 4]$

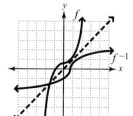

$-6\ -5\ -4\ -3\ -2\ -1\ \ 0\ \ 1\ \ 2\ \ 3\ \ 4\ \ 5\ \ 6$

21. $(-\infty, -2) \cup (1, \infty)$

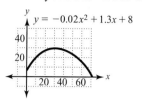

$-6\ -5\ -4\ -3\ -2\ -1\ 0\ 1\ 2\ 3\ 4\ 5\ 6$

22. 2.5 sec.   **23.** 12.36 ft. by 32.36 ft.   **24. a.** 29.125 m
b. 70.66 m   **c.**

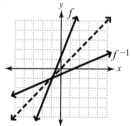

y $y = -0.02x^2 + 1.3x + 8$

25. a. $0 \leq w \leq 25$ **b.** $15 \leq l \leq 40$

Chapter 12

12.1 Exercises   **1.** No, $f \circ g = g \circ f$ if f and g are inverses
of each other and for a few other functions.   **3.** It is symmetric
with respect to the line $y = x$.   **5.** To prove that functions f and
g are inverses, you must show that $f[g(x)] = x$ and $g[f(x)] = x$.
7. 14   **9.** 3   **11.** 26   **13.** $3x^2 + 14$
15. $3\sqrt{x + 1} + 5$ **17.** $\sqrt{6}$
19. $(f \circ g)(x) = 6x + 6$; $(g \circ f)(x) = 6x - 2$
21. $(f \circ g)(x) = x^2 + 3$; $(g \circ f)(x) = x^2 + 4x + 5$
23. $(f \circ g)(x) = 9x^2 + 9x - 4$; $(g \circ f)(x) = 3x^2 + 9x - 12$
25. $(f \circ g)(x) = \sqrt{2x - 3}$; $(g \circ f)(x) = 2\sqrt{x + 2} - 5$
27. $(f \circ g)(x) = \dfrac{2x - 3}{x - 3}$; $(g \circ f)(x) = \dfrac{1 - 2x}{x + 1}$
29. Domain is $[0, \infty)$ and range is $[3, \infty)$.   **31.** yes   **33.** no
35. yes   **37.** yes   **39.** no   **41.** yes   **43.** no   **45.** yes   **47.** yes
49. Yes, since $(f \circ g)(x) = (g \circ f)(x) = x$.
51.

53.

55.

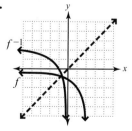

57.

59.

61.

63. yes **65.** no **67.** $f^{-1} = \{(2, -3), (-3, -1), (4, 0), (6, 4)\}$

69. $f^{-1} = \{(-2, 7), (2, 9), (1, -4), (3, 3)\}$

71. $f^{-1}(x) = x - 6$ **73.** $f^{-1}(x) = \dfrac{x - 3}{2}$

75. $f^{-1}(x) = \sqrt[3]{x - 2}$ **77.** $f^{-1}(x) = \dfrac{2 - 2x}{x}$

79. $f^{-1}(x) = \dfrac{3x + 2}{x - 1}$ **81.** $f^{-1}(x) = x^2 + 2, x \geq 2$

83. $f^{-1}(x) = \sqrt[3]{\dfrac{x - 4}{2}}$ **85.** $f^{-1}(x) = x^3 - 2$

87. $f^{-1}(x) = \dfrac{x^3}{16} - 2$ **89. a.** $y = \dfrac{x - 100}{0.05}$ **b.** x is salary, y is

sales. **c.** \$5000 sales. **91. a.** $y = \dfrac{6825 - x}{20}$ **b.** x is cost, y is

number of 36" fans. **c.** 80 36" fans. **93.** $(5, -4)$

95. $f^{-1}(x) = \dfrac{x - b}{a}$ **97.** No, the inverse of an even function is

not a function. If the graph of a function is symmetric with
respect to the y-axis, the function is not one-to-one.

Review Exercises
1. 8 **2.** 8 **3.** $\dfrac{1}{27}$ **4.** 16 **5.** Shift
the graph 3 units to the right, 1 unit up, and stretch by a factor
of 2. **6.** $32 = 2^5$

12.2 Exercises
1. Yes, the graphs are symmetric.
The point $(1, 2)$ on $f(x)$ corresponds to the point $(-1, 2)$ on
$g(x)$, etc. **3.** Domain is all real numbers, range is $(0, \infty)$.
5. Because the exponential function is one-to-one.

7.

9.

11.

13.

15.

17.

19.

21.

23.
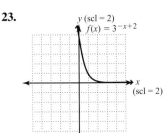

25. 3 **27.** $\dfrac{5}{3}$ **29.** $\dfrac{1}{2}$ **31.** -2

33. -2 **35.** 3 **37.** -4 **39.** $\dfrac{1}{2}$ **41.** -12

43.

45.
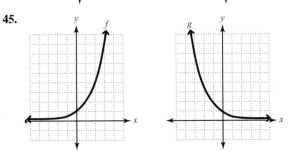

47. 6400 cells **49.** 6144 billion people **51.** \$25,870.70
53. 0.020 gm **55.** \$1046 **57.** 1.225 parts per million
59. 168.9 million transistors.

Review Exercises
1. $f^{-1}(x) = \dfrac{x - 4}{3}$

2.

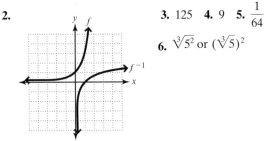

3. 125 **4.** 9 **5.** $\dfrac{1}{64}$

6. $\sqrt[3]{5^2}$ or $(\sqrt[3]{5})^2$

12.3 Exercises
1. $f^{-1}(x) = \log_3 x$, because logarithmic and exponential functions are inverses. **3.** $x > 0$
5. Logarithms are exponents because logarithms are inverses of exponential functions. **7.** $\log_2 32 = 5$ **9.** $\log_{10} 1000 = 3$
11. $\log_e x = 4$ **13.** $\log_5 \frac{1}{125} = -3$ **15.** $\log_{10} \frac{1}{100} = -2$
17. $\log_{625} 5 = \frac{1}{4}$ **19.** $\log_{1/4} \frac{1}{16} = 2$ **21.** $\log_7 \sqrt{7} = \frac{1}{2}$
23. $3^4 = 81$ **25.** $4^{-2} = \frac{1}{16}$ **27.** $10^2 = 100$ **29.** $e^5 = a$
31. $e^{-4} = \frac{1}{e^4}$ **33.** $\left(\frac{1}{8}\right)^2 = \frac{1}{64}$ **35.** $\left(\frac{1}{5}\right)^{-2} = 25$ **37.** $7^{1/2} = \sqrt{7}$
39. 8 **41.** $\frac{1}{25}$ **43.** 4 **45.** -1 **47.** 10 **49.** 2 **51.** $\frac{1}{4}$ **53.** 243
55. 2 **57.** -6
59.

61.

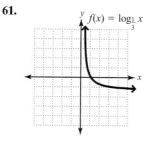

63. $\log_b b = 1$ because $b^1 = b$. **65.** Because $b^0 = 1$. **67.** 97%
69. After 4 days.

Review Exercises
1. x^6 **2.** x^3 **3.** x^{15} **4.** x^5
5. $x^{3/4}$ **6.** x^{-6}

12.4 Exercises
1. Because the exponential and logarithmic functions are inverses. **3.** $\log_b(x + y)$, there is no rule for the logarithm of a sum. **5.** 2 **7.** r **9.** $4x$ **11.** 5 **13.** a
15. $7x$ **17.** $\log_2 5 + \log_2 y$ **19.** $\log_a p + \log_a q$
21. $\log_4 m + \log_4 n + \log_4 p$ **23.** $\log_a x + \log_a(x - 5)$
25. $\log_3(40)$ **27.** $\log_4(16)$ **29.** $\log_a 7m$ **31.** $\log_4 ab$
33. $\log_a(2x^2 + 5x)$ **35.** $\log_4(x^2 + 4x + 3)$
37. $\log_2 7 - \log_2 9$ **39.** $\log_a x - \log_a 5$ **41.** $1 - \log_a b$
43. $\log_a x - \log_a(x - 3)$ **45.** $\log_4(2x - 3) - \log_4(4x + 5)$
47. $\log_6 8$ **49.** $\log_2 2 = 1$ **51.** $\log_a \frac{x}{3}$ **53.** $\log_4 \frac{p}{q}$ **55.** $\log_b \frac{x}{x - 4}$
57. $\log_x x = 1$ **59.** $3 \log_7 4$ **61.** $7 \log_a x$ **63.** $\frac{1}{2} \log_a 3$
65. $\frac{2}{3} \log_3 x$ **67.** $-2 \log_a y$ **69.** $\log_a 3 - 2 \log_a y$ **71.** $\log_3 5^4$
73. $\log_2 x^{-3}$ **75.** $\log_5 2$ **77.** $\log_a \sqrt[4]{x^3}$ **79.** $\log_a 4$
81. $\log_3 \frac{1}{\sqrt{x}}$ **83.** $3 \log_a x - 4 \log_a y$ **85.** $4 \log_3 a + 2 \log_3 b$
87. $\log_a x + \log_a y - \log_a z$ **89.** $2 \log_x a - \log_x b - 3 \log_x c$
91. $\frac{3}{2} \log_4 x - \frac{1}{2} \log_4 y$ **93.** $\log_a x + \frac{1}{2} \log_a y - \frac{3}{2} \log_a z$

95. $\log_3 \frac{1}{2}$ **97.** $\log_b x^4 y^3$ **99.** $\log_a \sqrt{\frac{5}{7}}$ **101.** $\log_a \sqrt[3]{(x^2 y)^2}$
103. $\log_b(3x^2 - 2)$ **105.** $\log_a \frac{(x - 2)^3}{(x + 1)^4}$ **107.** $\log_a \frac{x^2 z^4}{w^3 u^6}$

Review Exercises
1. 10^9 **2.** $10^{-2.5}$ **3.** $10^{1.6532} = 45$
4. $e^{-1.3863} = 0.25$ **5.** $\log_{10} 50 = 1.6990$ **6.** $13,468.55

12.5 Exercises
1. $b > 0, b \neq 1$. If $y = \log_b x$ then $b^y = x$.
If $b < 0$ and $y = \frac{1}{2}$ or any other fraction whose denominator is even, then x is imaginary. If $b = 1$, then $x = 1$ for all values of y and the graph is a vertical line. **3.** Positive. If $y = \ln x$, then $e^y = x$. If $x > 1$, then $e^y > 1$, which is true if $y > 0$. **5.** 0.91629
7. 1.8062 **9.** -2.1739 **11.** 2.3941 **13.** 4.1761 **15.** -5.7959
17. 2.2343 **19.** 4.3720 **21.** -3.3814 **23.** 5.6864 **25.** 0.4343
27. Error results, because the domain of $\log_a x$ is $(0, \infty)$, so log 0 is undefined. **29.** 2 **31.** -1 **33.** $\frac{1}{3}$ **35.** -3 **37.** 3 **39.** $\frac{1}{2}$
41. 90 dB **43.** 10^{-6} watts/m^2 **45.** 2.796 **47.** $10^{-3.5}$ moles/L
49. 22.9 or about 23 yr. **51.** The 1964 Alaska earthquake was about four times as severe as the 1906 San Francisco earthquake.
53. $10^{22.75}$ergs **55. a.** About 11,809 deaths. **b.** 1997
57. 6124 ft.

Review Exercises
1. $4x$ **2.** $0.05t$ **3.** $\log_5(x^2 + 3x)$
4. $\log_3 \frac{2x + 1}{x - 1}$ **5.** $3^2 = 2x + 5$ **6.** No, $\log_4(-3)$ is undefined.

12.6 Exercises
1. $m = n$ **3.** If $\log_b x = \log_b y$, then $x = y$. **5.** The solution $x = -5$ is rejected because substituting into the equation gives $\log(-5)$ and $\log(-3)$, which are both undefined. **7.** 3.1699 **9.** 1.0767 **11.** -1.5693 **13.** 2.8617
15. 12.4251 **17.** 7.1285 **19.** 0.5365 **21.** 107.2959
23. -73.1563 **25.** 6 **27.** 11 **29.** 6 **31.** 4 **33.** 4, -9 **35.** 7
37. No solution. **39.** $-6, 2$ **41.** 2 **43.** 10 **45.** 3 **47.** $\frac{3}{2}$
49. 3 **51.** 4 **53.** $\frac{5}{6}$ **55.** $\frac{5}{2}$ **57.** 6 **59.** 4 **61.** 9.5 yr.
63. a. $19,676.82 **b.** 9.3 yr. **65. a.** 7.18 billion **b.** About 8 yr., in 2010. **67. a.** 1.89 in. of mercury **b.** 21.8 mi. **69.** About 1173 colleges **71. a.** 5.58% **b.** 0.18 **73. a.** $82,243 **b.** 11.4 yr., about 2001. **75.** 11,714 **77.** 3276 **79. a.** 875 mosquitoes **b.** 75 days **81.** 1.7925 **83.** 0.5283 **85.** -2.3219 **87.** 0.3685

Review Exercises
1. $-1, 5, 15, 29$ **2.** $-1, \frac{1}{4}, -\frac{1}{7}, \frac{1}{10}$
3. 38 **4.** 312 **5.** 27.5 **6.** -84

Chapter 12 Review Exercises
1. true **2.** false **3.** false **4.** false **5.** true **6.** false **7.** 25 **8.** 167 **9.** -2
10. 14 **11.** $(f \circ g)(x) = 6x + 3$, $(g \circ f)(x) = 6x - 9$
12. $(f \circ g)(x) = 9x^2 - 42x + 53$, $(g \circ f)(x) = 3x^2 + 5$
13. $(f \circ g)(x) = \sqrt{2x - 4}$, $(g \circ f)(x) = 2\sqrt{x - 3} - 1$
14. $(f \circ g)(x) = \frac{4x - 4}{x - 4}$, $(g \circ f)(x) = \frac{3 - 3x}{x + 3}$ **15.** no

16. yes

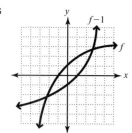

17. yes **18.** yes **19.** no

$f(x) = \log_b x$ and $g(x) = b^x$ are inverses. **62.** 30 dB **63.** 8
64. 6 **65.** $1 + \log_6 x$ **66.** $\log_4 x + \log_4(2x - 5)$
67. $\log_3 32$ **68.** $\log_5(3x^2 - 6x)$ **69.** $\log_b x - \log_b 5$
70. $\log_a(3x - 2) - \log_a(4x + 3)$ **71.** $\log_5 6$ **72.** $\log_2 \dfrac{x + 5}{2x - 3}$
73. $4 \log_3 7$ **74.** $\dfrac{1}{3} \log_a x$ **75.** $-4 \log_4 a$ **76.** $\dfrac{4}{5}$ **77.** $\log_a x^4$
78. $\log_a \sqrt[5]{y^3}$ **79.** $2 \log_a x + 3 \log_a y$ **80.** $4 \log_a c - 3 \log_a d$
81. $2 \log_a x + 3 \log_a y - 4 \log_a z$ **82.** $\dfrac{3}{2} - 2 \log_a b$
83. $\log_a b^2 c^4$ **84.** $\log_a \dfrac{4^3}{3^2}$ **85.** $\log_a \sqrt[4]{x^2 y^3}$
86. $\log_a(x + 5)^4(x - 3)^2$ **87.** 2.5132 **88.** -2.4559
89. -3.1466 **90.** 4.5218 **91.** -5 **92.** $\dfrac{1}{4}$ **93.** 85 dB
94. pH = 2.9 **95.** About 17 yr. **96.** The 7.8 earthquake is 10
times as severe. **97.** 1.5773 **98.** 0.5360 **99.** 1.3626
100. 4.9592 **101.** 0.5995 **102.** -536.4793 **103.** 13 **104.** $-4, 2$
105. 3 **106.** 20 **107.** 36 **108.** 8 **109.** $\dfrac{5}{2}$ **110.** 5 **111.** 6.1 yr.
112. a. \$12,254.71 **b.** 7.7 yr. **113. a.** 573 ants **b.** about 23
months **114. a.** 36,165 cases **b.** 1998 **115.** 1.9534
116. -2.5850

Chapter 12 Practice Test

1. $f[g(x)] = 9x^2 - 30x + 19$ **2. a.** $f^{-1}(x) = \dfrac{x + 3}{4}$

b. $f[f^{-1}(x)] = f\left[\dfrac{x + 3}{4}\right] = 4\left(\dfrac{x + 3}{4}\right) - 3 = x + 3 - 3 = x$

$f^{-1}[f(x)] = f^{-1}(4x + 3) = \dfrac{4x - 3 + 3}{4} = \dfrac{4x}{4} = x$

c. The graphs are symmetric about the line $y = x$.

20. no **21.** $f^{-1}(x) = \dfrac{x - 4}{5}$ **22.** $f^{-1}(x) = \sqrt[3]{x - 6}$
23. $f^{-1}(x) = \dfrac{4 - 5x}{x}$ **24.** $f^{-1}(x) = \dfrac{x^3 - 2}{3}$ **25.** $(-6, 4)$

26. $f^{-1}(b) = a$

27.

28.

3.

4.

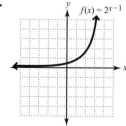

5. -10 **6. a.** \$42,143.63 **b.** About 8.2 yr. **7. a.** 0.68 gm
b. About 1.16 hr. **8. a.** 14.8 million people, 46.9 million
people **b.** 1989 **9.** $\left(\dfrac{1}{3}\right)^{-4} = 81$ **10.** -3 **11.** 5

29.

30.

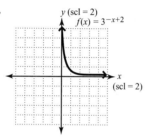

31. 4 **32.** $\dfrac{3}{2}$ **33.** -2 **34.** -2 **35.** 4 **36.** 2 **37.** 3 **38.** $-\dfrac{6}{7}$
39. 6277 cells **40.** \$40,353.57 **41.** 7.16 gm **42.** About 865
addicts. **43.** $\log_7 343 = 3$ **44.** $\log_4 \dfrac{1}{64} = -3$
45. $\log_{3/2} \dfrac{81}{16} = 4$ **46.** $\log_{11} \sqrt[3]{11} = \dfrac{1}{3}$ **47.** $9^2 = 81$
48. $\left(\dfrac{1}{5}\right)^{-3} = 125$ **49.** $a^4 = 16$ **50.** $e^b = c$ **51.** $\dfrac{1}{81}$ **52.** 4
53. 5 **54.** -2 **55.** 3 **56.** $\dfrac{1}{10}$ **57.** 2 **58.** 11

12.

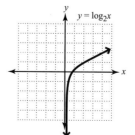

13. $4 \log_b x + 2 \log_b y - \log_b z$

59.

60.

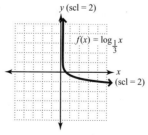

61. Domain is $(-\infty, \infty)$ and range is $(0, \infty)$ because

14. $\frac{5}{4}\log_b x - \frac{7}{4}\log_b y$ **15.** $\log_b \sqrt[4]{x^6 y^9}$ **16. a.** About 74 g.
b. About 247.6 days.   **17.** 4.6433   **18.** 3   **19.** 3
20. a. 307 cases   **b.** 2001

Chapter 13

Exercise Set 13.1
1. Parabola, because only one variable is squared.   **3.** Parabola opens to the right, since $a = 3$ and $3 > 0$.   **5.** Circle; center at $(-3, 2)$; radius of 4.

7. opens up; vertex $(1, 2)$;
axis of symmetry $x = 1$

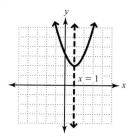

9. opens up; vertex $(2, -3)$;
axis of symmetry $x = 2$

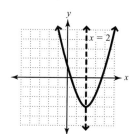

11. opens right; vertex $(-2, -2)$;
axis of symmetry $y = -2$

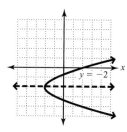

13. opens left; vertex $(3, 1)$;
axis of symmetry $y = 1$

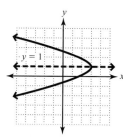

15. opens right; vertex $(-4, -2)$;
axis of symmetry $y = -2$

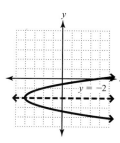

17. opens left; vertex $(-5, -2)$;
axis of symmetry $y = -2$

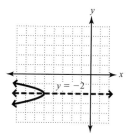

19. opens right; vertex $(-1, -2)$;
axis of symmetry $y = -2$

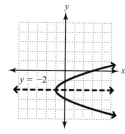

21. opens left; vertex $(4, 3)$;
axis of symmetry $y = 3$

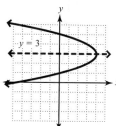

23. opens right; vertex $(-5, -2)$;
axis of symmetry $y = -2$

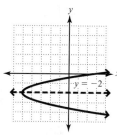

25. opens right; vertex $(0, 1)$;
axis of symmetry $y = 1$

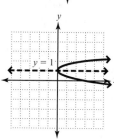

27. opens left; vertex $(7, 1)$;
axis of symmetry $y = 1$

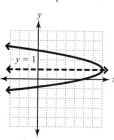

29. c   **31.** a   **33.** 5   **35.** 13   **37.** 13   **39.** $2\sqrt{5}$ **41.** $4\sqrt{5}$
43. $\sqrt{34}$ **45.** 5   **47.** $\sqrt{89}$

49. center (2,1); radius 2

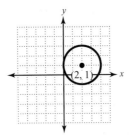

51. center (−3, −2); radius 9

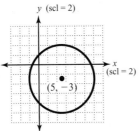

53. center (5, −3) ; radius 7

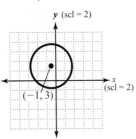

55. center (−1, 3) ; radius $3\sqrt{2}$

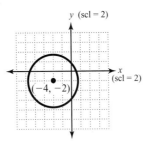

57. center (−4, −2) ; radius $4\sqrt{2}$

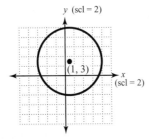

59. center (1, 3) ; radius 7

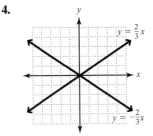

61. center (−5, 2) ; radius 8

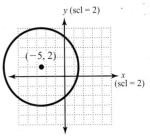

63. center (−7, 2) ; radius 2

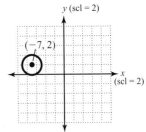

65. b   **67.** c

69.

71.

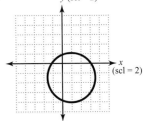

73. $(x − 8)^2 + (y + 1)^2 = 36$ **75.** $(x + 4)^2 + (y + 3)^2 = 25$
77. $(x − 6)^2 + (y + 2)^2 = 14$ **79.** $(x + 5)^2 + (y − 2)^2 = 45$
81. $(x − 2)^2 + (y − 4)^2 = 25$ **83.** $(x − 2)^2 + (y − 4)^2 = 169$
85. $(x − 2)^2 + (y + 5)^2 = 64$ **87. a.** 256 feet   **b.** 3 seconds
c. 7 seconds   **89.** $y = \frac{-2}{25}x^2 + 18$ or $y = −0.08x^2 + 18$
91. 3.5%   **93.** $x^2 + y^2 = 400$ **95.** $x^2 + (y − 110)^2 = 10,000$
97. $x^2 + y^2 = 43,560,000$

Review Exercises   **1.** ± 4   **2.** (4, 0), (−4, 0), (0, 3),
(0, −3)   **3.** (3, 0), (−3, 0), (0, 5), (0, −5)
4.

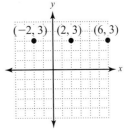

5. (−2, 3) and (6, 3)

6. $(5x + 3y)(5x − 3y)$

Exercise Set 13.2

1. An ellipse is the set of all points the sum of whose distances from two fixed points is constant. **3.** The equation of an ellipse is a sum; the equation of a hyperbola is a difference. **5.** The parabola has only one squared term.

7.

9.

11.

13.

15.

17.

19.

21.
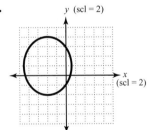

23. $\dfrac{x^2}{9} + \dfrac{y^2}{16} = 1$

25.

27.

29.

31.

33.
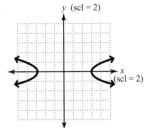

35. $\dfrac{x^2}{9} - \dfrac{y^2}{4} = 1$ **37.** c **39.** b **41.** ellipse **43.** circle

45. circle

47. ellipse

49. parabola

51. hyperbola

53. parabola

55. hyperbola

57. ellipse

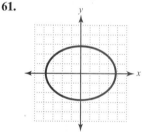

59. a. Yes, the sailboat will clear the bridge. The height of the bridge at the center is 20 feet and the boat's mast is only 18 feet tall. **b.** The bridge is 32 feet wide at the base of the arch.

61.

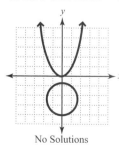

63. $\dfrac{x^2}{2.25} + \dfrac{y^2}{4} = 1$ **65.** 4 units

Review Exercises   **1.** $(2, -3)$ **2.** $(-3, 4)$ **3.** $(1, 2)$

4. $y = -3x^2 + 6$ **5.** $x = \pm 3$ **6.** $x = -4, \dfrac{2}{3}$

Exercise Set 13.3   **1.** 0, 1, 2, 3, or 4

a.

No Solutions

b.

One Solution

c.

Two Solutions

d.

Three Solutions

e.

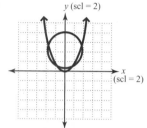

Four Solutions

3. Substitution, because the first equation has a linear term and is easy to solve for y.
5. This is a line and an ellipse, so there will be 0, 1, or 2 solutions.

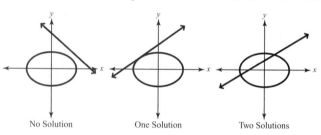

No Solution   One Solution   Two Solutions

7. no solution   **9.** $(-2, 0)$ and $(1, 9)$ **11.** $(-4, -3)$ and $(3, 4)$
13. $(3.6, -0.2)$ and $(2, 3)$ **15.** no solution   **17.** $(0, 3)$ and $(1, 5)$
19. $(3, 2)$ and $(2, 3)$ **21.** $(4, 2), (4, -2), (-4, -2)$ and $(-4, 2)$
23. $(3, 4), (-3, 4), (-3, -4), (3, -4)$ **25.** $(2, 1), (2, -1), (-2, -1), (-2, 1)$ **27.** $(-6, 0), (3, 3), (3, -3)$ **29.** no solution
31. $(2, 4), (-2, 4)$ **33.** $(4, 0), (-4, 0)$ **35.** $(1, 2), (-1, -2)$
37. no solution   **39.** $(2, 2), (2, -2), (-2, -2), (-2, 2)$
41. $(-5, \sqrt{7}), (-5, -\sqrt{7}), (2, 0)$

43. Answers will vary, but one possible choice is the system
$\begin{cases} x + y = 4 \\ x^2 + y^2 = 1 \end{cases}$.

45. $(5, 3), (5, -3), (-5, -3), (-5, 3)$ **47.** The computer keyboard is 8 inches by 18 inches.   **49.** At equilibrium there should be 400 chairs at \$72 per chair.

51.

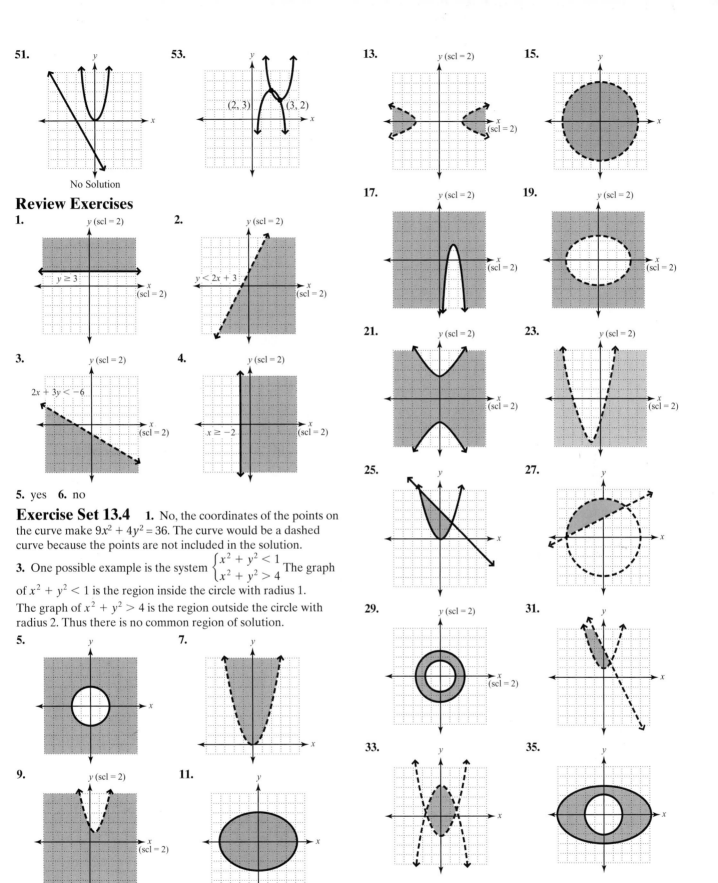

No Solution

Review Exercises

1.

2.

$y \geq 3$

$y < 2x + 3$

3.

4.

$2x + 3y < -6$

$x \geq -2$

5. yes   **6.** no

Exercise Set 13.4

1. No, the coordinates of the points on the curve make $9x^2 + 4y^2 = 36$. The curve would be a dashed curve because the points are not included in the solution.

3. One possible example is the system $\begin{cases} x^2 + y^2 < 1 \\ x^2 + y^2 > 4 \end{cases}$ The graph of $x^2 + y^2 < 1$ is the region inside the circle with radius 1. The graph of $x^2 + y^2 > 4$ is the region outside the circle with radius 2. Thus there is no common region of solution.

5.

7.

9.

11.

53.

(2, 3)   (3, 2)

13.

15.

17.

19.

21.

23.

25.

27.

29.

31.

33.

35.

37.

39.

41.

43.

45.

47.

49.

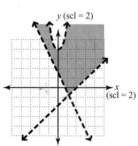

Review Exercises   **1.** 20   **2.** $a^2 + 3a + 2$ **3.** $x + 3$
4. $6x^2 - x - 2$ **5.** $x^2 - x - 1$ **6.** $4x^2 + 4x + 5$

Chapter 13 Review Exercises   **1.** false   **2.** true

3. false   **4.** true   **5.** true   **6.** false   **7.** ellipse; $(4, 0)$ and
$(-4, 0)$ **8.** 10   **9.** $(5, 4), (5, -4), (-5, -4), (-5, 4)$
10. $(-4, 1); r = 2$
11. opens up; vertex $(3, -5)$;
axis of symmetry $x = 3$

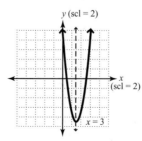

12. opens down; vertex $(-2, 3)$;
axis of symmetry $x = -2$

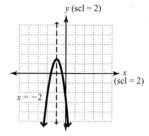

13. opens left; vertex $(4, 2)$;
axis of symmetry $y = 2$

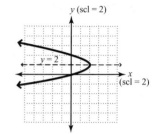

14. opens right; vertex $(-2, -3)$;
axis of symmetry $y = -3$

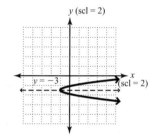

15. opens right; vertex $(-1, -3)$;
axis of symmetry $y = -3$

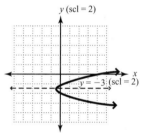

16. opens right; vertex $(-14, 2)$;
axis of symmetry $y = 2$

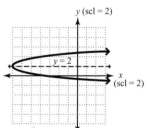

17. opens left; vertex $(4, -1)$;
axis of symmetry $y = -1$

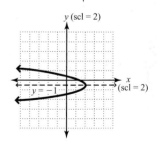

18. opens left; vertex $(3, -2)$;
axis of symmetry $y = -2$

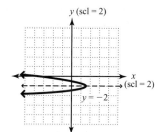

19. 5 **20.** $4\sqrt{5}$ **21.** 10 **22.** $(x + 6)^2 + (y - 8)^2 = 225$
23. a. $(3, -2)$ **b.** 5

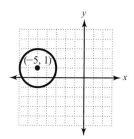

24. a. $(-5, 1)$ **b.** 2

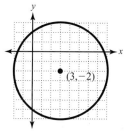

25. a. $(2, -4)$ **b.** 3

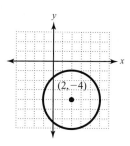

26. a. $(-5, -1)$ **b.** 2

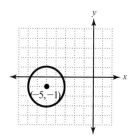

27. $(x - 6)^2 + (y + 8)^2 = 81$ **28.** $(x + 3)^2 + (y + 5)^2 = 100$
29. a. 144 feet **b.** 1 second **c.** 4 seconds **30.** $x^2 + y^2 = 400$

31.

32.

33.

34.

35.

36.

37.

38.

39.

40.

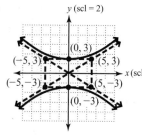

41. $\dfrac{x^2}{625} + \dfrac{y^2}{225} = 1$ **42.** $\dfrac{x^2}{1.625^2} + \dfrac{y^2}{4} = 1$ **43.** $(2, 5), (-1, -1)$
44. $(-1, -4), (4, 1)$ **45.** $(-73/37, -35/37), (1, 5)$ **46.** $(8, 0),$
$(-8, 0)$ **47.** $(3, 4), (3, -4), (-3, -4), (-3, 4)$ **48.** $(4, 2), (4, -2),$
$(-4, -2), (-4, 2)$ **49.** $(2, 2), (-2, 2)$ **50.** $(2, 3), (-2, 3)$
51. $(8, 5), (8, -5), (-8, -5), (-8, 5)$ **52.** 8 feet by 10 feet

53.

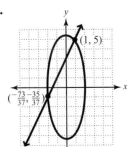

(1, 5)

$\left(-\frac{73}{37}, -\frac{35}{37}\right)$

54.

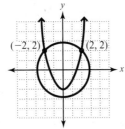

$(-2, 2)$ $(2, 2)$

55.

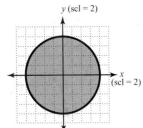

y (scl = 2)

x (scl = 2)

56.

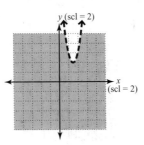

y (scl = 2)

x (scl = 2)

57.

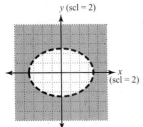

y (scl = 2)

x (scl = 2)

58.

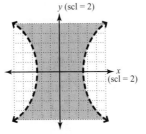

y (scl = 2)

x (scl = 2)

59.

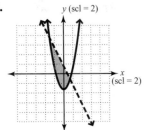

y (scl = 2)

x (scl = 2)

60.

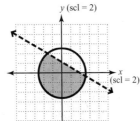

y (scl = 2)

x (scl = 2)

61.

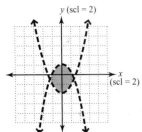

y (scl = 2)

x (scl = 2)

62.

y

x

63.

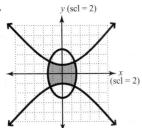

y (scl = 2)

x (scl = 2)

64.

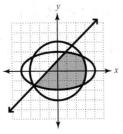

y

x

Chapter 13 Practice Test

1. opens up; vertex $(-1, -4)$; axis of symmetry $x = -1$

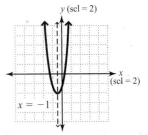

y (scl = 2)

x (scl = 2)

$x = -1$

2. opens left; vertex $(1, 3)$; axis of symmetry $y = 3$

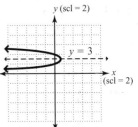

y (scl = 2)

$y = 3$

x (scl = 2)

3. opens right; vertex $(-7, -2)$; axis of symmetry $y = -2$

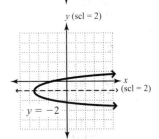

y (scl = 2)

x (scl = 2)

$y = -2$

4. $2\sqrt{5}$ **5. a.** $(-4, 3)$ **b.** 6

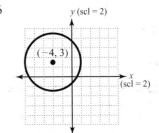

y (scl = 2)

$(-4, 3)$

x (scl = 2)

6. a. $(-2, 5)$ **b.** 3

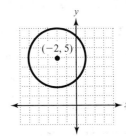

y

$(-2, 5)$

x

7. $(x - 2)^2 + (y + 4)^2 = 100$

8.

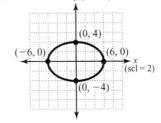

y (scl = 2)

$(0, 4)$

$(-6, 0)$ $(6, 0)$

x (scl = 2)

$(0, -4)$

9.

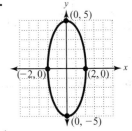

y

$(0, 5)$

$(-2, 0)$ $(2, 0)$

x

$(0, -5)$

10.

11.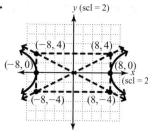

b. $\left\{m\mid 0 < m < \dfrac{3}{5}\right\}$

c. $\left(0, \dfrac{3}{5}\right)$

31. a. $x^2 + 2x$ **b.** $-x^2 + 2x + 2$ **c.** $2x^3 + x^2 - 2x - 1$
d. $\dfrac{2x + 1}{x^2 - 1}; x \neq \pm 1$ **e.** $2x^2 - 1$ **f.** $4x^2 + 4x$ **32.** yes

33. $y = \dfrac{4}{3}x - 12, 4x - 3y = 36$ **34.** $\log_5 125 = 3$

35. $5\log_5 x + \log_5 y$ **36.** $(x - 2)^2 + (y - 4)^2 = 13^2$

12. $(-5, 18), (1, 6)$ **13.** $(3, 5), (-3/5, -29/5)$ **14.** $(2, 3), (2, -3),$ $(-2, -3), (-2, 3)$ **15.** $(3, 2), (3, -2), (-3, -2), (-3, 2)$

16.

17.

18.

37.

38.

39.

40.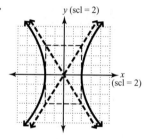

19. Height is 18, distance across the base is 12.

20. $\dfrac{x^2}{9.5^2} + y^2 = 1$

41.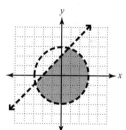

Chapters 1–13 Cumulative Review   **1.** true
2. true   **3.** true   **4.** true   **5.** false   **6.** false
7. $(x - h)^2 + (y - k)^2 = r^2$ **8.** i
9. $d = \sqrt{(x_2 - x_1)^2 + (y_2 - y_1)^2}$ **10.** $a_1 b_2 - a_2 b_1$ **11.** $9\dfrac{1}{16}$
12. $2i\sqrt{5}$ **13.** $\dfrac{8}{3x}$ **14.** $\dfrac{x}{3x^2 + 1}$ **15.** $-\dfrac{1}{4}$ **16.** $\dfrac{n^6}{125m^9}$
17. $(k^2 + 9)(k + 3)(k - 3)$ **18.** $(x + y)(a + b)$
19. $(2, -1)$ **20.** $\dfrac{1 \pm i\sqrt{3}}{4}$ **21.** $-4, \dfrac{2}{3}$ **22.** 52   **23.** $\pm 2, \pm i$
24. $-\dfrac{4}{3}$ **25.** 1.5   **26.** No solution.   **27.** $\dfrac{1}{9}$ **28.** $\dfrac{1}{8}$

29. a.

b. $\left\{x \mid x < -\dfrac{5}{3} \text{ or } x > 1\right\}$

c. $\left(-\infty, -\dfrac{5}{3}\right) \cup (1, \infty)$

30. a.

42. 7.3%   **43. a.** $h = \dfrac{V}{\pi r^2}$ **b.** $\dfrac{15}{r^2}$ **44.** 2.9 lbs. of tuna and
3.4 lbs. of shrimp   **45.** 0.857 hr. or approximately 51 min.
46. $1500 at 5%, $2500 at 6%, $1000 at 3%
47. $\dfrac{x^2}{1406.25} + \dfrac{y^2}{400} = 1$ **48.** 7.37   **49.** $16,909.08   **50.** 6.9

Appendix A Exercises

1. The set of natural numbers 1 to n.   **3.** A series is the sum of
the terms of a sequence.   **5.** arithmetic, 5   **7.** 3, 5, 7, 9, 41
9. 3, 6, 11, 18, 227   **11.** $\dfrac{1}{3}, \dfrac{1}{2}, \dfrac{3}{5}, \dfrac{2}{3}, \dfrac{11}{12}$ **13.** $-\dfrac{1}{2}, \dfrac{1}{5}, -\dfrac{1}{10}, \dfrac{1}{17}, -\dfrac{1}{226}$
15. 5   **17.** -3 **19.** 7   **21.** $a_n = 14 + (n - 1)(4), a_{14} = 66$
23. $a_n = -8 + (n - 1)(-3), a_{28} = -89$

25. $a_n = -5 + (n-1)(4), a_{34} = 127$ **27.** $-6, 1, 8, 15$
29. $-5, -8, -11, -14$ **31.** $7, 11, 15, 19$ **32.** $6, 13, 20, 27$
33. $a_1 = 11$ **35.** $d = 4$ **37.** $-8, -1, 6, 13$ **39.** $27, 24, 21, 18$
41. $1 + 4 + 9 + 16 + 25 + 36 = 91$ **43.** $-3 - 1 + 1 + 3 = 0$
45. $-1 + 8 + 23 = 30$ **47.** $9 + 13 + 17 + 21 = 60$ **49.** 1450
51. -429 **53.** 675 **55.** 120 **57.** 5050 **59.** 144 seats; 4480 seats
61. a. 16 posts   **b.** 325 posts   **63. a.** \$28,000 \$29,500 \$31,000
\$32,500 \$34,000   **b.** \$41,500   **c.** \$347,500

Appendix B Exercises

1. The common ratio, r.   **3.** Yes. The first n terms form a finite
geometric series.   **5.** $2 + 6 + 18 + 54 = 80$ **7. a.** 3 **b.** $a_9 = 6561$
c. $a_n = 1(3)^{n-1}$ **9. a.** -2 **b.** $a_{14} = -32,768$ **c.** $a_n = -2(-2)^{n-1}$
11. a. $\dfrac{1}{3}$ **b.** $a_{10} = \dfrac{1}{81}$ **c.** $a_n = 243\left(\dfrac{1}{3}\right)^{n-1}$ **13. a.** $-\dfrac{1}{4}$

b. $a_7 = \dfrac{1}{32}$ **c.** $a_n = 128\left(-\dfrac{1}{4}\right)^{n-1}$ **15. a.** $-4, 12, -36, 108, -324$

b. 8748 **17. a.** $243, 81, 27, 9, 3$ **b.** $\dfrac{1}{3}$ **19. a.** 2 **b.** $1, 2, 4, 8$

21. a. 2 **b.** $-6, -12, -24, -48$ **23. a.** $\dfrac{1}{4}$ **b.** $128, 32, 8, 2$

25. -6 **27.** $265,719$ **29.** 63.875 **31.** 102.4 **33.** 40.5

35. 9.375 **37.** Does not exist.   **39.** $\dfrac{4}{9}$ **41.** $\dfrac{17}{99}$ **43. a.** $520,$
$540.8, 562.432, 584.92928, 608.3264512$ **b.** 740.122142459
45. a. $100, 108, 117, 126$ **b.** 171 mink   **c.** $a_n = 100(1.08)^{n-1}$
47. a. \$0.01, \$0.02, \$0.04, \$0.08, \$0.16   **b.** $a_n = 0.01(2)^{n-1}$
c. \$5242.88   **d.** \$10,485.75   **49.** \$11,191.64

Appendix C Exercises

1. 13 **3.** 8 factorial   **5.** $5, 8$ **7.** 24 **9.** 144 **11.** 720 **13.** $\dfrac{1}{90}$
or 0.011 **15.** 10 **17.** 28 **19.** 35 **21.** 210 **23.** 1 **25.** 1

27. $a^5 + 5a^4b + 10a^3b^2 + 10a^2b^3 + 5ab^4 + b^5$
29. $x^4 - 4x^3y + 6x^2y^2 - 4xy^3 + y^4$ **31.** $8a^3 + 12a^2b + 6ab^2 + b^3$
33. $x^5 - 10x^4y + 40x^3y^2 - 80x^2y^3 + 80xy^4 - 32y^5$
35. $64m^6 + 576m^5n + 2160m^4n^2 + 4320m^3n^3 + 4860m^2n^4 + 2916mn^5 + 729n^6$
37. $81x^4 - 432x^3y + 864x^2y^2 - 768xy^3 + 256y^4$
39. $70x^4y^4$ **41.** $45a^8b^2$ **43.** $32,256x^4y^5$ **45.** $-22,680m^4n^3$

Appendix D Exercises

1. The number of ways the two tasks can be performed.
3. nCr because the order of selection does not matter.
5. nPr because the order matters.   **7.** 42 **9.** 720 **11.** 1
13. 15 **15.** 1 **17.** 1 **19.** 360 outfits   **21.** 120 ways
23. a. 6 **b.** 27 **25.** $17,576,000$ **27.** $10,000; 5040$
29. $39,916,800$ **31.** $1,906,884$ **33.** 816

Appendix E Exercises

1. The sample space is the set of all possible outcomes.
3. 0 and 1 **5.** The probability of an event is the ratio of the
number of ways the event can occur to the number of outcomes
in the sample space.   **7.** $\{a, e, i, o, u\}$
9. $\{HHH, HHT, HTH, THH, THT, HTT, TTH, TTT\}$
11. 90 **13.** 1820 **15.** $\{HTT, THT, TTH\}$ **17.** $(1, 3)\ (3, 1)\ (2, 2)$
19. $\dfrac{1}{6}$ **21.** $\dfrac{1}{2}$ **23.** $\dfrac{3}{4}$ **25.** $\dfrac{1}{12}$ **27.** $\dfrac{1}{12}$ **29.** $\dfrac{1}{13}$ **31.** $\dfrac{3}{13}$ **33.** $\dfrac{3}{8}$
35. $\dfrac{1}{3}$ **37.** 0 **39.** $\dfrac{1}{9}$ **41.** $\dfrac{10}{21}$ **43.** 0.018 **45.** 0.070

Powers and Roots

n	n^2	n^3	$\sqrt{n}$	$\sqrt[3]{n}$	$\sqrt{10n}$	n	n^2	n^3	$\sqrt{n}$	$\sqrt[3]{n}$	$\sqrt{10n}$
1	1	1	1.000	1.000	3.162	51	2,601	132,651	7.141	3.708	22.583
2	4	8	1.414	1.260	4.472	52	2,704	140,608	7.211	3.733	22.804
3	9	27	1.732	1.442	5.477	53	2,809	148,877	7.280	3.756	23.022
4	16	64	2.000	1.587	6.325	54	2,916	157,464	7.348	3.780	23.238
5	25	125	2.236	1.710	7.071	55	3,025	166,375	7.416	3.803	23.452
6	36	216	2.449	1.817	7.746	56	3,136	175,616	7.483	3.826	23.664
7	49	343	2.646	1.913	8.367	57	3,249	185,193	7.550	3.849	23.875
8	64	512	2.828	2.000	8.944	58	3,364	195,112	7.616	3.871	24.083
9	81	729	3.000	2.080	9.487	59	3,481	205,379	7.681	3.893	24.290
10	100	1,000	3.162	2.154	10.000	60	3,600	216,000	7.746	3.915	24.495
11	121	1,331	3.317	2.224	10.488	61	3,721	226,981	7.810	3.936	24.698
12	144	1,728	3.464	2.289	10.954	62	3,844	238,328	7.874	3.958	24.900
13	169	2,197	3.606	2.351	11.402	63	3,969	250,047	7.937	3.979	25.100
14	196	2,744	3.742	2.410	11.832	64	4,096	262,144	8.000	4.000	25.298
15	225	3,375	3.873	2.466	12.247	65	4,225	274,625	8.062	4.021	25.495
16	256	4,096	4.000	2.520	12.649	66	4,356	287,496	8.124	4.041	25.690
17	289	4,913	4.123	2.571	13.038	67	4,489	300,763	8.185	4.062	25.884
18	324	5,832	4.243	2.621	13.416	68	4,624	314,432	8.246	4.082	26.077
19	361	6,859	4.359	2.688	13.784	69	4,761	328,509	8.307	4.102	26.268
20	400	8,000	4.472	2.714	14.142	70	4,900	343,000	8.367	4.121	26.458
21	441	9,261	4.583	2.759	14.491	71	5,041	357,911	8.426	4.141	26.646
22	484	10,648	4.690	2.802	14.832	72	5,184	373,248	8.485	4.160	26.833
23	529	12,167	4.796	2.844	15.166	73	5,329	389,017	8.544	4.179	27.019
24	576	13,824	4.899	2.884	15.492	74	5,476	405,224	8.602	4.198	27.203
25	625	15,625	5.000	2.924	15.811	75	5,625	421,875	8.660	4.217	27.386
26	676	17,576	5.099	2.962	16.125	76	5,776	438,976	8.718	4.236	27.568
27	729	19,683	5.196	3.000	16.432	77	5,929	456,533	8.775	4.254	27.749
28	784	21,952	5.292	3.037	16.733	78	6,084	474,552	8.832	4.273	27.928
29	841	24,389	5.385	3.072	17.029	79	6,241	493,039	8.888	4.291	28.107
30	900	27,000	5.477	3.107	17.321	80	6,400	512,000	8.944	4.309	28.284
31	961	29,791	5.568	3.141	17.607	81	6,561	531,441	9.000	4.327	28.460
32	1,024	32,768	5.657	3.175	17.889	82	6,724	551,368	9.055	4.344	28.636
33	1,089	35,937	5.745	3.208	18.166	83	6,889	571,787	9.110	4.362	28.810
34	1,156	39,304	5.831	3.240	18.439	84	7,056	592,704	9.165	4.380	28.983
35	1,225	42,875	5.916	3.271	18.708	85	7,225	614,125	9.220	4.397	29.155
36	1,296	46,656	6.000	3.302	18.974	86	7,396	636,056	9.274	4.414	29.326
37	1,369	50,653	6.083	3.332	19.235	87	7,569	658,503	9.327	4.431	29.496
38	1,444	54,872	6.164	3.362	19.494	88	7,744	981,472	9.381	4.448	29.665
39	1,521	59,319	6.245	3.391	19.748	89	7,921	704,969	9.434	4.465	29.833
40	1,600	64,000	6.325	3.420	20.000	90	8,100	729,000	9.487	4.481	30.000
41	1,681	68,921	6.403	3.448	20.248	91	8,281	753,571	9.539	4.498	30.166
42	1,764	74,088	6.481	3.476	20.494	92	8,464	778,688	9.592	4.514	30.332
43	2,849	79,507	6.557	3.503	20.736	93	8,649	804,357	9.644	4.531	30.496
44	2,936	85,184	6.633	3.530	20.976	94	8,836	830,584	9.695	4.547	30.659
45	2,025	91,125	6.708	3.557	21.213	95	9,025	857,375	9.747	4.563	30.882
46	2,116	97,336	6.782	3.583	21.148	96	9,216	884,736	9.798	4.579	30.984
47	2,209	103,823	6.856	3.609	21.679	97	9,409	912,673	9.849	4.595	31.145
48	2,304	110,592	6.928	3.534	21.909	98	9,604	941,192	9.899	4.610	31.305
49	2,401	117,649	7.000	3.659	22.136	99	9,801	970,299	9.950	4.626	31.464
50	2,500	125,000	7.071	3.684	22.361	100	10,000	1,000,000	10.000	4.642	31.623

Linear Functions

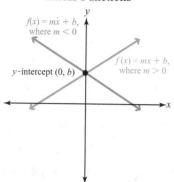

$f(x) = mx + b$, where $m < 0$

$f(x) = mx + b$, where $m > 0$

y-intercept $(0, b)$

Quadratic Functions

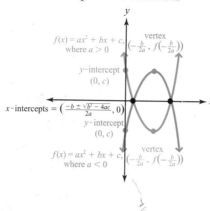

$f(x) = ax^2 + bx + c$, where $a > 0$

vertex $\left(-\frac{b}{2a}, f\left(-\frac{b}{2a}\right)\right)$

y-intercept $(0, c)$

x-intercepts $= \left(\frac{-b \pm \sqrt{b^2 - 4ac}}{2a}, 0\right)$

y-intercept $(0, c)$

$f(x) = ax^2 + bx + c$, where $a < 0$

vertex $\left(-\frac{b}{2a}, f\left(-\frac{b}{2a}\right)\right)$

Cubic Functions

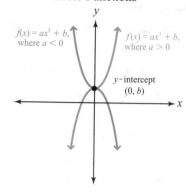

$f(x) = ax^3 + b$, where $a < 0$

$f(x) = ax^3 + b$, where $a > 0$

y-intercept $(0, b)$

Absolute Value Functions

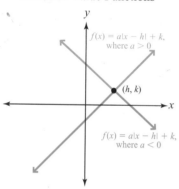

$f(x) = a|x - h| + k$, where $a > 0$

(h, k)

$f(x) = a|x - h| + k$, where $a < 0$

Radical Functions

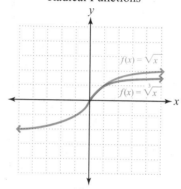

$f(x) = \sqrt{x}$

$f(x) = \sqrt[3]{x}$

Exponential and Logarithmic Functions

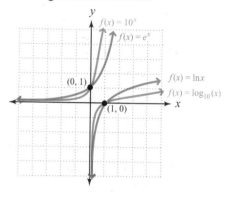

$f(x) = 10^x$

$f(x) = e^x$

$(0, 1)$

$f(x) = \ln x$

$f(x) = \log_{10}(x)$

$(1, 0)$

Conic Sections Parabola

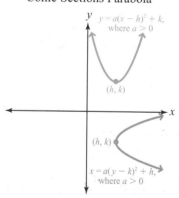

$y = a(x - h)^2 + k$, where $a > 0$

(h, k)

(h, k)

$x = a(y - k)^2 + h$, where $a > 0$

Circle

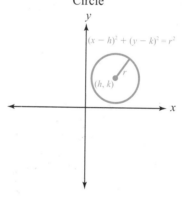

$(x - h)^2 + (y - k)^2 = r^2$

(h, k) r

Ellipse

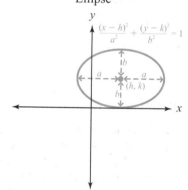

$\frac{(x - h)^2}{a^2} + \frac{(y - k)^2}{b^2} = 1$

b

a a

(h, k)

b

Hyperbola with center (0, 0)

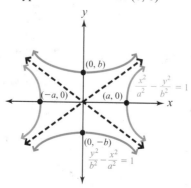

$(0, b)$

$(-a, 0)$

$(a, 0)$

$\frac{x^2}{a^2} - \frac{y^2}{b^2} = 1$

$(0, -b)$

$\frac{y^2}{b^2} - \frac{x^2}{a^2} = 1$